INTEGRALS CONTAINING $\sqrt{a + bu}$

14. $\displaystyle\int u\sqrt{a + bu}\,du = \frac{2}{15b^2}(3bu - 2a)(a + bu)^{3/2} + C$

15. $\displaystyle\int u^2\sqrt{a + bu}\,du = \frac{2}{105b^3}(15b^2u^2 - 12abu + 8a^2)(a + bu)^{3/2} + C$

16. $\displaystyle\int u^n\sqrt{a + bu}\,du = \frac{2u^n(a + bu)^{3/2}}{b(2n + 3)} - \frac{2an}{b(2n + 3)}\int u^{n-1}\sqrt{a + bu}\,du$

17. $\displaystyle\int \frac{u\,du}{\sqrt{a + bu}} = \frac{2}{3b^2}(bu - 2a)\sqrt{a + bu} + C$

18. $\displaystyle\int \frac{u^2\,du}{\sqrt{a + bu}} = \frac{2}{15b^3}(3b^2u^2 - 4abu + 8a^2)\sqrt{a + bu} + C$

19. $\displaystyle\int \frac{u^n\,du}{\sqrt{a + bu}} = \frac{2u^n\sqrt{a + bu}}{b(2n + 1)} - \frac{2an}{b(2n + 1)}\int \frac{u^{n-1}\,du}{\sqrt{a + bu}}$

20. $\displaystyle\int \frac{du}{u\sqrt{a + bu}} = \begin{cases} \dfrac{1}{\sqrt{a}}\ln\left|\dfrac{\sqrt{a + bu} - \sqrt{a}}{\sqrt{a + bu} + \sqrt{a}}\right| + C & (a > 0) \\[3mm] \dfrac{2}{\sqrt{-a}}\tan^{-1}\sqrt{\dfrac{a + bu}{-a}} + C & (a < 0) \end{cases}$

21. $\displaystyle\int \frac{du}{u^n\sqrt{a + bu}} = -\frac{\sqrt{a + bu}}{a(n - 1)u^{n-1}} - \frac{b(2n - 3)}{2a(n - 1)}\int \frac{du}{u^{n-1}\sqrt{a + bu}}$

22. $\displaystyle\int \frac{\sqrt{a + bu}\,du}{u} = 2\sqrt{a + bu} + a\int \frac{du}{u\sqrt{a + bu}}$

23. $\displaystyle\int \frac{\sqrt{a + bu}\,du}{u^n} = -\frac{(a + bu)^{3/2}}{a(n - 1)u^{n-1}} - \frac{b(2n - 5)}{2a(n - 1)}\int \frac{\sqrt{a + bu}\,du}{u^{n-1}}$

INTEGRALS CONTAINING $a^2 \pm u^2$ $(a > 0)$

24. $\displaystyle\int \frac{du}{a^2 + u^2} = \frac{1}{a}\tan^{-1}\frac{u}{a} + C$

25. $\displaystyle\int \frac{du}{a^2 - u^2} = \frac{1}{2a}\ln\left|\frac{u + a}{u - a}\right| + C$

26. $\displaystyle\int \frac{du}{u^2 - a^2} = \frac{1}{2a}\ln\left|\frac{u - a}{u + a}\right| + C$

INTEGRALS CONTAINING $\sqrt{u^2 \pm a^2}$ $(a > 0)$

27. $\displaystyle\int \frac{du}{\sqrt{u^2 \pm a^2}} = \ln\left|u + \sqrt{u^2 \pm a^2}\right| + C$

28. $\displaystyle\int \sqrt{u^2 \pm a^2}\,du = \frac{u}{2}\sqrt{u^2 \pm a^2} \pm \frac{a^2}{2}\ln\left|u + \sqrt{u^2 \pm a^2}\right| + C$

29. $\displaystyle\int u\sqrt{u^2 \pm a^2}\,du = \frac{1}{3}(u^2 \pm a^2)^{3/2} + C$

30. $\displaystyle\int u^2\sqrt{u^2 \pm a^2}\,du = \frac{u}{8}(2u^2 \pm a^2)\sqrt{u^2 \pm a^2} - \frac{a^4}{8}\ln\left|u + \sqrt{u^2 \pm a^2}\right| + C$

31. $\displaystyle\int \frac{\sqrt{u^2 + a^2}\,du}{u} = \sqrt{u^2 + a^2} - a\ln\left|\frac{a + \sqrt{u^2 + a^2}}{u}\right| + C$

32. $\displaystyle\int \frac{\sqrt{u^2 - a^2}\,du}{u} = \sqrt{u^2 - a^2} - a\sec^{-1}\frac{u}{a} + C$

33. $\displaystyle\int \frac{\sqrt{u^2 \pm a^2}\,du}{u^2} = -\frac{\sqrt{u^2 \pm a^2}}{u} + \ln\left|u + \sqrt{u^2 \pm a^2}\right| + C$

34. $\displaystyle\int \frac{u^2\,du}{\sqrt{u^2 \pm a^2}} = \frac{u}{2}\sqrt{u^2 \pm a^2} \mp \frac{a^2}{2}\ln\left|u + \sqrt{u^2 \pm a^2}\right| + C$

35. $\displaystyle\int \frac{du}{u\sqrt{u^2 + a^2}} = -\frac{1}{a}\ln\left|\frac{a + \sqrt{u^2 + a^2}}{u}\right| + C$

36. $\displaystyle\int \frac{du}{u\sqrt{u^2 - a^2}} = \frac{1}{a}\sec^{-1}\frac{u}{a} + C$

37. $\displaystyle\int \frac{du}{u^2\sqrt{u^2 \pm a^2}} = \mp\frac{\sqrt{u^2 \pm a^2}}{a^2 u} + C$

38. $\displaystyle\int (u^2 \pm a^2)^{3/2}\,du = \frac{u}{8}(2u^2 \pm 5a^2)\sqrt{u^2 \pm a^2} + \frac{3a^4}{8}\ln\left|u + \sqrt{u^2 \pm a^2}\right| + C$

39. $\displaystyle\int \frac{du}{(u^2 \pm a^2)^{3/2}} = \pm\frac{u}{a^2\sqrt{u^2 \pm a^2}} + C$

INTEGRALS CONTAINING $\sqrt{a^2 - u^2}$ $(a > 0)$

40. $\displaystyle\int \frac{du}{\sqrt{a^2 - u^2}} = \sin^{-1}\frac{u}{a} + C$

41. $\displaystyle\int \sqrt{a^2 - u^2}\,du = \frac{u}{2}\sqrt{a^2 - u^2} + \frac{a^2}{2}\sin^{-1}\frac{u}{a} + C$

42. $\displaystyle\int u^2\sqrt{a^2 - u^2}\,du = \frac{u}{8}(2u^2 - a^2)\sqrt{a^2 - u^2} + \frac{a^4}{8}\sin^{-1}\frac{u}{a} + C$

43. $\displaystyle\int \frac{\sqrt{a^2 - u^2}\,du}{u} = \sqrt{a^2 - u^2} - a\ln\left|\frac{a + \sqrt{a^2 - u^2}}{u}\right| + C$

44. $\displaystyle\int \frac{\sqrt{a^2 - u^2}\,du}{u^2} = -\frac{\sqrt{a^2 - u^2}}{u} - \sin^{-1}\frac{u}{a} + C$

45. $\displaystyle\int \frac{u^2\,du}{\sqrt{a^2 - u^2}} = -\frac{u}{2}\sqrt{a^2 - u^2} + \frac{a^2}{2}\sin^{-1}\frac{u}{a} + C$

46. $\displaystyle\int \frac{du}{u\sqrt{a^2 - u^2}} = -\frac{1}{a}\ln\left|\frac{a + \sqrt{a^2 - u^2}}{u}\right| + C$

47. $\displaystyle\int \frac{du}{u^2\sqrt{a^2 - u^2}} = -\frac{\sqrt{a^2 - u^2}}{a^2 u} + C$

48. $\displaystyle\int (a^2 - u^2)^{3/2}\,du = -\frac{u}{8}(2u^2 - 5a^2)\sqrt{a^2 - u^2} + \frac{3a^4}{8}\sin^{-1}\frac{u}{a} + C$

49. $\displaystyle\int \frac{du}{(a^2 - u^2)^{3/2}} = \frac{u}{a^2\sqrt{a^2 - u^2}} + C$

Calculus
with Analytic Geometry

Brief Edition

Calculus
with Analytic Geometry

Brief Edition

FOURTH EDITION

HOWARD ANTON
DREXEL UNIVERSITY

with contributions from
Albert Herr, **Drexel University**

JOHN WILEY & SONS, INC.
New York   Chichester   Brisbane   Toronto   Singapore

Acquisitions Editor: Robert Macek
Developmental Editor: Joan Carrafiello
Production Manager: Joseph Ford
Design Supervisor: Madelyn Lesure
Production Supervisor: Lucille Buonocore
Manufacturing Manager: Denis Clarke
Copy Editor: Lilian Brady
Photo Researcher: Jennifer Atkins
Photo Research Manager: Stella Kupferberg
Illustration: John Balbalis
Cover and Chapter Opening Art: Hudson River Studio
Text Design: Hudson River Studio
Electronic Illustration: Techsetters, Inc.
Compositor: General Graphic Services, Inc.

Anton, *Calculus*, 4th ed. Cover Photo Credits

Front cover (*clockwise from top center*): David Eugene Smith Collection/Columbia University; H. Josse/Art Resource; New York Public Library Picture Collection; Art Matrix/Cornell National Supercomputer Facility; David Eugene Smith Collection/Columbia University.

Back cover (*clockwise from top right*): New York Public Library Picture Collection; David Eugene Smith Collection/Columbia University; The Bettmann Archive; The Bettmann Archive; AIP Neils Bohr Library.

Library of Congress Cataloging in Publication Data:

Anton, Howard.
    Calculus with analytic geometry / Howard Anton, with contributions
from Albert Herr. — 4th ed., brief ed.
      p.   cm.
   Includes index.
   ISBN 0-471-54805-7
    1. Calculus. 2. Geometry, Analytic.   I. Herr, Albert.
II. Title.
QA303.A53 1992b
515'.15--dc20   91-29332
                                                        CIP

Printed and bound by Von Hoffmann Press, Inc.

10 9 8 7 6 5 4

About the Author

Howard Anton was born in Philadelphia, Pennsylvania. He obtained his B.A. from Lehigh University, his M.A. from the University of Illinois, and his Ph.D. from the Polytechnic Institute of Brooklyn, all in mathematics. In the early 1960s he worked for Burroughs Corporation and Avco Corporation at Cape Canaveral, Florida, where he was involved with missile tracking problems for the manned space program. In 1968 he joined the Mathematics Department at Drexel University, where he taught full time until 1983. Since that time he has been an adjunct professor at Drexel and has devoted the majority of his time to textbook writing, development of pedagogical software, and activities for mathematical associations.

He has published numerous research papers in Functional Analysis, Approximation Theory, and Topology, as well as pedagogical papers on applications of mathematics. He is best known for his textbooks in mathematics, which are among the most widely used in the world. There are currently nearly seventy versions of his books including translations into Spanish, Arabic, Portuguese, Italian, Indonesian, and Japanese.

Professor Anton was President of the EPADEL Section of the Mathematical Association of America and served on the Board of Governors of that organization. He currently lives in Cherry Hill, New Jersey with his wife, Pat, and his three children, Brian, David, and Lauren. He enjoys traveling and is an avid photographer.

To
My Wife Pat
My Children Brian, David, and Lauren
My Mother Shirley

Preface

I am gratified that the third edition of this text continued the tradition of its predecessors as the most widely used textbook in calculus and that after twelve years in print users and reviewers have continued to praise the clarity of the exposition.

We are now on the threshold of major changes in the way calculus and, indeed, all of mathematics will be taught. Fueled by rapid advances in technology and a reevaluation of traditional course content, we are entering a period of exploration in the teaching of calculus. This new edition reflects a clear, but cautious, commitment to the newer visions of calculus. My goal in this revision is to provide instructors with all of the *supplemental* resources required to begin experimenting with new ideas and new technology within the framework of the traditional course structure.

Users of the earlier editions will be able to ease comfortably into the new edition, but will sense a more contemporary philosophy in the exposition and exercises, with increased attention to numerical computations and estimation, as well as flexibility for varying the order and emphasis of topics. For instructors on the cutting edge of calculus reform, there is an extensive array of supplements that use new technology: symbolic algebra software, calculators, Hypercard stacks, and even CD-ROM.

Although there are many changes in this new edition, I remain committed to the philosophy that the heart and soul of a fine textbook is the clarity of its exposition; there are very few sections in this new edition that have not been polished and refined. My goal, as in earlier editions, is to *teach* the material in the clearest possible way with a level of rigor that is suitable for the mainstream calculus audience.

☐ FEATURES

About the Brief Edition

This text is designed for a standard course in the calculus of one variable. The first thirteen chapters are identical to those in *Calculus with Analytic Geometry*, Fourth Edition; the fourteenth chapter, which is concerned with second-order differential equations, is identical to Chapter 19 of that text.

Precalculus Material

Because of the vast amount of material to be covered, it is desirable to spend as little time as possible on precalculus topics. However, the reality is that freshmen have a wide variety of educational backgrounds and different levels of preparedness. Therefore, I have included an optional first chapter devoted to precalculus material. It is written in enough detail for the instructor to feel confident in moving quickly through these preliminaries.

Trigonometry

Deficiencies in trigonometry plague many students throughout the entire calculus sequence. Therefore, I have included a substantial trigonometry review in the appendix. It is more detailed than most such reviews because I feel that students will appreciate having this material readily available. The review is broken into two sections: the first to be mastered before reading Section 1.4 and the second before Section 2.8.

Rigor

The challenge of writing a good calculus textbook is to strike the right balance of rigor and clarity. It is my goal in this text to present precise mathematics to the fullest extent possible for the freshman audience. The theory is presented in a style tailored for beginners, but in those places where precision conflicts with clarity, the exposition is designed for clarity. However, I believe it to be of fundamental importance for the student to understand the distinction between a careful proof and an informal argument; thus, when the arguments are informal, I make this clear to the reader. Theory involving $\delta\epsilon$-arguments has been placed in optional sections so that it can be bypassed if desired.

Illustrations

This text is more heavily illustrated than most calculus books because illustrations play a special role in my philosophy of pedagogical exposition. Freshmen have great difficulty in reading mathematics and extracting concepts from mathematical formulas, yet they can often grasp a concept immediately when the right picture is presented. This is not surprising since mathematics is a language that must be learned. Until the language is mastered one cannot hope to understand the ideas that the language conveys. I use illustrations to help teach the language of mathematics by supporting the theorems and formulas with illustrations that help the reader understand the concepts embodied in the mathematical symbolism. In keeping with current trends, I have used modern four-color typography with a consistency of style and color selection throughout the text.

Exercises

The exercises in this new edition have been extensively modified and expanded. Each exercise set begins with routine drill problems and progresses gradually toward problems of greater difficulty. I have tried to construct well-balanced exercise sets with more variety than is available in most calculus texts. Each chapter ends with a set of supplementary exercises to help the student consolidate his or her mastery of the chapter. In addition, many of the exercise sets now contain problems requiring a calculator as well as so-called "spiral" problems (problems that use concepts from earlier chapters). Answers are given to odd-numbered problems, and at the beginning of each exercise set there is a list of those exercises that require a calculator; these are labeled with the icon $\boxed{C}$.

Order of Presentation

The topics are ordered in a fairly standard way with an early introduction of trigonometric functions (Chapter 3). Users should have no difficulty in permuting sections or chapters in any reasonable way.

Early Differential Equations Option

Of special note is the placement of the section on first-order linear and separable differential equations in the chapter on logarithmic and exponential functions (Chapter 7). This allows us to give some nice applications of logarithms and exponentials immediately and also helps meet the needs of those engineering and science students who will need this material in courses taken concurrently with calculus. This section can be omitted or deferred with no difficulty.

Early Logarithm Option

Instructors who want to teach logarithms early will have no trouble doing so. There are two possibilities: The basic material on logarithms (Sections 7.1–7.3) can be covered without modification at the end of Chapter 5 (before the chapter on applications of the definite integral); alternatively, Sections 7.1 to 7.3 can be covered immediately after Section 5.3 (before the definite integral) by omitting Definition 7.3.1 and treating integral formula (6) of Section 7.3 as a consequence of derivative formulas (1) and (5) of that section. These variations are appropriate when it is desirable to cover transcendental functions and the definite integral in the first semester.

Brief Techniques of Integration Option

There is a new introductory section on tables of integrals. By covering this section and utilizing the Table of Integrals printed on the endpapers of the text, instructors can reduce the time devoted to techniques of integration.

Reviewing and Class Testing

This edition is the outgrowth of twelve years of classroom use and input from hundreds of students and instructors who have written to me with constructive comments over the years. In addition, the exposition and exercise sets have been refined and polished in response to input from a team of professors who carefully critiqued every single line of the third edition as they used it in the classroom. It is my hope that this painstaking attention to pedagogical detail has resulted in a book that surpasses the earlier editions for clarity and accuracy.

☐ CHANGES FOR THE FOURTH EDITION

As stated earlier, the exposition has been improved and polished throughout the text. In addition, some of the more salient changes for this new edition are as follows:

- Major revision of the exercise sets: nearly 1000 new exercises with added emphasis on applications and the use of the calculator; many new higher-level exercises requiring multistep thinking; and many spiral exercises that build on material from earlier sections.

- A new section on using integral tables, which allows for a briefer presentation of techniques of integration.

- Major revision of the material on numerical integration with heavier emphasis on error estimation.

- More emphasis on understanding information conveyed by graphs, taking into account that the graphs themselves can be generated on a calculator or computer.

- Major revision of the material on logarithms for greater clarity and added flexibility for moving the material to an earlier position in the text if the curriculum requires it.

- Revision and addition of many examples to coordinate better with exercises and to reflect increased emphasis on numerical considerations.

HOWARD ANTON

Supplements

Graphing Calculator Supplement
The following supplement contains a collection of problems that are intended to be solved on a graphing calculator. The problems are not specific to a particular brand of calculator. The manual also provides an overview of the types of calculators available, general instructions for calculator use, and a discussion of the numerical pitfalls of roundoff and truncation error.

- *Discovering Calculus with Graphing Calculators*
  Joan McCarter, *Arizona State University*
  ISBN: 0-471-55609-2

Symbolic Algebra Supplements
The following supplements are collections of problems for the student to solve. Each contains a brief set of instructions for using the software as well as an extensive set of problems utilizing the capabilities of the software. The problems range from very basic to those involving real-world applications.

- *Discovering Calculus with DERIVE*
  Jerry Johnson and Benny Evans, *Oklahoma State University*
  ISBN: 0-471-55155-4

- *Discovering Calculus with MAPLE*
  Kent Harris, *Western Illinois University*
  ISBN: 0-471-55156-2

- *Discovering Calculus with MATHEMATICA*
  Bert Braden, Don Krug, Steve Wilkinson, *Northern Kentucky University*
  ISBN: 0-471-53969-4

Macintosh Hypercard Stacks
This supplement is a set of Hypercard 2.0 stacks that are designed primarily for lecture demonstrations. Each stack is self-contained, but all have a common interface. There is an initial set of six stacks (with more to be developed). The six initial stacks are concerned with limits, Newton's Method, the definition of the definite integral, convergence of Taylor polynomials, polar coordinates, and mathematics history.

- *HYPERCALCULUS*
  Chris Rorres and Loren Argabright, *Drexel University*
  ISBN: 0-471-57052-4

CD-ROM Version of Calculus for IBM Compatible Computers

This supplement is an electronic version of the entire text and the *Student's Solutions Manual* on compact disk for use with IBM compatible computers equipped with a CD-ROM drive. All text material and illustrations are stored on disk with an interconnecting network of hyperlinks that allows the student to access related items that do not appear in proximity in the text. A complete keyword glossary and step-by-step discussions of key concepts are also included.

- *CD-ROM Version of Anton Calculus/4E: An Electronic Study Environment*
  Developed by Electric Book Company
  ISBN: 0-471-55803-6

Linear Algebra Supplement

The following supplement gives a brief introduction to those aspects of linear algebra that are of immediate concern to the calculus student. The emphasis is on methods rather than proof.

- *Linear Algebra Supplement to Accompany Anton Calculus/4E*
  ISBN: 0-471-56893-7

Student Study Resources

The following supplement is a tutorial, review, and study aid for the student.

- *The Calculus Companion to Accompany Anton Calculus/4E, Vols. 1 and 2*
  William H. Barker and James E. Ward, *Bowdoin College*
  ISBN: 0-471-55139-2 (Volume 1); ISBN: 0-471-55138-4 (Volume 2)

The following supplement contains detailed solutions to all odd-numbered exercises.

- *Student's Solutions Manual to Accompany Anton Calculus Brief/4E*
  Albert Herr, *Drexel University*
  ISBN: 0-471-55140-6

Calculus Video

This $2\frac{1}{2}$-hour tutorial covers some of the central topics in calculus: limits, differentiation, Newton's Method, the Mean-Value Theorem, integration, infinite series, and Taylor polynomials. This video, which is not intended to replace lectures or text material, is provided as an aid for students who need additional work on key topics. The video consists of five 30-minute units, each of which is divided into segments of 10 or 15 minutes. The presentations make heavy use of television graphics and incorporate video versions of the Hypercalculus stacks.

- *Calculus Video: A Tutorial to Accompany Anton Calculus/4E*
  Dr. Joby Anthony, *University of Central Florida*
  ISBN: 0-471-57049-4

Resources for the Instructor

There is a resource package for the instructor that includes hard copy and electronic test banks and other materials. These can be obtained by writing on your institutional letterhead to Susan Elbe, Senior Marketing Manager, John Wiley & Sons, Inc., 605 Third Avenue, New York, N.Y., 10158-0012.

Acknowledgments

It has been my good fortune to have the advice and guidance of many talented people, whose knowledge and skills have enhanced this book in many ways. For their valuable help I thank:

Reviewers and Contributors to Earlier Editions
Edith Ainsworth, *University of Alabama*
David Armacost, *Amherst College*
Larry Bates, *University of Calgary*
Marilyn Blockus, *San Jose State University*
David Bolen, *Virginia Military Institute*
George W. Booth, *Brooklyn College*
Mark Bridger, *Northeastern University*
John Brothers, *Indiana University*
Robert C. Bueker, *Western Kentucky University*
Robert Bumcrot, *Hofstra University*
Chris Christensen, *Northern Kentucky University*
David Cohen, *University of California, Los Angeles*
Michael Cohen, *Hofstra University*
Robert Conley, *Precision Visuals*
A. L. Deal, *Virginia Military Institute*
Charles Denlinger, *Millersville State College*
Dennis DeTurck, *University of Pennsylvania*
Jacqueline Dewar, *Loyola Marymount University*
Irving Drooyan, *Los Angeles Pierce College*
Hugh B. Easler, *College of William and Mary*
Joseph M. Egar, *Cleveland State University*
Garret J. Etgen, *University of Houston*
James H. Fife, *University of Richmond*
Barbara Flajnik, *Virginia Military Institute*
Nicholas E. Frangos, *Hofstra University*
Katherine Franklin, *Los Angeles Pierce College*
Michael Frantz, *University of La Verne*
William R. Fuller, *Purdue University*
Raymond Greenwell, *Hofstra University*
Gary Grimes, *Mt. Hood Community College*
Jane Grossman, *University of Lowell*

Michael Grossman, *University of Lowell*
Douglas W. Hall, *Michigan State University*
Nancy A. Harrington, *University of Lowell*
Albert Herr, *Drexel University*
Peter Herron, *Suffolk County Community College*
Robert Higgins, *Quantics Corporation*
Louis F. Hoelzle, *Bucks County Community College*
Harvey B. Keynes, *University of Minnesota*
Paul Kumpel, *SUNY, Stony Brook*
Leo Lampone, *Quantics Corporation*
Bruce Landman, *Hofstra University*
Benjamin Levy, *Lexington H.S., Lexington, Mass.*
Phil Locke, *University of Maine, Orono*
Stanley M. Lukawecki, *Clemson University*
Nicholas Macri, *Temple University*
Melvin J. Maron, *University of Louisville*
Thomas McElligott, *University of Lowell*
Judith McKinney, *California State Polytechnic University, Pomona*
Joseph Meier, *Millersville State College*
David Nash, *VP Research, Autofacts, Inc.*
Mark A. Pinsky, *Northeastern University*
William H. Richardson, *Wichita State University*
David Sandell, *U.S. Coast Guard Academy*
Donald R. Sherbert, *University of Illinois*
Wolfe Snow, *Brooklyn College*
Norton Starr, *Amherst College*
Richard B. Thompson, *The University of Arizona*
William F. Trench, *Trinity University*
Walter W. Turner, *Western Michigan University*
Richard C. Vile, *Eastern Michigan University*
James Warner, *Precision Visuals*
Candice A. Weston, *University of Lowell*
Yihren Wu, *Hofstra University*
Richard Yuskaitis, *Precision Visuals*

Content Reviewers
The following people critiqued the previous edition and recommended many
of the changes that found their way into the new edition:
Ray Boersma, *Front Range Community College*
Terrance Cremeans, *Oakland Community College*
Tom Drouet, *East Los Angeles College*
Ken Dunn, *Dalhousie University*
Kent Harris, *Western Illinois University*
Maureen Kelly, *Northern Essex Community College*
Richard Nowakowski, *Dalhousie University*
Robert Phillips, *University of South Carolina at Aiken*
David Randall, *Oakland Community College*
George Shapiro, *Brooklyn College*
Ian Spatz, *Brooklyn College*

Accuracy Reviewers of the Fourth Edition

The following people worked with me in reading the manuscript, galley proofs, and page proofs for mathematical accuracy. Their perceptive comments have improved the text immeasurably:

Irl C. Bivens, *Davidson College*
Hannah Clavner, *Drexel University*
Garret J. Etgen, *University of Houston*
Daniel Flath, *University of South Alabama*
Susan L. Friedman, *Bernard M. Baruch College*, *CUNY*
Evelyn Weinstock, *Glassboro State College*

Problem Contributors

The following people contributed numerous new and imaginative problems to the text:

Irl C. Bivens, *Davidson College*
Daniel Bonar, *Denison University*
James Caristi, *Valparaiso University*
G. S. Gill, *Brigham Young University*
Albert Herr, *Drexel University*
Herbert Kasube, *Bradley University*
Phil Kavanaugh, *Illinois Wesleyan University*
John Lucas, *University of Wisconsin–Oshkosh*
Ron Moore, *Ryerson Polytechnical Institute*
Barbara Moses, *Bowling Green State University*
David Randall, *Oakland Community College*
Jean Springer, *Mount Royal College*
Peter Waterman, *Northern Illinois University*

Answers, Solutions, and Index

The following people assisted with the critically important job of preparing the index and obtaining answers for the text and solutions for the *Student's Solutions Manual*:

Harry N. Bixler, *Bernard M. Baruch College*, *CUNY*
Hannah Clavner, *Drexel University*
Michael Dagg
Stephen L. Davis, *Davidson College*
Susan L. Friedman, *Bernard M. Baruch College*, *CUNY*
Shirley Wakin, *University of New Haven*

Computer Illustrations

Many of the illustrations involving two-dimensional mathematical graphs were generated electronically (including the four-color separation) by using software developed by Techsetters, Inc. A number of people contributed to the development of the software:

Edward A. Burke, *Hudson River Studio*—color separation
John R. DiStefano (student), *Drexel University*—programmer
Theo Gray, *Wolfram Research*—technical assistance
Julie Varbalow (student), *Skidmore College*—testing

Some of the three-dimensional surfaces were generated by Mark Bridger of Northeastern University using the SURFS software package that he developed.

The Wiley Staff

There are so many people at Wiley who contributed in special ways to this project that it is impossible to mention them all. However, a word of appreciation is due to the people I worked with very closely: Joan Carrafiello, Lucille Buonocore, Suzanne Ingrao, Steve Kraham, Susan Elbe, and my editor Robert Macek.

Other Supplementary Materials

The following people provided additional material for tests and other supplements:

Pasquale Condo, *University of Lowell*
Maureen Kelley, *Northern Essex Community College*
Catherine H. Pirri, *Northern Essex Community College*

Special Contributions

I owe an enormous debt of gratitude to Albert Herr of Drexel University, who worked so closely with me on this edition that his name appears on the title page. Professor Herr is a gifted, award-winning teacher of mathematics with years of experience in the calculus classroom. His careful attention to technical detail and his dogged challenges to the clarity of virtually every line of exposition fostered hours of debate over pedagogy and helped bring this edition to a level of quality I could not have achieved alone. Professor Herr is also the source for many of the imaginative new exercises in this edition. I feel fortunate that Al was by my side in preparing this revision.

If I had the power to grant an honorary degree in mathematics, I would give it to my assistant, Mary Parker, who worked on virtually every aspect of this text—from the tedious task of xeroxing tens of thousands of pages to the technical tasks of accuracy checking and coordination. Her warm sense of humor and dedication to the quality of this book were a constant source of inspiration.

H. A.

Contents

Introduction

Calculus is the mathematical tool used to analyze changes in physical quantities. It was developed in the seventeenth century to study four major classes of scientific and mathematical problems of the time:

1. Find the tangent line to a curve at a point.

2. Find the length of a curve, the area of a region, and the volume of a solid.

3. Find the maximum or minimum value of a quantity—for example, the maximum and minimum distances of a planet from the sun, or the maximum range attainable for a projectile by varying its angle of fire.

4. Given a formula for the distance traveled by a body in any specified amount of time, find the velocity and acceleration of the body at any instant. Conversely, given a formula that specifies the acceleration or velocity at any instant, find the distance traveled by the body in a specified period of time.

These problems were attacked by the greatest minds of the seventeenth century, culminating in the crowning achievements of Gottfried Wilhelm Leibniz and Isaac Newton—the creation of calculus.

Gottfried Wilhelm Leibniz (1646–1716)

This gifted genius was one of the last people to have mastered most major fields of knowledge—an impossible accomplishment in our own era of specialization. He was an expert in law, religion, philosophy, literature, politics, geology, metaphysics, alchemy, history, and mathematics.

Leibniz was born in Leipzig, Germany. His father, a professor of moral philosophy at the University of Leipzig, died when Leibniz was six years old. The precocious boy then gained access to his father's library and began reading voraciously on a wide range of subjects, a habit that he maintained throughout his life. At age 15 he entered the University of Leipzig as a law student and by the age of 20 received a doctorate from the University of Altdorf. Subsequently, Leibniz followed a career in law and international politics, serving as counsel to kings and princes.

During his numerous foreign missions, Leibniz came in contact with outstanding mathematicians and scientists who stimulated his interest in mathematics—most notably, the physicist Christian Huygens. In mathematics Leibniz was self-taught, learning the subject by reading papers and journals. As a result of this fragmented mathematical education, Leibniz often duplicated the results of others, and this ultimately led to a raging conflict as to whether he or Isaac Newton should be regarded as the inventor of calculus. The argument over this question engulfed the scientific circles of England and Europe, with most scientists on the continent supporting Leibniz and those in England supporting Newton. The conflict was unfortunate, and both sides suffered in the end. The continent lost the benefit of Newton's discoveries in astronomy and physics for more than 50 years, and for a long period England became a second-rate country mathematically because its mathematicians were hampered by Newton's inferior calculus notation. It is of interest to note that Newton and Leibniz never went to the lengths of vituperation of their advocates—both were sincere admirers of each other's work. The fact is that both men invented calculus independently. Leibniz invented it 10 years after Newton, in 1685, but he published his results 20 years before Newton published his own work on the subject.

Leibniz never married. He was moderate in his habits, quick-tempered, but easily appeased, and charitable in his judgment of other people's work. In spite of his great achievements, Leibniz never received the honors showered on Newton, and he spent his final years as a lonely embittered man. At his funeral there was one mourner, his secretary. An eyewitness stated, "He was buried more like a robber than what he really was—an ornament of his country."

Isaac Newton (1642–1727)

Newton was born in the village of Woolsthorpe, England. His father died before he was born and his mother raised him on the family farm. As a youth he

Gottfried Leibniz
(Culver Pictures)

Isaac Newton
(Culver Pictures)

showed little evidence of his later brilliance, except for an unusual talent with mechanical devices—he apparently built a working water clock and a toy flour mill powered by a mouse. In 1661 he entered Trinity College in Cambridge with a deficiency in geometry. Fortunately, Newton caught the eye of Isaac Barrow, a gifted mathematician and teacher. Under Barrow's guidance Newton immersed himself in mathematics and science, but he graduated without any special distinction. Because the Plague was spreading rapidly through London, Newton returned to his home in Woolsthorpe and stayed there during the years of 1665 and 1666. In those two momentous years the entire framework of modern science was miraculously created in Newton's mind—he discovered calculus, recognized the underlying principles of planetary motion and gravity, and determined that "white" sunlight was composed of all colors, red to violet. For some reason he kept his discoveries to himself. In 1667 he returned to Cambridge to obtain his Master's degree and upon graduation became a teacher at Trinity. Then in 1669 Newton succeeded his teacher, Isaac Barrow, to the Lucasian chair of mathematics at Trinity, one of the most honored chairs of mathematics in the world. Thereafter, brilliant discoveries flowed from Newton steadily. He formulated the law of gravitation and used it to explain the motion of the moon, the planets, and the tides; he formulated basic theories of light, thermodynamics, and hydrodynamics; and he devised and constructed the first modern reflecting telescope.

Throughout his life Newton was hesitant to publish his major discoveries, revealing them only to a select circle of friends, perhaps because of a fear of criticism or controversy. In 1687, only after intense coaxing by the astronomer, Edmond Halley (Halley's comet), did Newton publish his masterpiece, *Philosophiae Naturalis Principia Mathematica* (The Mathematical Principles of Natural Philosophy). This work is generally considered to be the most important and influential scientific book ever written. In it Newton explained the workings of the solar system and formulated the basic laws of motion which to this day are fundamental in engineering and physics. However, not even the pleas of his friends could convince Newton to publish his discovery of calculus. Only after Leibniz published his results did Newton relent and publish his own work on calculus.

After 35 years as a professor, Newton suffered depression and a nervous breakdown. He gave up research in 1695 to accept a position as warden and later master of the London mint. During the 25 years that he worked at the mint, he did virtually no scientific or mathematical work. He was knighted in 1705 and on his death was buried in Westminster Abbey with all the honors his country could bestow. It is interesting to note that Newton was a learned theologian who viewed the primary value of his work to be its support of the existence of God. Throughout his life he worked passionately to date biblical events by relating them to astronomical phenomena. He was so consumed with this passion that he spent years searching the Book of Daniel for clues to the end of the world and the geography of hell.

Newton described his brilliant accomplishments as follows: "I seem to have been only like a boy playing on the seashore and diverting myself in now and then finding a smoother pebble or prettier shell than ordinary, whilst the great ocean of truth lay all undiscovered before me."

Calculus
with Analytic Geometry

Brief Edition

1
Coordinates, Graphs, Lines

René Descartes   (1596-1650)

■ 1.1   REAL NUMBERS, INTERVALS, AND INEQUALITIES (A REVIEW)

> *Because numbers are the foundation of all mathematics, it is important to be familiar with the various kinds of numbers and the differences between them. In this section we shall review the basic facts and terminology relating to real numbers.*

□ **CLASSIFICATION OF NUMBERS**

The simplest numbers are the *natural numbers:*

$$1, 2, 3, 4, 5, \ldots$$

The natural numbers form a subset of a larger class of numbers called the *integers:*

$$\ldots, -4, -3, -2, -1, 0, 1, 2, 3, 4, \ldots$$

These consist of the positive integers, the negative integers, and zero.

The integers in turn are a subset of a still larger class of numbers called the *rational numbers*. With the exception that division by zero is ruled out, the rational numbers are formed by taking ratios of integers. Examples are

$$\frac{2}{3}, \frac{7}{5}, \frac{6}{1}, \frac{0}{9}, -\frac{5}{2} \left(= \frac{-5}{2} = \frac{5}{-2} \right)$$

Observe that every integer is also a rational number because an integer p can be written as the ratio

$$p = \frac{p}{1}$$

The early Greeks believed that the size of every physical quantity could, in theory, be represented by a rational number. They reasoned that the size of a physical quantity must consist of a certain whole number of units plus some fraction m/n of an additional unit. This idea was shattered in the fifth century B.C. by Hippasus of Metapontum* who demonstrated the existence of *irrational numbers,* that is, numbers that cannot be expressed as the ratio of integers. Using geometric methods, he showed that the hypotenuse of the triangle in Figure 1.1.1 cannot be expressed as the ratio of integers, thereby proving that $\sqrt{2}$ is an irrational number. Other examples of irrational numbers are

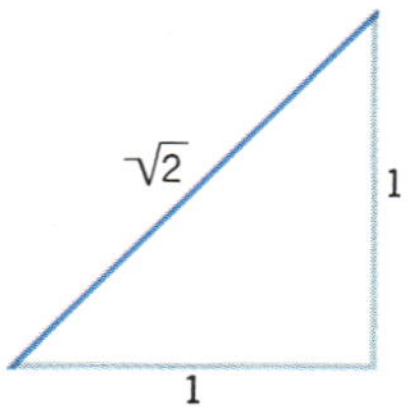

$$\sqrt{3}, \quad \sqrt{5}, \quad 1 + \sqrt{2}, \quad \sqrt[3]{7}, \quad \pi, \quad \cos 19°$$

Figure 1.1.1

---

* HIPPASUS OF METAPONTUM (circa 500 B.C.). A Greek Pythagorean philosopher. According to legend, Hippasus made his discovery at sea and was thrown overboard by fanatic Pythagoreans because his result contradicted their doctrine. The discovery of Hippasus is one of the most fundamental in the entire history of science.

The rational and irrational numbers together comprise a larger class of numbers, called *real numbers* or sometimes the *real number system*.

☐ **DIVISION BY ZERO**

In computations with real numbers, division by zero is never allowed because a relationship of the form $y = p/0$ would imply that

$$0 \cdot y = p$$

If p is different from zero, this equation is contradictory; and if p is equal to zero, this equation is satisfied by any number y, so the ratio $0/0$ does not have a unique value—a situation that is mathematically unsatisfactory. For these reasons such symbols as

$$\frac{p}{0} \quad \text{and} \quad \frac{0}{0}$$

are not assigned a value; they are said to be *undefined*.

☐ **COMPLEX NUMBERS**

Because the square of a real number cannot be negative, the equation

$$x^2 = -1$$

has no solutions in the real number system. In the eighteenth century mathematicians remedied this problem by inventing a new number, which they denoted by

$$i = \sqrt{-1}$$

and which they defined to have the property $i^2 = -1$. This, in turn, led to the development of the *complex numbers,* which are numbers of the form

$$a + bi$$

where a and b are real numbers. Some examples are

$$2 + 3i \qquad 3 - 4i \qquad 6i \qquad \tfrac{2}{3}$$
$$[a = 2, b = 3] \quad [a = 3, b = -4] \quad [a = 0, b = 6] \quad [a = \tfrac{2}{3}, b = 0]$$

Observe that every real number a is also a complex number because it can be written as

$$a = a + 0i$$

Thus, the real numbers are a subset of the complex numbers. Those complex numbers that are not real numbers are called *imaginary numbers*.

In this text, we shall be concerned primarily with real numbers; however, complex numbers will arise in the course of solving equations. For example, the solutions of the quadratic equation

$$ax^2 + bx + c = 0 \tag{1}$$

which are given by the *quadratic formula*

$$x = \frac{-b \pm \sqrt{b^2 - 4ac}}{2a}$$

are imaginary if the quantity $b^2 - 4ac$ inside the radical is negative. [Recall that $b^2 - 4ac$ is called the *discriminant* of (1).]

The hierarchy of numbers is summarized in Figure 1.1.2.

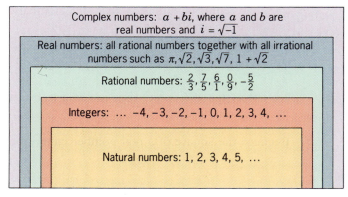

Complex numbers: $a + bi$, where a and b are real numbers and $i = \sqrt{-1}$

Real numbers: all rational numbers together with all irrational numbers such as $\pi, \sqrt{2}, \sqrt{3}, \sqrt{7}, 1 + \sqrt{2}$

Rational numbers: $\frac{2}{3}, \frac{7}{5}, \frac{6}{1}, \frac{0}{9}, -\frac{5}{2}$

Integers: ... $-4, -3, -2, -1, 0, 1, 2, 3, 4,$ ...

Natural numbers: $1, 2, 3, 4, 5,$ ...

Figure 1.1.2

☐ **DECIMAL REPRESENTATION OF REAL NUMBERS**

Rational and irrational numbers can be distinguished by their decimal representations. Rational numbers have decimals that are *repeating*, by which we mean that there is some point in the decimal representation at which the digits that follow consist of a fixed block of integers repeated over and over. For example,

$$\frac{4}{3} = 1.333\ldots \qquad \text{3 repeats}$$

$$\frac{3}{11} = .272727\ldots \qquad \text{27 repeats}$$

$$\frac{5}{7} = .714285714285714285\ldots \qquad \text{714285 repeats}$$

Repeating decimals that consist of zeros from some point on are called *terminating decimals*. Some examples are

$$\frac{1}{2} = .50000\ldots$$

$$\frac{12}{4} = 3.0000\ldots$$

$$\frac{8}{25} = .32000\ldots$$

It is usual to omit the repetitive zeros in terminating decimals. For example,

$$\frac{1}{2} = .5, \quad \frac{12}{4} = 3, \quad \frac{8}{25} = .32$$

Moreover, it has become common to denote repeating decimals by writing the repeating digits only once, but with a bar over them to indicate the repetition. For example,

$$\frac{4}{3} = 1.\bar{3}, \quad \frac{3}{11} = .\overline{27}, \quad \frac{5}{7} = .\overline{714285}$$

Rational numbers are represented by repeating decimals, and conversely every repeating decimal represents a rational number. Thus, the *irrational* numbers can be viewed as those real numbers that are represented by *nonrepeating* decimals. For example, the decimal

$$.101001000100001000001 \ldots$$

does not repeat because the number of zeros between the ones keeps growing. Thus, it represents an irrational number.

Irrational numbers cannot be represented with perfect accuracy in decimal notation. For example, π is only approximated by the decimal 3.14. Moreover, no matter how many decimal places we use, even if we compute π to 4800 places, as in Figure 1.1.3, we still have only an approximation to π.

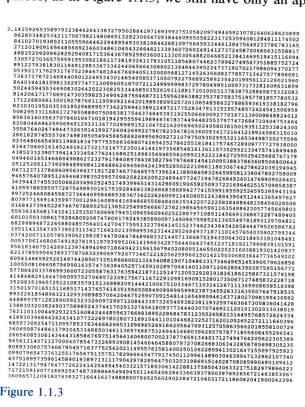

Figure 1.1.3

REMARK.   Beginning mathematics students are sometimes taught to approximate π by 22/7. Note, however, that

$$\frac{22}{7} = 3.\overline{142857}$$

is a rational number whose decimal representation begins to differ from π in the third decimal place.

COORDINATE LINES

In 1637 René Descartes* published a philosophical work called *Discourse on the Method of Rightly Conducting the Reason.* In the back of that book were three appendices that purported to show how the "method" could be applied to concrete examples. The first two appendices were minor works that endeavored to explain the behavior of lenses and the movement of shooting stars. The third appendix, however, was an inspired stroke of genius; it was described by the nineteenth century British philosopher John Stuart Mill as, "The greatest single step ever made in the progress of the exact sciences." In that appendix René Descartes linked together two branches of mathematics, algebra and geometry. Descartes' work evolved into a new subject called *analytic geometry;* it gave a way of describing algebraic formulas by means of geometric curves and, conversely, geometric curves by algebraic formulas.

In analytic geometry, the key step is to establish a correspondence between real numbers and points on a line. This is done by arbitrarily designating one of the two directions along the line to be the *positive direction* and the other to be the *negative direction.* The positive direction is usually marked with an arrowhead as in Figure 1.1.4; for horizontal lines the positive direction is generally taken to the right. A unit of measurement is then chosen and an arbitrary point, called the *origin,* is selected anywhere along the line. The line, the origin, the positive direction, and the unit of measurement define what is called a *coordinate line* or sometimes a *real line.* With each real number we can associate a point on the line as follows:

Figure 1.1.4

*RENÉ DESCARTES (1596–1650). Descartes, a French aristocrat, was the son of a government official. He graduated from the University of Poitiers with a law degree at age 20. After a brief probe into the pleasures of Paris he became a military engineer, first for the Dutch Prince of Nassau and then for the German Duke of Bavaria. It was during his service as a soldier that Descartes began to pursue mathematics seriously and develop his analytic geometry. After the wars, he returned to Paris where he stalked the city as an eccentric, wearing a sword in his belt and a plumed hat. He lived in leisure, seldom arose before 11 A.M., and dabbled in the study of human physiology, philosophy, glaciers, meteors, and rainbows. He eventually moved to Holland, where he published his *Discourse on the Method,* and finally to Sweden where he died while serving as tutor to Queen Christina. Descartes is regarded as a genius of the first magnitude. In addition to major contributions in mathematics and philosophy, he is considered, along with William Harvey, to be a founder of modern physiology.

- Associate with each positive number r the point that is a distance of r units in the positive direction from the origin.
- Associate with each negative number $-r$ the point that is a distance of r units in the negative direction from the origin.
- Associate the origin with the number 0.

The real number corresponding to a point on the line is called the ***coordinate*** of the point.

Example 1 In Figure 1.1.5 we have marked the locations of the points with coordinates -4, -3, -1.75, $-\frac{1}{2}$, $\sqrt{2}$, π, and 4. The locations of π and $\sqrt{2}$, which are approximate, were obtained from their decimal approximations, $\pi \approx 3.14$ and $\sqrt{2} \approx 1.41$. ◀

Figure 1.1.5

It is evident from the way in which real numbers and points on a coordinate line are related that each real number corresponds to a single point and each point corresponds to a single real number. To describe this fact we say that the real numbers and the points on a coordinate line are in ***one-to-one correspondence.***

☐ **ORDER PROPERTIES**

If we traverse a coordinate line in the positive direction, then the numbers increase in size. This is a reflection of the fact that the real numbers are ***ordered;*** that is, given any two numbers, a and b, exactly one of the following is true:

> a is less than b
> b is less than a
> a is equal to b

To describe the relative size of two real numbers, we use the order symbols $<$ (less than) and $\leq$ (less than or equal to), which are called ***inequalities.*** These symbols are defined as follows:

1.1.1 DEFINITION.   If a and b are real numbers, then

$a < b$ means $b - a$ is positive
$a \leq b$ means $a < b$ or $a = b$

The inequality $a < b$, which is read "*a is less than b,*" can also be written as $b > a$, which is read "*b is greater than a*"; and the inequality $a \leq b$, which is read "*a is less than or equal to b,*" can also be written as $b \geq a$, which is read "*b is greater than or equal to a.*"

In Table 1.1.1 we have summarized the geometric interpretations of the basic inequality symbols.

Table 1.1.1

INEQUALITY	GEOMETRIC INTERPRETATION	ILLUSTRATION
$a < b$ or $b > a$	a is to the left of b on a coordinate line.	
$a \leq b$ or $b \geq a$	a is to the left of b or coincides with b on a coordinate line.	
$0 < a$ or $a > 0$	a is to the right of the origin on a coordinate line.	
$a < 0$ or $0 > a$	a is to the left of the origin on a coordinate line.	

> **1.1.2 DEFINITION.**   If a, b, and c are real numbers, we shall write
>
> $$a < b < c$$
>
> when $a < b$ and $b < c$.

Geometrically, $a < b < c$ states that on a coordinate line, b is to the right of a and c is to the right of b (Figure 1.1.6).

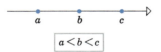

Figure 1.1.6

The symbol $a < b \leq c$ means $a < b$ and $b \leq c$. We leave it to the reader to deduce the meanings of such symbols as

$$a \leq b < c, \quad a \leq b \leq c, \quad \text{and} \quad a < b < c < d$$

Example 2 The following inequalities are all correct:

$$3 < 8, \quad -7 < 1.5, \quad -12 \leq -\pi, \quad 5 \leq 5, \quad 0 \leq 2 \leq 4$$
$$8 \geq 3, \quad 1.5 > -7, \quad -\pi > -12, \quad 5 \geq 5, \quad 3 > 0 > -1 \quad \blacktriangleleft$$

REMARK.   To distinguish verbally between numbers that satisfy $a \geq 0$ and those that satisfy $a > 0$, we shall call a **_nonnegative_** if $a \geq 0$ and **_positive_** if $a > 0$. Thus, a nonnegative number is either positive or zero.

The following properties of inequalities are frequently used in calculus. We omit the proofs.

1.1.3 THEOREM. *Let a, b, c, and d be real numbers.*

(a) *If $a < b$ and $b < c$, then $a < c$.*

(b) *If $a < b$, then $a + c < b + c$ and $a - c < b - c$.*

(c) *If $a < b$, then $ac < bc$ when c is positive and $ac > bc$ when c is negative.*

(d) *If $a < b$ and $c < d$, then $a + c < b + d$.*

(e) *If a and b are both positive or both negative and $a < b$, then $1/a > 1/b$.*

REMARK.   These five properties remain true if $<$ and $>$ are replaced by $\leq$ and $\geq$, respectively.

If we call the direction in which an inequality points its *sense*, then parts (*b*)–(*e*) of this theorem can be paraphrased informally as follows:

(b) *The sense of an inequality is unchanged if the same number is added to or subtracted from both sides.*

(c) *The sense of an inequality is unchanged if both sides are multiplied by the same positive number, but the sense is reversed if both sides are multiplied by the same negative number.*

(d) *Inequalities with the same sense can be added.*

(e) *If both sides of an inequality have the same sign, then the sense of the inequality is reversed by taking the reciprocal of each side.*

Example 3   The statements in Theorem 1.1.3 are illustrated in Table 1.1.2.

Table 1.1.2

STARTING INEQUALITY	OPERATION	RESULTING INEQUALITY
$-2 < 6$	Add 7 to both sides.	$5 < 13$
$-2 < 6$	Subtract 8 from both sides.	$-10 < -2$
$-2 < 6$	Multiply both sides by 3.	$-6 < 18$
$-2 < 6$	Multiply both sides by -3.	$6 > -18$
$3 < 7$	Multiply both sides by 4.	$12 < 28$
$3 < 7$	Multiply both sides by -4.	$-12 > -28$
$3 < 7$	Take reciprocals of both sides.	$\frac{1}{3} > \frac{1}{7}$
$-8 < -6$	Take reciprocals of both sides.	$-\frac{1}{8} > -\frac{1}{6}$
$4 < 5, -7 < 8$	Add corresponding sides.	$-3 < 13$

◀

☐ **INTERVALS**

We shall assume in the following discussion that you are familiar with the concept of a set and understand the meaning of such symbols as $a \in A, a \notin A$, $\varnothing$ (the empty set), $A \cap B, A \cup B, A = B$, and $A \subset B$. If this is not the case, then read the review of sets in Appendix A.

One way to specify a set is to list its members between braces. Thus, the set of all positive integers less than 5 can be written as

$$\{1, 2, 3, 4\}$$

and the set of all positive even integers can be written as

$$\{2, 4, 6, \ldots\}$$

where the dots are used to indicate that only some of the members are listed explicitly and the rest can be obtained by continuing the pattern.

When it is inconvenient or impossible to list the members, it is sufficient to define a set by stating a property common only to its members. For example,

- the set of all rational numbers
- the set of all real numbers x such that $2x^2 - 4x + 1 = 0$
- the set of all real numbers between 2 and 3

As an alternative to such verbal descriptions of sets, we can use the notation

$$\{x: \underline{\qquad}\}$$

which is read, "the set of all x such that _____." Where the line is placed, one would state a property that specifies the set. Thus,

$$\{x:x \text{ is a rational number}\}$$

is read, "the set of all x such that x is a rational number."

When it is clear that the members of a set are real numbers, we will omit the reference to this fact. Thus, the three sets marked above would be denoted by

$$\{x:x \text{ is rational}\}$$
$$\{x:2x^2 - 4x + 1 = 0\}$$
$$\{x:2 < x < 3\}$$

Of special interest are sets of real numbers called **intervals**. Geometrically, an interval is a line segment. For example, if $a < b$, then the **closed interval** from a to b is the set

$$\{x:a \le x \le b\}$$

and the **open interval** from a to b is the set

$$\{x:a < x < b\}$$

These sets are pictured in Figure 1.1.7.

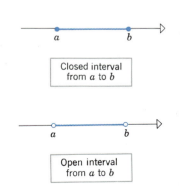

Closed interval from a to b

Open interval from a to b

Figure 1.1.7

A closed interval includes both its endpoints (indicated by solid dots in Figure 1.1.7), while an open interval does not include the endpoints (indicated by open dots in Figure 1.1.7). Closed and open intervals are usually denoted by the symbols $[a, b]$ and (a, b), respectively, so

$$[a, b] = \{x : a \leq x \leq b\}$$
$$(a, b) = \{x : a < x < b\}$$

A square bracket such as [or] indicates that the endpoint is included, while a rounded bracket such as (or) indicates that the endpoint is not included.

An interval can include one endpoint and not the other. Such intervals are called **half-open** (or sometimes **half-closed**). For example,

$$[a, b) = \{x : a \leq x < b\}$$
$$(a, b] = \{x : a < x \leq b\}$$

An interval can extend indefinitely in either the positive direction, the negative direction, or both. Some examples are shown in Figure 1.1.8. The intervals in that figure are denoted by

$$(a, +\infty) = \{x : x > a\}$$
$$[a, +\infty) = \{x : x \geq a\}$$
$$(-\infty, b) = \{x : x < b\}$$
$$(-\infty, b] = \{x : x \leq b\}$$
$$(-\infty, +\infty) = \{x : x \text{ is any real number}\}$$

In this notation the symbols $-\infty$ and $+\infty$ do not represent numbers; the $+\infty$ ("positive infinity") indicates that the interval extends indefinitely in the positive direction, and the $-\infty$ ("negative infinity") that it extends indefinitely in the negative direction. Intervals of infinite extent are called **infinite intervals,** and intervals of finite extent are called **finite intervals.**

Infinite intervals of the form $[a, +\infty)$ or $(-\infty, b]$ are considered to be **closed** because they contain their endpoint, and those of the form $(-\infty, a)$ or $(b, +\infty)$ are considered to be **open** because they do not. The interval $(-\infty, +\infty)$ has no endpoints; it is regarded to be both open and closed.

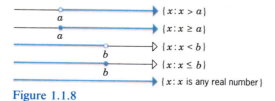

Figure 1.1.8

Example 4   Some specific intervals and their notations are shown in Table 1.1.3.

Table 1.1.3

INTERVAL NOTATION	SET NOTATION	GEOMETRIC PICTURE	CLASSIFICATION
$[-2, 5]$	$\{x: -2 \leq x \leq 5\}$		Finite; closed
$(-2, 5)$	$\{x: -2 < x < 5\}$		Finite; open
$[-2, 5)$	$\{x: -2 \leq x < 5\}$		Finite; $\begin{cases}\text{half-open}\\\text{half-closed}\end{cases}$
$(-2, 5]$	$\{x: -2 < x \leq 5\}$		Finite; $\begin{cases}\text{half-open}\\\text{half-closed}\end{cases}$
$(-\infty, 5]$	$\{x: x \leq 5\}$		Infinite; closed
$(-\infty, 5)$	$\{x: x < 5\}$		Infinite; open
$[-2, +\infty)$	$\{x: x \geq -2\}$		Infinite; closed
$(-2, +\infty)$	$\{x: x > -2\}$		Infinite; open

◀

UNIONS AND INTERSECTIONS OF INTERVALS

In many problems we shall be interested in finding points that lie in two or more intervals or points that lie in at least one of several possible intervals. For such problems it is helpful to use the notation of union and intersection of sets.

Example 5 Sketch the sets

(a) $(-4, 3] \cap (2, 6)$ (b) $[1, 2] \cap [3, 4]$

Solution (a). We can write

$$(-4, 3] \cap (2, 6) = \{x: -4 < x \leq 3 \quad \text{and} \quad 2 < x < 6\}$$

This set consists of all points common to the intervals $(-4, 3]$ and $(2, 6)$. This intersection is the interval $(2, 3]$ shown in Figure 1.1.9.

Figure 1.1.9

Solution (b). We can write

$$[1, 2] \cap [3, 4] = \{x: 1 \leq x \leq 2 \quad \text{and} \quad 3 \leq x \leq 4\}$$

There are no points common to the intervals $[1, 2]$ and $[3, 4]$, so $[1, 2] \cap [3, 4] = \varnothing$ and there is no sketch to be made.   ◀

Example 6 Sketch the sets

(a) $[-4, 1) \cup (3, 6)$ (b) $(-3, 2] \cup (1, 7]$

Solution. We can write

(a) $[-4, 1) \cup (3, 6) = \{x : -4 \leq x < 1 \quad \text{or} \quad 3 < x < 6\}$
(b) $(-3, 2] \cup (1, 7] = \{x : -3 < x \leq 2 \quad \text{or} \quad 1 < x \leq 7\}$

These sets are sketched in Figure 1.1.10. Observe that the set in part (b) can be expressed as a single interval, namely $(-3, 2] \cup (1, 7] = (-3, 7]$. ◄

Figure 1.1.10

REMARK.   There is a convention about the use of the word "or" in mathematics that should be noted. In mathematical parlance, the statement that "P is true or Q is true" is understood to allow for the possibility that P and Q are both true. This is called the *inclusive* interpretation of "or." Thus, if A and B are sets, it is correct to say that $A \cup B$ consists of all points that belong to A or B. In particular, for the sets in part (b) of the foregoing example, it is correct to say that the points in $(-3, 2] \cap (1, 7] = (1, 2]$ lie in $(-3, 2]$ or $(1, 7]$, even though these points lie in both intervals.

☐ **SOLVING INEQUALITIES**

A *solution* of an inequality in an unknown x is a value for x that makes the inequality a true statement. For example, $x = 1$ is a solution of the inequality $x < 5$, but $x = 7$ is not. The set of all solutions of an inequality is called its *solution set*. It can be shown that as long as one does not multiply both sides of an inequality by zero or an expression involving an unknown, then the operations in Theorem 1.1.3 will not change the solution set of the inequality. The process of finding the solution set of an inequality is called *solving* the inequality.

Example 7 Solve $3 + 7x \leq 2x - 9$.

Solution. We shall use the operations of Theorem 1.1.3 to isolate x on one side of the inequality.

$3 + 7x \leq 2x - 9$ | Given. |

$7x \leq 2x - 12$ | We added -3 to both sides. |

$5x \leq -12$ | We added $-2x$ to both sides. |

$x \leq -\frac{12}{5}$ | We multiplied both sides by $\frac{1}{5}$. |

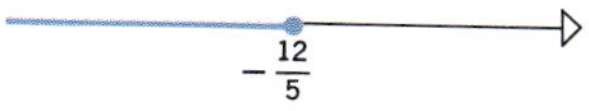

Because we have not multiplied by any expressions involving the unknown x, the last inequality has the same solution set as the first. Thus, the solution set is the interval $(-\infty, -\frac{12}{5}]$ shown in Figure 1.1.11. ◄

$-\frac{12}{5}$

Figure 1.1.11

Example 8 Solve $7 \leq 2 - 5x < 9$.

Solution. The given inequality is actually a combination of the two inequalities

$$7 \leq 2 - 5x \quad \text{and} \quad 2 - 5x < 9$$

We could solve the two inequalities separately, then determine the values of x that satisfy both by taking the intersection of the two solution sets. However, it is more efficient to work with the combined inequalities:

$$7 \leq 2 - 5x < 9 \quad \boxed{\text{Given.}}$$

$$5 \leq -5x < 7 \quad \boxed{\begin{array}{l}\text{We added } -2 \text{ to} \\ \text{each member.}\end{array}}$$

$$-1 \geq x > -\frac{7}{5} \quad \boxed{\begin{array}{l}\text{We multiplied by } -\frac{1}{5} \text{ and reversed} \\ \text{the sense of the inequalities.}\end{array}}$$

$$-\frac{7}{5} < x \leq -1 \quad \boxed{\begin{array}{l}\text{For clarity, we rewrote the inequalities} \\ \text{with the smaller number on the left.}\end{array}}$$

Figure 1.1.12

Thus, the solution set is the interval $(-\frac{7}{5}, -1]$ shown in Figure 1.1.12. ◀
Inequalities of the forms

$$ax^2 + bx + c > 0, \quad ax^2 + bx + c \geq 0$$

$$ax^2 + bx + c < 0, \quad ax^2 + bx + c \leq 0$$

can be solved by two procedures: the ***method of factoring*** and the ***method of test points***. The following example illustrates both methods.

Example 9 Solve $x^2 - 3x > 10$.

Solution (The Method of Factoring). By adding -10 to both sides, the inequality can be rewritten as

$$x^2 - 3x - 10 > 0$$

Factoring the left side yields

$$(x + 2)(x - 5) > 0 \tag{2}$$

The signs of the two factors on the left side are shown in the first two lines of Figure 1.1.13. These two lines and the following facts about signs

$$(-)(-) = (+) \quad (+)(-) = (-) \quad (+)(+) = (+)$$

yield the third line in the figure. It follows from the third line that (2) holds if $x < -2$ or $x > 5$, so the solution set is $(-\infty, -2) \cup (5, +\infty)$, which is shown at the bottom of Figure 1.1.13.

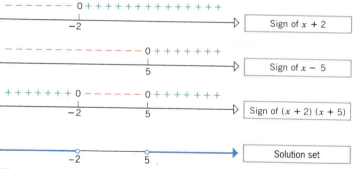

Figure 1.1.13

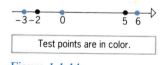

Test points are in color.

Figure 1.1.14

Solution (The Method of Test Points).   As above, we begin by rewriting the inequality as

$$x^2 - 3x - 10 > 0$$

Next we solve the *equation*

$$x^2 - 3x - 10 = 0$$

or in factored form

$$(x + 2)(x - 5) = 0$$

This yields the solutions $x = -2$ and $x = 5$, which we now plot on a coordinate line (Figure 1.1.14). These points divide the coordinate line into three intervals,

$$(-\infty, -2), \quad (-2, 5), \quad (5, +\infty)$$

on which the expression $x^2 - 3x - 10$ has constant sign. To determine those signs we shall choose an *arbitrary* point in each interval at which we shall test the sign. These are called *test points*. As shown in Figure 1.1.14, we shall use -3, 0, and 6 as our test points. The calculations can be organized as follows:

INTERVAL	TEST POINT	VALUE OF $x^2 - 3x - 10$	SIGN OF $x^2 - 3x - 10$ IN THE INTERVAL
$(-\infty, -2)$	-3	8	+
$(-2, 5)$	0	-10	$-$
$(5, +\infty)$	6	8	+

From this table we deduce that the solution set of $x^2 - 3x - 10 > 0$ is $(-\infty, -2) \cup (5, +\infty)$, which agrees with the result obtained by the method of factoring.   ◀

Example 10   Solve $\dfrac{2x - 5}{x - 2} < 1$.

Solution.   We could start by multiplying both sides by $x - 2$ to eliminate the fraction. However, this would require us to consider the cases $x - 2 > 0$ and $x - 2 < 0$ separately because the sense of the inequality would be reversed in the second case, but not the first. The following approach is simpler:

$$\frac{2x - 5}{x - 2} < 1 \qquad \boxed{\text{Given.}}$$

$$\frac{2x - 5}{x - 2} - 1 < 0 \qquad \boxed{\begin{array}{l}\text{We subtracted 1 from both}\\ \text{sides to obtain a 0 on the}\\ \text{right.}\end{array}}$$

$$\frac{(2x - 5) - (x - 2)}{x - 2} < 0 \qquad \boxed{\text{We combined terms.}}$$

$$\frac{x - 3}{x - 2} < 0 \qquad \boxed{\text{We simplified.}}$$

The signs of the numerator, denominator, and quotient on the left side of this inequality are shown in Figure 1.1.15. From that figure we see that the solution set consists of all real values of x such that $2 < x < 3$. This is the interval $(2, 3)$ shown at the bottom of the figure.   ◀

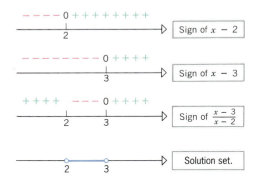

Figure 1.1.15

► **Exercise Set 1.1** $\boxed{C}$ *5, 6*

1. Among the terms, *integer*, *rational*, and *irrational*, which ones apply to the given number?

   (a) $-\frac{3}{4}$ (b) 0

   (c) $\frac{24}{8}$ (d) 0.25

   (e) $-\sqrt{16}$ (f) $2^{1/2}$

   (g) 0.020202 . . .   (h) 7.000 . . . .

2. Which of the terms, *integer*, *rational*, and *irrational*, apply to the given number?

   (a) 0.31311311131111 . . .   (b) 0.729999 . . .

   (c) 0.376237623762 . . .   (d) $17\frac{3}{5}$.

3. The repeating decimal 0.137137137 . . . can be expressed as a ratio of integers by writing

$$x = 0.137137137\ldots$$
$$1000x = 137.137137137\ldots$$

and subtracting to obtain $999x = 137$ or $x = \frac{137}{999}$. Use this idea, where needed, to express the following decimals as ratios of integers.

(a) $0.123123123\ldots$ (b) $12.7777\ldots$
(c) $38.07818181\ldots$ (d) $0.4296000\ldots.$

4. Show that the repeating decimal $0.99999\ldots$ represents the number 1. Since $1.000\ldots$ is also a decimal representation of 1, this problem shows that a real number can have two different decimal representations. [*Hint:* Use the technique of Exercise 3.]

5. The Rhind Papyrus, which is a fragment of Egyptian mathematical writing from about 1650 B.C., is one of the oldest known examples of written mathematics. It is stated in the papyrus that the area A of a circle is related to its diameter D by

$$A = \left(\frac{8}{9}D\right)^2$$

(a) What approximation to π were the Egyptians using?
(b) Use a calculator to determine if this approximation is better or worse than the approximation of $22/7$.

6. The following are all famous historical approximations to π:

$$\frac{333}{106}$$ Adrian Athoniszoon, c. 1583

$$\frac{355}{113}$$ Tsu Chung-Chi and others

$$\frac{63}{25}\left(\frac{17 + 15\sqrt{5}}{7 + 15\sqrt{5}}\right)$$ Ramanujan

$$\frac{22}{7}$$ Archimedes

$$\frac{223}{71}$$ Archimedes

(a) Use a calculator to order these approximations according to size.
(b) Which of these approximations is closest to but larger than π?
(c) Which of these approximations is closest to but smaller than π?
(d) Which of these approximations is most accurate?

7. In each line of the following table, check the blocks, if any, that describe a valid relationship between the real numbers a and b. The first line is already completed as an illustration.

a	b	$a < b$	$a \le b$	$a > b$	$a \ge b$	$a = b$
1	6	✓	✓			
6	1					
-3	5					
5	-3					
-4	-4					
0.25	$\frac{1}{3}$					
$-\frac{1}{4}$	$-\frac{3}{4}$					

8. In each line of the following table, check the blocks, if any, that describe a valid relationship between the real numbers a, b, and c.

a	b	c	$a < b < c$	$a \le b \le c$	$a < b \le c$	$a \le b < c$
-1	0	2				
2	4	-3				
$\frac{1}{2}$	$\frac{1}{2}$	$\frac{3}{4}$				
-5	-5	-5				
0.75	1.25	1.25				

9. Which of the following are always correct if $a \le b$?
(a) $a - 3 \le b - 3$ (b) $-a \le -b$
(c) $3 - a \le 3 - b$ (d) $6a \le 6b$
(e) $a^2 \le ab$ (f) $a^3 \le a^2b.$

10. Which of the following are always correct if $a \le b$ and $c \le d$?
(a) $a + 2c \le b + 2d$ (b) $a - 2c \le b - 2d$
(c) $a - 2c \ge b - 2d.$

11. For what values of a are the following inequalities valid?
(a) $a \le a$ (b) $a < a.$

12. If $a \le b$ and $b \le a$, what can you say about a and b?

13. (a) If $a < b$ is true, does it follow that $a \le b$ must also be true?
(b) If $a \le b$ is true, does it follow that $a < b$ must also be true?

14. In each part, list the elements in the set.
(a) $\{x:x^2 - 5x = 0\}$
(b) $\{x:x$ is an integer satisfying $-2 < x < 3\}.$

15. In each part, express the set in the notation $\{x:\underline{\qquad}\}$.
   (a) $\{1, 3, 5, 7, 9, \ldots\}$
   (b) the set of even integers
   (c) the set of irrational numbers
   (d) $\{7, 8, 9, 10\}$.

16. Let $A = \{1, 2, 3\}$. Which of the following are equal to A?
   (a) $\{0, 1, 2, 3\}$ (b) $\{3, 2, 1\}$
   (c) $\{x:(x - 3)(x^2 - 3x + 2) = 0\}$.

17. In Figure 1.1.16, let
   S = the set of points inside the square
   T = the set of points inside the triangle
   C = the set of points inside the circle
   and let a, b, and c be the points shown. Answer the following as true or false.
   (a) $T \subset C$ (b) $T \subset S$
   (c) $a \notin T$ (d) $a \notin S$
   (e) $b \in T$ and $b \in C$ (f) $a \in C$ or $a \in T$
   (g) $c \in T$ and $c \notin C$.

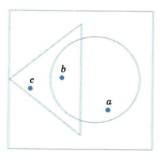

Figure 1.1.16

18. List all subsets of
   (a) $\{a_1, a_2, a_3\}$ (b) $\varnothing$.

19. In each part, sketch on a coordinate line all values of x that satisfy the stated condition.
   (a) $x \le 4$ (b) $x \ge -3$
   (c) $-1 \le x \le 7$ (d) $x^2 = 9$
   (e) $x^2 \le 9$ (f) $x^2 \ge 9$.

20. In parts (a)–(d), sketch on a coordinate line all values of x, if any, that satisfy the stated conditions.
   (a) $x > 4$ and $x \le 8$
   (b) $x \le 2$ or $x \ge 5$
   (c) $x > -2$ and $x \ge 3$
   (d) $x \le 5$ and $x > 7$.

21. Express in interval notation.
   (a) $\{x: x^2 \le 4\}$ (b) $\{x: x^2 > 4\}$.

22. In each part, sketch the set on a coordinate line.
   (a) $[-3, 2] \cup [1, 4]$ (b) $[4, 6] \cup [8, 11]$
   (c) $(-4, 0) \cup (-5, 1)$ (d) $[2, 4) \cup (4, 7)$
   (e) $(-2, 4) \cap (0, 5]$
   (f) $[1, 2.3) \cup (1.4, \sqrt{2})$
   (g) $(-\infty, -1) \cup (-3, +\infty)$
   (h) $(-\infty, 5) \cap [0, +\infty)$.

In Exercises 23–44, solve the inequality and sketch the solution on a coordinate line.

23. $3x - 2 < 8$.
24. $\frac{1}{5}x + 6 \ge 14$.
25. $4 + 5x \le 3x - 7$.
26. $2x - 1 > 11x + 9$.
27. $3 \le 4 - 2x < 7$.
28. $-2 \ge 3 - 8x \ge -11$.
29. $\dfrac{x}{x - 3} < 4$.
30. $\dfrac{x}{8 - x} \ge -2$.
31. $\dfrac{3x + 1}{x - 2} < 1$.
32. $\dfrac{\frac{1}{2}x - 3}{4 + x} > 1$.
33. $\dfrac{4}{2 - x} \le 1$.
34. $\dfrac{3}{x - 5} \le 2$.
35. $x^2 > 9$.
36. $x^2 \le 5$.
37. $(x - 4)(x + 2) > 0$.
38. $(x - 3)(x + 4) < 0$.
39. $x^2 - 9x + 20 \le 0$.
40. $2 - 3x + x^2 \ge 0$.
41. $\dfrac{2}{x} < \dfrac{3}{x - 4}$.
42. $\dfrac{1}{x + 1} \ge \dfrac{3}{x - 2}$.
43. $x^3 - x^2 - x - 2 > 0$.
44. $x^3 - 3x + 2 \le 0$.

In Exercises 45 and 46, find all values of x for which the given expression yields a real number.

45. $\sqrt{x^2 + x - 6}$.
46. $\sqrt{\dfrac{x + 2}{x - 1}}$.

47. Fahrenheit and Celsius temperatures are related by the formula $C = \frac{5}{9}(F - 32)$. If the temperature in degrees Celsius ranges over the interval $25 \le C \le 40$ on a certain day, what is the temperature range in degrees Fahrenheit that day?

48. Every integer is either even or odd. The even integers are those that are divisible by 2, so n is even if and only if $n = 2k$ for some integer k. Each odd integer is one unit larger than an even integer, so n is odd if and only if $n = 2k + 1$ for some integer k. Show:
   (a) If n is even, then so is n^2.
   (b) If n is odd, then so is n^2.

49. Prove the following results about sums of rational and irrational numbers:
   (a) rational + rational = rational
   (b) rational + irrational = irrational.

50. Prove the following results about products of rational and irrational numbers:
 (a) rational · rational = rational
 (b) rational · irrational = irrational (provided the rational factor is nonzero).

51. Show that the sum or product of two irrational numbers can be rational or irrational.

52. Classify the following as rational or irrational and justify your conclusion.
 (a) $3 + \pi$ (b) $\sqrt[3]{4}\sqrt{2}$
 (c) $\sqrt{8}\,\sqrt{2}$ (d) $\sqrt{\pi}$.
 (See Exercises 49 and 50.)

53. Prove: The average of two rational numbers is a rational number, but the average of two irrational numbers can be rational or irrational.

54. Can a rational number satisfy $10^x = 3$?

55. Solve: $8x^3 - 4x^2 - 2x + 1 < 0$.

56. Solve: $12x^3 - 20x^2 \geq -11x + 2$.

57. Prove: If a, b, c, and d are positive numbers such that $a < b$ and $c < d$, then $ac < bd$. (This result gives conditions under which inequalities can be ''multiplied together.'')

58. Show that rational numbers are represented by repeating decimals. [*Hint:* Examine the long division process that converts a ratio of integers into a decimal.]

1.2 ABSOLUTE VALUE

In this section we shall review the notion of absolute value. This concept plays an important role in algebraic computations involving radicals and in determining the distance between points on a coordinate line.

☐ **ABSOLUTE VALUE**

1.2.1 DEFINITION. The *absolute value* or *magnitude* of a real number a is denoted by $|a|$ and is defined by

$$|a| = \begin{cases} a & \text{if} \quad a \geq 0 \\ -a & \text{if} \quad a < 0 \end{cases}$$

Example 1

$$|5| = 5 \qquad \boxed{\text{Since } 5 > 0}$$
$$\left|-\tfrac{4}{7}\right| = -\left(-\tfrac{4}{7}\right) = \tfrac{4}{7} \qquad \boxed{\text{Since } -\tfrac{4}{7} < 0}$$
$$|0| = 0 \qquad \boxed{\text{Since } 0 \geq 0} \qquad \blacktriangleleft$$

REMARK. Note that the effect of taking the absolute value of a number is to strip away the minus sign if the number is negative and to leave the number unchanged if it is nonnegative. Thus, $|a| \geq 0$ for all values of a.

REMARK. Symbols such as $+a$ and $-a$ are deceptive, since it is tempting to conclude that $+a$ is positive and $-a$ is negative. However, this need not be so, since a itself can represent either a positive or negative number. In fact, if a itself is negative, then $-a$ is positive and $+a$ is negative.

Example 2   Solve $|x - 3| = 4$.

Solution.   Depending on whether $x - 3$ is positive or negative, the equation $|x - 3| = 4$ can be written as

$$x - 3 = 4 \quad \text{or} \quad -(x - 3) = 4$$

Solving these two equations gives $x = 7$ and $x = -1$.   ◄

Example 3   Solve $|3x - 2| = |5x + 4|$.

Solution.   Because two numbers with the same absolute value are either equal or differ only in sign, the given equation will be satisfied if either

$$3x - 2 = 5x + 4 \quad \text{or} \quad 3x - 2 = -(5x + 4)$$

Solving the first equation yields $x = -3$ and solving the second yields $x = -\frac{1}{4}$ (verify); thus, the given equation has the solutions $x = -3$ and $x = -\frac{1}{4}$.   ◄

☐ **RELATIONSHIP BETWEEN SQUARE ROOTS AND ABSOLUTE VALUES**

Recall that a number whose square is a is called a ***square root*** of a. In algebra it is learned that every positive real number a has two real square roots, one positive and one negative. The positive square root is denoted by $\sqrt{a}$. For example, the number 9 has two square roots, -3 and 3. Since 3 is the positive square root, we have $\sqrt{9} = 3$. In addition, we define $\sqrt{0} = 0$.

REMARK.   Readers who were previously taught to write $\sqrt{9} = \pm 3$ should stop doing so, because it is incorrect.

It is a common error to write $\sqrt{a^2} = a$. Although this equality is correct when a is nonnegative, it is false for negative a. For example, if $a = -4$, then

$$\sqrt{a^2} = \sqrt{(-4)^2} = \sqrt{16} = 4 \neq a$$

A result that is correct for all a is given in the following theorem.

1.2.2   THEOREM.   *For any real number a,*

$$\sqrt{a^2} = |a|$$

Proof.   Since $a^2 = (+a)^2 = (-a)^2$, the numbers $+a$ and $-a$ are square roots of a^2. If $a \geq 0$, then $+a$ is the nonnegative square root of a^2, and if $a < 0$, then $-a$ is the nonnegative square root of a^2. Since $\sqrt{a^2}$ denotes the nonnegative square root of a^2, we have

$$\sqrt{a^2} = +a \quad \text{if} \quad a \geq 0$$
$$\sqrt{a^2} = -a \quad \text{if} \quad a < 0$$

That is, $\sqrt{a^2} = |a|$.   ▮

□ **PROPERTIES OF ABSOLUTE VALUE**

The next theorem, which lists some of the basic properties of absolute value, is based on Theorem 1.2.2 and the following properties of square roots, which should be familiar from algebra:

$$\sqrt{ab} = \sqrt{a}\sqrt{b} \quad (a \geq 0, b \geq 0)$$

$$\sqrt{\frac{a}{b}} = \frac{\sqrt{a}}{\sqrt{b}} \quad (a \geq 0, b > 0)$$

1.2.3 THEOREM. *If a and b are real numbers, then*

(a) $|-a| = |a|$

(b) $|ab| = |a|\,|b|$

(c) $\left|\dfrac{a}{b}\right| = \dfrac{|a|}{|b|}$

We shall prove parts (*a*) and (*b*) only.

Proof (a). From Theorem 1.2.2

$$|-a| = \sqrt{(-a)^2} = \sqrt{a^2} = |a|$$

Proof (b). From Theorem 1.2.2

$$|ab| = \sqrt{(ab)^2} = \sqrt{a^2b^2} = \sqrt{a^2}\sqrt{b^2} = |a|\,|b| \quad ■$$

REMARK. In words, parts (*b*) and (*c*) of Theorem 1.2.3 state that *the absolute value of a product is the product of the absolute values, and the absolute value of a ratio is the ratio of the absolute values.*

The result in part (*b*) of Theorem 1.2.3 can be extended to three or more factors. More precisely, for any *n* real numbers, $a_1, a_2, \ldots, a_n$, it follows that

$$|a_1 a_2 \cdots a_n| = |a_1|\,|a_2| \cdots |a_n| \tag{1}$$

In the special case where $a_1, a_2, \ldots, a_n$ have the same value, *a*, it follows from (1) that

$$|a^n| = |a|^n \tag{2}$$

REMARK. In part (*c*) of Theorem 1.2.3 we did not explicitly state that $b \neq 0$, but this must be so since division by zero is not allowed. Whenever divisions occur in this text, it will be assumed that the denominator is not zero, even if we do not mention it explicitly.

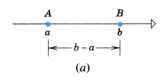

(a)

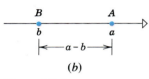

(b)

Figure 1.2.1

The notion of absolute value arises naturally in distance problems. On a co-ordinate line, let A and B be points with coordinates a and b. Because distance is nonnegative, the distance d between A and B is

$$d = b - a$$

when B is to the right of A (Figure 1.2.1a), and

$$d = a - b$$

when B is to the left of A (Figure 1.2.1b).
   In the first case, $b - a$ is nonnegative, so we can write

$$d = b - a = |b - a|$$

and in the second case, $b - a$ is nonpositive, so we can write

$$d = a - b = -(b - a) = |b - a|$$

Thus, regardless of whether B is to the right or left of A or A and B coincide, we have the following result:

1.2.4   THEOREM (*Distance Formula*).   *If A and B are points on a coordinate line with coordinates a and b, respectively, then the distance d between A and B is*

$$d = |b - a| \tag{3}$$

Formula (3) provides a useful geometric interpretation of some common mathematical expressions:

Table 1.2.1

EXPRESSION	GEOMETRIC INTERPRETATION ON A COORDINATE LINE
$|x - a|$	The distance between x and a
$|x + a|$	The distance between x and $-a$ (since $|x + a| = |x - (-a)|$)
$|x|$	The distance between x and the origin (since $|x| = |x - 0|$)

Inequalities of the forms

$$|x - a| < k \quad \text{and} \quad |x - a| > k$$

arise often, so we have summarized the key facts about them in Table 1.2.2.

REMARK.   In this table we can replace $<$ by $\leq$ and $>$ by $\geq$ if we replace the open dots by closed dots in the illustrations.

Example 4   Solve $|x - 3| < 4$.

Table 1.2.2

INEQUALITY $(k > 0)$	GEOMETRIC INTERPRETATION	FIGURE	ALTERNATIVE FORM OF THE INEQUALITY	SOLUTION SET
$\|x - a\| < k$	x is within k units of a.	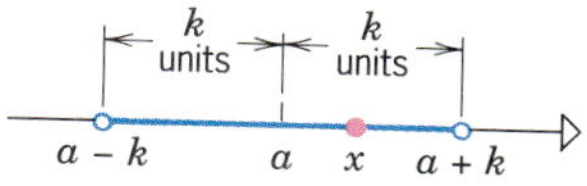 	$-k < x - a < k$	$a - k < x < a + k$
$\|x - a\| > k$	x is more than k units away from a.	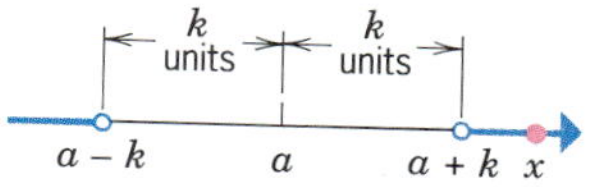 	$x - a < -k$ or $x - a > k$	$x < a - k$ or $x > a + k$

Solution. This inequality can be rewritten as

$$-4 < x - 3 < 4$$

from which we obtain

$$-4 + 3 < x < 4 + 3$$
$$-1 < x < 7$$

which can be written in interval notation as $(-1, 7)$. This solution set is shown in Figure 1.2.2. ◀

Figure 1.2.2

Example 5 Solve $|x + 4| \geq 2$.

Solution. The given inequality can be rewritten as

$$x + 4 \leq -2 \quad \text{or} \quad x + 4 \geq 2$$
$$x \leq -6 \quad \text{or} \quad x \geq -2$$

which can be written in interval notation as

$$(-\infty, -6] \cup [-2, +\infty)$$

This solution set is shown in Figure 1.2.3. ◀

Figure 1.2.3

REMARK. The problems in the last two examples were solved algebraically. However, we could also have proceeded geometrically. For example, the solution of $|x - 3| < 4$ consists of all x whose distance from 3 is less than 4 units

(Figure 1.2.2), and the solution of $|x + 4| \geq 2$ [which can be rewritten as $|x - (-4)| \geq 2$] consists of all x whose distance from -4 is 2 units or more (Figure 1.2.3).

Example 6   Solve $\dfrac{1}{|2x - 3|} > 5$.

Solution.   Observe first that $x = \frac{3}{2}$ is not a solution because this value of x results in a division by zero. Keeping this in mind, we shall begin by applying Theorem 1.1.3(e). Taking reciprocals and reversing the inequality yields

$$|2x - 3| < \tfrac{1}{5}$$

$$|2(x - \tfrac{3}{2})| < \tfrac{1}{5} \quad \boxed{\text{Factor out the coefficient of } x.}$$

$$|2|\,|x - \tfrac{3}{2}| < \tfrac{1}{5} \quad \boxed{\text{Theorem 1.2.3}(b)}$$

$$|x - \tfrac{3}{2}| < \tfrac{1}{10} \quad \boxed{\text{We multiplied both sides by } 1/|2| = 1/2.}$$

$$-\tfrac{1}{10} < x - \tfrac{3}{2} < \tfrac{1}{10} \quad \boxed{\text{Table 1.2.2}}$$

$$\tfrac{7}{5} < x < \tfrac{8}{5} \quad \boxed{\text{We added 3/2 throughout.}}$$

If, as noted above, we eliminate the value $x = \frac{3}{2}$ to avoid the division by zero, we see that the solution consists of all x that satisfy

$$\tfrac{7}{5} < x < \tfrac{3}{2} \quad \text{or} \quad \tfrac{3}{2} < x < \tfrac{8}{5}$$

Geometrically, the solution set consists of all x in the set $(\frac{7}{5}, \frac{3}{2}) \cup (\frac{3}{2}, \frac{8}{5})$ shown in Figure 1.2.4.   ◀

Figure 1.2.4

☐ **AN INEQUALITY FROM CALCULUS**

The inequality in the following example arises in calculus.

Example 7   Solve

$$0 < |x - a| < \delta \tag{4}$$

where a is any real number and δ (Greek "delta") is a positive real number.

Solution.   The solution set consists of all real values of x that satisfy the two inequalities

$$0 < |x - a| \quad \text{and} \quad |x - a| < \delta$$

The solutions of $|x - a| < \delta$ are those values of x such that

$$a - \delta < x < a + \delta$$

or in interval notation

$$(a - \delta, a + \delta) \tag{5}$$

The inequality $0 < |x - a|$ is satisfied by all real values of x except $x = a$

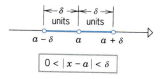

Figure 1.2.5

$$0 < |x - a| < \delta$$

(why?). Thus, the solution set of $0 < |x - a| < \delta$ is the interval in (5) with the point a removed. This is the set

$$(a - \delta, a) \cup (a, a + \delta)$$

shown in Figure 1.2.5. ◀

☐ **INEQUALITY PROPERTIES OF ABSOLUTE VALUE**

We conclude this section with two basic inequalities involving absolute values.

1.2.5 THEOREM. *If a is any real number, then*

$$-|a| \leq a \leq |a| \tag{6}$$

Proof. If $a \geq 0$, then we can write $-a \leq a \leq a$ and $|a| = a$ from which it follows that

$$-|a| \leq a \leq |a|$$

If $a < 0$, we can write $a \leq a \leq -a$ and $-a = |a|$ or $a = -|a|$ from which it again follows that

$$-|a| \leq a \leq |a|$$

Thus, (6) holds in all cases. ∎

It is *not* generally true that $|a + b| = |a| + |b|$. For example, if $a = 2$ and $b = -3$, then $a + b = -1$, so that

$$|a + b| = |-1| = 1$$

whereas

$$|a| + |b| = |2| + |-3| = 2 + 3 = 5$$

so $|a + b| \neq |a| + |b|$. It is true, however, that *the absolute value of a sum is always less than or equal to the sum of the absolute values.* This is the content of the following very important theorem, known as the ***triangle inequality***.

1.2.6 THEOREM (***Triangle Inequality***). *If a and b are any real numbers, then*

$$|a + b| \leq |a| + |b| \tag{7}$$

Proof. From Theorem 1.2.5

$$-|a| \leq a \leq |a| \quad \text{and} \quad -|b| \leq b \leq |b|$$

Adding these inequalities yields

$$-(|a| + |b|) \leq a + b \leq (|a| + |b|) \tag{8}$$

We now consider two cases: $a + b \geq 0$ and $a + b < 0$. In the former case $a + b = |a + b|$, so the right-hand inequality in (8) yields (7). In the second case $a + b = -|a + b|$, so the left-hand inequality in (8) can be written as

$$-(|a| + |b|) \leq -|a + b|$$

Multiplying both sides of this inequality by -1 yields (7). ∎

Various applications of the triangle inequality will occur throughout this text.

▶ Exercise Set 1.2

1. Compute $|x|$ if
   (a) $x = 7$
   (b) $x = -\sqrt{2}$
   (c) $x = k^2$
   (d) $x = -k^2$.

2. Rewrite $\sqrt{(x - 6)^2}$ without using a square root or absolute value sign.

In Exercises 3–10, find all values of x for which the given statement is true.

3. $|x - 3| = 3 - x$.
4. $|x + 2| = x + 2$.
5. $|x^2 + 9| = x^2 + 9$.
6. $|x^2 + 5x| = x^2 + 5x$.
7. $|3x^2 + 2x| = x|3x + 2|$.
8. $|6 - 2x| = 2|x - 3|$.
9. $\sqrt{(x + 5)^2} = x + 5$.
10. $\sqrt{(3x - 2)^2} = 2 - 3x$.
11. Verify $\sqrt{a^2} = |a|$ for $a = 7$ and $a = -7$.
12. Verify the inequalities $-|a| \leq a \leq |a|$ for $a = 2$ and $a = -5$.
13. Let A and B be points with coordinates a and b. In each part find the distance between A and B.
    (a) $a = 9, b = 7$
    (b) $a = 2, b = 3$
    (c) $a = -8, b = 6$
    (d) $a = \sqrt{2}, b = -3$
    (e) $a = -11, b = -4$
    (f) $a = 0, b = -5$.
14. Is the equality $\sqrt{a^4} = a^2$ valid for all values of a? Explain.
15. Let A and B be points with coordinates a and b. In each part, use the given information to find b.
    (a) $a = -3$, B is to the left of A, and $|b - a| = 6$
    (b) $a = -2$, B is to the right of A, and $|b - a| = 9$
    (c) $a = 5$, $|b - a| = 7$, and $b > 0$.
16. Let E and F be points with coordinates e and f. In each part, determine whether E is to the left or to the right of F on a coordinate line.
    (a) $f - e = 4$
    (b) $e - f = 4$
    (c) $f - e = -6$
    (d) $e - f = -7$.

In Exercises 17–24 solve for x.

17. $|6x - 2| = 7$.
18. $|3 + 2x| = 11$.
19. $|6x - 7| = |3 + 2x|$.
20. $|4x + 5| = |8x - 3|$.
21. $|9x| - 11 = x$.
22. $2x - 7 = |x + 1|$.
23. $\left|\dfrac{x + 5}{2 - x}\right| = 6$.
24. $\left|\dfrac{x - 3}{x + 4}\right| = 5$.

In Exercises 25–44, solve for x and express the solution in terms of intervals. For Exercises 37–44, use the fact that $|a| < |b|$ (or $\leq$) if and only if $a^2 < b^2$ (or $\leq$).

25. $|x + 6| < 3$.
26. $|7 - x| \leq 5$.
27. $|2x - 3| \leq 6$.
28. $|3x + 1| < 4$.
29. $|x + 2| > 1$.
30. $|\frac{1}{2}x - 1| \geq 2$.
31. $|5 - 2x| \geq 4$.
32. $|7x + 1| > 3$.
33. $\dfrac{1}{|x - 1|} < 2$.
34. $\dfrac{1}{|3x + 1|} \geq 5$.
35. $\dfrac{3}{|2x - 1|} \geq 4$.
36. $\dfrac{2}{|x + 3|} < 1$.
37. $|x + 3| < |x - 8|$.
38. $|3x| \leq |2x - 5|$.
39. $|4x| \geq |7 - 6x|$.
40. $|2x + 1| > |x - 5|$.
41. $\left|\dfrac{x - \frac{1}{2}}{x + \frac{1}{2}}\right| < 1$.
42. $\left|\dfrac{3 - 2x}{1 + x}\right| \leq 4$.
43. $\dfrac{1}{|x - 4|} < \dfrac{1}{|x + 7|}$.
44. $\dfrac{1}{|x - 3|} - \dfrac{1}{|x + 4|} \geq 0$.
45. For which values of x is
    $$\sqrt{(x^2 - 5x + 6)^2} = x^2 - 5x + 6?$$
46. Solve $3 \leq |x - 2| \leq 7$ for x.
47. Solve $|x - 3|^2 - 4|x - 3| = 12$ for x. [*Hint:* First let $u = |x - 3|$.]
48. Prove: If $|x + 2| < 2$, then $|3x - 2| < 14$.

49. Prove: If $|3x - 4| < 5$, then $|x - \frac{4}{3}| < \frac{8}{3}$.

50. Verify the triangle inequality $|a + b| \le |a| + |b|$ (Theorem 1.2.6) for
(a) $a = 3, b = 4$ (b) $a = -2, b = 6$
(c) $a = -7, b = -8$ (d) $a = -4, b = 4$.

51. Prove: $|a - b| \le |a| + |b|$.

52. Prove: $|a| - |b| \le |a - b|$.

53. Prove: $\big||a| - |b|\big| \le |a - b|$. [*Hint:* Use Exercise 52.]

54. Find the smallest value of M such that
$$\left|\frac{1}{x}\right| \le M \text{ for all } x \text{ in the interval } [2, 7]$$

55. Find the smallest value of M such that
$$\left|\frac{1}{x + 7}\right| \le M$$
for all x in the interval $(-4, 2)$.

56. Use the triangle inequality to find a value of M such that
$$|x^3 - 2x + 1| \le M$$
for all x in the interval $(-2, 3)$.

57. Find a value of M such that
$$\left|\frac{x + 3}{x - 3}\right| \le M$$
for all x in the interval $[-\frac{3}{4}, \frac{1}{4}]$.

▮ 1.3 COORDINATE PLANES AND GRAPHS

> *Just as points on a line can be placed in one-to-one correspondence with the real numbers, so points in the plane can be placed in one-to-one correspondence with pairs of real numbers. This fundamental observation will enable us to visualize algebraic equations as geometric curves and, conversely, to represent geometric curves by algebraic equations.*

☐ **RECTANGULAR COORDINATE SYSTEMS**

A *rectangular coordinate system* (also called a *Cartesian* coordinate system*) is a pair of perpendicular coordinate lines, called *coordinate axes*, which are placed so that they intersect at their origins. Usually, but not always, one of the lines is horizontal with its positive direction to the right, and the other is vertical with its positive direction up (Figure 1.3.1). As illustrated in this figure, the

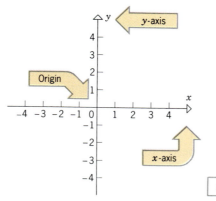

A rectangular coordinate system.

Figure 1.3.1

*See p. 6 for footnote on René Descartes.

intersection of the coordinate axes is called the *origin* of the coordinate system. The horizontal axis is usually called the *x-axis* and the vertical axis is called the *y-axis*. A plane in which a rectangular coordinate system has been introduced is called a *coordinate plane*.

Labeling the axes with letters x and y is a common convention, but any letters may be used. If the letters x and y are used to label the coordinate axes, then the resulting plane is called an *xy-plane*. In applications it is common to use letters other than x and y to label coordinate axes. Figure 1.3.2 shows a uv-plane and a ts-plane. The first letter in the name of the plane refers to the horizontal axis and the second to the vertical axis.

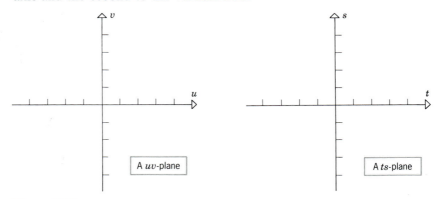

Figure 1.3.2

COORDINATES

We shall now show how to establish a one-to-one correspondence between points in a coordinate plane and "ordered pairs" of real numbers. By an *ordered pair* of real numbers we mean two real numbers in an assigned order. Thus, there is a "first number" and a "second number." The symbol (a, b) is used to denote the ordered pair of real numbers in which a is the first number and b is the second number. Because order matters, the ordered pairs $(2, -3)$ and $(-3, 2)$ are regarded to be different.

REMARK.   Recall that the symbol (a, b) also denotes the open interval between a and b. The appropriate interpretation will usually be clear from the context.

Every point P in a coordinate plane can be associated with a unique ordered pair of real numbers by drawing two lines through P, one perpendicular to the x-axis and the other perpendicular to the y-axis (Figure 1.3.3). If the first line intersects the x-axis at the point with coordinate a and the second line intersects the y-axis at the point with coordinate b, then we associate the ordered pair of real numbers (a, b) with the point P. The number a is called the *x-coordinate* or *abscissa* of P and the number b is called the *y-coordinate* or *ordinate* of P. We shall say that P has *coordinates* (a, b) and write $P(a, b)$ when we want to emphasize that the coordinates of P are (a, b).

It is evident that the construction process illustrated in Figure 1.3.3 associates a unique ordered pair of real numbers (a, b) with each point P in the plane. Conversely, if we *start* with an ordered pair of real numbers (a, b) and construct lines perpendicular to the x-axis and y-axis that pass through the points with coordinates a and b, respectively, then these lines intersect at a unique point P in the plane whose coordinates are (a, b). Thus, we have a one-to-one cor-

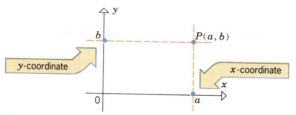

Figure 1.3.3

respondence between ordered pairs of real numbers and points in a coordinate plane.

To **plot** a point $P(a, b)$ means to locate the point with coordinates (a, b) in a coordinate plane. For example, in Figure 1.3.4 we have plotted the points

$$P(2, 5), \quad Q(-4, 3), \quad R(-5, -2), \quad \text{and} \quad S(4, -3)$$

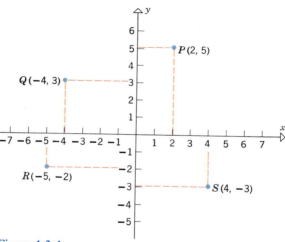

Figure 1.3.4

In a rectangular coordinate system the coordinate axes divide the plane into four regions called **quadrants**. These are numbered counterclockwise with roman numerals as shown in Figure 1.3.5. As illustrated in Figure 1.3.6, it is easy to determine the quadrant in which a given point lies from the signs of its coordinates: a point with two positive coordinates $(+, +)$ lies in Quadrant I, a point with a negative x-coordinate and a positive y-coordinate $(-, +)$ lies in Quadrant II, and so forth. Points with a zero x-coordinate lie on the y-axis and points with a zero y-coordinate lie on the x-axis.

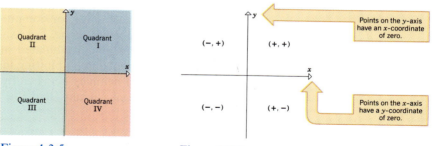

Figure 1.3.5   **Figure 1.3.6**

☐ **GRAPHS**

To see how rectangular coordinate systems enable us to describe equations geometrically, assume that we have constructed a rectangular coordinate system and are given an equation involving only two variables, x and y; for example,

$$5xy = 2, \quad x^2 + 2y^2 = 7, \quad \text{or} \quad y = \frac{1}{1-x}$$

We define a *solution* of such an equation to be an ordered pair of real numbers (a, b) such that the equation is satisfied when we substitute $x = a$ and $y = b$. The set of all solutions is called the *solution set* of the equation.

Example 1 The pair $(3, 2)$ is a solution of $6x - 4y = 10$ since the equation is satisfied when we substitute $x = 3$ and $y = 2$. However, the pair $(2, 0)$ is not a solution, since the equation is not satisfied when we substitute $x = 2$ and $y = 0$. ◄

We make the following definition.

1.3.1 DEFINITION. The *graph* of an equation in two variables x and y is the set of all points in the xy-plane whose coordinates are members of the solution set of the equation.

Example 2 Sketch the graph of $y = x^2$.

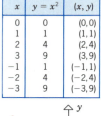

x	$y = x^2$	(x, y)
0	0	$(0, 0)$
1	1	$(1, 1)$
2	4	$(2, 4)$
3	9	$(3, 9)$
−1	1	$(−1, 1)$
−2	4	$(−2, 4)$
−3	9	$(−3, 9)$

Solution. The solution set of $y = x^2$ has infinitely many members, so that it is impossible to plot them all. However, some sample members of the solution set can be obtained by substituting some arbitrary x values into the right side of $y = x^2$ and solving for the associated values of y.

In Figure 1.3.7, some typical computations are given in the table. The accompanying curve was obtained by plotting the points in the table and connecting them with a smooth curve. This curve approximates the graph of $y = x^2$. ◄

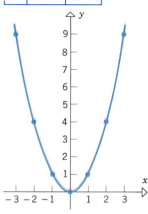

Figure 1.3.7

REMARK. It should be kept in mind that the curve in Figure 1.3.7 is only an *approximation* to the graph of $y = x^2$. When a graph is obtained by plotting points, whether by hand, calculator, or computer, there is no guarantee that the resulting curve has the correct shape. For example, the two curves in Figure 1.3.8 both pass through the points tabulated in Figure 1.3.7. Moreover, we cannot totally resolve the problem by plotting more points since we will never be sure how the true graph behaves *between* the plotted points. In general, it is only with techniques from calculus that the true shape of a graph can be ascertained.

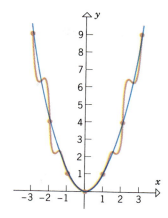

Figure 1.3.8

x	$y = x^3$	(x, y)
0	0	$(0,0)$
1	1	$(1,1)$
2	8	$(2,8)$
−1	−1	$(−1,−1)$
−2	−8	$(−2,−8)$

Example 3   Sketch the graph of $y = x^3$.

Solution.   A table of points and the graph are given in Figure 1.3.9.   ◀

Example 4   Sketch the graph of $y^2 - 2y - x = 0$.

Solution.   To calculate coordinates of points on the graph of an equation in x and y, it is desirable to have y expressed in terms of x or x in terms of y. In this case it is easier to express x in terms of y. We obtain

$$x = y^2 - 2y$$

Some sample members of the solution set can now be obtained by substituting arbitrary y values into the right side of the above equation and computing the associated x values (Figure 1.3.10).   ◀

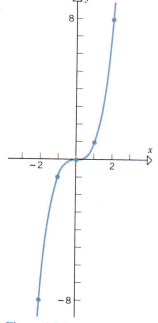

Figure 1.3.9

y	$x = y^2 - 2y$	(x, y)
−2	8	$(8, −2)$
−1	3	$(3, −1)$
0	0	$(0,0)$
1	−1	$(−1,1)$
2	0	$(0,2)$
3	3	$(3,3)$
4	8	$(8,4)$

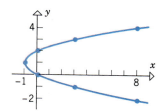

Figure 1.3.10

Example 5 Sketch the graph of $y = 1/x$.

Solution. Because $1/x$ is undefined when $x = 0$, we can plot only points for which $x \neq 0$. This forces a break (commonly called a *discontinuity*) in the graph at $x = 0$ (Figure 1.3.11). ◀

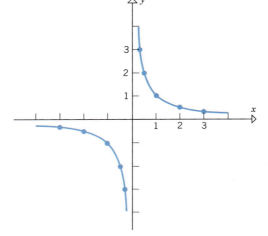

x	$y = 1/x$	(x, y)
$\frac{1}{3}$	3	$(\frac{1}{3}, 3)$
$\frac{1}{2}$	2	$(\frac{1}{2}, 2)$
1	1	$(1, 1)$
2	$\frac{1}{2}$	$(2, \frac{1}{2})$
3	$\frac{1}{3}$	$(3, \frac{1}{3})$
$-\frac{1}{3}$	-3	$(-\frac{1}{3}, -3)$
$-\frac{1}{2}$	-2	$(-\frac{1}{2}, -2)$
-1	-1	$(-1, -1)$
-2	$-\frac{1}{2}$	$(-2, -\frac{1}{2})$
-3	$-\frac{1}{3}$	$(-3, -\frac{1}{3})$

Figure 1.3.11

Example 6 Sketch the graph of $y = \sqrt{x}$.

Solution. If $x < 0$, then $\sqrt{x}$ is a complex number. Since the coordinates of points in the xy-plane are real numbers, we can only plot those points for which $x \geq 0$. With the help of a calculator we obtained the table and graph in Figure 1.3.12. ◀

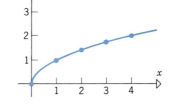

x	$y = \sqrt{x}$	(x, y)
0	0	$(0, 0)$
1	1	$(1, 1)$
2	$\sqrt{2}$	$(2, \sqrt{2}) \approx (2, 1.4)$
3	$\sqrt{3}$	$(3, \sqrt{3}) \approx (3, 1.7)$
4	2	$(4, 2)$

Figure 1.3.12

▢ INTERCEPTS

Points where a graph intersects the coordinate axes are of special interest in many problems. As illustrated in Figure 1.3.13, intersections of a graph with the x-axis have the form $(a, 0)$ and intersections with the y-axis have the form $(0, b)$. The number a is called an *x-intercept* of the graph and the number b a *y-intercept*.

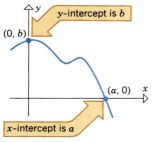

Figure 1.3.13

Example 7 Find all intercepts of

(a) $3x + 2y = 6$ (b) $x = y^2 - 2y$ (c) $y = 1/x$

Solution (a). To find the x-intercepts we set $y = 0$ and solve for x:

$$3x = 6 \quad \text{or} \quad x = 2$$

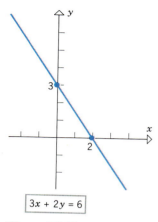

Figure 1.3.14

To find the y-intercepts we set $x = 0$ and solve for y:

$$2y = 6 \quad \text{or} \quad y = 3$$

As we shall see later, the graph of $3x + 2y = 6$ is the line shown in Figure 1.3.14.

Solution (b). To find the x-intercepts, set $y = 0$ and solve for x:

$$x = 0$$

Thus, $x = 0$ is the only x-intercept. To find the y-intercepts, set $x = 0$ and solve for y:

$$y^2 - 2y = 0$$
$$y(y - 2) = 0$$

so the y-intercepts are $y = 0$ and $y = 2$. The graph of the equation is shown in Figure 1.3.10.

Solution (c). To find the x-intercepts, set $y = 0$:

$$\frac{1}{x} = 0$$

This equation has no solutions (why?), so there are no x-intercepts. To find y-intercepts we would set $x = 0$ and solve for y. But, substituting $x = 0$ leads to a division by zero, which is not allowed, so there are no y-intercepts either. The graph of the equation is shown in Figure 1.3.11. ◀

☐ **SYMMETRY**

As illustrated in Figure 1.3.15, the points (x, y), $(-x, y)$, $(x, -y)$, and $(-x, -y)$ form the corners of a rectangle. For obvious reasons, the points (x, y) and $(x, -y)$ are said to be *symmetric about the x-axis,* the points (x, y) and $(-x, y)$ *symmetric about the y-axis,* and the points (x, y) and $(-x, -y)$ *symmetric about the origin.*

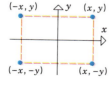
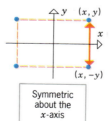
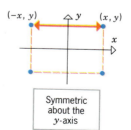
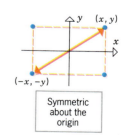

Figure 1.3.15

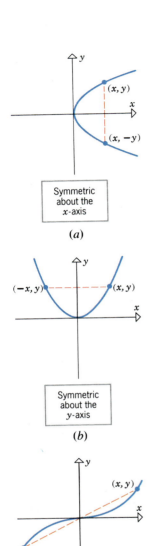

Symmetric
about the
x-axis

(a)

Symmetric
about the
y-axis

(b)

Symmetric
about the
origin

(c)

Figure 1.3.16

Analogously, we say that a graph in the xy-plane is

- *symmetric about the x-axis* if for each point (x, y) on the graph the point $(x, -y)$ is also on the graph (Figure 1.3.16a);
- *symmetric about the y-axis* if for each point (x, y) on the graph the point $(-x, y)$ is also on the graph (Figure 1.3.16b);
- *symmetric about the origin* if for each point (x, y) on the graph the point $(-x, -y)$ is also on the graph (Figure 1.3.16c).

Geometrically, a graph is symmetric about the x-axis if folding the plane along the x-axis makes the upper and lower portions of the graph align exactly, and it is symmetric about the y-axis if folding the plane along the y-axis makes the left and right portions of the graph align exactly. A graph is symmetric about the origin if rotating the graph 180° about the origin results in the original graph.

Symmetries can often be detected from the equation of a curve. For example, the graph of

$$y = x^3 \tag{1}$$

must be symmetric about the origin (see Figure 1.3.9) because for any point (x, y) whose coordinates satisfy (1), the coordinates of the point $(-x, -y)$ also satisfy (1), since substituting these values in (1) yields

$$-y = (-x)^3$$

which simplifies to (1). This discussion suggests the following symmetry tests.

1.3.2   THEOREM (*Symmetry Tests*).

(a)   *A plane curve is symmetric about the y-axis if replacing x by $-x$ in its equation produces an equivalent equation.*

(b)   *A plane curve is symmetric about the x-axis if replacing y by $-y$ in its equation produces an equivalent equation.*

(c)   *A plane curve is symmetric about the origin if replacing both x by $-x$ and y by $-y$ in its equation produces an equivalent equation.*

Example 8   Figure 1.3.7 suggests that the graph of $y = x^2$ is symmetric about the y-axis, but not about the x-axis or the origin. Use Theorem 1.3.2 to verify that this is so.

Solution.   To test for symmetry about the y-axis, we substitute $-x$ for x in $y = x^2$, which yields

$$y = (-x)^2$$

This is equivalent to the original equation since squaring the right side yields $y = x^2$. Thus, the graph is symmetric about the y-axis.

To test for symmetry about the x-axis, we substitute $-y$ for y in $y = x^2$, which yields

$$-y = x^2$$

This is not equivalent to $y = x^2$, so the graph is not symmetric about the x-axis.

To test for symmetry about the origin, we substitute $-x$ for x and $-y$ for y in $y = x^2$, which yields

$$-y = (-x)^2$$
$$-y = x^2$$

The last equation is not equivalent to $y = x^2$, so the graph is not symmetric about the origin. ◄

☐ **SYMMETRY AS A TOOL FOR GRAPHING**

By taking advantage of symmetries when they exist, the work required to obtain a graph can be reduced considerably.

Example 9 Sketch the graph of the equation $y = \frac{1}{8}x^4 - x^2$.

Solution. The graph is symmetric about the y-axis since substituting $-x$ for x yields

$$y = \frac{1}{8}(-x)^4 - (-x)^2$$

which simplifies to the original equation.

As a consequence of this symmetry, we need only calculate points on the graph that lie in the right half of the xy-plane ($x \geq 0$). The corresponding points in the left half of the xy-plane ($x \leq 0$) can be obtained with no additional computation by using the symmetry. The top part of the table in Figure 1.3.17 was obtained with the help of a calculator. The bottom part was obtained with no computation using the symmetry. ◄

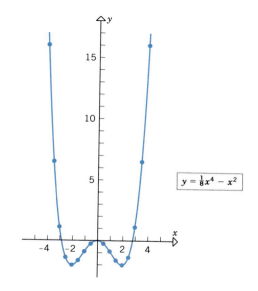

x	$y = \frac{1}{8}x^4 - x^2$
0	0
0.5	−0.24
1	−0.88
1.5	−1.62
2	−2
2.5	−1.37
3	1.12
3.5	6.51
4	16
−4	16
−3.5	6.51
−3	1.12
−2.5	−1.37
−2	−2
−1.5	−1.62
−1	−0.88
−0.5	−0.24

Figure 1.3.17

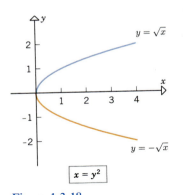

Figure 1.3.18

Example 10 Sketch the graph of $x = y^2$.

Solution. If we solve $x = y^2$ for y in terms of x, we obtain two solutions, $y = \sqrt{x}$ and $y = -\sqrt{x}$. The graph of $y = \sqrt{x}$ is the portion of the curve $x = y^2$ that lies above or touches the x-axis (since $y = \sqrt{x} \geq 0$), and the graph of $y = -\sqrt{x}$ is the portion that lies below or touches the x-axis (since $y = -\sqrt{x} \leq 0$). However, the curve $x = y^2$ is symmetric about the x-axis because substituting $-y$ for y yields $x = (-y)^2$, which is equivalent to the original equation. Thus, we need only graph $y = \sqrt{x}$ (see Figure 1.3.12) and then reflect it about the x-axis to complete the graph (Figure 1.3.18). ◀

☐ **GRAPHS WITH UNEQUAL SCALES**

In all of our graphs thus far we have used equal scales on the coordinate axes, that is, one unit along the x-axis had the same length as one unit along the y-axis. However, sometimes it is necessary to use unequal scales. For example, to graph $y = x^3$ for values of x between -10 and 10, we would have to allow for y values between $(-10)^3 = -1000$ and $10^3 = 1000$, which is difficult to do within the confines of a standard sheet of paper or printed page; the only solution is to use unequal scales. Although unequal scales distort the graph, the general shape is usually maintained, which is adequate for many purposes. Figure 1.3.19 shows two versions of the graph of $y = x^3$.

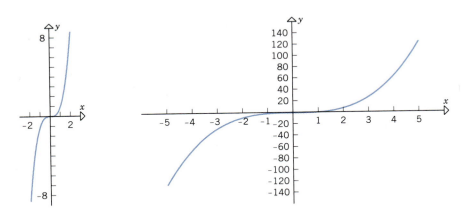

Figure 1.3.19

In applications where the variables in an equation represent physical quantities, there is usually no reason to use equal scales on the coordinate axes when plotting a graph. However, the units of measure should be indicated on the coordinate axes for clarity. For example, Figure 1.3.20 is a graph of T versus R, where T is the time (in thousands of days) for an object orbiting about the sun to make one revolution and R is the mean distance from the object to the sun (in billions of kilometers).

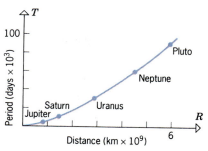

Figure 1.3.20

GRAPHING BY CALCULATOR AND COMPUTER

Laborious hand-plotting of graphs is rapidly becoming a thing of the past since computers and hand calculators with graphics capabilities can generate complex graphs in a matter of seconds. For example, Figure 1.3.21 shows several graphs that were generated on a computer using hundreds of plotted points. Although you may have to do a certain amount of hand-plotting as you work through this text, you should keep in mind that in practical applications computers are available to do such work. Thus, in this text the primary focus will be on learning to understand the kind of information that a graph conveys rather than on the numerical process of generating graphs.

The reader may find it instructive to find the symmetries, if any, of the curves in Figure 1.3.21 using Theorem 1.3.2, then verify that the results obtained agree with the figure.

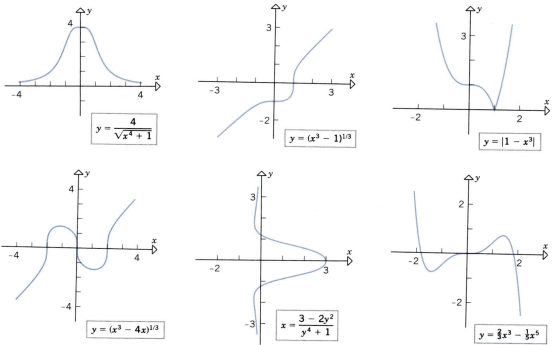

Figure 1.3.21

□ **A CATALOG OF**
BASIC GRAPHS

We conclude this section with a catalog of graphs that arise frequently (Figure 1.3.22). You will find it helpful to be familiar with the general shapes of these graphs without resorting to point-plotting.

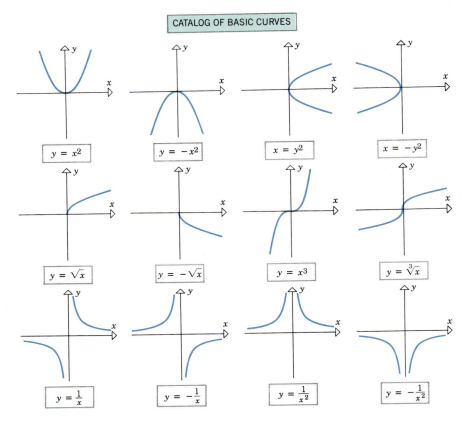

CATALOG OF BASIC CURVES

$y = x^2$ $y = -x^2$ $x = y^2$ $x = -y^2$

$y = \sqrt{x}$ $y = -\sqrt{x}$ $y = x^3$ $y = \sqrt[3]{x}$

$y = \dfrac{1}{x}$ $y = -\dfrac{1}{x}$ $y = \dfrac{1}{x^2}$ $y = -\dfrac{1}{x^2}$

Figure 1.3.22

▶ Exercise Set 1.3

1. Draw a rectangular coordinate system and locate the points
    (a) $(3, 4)$
    (b) $(-2, 5)$
    (c) $(-2.5, -3)$
    (d) $(1.7, -2)$
    (e) $(0, -6)$
    (f) $(4, 0)$.

In Exercises 2 and 3, draw a rectangular coordinate system and sketch the set of points whose coordinates (x, y) satisfy the given conditions.

2. (a) $x = 0$
    (b) $y = 0$
    (c) $y < 0$
    (d) $x \geq 1$ and $y \leq 2$

3. (a) $x = 2$
    (c) $x \geq 0$
    (e) $y \geq x$
    (e) $x = 3$
    (f) $|x| = 5$.
    (b) $y = -3$
    (d) $y = x$
    (f) $|x| \geq 1$.

In Exercises 4 and 5, the points lie on a horizontal or vertical line. Determine whether the line is horizontal or vertical.

4. (a) $A(9, 2)$, $B(7, 2)$
    (b) $A(2, -6)$, $B(3, -6)$
    (c) $A(6, 6)$, $B(6, 1)$.

5. (a) $A(-4, \sqrt{2})$, $B(-4, -3)$
(b) $A(0, -4)$, $B(3, -4)$
(c) $A(0, 0)$, $B(0, -5)$.

6. Find the fourth vertex of the rectangle, three of whose vertices are $(-1, 4)$, $(6, 4)$, and $(-1, 9)$.

7. In each part determine if the given ordered pair (x, y) is a solution of $x^2 - 2x + y = 4$.
(a) $(0, 4)$
(b) $(-3, 7)$
(c) $(\frac{1}{2}, \frac{19}{4})$
(d) $(1 + \sqrt{5 - t}, t)$.

8. Figure 1.3.23 shows only a portion of the graph of an equation in the variables x and y for $x \geq 0$ and $y \geq 0$. Sketch the complete graph if it is known to be
(a) symmetric about the x-axis
(b) symmetric about the y-axis
(c) symmetric about the origin.

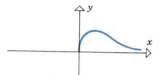

Figure 1.3.23

9. In parts (a)–(d) of Figure 1.3.24, the graphs of equations in the variables x and y are shown. In each case determine whether the graph is symmetric about the x-axis, y-axis, origin, or none of the preceding.

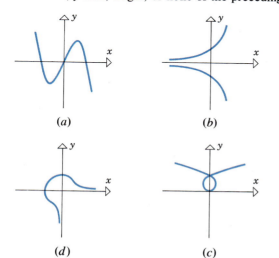

(a)   *(b)*

(d)   *(c)*

Figure 1.3.24

10. Use Theorem 1.3.2 to show that a graph which is symmetric about the x-axis and y-axis must be symmetric about the origin. Give an example to show that the converse is not true.

11. Which of the curves shown in Figure 1.3.22 have symmetry about
(a) the x-axis
(b) the y-axis
(c) the origin?

In Exercises 12 and 13, determine whether the graph is symmetric about the x-axis, the y-axis, or the origin.

12. (a) $x = 5y^2 + 9$
(b) $x^2 - 2y^2 = 3$
(c) $xy = 5$.

13. (a) $x^4 = 2y^3 + y$
(b) $y = \dfrac{x}{3 + x^2}$
(c) $y^2 = |x| - 5$.

In Exercises 14–23, sketch the graph of the equation. (A calculator will be helpful in some of these problems.)

14. $y = 2x - 3$.   **15.** $y = 6 - x$.
16. $y = 1 + x^2$.   **17.** $y = 4 - x^2$.
18. $y = -\sqrt{x + 1}$.   **19.** $y = \sqrt{x - 4}$.
20. $y = |x|$.   **21.** $y = |x - 3|$.
22. $xy = -1$.   **23.** $x^2y = 2$.

In Exercises 24 and 25, sketch the portion of the graph in the first quadrant, and use symmetry to complete the rest of the graph. (A hand calculator will be helpful.)

24. $9x^2 + 4y^2 = 36$.

25. $4x^2 + 16y^2 = 16$.

26. Sketch the graph of $y^2 = 3x$ and explain how this graph is related to the graphs of $y = \sqrt{3x}$ and $y = -\sqrt{3x}$.

27. Sketch the graph of $(x - y)(x + y) = 0$ and explain how it is related to the graphs of $x - y = 0$ and $x + y = 0$.

28. Graph $F = \frac{9}{5}C + 32$ in a CF-coordinate system.

29. Graph $u = 3v^2$ in a uv-coordinate system.

30. Graph $Y = 4X + 5$ in a YX-coordinate system.

◼ **1.4** SLOPE OF A LINE

In this section we shall discuss ways to measure the "steepness" or "slope" of a line in the plane. The ideas we develop here will be important when we discuss equations and graphs of straight lines.

This section requires a knowledge of the trigonometry material in Section I of Appendix B. Readers who need to review that material are advised to do so before starting this section.

In surveying, the *grade* or *slope* of a hill is defined to be the ratio of its *rise* to its *run* (Figure 1.4.1). We shall now show how the surveyor's notion of slope can be adapted to measure the steepness of a line in the *xy*-plane.

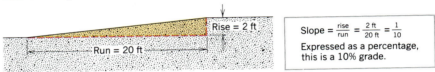

Slope $= \dfrac{\text{rise}}{\text{run}} = \dfrac{2\text{ ft}}{20\text{ ft}} = \dfrac{1}{10}$

Expressed as a percentage, this is a 10% grade.

Figure 1.4.1

Consider a particle moving left to right along a *nonvertical* line segment from a point $P_1(x_1, y_1)$ to a point $P_2(x_2, y_2)$. As shown in Figure 1.4.2, the particle moves $y_2 - y_1$ units in the *y*-direction as it travels $x_2 - x_1$ units in the positive *x*-direction. The vertical change $y_2 - y_1$ is called the **rise**, and the horizontal change $x_2 - x_1$ the **run**.

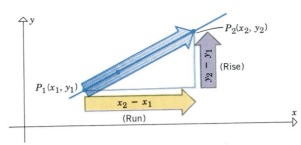

Figure 1.4.2

By analogy with the surveyor's notion of slope we make the following definition.

1.4.1 DEFINITION. If $P_1(x_1, y_1)$ and $P_2(x_2, y_2)$ are the endpoints of a nonvertical line segment, then the **slope** m of the line segment is defined by

$$m = \frac{\text{rise}}{\text{run}} = \frac{y_2 - y_1}{x_2 - x_1} \tag{1}$$

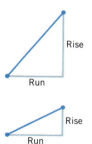

Figure 1.4.3

For two line segments with positive rise and the same run, the more steeply inclined segment will have the greater rise and therefore the greater slope (Figure 1.4.3). Thus, the slope m is a numerical measure of steepness.

REMARK.   Note that Definition 1.4.1 does not apply to vertical line segments. For such segments we would have $x_2 = x_1$, so (1) would involve a division by zero. The slope of a vertical line segment is **undefined**.

Example 1   In each part find the slope of the line segment that connects

(a)   the points $(6, 2)$ and $(9, 8)$
(b)   the points $(2, 9)$ and $(4, 3)$
(c)   the points $(-2, 7)$ and $(5, 7)$

Solution (a).   If we label $(6, 2)$ as P_1 and $(9, 8)$ as P_2, then from (1) the slope is

$$m = \frac{8 - 2}{9 - 6} = \frac{6}{3} = 2$$

Solution (b).   The slope of the line segment connecting $(2, 9)$ and $(4, 3)$ is

$$m = \frac{3 - 9}{4 - 2} = \frac{-6}{2} = -3$$

Solution (c).   The slope of the line segment connecting $(-2, 7)$ and $(5, 7)$ is

$$m = \frac{7 - 7}{5 - (-2)} = 0 \quad \blacktriangleleft$$

REMARK.   When applying Formula (1), it does not matter which point is used as P_1 and which is used as P_2. For example, had we reversed the roles of P_1 and P_2 in part (a) of Example 1 we would have obtained

$$m = \frac{2 - 8}{6 - 9} = \frac{-6}{-3} = 2$$

which is the same value obtained for the slope using the points in the original order.

It is clear geometrically that two line segments on the same straight line will be equally "steep," and consequently have the same slope. This is the content of the following theorem.

1.4.2   THEOREM.   *On a nonvertical line, all line segments have the same slope.*

Proof.   Let $P_1(x_1, y_1)$ and $P_2(x_2, y_2)$ be distinct points on a nonvertical line L and let $P_1'(x_1', y_1')$ and $P_2'(x_2', y_2')$ be another pair of distinct points on L. We shall show that the slope

$$m = \frac{y_2 - y_1}{x_2 - x_1}$$

of the line segment joining P_1 and P_2 is equal to the slope

$$m' = \frac{y_2' - y_1'}{x_2' - x_1'}$$

of the line segment joining P_1' and P_2'. (We shall assume the points are ordered as in Figure 1.4.4. The proofs of the remaining cases are similar and will be omitted for brevity.) The triangles P_1QP_2 and $P_1'Q'P_2'$ in Figure 1.4.4 are similar, so that the ratios of the lengths of corresponding sides are equal. Therefore,

$$\frac{y_2 - y_1}{x_2 - x_1} = \frac{y_2' - y_1'}{x_2' - x_1'}$$

or, equivalently, $m = m'$.   ▌

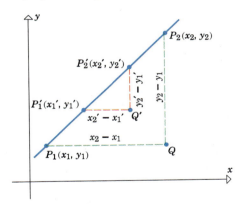

Figure 1.4.4

We define the *slope* of a nonvertical line L to be the common slope of all line segments on L. The slope of a vertical line is **undefined**. Speaking informally, some people say that a vertical line has **infinite slope**.

The significance of the numerical value of the slope can be seen by rewriting (1) as

$$y_2 - y_1 = m(x_2 - x_1) \tag{2}$$

Because $y_2 - y_1$ represents the rise and $x_2 - x_1$ the run, it follows from (2) that as we travel left to right along a nonvertical line, the rise is proportional to the run and the slope m is the constant of proportionality that relates the two. For this reason the slope m is called the *rate of change of y with respect to x* along the line. As we move from point to point along the graph, through increasing values of the x-coordinate, the slope measures the change in the y-coordinate for each unit change in x.

Example 2   In applied problems it is essential to include the units in the calculation of slope because changing the units can alter the slope of a line. For example, suppose that a uniform rod 40 cm long is thermally insulated around the lateral surface and the ends of the rod are placed in contact with objects held at constant temperatures of 27° C and 15° C (Figure 1.4.5a). After a period of time the graph of the temperature T at a distance x from the left-hand end of the rod will be the line segment shown in Figure 1.4.5b.

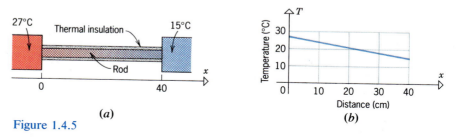

Figure 1.4.5

Using the temperatures at the ends of the rod to obtain the slope of the temperature distribution yields

$$m = \frac{15 - 27}{40 - 0} = \frac{-12}{40} = -0.3 \text{ °C/cm}$$

This slope measures the rate of change of temperature with distance along the rod. Since the negative sign indicates that the temperature is decreasing with distance along the rod, the slope $m = -0.3$ °C/cm can be interpreted to mean that the temperature is decreasing at a rate of 0.3 °C/cm along the rod. We can also describe this by saying that the *rate of change of T with respect to x is* -0.3 °C/cm. It should be noted that the calculated slope would be different if the temperature and distance were measured in different units (Exercise 10). ◄

Example 3 Figure 1.4.6 shows the three lines determined by the points in Example 1 and interprets the significance of their slopes. ◄

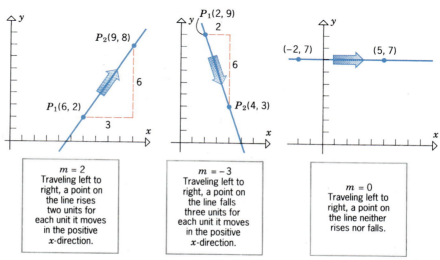

Figure 1.4.6

As illustrated in the last example, the slope of a line can be positive, negative, or zero. A positive slope means that the line is inclined upward to the right, a negative slope means that it is inclined downward to the right, and a zero slope means that the line is horizontal. An undefined slope means that the line is vertical. In Figure 1.4.7 we have sketched some lines with different slopes.

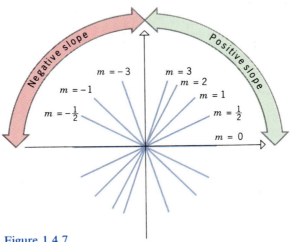

Figure 1.4.7

□ **ANGLE OF INCLINATION**   The slope of a line is related to the angle the line makes with the positive x-axis. To establish this relationship we need the following definition.

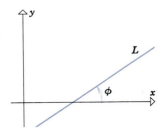

> **1.4.3 DEFINITION.**   For a line L not parallel to the x-axis, the *angle of inclination* is the smallest angle ϕ measured counterclockwise from the direction of the positive x-axis to L (Figure 1.4.8). For a line parallel to the x-axis, we take $\phi = 0$.

REMARK.   In degree measure the angle of inclination satisfies $0° \leq \phi < 180°$ and in radian measure it satisfies $0 \leq \phi < \pi$.

The following theorem relates the slope of a line to its angle of inclination.

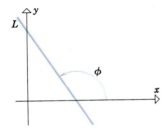

> **1.4.4 THEOREM.**   *For a nonvertical line, the slope m and angle of inclination ϕ are related by*
>
> $$m = \tan \phi \tag{3}$$

Figure 1.4.8

Proof.   If the line is horizontal, then $m = 0$ and $\phi = 0°$. Thus,

$$\tan \phi = \tan 0 = 0 = m$$

so (3) holds in this case. If the line is not horizontal, let $(x_0, 0)$ be its point of intersection with the x-axis, and construct a circle of radius 1 centered at this

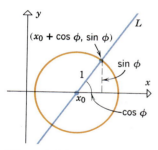

Figure 1.4.9

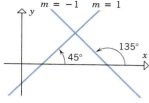

Figure 1.4.10

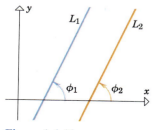

Figure 1.4.11

point. Because the angle of inclination of the line is ϕ, the line will intersect this circle at the point $(x_0 + \cos\phi, \sin\phi)$ (Figure 1.4.9).

Since the points $(x_0, 0)$ and $(x_0 + \cos\phi, \sin\phi)$ lie on L, we can use them to compute the slope m. This gives

$$m = \frac{\sin\phi - 0}{(x_0 + \cos\phi) - x_0} = \frac{\sin\phi}{\cos\phi} = \tan\phi$$

which proves (3). ∎

REMARK.   If the line L is parallel to the y-axis, then $\phi = \frac{1}{2}\pi$, so $\tan\phi$ in (3) is undefined. This agrees with the fact that the slope m is also undefined in this case.

Example 4   Find the angle of inclination for a line of slope $m = 1$ and also for a line of slope $m = -1$.

Solution.   If $m = 1$, then from (3), $\tan\phi = 1$ so that $\phi = \pi/4$ (or in degree measure $\phi = 45°$). If $m = -1$, then from (3), $\tan\phi = -1$. From this equality and the fact that $0 \le \phi < \pi$ we obtain $\phi = 3\pi/4$ or, in degree measure, $\phi = 135°$ (Figure 1.4.10).   ◀

☐ **PARALLEL AND PERPENDICULAR LINES**

As a consequence of Theorem 1.4.4, we obtain the following basic result.

1.4.5   THEOREM.   *Two nonvertical lines are parallel if and only if they have the same slope.*

Proof.   If L_1 and L_2 are nonvertical parallel lines, then their angles of inclination ϕ_1 and ϕ_2 are equal, since two parallel lines cut by a transversal have equal corresponding angles (Figure 1.4.11). Thus,

$$\text{slope } L_1 = \tan\phi_1 = \tan\phi_2 = \text{slope } L_2$$

Conversely, if L_1 and L_2 have the same slope m, then

$$m = \tan\phi_1 = \tan\phi_2 \qquad (4)$$

Because $0 \le \phi_1 < \pi$ and $0 \le \phi_2 < \pi$, it follows from (4) that ϕ_1 and ϕ_2 are equal. Thus, L_1 and L_2 are parallel (Figure 1.4.11). ∎

The next theorem shows how slopes can be used to determine whether two lines are perpendicular.

1.4.6   THEOREM.   *Two nonvertical lines are perpendicular if and only if the product of their slopes is -1; that is, lines with slopes m_1 and m_2 are perpendicular if and only if $m_1 m_2 = -1$ or equivalently*

$$m_2 = -\frac{1}{m_1} \qquad (5)$$

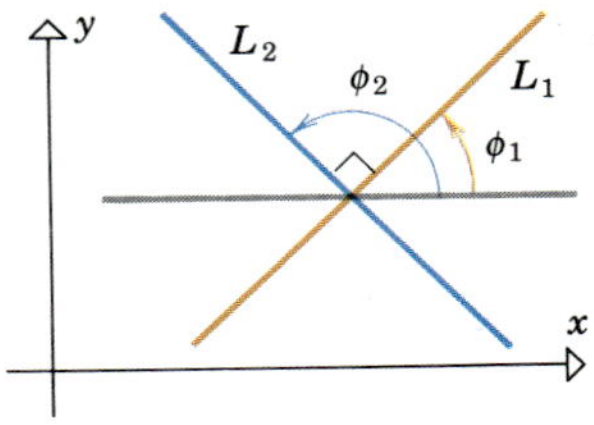

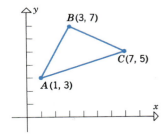

Figure 1.4.12

Proof. We shall prove first that if L_1 and L_2 are nonvertical perpendicular lines, then (5) holds. Assume that L_1 and L_2 have angles of inclination ϕ_1 and ϕ_2 and that $\phi_1 < \phi_2$ (Figure 1.4.12), so that

$$\phi_2 = \phi_1 + \frac{\pi}{2}$$

Thus,

$$m_2 = \tan \phi_2 = \tan\left(\phi_1 + \frac{\pi}{2}\right) = \frac{\sin\left(\phi_1 + \frac{\pi}{2}\right)}{\cos\left(\phi_1 + \frac{\pi}{2}\right)}$$

$$= \frac{\sin \phi_1 \cos \frac{\pi}{2} + \cos \phi_1 \sin \frac{\pi}{2}}{\cos \phi_1 \cos \frac{\pi}{2} - \sin \phi_1 \sin \frac{\pi}{2}}$$

$$= -\frac{\cos \phi_1}{\sin \phi_1} = -\frac{1}{\sin \phi_1/\cos \phi_1} = -\frac{1}{\tan \phi_1} = -\frac{1}{m_1}$$

which establishes (5). The proof of the converse is left as an exercise (Exercise 39). ∎

Example 5 Use slopes to show that the points $A(1, 3)$, $B(3, 7)$, and $C(7, 5)$ are vertices of a right triangle.

Solution. The line through A and B has slope

$$m_1 = \frac{7 - 3}{3 - 1} = 2$$

and the line through B and C has slope

$$m_2 = \frac{5 - 7}{7 - 3} = -\frac{1}{2}$$

Since $m_1 m_2 = -1$, the line through A and B is perpendicular to the line through B and C; thus, ABC is a right triangle (Figure 1.4.13). ◀

Figure 1.4.13

▶ **Exercise Set 1.4**   Ⓒ *10, 11, 12, 13, 22, 23, 34, 35, 36, 37*

---

1. Find the slope of the line through
(a) $(-1, 2)$ and $(3, 4)$
(b) $(5, 3)$ and $(7, 1)$
(c) $(4, \sqrt{2})$ and $(-3, \sqrt{2})$
(d) $(-2, -6)$ and $(-2, 12)$.

2. Find the slopes of the sides of the triangle with vertices $(-1, 2)$, $(6, 5)$, and $(2, 7)$.

3. Use slopes to determine whether the given points lie on the same line.
(a) $(1, 1)$, $(-2, -5)$, and $(0, -1)$
(b) $(-2, 4)$, $(0, 2)$, and $(1, 5)$.

4. Draw the line through $(4, 2)$ with slope
(a) $m = 3$
(b) $m = -2$
(c) $m = -\frac{3}{4}$.

5. Draw the line through $(-1, -2)$ with slope
  (a) $m = \frac{3}{5}$ (b) $m = -1$
  (c) $m = \sqrt{2}$.

6. List the lines in parts (a)–(d) of Figure 1.4.14 in the order of decreasing slope.

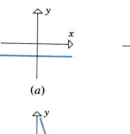
(a)   (b)

(c)   (d)

Figure 1.4.14

7. List the lines in parts (a)–(d) of Figure 1.4.15 in the order of increasing slope.

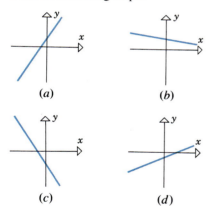
(a)   (b)

(c)   (d)

Figure 1.4.15

8. Find the slope of the lines in parts (a) and (b) of Figure 1.4.16.

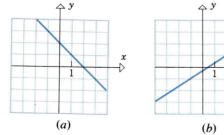
(a)   (b)

Figure 1.4.16

9. Find the slope of the lines in parts (a) and (b) of Figure 1.4.17.

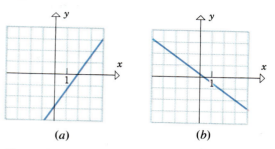
(a)   (b)

Figure 1.4.17

10. Find the slope of the temperature graph in Example 2 if temperature is in degrees Fahrenheit (°F) and distance is in feet. [*Note:* $T_F = \frac{9}{5}T + 32$, where T_F is the temperature in °F and T is the temperature in °C. Use the approximation 1 cm = 0.03281 ft.]

11. The force F (in newtons) needed to stretch a certain spring x mm beyond its natural length is related linearly to x as shown in Figure 1.4.18. The slope of the line is called the ***spring constant***. Estimate the spring constant in units of newtons/meter (N/m). [*Note:* 1 m = 1000 mm.]

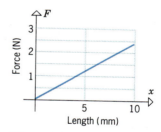
Figure 1.4.18

12. In the interval between 0° C and 100° C, the mass density ρ of mercury (g/cm³) varies linearly with the temperature T (°C), as shown in Figure 1.4.19. Estimate the slope of the line and interpret the slope as a rate of change using the correct units.

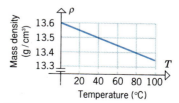
Figure 1.4.19

13. In the interval between $0°$ C and $40°$ C, the speed of sound in air (m/sec) varies linearly with the temperature T ($°$C), as shown in Figure 1.4.20. Estimate the slope of the line and interpret the slope as a rate of change using the correct units.

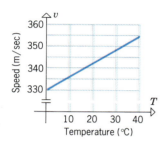

Figure 1.4.20

14. A particle, initially at $(7, 5)$, moves along a line of slope $m = -2$ to a new position (x, y).
    (a) Find y if $x = 9$.   (b) Find x if $y = 12$.

15. A particle, initially at $(1, 2)$, moves along a line of slope $m = 3$ to a new position (x, y).
    (a) Find y if $x = 5$.   (b) Find x if $y = -2$.

16. Given that the point $(k, 4)$ is on the line through $(1, 5)$ and $(2, -3)$, find k.

17. Let the point $(3, k)$ lie on the line of slope $m = 5$ through $(-2, 4)$; find k.

18. Find x and y if the line through $(0, 0)$ and (x, y) has slope $\frac{1}{2}$, and the line through (x, y) and $(7, 5)$ has slope 2.

19. Find x if the slope of the line through $(1, 2)$ and $(x, 0)$ is the negative of the slope of the line through $(4, 5)$ and $(x, 0)$.

In Exercises 20 and 21, find the slope of the line whose angle of inclination is given. (You should be able to solve these problems without a calculator.)

20. (a) $45°$ (b) $\dfrac{2\pi}{3}$ (c) $30°$.

21. (a) $\dfrac{\pi}{6}$ (b) $135°$ (c) $60°$.

In Exercises 22 and 23, use a calculator, where necessary, to find (to the nearest degree) the angle of inclination of a line with the given slope.

22. (a) $m = \frac{1}{2}$ (b) $m = -1$
    (c) $m = 2$ (d) $m = -57$.

23. (a) $m = -\frac{1}{2}$ (b) $m = 1$
    (c) $m = -2$ (d) $m = 57$.

24. Let L be a line with slope $m = 2$. Determine whether the given line L' is parallel to L, perpendicular to L, or neither.
    (a) L' is the line through $(2, 4)$ and $(4, 8)$.
    (b) L' is the line through $(2, 4)$ and $(6, 2)$.
    (c) L' is the line through $(1, 5)$ and $(2, -3)$.

25. Let L be a line with slope $m = -3$. Determine whether the given line L' is parallel to L, perpendicular to L, or neither.
    (a) L' is the line through $(1, 8)$ and $(2, 5)$.
    (b) L' is the line through $(6, 5)$ and $(3, 4)$.
    (c) L' is the line through $(1, 0)$ and $(-2, 1)$.

26. An equilateral triangle has one vertex at the origin, another on the x-axis, and the third in the first quadrant. Find the slopes of its sides.

27. Find the coordinates of all points P on the x-axis so that the line through $A(1, 2)$ and P is perpendicular to the line through $B(8, 3)$ and P.

28. Use slopes to show that $(3, 1)$, $(6, 3)$, and $(2, 9)$ are vertices of a right triangle.

29. Use slopes to show that $(3, -1)$, $(6, 4)$, $(-3, 2)$, and $(-6, -3)$ are vertices of a parallelogram.

30. Let $P(2, 3)$ be a point on the circle with center $(4, -1)$. Find the slope of the line that is tangent to the circle at P.

31. Given two intersecting lines, let L_2 be the line with the larger angle of inclination ϕ_2, and let L_1 be the line with the smaller angle of inclination ϕ_1. We define the **angle θ between L_1 and L_2** by
    $$\theta = \phi_2 - \phi_1$$
    (a) Prove geometrically that θ is the angle pictured in Figure 1.4.21. [*Remark:* The angle θ is the smallest positive angle through which L_1 can be rotated until it coincides with L_2.]
    (b) Prove: If L_1 and L_2 are not perpendicular, then
    $$\tan \theta = \frac{m_2 - m_1}{1 + m_1 m_2}$$

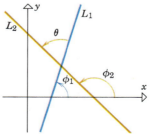

Figure 1.4.21

32. Use the result of Exercise 31 to find the tangent of the angle between the lines whose slopes are
(a) 1 and 3   (b) -2 and 4   (c) $-\frac{1}{2}$ and $-\frac{2}{3}$.

33. Use the result of Exercise 31 to find the tangent of the angle between the lines whose slopes are
(a) $\frac{4}{5}$ and $\frac{1}{3}$ (b) -0.7 and 5   (c) -6 and -2.

34. Use the result of Exercise 31 and a calculator to find, to the nearest degree, the angle between the lines whose slopes are given in Exercise 32.

35. Repeat the directions of Exercise 34 for the lines whose slopes are given in Exercise 33.

36. Use the result of Exercise 31 to find, to the nearest degree, the interior angles of the triangle whose vertices are $(4, -3)$, $(-3, -1)$, and $(6, 6)$.

37. Use the result of Exercise 31 to find the slope of the line that bisects the angle A of the triangle whose vertices are $A(0, 2)$, $B(-8, 8)$, and $C(8, 6)$.

38. Let L be a line of slope $m = -2$. Use the result of Exercise 31 to find the slope of a line K such that the angle between K and L is $45°$ (two solutions).

39. Complete the proof of Theorem 1.4.6 by showing: If L_1 and L_2 are two nonvertical lines whose slopes satisfy $m_1 m_2 = -1$, then L_1 and L_2 are perpendicular.

■ **1.5 EQUATIONS OF STRAIGHT LINES**

In this section we shall be concerned with recognizing those equations whose graphs are straight lines and finding equations for lines specified geometrically.

☐ **LINES PARALLEL TO THE COORDINATE AXES**

A line parallel to the y-axis intersects the x-axis at some point $(a, 0)$. This line consists precisely of those points whose x-coordinate is equal to a (Figure 1.5.1a). Similarly, a line parallel to the x-axis intersects the y-axis at some point $(0, b)$. This line consists precisely of those points whose y-coordinate is equal to b (Figure 1.5.1b). Thus, we have the following theorem.

1.5.1 THEOREM. *The vertical line through $(a, 0)$ and the horizontal line through $(0, b)$ are represented by the equations*

$$x = a \qquad \text{and} \qquad y = b$$

respectively.

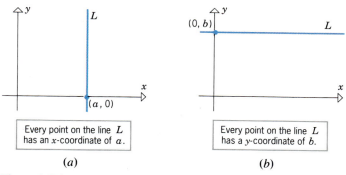

Every point on the line L has an x-coordinate of a.

(a)

Every point on the line L has a y-coordinate of b.

(b)

Figure 1.5.1

Example 1   The graph of $x = -5$ is the vertical line through $(-5, 0)$, and the graph of $y = 7$ is the horizontal line through $(0, 7)$ (Figure 1.5.2).   ◀

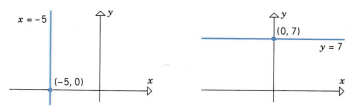

☐ **LINES DETERMINED BY POINT AND SLOPE**

There are infinitely many lines that pass through any given point in the plane. However, if we specify the slope of the line in addition to a point on it, then the point and the slope together determine a unique line (Figure 1.5.3).

Let us now consider how to find an equation of a nonvertical line L that passes through a point $P_1(x_1, y_1)$ and has slope m. If $P(x, y)$ is any point on L, different from P_1, then the slope m can be obtained from the points $P(x, y)$ and $P_1(x_1, y_1)$; this gives

$$m = \frac{y - y_1}{x - x_1}$$

which can be rewritten as

$$y - y_1 = m(x - x_1) \tag{1}$$

With the possible exception of (x_1, y_1), we have shown that every point on L satisfies (1). But $x = x_1$, $y = y_1$ satisfies (1), so that all points on L satisfy (1). We leave it as an exercise to show that every point satisfying (1) lies on L.

In summary, we have the following theorem.

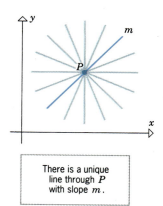

There is a unique line through P with slope m.

Figure 1.5.3

1.5.2   THEOREM.   *The line passing through $P_1(x_1, y_1)$ and having slope m is given by the equation*

$$y - y_1 = m(x - x_1) \tag{2}$$

*This is called the **point-slope form** of the line.*

Example 2   Find the point-slope form of the line through $(4, -3)$ with slope 5.

Solution.   Substituting the values $x_1 = 4$, $y_1 = -3$, and $m = 5$ in (2) yields the point-slope form $y + 3 = 5(x - 4)$.   ◀

☐ **LINES DETERMINED BY SLOPE AND *y*-INTERCEPT**

A nonvertical line crosses the y-axis at some point $(0, b)$. If we use this point in the point-slope form of its equation, we obtain

$$y - b = m(x - 0)$$

which we can rewrite as $y = mx + b$. To summarize:

> **1.5.3** THEOREM. *The line with y-intercept b and slope m is given by the equation*
>
> $$y = mx + b \tag{3}$$
>
> *This is called the **slope-intercept form** of the line.*

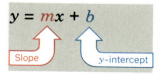

Figure 1.5.4

REMARK. Note that y is alone on one side of Equation (3). When the equation of a line is written in this way the slope of the line and its y-intercept can be determined by inspection of the equation: the slope is the coefficient of x and the y-intercept is the constant term (Figure 1.5.4).

Example 3

EQUATION	SLOPE	y-INTERCEPT
$y = 3x + 7$	$m = 3$	$b = 7$
$y = -x + \frac{1}{2}$	$m = -1$	$b = \frac{1}{2}$
$y = x$	$m = 1$	$b = 0$
$y = \sqrt{2}x - 8$	$m = \sqrt{2}$	$b = -8$
$y = 2$	$m = 0$	$b = 2$

◄

Example 4 Find the slope-intercept form of the equation of the line that satisfies the stated conditions:

(a)   slope is -9; crosses the y-axis at $(0, -4)$
(b)   slope is 1; passes through the origin
(c)   passes through $(5, -1)$; perpendicular to $y = 3x + 4$

Solution (a). From the given conditions $m = -9$ and $b = -4$, so (3) yields $y = -9x - 4$.

Solution (b). From the given conditions $m = 1$ and the line passes through $(0, 0)$, so $b = 0$. Thus, it follows from (3) that $y = x + 0$ or $y = x$.

Solution (c). The given line has slope 3, so the line to be determined will have slope $m = -\frac{1}{3}$. Substituting this slope and the given point in the point-slope form (2) and then simplifying yields

$$y - (-1) = -\tfrac{1}{3}(x - 5)$$

$$y = -\tfrac{1}{3}x + \tfrac{2}{3} \quad ◄$$

☐ **LINES DETERMINED BY TWO POINTS**

If $P_1(x_1, y_1)$ and $P_2(x_2, y_2)$ are distinct points on a nonvertical line, then the slope of the line is

$$m = \frac{y_2 - y_1}{x_2 - x_1}$$

Substituting this expression in (2) we obtain the following result.

1.5.4 THEOREM. *The nonvertical line determined by the distinct points $P_1(x_1, y_1)$ and $P_2(x_2, y_2)$ is given by the equation*

$$y - y_1 = \frac{y_2 - y_1}{x_2 - x_1}(x - x_1) \tag{4}$$

*This is called the **two-point form** of the line.*

Example 5 Find the slope-intercept form of the line passing through $(3, 4)$ and $(2, -5)$.

Solution. Letting $(x_1, y_1) = (3, 4)$ and $(x_2, y_2) = (2, -5)$ and substituting in (4), we obtain the two-point form

$$y - 4 = \frac{-5 - 4}{2 - 3}(x - 3)$$

which can be written as $y - 4 = 9(x - 3)$. Solving for y yields the slope-intercept form

$$y = 9x - 23 \quad \blacktriangleleft$$

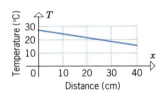

Figure 1.5.5

Example 6 Figure 1.5.5 shows the graph of the temperature distribution along the rod described in Example 2 of Section 1.4. Find the slope-intercept form of the equation for the graph.

Solution. The slope is $m = -0.3\ °C/cm$ and the intercept on the T-axis is $27°$, so

$$T = -0.3x + 27$$

Note, however, that this equation is not applicable for all real values of x. The value of x is limited by the length of the rod to lie in the interval $0 \le x \le 40$. Thus, the graph is actually a line segment rather than a complete line. We denote this restriction by writing the equation as

$$T = -0.3x + 27, \quad 0 \le x \le 40 \quad \blacktriangleleft$$

□ **THE GENERAL EQUATION OF A LINE**

An equation that is expressible in the form

$$Ax + By + C = 0 \tag{5}$$

where A, B, and C are constants and A and B are not both zero, is called a ***first-degree equation*** in x and y. For example,

$$4x + 6y - 5 = 0$$

is a first-degree equation in x and y since it has form (5) with

$$A = 4, \quad B = 6, \quad C = -5$$

In fact, all the equations of lines studied in this section are first-degree equations in x and y.

The following theorem states that the first-degree equations in x and y are precisely the equations whose graphs in the xy-plane are straight lines. (The proof is left as an exercise.)

1.5.5 THEOREM. *Every first-degree equation in x and y has a straight line as its graph and, conversely, every straight line can be represented by a first-degree equation in x and y.*

Because of this theorem, (5) is sometimes called the **general equation** of a line or a **linear equation** in x and y.

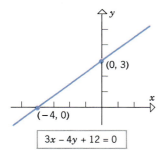

$3x - 4y + 12 = 0$

Figure 1.5.6

Example 7 Graph the equation $3x - 4y + 12 = 0$.

Solution. Since this is a linear equation in x and y, its graph is a straight line. Thus, to sketch the graph we need only plot any two points on the graph and draw the line through them. It is particularly convenient to plot the points where the line crosses the coordinate axes. These points are $(0, 3)$ and $(-4, 0)$ (verify), so the graph is the line shown in Figure 1.5.6. ◀

Example 8 If an object is thrown straight up near the surface of the earth at time $t = 0$ sec with an initial velocity of v_0, and if air resistance is ignored, then for $t \geq 0$ its velocity v is given by the linear equation

$$v = v_0 - gt \tag{6}$$

where g is a constant, called the **acceleration due to gravity,** whose value is approximately 9.8 m/sec^2. Suppose that $v_0 = 24.5$ m/sec.

(a) Graph velocity versus time for $t \geq 0$, making the t-axis horizontal and the v-axis vertical.

(b) Find the slope of the line and give an interpretation of the slope as a rate of change.

Solution (a). Substituting the given values of g and v_0 in (6) yields

$$v = 24.5 - 9.8t, \quad t \geq 0$$

It follows that $v = 24.5$ when $t = 0$ and that $v = 0$ when $t = 24.5/9.8 = 2.5$, which results in the graph shown in Figure 1.5.7.

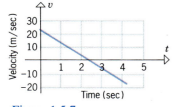

Figure 1.5.7

Solution (b). The slope is -9.8 m/sec^2, which can be interpreted to mean that the velocity is decreasing at the rate of 9.8 m/sec per second. ◀

▶ Exercise Set 1.5   Ⓒ 7, 8, 28, 29, 31, 32, 33

1. Graph the equations
   (a) $2x + 5y = 15$ (b) $x = 3$
   (c) $y = -2$ (d) $y = 2x - 7$.

2. Graph the equations
   (a) $\dfrac{x}{3} - \dfrac{y}{4} = 1$ (b) $x = -8$
   (c) $y = 0$ (d) $x = 3y + 2$.

3. Graph the equations
   (a) $y = 2x - 1$ (b) $y = 3$
   (c) $y = -2x$.

4. Graph the equations
   (a) $y = 2 - 3x$ (b) $y = \frac{1}{4}x$
   (c) $y = -\sqrt{3}$.

5. Find the slope and y-intercept of
   (a) $y = 3x + 2$ (b) $y = 3 - \frac{1}{4}x$
   (c) $3x + 5y = 8$ (d) $y = 1$
   (e) $\dfrac{x}{a} + \dfrac{y}{b} = 1$.

6. Find the slope and y-intercept of
   (a) $y = -4x + 2$ (b) $x = 3y + 2$
   (c) $\dfrac{x}{2} + \dfrac{y}{3} = 1$ (d) $y - 3 = 0$
   (e) $a_0 x + a_1 y = 0$ $(a_1 \neq 0)$.

7. To the nearest degree, find the angle of inclination of
   (a) $y = \sqrt{3}\,x + 2$ (b) $y + 2x + 5 = 0$.

8. To the nearest degree, find the angle of inclination of
   (a) $3y = 2 - \sqrt{3}\,x$ (b) $y - 4x + 7 = 0$.

In Exercises 9–22, find the slope-intercept form of the line satisfying the given conditions.

9. Slope $= -2$, y-intercept $= 4$.

10. $m = 5$, $b = -3$.

11. The line is parallel to $y = 4x - 2$ and its y-intercept is 7.

12. The line is parallel to $3x + 2y = 5$ and passes through $(-1, 2)$.

13. The line is perpendicular to $y = 5x + 9$ and has y-intercept 6.

14. The line is perpendicular to $x - 4y = 7$ and passes through $(3, -4)$.

15. The line passes through $(2, 4)$ and $(1, -7)$.

16. The line passes through $(-3, 6)$ and $(-2, 1)$.

17. The line has an angle of inclination of $\phi = \frac{1}{6}\pi$ and a y-intercept of -3.

18. The line has angle of inclination $\phi = \frac{2}{3}\pi$ and passes through the point $(1, 2)$.

19. The y-intercept is 2 and the x-intercept is -4.

20. The y-intercept is b and the x-intercept is a.

21. The line is perpendicular to the y-axis and passes through $(-4, 1)$.

22. The line is parallel to $y = -5$ and passes through $(-1, -8)$.

23. Find an equation for the line that passes through $(5, -2)$ and has angle of inclination $\phi = \frac{1}{2}\pi$.

24. Find an equation for the y-axis.

25. In each part, classify the lines as parallel, perpendicular, or neither.
   (a) $y = 4x - 7$ and $y = 4x + 9$
   (b) $y = 2x - 3$ and $y = 7 - \frac{1}{2}x$
   (c) $5x - 3y + 6 = 0$ and $10x - 6y + 7 = 0$
   (d) $Ax + By + C = 0$ and $Bx - Ay + D = 0$
   (e) $y - 2 = 4(x - 3)$ and $y - 7 = \frac{1}{4}(x - 3)$.

26. In each part, classify the lines as parallel, perpendicular, or neither.
   (a) $y = -5x + 1$ and $y = 3 - 5x$
   (b) $y - 1 = 2(x - 3)$ and $y - 4 = -\frac{1}{2}(x + 7)$
   (c) $4x + 5y + 7 = 0$ and $5x - 4y + 9 = 0$
   (d) $Ax + By + C = 0$ and $Ax + By + D = 0$
   (e) $y = \frac{1}{2}x$ and $x = \frac{1}{2}y$.

27. For what value of k will the line $3x + ky = 4$
   (a) have slope 2
   (b) have y-intercept 5
   (c) pass through the point $(-2, 4)$
   (d) be parallel to the line $2x - 5y = 1$
   (e) be perpendicular to the line $4x + 3y = 2$?

28. When light of suitably high frequency falls on a metal surface, electrons are emitted (Figure 1.5.8). In his explanation of this phenomenon, Albert Einstein showed in 1905 that the stopping voltage V ($V \geq 0$) needed to prevent emission is related to the frequency f of the light according to the equation

   $$V = kf - P$$

   where $k = 4.1 \times 10^{-15}$ volts/Hz and $P = 1.77$ volts for sodium metal.
   (a) Graph this equation in an fV-coordinate system for $f \leq 10^{15}$ Hz, and $V \geq 0$.

(b) What is the frequency below which electrons will not be emitted? [*Hint:* Electrons are not emitted if $V \leq 0$.]

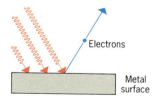

Figure 1.5.8

29. There are two common systems for measuring temperature, Celsius and Fahrenheit. Water freezes at 0° Celsius (0° C) and 32° Fahrenheit (32° F); it boils at 100° C and 212° F.

(a) Assuming that the Celsius temperature T_C and the Fahrenheit temperature T_F are related by a linear equation, find the equation.

(b) What is the slope of the line relating T_F and T_C if T_F is plotted on the horizontal axis?

(c) At what temperature is the Fahrenheit reading equal to the Celsius reading?

(d) Normal body temperature is 98.6° F. What is it in °C?

30. Thermometers are calibrated using the so-called "triple point" of water, which is 273.16 K on the Kelvin scale and 0.01° C on the Celsius scale (where 1° C = 1 K). There is a linear relationship between the temperature T_C in degrees Celsius and the temperature T_K in kelvins.

(a) Find an equation that relates T_C and T_K.

(b) Absolute zero (0 K on the Kelvin scale) is the temperature below which a body's temperature cannot be lowered. Express absolute zero in °C.

31. To the extent that water can be assumed to be incompressible, the pressure p in a body of water varies linearly with the distance h below the surface.

(a) Given that the pressure is 1 atmosphere (1 atm) at the surface and 5.9 atm at a depth of 50 m, find an equation that relates pressure to depth.

(b) At what depth is the pressure twice that at the surface?

32. A resistance thermometer is a device that determines temperature by measuring the resistance of a fine wire whose resistance varies with temperature. Suppose that the resistance R in ohms (Ω) varies linearly with the temperature T in °C and that $R = 123.4\ \Omega$ when $T = 20°\,C$ and that $R = 133.9\ \Omega$ when $T = 45°\,C$.

(a) Find an equation for R in terms of T.

(b) If R is measured experimentally as 128.6 Ω, what is the temperature?

33. Suppose that the mass of a spherical mothball decreases with time, due to evaporation, at a rate that is proportional to its surface area. Assuming that it always retains the shape of a sphere, it can be shown that the radius r of the sphere decreases linearly with the time t.

(a) If, at a certain instant, the radius is 0.80 mm and 4 days later it is 0.75 mm, find an equation for r (in millimeters) in terms of the elapsed time t (in days).

(b) How long will it take for the mothball to completely evaporate?

34. In physical problems linear equations involving variables other than x and y often arise. In parts (a)–(g) determine whether the equation is linear.

(a) $3\alpha - 2\beta = 5$

(b) $A = 2000(1 + 0.06t)$

(c) $A = \pi r^2$

(d) $E = mc^2$ (c constant)

(e) $V = C(1 - rt)$ (r and C constant)

(f) $V = \frac{1}{3}\pi r^2 h$ (r constant)

(g) $V = \frac{1}{3}\pi r^2 h$ (h constant).

35. A point moves in the xy-plane in such a way that at any time t its coordinates are given by $x = 5t + 2$ and $y = t - 3$. By expressing y in terms of x, show that the point moves along a straight line.

36. A point moves in the xy-plane in such a way that at any time t its coordinates are given by $x = 1 + 3t^2$ and $y = 2 - t^2$. By expressing y in terms of x, show that the point moves along a straight line path and specify the values of x for which the equation is valid.

37. Find the area of the triangle formed by the coordinate axes and the line through $(1, 4)$ and $(2, 1)$.

38. Draw the graph of $4x^2 - 9y^2 = 0$. [*Hint:* Factor.]

39. In each part, find the point of intersection of the lines.

(a) $2x + 3y = 5$ and $y = -1$

(b) $4x + 3y = -2$ and $5x - 2y = 9$.

40. In each part, find the point of intersection of the lines.

(a) $6x - 9y = 7$ and $x = -\frac{2}{3}$

(b) $6x - 2y = -3$ and $-8x + 3y = 5$.

41. Find the point on the graph of $y = x^2$ that is closest to the line $y = x - 2$. [*Hint:* First determine those lines that are parallel to $y = x - 2$ and intersect the graph of $y = x^2$.]

42. Prove: If (x, y) satisfies Equation (1), then the point $P(x, y)$ lies on the line with slope m passing through $P_1(x_1, y_1)$.

43. Prove Theorem 1.5.5.

■ 1.6 DISTANCE; CIRCLES; EQUATIONS OF THE FORM $y = ax^2 + bx + c$

In this section we shall derive a formula for the distance between two points in a coordinate plane, and we shall use that formula to study equations of circles. We shall also study equations of the form $y = ax^2 + bx + c$. Such equations will be considered in more detail later in the text, but they are introduced here to provide an immediate source of useful examples for subsequent sections.

☐ **DISTANCE BETWEEN TWO POINTS IN THE PLANE**

In Section 1.2 we showed that if A and B are points on a coordinate line, and if A and B have coordinates a and b, respectively, then the distance d between the points is $d = |b - a|$. It follows from this result that the distance d between two points $A(x_1, y)$ and $B(x_2, y)$ on any horizontal line in the xy-plane is

$$d = |x_2 - x_1|$$

(Figure 1.6.1) and the distance d between two points $A(x, y_1)$ and $B(x, y_2)$ on any vertical line in the xy-plane is

$$d = |y_2 - y_1|$$

(Figure 1.6.2).

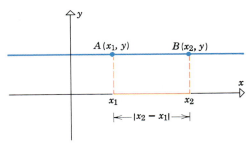

Figure 1.6.1

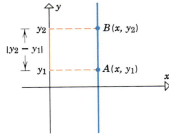

Figure 1.6.2

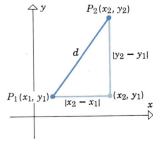

Figure 1.6.3

To find the distance d between any two points $P_1(x_1, y_1)$ and $P_2(x_2, y_2)$ in the xy-plane we apply the Theorem of Pythagoras to the triangle shown in Figure 1.6.3. This yields

$$d = \sqrt{|x_2 - x_1|^2 + |y_2 - y_1|^2}$$

But for every real number a we have $|a|^2 = a^2$ (since a and $|a|$ can differ only in sign). Thus, $|x_2 - x_1|^2 = (x_2 - x_1)^2$ and $|y_2 - y_1|^2 = (y_2 - y_1)^2$, which yields the following result.

1.6.1 THEOREM. *The distance d between two points (x_1, y_1) and (x_2, y_2) in a coordinate plane is given by*

$$d = \sqrt{(x_2 - x_1)^2 + (y_2 - y_1)^2}$$

(1)

It is assumed in this formula that equal scales are used on the coordinate axes.

Example 1 Find the distance between the points $(-2, 3)$ and $(1, 7)$.

Solution. If we let (x_1, y_1) be $(-2, 3)$ and let (x_2, y_2) be $(1, 7)$, then (1) yields

$$d = \sqrt{[1 - (-2)]^2 + [7 - 3]^2} = \sqrt{3^2 + 4^2} = \sqrt{25} = 5 \quad \blacktriangleleft$$

REMARK. When using (1) it does not matter which point is labeled (x_1, y_1) and which is labeled (x_2, y_2). Thus, in Example 1, if we had let (x_1, y_1) be the point $(1, 7)$ and (x_2, y_2) the point $(-2, 3)$, we would have obtained

$$d = \sqrt{[-2 - 1]^2 + [3 - 7]^2} = \sqrt{(-3)^2 + (-4)^2} = \sqrt{25} = 5$$

which is the same result we obtained with the opposite labeling.

REMARK. The distance between two points P_1 and P_2 in a coordinate plane is commonly denoted by $d(P_1, P_2)$, or $d(P_2, P_1)$.

Example 2 It can be shown that the converse of the Theorem of Pythagoras is true; that is, if the sides of a triangle satisfy the relationship $a^2 + b^2 = c^2$, then the triangle must be a right triangle. Use this result to show that the points $A(4, 6)$, $B(1, -3)$, $C(7, 5)$ are vertices of a right triangle.

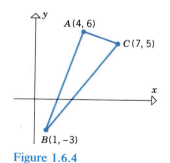

$A(4, 6)$

$C(7, 5)$

$B(1, -3)$

Figure 1.6.4

Solution. The points and the triangle are shown in Figure 1.6.4. From (1) the lengths of the sides of the triangles are

$$d(A, B) = \sqrt{(1 - 4)^2 + (-3 - 6)^2} = \sqrt{9 + 81} = \sqrt{90}$$
$$d(A, C) = \sqrt{(7 - 4)^2 + (5 - 6)^2} = \sqrt{9 + 1} = \sqrt{10}$$
$$d(B, C) = \sqrt{(7 - 1)^2 + [5 - (-3)]^2} = \sqrt{36 + 64} = \sqrt{100} = 10$$

Since

$$[d(A, B)]^2 + [d(A, C)]^2 = [d(B, C)]^2$$

it follows that $\triangle ABC$ is a right triangle with hypotenuse BC. $\blacktriangleleft$

☐ **THE MIDPOINT FORMULA**

It is often necessary to find the coordinates of the midpoint of a line segment joining two points in the plane. To derive the midpoint formula, we shall start with two points on a coordinate line. If we assume that the points have coordinates a and b and that $a \leq b$, then, as shown in Figure 1.6.5a, the distance between a and b is $b - a$, and the coordinate of the midpoint between a and b is

$$a + \tfrac{1}{2}(b - a) = \tfrac{1}{2}a + \tfrac{1}{2}b = \tfrac{1}{2}(a + b)$$

which is the arithmetic average of a and b. Had the points been labeled with $b \leq a$, the same formula would have resulted (verify). Therefore, *the midpoint of two points on a coordinate line is the arithmetic average of their coordinates, regardless of their relative positions.*

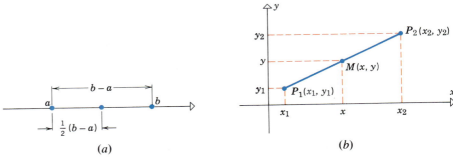

Figure 1.6.5

If we now let $P_1(x_1, y_1)$ and $P_2(x_2, y_2)$ be any two points in the plane and $M(x, y)$ the midpoint of the line segment joining them (Figure 1.6.5b), then it can be shown using plane geometry that x is the midpoint of x_1 and x_2 on the x-axis and y is the midpoint of y_1 and y_2 on the y-axis, so

$$x = \tfrac{1}{2}(x_1 + x_2) \quad \text{and} \quad y = \tfrac{1}{2}(y_1 + y_2)$$

Thus, we have the following result.

1.6.2 THEOREM (*The Midpoint Formula*). *The midpoint of the line segment joining two points* (x_1, y_1) *and* (x_2, y_2) *in a coordinate plane is*

$$\left(\tfrac{1}{2}(x_1 + x_2), \tfrac{1}{2}(y_1 + y_2)\right) \tag{2}$$

Example 3 Find the midpoint of the line segment joining $(3, -4)$ and $(7, 2)$.

Solution. From (2) the midpoint is

$$\left(\frac{3 + 7}{2}, \frac{-4 + 2}{2}\right) = (5, -1) \quad \blacktriangleleft$$

□ **CIRCLES**

If (x_0, y_0) is a fixed point in the plane, then the circle of radius r centered at (x_0, y_0) is the set of all points in the plane whose distance from (x_0, y_0) is r (Figure 1.6.6). Thus, a point (x, y) will lie on this circle if and only if

$$\sqrt{(x - x_0)^2 + (y - y_0)^2} = r$$

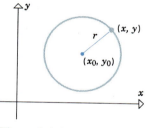

Figure 1.6.6

or equivalently

$$(x - x_0)^2 + (y - y_0)^2 = r^2 \qquad (3)$$

This is called the *standard form of the equation of a circle.*

Example 4 Find an equation for the circle of radius 4 centered at $(-5, 3)$.

Solution. From (3) with $x_0 = -5$, $y_0 = 3$, and $r = 4$ we obtain

$$(x + 5)^2 + (y - 3)^2 = 16$$

If desired, this equation can be written in an expanded form by squaring the terms, then simplifying:

$$(x^2 + 10x + 25) + (y^2 - 6y + 9) - 16 = 0$$
$$x^2 + y^2 + 10x - 6y + 18 = 0 \qquad \blacktriangleleft$$

Example 5 Find an equation for the circle with center $(1, -2)$ that passes through $(4, 2)$.

Solution. The radius r of the circle is the distance between $(4, 2)$ and $(1, -2)$, so

$$r = \sqrt{(1 - 4)^2 + (-2 - 2)^2} = 5$$

We now know the center and radius, so we can use (3) to obtain the equation

$$(x - 1)^2 + (y + 2)^2 = 25$$

which can also be written as

$$x^2 + y^2 - 2x + 4y - 20 = 0 \qquad \blacktriangleleft$$

☐ **FINDING THE CENTER AND RADIUS OF A CIRCLE**

When you encounter an equation of form (3), you will know immediately that its graph is a circle; its center and radius can then be found from the constants that appear in the equation:

$$\underbrace{(x - x_0)^2}_{\text{x-coordinate of the center}} + \underbrace{(y - y_0)^2}_{\text{y-coordinate of the center}} = \underbrace{r^2}_{\text{radius squared}}$$

Example 6

EQUATION OF A CIRCLE	CENTER (x_0, y_0)	RADIUS r
$(x - 2)^2 + (y - 5)^2 = 9$	$(2, 5)$	3
$(x + 7)^2 + (y + 1)^2 = 16$	$(-7, -1)$	4
$x^2 + y^2 = 25$	$(0, 0)$	5
$(x - 4)^2 + y^2 = 5$	$(4, 0)$	$\sqrt{5}$

$\blacktriangleleft$

The circle $x^2 + y^2 = 1$, which is centered at the origin and has radius 1, is of special importance; it is called the **unit circle** (Figure 1.6.7).

☐ **OTHER FORMS FOR THE EQUATION OF A CIRCLE**

An alternative version of Equation (3) can be obtained by squaring the terms and simplifying. This yields an equation of the form

$$x^2 + y^2 + dx + ey + f = 0 \tag{4}$$

where d, e, and f are constants. (See the final equations in Examples 4 and 5.)

Still another version of the equation of a circle can be obtained by multiplying both sides of (4) by a nonzero constant A. This yields an equation of the form

$$Ax^2 + Ay^2 + Dx + Ey + F = 0 \tag{5}$$

where A, D, E, and F are constants and $A \neq 0$.

If the equation of a circle is given by (4) or (5), then the center and radius can be found by first rewriting the equation in standard form, then reading off the center and radius from that equation. The following example shows how to do this using the technique of *completing the square*.

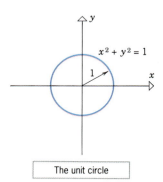

The unit circle

Figure 1.6.7

Example 7   Find the center and radius of the circle with equation

(a) $x^2 + y^2 - 8x + 2y + 8 = 0$

(b) $2x^2 + 2y^2 + 24x - 81 = 0$

Solution (a).   First, group the x-terms, group the y-terms, and take the constant to the right side:

$$(x^2 - 8x) + (y^2 + 2y) = -8$$

Next we want to add the appropriate constant within each set of parentheses to complete the square, and add the same constants to the right side to maintain equality. The appropriate constant is obtained by taking half the coefficient of the first-degree term and squaring it. This yields

$$(x^2 - 8x + 16) + (y^2 + 2y + 1) = -8 + 16 + 1$$

or

$$(x - 4)^2 + (y + 1)^2 = 9$$

Thus from (3), the circle has center $(4, -1)$ and radius 3.

Solution (b).   The given equation is of form (5). We shall first divide through by 2 (the coefficient of the squared terms) to reduce the equation to form (4). Then we shall proceed as in part (a) of this example. The computations are as follows:

$$x^2 + y^2 + 12x - \frac{81}{2} = 0 \qquad \boxed{\text{We divided through by 2.}}$$

$$(x^2 + 12x) + y^2 = \frac{81}{2}$$

$$(x^2 + 12x + 36) + y^2 = \frac{81}{2} + 36 \quad \boxed{\text{We completed the square.}}$$

$$(x + 6)^2 + y^2 = \frac{153}{2}$$

From (3), the circle has center $(-6, 0)$ and radius $\sqrt{\frac{153}{2}}$. ◀

☐ DEGENERATE CASES OF A CIRCLE

There is no guarantee that an equation of form (5) represents a circle. For example, suppose we divide both sides of (5) by A, then complete the squares to obtain

$$(x - x_0)^2 + (y - y_0)^2 = k$$

Depending on the value of k, the following situations occur:

- $(k > 0)$ The graph is a circle with center (x_0, y_0) and radius $\sqrt{k}$.
- $(k = 0)$ The only solution of the equation is $x = x_0$, $y = y_0$, so the graph is the single point (x_0, y_0).
- $(k < 0)$ The equation has no real solutions and consequently no graph.

Example 8 Describe the graphs of

(a) $(x - 1)^2 + (y + 4)^2 = -9$ (b) $(x - 1)^2 + (y + 4)^2 = 0$

Solution (a). There are no real values of x and y that will make the left side of the equation negative. Thus, the solution set of the equation is empty, and the equation has no graph.

Solution (b). The only values of x and y that will make the left side of the equation 0 are $x = 1$, $y = -4$. Thus, the graph of the equation is the single point $(1, -4)$. ◀

The following theorem summarizes our observations:

1.6.3 THEOREM. *An equation of the form*

$$Ax^2 + Ay^2 + Dx + Ey + F = 0 \tag{6}$$

where $A \neq 0$, represents a circle, or a point, or else has no graph.

REMARK. The last two cases in Theorem 1.6.3 are called ***degenerate cases***. In spite of the fact that these degenerate cases can occur, (6) is often called the ***general equation of a circle***.

☐ **THE GRAPH OF**
   $y = ax^2 + bx + c$

An equation of the form

$$y = ax^2 + bx + c \quad (a \neq 0) \tag{7}$$

is called a *quadratic equation in x*. Its graph is a curve called a *parabola*. If a is positive, the parabola will open up as shown in Figure 1.6.8*a*, and if a is negative, it will open down as shown in Figure 1.6.8*b*. In both cases the parabola is symmetric about a vertical line parallel to the *y*-axis. This line of symmetry cuts the parabola at a point called the *vertex*. The vertex is the low point on the curve if $a > 0$ and the high point if $a < 0$.

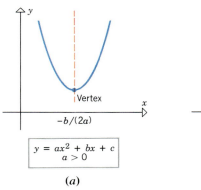

Figure 1.6.8

In the exercises (Exercise 78) we will help the reader show that the *x*-coordinate of the vertex is given by the formula

$$x = -\frac{b}{2a} \tag{8}$$

With the aid of this formula, a reasonably accurate graph of a quadratic equation in *x* can be obtained by plotting the vertex and two points on each side of it.

Example 9   Sketch the graph of

(a) $y = x^2 - 4x + 5$ (b) $y = -x^2 + 2x + 2$

Solution (a).   The equation is of form (7) with $a = 1$, $b = -4$, and $c = 5$, so by (8) the *x*-coordinate of the vertex is

$$x = -\frac{b}{2a} = 2$$

Using this value and two additional values on each side, we obtain the table and graph in Figure 1.6.9.

x	$y = x^2 - 4x + 5$
0	5
1	2
2	1
3	2
4	5

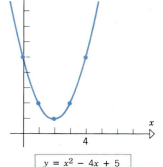

$y = x^2 - 4x + 5$

Figure 1.6.9

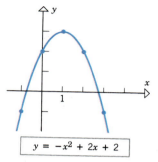

x	$y = -x^2 + 2x + 2$
-1	-1
0	2
1	3
2	2
3	-1

$y = -x^2 + 2x + 2$

Figure 1.6.10

Solution (b). The equation is of form (7) with $a = -1, b = 2$, and $c = 2$, so by (8) the x-coordinate of the vertex is

$$x = -\frac{b}{2a} = 1$$

Using this value and two additional values on each side, we obtain the table and graph in Figure 1.6.10. ◀

Often, the intercepts of a parabola $y = ax^2 + bx + c$ are important to know. The y-intercept, $y = c$, results immediately by setting $x = 0$. However, to obtain the x-intercepts, if any, we must set $y = 0$ and solve the resulting quadratic equation $ax^2 + bx + c = 0$.

Example 10 From Figure 1.6.9, we see that the parabola $y = x^2 - 4x + 5$ has no x-intercepts. This can be verified algebraically by setting $y = 0$ and solving for x. We obtain

$$x^2 - 4x + 5 = 0$$

By the quadratic formula

$$x = \frac{-b \pm \sqrt{b^2 - 4ac}}{2a} = \frac{4 \pm \sqrt{16 - 20}}{2} = 2 \pm \frac{\sqrt{-4}}{2}$$

so the solutions are not real. Thus, there are no x-intercepts. ◀

Example 11 It appears from Figure 1.6.10 that the x-intercepts of the parabola $y = -x^2 + 2x + 2$ lie somewhere in the open intervals $(-1, 0)$ and $(2, 3)$. To find these intercepts we set $y = 0$ to obtain

$$-x^2 + 2x + 2 = 0$$

Solving by the quadratic formula gives

$$x = \frac{-b \pm \sqrt{b^2 - 4ac}}{2a} = \frac{-2 \pm \sqrt{12}}{-2} = 1 \pm \sqrt{3}$$

Thus, the x-intercepts are

$$x = 1 + \sqrt{3} \approx 2.7 \quad \text{and} \quad x = 1 - \sqrt{3} \approx -.7 \quad ◀$$

Example 12 A ball is thrown straight up from the surface of the earth at time $t = 0$ sec with an initial velocity of 24.5 m/sec. If air resistance is ignored, it can be shown that the distance s (in meters) of the ball above the ground after t sec is given by

$$s = 24.5t - 4.9t^2 \tag{9}$$

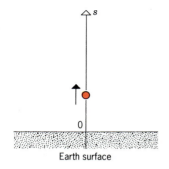

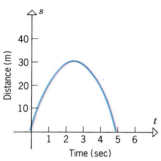

Figure 1.6.11

(a)   Graph s versus t, making the t-axis horizontal and the s-axis vertical.

(b)   How high does the ball rise above the ground?

Solution (a).   Equation (9) is of form (7) with $a = -4.9$, $b = 24.5$, and $c = 0$, so by (8) the t-coordinate of the vertex is

$$t = -\frac{b}{2a} = -\frac{24.5}{2(-4.9)} = 2.5 \text{ sec}$$

and consequently the s-coordinate of the vertex is

$$s = 24.5(2.5) - 4.9(2.5)^2 = 30.625 \text{ m}$$

The factored form of (9) is

$$s = 4.9t(5 - t)$$

so the graph has t-intercepts $t = 0$ and $t = 5$. From the vertex and the intercepts we obtain the graph shown in Figure 1.6.11.

Solution (b).   From the s-coordinate of the vertex we deduce that the ball rises 30.625 m above the ground.   ◀

REMARK.   If x and y are interchanged in (7), the resulting equation,

$$x = ay^2 + by + c$$

is called a *quadratic equation in y.* The graph of such an equation is a parabola with its line of symmetry parallel to the x-axis and its vertex at the point with y-coordinate $y = -b/(2a)$ (Figure 1.6.12). Some problems relating to such equations appear in the exercises.

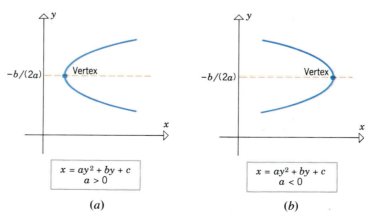

Figure 1.6.12

▶ Exercise Set 1.6

1. Where in this section did we use the fact that the same unit of measure was used on both coordinate axes?

In Exercises 2–5, find
(a) the distance between A and B
(b) the midpoint of the line segment joining A and B.

2. $A(2, 5)$, $B(-1, 1)$.

3. $A(7, 1)$, $B(1, 9)$.

4. $A(2, 0)$, $B(-3, 6)$.

5. $A(-2, -6)$, $B(-7, -4)$.

In Exercises 6–10, use the distance formula to solve the problem.

6. Prove that $(1, 1)$, $(-2, -8)$, and $(4, 10)$ lie on a straight line.

7. Prove that the triangle with vertices $(5, -2)$, $(6, 5)$, $(2, 2)$ is isosceles.

8. Prove that $(1, 3)$, $(4, 2)$, and $(-2, -6)$ are vertices of a right triangle and specify the vertex at which the right angle occurs.

9. Prove that $(0, -2)$, $(-4, 8)$, and $(3, 1)$ lie on a circle with center $(-2, 3)$.

10. Prove that for all values of t the point $(t, 2t - 6)$ is equidistant from $(0, 4)$ and $(8, 0)$.

11. Find k, given that $(2, k)$ is equidistant from $(3, 7)$ and $(9, 1)$.

12. Find x and y if $(4, -5)$ is the midpoint of the line segment joining $(-3, 2)$ and (x, y).

In Exercises 13 and 14, find an equation of the given line.

13. The line is the perpendicular bisector of the line segment joining $(2, 8)$ and $(-4, 6)$.

14. The line is the perpendicular bisector of the line segment joining $(5, -1)$ and $(4, 8)$.

15. Find the point on the line $4x - 2y + 3 = 0$ that is equidistant from $(3, 3)$ and $(7, -3)$. [*Hint:* First find an equation of the line that is the perpendicular bisector of the line segment joining $(3, 3)$ and $(7, -3)$.]

16. Find the distance from the point $(3, -2)$ to the line
(a) $y = 4$ (b) $x = -1$.

17. Find the distance from the point $(2, 1)$ to the line $4x - 3y + 10 = 0$. [*Hint:* Find the foot of the perpendicular dropped from the point to the line.]

18. Find the distance from the point $(8, 4)$ to the line $5x + 12y - 36 = 0$. [*Hint:* See the hint in Exercise 17.]

19. Use the method described in Exercise 17 to prove that the distance d from (x_0, y_0) to the line $Ax + By + C = 0$ is

$$d = \frac{|Ax_0 + By_0 + C|}{\sqrt{A^2 + B^2}}$$

20. Use the formula in Exercise 19 to solve Exercise 17.

21. Use the formula in Exercise 19 to solve Exercise 18.

22. Prove: For any triangle, the perpendicular bisectors of the sides meet at a point. [*Hint:* Position the triangle with one vertex on the y-axis and the opposite side on the x-axis, so that the vertices are $(0, a)$, $(b, 0)$, and $(c, 0)$.]

In Exercises 23 and 24, find the center and radius of each circle.

23. (a) $x^2 + y^2 = 25$
(b) $(x - 1)^2 + (y - 4)^2 = 16$
(c) $(x + 1)^2 + (y + 3)^2 = 5$
(d) $x^2 + (y + 2)^2 = 1$.

24. (a) $x^2 + y^2 = 9$
(b) $(x - 3)^2 + (y - 5)^2 = 36$
(c) $(x + 4)^2 + (y + 1)^2 = 8$
(d) $(x + 1)^2 + y^2 = 1$.

In Exercises 25–32, find the standard equation of the circle satisfying the given conditions.

25. Center $(3, -2)$; radius $= 4$.

26. Center $(1, 0)$; diameter $= \sqrt{8}$.

27. Center $(-4, 8)$; circle is tangent to the x-axis.

28. Center $(5, 8)$; circle is tangent to the y-axis.

29. Center $(-3, -4)$; circle passes through the origin.

30. Center $(4, -5)$; circle passes through $(1, 3)$.

31. A diameter has endpoints $(2, 0)$ and $(0, 2)$.

32. A diameter has endpoints $(6, 1)$ and $(-2, 3)$.

In Exercises 33–44, determine whether the equation represents a circle, a point, or no graph. If the equation represents a circle, find the center and radius.

33. $x^2 + y^2 - 2x - 4y - 11 = 0$.

34. $x^2 + y^2 + 8x + 8 = 0$.

35. $2x^2 + 2y^2 + 4x - 4y = 0$.

36. $6x^2 + 6y^2 - 6x + 6y = 3$.

37. $x^2 + y^2 + 2x + 2y + 2 = 0$.

38. $x^2 + y^2 - 4x - 6y + 13 = 0$.

39. $9x^2 + 9y^2 = 1$.

40. $\dfrac{x^2}{4} + \dfrac{y^2}{4} = 1$.

41. $x^2 + y^2 + 10y + 26 = 0$.

42. $x^2 + y^2 - 10x - 2y + 29 = 0$.

43. $16x^2 + 16y^2 + 40x + 16y - 7 = 0$.

44. $4x^2 + 4y^2 - 16x - 24y = 9$.

45. Find an equation of
(a) the bottom half of the circle $x^2 + y^2 = 16$
(b) the top half of the circle
$x^2 + y^2 + 2x - 4y + 1 = 0$.

46. Find an equation of
(a) the right half of the circle $x^2 + y^2 = 9$
(b) the left half of the circle
$x^2 + y^2 - 4x + 3 = 0$.

47. Graph
(a) $y = \sqrt{25 - x^2}$
(b) $y = \sqrt{5 + 4x - x^2}$.

48. Graph
(a) $x = -\sqrt{4 - y^2}$
(b) $x = 3 + \sqrt{4 - y^2}$.

49. Find an equation of the line that is tangent to the circle $x^2 + y^2 = 25$ at the point $(3, 4)$ on the circle.

50. Find an equation of the line that is tangent to the circle at the point P on the circle
(a) $x^2 + y^2 + 2x = 9$; $P(2, -1)$
(b) $x^2 + y^2 - 6x + 4y = 13$; $P(4, 3)$.

51. For the circle $x^2 + y^2 = 20$ and the point $P(-1, 2)$:
(a) Is P inside, outside, or on the circle?
(b) Find the largest and smallest distances between P and points on the circle.

52. Follow the directions of Exercise 51 for the circle $x^2 + y^2 - 2y - 4 = 0$ and the point $P(3, \frac{5}{2})$.

53. Referring to Figure 1.6.13, find the coordinates of the points T and T', where the lines L and L' are

tangent to the circle of radius 1 with center at the origin.

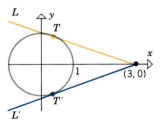

Figure 1.6.13

54. A point (x, y) moves so that its distance to $(2, 0)$ is $\sqrt{2}$ times its distance to $(0, 1)$.
(a) Show that the point moves along a circle.
(b) Find the center and radius.

55. A point (x, y) moves so that the sum of the squares of its distances from $(4, 1)$ and $(2, -5)$ is 45.
(a) Show that the point moves along a circle.
(b) Find the center and radius.

56. Find all values of c for which the system of equations
$$\begin{cases} x^2 - y^2 = 0 \\ (x - c)^2 + y^2 = 1 \end{cases}$$
has 0, 1, 2, 3, or 4 solutions. [*Hint:* Sketch a graph.]

In Exercises 57–70, graph the parabola and label the coordinates of the vertex and the intersections with the coordinate axes.

57. $y = x^2 + 2$.

58. $y = x^2 - 3$.

59. $y = x^2 + 2x - 3$.

60. $y = x^2 - 3x - 4$.

61. $y = -x^2 + 4x + 5$.

62. $y = -x^2 + x$.

63. $y = (x - 2)^2$.

64. $y = (3 + x)^2$.

65. $x^2 - 2x + y = 0$.

66. $x^2 + 8x + 8y = 0$.

67. $y = 3x^2 - 2x + 1$.

68. $y = x^2 + x + 2$.

69. $x = -y^2 + 2y + 2$.

70. $x = y^2 - 4y + 5$.

71. Find an equation of
(a) the right half of the parabola $y = 3 - x^2$
(b) the left half of the parabola $y = x^2 - 2x$.

72. Find an equation of
(a) the upper half of the parabola $x = y^2 - 5$
(b) the lower half of the parabola $x = y^2 - y - 2$.

73. Graph
(a) $y = \sqrt{x + 5}$
(b) $x = -\sqrt{4 - y}$.

74. Graph
(a) $y = 1 + \sqrt{4 - x}$
(b) $x = 3 + \sqrt{y}$.

75. If a ball is thrown straight up with an initial velocity of 32 ft/sec, then after t seconds the distance s above its starting height, in feet, is given by $s = 32t - 16t^2$.
  (a) Graph this equation in a ts-coordinate system (t-axis horizontal).
  (b) At what time t will the ball be at its highest point, and how high will it rise?

76. A rectangular field is to be enclosed with 500 ft of fencing along three sides and by a straight stream on the fourth side. Let x be the length of each side perpendicular to the stream, and let y be the length of the side parallel to the stream.
  (a) Express y in terms of x.
  (b) Find a formula for the area A of the field in terms of x.
  (c) What is the largest area that can be enclosed?

77. A rectangular plot of land is to be enclosed using two kinds of fencing. Two opposite sides will have heavy-duty fencing costing \$3/ft, while the other two sides will have standard fencing costing \$2/ft. A total of \$600 is available for the fencing. Let x be the length of each side with heavy-duty fencing, and let y be the length of each side with standard fencing.
  (a) Express y in terms of x.
  (b) Find a formula for the area A of the rectangular plot in terms of x.
  (c) What is the largest area that can be enclosed?

78. (a) By completing the square, show that the equation $y = ax^2 + bx + c$ can be rewritten as
$$y = a\left(x + \frac{b}{2a}\right)^2 + \left(c - \frac{b^2}{4a}\right)$$
  if $a \neq 0$.
  (b) Use the result in part (a) to show that the graph of $y = ax^2 + bx + c$ has its high point at $x = -b/(2a)$ if $a < 0$ and its low point there if $a > 0$.

▶ SUPPLEMENTARY EXERCISES

In Exercises 1–5, use interval notation to describe the set of all values of x (if any) that satisfy the given inequalities.

1. (a) $-3 < x \leq 5$ (b) $-1 < x^2 \leq 9$
  (c) $x^2 \geq \frac{1}{4}$.

2. (a) $|2x + 1| > 5$ (b) $|x^2 - 9| \geq 7$
  (c) $1 \leq |x| \leq 3$.

3. (a) $2x^2 - 5x > 3$ (b) $x^2 - 5x + 4 \leq 0$.

4. (a) $\dfrac{x}{1 - x} \geq 3$ (b) $\dfrac{2x + 3}{x} \geq x$.

5. (a) $\dfrac{|x| - 1}{|x| - 2} \leq 0$ (b) $|x - 1| \leq 2|x + 2|$.

6. Among the terms *integer, rational, irrational*, which ones apply to the given number?
  (a) $\sqrt{4/9}$ (b) 2^{-2}
  (c) $-4^{1/3}$ (d) 0.87
  (e) $-4^{1/2}$ (f) $0.1010010001\ldots$
  (g) 3.222 (h) $3/(-1)$.

7. (a) Find values of a and b such that $a < b$, but $a^2 > b^2$.
  (b) If $a < b$, what additional assumptions on a and b are required to ensure that $a^2 < b^2$?

8. Which of the following are true for all sets A and B?
  (a) $A \subset (A \cap B)$ (b) $(A \cap B) \subset A$
  (c) $\varnothing \subset A$ (d) $A \subset (A \cup B)$
  (e) $(A \cap B) \subset (A \cup B)$
  (f) Either $A \subset B$ or $B \subset A$
  (g) $A \in B$.

9. Prove: $|x| \leq \sqrt{x^2 + y^2}$ and $|y| \leq \sqrt{x^2 + y^2}$. Interpret this geometrically.

In Exercises 10–14, draw a rectangular coordinate system and sketch the set of points whose coordinates (x, y) satisfy the given conditions.

10. (a) $y = 0$ and $x > 0$
  (b) $2x - y \leq 3$.

11. (a) $xy = x^2$
  (b) $y(x - 1) = x^2 - 1$.

12. (a) $y = (x^3 - 1)/(x - 1)$
  (b) $y > x^2 - 9$.

13. (a) $y^2 - 6y + x^2 - 2x - 6 \geq 0$
  (b) $x + |y - 2| = 1$.

14. (a) $|x| + |y| = 4$ (b) $|x| - |y| = 4$.

15. Where does the parabola $y = x^2$ intersect the line $y - 2 = x$?

In Exercises 16–19, sketch the graph of the given equation.

16. $xy + 4 = 0$.

17. $y = |x - 2|$.

18. $y = \sqrt{4 - x^2}$.

19. $y = x(x - 2)$.

In Exercises 20–24, find the standard equation for the circle satisfying the given conditions.

20. The circle centered at $(3, -2)$ and tangent to the line $y = 1$.

21. The circle centered at $(1, 2)$ and passing through the point $(4, -2)$.

22. The circle centered on the line $x = 2$ and passing through the points $(1, 3)$ and $(3, -11)$.

23. The circle of radius 5 tangent to the lines $y = 7$ and $x = 6$.

24. The circle of radius 13 that passes through the origin and the point $(0, -24)$.

In Exercises 25–28, determine whether the equation represents a circle, a point, or has no graph. If it represents a circle, find the center and radius.

25. $x^2 + y^2 + 4x + 2y + 5 = 0$.

26. $4x^2 + 4y^2 - 4x + 8y + 1 = 0$.

27. $x^2 + y^2 - 3x + 2y + 4 = 0$.

28. $3x^2 + 3y^2 - 5x + 7y + 3 = 0$.

29. In each part, find an equation for the line through A and B, the distance between A and B, and the coordinates of the midpoint of the line segment joining A and B.
  (a) $A(3, 4)$, $B(-3, -4)$
  (b) $A(3, 4)$, $B(3, -4)$
  (c) $A(3, 4)$, $B(-3, 4)$
  (d) $A(3, 4)$, $B(4, 3)$.

30. Show that the point $(8, 1)$ is *not* on the line through the points $(-3, -2)$ and $(1, -1)$.

31. Where does the circle of radius 5 centered at the origin intersect the line of slope $-3/4$ through the origin?

32. Find the slope of the line whose angle of inclination is
  (a) $30°$ (b) $120°$ (c) $90°$.

In Exercises 33–35, find the slope-intercept form of the line satisfying the stated conditions.

33. The line through $(2, -3)$ and $(4, -3)$.

34. The line with x-intercept -2 and angle of inclination $\phi = 45°$.

35. The line parallel to $x + 2y = 3$ that passes through the origin.

36. Find an equation of the perpendicular bisector of the line segment joining $A(-2, -3)$ and $B(1, 1)$.

In Exercises 37–39, find equations of the lines L and L' and determine their point of intersection.

37. L passes through $(1, 0)$ and $(-1, 4)$.
  L' is perpendicular to L and has y-intercept -3.

38. L passes through $(-2, 0)$ and $(-2, 3)$.
  L' passes through $(-1, 4)$ and is perpendicular to L.

39. L has slope $2/5$ and passes through $(3, 1)$.
  L' has x-intercept $-8/3$ and y-intercept -4.

40. Consider the triangle with vertices $A(5, 2)$, $B(1, -3)$, and $C(-3, 4)$. Find the point-slope form of the line containing
  (a) the median from C to AB
  (b) the altitude from C to AB.

41. Use slopes to show that the points $(5, 6)$, $(-4, 3)$, $(-3, -2)$, and $(6, 1)$ are vertices of a parallelogram. Is it a rectangle?

42. For what value of k (if any) will the line $2x + ky = 3k$ satisfy the stated condition?
  (a) Have slope 3
  (b) Have y-intercept 3
  (c) Be parallel to the x-axis
  (d) Pass through $(1, 2)$.

2
Functions and Limits

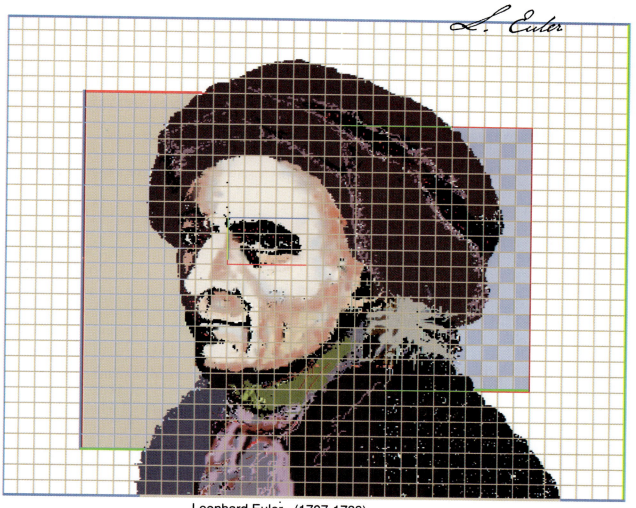

Leonhard Euler   (1707-1783)

■ 2.1 FUNCTIONS

In this section we shall define one of the most fundamental concepts in mathematics, the notion of a function. We shall discuss the notation used to describe functions and investigate some of their basic properties.

□ **THE CONCEPT OF A FUNCTION**

Historically, the term "function" was first used by Leibniz in 1673 to denote the dependence of one quantity on another. For example:

- The area of a circle depends on its radius r by the equation $A = \pi r^2$, so we say that "A is a function of r."
- The velocity v of a ball falling freely in the earth's gravitational field increases with time t until it hits the ground, so we say that "v is a function of t."
- In a bacteria culture, the number n of bacteria present after one hour of growth depends on the number N of bacteria present initially, so we say that "n is a function of N."

In general, if a quantity y depends on a quantity x in such a way that each value of x determines exactly one value of y, then we say that "y is a function of x." For example, the equation

$$y = 4x + 1$$

defines y as a function of x because each value assigned to x determines one value of y (Table 2.1.1).

Table 2.1.1

VALUE OF x	VALUE OF $y = 4x + 1$
$x = 2$	$y = 9$
$x = 1$	$y = 5$
$x = 0$	$y = 1$
$x = -\frac{1}{4}$	$y = 0$
$x = \sqrt{3}$	$y = 4\sqrt{3} + 1$

In order to describe functions without stating specific formulas, the Swiss mathematician, Leonhard Euler* (pronounced "oiler") conceived the idea of denoting functions by letters of the alphabet. For example, if we use the letter f to denote a function, then the equation

$$y = f(x) \tag{1}$$

(read, "y equals f of x") indicates that y is a function of x. The quantity x in (1) is called the *independent variable* of f and the quantity y the *dependent variable* of f. This terminology conveys the idea that we are free to assign values to the variable x, but that once a value is assigned to x, a unique value of y is determined. For example, in Table 2.1.1 we arbitrarily assigned values to x, then determined the corresponding y values from the equation $y = 4x + 1$.

REMARK.   It is important to understand that in (1), x and y may represent numerical quantities, but f itself does not represent a numerical quantity; it stands for a "relationship" between y and x.

Functions are commonly specified by formulas that do not explicitly involve a dependent variable. For example, the formula

$$f(x) = x^2$$

(2)

defines a function f that associates the number x^2 with the number x. Thus,

$$f(3) = 3^2 = 9 \qquad \boxed{f \text{ associates } 9 \text{ with } 3.}$$

$$f(-2) = (-2)^2 = 4 \qquad \boxed{f \text{ associates } 4 \text{ with } -2.}$$

$$f(0) = 0^2 = 0 \qquad \boxed{f \text{ associates } 0 \text{ with } 0.}$$

$$f(\sqrt{2}) = (\sqrt{2})^2 = 2 \qquad \boxed{f \text{ associates } 2 \text{ with } \sqrt{2}.}$$

Although Formula (2) does not involve a dependent variable, we can introduce one by letting $y = f(x)$ and rewriting (2) as

$$y = x^2$$

---

*LEONHARD EULER (1707–1783). Euler was probably the most prolific mathematician who ever lived. It has been said that, "Euler wrote mathematics as effortlessly as most men breathe." He was born in Basel, Switzerland, and was the son of a Protestant minister who had himself studied mathematics. Euler's genius developed early. He attended the University of Basel, where by age 16 he obtained both a Bachelor of Arts degree and a Master's degree in philosophy. While at Basel, Euler had the good fortune to be tutored one day a week in mathematics by a distinguished mathematician, Johann Bernoulli. At the urging of his father, Euler then began to study theology. The lure of mathematics was too great, however, and by age 18 Euler had begun to do mathematical research. Nevertheless, the influence of his father and his theological studies remained, and throughout his life Euler was a deeply religious, unaffected person. At various times Euler taught at St. Petersburg Academy of Sciences (in Russia), the University of Basel, and the Berlin Academy of Sciences. Euler's energy and capacity for work were virtually boundless. His collected works form more than 100 quarto-sized volumes and it is believed that much of his work has been lost. What is particularly astonishing is that Euler was blind for the last 17 years of his life, and this was one of his most productive periods! Euler's flawless memory was phenomenal. Early in his life he memorized the entire *Aeneid* by Virgil and at age 70 could not only recite the entire work, but could also state the first and last sentence on each page of the book from which he memorized the work. His ability to solve problems in his head was beyond belief. He worked out in his head major problems of lunar motion that baffled Isaac Newton and once did a complicated calculation in his head to settle an argument between two students whose computations differed in the fiftieth decimal place.

Following the development of calculus by Leibniz and Newton, results in mathematics developed rapidly in a disorganized way. Euler's genius gave coherence to the mathematical landscape. He was the first mathematician to bring the full power of calculus to bear on problems from physics. He made major contributions to virtually every branch of mathematics as well as to the theory of optics, planetary motion, electricity, magnetism, and general mechanics.

In some situations it is preferable to define a function without using a dependent variable, and in other situations a dependent variable is desirable.

Although f is the symbol most commonly used to denote a function, any symbol can be used. Thus,

$$y = F(x), \quad y = f_1(x), \quad y = g(x), \quad \text{and} \quad y = \phi(x)$$

all express the fact that y is a function of x. It is also not necessary to use the letters x and y for the independent and dependent variables. Any symbols can be used. For example, the equation

$$s = f(t)$$

states that the dependent variable s is a function of the independent variable t.

Example 1 If $f(x) = 3x - 4$, then

$$f(0) = 3 \cdot 0 - 4 = -4$$
$$f(1) = (3 \cdot 1) - 4 = -1$$
$$f(2) = (3 \cdot 2) - 4 = 2$$
$$f(-3) = [3 \cdot (-3)] - 4 = -13$$
$$f(\sqrt{5}) = (3 \cdot \sqrt{5}) - 4 = 3\sqrt{5} - 4 \quad \blacktriangleleft$$

Example 2 If $\phi(x) = \dfrac{1}{x^3 - 1}$, then

$$\phi(-1) = \frac{1}{(-1)^3 - 1} = \frac{1}{-2} = -\frac{1}{2}$$

$$\phi(0) = \frac{1}{0^3 - 1} = -1$$

$$\phi(\sqrt[3]{7}) = \frac{1}{(\sqrt[3]{7})^3 - 1} = \frac{1}{7 - 1} = \frac{1}{6}$$

$$\phi(5^{1/6}) = \frac{1}{(5^{1/6})^3 - 1} = \frac{1}{5^{3/6} - 1} = \frac{1}{\sqrt{5} - 1}$$

$$\phi(1) \text{ is undefined (division by zero)} \quad \blacktriangleleft$$

Example 3 If $F(x) = 2x^2 - 1$, then

$$F(4) = 2(4)^2 - 1 = 31$$
$$F(t) = 2t^2 - 1$$
$$F(k + 1) = 2(k + 1)^2 - 1 = 2k^2 + 4k + 1$$
$$F(k) + 1 = (2k^2 - 1) + 1 = 2k^2 \quad \blacktriangleleft$$

For reasons that will be discussed later, radian measure of angles is preferred over degree measure in calculus. Therefore, in trigonometric expressions such as $\sin x$, $\cos x$, $\tan x$, $\cot x$, $\sec x$, and $\csc x$ it is understood that x is measured in radians. When degree measure is called for, we shall make this clear by writing $\sin x°$, $\cos x°$, and so forth.

Example 4   If $f(x) = \sin x$, then

$$f(\pi/2) = \sin(\pi/2) = 1$$

$$f(1) = \sin(1) \approx .841471$$

> Use a calculator set to the radian mode to obtain this approximation.

$$f(0) = \sin(0) = 0 \quad \blacktriangleleft$$

Example 5   Find $g(2)$ if

    (a) $g(x) = x^2 - 3x + 5$ (b) $g(t) = t^2 - 3t + 5$

Solution (a).   Substitute 2 for x:

$$g(2) = 2^2 - 3 \cdot 2 + 5 = 3$$

Solution (b).   Substitute 2 for t:

$$g(2) = 2^2 - 3 \cdot 2 + 5 = 3 \quad \blacktriangleleft$$

Observe that the computations in both parts of the foregoing example are exactly the same. This is because

$$g(x) = x^2 - 3x + 5 \quad \text{and} \quad g(t) = t^2 - 3t + 5$$

really define the same function g; the two formulas differ only in the choice of the symbol used for the independent variable.

 DOMAIN

In many situations the independent variable of a function is not free to vary arbitrarily. Rather, it must be restricted to lie in some set, called the *domain* of the function. If the function is f and $y = f(x)$, then the domain of f can be viewed as the set of allowable values for the independent variable x.

In applications, the domain of a function is often determined by physical considerations; for example, suppose that a square with a side of length x cm is cut from each corner of a piece of cardboard that is 1 cm square, and let y be the area (in cm²) of the cardboard that remains (Figure 2.1.1).

By subtracting the areas of the four corner squares from the original area, it follows that

$$y = 1 - 4x^2 \tag{3}$$

However, x cannot be negative because it denotes a length, and its value cannot exceed $\frac{1}{2}$ (why?). Thus, x must satisfy the restriction $0 \leq x \leq \frac{1}{2}$. Therefore, even though it is not stated explicitly, the underlying physical meaning of x dictates that the domain of the function defined by Equation (3) is the set

$$\{x : 0 \leq x \leq \tfrac{1}{2}\} = [0, \tfrac{1}{2}] \tag{4}$$

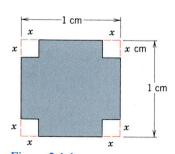

Figure 2.1.1

REMARK. A practical-minded engineer might argue that $x = 0$ and $x = \frac{1}{2}$ should be excluded from the domain because these values do not correspond to "physical" cuts in the cardboard (why?). On the other hand, a mathematician might argue that these values should be regarded as "theoretical" possibilities and thus included. This issue often arises when a physical problem is represented (or "modeled") mathematically. Usually, it will not matter which viewpoint you take as long as you are consistent within a problem and give an appropriate physical interpretation to any answers you obtain.

☐ **RANGE**

If x is a number in the domain of a function f, then the number $f(x)$ that f associates with x is called the ***value of f at x*** or the ***image of x under f***. Thus, if

$$f(x) = x^2$$

then the value of f at $x = 3$ is $f(3) = 3^2 = 9$, and the value of f at $x = -2$ is $f(-2) = (-2)^2 = 4$. Equivalently, if we introduce a dependent variable

$$y = x^2$$

then the value of y at $x = 3$ is $y = 3^2 = 9$, and the value of y at $x = -2$ is $y = (-2)^2 = 4$. The set of all possible values of $f(x)$ as x varies over the domain of f is called the ***range of f***. If $y = f(x)$, then the range of f can be viewed as the set of all possible values for the dependent variable y.

Example 6   We saw above that the function relating the variables x and y in Figure 2.1.1 is given by

$$y = 1 - 4x^2$$

where the domain of the function is the set

$$\{x : 0 \le x \le \tfrac{1}{2}\} = [0, \tfrac{1}{2}]$$

As x varies over this domain, the values of $y = 1 - 4x^2$ vary between $y = 0$ (when $x = \frac{1}{2}$) and $y = 1$ (when $x = 0$), so the range of the function is

$$\{y : 0 \le y \le 1\} = [0, 1] \quad \blacktriangleleft$$

☐ **DOMAINS DETERMINED BY PHYSICAL OR GEOMETRIC CONSIDERATIONS**

Restrictions on the independent variable that affect the domain of a function generally come about in one of three ways:

- Physical or geometric considerations.
- Natural restrictions that result from a formula used to define the function.
- Artificial restrictions imposed by a problem solver for one purpose or another.

We shall give examples of each of these.

Example 7   We saw in Example 6 that the function relating the variables x and y in Figure 2.1.1 is given by

$$y = 1 - 4x^2$$

where the domain of this function is $\{x : 0 \le x \le \frac{1}{2}\} = [0, \frac{1}{2}]$. This domain is a consequence of the physical restrictions on x in the problem. We would generally denote this function and its domain by writing

$$y = 1 - 4x^2, \quad 0 \le x \le \tfrac{1}{2} \quad \blacktriangleleft$$

Example 8   Assume that it costs 12 cents to manufacture a certain kind of computer chip, and let $f(x)$ be the cost in dollars of manufacturing x such chips. The formula for $f(x)$ is

$$f(x) = 0.12x \tag{5}$$

Physically, x must be a nonnegative integer, so the domain of the function f in (5) is the set $\{x : x = 0, 1, 2, \ldots\}$. To make the domain of f completely clear we would write

$$f(x) = 0.12x, \quad x = 0, 1, 2, \ldots \quad \blacktriangleleft$$

□ **THE NATURAL DOMAIN**

Not all functions studied in mathematics arise from physical or geometric problems—many are studied purely as mathematical objects. For such functions there are no physical or geometric restrictions on the independent variable. However, restrictions can arise from formulas used to define such functions.

Example 9   If

$$h(x) = \frac{1}{(x - 1)(x - 3)} \tag{6}$$

then the function h is undefined if $x = 1$ or $x = 3$ because division by zero is not allowed. However, for all other values of x, the function h is defined and has real values. Thus, the domain of h consists of all real numbers x except $x = 1$ and $x = 3$. In interval notation the domain is

$$(-\infty, 1) \cup (1, 3) \cup (3, +\infty)$$

Figure 2.1.2

(Figure 2.1.2). Although we could explicitly state the domain of h by writing its formula as

$$h(x) = \frac{1}{(x - 1)(x - 3)}, \quad x \ne 1, 3$$

we would not ordinarily do so; we would usually just write (6) and assume that the restrictions $x \ne 1, 3$ are evident from that formula.   ◀

In general, we make the following convention:

*If a function is defined by a formula and there is no domain explicitly stated, then it is understood that the domain consists of all real numbers for which the formula makes sense, and the function has a real value. This is called the **natural domain** of the function.*

Example 10   The formula

$$f(x) = x^2$$

makes sense and yields a real value for all real values of x. Thus, the natural domain of f is $\{x : -\infty < x < +\infty\} = (-\infty, +\infty)$.   ◄

Example 11   If

$$g(x) = 2 + \sqrt{x - 1}$$

then g is undefined for $x < 1$, because negative numbers do not have real square roots; the function g is defined and has real values otherwise, so its natural domain is $\{x : 1 \le x\} = [1, +\infty)$.   ◄

Example 12   If $f(x) = \tan x$, then $f(x)$ is undefined if

$$x = \pm \frac{\pi}{2},\ \pm \frac{3\pi}{2},\ \pm \frac{5\pi}{2}, \ldots$$

but has a real value otherwise Figure 2.1.3. Thus, the natural domain consists of all x except

$$x = \pm \frac{\pi}{2},\ \pm \frac{3\pi}{2},\ \pm \frac{5\pi}{2}, \ldots \qquad ◄$$

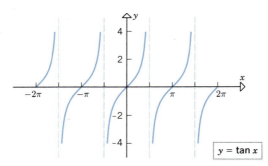

Figure 2.1.3

☐ **DOMAINS DETERMINED BY SPECIAL RESTRICTIONS**

Sometimes it is desirable to limit the domain of a function to some subset of its natural domain by imposing specific restrictions on the independent variable. Restrictions of this type are often used to focus attention on certain values of the independent variable that may be of special interest in a given problem or situation.

Example 13   The formula

$$f(x) = x^2, \quad 0 \le x \le 1$$

defines a function whose domain is the interval $[0, 1]$. Without the stated restriction the domain of f would be the natural domain, $(-\infty, +\infty)$.   ◄

It is a common procedure in algebra to simplify functions by canceling common factors in the numerator and denominator. However, the following example shows that this operation can alter the domain of a function.

Example 14   The function

$$h(x) = \frac{x^2 - 4}{x - 2} \tag{7}$$

has a real value everywhere except at $x = 2$, where we have a division by zero:

$$h(2) = \frac{2^2 - 4}{2 - 2} = \frac{0}{0}$$

Thus, the domain of h consists of all x except $x = 2$. However, if we rewrite (7) as

$$h(x) = \frac{(x - 2)(x + 2)}{(x - 2)} \tag{8}$$

then cancel the common factors, we obtain

$$h(x) = x + 2 \tag{9}$$

which *is* defined at $x = 2$ since

$$h(2) = 2 + 2 = 4$$

Thus, our algebraic simplification has altered the domain of the function. In order to cancel the factors in (8) and not alter the domain of h, we must restrict the domain in (9) and write

$$h(x) = x + 2, \quad x \neq 2 \quad \blacktriangleleft$$

REMARK.   In algebra the reader undoubtedly learned to simplify (8) by writing (9). In elementary problems this often causes no difficulty because the domain of the function is irrelevant to the solution of the problem. However, in more advanced problems the domain is often important, so the reader should learn to be precise and place the appropriate restrictions on the domain of the function after canceling factors.

☐ **TECHNIQUES FOR FINDING THE RANGE**

Often, the range of a function is evident by inspection.

Example 15   Find the range of

(a) $f(x) = x^2$ (b) $g(x) = 2 + \sqrt{x - 1}$

Solution (a).   Since no domain is stated explicitly, the domain of f is the natural domain $\{x : -\infty < x + \infty\} = (-\infty, +\infty)$ (see Example 10). To find the range of f it will help to introduce a dependent variable

$$y = x^2$$

As x varies over the domain of f, the corresponding y values (which must be nonnegative) vary over the set $\{y : 0 \leq y < +\infty\} = [0, +\infty)$. This is the range of f.

Solution (b).   Since no domain is stated explicitly, the domain of g is the natural domain $\{x : 1 \leq x\} = [1, +\infty)$. To determine the range of the function g, let $y = 2 + \sqrt{x - 1}$. As x varies over the interval $[1, +\infty)$, the value of $\sqrt{x - 1}$ varies over the interval $[0, +\infty)$, so the value of $y = 2 + \sqrt{x - 1}$ varies over the interval $[2, +\infty)$. This is the range of g.   ◀

The next example illustrates a technique that can sometimes be used to find a range of a function when it is not evident by inspection.

Example 16   Find the range of the function $f(x) = \dfrac{x + 1}{x - 1}$.

Solution.   The natural domain of f consists of all x, except $x = 1$. As in our earlier examples, let us introduce a dependent variable

$$y = \frac{x + 1}{x - 1}$$

The set of all possible y values is not at all evident from this equation. However, solving this equation for x in terms of y yields

$$(x - 1)y = x + 1$$
$$xy - y = x + 1$$
$$xy - x = y + 1$$
$$x(y - 1) = y + 1$$
$$x = \frac{y + 1}{y - 1}$$

It is now evident from the right side of this equation that $y = 1$ is not in the range, for otherwise we would have a division by zero. No other values of y are excluded by this equation, so that the range of the function f is $\{y : y \neq 1\} = (-\infty, 1) \cup (1, +\infty)$.   ◀

□ **FUNCTIONS DEFINED PIECEWISE**

The next example shows that functions must sometimes be defined by formulas that have been "pieced together."

Example 17   The cost of a taxicab ride in a certain metropolitan area is $1.75 for any ride up to and including one mile. After one mile the rider pays an additional amount at the rate of 50 cents per mile (or fraction, thereof). If $f(x)$ is the total cost in dollars for a ride of x miles, then the value of $f(x)$ is

$$f(x) = \begin{cases} 1.75, & 0 < x \leq 1 \\ 1.75 + 0.50(x - 1), & 1 < x \end{cases}$$

$1.75 for a ride up to and including one mile

$1.75 for the first mile plus a rate of $0.50 a mile for each mile, or fraction thereof, after the first   ◀

Example 18 From the definition of absolute value, the function $f(x) = |x|$ can be written in an equivalent piecewise form as

$$f(x) = \begin{cases} x, & x \geq 0 \\ -x, & x < 0 \end{cases} \quad \blacktriangleleft$$

☐ **REVERSING THE ROLES OF x AND y**

Sometimes it is desirable to reverse the roles of x and y by treating y as an independent variable and x as a dependent variable. For example, consider the equation

$$x = 4y^5 - 2y^3 + 7y - 5 \tag{10}$$

It is a straightforward task to calculate the value of x if a value of y is given. For example, if we let $y = 1$, then the corresponding value of x is

$$x = 4(1)^5 - 2(1)^3 + 7(1) - 5 = 4$$

However, it is a complicated task to calculate the value of y if a value of x is given. For example, if we let $x = 1$, then to find a corresponding value of y we would have to solve the equation

$$1 = 4y^5 - 2y^3 + 7y - 5$$

Thus, it is preferable to think of (10) as defining x as a function of y, where y is the independent variable and x the dependent variable.

In some equations we can express y as a function of x or x as a function of y with equal simplicity.

Example 19 The equation

$$3x + 2y = 6$$

can be solved for y to obtain

$$y = -\frac{3}{2}x + 3 \tag{11}$$

or it can be solved for x in terms of y to obtain

$$x = -\frac{2}{3}y + 2 \tag{12}$$

Equation (11) defines y as a function of x and Equation (12) defines x as a function of y. $\blacktriangleleft$

☐ **THE GENERAL CONCEPT OF A FUNCTION**

For all of the functions encountered in this section thus far, the domains and ranges have been sets of real numbers. However, later in this text we shall be interested in functions whose domains and ranges are other kinds of objects. For this purpose we make the following more general definition of a function.

2.1.1 DEFINITION. A *function* is a rule that assigns to each element in a nonempty set A one and only one element in a set B.

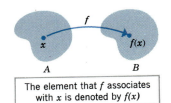

The element that f associates with x is denoted by $f(x)$

Figure 2.1.4

The set A in this definition is called the **domain** of the function. If x is an element in the domain of a function f, then the element in B that f associates with x is denoted by the symbol $f(x)$ and is called the **image of x under f** or the **value of f at x** (Figure 2.1.4). The set of all images of points in the domain is a subset of B called the **range** of f.

Example 20 Most computers and many calculators represent characters (numbers, letters, and special symbols) by a numerical code known as ASCII code, which stands for **American Standard Code for Information Interchange**. Some of the 256 ASCII codes and the characters assigned to them are shown in Figure 2.1.5.

In computer terminology, a **string** is any sequence of characters. Many computer programming languages have built-in functions whose domains are strings. For example, in the BASIC programming language an arbitrary string can be denoted by X$ and there is a function, denoted by "len," that calculates the number of characters in a string; thus,

len(X$) = number of characters in the string X$

For example, if

X$ = QWER*+%??25

then

len(X$) = 11

The domain of the function "len" is the set of all possible strings that the computer hardware will allow, and the range is the set of positive integers that represent the possible string lengths. ◀

ASCII value	Character	ASCII value	Character	ASCII value	Character	ASCII value	Character	ASCII value	Character
042	.	084	T	127	⌂	170	¬	213	⊨
043	+	085	U	128	Ç	171	½	214	┌
044	-	086	V	129	ü	172	¼	215	╪
045	-	087	W	130	é	173	¡	216	╪
046	.	088	X	131	â	174	«	217	┘
047	/	089	Y	132	ä	175	»	218	┌
048	0	090	Z	133	à	176	░	219	█
049	1	091	[	134	å	177	▒	220	▄
050	2	092	/	135	ç	178	▓	221	▌
051	3	093	]	136	ê	179	│	222	▐
052	4	094	<	137	ë	180	┤	223	▀
053	5	095	-	138	è	181	╡	224	α
054	6	096	,	139	ï	182	╢	225	β
055	7	097	a	140	î	183	╖	226	Γ
056	8	098	b	141	ì	184	╕	227	η
057	9	099	c	142	Ä	185	╣	228	Σ
058	:	100	d	143	Å	186	║	229	o
059	;	101	e	144	É	187	╗	230	μ

ASCII code

Figure 2.1.5

▶ Exercise Set 2.1   C 57, 58, 59, 60

1. Given that $f(x) = 3x^2 + 2$, find
   (a) $f(-2)$ (b) $f(4)$ (c) $f(0)$
   (d) $f(-\sqrt{3})$ (e) $f(a + 1)$ (f) $f(3t)$.

2. Given that $g(x) = \dfrac{x + 1}{x - 1}$, find
   (a) $g(2)$ (b) $g(-2)$ (c) $g(\frac{1}{4})$
   (d) $g(\pi)$ (e) $g(a - 1)$ (f) $g(2t + 1)$.

3. Given that
$$f(x) = \begin{cases} \dfrac{1}{x}, & x > 3 \\ \\ 2x, & x \le 3 \end{cases}$$
   find
   (a) $f(-4)$ (b) $f(4)$ (c) $f(0)$
   (d) $f(3)$ (e) $f(2.9)$ (f) $f(t^2 + 5)$.

4. Given that
$$g(x) = \begin{cases} \sqrt{x + 1}, & x \ge -1 \\ \\ 3, & x < -1 \end{cases}$$
   find
   (a) $g(0)$ (b) $g(-1.1)$ (c) $g(3)$
   (d) $g(-1)$ (e) $g(-\pi)$ (f) $g(t^2 - 1)$.

In Exercises 5–22, find the natural domain of the given function.

5. $f(x) = \dfrac{1}{x - 3}$.   6. $f(x) = \dfrac{1}{5x + 7}$.

7. $g(x) = \sqrt{x^2 - 3}$.   8. $g(x) = \sqrt{x^2 + 3}$.

9. $h(x) = \sqrt{\dfrac{x - 1}{x + 2}}$.   10. $h(x) = \sqrt{x - 3x^2}$.

11. $\phi(x) = \dfrac{x}{|x| + 1}$.   12. $\phi(x) = \sqrt{3 - \sqrt{x}}$.

13. $F(x) = \sqrt{x - 5} + \sqrt{8 - x}$.

14. $F(x) = 3\sqrt{x} - \sqrt{x^2 - 4}$.

15. $G(x) = \sqrt{x^2 - 2x + 5}$.

16. $G(x) = \sqrt{\dfrac{x^2 - 4}{x - 4}}$.

17. $f(x) = \dfrac{x}{|x|}$.   18. $f(x) = \dfrac{x^2 - 1}{x + 1}$.

19. $g(x) = \sin \sqrt{x}$.   20. $g(x) = \cos \dfrac{1}{x}$.

21. $h(x) = \dfrac{1}{1 - \sin x}$.   22. $h(x) = \dfrac{3}{2 - \cos x}$.

In Exercises 23–36, find the natural domain and the range of the given function.

23. $f(x) = \sqrt{3 - x}$.   24. $f(x) = \sqrt{3x - 2}$.

25. $g(x) = \sqrt{4 - x^2}$.   26. $g(x) = \sqrt{9 - 4x^2}$.

27. $h(x) = 3 + \sqrt{x}$.   28. $h(x) = \dfrac{1}{3 + \sqrt{x}}$.

29. $F(x) = x^2 + 3$.   30. $F(x) = \dfrac{2}{x^2 + 3}$.

31. $G(x) = x^3 + 2$.   32. $G(x) = \dfrac{3}{x}$.

33. $H(x) = 3 \sin x$.   34. $H(x) = \sin^2 \sqrt{x}$.

35. $\phi(x) = 2 + \cos x$.   36. $\phi(x) = \dfrac{5}{3 - \cos 2x}$.

In Exercises 37–40, express the given function in piecewise form without using absolute values.

37. $f(x) = |x| + 3x + 1$.

38. $f(x) = 3 + |2x - 5|$.

39. $g(x) = |x| + |x - 1|$.

40. $g(x) = 3|x - 2| - |x + 1|$.

In Exercises 41–48, find all values of x for which $f(x) = a$.

41. $f(x) = \sqrt{3x - 2}$; $a = 6$.

42. $f(x) = \dfrac{1}{x + 3}$; $a = 5$.

43. $f(x) = x^2 + 5$; $a = 7$.

44. $f(x) = \dfrac{x}{x^2 + 3}$; $a = \frac{1}{4}$.

45. $f(x) = \cos x$; $a = 1$.

46. $f(x) = \sin \dfrac{1}{x}$; $a = 1$.

47. $f(x) = \sin \sqrt{x}$; $a = \frac{1}{2}$.

48. $f(x) = 3 \tan x$; $a = 3$.

49. Express the area A of a circle as a function of its circumference C.

50. Express the area A of an equilateral triangle as a function of
    (a) the length s of each side
    (b) the altitude h.

51. Express the total surface area S of a cube as a function of
(a) the length x of an edge
(b) the volume V of the cube.

52. Express the total surface area S of a right-circular cylinder with given volume V as a function of its radius r.

53. An open box is to be made from an 8 in. × 15 in. piece of sheet metal by cutting out squares with sides of length x from each of the four corners and bending up the sides (Figure 2.1.6). Express the volume V of the box as a function of x, and state the domain of the function.

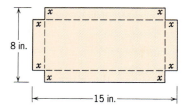

Figure 2.1.6

54. A camera is mounted at a point 3000 ft from the base of a rocket launching pad. The rocket rises vertically when launched. Express the distance x traveled by the rocket as a function of the camera elevation angle θ (Figure 2.1.7).

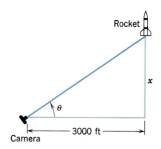

Rocket
x
θ
3000 ft
Camera

Figure 2.1.7

55. A pendulum of constant length L makes an angle θ with its vertical position (Figure 2.1.8). Express the height h as a function of the angle θ.

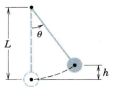

L
θ
h

Figure 2.1.8

56. Express the length L of a chord of a circle with radius 10 cm as a function of the central angle θ (Figure 2.1.9).

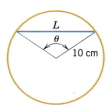

L
θ
10 cm

Figure 2.1.9

57. For a given outside temperature T and wind speed v, the Windchill Index (WCI) is the equivalent temperature that exposed skin would feel with a wind speed of 4 mi/hr. An empirical formula for the WCI (based on experience and observation) is

$$\text{WCI} = \begin{cases} T, & 0 \le v \le 4 \\ 91.4 + (91.4 - T)(0.0203v - 0.304\sqrt{v} - 0.474), & 4 < v < 45 \\ 1.6T - 55, & v \ge 45 \end{cases}$$

where T is the air temperature in °F, v is the wind speed in mi/hr, and WCI is the equivalent temperature in °F. Find the WCI to the nearest degree if the air temperature is 25° F and
(a) $v = 3$ mi/hr
(b) $v = 15$ mi/hr
(c) $v = 46$ mi/hr.

[Adapted from UMAP Module 658, *Windchill*, W. Bosch and L. Cobb, COMAP, Arlington, MA.]

In Exercises 58–60, use the formula for the Windchill Index described in Exercise 57.

58. Find the air temperature to the nearest degree if the WCI is reported as −60° F with a wind speed of 48 mi/hr.

59. Find the air temperature to the nearest degree if the WCI is reported as −10° F with a wind speed of 8 mi/hr.

60. Find the wind speed to the nearest mile per hour if the WCI is reported as $-15°$ F with an air temperature of $20°$ F.

61. Criticize the following statement: The function

$$\frac{1 - (1/x)}{1 + (1/x)}$$

can be simplified by multiplying numerator and denominator by x to obtain

$$\frac{1 - (1/x)}{1 + (1/x)} = \frac{x - 1}{x + 1}$$

How would you rewrite the statement to make it accurate?

In Exercises 62–68, simplify the function by canceling factors, and state the natural domain of the simplified function.

62. $f(x) = \dfrac{x^2 - 4}{x + 2}$.

63. $f(x) = \dfrac{(x + 2)(x^2 - 1)}{(x + 2)(x + 1)}$.

64. $f(x) = \dfrac{x^2 + x}{x}$.

65. $f(x) = \dfrac{x + 1 + \sqrt{x + 1}}{\sqrt{x + 1}}$.

66. $f(x) = \dfrac{x^2 - 9}{x - 3}$.

67. $f(x) = \dfrac{x^3 + 2x^2 - 3x}{(x - 1)(x + 3)}$.

68. $f(x) = \dfrac{x + \sqrt{x}}{\sqrt{x}}$.

■ 2.2 OPERATIONS ON FUNCTIONS

Just as numbers can be added, subtracted, multiplied, and divided to produce other numbers, so functions can be added, subtracted, multiplied, and divided to produce other functions. In this section we shall discuss these operations and some others that have no analogs in ordinary arithmetic.

☐ ARITHMETIC
OPERATIONS ON
FUNCTIONS

Functions are added, subtracted, multiplied, and divided in a very natural way. For example, if

$$f(x) = x \quad \text{and} \quad g(x) = x^2$$

then

$$f(x) + g(x) = x + x^2$$

This formula defines a new function which we call the **sum** of f and g and denote by $f + g$. Thus,

$$(f + g)(x) = f(x) + g(x) = x + x^2$$

Similarly,

$$(f - g)(x) = f(x) - g(x) = x - x^2$$
$$(f \cdot g)(x) = f(x) \cdot g(x) = x \cdot x^2 = x^3$$
$$(f/g)(x) = f(x)/g(x) = \frac{x}{x^2} = \frac{1}{x}$$

In general, we make the following definition.

> **2.2.1** DEFINITION.   Given functions f and g, their **sum** $f + g$, **difference** $f - g$, **product** $f \cdot g$, and **quotient** f/g are defined by
>
> $$(f + g)(x) = f(x) + g(x)$$
> $$(f - g)(x) = f(x) - g(x)$$
> $$(f \cdot g)(x) = f(x) \cdot g(x)$$
> $$(f/g)(x) = f(x)/g(x)$$
>
> For the functions $f + g$, $f - g$, and $f \cdot g$ the domain is defined to be the intersection of the domains of f and g, and for f/g the domain is this intersection with the points where $g(x) = 0$ excluded.

Example 1   Let

$$f(x) = 1 + \sqrt{x - 2} \quad \text{and} \quad g(x) = x - 1$$

Find $(f + g)(x)$, $(f - g)(x)$, $(f \cdot g)(x)$, and $(f/g)(x)$.

Solution.   First, we find formulas for the functions, then we find the domains. The formulas are

$$(f + g)(x) = f(x) + g(x) = (1 + \sqrt{x - 2}) + (x - 1) = x + \sqrt{x - 2} \quad (1)$$

$$(f - g)(x) = f(x) - g(x) = (1 + \sqrt{x - 2}) - (x - 1) = 2 - x + \sqrt{x - 2} \quad (2)$$

$$(f \cdot g)(x) = f(x) \cdot g(x) = (1 + \sqrt{x - 2})(x - 1) \quad (3)$$

$$(f/g)(x) = f(x)/g(x) = \frac{1 + \sqrt{x - 2}}{x - 1} \quad (4)$$

The domain of f is the interval $[2, +\infty)$ and the domain of g is the interval $(-\infty, +\infty)$. Thus, the domains of $f + g$, $f - g$, and $f \cdot g$ are the intersections of these two intervals, which is $[2, +\infty)$. But this is precisely the natural domain of (1), (2), and (3), so there is no need to explicitly state the domains. Similarly, the domain of f/g consists of all x in $[2, +\infty)$. Again this is the natural domain in (4).   ◀

In the foregoing example the formulas for $(f + g)(x)$, $(f - g)(x)$, $(f \cdot g)(x)$, and $(f/g)(x)$ determined the correct domains for the functions $f + g$, $f - g$, $f \cdot g$, and f/g. However, the following example shows that this is not always the case.

Example 2   Let

$$f(x) = 3\sqrt{x} \quad \text{and} \quad g(x) = \sqrt{x}$$

Find $(f \cdot g)(x)$.

Solution.   From Definition 2.2.1,

$$(f \cdot g)(x) = f(x) \cdot g(x) = (3\sqrt{x}) \cdot (\sqrt{x}) = 3x$$

The natural domain of the function $3x$ is $(-\infty, +\infty)$; however, this is not the correct domain of $f \cdot g$. To see why, observe that both f and g have domain $[0, +\infty)$, so that by definition $f \cdot g$ also has domain $[0, +\infty)$, since this is the intersection of the domains of f and g. Thus, the correct formula for $(f \cdot g)(x)$ is

$$(f \cdot g)(x) = 3x, \quad x \geq 0 \quad \blacktriangleleft$$

REMARK. In light of the foregoing example, we recommend that the reader develop the habit of indicating the appropriate restrictions on the domain after performing any operations on functions.

Sometimes we shall write f^2 to denote the product $f \cdot f$. For example, if $f(x) = 8x$, then

$$f^2(x) = (f \cdot f)(x) = f(x) \cdot f(x) = (8x) \cdot (8x) = 64x^2$$

Similarly,

$$f^3 = f \cdot f^2, \quad f^4 = f \cdot f^3, \quad f^5 = f \cdot f^4, \ldots$$

This notation is especially common with trigonometric functions. For example, $(\sin x)^2$ is generally written as $\sin^2 x$.

COMPOSITION OF FUNCTIONS

We shall now consider an operation on functions, called *composition*, which has no direct analog in ordinary arithmetic. Informally stated, the operation of composition is performed by substituting some function for the independent variable of another function. For example, suppose that

$$f(x) = x^2 \quad \text{and} \quad g(x) = x + 1$$

If we substitute $g(x)$ for x in the formula for f we obtain a new function

$$f(g(x)) = (g(x))^2 = (x + 1)^2$$

which we denote by $f \circ g$. Thus,

$$(f \circ g)(x) = f(g(x)) = (g(x))^2 = (x + 1)^2$$

In general, we make the following definition.

2.2.2 DEFINITION. Given functions f and g, the **composition of f with g**, denoted by $f \circ g$, is the function defined by

$$(f \circ g)(x) = f(g(x))$$

The domain of $f \circ g$ is defined to consist of all x in the domain of g for which $g(x)$ is in the domain of f.

REMARK. Although the domain of $f \circ g$ may seem complicated at first glance, it is quite natural: To compute $f(g(x))$ one needs x in the domain of g to compute $g(x)$, then one needs $g(x)$ in the domain of f to compute $f(g(x))$.

Example 3   Find $(f \circ g)(x)$ if $f(x) = x^2 + 3$ and $g(x) = \sqrt{x}$.

Solution.   The formula for $f(g(x))$ is

$$f(g(x)) = [g(x)]^2 + 3 = (\sqrt{x})^2 + 3 = x + 3$$

Since the domain of g is $[0, +\infty)$ and the domain of f is $(-\infty, +\infty)$, the domain of $f \circ g$ consists of all x in $[0, +\infty)$ such that $g(x) = \sqrt{x}$ lies in $(-\infty, +\infty)$; thus, the domain of $f \circ g$ is $[0, +\infty)$. Therefore,

$$(f \circ g)(x) = x + 3, \quad x \geq 0 \quad \blacktriangleleft$$

Example 4   Let $f(x) = x - 1$ and $g(x) = \sqrt{x}$. Find

(a) $(f \circ g)(x)$ (b) $(g \circ f)(x)$

Solution (a).   The formula for $f(g(x))$ is

$$f(g(x)) = g(x) - 1 = \sqrt{x} - 1$$

Since the domain of g is $[0, +\infty)$ and the domain of f is $(-\infty, +\infty)$, the domain of $f \circ g$ consists of all x in $[0, +\infty)$ such that $g(x) = \sqrt{x}$ lies in $(-\infty, +\infty)$, that is, all x in $[0, +\infty)$. Therefore,

$$(f \circ g)(x) = \sqrt{x} - 1$$

In this formula there is no need to indicate that the domain is $[0, +\infty)$ since this is the natural domain of $\sqrt{x} - 1$.

Solution (b).   The formula for $g(f(x))$ is

$$g(f(x)) = \sqrt{f(x)} = \sqrt{x - 1}$$

The domain of $g \circ f$ consists of all x in the domain of f such that $f(x)$ lies in the domain of g. Since the domain of f is $(-\infty, +\infty)$ and the domain of g is $[0, +\infty)$, the domain of $g \circ f$ consists of all x in $(-\infty, +\infty)$ such that $f(x) = x - 1$ lies in $[0, +\infty)$; thus, the domain is $[1, +\infty)$. Therefore,

$$(g \circ f)(x) = \sqrt{x - 1}$$

[As in part (a), there is no need to indicate specifically that the domain is $[1, +\infty)$ since this is the natural domain of $\sqrt{x - 1}$.] $\blacktriangleleft$

REMARK.   Note that $f \circ g$ and $g \circ f$ are different functions in the foregoing example. Thus, the order in which functions are composed can make a difference in the end result. In fact, it is relatively rare that $f \circ g$ and $g \circ f$ are the same.

☐ **EXPRESSING A FUNCTION AS A COMPOSITION**

Many problems in mathematics are attacked by "decomposing" functions into compositions of simpler functions. For example, consider the function h given by

$$h(x) = (x + 1)^2$$

To evaluate $h(x)$ for a given value of x, we would first compute $x + 1$ and then square the result. These two operations are performed by the functions

$$g(x) = x + 1 \quad \text{and} \quad f(x) = x^2$$

We can express h in terms of f and g by writing

$$h(x) = (x + 1)^2 = [g(x)]^2 = f(g(x))$$

so we have succeeded in expressing h as the composition $h = f \circ g$.

   The thought process in this example suggests a general procedure for decomposing a function h into a composition $h = f \circ g$:

- Think about how you would evaluate $h(x)$ for a specific value of x, trying to break the evaluation into two steps performed in succession.
- The first operation in the evaluation will determine a function g and the second a function f.
- The formula for h can then be written as $h(x) = f(g(x))$.

   For descriptive purposes, we shall refer to g as the "inside function" and f as the "outside function" in the expression $f(g(x))$. The inside function performs the first operation and the outside function performs the second.

Example 5   Express $h(x) = (x - 4)^5$ as a composition of two functions.

Solution.   To evaluate $h(x)$ for a given value of x we would first compute $x - 4$ and then raise the result to the fifth power. Therefore, the inside function (first operation) is

$$g(x) = x - 4$$

and the outside function (second operation) is

$$f(x) = x^5$$

so $h(x) = f(g(x))$. As a check,

$$f(g(x)) = [g(x)]^5 = (x - 4)^5 = h(x) \quad \blacktriangleleft$$

Example 6   Express $\sin(x^3)$ as a composition of two functions.

Solution.   To evaluate $\sin(x^3)$, we would first compute x^3 and then take the sine, so $g(x) = x^3$ is the inside function and $f(x) = \sin x$ the outside function. Therefore,

$$\sin(x^3) = f(g(x)) \quad \boxed{\text{where } g(x) = x^3 \text{ and } f(x) = \sin x} \quad \blacktriangleleft$$

Example 7   The following table gives some more examples of decomposing functions into compositions.

Table 2.2.1

FUNCTION	$g(x)$ INSIDE	$f(x)$ OUTSIDE	COMPOSITION
$(x^2 + 1)^{10}$	$x^2 + 1$	x^{10}	$(x^2 + 1)^{10} = f(g(x))$
$\sin^3 x$	$\sin x$	x^3	$\sin^3 x = f(g(x))$
$\tan(x^5)$	x^5	$\tan x$	$\tan(x^5) = f(g(x))$
$\sqrt{4 - 3x}$	$4 - 3x$	$\sqrt{x}$	$\sqrt{4 - 3x} = f(g(x))$
$8 + \sqrt{x}$	$\sqrt{x}$	$8 + x$	$8 + \sqrt{x} = f(g(x))$
$\dfrac{1}{x + 1}$	$x + 1$	$\dfrac{1}{x}$	$\dfrac{1}{x + 1} = f(g(x))$

◄

REMARK.   It should be noted that there is always more than one way to express a function as a composition. For example, here are two ways to express $(x^2 + 1)^{10}$ as a composition that differ from that in Table 2.2.1:

$$(x^2 + 1)^{10} = [(x^2 + 1)^2]^5 = f(g(x)) \qquad \boxed{g(x) = (x^2 + 1)^2 \text{ and } f(x) = x^5}$$

$$(x^2 + 1)^{10} = [(x^2 + 1)^3]^{10/3} = f(g(x)) \qquad \boxed{g(x) = (x^2 + 1)^3 \text{ and } f(x) = x^{10/3}}$$

☐ **AN EXAMPLE FROM CALCULUS**

Calculations of the type in the next example will arise in a variety of calculus problems later in the text.

Example 8   Let $f(x) = x^2$, and let h be any nonzero real number. Find

$$\frac{f(x + h) - f(x)}{h}$$

and simplify as much as possible.

Solution.

$$\frac{f(x + h) - f(x)}{h} = \frac{(x + h)^2 - x^2}{h}$$

$$= \frac{x^2 + 2xh + h^2 - x^2}{h}$$

$$= \frac{2xh + h^2}{h}$$

$$= \frac{h(2x + h)}{h}$$

Canceling the common factor h and being careful to emphasize the restriction on h, we obtain

$$\frac{f(x + h) - f(x)}{h} = 2x + h, \quad h \neq 0 \qquad ◄$$

□ **CLASSIFICATION OF FUNCTIONS**

We shall conclude this section by discussing some of the important categories of functions that will occur in this text.

The simplest of all functions are those that assign the same value to every member of the domain. These are called **constant functions**. For example, if f is the constant function defined by $f(x) = 3$, then

$$f(-1) = 3, \quad f(0) = 3, \quad f(\sqrt{2}) = 3, \quad f(9) = 3$$

and so forth.

A function of the form cx^n, where c is a constant and n is a nonnegative integer, is called a **monomial in x**. Examples are:

$$2x^3, \quad \pi x^7, \quad 4x^0 \,(= 4), \quad -6x, \quad x^{17}$$

The functions $4x^{1/2}$ and x^{-3} are *not* monomials because the powers of x are not nonnegative integers. A function that is expressible as the sum of finitely many monomials in x is called a **polynomial in x**. Examples are:

$$x^3 + 4x + 7, \quad 3 - 2x^3 + x^{17}, \quad 9, \quad 17 - \tfrac{2}{3}x, \quad x^5$$

The function $(x^2 - 4)^3$ is also a polynomial because it can be expressed as a sum of monomials by performing the cubing operation. Depending on whether one wants the powers written in ascending or descending order, the general formula for a polynomial in x is

$$f(x) = a_0 + a_1 x + a_2 x^2 + \cdots + a_n x^n$$

or

$$f(x) = a_n x^n + a_{n-1} x^{n-1} + \cdots + a_1 x + a_0$$

where n is a nonnegative integer and $a_0, a_1, a_2, \ldots, a_n$ are all constants. The highest power of x that occurs in a nonconstant polynomial is called the **degree** of the polynomial. Thus, $x^3 + 4x + 7$ has degree 3 and $17 - \tfrac{2}{3}x$ has degree 1. A nonzero constant c has degree zero (it can be written as $c = cx^0$). The constant zero is not assigned a degree. First-, second-, and third-degree polynomials are called **linear, quadratic,** and **cubic,** respectively. These have the following forms:

DESCRIPTION	GENERAL FORMULA
Linear polynomial	$a_0 + a_1 x \quad (a_1 \neq 0)$
Quadratic polynomial	$a_0 + a_1 x + a_2 x^2 \quad (a_2 \neq 0)$
Cubic polynomial	$a_0 + a_1 x + a_2 x^2 + a_3 x^3 \quad (a_3 \neq 0)$

A function that is expressible as a ratio of two polynomials is called a **rational function**. Examples are

$$\frac{x^5 - 2x^2 + 1}{x^2 - 4}, \quad \frac{x}{x + 1}, \quad \frac{1}{x^5}$$

In general, f is a rational function if it is expressible in the form

$$f(x) = \frac{a_0 + a_1 x + \cdots + a_n x^n}{b_0 + b_1 x + \cdots + b_m x^m}$$

The natural domain of f consists of all x where the denominator differs from zero.

The rational functions are part of a broader class of functions called *explicit algebraic functions*. These are functions that can be evaluated using finitely many additions, subtractions, multiplications, divisions, and root extractions. For example,

$$f(x) = x^{2/3} = (\sqrt[3]{x})^2 \quad \text{and} \quad g(x) = \frac{(x - 3)\sqrt[4]{x}}{x^5 + \sqrt{x^2 + 1}}$$

are all explicit algebraic functions.

All remaining functions fall into two categories, *implicit algebraic functions* and *transcendental functions*. We shall not define these terms, but instead refer the interested reader to a classic book in calculus, G. H. Hardy,* *A Course of Pure Mathematics*, Cambridge Press, 1958 (10th edition). Among the transcendental functions are those that involve trigonometric expressions, exponentials, and logarithms.

▶ Exercise Set 2.2

1. Let $f(x) = x^2 + 1$. Find
   (a) $f(t)$ (b) $f(t + 2)$ (c) $f(x + 2)$
   (d) $f\left(\dfrac{1}{x}\right)$ (e) $f(x + h)$ (f) $f(-x)$
   (g) $f(\sqrt{x})$ (h) $f(3x)$.

2. Let $g(x) = \sqrt{x}$. Find
   (a) $g(5s + 2)$ (b) $g(\sqrt{x} + 2)$ (c) $3g(5x)$
   (d) $\dfrac{1}{g(x)}$ (e) $g(g(x))$ (f) $g^2(x)$
   (g) $g\left(\dfrac{1}{\sqrt{x}}\right)$ (h) $g((x - 1)^2)$.

3. Given that $f(-1) = 4$, $f(2) = 5$, $g(-1) = 3$, and $g(2) = -1$, find
   (a) $(f - g)(-1)$ (b) $(f \cdot g)(-1)$
   (c) $(f/g)(2)$ (d) $(f \circ g)(2)$.

4. Let $f(x) = \dfrac{3}{x}$. Find
   (a) $f\left(\dfrac{1}{x}\right) + \dfrac{1}{f(x)}$
   (b) $f(x^2) - f^2(x)$.

In Exercises 5–12, find formulas for the following functions, making sure to specify the domain in each case.
   (a) $(f + g)(x)$ (b) $(f - g)(x)$
   (c) $(f \cdot g)(x)$ (d) $(f/g)(x)$
   (e) $(f \circ g)(x)$ (f) $(g \circ f)(x)$

5. $f(x) = 2x$, $g(x) = x^2 + 1$.

6. $f(x) = 3x - 2$, $g(x) = |x|$.

7. $f(x) = \sqrt{x + 1}$, $g(x) = x - 2$.

---

*G. H. HARDY (1877–1947). Hardy was a world-renowned British mathematician. He taught at Cambridge and Oxford universities and was a prolific researcher who produced over 300 research papers. He received numerous medals and honorary degrees for his accomplishments. Hardy's book, *A Course of Pure Mathematics*, which gave the first rigorous English exposition of functions and limits for the college undergraduate, had a great impact on university teaching. Hardy had a rebellious spirit; he once listed among his most ardent wishes: (1) to prove the Riemann hypothesis (a famous unsolved mathematical problem), (2) to make a brilliant play in a crucial cricket match, (3) to prove the nonexistence of God, and (4) to murder Mussolini.

8. $f(x) = \dfrac{x}{1 + x^2}$, $g(x) = \dfrac{1}{x}$.

9. $f(x) = \sqrt{x - 2}$, $g(x) = \sqrt{x - 3}$.

10. $f(x) = x^3$, $g(x) = \dfrac{1}{\sqrt[3]{x}}$.

11. $f(x) = \sqrt{1 - x^2}$, $g(x) = \sin 3x$.

12. $f(x) = \sin^2 x$, $g(x) = \cos x$.

13. Let

$$f(x) = \frac{4}{x^2 + 5} \quad \text{and} \quad g(x) = \sqrt{x}$$

Find $(f \circ g)(x)$ and $(g \circ f)(x)$.

14. Let $f(x) = \sqrt{2x - 10}$ and $g(x) = 8x^2 + 5$. Find $(f \circ g)(x)$ and $(g \circ f)(x)$.

15. Let h be defined by $h(x) = 2x - 5$. Find
    (a) $h \circ h$ (b) h^2.

16. Let $g(x) = x^3$ and
$$f(x) = \begin{cases} 5x, & x \le 0 \\ -x, & 0 < x \le 8 \\ \sqrt{x}, & x > 8 \end{cases}$$

    Find $(f \circ g)(x)$.

17. Let $f(x) = \dfrac{1}{x}$.

    (a) If $g(x) = x^2 + 1$, show that $f \circ g$ is defined for all x even though f is not defined when $x = 0$.
    (b) Find another function g such that $f \circ g$ is defined for all x?
    (c) What property must a function g have in order for $f \circ g$ to be defined for all x?

18. Is it always true that $f \circ g = g \circ f$? Is it ever true that $f \circ g = g \circ f$?

19. Prove or disprove: For any three functions f, g, and h, $f \circ (g \circ h) = (f \circ g) \circ h$.

In Exercises 20–31, express f as a composition of two functions; that is, find functions g and h such that $f = g \circ h$. (Each exercise has more than one correct solution.)

20. $f(x) = x^2 + 1$.

21. $f(x) = \sqrt{x + 2}$.

22. $f(x) = \dfrac{1}{x - 3}$.

23. $f(x) = (x - 5)^7$.

24. $f(x) = a + bx$.

25. $f(x) = |x^2 - 3x + 5|$.

26. $f(x) = 3 \sin(x^2)$.

27. $f(x) = \sin^2 x$.

28. $f(x) = \cos^3 2x$.

29. $f(x) = \dfrac{3}{5 + \cos x}$.

30. $f(x) = 3 \sin^2 x + 4 \sin x$.

31. $f(x) = \dfrac{\tan x}{3 + \tan x}$.

32. Find functions f, g, and h such that
$$f(g(h(x))) = \sin \sqrt{x^2 + 3x + 7}$$

33. Find functions f, g, and h such that
$$f(g(h(x))) = \sqrt{3 - \sin^2 x}$$

34. Find $(f \circ f)(\pi)$ if $f(x) = \begin{cases} 1 & \text{if } x \text{ is rational} \\ 0 & \text{if } x \text{ is irrational.} \end{cases}$

35. Find $f(x)$ if $f(x + 1) = x^2 + 3x + 5$. [*Hint:* Let $u = x + 1$ and find $f(u)$.]

36. Find $f(x)$ if $f(3x) = \dfrac{x}{x^2 + 1}$. [*Hint:* Let $u = 3x$ and find $f(u)$.]

37. Given that $f(x) = 0$ only for $x = -1$ and $x = 2$ and that $g(x) = 2x - 1$, find x such that $(f \circ g)(x) = 0$.

38. Find $g(x)$ if $f(x) = 2x - 1$ and $(f \circ g)(x) = x^2$.

39. Find $g(x)$ if $f(x) = \sqrt{x + 5}$ and $(f \circ g)(x) = 3|x|$.

40. Let $f(x) = x^2$, $g(x) = \sin x$, and $h(x) = \cos x$.
    (a) Find the exact numerical value of
    $$f(g(0.3)) + f(h(0.3))$$
    (b) Show that $f(h(x)) - f(g(x)) = h(2x)$.

41. If $f(x + y) = f(x) - f(y)$ for all real x and y, show that $f(x) = 0$ for all real x.

42. If $f(-x) = -f(x)$ for all real x, show that $f(0) = 0$.

For the functions in Exercises 43–46, state which of the following term(s) apply: *monomial, polynomial, rational function, explicit algebraic function.*

43. (a) $3x^7$ (b) $4x^{1/7}$
    (c) $\dfrac{1}{5x^6}$ (d) $2x^3 - 1$.

44. (a) $2x^{1/3} + 1$ (b) x^{-2}
    (c) $x^{-1/2}$ (d) $(x - 3)^{12}$.

45. (a) $x^2 \sqrt{x^3 - 3}$ (b) $\dfrac{x + 1}{x + 2}$
    (c) $\pi^{-2} + 1$ (d) $|x|$.

46. (a) $\sqrt{x^2 + \sqrt{x}}$ (b) $\dfrac{x^3 - 2x + 1}{x^2 - 9}$
    (c) $\sqrt{\pi - 3}$ (d) $|x - 2|$.

■ **2.3** GRAPHS OF FUNCTIONS

In this section we shall show how to represent functions geometrically by graphs. Such graphs provide a useful way of visualizing the behavior of a function. We shall also develop some basic techniques for using graphs of simple functions to construct graphs of more complicated functions.

☐ **DEFINITION OF THE GRAPH OF A FUNCTION**

We begin with the following definition.

2.3.1 DEFINITION. The ***graph in the xy-plane of a function f*** is defined to be the graph of the equation $y = f(x)$.

Example 1 Sketch the graph of $f(x) = x + 2$.

Solution. By definition, the graph in the xy-plane of f is the graph of the equation $y = x + 2$; this is a line with slope 1 and y-intercept 2 (Figure 2.3.1). ◄

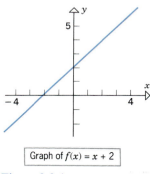

Graph of $f(x) = x + 2$

Figure 2.3.1

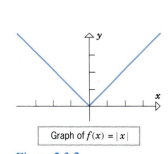

Graph of $f(x) = |x|$

Figure 2.3.2

Example 2 Sketch the graph of $f(x) = |x|$.

Solution. By definition, the graph in the xy-plane of f is the graph of $y = |x|$, or equivalently

$$y = \begin{cases} x, & x \geq 0 \\ -x, & x < 0 \end{cases}$$

The graph coincides with the line $y = x$ for $x \geq 0$ and with the line $y = -x$ for $x < 0$ (Figure 2.3.2). ◄

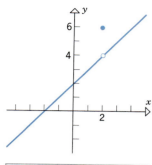

Graph of $\phi(x) = \begin{cases} x + 2, & x \ne 2 \\ 6, & x = 2 \end{cases}$

Figure 2.3.3

Example 3   Sketch the graph of

$$\phi(x) = \begin{cases} x + 2, & x \ne 2 \\ 6, & x = 2 \end{cases}$$

Solution.   The function ϕ is identical to the function f in Example 1, except at $x = 2$, where we have

$$\phi(2) = 6 \quad \text{and} \quad f(2) = 4$$

Thus, the graph of ϕ is identical to the graph of f shown in Figure 2.3.1, except that the graph of ϕ has a point separated from the line at $x = 2$ (Figure 2.3.3).   ◄

Example 4   Sketch the graph of

$$h(x) = \frac{x^2 - 4}{x - 2} \tag{1}$$

Solution.   As noted in Example 14 of Section 2.1, this function can be rewritten as

$$h(x) = x + 2, \quad x \ne 2$$

Thus, the function h in (1) is identical to the function f in Example 1, except that h is undefined at $x = 2$. It follows that the graph of h is identical to the graph of f in Figure 2.3.1, except that the graph of h has a hole in it above $x = 2$ (Figure 2.3.4).   ◄

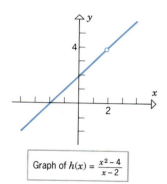

Graph of $h(x) = \frac{x^2 - 4}{x - 2}$

Figure 2.3.4

Example 5   Sketch the graph of

$$g(x) = \begin{cases} 1, & x \le 2 \\ x + 2, & x > 2 \end{cases}$$

Solution.   By definition, the graph in the xy-plane of g is the graph of

$$y = \begin{cases} 1, & x \le 2 \\ x + 2, & x > 2 \end{cases}$$

Thus, for $x \le 2$, we have $y = 1$, and for $x > 2$ we have $y = x + 2$. The graph of $y = 1$ is a horizontal line, and the graph of $y = x + 2$ is the straight line graphed in Example 1. The graph of g is shown in Figure 2.3.5. In that figure we used the heavy dot and open circle above $x = 2$ to emphasize that the value $g(2) = 1$ lies on the horizontal line and not on the inclined line.   ◄

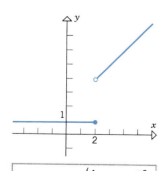

Graph of $g(x) = \begin{cases} 1, & x \le 2 \\ x + 2, & x > 2 \end{cases}$

Figure 2.3.5

☐ **GRAPHING FUNCTIONS**
   BY TRANSLATION

Once you have obtained the graph of an equation $y = f(x)$, there are some special techniques that can be used to obtain the graphs of the closely related equations

$$y = f(x + c), \quad y = f(x) + c$$
$$y = f(x - c), \quad y = f(x) - c$$

where c is any positive constant.

   If a positive constant is added to or subtracted from $f(x)$, the geometric effect is to translate the graph of the function f parallel to the vertical axis; addition translates the graph in the positive direction and subtraction translates it in the negative direction. This is illustrated in the top two lines of Table 2.3.1. Similarly, if a positive constant is added to or subtracted from the independent variable of a function, the geometric effect is to translate the graph of the function parallel to the horizontal axis; subtraction translates the graph in the positive direction, and addition translates it in the negative direction. This is illustrated in the bottom two lines of Table 2.3.1.

Before proceeding to the following examples, it will be helpful to review the graphs in Figure 1.3.22.

Example 6   Sketch the graph of

   (a) $y = x^3 - 4$ (b) $y = x^3 + 4$

Solution.   The graph of the equation $y = x^3 - 4$ can be obtained by translating the graph of $y = x^3$ down 4 units, and the graph of $y = x^3 + 4$ by translating the graph of $y = x^3$ up 4 units (Figure 2.3.6).   ◄

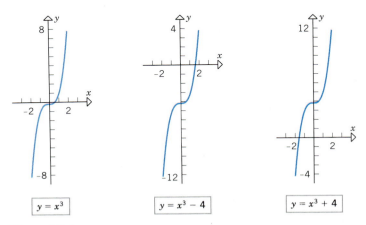

Figure 2.3.6

Example 7   Sketch the graph of

   (a) $y = \sqrt{x - 3}$ (b) $y = \sqrt{x + 3}$

Table 2.3.1

OPERATION ON $y = f(x)$	NEW EQUATION	GEOMETRIC EFFECT	EXAMPLE
Add a positive constant c to $f(x)$	$y = f(x) + c$	Translates the graph of $y = f(x)$ up c units	
Subtract a positive constant c from $f(x)$	$y = f(x) - c$	Translates the graph of $y = f(x)$ down c units	
Add a positive constant c to x	$y = f(x + c)$	Translates the graph of $y = f(x)$ left c units	
Subtract a positive constant c from x	$y = f(x - c)$	Translates the graph of $y = f(x)$ right c units	

Solution. The graph of the equation $y = \sqrt{x - 3}$ can be obtained by translating the graph of $y = \sqrt{x}$ right 3 units, and the graph of $y = \sqrt{x + 3}$ by translating the graph of $y = \sqrt{x}$ left 3 units (Figure 2.3.7). ◄

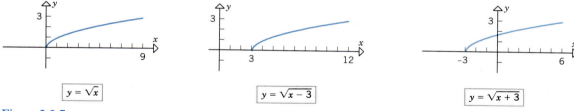

Figure 2.3.7

Example 8   Sketch the graph of $y = |x - 3| + 2$.

Solution.   The graph can be obtained by two translations: first translate the graph of $y = |x|$ right 3 units to obtain the graph of $y = |x - 3|$, then translate this graph up 2 units to obtain the graph of $y = |x - 3| + 2$ (Figure 2.3.8).   ◄

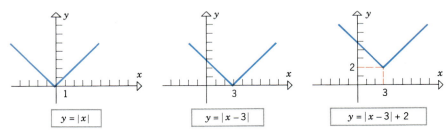

Figure 2.3.8

REMARK.   The graph in the foregoing example could also have been obtained by performing the translations in the opposite order: first translating the graph of $y = |x|$ up 2 units to obtain the graph of $y = |x| + 2$, then translating this graph right 3 units to obtain the graph of $y = |x - 3| + 2$.

Example 9   Sketch the graph of $y = x^2 - 4x + 5$.

Solution.   Complete the square on the first two terms:

$$y = (x^2 - 4x + 4) - 4 + 5 = (x - 2)^2 + 1$$

In this form we see that the graph can be obtained by translating the graph of $y = x^2$ right 2 units because of the $x - 2$, and up 1 unit because of the $+1$ (Figure 2.3.9).

Alternative Solution.   Use the procedure in Example 9 of Section 1.6.   ◄

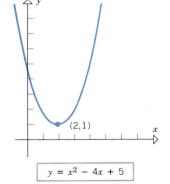

Figure 2.3.9

☐ **REFLECTIONS**

In Section 1.3 (Figure 1.3.16) we observed that $(-x, y)$ is the reflection of (x, y) about the y-axis, and $(x, -y)$ the reflection of (x, y) about the x-axis. Thus, replacing x by $-x$ in an equation reflects its graph about the y-axis and replacing y by $-y$ reflects its graph about the x-axis. In particular,

• the graphs of $y = f(x)$ and $y = f(-x)$ are reflections of one another about the y-axis;

• the graphs of $y = f(x)$ and $-y = f(x)$ [or equivalently $y = -f(x)$] are reflections of one another about the x-axis.

Examples are given in Table 2.3.2.

Table 2.3.2

OPERATION ON $y = f(x)$	NEW EQUATION	GEOMETRIC EFFECT	EXAMPLE
Replace x by $-x$	$y = f(-x)$	Reflects the graph of $y = f(x)$ about the y-axis	
Multiply $f(x)$ by -1	$y = -f(x)$	Reflects the graph of $y = f(x)$ about the x-axis	

Example 10   Sketch the graph of $y = \sqrt[3]{2 - x}$.

Solution.   The graph can be obtained by a reflection and a translation: first reflect the graph of $y = \sqrt[3]{x}$ about the y-axis to obtain the graph of $y = \sqrt[3]{-x}$, then translate this graph right 2 units to obtain the graph of the equation $y = \sqrt[3]{-(x - 2)} = \sqrt[3]{2 - x}$ (Figure 2.3.10).   ◀

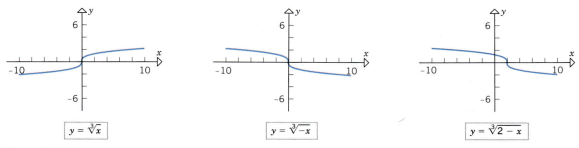

Figure 2.3.10

Example 11   Sketch the graph of $y = 4 - |x - 2|$.

Solution.   The graph can be obtained by a reflection and two translations: first reflect the graph of $y = |x|$ about the x-axis to obtain the graph of $y = -|x|$; then translate this graph right 2 units to obtain the graph of $y = -|x - 2|$; and then translate this graph up 4 units to obtain the graph of the equation $y = -|x - 2| + 4 = 4 - |x - 2|$ (Figure 2.3.11).   ◀

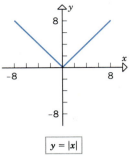

$$y = |x|$$

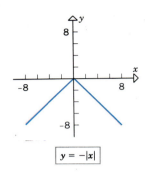

$$y = -|x|$$

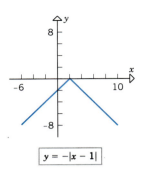

$$y = -|x - 1|$$

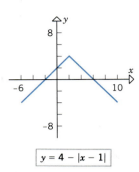

$$y = 4 - |x - 1|$$

Figure 2.3.11

SCALING

Translations and reflections are called ***rigid transformations*** because they do not change the shape of a graph, only its position. We shall now discuss a transformation, called ***scaling,*** that actually changes the shape of a graph.

If $f(x)$ is multiplied by a *positive* constant c, the geometric effect is to compress or stretch the graph of $y = f(x)$ in the y-direction; a compression occurs if $0 < c < 1$ and a stretching if $c > 1$. This operation is called ***vertical scaling by a factor of c*** (Figure 2.3.12a). Similarly, if x is multiplied by a *positive* constant c, the geometric effect is to compress or stretch the graph of $y = f(x)$ in the x-direction; a compression occurs if $c > 1$ and a stretching if $0 < c < 1$. This operation is called ***horizontal scaling by a factor of c*** (Figure 2.3.12b).

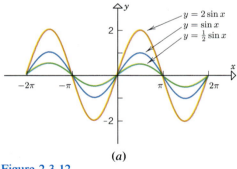

(a)

(b)

Figure 2.3.12

VERTICAL LINE TEST

In all of the examples given in this section so far, we *started* with a function f and by one method or another we *found* the graph of the equation $y = f(x)$. By reversing the situation and starting with a graph, we are led to the following interesting question.

> **2.3.2 PROBLEM.** *If we draw an arbitrary curve in the xy-plane, must that curve necessarily be the graph of $y = f(x)$ for some function f?*

The answer to this question is no! There exist curves in the plane that are not graphs of any functions. To see why this is so, consider the curve in Figure

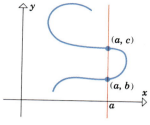

This curve is not the graph of $y = f(x)$ for any function f.

Figure 2.3.13

2.3.13 and the vertical line that intersects it at the two distinct points (a, b) and (a, c). This curve cannot be the graph of

$$y = f(x) \tag{2}$$

for any function f. For if it were, then (a, b) and (a, c), being points on the curve, would both have coordinates satisfying (2). Thus, we would have

$$b = f(a) \quad \text{and} \quad c = f(a)$$

But this is impossible since f cannot assign two different values to a. Therefore, there is no function of x whose graph is the curve in Figure 2.3.13.

This discussion illustrates the following general result, which we shall call the **vertical line test**.

2.3.3   THE VERTICAL LINE TEST.   *A curve in the xy-plane is the graph of* $y = f(x)$ *for some function f if and only if no vertical line intersects the curve more than once.*

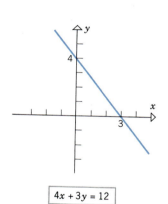

$4x + 3y = 12$

Figure 2.3.14

Example 12   Is the graph of the equation

$$4x + 3y = 12 \tag{3}$$

the graph of $y = f(x)$ for some function f?

Solution.   We can answer this question by rewriting the equation in the equivalent form

$$y = 4 - \frac{4}{3}x$$

from which it is evident that the graph of (3) is also the graph of the function

$$f(x) = 4 - \frac{4}{3}x$$

Observe that the graph of this function, which is shown in Figure 2.3.14, passes the vertical line test.   ◄

Example 13   Is the graph of the circle

$$x^2 + y^2 = 25 \tag{4}$$

also the graph of $y = f(x)$ for some function f?

Solution.   Since some vertical lines intersect the circle more than once (see Figure 2.3.15), the circle is not the graph of any function of x.

We can also deduce this result algebraically by observing that (4) can be written as

$$y = \pm\sqrt{25 - x^2} \tag{5}$$

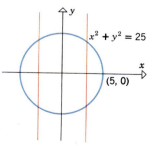

$x^2 + y^2 = 25$

$(5, 0)$

Figure 2.3.15

But the right side of (5) is not a function of x since it is "multiple-valued", that is, each value of x results in more than one value of y. Thus, (4) is not equivalent to an equation of the form $y = f(x)$.   ◄

REVERSING THE ROLES OF x AND y

Sometimes the roles of x and y will be reversed, and we will be interested in knowing whether a given curve is the graph in the xy-plane of an equation

$$x = g(y)$$

for some function g. Because g cannot assign two different x values to the same y, the graph of such an equation cannot be cut twice by any horizontal line (Figure 2.3.16). Thus, we have the following analog of the vertical line test.

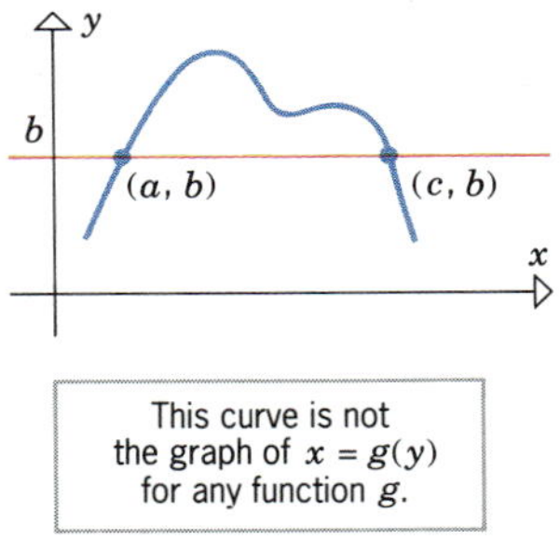

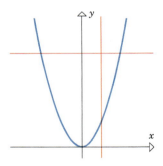

This curve is not the graph of $x = g(y)$ for any function g.

Figure 2.3.16

> **2.3.4** THE HORIZONTAL LINE TEST.   *A curve in the xy-plane is the graph of $x = g(y)$ for some function g if and only if no horizontal line intersects the curve more than once.*

Example 14   The graph of the equation

$$y = x^2 \tag{6}$$

(Figure 2.3.17) passes the vertical line test, but not the horizontal line test, so it is the graph of a function of x [namely, $f(x) = x^2$], but it is not the graph of any function of y. This can also be seen algebraically by solving (6) for x in terms of y. This yields

$$x = \pm\sqrt{y} \tag{7}$$

The right side of this equation is "multiple-valued", so it is not the formula for a function of y.   ◄

The graph of $y = x^2$ passes the vertical line test but not the horizontal line test.

Figure 2.3.17

REMARK.   When graphing equations of the form $y = f(x)$, we have been keeping the x-axis horizontal and the y-axis vertical. Thus, it would seem natural to graph equations of the form $x = g(y)$ with the y-axis horizontal and the x-axis vertical. However, there are many situations in which one wants to graph $y = f(x)$ and $x = g(y)$ in the same coordinate system, so it is common to keep the x-axis horizontal and the y-axis vertical for equations of both types.

IMPLICITLY DEFINED FUNCTIONS

An equation of the form $y = f(x)$ is said to **define y explicitly as a function of x** (the function being f), and an equation of the form $x = g(y)$ is said to **define x explicitly as a function of y** (the function being g). For example,

$$y = 5x^2 \sin x$$

defines y explicitly as a function of x and

$$x = (7y^3 - 2y)^{3/2}$$

defines x explicitly as a function of y.

An equation that is not of the form $y = f(x)$ but whose graph in the xy-plane is cut at most once by any vertical line is said to **define y implicitly as a function of x,** and an equation that is not of the form $x = g(y)$ but whose graph in the xy-plane is cut at most once by any horizontal line is said to **define x implicitly as a function of y.**

Sometimes one can determine whether an equation defines a function implicitly by solving the equation for y in terms of x or x in terms of y. For example, the equation

$$y^3 - 3y^2 + 3y - x^2 = 1 \tag{8}$$

can be solved for y in terms of x as

$$y^3 - 3y^2 + 3y - 1 = x^2$$
$$(y - 1)^3 = x^2$$
$$y = x^{2/3} + 1 \tag{9}$$

and for x in terms of y as

$$x^2 = y^3 - 3y^2 + 3y - 1$$
$$x = \pm\sqrt{y^3 - 3y^2 + 3y - 1} \tag{10}$$

It follows from (9) that (8) defines y implicitly as a function of x and from (10) that (8) does not define x implicitly as a function of y.

☐ **SOME COMPLICATIONS**

Some equations in x and y cannot be solved for x in terms of y or y in terms of x. This can occur either because the algebra required is too complicated or because it is mathematically impossible to do so. For example, one cannot determine by simple algebra whether the equation

$$1 + xy^3 - \sin(x^2y) = 0$$

implicitly defines y as a function of x, or x as a function of y, or neither. There are theorems studied in advanced courses that can sometimes be used to ascertain whether an equation defines a function implicitly. However, even if it can be shown that the equation does define a function implicitly, there may be no way to produce an explicit formula for the function, so all of its properties would have to be developed indirectly from the given equation. This can be quite complicated in some cases.

☐ **FUNCTIONS DEFINED IN A NEIGHBORHOOD OF A POINT**

Sometimes a curve, when considered in its entirety, is not the graph of a function, but some smaller portion of the curve is. As an illustration, we saw in Example 13 that the circle $x^2 + y^2 = 25$ is not the graph of $y = f(x)$ for any function f. However, the upper and lower semicircles pass the vertical line test, so they are each graphs of functions. From (5) the upper semicircle has the equation $y = \sqrt{25 - x^2}$, so it is the graph of the function $f_1(x) = \sqrt{25 - x^2}$, and the lower semicircle has the equation $y = -\sqrt{25 - x^2}$, so it is the graph of the function $f_2(x) = -\sqrt{25 - x^2}$ (Figure 2.3.18a). Similarly, solving $x^2 + y^2 = 25$ for x in terms of y we obtain $x = \pm\sqrt{25 - y^2}$ from which

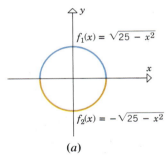

$f_1(x) = \sqrt{25 - x^2}$

$f_2(x) = -\sqrt{25 - x^2}$

(a)

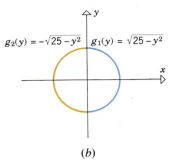

$g_2(y) = -\sqrt{25 - y^2}$ | $g_1(y) = \sqrt{25 - y^2}$

(b)

Figure 2.3.18

it follows that the right semicircle is the graph of $g_1(y) = \sqrt{25 - y^2}$ and the left semicircle is the graph of $g_2(y) = -\sqrt{25 - y^2}$ (Figure 2.3.18b).

To make the foregoing ideas more precise, let C denote the graph of an equation in the xy-plane. We will say that the equation *defines y implicitly as a function of x in a neighborhood of* (x_0, y_0) if there is some rectangle centered at (x_0, y_0), with sides parallel to the coordinate axes, such that the portion of C contained within the rectangle is cut at most once by any vertical line. Similarly, we will say that the equation *defines x implicitly as a function of y in a neighborhood of* (x_0, y_0) if there is some rectangle centered at (x_0, y_0), with sides parallel to the coordinate axes, such that the portion of C contained within the rectangle is cut at most once by any horizontal line.

Example 15 The equation of the unit circle $x^2 + y^2 = 1$ does not implicitly define y as a function of x in a neighborhood of the point $P(1, 0)$, because the portion of the circle within any rectangle centered at P is cut twice by some vertical line (Figure 2.3.19a). Similarly, the equation does not implicitly define x as a function of y in a neighborhood of the point $Q(0, 1)$, because the portion of the circle within any rectangle centered at Q is cut twice by some horizontal line (Figure 2.3.19b). However, the equation implicitly defines both y as a function of x and x as a function of y in a neighborhood of the point $(1/\sqrt{2}, 1/\sqrt{2})$ (Figure 2.3.19c). ◀

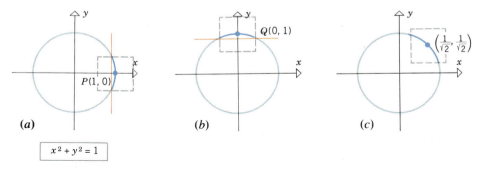

(a)   (b)   (c)

$x^2 + y^2 = 1$

Figure 2.3.19

▶ **Exercise Set 2.3**   [C] *53, 54, 55*

1. Consider the function f graphed in Figure 2.3.20. In each part, find all values of x satisfying the given condition.
   (a) $f(x) = 0$ (b) $f(x) = 3$
   (c) $f(x) \geq 0$ (d) $f(x) \leq 0$.

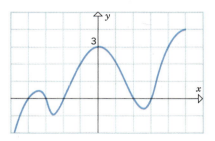

Figure 2.3.20

In Exercises 2–29, sketch the graph of the function.

2. $f(x) = 2x + 1$.   3. $f(x) = 3x - 2$.

4. $G(x) = x$, $1 \leq x \leq 2$.

5. $G(x) = x - 2$, $-1 \leq x \leq 1$.

6. $h(x) = x^2 - 3$.

7. $h(x) = (x - 2)^2$.

8. $F(x) = \sqrt{x + 1}$.

9. $F(x) = \sqrt{3 - x}$.

10. $f(x) = \sqrt{4 - x^2}$.

11. $f(x) = 2 + \sqrt{9 - x^2}$.

12. $g(x) = \sqrt{4x - x^2}$.

13. $g(x) = \sqrt{7 - 6x - x^2}$.

14. $f(x) = 2 \sin x$.   **15.** $f(x) = 3 \sin 2x$.

16. $f(x) = \cos 2x$.   **17.** $f(x) = 1 + \cos x$.

18. $f(x) = \dfrac{x^2 - 4}{x + 2}$.   **19.** $f(x) = \dfrac{x^2 + 2x}{x}$.

20. $g(x) = \dfrac{x^3 - x^2}{x - 1}$.   **21.** $g(x) = \dfrac{x - x^3}{x}$.

22. $\phi(x) = \dfrac{x}{|x|}$.   **23.** $\phi(x) = \dfrac{|x - 2|}{x - 2}$.

24. $g(x) = \begin{cases} x^2, & x \neq 4 \\ 0, & x = 4. \end{cases}$

25. $g(x) = \begin{cases} x - 1, & x \neq 1 \\ 3, & x = 1. \end{cases}$

26. $f(x) = \begin{cases} x + 2, & x \leq 3 \\ x + 4, & x > 3. \end{cases}$

27. $f(x) = \begin{cases} x^2, & x > 1 \\ 2, & x \leq 1. \end{cases}$

28. $h(x) = \begin{cases} 1, & 0 < x \leq 1 \\ 3, & 1 < x \leq 2 \\ -1, & 2 < x \leq 3 \\ 0, & \text{elsewhere.} \end{cases}$

29. $h(x) = \begin{cases} -2, & -2 \leq x < -1 \\ 1, & -1 \leq x < 0 \\ 2, & 0 \leq x < 1 \\ 0, & \text{elsewhere.} \end{cases}$

In Exercises 30–33, express the function in piecewise form without using absolute values and sketch its graph.

30. $f(x) = |x - 3| - x$.

31. $f(x) = 2x + |2 - x|$.

32. $g(x) = |x| + |x - 3|$.

33. $g(x) = |x - 5| - |x - 3|$.

34. Let $f(x) = x^2 - 1$. Sketch the graph of
(a) $y = \frac{1}{2}(f(x) + |f(x)|)$
(b) $y = \frac{1}{2}(f(x) - |f(x)|)$.
[*Hint:* Express the equations in piecewise form.]

35. Given that $f(x) = |x|$ and $g(x) = x$, sketch the graph of the equation.
(a) $y = (f + g)(x)$ (b) $y = (f - g)(x)$
(c) $y = (f \cdot g)(x)$ (d) $y = (f/g)(x)$.

36. A 10-foot ladder leans against a wall. The base of the ladder is x feet from the wall. Express the distance from the top of the ladder to the ground as a function of x, and sketch the graph of the function.

37. Express the area A enclosed by the graph of
$$f(x) = \begin{cases} 2x, & 0 \leq x \leq 1 \\ 2, & x > 1 \end{cases}$$
the x-axis, and the vertical line at x $(x \geq 0)$ as a function of x.

38. Find a formula for the function f graphed in Figure 2.3.21.

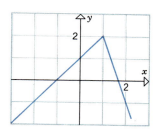

Figure 2.3.21

39. Find a formula for the function g graphed in Figure 2.3.22.

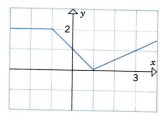

Figure 2.3.22

40. Use Table 2.3.1 and the graph of $y = |x|$ to graph the following:
(a) $y = |x - 4|$ (b) $y = |x| + 4$
(c) $y = |x - 4| + 4$ (d) $y = |x + 5| - 2$.

41. Use Table 2.3.1 and the graph of $y = \sqrt{x}$ to graph the following:
 (a) $y = \sqrt{x - 3}$ (b) $y = \sqrt{x} + 3$
 (c) $y = \sqrt{x - 3} + 3$ (d) $y = \sqrt{x + 1} - 2$.

42. A function f with domain $[-1, 3]$ has the graph shown in Figure 2.3.23. Use this graph to obtain the graphs of the equations
 (a) $y = f(x + 1)$ (b) $y = f(2x)$
 (c) $y = f(-x)$ (d) $y = -f(x)$.

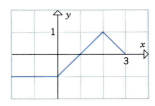

Figure 2.3.23

43. The equation $y = |f(x)|$ can be written as

$$y = \begin{cases} f(x), & f(x) \geq 0 \\ -f(x), & f(x) < 0 \end{cases}$$

which shows that the graph of $y = |f(x)|$ can be obtained from the graph of $y = f(x)$ by retaining the portion that lies on or above the x-axis and reflecting about the x-axis the portion that lies below the x-axis. Use this method to obtain the graph of $y = |2x - 3|$ from the graph of $y = 2x - 3$.

In Exercises 44 and 45, use the method described in Exercise 43.

44. Sketch the graph of $y = |1 - x^2|$.

45. Sketch the graph of
 (a) $f(x) = |\cos x|$
 (b) $f(x) = \cos x + |\cos x|$.

46. The **greatest integer function**, $[x]$, is defined to be the greatest integer that is less than or equal to x. For example, $[2.7] = 2$, $[-2.3] = -3$, and $[4] = 4$. Sketch the graph of
 (a) $f(x) = [x]$ (b) $f(x) = [x^2]$
 (c) $f(x) = [x]^2$ (d) $f(x) = [\sin x]$.

47. A function f is called **even** if $f(-x) = f(x)$ for each x in the domain of f and **odd** if $f(-x) = -f(x)$ for each such x. In each part, classify the function as even, odd, or neither.
 (a) $f(x) = x^2$ (b) $f(x) = x^3$
 (c) $f(x) = |x|$ (d) $f(x) = x + 1$
 (e) $f(x) = \dfrac{x^5 - x}{1 + x^2}$ (f) $f(x) = 2$.

48. In Figure 2.3.24 we have sketched part of the graph of a function f. Complete the graph assuming
 (a) f is an even function
 (b) f is an odd function.
 [See Exercise 47 for terminology.]

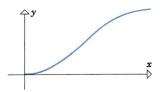

Figure 2.3.24

49. Classify the functions graphed in Figure 2.3.25 as even, odd, or neither. (See Exercise 47 for the definitions of even and odd functions.)

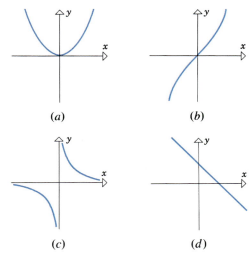

Figure 2.3.25

50. Can a function be both even and odd? [See Exercise 47 for terminology.]

51. Prove that the product of
 (a)   two even functions is an even function
 (b)   two odd functions is an even function
 (c)   an even and an odd function is an odd function. [See Exercise 47 for terminology.]

52. Let a be a constant and suppose

$$f(a - x) = f(a + x)$$

for all x. What geometric property must the graph of f have?

53. There are various numerical methods to obtain approximate solutions of equations of the form $f(x) = 0$. One such method requires that the equation be expressed as $x = g(x)$, so that a solution $x = c$ can be interpreted as the value of x where the line $y = x$ intersects the curve $y = g(x)$ (Figure 2.3.26). If x_1 is an initial estimate of c and the graph of $y = g(x)$ is not too steep in the vicinity of c, then a better approximation x_2 can be obtained from $x_2 = g(x_1)$, as shown in the figure. An even better value x_3 is obtained from $x_3 = g(x_2)$, and so on. Thus, the formula

$$x_{n+1} = g(x_n) \quad n = 1, 2, 3, \ldots$$

generates successive approximations $x_2, x_3, x_4, \ldots$ that get closer and closer to c.

(a) It can be shown that the equation $x^3 - x - 1 = 0$ has only one real solution. Show that the equation can be written as $x = g(x)$, where $g(x) = \sqrt[3]{x + 1}$. Sketch the graphs of $y = x$ and $y = \sqrt[3]{x + 1}$ in the same coordinate system for $-1 \le x \le 3$, and obtain an initial estimate x_1 for the solution. Use a calculator to find successive approximations $x_2, x_3, \ldots$ until there is no change in the value displayed by your calculator.

(b) Show that this method will not always work by writing $x^3 - x - 1 = 0$ in the form $x = g(x)$, where $g(x) = x^3 - 1$, and using the same initial value x_1 as in part (a). Sketch the graphs of $y = x$ and $y = x^3 - 1$ to see what goes wrong.

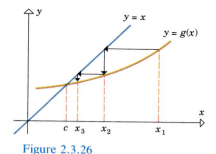

Figure 2.3.26

In Exercises 54 and 55, use the method of Exercise 53 to obtain an approximate solution of the given equation.

54. $x^5 - x - 2 = 0$.

55. $x - \cos x = 0$. [*Note:* Be sure that your calculator is set in the radian mode.]

In Exercises 56 and 57, determine whether the equation defines y as a function of x, or x as a function of y, or both, or neither.

56. (a) $3x - 4y = 12$ (b) $xy^2 = 1$
    (c) $x^2 + y^2 = 1$ (d) $\dfrac{1 + x}{1 - y} = 2$.

57. (a) $4x + 2y = -8$ (b) $x^2 y^3 = 1$
    (c) $3x^2 + 4y^2 = 12$ (d) $\dfrac{xy}{1 - xy} = 1$.

58. In each part express x explicitly as a function of y.
    (a) $xy - x = 1$ (b) $y = \dfrac{x}{1 + x}$
    (c) $x^2 + 2xy + y^2 = 0$.

59. In each part express y explicitly as a function of x.
    (a) $x^2 y - 1 = 0$ (b) $x = \dfrac{1 - y}{1 + y}$
    (c) $y^2 + 2xy + x^2 = 0$.

60. Show that $y^2 + 3xy + x^2 = 0$ does not define y implicitly as a function of x.

61. Show that $y^2 + 4xy + 1 = 0$ does not define y implicitly as a function of x.

62. Find two functions, the union of whose graphs is the graph of the equation in Exercise 61.

63. In Figure 2.3.27, determine whether the curve is the graph of a function of x, a function of y, both, or neither.

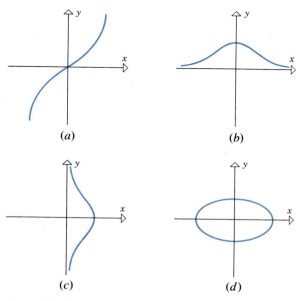

Figure 2.3.27

■ 2.4 LIMITS (AN INTUITIVE INTRODUCTION)

> *The development of calculus was stimulated in large part by two geometric problems: finding areas of plane regions and finding tangent lines to curves. In this section we shall show that both of these problems are closely related to a fundamental concept of calculus known as a "limit."*

□ **THE TANGENT LINE AND AREA PROBLEMS**

Calculus centers around the following two fundamental problems:

> **THE TANGENT PROBLEM.**   *Given a function f and a point $P(x_0, y_0)$ on its graph, find an equation of the line tangent to the graph at P* (Figure 2.4.1).

> **THE AREA PROBLEM.**   *Given a function f, find the area between the graph of f and an interval [a, b] on the x-axis* (Figure 2.4.2).

Traditionally, that portion of calculus arising from the tangent problem is called ***differential calculus*** and that arising from the area problem is called ***integral calculus***. As we shall see, however, the tangent and area problems are closely related, so that the distinction between differential calculus and integral calculus is often hard to discern.

In order to solve the tangent and area problems it is necessary to have a more precise understanding of the concepts of "tangent line" and "area." As we shall explain in this section, both of these concepts rest on a more fundamental concept, known as a "*limit*."

□ **TANGENT LINES AND LIMITS**

In plane geometry, a line is called ***tangent*** to a circle if it meets the circle at precisely one point (Figure 2.4.3a). However, this definition is not satisfactory for other kinds of curves. In Figure 2.4.3b the line meets the curve exactly once, yet is not a tangent, and in Figure 2.4.3c the line is tangent yet meets the curve more than once.

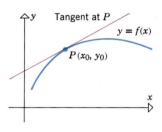

Figure 2.4.1

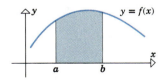

Figure 2.4.2

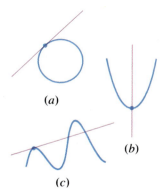

Figure 2.4.3

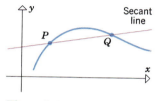

Figure 2.4.4

To define the concept of a tangent line so that it applies to curves other than circles, we must view tangent lines another way. Consider a point P on a curve in the xy-plane. If Q is any point on the curve different from P, the line through P and Q is called a **secant line** for the curve (Figure 2.4.4). Intuition suggests that if we move the point Q along the curve toward P, the secant line will rotate toward a "limiting" position (Figure 2.4.5). The line T occupying this limiting position we consider to be the **tangent line** at P.

As suggested by Figure 2.4.6, this new concept of a tangent line coincides with the traditional concept, when applied to circles.

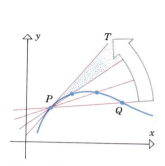

Figure 2.4.5

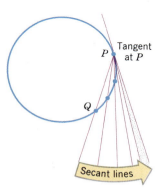

Figure 2.4.6

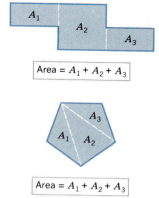

Figure 2.4.7

☐ **AREA AS A LIMIT**

Just as the general notion of a tangent line leads to the concept of a "limit," so does the general notion of area. Areas of some plane regions can be calculated by subdividing them into a *finite* number of rectangles or triangles, then adding the areas of the constituent parts (Figure 2.4.7). However, for many regions a more general approach is needed. Consider the shaded region in Figure 2.4.8a. We can *approximate* the area of this region by inscribing rectangles of equal width under the curve and adding the areas of these rectangles (Figure 2.4.8b). Moreover, intuition suggests that if we repeat the process using more and more rectangles, then the rectangles will tend to fill in the gaps under the curve and our approximations will "approach" the exact area under the curve as a "limiting value" (Figure 2.4.8c).

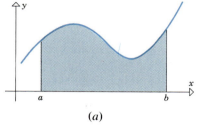

(a)

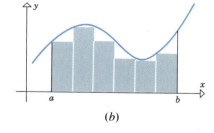

(b)

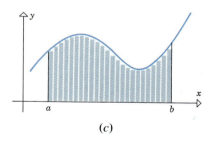

(c)

Figure 2.4.8

□ **LIMITS**

In the foregoing discussion, we saw that the concepts of a tangent line and the area of a plane region ultimately rest on the notion of a "limit." In this section and the next few we shall investigate the notion of limit in more detail. Our development of limits in this text proceeds in three stages:

- First we discuss limits intuitively.
- Then we discuss methods for computing limits.
- Finally, we give a precise mathematical discussion of limits.

Limits can be used to describe how a function behaves as the independent variable moves toward a certain value. For example, consider the function

$$f(x) = \frac{\sin x}{x}$$

Table 2.4.1

x (RADIANS)	$\dfrac{\sin x}{x}$
1.0	0.84147
0.9	0.87036
0.8	0.89670
0.7	0.92031
0.6	0.94107
0.5	0.95885
0.4	0.97355
0.3	0.98507
0.2	0.99335
0.1	0.99833
0.01	0.99998

where x is in radians. Although this function is not defined at $x = 0$, it still makes sense to ask what happens to the values of $f(x)$ as x moves along the x-axis toward $x = 0$. To answer this question, we used a calculator set in the radian mode to obtain values of $f(x)$ at a succession of points moving along the *positive* x-axis toward $x = 0$. The results, which appear in Table 2.4.1, suggest that the values of $f(x)$ approach 1. We call the number 1 the *limit* of

$$\frac{\sin x}{x}$$

as x approaches 0 from the right side, and we write

$$\lim_{x \to 0^+} \frac{\sin x}{x} = 1 \tag{1}$$

Table 2.4.2

x (RADIANS)	$\dfrac{\sin x}{x}$
−1.0	0.84147
−0.9	0.87036
−0.8	0.89670
−0.7	0.92031
−0.6	0.94107
−0.5	0.95885
−0.4	0.97355
−0.3	0.98507
−0.2	0.99335
−0.1	0.99833
−0.01	0.99998

In this expression, "lim" tells us that we are computing a limit; the symbol $x \to 0$ tells us that we are letting x approach 0; and the label "+" on $x \to 0$ tells us that x is approaching zero from the *right* side. We can also ask what happens to the values of

$$f(x) = \frac{\sin x}{x}$$

as x approaches 0 from the left side. From Table 2.4.2 [or from the fact that $f(-x) = f(x)$] it is evident that the values of $f(x)$ again approach 1. We denote this by writing

$$\lim_{x \to 0^-} \frac{\sin x}{x} = 1 \tag{2}$$

In (2), the label "−" on $x \to 0$ tells us that we are computing the limit as x approaches zero from the *left* side.

REMARK.   It is important to keep in mind that the limits given in (1) and (2) are really just guesses based on numerical evidence. It is conceivable that if we continue the calculator computations in Tables 2.4.1 and 2.4.2 for values of x still closer to 0, the pattern of values for $(\sin x)/x$ might change and approach some value different from 1 or perhaps approach no number at all. Moreover, no matter how much we enlarge our tables, this problem will persist since we can never be certain what happens for values of x not included in the table. To be sure that the limits in (1) and (2) are really correct, we need mathematical proof, and to give a mathematical proof we need a precise mathematical definition of a limit. These questions will be taken up later. However, in this section we shall rely completely on our intuition to obtain limits. Our main objective at this time is to introduce the mathematical notation associated with limits and to develop our intuitive understanding of limits geometrically.

□ **LIMIT NOTATION**

2.4.1 NOTATION. If the value of $f(x)$ approaches the number L_1 as x approaches x_0 from the right side, we write

$$\lim_{x \to x_0^+} f(x) = L_1 \tag{3}$$

which is read, "the limit of $f(x)$ as x approaches x_0 from the right is equal to L_1."

2.4.2 NOTATION. If the value of $f(x)$ approaches the number L_2 as x approaches x_0 from the left side, we write

$$\lim_{x \to x_0^-} f(x) = L_2 \tag{4}$$

which is read, "the limit of $f(x)$ as x approaches x_0 from the left is equal to L_2."

2.4.3 NOTATION. If the limit from the left side is the same as the limit from the right side, that is, if

$$\lim_{x \to x_0^-} f(x) = \lim_{x \to x_0^+} f(x) = L$$

then we write

$$\lim_{x \to x_0} f(x) = L \tag{5}$$

which is read, "the limit of $f(x)$ as x approaches x_0 is equal to L."

The expressions

$$\lim_{x \to x_0^-} f(x) \quad \text{and} \quad \lim_{x \to x_0^+} f(x)$$

are called the *one-sided limits of $f(x)$ at x_0* and the expression

$$\lim_{x \to x_0} f(x)$$

is called the *two-sided limit of $f(x)$ at x_0.*

Example 1 From (1) and (2)

$$\lim_{x \to 0^+} \frac{\sin x}{x} = \lim_{x \to 0^-} \frac{\sin x}{x} = 1$$

Therefore,

$$\lim_{x \to 0} \frac{\sin x}{x} = 1 \qquad \blacktriangleleft$$

The following examples illustrate limits geometrically.

Example 2 Let f be the function whose graph is shown in Figure 2.4.9. As x approaches 2 from the left, $f(x)$ approaches 1, so that

$$\lim_{x \to 2^-} f(x) = 1$$

As x approaches 2 from the right, $f(x)$ approaches 3, so that

$$\lim_{x \to 2^+} f(x) = 3$$

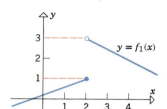

Figure 2.4.9

Since $f(2) = 1.5$, this example shows that the value of a function at a point, and the left- and right-hand limits there, can all be different. $\qquad \blacktriangleleft$

Example 3 The functions f_1 and f_2 whose graphs are shown in Figure 2.4.10 are identical to the function f in Example 2, except at the point $x = 2$; there we have $f(2) = 1.5$ and $f_1(2) = 1$, while $f_2(2)$ is undefined. However, even though the functions f, f_1, and f_2 behave differently *at* $x = 2$, their limits as x approaches 2 are the same, that is,

$$\lim_{x \to 2^-} f(x) = \lim_{x \to 2^-} f_1(x) = \lim_{x \to 2^-} f_2(x) = 1$$

and

$$\lim_{x \to 2^+} f(x) = \lim_{x \to 2^+} f_1(x) = \lim_{x \to 2^+} f_2(x) = 3 \qquad \blacktriangleleft$$

The foregoing example illustrates a general principle about limits that can be stated loosely as follows:

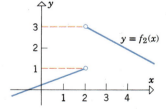

Figure 2.4.10

> The limit of a function as the independent variable approaches a point does not depend on the value of the function *at* the point.

Thus, if we alter the value of a function f only at $x = x_0$, then we do not affect

$$\lim_{x \to x_0^-} f(x), \quad \lim_{x \to x_0^+} f(x), \quad \text{or} \quad \lim_{x \to x_0} f(x)$$

Function f_2 in Example 3 illustrates another idea: namely, that the left- and right-hand limits can have values at a point where the function itself is undefined.

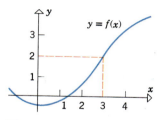

Figure 2.4.11

Example 4 Let f be the function whose graph is shown in Figure 2.4.11. From the graph we see that

$$\lim_{x \to 3^-} f(x) = 2 \quad \text{and} \quad \lim_{x \to 3^+} f(x) = 2$$

so that

$$\lim_{x \to 3} f(x) = 2$$

The graph also shows that $f(3) = 2$. Thus, the value of a function at a point and the one-sided and two-sided limits as x approaches that point may be equal. ◀

☐ **EXISTENCE OF LIMITS**

There is, in general, no guarantee that a function $f(x)$ actually has a limit as $x \to x_0^+$, $x \to x_0^-$, or $x \to x_0$. If there is no limit, then we say that *the limit does not exist*. Here are some examples.

Example 5 Let f be the function graphed in Figure 2.4.12a. As x approaches x_0 from the left side or the right side, the value of $f(x)$ gets larger and larger without bound and consequently approaches no fixed finite value. Thus, the limits

$$\lim_{x \to x_0^-} f(x) \quad \text{and} \quad \lim_{x \to x_0^+} f(x)$$

do not exist. To indicate that these limits fail to exist because $f(x)$ is increasing without bound, we write

$$\lim_{x \to x_0^-} f(x) = +\infty \quad \text{and} \quad \lim_{x \to x_0^+} f(x) = +\infty$$

These two expressions are commonly expressed in the form of a two-sided limit by writing

$$\lim_{x \to x_0} f(x) = +\infty$$

but again, this is just a notation for a type of limit that fails to exist. ◀

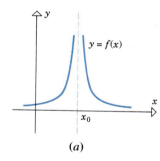

Figure 2.4.12

Example 6 Let f be the function graphed in Figure 2.4.12b. As x approaches x_0 from the left side or the right side, the value of $f(x)$ gets smaller and smaller without bound and consequently approaches no fixed finite value. Thus, the limits

$$\lim_{x \to x_0^-} f(x) \quad \text{and} \quad \lim_{x \to x_0^+} f(x)$$

do not exist. To indicate that these limits fail to exist because $f(x)$ is decreasing without bound, we write

$$\lim_{x \to x_0^-} f(x) = -\infty \quad \text{and} \quad \lim_{x \to x_0^+} f(x) = -\infty$$

These two expressions are commonly expressed in the form of a two-sided limit by writing

$$\lim_{x \to x_0} f(x) = -\infty \quad \blacktriangleleft$$

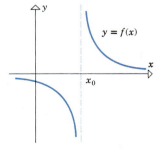

Figure 2.4.13

Example 7   Let f be the function whose graph is shown in Figure 2.4.13. We have

$$\lim_{x \to x_0^-} f(x) = -\infty \quad \text{and} \quad \lim_{x \to x_0^+} f(x) = +\infty$$

However, because these two expressions have opposite signs, there is no special notation for combining them into a two-sided form; we simply write

$$\lim_{x \to x_0} f(x) \; does \; not \; exist \quad \blacktriangleleft$$

REMARK.   Keep in mind that the symbols $+\infty$ and $-\infty$, as used in the preceding examples, are simply descriptions of limits that fail to exist. These symbols do not represent real numbers and consequently they cannot be manipulated using rules of algebra. For example, it is not correct to write $+\infty - \infty = 0$.

Example 8   Let f be the function whose graph is shown in Figure 2.4.14. This graph is intended to convey the idea that as x approaches 6 from the right side, $f(x)$ oscillates between -3 and 3 infinitely often and with increasing frequency. Since $f(x)$ approaches no fixed value as x approaches 6 from the right, we say that

$$\lim_{x \to 6^+} f(x) \; does \; not \; exist$$

On the other hand, as x approaches 6 from the left side, $f(x)$ approaches 1, so that

$$\lim_{x \to 6^-} f(x) = 1$$

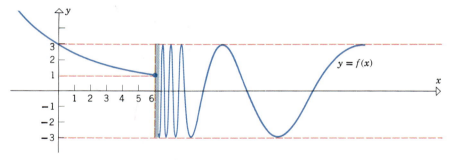

Figure 2.4.14

Recall from 2.4.3 that in order to write $\lim\limits_{x \to x_0} f(x) = L$ we must have

$$\lim_{x \to x_0^+} f(x) = L \quad \text{and} \quad \lim_{x \to x_0^-} f(x) = L$$

If either of these one-sided limits does not exist, then the two-sided limit does not exist. Thus, in this example

$$\lim_{x \to 6} f(x) \text{ does not exist}$$

because $\lim\limits_{x \to 6^+} f(x)$ does not exist.   ◀

□ **LIMITS AT INFINITY**   So far we have used limits to describe how a function behaves as the independent variable approaches a fixed point on the x-axis. However, limits can also be used to describe how a function behaves as the independent variable moves "indefinitely far" from the origin along the x-axis. If x is allowed to increase without bound, then we write $x \to +\infty$ (read, "x approaches plus infinity"), and if x is allowed to decrease without bound, then we write $x \to -\infty$ (read, "x approaches minus infinity").

Example 9   Let f be the function with the graph shown in Figure 2.4.15. As $x \to +\infty$, the graph of f approaches the line $y = 4$ so that the value of $f(x)$ approaches 4. We denote this by writing

$$\lim_{x \to +\infty} f(x) = 4$$

As $x \to -\infty$, the graph of f tends toward the line $y = -1$, and so the value of $f(x)$ approaches -1. We denote this by writing

$$\lim_{x \to -\infty} f(x) = -1 \quad ◀$$

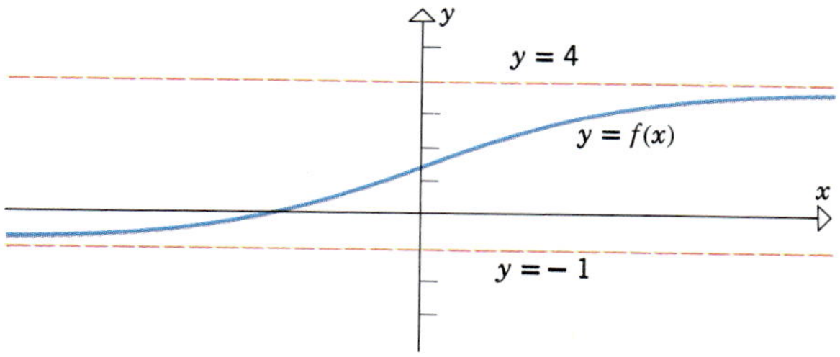

Figure 2.4.15

Example 10   Let f be the function with the graph shown in Figure 2.4.16. As x approaches $-\infty$, the value of $f(x)$ increases without bound and consequently approaches no fixed value. Thus,

$$\lim_{x \to -\infty} f(x)$$

does not exist. In this case, we would write

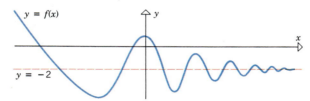

Figure 2.4.16

$$\lim_{x \to -\infty} f(x) = +\infty$$

to indicate that the limit fails to exist because $f(x)$ is increasing without bound.

As x approaches $+\infty$, the graph of f oscillates but approaches the line $y = -2$. Thus, the value of $f(x)$ approaches -2 as x approaches $+\infty$ and we write

$$\lim_{x \to +\infty} f(x) = -2 \qquad \blacktriangleleft$$

Example 11   Let f be the function with the graph shown in Figure 2.4.17. As x approaches $-\infty$, the value of $f(x)$ decreases without bound and consequently approaches no fixed finite value. Thus,

$$\lim_{x \to -\infty} f(x)$$

does not exist. In this case, we would write

$$\lim_{x \to -\infty} f(x) = -\infty$$

to indicate that the limit fails to exist because $f(x)$ is decreasing without bound.

As x approaches $+\infty$, the value of $f(x)$ oscillates between -3 and $+3$ and consequently approaches no fixed value. Thus,

$$\lim_{x \to +\infty} f(x)$$

does not exist. There is no special notation used to describe limits that fail to exist because of oscillation.   ◀

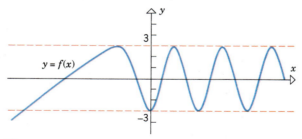

Figure 2.4.17

▶ Exercise Set 2.4

1. For the function f graphed below, find
  (a) $\lim\limits_{x\to 3^-} f(x)$ (b) $\lim\limits_{x\to 3^+} f(x)$
  (c) $\lim\limits_{x\to 3} f(x)$ (d) $f(3)$
  (e) $\lim\limits_{x\to -\infty} f(x)$ (f) $\lim\limits_{x\to +\infty} f(x)$.

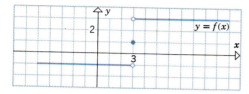

2. For the function f graphed below, find
  (a) $\lim\limits_{x\to 2^-} f(x)$ (b) $\lim\limits_{x\to 2^+} f(x)$
  (c) $\lim\limits_{x\to 2} f(x)$ (d) $f(2)$
  (e) $\lim\limits_{x\to -\infty} f(x)$ (f) $\lim\limits_{x\to +\infty} f(x)$.

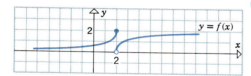

3. For the function g graphed below, find
  (a) $\lim\limits_{x\to 4^-} g(x)$ (b) $\lim\limits_{x\to 4^+} g(x)$
  (c) $\lim\limits_{x\to 4} g(x)$ (d) $g(4)$
  (e) $\lim\limits_{x\to -\infty} g(x)$ (f) $\lim\limits_{x\to +\infty} g(x)$.

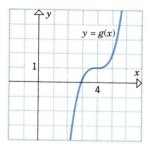

4. For the function g in the following graph, find
  (a) $\lim\limits_{x\to 0^-} g(x)$ (b) $\lim\limits_{x\to 0^+} g(x)$
  (c) $\lim\limits_{x\to 0} g(x)$ (d) $g(0)$
  (e) $\lim\limits_{x\to -\infty} g(x)$ (f) $\lim\limits_{x\to +\infty} g(x)$.

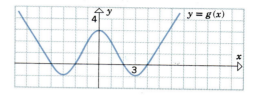

5. For the function F graphed below, find
  (a) $\lim\limits_{x\to -2^-} F(x)$ (b) $\lim\limits_{x\to -2^+} F(x)$
  (c) $\lim\limits_{x\to -2} F(x)$ (d) $F(-2)$
  (e) $\lim\limits_{x\to -\infty} F(x)$ (f) $\lim\limits_{x\to +\infty} F(x)$.

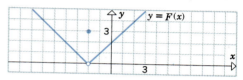

6. For the function F graphed below, find
  (a) $\lim\limits_{x\to 3^-} F(x)$ (b) $\lim\limits_{x\to 3^+} F(x)$
  (c) $\lim\limits_{x\to 3} F(x)$ (d) $F(3)$
  (e) $\lim\limits_{x\to -\infty} F(x)$ (f) $\lim\limits_{x\to +\infty} F(x)$.

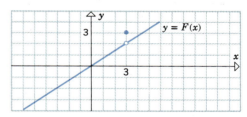

7. For the function ϕ graphed below, find
  (a) $\lim\limits_{x\to -2^-} \phi(x)$ (b) $\lim\limits_{x\to -2^+} \phi(x)$
  (c) $\lim\limits_{x\to -2} \phi(x)$ (d) $\phi(-2)$
  (e) $\lim\limits_{x\to -\infty} \phi(x)$ (f) $\lim\limits_{x\to +\infty} \phi(x)$.

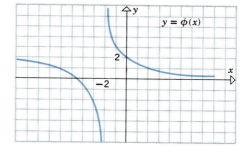

8. For the function ϕ graphed below, find

(a) $\lim\limits_{x \to 4^-} \phi(x)$ (b) $\lim\limits_{x \to 4^+} \phi(x)$

(c) $\lim\limits_{x \to 4} \phi(x)$ (d) $\phi(4)$

(e) $\lim\limits_{x \to -\infty} \phi(x)$ (f) $\lim\limits_{x \to +\infty} \phi(x)$.

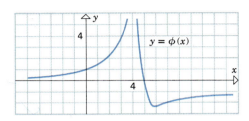

9. For the function f graphed below, find

(a) $\lim\limits_{x \to 3^-} f(x)$ (b) $\lim\limits_{x \to 3^+} f(x)$

(c) $\lim\limits_{x \to 3} f(x)$ (d) $f(3)$

(e) $\lim\limits_{x \to -\infty} f(x)$ (f) $\lim\limits_{x \to +\infty} f(x)$.

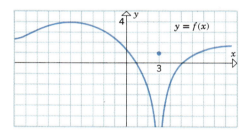

10. For the function f graphed below, find

(a) $\lim\limits_{x \to 0^-} f(x)$ (b) $\lim\limits_{x \to 0^+} f(x)$

(c) $\lim\limits_{x \to 0} f(x)$ (d) $f(0)$

(e) $\lim\limits_{x \to -\infty} f(x)$ (f) $\lim\limits_{x \to +\infty} f(x)$.

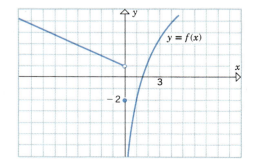

11. For the function G graphed below, find

(a) $\lim\limits_{x \to 0^-} G(x)$ (b) $\lim\limits_{x \to 0^+} G(x)$

(c) $\lim\limits_{x \to 0} G(x)$ (d) $G(0)$

(e) $\lim\limits_{x \to -\infty} G(x)$ (f) $\lim\limits_{x \to +\infty} G(x)$.

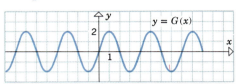

12. For the function G graphed below, find

(a) $\lim\limits_{x \to 0^-} G(x)$ (b) $\lim\limits_{x \to 0^+} G(x)$

(c) $\lim\limits_{x \to 0} G(x)$ (d) $G(0)$

(e) $\lim\limits_{x \to -\infty} G(x)$ (f) $\lim\limits_{x \to +\infty} G(x)$.

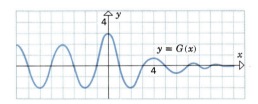

13. Consider the function g graphed below. For what values of x_0 does $\lim\limits_{x \to x_0} g(x)$ exist?

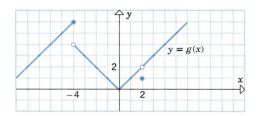

14. Consider the function f graphed below. For what values of x_0 does $\lim\limits_{x \to x_0} f(x)$ exist?

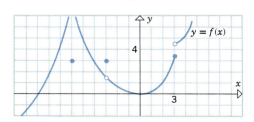

■ **2.5** LIMITS (COMPUTATIONAL TECHNIQUES)

In the last section we concentrated on the graphical interpretation of limits. In this section we shall discuss techniques for finding limits directly from the formula for a function. Again our results will be based on intuition. Mathematical proofs of the results here will have to wait until we have a precise mathematical definition of a limit.

□ **SOME BASIC LIMITS**

We will start by obtaining some simple limits that form the building blocks for more complex limits.

Since a constant function $f(x) = k$ has the same value k for all values of x, it follows that as x approaches any real number a the value of $f(x)$ remains fixed at k, so

$$\lim_{x \to a} k = k \tag{1}$$

(Figure 2.5.1). For the same reason, we also have

$$\lim_{x \to +\infty} k = \lim_{x \to -\infty} k = k \tag{2}$$

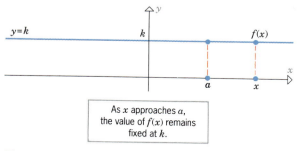

As x approaches a, the value of $f(x)$ remains fixed at k.

Figure 2.5.1

Example 1 It follows from (1) and (2) that

$$\lim_{x \to 2} 3 = 3, \quad \lim_{x \to -2} 3 = 3, \quad \lim_{x \to -\infty} 3 = 3, \quad \lim_{x \to +\infty} 3 = 3 \quad ◀$$

As x approaches any real number a, the value of the function $f(x) = x$ also approaches a, since x and $f(x)$ are equal (Figure 2.5.2). Thus,

$$\lim_{x \to a} x = a \tag{3}$$

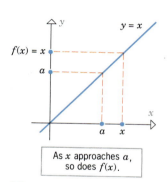

As x approaches a, so does $f(x)$.

Figure 2.5.2

For the same reason, we also have

$$\lim_{x \to +\infty} x = +\infty, \quad \lim_{x \to -\infty} x = -\infty \tag{4}$$

Example 2   It follows from (3) that

$$\lim_{x \to 5} x = 5, \quad \lim_{x \to 0} x = 0, \quad \lim_{x \to -2} x = -2 \quad \blacktriangleleft$$

The calculations in Table 2.5.1 suggest the following facts about the function $f(x) = 1/x$:

- As x approaches 0 from the right, the value of $1/x$ increases without bound.
- As x approaches 0 from the left, the value of $1/x$ decreases without bound.
- As $x \to +\infty$, the value of $1/x$ decreases toward 0.
- As $x \to -\infty$, the value of $1/x$ increases toward 0.

These four properties of $f(x) = 1/x$ are described by the following limits:

$$\lim_{x \to 0^+} \frac{1}{x} = +\infty \qquad \lim_{x \to 0^-} \frac{1}{x} = -\infty \tag{5a}$$

$$\lim_{x \to +\infty} \frac{1}{x} = 0 \qquad \lim_{x \to -\infty} \frac{1}{x} = 0 \tag{5b}$$

Table 2.5.1

	VALUES						CONCLUSION
x	1	.1	.01	.001	.0001	...	As $x \to 0^+$ the value of $1/x$ increases without bound.
$1/x$	1	10	100	1000	10,000	...	
x	−1	−.1	−.01	−.001	−.0001	...	As $x \to 0^-$ the value of $1/x$ decreases without bound.
$1/x$	−1	−10	−100	−1000	−10,000	...	
x	1	10	100	1000	10,000	...	As $x \to +\infty$ the value of $1/x$ decreases toward zero.
$1/x$	1	.1	.01	.001	.0001	...	
x	−1	−10	−100	−1000	−10,000	...	As $x \to -\infty$ the value of $1/x$ increases toward zero.
$1/x$	−1	−.1	−.01	−.001	−.0001	...	

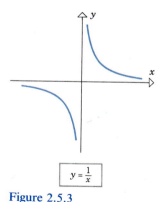

Figure 2.5.3

$y = \frac{1}{x}$

These limits should also be evident geometrically from the graph of $f(x) = 1/x$ shown in Figure 2.5.3.

REMARK. Note that

$$\lim_{x \to 0} \frac{1}{x} \text{ does not exist}$$

since the one-sided limits as x approaches zero do not exist.

The limits studied thus far, together with the fundamental properties of limits given in the following theorem, can be used to help solve more complicated limit problems. Parts of this theorem are proved in Appendix C.

2.5.1 THEOREM. *Let* lim *stand for one of the limits* $\lim_{x \to a}$, $\lim_{x \to a^-}$, $\lim_{x \to a^+}$, $\lim_{x \to +\infty}$, *or* $\lim_{x \to -\infty}$. *If* $L_1 = \lim f(x)$ *and* $L_2 = \lim g(x)$ *both exist, then*

(a) $\lim [f(x) + g(x)] = \lim f(x) + \lim g(x) = L_1 + L_2$

(b) $\lim [f(x) - g(x)] = \lim f(x) - \lim g(x) = L_1 - L_2$

(c) $\lim [f(x)g(x)] = \lim f(x) \lim g(x) = L_1 L_2$

(d) $\lim \dfrac{f(x)}{g(x)} = \dfrac{\lim f(x)}{\lim g(x)} = \dfrac{L_1}{L_2}$ *if* $L_2 \neq 0$

(e) $\lim \sqrt[n]{f(x)} = \sqrt[n]{\lim f(x)} = \sqrt[n]{L_1}$ *provided* $L_1 \geq 0$ *if* n *is even.*

In words, this theorem states:

(a) *The limit of a sum is the sum of the limits.*
(b) *The limit of a difference is the difference of the limits.*
(c) *The limit of a product is the product of the limits.*
(d) *The limit of a quotient is the quotient of the limits provided the limit of the denominator is nonzero.*
(e) *The limit of an nth root is the nth root of the limit.*

REMARK. Although results (a) and (c) are stated for two functions f and g, these results hold as well for any finite number of functions; that is, if

$$\lim f_1(x), \quad \lim f_2(x), \quad \ldots, \lim f_n(x)$$

all exist, then

$$\lim [f_1(x) + f_2(x) + \cdots + f_n(x)] = \lim f_1(x) + \lim f_2(x) + \cdots + \lim f_n(x) \quad (6)$$

and

$$\lim [f_1(x)f_2(x) \cdots f_n(x)] = \lim f_1(x) \lim f_2(x) \cdots \lim f_n(x) \quad (7)$$

In particular, if $f_1, f_2, \ldots, f_n$ are all the same function f, then (7) reduces to

$$\lim [f(x)]^n = [\lim f(x)]^n \qquad (8)$$

From (8) we obtain the useful result

$$\lim_{x \to a} x^n = [\lim_{x \to a} x]^n = a^n \qquad (9)$$

For example,

$$\lim_{x \to 3} x^4 = 3^4 = 81$$

Another useful result follows from part (c) of Theorem 2.5.1 in the special case where one of the factors is a constant k:

$$\lim kf(x) = \lim k \lim f(x) = k \lim f(x) \qquad (10)$$

In words, the first and last expressions in (10) state:

A constant factor can be moved through a limit sign.

☐ **LIMITS OF POLYNOMIALS**

Example 3   Find $\lim\limits_{x \to 5} (x^2 - 4x + 3)$ and justify each step.

Solution.

$$\lim_{x \to 5} (x^2 - 4x + 3) = \lim_{x \to 5} x^2 - \lim_{x \to 5} 4x + \lim_{x \to 5} 3 \qquad \boxed{\text{Theorem 2.5.1}(a), (b)}$$

$$= \lim_{x \to 5} x^2 - 4 \lim_{x \to 5} x + \lim_{x \to 5} 3 \qquad \boxed{\text{Equation (10)}}$$

$$= 5^2 - 4(5) + 3 \qquad \boxed{\text{Equation (9)}}$$

$$= 8 \qquad \blacktriangleleft$$

The following theorem greatly simplifies the computation of limits of polynomials. It shows that the limit of a polynomial $p(x)$ as x approaches a can be obtained by evaluating the polynomial at a.

2.5.2   THEOREM.   *For any polynomial*

$$p(x) = c_0 + c_1 x + \cdots + c_n x^n$$

and any real number a,

$$\lim_{x \to a} p(x) = c_0 + c_1 a + \cdots + c_n a^n = p(a)$$

Proof.

$$\lim_{x \to a} p(x) = \lim_{x \to a} (c_0 + c_1 x + \cdots + c_n x^n)$$

$$= \lim_{x \to a} c_0 + \lim_{x \to a} c_1 x + \cdots + \lim_{x \to a} c_n x^n$$

$$= \lim_{x \to a} c_0 + c_1 \lim_{x \to a} x + \cdots + c_n \lim_{x \to a} x^n$$

$$= c_0 + c_1 a + \cdots + c_n a^n = p(a) \quad \blacksquare$$

Example 4   If we apply Theorem 2.5.2 to the limit problem in Example 3, we can bypass the intermediate steps and write immediately

$$\lim_{x \to 5} (x^2 - 4x + 3) = 5^2 - 4(5) + 3 = 8 \quad \blacktriangleleft$$

☐ **LIMITS OF RATIONAL FUNCTIONS**

Recall that a rational function is the ratio of two polynomials. Theorem 2.5.2 and Theorem 2.5.1(*d*) can often be used in combination to compute limits of rational functions.

Example 5   Find $\lim\limits_{x \to 2} \dfrac{5x^3 + 4}{x - 3}$.

Solution.

$$\lim_{x \to 2} \frac{5x^3 + 4}{x - 3} = \frac{\lim\limits_{x \to 2} (5x^3 + 4)}{\lim\limits_{x \to 2} (x - 3)} = \frac{5 \cdot 2^3 + 4}{2 - 3} = -44 \quad \blacktriangleleft$$

The method of the foregoing example will not work if the limit of the denominator is zero, since Theorem 2.5.1(*d*) is not applicable in this situation. However, if the limit of a rational function is being calculated as x approaches a real number a, and if the numerator and denominator *both* approach zero as x approaches a, then the numerator and denominator will have a common factor of $x - a$ and the limit can often be obtained by first canceling the common factors. The following example illustrates this technique.

Example 6   Find $\lim\limits_{x \to 2} \dfrac{x^2 - 4}{x - 2}$.

Solution.   The numerator and denominator both have a limit of zero as x approaches 2, so they share a common factor of $x - 2$. The limit can be obtained as follows:

$$\lim_{x \to 2} \frac{x^2 - 4}{x - 2} = \lim_{x \to 2} \frac{(x - 2)(x + 2)}{x - 2} = \lim_{x \to 2} (x + 2) = 4 \quad \blacktriangleleft$$

REMARK.   Although correct, the second equality in the foregoing calculation needs some justification. We pointed out in Example 14 of Section 2.1 that to cancel the $x - 2$ factors from

$$h(x) = \frac{x^2 - 4}{x - 2} = \frac{(x - 2)(x + 2)}{(x - 2)}$$

we must write

$$h(x) = x + 2, \quad x \neq 2$$

in order not to alter the domain of h. However, the functions

$$f(x) = x + 2 \quad \text{and} \quad h(x) = x + 2, \quad x \neq 2$$

differ only at $x = 2$, so that their limits as x approaches 2 are the same. Thus,

$$\lim_{x \to 2} \frac{(x - 2)(x + 2)}{x - 2} = \lim_{x \to 2} (x + 2)$$

even though

$$\frac{(x - 2)(x + 2)}{x - 2} \quad \text{and} \quad x + 2$$

are different functions.

Example 7   Find

(a) $\displaystyle\lim_{x \to 3} \frac{x^2 - 6x + 9}{x - 3}$ (b) $\displaystyle\lim_{x \to -4} \frac{2x + 8}{x^2 + x - 12}$

Solution (a).   The numerator and denominator both have a limit of zero as x approaches 3, so there is a common factor of $x - 3$. We proceed as follows:

$$\lim_{x \to 3} \frac{x^2 - 6x + 9}{x - 3} = \lim_{x \to 3} \frac{(x - 3)^2}{x - 3} = \lim_{x \to 3} (x - 3) = 0$$

Solution (b).   The numerator and denominator both have a limit of zero as x approaches -4, so there is a common factor of $x - (-4) = x + 4$. We proceed as follows:

$$\lim_{x \to -4} \frac{2x + 8}{x^2 + x - 12} = \lim_{x \to -4} \frac{2(x + 4)}{(x + 4)(x - 3)}$$

$$= \lim_{x \to -4} \frac{2}{x - 3} = -\frac{2}{7} \quad \blacktriangleleft$$

If the denominator of a rational function has a limit of zero, and the numerator has a limit that is *not* zero, then the limit of the rational function itself does not exist. The simplest examples of such rational functions are of the form

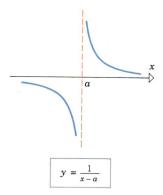

$$y = \frac{1}{x-a}$$

Figure 2.5.4

$$f(x) = \frac{1}{x-a}$$

where a is any real number. The graphs of these functions are horizontal translations of the graph of $f(x) = 1/x$ (Figures 2.5.3 and 2.5.4).

It is evident geometrically that the three limits

$$\lim_{x \to a^+} \frac{1}{x-a}, \quad \lim_{x \to a^-} \frac{1}{x-a}, \quad \lim_{x \to a} \frac{1}{x-a}$$

do not exist.

These results illustrate the three possibilities that can occur for rational functions when the limit of the denominator is zero and the limit of the numerator is nonzero:

- The limit may be $+\infty$.
- The limit may be $-\infty$.
- The limit may be $+\infty$ from one side and $-\infty$ from the other.

Example 8 Find

(a) $\displaystyle\lim_{x \to 4^+} \frac{2-x}{(x-4)(x+2)}$ (b) $\displaystyle\lim_{x \to 4^-} \frac{2-x}{(x-4)(x+2)}$ (c) $\displaystyle\lim_{x \to 4} \frac{2-x}{(x-4)(x+2)}$

Solution. In all three parts the limit of the numerator is -2 and the limit of the denominator is 0, so the limit of the ratio does not exist. To be more specific, we need to analyze the sign of the ratio. Using either of the methods illustrated in Example 9 of Section 1.1, the reader should be able to obtain the result in Figure 2.5.5. It follows from this figure that as x approaches 4 from the right, the ratio is always negative, so

$$\lim_{x \to 4^+} \frac{2-x}{(x-4)(x+2)} = -\infty \tag{11}$$

It is also evident from the figure that as x approaches 4 from the left, the sign of the ratio is eventually positive (after x exceeds 2), so

$$\lim_{x \to 4^-} \frac{2-x}{(x-4)(x+2)} = +\infty \tag{12}$$

From (11) and (12)

$$\lim_{x \to 4} \frac{2-x}{(x-4)(x+2)}$$

does not exist. ◀

Figure 2.5.5

In Figure 2.5.6 we have graphed the polynomials of the form x^n for $n = 1, 2, 3,$ and 4; and below each figure we have indicated the limits as $x \to +\infty$ and $x \to -\infty$. The results in the figure are special cases of the following general results:

$$\lim_{x \to +\infty} x^n = +\infty, \quad n = 1, 2, 3, \ldots \tag{13}$$

$$\lim_{x \to -\infty} x^n = \begin{cases} +\infty, & n = 2, 4, 6, \ldots \\ -\infty, & n = 1, 3, 5, \ldots \end{cases} \tag{14}$$

Multiplying x^n by a positive real number does not affect limits (13) and (14), but multiplying by a negative real number reverses the signs.

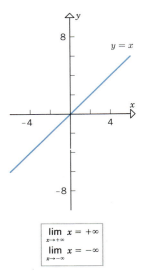
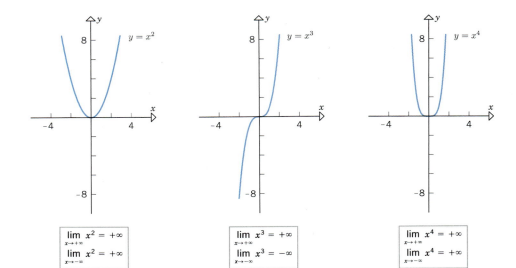

Figure 2.5.6

Example 9

$$\lim_{x \to +\infty} 2x^5 = +\infty, \quad \lim_{x \to -\infty} 2x^5 = -\infty$$

$$\lim_{x \to +\infty} -7x^6 = -\infty, \quad \lim_{x \to -\infty} -7x^6 = -\infty \quad \blacktriangleleft$$

In Table 2.5.1 we obtained the limits

$$\lim_{x \to +\infty} \frac{1}{x} = \lim_{x \to -\infty} \frac{1}{x} = 0$$

Consequently, for any positive integer n,

$$\lim_{x \to +\infty} \frac{1}{x^n} = \left(\lim_{x \to +\infty} \frac{1}{x} \right)^n = 0 \tag{15}$$

and

$$\lim_{x \to -\infty} \frac{1}{x^n} = \left(\lim_{x \to -\infty} \frac{1}{x} \right)^n = 0 \tag{16}$$

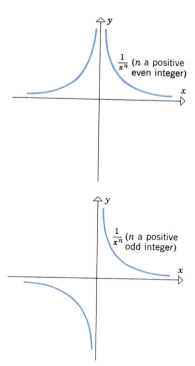

$\frac{1}{x^n}$ (n a positive even integer)

$\frac{1}{x^n}$ (n a positive odd integer)

Figure 2.5.7

Results (15) and (16) are also evident from the graph of $1/x^n$, which has one of the two general shapes shown in Figure 2.5.7, depending on whether n is even or odd.

There is an important principle about limits of polynomials which, expressed informally, states that *a polynomial behaves like its term of highest degree as $x \to +\infty$ or $x \to -\infty$*. More precisely, if $c_n \neq 0$, then

$$\lim_{x \to +\infty} (c_0 + c_1 x + \cdots + c_n x^n) = \lim_{x \to +\infty} c_n x^n \tag{17}$$

$$\lim_{x \to -\infty} (c_0 + c_1 x + \cdots + c_n x^n) = \lim_{x \to -\infty} c_n x^n \tag{18}$$

We can motivate these results by factoring out the highest power of x from the polynomial and examining the limit of the factored expression. Thus,

$$c_0 + c_1 x + \cdots + c_n x^n = x^n \left(\frac{c_0}{x^n} + \frac{c_1}{x^{n-1}} + \cdots + c_n \right)$$

As $x \to +\infty$ or $x \to -\infty$, it follows from (15) and (16) that all of the terms with positive powers of x in the denominator approach 0, so (17) and (18) are certainly plausible.

Example 10

$$\lim_{x \to +\infty} (7x^5 - 4x^3 + 2x - 9) = \lim_{x \to +\infty} 7x^5 = +\infty$$

$$\lim_{x \to -\infty} (-4x^8 + 17x^3 - 5x + 1) = \lim_{x \to -\infty} -4x^8 = -\infty \qquad \blacktriangleleft$$

◻ **LIMITS OF RATIONAL FUNCTIONS AS $x \to +\infty$ OR $x \to -\infty$**

If we divide the numerator and denominator of a rational function by the highest power of x that occurs in the function, then all the powers of x become constants or powers of $1/x$. The following examples show how this observation together with (15) and (16) can be used to find limits of rational functions as $x \to +\infty$ or $x \to -\infty$.

Example 11   Find $\lim\limits_{x \to +\infty} \dfrac{3x + 5}{6x - 8}$.

Solution.   Divide the numerator and denominator by the highest power of x that occurs; this is $x^1 = x$. We obtain

$$\lim_{x \to +\infty} \frac{3x + 5}{6x - 8} = \lim_{x \to +\infty} \frac{3 + 5/x}{6 - 8/x} = \frac{\lim\limits_{x \to +\infty} (3 + 5/x)}{\lim\limits_{x \to +\infty} (6 - 8/x)}$$

$$= \frac{\lim\limits_{x \to +\infty} 3 + \lim\limits_{x \to +\infty} 5/x}{\lim\limits_{x \to +\infty} 6 - \lim\limits_{x \to +\infty} 8/x} = \frac{3 + 5 \lim\limits_{x \to +\infty} 1/x}{6 - 8 \lim\limits_{x \to +\infty} 1/x}$$

$$= \frac{3 + (5 \cdot 0)}{6 - (8 \cdot 0)} = \frac{1}{2} \quad \blacktriangleleft$$

Example 12   Find $\lim\limits_{x \to -\infty} \dfrac{4x^2 - x}{2x^3 - 5}$.

Solution.   Divide the numerator and denominator by the highest power of x that occurs, namely x^3. We obtain

$$\lim_{x \to -\infty} \frac{4x^2 - x}{2x^3 - 5} = \lim_{x \to -\infty} \frac{4/x - 1/x^2}{2 - 5/x^3} = \frac{\lim\limits_{x \to -\infty} (4/x - 1/x^2)}{\lim\limits_{x \to -\infty} (2 - 5/x^3)}$$

$$= \frac{\lim\limits_{x \to -\infty} 4/x - \lim\limits_{x \to -\infty} 1/x^2}{\lim\limits_{x \to -\infty} 2 - \lim\limits_{x \to -\infty} 5/x^3} = \frac{4 \lim\limits_{x \to -\infty} 1/x - \lim\limits_{x \to -\infty} 1/x^2}{2 - 5 \lim\limits_{x \to -\infty} 1/x^3}$$

$$= \frac{(4 \cdot 0) - 0}{2 - (5 \cdot 0)} = \frac{0}{2} = 0 \quad \blacktriangleleft$$

☐ **A QUICK METHOD FOR FINDING LIMITS OF RATIONAL FUNCTIONS AS $x \to +\infty$ OR $x \to -\infty$**

It can be shown (Exercise 76) that the limit of a *rational function* as $x \to +\infty$ or $x \to -\infty$ is unaffected if all but the highest degree terms in the numerator and denominator are discarded; that is, if $c_n \neq 0$ and $d_m \neq 0$, then

$$\lim_{x \to +\infty} \frac{c_0 + c_1 x + \cdots + c_n x^n}{d_0 + d_1 x + \cdots + d_m x^m} = \lim_{x \to +\infty} \frac{c_n x^n}{d_m x^m} \qquad (19)$$

and

$$\lim_{x \to -\infty} \frac{c_0 + c_1 x + \cdots + c_n x^n}{d_0 + d_1 x + \cdots + d_m x^m} = \lim_{x \to -\infty} \frac{c_n x^n}{d_m x^m} \qquad (20)$$

Example 13   Use Formulas (19) and (20) to find

(a) $\lim\limits_{x \to +\infty} \dfrac{3x + 5}{6x - 8}$ (b) $\lim\limits_{x \to -\infty} \dfrac{4x^2 - x}{2x^3 - 5}$ (c) $\lim\limits_{x \to +\infty} \dfrac{3 - 2x^4}{x + 1}$

Solution (a).

$$\lim_{x \to +\infty} \frac{3x + 5}{6x - 8} = \lim_{x \to +\infty} \frac{3x}{6x} = \lim_{x \to +\infty} \frac{1}{2} = \frac{1}{2}$$

which agrees with the result obtained in Example 11.

Solution (b).

$$\lim_{x \to -\infty} \frac{4x^2 - x}{2x^3 - 5} = \lim_{x \to -\infty} \frac{4x^2}{2x^3} = \lim_{x \to -\infty} \frac{2}{x} = 0$$

which agrees with the result obtained in Example 12.

Solution (c).

$$\lim_{x \to +\infty} \frac{3 - 2x^4}{x + 1} = \lim_{x \to +\infty} \frac{-2x^4}{x} = \lim_{x \to +\infty} -2x^3 = -\infty \quad \blacktriangleleft$$

REMARK. We emphasize that Formulas (19) and (20) are only applicable if $x \to +\infty$ or $x \to -\infty$; they are not applicable to limits in which x approaches a *finite* number a.

☐ **LIMITS INVOLVING RADICALS**

Example 14 Find $\displaystyle\lim_{x \to +\infty} \sqrt[3]{\frac{3x + 5}{6x - 8}}$.

Solution.

$$\lim_{x \to +\infty} \sqrt[3]{\frac{3x + 5}{6x - 8}} = \sqrt[3]{\lim_{x \to +\infty} \frac{3x + 5}{6x - 8}} = \sqrt[3]{\frac{1}{2}} \quad \blacktriangleleft$$

Theorem 2.5.1(e)   Example 11

Example 15 Find

(a) $\displaystyle\lim_{x \to +\infty} \frac{\sqrt{x^2 + 2}}{3x - 6}$ (b) $\displaystyle\lim_{x \to -\infty} \frac{\sqrt{x^2 + 2}}{3x - 6}$

In both parts it would be helpful to manipulate the function so that the powers of x become powers of $1/x$. This can be achieved in both cases by dividing the numerator and denominator by $|x|$ and using the fact that

$$\sqrt{x^2} = |x|$$

Solution (a). As $x \to +\infty$, the values of x are eventually positive, so we can replace $|x|$ by x where desirable. We obtain

$$\lim_{x \to +\infty} \frac{\sqrt{x^2 + 2}}{3x - 6} = \lim_{x \to +\infty} \frac{\sqrt{x^2 + 2}/|x|}{(3x - 6)/|x|} = \lim_{x \to +\infty} \frac{\sqrt{x^2 + 2}/\sqrt{x^2}}{(3x - 6)/x}$$

$$= \lim_{x \to +\infty} \frac{\sqrt{1 + 2/x^2}}{3 - 6/x} = \frac{\displaystyle\lim_{x \to +\infty} \sqrt{1 + 2/x^2}}{\displaystyle\lim_{x \to +\infty} (3 - 6/x)}$$

$$= \frac{\sqrt{\lim_{x \to +\infty} (1 + 2/x^2)}}{\lim_{x \to +\infty} (3 - 6/x)} = \frac{\sqrt{\lim_{x \to +\infty} 1 + 2 \lim_{x \to +\infty} 1/x^2}}{\lim_{x \to +\infty} 3 - 6 \lim_{x \to +\infty} 1/x}$$

$$= \frac{\sqrt{1 + (2 \cdot 0)}}{3 - (6 \cdot 0)} = \frac{1}{3}$$

Solution (b).   As $x \to -\infty$, the values of x are eventually negative, so we can replace $|x|$ by $-x$ where desirable. We obtain

$$\lim_{x \to -\infty} \frac{\sqrt{x^2 + 2}}{3x - 6} = \lim_{x \to -\infty} \frac{\sqrt{x^2 + 2}/|x|}{(3x - 6)/|x|} = \lim_{x \to -\infty} \frac{\sqrt{x^2 + 2}/\sqrt{x^2}}{(3x - 6)/(-x)}$$

$$= \lim_{x \to -\infty} \frac{\sqrt{1 + 2/x^2}}{(6/x) - 3} = -\frac{1}{3} \qquad \blacktriangleleft$$

☐ **LIMITS OF PIECEWISE DEFINED FUNCTIONS**

For functions that are defined piecewise, a two-sided limit at a point where the formula for the function changes is best obtained by first finding the one-sided limits at the point.

Example 16   Find $\lim_{x \to 3} f(x)$ for

$$f(x) = \begin{cases} x^2 - 5, & x \le 3 \\ \\ \sqrt{x + 13}, & x > 3 \end{cases}$$

Solution.   As x approaches 3 from the left, the formula for f is

$$f(x) = x^2 - 5$$

so that

$$\lim_{x \to 3^-} f(x) = \lim_{x \to 3^-} (x^2 - 5) = 3^2 - 5 = 4$$

As x approaches 3 from the right, the formula for f is

$$f(x) = \sqrt{x + 13}$$

so that

$$\lim_{x \to 3^+} f(x) = \lim_{x \to 3^+} \sqrt{x + 13} = \sqrt{\lim_{x \to 3^+} (x + 13)} = \sqrt{16} = 4$$

Therefore,

$$\lim_{x \to 3} f(x) = 4$$

since the one-sided limits are equal. $\blacktriangleleft$

► Exercise Set 2.5

Find the limits in Exercises 1–56.

1. $\lim\limits_{x \to 8} 7$.

2. $\lim\limits_{x \to -\infty} (-3)$.

3. $\lim\limits_{x \to 0^+} \pi$.

4. $\lim\limits_{x \to -2} 3x$.

5. $\lim\limits_{y \to 3^+} 12y$.

6. $\lim\limits_{h \to +\infty} (-2h)$.

7. $\lim\limits_{x \to 5} \sqrt{x^3 - 3x - 1}$.

8. $\lim\limits_{x \to 0^-} (x^4 + 12x^3 - 17x + 2)$.

9. $\lim\limits_{y \to -1} (y^6 - 12y + 1)$.

10. $\lim\limits_{x \to 3} \dfrac{x^2 - 2x}{x + 1}$.

11. $\lim\limits_{y \to 2^-} \dfrac{(y - 1)(y - 2)}{y + 1}$.

12. $\lim\limits_{x \to 0} \dfrac{6x - 9}{x^3 - 12x + 3}$.

13. $\lim\limits_{x \to 4} \dfrac{x^2 - 16}{x - 4}$.

14. $\lim\limits_{t \to -2} \dfrac{t^3 + 8}{t + 2}$.

15. $\lim\limits_{x \to 1^+} \dfrac{x^4 - 1}{x - 1}$.

16. $\lim\limits_{x \to 2} \dfrac{x^2 - 4x + 4}{x^2 + x - 6}$.

17. $\lim\limits_{x \to -1} \dfrac{x^2 + 6x + 5}{x^2 - 3x - 4}$.

18. $\lim\limits_{t \to 1} \dfrac{t^3 + t^2 - 5t + 3}{t^3 - 3t + 2}$.

19. $\lim\limits_{x \to +\infty} \dfrac{3x + 1}{2x - 5}$.

20. $\lim\limits_{x \to +\infty} \dfrac{1}{x - 12}$.

21. $\lim\limits_{y \to -\infty} \dfrac{3}{y + 4}$.

22. $\lim\limits_{x \to +\infty} \dfrac{5x^2 + 7}{3x^2 - x}$.

23. $\lim\limits_{x \to -\infty} \dfrac{x - 2}{x^2 + 2x + 1}$.

24. $\lim\limits_{s \to +\infty} \sqrt[3]{\dfrac{3s^7 - 4s^5}{2s^7 + 1}}$.

25. $\lim\limits_{x \to -\infty} \dfrac{\sqrt{5x^2 - 2}}{x + 3}$.

26. $\lim\limits_{x \to +\infty} \dfrac{\sqrt{5x^2 - 2}}{x + 3}$.

27. $\lim\limits_{y \to -\infty} \dfrac{2 - y}{\sqrt{7 + 6y^2}}$.

28. $\lim\limits_{y \to +\infty} \dfrac{2 - y}{\sqrt{7 + 6y^2}}$.

29. $\lim\limits_{x \to -\infty} \dfrac{\sqrt{3x^4 + x}}{x^2 - 8}$.

30. $\lim\limits_{x \to +\infty} \dfrac{\sqrt{3x^4 + x}}{x^2 - 8}$.

31. $\lim\limits_{x \to 3^+} \dfrac{x}{x - 3}$.

32. $\lim\limits_{x \to 3^-} \dfrac{x}{x - 3}$.

33. $\lim\limits_{x \to 3} \dfrac{x}{x - 3}$.

34. $\lim\limits_{x \to 2^+} \dfrac{x}{x^2 - 4}$.

35. $\lim\limits_{x \to 2^-} \dfrac{x}{x^2 - 4}$.

36. $\lim\limits_{x \to 2} \dfrac{x}{x^2 - 4}$.

37. $\lim\limits_{y \to 6^+} \dfrac{y + 6}{y^2 - 36}$.

38. $\lim\limits_{y \to 6^-} \dfrac{y + 6}{y^2 - 36}$.

39. $\lim\limits_{y \to 6} \dfrac{y + 6}{y^2 - 36}$.

40. $\lim\limits_{x \to 4^+} \dfrac{3 - x}{x^2 - 2x - 8}$.

41. $\lim\limits_{x \to 4^-} \dfrac{3 - x}{x^2 - 2x - 8}$.

42. $\lim\limits_{x \to 4} \dfrac{3 - x}{x^2 - 2x - 8}$.

43. $\lim\limits_{x \to +\infty} \dfrac{7 - 6x^5}{x + 3}$.

44. $\lim\limits_{t \to -\infty} \dfrac{5 - 2t^3}{t^2 + 1}$.

45. $\lim\limits_{t \to +\infty} \dfrac{6 - t^3}{7t^3 + 3}$.

46. $\lim\limits_{x \to 0^+} \dfrac{x}{|x|}$.

47. $\lim\limits_{x \to 0^-} \dfrac{x}{|x|}$.

48. $\lim\limits_{x \to 3^-} \dfrac{1}{|x - 3|}$.

49. $\lim\limits_{x \to 9} \dfrac{x - 9}{\sqrt{x} - 3}$.

50. $\lim\limits_{y \to 4} \dfrac{4 - y}{2 - \sqrt{y}}$.

51. $\lim\limits_{x \to +\infty} \sqrt{x}$.

52. $\lim\limits_{x \to -\infty} \sqrt{5 - x}$.

53. $\lim\limits_{x \to -\infty} (3 - x)$.

54. $\lim\limits_{x \to -\infty} (3 - x^2)$.

55. $\lim\limits_{x \to +\infty} (1 + 2x - 3x^5)$.

56. $\lim\limits_{x \to +\infty} (2x^3 - 100x + 5)$.

57. Find
$$\lim\limits_{x \to a} \dfrac{x}{x + a}$$
where a is an arbitrary constant. [*Hint:* Consider the cases, $a \neq 0$ and $a = 0$ separately.]

58. Let $f(x) = \dfrac{x^3 - 1}{x - 1}$.
   (a) Find $\lim\limits_{x \to 1} f(x)$.
   (b) Sketch the graph of $y = f(x)$.

59. Let
$$f(x) = \begin{cases} x - 1, & x \leq 3 \\ 3x - 7, & x > 3 \end{cases}$$
Find
   (a) $\lim\limits_{x \to 3^-} f(x)$ (b) $\lim\limits_{x \to 3^+} f(x)$
   (c) $\lim\limits_{x \to 3} f(x)$.

60. Let
$$g(t) = \begin{cases} t^2, & t \geq 0 \\ t - 2, & t < 0 \end{cases}$$
Find
(a) $\lim\limits_{t \to 0^-} g(t)$ (b) $\lim\limits_{t \to 0^+} g(t)$ (c) $\lim\limits_{t \to 0} g(t)$.

61. Find $\lim\limits_{x \to 3} h(x)$ given that
$$h(x) = \begin{cases} x^2 - 2x + 1, & x \neq 3 \\ 7, & x = 3 \end{cases}$$

62. Let
$$F(x) = \begin{cases} \dfrac{x^2 - 9}{x + 3}, & x \neq -3 \\ k, & x = -3 \end{cases}$$
(a) Find k so that $F(-3) = \lim\limits_{x \to -3} F(x)$.

(b) With k assigned the value $\lim\limits_{x \to -3} F(x)$, show that $F(x)$ can be expressed as a polynomial.

63. (a) Explain why the following calculation is incorrect.
$$\lim_{x \to 0^+} \left(\frac{1}{x} - \frac{1}{x^2} \right) = \lim_{x \to 0^+} \frac{1}{x} - \lim_{x \to 0^+} \frac{1}{x^2}$$
$$= +\infty - (+\infty) = 0$$

(b) Show that $\lim\limits_{x \to 0^+} \left(\dfrac{1}{x} - \dfrac{1}{x^2} \right) = -\infty$.

64. Find $\lim\limits_{x \to 0^-} \left(\dfrac{1}{x} + \dfrac{1}{x^2} \right)$.

In Exercises 65–68, first rationalize the numerator, then find the limit.

65. $\lim\limits_{x \to 0} \dfrac{\sqrt{x + 4} - 2}{x}$.

66. $\lim\limits_{x \to 0} \dfrac{\sqrt{x^2 + 4} - 2}{x}$.

67. $\lim\limits_{x \to 0} \dfrac{\sqrt{5x + 9} - 3}{x}$.

68. $\lim\limits_{x \to 3} \dfrac{1 - \sqrt{x - 2}}{x - 3}$.

Find the limits in Exercises 69–74.

69. $\lim\limits_{x \to +\infty} (\sqrt{x^2 + 3} - x)$.

70. $\lim\limits_{x \to +\infty} (\sqrt{2x^2 + 5} - x)$.

71. $\lim\limits_{x \to +\infty} (\sqrt{x^2 + 5x} - x)$.

72. $\lim\limits_{x \to +\infty} (\sqrt{x^2 - 3x} - x)$.

73. $\lim\limits_{x \to +\infty} (\sqrt{x^2 + ax} - x)$.

74. $\lim\limits_{x \to +\infty} (\sqrt{x^2 + ax} - \sqrt{x^2 + bx})$.

75. Let $r(x)$ be a rational function. Under what conditions is it true that $\lim\limits_{x \to a} r(x) = r(a)$?

76. Find
$$\lim_{x \to +\infty} \frac{c_0 + c_1 x + \cdots + c_n x^n}{d_0 + d_1 x + \cdots + d_m x^m}$$
where $c_n \neq 0$ and $d_m \neq 0$. [*Hint:* Your answer will depend on whether $m < n$, $m = n$, or $m > n$.]

2.6   LIMITS: A RIGOROUS APPROACH (OPTIONAL)

This section gives a rigorous treatment of two-sided limits. Readers interested in pursuing the theory of one-sided and infinite limits can turn to Appendix C, where we give a rigorous treatment of these topics and also prove some of the basic limit theorems.

□ **DEFINITION OF A LIMIT**   Let a and L be real numbers. When we write

$$\lim_{x \to a} f(x) = L \tag{1}$$

we mean intuitively that the value of $f(x)$ approaches L as x approaches a from either side. To capture this concept mathematically, we must clarify the meanings of the phrases "x approaches a from either side" and "$f(x)$ approaches L."

The limit in (1) is intended to describe the behavior of f when x is near but *different from a*. The value of f at a is irrelevant to the limit. In fact, we have seen examples such as

$$\lim_{x \to 0} \frac{\sin x}{x} = 1$$

where the function f is not even defined at the point a. Thus, when we say "x approaches a from either side," we mean that x assumes values arbitrarily close to a, on one side or the other, but does not actually take on the value a. On the other hand, when we say that "$f(x)$ approaches L," we want to allow the possibility that $f(x)$ may actually take on the value L.

Let us see if we can phrase these ideas more precisely. To say that $f(x)$ becomes arbitrarily close to L as x approaches a means that if we pick any positive number ϵ (no matter how small), eventually the difference between $f(x)$ and L will be less than ϵ as x approaches a. That is, eventually $f(x)$ will satisfy the condition

$$|f(x) - L| < \epsilon$$

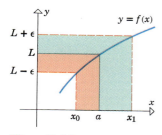

Figure 2.6.1

Figure 2.6.1 illustrates this idea. For the curve in that figure we have

$$\lim_{x \to a} f(x) = L$$

where a and L are the points shown. We picked an arbitrary positive number ϵ and indicated the interval

$$(L - \epsilon, L + \epsilon)$$

on the y-axis. As x approaches a, x will eventually lie between the points x_0 and x_1. When this occurs, $f(x)$ will lie in the interval $(L - \epsilon, L + \epsilon)$, that is, $f(x)$ will satisfy

$$L - \epsilon < f(x) < L + \epsilon$$

or equivalently

$$|f(x) - L| < \epsilon \tag{2}$$

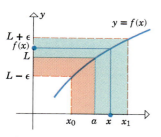

Figure 2.6.2

(Figure 2.6.2). It is evident geometrically that no matter how small we make ϵ, eventually $f(x)$ will satisfy (2) as x approaches a.

For the curve in Figure 2.6.1, we have

$$f(a) = \lim_{x \to a} f(x) = L$$

But the equality of $f(a)$ and L is irrelevant to the limit discussion. As x approaches a, x is not allowed to take on the value a. Thus, if $f(a)$ is undefined, as in Figure 2.6.3, or if $f(a)$ is defined, but unequal to L, as in Figure 2.6.4, it is still true that for an arbitrary positive value of ϵ, $f(x)$ will eventually satisfy

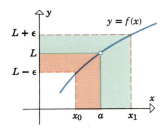

Figure 2.6.3

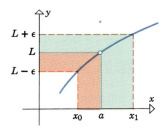

Figure 2.6.4

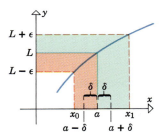

Figure 2.6.5

(2) as x approaches a. In brief, the condition $|f(x) - L| < \epsilon$ need only be satisfied for x in the set $(x_0, a) \cup (a, x_1)$.

Before we can give a precise definition of limit, we need a preliminary observation. If the condition

$$|f(x) - L| < \epsilon \tag{3}$$

holds for all x in the set

$$(x_0, a) \cup (a, x_1) \tag{4}$$

then this condition holds as well for all x in any *subset* of (4). In particular, if we let δ be any positive number that is smaller than both

$$a - x_0 \quad \text{and} \quad x_1 - a$$

then

$$(a - \delta, a) \cup (a, a + \delta) \tag{5}$$

will be a subset of (4) (Figure 2.6.5) and condition (3) will hold for all x in this subset.

Since the set in (5) consists of all x satisfying

$$0 < |x - a| < \delta$$

we are led to the following definition.

2.6.1   DEFINITION.   Let $f(x)$ be defined for all x in some open interval containing the number a, with the possible exception that $f(x)$ may or may not be defined at a. We shall write

$$\lim_{x \to a} f(x) = L$$

if, given any number $\epsilon > 0$, we can find a number $\delta > 0$ such that $f(x)$ satisfies

$$|f(x) - L| < \epsilon$$

whenever x satisfies

$$0 < |x - a| < \delta$$

Example 1   Use Definition 2.6.1 to prove

$$\lim_{x \to 2} (3x - 5) = 1$$

Solution.   We must show that given any positive number ϵ, we can find a positive number δ such that $f(x) = 3x - 5$ satisfies

$$|(3x - 5) - 1| < \epsilon \tag{6}$$

whenever x satisfies

$$0 < |x - 2| < \delta \qquad (7)$$

To find δ, we can rewrite (6) as

$$|3x - 6| < \epsilon$$

or equivalently

$$3|x - 2| < \epsilon$$

or

$$|x - 2| < \frac{\epsilon}{3} \qquad (8)$$

We must choose δ so that (8) is satisfied whenever (7) is satisfied. We can do this by taking

$$\delta = \frac{\epsilon}{3}$$

To prove that this choice for δ works, suppose that x satisfies (7). Since we are letting $\delta = \epsilon/3$, it follows from (7) that x satisfies

$$0 < |x - 2| < \frac{\epsilon}{3} \qquad (9)$$

Thus, (8) is satisfied since it is just the right-hand inequality in (9). This proves that $\lim_{x \to 2} (3x - 5) = 1$. ◀

As illustrated in Example 1, a limit proof proceeds as follows: We *assume* that an arbitrary positive number ϵ is given to us. Then we try to *find* a positive number δ such that the conditions in the definition are fulfilled. The idea is to find a formula for δ in terms of ϵ so that no matter what ϵ is chosen, we automatically obtain from the formula the required δ. (In Example 1, the formula was $\delta = \epsilon/3$.)

Example 1 is about as easy as a limit proof can get; most limit proofs require a little more algebraic and logical ingenuity. The reader who finds ''δ-ε'' discussions hard going should not become discouraged; the concepts and techniques are intrinsically difficult. In fact, a precise understanding of limits evaded the finest mathematical minds for centuries.

□ **THE VALUE OF δ IS NOT UNIQUE**

Note that the value of δ in Definition 2.6.1 is not unique. Once one value of δ is found that fulfills the requirements of the definition, any *smaller* positive value for δ will also fulfill these requirements. To see why this is so, assume that we have found a value of δ such that

$$|f(x) - L| < \epsilon \qquad (10)$$

whenever x satisfies

$$0 < |x - a| < \delta \qquad (11)$$

Let δ_1 be any positive number smaller than δ. If x satisfies

$$0 < |x - a| < \delta_1 \tag{12}$$

then we have

$$0 < |x - a| < \delta_1 < \delta$$

Thus, x satisfies (11) and consequently $f(x)$ satisfies (10). Therefore, δ_1 meets the requirements of Definition 2.6.1 also. For example, we found in Example 1 that $\delta = \epsilon/3$ fulfills the requirements of Definition 2.6.1. Consequently, any smaller value for δ, such as $\delta = \epsilon/4$, $\delta = \epsilon/5$, or $\delta = \epsilon/6$, does also.

Example 2   Prove that $\lim\limits_{x \to 3} x^2 = 9$.

Solution.   We must show that given any positive number ϵ, we can find a positive number δ such that

$$|x^2 - 9| < \epsilon \tag{13}$$

whenever x satisfies

$$0 < |x - 3| < \delta \tag{14}$$

Because $|x - 3|$ occurs in (14), it will be helpful to rewrite (13) so that $|x - 3|$ appears as a factor on the left side. Therefore, we shall rewrite (13) as

$$|x + 3|\,|x - 3| < \epsilon \tag{15}$$

If we can somehow ensure that when x satisfies (14) the factor $|x + 3|$ remains less than some positive constant, say,

$$|x + 3| < k \tag{16}$$

then by choosing

$$\delta = \frac{\epsilon}{k} \tag{17}$$

it will follow from (14) that

$$0 < |x - 3| < \frac{\epsilon}{k}$$

or

$$0 < k|x - 3| < \epsilon \tag{18}$$

From (16) and the right-hand inequality in (18) we shall then have

$$|x + 3|\,|x - 3| < k|x - 3| < \epsilon$$

so that (15) will be satisfied, and the proof will be complete. As we shall now explain, condition (16) can be satisfied by restricting the size of δ. We remarked in the discussion preceding this example that δ is not unique; once a value of δ is found, any *smaller* positive value for δ can also be used. At this point we do not have a formula relating δ and ϵ. However, let us agree in advance that if, for a given ϵ, δ turns out to exceed 1, we shall choose the smaller value $\delta = 1$ instead. Thus, we can assume in our discussion that δ satisfies

$$0 < \delta \leq 1 \tag{19}$$

Suppose that x satisfies (14). From (19) and the right side of (14) we obtain

$$|x - 3| < \delta \leq 1$$

so that

$$|x - 3| < 1$$

or equivalently

$$2 < x < 4$$

so

$$5 < x + 3 < 7$$

Therefore,

$$|x + 3| < 7$$

Comparing the last inequality to (16) suggests $k = 7$; and from (17)

$$\delta = \frac{\epsilon}{k} = \frac{\epsilon}{7}$$

In summary, given $\epsilon > 0$, we choose

$$\delta = \frac{\epsilon}{7}$$

provided $\epsilon/7$ does not exceed 1. If $\epsilon/7$ exceeds 1, we choose

$$\delta = 1$$

In other words, we take δ to be the minimum of the numbers $\epsilon/7$ and 1. This is sometimes written as

$$\delta = \min\left(\frac{\epsilon}{7}, 1\right) \quad \blacktriangleleft$$

REMARK. In the foregoing example the reader may have wondered how we knew to make the restriction $\delta \leq 1$ as opposed to some other restriction such as $\delta \leq \frac{1}{2}$ or $\delta \leq 5$. Actually, our selection was completely arbitrary; any other restriction of the form $\delta \leq c$ would have worked equally well (Exercise 29).

Example 3   Prove that $\lim\limits_{x \to 1/2} \dfrac{1}{x} = 2$.

Solution.   We must show that given $\epsilon > 0$, there exists a $\delta > 0$ such that

$$\left| \frac{1}{x} - 2 \right| < \epsilon \tag{20}$$

whenever x satisfies

$$0 < \left| x - \frac{1}{2} \right| < \delta \tag{21}$$

Because $|x - \tfrac{1}{2}|$ occurs in (21), it will be helpful to rewrite (20) so that $|x - \tfrac{1}{2}|$ appears as a factor on the left side. We do this as follows:

$$\left| \frac{1}{x} - 2 \right| = \left| \frac{2}{x} \left(\frac{1}{2} - x \right) \right| = \left| \frac{2}{x} \right| \left| \frac{1}{2} - x \right| = \left| \frac{2}{x} \right| \left| x - \frac{1}{2} \right|$$

Thus, (20) is equivalent to

$$\left| \frac{2}{x} \right| \left| x - \frac{1}{2} \right| < \epsilon \tag{22}$$

From here on, the procedure is similar to that used in the last example. By restricting δ we shall try to make the factor $|2/x|$ satisfy

$$\left| \frac{2}{x} \right| < k \tag{23}$$

for some constant k when x satisfies (21). Then on choosing

$$\delta = \frac{\epsilon}{k} \tag{24}$$

we shall have, from (21),

$$0 < \left| x - \frac{1}{2} \right| < \frac{\epsilon}{k}$$

or

$$0 < k \left| x - \frac{1}{2} \right| < \epsilon \tag{25}$$

From (23) and the right-hand inequality in (25) it will then follow that

$$\left| \frac{2}{x} \right| \left| x - \frac{1}{2} \right| < k \left| x - \frac{1}{2} \right| < \epsilon$$

so that (22) will hold and the proof will be complete. Therefore, to finish the proof, we must show that δ can be restricted so that (23) holds for some k when x satisfies (21).

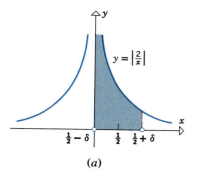

(a)

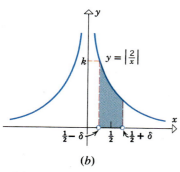

(b)

Figure 2.6.6

In Figure 2.6.6 we have sketched the graph of $|2/x|$ and marked the set of x values satisfying (21). In part (a) of the figure we took $\delta = \frac{1}{2}$ and in part (b) of the figure we took $\delta < \frac{1}{2}$. Part (a) of this figure makes it clear that if $\delta = \frac{1}{2}$ (or $\delta > \frac{1}{2}$), then the values of $|2/x|$ will have no upper bound, making it impossible to satisfy (23). However, part (b) shows that when $\delta < \frac{1}{2}$, there is an upper bound k to the values of $|2/x|$. Therefore, the arbitrary restriction

$$\delta \leq \frac{1}{4} \tag{26}$$

will be satisfactory to assure that $|2/x|$ has an upper bound.

Assume that x satisfies (21). From (26) and the right side of (21), we obtain

$$\left| x - \frac{1}{2} \right| < \delta \leq \frac{1}{4}$$

so that

$$\left| x - \frac{1}{2} \right| < \frac{1}{4}$$

or equivalently

$$\frac{1}{4} < x < \frac{3}{4} \tag{27}$$

To estimate the size of $|2/x|$ we take reciprocals in (27) and multiply through by 2, which gives

$$8 > \frac{2}{x} > \frac{8}{3}$$

Thus,

$$\left| \frac{2}{x} \right| < 8$$

Comparing the last inequality to (23) suggests $k = 8$; and from (24)

$$\delta = \frac{\epsilon}{k} = \frac{\epsilon}{8}$$

In summary, given $\epsilon > 0$, we can choose

$$\delta = \frac{\epsilon}{8}$$

provided this choice does not violate $\delta \leq \frac{1}{4}$. If it does, we choose the value $\delta = \frac{1}{4}$ instead. In other words,

$$\delta = \min\left(\frac{\epsilon}{8}, \frac{1}{4} \right) \quad \blacktriangleleft$$

The following example illustrates a technique for proving that a two-sided limit does not exist in the case where the one-sided limits exist but are not equal.

Example 4   The graph of

$$f(x) = \begin{cases} 1, & x > 0 \\ -1, & x < 0 \end{cases}$$

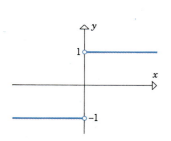

Figure 2.6.7

is shown in Figure 2.6.7. Because the one-sided limits

$$\lim_{x \to 0^+} f(x) = 1 \quad \text{and} \quad \lim_{x \to 0^-} f(x) = -1$$

are not equal, the two-sided limit

$$\lim_{x \to 0} f(x)$$

does not exist. Prove that this is so using Definition 2.6.1.

Solution.   We shall assume that there is a limit and obtain a contradiction. Assume that there is a number L such that

$$\lim_{x \to 0} f(x) = L$$

Then given any $\epsilon > 0$, there exists a $\delta > 0$ such that

$$|f(x) - L| < \epsilon \quad \text{whenever} \quad 0 < |x - 0| < \delta$$

In particular, if we take $\epsilon = 1$, there is a $\delta > 0$ such that

$$|f(x) - L| < 1$$

whenever

$$0 < |x - 0| < \delta \tag{28}$$

But $x = \delta/2$ and $x = -\delta/2$ both satisfy (28) so that

$$\left| f\left(\frac{\delta}{2}\right) - L \right| < 1 \quad \text{and} \quad \left| f\left(-\frac{\delta}{2}\right) - L \right| < 1 \tag{29}$$

However, $\delta/2$ is positive and $-\delta/2$ is negative, so

$$f\left(\frac{\delta}{2}\right) = 1 \quad \text{and} \quad f\left(-\frac{\delta}{2}\right) = -1$$

Thus, (29) states that

$$|1 - L| < 1 \quad \text{and} \quad |-1 - L| < 1$$

or equivalently

$$0 < L < 2 \quad \text{and} \quad -2 < L < 0$$

But this is a contradiction, since no number L can satisfy both of these conditions. ◄

▶ Exercise Set 2.6

In Exercises 1–10, we are told that $\lim_{x \to a} f(x) = L$ and we are given a value of ϵ. In each exercise, find a number δ such that $|f(x) - L| < \epsilon$ whenever $0 < |x - a| < \delta$.

1. $\lim_{x \to 4} 2x = 8; \epsilon = 0.1$.

2. $\lim_{x \to -2} \frac{1}{2}x = -1; \epsilon = 0.1$.

3. $\lim_{x \to -1} (7x + 5) = -2; \epsilon = 0.01$.

4. $\lim_{x \to 3} (5x - 2) = 13; \epsilon = 0.01$.

5. $\lim_{x \to 2} \frac{x^2 - 4}{x - 2} = 4; \epsilon = 0.05$.

6. $\lim_{x \to -1} \frac{x^2 - 1}{x + 1} = -2; \epsilon = 0.05$.

7. $\lim_{x \to 4} x^2 = 16; \epsilon = 0.001$.

8. $\lim_{x \to 9} \sqrt{x} = 3; \epsilon = 0.001$.

9. $\lim_{x \to 5} \frac{1}{x} = \frac{1}{5}; \epsilon = 0.05$.

10. $\lim_{x \to 0} |x| = 0; \epsilon = 0.05$.

In Exercises 11–24, use Definition 2.6.1 to prove that the given limit statement is correct.

11. $\lim_{x \to 5} 3x = 15$.

12. $\lim_{x \to 3} (4x - 5) = 7$.

13. $\lim_{x \to 2} (2x - 7) = -3$.

14. $\lim_{x \to -1} (2 - 3x) = 5$.

15. $\lim_{x \to 0} \frac{x^2 + x}{x} = 1$.

16. $\lim_{x \to -3} \frac{x^2 - 9}{x + 3} = -6$.

17. $\lim_{x \to 1} 2x^2 = 2$.

18. $\lim_{x \to 3} (x^2 - 5) = 4$.

19. $\lim_{x \to 1/3} \frac{1}{x} = 3$.

20. $\lim_{x \to -2} \frac{1}{x + 1} = -1$.

21. $\lim_{x \to 4} \sqrt{x} = 2$.

22. $\lim_{x \to 6} \sqrt{x + 3} = 3$.

23. $\lim_{x \to 1} f(x) = 3$, where $f(x) = \begin{cases} x + 2, & x \neq 1 \\ 10, & x = 1. \end{cases}$

24. $\lim_{x \to 2} (x^2 + 3x - 1) = 9$.

25. Let $f(x) = \begin{cases} \frac{1}{8}, & x > 0 \\ -\frac{1}{8}, & x < 0. \end{cases}$

Use the method of Example 4 to prove that $\lim_{x \to 0} f(x)$ does not exist.

26. Let $g(x) = \begin{cases} 1 + x, & x > 0 \\ x - 1, & x < 0. \end{cases}$

Prove that $\lim_{x \to 0} g(x)$ does not exist.

27. Prove that $\lim_{x \to 1} \frac{1}{x - 1}$ does not exist.

28. (a) In Definition 2.6.1 there is a condition requiring that $f(x)$ be defined for every x in an open interval containing a, except possibly at a itself. What is the purpose of this requirement?

(b) Why is $\lim_{x \to 0} \sqrt{x} = 0$ an incorrect statement?

(c) Is $\lim_{x \to 0.01} \sqrt{x} = 0.1$ a correct statement?

29. Prove the result in Example 2 under the assumption that $\delta \leq 2$ rather than $\delta \leq 1$.

■ 2.7 CONTINUITY

A moving physical object cannot vanish at some point and reappear someplace else to continue its motion. Thus, we perceive the path of a moving object as a single, unbroken curve without gaps, jumps, or holes. Such curves can be described as "continuous." In this section we shall express this intuitive idea mathematically and develop some properties of continuous curves.

☐ **DEFINITION OF CONTINUITY**

Before we give any formal definitions, let us consider some of the ways in which curves can be "discontinuous." In Figure 2.7.1 we have graphed some curves which, because of their behavior at the point c, are not continuous. The curve in Figure 2.7.1*a* has a hole at the point c because the function f is undefined there. For the curves in Figures 2.7.1*b* and 2.7.1*c*, the function f is defined at c, but

$$\lim_{x \to c} f(x) \tag{1}$$

does not exist, thereby causing a break in the graph. For the curve in Figure 2.7.1*d*, the function f is defined at c, and the limit in (1) exists, yet the graph still has a break at the point c because

$$\lim_{x \to c} f(x) \neq f(c)$$

Based on this discussion, we see that there is a break or discontinuity in the graph of $y = f(x)$ at a point $x = c$ if any of the following conditions occur:

- The function f is undefined at c.
- The limit $\lim_{x \to c} f(x)$ does not exist.
- The function f is defined at c and the limit $\lim_{x \to c} f(x)$ exists, but the value of the function at c and the value of the limit at c are different.

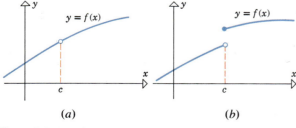

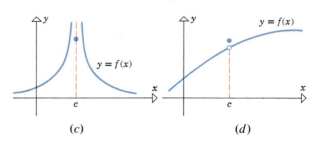

(a)   (b)   (c)   (d)

Figure 2.7.1

This suggests the following definition.

2.7.1 DEFINITION.   A function f is said to be *continuous at a point c* if the following conditions are satisfied:

1. $f(c)$ is defined.

2. $\lim\limits_{x \to c} f(x)$ exists.

3. $\lim\limits_{x \to c} f(x) = f(c)$.

If one or more of the conditions in this definition fails to hold, then f is called *discontinuous at c* and c is called a *point of discontinuity* of f. If f is continuous at all points of an open interval (a, b), then f is said to be *continuous on (a, b)*. A function that is continuous on $(-\infty, +\infty)$ is said to be *continuous everywhere* or simply *continuous*.

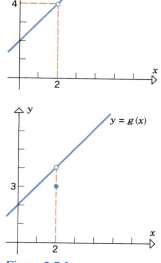

Figure 2.7.2

Example 1   Let

$$f(x) = \frac{x^2 - 4}{x - 2} \quad \text{and} \quad g(x) = \begin{cases} \dfrac{x^2 - 4}{x - 2}, & x \neq 2 \\ 3, & x = 2 \end{cases}$$

Both f and g are discontinuous at 2 (Figure 2.7.2), the function f because $f(2)$ is undefined, and the function g because $g(2) = 3$, and

$$\lim_{x \to 2} g(x) = \lim_{x \to 2} \frac{x^2 - 4}{x - 2} = \lim_{x \to 2} (x + 2) = 4$$

so that

$$\lim_{x \to 2} g(x) \neq g(2) \qquad \blacktriangleleft$$

REMARK.   Some authors define a function to be continuous at c if condition 3 in Definition 2.7.1 holds. This is really equivalent to our definition because if condition 3 holds, then 1 and 2 hold automatically. (Why?) We have stated the three conditions for clarity. However, when we want to show that a function is continuous at a point, we shall just show that condition 3 holds.

Example 2   Show that $f(x) = x^2 - 2x + 1$ is a continuous function.

Solution.   We must show that the third condition in Definition 2.7.1 holds for all real numbers c. But, by Theorem 2.5.2

$$\lim_{x \to c} f(x) = \lim_{x \to c} (x^2 - 2x + 1) = c^2 - 2c + 1 = f(c)$$

which shows that the third condition holds. $\blacktriangleleft$

The foregoing example is a special case of the following general result.

☐ **CONTINUITY OF POLYNOMIALS**

2.7.2   THEOREM.   *Polynomials are continuous functions.*

Proof.   If p is a polynomial and c is any real number, then by Theorem 2.5.2

$$\lim_{x \to c} p(x) = p(c)$$

which proves the continuity of p at c. Since c is an arbitrary real number, p is continuous everywhere.   ∎

Example 3   Show that $f(x) = |x|$ is a continuous function.

Solution.   We can write $f(x)$ as

$$f(x) = |x| = \begin{cases} x & \text{if } x > 0 \\ 0 & \text{if } x = 0 \\ -x & \text{if } x < 0 \end{cases}$$

It follows from Theorem 2.7.2 that $f(x) = |x|$ is continuous if $x > 0$ or $x < 0$ because $|x|$ is identical to the polynomial x in the former case and identical to the polynomial $-x$ in the latter. Thus, $x = 0$ is the only point that remains to be considered. At this point we have $f(0) = |0| = 0$, so it remains to show that

$$\lim_{x \to 0} f(x) = \lim_{x \to 0} |x| = 0 \tag{2}$$

Because the formula for f changes at 0, it will be helpful to consider the one-sided limits at 0 rather than the two-sided limit. We obtain

$$\lim_{x \to 0^+} |x| = \lim_{x \to 0^+} x = 0$$

$$\lim_{x \to 0^-} |x| = \lim_{x \to 0^-} (-x) = 0$$

Thus, (2) holds and $|x|$ is continuous at $x = 0$.   ◄

☐ **SOME PROPERTIES OF CONTINUOUS FUNCTIONS**

The following basic result is an immediate consequence of Theorem 2.5.1.

2.7.3   THEOREM.   *If the functions f and g are continuous at c, then*

(a)   *$f + g$ is continuous at c;*
(b)   *$f - g$ is continuous at c;*
(c)   *$f \cdot g$ is continuous at c;*
(d)   *f/g is continuous at c if $g(c) \neq 0$ and is discontinuous at c if $g(c) = 0$.*

We shall prove part (*d*) and leave the remaining proofs as exercises.

Proof (d). If $g(c) = 0$, then f/g is discontinuous at c because $f(c)/g(c)$ is undefined.

Assume that $g(c) \neq 0$. We must show that

$$\lim_{x \to c} \frac{f(x)}{g(x)} = \frac{f(c)}{g(c)} \tag{3}$$

Since f and g are continuous at c,

$$\lim_{x \to c} f(x) = f(c) \quad \text{and} \quad \lim_{x \to c} g(x) = g(c)$$

Thus, by Theorem 2.5.1(*d*)

$$\lim_{x \to c} \frac{f(x)}{g(x)} = \frac{\displaystyle\lim_{x \to c} f(x)}{\displaystyle\lim_{x \to c} g(x)} = \frac{f(c)}{g(c)}$$

which proves (3). ∎

CONTINUITY OF RATIONAL FUNCTIONS

Example 4 Where is $h(x) = \dfrac{x^2 - 9}{x^2 - 5x + 6}$ continuous?

Solution. The numerator and denominator of h are polynomials and therefore are continuous everywhere by Theorem 2.7.2. Thus, by Theorem 2.7.3(*d*), the ratio is continuous everywhere except at the points where the denominator is zero. Since the solutions of

$$x^2 - 5x + 6 = 0$$

are $x = 2$ and $x = 3$, $h(x)$ is continuous everywhere except at these two points. ◀

The result in the foregoing example is a special case of the following general theorem whose proof is left as an exercise.

> **2.7.4 THEOREM.** *A rational function is continuous everywhere except at the points where the denominator is zero.*

CONTINUITY OF COMPOSITIONS

The next theorem is useful for calculating limits of compositions of functions.

> **2.7.5 THEOREM.** *Let* lim *stand for one of the limits* $\lim_{x \to c}$, $\lim_{x \to c^-}$, $\lim_{x \to c^+}$, $\lim_{x \to +\infty}$, *or* $\lim_{x \to -\infty}$. *If* $\lim g(x) = L$ *and if the function* f *is continuous at* L, *then* $\lim f(g(x)) = f(L)$. *That is,* $\lim f(g(x)) = f(\lim g(x))$.

For those who have read Section 2.6, the optional proof of this result appears in Appendix C.

REMARK.   In words, this theorem states that the limit symbol can be moved through a function sign provided the limit of the expression inside the function sign exists and the function is continuous at this limit.

Example 5   We saw in Example 3 that $|x|$ is continuous everywhere, so Theorem 2.7.5 implies that

$$\lim |g(x)| = |\lim g(x)|$$

if $\lim g(x)$ exists. For example,

$$\lim_{x \to 3} |5 - x^2| = |\lim_{x \to 3} (5 - x^2)| = |-4| = 4 \quad \blacktriangleleft$$

The following consequence of Theorem 2.7.5 tells us that compositions of continuous functions are themselves continuous.

2.7.6   THEOREM.   *If the function g is continuous at the point c and the function f is continuous at the point g(c), then the composition f ∘ g is continuous at c.*

Proof.   We must show that $f \circ g$ satisfies the third condition of Definition 2.7.1 at c. But this is so since we can write

$$\lim_{x \to c} (f \circ g)(x) = \lim_{x \to c} f(g(x)) = f(\lim_{x \to c} g(x)) = f(g(c)) = (f \circ g)(c)$$

| Theorem 2.7.5 | g is continuous at c. | ∎

Example 6   The function $|5 - x^2|$ is continuous because it is the composition of $|x|$ with $5 - x^2$, both of which are continuous.   ◀

◻ CONTINUITY FROM THE LEFT AND RIGHT

Figure 2.7.3 shows the graphs of three functions defined only on a closed interval $[a, b]$. Obviously the function shown in Figure 2.7.3a should be regarded as discontinuous at the left-hand endpoint a, the function in Figure 2.7.3b should be regarded as discontinuous at the right-hand endpoint b, and the function in Figure 2.7.3c should be regarded as continuous at both end-

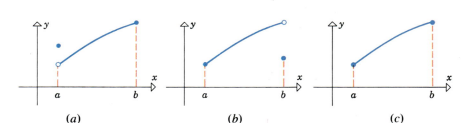

(a)   (b)   (c)

Figure 2.7.3

points. However, the definition of continuity (Definition 2.7.1) does not apply at the endpoints, since the two-sided limits appearing in parts 2 and 3 of the definition make no sense. At the left-hand endpoint the only sensible limit is the one-sided limit

$$\lim_{x \to a^+} f(x)$$

and at the right-hand endpoint the only sensible limit is the one-sided limit

$$\lim_{x \to b^-} f(x)$$

This motivates the following definitions.

2.7.7 DEFINITION. A function f is called **continuous from the left at the point c** if the conditions in the left column below are satisfied, and is called **continuous from the right at the point c** if the conditions in the right column are satisfied.

1. $f(c)$ is defined.	1'. $f(c)$ is defined.
2. $\lim_{x \to c^-} f(x)$ exists.	2'. $\lim_{x \to c^+} f(x)$ exists.
3. $\lim_{x \to c^-} f(x) = f(c)$.	3'. $\lim_{x \to c^+} f(x) = f(c)$.

2.7.8 DEFINITION. A function f is said to be **continuous on a closed interval** $[a, b]$ if the following conditions are satisfied:

1. f is continuous on (a, b).
2. f is continuous from the right at a.
3. f is continuous from the left at b.

The reader should have no trouble deducing the appropriate definitions of continuity on intervals of the form $[a, +\infty)$, $(-\infty, b]$, $[a, b)$, and $(a, b]$.

Example 7 If f denotes the function graphed in Figure 2.7.3a, then

$$f(a) \neq \lim_{x \to a^+} f(x) \quad \text{and} \quad f(b) = \lim_{x \to b^-} f(x)$$

Thus, f is continuous from the left at b, but is not continuous from the right at a. ◄

Example 8 Show that $f(x) = \sqrt{9 - x^2}$ is continuous on the closed interval $[-3, 3]$.

Solution. Observe that the domain of f is the interval $[-3, 3]$. For c in the interval $(-3, 3)$ we have by Theorem 2.5.1(e)

$$\lim_{x \to c} f(x) = \lim_{x \to c} \sqrt{9 - x^2} = \sqrt{\lim_{x \to c} (9 - x^2)} = \sqrt{9 - c^2} = f(c)$$

so that f is continuous on $(-3, 3)$. Also,

$$\lim_{x \to 3^-} f(x) = \lim_{x \to 3^-} \sqrt{9 - x^2} = \sqrt{\lim_{x \to 3^-} (9 - x^2)} = 0 = f(3)$$

and

$$\lim_{x \to -3^+} f(x) = \lim_{x \to -3^+} \sqrt{9 - x^2} = \sqrt{\lim_{x \to -3^+} (9 - x^2)} = 0 = f(-3)$$

so that f is continuous on $[-3, 3]$.   ◀

☐ THE INTERMEDIATE VALUE THEOREM

If f is continuous on a closed interval $[a, b]$, and we draw a horizontal line crossing the y-axis between the numbers $f(a)$ and $f(b)$ (Figure 2.7.4), then it is geometrically obvious that this line will cross the curve $y = f(x)$ at least once over the interval $[a, b]$. Stated another way, a continuous function must assume every possible value between $f(a)$ and $f(b)$ as x varies from a to b. This idea is formalized in the following theorem.

2.7.9   THEOREM (*Intermediate Value Theorem*).   *If f is continuous on a closed interval $[a, b]$ and C is any number between $f(a)$ and $f(b)$, inclusive, then there is at least one number x in the interval $[a, b]$ such that $f(x) = C$.*

(See Figure 2.7.4.) This theorem is intuitively obvious, but its proof depends on a mathematically precise development of the real number system which is beyond the scope of this text. The proof can be found in most advanced calculus texts.

The following useful result is an immediate consequence of the Intermediate Value Theorem.

2.7.10   THEOREM.   *If f is continuous on $[a, b]$, and if $f(a)$ and $f(b)$ have opposite signs, then there is at least one solution of the equation $f(x) = 0$ in the interval (a, b).*

The proof is left as an exercise, but the result is illustrated in Figure 2.7.5 in the case where $f(a) > 0$ and $f(b) < 0$.

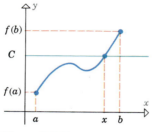

Figure 2.7.4

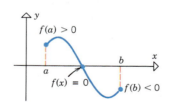

Figure 2.7.5

Theorem 2.7.10 is the basis for various numerical techniques for approximating solutions of equations of the form $f(x) = 0$.

Example 9 The equation

$$x^3 - x - 1 = 0$$

cannot be solved readily by factoring because the left side has no simple factors. However, from Figure 2.7.6, which was generated on a microcomputer, we see that there is a solution in the interval $(1, 2)$. This can also be seen from Theorem 2.7.10 by letting $f(x) = x^3 - x - 1$, and noting that $f(1) = -1$ and $f(2) = 5$ have opposite signs.

To pinpoint the solution more exactly, we can divide the interval $(1, 2)$ into 10 equal parts and evaluate $f(x)$ at each point of subdivision using a calculator (Table 2.7.1). From the table we see that $f(1.3)$ and $f(1.4)$ have opposite signs, so there is a solution in the subinterval $(1.3, 1.4)$. If we use the midpoint 1.35 as an approximation to the solution, then our error is at most half the subinterval length or .05. If desired, we could reduce the error to at most .005 by subdividing the subinterval $(1.3, 1.4)$ into 10 parts and repeating the sign analysis process. With sufficiently many subdivisions, the solution can be determined to any degree of accuracy. ◄

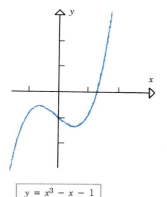

$$y = x^3 - x - 1$$

Figure 2.7.6

Table 2.7.1

x	1	1.1	1.2	1.3	1.4	1.5	1.6	1.7	1.8	1.9	2
$f(x)$	-1	$-.77$	$-.47$	$-.10$	.34	.88	1.5	2.2	3.0	3.9	5

We conclude this section by stating an alternative form of the continuity definition that will be useful in later sections.

If we let $h = x - c$ in Definition 2.7.1, then the condition that $x \to c$ is equivalent to $h \to 0$. Thus, Definition 2.7.1 can be restated as follows.

☐ **ALTERNATIVE DEFINITION OF CONTINUITY**

2.7.11 **DEFINITION.** A function f is said to be **continuous at a point c** if the following conditions are satisfied:

1. $f(c)$ is defined.
2. $\lim\limits_{h \to 0} f(c + h)$ exists.
3. $\lim\limits_{h \to 0} f(c + h) = f(c)$.

REMARK. As with Definition 2.7.1, conditions 1 and 2 hold automatically if condition 3 holds. Thus, to prove that a function is continuous at a point, it is only necessary to show that condition 3 holds.

▶ Exercise Set 2.7 Ⓒ 32, 33, 34, 35

In Exercises 1–4, let f be the function whose graph is shown. On which of the following intervals, if any, is f continuous?

(a) $[1, 3]$ (b) $(1, 3)$ (c) $[1, 2]$
(d) $(1, 2)$ (e) $[2, 3]$ (f) $(2, 3)$.

On those intervals where f is discontinuous, state where the discontinuities occur.

1.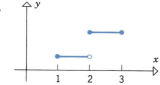

2.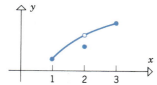

3.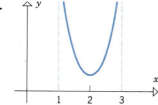

4.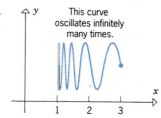
This curve oscillates infinitely many times.

In Exercises 5–16, find the points of discontinuity, if any.

5. $f(x) = x^3 - 2x + 3$.

6. $f(x) = (x - 5)^{17}$.

7. $f(x) = \dfrac{x}{x^2 + 1}$.

8. $f(x) = \dfrac{x}{x^2 - 1}$.

9. $f(x) = \dfrac{x - 4}{x^2 - 16}$.

10. $f(x) = \dfrac{3x + 1}{x^2 + 7x - 2}$.

11. $f(x) = \dfrac{x}{|x| - 3}$.

12. $f(x) = \dfrac{5}{x} + \dfrac{2x}{x + 4}$.

13. $f(x) = |x^3 - 2x^2|$.

14. $f(x) = \dfrac{x + 3}{|x^2 + 3x|}$.

15. $f(x) = \begin{cases} 2x + 3, & x \le 4 \\ 7 + \dfrac{16}{x}, & x > 4. \end{cases}$

16. $f(x) = \begin{cases} \dfrac{3}{x - 1}, & x \ne 1 \\ 3, & x = 1. \end{cases}$

17. Find a value for the constant k, if possible, that will make the function continuous.

(a) $f(x) = \begin{cases} 7x - 2, & x \le 1 \\ kx^2, & x > 1 \end{cases}$

(b) $f(x) = \begin{cases} kx^2, & x \le 2 \\ 2x + k, & x > 2. \end{cases}$

18. On which of the following intervals is
$$f(x) = \frac{1}{\sqrt{x - 2}}$$
continuous?

(a) $[2, +\infty)$ (b) $(-\infty, +\infty)$
(c) $(2, +\infty)$ (d) $[1, 2)$.

19. (a) Prove that $f(x) = \sqrt{x}$ is continuous on $[0, +\infty)$.

(b) Prove that if $g(x)$ is continuous and nonnegative, then $\sqrt{g(x)}$ is continuous.

20. Prove that if $g(x)$ is continuous, then $|g(x)|$ is continuous.

21. Find all points of discontinuity of the greatest integer function, $f(x) = [x]$. (See Exercise 46, Section 2.3, for the definition of $[x]$.)

22. A function f is said to have a ***removable discontinuity*** at $x = c$ if the limit
$$\lim_{x \to c} f(x)$$
exists, but
$$f(c) \ne \lim_{x \to c} f(x)$$
either because $f(c)$ is undefined or the value of $f(c)$ differs from the value of the limit.

(a) Show that the following functions have removable discontinuities at $x = 1$ and sketch their graphs.

$$f(x) = \frac{x^2 - 1}{x - 1} \quad \text{and} \quad g(x) = \begin{cases} 1, & x > 1 \\ 0, & x = 1 \\ 1, & x < 1 \end{cases}$$

(b) The terminology "removable discontinuity" is used because a function with a removable discontinuity at $x = c$ can be made continuous at $x = c$ by defining (or redefining) the value of the function at $x = c$ appropriately. Do this for the functions f and g in part (a).

In Exercises 23 and 24, find all points of discontinuity for the functions, and determine whether the discontinuities are removable or not. [See Exercise 22 for terminology.]

23. (a) $f(x) = \dfrac{|x|}{x}$ (b) $f(x) = \dfrac{x^2 + 3x}{x + 3}$

(c) $f(x) = \dfrac{x - 2}{|x| - 2}.$

24. (a) $f(x) = \dfrac{x^2 - 4}{x^3 - 8}$ (b) $f(x) = \begin{cases} 2x - 3, & x \leq 2 \\ x^2, & x > 2 \end{cases}$

(c) $f(x) = \begin{cases} 3x^2 + 5, & x \neq 1 \\ 6, & x = 1. \end{cases}$

25. Prove:
   (a) part (a) of Theorem 2.7.3
   (b) part (b) of Theorem 2.7.3
   (c) part (c) of Theorem 2.7.3.

26. Prove Theorem 2.7.4.

27. Let f and g be discontinuous at c. Give examples to show:
   (a) $f + g$ can be continuous or discontinuous at c
   (b) $f \cdot g$ can be continuous or discontinuous at c.

28. **(For students who have read Section 2.6.)** Let f be defined at c. Prove that f is continuous at c if and only if, given $\epsilon > 0$, there exists a $\delta > 0$ such that $|f(x) - f(c)| < \epsilon$ whenever $|x - c| < \delta$.

29. Use Theorem 2.7.9 to prove Theorem 2.7.10.

In Exercises 30 and 31, show that the equation has at least one solution in the given interval.

30. $x^3 - 4x + 1 = 0$; $[1, 2]$.

31. $x^3 + x^2 - 2x = 1$; $[-1, 1]$.

32. Figure 2.7.7 shows the graph of $y = x^4 + x - 1$ generated on a microcomputer. Use the method of Example 9 to approximate the real solutions of $x^4 + x - 1 = 0$ with an error of at most .05.

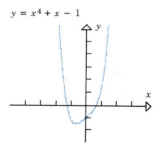

$y = x^4 + x - 1$

Figure 2.7.7

33. Figure 2.7.8 shows the graph of $y = 5 - x - x^4$ generated on a microcomputer. Use the method of Example 9 to approximate the real solutions of $5 - x - x^4 = 0$ with an error of at most .05.

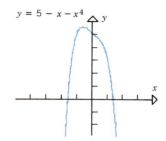

$y = 5 - x - x^4$

Figure 2.7.8

34. For the equation $x^3 - x - 1 = 0$ discussed in Example 9, show that the real solution is approximately 1.325, with an error of at most .005.

35. Use the fact that $\sqrt{5}$ is a solution of $x^2 - 5 = 0$ to approximate $\sqrt{5}$ with an error of at most
   (a) .05   (b) .005.

36. Prove: If f and g are continuous on $[a, b]$, and $f(a) > g(a)$, $f(b) < g(b)$, then there is at least one solution of the equation $f(x) = g(x)$ in (a, b). [*Hint:* Consider $f(x) - g(x)$.]

37. Construct an example of a function f that is defined at every point in a closed interval, and whose values at the endpoints have opposite signs, but for which the equation $f(x) = 0$ has no solution in the interval.

38. Prove that if a and b are positive, then the equation

$$\frac{a}{x - 1} + \frac{b}{x - 3} = 0$$

has at least one solution in the interval $(1, 3)$.

39. Prove: If $p(x)$ is a polynomial of odd degree, then the equation $p(x) = 0$ has at least one real solution.

■ 2.8 LIMITS AND CONTINUITY OF TRIGONOMETRIC FUNCTIONS

In this section we shall derive some basic limits involving trigonometric functions and study the continuity of these functions.

This section requires a knowledge of the trigonometry material in Section II of Appendix B. Readers who need to review that material are advised to do so before starting this section.

☐ **AREA OF A SECTOR**

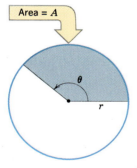

Area = A

Figure 2.8.1

Before proceeding to the main results in this section, it will be helpful to review the formula for the area of a sector.

Let A denote the area of a sector with radius r and a central angle of θ radians (Figure 2.8.1). If $\theta = 2\pi$, the sector is the entire circle and the area is πr^2. For a general sector the area A is proportional to the central angle θ, so we can write

$$\frac{\text{area of the sector}}{\text{area of the circle}} = \frac{\text{central angle of the sector}}{\text{central angle of the circle}}$$

This yields

$$\frac{A}{\pi r^2} = \frac{\theta}{2\pi}$$

from which we obtain the formula

$$A = \frac{1}{2} r^2 \theta \tag{1}$$

REMARK. It is important to keep in mind that the angle θ in the foregoing formula is measured in radians. As we progress through this text you will encounter many calculus formulas that build on (1). All such formulas will also require that angles be measured in radians.

☐ **CONTINUITY OF SINE AND COSINE**

In trigonometry the graphs of $y = \sin x$ and $y = \cos x$ are drawn as continuous curves (Figures 2.8.2 and 2.8.3). Our first objective in this section is to demonstrate that $\sin x$ and $\cos x$ are indeed continuous functions. For this purpose, consider Figure 2.8.4, which shows an angle of h (radians) drawn in standard position with its terminal side intersecting the unit circle at the point $P(\cos h, \sin h)$. It is evident geometrically that as h approaches 0, the point $P(\cos h, \sin h)$ moves toward the point $B(1, 0)$. (Although h was drawn as a positive angle,

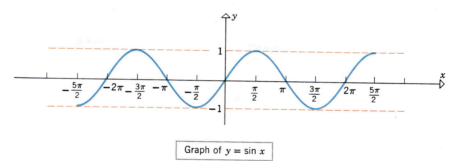

Graph of $y = \sin x$

Figure 2.8.2

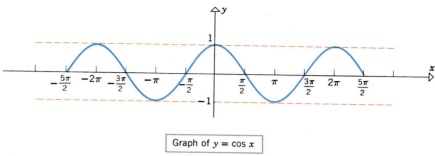

Graph of $y = \cos x$

Figure 2.8.3

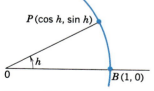

Figure 2.8.4

the same conclusion holds for negative h.) This suggests that

$$\lim_{h \to 0} \cos h = 1 \tag{2a}$$

$$\lim_{h \to 0} \sin h = 0 \tag{2b}$$

These limits tell us that the functions $\sin x$ and $\cos x$ are continuous at the origin. The following theorem shows that they are actually continuous everywhere.

2.8.1 THEOREM.   *The functions $\sin x$ and $\cos x$ are continuous.*

Proof.   We shall prove the result for $\sin x$; the proof for $\cos x$ is similar and will be left as an exercise. By the remark following Definition 2.7.11, it suffices to show that the condition

$$\lim_{h \to 0} \sin(c + h) = \sin c \tag{3}$$

is satisfied. Using the addition formula for the sine function we obtain

$$\lim_{h \to 0} \sin (c + h) = \lim_{h \to 0} [\sin c \cos h + \cos c \sin h]$$

$$= \lim_{h \to 0} [\sin c \cos h] + \lim_{h \to 0} [\cos c \sin h] \qquad (4)$$

The expressions $\sin c$ and $\cos c$ in (4) do not involve h, so they remain *constant* as $h \to 0$. This allows us to move these expressions through the limit signs and write

$$\lim_{h \to 0} \sin (c + h) = \sin c \lim_{h \to 0} \cos h + \cos c \lim_{h \to 0} \sin h$$

$$= (\sin c)(1) + (\cos c)(0) = \sin c$$

which proves (3). ∎

☐ CONTINUITY OF OTHER TRIGONOMETRIC FUNCTIONS

The continuity properties of $\tan x$, $\cot x$, $\sec x$, and $\csc x$ can be deduced by expressing these functions in terms of $\sin x$ and $\cos x$. For example, $\tan x = \sin x / \cos x$, so by part (*d*) of Theorem 2.7.3 the function $\tan x$ is continuous everywhere except at the points where $\cos x = 0$. These points of discontinuity are

$$x = \pm \frac{\pi}{2}, \pm \frac{3\pi}{2}, \pm \frac{5\pi}{2}, \dots$$

(see Figure B.28 in Appendix B).

Example 1   Since the functions $\sin x$ and $\cos x$ are continuous everywhere, it follows from Theorem 2.7.5 that we can write

$$\lim [\sin (g(x))] = \sin [\lim g(x)]$$

and

$$\lim [\cos (g(x))] = \cos [\lim g(x)]$$

if $\lim g(x)$ exists. For example,

$$\lim_{x \to \pi} \left[\sin \left(\frac{x^2}{\pi + x} \right) \right] = \sin \left[\lim_{x \to \pi} \left(\frac{x^2}{\pi + x} \right) \right] = \sin \frac{\pi}{2} = 1$$

$$\lim_{x \to +\infty} \left[\cos \left(\frac{\pi x^2 + 1}{x^2 + 3} \right) \right] = \cos \left[\lim_{x \to +\infty} \left(\frac{\pi x^2 + 1}{x^2 + 3} \right) \right]$$

$$= \cos \left[\lim_{x \to +\infty} \left(\frac{\pi + (1/x^2)}{1 + (3/x^2)} \right) \right] = \cos \pi = -1 \qquad ◀$$

Our next objective is to establish two fundamental limits:

$$\lim_{h \to 0} \frac{\sin h}{h} = 1 \quad \text{and} \quad \lim_{h \to 0} \frac{1 - \cos h}{h} = 0 \qquad (5)$$

These limits are not at all obvious. For example, in the limit

$$\lim_{h \to 0} \frac{\sin h}{h}$$

the numerator and denominator both approach zero as $h \to 0$. As a result, there are two conflicting influences on the ratio. The numerator approaching 0 drives the magnitude of the ratio toward zero, while the denominator approaching 0 drives the magnitude of the ratio toward $+\infty$. The precise way in which these influences offset one another determines whether the limit exists and what its value is. In a limit problem where the numerator and denominator both approach 0, it is often possible to obtain the limit by some algebraic manipulations. Unfortunately, no such algebraic manipulations will work for the limits in (5); we must develop other techniques. However, the computer-generated graphs in Figures 2.8.5 and 2.8.6 suggest that these limits are correct.

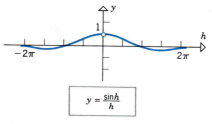

Figure 2.8.5   Figure 2.8.6

□ **OBTAINING LIMITS BY SQUEEZING**

When it is difficult to find the limit of a function directly, it is sometimes possible to obtain the limit indirectly by "squeezing" the function between simpler functions whose limits are known. For example, suppose we are unable to calculate

$$\lim_{x \to a} f(x)$$

directly, but we are able to find two functions, g and h, that have the same limit L as $x \to a$ and such that f is "squeezed" between g and h by means of the inequalities

$$g(x) \le f(x) \le h(x)$$

It is evident geometrically that $f(x)$ must also approach L as $x \to a$ because the graph of f lies between the graphs of g and h (Figure 2.8.7).

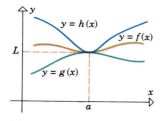

Figure 2.8.7

This idea is formalized in the following theorem, which is called the *Squeezing Theorem* or sometimes the *Pinching Theorem*. We omit the proof.

2.8.2   THEOREM (*The Squeezing Theorem*).   *Let f, g, and h be functions satisfying*

$$g(x) \le f(x) \le h(x)$$

for all x in some open interval containing the point a, with the possible exception that the inequalities need not hold at a. If g and h have the same limit as x approaches a, say

$$\lim_{x \to a} g(x) = \lim_{x \to a} h(x) = L$$

then f also has this limit as x approaches a, that is,

$$\lim_{x \to a} f(x) = L$$

REMARK.   The Squeezing Theorem remains true if $\lim_{x \to a}$ is replaced by $\lim_{x \to a^+}$ or $\lim_{x \to a^-}$. Moreover, for $\lim_{x \to a^+}$ the condition $g(x) \le f(x) \le h(x)$ need only hold on an open interval extending to the right from a, and for $\lim_{x \to a^-}$ the condition need only hold on an open interval extending to the left from a. The Squeezing Theorem can also be extended to limits of the form $\lim_{x \to -\infty}$ and $\lim_{x \to +\infty}$.

Example 2   Use the Squeezing Theorem to evaluate the limit

$$\lim_{x \to 0} x^2 \sin^2 \frac{1}{x}$$

Solution.   If $x \ne 0$, we can write

$$0 \le \sin^2 \frac{1}{x} \le 1$$

Multiplying through by x^2 yields

$$0 \le x^2 \sin^2 \frac{1}{x} \le x^2$$

if $x \ne 0$. But

$$\lim_{x \to 0} 0 = \lim_{x \to 0} x^2 = 0$$

so by the Squeezing Theorem

$$\lim_{x \to 0} x^2 \sin^2 \frac{1}{x} = 0 \qquad \blacktriangleleft$$

SOME IMPORTANT LIMITS OF TRIGONOMETRIC FUNCTIONS

We are now ready to prove the validity of the limits in (5).

2.8.3   THEOREM.

$$\lim_{h \to 0} \frac{\sin h}{h} = 1 \qquad (6)$$

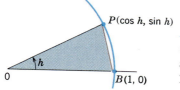

Proof.   Assume that h satisfies $0 < h < \pi/2$ and construct the angle h in standard position. The terminal side of h intersects the unit circle at $P(\cos h, \sin h)$ and intersects the vertical line through $B(1,0)$ at the point $Q(1, \tan h)$ (Figure 2.8.8). From the figure we obtain

$$0 < \text{area of } \triangle\, OBP < \text{area of sector } OBP < \text{area of } \triangle\, OBQ$$

But

$$\text{area } \triangle\, OBP = \tfrac{1}{2} \text{ base} \cdot \text{altitude} = \tfrac{1}{2} \cdot 1 \cdot \sin h = \tfrac{1}{2} \sin h$$
$$\text{area sector } OBP = \tfrac{1}{2}(1)^2 \cdot h = \tfrac{1}{2} h$$
$$\text{area } \triangle\, OBQ = \tfrac{1}{2} \text{ base} \cdot \text{altitude} = \tfrac{1}{2} \cdot 1 \cdot \tan h = \tfrac{1}{2} \tan h$$

Therefore,

$$0 < \tfrac{1}{2} \sin h < \tfrac{1}{2} h < \tfrac{1}{2} \tan h$$

Multiplying through by $2/\sin h$ yields

$$1 < \frac{h}{\sin h} < \frac{1}{\cos h}$$

and taking reciprocals yields

$$\cos h < \frac{\sin h}{h} < 1 \qquad (7)$$

We have derived (7) under the assumption that $0 < h < \pi/2$. However, the inequalities in (7) are also valid if $-\pi/2 < h < 0$ (Exercise 49) so that (7) holds for all h in the interval $(-\pi/2, \pi/2)$ except $h = 0$.
  From (2a) we have

$$\lim_{h \to 0} \cos h = 1$$

and since

$$\lim_{h \to 0} 1 = 1$$

(6) follows by applying the Squeezing Theorem to (7). ▌

Figure 2.8.8

2.8.4   COROLLARY.

$$\lim_{h \to 0} \frac{1 - \cos h}{h} = 0$$

Proof.   With the help of limits (2a), (2b), and the trigonometric identity $\sin^2 h = 1 - \cos^2 h$ we obtain

$$\lim_{h \to 0} \frac{1 - \cos h}{h} = \lim_{h \to 0} \left[\frac{1 - \cos h}{h} \cdot \frac{1 + \cos h}{1 + \cos h} \right] = \lim_{h \to 0} \frac{\sin^2 h}{h(1 + \cos h)}$$

$$= \left(\lim_{h \to 0} \frac{\sin h}{h} \right) \left(\lim_{h \to 0} \frac{\sin h}{1 + \cos h} \right) = (1) \left(\frac{0}{1 + 1} \right) = 0 \quad \blacksquare$$

The following examples illustrate how the limits obtained in this section can be used to calculate other limits involving trigonometric functions.

Example 3   Find $\lim_{x \to 0} \dfrac{\tan x}{x}$.

Solution.

$$\lim_{x \to 0} \frac{\tan x}{x} = \lim_{x \to 0} \left(\frac{\sin x}{x} \cdot \frac{1}{\cos x} \right) = (1)(1) = 1 \quad \blacktriangleleft$$

Example 4   Find $\lim_{\theta \to 0} \dfrac{\sin 2\theta}{\theta}$.

Solution.   The trick is to multiply and divide by 2, so that the resulting denominator is the same as the expression inside the sine function in the numerator. We obtain

$$\lim_{\theta \to 0} \frac{\sin 2\theta}{\theta} = \lim_{\theta \to 0} 2 \cdot \frac{\sin 2\theta}{2\theta} = 2 \lim_{\theta \to 0} \frac{\sin 2\theta}{2\theta}$$

Now make the substitution $t = 2\theta$ and use the fact that $t \to 0$ as $\theta \to 0$. This yields

$$\lim_{\theta \to 0} \frac{\sin 2\theta}{\theta} = 2 \lim_{\theta \to 0} \frac{\sin 2\theta}{2\theta} = 2 \lim_{t \to 0} \frac{\sin t}{t} = 2(1) = 2 \quad \blacktriangleleft$$

Example 5   Find $\lim_{x \to 0} \dfrac{\sin 3x}{\sin 5x}$.

Solution.

$$\lim_{x \to 0} \frac{\sin 3x}{\sin 5x} = \lim_{x \to 0} \frac{\dfrac{\sin 3x}{x}}{\dfrac{\sin 5x}{x}} = \lim_{x \to 0} \frac{3 \cdot \dfrac{\sin 3x}{3x}}{5 \cdot \dfrac{\sin 5x}{5x}}$$

$$= \frac{3 \cdot 1}{5 \cdot 1} = \frac{3}{5} \quad \blacktriangleleft$$

☐ **LIMITS OF** $\sin x$ **AND** $\cos x$ **AS** $x \to +\infty$ **OR** $x \to -\infty$

As $x \to +\infty$ or $x \to -\infty$, the values of $\sin x$ and $\cos x$ oscillate repeatedly between -1 and 1 without approaching any fixed real value. Thus, the limits

$$\lim_{x \to +\infty} \sin x, \qquad \lim_{x \to -\infty} \sin x, \qquad \lim_{x \to +\infty} \cos x, \qquad \lim_{x \to -\infty} \cos x$$

do not exist. We shall say that they *fail to exist due to oscillation*.

▶ **Exercise Set 2.8**   Ⓒ 51, 52, 53, 54

In Exercises 1–10, find the points of discontinuity, if any.

1. $f(x) = \sin(x^2 - 2)$.   **2.** $f(x) = \cos\left(\dfrac{x}{x - \pi}\right)$.

3. $f(x) = \cot x$.   **4.** $f(x) = \sec x$.

5. $f(x) = \csc x$.   **6.** $f(x) = \dfrac{1}{1 + \sin^2 x}$.

7. $f(x) = |\cos x|$.   **8.** $f(x) = \sqrt{2 + \tan^2 x}$.

9. $f(x) = \dfrac{1}{1 - 2 \sin x}$.   **10.** $f(x) = \dfrac{3}{5 + 2 \cos x}$.

11. Prove that $\sin(g(x))$ is continuous at every point where $g(x)$ is continuous.

12. Use Theorem 2.7.6 to prove that the following functions are continuous.
   (a) $\sin(x^3 + 7x + 1)$ (b) $|\sin x|$
   (c) $\cos^3(x + 1)$ (d) $\sqrt{3 + \sin 2x}$.

Find the limits in Exercises 13–35.

13. $\lim\limits_{x \to +\infty} \cos\left(\dfrac{1}{x}\right)$.   **14.** $\lim\limits_{x \to +\infty} \sin\left(\dfrac{2}{x}\right)$.

15. $\lim\limits_{x \to +\infty} \sin\left(\dfrac{\pi x}{2 - 3x}\right)$.   **16.** $\lim\limits_{h \to 0} \dfrac{\sin h}{2h}$.

17. $\lim\limits_{\theta \to 0} \dfrac{\sin 3\theta}{\theta}$.   **18.** $\lim\limits_{\theta \to 0^+} \dfrac{\sin \theta}{\theta^2}$.

19. $\lim\limits_{x \to 0^-} \dfrac{\sin x}{|x|}$.   **20.** $\lim\limits_{x \to 0} \dfrac{\sin^2 x}{3x^2}$.

21. $\lim\limits_{x \to 0^+} \dfrac{\sin x}{5\sqrt{x}}$.   **22.** $\lim\limits_{x \to 0} \dfrac{\sin 6x}{\sin 8x}$.

23. $\lim\limits_{x \to 0} \dfrac{\tan 7x}{\sin 3x}$.   **24.** $\lim\limits_{\theta \to 0} \dfrac{\sin^2 \theta}{\theta}$.

25. $\lim\limits_{h \to 0} \dfrac{h}{\tan h}$.   **26.** $\lim\limits_{h \to 0} \dfrac{\sin h}{1 - \cos h}$.

27. $\lim\limits_{\theta \to 0} \dfrac{\theta^2}{1 - \cos \theta}$.   **28.** $\lim\limits_{x \to 0} \dfrac{x}{\cos\left(\frac{1}{2}\pi - x\right)}$.

29. $\lim\limits_{\theta \to 0} \dfrac{\theta}{\cos \theta}$.   **30.** $\lim\limits_{t \to 0} \dfrac{t^2}{1 - \cos^2 t}$.

31. $\lim\limits_{h \to 0} \dfrac{1 - \cos 5h}{\cos 7h - 1}$.   **32.** $\lim\limits_{x \to 0^+} \sin\left(\dfrac{1}{x}\right)$.

33. $\lim\limits_{x \to 0^+} \cos\left(\dfrac{1}{x}\right)$.   **34.** $\lim\limits_{x \to 0} \dfrac{x^2 - 3 \sin x}{x}$.

35. $\lim\limits_{x \to 0} \dfrac{2x + \sin x}{x}$.

36. Find a value for the constant k so that

$$f(x) = \begin{cases} \dfrac{\sin 3x}{x}, & x \neq 0 \\ k, & x = 0 \end{cases}$$

will be continuous at $x = 0$.

37. Find a nonzero value for the constant k so that

$$f(x) = \begin{cases} \dfrac{\tan kx}{x}, & x < 0 \\[2ex] 3x + 2k^2, & x \geq 0 \end{cases}$$

will be continuous at $x = 0$.

38. Is

$$f(x) = \begin{cases} \dfrac{\sin x}{|x|}, & x \neq 0 \\[2ex] 1, & x = 0 \end{cases}$$

continuous at $x = 0$?

39. In each part, find the limit by making the indicated substitution.

(a) $\displaystyle\lim_{x \to +\infty} x \sin\frac{1}{x}$. $\left[\text{Let } t = \dfrac{1}{x}. \right]$

(b) $\displaystyle\lim_{x \to -\infty} x \left(1 - \cos\frac{1}{x} \right)$. $\left[\text{Let } t = \dfrac{1}{x}. \right]$

(c) $\displaystyle\lim_{x \to \pi} \frac{\pi - x}{\sin x}$. [Let $t = \pi - x$.]

40. Find $\displaystyle\lim_{x \to 2} \frac{\cos (\pi/x)}{x - 2}$. $\left[\text{Hint: Let } t = \dfrac{\pi}{2} - \dfrac{\pi}{x}. \right]$

41. Find $\displaystyle\lim_{x \to 1} \frac{\sin (\pi x)}{x - 1}$.

42. Find $\displaystyle\lim_{x \to \pi/4} \frac{\tan x - 1}{x - \pi/4}$.

43. Use the Squeezing Theorem to find the following limits.

(a) $\displaystyle\lim_{x \to +\infty} \frac{\sin x}{x}$ (b) $\displaystyle\lim_{x \to +\infty} \frac{\cos x}{x}$.

44. Use the Squeezing Theorem to find

$$\lim_{x \to 0^+} x \sin (1/x).$$

45. Let f be a function that satisfies

$$1 - x^2 \leq f(x) \leq \cos x$$

for all x in $(-\pi/2, \pi/2)$. Does $\displaystyle\lim_{x \to 0} f(x)$ exist? If so, find the limit. If not, explain why.

46. Let

$$f(x) = \begin{cases} 1 & \text{if } x \text{ is a rational number} \\ 0 & \text{if } x \text{ is an irrational number} \end{cases}$$

Use the Squeezing Theorem to prove $\displaystyle\lim_{x \to 0} xf(x) = 0$.

47. Prove: If there are constants L and M such that

$$L \leq f(x) \leq M$$

for all x in some open interval containing 0, with the possible exception that the inequalities may not hold at 0, then

$$\lim_{x \to 0} xf(x) = 0$$

48. Draw pictures analogous to Figure 2.8.7 that illustrate the Squeezing Theorem for limits of the form $\displaystyle\lim_{x \to +\infty} f(x)$ and $\displaystyle\lim_{x \to -\infty} f(x)$.

49. Prove that (7) holds if $-\pi/2 < h < 0$ by substituting $-h$ for h in (7).

50. Prove: If θ is in degrees, then

$$\lim_{\theta \to 0} \frac{\sin \theta}{\theta} = \frac{\pi}{180}$$

51. It follows from Theorem 2.8.3 that if θ is small (near zero) and measured in radians, then one should expect the approximation

$$\sin \theta \approx \theta$$

to be good.
(a) Find $\sin 10°$ using a calculator.
(b) Find $\sin 10°$ using the approximation above.

52. (a) Use the approximation of $\sin \theta$ that is given in Exercise 51 and the trigonometric identity $\cos 2\alpha = 1 - 2 \sin^2 \alpha$ with $\alpha = \theta/2$ to show that if θ is small (near zero) and measured in radians, then one should expect the approximation

$$\cos \theta \approx 1 - \frac{1}{2} \theta^2$$

to be good.
(b) Find $\cos 10°$ using a calculator.
(c) Find $\cos 10°$ using the approximation above.

53. It follows from Example 3 that if θ is small (near zero) and measured in radians, then one should expect the approximation

$$\tan \theta \approx \theta$$

to be good.
(a) Find $\tan 5°$ using a calculator.
(b) Find $\tan 5°$ using the approximation above.

54. Referring to Figure 2.8.9, suppose that the angle of elevation of the top of a building, as measured from a point L ft from its base, is found to be α degrees.

(a) Use the relationship $h = L \tan \alpha$ to calculate the height of a building for which $L = 500$ ft and $\alpha = 6°$.

(b) Show that if L is large compared to the building height h, then one should expect good results in approximating h by

$$h \approx \frac{\pi L \alpha}{180}$$

(c) Use the result in part (b) to approximate the building height h in part (a).

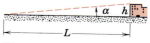

Figure 2.8.9

In Exercises 55 and 56, show that the equation has at least one solution in the given interval.

55. $x = \cos x$; $[0, \pi/2]$.

56. $x + \sin x = 1$; $[0, \pi/6]$.

▶ SUPPLEMENTARY EXERCISES

In Exercises 1–5, find the natural domain of f and then evaluate f (if defined) at the given values of x.

1. $f(x) = \sqrt{4 - x^2}$; $x = -\sqrt{2}, 0, \sqrt{3}$.

2. $f(x) = 1/\sqrt{(x - 1)^3}$; $x = 0, 1, 2$.

3. $f(x) = (x - 1)/(x^2 + x - 2)$; $x = 0, 1, 2$.

4. $f(x) = \sqrt{|x| - 2}$; $x = -3, 0, 2$.

5. $f(x) = \begin{cases} x^2 - 1, & x \le 2 \\ \sqrt{x - 1}, & x > 2 \end{cases}$; $x = 0, 2, 4$.

In Exercises 6 and 7, find
 (a) $f(x^2) - (f(x))^2$
 (b) $f(x + 3) - [f(x) + f(3)]$
 (c) $f(1/x) - 1/f(x)$
 (d) $(f \circ f)(x)$.

6. $f(x) = \sqrt{3 - x}$.

7. $f(x) = \dfrac{3 - x}{x}$.

In Exercises 8–15, sketch the graph of f and find its domain and range.

8. $f(x) = (x - 2)^2$.

9. $f(x) = -\pi$.

10. $f(x) = |2 - 4x|$.

11. $f(x) = \dfrac{x^2 - 4}{2x + 4}$.

12. $f(x) = \sqrt{-2x}$.

13. $f(x) = -\sqrt{3x + 1}$.

14. $f(x) = 2 - |x|$.

15. $f(x) = \dfrac{2x - 4}{x^2 - 4}$.

16. In each part, complete the square, and then find the range of f.

(a) $f(x) = x^2 - 5x + 6$

(b) $f(x) = -3x^2 + 12x - 7$.

17. Express $f(x)$ as a composite function $(g \circ h)(x)$ in two different ways.

(a) $f(x) = x^6 + 3$

(b) $f(x) = \sqrt{x^2 + 1}$

(c) $f(x) = \sin(3x + 2)$.

18. Find $\displaystyle\lim_{x \to k} \frac{x^3 - kx^2}{x^2 - k^2}$, where k is a constant.

In Exercises 19 and 20, sketch the graph of f and find the indicated limits of $f(x)$ (if they exist).

19.

$$f(x) = \begin{cases} 1/x, & x < 0 \\ x^2, & 0 \le x < 1 \\ 2, & x = 1 \\ 2 - x, & x > 1 \end{cases}$$

(a) as $x \to -1$ (b) as $x \to 0$

(c) as $x \to 1$ (d) as $x \to 0^+$

(e) as $x \to 0^-$ (f) as $x \to 2^+$

(g) as $x \to -\infty$ (h) as $x \to +\infty$.

20.

$$f(x) = \begin{cases} 2, & x \le -1 \\ -x, & -1 < x < 0 \\ x/(2 - x), & 0 < x < 2 \\ 1, & x \ge 2 \end{cases}$$

(a) as $x \to -1^+$ (b) as $x \to -1^-$

(c) as $x \to -1$ (d) as $x \to 0$

(e) as $x \to 2^+$ (f) as $x \to 2^-$

(g) as $x \to 2$ (h) as $x \to -\infty$.

In Exercises 21–24, find $\lim\limits_{x \to a} f(x)$ (if it exists).

21. $f(x) = \sqrt{2 - x}$;
$a = -2, 1, 2^-, 2^+, -\infty, +\infty$.

22. $f(x) = \begin{cases} (x - 2)/|x - 2|, & x \neq 2 \\ 0, & x = 2 \end{cases}$
$a = 0, 2^-, 2^+, 2, -\infty, +\infty$.

23. $f(x) = (x^2 - 25)/(x - 5)$;
$a = 0, 5^+, -5^-, 5, -5, -\infty, +\infty$.

24. $f(x) = (x + 5)/(x^2 - 25)$;
$a = 0, 5^+, -5^-, -5, 5, -\infty, +\infty$.

In Exercises 25–32, find the indicated limit if it exists.

25. $\lim\limits_{x \to 0} \dfrac{\tan ax}{\sin bx}$ $(a \neq 0, b \neq 0)$.

26. $\lim\limits_{x \to 0} \dfrac{\sin 3x}{\tan 3x}$.

27. $\lim\limits_{\theta \to 0} \dfrac{\sin 2\theta}{\theta^2}$.

28. $\lim\limits_{x \to 0} \dfrac{x \sin x}{1 - \cos x}$.

29. $\lim\limits_{x \to 0^+} \dfrac{\sin x}{\sqrt{x}}$.

30. $\lim\limits_{x \to 0} \dfrac{\sin^2(kx)}{x^2}$, $k \neq 0$.

31. $\lim\limits_{x \to 0} \dfrac{3x - \sin(kx)}{x}$, $k \neq 0$.

32. $\lim\limits_{x \to +\infty} \dfrac{2x + x \sin 3x}{5x^2 - 2x + 1}$.

3
Differentiation

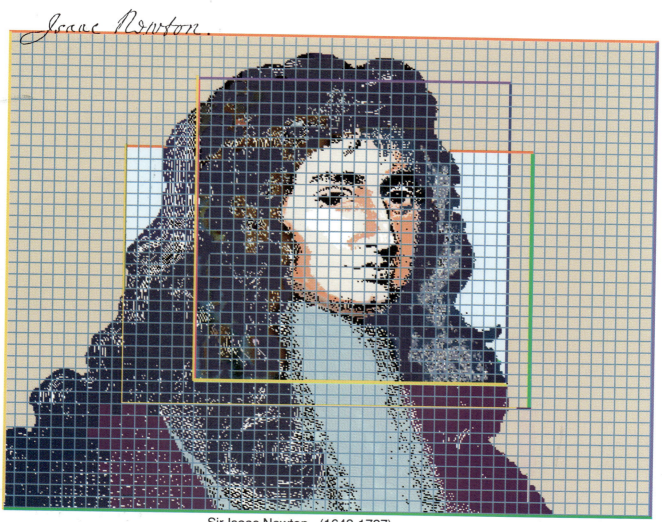

Sir Isaac Newton   (1642-1727)

◼ 3.1 TANGENT LINES AND RATES OF CHANGE

> *Many physical phenomena involve changing quantities—the speed of a rocket, the inflation of currency, the number of bacteria in a culture, the shock intensity of an earthquake, the voltage of an electrical signal, and so forth. In this section we shall establish a basic relationship between tangent lines and rates of change.*

□ **DEFINITION OF A TANGENT LINE**

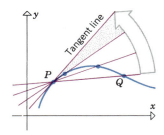

Figure 3.1.1

In Section 2.4 we observed informally that if a secant line is drawn between two points P and Q on a curve, and Q is allowed to move along the curve toward P, then we can expect the secant line to rotate toward a "limiting position" which can be regarded as the tangent line to the curve at the point P (Figure 3.1.1). Our first objective in this section is to make this informal idea mathematically precise.

For the moment we shall only consider lines tangent to curves of the form $y = f(x)$. If $P(x_0, y_0)$ and $Q(x_1, y_1)$ are distinct points on such a curve, then the secant line connecting P and Q has slope

$$m_{\text{sec}} = \frac{f(x_1) - f(x_0)}{x_1 - x_0} \tag{1}$$

(See Figure 3.1.2a.) If we let x_1 approach x_0, then Q will approach P along the graph of f, and the secant line through P and Q will approach the tangent line at P. Thus, the slope m_{sec} of the secant line approaches the slope m_{tan} of the tangent line as x_1 approaches x_0. Therefore, from (1)

$$m_{\text{tan}} = \lim_{x_1 \to x_0} \frac{f(x_1) - f(x_0)}{x_1 - x_0} \tag{2}$$

For many purposes it is desirable to rewrite this expression in an alternative form by letting

$$h = x_1 - x_0$$

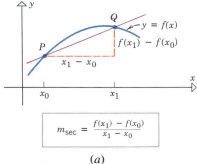

$$m_{\text{sec}} = \frac{f(x_1) - f(x_0)}{x_1 - x_0}$$

(a)

$$m_{\text{sec}} = \frac{f(x_0+h) - f(x_0)}{h}$$

(b)

Figure 3.1.2

Thus (see Figure 3.1.2b), $x_1 = x_0 + h$ and $h \to 0$ as $x_1 \to x_0$, so (2) can be rewritten as

$$m_{\tan} = \lim_{h \to 0} \frac{f(x_0 + h) - f(x_0)}{h}$$

Motivated by the foregoing discussion, we make the following definition.

3.1.1 DEFINITION. If $P(x_0, y_0)$ is a point on the graph of a function f, then the *tangent line* to the graph of f at P is defined to be the line through P with slope

$$m_{\tan} = \lim_{h \to 0} \frac{f(x_0 + h) - f(x_0)}{h} \tag{3}$$

provided this limit exists.

For brevity, the tangent line at $P(x_0, y_0)$ is often called the *tangent line at x_0.* It follows from the foregoing definition that the point-slope form of the equation of the tangent line at x_0 is

$$y - y_0 = m_{\tan}(x - x_0) \tag{4}$$

REMARK. It is possible that the limit in (3) may not exist, in which case $m_{\tan}$ will be undefined. We shall consider the geometric significance of this in later sections.

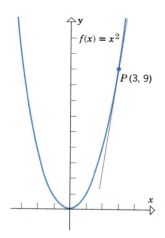

Figure 3.1.3

Example 1 Find the slope and an equation of the tangent line to the graph of $f(x) = x^2$ at the point $P(3, 9)$. (See Figure 3.1.3.)

Solution. We have $x_0 = 3$ and $y_0 = 9$, so from (3)

$$m_{\tan} = \lim_{h \to 0} \frac{f(3 + h) - f(3)}{h} = \lim_{h \to 0} \frac{(3 + h)^2 - 9}{h}$$

$$= \lim_{h \to 0} \frac{(9 + 6h + h^2) - 9}{h} = \lim_{h \to 0} \frac{6h + h^2}{h}$$

$$= \lim_{h \to 0} (6 + h) = 6$$

Thus, from (4) the point-slope form of the tangent line is

$$y - 9 = 6(x - 3)$$

and the slope-intercept form is $y = 6x - 9$. ◀

In later problems we shall want to know how the slope of the tangent line varies from point to point along a curve $y = f(x)$. For this purpose we shall need a formula that produces the slope of the tangent line at an arbitrary point

$P(x, y)$ on the curve. Such a formula can be obtained by replacing the constant x_0 in (3) by the variable x. This yields

$$m_{\tan} = \lim_{h \to 0} \frac{f(x + h) - f(x)}{h} \tag{5}$$

The limit in (5) occurs so frequently that it has a special notation. We denote it by

$$f'(x) = \lim_{h \to 0} \frac{f(x + h) - f(x)}{h} \tag{6}$$

where $f'(x)$ is read, "f prime of x." This notation emphasizes the fact that the slope of the tangent line at x is a function of x that is "derived" from the function f.

Example 2   Let $f(x) = x^2 + 1$.

(a)   Find $f'(x)$.

(b)   Use the result in part (a) to find the slope of the tangent line to $y = x^2 + 1$ at $x = 2$, $x = 0$, and $x = -2$.

Solution (a).   From (6)

$$f'(x) = \lim_{h \to 0} \frac{f(x + h) - f(x)}{h} = \lim_{h \to 0} \frac{[(x + h)^2 + 1] - [x^2 + 1]}{h}$$

$$= \lim_{h \to 0} \frac{x^2 + 2xh + h^2 + 1 - x^2 - 1}{h} = \lim_{h \to 0} \frac{2xh + h^2}{h}$$

$$= \lim_{h \to 0} (2x + h) = 2x$$

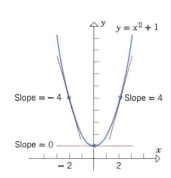

Figure 3.1.4

Solution (b).   From part (a) the slope of the tangent line at any point x is

$$f'(x) = 2x$$

Thus, at $x = 2$, $x = 0$, and $x = -2$ the slopes are

$$f'(2) = 2(2) = 4, \qquad f'(0) = 2(0) = 0, \qquad \text{and} \qquad f'(-2) = 2(-2) = -4$$

respectively (Figure 3.1.4).   ◀

Example 3   It is obvious geometrically that at each point on a line $y = mx + b$, the tangent line coincides with the line itself and thus has slope m (Figure 3.1.5). Therefore, if $f(x) = mx + b$, we should anticipate that $f'(x) = m$ for all x. The following computation shows that this is so.

$$f'(x) = \lim_{h \to 0} \frac{f(x + h) - f(x)}{h} = \lim_{h \to 0} \frac{[m(x + h) + b] - (mx + b)}{h}$$

$$= \lim_{h \to 0} \frac{mx + mh + b - mx - b}{h} = \lim_{h \to 0} \frac{mh}{h} = \lim_{h \to 0} m = m \qquad ◀$$

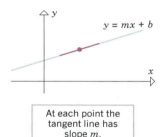

At each point the tangent line has slope m.

Figure 3.1.5

□ AVERAGE AND INSTANTANEOUS VELOCITY

While tangent lines are of interest as a matter of pure geometry, much of the impetus for studying them arose in the seventeenth century when scientists recognized that many problems involving objects moving with varying velocity could be reduced to problems involving tangents. To see why this is so, we need to examine critically the meaning of the word "velocity."

If a car travels 75 miles over a straight road in a 3-hour period, then we say that the average velocity of the car is 25 miles per hour (25 mi/hr). More generally, the *average velocity* of an object moving in *one direction* along a line is

$$\text{average velocity} = \frac{\text{distance traveled}}{\text{time elapsed}}$$

Obviously, if an object travels with an average velocity of 25 mi/hr during a 3-hour trip, it need not travel at a fixed velocity of 25 mi/hr; sometimes it may speed up and sometimes it may slow down.

Although average velocity is useful for some purposes, it is not always significant in physical problems. For example, if a moving car strikes a tree, the damage is not determined by the average velocity up to the time of impact, but rather by the *instantaneous velocity* at the precise moment of impact.

A clear understanding of instantaneous velocity evaded scientists until the advent of calculus in the seventeenth century. The subtlety of this concept was nicely described by Morris Kline* who wrote,

> *In contrasting average velocity with instantaneous velocity we implicitly utilize a distinction between interval and instant.... An average velocity is one that concerns what happens over an interval of time—3 hours, 5 seconds, one-half second, and so forth. The interval may be small or large, but it does represent the passage of a definite amount of time. We use the word instant, however, to state the fact that something happens so fast that no time elapses. The event is momentary. When we say, for example, that it is 3 o'clock, we refer to an instant, a precise moment. If the lapse of time is pictured by length along a line, then an interval (of time) is represented by a line segment, whereas an instant corresponds to a point. The notion of an instant, although it is used in everyday life, is strictly a mathematical idealization.*
>
> *Our ways of thinking about real events cause us to speak in terms of instants and velocity at an instant, but closer examination shows that the concept of velocity at an instant presents difficulties. Average velocity, which is simply the distance traveled during some interval of time divided by that amount of time, is easily calculated. Suppose, however, that we try to carry over this process to instantaneous velocity. The distance an automobile travels in one instant is 0 and the time that elapses during one instant is also 0. Hence the distance divided by the time is 0/0, which is meaningless. Thus, although instantaneous velocity is a physical reality, there seems to be a difficulty in calculating it, and unless we can calculate it, we cannot work with it mathematically.*

---

*MORRIS KLINE (1908–   ). American mathematician, scholar, and educator. Kline has made numerous contributions to mathematical thought, written extensively on education, especially mathematics education, and has taught, lectured, and served as a consultant throughout his very active career. He is the author of many popular books including *Mathematical Thought from Ancient to Modern Times* and *Why Johnny Can't Add: The Failure of the New Mathematics*. I wish to thank him for permission to use the above quotation, which is taken from *Calculus: An Intuitive and Physical Approach*. Wiley, New York, 1977, p. 17.

In order to understand the concept of instantaneous velocity and work with it mathematically we must think in terms of approximating the instantaneous velocity by the average velocity. For example, suppose we are interested in the instantaneous velocity of a car at a certain instant of time, say, exactly 5 sec after it starts to move along a straight road. Although the velocity of the car may be changing, it is evident that over a short interval of time, say, 0.1 sec, the velocity does not vary much. Thus, we make only a small error if we approximate the instantaneous velocity after 5 sec by the *average velocity* over the *interval* from 5 to 5.1 sec. This average velocity can be calculated by measuring the distance traveled during the time interval between 5 and 5.1 sec and then dividing by the time elapsed, which is 0.1 sec. Thus, even though the instantaneous velocity after 5 sec cannot be calculated directly, it can be approximated by an average velocity that can be determined from physical measurements.

Now let us look at these ideas geometrically. Suppose that the car is moving along a straight road and a coordinate line is introduced along the road with its positive direction in the direction of motion of the car. Imagine that a clock is keeping track of the elapsed time, starting with $t = 0$ initially, and that after t seconds have elapsed the car is at a distance of s units from the origin (Figure 3.1.6).

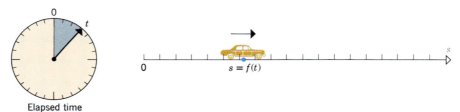

Elapsed time

Figure 3.1.6

Because s changes with t, the position coordinate s is some function of t, say $s = f(t)$. If we graph this function with the t-axis horizontal and the s-axis vertical, we obtain a *position versus time curve* (Figure 3.1.7). Using this curve, we can interpret average velocity and instantaneous velocity geometrically. For example, suppose we are interested in the average velocity of the car over the time interval from t_0 to t_1. At time t_0 the car is at some distance s_0 from the origin and at time t_1 it is at some distance s_1. Over the time interval from t_0 to t_1 the distance traveled is $s_1 - s_0$ and the time elapsed is $t_1 - t_0$ so that the average velocity of the car during the interval is given by

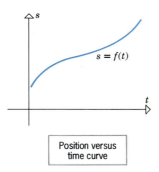

Position versus time curve

Figure 3.1.7

$$\begin{bmatrix} \text{average velocity} \\ \text{from times} \\ t_0 \text{ to } t_1 \end{bmatrix} = \frac{s_1 - s_0}{t_1 - t_0} \tag{7}$$

However, the points (t_0, s_0) and (t_1, s_1) lie on the position versus time curve, so that expression (7) is also the slope of the secant line connecting these points (Figure 3.1.8). Thus, we conclude:

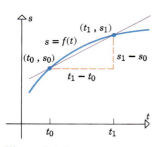

Figure 3.1.8

> ### Geometric Interpretation of Average Velocity
> The average velocity of the car between time t_0 and time t_1 is represented geometrically by the slope of the secant line connecting (t_0, s_0) and (t_1, s_1) on the position versus time curve.

If we choose t_1 close to t_0, then the average velocity between times t_0 and t_1 closely approximates the instantaneous velocity at time t_0; moreover, as we move t_1 closer and closer to t_0, intuition suggests that these approximations will approach the exact value of the instantaneous velocity at t_0. However, as t_1 moves toward t_0, the point (t_1, s_1) moves toward (t_0, s_0) on the position versus time curve (Figure 3.1.9), so that the slope of the secant line between (t_0, s_0) and (t_1, s_1) approaches the slope of the tangent line at (t_0, s_0). Thus, we conclude:

> *Geometric Interpretation of Instantaneous Velocity*
> The instantaneous velocity of the car at time t_0 is represented geometrically by the slope of the tangent line at (t_0, s_0) on the position versus time curve.

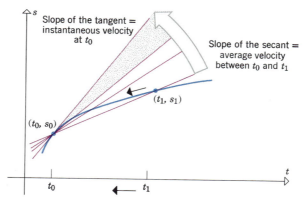

Figure 3.1.9

□ **AVERAGE AND INSTANTANEOUS RATES OF CHANGE**

Velocity can be viewed as a *rate of change*—the rate of change of position with time, or in algebraic terms, the rate of change of s with t. Rates of change occur in many applications. For example:

- A microbiologist might be interested in the rate at which the number of bacteria in a colony changes with time.
- An engineer might be interested in the rate at which the length of a metal rod changes with temperature.
- An economist might be interested in the rate at which production cost changes with the quantity of a product that is manufactured.
- A medical researcher might be interested in the rate at which the radius of an artery changes with the concentration of alcohol in the bloodstream.

In general, if x and y are any quantities related by an equation $y = f(x)$, we can consider the rate at which y varies with x. As with velocity, we distinguish between an average rate of change represented by the slope of a secant line and an instantaneous rate of change represented by the slope of the tangent line. More precisely, we make the following definitions:

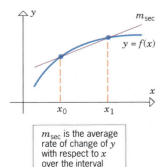

m_{sec} is the average rate of change of y with respect to x over the interval $[x_0, x_1]$.

(a)

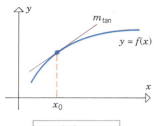

m_{tan} is the instantaneous rate of change of y with respect to x at the point x_0.

(b)

Figure 3.1.10

3.1.2   DEFINITION.   If $y = f(x)$, then the **average rate of change of y with respect to x over the interval $[x_0, x_1]$** is the slope m_{sec} of the secant line joining the points $(x_0, f(x_0))$ and $(x_1, f(x_1))$ on the graph of f (Figure 3.1.10a).

3.1.3   DEFINITION.   If $y = f(x)$, then the **instantaneous rate of change of y with respect to x at the point x_0,** is the slope m_{tan} of the tangent line to the graph of f at the point x_0 (Figure 3.1.10b).

If $y = f(x)$, then the average and instantaneous rates of change of y with respect to x are given by the following formulas (Figure 3.1.11):

$$m_{sec} = \frac{f(x_1) - f(x_0)}{x_1 - x_0} \tag{8}$$

$$m_{tan} = f'(x_0) = \lim_{x_1 \to x_0} \frac{f(x_1) - f(x_0)}{x_1 - x_0} \tag{9}$$

Consider the curve $y = f(x)$ in Figure 3.1.12. Traveling in the positive direction over the interval $[1, 9]$, the curve increases slowly at first, then more rapidly, then more slowly again. However, at the end of the trip y has increased by 6 units and x by 8 units, so the average rate of change of y with respect to x over the interval is

$$m_{sec} = \frac{6}{8} = \frac{3}{4}$$

Stated informally, "on the average" y increases 3/4 of a unit for each 1-unit increase in x over the interval $[1, 9]$.

The slopes of the tangent lines at $x = 5$ and $x = 7$ can be calculated using any two points on those lines. If we denote the slopes of those lines by m_1 and m_2, respectively, then m_1 can be calculated using the points $(3, 0)$ and $(6, 8)$; and m_2 can be calculated using the points $(0, 4)$ and $(9, 9)$. This yields

$$m_1 = \frac{8 - 0}{6 - 3} = \frac{8}{3} \quad \text{and} \quad m_2 = \frac{9 - 4}{9 - 0} = \frac{5}{9}$$

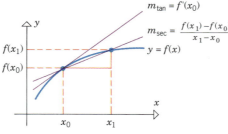

Figure 3.1.11

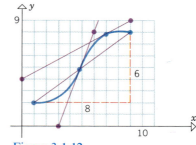

Figure 3.1.12

Note that the rate of change of y with respect to x is greater at $x = 5$ than at $x = 7$. Moreover, the instantaneous rate of change at $x = 5$ is greater than the average rate of change over the interval $[1, 9]$, and the instantaneous rate of change at $x = 7$ is less than the average rate of change over the interval $[1, 9]$. These results can be visualized from Figure 3.1.12 by comparing the inclinations of the tangent lines and the secant line.   ◀

Example 5   Suppose that $y = x^2 + 1$.

(a)   Find the average rate of change of y with respect to x over the interval $[3, 5]$.

(b)   Find the instantaneous rate of change of y with respect to x at the point $x = -4$.

Solution (a).   We will apply Formula (8) with $f(x) = x^2 + 1$, $x_0 = 3$, and $x_1 = 5$. This yields

$$m_{\text{sec}} = \frac{f(x_1) - f(x_0)}{x_1 - x_0} = \frac{f(5) - f(3)}{5 - 3} = \frac{26 - 10}{5 - 3} = 8$$

Thus, on the average, y increases 8 units per unit increase in x over the interval $[3, 5]$.

Solution (b).   We will apply Formula (9) with $f(x) = x^2 + 1$ and $x_0 = -4$. In Example 2 we found $f'(x) = 2x$, so

$$m_{\text{tan}} = f'(x_0) = f'(-4) = 2(-4) = -8$$

Because the instantaneous rate of change is negative, y is *decreasing* at the point $x = -4$; it is decreasing at a rate of 8 units per unit increase in x.   ◀

☐ **RATES OF CHANGE IN APPLICATIONS**

In applied problems, average and instantaneous rates of change must be accompanied by appropriate units.

Example 6   The limiting factor in athletic endurance is cardiac output. Figure 3.1.13 shows a stress-test graph of cardiac output (measured in liters/min of blood) versus work load (measured in $kg \cdot m/min$). This graph illustrates the known medical fact that cardiac output increases with the work load, but after reaching a peak value it begins to decrease at very high work loads.

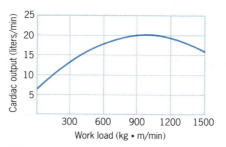

Figure 3.1.13

(a) Estimate the average rate of change of cardiac output with respect to work load as the work load increases from 300 to 1200 kg · m/min.

(b) Estimate the instantaneous rate of change of cardiac output with respect to work load at the point where the work load is 300 kg · m/min.

Solution (a).   Using the estimated points $(300, 13.5)$ and $(1200, 19)$, the slope of the secant line indicated in Figure 3.1.14a is

$$m_{\text{sec}} \approx \frac{19 - 13}{1200 - 300} \approx 0.0066 \; \frac{\text{liters/min}}{\text{kg} \cdot \text{m/min}}$$

On simplifying the expression for the units we conclude that the average rate of change of cardiac output with respect to work load is

$$0.0066 \; \frac{\text{liters}}{\text{kg} \cdot \text{m}}$$

Solution (b).   Using the estimated tangent line in Figure 3.1.14b and the estimated points $(0, 7)$ and $(900, 25)$ on this tangent line, we obtain

$$m_{\text{tan}} \approx \frac{25 - 7}{900 - 0} \approx 0.02 \; \frac{\text{liters/min}}{\text{kg} \cdot \text{m/min}}$$

Thus, the instantaneous rate of change of cardiac output with respect to work load is

$$0.018 \; \frac{\text{liters}}{\text{kg} \cdot \text{m}} \quad \blacktriangleleft$$

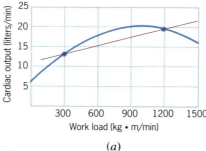

(a)

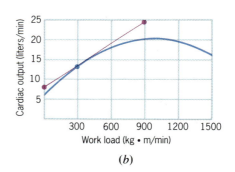

(b)

Figure 3.1.14

▶ Exercise Set 3.1   ⒸC 15, 16, 17

1. Let $f(x) = \frac{1}{2}x^2$.

(a) Find the slope of the secant line between those points on the graph of f for which $x = 3$ and $x = 4$.

(b) Use the method of Example 1 to find the slope and equation for the tangent to the graph of f at the point where $x = 3$.

(c) Sketch the graph of f together with the secant and tangent lines from parts (a) and (b).

2. Let $f(x) = x^3$.

(a) Find the slope of the secant line between those points on the graph of f for which $x = 1$ and $x = 2$.

(b) Use the method of Example 1 to find the slope

and equation for the tangent to the graph of f at the point where $x = 1$.

(c) Sketch the graph of f together with the secant and tangent lines from parts (a) and (b).

3. Let $f(x) = 1/x$.

(a) Find the slope of the secant line between those points on the graph of f for which $x = 2$ and $x = 3$.

(b) Use the method of Example 1 to find the slope and equation for the tangent to the graph of f at the point where $x = 2$.

(c) Sketch the graph of f together with the secant and tangent lines from parts (a) and (b).

4. Let $f(x) = 1/x^2$.

(a) Find the slope of the secant line between those points on the graph of f for which $x = 1$ and $x = 2$.

(b) Use the method of Example 1 to find the slope and equation for the tangent to the graph of f at the point where $x = 1$.

(c) Sketch the graph of f together with the secant and tangent lines from parts (a) and (b).

5. Let $f(x) = x^3$.

(a) Use the method of Example 1 to show that the slope of the tangent to the graph of f at the point where $x = x_0$ is $3x_0^2$.

(b) Use the result in part (a) to find the equation of the tangent to the graph of f at the point where $x = 5$.

(c) Use the result in part (a) to find the equation of the tangent to the graph of f at the point where $x = x_0$.

6. Let $f(x) = 1/x$.

(a) Use the method of Example 1 to show that the slope of the tangent to the graph of f at the point where $x = x_0$ is $-1/x_0^2$.

(b) Use the result in part (a) to find the equation of the tangent to the graph of f at the point where $x = -7$.

(c) Use the result in part (a) to find the equation of the tangent to the graph of f at the point where $x = x_0$.

7. Let $f(x) = x^2 + x$.

(a) Use the method of Example 1 to find the slope of the tangent to the graph of f at the point where $x = x_0$.

(b) Use the result in part (a) to find the equation of the tangent to the graph of f at the point where $x = 2$.

(c) Use the result in part (a) to find the equation of the tangent to the graph of f at the point where $x = x_0$.

8. Follow the directions of Exercise 7 for the function $f(x) = x^2 + 3x + 2$.

9. Figure 3.1.15 shows the position versus time curve for an elevator that moves upward a distance of 60 meters and then discharges its passengers.

(a) Estimate the instantaneous velocity of the elevator at $t = 10$ seconds.

(b) Sketch a velocity versus time curve for the motion of the elevator for $0 \leq t \leq 20$.

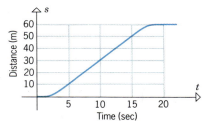

Figure 3.1.15

10. Figure 3.1.16 shows the position versus time curve for a certain particle moving along a straight line. Estimate each of the following from the graph:

(a) the average velocity over the interval $0 \leq t \leq 3$

(b) the values of t at which the instantaneous velocity is zero

(c) the values of t at which the instantaneous velocity is either a maximum or a minimum

(d) the instantaneous velocity when $t = 3$ seconds.

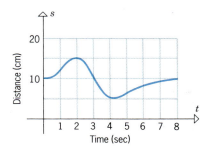

Figure 3.1.16

11. Figure 3.1.17 shows the position versus time curve for a certain particle moving on a straight line.
   (a) Is the particle moving faster at time t_0 or time t_2? Explain.
   (b) At the origin, the tangent is horizontal. What does this tell us about the initial velocity of the particle?
   (c) Is the particle speeding up or slowing down in the interval $[t_0, t_1]$? Explain.
   (d) Is the particle speeding up or slowing down in the interval $[t_1, t_2]$? Explain.

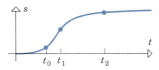

Figure 3.1.17

12. An automobile, initially at rest, begins to move along a straight track. The velocity increases steadily until suddenly the driver sees a concrete barrier in the road and applies the brakes sharply at time t_0. The car decelerates rapidly, but it is too late—the car crashes into the barrier at time t_1 and instantaneously comes to rest. Sketch a position versus time curve that might represent the motion of the car.

13. If a particle moves at constant velocity, what can you say about its position versus time curve?

14. Figure 3.1.18 shows the position versus time curves of four different particles moving on a straight line. For each particle, determine whether its instantaneous velocity is increasing or decreasing with time.

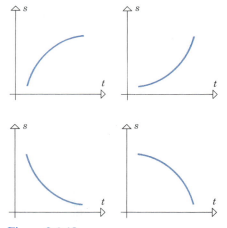

Figure 3.1.18

15. Suppose that the outside temperature versus time curve over a 24-hour period is as shown in Figure 3.1.19.
   (a) Estimate the maximum temperature and the time at which it occurs.
   (b) The temperature rise is fairly linear from 8 A.M. to 2 P.M. Estimate the rate at which the temperature is increasing during this time period.
   (c) Estimate the time at which the temperature is decreasing most rapidly. Estimate the instantaneous rate of change of temperature with respect to time at this instant.

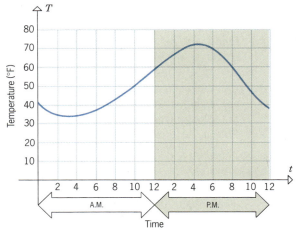

Figure 3.1.19

16. Figure 3.1.20 shows the graph of the pressure p (in atmospheres) versus the volume V (in liters) of 1 mole of an ideal gas at a constant temperature of 300 K (kelvin). Use the tangent lines shown in the figure to estimate the rate of change of pressure with respect to volume at the points where $V = 10$ L and $V = 25$ L.

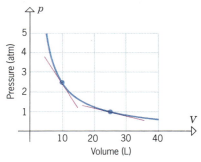

Figure 3.1.20

17. Figure 3.1.21 shows the graph of the height h (in centimeters) versus the age t (in years) of an individual from sometime after birth to age 20.

(a) When is the growth rate greatest?

(b) Estimate the growth rate at age 5.

(c) At approximately what age between 10 and 20 is the growth rate greatest? Estimate the growth rate at this age.

(d) Draw a rough graph of the growth rate versus age.

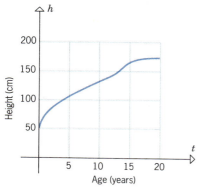

Figure 3.1.21

In Exercises 18–21, use the methods of Examples 1 and 5 with $s = f(t)$ to find the average and instantaneous velocity.

18. A rock is dropped from a height of 576 ft and falls toward earth in a straight line. In t sec the rock drops a distance of $s = 16t^2$ ft.

(a) How many seconds after release does the rock hit the ground?

(b) What is the average velocity of the rock during the time it is falling?

(c) What is the average velocity of the rock for the first 3 sec?

(d) What is the instantaneous velocity of the rock when it hits the ground?

19. During the first 40 sec of a rocket flight, the rocket is propelled straight up so that in t sec it reaches a height of $s = 5t^3$ ft.

(a) How high does the rocket travel in 40 sec?

(b) What is the average velocity of the rocket during the first 40 sec?

(c) What is the average velocity of the rocket during the first 135 ft of its flight?

(d) What is the instantaneous velocity of the rocket at the end of 40 sec?

20. A particle moves on a line away from its initial position so that after t hr it is $s = 3t^2 + t$ mi from its initial position.

(a) Find the average velocity of the particle over the interval $[1, 3]$.

(b) Find the instantaneous velocity at $t = 1$.

21. A particle moves in one direction along a straight line so that after t min its distance is $s = 6t^4$ ft from the origin.

(a) Find the average velocity of the particle over the interval $[2, 4]$.

(b) Find the instantaneous velocity at $t = 2$.

22. A car is traveling on a straight road that is 120 mi long. For the first 100 mi the car travels at an average velocity of 50 mi/hr. Show that no matter how fast the car travels for the final 20 mi it cannot bring the average velocity up to 60 mi/hr for the entire trip.

23. Let $y = 2x^2 - 1$.

(a) Find the average rate at which y changes with x over the interval $[1, 4]$.

(b) Find the instantaneous rate at which y changes with x at the point $x = 1$.

24. Let $y = \dfrac{1}{x^2 + 1}$.

(a) Find the average rate at which y changes with x over the interval $[-1, 2]$.

(b) Find the instantaneous rate at which y changes with x at the point $x = -1$.

25. Use the formula $A = \pi r^2$ for the area of a circle to find

(a) the average rate at which the area of a circle changes with r as the radius increases from $r = 1$ to $r = 2$

(b) the instantaneous rate at which the area changes with r when $r = 2$.

26. Use the formula $V = l^3$ for the volume of a cube of side l to find

(a) the average rate at which the volume of a cube changes with l as l increases from $l = 2$ to $l = 4$

(b) the instantaneous rate at which the volume of a cube changes with l when $l = 5$.

27. Let $f(x) = x^2$. If we approximate the slope of the tangent line at the point $(x_0, f(x_0))$ by the slope of the secant line between $(x_0, f(x_0))$ and $(x_1, f(x_1))$, show that the error is $|x_1 - x_0|$. (By error we mean $|m_{\text{tan}} - m_{\text{sec}}|$, where m_{tan} is the slope of the tangent line and m_{sec} the slope of the secant line.)

■ 3.2 THE DERIVATIVE

The "slope-producing" function f' is of fundamental importance in mathematics. In this section we shall study some of the basic properties of this function.

□ **DEFINITION OF THE DERIVATIVE**

The function f' defined in the previous section is so important that it has its own name.

3.2.1 DEFINITION. The function f' defined by the formula

$$f'(x) = \lim_{h \to 0} \frac{f(x + h) - f(x)}{h} \tag{1}$$

is called the ***derivative with respect to x*** of the function f. The domain of f' consists of all x for which the limit exists.

Based on our discussion in the previous section, the derivative of a function f can be interpreted two ways:

Geometric Interpretation of the Derivative
f' is the function whose value at x is the slope of the tangent line to the graph of f at x.

Rate of Change Interpretation of the Derivative
If $y = f(x)$, then f' is the function whose value at x is the instantaneous rate of change of y with respect to x at the point x.

In Example 2 of Section 3.1, we found the derivative of $f(x) = x^2 + 1$ to be $f'(x) = 2x$. The graphs of f' and f are shown in Figure 3.2.1. As illustrated in the figure, the points

$$(-2, -4), \quad (0, 0), \quad (2, 4)$$

lie on the graph of f'. Thus, the y-coordinates of these points, namely, $y = -4$, 0, and 4, represent the slopes of the tangent lines to the graph of f at the corresponding x-coordinates, $x = -2$, 0, and 2.

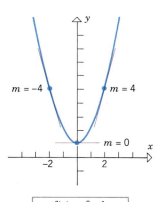

$f'(x) = 2x$

$f(x) = x^2 + 1$

Figure 3.2.1

Example 1 Find the derivative of $f(x) = \sqrt{x}$.

Solution. From Definition 3.2.1,

$$f'(x) = \lim_{h \to 0} \frac{f(x + h) - f(x)}{h} = \lim_{h \to 0} \frac{\sqrt{x + h} - \sqrt{x}}{h}$$

$$= \lim_{h \to 0} \frac{(\sqrt{x + h} - \sqrt{x})(\sqrt{x + h} + \sqrt{x})}{h(\sqrt{x + h} + \sqrt{x})} = \lim_{h \to 0} \frac{(x + h) - x}{h(\sqrt{x + h} + \sqrt{x})}$$

$$= \lim_{h \to 0} \frac{h}{h(\sqrt{x + h} + \sqrt{x})} = \lim_{h \to 0} \frac{1}{\sqrt{x + h} + \sqrt{x}}$$

$$= \frac{1}{\sqrt{x} + \sqrt{x}} = \frac{1}{2\sqrt{x}} \qquad \blacktriangleleft$$

The graphs of the functions

$$f(x) = \sqrt{x} \quad \text{and} \quad f'(x) = \frac{1}{2\sqrt{x}}$$

from the foregoing example are shown in Figure 3.2.2. Observe that

$$\lim_{x \to 0^+} \frac{1}{2\sqrt{x}} = +\infty$$

This implies that the slopes of the tangent lines to the graph of f increase without bound as x approaches zero from the right. (Can you see that this is so from the graph of f?)

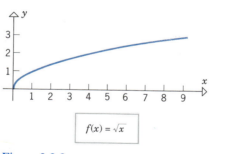

$f(x) = \sqrt{x}$

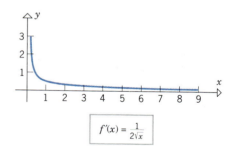

$f'(x) = \frac{1}{2\sqrt{x}}$

Figure 3.2.2

Example 2 Let $y = \sqrt{x}$, and use the result in Example 1 to find

(a) the slope of the tangent line to the graph of this equation at $x = 9$;

(b) the instantaneous rate of change of y with respect to x at $x = 5$.

Solution. If we let $f(x) = \sqrt{x}$, then $f'(x) = 1/(2\sqrt{x})$, so the slope of the tangent line at $x = 9$ is $f'(9) = 1/(2\sqrt{9}) = \frac{1}{6}$, and the instantaneous rate of change of y with respect to x at $x = 5$ is $f'(5) = 1/(2\sqrt{5}) = \sqrt{5}/10$. $\blacktriangleleft$

Example 3 Find the derivative of $f(x) = \dfrac{x}{x - 9}$.

Solution.

$$f'(x) = \lim_{h \to 0} \frac{f(x + h) - f(x)}{h}$$

$$= \lim_{h \to 0} \frac{1}{h} \left[\frac{x + h}{x + h - 9} - \frac{x}{x - 9} \right]$$

$$= \lim_{h \to 0} \frac{1}{h} \left[\frac{(x + h)(x - 9) - x(x + h - 9)}{(x + h - 9)(x - 9)} \right]$$

$$= \lim_{h \to 0} \frac{(x^2 - 9x + hx - 9h) - (x^2 + hx - 9x)}{h(x + h - 9)(x - 9)}$$

$$= \lim_{h \to 0} \frac{-9h}{h(x + h - 9)(x - 9)}$$

$$= \lim_{h \to 0} \frac{-9}{(x + h - 9)(x - 9)} = -\frac{9}{(x - 9)^2} \quad \blacktriangleleft$$

☐ **DERIVATIVE NOTATION**

In the early days of calculus there was vitriolic debate over the proper notation for the derivative. The English mathematicians advocated the notation of Isaac Newton and the Continental mathematicians advocated the notation of Leibniz. Today, there is a variety of different notations in common use, all of which have useful purposes. We now discuss some of these.

The process of finding a derivative is called *differentiation.* It is often useful to think of differentiation as an operation that, when applied to a function f, produces a new function f'. In the case where the independent variable is x, the differentiation operation is often denoted by the symbol

$$\frac{d}{dx} [f(x)]$$

which is read, "*the derivative of f(x) with respect to x.*" Thus,

$$\frac{d}{dx} [f(x)] = f'(x) \tag{2}$$

With this notation, the results in Examples 2 and 3 can be written as

$$\frac{d}{dx} [\sqrt{x}] = \frac{1}{2\sqrt{x}} \quad \text{and} \quad \frac{d}{dx} \left[\frac{x}{x - 9} \right] = -\frac{9}{(x - 9)^2} \tag{3}$$

If there is a dependent variable $y = f(x)$ then (2) can be written as

$$\frac{d}{dx} [y] = f'(x) \tag{4}$$

However, it is usual to omit the brackets on the left side and write more simply

$$\frac{dy}{dx} = f'(x) \tag{5}$$

For example, if $y = \sqrt{x}$, then the first result in (3) can be written as

$$\frac{dy}{dx} = \frac{1}{2\sqrt{x}} \tag{6}$$

REMARK.   Later, the symbols dy and dx will be defined separately. However, for the time being, dy/dx should not be regarded as a ratio; rather, it should be considered as a single symbol denoting the derivative.

If $y = f(x)$, then the value of the derivative of a specific point $x = x_0$ can be denoted as

$$\frac{d}{dx}[f(x)]\bigg|_{x=x_0} \quad \text{or} \quad \frac{dy}{dx}\bigg|_{x=x_0}$$

With these notations, it follows from (2) and (5) that

$$\frac{d}{dx}[f(x)]\bigg|_{x=x_0} = f'(x_0) \tag{7}$$

$$\frac{dy}{dx}\bigg|_{x=x_0} = f'(x_0) \tag{8}$$

For example, if $f(x) = \sqrt{x}$, then from (3) and (7)

$$\frac{d}{dx}[\sqrt{x}]\bigg|_{x=x_0} = \frac{1}{2\sqrt{x_0}}$$

and if $y = \dfrac{x}{x-9}$ then from (3) and (8)

$$\frac{dy}{dx}\bigg|_{x=3} = -\frac{9}{(3-9)^2} = -\frac{1}{4}$$

If the independent variable is not x, then appropriate adjustments in notation have to be made. For example, if $y = f(u)$, then (2) and (5) would be written as

$$\frac{d}{du}[f(u)] = f'(u) \quad \text{and} \quad \frac{dy}{du} = f'(u)$$

In particular, adjusting the notation in (3) yields

$$\frac{d}{du}[\sqrt{u}] = \frac{1}{2\sqrt{u}} \quad \text{and} \quad \frac{d}{du}\left[\frac{u}{u-9}\right] = -\frac{9}{(u-9)^2}$$

From the second of these formulas it would follow that

$$\frac{d}{du}\left[\frac{u}{u-9}\right]\bigg|_{u=10} = -\frac{9}{(10-9)^2} = -9$$

The variety of notations for derivatives can be confusing at first, but they are all used, so it is important to be familiar with each of them. To compound the problem, if y is the dependent variable for a function, and if the independent variable is self-evident from the problem, then it is common to denote the derivative simply as y'. Thus, y' would mean dy/dx if the independent variable is x or dy/du if the independent variable is u.

REMARK.   Some writers use the notation $D_x[\ \ ]$ to denote the differentiation operation. We shall not use this notation, however.

☐ **EXISTENCE OF DERIVATIVES**

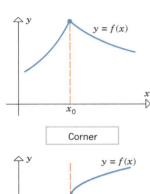

Corner

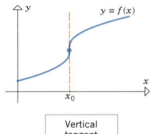

Vertical tangent

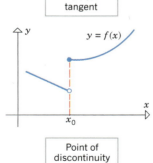

Point of discontinuity

Figure 3.2.3

The derivative of a function f is defined at those points where the limit in (1) exists. If x_0 is such a point, then we say that *f is differentiable at x_0 or f has a derivative at x_0*. Stated another way, the domain of f' consists of those points where f is differentiable. We say that f is *differentiable on an open interval* (a, b) if it is differentiable at each point in (a, b), and we say that f is a *differentiable function* if it is differentiable on $(-\infty, +\infty)$. At points where f is not differentiable we say that *the derivative of f does not exist*. Informally speaking, the most commonly encountered points of nondifferentiability can be classified as

- corners
- vertical tangents
- points of discontinuity

(see Figure 3.2.3).

The graph of a function f has a "corner" at a point $P(x_0, f(x_0))$ if f is continuous at P and the limiting position for the secant line joining P and Q depends on whether Q approaches P from the left or the right (Figure 3.2.4). At a corner a tangent line does not exist because the slopes of the secant lines do not have a (two-sided) limit.

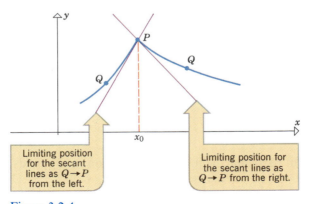

Limiting position for the secant lines as $Q \to P$ from the left.

Limiting position for the secant lines as $Q \to P$ from the right.

Figure 3.2.4

If the slope of the secant line joining P and Q tends toward $+\infty$ or $-\infty$ as Q approaches P along the graph of f, then f is not differentiable at x_0. Geometrically, such points occur where the secant lines tend toward a vertical limiting position (Figure 3.2.5).

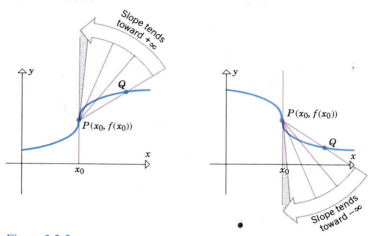

Figure 3.2.5

The limiting behavior of secant lines can be very strange at points of discontinuity. For example, in Figure 3.2.6 the slope of the secant line tends toward $+\infty$ as x approaches x_0 from the left and tends toward 0 as x approaches x_0 from the right. Since the slopes of the secant lines do not have a two-sided limit, the function graphed is not differentiable at x_0.

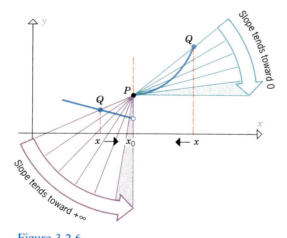

Figure 3.2.6

□ **RELATIONSHIP BETWEEN DIFFERENTIABILITY AND CONTINUITY**

The following major theorem shows that a function must be continuous at each point where it is differentiable.

3.2.2   THEOREM.   *If f is differentiable at a point x_0, then f is also continuous at x_0.*

Proof. We shall use Definition 2.7.11 to prove the continuity. We must show that $\lim_{h \to 0} f(x_0 + h) = f(x_0)$ or equivalently $\lim_{h \to 0} [f(x_0 + h) - f(x_0)] = 0$. But

$$\lim_{h \to 0} [f(x_0 + h) - f(x_0)] = \lim_{h \to 0} \left[\frac{f(x_0 + h) - f(x_0)}{h} \cdot h \right]$$

$$= \lim_{h \to 0} \left[\frac{f(x_0 + h) - f(x_0)}{h} \right] \cdot \lim_{h \to 0} h$$

$$= f'(x_0) \cdot 0 = 0 \quad \blacksquare$$

REMARK. It follows from the foregoing theorem that a function cannot be differentiable at a point of discontinuity.

Theorem 3.2.2 shows that differentiability at a point implies continuity at that point. The converse, however, is false—*a function may be continuous at a point but not differentiable there.* Whenever the graph of a function has a corner at a point, but no break or gap there, we have a point where the function is continuous, but not differentiable.

Example 4 The function $f(x) = |x|$ is continuous for all x and consequently is continuous at $x = 0$ (Example 3, Section 2.7).

(a) Show that $f(x) = |x|$ is not differentiable at $x = 0$.
(b) Find $f'(x)$.

Solution (a). This result is evident geometrically, since the graph of $|x|$ has a corner at $x = 0$ (Figure 3.2.7). However, an analytic argument can be given as follows. From Definition 3.2.1,

$$f'(0) = \lim_{h \to 0} \frac{f(0 + h) - f(0)}{h} = \lim_{h \to 0} \frac{f(h) - f(0)}{h} = \lim_{h \to 0} \frac{|h| - |0|}{h} = \lim_{h \to 0} \frac{|h|}{h}$$

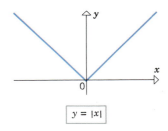

$y = |x|$

Figure 3.2.7

But

$$\frac{|h|}{h} = \begin{cases} 1, & h > 0 \\ -1, & h < 0 \end{cases}$$

so that

$$\lim_{h \to 0^-} \frac{|h|}{h} = -1 \quad \text{and} \quad \lim_{h \to 0^+} \frac{|h|}{h} = 1$$

Thus,

$$f'(0) = \lim_{h \to 0} \frac{|h|}{h}$$

does not exist because the one-sided limits are not equal. Consequently, $f(x) = |x|$ is not differentiable at $x = 0$.

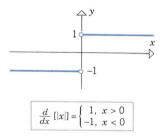

$$\frac{d}{dx}[|x|] = \begin{cases} 1, & x > 0 \\ -1, & x < 0 \end{cases}$$

Figure 3.2.8

Solution (b).   If $x > 0$, then $f(x) = |x| = x$, so $f'(x) = 1$, and if $x < 0$, then $f(x) = |x| = -x$, so $f'(x) = -1$. Combining these results into a single formula expressed piecewise, we obtain

$$f'(x) = \frac{d}{dx}[|x|] = \begin{cases} 1, & x > 0 \\ -1, & x < 0 \end{cases}$$

The graph of f' is shown in Figure 3.2.8. Observe that f' is not a continuous function, so this example shows that a continuous function can have a derivative that is not continuous.   ◄

The relationship between continuity and differentiability was of great historical significance in the development of calculus. In the early nineteenth century mathematicians believed that the graph of a continuous function could not have too many points of nondifferentiability bunched up. They felt that if a continuous function had many points of nondifferentiability, these points, like the tips of a sawblade, would have to be separated from each other and joined by smooth curve segments (Figure 3.2.9). This misconception was shattered by a series of discoveries beginning in 1834. In that year a Bohemian priest, philosopher, and mathematician named Bernhard Bolzano* discovered a procedure for constructing a continuous function that is not differentiable at any point. Later, in 1860, the great German mathematician, Karl Weierstrass** (biography on p. 182) produced the first formula for such a function. The graphs of such functions are impossible to draw; it is as if the corners are so numerous that any segment of the curve, when suitably enlarged, reveals more corners. The discovery of these pathological functions was important in that it made mathematicians distrustful of their geometric intuition and more reliant on precise mathematical proof. However, they remained only mathematical curiosities until the early 1980s, when applications of them began to emerge. During the past 10 years they have started to play a fundamental role in the study of geometric objects called **fractals**. Fractals have revealed an order to natural phenomena that were previously dismissed as random and chaotic.

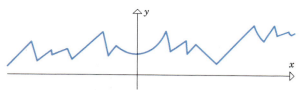

Figure 3.2.9

---

*BERNHARD BOLZANO (1781–1848). Bolzano, the son of an art dealer, was born in Prague, Bohemia (Czechoslovakia). He was educated at the University of Prague, and eventually won enough mathematical fame to be recommended for a mathematics chair there. However, Bolzano became an ordained Roman Catholic priest, and in 1805 he was appointed to a chair of Philosophy at the University of Prague. Bolzano was a man of great human compassion; he spoke out for educational reform, he voiced the right of individual conscience over government demands, and he lectured on the absurdity of war and militarism. His views so disenchanted Emperor Franz I of Austria that the emperor pressed the Archbishop of Prague to have Bolzano recant his statements. Bolzano refused and was then forced to retire in 1824 on a small pension. Bolzano's main contribution to mathematics was philosophical. His work helped convince mathematicians that sound mathematics must ultimately rest on rigorous proof rather than intuition. In addition to his work in mathematics, Bolzano investigated problems concerning space, force, and wave propagation.

DERIVATIVES AT THE ENDPOINTS OF AN INTERVAL

If a function f is defined on a closed interval $[a, b]$ and is not defined outside of that interval, then the derivative $f'(x)$ is not defined at the endpoints a and b because

$$f'(x) = \lim_{h \to 0} \frac{f(x + h) - f(x)}{h}$$

is a two-sided limit and only one-sided limits make sense at the endpoints. To deal with this situation, we define *derivatives from the left and right*. These are denoted by f'_- and f'_+, respectively, and defined by

$$f'_-(x) = \lim_{h \to 0^-} \frac{f(x + h) - f(x)}{h} \quad \text{and} \quad f'_+(x) = \lim_{h \to 0^+} \frac{f(x + h) - f(x)}{h}$$

At points where $f'_+(x)$ exists we say that the function f is *differentiable from the right* and at points where $f'_-(x)$ exists we say that f is *differentiable from the left*. Geometrically, $f'_+(x)$ is the limit of the slopes of the secant lines approaching x from the right and $f'_-(x)$ is the limit of the slopes of the secant lines approaching x from the left (Figure 3.2.10).

It can be proved that a function f is continuous from the left at those points where it is differentiable from the left and continuous from the right at those points where it is differentiable from the right. (See Definition 2.7.7.)

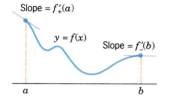

Slope = $f'_+(a)$

$y = f(x)$

Slope = $f'_-(b)$

a b

Figure 3.2.10

3.2.3   DEFINITION.   A function f is said to be *differentiable on a closed interval* $[a, b]$ if the following conditions are satisfied:

1. f is differentiable on (a, b).
2. f is differentiable from the right at a.
3. f is differentiable from the left at b.

The reader should have no trouble formulating the appropriate definitions of differentiability on intervals of the form $[a, +\infty)$, $(-\infty, b]$, $[a, b)$, and $(a, b]$.

**KARL WEIERSTRASS (1815–1897). Weierstrass, the son of a customs officer, was born in Ostenfelde, Germany. As a youth Weierstrass showed outstanding skills in languages and mathematics. However, at the urging of his dominant father, Weierstrass entered the law and commerce program at the University of Bonn. To the chagrin of his family, the rugged and congenial young man concentrated instead on fencing and beer drinking. Four years later he returned home without a degree. In 1839 Weierstrass entered the Academy of Münster to study for a career in secondary education, and he met and studied under an excellent mathematician named Christof Gudermann. Gudermann's ideas greatly influenced the work of Weierstrass. After receiving his teaching certificate, Weierstrass spent the next 15 years in secondary education teaching German, geography, and mathematics. In addition, he taught handwriting to small children. During this period much of Weierstrass's mathematical work was ignored because he was a secondary schoolteacher and not a college professor. Then, in 1854, he published a paper of major importance which created a sensation in the mathematics world and catapulted him to international fame overnight. He was immediately given an honorary Doctorate at the University of Königsberg and began a new career in college teaching at the University of Berlin in 1856. In 1859 the strain of his mathematical research caused a temporary nervous breakdown and led to spells of dizziness that plagued him for the rest of his life. Weierstrass was a brilliant teacher and his classes overflowed with multitudes of auditors. In spite of his fame, he never lost his early beer-drinking congeniality and was always in the company of students, both ordinary and brilliant. Weierstrass was acknowledged as the leading mathematical analyst in the world. He and his students opened the door to the modern school of mathematical analysis.

▶ Exercise Set 3.2   Ⓒ 27, 28, 29, 30

In Exercises 1–12, use Definition 3.2.1 to find $f'(x)$.

1. $f(x) = 3x^2$.

2. $f(x) = x^2 - x$.

3. $f(x) = x^3$.

4. $f(x) = 2x^3 + 1$.

5. $f(x) = \sqrt{x + 1}$.

6. $f(x) = x^4$.

7. $f(x) = \dfrac{1}{x}$.

8. $f(x) = \dfrac{1}{x^2}$.

9. $f(x) = ax^2 + b$ $(a, b \text{ constants})$.

10. $f(x) = \dfrac{1}{x + 1}$.

11. $f(x) = \dfrac{1}{\sqrt{x}}$.

12. $f(x) = x^{1/3}$.

In Exercises 13–18, find $f'(a)$ and the equation of the tangent to the graph of f at the point where $x = a$.

13. f is the function in Exercise 1; $a = 3$.

14. f is the function in Exercise 2; $a = 2$.

15. f is the function in Exercise 3; $a = 0$.

16. f is the function in Exercise 4; $a = -1$.

17. f is the function in Exercise 5; $a = 8$.

18. f is the function in Exercise 6; $a = -2$.

19. Let $y = 4x^2 + 2$. Find

(a) $\dfrac{dy}{dx}$ (b) $\dfrac{dy}{dx}\Big|_{x=1}$.

20. Let $y = \dfrac{5}{x} + 1$. Find

(a) $\dfrac{dy}{dx}$ (b) $\dfrac{dy}{dx}\Big|_{x=-2}$.

In Exercises 21–24, use Definition 3.2.1 (with the appropriate change in notation) to obtain the derivative requested.

21. Find $f'(t)$ if $f(t) = 4t^2 + t$.

22. Find $g'(u)$ if $g(u) = 5u + 3$.

23. Find $\dfrac{dA}{d\lambda}$ if $A = 3\lambda^2 - \lambda$.

24. Find $\dfrac{dV}{dr}$ if $V = \frac{4}{3}\pi r^3$.

25. Match the graphs of the functions shown in (a)–(f) with the graphs of their derivatives in (A)–(F).

(a)
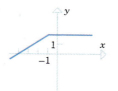

(b)
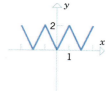

(c)
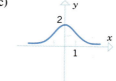

(d)
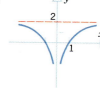

(e)

(f)
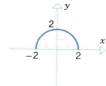

(A)

(B)

(C)
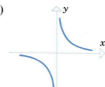

(D)
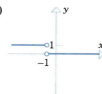

(E)
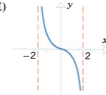

(F)
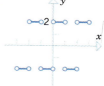

26. Use the graph of $y = f(x)$ shown in Figure 3.2.11 to estimate the value of $f'(1)$, $f'(3)$, $f'(5)$, and $f'(6)$.

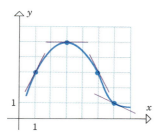

Figure 3.2.11

27. Biological systems in which the exchange of substances occur are sometimes described mathematically by using a *compartment model*. Figure 3.2.12a shows two compartments of equal size separated by a permeable membrane, the left containing a red dye dissolved in water and the right just water. The compartment model states that over time the dye will diffuse from the compartment with higher concentration to the compartment with lower concentration at a rate that is proportional to the difference in concentrations until the concentration of dye is the same in both compartments (Figures 3.2.12b and 3.2.12c). Figure 3.2.13 shows the graph of the concentration C versus the time t for the dye in the right compartment in the case where the left compartment has a concentration of 1.6 mol/L (moles per liter) and the right compartment contains only water at time $t = 0$. Estimate the values of

$$dC/dt|_{t=0} \quad \text{and} \quad dC/dt|_{t=15}$$

[Adapted from *Compartment Models in Biology*, by Ron Barnes, UMAP Module 676, COMAP, Inc.]

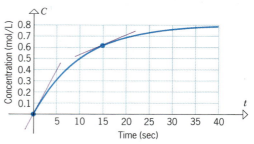

Figure 3.2.13

28. According to *The World Almanac and Book of Facts* (1987), the estimated world population, N, in millions for the years 1850, 1900, 1950, and 1985 was, 1175, 1600, 2490, and 4843, respectively. Although the increase in population is not a continuous function of the time t, we can apply the ideas in this section if we are willing to approximate the graph of N versus t by a continuous curve, as shown in Figure 3.2.14.

(a) Use the estimated tangent line shown in the figure at the point where $t = 1950$ to approximate the value of dN/dt there. Describe your result as a rate of change.

(b) At any instant, the **growth rate** is defined as

$$\frac{dN/dt}{N}$$

Use your answer to part (a) to approximate the growth rate in 1950. Express the result as a percentage and include the proper units.

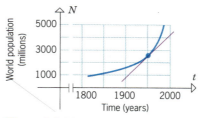

Figure 3.2.14

29. It is a fact that when a flexible rope is wrapped around a rough cylinder, a small force of magnitude F_0 at one end can resist a large force of magnitude F at the other end (Figure 3.2.15a). The size of F depends on the angle θ through which the rope is wrapped around the cylinder (Figure 3.2.15b). For example, Figure 3.2.16 shows the graph of F (in pounds) versus θ (in radians), where F is the magnitude of the force that can be resisted by a force

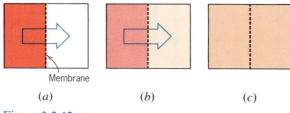

Membrane

(a)   (b)   (c)

Figure 3.2.12

with magnitude $F_0 = 10$ lb for a certain rope and cylinder.

(a) Estimate the values of F and $dF/d\theta$ when $\theta = 10$ radians.

(b) It can be shown that F satisfies the equation $dF/d\theta = \mu F$, where the constant μ is called the **coefficient of friction.** Use the results in part (a) to estimate the value of μ.

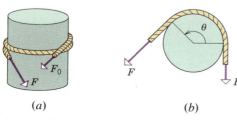

(a)   (b)

Figure 3.2.15

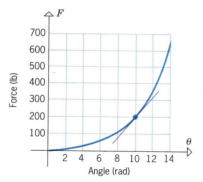

Figure 3.2.16

30. According to **Newton's law of cooling**, the rate of change of an object's temperature is proportional to the difference between the temperature of the object and that of the surrounding medium. Figure 3.2.17 shows the graph of the temperature T (in degrees Fahrenheit) versus time t (in minutes) for a cup of coffee, initially with a temperature of 200° F, that is allowed to cool in a room with a constant temperature of 75° F.

(a) Estimate T and dT/dt when $t = 10$ min.

(b) Newton's law of cooling can be expressed as

$$\frac{dT}{dt} = k(T - T_0)$$

where k is the constant of proportionality and T_0 is the temperature (assumed constant) of the

surrounding medium. Use the results in part (a) to estimate the value of k.

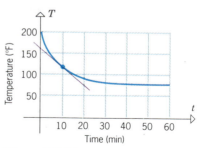

Figure 3.2.17

In Exercises 31–38, sketch the graph of the derivative of the function whose graph is shown.

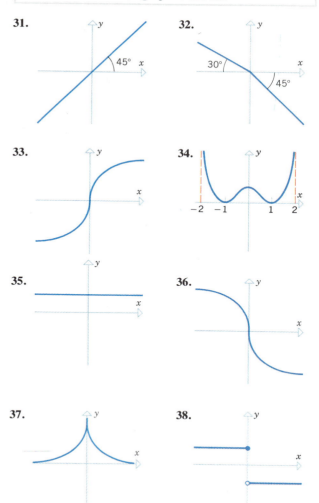

31.   32.   33.   34.   35.   36.   37.   38.

39. Show that $f(x) = \sqrt[3]{x}$ is continuous at $x = 0$ but not differentiable at $x = 0$. Sketch the graph of f.

40. Show that $f(x) = \sqrt[3]{(x - 2)^2}$ is continuous at $x = 2$ but not differentiable at $x = 2$. Sketch the graph of f.

41. Show that

$$f(x) = \begin{cases} x^2 + 1, & x \le 1 \\ 2x, & x > 1 \end{cases}$$

is continuous and differentiable at $x = 1$. Sketch the graph of f.

42. Show that

$$f(x) = \begin{cases} x^2 + 2, & x \le 1 \\ x + 2, & x > 1 \end{cases}$$

is continuous but not differentiable at $x = 1$. Sketch the graph of f.

43. Suppose that the function f is differentiable at $x = 1$ and $\lim\limits_{h \to 0} \dfrac{f(1 + h)}{h} = 5$. Find $f(1)$ and $f'(1)$.

44. Suppose that f is a differentiable function with the property that $f(x + y) = f(x) + f(y) + 5xy$ and $\lim\limits_{h \to 0} \dfrac{f(h)}{h} = 3$. Find $f(0)$ and $f'(x)$.

45. Suppose that the function f has the property $f(x + y) = f(x)f(y)$ for all values of x and y and that $f(0) = f'(0) = 1$. Show that f is differentiable and $f'(x) = f(x)$. [*Hint:* Start by expressing $f'(x)$ as a limit.]

■ 3.3 TECHNIQUES OF DIFFERENTIATION

Up to now we have obtained derivatives directly from the definition. In this section we shall develop some theorems and formulas that provide more efficient methods.

□ **DERIVATIVE OF A CONSTANT**

> **3.3.1 THEOREM.** *If f is a constant function, say $f(x) = c$ for all x, then $f'(x) = 0$.*

Proof.

$$f'(x) = \lim_{h \to 0} \frac{f(x + h) - f(x)}{h} = \lim_{h \to 0} \frac{c - c}{h} = \lim_{h \to 0} 0 = 0 \quad ■$$

This result is obvious geometrically. If $f(x) = c$ is a constant function, then the graph of f is a line parallel to the x-axis; consequently, the tangent line at each point is horizontal. Therefore, $f'(x) = m_{\text{tan}} = 0$, since a horizontal line has slope 0.

In another notation, Theorem 3.3.1 states

$$\frac{d}{dx}[c] = 0$$

Example 1   If $f(x) = 5$ for all x, then $f'(x) = 0$ for all x; that is,

$$\frac{d}{dx}[5] = 0 \quad ◀$$

☐ **DERIVATIVE OF A POWER**

3.3.2 THEOREM (*The Power Rule*). *If n is a positive integer, then for every real value of x*

$$\frac{d}{dx}[x^n] = nx^{n-1}$$

Proof.   Let $f(x) = x^n$. Then

$$f'(x) = \lim_{h \to 0} \frac{f(x+h) - f(x)}{h} = \lim_{h \to 0} \frac{(x+h)^n - x^n}{h}$$

Expanding $(x + h)^n$ by the binomial theorem, we obtain

$$f'(x) = \lim_{h \to 0} \frac{\left[x^n + nx^{n-1}h + \frac{n(n-1)}{2!} x^{n-2}h^2 + \cdots + nxh^{n-1} + h^n \right] - x^n}{h}$$

$$= \lim_{h \to 0} \frac{nx^{n-1}h + \frac{n(n-1)}{2!} x^{n-2}h^2 + \cdots + nxh^{n-1} + h^n}{h}$$

Canceling a factor of h we obtain

$$f'(x) = \lim_{h \to 0} \left[nx^{n-1} + \frac{n(n-1)}{2!} x^{n-2}h + \cdots + nxh^{n-2} + h^{n-1} \right]$$

Every term but the first has a factor of h, and so every term but the first approaches zero as $h \to 0$. Therefore,

$$f'(x) = nx^{n-1} \quad \blacksquare$$

REMARK.   In words, *to differentiate x to a positive integer power, take the power and multiply it by x to the next lower integer power.*

Example 2

$$\frac{d}{dx}[x^5] = 5x^4, \quad \frac{d}{dx}[x] = 1 \cdot x^0 = 1, \quad \frac{d}{dx}[x^{12}] = 12x^{11} \quad \blacktriangleleft$$

☐ **DERIVATIVE OF A CONSTANT TIMES A FUNCTION**

3.3.3 THEOREM.   *Let c be a constant. If f is differentiable at x, then so is cf, and*

$$\frac{d}{dx}[cf(x)] = c\frac{d}{dx}[f(x)]$$

Proof.

$$\frac{d}{dx}[cf(x)] = \lim_{h \to 0} \frac{cf(x+h) - cf(x)}{h} = \lim_{h \to 0} c\left[\frac{f(x+h) - f(x)}{h}\right]$$

$$= c \lim_{h \to 0} \frac{f(x+h) - f(x)}{h} = c \frac{d}{dx}[f(x)] \quad \blacksquare$$

A constant factor can be moved through a limit sign.

In function notation, Theorem 3.3.3 states

$$(cf)' = cf'$$

REMARK.   In words, *a constant factor can be moved through a derivative sign.*

Example 3

$$\frac{d}{dx}[4x^8] = 4\frac{d}{dx}[x^8] = 4[8x^7] = 32x^7$$

$$\frac{d}{dx}[-x^{12}] = (-1)\frac{d}{dx}[x^{12}] = -12x^{11} \quad \blacktriangleleft$$

☐ **DERIVATIVES OF SUMS AND DIFFERENCES**

3.3.4   THEOREM.   *If f and g are differentiable at x, then so is f + g, and*

$$\frac{d}{dx}[f(x) + g(x)] = \frac{d}{dx}[f(x)] + \frac{d}{dx}[g(x)]$$

Proof.

$$\frac{d}{dx}[f(x) + g(x)] = \lim_{h \to 0} \frac{[f(x+h) + g(x+h)] - [f(x) + g(x)]}{h}$$

$$= \lim_{h \to 0} \frac{[f(x+h) - f(x)] + [g(x+h) - g(x)]}{h}$$

$$= \lim_{h \to 0} \frac{f(x+h) - f(x)}{h} + \lim_{h \to 0} \frac{g(x+h) - g(x)}{h}$$

The limit of a sum is the sum of the limits.

$$= \frac{d}{dx}[f(x)] + \frac{d}{dx}[g(x)] \quad \blacksquare$$

Theorem 3.3.4 can be written in function notation as

$$(f + g)' = f' + g'$$

By writing $f - g = f + (-1)g$ and then applying Theorems 3.3.3 and 3.3.4 it follows that

$$\frac{d}{dx}[f(x) - g(x)] = \frac{d}{dx}[f(x)] - \frac{d}{dx}[g(x)]$$

or in function notation

$$(f - g)' = f' - g'$$

REMARK.   In words, *the derivative of a sum equals the sum of the derivatives, and the derivative of a difference equals the difference of the derivatives.*

Example 4

$$\frac{d}{dx}[x^4 + x^2] = \frac{d}{dx}[x^4] + \frac{d}{dx}[x^2] = 4x^3 + 2x$$

$$\frac{d}{dx}[x^4 - x^2] = \frac{d}{dx}[x^4] - \frac{d}{dx}[x^2] = 4x^3 - 2x$$

$$\frac{d}{dx}[6x^{11} + 9] = \frac{d}{dx}[6x^{11}] + \frac{d}{dx}[9] = 66x^{10} + 0 = 66x^{10} \quad \blacktriangleleft$$

The result in Theorem 3.3.4 can be extended to any finite number of functions. More precisely, if the functions $f_1, f_2, \ldots, f_n$ are all differentiable at x, then their sum is differentiable at x and

$$\frac{d}{dx}[f_1(x) + f_2(x) + \cdots + f_n(x)] = \frac{d}{dx}[f_1(x)] + \frac{d}{dx}[f_2(x)] + \cdots + \frac{d}{dx}[f_n(x)]$$

Example 5

$$\frac{d}{dx}[3x^8 - 2x^5 + 6x + 1] = \frac{d}{dx}[3x^8] + \frac{d}{dx}[-2x^5] + \frac{d}{dx}[6x] + \frac{d}{dx}[1]$$

$$= 24x^7 - 10x^4 + 6 \quad \blacktriangleleft$$

☐ **DERIVATIVE OF A PRODUCT**

3.3.5 THEOREM (*The Product Rule*).   *If f and g are differentiable at x, then so is the product $f \cdot g$, and*

$$\frac{d}{dx}[f(x)g(x)] = f(x)\frac{d}{dx}[g(x)] + g(x)\frac{d}{dx}[f(x)]$$

Proof.

$$\frac{d}{dx}\left[f(x)g(x)\right] = \lim_{h \to 0} \frac{f(x + h) \cdot g(x + h) - f(x) \cdot g(x)}{h}$$

If we add and subtract $f(x + h) \cdot g(x)$ in the numerator, we obtain

$$\frac{d}{dx}\left[f(x)g(x)\right] = \lim_{h \to 0} \frac{f(x + h)g(x + h) - f(x + h)g(x) + f(x + h)g(x) - f(x)g(x)}{h}$$

$$= \lim_{h \to 0} \left[f(x + h) \cdot \frac{g(x + h) - g(x)}{h} + g(x) \cdot \frac{f(x + h) - f(x)}{h}\right]$$

$$= \lim_{h \to 0} f(x + h) \cdot \lim_{h \to 0} \frac{g(x + h) - g(x)}{h} + \lim_{h \to 0} g(x) \cdot \lim_{h \to 0} \frac{f(x + h) - f(x)}{h}$$

$$= [\lim_{h \to 0} f(x + h)] \frac{d}{dx}[g(x)] + [\lim_{h \to 0} g(x)] \frac{d}{dx}[f(x)] \qquad (1)$$

But

$$\lim_{h \to 0} g(x) = g(x) \qquad (2)$$

because $g(x)$ does not involve h and thus remains *constant* as $h \to 0$. Also, it follows from Definition 2.7.11 that

$$\lim_{h \to 0} f(x + h) = f(x) \qquad (3)$$

because f is assumed to be differentiable at x and is therefore continuous at x by Theorem 3.2.2. Substituting (2) and (3) into (1) yields

$$\frac{d}{dx}\left[f(x)g(x)\right] = f(x) \frac{d}{dx}[g(x)] + g(x) \frac{d}{dx}[f(x)] \qquad ∎$$

The product rule can be written in function notation as

$$(f \cdot g)' = f \cdot g' + g \cdot f'$$

REMARK.   In words, *the derivative of a product of two functions is the first function times the derivative of the second plus the second function times the derivative of the first.*

WARNING.   Note that it is *not* true in general that $(f \cdot g)' = f' \cdot g'$; that is, the derivative of a product is *not* generally the product of the derivatives!

Example 6   Find dy/dx if $y = (4x^2 - 1)(7x^3 + x)$.

Solution.   There are two methods that can be used to find dy/dx: We can either use the product rule or we can multiply out the factors in y and then differentiate. We shall give both methods.

Method I. (*Using the Product Rule.*)

$$\frac{dy}{dx} = \frac{d}{dx}\left[(4x^2 - 1)(7x^3 + x)\right]$$

$$= (4x^2 - 1)\frac{d}{dx}\left[7x^3 + x\right] + (7x^3 + x)\frac{d}{dx}\left[4x^2 - 1\right]$$

$$= (4x^2 - 1)(21x^2 + 1) + (7x^3 + x)(8x) = 140x^4 - 9x^2 - 1$$

Method II. (*Multiplying First.*)

$$y = (4x^2 - 1)(7x^3 + x) = 28x^5 - 3x^3 - x$$

Thus,

$$\frac{dy}{dx} = \frac{d}{dx}\left[28x^5 - 3x^3 - x\right] = 140x^4 - 9x^2 - 1$$

which agrees with the result obtained using the product rule.   ◀

□ **DERIVATIVE OF A QUOTIENT**

> **3.3.6** **THEOREM** (*The Quotient Rule*).   *If f and g are differentiable at x and* $g(x) \neq 0$, *then f/g is differentiable at x and*
>
> $$\frac{d}{dx}\left[\frac{f(x)}{g(x)}\right] = \frac{g(x)\dfrac{d}{dx}\left[f(x)\right] - f(x)\dfrac{d}{dx}\left[g(x)\right]}{[g(x)]^2}$$

Proof.

$$\frac{d}{dx}\left[\frac{f(x)}{g(x)}\right] = \lim_{h \to 0}\frac{\dfrac{f(x+h)}{g(x+h)} - \dfrac{f(x)}{g(x)}}{h}$$

$$= \lim_{h \to 0}\frac{f(x+h)\cdot g(x) - f(x)\cdot g(x+h)}{h \cdot g(x) \cdot g(x+h)}$$

Adding and subtracting $f(x) \cdot g(x)$ in the numerator yields

$$\frac{d}{dx}\left[\frac{f(x)}{g(x)}\right] = \lim_{h \to 0}\frac{f(x+h)\cdot g(x) - f(x)\cdot g(x) - f(x)\cdot g(x+h) + f(x)\cdot g(x)}{h \cdot g(x) \cdot g(x+h)}$$

$$= \lim_{h \to 0}\frac{\left[g(x) \cdot \dfrac{f(x+h) - f(x)}{h}\right] - \left[f(x)\cdot\dfrac{g(x+h) - g(x)}{h}\right]}{g(x)\cdot g(x+h)}$$

$$= \frac{\displaystyle\lim_{h \to 0} g(x) \cdot \lim_{h \to 0}\frac{f(x+h) - f(x)}{h} - \lim_{h \to 0} f(x) \cdot \lim_{h \to 0}\frac{g(x+h) - g(x)}{h}}{\displaystyle\lim_{h \to 0} g(x) \cdot \lim_{h \to 0} g(x+h)}$$

$$= \frac{[\lim_{h \to 0} g(x)] \cdot \dfrac{d}{dx}[f(x)] - [\lim_{h \to 0} f(x)] \cdot \dfrac{d}{dx}[g(x)]}{\lim_{h \to 0} g(x) \cdot \lim_{h \to 0} g(x + h)} \tag{4}$$

Since the expressions $f(x)$ and $g(x)$ do not involve h,

$$\lim_{h \to 0} g(x) = g(x) \quad \text{and} \quad \lim_{h \to 0} f(x) = f(x) \tag{5}$$

Because g is assumed to be differentiable at x, it is continuous at x, and so by Definition 2.7.11

$$\lim_{h \to 0} g(x + h) = g(x) \tag{6}$$

Substituting (5) and (6) into (4) yields

$$\frac{d}{dx}\left[\frac{f(x)}{g(x)}\right] = \frac{g(x)\dfrac{d}{dx}[f(x)] - f(x)\dfrac{d}{dx}[g(x)]}{[g(x)]^2} \qquad \blacksquare$$

The quotient rule can be written in function notation as

$$\left(\frac{f}{g}\right)' = \frac{g \cdot f' - f \cdot g'}{g^2}$$

REMARK.   In words, *the derivative of a quotient of two functions is the denominator times the derivative of the numerator minus the numerator times the derivative of the denominator, all divided by the denominator squared.*

WARNING.   Note that it is *not* generally true that $(f/g)' = f'/g'$; that is, the derivative of a quotient is *not* generally the quotient of the derivatives.

Example 7   Let $y = \dfrac{x^2 - 1}{x^4 + 1}$.

(a)   Find dy/dx.

(b)   At which points does the graph of the equation have a horizontal tangent line?

Solution (a).

$$\frac{dy}{dx} = \frac{d}{dx}\left[\frac{x^2 - 1}{x^4 + 1}\right] = \frac{(x^4 + 1)\dfrac{d}{dx}[x^2 - 1] - (x^2 - 1)\dfrac{d}{dx}[x^4 + 1]}{(x^4 + 1)^2}$$

$$= \frac{(x^4 + 1)(2x) - (x^2 - 1)(4x^3)}{(x^4 + 1)^2} \qquad \boxed{\begin{array}{l}\text{The differentiation is complete.}\\ \text{The rest is simplification.}\end{array}}$$

$$= \frac{-2x^5 + 4x^3 + 2x}{(x^4 + 1)^2} = -\frac{2x(x^4 - 2x^2 - 1)}{(x^4 + 1)^2}$$

Solution (b).   Since dy/dx can be interpreted as the slope of the tangent line to the graph, and since a horizontal tangent line has slope 0, the points where the tangent line is horizontal can be found by solving the equation

$$\frac{dy}{dx} = 0$$

or from part (a)

$$-\frac{2x(x^4 - 2x^2 - 1)}{(x^4 + 1)^2} = 0$$

The solutions of this equation are the values of x for which the numerator is 0:

$$2x(x^4 - 2x^2 - 1) = 0$$

The first factor yields the solution $x = 0$. Other solutions can be found by solving the equation

$$x^4 - 2x^2 - 1 = 0$$

This can be treated as a quadratic equation in x^2 and solved by the quadratic formula. This yields

$$x^2 = \frac{2 \pm \sqrt{8}}{2} = 1 \pm \sqrt{2}$$

The minus sign yields imaginary values for x (why?); these solutions are not relevant to our problem, so we ignore them. The plus sign yields the solutions

$$x = \pm\sqrt{1 + \sqrt{2}}$$

In summary, horizontal tangent lines occur at

$$x = 0, \quad x = \sqrt{1 + \sqrt{2}} \approx 1.55, \quad \text{and} \quad x = -\sqrt{1 + \sqrt{2}} \approx -1.55$$

These results are consistent with the graph of the given equation, which we generated for Figure 3.3.1 using a graphing program.   ◀

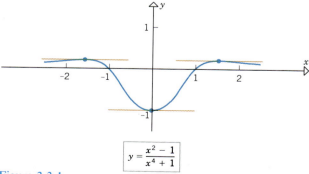

$$y = \frac{x^2 - 1}{x^4 + 1}$$

Figure 3.3.1

DERIVATIVE OF A RECIPROCAL

The special case of Theorem 3.3.6 in which f is the constant function 1 is of interest in its own right. We leave it as an exercise for the reader to deduce the following result from Theorem 3.3.6.

3.3.7 THEOREM (*The Reciprocal Rule*). *If g is differentiable at x and $g(x) \neq 0$, then $1/g$ is differentiable at x and*

$$\frac{d}{dx}\left[\frac{1}{g(x)}\right] = -\frac{\dfrac{d}{dx}[g(x)]}{[g(x)]^2}$$

The reciprocal rule can be written in function notation as

$$\left(\frac{1}{g}\right)' = -\frac{g'}{g^2}$$

REMARK. In words, *the derivative of the reciprocal of a function is the negative of the derivative of the function divided by the function squared.*

Example 8

$$\frac{d}{dx}\left[\frac{1}{x}\right] = -\frac{\dfrac{d}{dx}[x]}{x^2} = -\frac{1}{x^2}$$

$$\frac{d}{dx}\left[\frac{1}{x^3 + 2x - 3}\right] = -\frac{\dfrac{d}{dx}[x^3 + 2x - 3]}{(x^3 + 2x - 3)^2} = -\frac{3x^2 + 2}{(x^3 + 2x - 3)^2} \quad \blacktriangleleft$$

REMARK. The computations in the foregoing example could have been done using the quotient rule, but this would have been more work. Where it applies, the reciprocal rule is preferable to the quotient rule.

THE POWER RULE FOR INTEGER EXPONENTS

In Theorem 3.3.2 we established the formula

$$\frac{d}{dx}[x^n] = nx^{n-1}$$

for *positive* integer values of n. The following theorem extends this formula so that it applies to *all* integer values of n.

3.3.8 THEOREM. *If n is any integer, then*

$$\frac{d}{dx}[x^n] = nx^{n-1} \tag{7}$$

Proof. The result has already been established in the case where $n > 0$. If $n < 0$, then let $m = -n$ and let

$$f(x) = x^{-m} = \frac{1}{x^m}$$

From Theorem 3.3.7,

$$f'(x) = \frac{d}{dx}\left[\frac{1}{x^m}\right] = -\frac{\frac{d}{dx}[x^m]}{(x^m)^2}$$

Since $n < 0$, it follows that $m > 0$, so x^m can be differentiated using Theorem 3.3.2. Thus,

$$f'(x) = -\frac{mx^{m-1}}{x^{2m}} = -mx^{m-1-2m} = -mx^{-m-1} = nx^{n-1}$$

which proves (7). In the case $n = 0$ Formula (7) reduces to

$$\frac{d}{dx}[1] = 0 \cdot x^{-1} = 0$$

which is correct by Theorem 3.3.1.   ∎

Example 9

$$\frac{d}{dx}[x^{-9}] = -9x^{-9-1} = -9x^{-10}$$

$$\frac{d}{dx}\left[\frac{1}{x}\right] = \frac{d}{dx}[x^{-1}] = (-1)x^{-1-1} = -x^{-2} = -\frac{1}{x^2}$$

Note that the last result agrees with that obtained in Example 8.   ◀

In Example 1 of Section 3.2 we showed that

$$\frac{d}{dx}[\sqrt{x}] = \frac{1}{2\sqrt{x}} \tag{8}$$

If we write this result in exponential notation, we obtain

$$\frac{d}{dx}[x^{1/2}] = \frac{1}{2x^{1/2}} = \frac{1}{2}x^{-1/2}$$

which shows that (7) holds for the rational exponent $n = \frac{1}{2}$. Later, we shall show that (7) actually holds for all rational exponents.

☐ **HIGHER DERIVATIVES**   If the derivative f' of a function f is itself differentiable, then the derivative of f' is denoted by f'' and is called the *second derivative* of f. As long as we have differentiability, we can continue the process of differentiating derivatives to obtain third, fourth, fifth, and even higher derivatives of f. The successive derivatives of f are denoted by

f'	The first derivative of f
$f'' = (f')'$	The second derivative of f
$f''' = (f'')'$	The third derivative of f
$f^{(4)} = (f''')'$	The fourth derivative of f
$f^{(5)} = (f^{(4)})'$	The fifth derivative of f

$$\vdots \qquad \vdots \qquad \qquad \vdots$$

Beyond the third derivative, it is too clumsy to continue using primes, so we switch from primes to integers in parentheses to denote the **order** of the derivative. In this notation it is easy to denote a derivative of arbitrary order by writing

$f^{(n)}$	The nth derivative of f

The significance of the derivatives of order 2 and higher will be discussed later.

Example 10   If $f(x) = 3x^4 - 2x^3 + x^2 - 4x + 2$, then

$$f'(x) = 12x^3 - 6x^2 + 2x - 4$$
$$f''(x) = 36x^2 - 12x + 2$$
$$f'''(x) = 72x - 12$$
$$f^{(4)}(x) = 72$$
$$f^{(5)}(x) = 0$$

$$\vdots$$

$$f^{(n)}(x) = 0 \quad (n \geq 5) \qquad \blacktriangleleft$$

Successive derivatives can also be denoted as follows:

$$f'(x) = \frac{d}{dx}[f(x)]$$

$$f''(x) = \frac{d}{dx}\left[\frac{d}{dx}[f(x)]\right] = \frac{d^2}{dx^2}[f(x)]$$

$$f'''(x) = \frac{d}{dx}\left[\frac{d}{dx}\left[\frac{d}{dx}[f(x)]\right]\right] = \frac{d^3}{dx^3}[f(x)]$$

$$\vdots \qquad\qquad\qquad \vdots$$

In general, we write

$$f^{(n)}(x) = \frac{d^n}{dx^n}[f(x)]$$

which is read, *the nth derivative of f with respect to x.*
When a dependent variable is involved, say,

$$y = f(x)$$

then successive derivatives can be denoted by writing

$$\frac{dy}{dx}, \quad \frac{d^2y}{dx^2}, \quad \frac{d^3y}{dx^3}, \quad \frac{d^4y}{dx^4}, \dots, \frac{d^ny}{dx^n}, \dots$$

or more briefly

$$y', \quad y'', \quad y''', \quad y^{(4)}, \dots, y^{(n)}, \dots$$

The following symbols denote values of derivatives at a particular point x_0; their meanings should be self-evident.

$$y''(x_0), \quad f^{(4)}(x_0), \quad \frac{d^3y}{dx^3}\bigg|_{x=x_0}, \quad \frac{d^2}{dx^2}[x^7 - x]\bigg|_{x=x_0}$$

Example 11

$$\frac{d^2}{dx^2}[x^5] = \frac{d}{dx}\left[\frac{d}{dx}(x^5)\right] = \frac{d}{dx}[5x^4] = 20x^3$$

Thus,

$$\frac{d^2}{dx^2}[x^5]\bigg|_{x=2} = 20 \cdot 8 = 160 \quad \blacktriangleleft$$

▶ Exercise Set 3.3

In Exercises 1–28, use the results of this section to find dy/dx.

1. $y = 4x^7$.

2. $y = -3x^{12}$.

3. $y = 3x^8 + 2x + 1$.

4. $y = \frac{1}{2}(x^4 + 7)$.

5. $y = \pi^3$.

6. $y = \sqrt{2}x + \frac{1}{\sqrt{2}}$.

7. $y = -\frac{1}{3}(x^7 + 2x - 9)$.

8. $y = \frac{x^2 + 1}{5}$.

9. $y = ax^3 + bx^2 + cx + d$ $(a, b, c, d$ constant$)$.

10. $y = \frac{1}{a}\left(x^2 + \frac{1}{b}x + c\right)$ $(a, b, c$ constant$)$.

11. $y = -3x^{-8} + 2\sqrt{x}$.

12. $y = 7x^{-6} - 5\sqrt{x}$.

13. $y = x^{-3} + \frac{1}{x^7}$.

14. $y = \sqrt{x} + \frac{1}{x}$.

15. $y = (3x^2 + 6)(2x - \frac{1}{4})$.

16. $y = (2 - x - 3x^3)(7 + x^5)$.

17. $y = (x^3 + 7x^2 - 8)(2x^{-3} + x^{-4})$.

18. $y = \left(\frac{1}{x} + \frac{1}{x^2}\right)(3x^3 + 27)$.

19. $y = (3x^2 + 1)^2$.

20. $y = (x^5 + 2x)^2$.

21. $y = \dfrac{1}{5x - 3}$.

22. $y = \dfrac{3}{\sqrt{x} + 2}$.

23. $y = \dfrac{3x}{2x + 1}$.

24. $y = \dfrac{x^2 + 1}{3x}$.

25. $y = \dfrac{2x - 1}{x + 3}$.

26. $y = \dfrac{4x + 1}{x^2 - 5}$.

27. $y = \left(\dfrac{3x + 2}{x}\right)(x^{-5} + 1)$.

28. $y = (2x^7 - x^2)\left(\dfrac{x - 1}{x + 1}\right)$.

29. If $f(4) = 3$ and $f'(4) = -5$, find $g'(4)$.
    (a) $g(x) = \sqrt{x}\, f(x)$
    (b) $g(x) = \dfrac{f(x)}{x}$.

30. If $f(3) = -2$ and $f'(3) = 4$, find $g'(3)$.
    (a) $g(x) = 3x^2 - 5f(x)$
    (b) $g(x) = \dfrac{2x + 1}{f(x)}$.

In Exercises 31–36, the functions involve independent variables other than x. Use the results in this section to find the indicated derivative.

---

31. Find $\dfrac{d}{dt}[16t^2]$.

32. $c = 2\pi r$; find $\dfrac{dc}{dr}$.

33. $V(r) = \pi r^3$; find $V'(r)$.

34. Find $\dfrac{d}{d\alpha}[2\alpha^{-1} + \alpha]$.

35. $s = \dfrac{t}{t^3 + 7}$; find $\dfrac{ds}{dt}$.

36. Find $\dfrac{d}{d\lambda}\left[\dfrac{\lambda\lambda_0 + \lambda^6}{2 - \lambda_0}\right]$　(λ_0 is constant).

37. Newton's law of gravitation states that the magnitude F of the force exerted by a point with mass M on a point with mass m is

$$F = \frac{GmM}{r^2}$$

where G is a constant and r is the distance between the bodies. Assuming that the points are moving, find a formula for the instantaneous rate of change of F with respect to r.

38. The volume of a sphere is $V = \frac{4}{3}\pi r^3$. Assuming that the radius is changing, find a formula for the instantaneous rate of change of V with respect to r.

39. Find d^2y/dx^2.
    (a) $y = 7x^3 - 5x^2 + x$
    (b) $y = 12x^2 - 2x + 3$
    (c) $y = \dfrac{x + 1}{x}$
    (d) $y = (5x^2 - 3)(7x^3 + x)$.

40. Find y''.
    (a) $y = 4x^7 - 5x^3 + 2x$　(b) $y = 3x + 2$
    (c) $y = \dfrac{3x - 2}{5x}$　(d) $y = (x^3 - 5)(2x + 3)$.

41. Find y'''.
    (a) $y = x^{-5} + x^5$　(b) $y = 1/x$
    (c) $y = ax^3 + bx + c$　(a, b, c constant).

42. Find $\dfrac{d^3y}{dx^3}$.
    (a) $y = 5x^2 - 4x + 7$
    (b) $y = 3x^{-2} + 4x^{-1} + x$
    (c) $y = ax^4 + bx^2 + c$　(a, b, c constant).

43. Find
    (a) $f'''(2)$, where $f(x) = 3x^2 - 2$
    (b) $\left.\dfrac{d^2y}{dx^2}\right|_{x=1}$, where $y = 6x^5 - 4x^2$
    (c) $\left.\dfrac{d^4}{dx^4}[x^{-3}]\right|_{x=1}$

44. Find
    (a) $y'''(0)$, where $y = 4x^4 + 2x^3 + 3$
    (b) $\left.\dfrac{d^4y}{dx^4}\right|_{x=1}$, where $y = \dfrac{6}{x^4}$.

45. Show that $y = x^3 + 3x + 1$ satisfies the equation $y''' + xy'' - 2y' = 0$.

46. Show that if $x \neq 0$, then $y = 1/x$ satisfies the equation $x^3y'' + x^2y' - xy = 0$.

47. Find a general formula for $F''(x)$ if $F(x) = xf(x)$ and f and f' are differentiable at x.

48. In the temperature range between $0°\,C$ and $700°\,C$ the resistance R [in ohms (Ω)] of a certain platinum resistance thermometer is given by

$$R = 10 + 0.04124T - 1.779 \times 10^{-5}T^2$$

where T is the temperature in degrees Celsius. Where in the interval from $0°\,C$ to $700°\,C$ is the resistance of the thermometer most sensitive and least sensitive to temperature changes? [*Hint:* Consider the size of dR/dT in the interval $0 \leq T \leq 700$.]

49. At which point(s) does the graph of the equation $y = \frac{1}{3}x^3 - \frac{3}{2}x^2 + 2x$ have a horizontal tangent line?

50. At which point(s) does the graph of $y = \dfrac{x}{x^2 + 9}$ have a horizontal tangent line?

51. Find an equation of the tangent line to the graph of $y = f(x)$ at the point where $x = -3$ if $f(-3) = 2$ and $f'(-3) = 5$.

52. Find an equation for the line that is tangent to $y = (1 - x)/(1 + x)$ at the point where $x = 2$.

53. Find the values of a and b if the tangent to $y = ax^2 + bx$ at $(1, 5)$ has slope $m_{\tan} = 8$.

54. Find the values of a and b if the tangent to

$$y = \frac{a}{x^2} + b$$

at $(2, 4)$ has slope $m_{\tan} = -2$.

55. Find a function $y = ax^2 + bx + c$ whose graph has an x-intercept of 1, a y-intercept of -2, and a tangent line with a slope of -1 at the y-intercept.

56. Find k if the curve $y = x^2 + k$ is tangent to the line $y = 2x$.

57. Find the x-coordinate of the point on the graph of $y = x^2$ where the tangent line is parallel to the secant line that cuts the curve at $x = -1$ and $x = 2$.

58. Find the x-coordinate of the point on the graph of $y = \sqrt{x}$ where the tangent line is parallel to the secant line that cuts the curve at $x = 1$ and $x = 4$.

59. Find the x-coordinate of all points on the graph of $y = 1 - x^2$ at which the tangent line passes through the point $(2, 0)$.

60. Show that any two tangent lines to the parabola $y = ax^2$, $a \neq 0$, intersect at a point that is on the vertical line halfway between the points of tangency.

61. Suppose that L is the tangent line at $x = x_0$ to the graph of the cubic equation $y = ax^3 + bx$. Find the x-coordinate of the point where L intersects the graph a second time.

62. Show that the segment of the tangent line to the graph of $y = 1/x$ that is cut off by the coordinate axes is bisected by the point of tangency.

63. Show that the triangle that is formed by any tangent line to the graph of $y = 1/x$, $x > 0$, and the coordinate axes has an area of 2 square units.

64. Find conditions on a, b, c, and d so that the graph of the polynomial $f(x) = ax^3 + bx^2 + cx + d$ has
(a) exactly two horizontal tangents
(b) exactly one horizontal tangent
(c) no horizontal tangents.

65. Prove: If the graphs of $y = f(x)$ and $y = g(x)$ have parallel tangent lines at $x = c$, then the graph of $y = f(x) - g(x)$ has a horizontal tangent line at $x = c$.

66. (a) Let the functions f, g, and h be differentiable at x. By applying Theorem 3.3.5 twice, show that the product $f \cdot g \cdot h$ is differentiable at x and

$$(f \cdot g \cdot h)'(x) =$$
$$f(x)g(x)h'(x) + f(x)g'(x)h(x) + f'(x)g(x)h(x)$$

(b) State a formula for differentiating a product of n functions.

67. Use the results of Exercise 66 to find

(a) $\dfrac{d}{dx}\left[(2x + 1)\left(1 + \dfrac{1}{x}\right)(x^{-3} + 7)\right]$

(b) $\dfrac{d}{dx}[x^{-5}(x^2 + 2x)(4 - 3x)(2x^9 + 1)]$

(c) $\dfrac{d}{dx}[(x^7 + 2x - 3)^3]$

(d) $\dfrac{d}{dx}[(x^2 + 1)^{50}]$.

68. Prove: If the function f is differentiable at x, then $\dfrac{d}{dx}[f^2(x)] = 2f(x)f'(x)$. [*Hint:* Use the product rule.]

69. Use the result obtained in Exercise 68 to find $\dfrac{d}{dx}(2x^3 - 5x^2 + 7x - 2)^2$.

70. Let $f(x) = \sqrt{x}$. Assuming that f is differentiable at x, use the result of Exercise 68 to show that $f'(x) = 1/(2\sqrt{x})$.

In Exercises 71–75, you will have to determine whether a function f is differentiable at a point x_0 where the formula for f changes. Use the following result:

Theorem. *Let f be continuous at x_0 and suppose that*

$$\lim_{x \to x_0^+} f'(x) \quad \text{and} \quad \lim_{x \to x_0^-} f'(x)$$

exist. Then f is differentiable at x_0 if and only if these limits are equal. Moreover, in the case of equality

$$f'(x_0) = \lim_{x \to x_0^+} f'(x) = \lim_{x \to x_0^-} f'(x)$$

71. Let

$$f(x) = \begin{cases} x^2, & x \leq 1 \\ \sqrt{x}, & x > 1 \end{cases}$$

Determine whether f is differentiable at $x = 1$. If so, find the value of the derivative there.

72. Let

$$f(x) = \begin{cases} x^3 + \frac{1}{16}, & x < \frac{1}{2} \\ \frac{3}{4}x^2, & x \geq \frac{1}{2} \end{cases}$$

Determine whether f is differentiable at $x = \frac{1}{2}$. If so, find the value of the derivative there.

73. Let

$$f(x) = \begin{cases} 3x^2, & x \leq 1 \\ ax + b, & x > 1 \end{cases}$$

Find the values of a and b so that f will be differentiable at $x = 1$.

74. (a) Let

$$f(x) = \begin{cases} x^2, & x \leq 0 \\ x^2 + 1, & x > 0 \end{cases}$$

Show that

$$\lim_{x \to 0^-} f'(x) = \lim_{x \to 0^+} f'(x)$$

but that $f'(0)$ does not exist.

(b) Let

$$f(x) = \begin{cases} x^2, & x \leq 0 \\ x^3, & x > 0 \end{cases}$$

Show that $f'(0)$ exists but $f''(0)$ does not.

75. Find all points where f fails to be differentiable. Justify your answer.
   (a) $f(x) = |3x - 2|$
   (b) $f(x) = |x^2 - 4|$.

76. Prove: If f is differentiable at x and $f(x) \neq 0$, then $1/f(x)$ is differentiable at x and

$$\frac{d}{dx}\left[\frac{1}{f(x)}\right] = -\frac{f'(x)}{[f(x)]^2}$$

77. (a) Find $f^{(n)}(x)$ if $f(x) = x^n$.
   (b) Find $f^{(n)}(x)$ if $f(x) = x^k$ and $n > k$, where k is a positive integer.
   (c) Find $f^{(n)}(x)$ if

$$f(x) = a_0 + a_1 x + a_2 x^2 + \cdots + a_n x^n$$

78. In each part compute f', f'', f''' and then state the formula for $f^{(n)}$.
   (a) $f(x) = 1/x$

(b) $f(x) = 1/x^2$.

[*Hint:* The expression $(-1)^n$ has a value of 1 if n is even and -1 if n is odd. Use this expression in your answer.]

79. (a) Prove:

$$\frac{d^2}{dx^2}[cf(x)] = c\frac{d^2}{dx^2}[f(x)]$$

$$\frac{d^2}{dx^2}[f(x) + g(x)] = \frac{d^2}{dx^2}[f(x)] + \frac{d^2}{dx^2}[g(x)]$$

(b) Do the results in part (a) generalize to nth derivatives? Justify your answer.

80. Prove:

$$(f \cdot g)''(x) = f''(x)g(x) + 2f'(x)g'(x) + f(x)g''(x)$$

81. (a) In our proof of Theorem 3.3.5, we used the fact that

$$\lim_{h \to 0} f(x + h) = f(x)$$

[see Equation (3)]. The argument given used the hypothesis that f is differentiable at x. Find the fallacy in the following proof that makes no assumptions about f: As h approaches 0, the quantity $x + h$ approaches x; conse٬uently

$$\lim_{h \to 0} f(x + h) = f(x)$$

(b) Let

$$f(x) = \begin{cases} x, & x \neq 1 \\ 3, & x = 1 \end{cases}$$

Show that

$$\lim_{h \to 0} f(x + h) \neq f(x)$$

when $x = 1$.

82. Let $f(x) = x^8 - 2x + 3$ and $x_0 = 2$; find

$$\lim_{h \to 0} \frac{f'(x_0 + h) - f'(x_0)}{h}$$

83. (a) Prove: If $f''(x)$ exists for each x in (a, b), then both f and f' are continuous on (a, b).
   (b) What can be said about the continuity of f and its derivatives if $f^{(n)}(x)$ exists for each x in (a, b)?

■ **3.4 DERIVATIVES OF TRIGONOMETRIC FUNCTIONS**

The main objective of this section is to obtain formulas for the derivatives of trigonometric functions.

Recall that in the expressions $\sin x$, $\cos x$, $\tan x$, $\cot x$, $\sec x$, and $\csc x$ it is understood that x is measured in radians. Also, we remind the reader of the limits

$$\lim_{h \to 0} \frac{\sin h}{h} = 1 \quad \text{and} \quad \lim_{h \to 0} \frac{1 - \cos h}{h} = 0$$

derived in Section 2.8.

Let us first consider the problem of differentiating $\sin x$. Let x be any real number. From the definition of a derivative,

$$\frac{d}{dx}[\sin x] = \lim_{h \to 0} \frac{\sin(x + h) - \sin x}{h}$$

$$= \lim_{h \to 0} \frac{\sin x \cos h + \cos x \sin h - \sin x}{h}$$

$$= \lim_{h \to 0} \left[\sin x \left(\frac{\cos h - 1}{h} \right) + \cos x \left(\frac{\sin h}{h} \right) \right]$$

$$= \lim_{h \to 0} \left[\cos x \left(\frac{\sin h}{h} \right) - \sin x \left(\frac{1 - \cos h}{h} \right) \right]$$

Since $\sin x$ and $\cos x$ do not involve h, they remain constant as $h \to 0$; thus,

$$\lim_{h \to 0} (\sin x) = \sin x \quad \text{and} \quad \lim_{h \to 0} (\cos x) = \cos x$$

Consequently,

$$\frac{d}{dx}[\sin x] = \cos x \cdot \lim_{h \to 0} \left(\frac{\sin h}{h} \right) - \sin x \cdot \lim_{h \to 0} \left(\frac{1 - \cos h}{h} \right)$$

$$= \cos x \cdot (1) - \sin x \cdot (0) = \cos x$$

Thus, we have shown that

$$\frac{d}{dx}[\sin x] = \cos x \tag{1}$$

The derivative of $\cos x$ is obtained similarly:

$$\frac{d}{dx}[\cos x] = \lim_{h \to 0} \frac{\cos(x + h) - \cos x}{h}$$

$$= \lim_{h \to 0} \frac{\cos x \cos h - \sin x \sin h - \cos x}{h}$$

$$= \lim_{h \to 0} \left[\cos x \cdot \left(\frac{\cos h - 1}{h} \right) - \sin x \cdot \left(\frac{\sin h}{h} \right) \right]$$

$$= -\cos x \cdot \lim_{h \to 0} \left(\frac{1 - \cos h}{h} \right) - \sin x \cdot \lim_{h \to 0} \left(\frac{\sin h}{h} \right)$$

$$= (-\cos x)(0) - (\sin x)(1) = -\sin x$$

Thus, we have shown that

$$\frac{d}{dx} [\cos x] = -\sin x \tag{2}$$

The derivatives of the remaining trigonometric functions can be obtained using the relationships

$$\tan x = \frac{\sin x}{\cos x} \qquad \cot x = \frac{\cos x}{\sin x} \qquad \sec x = \frac{1}{\cos x} \qquad \csc x = \frac{1}{\sin x}$$

For example,

$$\frac{d}{dx} [\tan x] = \frac{d}{dx} \left[\frac{\sin x}{\cos x} \right]$$

$$= \frac{\cos x \cdot \dfrac{d}{dx} [\sin x] - \sin x \cdot \dfrac{d}{dx} [\cos x]}{\cos^2 x}$$

$$= \frac{\cos x \cdot \cos x - \sin x \cdot (-\sin x)}{\cos^2 x}$$

$$= \frac{\cos^2 x + \sin^2 x}{\cos^2 x} = \frac{1}{\cos^2 x} = \sec^2 x$$

Thus,

$$\frac{d}{dx} [\tan x] = \sec^2 x \tag{3}$$

We leave the remaining formulas for the exercises:

$$\frac{d}{dx} [\cot x] = -\csc^2 x \tag{4}$$

$$\frac{d}{dx} [\sec x] = \sec x \tan x \tag{5}$$

$$\frac{d}{dx} [\csc x] = -\csc x \cot x \tag{6}$$

REMARK.   The derivative formulas for the trigonometric functions should be memorized. An easy way of doing this is discussed in Exercise 38. Moreover, we emphasize that in all of the derivative formulas for the trigonometric functions, x is measured in radians.

Example 1   Find $f'(x)$ if $f(x) = x^2 \tan x$.

Solution.   Using the product rule and Formula (3), we obtain

$$f'(x) = x^2 \cdot \frac{d}{dx} [\tan x] + \tan x \cdot \frac{d}{dx} [x^2] = x^2 \sec^2 x + 2x \tan x \quad \blacktriangleleft$$

Example 2   Find dy/dx if $y = \dfrac{\sin x}{1 + \cos x}$.

Solution.   Using the quotient rule together with Formulas (1) and (2) we obtain

$$\frac{dy}{dx} = \frac{(1 + \cos x) \cdot \dfrac{d}{dx} [\sin x] - \sin x \cdot \dfrac{d}{dx} [1 + \cos x]}{(1 + \cos x)^2}$$

$$= \frac{(1 + \cos x)(\cos x) - (\sin x)(-\sin x)}{(1 + \cos x)^2}$$

$$= \frac{\cos x + \cos^2 x + \sin^2 x}{(1 + \cos x)^2} = \frac{\cos x + 1}{(1 + \cos x)^2} = \frac{1}{1 + \cos x} \quad \blacktriangleleft$$

Example 3   Find $y''(\pi/4)$ if $y(x) = \sec x$.

Solution.

$$y'(x) = \sec x \tan x$$

$$y''(x) = \sec x \cdot \frac{d}{dx} [\tan x] + \tan x \cdot \frac{d}{dx} [\sec x]$$

$$= \sec x \cdot \sec^2 x + \tan x \cdot \sec x \tan x$$

$$= \sec^3 x + \sec x \tan^2 x$$

Thus,

$$y''(\pi/4) = \sec^3 (\pi/4) + \sec (\pi/4) \tan^2 (\pi/4)$$
$$= (\sqrt{2})^3 + (\sqrt{2})(1)^2 = 3\sqrt{2} \quad \blacktriangleleft$$

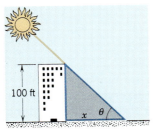

Figure 3.4.1

Example 4 Suppose that the rising sun passes directly over a building that is 100 feet high, and let θ be the sun's angle of elevation (Figure 3.4.1). Find the rate at which the length x of the building's shadow is changing with respect to θ when $\theta = 45°$. Express the answer in units of feet/degree.

Solution. The variables x and θ are related by the equation $\tan \theta = 100/x$, or equivalently,

$$x = 100 \cot \theta \tag{7}$$

If we assume for the moment that θ is measured in radians, then Formula (4) is applicable, and (7) yields

$$\frac{dx}{d\theta} = -100 \csc^2 \theta$$

which is the rate of change of shadow length with respect to the elevation angle θ in units of feet/radian. When $\theta = 45°$ (or equivalently, $\theta = \pi/4$ radians), we obtain

$$\left. \frac{dx}{d\theta} \right|_{\theta = \pi/4} = -100 \csc^2 (\pi/4) = -200 \text{ feet/radian}$$

Converting radians to degrees yields

$$-200 \frac{\text{feet}}{\text{radian}} \cdot \frac{\pi}{180} \frac{\text{radians}}{\text{degree}} = -\frac{10}{9} \pi \approx -3.49 \text{ feet/degree}$$

Thus, when $\theta = 45°$, the shadow length is decreasing (because of the minus sign) at an approximate rate of 3.49 feet/degree increase in elevation. ◄

► Exercise Set 3.4   ⃞C 29, 30, 31, 32

In Exercises 1–18, find $f'(x)$.

1. $f(x) = 2 \cos x - 3 \sin x$.

2. $f(x) = \sin x \cos x$.

3. $f(x) = \dfrac{\sin x}{x}$.

4. $f(x) = x^2 \cos x$.

5. $f(x) = x^3 \sin x - 5 \cos x$.

6. $f(x) = \dfrac{\cos x}{x \sin x}$.

7. $f(x) = \sec x - \sqrt{2} \tan x$.

8. $f(x) = (x^2 + 1) \sec x$.

9. $f(x) = \sec x \tan x$.

10. $f(x) = \dfrac{\sec x}{1 + \tan x}$.

11. $f(x) = x - 4 \csc x + 2 \cot x$.

12. $f(x) = \csc x \cot x$.

13. $f(x) = \dfrac{\cot x}{1 + \csc x}$.

14. $f(x) = \dfrac{\csc x}{\tan x}$.

15. $f(x) = \sin^2 x + \cos^2 x$.

16. $f(x) = \dfrac{1}{\cot x}$.

17. $f(x) = \dfrac{\sin x \sec x}{1 + x \tan x}$.

18. $f(x) = \dfrac{(x^2 + 1) \cot x}{3 - \cos x \csc x}$.

In Exercises 19–23, find d^2y/dx^2.

19. $y = x \cos x$.

20. $y = \csc x$.

21. $y = x \sin x - 3 \cos x$.

22. $y = x^2 \cos x + 4 \sin x$.

23. $y = \sin x \cos x$.

In Exercises 24 and 25, find all points where the graph of f has a horizontal tangent line.

24. (a) $f(x) = \sin x$ (b) $f(x) = \tan x$
    (c) $f(x) = \sec x$.

25. (a) $f(x) = \cos x$ (b) $f(x) = \cot x$
    (c) $f(x) = \csc x$.

26. Find the equation of the line tangent to the graph of $\sin x$ at the point where
    (a) $x = 0$ (b) $x = \pi$ (c) $x = \dfrac{\pi}{4}$.

27. Find the equation of the line tangent to the graph of $\tan x$ at the point where
    (a) $x = 0$ (b) $x = \pi/4$ (c) $x = -\pi/4$.

28. (a) Show that $y = \cos x$ and $y = \sin x$ are solutions of the equation $y'' + y = 0$.
    (b) Show that $y = A \sin x + B \cos x$ is a solution for all constants A and B.

29. A 10-foot ladder leans against a wall at an angle θ with the horizontal, as shown in Figure 3.4.2. The top of the ladder is x feet above the ground. If the bottom of the ladder is pushed toward the wall, find the rate at which x changes with respect to θ when $\theta = 60°$. Express the answer in units of feet/degree.

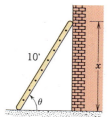

10'

x

θ

Figure 3.4.2

30. An airplane is flying on a horizontal path at a height of 3800 feet, as shown in Figure 3.4.3. At what rate is the distance s between the airplane and the fixed point P changing with respect to θ when $\theta = 30°$? Express the answer in units of feet/degree.

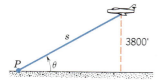

s

3800'

P θ

Figure 3.4.3

31. A searchlight is located 50 meters from a straight wall (Figure 3.4.4). Find the rate at which the distance D is changing with θ when $\theta = 45°$. Express the answer in units of meters/degree.

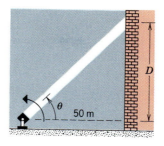

D

θ 50 m

Figure 3.4.4

32. An Earth-observing satellite can see only a portion of the earth's surface. The satellite has horizon sensors that can detect the angle θ shown in Figure 3.4.5. Let r be the radius of the earth (assumed spherical) and h the distance of the satellite from the earth's surface.
    (a) Show that $h = r(\csc\theta - 1)$.
    (b) Using $r = 6378$ kilometers, and assuming that the satellite is getting closer to the earth, find the rate at which h is changing with respect to θ when $\theta = 30°$. Express the answer in units of kilometers/degree. [Adapted from *Space Mathematics*, NASA, 1985.]

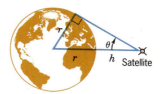

r

θ

r h Satellite

Earth   Figure 3.4.5

33. In each part, determine where f is differentiable.
    (a) $f(x) = \sin x$ (b) $f(x) = \cos x$
    (c) $f(x) = \tan x$ (d) $f(x) = \cot x$
    (e) $f(x) = \sec x$ (f) $f(x) = \csc x$
    (g) $f(x) = \dfrac{1}{1 + \cos x}$ (h) $f(x) = \dfrac{1}{\sin x \cos x}$
    (i) $f(x) = \dfrac{\cos x}{2 - \sin x}$.

34. Derive the formulas
    (a) $\dfrac{d}{dx}[\cot x] = -\csc^2 x$
    (b) $\dfrac{d}{dx}[\sec x] = \sec x \tan x$
    (c) $\dfrac{d}{dx}[\csc x] = -\csc x \cot x$.

35. Let $f(x) = \cos x$. Find all positive integers n for which $f^{(n)}(x) = \sin x$.

36. (a)   Show that $\lim\limits_{h \to 0} \dfrac{\tan h}{h} = 1$.

(b)   Use the result in part (a) to help derive the formula for the derivative of $\tan x$ directly from the definition of a derivative.

37. Without using any trigonometric identities, find

$$\lim_{x \to 0} \frac{\tan(x + y) - \tan y}{x}$$

38. Let us agree to call the functions $\cos x$, $\cot x$, and $\csc x$ the **cofunctions** of $\sin x$, $\tan x$, and $\sec x$, respectively. Convince yourself that the derivative of any cofunction can be obtained from the derivative of the corresponding function by introducing a minus sign and replacing each function in the derivative by its cofunction. Memorize the derivatives of $\sin x$, $\tan x$, and $\sec x$ and then use the above observation to deduce the derivatives of the cofunctions.

39. The derivative formulas for $\sin x$, $\cos x$, $\tan x$, $\cot x$, $\sec x$, and $\csc x$ were obtained under the assumption that x is measured in radians. This exercise shows that different (more complicated) formulas result if x is measured in degrees. Prove that if h and x are degree measures, then

(a) $\lim\limits_{h \to 0} \dfrac{\cos h - 1}{h} = 0$ (b) $\lim\limits_{h \to 0} \dfrac{\sin h}{h} = \dfrac{\pi}{180}$

(c) $\dfrac{d}{dx}[\sin x] = \dfrac{\pi}{180} \cos x$.

■ 3.5 THE CHAIN RULE

In this section we shall derive a formula that expresses the derivative of a composition $f \circ g$ in terms of the derivatives of f and g. This formula will enable us to differentiate complicated functions using known derivatives of simpler functions.

□ **DERIVATIVES OF COMPOSITIONS**

3.5.1 PROBLEM.   *If we know the derivatives of f and g, how can we use this information to find the derivative of the composition $f \circ g$?*

The key to solving this problem is to introduce dependent variables

$$y = (f \circ g)(x) = f(g(x)) \quad \text{and} \quad u = g(x)$$

so that $y = f(u)$. We are interested in using the known derivatives

$$\frac{dy}{du} = f'(u) \quad \text{and} \quad \frac{du}{dx} = g'(x)$$

to find the unknown derivative

$$\frac{dy}{dx} = \frac{d}{dx}[f(g(x))]$$

Stated another way, we are interested in using the known rates of change dy/du and du/dx to find the unknown rate of change dy/dx. But intuition suggests that rates of change multiply. For example, if y changes at 4 times the rate of u and u changes at 2 times the rate of x, then y changes at $4 \times 2 = 8$ times the rate

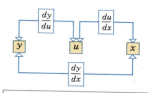

Rates of change multiply:
$$\frac{dy}{dx} = \frac{dy}{du} \cdot \frac{du}{dx}$$

Figure 3.5.1

of x. Thus, Figure 3.5.1 suggests that

$$\frac{dy}{dx} = \frac{dy}{du} \cdot \frac{du}{dx}$$

These ideas are formalized in the following theorem.

3.5.2 THEOREM (*The Chain Rule*). *If g is differentiable at the point x and f is differentiable at the point $g(x)$, then the composition $f \circ g$ is differentiable at the point x. Moreover, if*

$$y = f(g(x)) \quad and \quad u = g(x)$$

then $y = f(u)$ and

$$\frac{dy}{dx} = \frac{dy}{du} \cdot \frac{du}{dx} \tag{1}$$

The proof of this result, which is optional, is given in Section III of Appendix C.

Example 1 Find dy/dx if $y = 4 \cos (x^3)$.

Solution. Let $u = x^3$ so that

$$y = 4 \cos u$$

By the chain rule,

$$\frac{dy}{dx} = \frac{dy}{du} \cdot \frac{du}{dx} = \frac{d}{du} [4 \cos u] \cdot \frac{d}{dx} [x^3]$$

$$= (-4 \sin u) \cdot (3x^2) = (-4 \sin (x^3)) \cdot (3x^2) = -12x^2 \sin (x^3) \quad \blacktriangleleft$$

REMARK. Formula (1) is easy to remember because the left side is exactly what results if we "cancel" the du's on the right side. This "canceling" device provides a good way to remember the chain rule when variables other than x, y, and u are used.

Example 2 Find dw/dt if

$$w = \tan x \quad and \quad x = 4t^3 + t$$

Solution. In this case, the chain rule takes the form

$$\frac{dw}{dt} = \frac{dw}{dx} \cdot \frac{dx}{dt} = \frac{d}{dx} [\tan x] \cdot \frac{d}{dt} [4t^3 + t]$$

$$= (\sec^2 x)(12t^2 + 1) = (12t^2 + 1) \sec^2 (4t^3 + t) \quad \blacktriangleleft$$

☐ **GENERALIZED**
DERIVATIVE
FORMULAS

Although Formula (1) is useful, it is sometimes unwieldy because it involves so many dependent variables. A simpler version of the chain rule can be obtained by noting that $y = f(u)$ in (1), so

$$\frac{dy}{dx} = \frac{d}{dx}[f(u)] \quad \text{and} \quad \frac{dy}{du} = f'(u)$$

Substituting these expressions in (1) yields the following alternative form of the chain rule.

$$\frac{d}{dx}[f(u)] = f'(u)\frac{du}{dx} \tag{2}$$

This very powerful formula vastly extends our differentiation capabilities. For example, to differentiate the function

$$f(x) = (x^2 - x + 1)^{23} \tag{3}$$

we can let $u = x^2 - x + 1$, so (3) becomes $f(u) = u^{23}$, then apply (2) to obtain

$$\frac{d}{dx}[(x^2 - x + 1)^{23}] = \frac{d}{dx}[u^{23}] = \underbrace{23u^{22}}_{f'(u)}\frac{du}{dx}$$

$$= 23(x^2 - x + 1)^{22}\frac{d}{dx}[x^2 - x + 1]$$

$$= 23(x^2 - x + 1)^{22} \cdot (2x - 1)$$

More generally, if u were any other differentiable function of x, the pattern of computations would be virtually the same. For example, if $u = \cos x$, then

$$\frac{d}{dx}[\cos^{23} x] = \frac{d}{dx}[u^{23}] = 23u^{22}\frac{du}{dx} = 23\cos^{22} x \frac{d}{dx}[\cos x]$$

$$= 23\cos^{22} x \cdot (-\sin x) = -23\sin x \cos^{22} x$$

☐ **GENERALIZED**
DERIVATIVE
FORMULAS

In both of the preceding computations, the chain rule took the form

$$\frac{d}{dx}[u^{23}] = 23u^{22}\frac{du}{dx} \tag{4}$$

This formula is a generalization of the more basic formula

$$\frac{d}{dx}[x^{23}] = 23x^{22} \tag{5}$$

In fact, in the special case where $u = x$, Formula (4) reduces to (5) since

$$\frac{d}{dx}[u^{23}] = 23u^{22}\frac{du}{dx} = 23x^{22}\frac{d[x]}{dx} = 23x^{22}$$

Table 3.5.1 contains a list of *generalized derivative formulas* that are conse-quences of (2).

Table 3.5.1

GENERALIZED DERIVATIVE FORMULAS

$$\frac{d}{dx}[u^n] = nu^{n-1}\frac{du}{dx} \quad (n \text{ an integer}) \qquad \frac{d}{dx}[\sqrt{u}] = \frac{1}{2\sqrt{u}}\frac{du}{dx}$$

$$\frac{d}{dx}[\sin u] = \cos u \frac{du}{dx} \qquad \frac{d}{dx}[\cos u) = -\sin u \frac{du}{dx}$$

$$\frac{d}{dx}[\tan u] = \sec^2 u \frac{du}{dx} \qquad \frac{d}{dx}[\cot u] = -\csc^2 u \frac{du}{dx}$$

$$\frac{d}{dx}[\sec u] = \sec u \tan u \frac{du}{dx} \qquad \frac{d}{dx}[\csc u] = -\csc u \cot u \frac{du}{dx}$$

Example 3   Find

(a) $\dfrac{d}{dx}[\sin(2x)]$ (b) $\dfrac{d}{dx}[\tan(x^2 + 1)]$

Solution (a).   Taking $u = 2x$ in the generalized derivative formula for $\sin u$ yields

$$\frac{d}{dx}[\sin(2x)] = \frac{d}{dx}[\sin u] = \cos u \frac{du}{dx} = \cos 2x \cdot \frac{d}{dx}[2x]$$

$$= \cos 2x \cdot 2 = 2\cos 2x$$

Solution (b).   Taking $u = x^2 + 1$ in the generalized derivative formula for $\tan u$ yields

$$\frac{d}{dx}[\tan(x^2 + 1)] = \frac{d}{dx}[\tan u] = \sec^2 u \frac{du}{dx}$$

$$= \sec^2(x^2 + 1) \cdot \frac{d}{dx}[x^2 + 1] = \sec^2(x^2 + 1) \cdot 2x$$

$$= 2x \sec^2(x^2 + 1) \qquad \blacktriangleleft$$

Example 4   Find

(a) $\dfrac{d}{dx}[\sqrt{x^3 + \csc x}]$ (b) $\dfrac{d}{dx}[(1 + x^5 \cot x)^{-8}]$

Solution (a).   Taking $u = x^3 + \csc x$ in the generalized derivative formula for $\sqrt{u}$ yields

$$\frac{d}{dx}[\sqrt{x^3 + \csc x}] = \frac{d}{dx}[\sqrt{u}] = \frac{1}{2\sqrt{u}}\frac{du}{dx} = \frac{1}{2\sqrt{x^3 + \csc x}} \cdot \frac{d}{dx}[x^3 + \csc x]$$

$$= \frac{1}{2\sqrt{x^3 + \csc x}} \cdot (3x^2 - \csc x \cot x) = \frac{3x^2 - \csc x \cot x}{2\sqrt{x^3 + \csc x}}$$

Solution (b). Taking $u = 1 + x^5 \cot x$ in the generalized derivative formula for u^{-8} yields

$$\frac{d}{dx}[(1 + x^5 \cot x)^{-8}] = \frac{d}{dx}[u^{-8}] = -8u^{-9}\frac{du}{dx}$$

$$= -8(1 + x^5 \cot x)^{-9} \cdot \frac{d}{dx}[1 + x^5 \cot x]$$

$$= -8(1 + x^5 \cot x)^{-9} \cdot (x^5(-\csc^2 x) + 5x^4 \cot x)$$

$$= (8x^5 \csc^2 x - 40x^4 \cot x)(1 + x^5 \cot x)^{-9} \quad \blacktriangleleft$$

As the following example shows, it is sometimes necessary to apply the chain rule more than once to find a derivative.

Example 5 Find

(a) $\dfrac{d}{dx}[\cos^2 \pi x]$ (b) $\dfrac{d}{dx}[\sin(\sqrt{1 + \cos x})]$

Solution (a). Taking $u = \cos \pi x$ in the generalized derivative formula for u^2 yields

$$\frac{d}{dx}[\cos^2 \pi x] = \frac{d}{dx}[u^2] = 2u\frac{du}{dx}$$

$$= 2 \cos \pi x \cdot \frac{d}{dx}[\cos \pi x]$$

$$= 2 \cos \pi x \cdot (-\pi \sin \pi x)$$

> We used the generalized derivative formula for $\cos u$ with $u = \pi x$.

$$= -2\pi \sin \pi x \cos \pi x$$

Solution (b). Taking $u = \sqrt{1 + \cos x}$ in the generalized derivative formula for $\sin u$ yields

$$\frac{d}{dx}[\sin(\sqrt{1 + \cos x})] = \frac{d}{dx}[\sin u] = \cos u \frac{du}{dx}$$

$$= \cos(\sqrt{1 + \cos x}) \cdot \frac{d}{dx}[\sqrt{1 + \cos x}]$$

> We used the generalized derivative formula for $\sqrt{u}$ with $u = 1 + \cos x$.

$$= \cos(\sqrt{1 + \cos x}) \cdot \frac{-\sin x}{2\sqrt{1 + \cos x}}$$

$$= -\frac{\sin x \cos(\sqrt{1 + \cos x})}{2\sqrt{1 + \cos x}} \quad \blacktriangleleft$$

Example 6 Find $d\mu/dt$ if $\mu = t^2 \sec \sqrt{\omega t}$, where ω is a constant.

Solution. Because the independent variable is t rather than x, appropriate adjustments in notation have to be made.

$$\frac{d\mu}{dt} = \frac{d}{dt} [t^2 \sec \sqrt{\omega t}] = t^2 \frac{d}{dt} [\sec \sqrt{\omega t}] + (\sec \sqrt{\omega t}) \cdot 2t$$

$$= t^2 \sec \sqrt{\omega t} \tan \sqrt{\omega t} \frac{d}{dt} [\sqrt{\omega t}] + 2t \sec \sqrt{\omega t} \qquad \boxed{\begin{array}{l} \text{We used the generalized} \\ \text{derivative formula for} \\ \sec u \text{ with } u = \sqrt{\omega t}. \end{array}}$$

$$= t^2 \sec \sqrt{\omega t} \tan \sqrt{\omega t} \frac{\omega}{2\sqrt{\omega t}} + 2t \sec \sqrt{\omega t} \qquad \boxed{\begin{array}{l} \text{We used the generalized} \\ \text{derivative formula for} \\ \sqrt{u} \text{ with } u = \omega t. \end{array}} \qquad ◀$$

☐ **AN ALTERNATIVE APPROACH TO USING THE CHAIN RULE**

As you become more comfortable with using the chain rule, you may want to dispense with actually writing out the expression for u in your computations. To accomplish this, it is helpful to express Formula (2) in words. If we call u the "inside function" and f the "outside function" in the composition $f(u)$, then (2) states:

> *The derivative of $f(u)$ is the derivative of the outside function evaluated at the inside function times the derivative of the inside function.*

For example,

$$\frac{d}{dx} [\cos (x^2 + 9)] = \underbrace{-\sin (x^2 + 9)}_{\substack{\text{Derivative of} \\ \text{the outside} \\ \text{evaluated at} \\ \text{the inside}}} \cdot \underbrace{2x}_{\substack{\text{Derivative} \\ \text{of the inside}}}$$

$$\frac{d}{dx} [\tan^2 x] = \underbrace{(2 \tan x)}_{\substack{\text{Derivative of} \\ \text{the outside} \\ \text{evaluated at} \\ \text{the inside}}} \cdot \underbrace{(\sec^2 x)}_{\substack{\text{Derivative} \\ \text{of the inside}}} = 2 \tan x \sec^2 x$$

In general, if $f(g(x))$ is a composition of functions in which the inside function g and the outside function f are differentiable, then

$$\frac{d}{dx} [f(g(x))] = \underbrace{f'(g(x))}_{\substack{\text{Derivative of} \\ \text{the outside} \\ \text{evaluated at} \\ \text{the inside}}} \cdot \underbrace{g'(x)}_{\substack{\text{Derivative} \\ \text{of the inside}}} \qquad (6)$$

▶ **Exercise Set 3.5**   ☐ 53, 54

In Exercises 1–39, find $f'(x)$.

1. $f(x) = (x^3 + 2x)^{37}$.

2. $f(x) = (3x^2 + 2x - 1)^6$.

3. $f(x) = \left(x^3 - \frac{7}{x} \right)^{-2}$.

4. $f(x) = \dfrac{1}{(x^5 - x + 1)^9}$.

5. $f(x) = \dfrac{4}{(3x^2 - 2x + 1)^3}$.

6. $f(x) = \sqrt{x^3 - 2x + 5}$.

7. $f(x) = \sqrt{4 + 3\sqrt{x}}$.

8. $f(x) = \sin^3 x$.

9. $f(x) = \sin (x^3)$.

10. $f(x) = \cos^2 (3\sqrt{x})$.

11. $f(x) = \tan (4x^2)$.

12. $f(x) = 3 \cot^4 x$.

13. $f(x) = 4 \cos^5 x$.

14. $f(x) = \csc(x^3)$.

15. $f(x) = \sin\left(\dfrac{1}{x^2}\right)$.

16. $f(x) = \tan^4(x^3)$.

17. $f(x) = 2 \sec^2(x^7)$.

18. $f(x) = \cos^3\left(\dfrac{x}{x+1}\right)$.

19. $f(x) = \sqrt{\cos(5x)}$.

20. $f(x) = \sqrt{3x - \sin^2(4x)}$.

21. $f(x) = [x + \csc(x^3 + 3)]^{-3}$.

22. $f(x) = [x^4 - \sec(4x^2 - 2)]^{-4}$.

23. $f(x) = x^2\sqrt{5 - x^2}$.

24. $f(x) = \dfrac{x}{\sqrt{1 - x^2}}$.

25. $f(x) = x^3 \sin^2(5x)$.

26. $f(x) = \sqrt{x} \tan^3(\sqrt{x})$.

27. $f(x) = x^5 \sec(1/x)$.

28. $f(x) = \dfrac{\sin x}{\sec(3x + 1)}$.

29. $f(x) = \cos(\cos x)$.

30. $f(x) = \sin(\tan 3x)$.

31. $f(x) = \cos^3(\sin 2x)$.

32. $f(x) = \dfrac{1 + \csc(x^2)}{1 - \cot(x^2)}$.

33. $f(x) = (5x + 8)^{13}(x^3 + 7x)^{12}$.

34. $f(x) = (2x - 5)^2(x^2 + 4)^3$.

35. $f(x) = \left(\dfrac{x - 5}{2x + 1}\right)^3$.

36. $f(x) = \left(\dfrac{1 + x^2}{1 - x^2}\right)^{17}$.

37. $f(x) = \dfrac{(2x + 3)^3}{(4x^2 - 1)^8}$.

38. $f(x) = [1 + \sin^3(x^5)]^{12}$.

39. $f(x) = [x \sin 2x + \tan^4(x^7)]^5$.

In Exercises 40–42, find d^2y/dx^2.

40. $y = \sin(3x^2)$.

41. $y = x \cos(5x) - \sin^2 x$.

42. $y = x \tan\left(\dfrac{1}{x}\right)$.

In Exercises 43–46, find an equation for the tangent to the graph at the specified point.

43. $y = x \cos 3x$, $x = \pi$.

44. $y = \sin(1 + x^3)$, $x = -3$.

45. $y = \sec^3\left(\dfrac{\pi}{2} - x\right)$, $x = -\dfrac{\pi}{2}$.

46. $y = \left(x - \dfrac{1}{x}\right)^3$, $x = 2$.

In Exercises 47–50, find the indicated derivative.

47. $y = \cot^3(\pi - \theta)$; find $\dfrac{dy}{d\theta}$.

48. $\lambda = \left(\dfrac{au + b}{cu + d}\right)^6$; find $\dfrac{d\lambda}{du}$ (a, b, c, d constants).

49. $\dfrac{d}{d\omega}[a \cos^2 \pi\omega + b \sin^2 \pi\omega]$ (a, b constants).

50. $x = \csc^2\left(\dfrac{\pi}{3} - y\right)$; find $\dfrac{dx}{dy}$.

51. If an object suspended from a spring is displaced vertically from its equilibrium position by a small amount and released, and if the air resistance and the mass of the spring are ignored, then the resulting oscillation of the object is called **simple harmonic motion**. For such motion the displacement y from equilibrium in terms of time t is given by

$$y = A \cos \omega t$$

where A is the initial displacement at time $t = 0$, and ω is a constant that depends on the mass of the object and the stiffness of the spring (Figure 3.5.2). The constant $|A|$ is called the **amplitude** of the motion and ω the **angular frequency**.

(a) Show that

$$\dfrac{d^2y}{dt^2} = -\omega^2 y$$

(b) The **period** T is the time required to make one complete oscillation. Show that $T = 2\pi/\omega$.

(c) The **frequency** f of the vibration is the number of oscillations per unit time. Find f in terms of the period T.

(d) Find the amplitude, period, and frequency of an object that is executing simple harmonic motion given by

$$y = 0.6 \cos 15t$$

where t is in seconds and y is in centimeters.

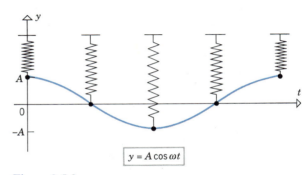

Figure 3.5.2

52. Find the value of the constant A so that $y = A \sin 3t$ satisfies the equation

$$\frac{d^2y}{dt^2} + 2y = 4 \sin 3t$$

53. Figure 3.5.3 shows the graph of atmospheric pressure p (lb/in²) versus the altitude h (mi) above sea level.
   (a) From the graph and the tangent line at $h = 2$ shown on the graph, estimate the values of p and dp/dh at an altitude of 2 mi.
   (b) If the altitude of a space vehicle is increasing at the rate of 0.3 mi/sec at the instant when it is 2 mi above sea level, how fast is the pressure changing with time at this instant?

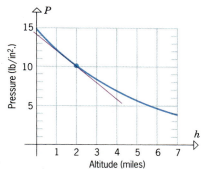

Figure 3.5.3

54. The force F (in pounds) acting at an angle θ with the horizontal that is needed to drag a crate weighing W lb along a horizontal surface at a constant velocity is given by

$$F = \frac{\mu W}{\cos \theta + \mu \sin \theta}$$

where μ is a constant called the **coefficient of sliding friction** between the crate and the surface (Figure 3.5.4). Suppose that the crate weighs 150 lb and that $\mu = 0.3$.
   (a) Find $dF/d\theta$ when $\theta = 30°$. Express the answer in units of pounds/degree.
   (b) Find dF/dt when $\theta = 30°$ if θ is decreasing at the rate of 0.5°/sec at this instant.

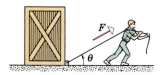

Figure 3.5.4

In Exercises 55–58, find the indicated derivative in terms of x, y, and dy/dx, assuming that y is a differentiable function of x.

55. $\dfrac{d}{dx}(x^2y^3)$.

56. $\dfrac{d}{dx}\left(\dfrac{x}{y^2}\right)$.

57. $\dfrac{d}{dx}[\sin(xy)]$.

58. $\dfrac{d}{dx}(\sqrt{x^2 + y^2})$.

In Exercises 59–62, find the indicated derivative in terms of x, y, dx/dt, and dy/dt, assuming that x and y are differentiable functions of t.

59. $\dfrac{d}{dt}(x^2 + y^2)$.

60. $\dfrac{d}{dt}[\tan(xy^3)]$.

61. $\dfrac{d}{dt}(x^2\sqrt{y})$.

62. $\dfrac{d}{dt}\left(\dfrac{y^2}{x}\right)$.

63. Recall that

$$\frac{d}{dx}(|x|) = \begin{cases} 1, & x > 0 \\ -1, & x < 0 \end{cases}$$

Use this result and the chain rule to find

$$\frac{d}{dx}(|\sin x|)$$

for nonzero x in the interval $(-\pi, \pi)$.

64. Use the derivative formula for $\sin x$ and the identity

$$\cos x = \sin\left(\frac{\pi}{2} - x\right)$$

to obtain the derivative formula for $\cos x$.

65. Let

$$f(x) = \begin{cases} x \sin \dfrac{1}{x}, & x \neq 0 \\ 0, & x = 0 \end{cases}$$

   (a) Find $f'(x)$ for $x \neq 0$.
   (b) Show that f is continuous at $x = 0$.
   (c) Use Definition 3.2.1 to show that $f'(0)$ does not exist.

66. Let

$$f(x) = \begin{cases} x^2 \sin \dfrac{1}{x}, & x \neq 0 \\ 0, & x = 0 \end{cases}$$

   (a) Find $f'(x)$ for $x \neq 0$.
   (b) Show that f is continuous at $x = 0$.
   (c) Use Definition 3.2.1 to find $f'(0)$.
   (d) Show that f' is not continuous at $x = 0$.

67. Given the following table of values, find the indicated derivatives in parts (a) and (b).

x	$f(x)$	$f'(x)$
2	1	7
8	5	−3

(a) $g'(2)$, where $g(x) = [f(x)]^3$
(b) $h'(2)$, where $h(x) = f(x^3)$.

68. Given the following table of values, find the indicated derivatives in parts (a) and (b).

x	$f(x)$	$f'(x)$	$g(x)$	$g'(x)$
−1	2	3	2	−3
2	0	4	1	−5

(a) $F'(-1)$, where $F(x) = f(g(x))$
(b) $G'(-1)$, where $G(x) = g(f(x))$.

69. Given that $f'(0) = 2$, $g(0) = 0$, and $g'(0) = 3$, find $(f \circ g)'(0)$.

70. Given that $f'(x) = \sqrt{3x + 4}$ and $g(x) = x^2 - 1$, find $F'(x)$ if $F(x) = f(g(x))$.

71. Given that $f'(x) = \dfrac{x}{x^2 + 1}$ and $g(x) = \sqrt{3x - 1}$, find $F'(x)$ if $F(x) = f(g(x))$.

72. Find $f'(x^2)$ if $\dfrac{d}{dx}[f(x^2)] = x^2$.

73. Find $\dfrac{d}{dx}[f(x)]$ if $\dfrac{d}{dx}[f(3x)] = 6x$.

74. A function f is **even** if $f(-x) = f(x)$; **odd** if $f(-x) = -f(x)$, for all x in the domain of f. Assuming that f is differentiable, prove:
(a) f' is odd if f is even
(b) f' is even if f is odd.

75. Find a formula for

$$\frac{d}{dx}[f(g(h(x)))]$$

76. Let $y = f_1(u)$, $u = f_2(v)$, $v = f_3(w)$, and $w = f_4(x)$. Express dy/dx in terms of dy/du, dw/dx, du/dv, and dv/dw.

■ **3.6** IMPLICIT DIFFERENTIATION

> *In the previous sections of this chapter we showed how to differentiate functions defined explicitly by equations of the form $y = f(x)$. In this section we shall learn to find derivatives of functions that are defined implicitly.*

☐ **THE METHOD OF IMPLICIT DIFFERENTIATION**

Consider the equation

$$xy = 1 \tag{1}$$

One way to obtain dy/dx is to rewrite this equation as

$$y = \frac{1}{x} \tag{2}$$

from which it follows that

$$\frac{dy}{dx} = \frac{d}{dx}\left[\frac{1}{x}\right] = -\frac{1}{x^2}$$

However, there is another possibility. We can differentiate both sides of (1) *before* solving for y in terms of x, treating y as a (temporarily unspecified) differentiable function of x. With this approach we obtain

$$\frac{d}{dx}[xy] = \frac{d}{dx}[1]$$

$$x\frac{d}{dx}[y] + y\frac{d}{dx}[x] = 0$$

$$x\frac{dy}{dx} + y = 0$$

$$\frac{dy}{dx} = -\frac{y}{x}$$

If we now substitute (2) into the last expression, we obtain

$$\frac{dy}{dx} = -\frac{1}{x^2}$$

which agrees with the previous computation. This second method of obtaining derivatives is called *implicit differentiation.* It is especially useful when it is ·inconvenient or impossible to solve explicitly for y in terms of x.

Example 1   By implicit differentiation find dy/dx if $5y^2 + \sin y = x^2$.

Solution.   Differentiating both sides with respect to x and treating y as a differentiable function of x, we obtain

$$\frac{d}{dx}[5y^2 + \sin y] = \frac{d}{dx}[x^2]$$

$$5\frac{d}{dx}[y^2] + \frac{d}{dx}[\sin y] = 2x$$

$$5\left(2y\frac{dy}{dx}\right) + (\cos y)\frac{dy}{dx} = 2x \qquad \boxed{\begin{array}{l}\text{The chain rule was}\\ \text{used here because}\\ y \text{ is a function of } x.\end{array}}$$

$$10y\frac{dy}{dx} + (\cos y)\frac{dy}{dx} = 2x$$

Solving for dy/dx, we obtain

$$\frac{dy}{dx} = \frac{2x}{10y + \cos y} \tag{3}$$

Note that this formula for dy/dx involves both the variables x and y. In order to obtain a formula involving x alone we would have to solve the original equation for y in terms of x and substitute in (3). However, it is impossible to do this, so the formula for dy/dx must be left in terms of x and y. ◀

In problems where dy/dx is used to calculate the slope of a tangent line at a point on a curve, the x and y coordinates of the point are often both known, so that there is no problem using a formula for dy/dx that involves both x and y.

Example 2   Find the slope of the tangent line at $(4, 0)$ to the graph of

$$7y^4 + x^3y + x = 4 \tag{4}$$

[Note that $x = 4$, $y = 0$ satisfies the equation so that $(4, 0)$ is actually a point on the graph.]

Solution.   It is difficult to solve (4) for y in terms of x, so we shall differentiate implicitly. We obtain

$$\frac{d}{dx}[7y^4 + x^3y + x] = \frac{d}{dx}[4]$$

$$\frac{d}{dx}[7y^4] + \frac{d}{dx}[x^3y] + \frac{d}{dx}[x] = 0$$

$$\frac{d}{dx}[7y^4] + \left(x^3\frac{dy}{dx} + y\frac{d}{dx}[x^3]\right) + \frac{d}{dx}[x] = 0$$

$$28y^3\frac{dy}{dx} + x^3\frac{dy}{dx} + 3yx^2 + 1 = 0 \qquad \boxed{\text{Note the use of the chain rule in the first term.}}$$

Solving for dy/dx yields

$$\frac{dy}{dx} = -\frac{3yx^2 + 1}{28y^3 + x^3} \tag{5}$$

At the point $(4, 0)$ we have $x = 4$ and $y = 0$, so (5) yields

$$m_{\tan} = \frac{dy}{dx}\bigg|_{\substack{x=4 \\ y=0}} = -\frac{1}{64} \qquad \blacktriangleleft$$

Example 3   Find the slope of the tangent line at $(2, -1)$ to $y^2 - x + 1 = 0$ (Figure 3.6.1).

Solution.   Differentiating implicitly yields

$$\frac{d}{dx}[y^2 - x + 1] = \frac{d}{dx}[0]$$

$$\frac{d}{dx}[y^2] - \frac{d}{dx}[x] + \frac{d}{dx}[1] = \frac{d}{dx}[0]$$

$$2y\frac{dy}{dx} - 1 = 0$$

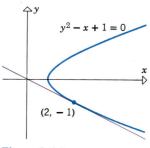

$y^2 - x + 1 = 0$

$(2, -1)$

Figure 3.6.1

$$\frac{dy}{dx} = \frac{1}{2y}$$

At $(2, -1)$ we have $y = -1$, so the slope of the tangent line there is

$$m_{\text{tan}} = \frac{dy}{dx}\bigg|_{y=-1} = -\frac{1}{2}$$

Alternative Solution.   If we solve $y^2 - x + 1 = 0$ for y in terms of x, we obtain

$$y = \sqrt{x - 1} \quad \text{and} \quad y = -\sqrt{x - 1}$$

As shown in Figure 3.6.2, the graph of the first equation is the upper half of the curve $y^2 - x + 1 = 0$ (since $y \geq 0$), and the graph of the second is the lower half (since $y \leq 0$).

Since the point $(2, -1)$ lies on the lower half of the curve, the slope m_{tan} of the tangent line at this point can be obtained by evaluating the derivative of $y = -\sqrt{x - 1}$ when $x = 2$; thus,

$$\frac{dy}{dx} = \frac{d}{dx}[-\sqrt{x-1}] = -\frac{1}{2\sqrt{x-1}} \cdot \frac{d}{dx}[x-1] = -\frac{1}{2\sqrt{x-1}}$$

and

$$m_{\text{tan}} = \frac{dy}{dx}\bigg|_{x=2} = -\frac{1}{2}$$

which agrees with the solution obtained by differentiating implicitly.   ◀

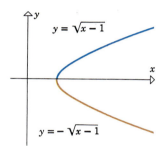

$y = \sqrt{x-1}$

$y = -\sqrt{x-1}$

Figure 3.6.2

Example 4   Use implicit differentiation to find d^2y/dx^2 if $4x^2 - 2y^2 = 9$.

Solution.   Differentiating both sides of $4x^2 - 2y^2 = 9$ implicitly yields

$$8x - 4y\frac{dy}{dx} = 0$$

from which we obtain

$$\frac{dy}{dx} = \frac{2x}{y} \tag{6}$$

Differentiating both sides of (6) implicitly yields

$$\frac{d^2y}{dx^2} = \frac{(y)(2) - (2x)(dy/dx)}{y^2} \tag{7}$$

Substituting (6) into (7) and simplifying, we obtain

$$\frac{d^2y}{dx^2} = \frac{2y - 2x(2x/y)}{y^2} = \frac{2y^2 - 4x^2}{y^3}$$

Finally, using the original equation to simplify further, we obtain

$$\frac{d^2y}{dx^2} = \frac{(-9)}{y^3} = -\frac{9}{y^3} \quad \blacktriangleleft$$

□ **DERIVATIVES OF RATIONAL POWERS OF x**

In Section 3.3 we showed that the formula

$$\frac{d}{dx}[x^n] = nx^{n-1} \tag{8}$$

holds for integer values of n and for $n = \frac{1}{2}$. By using implicit differentiation we shall show that this formula holds for any rational power of x. More precisely, we shall show that if r is a rational number, then

$$\frac{d}{dx}[x^r] = rx^{r-1} \tag{9}$$

whenever x^r and x^{r-1} are defined. In our computations we shall assume that x^r is differentiable; the justification for this assumption will be considered later.

Let $y = x^r$. Since r is a rational number, it can be expressed as a ratio of integers $r = m/n$. Thus, $y = x^r = x^{m/n}$ can be written as

$$y^n = x^m \qquad \text{so that} \qquad \frac{d}{dx}[y^n] = \frac{d}{dx}[x^m]$$

On differentiating implicitly with respect to x and using (8), we obtain

$$ny^{n-1}\frac{dy}{dx} = mx^{m-1} \tag{10}$$

But

$$y^{n-1} = [x^{m/n}]^{n-1} = x^{m-(m/n)}$$

Thus, (10) can be written as

$$nx^{m-(m/n)}\frac{dy}{dx} = mx^{m-1}$$

so that

$$\frac{dy}{dx} = \frac{m}{n}x^{(m/n)-1} = rx^{r-1}$$

which establishes (9).

Example 5 From (9)

$$\frac{d}{dx}[x^{4/5}] = \frac{4}{5}x^{(4/5)-1} = \frac{4}{5}x^{-1/5}$$

$$\frac{d}{dx}[x^{-7/8}] = -\frac{7}{8}x^{(-7/8)-1} = -\frac{7}{8}x^{-15/8}$$

$$\frac{d}{dx}[\sqrt[3]{x}] = \frac{d}{dx}[x^{1/3}] = \frac{1}{3}x^{-2/3} = \frac{1}{3\sqrt[3]{x^2}} \qquad \blacktriangleleft$$

If u is a differentiable function of x, and r is a rational number, then the chain rule yields the following generalization of (9):

$$\frac{d}{dx}[u^r] = ru^{r-1} \cdot \frac{du}{dx} \qquad (11)$$

Example 6

$$\frac{d}{dx}[x^2 - x + 2]^{3/4} = \frac{3}{4}(x^2 - x + 2)^{-1/4} \cdot \frac{d}{dx}[x^2 - x + 2]$$

$$= \frac{3}{4}(x^2 - x + 2)^{-1/4}(2x - 1)$$

$$\frac{d}{dx}[(\sec \pi x)^{-4/5}] = -\frac{4}{5}(\sec \pi x)^{-9/5} \cdot \frac{d}{dx}[\sec \pi x]$$

$$= -\frac{4}{5}(\sec \pi x)^{-9/5} \cdot \sec \pi x \tan \pi x \cdot \pi$$

$$= -\frac{4\pi}{5}(\sec \pi x)^{-4/5} \tan \pi x \qquad \blacktriangleleft$$

☐ **DIFFERENTIABILITY OF IMPLICIT FUNCTIONS**

We conclude this section with some observations about the mathematical assumptions that underlie the method of implicit differentiation.

When differentiating implicitly, it is assumed that y represents a differentiable function of x. If this is not so, the resulting calculations may be nonsense. For example, if we implicitly differentiate the equation

$$x^2 + y^2 + 1 = 0 \qquad (12)$$

we obtain

$$2x + 2y\frac{dy}{dx} = 0$$

or

$$\frac{dy}{dx} = -\frac{x}{y} \qquad (13)$$

However, the derivative in (13) is meaningless because no function satisfies (12). (The left side of the equation is always greater than zero.) Unfortunately, it can be difficult to determine whether an equation defines y as a function of x and if so, whether the function is differentiable. We leave such questions for a course in advanced calculus. However, the following example should clarify the basic idea.

Example 7 The equation of the unit circle $x^2 + y^2 = 1$ does not define y as a function of x in a neighborhood of the point $P(1, 0)$, because the portion of the circle within any rectangle centered at P is cut twice by some vertical line (Figure 3.6.3a). The equation does, however, define y as a differentiable function of x in a neighborhood of the point $(1/\sqrt{2}, 1/\sqrt{2})$ (Figure 3.6.3b).

The graph of the equation $x^{2/3} + y^{2/3} = 1$, called a *four-cusped hypocycloid* (Figure 3.6.3c), defines y as a function of x in a neighborhood of the point $(0, 1)$; however, the function is not differentiable at $x = 0$, because of the corner or "cusp" at $(0, 1)$. ◀

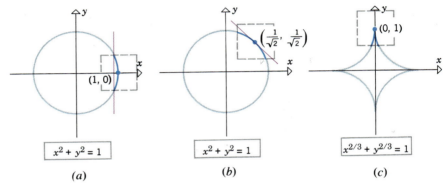

Figure 3.6.3

▶ Exercise Set 3.6

In Exercises 1–10, find dy/dx.

1. $y = \sqrt[3]{2x - 5}$.

2. $y = \sqrt[3]{2 + \tan(x^2)}$.

3. $y = \left(\dfrac{x - 1}{x + 2}\right)^{3/2}$.

4. $y = \sqrt{\dfrac{x^2 + 1}{x^2 - 5}}$.

5. $y = x^3(5x^2 + 1)^{-2/3}$.

6. $y = \dfrac{(3 - 2x)^{4/3}}{x^2}$.

7. $y = [\sin(3/x)]^{5/2}$.

8. $y = [\cos(x^3)]^{-1/2}$.

9. $y = \tan[(2x - 1)^{-1/3}]$.

10. $y = [\tan(2x - 1)]^{-1/3}$.

In Exercises 11 and 12, find all rational values of r so that $y = x^r$ satisfies the given equation.

11. $3x^2 y'' + 4xy' - 2y = 0$.

12. $16x^2 y'' + 24xy' + y = 0$.

In Exercises 13–29, find dy/dx by implicit differentiation.

13. $x^2 + y^2 = 100$.

14. $x^3 - y^3 = 6xy$.

15. $x^2 y + 3xy^3 - x = 3$.

16. $x^3 y^2 - 5x^2 y + x = 1$.

17. $\dfrac{1}{y} + \dfrac{1}{x} = 1$.

18. $x^2 = \dfrac{x + y}{x - y}$.

19. $\sqrt{x} + \sqrt{y} = 8$.

20. $\sqrt{xy} + 1 = y$.

21. $(x^2 + 3y^2)^{35} = x$.

22. $xy^{2/3} + yx^{2/3} = x^2$.

23. $3xy = (x^3 + y^2)^{3/2}$.

24. $\cos xy = y$.

25. $\sin(x^2 y^2) = x$.

26. $x^2 = \dfrac{\cot y}{1 + \csc y}$.

27. $\tan^3(xy^2 + y) = x$.

28. $\dfrac{xy^3}{1 + \sec y} = 1 + y^4$.

29. $\sqrt{1 + \sin^3(xy^2)} = y$.

In Exercises 30–34, use implicit differentiation to find the slope of the tangent line to the given curve at the specified point.

30. $x^2y - 5xy^2 + 6 = 0$; $(3, 1)$.

31. $x^3y + y^3x = 10$; $(1, 2)$.

32. $\sin xy = y$; $(\pi/2, 1)$.

33. $x^{2/3} - y^{2/3} - y = 1$; $(1, -1)$.

34. $\sqrt{3 + \tan xy} - 2 = 0$; $(\pi/12, 3)$.

In Exercises 35–39, find the value of dy/dx at the given point in two ways: first by solving for y in terms of x and then by implicit differentiation. (See Example 3.)

35. $xy = 8$; $(2, 4)$.

36. $y^2 - x + 1 = 0$; $(10, 3)$.

37. $x^2 + y^2 = 1$; $(1/\sqrt{2}, -1/\sqrt{2})$.

38. $\dfrac{1 - y}{1 + y} = x$; $(0, 1)$.

39. $y^2 - 3xy + 2x^2 = 4$; $(3, 2)$.

In Exercises 40–45, find d^2y/dx^2 by implicit differentiation.

40. $3x^2 - 4y^2 = 7$.

41. $x^3 + y^3 = 1$.

42. $x^3y^3 - 4 = 0$.

43. $2xy - y^2 = 3$.

44. $y + \sin y = x$.

45. $x \cos y = y$.

In Exercises 46–49, use implicit differentiation to find the specified derivative.

46. $\sqrt{u} + \sqrt{v} = 5$; du/dv.

47. $a^4 - t^4 = 6a^2t$; da/dt.

48. $y = \sin x$; dx/dy.

49. $a^2\omega^2 + b^2\lambda^2 = 1$ (a, b constants); $d\omega/d\lambda$.

50. Use implicit differentiation to show that the equation of the tangent to the curve $y^2 = kx$ at (x_0, y_0) is
$$y_0 y = \tfrac{1}{2}k(x + x_0)$$

51. Find dy/dx if
$$2y^3t + t^3y = 1 \quad \text{and} \quad \frac{dt}{dx} = \frac{1}{\cos t}$$

In Exercises 52 and 53, find dy/dt in terms of x, y, and dx/dt, assuming that x and y are differentiable functions of the variable t. [*Hint:* Differentiate both sides of the given equation with respect to t.]

52. $x^3y^2 + y = 3$.

53. $xy^2 = \sin 3x$.

54. At what point(s) is the tangent line to the curve $y^2 = 2x^3$ perpendicular to the line $4x - 3y + 1 = 0$?

55. Find the values of a and b for the curve $x^2y + ay^2 = b$ if the point $(1, 1)$ is on its graph and the tangent line at $(1, 1)$ has the equation $4x + 3y = 7$.

56. Find the coordinates of the point in the first quadrant at which the tangent line to the curve $x^3 - xy + y^3 = 0$ is parallel to the x-axis.

57. Find equations for two lines through the origin that are tangent to the curve $x^2 - 4x + y^2 + 3 = 0$.

58. (a) Use implicit differentiation to find the slope of the tangent to the four-cusped hypocycloid $x^{2/3} + y^{2/3} = 1$ at $P(-\tfrac{1}{4}\sqrt{2}, \tfrac{1}{4}\sqrt{2})$. (See Figure 3.6.3c.)

(b) Do part (a) by solving explicitly for y as a function of x and then differentiating.

(c) Trace the curve in Figure 3.6.3c and sketch the tangent line at P.

(d) Where does $x^{2/3} + y^{2/3} = 1$ *not* define y as an implicit function of x?

59. The graph of $8(x^2 + y^2)^2 = 100(x^2 - y^2)$, shown in Figure 3.6.4, is called a **lemniscate**.

(a) Where does this equation *not* define y as an implicit function of x?

(b) Use implicit differentiation to find the equation of the tangent to the curve at the point $(3, 1)$.

(c) Trace the curve in Figure 3.6.4 and sketch the tangent at $(3, 1)$.

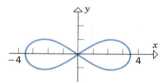

Figure 3.6.4

60. (a) Show that $f(x) = x^{4/3}$ is differentiable at 0, but not twice differentiable at 0.

(b) Show that $f(x) = x^{7/3}$ is twice differentiable at 0, but not three times differentiable at 0.

(c) Find an exponent k such that $f(x) = x^k$ is $(n - 1)$ times differentiable at 0, but not n times differentiable at 0.

■ 3.7 Δ-NOTATION; DIFFERENTIALS

Often, the key to solving or understanding a problem is to introduce the right notation. In this section, we shall discuss some powerful notational tools.

☐ **INCREMENTS**

If the value of a variable changes from one number to another, then the final value minus the initial value is called an ***increment*** in the variable. It is traditional in calculus to denote an increment in a variable x by the symbol Δx ("delta x"). In this notation, "Δx" is not a product of "Δ" and "x." Rather, Δx should be viewed as a single entity representing the *change* in the value of x. Similarly, Δy, Δt, and $\Delta \theta$ denote increments in the variables y, t, and θ.

If $y = f(x)$, and x changes from an initial value x_0 to a final value x_1, then there is a corresponding change in the value of y from $y_0 = f(x_0)$ to $y_1 = f(x_1)$. Stated another way, the increment

$$\Delta x = x_1 - x_0 \tag{1}$$

produces a corresponding increment

$$\Delta y = y_1 - y_0 = f(x_1) - f(x_0) \tag{2}$$

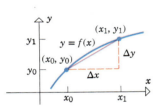

Figure 3.7.1

in y (Figure 3.7.1). Increments can be either positive, negative, or zero depending on the relative positions of the initial and final points. In Figure 3.7.1, Δx is positive since the final point x_1 is to the right of the initial point x_0. If x_1 were to the left of x_0, then Δx would be negative. Relation (1) can be rewritten as $x_1 = x_0 + \Delta x$, which states that the final value of x is the initial value of x plus the increment in x. With this notation, (2) can be rewritten as

$$\Delta y = f(x_0 + \Delta x) - f(x_0) \tag{3}$$

Sometimes it is convenient to drop the zero subscript on x_0 and use x to denote the initial value as well as the name of the variable. When this is done, $x + \Delta x$ represents the final value of the variable. Similarly, the initial and final values of y may be denoted by y and $y + \Delta y$ rather than y_0 and y_1 (Figure 3.7.2). Moreover, (3) becomes

$$\Delta y = f(x + \Delta x) - f(x) \tag{4}$$

The Δ-notation provides a compact way of rewriting the derivative definition,

$$\frac{dy}{dx} = \lim_{h \to 0} \frac{f(x + h) - f(x)}{h} \tag{5}$$

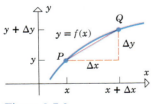

Figure 3.7.2

Referring to Figure 3.7.2, the slope of the secant line joining the points P and Q is $\Delta y/\Delta x$. As $\Delta x \to 0$, Q approaches P along the curve $y = f(x)$, so that the slope of the secant line approaches the slope of the tangent line at P; that is,

$$\frac{dy}{dx} = \lim_{\Delta x \to 0} \frac{\Delta y}{\Delta x} \tag{6}$$

This result is expected since $\Delta y = f(x + \Delta x) - f(x)$, so (6) can be written as

$$\frac{dy}{dx} = \lim_{\Delta x \to 0} \frac{f(x + \Delta x) - f(x)}{\Delta x} \tag{7}$$

which is simply a restatement of (5) with Δx used in place of h.

If it is undesirable to introduce a dependent variable y, then we write

$$\Delta f = f(x + \Delta x) - f(x)$$

and (6) as

$$f'(x) = \lim_{\Delta x \to 0} \frac{\Delta f}{\Delta x} \tag{8}$$

Expressions (5), (6), (7), and (8) provide four ways of writing the definition of a derivative. All are commonly used.

□ **DIFFERENTIALS**

Up to now we have been viewing the expression dy/dx as a single symbol for the derivative. The symbols "dy" and "dx" in this expression, which are called *differentials,* have had no meaning by themselves. We shall now show how to interpret these symbols so that dy/dx can be viewed as the ratio of dy and dx.

Regard x as fixed and *define dx* to be an independent variable that can be assigned an arbitrary value. If f is differentiable at x, then we *define dy* by the formula

$$dy = f'(x)\, dx \tag{9}$$

If $dx \neq 0$, then we can divide both sides of (9) by dx to obtain

$$\frac{dy}{dx} = f'(x)$$

Thus, we have achieved our goal of defining dy and dx so that their ratio is $f'(x)$.

Because

$$\frac{dy}{dx} = f'(x) = m_{\tan}$$

where $m_{\tan}$ is the slope of the tangent to $y = f(x)$ at x, the differentials dy and dx can be viewed as a corresponding rise and run of this tangent line (Figure 3.7.3).

It is important to understand the distinction between the increment Δy and the differential dy. To see the difference, let us assign the independent variables dx and Δx the same value, so $dx = \Delta x$. Then Δy represents the change in y

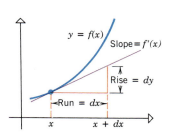

Figure 3.7.3

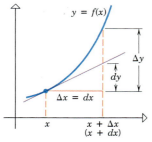

Figure 3.7.4

that occurs when we start at x and travel *along the curve* $y = f(x)$ until we have moved $\Delta x (= dx)$ units in the x-direction, while dy represents the change in y that occurs if we start at x and travel *along the tangent* line until we have moved $dx (= \Delta x)$ units in the x-direction (Figure 3.7.4).

Example 1   If $y = x^2$, then the relation $dy/dx = 2x$ can be written in the *differential form*

$$dy = 2x\, dx$$

When $x = 3$, this becomes

$$dy = 6\, dx$$

This tells us that if we travel along the tangent to the curve $y = x^2$ at $x = 3$, then a change of dx units in x produces a change of $6\, dx$ units in y. For example, if the change in x is $dx = 4$, then the change in y along the tangent is

$$dy = 6(4) = 24 \quad \text{units} \qquad \blacktriangleleft$$

Example 2   Let $y = \sqrt{x}$. Find dy and Δy if $x = 4$ and $dx = \Delta x = 3$. Then make a sketch of $y = \sqrt{x}$, showing dy and Δy in the picture.

Solution.   From (4) with $f(x) = \sqrt{x}$,

$$\Delta y = \sqrt{x + \Delta x} - \sqrt{x} = \sqrt{7} - \sqrt{4} \approx .65$$

If $y = \sqrt{x}$, then

$$\frac{dy}{dx} = \frac{1}{2\sqrt{x}} \quad \text{so} \quad dy = \frac{1}{2\sqrt{x}}\, dx = \frac{1}{2\sqrt{4}}(3) = \frac{3}{4} = .75$$

Figure 3.7.5 shows the curve $y = \sqrt{x}$ together with dy and Δy. $\blacktriangleleft$

☐ **TANGENT LINE**
APPROXIMATIONS

Figure 3.7.6 suggests that if f is differentiable at x_0, then the tangent line to the curve $y = f(x)$ at x_0 is a reasonably good approximation to the curve $y = f(x)$ for values of x near x_0. Since the tangent line passes through the point $(x_0, f(x_0))$ and has slope $f'(x_0)$, the point-slope form of its equation is

$$y - f(x_0) = f'(x_0)(x - x_0) \qquad \text{or} \qquad y = f(x_0) + f'(x_0)(x - x_0)$$

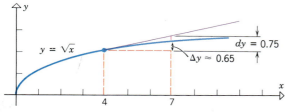

Figure 3.7.5

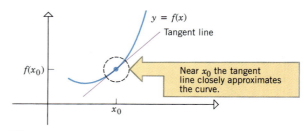

Figure 3.7.6

For values of x close to x_0, the height y of this tangent line will closely approximate the height $f(x)$ of the curve, which yields the approximation

$$f(x) \approx f(x_0) + f'(x_0)(x - x_0) \qquad (10)$$

Right side = height of the tangent line

Left side = height of the curve

for x near x_0. If we let $\Delta x = x - x_0$, so that $x = x_0 + \Delta x$, then (10) can be written in the alternate form

$$f(x_0 + \Delta x) \approx f(x_0) + f'(x_0)\,\Delta x \qquad (11)$$

which is a good approximation when Δx is near zero. This result is called the *linear approximation of f near x_0.*

In the event that $f(x_0 + \Delta x)$ is tedious to calculate, but $f(x_0)$ and $f'(x_0)$ are not, then this formula enables us to use the values of $f(x_0)$ and $f'(x_0)$ to approximate $f(x_0 + \Delta x)$.

Example 3 Use (11) to approximate $\sqrt{1.1}$.

Solution. If we let $f(x) = \sqrt{x}$, then the problem is to approximate $f(1.1)$. But $f'(x) = 1/(2\sqrt{x})$, so $f(1)$ and $f'(1)$ are easy to compute. Thus, we shall apply (11) with

$$x_0 = 1 \quad \text{and} \quad x_0 + \Delta x = 1.1$$

It follows that $\Delta x = .1$, so (11) yields

$$f(1.1) \approx f(1) + f'(1)(.1)$$

or

$$\sqrt{1.1} \approx \sqrt{1} + \frac{1}{2\sqrt{1}}(.1) = 1.05 \qquad \blacktriangleleft$$

Example 4 Use (11) to approximate $\cos 62°$.

Solution. We shall take advantage of the fact that $62°$ is close to $60°$, at which point the trigonometric functions are easy to evaluate. However, to apply (11), we must first convert to radian measure because the derivative formulas for the trigonometric functions are based on the assumption that x is in radians. If we let $f(x) = \cos x$ and note that $62° = 31\pi/90$ radians and $60° = \pi/3$ radians, then the problem is to approximate $f(31\pi/90)$ using the known values of $f(\pi/3)$ and $f'(\pi/3)$. Thus, we will apply (11) with

$$x_0 + \Delta x = \frac{31\pi}{90} \quad \text{and} \quad x_0 = \frac{\pi}{3}\,(= 60°)$$

It follows that $\Delta x = \pi/90$, so (11) yields

$$f\left(\frac{31\pi}{90}\right) \approx f\left(\frac{\pi}{3}\right) + f'\left(\frac{\pi}{3}\right)\left(\frac{\pi}{90}\right)$$

$$\cos\frac{31\pi}{90} \approx \cos\frac{\pi}{3} - \left(\sin\frac{\pi}{3}\right)\left(\frac{\pi}{90}\right)$$

$$\cos\frac{31\pi}{90} \approx \frac{1}{2} - \frac{\sqrt{3}}{2}\left(\frac{\pi}{90}\right) \approx 0.5 - 0.0302300 = 0.4697700$$

To seven digits, the value of $\cos 62°$ is 0.4694716, so that the error is less than 0.0003. ◄

Formula (11) has a useful alternative form, which we shall now derive. Subtract $f(x_0)$ from both sides to obtain

$$f(x_0 + \Delta x) - f(x_0) \approx f'(x_0)\,\Delta x$$

and use (3) to rewrite this as

$$\Delta y \approx f'(x_0)\,\Delta x$$

If we drop the subscript on x_0 and assume that $\Delta x = dx$, then from (9) this can be rewritten as

$$\Delta y \approx dy \tag{12}$$

This formula has applications in the study of **error propagation.** Suppose a researcher measures a physical quantity. Because of limitations in the instrumentation and other factors, the researcher will not usually obtain the exact value x of the quantity, but rather will obtain $x + \Delta x$, where Δx is a measurement error. This recorded value may then be used to calculate some other quantity y. In this way the measurement error Δx propagates to produce an error Δy in the calculated value of y.

Example 5   The radius of a sphere is measured to be 50 in. with a possible measurement error of ± 0.02 in. Estimate the possible error in the computed volume of the sphere.

Solution.   The volume of the sphere is

$$V = \tfrac{4}{3}\pi r^3 \tag{13}$$

We are given that the error in the radius is $\Delta r = \pm 0.02$, and we want to find the error ΔV in V. If we consider Δr to be small and we let $dr = \Delta r$, then ΔV can be approximated by dV. Thus, from (13),

$$\Delta V \approx dV = 4\pi r^2\, dr \tag{14}$$

Substituting $r = 50$ and $dr = \pm 0.02$ in (14), we obtain

$$\Delta V \approx 4\pi(2500)(\pm 0.02) \approx \pm 628.32$$

Therefore, the possible error in the calculated volume is approximately ± 628.32 cubic inches (in³).   ◄

REMARK.   In (14), r represents the exact value of the radius. Since the exact value of r was unknown, we substituted the measured value $r = 50$ to obtain ΔV. This is reasonable since the error Δr was assumed to be small.

If the exact value of a quantity is q and a measurement or calculation results in an error Δq, then $\Delta q/q$ is called the **relative error** in the measurement or calculation; when expressed as a percentage, $\Delta q/q$ is called the **percentage error**. As a practical matter, the exact value q is usually unknown, so that the measured or calculated value of q is used instead; and the relative error is approximated by dq/q.

Example 6   For the sphere in Example 5,

$$\text{relative error in } r \approx \frac{dr}{r} = \frac{\pm 0.02}{50} = \pm 0.0004$$

$$\text{relative error in } V \approx \frac{dV}{V} = \frac{4\pi r^2 \, dr}{\frac{4}{3}\pi r^3} = 3\frac{dr}{r} = \pm 0.0012$$

Thus, the percentage error in the calculated radius is approximately $\pm 0.04\%$, and the percentage error in the calculated volume is approximately $\pm 0.12\%$.

◄

Example 7   The side of a square is measured with a possible percentage error of $\pm 5\%$. Use differentials to estimate the possible percentage error in the area of the square.

Solution.   The area of a square with side x is given by

$$A = x^2 \tag{15}$$

and the relative errors in A and x are approximately dA/A and dx/x, respectively. From (15),

$$dA = 2x \, dx$$

so

$$\frac{dA}{A} = \frac{2x \, dx}{A} = \frac{2x \, dx}{x^2} = 2\frac{dx}{x}$$

We are given that $dx/x \approx \pm 0.05$, so that

$$\frac{dA}{A} \approx \pm 2(0.05) = \pm 0.1$$

Thus, there is a possible percentage error of $\pm 10\%$ in the area of the square.   ◄

DIFFERENTIAL FORMULAS

Just as

$$\frac{d}{dx}[\ \ ]$$

denotes the derivative of the expression inside the brackets, so

$$d[\ \ ]$$

denotes the differential of the expression in the brackets. For example,

$$d[x^3] = 3x^2\, dx$$
$$d[f(x)] = f'(x)\, dx$$

The basic rules of differentiation can be expressed in terms of differentials. In the following table, the differential formulas on the right result when the derivative formulas on the left are multiplied through by dx.

DERIVATIVE FORMULA	DIFFERENTIAL FORMULA
$\dfrac{d}{dx}[c] = 0$	$d[c] = 0$
$\dfrac{d}{dx}[cf] = c\dfrac{df}{dx}$	$d[cf] = c\, df$
$\dfrac{d}{dx}[f + g] = \dfrac{df}{dx} + \dfrac{dg}{dx}$	$d[f + g] = df + dg$
$\dfrac{d}{dx}[fg] = f\dfrac{dg}{dx} + g\dfrac{df}{dx}$	$d[fg] = f\, dg + g\, df$
$\dfrac{d}{dx}\left[\dfrac{f}{g}\right] = \dfrac{g\dfrac{df}{dx} - f\dfrac{dg}{dx}}{g^2}$	$d\left[\dfrac{f}{g}\right] = \dfrac{g\, df - f\, dg}{g^2}$

Example 8 Find dy if $y = x \sin x$.

Solution.

$$dy = d[x \sin x] = x\, d[\sin x] + (\sin x)\, d[x]$$
$$= x(\cos x)\, dx + (\sin x)\, dx \quad \boxed{\text{Since } d[x] = 1 \cdot dx = dx}$$
$$= (x \cos x + \sin x)\, dx \quad \blacktriangleleft$$

▶ **Exercise Set 3.7** Ⓒ *19, 21, 24, 25, 26, 27, 28, 35, 36, 45, 46, 47, 48*

1. Let $y = x^2$.
   (a) Find Δy if $\Delta x = 1$ and the initial value of x is $x = 2$.
   (b) Find dy if $dx = 1$ and the initial value of x is $x = 2$.
   (c) Make a sketch of $y = x^2$ and show Δy and dy in the picture.

2. Repeat Exercise 1 with $x = 2$ as the initial value again, but $\Delta x = -1$ and $dx = -1$.

3. Let $y = 1/x$.
   (a) Find Δy if $\Delta x = 0.5$ and the initial value of x is $x = 1$.
   (b) Find dy if $dx = 0.5$ and the initial value of x is $x = 1$.

(c) Make a sketch of $y = 1/x$ and show Δy and dy in the picture.

4. Repeat Exercise 3 with $x = 1$ as the initial value again, but $\Delta x = -0.5$ and $dx = -0.5$.

In Exercises 5–8, find general formulas for dy and Δy.

5. $y = x^3$.

6. $y = 8x - 4$.

7. $y = x^2 - 2x + 1$.

8. $y = \sin x$.

In Exercises 9–12, find dy.

9. $y = 4x^3 - 7x^2 + 2x - 1$.

10. $y = \dfrac{1}{x^3 - 1}$.

11. $y = x \cos x$.

12. $y = \dfrac{1 - x^3}{2 - x}$.

In Exercises 13–16, find the limit by expressing it as a derivative.

13. $\lim\limits_{\Delta x \to 0} \dfrac{(x + \Delta x)^2 - x^2}{\Delta x}$.

14. $\lim\limits_{\Delta x \to 0} \dfrac{(3 + \Delta x)^2 - 3^2}{\Delta x}$.

15. $\lim\limits_{\Delta x \to 0} \dfrac{\sin(\pi + \Delta x) - \sin \pi}{\Delta x}$.

16. $\lim\limits_{\Delta x \to 0} \dfrac{5(2 + \Delta x)^4 - 5(2)^4}{\Delta x}$.

In Exercises 17–28, use a linear approximation [Formula (11)] to estimate the value of the given quantity.

17. $(3.02)^4$.

18. $(1.97)^3$.

19. $\sqrt{65}$.

20. $\sqrt{24}$.

21. $\sqrt{80.9}$.

22. $\sqrt{36.03}$.

23. $\sqrt[3]{8.06}$.

24. $\sqrt[3]{63.7}$.

25. $\cos 31°$.

26. $\sin 59°$.

27. $\sin 44°$.

28. $\tan 61°$.

In Exercises 29–32, use dy to approximate Δy when x changes as indicated.

29. $y = \sqrt{3x - 2}$; from $x = 2$ to $x = 2.03$.

30. $y = \sqrt{x^2 + 8}$; from $x = 1$ to $x = 0.97$.

31. $y = \dfrac{x}{x^2 + 1}$; from $x = 2$ to $x = 1.96$.

32. $y = x\sqrt{8x + 1}$; from $x = 3$ to $x = 3.05$.

33. The side of a square is measured to be 10 ft, with a possible error of ± 0.1 ft.
    (a) Use differentials to estimate the error in the calculated area.
    (b) Estimate the percentage errors in the side and the area.

34. The side of a cube is measured to be 25 cm, with a possible error of ± 1 cm.
    (a) Use differentials to estimate the error in the calculated volume.
    (b) Estimate the percentage errors in the side and volume.

35. The hypotenuse of a right triangle is known to be 10 in. exactly, and one of the acute angles is measured to be 30°, with a possible error of $\pm 1°$.
    (a) Use differentials to estimate the errors in the sides opposite and adjacent to the measured angle.
    (b) Estimate the percentage errors in the sides.

36. One side of a right triangle is known to be 25 cm exactly. The angle opposite to this side is measured to be 60°, with a possible error of $\pm 0.5°$.
    (a) Use differentials to estimate the errors in the adjacent side and the hypotenuse.
    (b) Estimate the percentage errors in the adjacent side and hypotenuse.

37. The electrical resistance R of a certain wire is given by $R = k/r^2$, where k is a constant and r is the radius of the wire. Assuming that the radius r has a possible error of $\pm 5\%$, use differentials to estimate the percentage error in R. (Assume k is exact.)

38. The side of a square is measured with a possible percentage error of $\pm 1\%$. Use differentials to estimate the percentage error in the area.

39. The side of a cube is measured with a possible percentage error of $\pm 2\%$. Use differentials to estimate the percentage error in the volume.

40. The volume of a sphere is to be computed from a measured value of its radius. Estimate the maximum permissible percentage error in the measurement if the percentage error in the volume must be kept within $\pm 3\%$. ($V = \frac{4}{3}\pi r^3$ is the volume of a sphere of radius r.)

41. The area of a circle is to be computed from a measured value of its diameter. Estimate the maximum permissible percentage error in the measurement if the percentage error in the area must be kept within $\pm 1\%$.

42. A steel cube with 1-in. sides is coated with 0.01 in. of copper. Use differentials to estimate the volume of copper in the coating. [*Hint:* Let ΔV be the change in the volume of the cube.]

43. A metal rod 15 cm long and 5 cm in diameter is to be covered (except for the ends) with insulation that is 0.001 cm thick. Use differentials to estimate the volume of insulation. [*Hint:* Let ΔV be the change in volume of the rod.]

44. The time required for one complete oscillation of a pendulum is called its *period.* If the length L of the pendulum is measured in feet and the period P in seconds, then the period is given by $P = 2\pi\sqrt{L/g}$, where g is a constant. Use differentials to show that the percentage error in P is approximately half the percentage error in L.

45. If the temperature T of a metal rod of length L is changed by an amount ΔT, then the length will change by the amount $\Delta L = \alpha L \Delta T$, where α is called the *coefficient of linear expansion.* For moderate changes in temperature α is taken as constant.
 (a) Suppose that a rod 40 cm long at $20°$ C is found to be 40.006 cm long when the temperature is raised to $30°$ C. Find α.
 (b) If an aluminum pole is 180 cm long at $15°$C, how long is the pole if the temperature is raised to $40°$ C? [Take $\alpha = 2.3 \times 10^{-5}/°$C.]

46. If the temperature T of a solid or liquid of volume V is changed by an amount ΔT, then the volume will change by the amount $\Delta V = \beta V \Delta T$, where β is called the *coefficient of volume expansion.* For moderate changes in temperature β is taken as constant. Suppose that a tank truck loads 4000 gallons of ethyl alcohol at a temperature of $35°$ C and delivers its load sometime later at a temperature of $15°$ C. Using $\beta = 7.5 \times 10^{-4}/°$C for ethyl alcohol, find the number of gallons delivered.

47. Use differentials to show that $(1 + x)^{-1} \approx 1 - x$ when x is near zero. Calculate, to 4 decimal places, the values of $(1 + x)^{-1}$ and $1 - x$ for $x = 0.1$ and $x = -0.02$.

48. Use differentials to show that $\sqrt{1 + x} \approx 1 + \frac{1}{2}x$ when x is near zero. Calculate, to 4 decimal places, the values of $\sqrt{1 + x}$ and $1 + \frac{1}{2}x$ for $x = 0.1$ and $x = -0.02$.

49. Let $y = x^k$, where k is a positive rational number. Show that the percentage error in y is approximately k times the percentage error in x.

SUPPLEMENTARY EXERCISES

In Exercises 1–4, use Definition 3.2.1 to find $f'(x)$.

1. $f(x) = kx$ (k constant).

2. $f(x) = (x - a)^2$ (a constant).

3. $f(x) = \sqrt{9 - 4x}$.

4. $f(x) = \dfrac{x}{x + 1}$.

5. Use Definition 3.2.1 to find $\dfrac{d}{dx}[|x|^3]\Big|_{x=0}$.

6. Suppose $f(x) = \begin{cases} x^2 - 1, & x \le 1 \\ k(x - 1), & x > 1. \end{cases}$
 For what values of k is f
 (a) continuous
 (b) differentiable?

7. Suppose $f(3) = -1$ and $f'(3) = 5$. Find an equation for the tangent line to the graph of f at $x = 3$.

8. Let $f(x) = x^2$. Show that for any distinct values of a and b, the slope of the tangent line to $y = f(x)$ at $x = \frac{1}{2}(a + b)$ is equal to the slope of the secant line through the points (a, a^2) and (b, b^2).

9. Given the following table of values at $x = 1$ and $x = -2$, find the indicated derivatives in parts (a)–(l).

x	$f(x)$	$f'(x)$	$g(x)$	$g'(x)$
1	1	3	-2	-1
-2	-2	-5	1	7

 (a) $\dfrac{d}{dx}[f^2(x) - 3g(x^2)]\Big|_{x=1}$

 (b) $\dfrac{d}{dx}[f(x)g(x)]\Big|_{x=1}$

(c) $\dfrac{d}{dx}\left[\dfrac{f(x)}{g(x)}\right]\Bigg|_{x=-2}$

(d) $\dfrac{d}{dx}\left[\dfrac{g(x)}{f(x)}\right]\Bigg|_{x=-2}$

(e) $\dfrac{d}{dx}[f(g(x))]\Bigg|_{x=1}$

(f) $\dfrac{d}{dx}[f(g(x))]\Bigg|_{x=-2}$

(g) $\dfrac{d}{dx}[g(f(x))]\Bigg|_{x=-2}$

(h) $\dfrac{d}{dx}[g(g(x))]\Bigg|_{x=-2}$

(i) $\dfrac{d}{dx}[f(g(4-6x))]\Bigg|_{x=1}$

(j) $\dfrac{d}{dx}[g^{3}(x)]\Bigg|_{x=1}$

(k) $\dfrac{d}{dx}[\sqrt{f(x)}]\Bigg|_{x=1}$

(l) $\dfrac{d}{dx}[f(-\tfrac{1}{2}x)]\Bigg|_{x=-2}$

In Exercises 10–15, find $f'(x)$ and determine those values of x for which $f'(x) = 0$.

10. $f(x) = (2x + 7)^6(x - 2)^5$.

11. $f(x) = \dfrac{(x - 3)^4}{x^2 + 2x}$.

12. $f(x) = \sqrt{3x + 1}\,(x - 1)^2$.

13. $f(x) = \left(\dfrac{3x + 1}{x^2}\right)^3$.

14. $f(x) = \dfrac{3(5x - 1)^{1/3}}{3x - 5}$.

15. $f(x) = \sqrt{x}\,\sqrt[3]{x^2 + x + 1}$.

16. Suppose that $f'(x) = 1/x$ for all $x \neq 0$.
   (a) Use the chain rule to show that for any non-zero constant a, $d(f(ax))/dx = d(f(x))/dx$.
   (b) If $y = f(\sin x)$ and $v = f(1/x)$, find dy/dx and dv/dx.

In Exercises 17–26, find the indicated derivatives.

17. $\dfrac{d}{dx}\left(\dfrac{\sqrt{2}}{x^2} - \dfrac{2}{5x}\right)$.

18. $\dfrac{dy}{dx}$ if $y = \dfrac{3x^2 + 7}{x^2 - 1}$.

19. $\dfrac{dz}{dr}\Bigg|_{r=\pi/6}$ if $z = 4\sin^2 r\cos^2 r$.

20. $g'(2)$ if $g(x) = 1/\sqrt{2x}$.

21. $\dfrac{du}{dx}$ if $u = \left(\dfrac{x}{x - 1}\right)^{-2}$.

22. $\dfrac{dw}{dv}$ if $w = \sqrt[5]{v^3} - \sqrt[4]{v}$.

23. $d(\sec^2 x - \tan^2 x)/dx$.

24. $\dfrac{dy}{dx}\Bigg|_{x=\pi/4}$ if $y = \tan t$ and $t = \cos(2x)$.

25. $F'(x)$ if $F(x) = \dfrac{(1/x) + 2x}{\tfrac{1}{2}(1/x^2) + 1}$.

26. $\Phi'(x)$ if $\Phi(x) = \dfrac{x^2 - 4x}{5\sqrt{x}}$.

27. Find all values of x for which the tangent to $y = x - (1/x)$ is parallel to the line $2x - y = 5$.

28. Find all values of x for which the tangent to $y = 2x^3 - x^2$ is perpendicular to the line $x + 4y = 10$.

29. Find all values of x for which the tangent to $y = (x + 2)^2$ passes through the origin.

30. Find all values of x for which the tangent to $y = x - \sin 2x$ is horizontal.

31. Find all values of x for which the tangent to $y = 3x - \tan x$ is parallel to the line $y - x = 2$.

In Exercises 32–34, find Δx, Δy, and dy.

32. $y = 1/(x - 1)$; x decreases from 2 to 1.5.

33. $y = \tan x$; x increases from $-\pi/4$ to 0.

34. $y = \sqrt{25 - x^2}$; x increases from 0 to 3.

35. Use a differential to approximate
   (a) $\sqrt[3]{-8.25}$ (b) $\cot 46°$.

36. The area of a right triangle with a hypotenuse of H is calculated using the formula $A = \tfrac{1}{4}H^2\sin 2\theta$, where θ is one of the acute angles. Use differentials to approximate the error in calculating A if $H = 4$ cm (exactly) and $\theta = 30° \pm 15'$.

37. A 12-foot ladder leaning against a wall makes an angle θ with the floor. If the top of the ladder is h feet up the wall, express h in terms of θ and then use dh to estimate the change in h if θ changes from 60° to 59°.

38. Let V and S denote the volume and surface area of a cube. Find the rate of change of V with respect to S.

39. The amount of water in a tank t minutes after it has started to drain is given by $W = 100(t - 15)^2$ gal.
  (a) At what rate is the water running out at the end of 5 min?
  (b) What is the average rate at which the water flows out during the first 5 min?

In Exercises 40–42, find dy/dx by implicit differentiation and use it to find the equation of the tangent line at the indicated points.

40. $(x + y)^3 + 3xy = -7$; $(-2, 1)$.

41. $xy^2 = \sin(x + 2y)$; $(0, 0)$.

42. $(x + y)^3 - 5x + y = 1$; at the point where the curve intersects the line $x + y = 1$.

43. Show that the curves whose equations are $y^2 = x^3$ and $2x^2 + 3y^2 = 5$ intersect at the point $(1, 1)$ and that their tangent lines are perpendicular there.

44. Show that for any point $P_0(x_0, y_0)$ on the circle $x^2 + y^2 = r^2$, the tangent line at P_0 is perpendicular to the radial line from the origin to P_0.

45. Verify that the function $y = \cos x - 3 \sin x$ satisfies $y''' + y'' + y' + y = 0$.

46. Find d^2y/dx^2 implicitly if
  (a) $y^3 + 3x^2 = 4y$ (b) $\sin y + \cos x = 1$.

4

Applications of Differentiation

Pierre de Fermat   (1601-1665)

■ 4.1 RELATED RATES

*In this section we shall study **related rates** problems. In such problems one tries to find the rate at which some quantity is changing by relating it to other quantities whose rates of change are known.*

Example 1   Assume that oil spilled from a ruptured tanker spreads in a circular pattern whose radius increases at a constant rate of 2 ft/sec. How fast is the area of the spill increasing when the radius of the spill is 60 ft?

Solution.   Let

t = number of seconds elapsed from the time of the spill
r = radius of the spill in feet after t seconds
A = area of the spill in square feet after t seconds

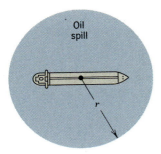

Figure 4.1.1

(See Figure 4.1.1.) At each instant the rate at which the radius is increasing with time is dr/dt, and the rate at which the area is increasing with time is dA/dt. We want to find

$$\left. \frac{dA}{dt} \right|_{r=60}$$

which is the rate at which the area of the spill is increasing at the instant when $r = 60$.

Since the radius increases at the constant rate of 2 ft/sec, we know

$$\frac{dr}{dt} = 2 \tag{1}$$

for all t. From the formula for the area of a circle we obtain

$$A = \pi r^2 \tag{2}$$

Because A and r are functions of t, we can differentiate both sides of (2) with respect to t to obtain

$$\frac{dA}{dt} = 2\pi r \frac{dr}{dt}$$

Substituting (1) yields

$$\frac{dA}{dt} = 2\pi r(2) = 4\pi r$$

Thus, when $r = 60$ the area of the spill is increasing at the rate

$$\frac{dA}{dt}\bigg|_{r=60} = 4\pi(60) = 240\pi \text{ ft}^2/\text{sec}$$

or approximately 754 ft²/sec. ◀

With only minor variations, the method used in Example 1 can be used to solve a variety of related rates problems. The method consists of four steps:

Step 1. Draw a figure and label the quantities that vary.

Step 2. Find an equation relating the quantity with the unknown rate of change to quantities whose rates of change are known.

Step 3. Differentiate both sides of this equation with respect to time and solve for the derivative that will give the unknown rate of change.

Step 4. Evaluate this derivative at the appropriate point.

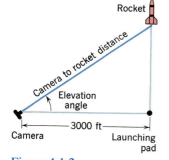

Figure 4.1.2

In Figure 4.1.2 we have shown a camera mounted at a point 3000 ft from the base of a rocket launching pad. Let us assume that the rocket rises vertically and the camera is to take a series of photographs of the rocket. Because the rocket will be rising, the elevation angle of the camera will have to vary at just the right rate to keep the rocket in sight. Moreover, because the camera-to-rocket distance will be changing constantly, the camera focusing mechanism will also have to vary at just the right rate to keep the picture sharp. Our next two examples are concerned with these problems.

Example 2 If the rocket shown in Figure 4.1.2 is rising vertically at 880 ft/sec when it is 4000 ft up, how fast is the camera-to-rocket distance changing at that instant?

Solution. Let

t = number of seconds elapsed from the time of launch
y = camera-to-rocket distance in feet after t seconds
x = height of the rocket in feet after t seconds

At each instant the rate at which the camera-to-rocket distance is changing is dy/dt, and the rate at which the rocket is rising is dx/dt. We want to find

$$\frac{dy}{dt}\bigg|_{x=4000}$$

which is the rate the camera-to-rocket distance is changing at the instant when $x = 4000$.

From the Theorem of Pythagoras and Figure 4.1.3 we have

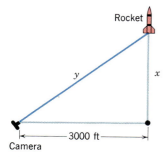

Figure 4.1.3

$$y^2 = x^2 + (3000)^2 \tag{3}$$

Because x and y are functions of t, we can differentiate both sides of (3) with respect to t to obtain

$$2y\frac{dy}{dt} = 2x\frac{dx}{dt} \quad \text{or} \quad \frac{dy}{dt} = \frac{x}{y}\frac{dx}{dt} \tag{4}$$

Note the use of the chain rule here.

When $x = 4000$, it follows from (3) that $y = 5000$; moreover, we are given that $dx/dt = 880$ when $x = 4000$ so that from (4)

$$\left.\frac{dy}{dt}\right|_{x=4000} = \frac{4000}{5000}(880) = 704 \text{ ft/sec} \quad \blacktriangleleft$$

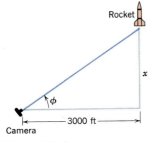

Rocket

x

ϕ

3000 ft

Camera

Figure 4.1.4

Example 3 If the rocket shown in Figure 4.1.4 is rising vertically at 880 ft/sec when it is 4000 ft up, how fast must the camera elevation angle change at that instant to keep the rocket in sight?

Solution. Let

t = number of seconds elapsed from the time of launch
ϕ = camera elevation angle in radians after t seconds
x = height of the rocket in feet after t seconds

(See Figure 4.1.4.) At each instant the rate at which the camera elevation angle must change is $d\phi/dt$, and the rate at which the rocket is rising is dx/dt. We want to find

$$\left.\frac{d\phi}{dt}\right|_{x=4000}$$

which is the rate the camera elevation angle must change at the instant when $x = 4000$.

From Figure 4.1.4 we see that

$$\tan\phi = \frac{x}{3000} \tag{5}$$

Because ϕ and x are functions of t, we can differentiate both sides of (5) with respect to t to obtain

$$(\sec^2\phi)\frac{d\phi}{dt} = \frac{1}{3000}\frac{dx}{dt} \quad \text{or} \quad \frac{d\phi}{dt} = \frac{1}{3000\sec^2\phi}\frac{dx}{dt} \tag{6}$$

When $x = 4000$, it follows that

$$\sec\phi = \frac{5000}{3000} = \frac{5}{3}$$

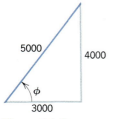

5000

4000

ϕ

3000

Figure 4.1.5

(See Figure 4.1.5.) Moreover, we are given that $dx/dt = 880$ when $x = 4000$, so that from (6)

$$\frac{d\phi}{dt}\Bigg|_{x=4000} = \frac{1}{3000(\frac{5}{3})^2} \cdot 880 = \frac{66}{625} \approx 0.11 \text{ radian/sec}$$

If desired, this rate of change can be expressed in degrees per second. The computations are as follows:

$$\frac{d\phi}{dt}\Bigg|_{x=4000} = \frac{66}{625}\frac{\text{radian}}{\text{sec}} \cdot \frac{180}{\pi}\frac{\text{degrees}}{\text{radian}} \approx 6.05 \text{ degrees/sec} \quad \blacktriangleleft$$

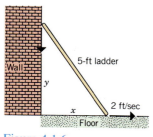

Figure 4.1.6

Example 4 A 5-ft ladder, leaning against a wall (Figure 4.1.6), slips so that its base moves away from the wall at a rate of 2 ft/sec. How fast will the top of the ladder be moving down the wall when the base is 4 ft from the wall?

Solution. Let

t = number of seconds after the ladder starts to slip
x = distance in feet from the base of the ladder to the wall
y = distance in feet from the top of the ladder to the floor

(See Figure 4.1.6.) At each instant the rate at which the base moves is dx/dt, and the rate at which the top moves is dy/dt. We want to find

$$\frac{dy}{dt}\Bigg|_{x=4}$$

which is the rate at which the top is moving at the instant the base is 4 ft from the wall.

From the Theorem of Pythagoras we have

$$x^2 + y^2 = 25 \tag{7}$$

Differentiating both sides of this equation with respect to t yields

$$2x\frac{dx}{dt} + 2y\frac{dy}{dt} = 0$$

or

$$\frac{dy}{dt} = -\frac{x}{y}\frac{dx}{dt} \tag{8}$$

When $x = 4$, it follows from (7) that $y = 3$; moreover, we are given that $dx/dt = 2$, so that (8) yields

$$\frac{dy}{dt}\Bigg|_{x=4} = -\frac{4}{3}(2) = -\frac{8}{3} \text{ ft/sec}$$

The negative sign in the answer tells us that y is decreasing, which makes sense physically, since the top of the ladder is moving *down* the wall. $\quad \blacktriangleleft$

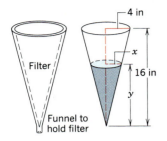

Figure 4.1.7

Example 5 Suppose a liquid is to be cleared of sediment by pouring it through a cone-shaped filter. Assume that the height of the cone is 16 in. and the radius at the base of the cone is 4 in. (Figure 4.1.7). If the liquid is flowing out of the cone at a rate of 2 in³/min when the level is 8 in. deep, how fast is the depth of the liquid decreasing at that instant?

Solution. Let

t = time elapsed from the initial observation (min)
V = volume of liquid in the cone at time t (in³)
y = depth of the liquid in the cone at time t (in.)
x = radius of the liquid surface at time t (in.)

(See Figure 4.1.7b.) At each instant the rate at which the volume of liquid is changing is dV/dt, and the rate at which the depth is changing is dy/dt. We want to find

$$\frac{dy}{dt}\bigg|_{y=8}$$

which is the rate at which the depth is changing at the instant the depth is 8 in.

From the formula for the volume of a cone, the volume V and the depth y are related by

$$V = \tfrac{1}{3}\pi x^2 y \tag{9}$$

Since we are given that the volume of liquid is decreasing at a rate of 2 in³/min when $y = 8$, we have

$$\frac{dV}{dt}\bigg|_{y=8} = -2 \tag{10}$$

(We must use a minus sign here because V *decreases* as t increases.) If we differentiate both sides of (9) with respect to t, the right side will involve the quantity dx/dt. Since we have no direct information about dx/dt, it is desirable to eliminate x from (9) before differentiating. This can be done using similar triangles. From Figure 4.1.7b, we see that

$$\frac{x}{y} \doteq \frac{4}{16} \quad \text{or} \quad x = \frac{1}{4}y$$

Substituting this expression in (9) gives

$$V = \frac{\pi}{48} y^3 \tag{11}$$

Differentiating both sides of (11) with respect to t we obtain

$$\frac{dV}{dt} = \frac{\pi}{48}\left(3y^2 \frac{dy}{dt}\right)$$

or

$$\frac{dy}{dt} = \frac{16}{\pi y^2} \frac{dV}{dt}$$

Letting $y = 8$ and substituting (10) into this equation yields

$$\left. \frac{dy}{dt} \right|_{y=8} = \frac{16}{\pi(8)^2}(-2) = -\frac{1}{2\pi} \approx -0.16 \text{ in/min}$$

Thus, when $y = 8$ in., the depth of the liquid is decreasing at an approximate rate of 0.16 in/min. (The minus sign simply means that the depth y is decreasing as t is increasing.) ◄

▶ Exercise Set 4.1 $\boxed{C}$ 19, 20

1. Let A be the area of a square whose sides have length x, and assume that x varies with time.
   (a) How are dA/dt and dx/dt related?
   (b) At a certain instant the sides are 3 ft long and growing at a rate of 2 ft/min. How fast is the area growing at that instant?

2. Let A be the area of a circle whose radius is r, and assume that r varies with time.
   (a) How are dA/dt and dr/dt related?
   (b) At a certain instant the radius is 5 in. and is increasing at a rate of 2 in/sec. How fast is the area of the circle increasing at that instant?

3. Let V be the volume of a cylinder having height h and radius r, and assume that h and r vary with time.
   (a) How are dV/dt, dh/dt, and dr/dt related?
   (b) At a certain instant, the height is 6 in. and increasing at 1 in/sec, while the radius is 10 in. and decreasing at 1 in/sec. How fast is the volume changing at that instant? Is the volume increasing or decreasing at that instant?

4. Let l be the length of a diagonal of a rectangle whose sides have lengths x and y, and assume that x and y vary with time.
   (a) How are dl/dt, dx/dt, and dy/dt related?
   (b) If x increases at a constant rate of $\frac{1}{2}$ ft/sec and y decreases at a constant rate of $\frac{1}{4}$ ft/sec, how fast is the size of the diagonal changing when $x = 3$ ft and $y = 4$ ft? Is the diagonal increasing or decreasing at that instant?

5. Let θ (in radians) be an acute angle in a right triangle, and let x and y, respectively, be the lengths of the sides adjacent and opposite θ. Suppose also that x and y vary with time.
   (a) How are $d\theta/dt$, dx/dt, and dy/dt related?
   (b) At a certain instant, $x = 2$ units and is increasing at 1 unit/sec, while $y = 2$ units and is decreasing at $\frac{1}{4}$ unit/sec. How fast is θ changing at that instant? Is θ increasing or decreasing at that instant?

6. Suppose that $z = x^3 y^2$, where both x and y are changing with time. At a certain instant when $x = 1$ and $y = 2$, x is decreasing at the rate of 2 units/sec, and y is increasing at the rate of 3 units/sec. How fast is z changing at this instant? Is z increasing or decreasing?

7. The minute hand of a certain clock is 4 in. long. Starting from the moment when the hand is pointing straight up, how fast is the area of the sector that is swept out by the hand increasing at any instant during the next revolution of the hand?

8. A stone dropped into a still pond sends out a circular ripple whose radius increases at a constant rate of 3 ft/sec. How rapidly is the area enclosed by the ripple increasing at the end of 10 sec?

9. Oil spilled from a ruptured tanker spreads in a circle whose area increases at a constant rate of 6 mi²/hr. How fast is the radius of the spill increasing when the area is 9 mi²?

10. A spherical balloon is inflated so that its volume is increasing at the rate of 3 ft³/min. How fast is the diameter of the balloon increasing when the radius is 1 ft?

11. A spherical balloon is to be deflated so that its radius decreases at a constant rate of 15 cm/min. At what rate must air be removed when the radius is 9 cm?

12. A 17-ft ladder is leaning against a wall. If the bottom of the ladder is pulled along the ground away from the wall at a constant rate of 5 ft/sec, how fast will the top of the ladder be moving down the wall when it is 8 ft above the ground?

13. A 13-ft ladder is leaning against a wall. If the top of the ladder slips down the wall at a rate of 2 ft/sec, how fast will the foot be moving away from the wall when the top is 5 ft above the ground?

14. A 10-ft plank is leaning against a wall. If at a certain instant the bottom of the plank is 2 ft from the wall and is being pushed toward the wall at the rate of 6 in/sec, how fast is the acute angle that the plank makes with the ground increasing?

15. At a certain instant each edge of a cube is 5 in. long and the volume is increasing at the rate of 2 in^3/min. How fast is the surface area of the cube increasing?

16. A rocket, rising vertically, is tracked by a radar station that is on the ground 5 mi from the launchpad. How fast is the rocket rising when it is 4 mi high and its distance from the radar station is increasing at a rate of 2000 mi/hr?

17. For the camera and rocket shown in Figure 4.1.2, at what rate is the elevation angle changing when the rocket is 3000 ft up and rising vertically at 500 ft/sec?

18. For the camera and rocket shown in Figure 4.1.2, at what rate is the rocket rising when the elevation angle is $\pi/4$ radians and increasing at a rate of 0.2 radian/sec?

19. A satellite is in an elliptical orbit around Earth. Its distance r (in miles) from the center of Earth is given by
$$r = \frac{4995}{1 + 0.12 \cos \theta}$$
where θ is the angle measured from the point on the orbit nearest the surface of Earth (Figure 4.1.8).
(a) Find the altitude of the satellite at *perigee* (the point nearest the surface of Earth) and at *apogee* (the point farthest from the surface of Earth). Use 3960 mi as the radius of Earth.
(b) At the instant when θ is 120°; the angle θ is increasing at the rate of 2.7°/min. Find the altitude of the satellite and the rate at which the

altitude is changing at this instant. Express the rate in units of mi/min.

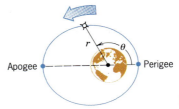

Apogee •------------• Perigee

Figure 4.1.8

20. An aircraft is flying horizontally at a constant height of 4000 ft above a fixed observation point (Figure 4.1.9). At a certain instant the angle of elevation θ is 30° and decreasing, and the speed of the aircraft is 300 mi/hr.
(a) How fast is θ decreasing at this instant? Express the result in units of degrees/sec.
(b) How fast is the distance between the aircraft and the observation point changing at this instant? Express the result in units of ft/sec. Use 1 mi = 5280 ft.

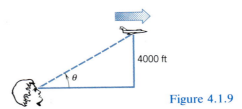

4000 ft

Figure 4.1.9

21. A conical water tank with vertex down has a radius of 10 ft at the top and is 24 ft high. If water flows into the tank at a rate of 20 ft^3/min, how fast is the depth of the water increasing when the water is 16 ft deep?

22. Grain pouring from a chute at the rate of 8 ft^3/min forms a conical pile whose altitude is always twice its radius. How fast is the altitude of the pile increasing at the instant when the pile is 6 ft high?

23. Sand pouring from a chute forms a conical pile whose height is always equal to the diameter. If the height increases at a constant rate of 5 ft/min, at what rate is sand pouring from the chute when the pile is 10 ft high?

24. Wheat is poured through a chute at the rate of 10 ft^3/min, and falls in a conical pile whose bottom radius is always half the altitude. How fast will the circumference of the base be increasing when the pile is 8 ft high?

25. An aircraft is climbing at a 30° angle to the horizontal. How fast is the aircraft gaining altitude if its speed is 500 mi/hr?

26. A boat is pulled into a dock by means of a rope attached to a pulley on the dock (Figure 4.1.10). The rope is attached to the bow of the boat at a point 10 ft below the pulley. If the rope is pulled through the pulley at a rate of 20 ft/min, at what rate will the boat be approaching the dock when 125 ft of rope is out?

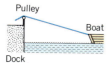

Figure 4.1.10

27. For the boat in Exercise 26, how fast must the rope be pulled if we want the boat to approach the dock at a rate of 12 ft/min at the instant when 125 ft of rope is out?

28. A man 6 ft tall is walking at the rate of 3 ft/sec toward a streetlight 18 ft high (Figure 4.1.11).
(a) At what rate is his shadow length changing?
(b) How fast is the tip of his shadow moving?

Figure 4.1.11

29. A beacon that makes one revolution every 10 sec is located on a ship 4 kilometers from a straight shoreline. How fast is the beam moving along the shoreline when it makes an angle of 45° with the shore?

30. An aircraft is flying at a constant altitude with a constant speed of 600 mi/hr. An antiaircraft missile is fired on a straight line perpendicular to the flight path of the aircraft so that it will hit the aircraft at a point *P* (Figure 4.1.12). At the instant the aircraft is 2 mi from the impact point *P* the missile is 4 mi from *P* and flying at 1200 mi/hr. At that instant, how rapidly is the distance between missile and aircraft decreasing?

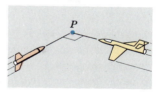

Figure 4.1.12

31. Solve Exercise 30 under the assumption that the angle between the flight paths is 120° instead of the assumption that the paths are perpendicular. [*Hint:* Use the law of cosines.]

32. A police helicopter is flying due north at 100 mi/hr, and at a constant altitude of $\frac{1}{2}$ mi. Below, a car is traveling west on a highway at 75 mi/hr. At the moment the helicopter crosses over the highway the car is 2 mi east of the helicopter.
(a) How fast is the distance between the car and helicopter changing at the moment the helicopter crosses the highway?
(b) Is the distance between car and helicopter increasing or decreasing at that moment?

33. A particle is moving along the curve whose equation is
$$\frac{xy^3}{1 + y^2} = \frac{8}{5}$$
Assume that the *x*-coordinate is increasing at the rate of 6 units/sec when the particle is at the point (1, 2).
(a) At what rate is the *y*-coordinate of the point changing at that instant?
(b) Is the particle rising or falling at that instant?

34. A point *P* is moving along the curve whose equation is $y = \sqrt{x^3 + 17}$. When *P* is at (2, 5), *y* is increasing at the rate of 2 units/sec. How fast is *x* changing?

35. A point *P* is moving along the line whose equation is $y = 2x$. How fast is the distance between *P* and the point (3, 0) changing at the instant when *P* is at (3, 6) if *x* is decreasing at the rate of 2 units/sec at that instant?

36. A point *P* is moving along the curve whose equation is $y = \sqrt{x}$. Suppose that *x* is increasing at the rate of 4 units/sec when $x = 3$.
(a) How fast is the distance between *P* and the point (2, 0) changing at this instant?
(b) How fast is the angle of inclination of the line segment from *P* to (2, 0) changing at this instant?

37. A particle is moving along the curve $y = x^2$. Find all values of x at which the rate of change of y with respect to time is three times that of x. [Assume that dx/dt is never zero.]

38. A particle is moving along the curve $16x^2 + 9y^2 = 144$. Find all points (x, y) at which the rates of change of x and y with respect to time are equal. [Assume that dx/dt and dy/dt are never both zero at the same point.]

39. The *thin lens equation* in physics is

$$\frac{1}{s} + \frac{1}{S} = \frac{1}{f}$$

where s is the object distance from the lens, S is the image distance from the lens, and f is the focal length of the lens. Suppose that a certain lens has a focal length of 6 cm and that an object is moving toward the lens at the rate of 2 cm/sec. How fast is the image distance changing at the instant when the object is 10 cm from the lens? Is the image moving away from the lens or toward the lens?

40. Water is stored in a cone-shaped reservoir (vertex down). Assuming that the water evaporates at a rate proportional to the surface area exposed to the air, show the depth of the water will decrease at a con-stant rate that does not depend on the dimensions of the reservoir.

41. A meteorite enters the earth's atmosphere and burns up at a rate that, at each instant, is proportional to its surface area. Assuming that the meteorite is always spherical, show that the radius decreases at a constant rate.

42. On a certain clock the minute hand is 4 in. long and the hour hand is 3 in. long. How fast is the distance between the tips of the hands changing at 9 o'clock?

43. Coffee is poured at a uniform rate of 2 cm³/sec into a cup whose inside is shaped like a truncated cone (Figure 4.1.13). If the upper and lower radii of the cup are 4 cm and 2 cm and the height of the cup is 6 cm, how fast will the coffee level be rising when the coffee is halfway up? [*Hint:* Extend the cup downward to form a cone.]

Figure 4.1.13

4.2 INTERVALS OF INCREASE AND DECREASE; CONCAVITY

Although point-plotting is useful for determining the general shape of a graph, it only provides an approximation because no matter how many points are plotted, we can only guess at the shape of the graph between those points. In this section we shall show how the derivative can be used to resolve such ambiguities.

INCREASING AND DECREASING FUNCTIONS

The terms *increasing*, *decreasing*, and *constant* are used to describe the behavior of a function as we travel left to right along its graph. For example, the function graphed in Figure 4.2.1 can be described as increasing on the interval $(-\infty, 0]$, decreasing on the interval $[0, 2]$, increasing again on the interval $[2, 4]$, and constant on the interval $[4, +\infty)$.

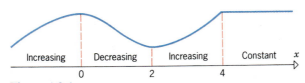

Figure 4.2.1

The following definition, which is illustrated in Figure 4.2.2, expresses these intuitive ideas precisely.

4.2.1 DEFINITION. Let f be defined on an interval, and let x_1 and x_2 denote points in that interval.

(a) f is ***increasing*** on the interval if $f(x_1) < f(x_2)$ whenever $x_1 < x_2$.
(b) f is ***decreasing*** on the interval if $f(x_2) < f(x_1)$ whenever $x_1 < x_2$.
(c) f is ***constant*** on the interval if $f(x_1) = f(x_2)$ for all points x_1 and x_2.

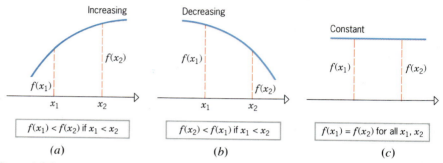

Figure 4.2.2

Figure 4.2.3 suggests that a function f is increasing on any interval where its graph has tangent lines with positive slope, is decreasing on any interval where its graph has tangent lines with negative slope, and is constant on any interval where its graph has tangent lines with zero slope. This intuitive observation suggests the following important theorem. (The proof, which requires results not yet discussed, is given in Section 4.10.)

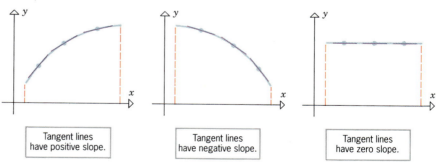

Figure 4.2.3

4.2.2 THEOREM. *Let f be a function that is continuous on a closed interval $[a, b]$ and differentiable on the open interval (a, b).*

(a) *If $f'(x) > 0$ for every value of x in (a, b), then f is increasing on $[a, b]$.*
(b) *If $f'(x) < 0$ for every value of x in (a, b), then f is decreasing on $[a, b]$.*
(c) *If $f'(x) = 0$ for every value of x in (a, b), then f is constant on $[a, b]$.*

REMARK.   This theorem is stated in a form that is useful for investigating whether a function is increasing, decreasing, or constant on a closed interval $[a, b]$. However, there are a number of variations of the theorem that are applicable to other kinds of intervals. For example, the theorem remains true if the closed interval $[a, b]$ is replaced by a finite or infinite interval (a, b), $[a, b)$, or $(a, b]$, provided f is differentiable on (a, b) and continuous on the entire interval. To take this idea one step further, the theorem also remains true if the closed interval $[a, b]$ is replaced by any interval I on which f is differentiable, because the continuity required in the hypothesis follows from the assumption of differentiability.

Example 1   Find the intervals on which the following functions are increasing and the intervals on which they are decreasing.

(a) $f(x) = x^2 - 4x + 3$ (b) $f(x) = x^3$

Solution (a).   Differentiating f we obtain

$$f'(x) = 2x - 4 = 2(x - 2)$$

It follows that

$$f'(x) < 0 \quad \text{if} \quad -\infty < x < 2$$
$$f'(x) > 0 \quad \text{if} \quad 2 < x < +\infty$$

Since f is continuous at $x = 2$, it follows from Theorem 4.2.2 and the subsequent remark that

$$f \text{ is decreasing} \quad \text{if} \quad -\infty < x \leq 2$$
$$f \text{ is increasing} \quad \text{if} \quad 2 \leq x < +\infty$$

Observe that $f'(x) = 0$ at the *point* $x = 2$, but there is no *interval* on which $f'(x) = 0$, so the function is not constant on any interval. For this function the point $x = 2$ marks the place on the x-axis where a transition from decrease to increase occurs. The graph of f is shown in Figure 4.2.4.

Solution (b).   Differentiating f we obtain $f'(x) = 3x^2$. Thus,

$$f'(x) > 0 \quad \text{if} \quad -\infty < x < 0$$
$$f'(x) > 0 \quad \text{if} \quad 0 < x < +\infty$$

Since f is continuous at $x = 0$,

$$f \text{ is increasing} \quad \text{if} \quad -\infty < x \leq 0$$
$$f \text{ is increasing} \quad \text{if} \quad 0 \leq x < +\infty$$

Combining these results, it follows that f is increasing on the interval $(-\infty, +\infty)$. (See Exercise 43.) Observe that $f'(x) = 0$ at the point $x = 0$, but unlike the function in part (a), this point does not mark a transition between increase and decrease. The graph of f is shown in Figure 4.2.5.   ◀

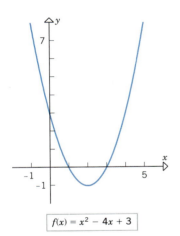

$f(x) = x^2 - 4x + 3$

Figure 4.2.4

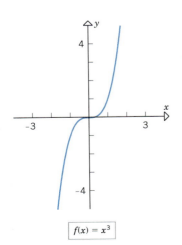

$f(x) = x^3$

Figure 4.2.5

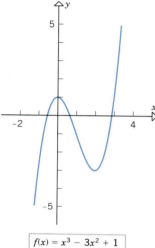

Figure 4.2.6

Figure 4.2.7

Example 2 Find the intervals on which the function

$$f(x) = x^3 - 3x^2 + 1$$

is increasing and the intervals on which it is decreasing.

Solution. Differentiating f we obtain

$$f'(x) = 3x^2 - 6x = 3x(x - 2)$$

From the sign analysis in Figure 4.2.6 we see that

$$f'(x) > 0 \quad \text{if} \quad -\infty < x < 0$$
$$f'(x) < 0 \quad \text{if} \quad 0 < x < 2$$
$$f'(x) > 0 \quad \text{if} \quad 2 < x < +\infty$$

Since f is continuous at $x = 0$ and $x = 2$, it follows that

$$f \text{ is increasing} \quad \text{if} \quad -\infty < x \le 0$$
$$f \text{ is decreasing} \quad \text{if} \quad 0 \le x \le 2$$
$$f \text{ is increasing} \quad \text{if} \quad 2 \le x < +\infty$$

The graph of f is shown in Figure 4.2.7. ◀

☐ **CONCAVITY**

Although the derivative of a function f can tell us where the graph of f is increasing or decreasing, it does not reveal where the graph has a downward curvature (Figure 4.2.8*a*) or an upward curvature (Figure 4.2.8*b*). To investigate this question, we must study the behavior of the tangent lines shown in the figure.

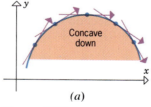

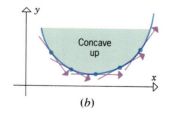

Figure 4.2.8

The curve in part (*a*) lies below its tangent lines and is called *concave down*. As we travel left to right along this curve, the tangent lines rotate clockwise so that their slopes *decrease*. In contrast, the curve in part (*b*) lies above its tangent lines and is called *concave up*. As we travel left to right, its tangent lines rotate counterclockwise so that their slopes *increase*. Since f' is the slope of a tangent line to the graph of f, we are led to the following definition.

4.2.3 DEFINITION. Let f be differentiable on an interval.

(a) f is called *concave up* on the interval if f' is increasing on the interval.

(b) f is called *concave down* on the interval if f' is decreasing on the interval.

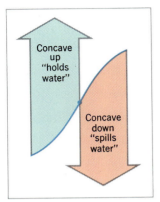

Figure 4.2.9

REMARK.   Informally, a curve that is concave up "holds water" and one that is concave down "spills water" (Figure 4.2.9).

REMARK.   Definition 4.2.3 requires the function f to be differentiable. There are more general definitions of concavity that do not require differentiability. However, we shall have no need for that added generality in this text.

Since f'' is the derivative of f', it follows from Theorem 4.2.2 that f' is increasing on an open interval (a, b) if $f''(x) > 0$ for all x in (a, b), and f' is decreasing on (a, b) if $f''(x) < 0$ for all x in (a, b). Thus, we have the following result.

4.2.4   THEOREM.

(a)   If $f''(x) > 0$ on an open interval (a, b), then f is concave up on (a, b).

(b)   If $f''(x) < 0$ on an open interval (a, b), then f is concave down on (a, b).

Example 3   Find open intervals on which the following functions are concave up and open intervals on which they are concave down.

(a) $f(x) = x^2 - 4x + 3$ (b) $f(x) = x^3$ (c) $f(x) = x^3 - 3x^2 + 1$

Solution (a).   Calculating the second derivative we obtain

$$f'(x) = 2x - 4$$
$$f''(x) = 2$$

Since $f''(x) > 0$ for all x, the function f is concave up on $(-\infty, +\infty)$. This is consistent with Figure 4.2.4.

Solution (b).   Calculating the second derivative we obtain

$$f'(x) = 3x^2$$
$$f''(x) = 6x$$

Since $f''(x) < 0$ if $x < 0$ and $f''(x) > 0$ if $x > 0$, the function f is concave down on $(-\infty, 0)$ and concave up on $(0, +\infty)$. This is consistent with Figure 4.2.5.

Solution (c).   Calculating the second derivative we obtain

$$f'(x) = 3x^2 - 6x$$
$$f''(x) = 6x - 6$$

Thus, the curve is concave up where

$$6x - 6 > 0 \tag{1}$$

and concave down where

$$6x - 6 < 0 \tag{2}$$

Since (1) holds if $x > 1$ and (2) holds if $x < 1$, we conclude that

f is concave up on $(1, +\infty)$

f is concave down on $(-\infty, 1)$

This is consistent with Figure 4.2.7.   ◄

Points where a graph changes from concave up to concave down or vice versa are of special interest. We make the following definition.

4.2.5 DEFINITION. If f is continuous on an open interval containing x_0, and if f changes the direction of its concavity at x_0, then the point $(x_0, f(x_0))$ on the graph of f is called an *inflection point* of f, and we say that f has an *inflection point at x_0*. (See Figure 4.2.10.)

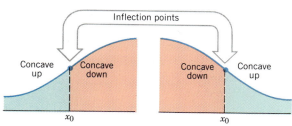

Figure 4.2.10

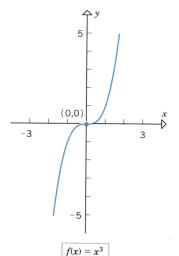

$f(x) = x^3$

Figure 4.2.11

Example 4   Find the inflection points of

(a) $f(x) = x^2 - 4x + 3$ (b) $f(x) = x^3$ (c) $f(x) = x^3 - 3x^2 + 1$

Solution (a).   From Example 3, the function f is concave up on the entire interval $(-\infty, +\infty)$. Since no changes in the direction of concavity occur, there are no inflection points.

Solution (b).   From Example 3, the function f changes from concave down to concave up at $x = 0$, so an inflection point occurs at this point. Since $f(0) = 0$, the inflection point is $(0, 0)$. (See Figure 4.2.11.)

Solution (c).   From Example 3, the function f changes from concave down to concave up at $x = 1$, so an inflection point occurs at this point. Since $f(1) = -1$, the inflection point is $(1, -1)$. (See Figure 4.2.12.)

Theorem 4.2.4 is not applicable on intervals in which there are points where $f''(x) = 0$. For such intervals the direction of concavity is best determined by using Definition 4.2.3 directly.

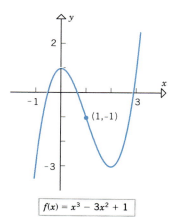

$f(x) = x^3 - 3x^2 + 1$

Figure 4.2.12

Example 5   Determine where the function $f(x) = x^4$ is concave up and where it is concave down.

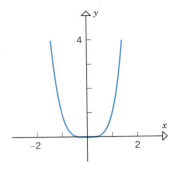

Figure 4.2.13

Solution. Differentiating yields

$$f'(x) = 4x^3 \quad \text{and} \quad f''(x) = 12x^2$$

Since $f''(x) > 0$ if $x < 0$, and $f''(x) > 0$ if $x > 0$, it follows from Theorem 4.2.4 that f is concave up on the open interval $(-\infty, 0)$ and concave up on the open interval $(0, +\infty)$. However, Theorem 4.2.4 is not applicable on the *entire* interval $(-\infty, +\infty)$ because $f''(0) = 0$. But $f'(x) = 4x^3$ is increasing on $(-\infty, +\infty)$ so it follows from Definition 4.2.3 that f is concave up on $(-\infty, +\infty)$. (See Figure 4.2.13.) ◄

Exercise Set 4.2

Exercises 1 and 2 refer to the function graphed in Figure 4.2.14.

1. Find the largest *open* intervals over which the function is
(a) increasing
(b) decreasing
(c) concave up
(d) concave down.

2. Find all values of x where the function has an inflection point.

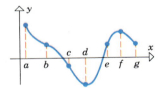

Figure 4.2.14

3. Determine the sign of dy/dx and d^2y/dx^2 at each of the points A, B, and C in Figure 4.2.15.

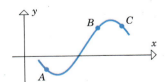

Figure 4.2.15

4. The graph of the *derivative* f' of a function f is shown in Figure 4.2.16. In parts (a)–(f), replace the question mark with $<$, $=$, or $>$.
(a) $f(0)$? $f(1)$
(b) $f(1)$? $f(2)$
(c) $f'(0)$? 0
(d) $f'(1)$? 0
(e) $f''(0)$? 0
(f) $f''(2)$? 0.

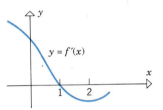

Figure 4.2.16

In Exercises 5–20, find the largest intervals on which f is (a) increasing, (b) decreasing; find the largest open interval on which f is (c) concave up, (d) concave down; and (e) find the x-coordinates of all inflection points.

5. $f(x) = x^2 - 5x + 6$.

6. $f(x) = 4 - 3x - x^2$.

7. $f(x) = (x + 2)^3$.

8. $f(x) = 5 + 12x - x^3$.

9. $f(x) = 3x^3 - 4x + 3$.

10. $f(x) = x^4 - 8x^2 + 16$.

11. $f(x) = 3x^4 - 4x^3$.

12. $f(x) = \dfrac{x}{x^2 + 2}$.

13. $f(x) = \cos x$, $0 < x < 2\pi$.

14. $f(x) = \sin^2 2x$, $0 < x < \pi$.

15. $f(x) = \tan x$, $-\pi/2 < x < \pi/2$.

16. $f(x) = x^{2/3}$.

17. $f(x) = \sqrt[3]{x + 2}$.

18. $f(x) = x^{4/3} - x^{1/3}$.

19. $f(x) = x^{1/3}(x + 4)$.

20. $f(x) = \begin{cases} \frac{1}{2}x^2, & x \le 0 \\ -x^2, & x > 0. \end{cases}$

[*Hint:* Refer to the theorem preceding Exercise 71, Section 3.3.]

21. Determine where $f(x) = x + \sin x$ is increasing.

22. Show that

$$\frac{a}{a^2 + 9} > \frac{b}{b^2 + 9}$$

if $3 \le a < b$. [*Hint:* Show that $x/(x^2 + 9)$ is decreasing on $[3, +\infty)$.]

23. Show that $x + 1/x > 2$ if $x > 1$. [*Hint:* Show that $f(x) = x + 1/x$ is increasing on $[1, +\infty)$.]

24. Show that $x < \tan x$ if $0 < x < \pi/2$. [*Hint:* Show that $f(x) = \tan x - x$ is increasing on $[0, \pi/2)$.]

25. Show that $\sqrt[3]{1 + x} < 1 + \frac{1}{3}x$ if $x > 0$. [*Hint:* Show that $f(x) = 1 + \frac{1}{3}x - \sqrt[3]{1 + x}$ is increasing on $[0, +\infty)$.]

26. In each part sketch a continuous curve $y = f(x)$ with the stated properties.
(a) $f(2) = 4$, $f'(2) = 0$, $f''(x) < 0$ for all x
(b) $f(2) = 4$, $f'(2) = 0$, $f''(x) > 0$ for $x < 2$, $f''(x) < 0$ for $x > 2$
(c) $f(2) = 4$, $f''(x) > 0$ for $x \ne 2$ and

$$\lim_{x \to 2^+} f'(x) = -\infty, \quad \lim_{x \to 2^-} f'(x) = +\infty$$

27. In each part sketch a continuous curve $y = f(x)$ with the stated properties.
(a) $f(2) = 4$, $f'(2) = 0$, $f''(x) > 0$ for all x
(b) $f(2) = 4$, $f'(2) = 0$, $f''(x) < 0$ for $x < 2$, $f''(x) > 0$ for $x > 2$
(c) $f(2) = 4$, $f''(x) < 0$ for $x \ne 2$ and

$$\lim_{x \to 2^+} f'(x) = +\infty, \quad \lim_{x \to 2^-} f'(x) = -\infty$$

28. In parts (a)–(c) sketch a continuous curve $y = f(x)$ with the stated properties.
(a) $f(2) = 4$, $f'(2) = 1$, $f''(x) < 0$ for $x < 2$, $f''(x) > 0$ for $x > 2$
(b) $f(2) = 4$, $f''(x) > 0$ for $x < 2$, $f''(x) < 0$ for $x > 2$, and

$$\lim_{x \to 2^-} f'(x) = +\infty, \quad \lim_{x \to 2^+} f'(x) = +\infty$$

(c) $f(2) = 4$, $f''(x) < 0$ for $x \ne 2$, and

$$\lim_{x \to 2^-} f'(x) = 1, \quad \lim_{x \to 2^+} f'(x) = -1$$

29. In each part sketch a continuous curve $y = f(x)$ with the stated properties.
(a) $f(2) = 4$, $f''(x) < 0$ if $x \ne 2$, and

$$\lim_{x \to 2^-} f'(x) = 0, \quad \lim_{x \to 2^+} f'(x) = +\infty$$

(b) $f(0) = 0$, $f(2) = 4$, $f''(x) = 0$ for all x.

30. Find the inflection points, if any, of the function $f(x) = (x - a)^3$, where a is constant.

31. Find the inflection points, if any, of the function $f(x) = (x - a)^4$, where a is constant.

32. Find the inflection points, if any, of the function $f(x) = ax^2 + bx + c$ $(a \ne 0)$.

In Exercises 33–38, use Definition 4.2.1 to prove the given statement.

33. $f(x) = x^2$ is increasing on $[0, +\infty)$.

34. $f(x) = x^2 - 2x$ is decreasing on $(-\infty, 1]$.

35. $f(x) = \sqrt{x}$ is increasing on $[0, +\infty)$.

36. $f(x) = 1/x$ is decreasing on $(0, +\infty)$.

37. If functions f and g are increasing on an interval I, then $f + g$ is increasing on I.

38. If functions f and g are positive and increasing on an interval I, then their product $f \cdot g$ is increasing on I.

39. Give an example of functions f and g that are increasing on an interval I, but $f - g$ is decreasing on I.

40. For the general cubic polynomial

$$f(x) = ax^3 + bx^2 + cx + d \ (a \ne 0)$$

find conditions on a, b, c, and d to assure that f is always increasing or always decreasing on $(-\infty, +\infty)$.

41. Prove that a third-degree polynomial

$$f(x) = ax^3 + bx^2 + cx + d \ (a \ne 0)$$

has exactly one inflection point.

42. Prove that an nth-degree polynomial

$$f(x) = a_0x^n + a_1x^{n-1} + \cdots + a_n \ (a_0 \ne 0)$$

has at most $n - 2$ inflection points.

43. Prove:
(a) If f is increasing on the intervals $(a, c]$ and $[c, b)$, then f is increasing on (a, b).
(b) If f is decreasing on the intervals $(a, c]$ and $[c, b)$, then f is decreasing on (a, b).

■ 4.3 RELATIVE EXTREMA; FIRST AND SECOND DERIVATIVE TESTS

In this section we shall develop techniques for finding the highest and lowest points on the graph of a function or, equivalently, the largest and smallest values of the function. The methods that we develop here will have important applications in later sections.

□ **RELATIVE MAXIMA AND MINIMA**

The graphs of many functions form hills and valleys. The tops of the hills are called *relative maxima* and the bottoms of the valleys are called *relative minima* (Figure 4.3.1). Just as the top of a hill on the earth's terrain need not be the highest point on earth, so a relative maximum need not be the highest point on the entire graph. However, relative maxima, like tops of hills, are the high points in their *immediate vicinity,* and relative minima, like valley bottoms, are the low points in their *immediate vicinity*. These ideas are captured in the following definitions.

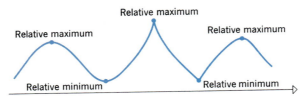

Figure 4.3.1

4.3.1 DEFINITION.   A function f is said to have a **relative maximum** at x_0 if $f(x_0) \geq f(x)$ for all x in some open interval containing x_0.

4.3.2 DEFINITION.   A function f is said to have a **relative minimum** at x_0 if $f(x_0) \leq f(x)$ for all x in some open interval containing x_0.

4.3.3 DEFINITION.   A function f is said to have a **relative extremum** at x_0 if it has either a relative maximum or a relative minimum at x_0.

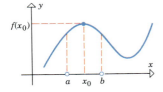

Figure 4.3.2

Example 1   The function f graphed in Figure 4.3.2 has a relative maximum at x_0 because there is an open interval containing x_0 on which $f(x_0) \geq f(x)$ holds. [See interval (a, b) in the figure.]   ◄

□ **CRITICAL POINTS**

Relative extrema can be viewed as the transition points that separate the regions where a graph is increasing from those where it is decreasing. As suggested by Figure 4.3.3, the relative extrema of a function f occur either at points where the graph of f has a horizontal tangent or at points where f is not differentiable.

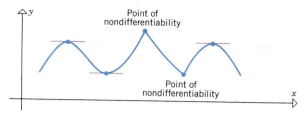

Figure 4.3.3

This is the content of the following theorem (whose proof is given at the end of this section).

> **4.3.4   THEOREM.**   *If f has a relative extremum at x_0, then either $f'(x_0) = 0$ or f is not differentiable at x_0.*

Theorem 4.3.4 is of such importance that there is some terminology associated with it.

> **4.3.5   DEFINITION.**   A ***critical point*** for a function f is any value of x in the domain of f at which $f'(x) = 0$ or at which f is not differentiable; the critical points where $f'(x) = 0$ are called ***stationary points*** of f.

With the terminology of this definition, Theorem 4.3.4 states:

> *The relative extrema of a function, if any, occur at critical points.*

Example 2   For each of the functions graphed in Figure 4.3.4, x_0 is a critical point. In parts (a), (b), (c), and (d), x_0 is a stationary point because there is a horizontal tangent at x_0, and in the remaining parts x_0 is a critical point, but not a stationary point, because the derivative does not exist there.   ◄

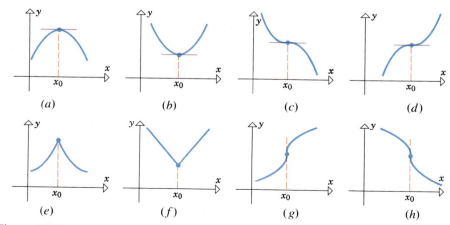

Figure 4.3.4

☐ **FIRST AND SECOND DERIVATIVE TESTS**

Although the relative extrema of a function must occur at critical points, a relative extremum need not occur at *every* critical point. For example, in parts (c), (d), (g), and (h) of Figure 4.3.4, there is a critical point but no relative extremum at x_0. However, what distinguishes these four cases from the rest is that there is no sign change in the first derivative at x_0. To be specific,

- a relative maximum occurs at those critical points where the sign of the first derivative changes from positive to negative moving in the positive x-direction (Figure 4.3.4a, e);
- a relative minimum occurs at those critical points where the sign of the first derivative changes from negative to positive moving in the positive x-direction (Figure 4.3.4b, f);
- no relative extremum occurs at those critical points where the sign of the first derivative does not change moving in the positive x-direction (Figure 4.3.4c, d, g, h).

These ideas are stated more precisely in the following theorem whose proof is given in Appendix C.

4.3.6   THEOREM (*First Derivative Test*).   *Suppose f is continuous at a critical point x_0.*

(a)   *If $f'(x) > 0$ on an open interval extending left from x_0 and $f'(x) < 0$ on an open interval extending right from x_0, then f has a relative maximum at x_0.*

(b)   *If $f'(x) < 0$ on an open interval extending left from x_0 and $f'(x) > 0$ on an open interval extending right from x_0, then f has a relative minimum at x_0.*

(c)   *If $f'(x)$ has the same sign [either $f'(x) > 0$ or $f'(x) < 0$] on an open interval extending left from x_0 and on an open interval extending right from x_0, then f does not have a relative extremum at x_0.*

It follows from this theorem that:

The relative extrema of a continuous nonconstant function f, if any, occur at those critical points where f' changes sign.

Example 3   Locate the relative extrema of $f(x) = 3x^{5/3} - 15x^{2/3}$.

Solution.

$$f'(x) = 5x^{2/3} - 10x^{-1/3} = 5x^{-1/3}(x - 2)$$

Since $f'(x)$ does not exist when $x = 0$, and $f'(x) = 0$ when $x = 2$, the critical points of f are $x = 0$ and $x = 2$.

As shown in Figure 4.3.5a, the sign of f' changes from positive to negative at 0 and from negative to positive at 2. Thus, there is a relative maximum at 0 and a relative minimum at 2. Although not needed for the solution, the graph of f is shown in Figure 4.3.5b.   ◄

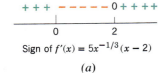

Sign of $f'(x) = 5x^{-1/3}(x - 2)$

(a)

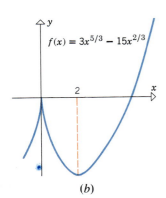

$f(x) = 3x^{5/3} - 15x^{2/3}$

(b)

Figure 4.3.5

Example 4   Locate the relative extrema of $f(x) = x^3 - 3x^2 + 3x - 1$.

Solution.   Since f is differentiable everywhere, the only possible critical points are stationary points. Differentiating yields

$$f'(x) = 3x^2 - 6x + 3 = 3(x - 1)^2$$

Solving $f'(x) = 0$ yields $x = 1$ as the only critical point. Since $3(x - 1)^2 \geq 0$ for all x, $f'(x)$ does not change sign at $x = 1$, so f does not have a relative extremum at $x = 1$. Thus, f has no relative extrema.   ◄

There is another test for relative extrema that is often easier to apply than the first derivative test. It is based on the geometric observation that a function has a relative maximum where its graph has a horizontal tangent line and is concave down (Figure 4.3.6), and has a relative minimum where its graph has a horizontal tangent line and is concave up (Figure 4.3.6).

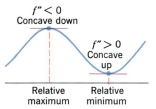

$f'' < 0$
Concave down

$f'' > 0$
Concave up

Relative maximum   Relative minimum

Figure 4.3.6

4.3.7   THEOREM (*Second Derivative Test*).   *Suppose f is twice differentiable at a stationary point x_0.*

(a)   If $f''(x_0) > 0$, then f has a relative minimum at x_0.
(b)   If $f''(x_0) < 0$, then f has a relative maximum at x_0.

A formal proof of this result is given in Appendix C.

Example 5   Locate and describe the relative extrema of $f(x) = x^4 - 2x^2$.

Solution.

$$f'(x) = 4x^3 - 4x = 4x(x - 1)(x + 1)$$
$$f''(x) = 12x^2 - 4$$

Solving $f'(x) = 0$ yields the stationary points $x = 0$, $x = 1$, and $x = -1$. Since

$$f''(0) = -4 < 0$$
$$f''(1) = 8 > 0$$
$$f''(-1) = 8 > 0$$

there is a relative maximum at $x = 0$ and there are relative minima at $x = 1$ and $x = -1$.   ◄

REMARK.   If f is not twice differentiable at the critical point x_0, or if $f''(x_0) = 0$, then we must either rely on the first derivative test or devise more imaginative techniques appropriate to the problem.

REMARK.   In some problems the first derivative test is easier to apply, and in others the second derivative test is easier. Thus, in each problem some thought should be given to determining the test that involves the least amount of computation.

■ OPTIONAL

Proof of Theorem 4.3.4. We shall prove this theorem in the case where there is a relative maximum at x_0. The proof in the case of a relative minimum is similar and is left to the reader.

There are two possibilities—either f is differentiable at x_0 or it is not. If it is not, then x_0 is a critical point for f and we are done. If f is differentiable at x_0, then we must show that $f'(x_0) = 0$. We shall do this by showing that $f'(x_0) \geq 0$ and $f'(x_0) \leq 0$, from which it follows that $f'(x_0) = 0$. From the definition of a derivative we have

$$f'(x_0) = \lim_{h \to 0} \frac{f(x_0 + h) - f(x_0)}{h}$$

so that

$$f'(x_0) = \lim_{h \to 0^+} \frac{f(x_0 + h) - f(x_0)}{h} \tag{1}$$

and

$$f'(x_0) = \lim_{h \to 0^-} \frac{f(x_0 + h) - f(x_0)}{h} \tag{2}$$

Because f has a relative maximum at x_0, there is an open interval (a, b) containing x_0 in which $f(x) \leq f(x_0)$ for all x in (a, b).

Assume that h is sufficiently small so that $x_0 + h$ lies in the interval (a, b). Thus,

$$f(x_0 + h) \leq f(x_0) \qquad \text{or equivalently} \qquad f(x_0 + h) - f(x_0) \leq 0$$

Thus, if h is negative,

$$\frac{f(x_0 + h) - f(x_0)}{h} \geq 0 \tag{3}$$

and if h is positive,

$$\frac{f(x_0 + h) - f(x_0)}{h} \leq 0 \tag{4}$$

But an expression that never assumes negative values cannot approach a negative limit and an expression that never assumes positive values cannot approach a positive limit, so that

$$f'(x_0) = \lim_{h \to 0^-} \frac{f(x_0 + h) - f(x_0)}{h} \geq 0 \quad \boxed{\text{From (2) and (3)}}$$

and

$$f'(x_0) = \lim_{h \to 0^+} \frac{f(x_0 + h) - f(x_0)}{h} \leq 0 \quad \boxed{\text{From (1) and (4)}}$$

Since $f'(x_0) \geq 0$ and $f'(x_0) \leq 0$, it must be that $f'(x_0) = 0$. ■

▶ Exercise Set 4.3

In Exercises 1–16, locate the critical points and classify them as stationary points or points of nondifferentiability.

1. $f(x) = x^2 - 5x + 6$.

2. $f(x) = 4x^2 + 2x - 5$.

3. $f(x) = x^3 + 3x^2 - 9x + 1$.

4. $f(x) = 2x^3 - 6x + 7$.

5. $f(x) = x^4 - 6x^2 - 3$.

6. $f(x) = 3x^4 - 4x^3$.   **7.** $f(x) = \dfrac{x}{x^2 + 2}$.

8. $f(x) = \dfrac{x^2 - 3}{x^2 + 1}$.   **9.** $f(x) = x^{2/3}$.

10. $f(x) = \sqrt[3]{x + 2}$.   **11.** $f(x) = \cos 3x$.

12. $f(x) = x \tan x$, $-\pi/2 < x < \pi/2$.

13. $f(x) = \sin^2 2x$, $0 < x < 2\pi$.

14. $f(x) = |\sin x|$.   **15.** $f(x) = x^{1/3}(x + 4)$.

16. $f(x) = x^{4/3} - 6x^{1/3}$.

17. Figure 4.3.7 shows the graph of the derivative f' of a function f. Find all values of x, if any, where f has

(a) relative minima   (b) relative maxima

(c) inflection points.

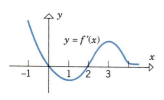

Figure 4.3.7   Figure 4.3.8

18. Repeat Exercise 17 using Figure 4.3.8.

In Exercises 19–22, the derivative of a continuous function is given. Find all critical points, and determine whether a relative maximum, relative minimum, or neither occurs there.

19. $f'(x) = x^3(x^2 - 5)$.

20. $f'(x) = x^2(2x + 1)(x - 1)$.

21. $f'(x) = \dfrac{9 - 4x^2}{\sqrt[3]{x + 1}}$.

22. $f'(x) = 2\sin^3 x - \sin^2 x$, $0 < x < 2\pi$.

In Exercises 23–26, find the relative extrema using (a) the first derivative test; (b) the second derivative test.

23. $f(x) = 1 - 4x - x^2$.

24. $f(x) = 2x^3 - 9x^2 + 12x$.

25. $f(x) = \sin^2 x$, $0 < x < 2\pi$.

26. $f(x) = \frac{1}{2}x - \sin x$, $0 < x < 2\pi$.

In Exercises 27–44, use any method to find the relative extrema.

27. $f(x) = x^3 + 5x - 2$.

28. $f(x) = x^4 - 2x^2 + 7$.

29. $f(x) = x(x - 1)^2$.

30. $f(x) = x^4 + 2x^3$.

31. $f(x) = 2x^2 - x^4$.   **32.** $f(x) = (2x - 1)^5$.

33. $f(x) = x^{4/5}$.   **34.** $f(x) = 2x + x^{2/3}$.

35. $f(x) = \dfrac{x^2}{x^2 + 1}$.   **36.** $f(x) = \dfrac{x}{x + 2}$.

37. $f(x) = |x^2 - 4|$.

38. $f(x) = \begin{cases} 9 - x, & x \le 3 \\ x^2 - 3, & x > 3. \end{cases}$

39. $f(x) = \cos^2 x$.

40. $f(x) = \sqrt{3}x + 2\sin x$, $0 < x < 2\pi$.

41. $f(x) = \tan(x^2 + 1)$.

42. $f(x) = \dfrac{\sin x}{2 + \cos x}$, $0 < x < 2\pi$.

43. $f(x) = |\sin 2x|$, $0 < x < 2\pi$.

44. $f(x) = \cos 4x + 2\sin 2x$, $0 < x < \pi$.

45. Find a value of k so that $x^2 + \dfrac{k}{x}$ will have a relative extremum at $x = 3$.

46. Find a value of k so that $\dfrac{x}{x^2 + k}$ will have a relative extremum at $x = 2.5$.

47. Recall that the second derivative test (Theorem 4.3.7) does not apply if $f''(x_0) = 0$. Give examples to show that a function f can have a relative maximum at x_0, a relative minimum at x_0, or neither if $f''(x_0) = 0$. [*Hint:* Try functions of the form $f(x) = x^n$.]

48. Let h and g have relative maxima at x_0. Prove or disprove:

(a) $h + g$ has a relative maximum at x_0

(b) $h - g$ has a relative maximum at x_0.

49. Sketch some curves that show that the three parts of the first derivative test (Theorem 4.3.6) can be false without the assumption that f is continuous at x_0.

50. Prove: If f is continuous on an open interval I that contains an inflection point at x_0, and if f' exists everywhere in I except perhaps at x_0, then either $f''(x_0) = 0$ or f' is not differentiable at x_0. [*Hint:* Modify the proof of Theorem 4.3.4 using f' and f'' in place of f and f', respectively.]

■ 4.4 GRAPHS OF POLYNOMIALS AND RATIONAL FUNCTIONS

Graphs of functions are used for many purposes in mathematics and science. In applications where numerical values are read from graphs, extremely accurate point-plotting is required. Such graphs are best generated on a computer. In many mathematical applications, however, accurately plotted points are not needed; what is required is the precise location of the key features such as critical points, inflection points, intervals of increase and decrease, points of discontinuity, and so forth. In this section we shall use the tools developed in the preceding section to investigate the key features of the graphs of polynomials and rational functions.

□ **GRAPHS OF POLYNOMIALS**

Polynomials are among the simplest functions to graph; they are continuous functions, so their graphs have no breaks or holes, and they are differentiable functions, so their graphs have no sharp corners. The following steps are usually sufficient to obtain the main elements of the graph.

> ***A Method for Graphing a Polynomial P(x)***
>
> **Step 1.** Calculate $P'(x)$ and $P''(x)$.
>
> **Step 2.** From $P'(x)$ determine the stationary points and the intervals where P is increasing and decreasing.
>
> **Step 3.** From $P''(x)$ determine the inflection points and the intervals where P is concave up and concave down.
>
> **Step 4.** Plot the intersection with the y-axis, the stationary points, the inflection points, and, if possible, the intersections with the x-axis. Finally, plot any additional points needed to obtain the accuracy desired in the graph.

Example 1 Sketch the graph of $y = x^3 - 3x + 2$.

Solution.

$$\frac{dy}{dx} = 3x^2 - 3 = 3(x - 1)(x + 1) \qquad \text{and} \qquad \frac{d^2y}{dx^2} = 6x$$

Figure 4.4.1 shows a convenient way of organizing the analysis. Part (*a*) in the figure shows the sign pattern of dy/dx, which can be obtained using either the method of factoring or the method of test points. (See Example 9 of Section 1.1.) The upward arrows indicate intervals where the graph is increasing, the downward arrows indicate intervals where the graph is decreasing, and the horizontal arrows mark the points where there is a horizontal tangent line. Part (*b*) of the figure shows the sign pattern of d^2y/dx^2. In part (*c*) of the figure the information in the previous parts is combined to produce the general graph shape.

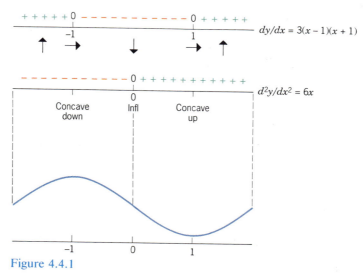

Figure 4.4.1

All that remains is to refine the sketch by plotting the stationary points, inflection points, and intersections with the coordinate axes. In this case, we will also plot the point of $x = 2$ for some extra precision (Figure 4.4.2). As a final note, the graph in Figure 4.4.2 was drawn as though it increases without bound as $x \to +\infty$ and decreases without bound as $x \to -\infty$. This is consistent with the limits

$$\lim_{x \to +\infty} (x^3 - 3x + 2) = +\infty \qquad \text{and} \qquad \lim_{x \to -\infty} (x^3 - 3x + 2) = -\infty$$

[See Formulas (13), (14), (17), (18), and Example 9 in Section 2.5.] ◀

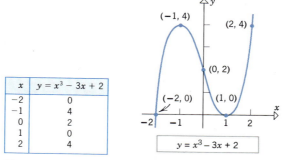

x	$y = x^3 - 3x + 2$
-2	0
-1	4
0	2
1	0
2	4

Figure 4.4.2

☐ **GRAPHS OF RATIONAL**
   FUNCTIONS

Recall that if $P(x)$ and $Q(x)$ are polynomials, then their ratio

$$f(x) = \frac{P(x)}{Q(x)}$$

is called a rational function of x. Graphing rational functions is complicated by the fact that discontinuities occur at points where $Q(x) = 0$. In Figure 4.4.3 we have sketched the graph of

$$f(x) = \frac{x}{x - 2}$$

This figure illustrates most of the characteristics that are typical of rational functions. At $x = 2$ the denominator of $f(x)$ is zero, so a discontinuity occurs in the graph at this point. Approaching the point $x = 2$, we have

$$\lim_{x \to 2^+} \frac{x}{x - 2} = +\infty \quad \text{and} \quad \lim_{x \to 2^-} \frac{x}{x - 2} = -\infty$$

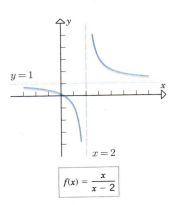

$$f(x) = \frac{x}{x - 2}$$

Figure 4.4.3

The line $x = 2$ is called a *vertical asymptote* for the graph. Also

$$\lim_{x \to +\infty} \frac{x}{x - 2} = 1 \quad \text{and} \quad \lim_{x \to -\infty} \frac{x}{x - 2} = 1$$

so that the graph approaches the line $y = 1$ as $x \to +\infty$ and as $x \to -\infty$. The line $y = 1$ is called a *horizontal asymptote* for the graph. More precisely, we make the following definition.

4.4.1   DEFINITION.   A line $x = x_0$ is called a *vertical asymptote* for the graph of a function f if $f(x) \to +\infty$ or $f(x) \to -\infty$ as x approaches x_0 from the right or from the left. A line $y = L$ is called a *horizontal asymptote* for the graph of f if

$$\lim_{x \to +\infty} f(x) = L \quad \text{or} \quad \lim_{x \to -\infty} f(x) = L$$

For rational functions whose numerator and denominator have no common factors, vertical asymptotes occur at points where the denominator is zero.

REMARK.   It is possible for the graph of a rational function to intersect a horizontal asymptote (Exercises 25–28), but it can not intersect a vertical asymptote.

Example 2   Find all vertical and horizontal asymptotes of

(a) $f(x) = \dfrac{x^2 + 2x}{x^2 - 1}$ (b) $f(x) = \dfrac{x^3 + 1}{x - 2}$

Solution (a). The vertical asymptotes occur at the points where $x^2 - 1 = 0$; these are the points $x = -1$ and $x = 1$. Since

$$\lim_{x \to +\infty} f(x) = \lim_{x \to +\infty} \frac{x^2 + 2x}{x^2 - 1} = \lim_{x \to +\infty} \frac{x^2}{x^2} = 1$$

$$\lim_{x \to -\infty} f(x) = \lim_{x \to -\infty} \frac{x^2 + 2x}{x^2 - 1} = \lim_{x \to -\infty} \frac{x^2}{x^2} = 1$$

it follows that $y = 1$ is a horizontal asymptote. (A computer-generated graph of f is shown in Figure 4.4.4a.)

Solution (b). The denominator is zero at $x = 2$, so a vertical asymptote occurs at this point. Since

$$\lim_{x \to +\infty} f(x) = \lim_{x \to +\infty} \frac{x^3 + 1}{x - 2} = \lim_{x \to +\infty} \frac{x^3}{x} = \lim_{x \to +\infty} x^2 = +\infty$$

$$\lim_{x \to -\infty} f(x) = \lim_{x \to -\infty} \frac{x^3 + 1}{x - 2} = \lim_{x \to -\infty} \frac{x^3}{x} = \lim_{x \to -\infty} x^2 = +\infty$$

there are no horizontal asymptotes. (A computer-generated graph of f is shown in Figure 4.4.4b.) ◄

For graphs of rational functions, the properties of interest are

1. symmetries
2. x-intercepts
3. y-intercept
4. asymptotes
5. intervals of increase and decrease
6. stationary points
7. concavity and inflection points

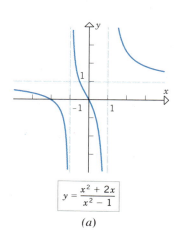

$$y = \frac{x^2 + 2x}{x^2 - 1}$$

(a)

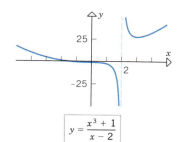

$$y = \frac{x^3 + 1}{x - 2}$$

(b)

Figure 4.4.4

One procedure for organizing the computations is to find the symmetries, the intercepts, and the asymptotes first. Once this is done, a preliminary sketch of the graph can be made by evaluating the function at some well-chosen points. This preliminary sketch can then be refined by calculating the first and second derivatives and using them to incorporate the missing information about stationary points, inflection points, and so forth.

Example 3 Sketch the graph of $y = \dfrac{2x^2 - 8}{x^2 - 16}$.

Solution.

Symmetries: Replacing x by $-x$ does not change the equation, so the graph is symmetric about the y-axis.

x-intercepts: Setting $y = 0$ yields the x-intercepts $x = -2$ and $x = 2$.

y-intercept: Setting $x = 0$ yields the y-intercept $y = 1/2$.

Vertical asymptotes: Setting $x^2 - 16 = 0$ yields the vertical asymptotes $x = -4$ and $x = 4$.

Horizontal asymptotes: The limits

$$\lim_{x \to +\infty} \frac{2x^2 - 8}{x^2 - 16} = \lim_{x \to +\infty} \frac{2x^2}{x^2} = 2$$

$$\lim_{x \to -\infty} \frac{2x^2 - 8}{x^2 - 16} = \lim_{x \to -\infty} \frac{2x^2}{x^2} = 2$$

yield the horizontal asymptote $y = 2$.

The set of points where x-intercepts or vertical asymptotes occur, namely $\{-4, -2, 2, 4\}$, divides the x-axis into the open intervals

$$(-\infty, -4), \quad (-4, -2), \quad (-2, 2), \quad (2, 4), \quad (4, +\infty)$$

Over each of these intervals, y cannot change sign (why?). We can find the sign of y on each interval by choosing an arbitrary test point in the interval and evaluating $y = f(x)$ at the test points (Table 4.4.1).

Table 4.4.1

INTERVAL	TEST POINT	$y = \dfrac{2x^2 - 8}{x^2 - 16}$	SIGN OF y
$(-\infty, -4)$	$x = -5$	$y = 14/3$	$+$
$(-4, -2)$	$x = -3$	$y = -10/7$	$-$
$(-2, 2)$	$x = 0$	$y = 1/2$	$+$
$(2, 4)$	$x = 3$	$y = -10/7$	$-$
$(4, +\infty)$	$x = 5$	$y = 14/3$	$+$

Using the information in the table, we can draw small curve segments at the intercepts and the test points, using the sign of y to position the curve segment properly above or below the x-axis; these segments allow us to make an educated guess about the shape of the curve near the vertical and horizontal asymptotes, and we can draw curve segments near the asymptotes as well (Figure 4.4.5a). The curve segments suggest the rough shape of the graph. Of course, we may have missed some stationary points and the concavity may be incorrect in places, but this can now be fixed by using information from the first and second derivatives.

Derivatives:

$$\frac{dy}{dx} = \frac{(x^2 - 16)(4x) - (2x^2 - 8)(2x)}{(x^2 - 16)^2} = -\frac{48x}{(x^2 - 16)^2}$$

$$\frac{d^2y}{dx^2} = \frac{48(16 + 3x^2)}{(x^2 - 16)^3} \quad \text{(verify)}$$

Intervals of increase and decrease: A sign analysis of dy/dx yields

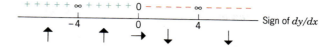

Thus, the graph is increasing on the intervals $(-\infty, -4)$, $(-4, -2)$, and $(-2, 0)$; and it is decreasing on the intervals $(0, 2)$, $(2, 4)$, and $(4, +\infty)$. There is a stationary point at $x = 0$.

Concavity: A sign analysis of d^2y/dx^2 yields

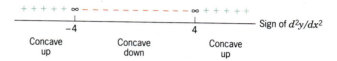

There are changes in concavity at the vertical asymptotes, $x = -4$ and $x = 4$, but there are no inflection points.

Using the information supplied by the derivatives to refine the rough sketch in Figure 4.4.5*a* yields the graph in Figure 4.4.5*b*. ◀

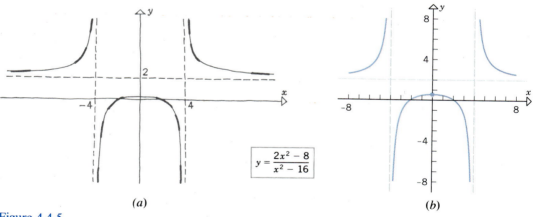

$$y = \frac{2x^2 - 8}{x^2 - 16}$$

(*a*)   (*b*)

Figure 4.4.5

Example 4   Sketch the graph of $y = \dfrac{x^2 - 1}{x^3}$.

Solution.

Symmetries: Replacing x by $-x$ and y by $-y$ yields an equation that simplifies back to the original equation, so the graph is symmetric about the origin.

x-intercepts: Setting $y = 0$ yields the x-intercepts $x = -1$ and $x = 1$.

y-intercept: Setting $x = 0$ leads to a division by zero, so that there is no y-intercept.

Vertical asymptotes: Setting $x^3 = 0$ yields the vertical asymptote $x = 0$.

Horizontal asymptotes: The limits

$$\lim_{x \to +\infty} \frac{x^2 - 1}{x^3} = \lim_{x \to +\infty} \frac{x^2}{x^3} = \lim_{x \to +\infty} \frac{1}{x} = 0$$

$$\lim_{x \to -\infty} \frac{x^2 - 1}{x^3} = \lim_{x \to -\infty} \frac{x^2}{x^3} = \lim_{x \to +\infty} \frac{1}{x} = 0$$

yield the horizontal asymptote $y = 0$.

The set of points where x-intercepts or vertical asymptotes occur, namely $\{-1, 0, 1\}$, divides the x-axis into the open intervals

$$(-\infty, -1), \quad (-1, 0), \quad (0, 1), \quad (1, +\infty)$$

Choosing a test point in each interval and finding the sign of y at the test points yields Table 4.4.2. From the information in the table we obtain the rough sketch of the graph in Figure 4.4.6a.

Table 4.4.2

INTERVAL	TEST POINT	$y = \dfrac{x^2 - 1}{x^3}$	SIGN OF y
$(-\infty, -1)$	$x = -2$	$y = -3/8$	$-$
$(-1, 0)$	$x = -1/2$	$y = 6$	$+$
$(0, 1)$	$x = 1/2$	$y = -6$	$-$
$(1, +\infty)$	$x = 2$	$y = 3/8$	$+$

We now refine the rough sketch.

Derivatives:

$$\frac{dy}{dx} = \frac{x^3(2x) - (x^2 - 1)(3x^2)}{(x^3)^2} = \frac{3 - x^2}{x^4}$$

$$\frac{d^2y}{dx^2} = \frac{x^4(-2x) - (3 - x^2)(4x^3)}{(x^4)^2} = \frac{2(x^2 - 6)}{x^5}$$

Intervals of increase and decrease:

This analysis reveals stationary points at $x = -\sqrt{3}$ and $x = \sqrt{3}$.

Concavity:

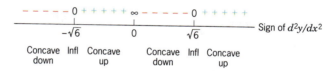

This analysis reveals a change in concavity at the vertical asymptote $x = 0$ and inflection points at $x = -\sqrt{6}$ and $x = \sqrt{6}$.

The final graph with the stationary points and inflection points plotted is shown in Figure 4.4.6*b*. ◄

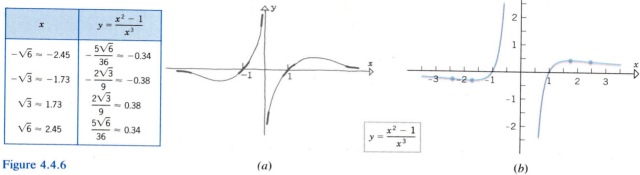

x	$y = \dfrac{x^2 - 1}{x^3}$
$-\sqrt{6} \approx -2.45$	$-\dfrac{5\sqrt{6}}{36} \approx -0.34$
$-\sqrt{3} \approx -1.73$	$-\dfrac{2\sqrt{3}}{9} \approx -0.38$
$\sqrt{3} \approx 1.73$	$\dfrac{2\sqrt{3}}{9} \approx 0.38$
$\sqrt{6} \approx 2.45$	$\dfrac{5\sqrt{6}}{36} \approx 0.34$

Figure 4.4.6

(a)

$$y = \frac{x^2 - 1}{x^3}$$

(b)

◻ COMPUTER-GENERATED GRAPHS

Although computers and graphing calculators can generate graphs with great speed and accuracy, the fact that a computer or calculator display can reveal only a finite portion of an entire graph can lead to problems. For example, it is evident that the graph of the polynomial equation $y = (x - 100)^3$ has an inflection point at $x = 100$, since the graph can be obtained by translating the graph of $y = x^3$ in the positive x-direction 100 units (Figure 4.4.7*a*). However, if one were to program a computer or calculator to graph this equation over the interval $[-40, 40]$, for example, the inflection point, which is the key feature of the graph, would be completely missed. For example, Figure 4.4.7*b* was generated using such a graphing program. Not only was the inflection point missed but the program was unable to show the y-intercept because of the wide vertical scale. This may create the false impression that the graph has a vertical asymptote at $x = 0$. In short, an intelligent use of computers and calculators for graphing requires a good understanding of the basic principles of graphing that result from the methods of calculus.

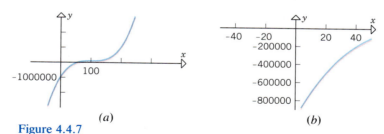

(a)

(b)

Figure 4.4.7

▶ Exercise Set 4.4   C 59, 60

In Exercises 1–18, use the techniques illustrated in this section to sketch the graph of the given polynomial. Plot the stationary points and the inflection points.

1. $x^2 - 2x - 3$.   **2.** $1 + x - x^2$.

3. $x^3 - 3x + 1$.   **4.** $2x^3 - 6x + 4$.

5. $x^3 + 3x^2 + 5$.   **6.** $x^2 - x^3$.

7. $2x^3 - 3x^2 + 12x + 9$.   **8.** $x^3 - 3x^2 + 3$.

9. $(x - 1)^4$.   **10.** $(x - 1)^5$.

11. $x^4 + 2x^3 - 1$.   **12.** $x^4 - 2x^2 - 12$.

13. $x^4 - 3x^3 + 3x^2 + 1$.   **14.** $x^5 - 4x^4 + 4x^3$.

15. $3x^5 - 5x^3$.   **16.** $3x^4 + 4x^3$.

17. $x(x - 1)^3$.   **18.** $x^5 + 5x^4$.

In Exercises 19–24, find equations of all vertical and horizontal asymptotes for the graph of the given rational function.

19. $\dfrac{3x}{x - 2}$.   **20.** $\dfrac{4x + 1}{3x + 2}$.   **21.** $\dfrac{x^3}{x^2 - 5}$.

22. $\dfrac{3 - x}{x^2 + 1}$.   **23.** $\dfrac{x^2}{x^2 - 2x - 3}$.   **24.** $\dfrac{2x^2 + 1}{3x^2 + 6x}$.

In Exercises 25–28, find all values of x where the graph of the given rational function crosses its horizontal asymptote.

25. $\dfrac{x^2}{x^2 + 2x + 5}$.   **26.** $\dfrac{x^2 - 3x + 2}{x^2}$.

27. $\dfrac{x^2 + 1}{2x^2 - 6x}$.   **28.** $\dfrac{25 - 9x^2}{x^3}$.

In Exercises 29–47, use the techniques illustrated in this section to sketch the graph of the given rational function. Show any horizontal and vertical asymptotes, and plot the stationary points and, if reasonable, the inflection points.

29. $\dfrac{2x}{x - 3}$.   **30.** $\dfrac{x}{x^2 - 1}$.   **31.** $\dfrac{x^2}{x^2 - 1}$.

32. $\dfrac{1}{(x - 1)^2}$.   **33.** $\dfrac{x}{1 + x^2}$.   **34.** $1 - \dfrac{1}{x}$.

35. $\dfrac{x - 1}{x - 2}$.   **36.** $\dfrac{1}{x^2 + 1}$.   **37.** $x^2 - \dfrac{1}{x}$.

38. $\dfrac{2x^2 - 1}{x^2}$.   **39.** $\dfrac{1 - x}{x^2}$.   **40.** $\dfrac{8}{4 - x^2}$.

41. $\dfrac{x - 1}{x^2 - 4}$.   **42.** $\dfrac{8(x - 2)}{x^2}$.   **43.** $\dfrac{(x - 1)^2}{x^2}$.

44. $2 + \dfrac{3}{x} - \dfrac{1}{x^3}$.   **45.** $3 - \dfrac{4}{x} - \dfrac{4}{x^2}$.   **46.** $\dfrac{x^2 - 1}{x^2 + 1}$.

47. $\dfrac{x^3 - 1}{x^3 + 1}$.

48. **(Oblique Asymptotes)** If a rational function $P(x)/Q(x)$ is such that the degree of the numerator exceeds the degree of the denominator by *one*, then the graph of $P(x)/Q(x)$ will have an *oblique asymptote,* that is, an asymptote that is neither vertical nor horizontal. To see why, we perform the division of $P(x)$ by $Q(x)$ to obtain

$$\frac{P(x)}{Q(x)} = (ax + b) + \frac{R(x)}{Q(x)}$$

where $ax + b$ is the quotient and $R(x)$ is the remainder. Use the fact that the degree of the remainder $R(x)$ is less than the degree of the divisor $Q(x)$ to help prove

$$\lim_{x \to +\infty} \left[\frac{P(x)}{Q(x)} - (ax + b) \right] = 0$$

$$\lim_{x \to -\infty} \left[\frac{P(x)}{Q(x)} - (ax + b) \right] = 0$$

These results tell us that the graph of the equation $y = P(x)/Q(x)$ "approaches" the line (an oblique asymptote) $y = ax + b$ as $x \to +\infty$ or $x \to -\infty$ (Figure 4.4.8).

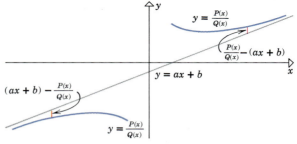

Figure 4.4.8

In Exercises 49–53, sketch the graph of the rational function. Show all vertical, horizontal, and oblique asymptotes (see Exercise 48).

49. $\dfrac{x^2 - 2}{x}$.   **50.** $\dfrac{x^2 - 2x - 3}{x + 2}$.

51. $\dfrac{(x-2)^3}{x^2}$.

52. $\dfrac{4-x^3}{x^2}$.

53. $x + 1 - \dfrac{1}{x} - \dfrac{1}{x^2}$.

54. Find all values of x where the graph of

$$y = \dfrac{2x^3 - 3x + 4}{x^2}$$

crosses its oblique asymptote. [See Exercise 48.]

55. Let $f(x) = x^2 + 1/x$. Show that the graph of $y = f(x)$ approaches the curve $y = x^2$ "asymptotically" in the sense that

$$\lim_{x \to +\infty} [f(x) - x^2] = 0 \quad \text{and} \quad \lim_{x \to -\infty} [f(x) - x^2] = 0$$

Sketch the graph of $y = f(x)$ showing this asymptotic behavior.

56. Let $f(x) = 3 - x^2 + 2/x$. Show that $y = f(x)$ approaches the curve $y = 3 - x^2$ asymptotically in the sense described in Exercise 55. Sketch the graph of $y = f(x)$ showing this asymptotic behavior.

57. A rectangular plot of land is to be fenced off so that the area enclosed will be 400 ft^2. Let L be the length of fencing needed and x the length of one side of the rectangle. Show that $L = 2x + 800/x$ for $x > 0$, and sketch the graph of L versus x for $x > 0$.

58. A box with a square base and open top is to be made from sheet metal so that its volume is 500 in^3. Let S be the area of the surface of the box and x the length of a side of the square base. Show that $S = x^2 + 2000/x$ for $x > 0$, and sketch the graph of S versus x for $x > 0$.

59. Figure 4.4.9 shows a computer-generated graph of the polynomial $y = 0.1x^5(x - 1)$ using a viewing window of $-2 \le x \le 2.5$ and $-1 \le y \le 5$. Show that the choice of the vertical scale caused the computer to miss an important feature of the graph. Find the feature that was missed and make your own sketch of the graph that shows the missing feature.

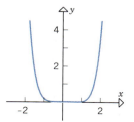

Figure 4.4.9

60. Figure 4.4.10 shows a computer-generated graph of the polynomial $y = 0.1x^5(x + 1)^2$ using a viewing window of $-2 \le x \le 1.5$ and $-0.2 \le y \le 0.2$. Show that the choice of the vertical scale caused the computer to miss an important feature of the graph. Find the feature that was missed and make your own sketch of the graph that shows the missing feature.

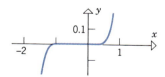

Figure 4.4.10

▮ 4.5 OTHER GRAPHING PROBLEMS

In this section we shall examine the graphs of some continuous functions whose derivatives are not continuous. Such functions have characteristics not found in the graphs of polynomials and rational functions.

☐ **VERTICAL TANGENT LINES AND CUSPS**

4.5.1 DEFINITION. The graph of a function f is said to have a ***vertical tangent line*** at x_0 if f is continuous at x_0 and $|f'(x)|$ approaches $+\infty$ as $x \to x_0$.

Four common ways in which vertical tangent lines occur are illustrated in Figure 4.5.1. The curve segments in parts (c) and (d) of Figure 4.5.1 are called *cusps*. To be precise, we make the following definition.

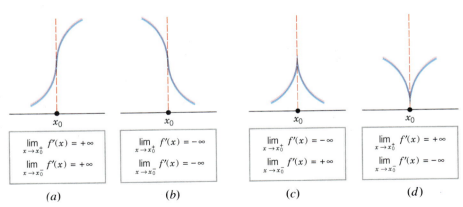

$$\lim_{x \to x_0^+} f'(x) = +\infty$$
$$\lim_{x \to x_0^-} f'(x) = +\infty$$

$$(a)$$

$$\lim_{x \to x_0^+} f'(x) = -\infty$$
$$\lim_{x \to x_0^-} f'(x) = -\infty$$

$$(b)$$

$$\lim_{x \to x_0^+} f'(x) = -\infty$$
$$\lim_{x \to x_0^-} f'(x) = +\infty$$

$$(c)$$

$$\lim_{x \to x_0^+} f'(x) = +\infty$$
$$\lim_{x \to x_0^-} f'(x) = -\infty$$

$$(d)$$

Figure 4.5.1

4.5.2 DEFINITION. The graph of a function f is said to have a **_cusp_** at x_0 if f is continuous at x_0 and $f'(x) \to +\infty$ as x approaches x_0 from one side, while $f'(x) \to -\infty$ as x approaches x_0 from the other side.

Example 1 Sketch the graph of $y = \sqrt[3]{x}$.

Solution. Let $f(x) = \sqrt[3]{x}$.

Symmetries: The graph is symmetric about the origin (verify).
x-intercepts: Setting $y = 0$ yields the *x*-intercept $x = 0$.
y-intercepts: Setting $x = 0$ yields the *y*-intercept $y = 0$.
Vertical asymptotes: None, since $f(x) = \sqrt[3]{x}$ is a continuous function.
Horizontal asymptotes: None, since

$$\lim_{x \to +\infty} \sqrt[3]{x} = +\infty \quad \text{and} \quad \lim_{x \to -\infty} \sqrt[3]{x} = -\infty$$

Derivatives:

$$\frac{dy}{dx} = f'(x) = \frac{1}{3} x^{-2/3} = \frac{1}{3x^{2/3}}$$

$$\frac{d^2y}{dx^2} = f''(x) = \frac{1}{3}\left(-\frac{2}{3}\right) x^{-5/3} = -\frac{2}{9x^{5/3}}$$

Vertical tangent lines: There is a vertical tangent line at $x = 0$ of the type in Figure 4.5.1*a* since $f(x) = \sqrt[3]{x}$ is continuous at $x = 0$ and

$$\lim_{x \to 0^+} f'(x) = \lim_{x \to 0^+} \frac{1}{3x^{2/3}} = +\infty \quad \text{and} \quad \lim_{x \to 0^-} f'(x) = \lim_{x \to 0^-} \frac{1}{3x^{2/3}} = +\infty$$

Intervals of increase and decrease; *concavity*: Combining the foregoing information with the following sign analysis of the first and second derivatives yields the graph in Figure 4.5.2. The points in the table accompanying the figure were plotted for some added accuracy. ◄

x	$y = \sqrt[3]{x}$
-8	-2
-1	-1
1	1
8	2

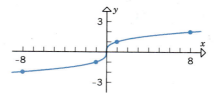

Figure 4.5.2

Example 2 Sketch the graph of $y = (x - 4)^{2/3}$.

Solution. Let $f(x) = (x - 4)^{2/3}$.

Symmetries: There are no symmetries about the coordinate axes or the origin (verify).

x-intercepts: Setting $y = 0$ yields the x-intercept $x = 4$.

y-intercepts: Setting $x = 0$ yields the y-intercept $y = \sqrt[3]{16}$.

Vertical asymptotes: None, since $f(x) = (x - 4)^{2/3}$ is a continuous function.

Horizontal asymptotes: None, since

$$\lim_{x \to +\infty} (x - 4)^{2/3} = +\infty$$

$$\lim_{x \to -\infty} (x - 4)^{2/3} = +\infty$$

Derivatives:

$$\frac{dy}{dx} = f'(x) = \frac{2}{3}(x - 4)^{-1/3} = \frac{2}{3(x - 4)^{1/3}}$$

$$\frac{d^2y}{dx^2} = f''(x) = -\frac{2}{9}(x - 4)^{-4/3} = -\frac{2}{9(x - 4)^{4/3}}$$

Vertical tangent lines: There is a vertical tangent line and cusp at $x = 4$ of the type in Figure 4.5.1d since $f(x) = (x - 4)^{2/3}$ is continuous at $x = 4$ and

$$\lim_{x \to 4^+} f'(x) = \lim_{x \to 4^+} \frac{2}{3(x - 4)^{1/3}} = +\infty$$

$$\lim_{x \to 4^-} f'(x) = \lim_{x \to 4^-} \frac{2}{3(x - 4)^{1/3}} = -\infty$$

Intervals of increase and decrease; *concavity*: Combining the foregoing information with the following sign analysis of the first and second derivatives yields the graph in Figure 4.5.3. The points in the table accompanying the figure were plotted for some added accuracy. ◀

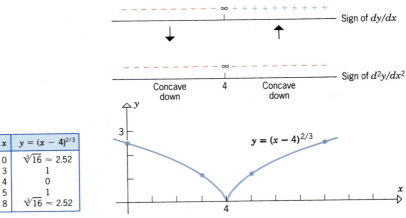

x	$y = (x - 4)^{2/3}$
0	$\sqrt[3]{16} \approx 2.52$
3	1
4	0
5	1
8	$\sqrt[3]{16} \approx 2.52$

Figure 4.5.3

Example 3 Sketch the graph of $y = 6x^{1/3} + 3x^{4/3}$.

Solution. Let $f(x) = 6x^{1/3} + 3x^{4/3} = 3x^{1/3}(2 + x)$.

Symmetries: There are no symmetries about the coordinate axes or the origin (verify).

x-intercepts: Setting $y = 3x^{1/3}(2 + x) = 0$ yields the x-intercepts $x = 0$ and $x = -2$.

y-intercept: Setting $x = 0$ yields the y-intercept $y = 0$.

Vertical asymptotes: None, since $f(x) = 6x^{1/3} + 3x^{4/3}$ is continuous.

Horizontal asymptotes: None, since

$$\lim_{x \to +\infty} (6x^{1/3} + 3x^{4/3}) = \lim_{x \to +\infty} 3x^{1/3}(2 + x) = +\infty$$

$$\lim_{x \to -\infty} (6x^{1/3} + 3x^{4/3}) = \lim_{x \to -\infty} 3x^{1/3}(2 + x) = +\infty$$

Derivatives:

$$\frac{dy}{dx} = f'(x) = 2x^{-2/3} + 4x^{1/3} = 2x^{-2/3}(1 + 2x) = \frac{2(2x + 1)}{x^{2/3}}$$

$$\frac{d^2y}{dx^2} = f''(x) = -\frac{4}{3}x^{-5/3} + \frac{4}{3}x^{-2/3} = \frac{4}{3}x^{-5/3}(-1 + x) = \frac{4(x - 1)}{3x^{5/3}}$$

Vertical tangent lines: There is a vertical tangent line at $x = 0$ of the type in Figure 4.5.1*a* since

$$\lim_{x \to 0^+} f'(x) = \lim_{x \to 0^+} \frac{2(2x + 1)}{x^{2/3}} = +\infty$$

$$\lim_{x \to 0^-} f'(x) = \lim_{x \to 0^-} \frac{2(2x + 1)}{x^{2/3}} = +\infty$$

Intervals of increase and decrease; concavity: Combining the foregoing information with the following sign analysis of the first and second derivatives yields the graph in Figure 4.5.4. In addition to the stationary point at $x = -\frac{1}{2}$ and the inflection points at $x = 0$ and $x = 1$, we plotted two additional points for some extra accuracy.   ◀

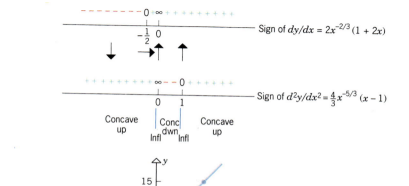

x	$y = 6x^{1/3} + 3x^{4/3}$
-3	$\approx$ 4.3
$-\frac{1}{2}$	$\approx$ -3.6
1	9
2	$\approx$ 15.1

Figure 4.5.4

REMARK.   The inflection at $x = 1$ is so subtle that it is hardly perceptible on the graph in Figure 4.5.4, yet we know it is there from our analysis of the second derivative.

▶ Exercise Set 4.5   C 14, 17, 18, 24

In Exercises 1–24, use the techniques illustrated in this section to sketch the graph of the given function.

1. $(x - 2)^{1/3}$.

2. $x^{1/4}$.

3. $x^{1/5}$.

4. $x^{2/5}$.

5. $x^{4/3}$.

6. $x^{-1/3}$.

7. $1 - x^{2/3}$.

8. $\sqrt{x + 2}$.

9. $\sqrt{x^2 - 1}$.

10. $\sqrt[3]{x^2 - 4}$.

11. $2x + 3x^{2/3}$.

12. $4x - 3x^{4/3}$.

13. $x\sqrt{3 - x}$.

14. $4x^{1/3} - x^{4/3}$.

15. $\dfrac{8(\sqrt{x} - 1)}{x}$.

16. $\dfrac{1 + \sqrt{x}}{1 - \sqrt{x}}$.

17. $\dfrac{\sqrt{x}}{x - 3}$.

18. $x^{2/3}(x - 5)$.

19. $x - \cos x$.

20. $x + \tan x$.

21. $\sin x + \cos x$.

22. $\sqrt{3} \cos x + \sin x$.

23. $\sin^2 x, \ 0 \le x \le 2\pi$.

24. $x \tan x, \ -\dfrac{\pi}{2} < x < \dfrac{\pi}{2}$.

◼ 4.6   MAXIMUM AND MINIMUM VALUES OF A FUNCTION

*Problems concerned with finding the "best" way to perform a task are called **optimization** problems. A large class of optimization problems can be reduced to finding the largest or smallest value of a function and determining where this value occurs. In this section we shall develop some mathematical tools for solving such problems.*

☐ **ABSOLUTE EXTREMA**

If we imagine the graph of a function f to be a two-dimensional profile of a mountain range (Figure 4.6.1), then the tops of the mountains correspond to the relative maxima and the bottoms of the valleys to the relative minima. Geologically, these are the high and low points of the terrain in their *immediate vicinity*. However, just as a geologist might be interested in finding the highest mountain and deepest valley in the entire mountain range (Figure 4.6.1), so a mathematician might be interested in finding the largest and smallest values of a function over its *entire domain*. This leads us to the following definitions.

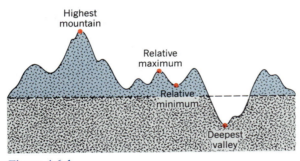

Figure 4.6.1

4.6.1   DEFINITION. If $f(x_0) \geq f(x)$ for all x in the domain of f, then $f(x_0)$ is called the **maximum value** or **absolute maximum value** of f.

4.6.2   DEFINITION. If $f(x_0) \leq f(x)$ for all x in the domain of f, then $f(x_0)$ is called the **minimum value** or **absolute minimum value** of f.

4.6.3   DEFINITION. A number that is either the maximum or the minimum value of a function f is called an **extreme value** or **absolute extreme value** of f. Sometimes the terms **extremum** or **absolute extremum** are also used.

Often we shall be concerned with the extreme values of f on some specified interval, rather than on the entire domain of f. The meaning of such terms as **maximum value of f on $[a, b]$** or **minimum value of f on (a, b)** should be clear.

Example 1   The function f graphed in Figure 4.6.2 has no maximum value. Its minimum value is 2, and the minimum occurs where $x = 3$.   ◄

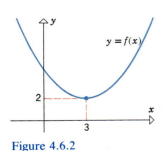

Figure 4.6.2

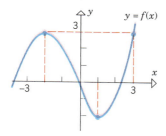

Figure 4.6.3

Example 2   The function graphed in Figure 4.6.3 has neither a maximum nor a minimum value. However, on the interval $[-3, 3]$ it has both. The maximum value on $[-3, 3]$ is $f(x) = 3$, which occurs at $x = -2$ and $x = 3$. The minimum value on $[-3, 3]$ is $f(x) = -2$, which occurs at $x = 1$.   ◀

Example 3   The function $f(x) = 2x + 1$ graphed in Figure 4.6.4 has a minimum, but no maximum value on $[0, 3)$. The minimum value is 1, and this occurs at $x = 0$. The reason why $f(x)$ has no maximum value is subtle, but important to understand. If we had been considering the interval $[0, 3]$ rather than $[0, 3)$, then $f(x)$ would have had a maximum value of 7 occurring at $x = 3$. However, the point $x = 3$ does not lie in the interval $[0, 3)$, so 7 is *not* the maximum value on $[0, 3)$. But no number *less* than 7 can be the maximum either, for if M is any number less than 7, there are values of x in $[0, 3)$ where $f(x) > M$ (see Figure 4.6.5). Thus, $f(x) = 2x + 1$ has no maximum value on $[0, 3)$.   ◀

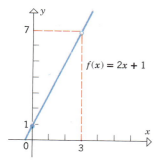

Figure 4.6.4

Given a function f, there are various questions we can ask about its maximum and minimum values; for example:

•   Does $f(x)$ have a maximum value?
•   If $f(x)$ has a maximum value, what is it?
•   If $f(x)$ has a maximum value, where does it occur?

These same questions could, of course, be asked about the minimum value of f. We shall now obtain some results that will help us to answer such questions.

The following theorem gives conditions under which the existence of maximum and minimum values of a function is ensured.

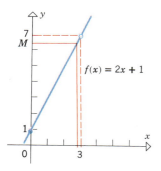

Figure 4.6.5

4.6.4   **THEOREM** (*Extreme-Value Theorem*).   *If a function f is continuous on a closed interval $[a, b]$, then f has both a maximum value and a minimum value on $[a, b]$.*

The proof of this theorem is surprisingly difficult, and will be omitted. However, the result is intuitively obvious if we imagine a particle moving along the graph of a continuous function over a closed interval $[a, b]$; during the trip the particle will have to pass through a highest point and a lowest point (Figure 4.6.6).

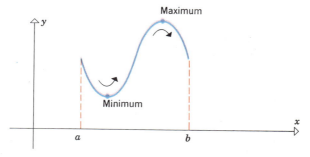

Figure 4.6.6

In the Extreme-Value Theorem the hypotheses that f is continuous and that the interval is closed are essential. If either hypothesis is violated, the existence of maximum or minimum values cannot be guaranteed; this is shown in the next two examples.

Example 4   Because $f(x) = 2x + 1$ is a polynomial, it is continuous everywhere. In particular, it is continuous on the interval $[0, 3)$. However, as shown in Example 3, the function f does not have a maximum value on $[0, 3)$. Thus, the closed-interval hypothesis in Theorem 4.6.4 is essential.   ◄

Example 5   The function f graphed in Figure 4.6.7 is defined everywhere on the closed interval $[1, 9]$, yet it has neither a maximum nor minimum value on the interval. Since f has a point of discontinuity on the interval $[1, 9]$, this example shows that the continuity hypothesis in Theorem 4.6.4 is essential.   ◄

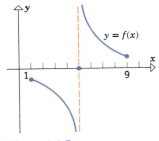

Figure 4.6.7

The Extreme-Value Theorem is an example of what mathematicians call an *existence theorem;* it states conditions under which something exists, in this case maximum and minimum values for a function f. However, finding these maximum and minimum values is a separate problem.

Example 6   The polynomial $f(x) = 2x^3 - 15x^2 + 36x$ is continuous everywhere. In particular, it is continuous on the closed interval $[1, 5]$. Thus, the Extreme-Value Theorem tells us that f has both maximum and minimum values on $[1, 5]$. However, this theorem does not tell us what the maximum and minimum values are or where they occur.   ◄

The following theorem is the key tool for finding the extreme values of a function.

> **4.6.5**   THEOREM.   *If a function f has an extreme value (either a maximum or a minimum) on an open interval (a, b), then the extreme value occurs at a critical point of f.*

Proof.   If f has a maximum value on (a, b) at x_0, then $f(x_0)$ is the largest value of f on (a, b) and therefore the largest value of f in the immediate vicinity of x_0. Thus, f has a relative maximum at x_0 and by Theorem 4.3.4, x_0 is a critical point for f. The proof in the case of a minimum value is similar.   ∎

□ FINDING ABSOLUTE EXTREMA

This theorem can be used to locate the extreme values of a continuous function f on a *closed* interval $[a, b]$. For example, consider the possible locations of the maximum. Either the maximum occurs at an endpoint (Figure 4.6.8a) or it occurs in the open interval (a, b), in which case it occurs at a critical point (Figures 4.6.8b and 4.6.8c). This suggests the following procedure:

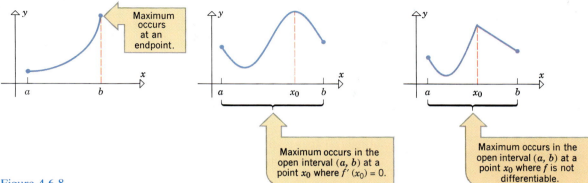

Figure 4.6.8

> **How to Find the Extreme Values of a Continuous Function f on a Closed Interval $[a, b]$**
>
> **Step 1.** Find the critical points of f in (a, b).
>
> **Step 2.** Evaluate f at all the critical points and at the endpoints a and b.
>
> **Step 3.** The largest of the values in Step 2 is the maximum value of f on $[a, b]$ and the smallest value is the minimum.

Example 7 Find the maximum value and minimum value of the function $f(x) = 2x^3 - 15x^2 + 36x$ on the interval $[1, 5]$, and determine where the maximum and minimum occur.

Solution. Since polynomials are differentiable everywhere, f is differentiable everywhere on the interval $(1, 5)$. Thus, if an extreme value occurs on the open interval $(1, 5)$, it must occur at a point where the derivative is zero. Since

$$f'(x) = 6x^2 - 30x + 36$$

the equation $f'(x) = 0$ becomes $6x^2 - 30x + 36 = 0$, which simplifies to

$$x^2 - 5x + 6 = 0 \quad \text{or} \quad (x - 3)(x - 2) = 0$$

Thus, there are two points in the interval $(1, 5)$ where $f'(x) = 0$, namely $x = 2$ and $x = 3$. Evaluating f at these points and the endpoints, we have

$$f(1) = 2(1)^3 - 15(1)^2 + 36(1) = 23$$
$$f(2) = 2(2)^3 - 15(2)^2 + 36(2) = 28$$
$$f(3) = 2(3)^3 - 15(3)^2 + 36(3) = 27$$
$$f(5) = 2(5)^3 - 15(5)^2 + 36(5) = 55$$

Thus, the minimum value is 23 and the maximum value is 55. The minimum occurs at $x = 1$ and the maximum occurs at $x = 5$. ◀

Example 8   Find the extreme values of $f(x) = 6x^{4/3} - 3x^{1/3}$ on the interval $[-1, 1]$ and determine where these values occur.

Solution.   Differentiating, we obtain

$$f'(x) = 8x^{1/3} - x^{-2/3} = x^{-2/3}(8x - 1) = \frac{8x - 1}{x^{2/3}}$$

Thus, $f'(x) = 0$ at $x = \frac{1}{8}$ and $f'(x)$ does not exist where $x = 0$. It follows that the critical points of f are $x = 0$, $x = \frac{1}{8}$, both of which lie in the interval $[-1, 1]$. Evaluating f at these critical points and the endpoints we obtain Table 4.6.1. Thus, the minimum value of f on $[-1, 1]$ is $-\frac{9}{8}$, which occurs at $x = \frac{1}{8}$, and the maximum value of f on $[-1, 1]$ is 9, which occurs at $x = -1$.   ◀

Table 4.6.1

x	-1	0	$\frac{1}{8}$	1
$f(x)$	9	0	$-\frac{9}{8}$	3

For a continuous function f on a closed interval $[a, b]$, the Extreme-Value Theorem ensures that f has both a maximum and minimum value; and the procedure for finding these values is mechanical—we simply evaluate f at the critical points and the endpoints. For a continuous function on an open interval, half-open interval, or infinite interval, the problem of finding maximum and minimum values is more complicated for two reasons:

- The function may not have a maximum or minimum value on such an interval.

- If the function does have maximum or minimum values, it often requires some ingenuity to find them.

It is often possible to determine whether a function has extrema by graphing it. Once it is determined that extrema exist, Theorem 4.6.5 can be applied to help locate them.

Example 9   Determine whether the function $f(x) = 3x^4 + 4x^3$ has maximum or minimum values on $(-\infty, +\infty)$ and, if so, find them.

Solution.   Using the methods of Section 4.4, the reader should be able to obtain the graph of f shown in Figure 4.6.9. From this graph we see that the function has a minimum value but no maximum. By Theorem 4.6.5 this minimum must occur at a critical point. To locate the critical points, we set $f'(x)$ equal to zero to obtain

$$12x^3 + 12x^2 = 0 \qquad \text{or} \qquad 12x^2(x + 1) = 0$$

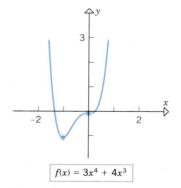

$f(x) = 3x^4 + 4x^3$

Figure 4.6.9

Thus, the critical points are $x = 0$ and $x = -1$. At $x = 0$ we have an inflection point, and at $x = -1$ we have the desired minimum. Substituting $x = -1$ in $f(x) = 3x^4 + 4x^3$ yields $f(-1) = -1$, which is the minimum value of f on $(-\infty, +\infty)$.   ◀

If f is a *continuous* function such that

$$\lim_{x \to +\infty} f(x) = \pm\infty \qquad \text{and} \qquad \lim_{x \to -\infty} f(x) = \pm\infty$$

then Table 4.6.2 shows how to ascertain if f has any extrema without actually constructing its graph.

Example 10 Find the maximum and minimum values, if any, of the function $f(x) = x^4 + 2x^3 - 1$ on $(-\infty, +\infty)$.

Solution. Because f is a polynomial it is continuous on $(-\infty, +\infty)$. Moreover,

$$\lim_{x \to +\infty} (x^4 + 2x^3 - 1) = +\infty \qquad \text{and} \qquad \lim_{x \to -\infty} (x^4 + 2x^3 - 1) = +\infty$$

so f has a minimum but no maximum on $(-\infty, +\infty)$. By Theorem 4.6.5 the minimum occurs at a critical point. Thus, if we evaluate f at each of its critical points, we shall be able to determine where the minimum occurs and find the minimum value. But,

$$f'(x) = 4x^3 + 6x^2 = 2x^2(2x + 3)$$

Table 4.6.2

$\lim_{x \to -\infty} f(x)$	$\lim_{x \to +\infty} f(x)$	CONCLUSION (if f is continuous)	GRAPH
$+\infty$	$+\infty$	f has a minimum but no maximum on $(-\infty, +\infty)$.	
$-\infty$	$-\infty$	f has a maximum but no minimum on $(-\infty, +\infty)$.	
$-\infty$	$+\infty$	f has neither a maximum nor a minimum on $(-\infty, +\infty)$.	
$+\infty$	$-\infty$	f has neither a maximum nor a minimum on $(-\infty, +\infty)$.	

so $f'(x) = 0$ yields the critical points $x = 0$ and $x = -\frac{3}{2}$. Evaluating f at these points yields

$$f(0) = -1 \quad \text{and} \quad f(-\tfrac{3}{2}) = -\tfrac{43}{16}$$

so the minimum value is $-\frac{43}{16}$ and this occurs at $x = -\frac{3}{2}$.   ◀

There is a variation of the results in Table 4.6.2 that is useful for finding extrema on an open interval. If f is continuous on an open interval (a, b) and

$$\lim_{x \to a^+} f(x) = \pm\infty \quad \text{and} \quad \lim_{x \to b^-} f(x) = \pm\infty$$

then Table 4.6.3 shows how to ascertain if f has any extrema on (a, b).

We leave it for the reader to give analogs of the results in Tables 4.6.2 and 4.6.3 for intervals of the form $(-\infty, b)$ and $(a, +\infty)$.

Table 4.6.3

$\lim\limits_{x \to a^+} f(x)$	$\lim\limits_{x \to b^-} f(x)$	CONCLUSION (IF f IS CONTINUOUS ON (a, b))	GRAPH
$+\infty$	$+\infty$	f has a minimum but no maximum on (a, b).	
$-\infty$	$-\infty$	f has a maximum but no minimum on (a, b).	
$-\infty$	$+\infty$	f has neither a maximum nor a minimum on (a, b).	
$+\infty$	$-\infty$	f has neither a maximum nor a minimum on (a, b).	

Example 11   Find the maximum and minimum values, if any, of

$$f(x) = \frac{1}{x^2 - x}$$

on the open interval $(0, 1)$.

Solution.   We have

$$\lim_{x \to 0^+} f(x) = \lim_{x \to 0^+} \frac{1}{x^2 - x} = \lim_{x \to 0^+} \frac{1}{x(x - 1)} = -\infty$$

$$\lim_{x \to 1^-} f(x) = \lim_{x \to 1^-} \frac{1}{x^2 - x} = \lim_{x \to 1^-} \frac{1}{x(x - 1)} = -\infty$$

so f has a maximum, but no minimum on $(0, 1)$. By Theorem 4.6.5 this maximum occurs at a critical point of f. We have

$$f'(x) = -\frac{2x - 1}{(x^2 - x)^2}$$

so $f'(x) = 0$ if $x = \frac{1}{2}$. Thus, the maximum value of f on $(0, 1)$ occurs at $x = \frac{1}{2}$ and this maximum value is

$$f(\tfrac{1}{2}) = \frac{1}{(\frac{1}{2})^2 - \frac{1}{2}} = -4 \quad \blacktriangleleft$$

Example 12   The function $f(x) = \tan x$ has neither a maximum nor a minimum on $(-\pi/2, \pi/2)$ because $\tan x$ is continuous on this interval and

$$\lim_{x \to -\pi/2^+} \tan x = -\infty, \quad \lim_{x \to \pi/2^-} \tan x = +\infty$$

(See Figure B.28 in Appendix B.) $\blacktriangleleft$

The following theorem can sometimes be helpful in situations where our previous methods do not apply.

4.6.6   THEOREM.   *Let f be continuous on an interval I and assume that f has exactly one relative extremum on I, say at x_0.*

(a)   *If f has a relative minimum at x_0, then $f(x_0)$ is the minimum value of f on the interval I.*

(b)   *If f has a relative maximum at x_0, then $f(x_0)$ is the maximum value of f on the interval I.*

Although we shall omit the proof, the result is easy to visualize. For example, if f has a relative maximum at x_0 and if $f(x_0)$ is *not* the maximum value of f on I, then the graph of f must make an upward turn somewhere on the interval I, thereby introducing a second relative extremum (Figure 4.6.10). Thus, if there is only one relative extremum on I, an absolute maximum value for f must also occur at x_0.

Where applicable, Theorem 4.6.6 reduces the problem of finding extrema to one of finding relative extrema.

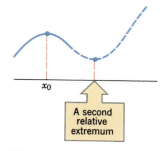

x_0

A second relative extremum

Figure 4.6.10

Example 13   Find the maximum and minimum values, if any, of the function $f(x) = x^3 - 3x^2 + 4$ on $(0, +\infty)$.

Solution. We have

$$\lim_{x \to 0^+} (x^3 - 3x^2 + 4) = 4$$

$$\lim_{x \to +\infty} (x^3 - 3x^2 + 4) = +\infty$$

Because the first limit is finite, Tables 4.6.2 and 4.6.3 do not apply. Moreover, the second limit tells us that f definitely has no maximum on $(0, +\infty)$. Thus, there are two possibilities: either f has no extreme values on $(0, +\infty)$ or f has a minimum on $(0, +\infty)$. To determine which is the case, we shall look for critical points. We have

$$f'(x) = 3x^2 - 6x = 3x(x - 2)$$

so $f'(x) = 0$ yields the critical points $x = 0$ and $x = 2$. However, $x = 2$ is the only critical point in the interval $(0, +\infty)$, so we shall try to apply Theorem 4.6.6. Since

$$f''(x) = 6x - 6$$

we have $f''(2) = 6 > 0$. Thus, a relative minimum occurs at $x = 2$ by the second derivative test and consequently a minimum value for f occurs at $x = 2$ by Theorem 4.6.6. This minimum value is $f(2) = 0$. ◀

▶ Exercise Set 4.6

In Exercises 1–14, find the maximum and minimum values of f on the given closed interval and state where these values occur.

1. $f(x) = 4x^2 - 4x + 1$; $[0, 1]$.

2. $f(x) = 8x - x^2$; $[0, 6]$.

3. $f(x) = (x - 1)^3$; $[0, 4]$.

4. $f(x) = 2x^3 - 3x^2 - 12x$; $[-2, 3]$.

5. $f(x) = \dfrac{3x}{\sqrt{4x^2 + 1}}$; $[-1, 1]$.

6. $f(x) = \dfrac{x}{x^2 + 2}$; $[-1, 4]$.

7. $f(x) = x^{2/3}(20 - x)$; $[-1, 20]$.

8. $f(x) = (x^2 + x)^{2/3}$; $[-2, 3]$.

9. $f(x) = x - \tan x$; $[-\pi/4, \pi/4]$.

10. $f(x) = \sin x - \cos x$; $[0, \pi]$.

11. $f(x) = 2 \sec x - \tan x$; $[0, \pi/4]$.

12. $f(x) = \sin^2 x + \cos x$; $[-\pi, \pi]$.

13. $f(x) = 1 + |9 - x^2|$; $[-5, 1]$.

14. $f(x) = |6 - 4x|$; $[-3, 3]$.

In Exercises 15–26, find the maximum and minimum values of f on the given interval, if they exist.

15. $f(x) = x^2 - 3x - 1$; $(-\infty, +\infty)$.

16. $f(x) = 3 - 4x - 2x^2$; $(-\infty, +\infty)$.

17. $f(x) = 4x^3 - 3x^4$; $(-\infty, +\infty)$.

18. $f(x) = x^4 + 4x$; $(-\infty, +\infty)$.

19. $f(x) = (x^2 - 1)^2$; $(-\infty, +\infty)$.

20. $f(x) = (x - 1)^2(x + 2)^2$; $(-\infty, +\infty)$.

21. $f(x) = x^3 - 3x - 2$; $(-\infty, +\infty)$.

22. $f(x) = x^3 - 9x + 1$; $(-\infty, +\infty)$.

23. $f(x) = 1 + \dfrac{1}{x}$; $(0, +\infty)$.

24. $f(x) = \dfrac{x}{x^2 + 1}$; $[0, +\infty)$.

25. $f(x) = \dfrac{x^2}{x + 1}$; $(-5, -1)$.

26. $f(x) = \dfrac{x + 3}{x - 3}$; $[-5, 5]$.

27. Find the maximum and minimum values of
$$f(x) = 2 \sin 2x + \sin 4x$$
and state where the maximum and minimum values

occur. [*Hint:* Since f is periodic, this problem can be solved by first finding the maximum and minimum values on an appropriate closed interval.]

28. Find the maximum and minimum values of

$$f(x) = 3\cos\frac{x}{3} + 2\cos\frac{x}{2}$$

[See the hint in the previous problem.]

29. Find the maximum and minimum values of the function $f(x) = \sin(\cos x)$ on $[0, 2\pi]$.

30. Find the maximum and minimum values of the function $f(x) = \cos(\sin x)$ on $[0, \pi]$.

31. Find the maximum and minimum values of

$$f(x) = \begin{cases} 4x - 2, & x < 1 \\ (x - 2)(x - 3), & x \ge 1 \end{cases}$$

on $[\frac{1}{2}, \frac{7}{2}]$.

32. Let $f(x) = x^2 + px + q$. Find values of p and q such that $f(1) = 3$ is an extreme value of f on $[0, 2]$. Is this value a maximum or minimum?

33. Let $f(x) = (x - a)^p$, where p is an integer greater than 1 and a is any real number. Find the relative extrema of f if
(a) p is even   (b) p is odd.

34. What is the smallest possible slope for a tangent to $y = x^3 - 3x^2 + 5x$?

35. (a) Show that

$$f(x) = \frac{64}{\sin x} + \frac{27}{\cos x}$$

has a minimum value, but no maximum value on the interval $(0, \pi/2)$.
(b) Find the minimum value.

36. Prove that the minimum value of

$$f(x) = x^2 + \frac{16x^2}{(8 - x)^2}, \quad x > 8$$

occurs at $x = 4(2 + \sqrt[3]{2})$.

37. Find the maximum value of the function $\sin^2\theta\cos\theta$ for $0 \le \theta \le \pi/2$.

38. Prove: For every positive value of t the function $f(x) = x + t/x$ has a minimum value but no maximum value on $(0, +\infty)$.

39. Find a_0, a_1, and a_2 such that the graph of $f(x) = a_0 + a_1x + a_2x^2$ passes through the point $(0, 9)$ and f has a minimum value of 1 at $x = 2$.

40. Prove that $\sin x \le x$ for all x in the interval $[0, 2\pi]$. [*Hint:* Find the minimum value of $x - \sin x$ on the interval.]

41. Prove that

$$1 - \frac{x^2}{2} \le \cos x$$

for all x in the interval $[0, 2\pi]$. [See the hint in the previous problem.]

42. (a) Sketch the graph of a function f that is continuous on the open interval $(-1, 1)$ and has both maximum and minimum values on the interval.
(b) Sketch the graph of a function f that is defined everywhere on the open interval $(-1, 1)$ and is not continuous everywhere on $(-1, 1)$, yet has both maximum and minimum values on $(-1, 1)$.
(c) Do the functions in parts (a) and (b) violate the Extreme-Value Theorem? Explain.

43. Prove: If $ax^2 + bx + c = 0$ has two distinct real roots, then the midpoint between these roots is a stationary point for $f(x) = ax^2 + bx + c$.

44. Prove Theorem 4.6.5 in the case where the extreme value is a minimum.

45. Let $f(x) = ax^2 + bx + c$, where $a > 0$. Prove that $f(x) \ge 0$ for all x if and only if $b^2 - 4ac \le 0$. [*Hint:* Find the minimum of $f(x)$.]

46. Assume that the function f is differentiable on an open interval (a, b) and L is a nonvertical line that does not intersect the curve $y = f(x)$ over the interval (a, b). Show that if there is a point on this curve over the interval (a, b) that is closest to or farthest from L, then the tangent line at this point is parallel to L. [*Hint:* Let $y = mx + b$ be the equation of L, and assume without proof that the perpendicular distance between a point on $y = f(x)$ and L is largest (smallest) at the same place where the vertical distance between the curve and the line is largest (smallest). (See Figure 4.6.11.)]

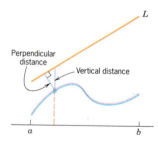

Figure 4.6.11

47. Use the result in Exercise 46 to find the coordinates of all points on the graph of $y = -x^2$ closest to and farthest from the line $y = 2 - x$ for $-1 \leq x \leq \frac{3}{2}$.

48. Use the result in Exercise 46 to find the coordinates of all points on the graph of $y = x^3$ closest to and farthest from the line $y = \frac{1}{3}x - 1$ for $-1 \leq x \leq 1$.

■ 4.7 APPLIED MAXIMUM AND MINIMUM PROBLEMS

> *In this section we shall show how the methods developed in the previous section can be used to solve some applied optimization problems.*

Example 1   Find the dimensions of a rectangle with perimeter 100 ft whose area is as large as possible.

Solution.   Let

x = length of the rectangle (ft)
y = width of the rectangle (ft)
A = area of the rectangle (ft^2)

Then

$$A = xy \tag{1}$$

Since the perimeter of the rectangle is 100 ft, the variables x and y are related by the equation

$$2x + 2y = 100 \quad \text{or} \quad y = 50 - x \tag{2}$$

(See Figure 4.7.1.) Substituting (2) in (1) yields

$$A = x(50 - x) = 50x - x^2 \tag{3}$$

Because x represents a length it cannot be negative, and because the two sides of length x cannot have a combined length exceeding the total perimeter of 100 ft, the variable x must satisfy

$$0 \leq x \leq 50 \tag{4}$$

Thus, we have reduced the problem to that of finding the value (or values) of x in $[0, 50]$, for which A is maximum. Since A is a polynomial in x, it is continuous on $[0, 50]$, and so the maximum must occur at an endpoint of this interval or at a critical point.

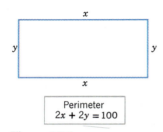

Perimeter
$2x + 2y = 100$

Figure 4.7.1

From (3) we obtain

$$\frac{dA}{dx} = 50 - 2x$$

Setting $dA/dx = 0$ we obtain

$$50 - 2x = 0$$

or $x = 25$. Thus, the maximum occurs at one of the points

$$x = 0, \quad x = 25, \quad x = 50$$

Table 4.7.1

x	0	25	50
A	0	625	0

Substituting these values in (3) yields Table 4.7.1, which tells us that the maximum area of 625 ft^2 occurs at $x = 25$. From (2) the corresponding value of y is $y = 25$, so the rectangle of perimeter 100 ft with greatest area is a square with sides of length 25 ft. ◀

REMARK. In (4) we included $x = 0$ and $x = 50$ as possible values for x. Because $x = 50$ implies $y = 0$ from (2), these x values correspond to rectangles with two sides of length zero. One can argue that these x values should not be allowed because a "true" rectangle cannot have sides of length zero.

Actually, it is just a matter of the assumptions we choose to make. If we view Example 1 as a purely mathematical problem, then there is nothing wrong with allowing sides of length zero. However, if we view it as a practical problem in which the rectangle is to be constructed from physical material, we would not want to allow $x = 0$ or $x = 50$ and (4) should be replaced by $0 < x < 50$. In this case we no longer have a closed interval to work with and there is no guarantee that a maximum exists. We shall consider this issue in more detail in the next section. In this section we shall consider only optimization problems over *finite closed intervals*.

Example 1 illustrates the following five-step procedure that can be used for solving many applied maximum and minimum problems.

Step 1. Draw an appropriate figure and label the quantities relevant to the problem.

Step 2. Find a formula for the quantity to be maximized or minimized.

Step 3. Using the conditions stated in the problem to eliminate variables, express the quantity to be maximized or minimized as a function of one variable.

Step 4. Find the interval of possible values for this variable from the physical restrictions in the problem.

Step 5. If applicable, use the techniques of the preceding section to obtain the maximum or minimum.

Example 2 An open box is to be made from a 16-in by 30-in piece of cardboard by cutting out squares of equal size from the four corners and bending up the sides (Figure 4.7.2). What size should the squares be to obtain a box with largest possible volume?

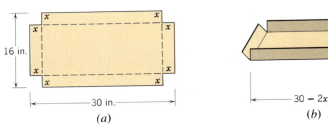

(a)   (b)

Figure 4.7.2

Solution. Let

$$x = \text{length (in inches) of the sides of the squares to be cut out}$$
$$V = \text{volume (in cubic inches) of the resulting box}$$

Because we are removing a square of side x from each corner, the resulting box will have dimensions $16 - 2x$ by $30 - 2x$ by x (Figure 4.7.2b). Since the volume of a box is the product of its dimensions, we have

$$V = (16 - 2x)(30 - 2x)x = 480x - 92x^2 + 4x^3 \tag{5}$$

The variable x in this expression is subject to certain restrictions. Because x represents a length it cannot be negative and because the width of the cardboard is 16 in., we cannot cut out squares whose sides are more than 8 in. long. Thus, the variable x in (5) must satisfy

$$0 \le x \le 8$$

We have thus reduced our problem to finding the value (or values) of x in the interval $[0, 8]$ for which (5) is maximum. Since the right side of (5) is a polynomial in x, it is continuous on the closed interval $[0, 8]$ and consequently we can use the methods developed in the preceding section to find the maximum. From (5) we obtain

$$\frac{dV}{dx} = 480 - 184x + 12x^2 = 4(120 - 46x + 3x^2)$$

Setting $dV/dx = 0$ yields

$$120 - 46x + 3x^2 = 0$$

which can be solved by the quadratic formula to obtain the critical points

$$x = \tfrac{10}{3} \quad \text{and} \quad x = 12$$

Since $x = 12$ falls outside the interval $[0, 8]$, the maximum value of V must occur either at the critical point $x = \tfrac{10}{3}$ or at one of the endpoints $x = 0$, $x = 8$. Substituting these values in (5) yields Table 4.7.2, which tells us that the greatest possible volume $V = \tfrac{19600}{27}$ in$^3 \approx 726$ in^3 occurs when we cut out squares whose sides have length $\tfrac{10}{3}$ in. ◄

Table 4.7.2

x	0	$\tfrac{10}{3}$	8
V	0	$\tfrac{19600}{27}$	0

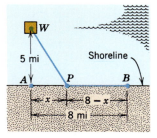

Figure 4.7.3

Example 3 An offshore oil well is located in the ocean at a point W, which is 5 mi from the closest shorepoint A on a straight shoreline (Figure 4.7.3). The oil is to be piped to a shorepoint B that is 8 mi from A by piping it on a straight line under water from W to some shorepoint P between A and B and then on to B via a pipe along the shoreline. If the cost of laying pipe is $100,000 per mile under water and $75,000 per mile over land, where should the point P be located to minimize the cost of laying the pipe?

REMARK.   Since the shortest distance between two points is a straight line, a pipeline directly from W to B uses the least amount of pipe. However, the pipe, being completely under water, would be expensive to lay. At the other extreme, a pipeline from W to A to B uses the least amount of expensive underwater pipe, but uses the greatest total amount of pipe. Thus, it seems plausible that by piping to some point P between A and B, one might incur less total cost than by piping to either of the extreme locations.

Solution.   Let

$$x = \text{distance (in miles) between } A \text{ and } P$$
$$c = \text{cost (in thousands of dollars) for the entire pipeline}$$

From Figure 4.7.3 the length of pipe under water is the distance between W and P. By the Theorem of Pythagoras, that length is

$$\sqrt{x^2 + 25} \text{ mi} \tag{6}$$

Also from Figure 4.7.3, the length of pipe over land is the distance between P and B, which is

$$8 - x \text{ mi} \tag{7}$$

From (6) and (7) it follows that the total cost c (in thousands of dollars) for the pipeline is

$$c = 100\sqrt{x^2 + 25} + 75(8 - x) \tag{8}$$

Because the distance between A and B is 8 mi, the distance x between A and P must satisfy

$$0 \leq x \leq 8$$

We have thus reduced our problem to finding the value (or values) of x in the interval $[0, 8]$ for which (8) is a minimum. Since c is a continuous function of x on the closed interval $[0, 8]$, we can use the methods developed in the preceding section to find the minimum.

From (8) we obtain

$$\frac{dc}{dx} = \frac{100x}{\sqrt{x^2 + 25}} - 75 = 25\left(\frac{4x}{\sqrt{x^2 + 25}} - 3\right)$$

Setting $dc/dx = 0$ yields

$$\frac{4x}{\sqrt{x^2 + 25}} - 3 = 0 \tag{9}$$

or

$$4x = 3\sqrt{x^2 + 25}$$

$$16x^2 = 9(x^2 + 25)$$

$$7x^2 = 225$$

$$x = \pm \frac{15}{\sqrt{7}}$$

The number $-15/\sqrt{7}$ is not a solution of (9) and must be discarded, leaving $x = 15/\sqrt{7}$ as the only critical point. Since this point lies in the interval $[0, 8]$, the minimum must occur at one of the points

$$x = 0, \quad x = 15/\sqrt{7}, \quad x = 8$$

Substituting these values in (8) yields Table 4.7.3, which tells us that the least possible cost for the pipeline, approximately $c = 930.719 = \$930,719$ occurs if the point P is located at a distance of $15/\sqrt{7} \approx 5.67$ mi from A.   ◀

Table 4.7.3

x	0	$15/\sqrt{7}$	8
c	1100	$100\sqrt{\dfrac{225}{7} + 25} + 75\left(8 - \dfrac{15}{\sqrt{7}}\right)$ $= 600 + 125\sqrt{7} \approx 930.719$	$100\sqrt{89} \approx 943.398$

Example 4   Find the radius and height of the right-circular cylinder of largest volume that can be inscribed in a right-circular cone with radius 6 in. and height 10 in. (Figure 4.7.4a).

Solution.   Let

$r =$ radius (in inches) of the cylinder
$h =$ height (in inches) of the cylinder
$V =$ volume (in cubic inches) of the cylinder

The formula for the volume of the inscribed cylinder is

$$V = \pi r^2 h \tag{10}$$

To eliminate one of the variables in (10) we need a relationship between r and h. Using similar triangles (Figure 4.7.4b) we obtain

$$\frac{10 - h}{r} = \frac{10}{6} \quad \text{or} \quad h = 10 - \tfrac{5}{3}r \tag{11}$$

Substituting (11) into (10) we obtain

$$V = \pi r^2(10 - \tfrac{5}{3}r) = 10\pi r^2 - \tfrac{5}{3}\pi r^3 \tag{12}$$

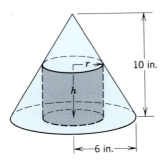

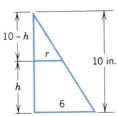

Figure 4.7.4

which expresses V in terms of r alone. Because r represents a radius it cannot be negative, and because the radius of the inscribed cylinder cannot exceed the radius of the cone, the variable r must satisfy

$$0 \le r \le 6$$

Thus, we have reduced the problem to that of finding the value (or values) of r in $[0, 6]$ for which (12) is a maximum. Since V is a continuous function of r on $[0, 6]$, the methods developed in the preceding section apply.

From (12) we obtain

$$\frac{dV}{dr} = 20\pi r - 5\pi r^2 = 5\pi r(4 - r)$$

Setting $dV/dr = 0$ gives

$$5\pi r(4 - r) = 0$$

so $r = 0$ and $r = 4$ are critical points. Since these lie in the interval $[0, 6]$, the maximum must occur at one of the points

$$r = 0, \quad r = 4, \quad r = 6$$

Table 4.7.4

r	0	4	6
V	0	$\frac{160}{3}\pi$	0

Substituting these values in (12) yields Table 4.7.4, which tells us that the maximum volume $V = \frac{160}{3}\pi \approx 168$ in³ occurs when the inscribed cylinder has radius 4 in. When $r = 4$ it follows from (11) that $h = \frac{10}{3}$. Thus, the inscribed cylinder of largest volume has radius $r = 4$ in. and height $h = \frac{10}{3}$ in.   ◀

□ **MARGINAL ANALYSIS (AN APPLICATION TO ECONOMICS)**

Economists and business people are interested in how changes in such variables as inventory, production, supply, advertising, and price affect other variables such as profit, revenue, demand, inflation, and employment. Such problems are studied using *marginal analysis*. The word, *marginal*, is the economist's term for a rate of change or derivative.

Three functions of importance to an economist or a manufacturer are

$C(x) =$ total cost of producing x units of a product during some time period

$R(x) =$ total revenue received from selling x units of the product during the time period

$P(x) =$ total profit obtained by selling x units of the product during the time period

These are called, respectively, the **cost function, revenue function,** and **profit function.** If all units produced are sold, then these are related by

$$P(x) = R(x) - C(x) \tag{13}$$
[profit] = [revenue] − [cost]

The derivatives $C'(x)$, $R'(x)$, and $P'(x)$ are called the **marginal cost, marginal revenue,** and **marginal profit.** If all units produced are sold, then a relationship between them is obtained by differentiating (13),

$$P'(x) = R'(x) - C'(x)$$

$$\begin{bmatrix} \text{marginal} \\ \text{profit} \end{bmatrix} = \begin{bmatrix} \text{marginal} \\ \text{revenue} \end{bmatrix} - \begin{bmatrix} \text{marginal} \\ \text{cost} \end{bmatrix}$$

The quantities $P'(x)$, $R'(x)$, and $C'(x)$ represent the instantaneous rates of change of profit, revenue, and cost with respect to x, where x is the amount of the product produced and sold.

In practice, $C'(x)$ is frequently interpreted as the cost of manufacturing the $(x + 1)$-st unit. Although this is not exact, it is usually a good approximation. The justification for this interpretation is based on the fact that x is usually large, so $\Delta x = 1$ can be considered close to zero by comparison. Thus,

$$C'(x) = \lim_{\Delta x \to 0} \frac{C(x + \Delta x) - C(x)}{\Delta x}$$

$$\approx \frac{C(x + 1) - C(x)}{1} = C(x + 1) - C(x)$$

Since $C(x + 1)$ is the cost of producing $x + 1$ units and $C(x)$ is the cost of producing x units, it follows that $C'(x) \approx C(x + 1) - C(x)$ is the approximate cost of producing the $(x + 1)$-th unit. Similarly, $R'(x)$ is the approximate revenue received from selling the $(x + 1)$-th unit and $P'(x)$ is the approximate profit from manufacturing and selling the $(x + 1)$-th unit.

The total cost $C(x)$ of producing x units can be expressed as a sum

$$C(x) = a + M(x) \tag{14}$$

where a is a constant, called **overhead,** and $M(x)$ is a function representing **manufacturing cost.** The overhead, which includes such fixed costs as rent and insurance, does not depend on x; it must be paid even if nothing is produced. On the other hand, the manufacturing cost $M(x)$, which includes such items as cost of materials and labor, depends on the number of items manufactured. It is shown in economics that with suitable simplifying assumptions, $M(x)$ can be expressed in the form

$$M(x) = bx + cx^2$$

where b and c are constants. Substituting this in (14) yields

$$C(x) = a + bx + cx^2 \tag{15}$$

Example 5   A paint manufacturer determines that the total cost in dollars of producing x gallons of paint per day is

$$C(x) = 5000 + x + 0.001x^2$$

(a)   Find the marginal cost when the production level is 500 gal per day.

(b)   Use the marginal cost to approximate the cost of producing the 501st gallon.

(c)   Find the exact cost of producing the 501st gallon.

Solution.

(a) The marginal cost is $C'(x) = 1 + 0.002x$, so

$$C'(500) = 1 + (0.002)(500) = 2$$

(b) Since $C'(500) = 2$, the cost of producing the 501st gallon is approximately $2.00.

(c) The total cost of producing 501 gal is

$$C(501) = 5000 + 501 + 0.001(501)^2 = 5752.001 \text{ dollars}$$

and the total cost of producing 500 gal is

$$C(500) = 5000 + 500 + 0.001(500)^2 = 5750 \text{ dollars}$$

so the exact cost of producing the 501st gallon is

$$C(501) - C(500) = 2.001 \text{ dollars} \qquad \blacktriangleleft$$

If a manufacturing firm can sell all the items it produces for p dollars apiece, then its total revenue $R(x)$ (in dollars) will be

$$R(x) = px \tag{16}$$

and its total profit $P(x)$ (in dollars) will be

$$P(x) = [\text{total revenue}] - [\text{total cost}] = R(x) - C(x) = px - C(x)$$

Thus, if the cost function is given by (15),

$$P(x) = px - (a + bx + cx^2) \tag{17}$$

Depending on such factors as number of employees, amount of machinery available, economic conditions, and competition, there will be some upper limit l on the number of items a manufacturer is capable of producing and selling. Thus, during a fixed time period the variable x in (17) will satisfy

$$0 \leq x \leq l$$

By determining the value or values of x in $[0, l]$ that maximize (17), the firm can determine how many units of its product must be manufactured and sold to yield the greatest profit. This is illustrated in the following numerical example.

Example 6   A liquid form of penicillin manufactured by a pharmaceutical firm is sold in bulk at a price of $200 per unit. If the total production cost (in dollars) for x units is

$$C(x) = 500,000 + 80x + 0.003x^2$$

and if the production capacity of the firm is at most 30,000 units in a specified time, how many units of penicillin must be manufactured and sold in that time to maximize the profit?

Solution.   Since the total revenue for selling x units is $R(x) = 200x$, the profit $P(x)$ on x units will be

$$P(x) = R(x) - C(x) = 200x - (500{,}000 + 80x + 0.003x^2) \qquad (18)$$

Since the production capacity is at most 30,000 units, x must lie in the interval $[0, 30{,}000]$. From (18)

$$\frac{dP}{dx} = 200 - (80 + 0.006x) = 120 - 0.006x$$

Setting $dP/dx = 0$ gives

$$120 - 0.006x = 0 \qquad \text{or} \qquad x = 20{,}000$$

Since this critical point lies in the interval $[0, 30{,}000]$, the maximum profit must occur at one of the points

$$x = 0, \quad x = 20{,}000, \quad \text{or} \quad x = 30{,}000$$

Substituting these values in (18) yields Table 4.7.5, which tells us that the maximum profit $P = \$700{,}000$ occurs when $x = 20{,}000$ units are manufactured and sold in the specified time.   ◀

Table 4.7.5

x	0	20,000	30,000
$P(x)$	$-500{,}000$	700,000	400,000

▶ Exercise Set 4.7

1. Express the number 10 as a sum of two nonnegative numbers whose product is as large as possible.

2. How should two nonnegative numbers be chosen so that their sum is 1 and the sum of their squares is
   (a) as large as possible
   (b) as small as possible?

3. Find a number in the closed interval $[\frac{1}{2}, \frac{3}{2}]$ such that the sum of the number and its reciprocal is
   (a) as small as possible
   (b) as large as possible.

4. A rectangular field is to be bounded by a fence on three sides and by a straight stream on the fourth side. Find the dimensions of the field with maximum area that can be enclosed with 1000 feet of fence.

5. A rectangular plot of land is to be fenced in using two kinds of fencing. Two opposite sides will use heavy-duty fencing selling for $3 a foot, while the remaining two sides will use standard fencing selling for $2 a foot. What are the dimensions of the rectangular plot of greatest area that can be fenced in at a cost of $6000?

6. Show that among all rectangles with perimeter p, the square with side $p/4$ has the maximum area.

7. Find the dimensions of the rectangle with maximum area that can be inscribed in a circle of radius 10.

8. A rectangle is to be inscribed in a right triangle having sides of length 6 in., 8 in., and 10 in. Find the dimensions of the rectangle with greatest area as-

suming the rectangle is positioned as in Figure 4.7.5a.

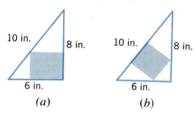

10 in. 8 in. 6 in. (a)   10 in. 8 in. 6 in. (b)   **Figure 4.7.5**

9. Solve the problem in Exercise 8 assuming the rectangle is positioned as in Figure 4.7.5b.

10. A rectangle has its two lower corners on the x-axis and its two upper corners on the curve $y = 16 - x^2$. For all such rectangles, what are the dimensions of the one with largest area?

11. A sheet of cardboard 12 in. square is used to make an open box by cutting squares of equal size from the four corners and folding up the sides. What size squares should be cut to obtain a box with largest possible volume?

12. A square sheet of cardboard of side k is used to make an open box by cutting squares of equal size from the four corners and folding up the sides. What size squares should be cut from the corners to obtain a box with largest possible volume?

13. An open box is to be made from a 3-ft by 8-ft rectangular piece of sheet metal by cutting out squares of equal size from the four corners and bending up the sides. Find the maximum volume that the box can have.

14. Show that among all isosceles triangles whose equal sides have length L, the triangle of largest area is a right triangle.

15. A wire of length 12 in. can be bent into a circle, bent into a square, or cut into two pieces to make both a circle and a square. How much wire should be used for the circle if the total area enclosed by the figure(s) is to be
    (a) a maximum   (b) a minimum?

16. Assume that the operating cost of a certain truck (excluding driver's wages) is $12 + x/6$ cents per mile when the truck travels at x mi/hr. If the driver earns $6 per hour, what is the most economical speed to operate the truck on a 400-mi turnpike where the minimum speed is 40 mi/hr and the maximum speed is 70 mi/hr?

17. Find the lengths of the sides of the isosceles triangle with perimeter 12 and maximum area.

18. A trapezoid is inscribed in a semicircle of radius 2 so that one side is along the diameter (Figure 4.7.6). Find the maximum possible area for the trapezoid. [*Hint:* Express the area of the trapezoid in terms of θ.]

θ

$\leftarrow$ 2 $\rightarrow$ **Figure 4.7.6**

19. A box-shaped wire frame consists of two identical wire squares whose vertices are connected by four straight wires of equal length (Figure 4.7.7). If the frame is to be made from a wire of length L, what should the dimensions be to obtain a box of greatest volume?

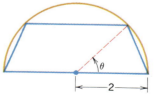

Figure 4.7.7

20. If air resistance is neglected, then the range R of a cannonball fired from a cannon whose barrel makes an angle θ with the horizontal is $R = (v_0^2/g)\sin 2\theta$, where the constants v_0 and g are the initial velocity of the cannonball and the acceleration due to gravity. Show that the maximum range is achieved when $\theta = 45°$.

21. (a) A chemical manufacturer sells sulfuric acid in bulk at a price of $100 per unit. If the daily total production cost in dollars for x units is
$$C(x) = 100{,}000 + 50x + 0.0025x^2$$
and if the daily production capacity is at most 7000 units, how many units of sulfuric acid must be manufactured and sold daily to maximize the profit?
    (b) Would it benefit the manufacturer to expand the daily production capacity?

22. Find the dimensions of the rectangle of greatest area that can be inscribed in a semicircle of radius R.

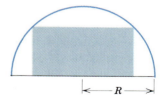

$\leftarrow$ R $\rightarrow$

23. Find the height and radius of the cone of slant height *L* whose volume is as large as possible.

24. Find the dimensions of the right-circular cylinder of largest volume that can be inscribed in a sphere of radius *R*.

25. Find the dimensions of the right-circular cylinder of greatest surface area that can be inscribed in a sphere of radius *R*.

26. Show that the right-circular cylinder of greatest volume that can be inscribed in a right-circular cone has volume that is $\frac{4}{9}$ the volume of the cone (Figure 4.7.8).

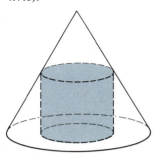

Figure 4.7.8

27. A drainage channel is to be made so that its cross section is a trapezoid with equally sloping sides (Figure 4.7.9). If the sides and bottom all have a length of 5 ft, how should the angle θ $(0 \le \theta \le \pi/2)$ be chosen to yield the greatest cross-sectional area?

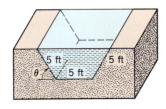

Figure 4.7.9

28. A commercial cattle ranch currently allows 20 steers per acre of grazing land; on the average its steers weigh 2000 lb at market. Estimates by the Agriculture Department indicate that the average market weight per steer will be reduced by 50 lb for each additional steer added per acre of grazing land. How many steers per acre should be allowed in order for the ranch to get the largest possible total market weight for its cattle?

29. A cone is made from a circular sheet of radius *R* by cutting out a sector and gluing the cut edges of the remaining piece together (Figure 4.7.10). What is the maximum volume attainable for the cone?

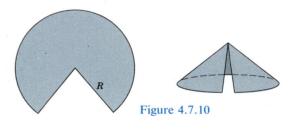

Figure 4.7.10

30. A man is on the bank of a river that is 1 mi wide. He wants to travel to a town on the opposite bank, but 1 mi upstream. He intends to row on a straight line to some point *P* on the opposite bank and then walk the remaining distance along the bank (Figure 4.7.11). To what point should he row in order to reach his destination in the least time if
(a) he can walk 5 mi/hr and row 3 mi/hr
(b) he can walk 5 mi/hr and row 4 mi/hr?

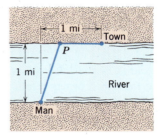

Figure 4.7.11

31. The total cost of producing *x* units of a commodity per week is

$$C(x) = 200 + 4x + 0.1x^2$$

(a) Find the marginal cost when the production level
(b) Use the marginal cost to approximate the cost of producing the 101st unit.
(c) Find the exact cost of producing the 101st unit.
(d) Assuming that the commodity is sold for $10 per unit, find the marginal revenue and marginal profit functions.

32. A firm determines that *x* units of its product can be sold daily at *p* dollars per unit, where

$$x = 1000 - p$$

The cost of producing *x* units per day is

$$C(x) = 3000 + 20x$$

(a) Find the revenue function *R(x)*.
(b) Find the profit function *P(x)*.

(c) Assuming that the production capacity is at most 500 units per day, determine how many units the company must produce and sell each day to maximize the profit.

(d) Find the maximum profit.

(e) What price per unit must be charged to obtain the maximum profit?

33. Prove that $(1, 0)$ is the closest point on the curve $x^2 + y^2 = 1$ to $(2, 0)$.

34. Find all points on the curve $y = \sqrt{x}$ for $0 \leq x \leq 3$ that are closest to, and at the greatest distance from, the point $(2, 0)$.

35. A church window consisting of a rectangle topped by a semicircle is to have a perimeter p. Find the radius of the semicircle if the area of the window is to be maximum.

36. A triangle is inscribed in a segment of the parabola $y = kx^2$ ($k > 0$), as shown in Figure 4.7.12. Show that the area of the triangle is greatest when $x = (a + b)/2$, and find the maximum area. [*Hint:* First show that the area of the triangle is $(b - a) \cdot D/2$, where D is the vertical distance between the line and the point (x, kx^2).]

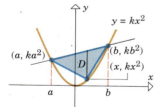

Figure 4.7.12

37. Suppose that the sum of the surface areas of a sphere and a cube is a constant.

(a) Show that the sum of their volumes is smallest when the diameter of the sphere is equal to the length of an edge of the cube.

(b) When will the sum of their volumes be greatest?

38. We are given a straight fence 100 ft long and we wish to use 200 ft of additional fence to form a rectangular enclosure whose boundary contains all of the original fence. How should we lay out the additional fence to obtain an enclosure of maximum area? [This problem, due to V. L. Klee, Jr., appeared in *Selected Papers on Calculus,* Mathematical Association of America, 1969, p. 246.]

39. (a) Find the smallest value of M such that
$$|x^2 - 3x + 2| \leq M$$
for all x in the interval $[1, \tfrac{5}{2}]$.

(b) Find the largest value of m such that
$$|x^2 - 3x + 2| \geq m$$
for all x in the interval $[\tfrac{3}{2}, \tfrac{7}{4}]$.

40. At what point(s) in the interval $[0, \pi]$ is the vertical distance between the graphs of $y = \tfrac{1}{2}x$ and $y = \sin x$ largest?

41. Fermat's* (biography on p. 292) principle in optics states that light traveling from one point to another follows that path for which the total travel time is minimum. In a uniform medium, the paths of "minimum time" and "shortest distance" turn out to be the same, so that light, if unobstructed, travels along a straight line. Assume that we have a light source, a flat mirror, and an observer in a uniform medium. If a light ray leaves the source, bounces off the mirror, and travels on to the observer, then its path will consist of two line segments, as shown in Figure 4.7.13. According to Fermat's principle, the path will be such that the total travel time t is minimum or, since the medium is uniform, the path will be such that the total distance traveled from A to P to B is as small as possible. Assuming that the minimum occurs when $dt/dx = 0$, show that the light ray will strike the mirror at the point P where the "angle of incidence" θ_1 equals the "angle of reflection" θ_2.

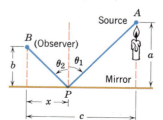

Figure 4.7.13

42. Fermat's principle (Exercise 41) also explains why light rays traveling between air and water undergo bending (refraction). Imagine we have two uniform media (such as air and water) and a light ray traveling from a source A in one medium to an observer B in the other medium (Figure 4.7.14). It is known that light travels at a constant speed in a uniform medium, but more slowly in a dense medium (such as water) than in a thin medium (such as air). Consequently, the path of shortest time from A to B is not necessarily a straight line, but rather some broken line path A to P to B allowing the light to take greatest advantage of its higher speed through the thin medium. Snell's† law of refraction states that the path of the light ray will be such that

$$\frac{\sin \theta_1}{v_1} = \frac{\sin \theta_2}{v_2}$$

where v_1 is the speed of light in the first medium, v_2 is the speed of light in the second medium, and θ_1 and θ_2 are the angles shown in Figure 4.7.14. Show that this follows from the assumption that the path of minimum time occurs when $dt/dx = 0$.

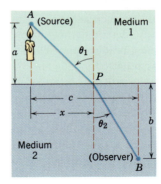

Figure 4.7.14

*PIERRE DE FERMAT (1601–1665). Fermat, the son of a successful French leather merchant, was a lawyer who practiced mathematics as a hobby. He received a Bachelor of Civil Laws degree from the University of Orleans in 1631 and subsequently held various government positions, including a post as councillor to the Toulouse parliament. Although he was apparently financially successful, confidential documents of that time suggest that his performance in office and as a lawyer was poor, perhaps because he devoted so much time to mathematics. Throughout his life, Fermat fought all efforts to have his mathematical results published. He had the unfortunate habit of scribbling his work in the margins of books and often sent his results to friends without keeping copies for himself. As a result, he never received credit for many major achievements until his name was raised from obscurity in the mid-nineteenth century. It is now known that Fermat, simultaneously and independently of Descartes, developed analytic geometry. Unfortunately, Descartes and Fermat argued bitterly over various problems so that there was never any real cooperation between these two great geniuses.

Fermat solved many fundamental calculus problems. He obtained the first procedure for differentiating polynomials, and solved many important maximization, minimization, area, and tangent problems. His work served to inspire Isaac Newton. Fermat is best known for his work in number theory, the study of properties and relationships between whole numbers. He was the first mathematician to make substantial contributions to this field after the ancient Greek mathematician Diophantus. Unfortunately, none of Fermat's contemporaries appreciated his work in this area, a fact that eventually pushed Fermat into isolation and obscurity in later life.

In addition to his work in calculus and number theory, Fermat was one of the founders of probability theory and made major contributions to the theory of optics. Outside mathematics, Fermat was a classical scholar of some note, was fluent in French, Italian, Spanish, Latin, and Greek, and he composed a considerable amount of Latin poetry.

One of the great mysteries of mathematics is shrouded in Fermat's work in number theory. In the margin of a book by Diophantus, Fermat scribbled that for integer values of n greater than 2, the equation $x^n + y^n = z^n$ has no nonzero integer solutions for x, y, and z. He stated, "I have discovered a truly marvelous proof of this, which however the margin is not large enough to contain." For the past 300 years the greatest mathematical geniuses have been unable to prove this result, even though it seems to be true. This result is now known as "Fermat's last theorem." In 1908 a prize of 100,000 German marks was offered for its solution; it has never been won. Whether or not Fermat really proved this theorem is a mystery to this day.

NOTE. A major breakthrough in Fermat's last theorem was recently made by a young West German mathematician named Gerd Faltings (see *Newsweek*, Aug. 1, 1983, p. 66).

†WILLEBRORD VAN ROIJEN SNELL (1591–1626). Dutch mathematician. Snell, who succeeded his father to the post of Professor of Mathematics at the University of Leiden in 1613, is most famous for the result on light refraction that bears his name. Although this phenomenon was studied as far back as the ancient Greek astronomer Ptolemy, until Snell's work the relationship was incorrectly thought to be $\theta_1/v_1 = \theta_2/v_2$. Snell's law was published by Descartes in 1638 without giving proper credit to Snell. Snell also discovered a method for determining distances by triangulation that founded the modern technique of mapmaking.

43. A farmer wants to walk at a constant rate from her barn to a straight river, fill her pail, and carry it to her house in the least time.

    (a) Explain how this problem relates to Fermat's principle and the light-reflection problem in Exercise 41.

    (b) Use the result of Exercise 41 to describe geometrically the best path for the farmer to take.

    (c) Use part (b) to determine where the farmer should fill her pail if her house and barn are located as in Figure 4.7.15.

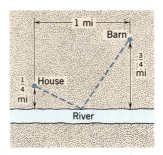

Figure 4.7.15

4.8 MORE APPLIED MAXIMUM AND MINIMUM PROBLEMS

In the preceding section we discussed optimization problems that reduced to maximizing or minimizing a continuous function over a closed interval [a, b]. For such problems, the Extreme-Value Theorem guarantees that a solution exists. In this section we shall consider optimization problems that reduce to maximizing or minimizing continuous functions over open intervals or infinite intervals. For such problems, there is no general guarantee that a solution exists. Thus, part of the problem is to determine whether a solution exists and then to find it if it does.

Example 1   A closed cylindrical can is to hold 1 liter (1000 cm³) of liquid. How should we choose the height and radius to minimize the amount of material needed to manufacture the can?

Solution.   Let

    h = height (in cm) of the can
    r = radius (in cm) of the can
    S = surface area (in cm²) of the can

Assuming there is no waste or overlap, the amount of material needed for manufacture will be the same as the surface area of the can. Since the can can be made from two circular disks of radius r and a rectangular sheet with dimensions h by $2\pi r$ (Figure 4.8.1), the surface area will be

$$S = 2\pi r^2 + 2\pi rh \tag{1}$$

We shall now eliminate one of the variables in (1) so that S will be expressed as a function of one variable. Since the volume of the can is 1000 cm³, it follows from the formula $V = \pi r^2 h$ for the volume of a cylinder that

$$1000 = \pi r^2 h \qquad \text{or} \qquad h = \frac{1000}{\pi r^2} \tag{2–3}$$

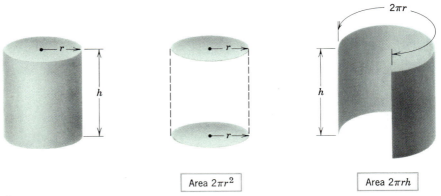

Area $2\pi r^2$ Area $2\pi rh$

Figure 4.8.1

Substituting (3) in (1) yields

$$S = 2\pi r^2 + \frac{2000}{r} \tag{4}$$

Because r represents a radius it must be positive;* thus the variable r in (4) must satisfy

$$0 < r < +\infty$$

We have now reduced the problem to that of finding a value of r in $(0, +\infty)$ for which (4) is minimum (provided such a value exists). But S is a continuous function of r on $(0, +\infty)$ with

$$\lim_{r \to 0^+} \left(2\pi r^2 + \frac{2000}{r}\right) = +\infty \quad \text{and} \quad \lim_{r \to +\infty} \left(2\pi r^2 + \frac{2000}{r}\right) = +\infty$$

so (4) has a minimum, but no maximum on $(0, +\infty)$. This minimum must occur at a critical point, so we calculate

$$\frac{dS}{dr} = 4\pi r - \frac{2000}{r^2} \tag{5}$$

Setting $dS/dr = 0$ gives

$$4\pi r - \frac{2000}{r^2} = 0 \quad \text{or} \quad r = \frac{10}{\sqrt[3]{2\pi}} \tag{6}$$

Since (6) is the only critical point in the interval $(0, +\infty)$, this value of r yields the minimum value of S. From (3) the value of h corresponding to this r is

$$h = \frac{1000}{\pi(10/\sqrt[3]{2\pi})^2} = \frac{20}{\sqrt[3]{2\pi}}$$

---

* The value $r = 0$ must be excluded, otherwise (2) cannot be satisfied.

Second Solution. The conclusion that a minimum occurs at the value of r in (6) can be deduced from Theorem 4.6.6 and the second derivative test by noting that

$$\frac{d^2S}{dr^2} = 4\pi + \frac{4000}{r^3}$$

so that

$$\left. \frac{d^2S}{dr^2} \right|_{r=10/\sqrt[3]{2\pi}} = 4\pi + \frac{4000}{(10/\sqrt[3]{2\pi})^3} = 12\pi$$

Since the second derivative is positive, a relative minimum, and therefore a minimum, occurs at the critical point $r = 10/\sqrt[3]{2\pi} \approx 5.4$.

Third Solution. Using the methods of Section 4.4, the reader should be able to obtain the graph in Figure 4.8.2. From this graph we see that S has a minimum, and this minimum occurs at the critical point $r = 10/\sqrt[3]{2\pi}$ calculated in our first solution of this problem [see (6)]. ◀

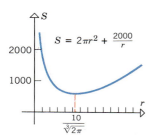

$$S = 2\pi r^2 + \frac{2000}{r}$$

Figure 4.8.2

REMARK. Note that S has no maximum on $(0, +\infty)$. Thus, had we asked for the dimensions of the can requiring the maximum amount of material for its manufacture, there would have been no solution to the problem. Optimization problems with no solution are sometimes called *ill posed*.

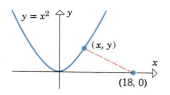

$y = x^2$

(x, y)

$(18, 0)$

Figure 4.8.3

Example 2 Find a point on the curve $y = x^2$ that is closest to the point $(18, 0)$.

Solution. The distance L between $(18, 0)$ and an arbitrary point (x, y) on the curve $y = x^2$ (Figure 4.8.3) is given by

$$L = \sqrt{(x - 18)^2 + (y - 0)^2}$$

Since (x, y) lies on the curve, x and y satisfy $y = x^2$; thus,

$$L = \sqrt{(x - 18)^2 + x^4} \tag{7}$$

Because there are no restrictions on x, the problem reduces to finding a value of x in $(-\infty, +\infty)$ for which (7) is minimum, provided such a value exists.

In problems of minimizing or maximizing a distance, there is a trick that is helpful for simplifying the computations. It is based on the observation that the distance and the square of the distance have their maximum or minimum at the same point (see Exercise 28). Thus, the minimum value of L in (7) and the minimum value of

$$S = L^2 = (x - 18)^2 + x^4 \tag{8}$$

occur at the same x value.

From (8),

$$\frac{dS}{dx} = 2(x - 18) + 4x^3 = 4x^3 + 2x - 36 \tag{9}$$

so that the critical points satisfy $4x^3 + 2x - 36 = 0$ or equivalently

$$2x^3 + x - 18 = 0 \tag{10}$$

To solve for x we shall begin by searching for integer solutions. This task can be simplified by using the fact that all integer solutions, if there are any, to a polynomial equation with integer coefficients

$$a_n x^n + \cdots + a_1 x + a_0 = 0$$

must be divisors of the constant term a_0. This result is usually proved in algebra courses. Thus, the only possible integer solutions of (10) are the divisors of -18: $\pm 1, \pm 2, \pm 3, \pm 6, \pm 9, \pm 18$. Successively substituting these values in (10) we find that $x = 2$ is a solution; therefore, $x - 2$ is a factor of the left side of (10). After dividing by the factor $x - 2$ we can rewrite (10) as

$$(x - 2)(2x^2 + 4x + 9) = 0$$

Thus, the remaining solutions of (10) satisfy the quadratic equation

$$2x^2 + 4x + 9 = 0$$

But these solutions are imaginary numbers (use the quadratic formula) so that $x = 2$ is the only real solution of (10) and consequently the only critical point of S. To determine the nature of this critical point we shall use the second derivative test. From (9),

$$\frac{d^2 S}{dx^2} = 12x^2 + 2 \quad \text{so} \quad \frac{d^2 S}{dx^2}\bigg|_{x=2} = 50 > 0$$

which shows that a relative minimum occurs at $x = 2$. Since $x = 2$ is the only relative extremum for L, it follows from Theorem 4.6.6 that an absolute minimum value of L also occurs at $x = 2$. Thus, the point on the curve $y = x^2$ closest to $(18, 0)$ is

$$(x, y) = (x, x^2) = (2, 4) \quad \blacktriangleleft$$

▶ Exercise Set 4.8

1. Find two numbers whose sum is 20 and whose product is
   (a) maximum   (b) minimum.

2. Find the dimensions of the rectangle of area A whose perimeter is
   (a) minimum   (b) maximum.

3. A rectangular area of 3200 ft^2 is to be fenced off. Two opposite sides will use fencing costing $1 per foot and the remaining sides will use fencing costing $2 per foot. Find the dimensions of the rectangle of least cost.

4. A closed rectangular container with a square base is to have a volume of 2250 in^3. The material for the top and bottom of the container will cost $2 per in^2, and the material for the sides will cost $3 per in^2. Find the dimensions of the container of least cost.

5. A closed rectangular container with a square base is to have a volume of 2000 cm^3. It costs twice as much per square centimeter for the top and bottom as it does for the sides. Find the dimensions of the container of least cost.

6. A container with square base, vertical sides, and open top is to be made from 1000 ft² of material. Find the dimensions of the container with greatest volume.

7. A rectangular container with two square sides and an open top is to have a volume of V cubic units. Find the dimensions of the container of minimum surface area.

8. A closed cylindrical can is to have a surface area of S square units. Show that the can of maximum volume is achieved when the height is equal to the diameter of the base.

9. A cylindrical can, open at the top, is to hold 500 cm³ of liquid. Find the height and radius that minimize the amount of material needed to manufacture the can.

10. A rectangular sheet of paper is to contain 72 in² of printed matter with 2-in margins at top and bottom and 1-in margins on each side. What dimensions for the sheet will use the least paper?

11. A cone-shaped paper drinking cup is to hold 10 cm³ of water. Find the height and radius of the cup that will require the least amount of paper.

12. Find the x-coordinate of the point P on the parabola $y = 1 - x^2$ for $0 < x \leq 1$ where the triangle that is enclosed by the tangent line at P and the coordinate axes has the smallest area.

13. Find all points on the curve $x^2 - y^2 = 1$ closest to $(0, 2)$.

14. Find a point on the curve $x = 2y^2$ closest to $(0, 9)$.

15. Suppose a line L of variable slope passes through $(1, 3)$ and intersects the coordinate axes at the points $(a, 0)$ and $(0, b)$ where a and b are positive. Find the slope of L for which the area of the triangle with vertices $(a, 0)$, $(0, b)$, and $(0, 0)$ is
(a) maximum   (b) minimum.

16. In a certain chemical manufacturing process, the daily weight y of defective chemical output depends on the total weight x of all output according to the empirical formula

$$y = 0.01x + 0.00003x^2$$

where x and y are in pounds. If the profit is $100 per pound of nondefective chemical produced and the loss is $20 per pound of defective chemical produced, how many pounds of chemical should be produced daily to maximize the profit?

17. Two particles, A and B, are in motion in the xy-plane. Their coordinates at each instant of time t $(t \geq 0)$ are given by $x_A = t$, $y_A = 2t$, $x_B = 1 - t$, and $y_B = t$. Find the minimum distance between A and B.

18. Follow the directions of Exercise 17, with $x_A = t$, $y_A = t^2$, $x_B = 2t$, and $y_B = 2$.

19. Find the coordinates of the point P on the curve $y = 1/x^2$ $(x > 0)$ where the segment of the tangent line at P that is cut off by the coordinate axes will have its shortest length.

20. A lamp is suspended above the center of a round table of radius r. How high above the table should the lamp be placed to achieve maximum illumination at the edge of the table? [Assume that the illumination I is directly proportional to the cosine of the angle of incidence ϕ of the light rays and inversely proportional to the square of the distance l from the light source (Figure 4.8.4).]

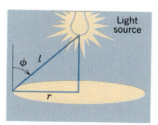

Figure 4.8.4

21. Where on the curve $y = (1 + x^2)^{-1}$ does the tangent line have the greatest slope?

22. Find the dimensions of the isosceles triangle of least area that can be circumscribed about a circle of radius R.

23. Find the height and radius of the right-circular cone with least volume that can be circumscribed about a sphere of radius R.

24. If an unknown physical quantity x is measured n times, the measurements $x_1, x_2, \ldots, x_n$ often vary because of uncontrollable factors such as temperature, atmospheric pressure, and so forth. Thus, a scientist is often faced with the problem of using n different observed measurements to obtain an estimate $\bar{x}$ of an unknown quantity x. One method for making such an estimate is based on the **least squares principle,** which states that the estimate $\bar{x}$ should be

chosen to minimize

$$s = (x_1 - \bar{x})^2 + (x_2 - \bar{x})^2 + \cdots + (x_n - \bar{x})^2$$

which is the sum of the squares of the deviations between the estimate $\bar{x}$ and the measured values. Show that the estimate resulting from the least squares principle is

$$\bar{x} = \frac{1}{n}(x_1 + x_2 + \cdots + x_n)$$

that is, $\bar{x}$ is the arithmetic average of the observed values.

25. A pipe of negligible diameter is to be carried horizontally around a corner from a hallway 8 ft wide into a hallway 4 ft wide (Figure 4.8.5). What is the maximum length that the pipe can have? [An interesting discussion of this problem in the case where the diameter of the pipe is not neglected is given by Norman Miller in the *American Mathematical Monthly*, vol. 56, 1949, pp. 177–179.]

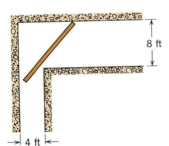

8 ft

4 ft

Figure 4.8.5

26. A plank is used to reach over a fence 8 ft high to support a wall that is 1 ft behind the fence (Figure 4.8.6). What is the length of the shortest plank that can be used? [*Hint:* Express the length of the plank in terms of the angle θ shown in the figure.]

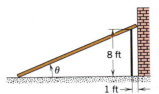

8 ft

θ

1 ft

Figure 4.8.6

27. The intensity of a point light source is directly proportional to the strength of the source and inversely proportional to the square of the distance from the source. Two point light sources with strengths of S and $8S$ are separated by a distance of 90 cm. Where on the line segment between the two sources is the intensity a minimum?

28. Prove: If $f(x) \geq 0$ on an interval I and if $f(x)$ has a maximum value on I at x_0, then $\sqrt{f(x)}$ also has a maximum value at x_0. Similarly, for minimum values. [*Hint:* Use the fact that $\sqrt{x}$ is an increasing function on the interval $[0, +\infty)$.]

29. Prove: If $P(x_1, y_1)$ is a fixed point, then the minimum distance between P and a point on the line $ax + by + c = 0$ is

$$\frac{|ax_1 + by_1 + c|}{\sqrt{a^2 + b^2}}$$

4.9 NEWTON'S METHOD

In Section 2.7 we showed how to approximate a solution of an equation $f(x) = 0$ to any degree of accuracy using the Intermediate Value Theorem. In this section we shall study a technique, called Newton's Method, that is generally more efficient.

In beginning algebra one learns that the solution of a first-degree equation $ax + b = 0$ is given by the formula $x = -b/a$, and the solutions of a second-degree equation $ax^2 + bx + c = 0$ are given by the quadratic formula. Formulas also exist for the solutions of all third- and fourth-degree equations though they are too complicated to be of practical use. In 1826 it was shown by the

Norwegian mathematician Niels Henrik Abel* that it is impossible to construct a formula of the same type for the solutions of a *general* fifth-degree equation or higher. Thus, for a *specific* fifth-degree polynomial equation such as

$$x^5 - 9x^4 + 2x^3 - 5x^2 + 17x - 8 = 0$$

it may be difficult or impossible to find exact values for all of the solutions. Similar difficulties occur for trigonometric equations such as

$$x - \cos x = 0$$

as well as equations of other types. For such equations the solutions are generally approximated in some way, often by the method we shall now discuss.

☐ **NEWTON'S METHOD**

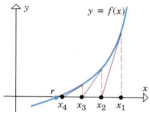

Figure 4.9.1

We note first that the solutions of $f(x) = 0$ are the values of x where the graph of f crosses the x-axis. Suppose that $x = r$ is the solution we are seeking. Even if we cannot find the value of r exactly, it is usually possible to approximate it by graphing f and applying Theorem 2.7.10 to estimate where the graph crosses the x-axis. If we let x_1 denote our initial approximation to r, then we can generally improve on this approximation by moving along the tangent line to $y = f(x)$ at x_1 until we meet the x-axis at a point x_2 (Figure 4.9.1). Usually,

*NIELS HENRIK ABEL (1802–1829). Norwegian mathematician. Abel was the son of a poor Lutheran minister and a remarkably beautiful mother from whom he inherited strikingly good looks. In his brief life of 26 years Abel lived in virtual poverty and suffered a succession of adversities; yet he managed to prove major results that altered the mathematical landscape forever. At the age of thirteen he was sent away from home to a school whose better days had long passed. By a stroke of luck the school had just hired a teacher named Bernt Michael Holmboe, who quickly discovered that Abel had extraordinary mathematical ability. Together, they studied the calculus texts of Euler and works of Newton and the later French mathematicians. By the time he graduated, Abel was familiar with most of the great mathematical literature. In 1820 his father died, leaving the family in dire financial straits. Abel was able to enter the University of Christiania in Oslo only because he was granted a free room and several professors supported him directly from their salaries. The University had no advanced courses in mathematics, so Abel took a preliminary degree in 1822 and then continued to study mathematics on his own. In 1824 he published at his own expense the proof that it is impossible to solve the general fifth-degree polynomial equation algebraically. With the hope that this landmark paper would lead to his recognition and acceptance by the European mathematical community, Abel sent the paper to the great German mathematician Gauss, who casually declared it to be a ''monstrosity'' and tossed it aside. However, in 1826 Abel's paper on the fifth-degree equation and other work was published in the first issue of a new journal, founded by his friend, Leopold Crelle. In the summer of 1826 he completed a landmark work on transcendental functions, which he submitted to the French Academy of Sciences in the hope of establishing himself as a major mathematician, for many young mathematicians had gained quick distinction by having their work accepted by the Academy. However, Abel waited in vain because the paper was either ignored or misplaced by one of the referees, and it did not surface again until two years after his death. That paper was later described by one major mathematician as ''. . . the most important mathematical discovery that has been made in our century. . . .'' After submitting his paper, Abel returned to Norway, ill with tuberculosis and in heavy debt. While eking out a meager living as a tutor, he continued to produce great work and his fame spread. Soon great efforts were being made to secure a suitable mathematical position for him. Fearing that his great work had been lost by the Academy, he mailed a proof of the main result to Crelle in January of 1829. In April he suffered a violent hemorrhage and died. Two days later Crelle wrote to inform him that an appointment had been secured for him in Berlin and his days of poverty were over! Abel's great paper was finally published by the Academy twelve years after his death.

x_2 will be closer to r than x_1. To improve the approximation further, we can repeat the process by moving along the tangent line to $y = f(x)$ at x_2 until we meet the x-axis at a point x_3. Continuing in this way we can generate a succession of values $x_1, x_2, x_3, x_4, \ldots$ that will usually get closer and closer to r. This procedure for approximating r is called **Newton's Method.**

To implement Newton's Method analytically, we must derive a formula that will tell us how to calculate each improved approximation from the preceding approximation. For this purpose, we note that the point-slope form of the tangent line to $y = f(x)$ at the initial approximation x_1 is

$$y - f(x_1) = f'(x_1)(x - x_1) \tag{1}$$

If $f'(x_1) \neq 0$, then this line is not parallel to the x-axis and consequently it crosses the x-axis at some point $(x_2, 0)$. Substituting the coordinates of this point in (1) yields

$$-f(x_1) = f'(x_1)(x_2 - x_1)$$

Solving for x_2 we obtain

$$x_2 - x_1 = -\frac{f(x_1)}{f'(x_1)}$$

or

$$x_2 = x_1 - \frac{f(x_1)}{f'(x_1)} \tag{2}$$

The next approximation can be obtained more easily. If we view x_2 as the starting approximation and x_3 the new approximation, we can simply apply (2) with x_2 in place of x_1 and x_3 in place of x_2. This yields

$$x_3 = x_2 - \frac{f(x_2)}{f'(x_2)} \tag{3}$$

provided $f'(x_2) \neq 0$. In general, if x_n is the nth approximation, then it is evident from the pattern in (2) and (3) that the improved approximation x_{n+1} is given by

Newton's Method

$$x_{n+1} = x_n - \frac{f(x_n)}{f'(x_n)} \quad n = 1, 2, 3, \ldots \tag{4}$$

Example 1   Use Newton's Method to approximate the real solutions of

$$x^3 - x - 1 = 0$$

Solution.   Let $f(x) = x^3 - x - 1$, so $f'(x) = 3x^2 - 1$ and (4) becomes

$$x_{n+1} = x_n - \frac{x_n^3 - x_n - 1}{3x_n^2 - 1} \tag{5}$$

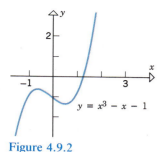

Figure 4.9.2

From the graph of f in Figure 4.9.2, we see that the given equation has only one real solution. This solution lies between 1 and 2 because $f(1) = -1 < 0$ and $f(2) = 5 > 0$. We shall use $x_1 = 1.5$ as our first approximation ($x_1 = 1$ or $x_1 = 2$ would also be reasonable choices).

Letting $n = 1$ in (5) and substituting $x_1 = 1.5$ yields

$$x_2 = 1.5 - \frac{(1.5)^3 - 1.5 - 1}{3(1.5)^2 - 1} = 1.34782609$$

(We used a calculator that displays nine digits.) Next, we let $n = 2$ in (5) and substitute $x_2 = 1.34782609$ to obtain

$$x_3 = 1.34782609 - \frac{(1.34782609)^3 - (1.34782609) - 1}{3(1.34782609)^2 - 1} = 1.32520040$$

If we continue this process until two identical approximations are generated in succession, we obtain

$$x_1 = 1.5$$
$$x_2 = 1.34782609$$
$$x_3 = 1.32520040$$
$$x_4 = 1.32471817$$
$$x_5 = 1.32471796$$
$$x_6 = 1.32471796$$

At this stage there is no need to continue further because we have reached the accuracy limit of our calculator, and all subsequent approximations that the calculator generates will be the same. Thus, the solution is approximately $x \approx 1.32471796$. ◀

Example 2 It is evident from Figure 4.9.3 that if x is in radians, then the equation

$$\cos x = x$$

has a solution between 0 and 1. Use Newton's Method to approximate it.

Solution. Rewrite the equation as

$$x - \cos x = 0$$

and apply (4) with $f(x) = x - \cos x$. Since $f'(x) = 1 + \sin x$, (4) becomes

$$x_{n+1} = x_n - \frac{x_n - \cos x_n}{1 + \sin x_n} \tag{6}$$

From Figure 4.9.3, the solution seems closer to $x = 1$ than $x = 0$, so we shall use $x_1 = 1$ (radian) as our initial approximation. Letting $n = 1$ in (6) and substituting $x_1 = 1$ yields

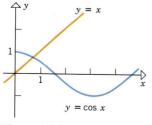

Figure 4.9.3

$$x_2 = 1 - \frac{1 - \cos 1}{1 + \sin 1} = .750363868$$

Next, letting $n = 2$ in (6) and substituting this value of x_2 yields

$$x_3 = .750363868 - \frac{.750363868 - \cos(.750363868)}{1 + \sin(.750363868)} = .739112891$$

If we continue this process until two identical approximations are generated in succession, we obtain

$$x_1 = 1$$
$$x_2 = .750363868$$
$$x_3 = .739112891$$
$$x_4 = .739085133$$
$$x_5 = .739085133$$

Thus, to the accuracy limit of our calculator, the solution of the equation $\cos x = x$ is $x \approx .739085133$. ◀

☐ **SOME DIFFICULTIES WITH NEWTON'S METHOD**

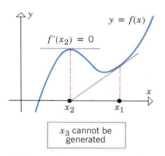

Figure 4.9.4

Newton's Method does not always work. For example, if $f'(x_n) = 0$ for some n, then (4) involves a division by zero, making it impossible to generate x_{n+1}. However, this is to be expected because the tangent line to $y = f(x)$ is parallel to the x-axis where $f'(x_n) = 0$, and so this tangent line does not cross the x-axis to generate the next approximation (Figure 4.9.4).

Sometimes the "approximations" produced by Newton's Method do not converge to a solution. For example, consider the equation

$$x^{1/3} = 0$$

which has $x = 0$ as its only solution, and try to approximate this solution by Newton's Method with a starting value of $x_0 = 1$. Letting $f(x) = x^{1/3}$, Formula (4) becomes

$$x_{n+1} = x_n - \frac{(x_n)^{1/3}}{\frac{1}{3}(x_n)^{-2/3}} = x_n - 3x_n = -2x_n$$

Beginning with $x_1 = 1$, the successive values generated by this formula are

$$x_1 = 1, \quad x_2 = -2, \quad x_3 = 4, \quad x_4 = -8, \ldots$$

which obviously don't converge to $x = 0$. Figure 4.9.5 illustrates what is happening geometrically.

For a statement of conditions under which Newton's Method works and a discussion of error questions, the reader should consult a book on numerical analysis, for example, Peter Henrici, *Elements of Numerical Analysis,* Wiley, New York, 1982. In situations where Newton's method fails, the method discussed in Section 2.7 can be used; but when Newton's Method works, it is the method of choice because it converges more rapidly.

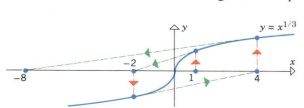

Figure 4.9.5

▶ Exercise Set 4.9   C *All*

In this exercise set, use a calculator, and keep as many decimal places as it can display.

1. Approximate $\sqrt{2}$ by applying Newton's Method to the equation $x^2 - 2 = 0$.

2. Approximate $\sqrt{7}$ by applying Newton's Method to the equation $x^2 - 7 = 0$.

3. Approximate $\sqrt[3]{6}$ by applying Newton's Method to the equation $x^3 - 6 = 0$.

In Exercises 4–7, the equation has one real solution. Approximate it by Newton's Method.

4. $x^3 + x - 1 = 0$.   **5.** $x^3 - x + 3 = 0$.

6. $x^5 - x + 1 = 0$.   **7.** $x^5 + x^4 - 5 = 0$.

In Exercises 8–15, the equation has one solution satisfying the given conditions. Approximate it by Newton's Method.

8. $2x^2 + 4x - 3 = 0$; $x < 0$.

9. $2x^2 + 4x - 3 = 0$; $x > 0$.

10. $x^4 + x - 3 = 0$; $x > 0$.

11. $x^4 + x - 3 = 0$; $x < 0$.

12. $x^5 - 5x^3 - 2 = 0$; $x > 0$.

13. $2 \sin x = x$; $x > 0$.

14. $\sin x = x^2$; $x > 0$.

15. $x - \tan x = 0$; $\pi/2 < x < 3\pi/2$.

16. Many calculators compute reciprocals using the approximation $1/a \approx x_{n+1}$ where
$$x_{n+1} = x_n(2 - ax_n) \qquad n = 1, 2, 3, \ldots$$
and x_1 is an initial approximation to $1/a$. This formula makes it possible to use multiplications and subtractions (which can be done quickly) to perform divisions that would be slow to obtain directly.
(a) Apply Newton's Method to the function $f(x) = (1/x) - a$ to derive this approximation.
(b) Use the formula to approximate $\frac{1}{17}$.

17. The *mechanic's rule* for approximating square roots states that $\sqrt{a} \approx x_{n+1}$ where
$$x_{n+1} = \frac{1}{2}\left(x_n + \frac{a}{x_n}\right) \qquad n = 1, 2, 3, \ldots$$
and x_1 is any positive approximation to $\sqrt{a}$.
(a) Apply Newton's Method to the function $f(x) = x^2 - a$ to derive the mechanic's rule.
(b) Use the mechanic's rule to approximate $\sqrt{10}$.

In Exercises 18–22, find the x-coordinates of all points of intersection of the given curves. Use Newton's Method, if needed.

18. $y = x^2 + 1$ and $y = x^3$.

19. $y = x^3$ and $y = \frac{1}{2}x - 1$.

20. $y = \frac{1}{4}x^2$ and $y = \dfrac{2x}{x^2 + 1}$.

21. $y = x^2$ and $y = \sqrt{2x + 1}$.

22. $y = \frac{1}{8}x^3 + 1$ and $y = \cos 2x$.

In Exercises 23 and 24, use Newton's Method to approximate all real values of y satisfying the given equation for the indicated value of x.

23. $xy^4 + x^3 y = 1$; $x = 1$.

24. $xy - \cos\left(\frac{1}{2}xy\right) = 0$; $x = 2$.

In Exercises 25–30, use Newton's Method.

25. Find the minimum value of
$$f(x) = \frac{1}{4}x^4 + x^2 + 5x$$

26. Find the maximum value of $f(x) = x \sin x$ on the interval $[0, \pi]$.

27. Find the coordinates of the point on the parabola $y = x^2$ that is nearest the point $(1, 0)$.

28. Find the dimensions of the rectangle of largest area that can be inscribed under the curve $y = \cos x$ for $0 \le x \le \pi/2$, as shown in Figure 4.9.6.

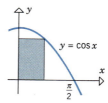

Figure 4.9.6

29. Find the central angle θ of a circle that subtends an arc whose length is 1.5 times the length of its chord (Figure 4.9.7). Give θ to the nearest degree. [*Hint:* Recall that $r\theta$ is the length of arc subtended by θ, where θ is in radians and r is the radius of the circle.]

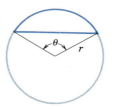

Figure 4.9.7

30. A *segment* of a circle is the region enclosed by an arc and its chord (Figure 4.9.8). If r is the radius of the circle and θ the angle subtended at the center of the circle, then it can be shown that the area A of the segment is $A = \frac{1}{2}r^2(\theta - \sin\theta)$, where θ is in radians. Find the value of θ for which the area of the segment is one-fourth the area of the circle. Give θ to the nearest degree.

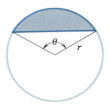

Figure 4.9.8

4.10   ROLLE'S THEOREM; MEAN-VALUE THEOREM

In this section we shall discuss a result called the Mean-Value Theorem. There are so many major consequences of this theorem that it is regarded as one of the most fundamental results in calculus.

□ **ROLLE'S THEOREM**

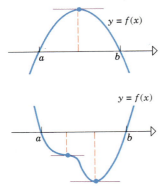

Figure 4.10.1

We shall begin with a special case of the Mean-Value Theorem, called **Rolle's* Theorem**. Geometrically, Rolle's Theorem states that between any two points, a and b, where a "well-behaved" curve $y = f(x)$ crosses the x-axis, there must be at least one place where the tangent line to the curve is horizontal (Figure 4.10.1).

The precise statement of this result is as follows.

> **4.10.1   THEOREM** (*Rolle's Theorem*).   *Let f be differentiable on (a, b) and continuous on $[a, b]$. If $f(a) = f(b) = 0$, then there is at least one point c in (a, b) where $f'(c) = 0$.*

*MICHEL ROLLE (1652–1719). French mathematician. Rolle, the son of a shopkeeper, received only an elementary education. He married early and as a young man struggled hard to support his family on the meager wages of a transcriber for notaries and attorneys. In spite of his financial problems and minimal education, Rolle studied algebra and Diophantine analysis (a branch of number theory) on his own. Rolle's fortune changed dramatically in 1682 when he published an elegant solution of a difficult, unsolved problem in Diophantine analysis. The public recognition of his achievement led to a patronage under minister Louvois, a job as an elementary mathematics teacher, and eventually to a short-term administrative post in the Ministry of War. In 1685 he joined the Académie des Sciences in a low-level position for which he received no regular salary until 1699. He stayed there until he died of apoplexy in 1719.

While Rolle's forté was always Diophantine analysis, his most important work was a book on the algebra of equations, called *Traité d'algèbre,* published in 1690. In that book Rolle firmly established the notation $\sqrt[n]{a}$ [earlier written as $\sqrt{(n)}\,a$] for the nth root of a, and proved a polynomial version of the theorem that today bears his name. (Rolle's Theorem was named by Giusto Bellavitis in 1846.) Ironically, Rolle was one of the most vocal early antagonists of calculus. He strove intently to demonstrate that it gave erroneous results and was based on unsound reasoning. He quarreled so vigorously on the subject that the Académie des Sciences was forced to intervene on several occasions. Among his several achievements, Rolle helped advance the currently accepted size order for negative numbers. Descartes, for example, viewed -2 as smaller than -5. Rolle preceded most of his contemporaries by adopting the current convention in 1691.*

Proof. Either $f(x)$ is equal to zero for all x in $[a, b]$ or it is not. If it is, then $f'(x) = 0$ for all x in (a, b), since f is constant on (a, b). Thus, for any c in (a, b)

$$f'(c) = 0$$

If $f(x)$ is not equal to zero for all x in $[a, b]$, then there must be a point x in (a, b) where $f(x) > 0$ or $f(x) < 0$. We shall consider the first case and leave the second as an exercise.

Since f is continuous on $[a, b]$, it follows from the Extreme-Value Theorem that f has a maximum value at some point c in $[a, b]$. Since $f(a) = f(b) = 0$ and $f(x) > 0$ at some point in (a, b), the point c cannot be an endpoint; it must lie in (a, b). By hypothesis, f is differentiable everywhere on (a, b). In particular, it is differentiable at c so that $f'(c) = 0$ by Theorem 4.6.5. ∎

Example 1 The function

$$f(x) = \sin x$$

is both continuous and differentiable everywhere, hence is continuous on $[0, 2\pi]$ and differentiable on $(0, 2\pi)$. Moreover,

$$f(0) = \sin 0 = 0 \quad \text{and} \quad f(2\pi) = \sin 2\pi = 0$$

so that f satisfies the hypotheses of Rolle's Theorem on the interval $[0, 2\pi]$. Since $f'(c) = \cos c$, Rolle's Theorem guarantees that there is at least one point c in $(0, 2\pi)$ such that

$$\cos c = 0 \tag{1}$$

Solving (1) yields two such values for c, namely $c_1 = \pi/2$ and $c_2 = 3\pi/2$ (Figure 4.10.2). ◀

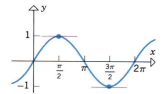

Figure 4.10.2

REMARK. In the foregoing example we were able to find c because (1) was easy to solve. However, frequently the equation $f'(c) = 0$ is sufficiently complicated that an exact value for c cannot be obtained.

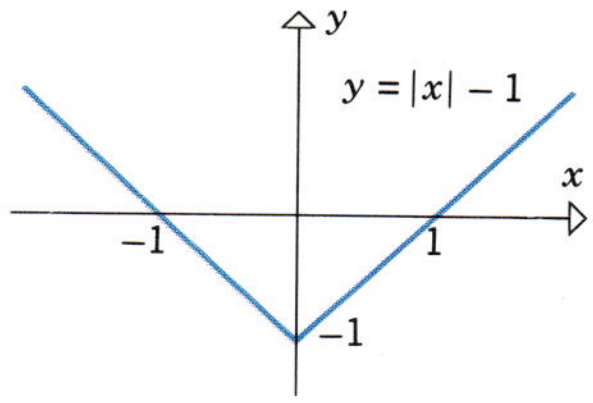

Example 2 The function

$$f(x) = |x| - 1$$

has the property that $f(-1) = 0$ and $f(1) = 0$, yet there is no point in the interval $(-1, 1)$ where the graph of f has a horizontal tangent line (Figure 4.10.3). This does *not* contradict Rolle's Theorem because the function f is not differentiable at every point of the interval $(-1, 1)$. ◀

Figure 4.10.3

☐ **THE MEAN-VALUE THEOREM**

Rolle's Theorem is a special case of the **Mean-Value Theorem,** which states that between any two points A and B on a "well-behaved" curve $y = f(x)$, there must be at least one place where the tangent line to the curve is parallel to the secant line joining A and B (Figure 4.10.4).

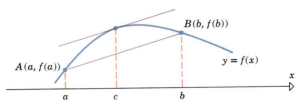

Figure 4.10.4

Noting that the slope of the secant line joining $A(a, f(a))$ and $B(b, f(b))$ is

$$\frac{f(b) - f(a)}{b - a}$$

and the slope of the tangent at c is $f'(c)$, the Mean-Value Theorem can be stated precisely as follows.

4.10.2   THEOREM (*Mean-Value Theorem*).   *Let f be differentiable on (a, b) and continuous on $[a, b]$. Then there is at least one point c in (a, b) where*

$$f'(c) = \frac{f(b) - f(a)}{b - a} \tag{2}$$

Motivation for the Proof of Theorem 4.10.2.   Figure 4.10.4 suggests that (2) will hold (i.e., the tangent line will be parallel to the secant line) at a point c where the vertical distance between the curve and the secant line is maximum. Thus, to prove the Mean-Value Theorem it is natural to begin by looking for a formula for the vertical distance $v(x)$ between the curve $y = f(x)$ and the secant line joining $(a, f(a))$ and $(b, f(b))$.

Proof of Theorem 4.10.2.   Since the two-point form of the equation of the secant line joining $(a, f(a))$ and $(b, f(b))$ is

$$y - f(a) = \frac{f(b) - f(a)}{b - a}(x - a)$$

or equivalently

$$y = \frac{f(b) - f(a)}{b - a}(x - a) + f(a)$$

the difference $v(x)$ between the height of the graph of f and the height of the secant line is

$$v(x) = f(x) - \left[\frac{f(b) - f(a)}{b - a}(x - a) + f(a)\right] \tag{3}$$

Since $f(x)$ is continuous on $[a, b]$ and differentiable on (a, b), so is $v(x)$. Moreover,

$$v(a) = 0 \quad \text{and} \quad v(b) = 0$$

so that $v(x)$ satisfies the hypotheses of Rolle's Theorem on the interval $[a, b]$.

Thus, there is a point c in (a, b) such that $v'(c) = 0$. But from Equation (3)

$$v'(x) = f'(x) - \frac{f(b) - f(a)}{b - a}$$

so

$$v'(c) = f'(c) - \frac{f(b) - f(a)}{b - a}$$

Thus, at the point c in (a, b), where $v'(c) = 0$, we have

$$f'(c) = \frac{f(b) - f(a)}{b - a} \qquad \blacksquare$$

Example 3   Let $f(x) = x^3 + 1$. Show that f satisfies the hypotheses of the Mean-Value Theorem on the interval $[1, 2]$ and find all values of c in this interval whose existence is guaranteed by the theorem.

Solution.   Because f is a polynomial, f is continuous and differentiable everywhere, hence is continuous on $[1, 2]$ and differentiable on $(1, 2)$. Thus, the hypotheses of the Mean-Value Theorem are satisfied with $a = 1$ and $b = 2$. But

$$f(a) = f(1) = 2, \quad f(b) = f(2) = 9$$

$$f'(x) = 3x^2, \qquad f'(c) = 3c^2$$

so that the equation

$$f'(c) = \frac{f(b) - f(a)}{b - a}$$

becomes $3c^2 = 7$, which has two solutions

$$c = \sqrt{7/3} \quad \text{and} \quad c = -\sqrt{7/3}$$

Only the first is in the interval $(1, 2)$, so $c = \sqrt{7/3}$ is the number whose existence is guaranteed by the Mean-Value Theorem.   ◀

□ **CONSEQUENCES OF THE MEAN-VALUE THEOREM**

We stated earlier that the Mean-Value Theorem is the starting point for many important results in calculus. We shall conclude this section by using it to prove Theorem 4.2.2, which was one of our main graphing tools.

4.2.2   THEOREM (*Revisited*).   *Let f be a function that is continuous on a closed interval $[a, b]$ and differentiable on the open interval (a, b).*

(a)   *If $f'(x) > 0$ for every value of x in (a, b), then f is increasing on $[a, b]$.*
(b)   *If $f'(x) < 0$ for every value of x in (a, b), then f is decreasing on $[a, b]$.*
(c)   *If $f'(x) = 0$ for every value of x in (a, b), then f is constant on $[a, b]$.*

Proof (a). Let x_1 and x_2 be points in $[a, b]$ such that $x_1 < x_2$. We must show that $f(x_1) < f(x_2)$. Because the hypotheses of the Mean-Value Theorem are satisfied on the entire interval $[a, b]$, they are satisfied on the subinterval $[x_1, x_2]$. Thus, there is some point c in the open interval (x_1, x_2) such that

$$f'(c) = \frac{f(x_2) - f(x_1)}{x_2 - x_1}$$

or equivalently

$$f(x_2) - f(x_1) = f'(c)(x_2 - x_1) \tag{4}$$

Since c is in the open interval (x_1, x_2), it follows that $a < c < b$; thus, $f'(c) > 0$. But $x_2 - x_1 > 0$ since we assumed that $x_1 < x_2$. Thus, it follows from (4) that $f(x_2) - f(x_1) > 0$ or, equivalently, $f(x_1) < f(x_2)$, which is what we were to prove. The proofs of parts (b) and (c) are similar and are left as exercises. ∎

We know from earlier work that the derivative of a constant is zero. Part (c) of Theorem 4.2.2 is a converse; it states that a function whose derivative is zero at every point of an interval must be constant on that interval. The following useful theorem is a consequence of this result.

> **4.10.3 THEOREM.** *If f and g are continuous on a closed interval $[a, b]$, and if $f'(x) = g'(x)$ for all x in the open interval (a, b), then f and g differ by a constant on $[a, b]$; that is, there is a constant k such that $f(x) - g(x) = k$ for all x in $[a, b]$.*

Proof. Let $h(x) = f(x) - g(x)$. Then for every x in (a, b)

$$h'(x) = f'(x) - g'(x) = 0$$

Thus, $h(x) = f(x) - g(x)$ is constant on $[a, b]$ by Theorem 4.2.2(c). ∎

REMARK. This theorem remains true if the closed interval $[a, b]$ is replaced by a finite or infinite interval (a, b), $[a, b)$, or $(a, b]$, provided f and g are differentiable on (a, b) and continuous on the entire interval.

Theorem 4.10.3 has a useful geometric interpretation. It implies that two functions with the same derivative at each point of an interval have graphs that are vertical translations of one another over the interval (Figure 4.10.5).

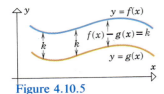

Figure 4.10.5

▶ Exercise Set 4.10

In Exercises 1–6, verify that the hypotheses of Rolle's Theorem are satisfied on the given interval and find all values of c that satisfy the conclusion of the theorem.

1. $f(x) = x^2 - 6x + 8$; $[2, 4]$.

2. $f(x) = x^3 - 3x^2 + 2x$; $[0, 2]$.

3. $f(x) = \cos x$; $\left[\dfrac{\pi}{2}, \dfrac{3\pi}{2}\right]$.

4. $f(x) = \dfrac{x^2 - 1}{x - 2}$; $[-1, 1]$.

5. $f(x) = \tfrac{1}{2}x - \sqrt{x}$; $[0, 4]$.

6. $f(x) = \dfrac{1}{x^2} - \dfrac{4}{3x} + \dfrac{1}{3};\ [1, 3].$

In Exercises 7–12, verify that the hypotheses of the Mean-Value Theorem are satisfied on the given interval and find all values of c that satisfy the conclusion of the theorem.

7. $f(x) = x^2 + x;\ [-4, 6].$

8. $f(x) = x^3 + x - 4;\ [-1, 2].$

9. $f(x) = \sqrt{x + 1};\ [0, 3].$

10. $f(x) = x + \dfrac{1}{x};\ [3, 4].$

11. $f(x) = \sqrt{25 - x^2};\ [-5, 3].$

12. $f(x) = \dfrac{1}{x - 1};\ [2, 5].$

13. Let $f(x) = \tan x.$
(a) Show that there is no point c in $(0, \pi)$ such that $f'(c) = 0$, even though $f(0) = f(\pi) = 0$.
(b) Explain why the result in part (a) does not violate Rolle's Theorem.

14. Let $f(x) = x^{2/3}$, $a = -1$, and $b = 8$.
(a) Show that there is no point c in (a, b) such that
$$f'(c) = \frac{f(b) - f(a)}{b - a}$$
(b) Explain why the result in part (a) does not violate the Mean-Value Theorem.

In Exercises 15–18, use the Mean-Value Theorem to prove the given statement.

15. $|\sin x - \sin y| \le |x - y|$ for all real values of x and y.

16. $|\tan x + \tan y| \ge |x + y|$ for all real values of x and y in the interval $(-\pi/2, \pi/2)$.

17. If $0 < x < y$, then $\sqrt{xy} < \frac{1}{2}(x + y).$ $\left[\textit{Hint:} \text{First show that } \sqrt{y} - \sqrt{x} < \dfrac{y - x}{2\sqrt{x}}.\right]$

18. If $f(1) = 0$ and $f'(x) = 1/x$ for all x in $(0, +\infty)$, then $f(x) \le x - 1$ for all x in $(0, +\infty)$. [*Hint:* Consider the cases $x > 1$, $0 < x < 1$, and $x = 1$ separately.]

19. Prove: If $f(x) = a_2x^2 + a_1x + a_0$ $(a_2 \ne 0)$, then the number c whose existence is guaranteed by the Mean-Value Theorem occurs at the midpoint of the interval $[a, b]$.

In Exercises 20–26, use Rolle's Theorem to prove the given statement.

20. The equation $6x^5 - 4x + 1 = 0$ has at least one solution in the interval $(0, 1)$.
$$\left[\textit{Hint:} \frac{d}{dx}(x^6 - 2x^2 + x) = 6x^5 - 4x + 1.\right]$$

21. The equation $3ax^2 + 2bx = a + b$ has at least one solution in the interval $(0, 1)$. [*Hint:* Consider the function $f(x) = ax^3 + bx^2 - (a + b)x$.]

22. If $f'(x)$ exists and $f'(x) \ne 0$ for all x in some open interval I, then $f(x) = 0$ can have at most one solution in I.

23. $x^3 + 4x - 1 = 0$ must have fewer than two distinct real solutions.

24. If $b^2 - 3ac < 0$, then $ax^3 + bx^2 + cx + d = 0$ $(a \ne 0)$ can have at most one real solution.

25. If $f(x) = x^5 + ax^4 + bx^3 + cx^2 + dx + e$, where $2a^2 < 5b$, then the equation $f(x) = 0$ cannot have more than three distinct real solutions. [*Hint:* Consider $f'''(x)$.]

26. If $f(x) = x^4 - 7x + 2$, then the equation $f(x) = 0$ has exactly two distinct real solutions. [*Hint:* First find $f(0)$, $f(1)$, and $f(2)$ to show that there are at least two distinct real solutions.]

In Exercises 27 and 28, use part (c) of Theorem 4.2.2 to prove the given statement.

27. If f and g are functions for which $f'(x) = g(x)$ and $g'(x) = -f(x)$ for all x, then $f^2(x) + g^2(x)$ is a constant.

28. If f and g are functions for which $f'(x) = g(x)$ and $g'(x) = f(x)$ for all x, then $f^2(x) - g^2(x)$ is a constant.

In Exercises 29–32, use Theorem 4.10.3.

29. Let $f(x) = (x - 1)^3$ and $g(x) = (x^2 + 3)(x - 3)$. Show that $f(x) - g(x)$ is a constant. Find the constant.

30. Let $f(x) = \dfrac{x + 2}{3 - x}$ and $g(x) = -\dfrac{5}{x - 3}$ for $x > 3$. Show that $f(x) - g(x)$ is a constant. Find the constant.

31. Let $g(x) = x^3 - 4x + 6$. Find $f(x)$ so that $f'(x) = g'(x)$ and $f(1) = 2$.

32. Let $g(x) = \sqrt{x^2 + 7}$. Find $f(x)$ so that $f'(x) = g'(x)$ and $f(-3) = 1$.

33. Let f and g be continuous on $[a, b]$ and differentiable on (a, b). Prove: If $f(a) = g(a)$ and $f(b) = g(b)$, then there is a point c in (a, b) where $f'(c) = g'(c)$.

34. Prove part (b) of Theorem 4.2.2.

35. Prove part (c) of Theorem 4.2.2.

36. (a) Prove: If $f''(x) > 0$ for all x in (a, b), then $f'(x) = 0$ at most once in (a, b).
   (b) Give a geometric interpretation of the result in part (a).

37. Prove: If f is continuous on $[a, b]$ and differentiable on (a, b), and if $f(a) = f(b)$, then there is a point c in (a, b) where $f'(c) = 0$.

38. Let $s = f(t)$ be the position versus time curve for a particle moving in the positive direction along a coordinate line. Prove: If f satisfies the hypotheses of the Mean-Value Theorem on a time interval $[a, b]$, then there will be an instant t_0 in (a, b) where the instantaneous velocity at time t_0 equals the average velocity over the interval $[a, b]$.

39. Complete the proof of Rolle's Theorem by considering the case where $f(x) < 0$ at some point x in (a, b).

40. Use the Mean-Value Theorem to prove
$$1.71 < \sqrt{3} < 1.75$$
[*Hint:* Let $f(x) = \sqrt{x}$, $a = 3$, and $b = 4$ in the Mean-Value Theorem.]

41. An automobile starts from rest and travels 4 miles along a straight road in 5 minutes. Use the Mean-Value Theorem to show that at some instant during the trip its velocity was exactly 48 mi/hr.

42. At 11 A.M. on a certain morning the outside temperature was 76° F. At 11 P.M. that evening it had dropped to 52° F.
   (a) Use the Mean-Value Theorem to show that at some instant during this period the temperature was decreasing at the rate of 2° F/hr.
   (b) Suppose you know that the temperature reached a high of 88° F sometime between 11 A.M. and 11 P.M. Show that at some instant during this period the temperature was decreasing at a rate greater than 3° F/hr.

■ **4.11   MOTION ALONG A LINE (RECTILINEAR MOTION)**

*In this section we shall study the motion of a particle along a line, sometimes called **rectilinear motion**. Examples are a piston moving up and down in a cylinder, a rock tossed straight up and returning to earth straight down, and the back and forth vibrations of a spring. In later sections we shall study motion along curves in two- and three-dimensional space.*

☐ **VELOCITY AND ACCELERATION**

To study the motion of a particle along a line, it is usually desirable to coordinatize the line. We do this by selecting an arbitrary origin, positive direction, and unit of length. We also choose a unit for measuring time and let t be the amount of time elapsed from some arbitrary initial observation. Thus, at the initial observation we have $t = 0$.

As the particle moves along the coordinate line its coordinate s will vary as a function of time t. This function, denoted by $s(t)$, is called the **position function** of the particle. The rate at which the particle's coordinate changes with time is called the *velocity* of the particle, and the rate at which the velocity changes with time is called the *acceleration* of the particle. More precisely, we make the following definition.

4.11.1 DEFINITION. If $s(t)$ is the position function of a particle moving on a coordinate line, then the ***instantaneous velocity*** at time t is defined by

$$v(t) = s'(t) = \frac{ds}{dt} \tag{1}$$

and the ***instantaneous acceleration*** at time t is defined by

$$a(t) = v'(t) = \frac{dv}{dt} \tag{2}$$

Since $v(t) = s'(t) = ds/dt$, the instantaneous acceleration can also be written as

$$a(t) = s''(t) = \frac{d^2s}{dt^2} \tag{3}$$

In everyday language the terms "speed" and "velocity" are often used interchangeably; however, in mathematics and science there is a distinction: The ***instantaneous speed*** of a particle moving on a coordinate line is defined to be the absolute value of the velocity. Thus,

$$\begin{bmatrix} \text{instantaneous} \\ \text{speed} \end{bmatrix} = |v(t)| = |s'(t)| = \left| \frac{ds}{dt} \right| \tag{4}$$

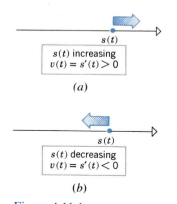

$s(t)$

$s(t)$ increasing
$v(t) = s'(t) > 0$

(a)

$s(t)$

$s(t)$ decreasing
$v(t) = s'(t) < 0$

(b)

Figure 4.11.1

If $v(t) > 0$ at a given time t, then the coordinate s of the particle is increasing at that instant, which means that the particle is moving in the positive direction along the line; similarly, if $v(t) < 0$, then s is decreasing at time t and the particle is moving in the negative direction (Figures 4.11.1a and 4.11.1b). The speed of a particle is always nonnegative; it tells how fast the particle is moving, but provides no information about the direction of motion.

□ **INTERPRETING THE SIGN OF ACCELERATION**

We will say that a particle is ***speeding up*** when its instantaneous speed is increasing and ***slowing down*** when its instantaneous speed is decreasing. It is important to understand that a positive acceleration does not necessarily mean that the particle is speeding up and a negative acceleration does not necessarily mean that it is slowing down; the sign of the velocity also plays a role. We leave it as an exercise to show that a particle is speeding up when the velocity and acceleration have the same sign and is slowing down when they have opposite signs. Thus, a particle with negative velocity is speeding up when its acceleration is negative and slowing down when its acceleration is positive. A particle with positive velocity is speeding up when its acceleration is positive and is slowing down when it is negative.

☐ **THE POSITION VERSUS TIME CURVE**

The graph of $s(t)$ in a *ts*-coordinate system is called the ***position versus time curve***; most of the significant information about the motion of a particle on a coordinate line can be obtained from that curve:

- Where $s(t) > 0$, the particle is on the positive side of the origin on the line of motion.
- Where $s(t) < 0$, the particle is on the negative side of the origin on the line of motion.
- Where a tangent line has positive slope we have $v(t) = s'(t) > 0$, so the particle is moving in the positive direction along its line of motion.
- Where a tangent line has negative slope we have $v(t) = s'(t) < 0$, so the particle is moving in the negative direction along its line of motion.
- Where a tangent line is horizontal we have $v(t) = s'(t) = 0$, so the particle is momentarily stopped.
- Where the graph is concave up we have $a(t) = v'(t) = s''(t) > 0$, so the velocity is increasing.
- Where the graph is concave down we have $a(t) = v'(t) = s''(t) < 0$, so the velocity is decreasing.

Example 1   The following table illustrates the above observations.

POSITION VERSUS TIME CURVE	CHARACTERISTICS AT $t = t_0$	BEHAVIOR OF PARTICLE AT TIME $t = t_0$
	$s(t_0) > 0$ Tangent line has positive slope. Curve is concave down.	Particle is on the positive side of the origin. Particle is moving in the positive direction. Velocity is decreasing. Particle is slowing down.
	$s(t_0) > 0$ Tangent line has negative slope. Curve is concave down.	Particle is on the positive side of the origin. Particle is moving in the negative direction. Velocity is decreasing. Particle is speeding up.
	$s(t_0) < 0$ Tangent line has negative slope. Curve is concave up.	Particle is on the negative side of the origin. Particle is moving in the negative direction. Velocity is increasing. Particle is slowing down.
	$s(t_0) > 0$ Tangent line has zero slope. Curve is concave down.	Particle is on the positive side of the origin. Particle is momentarily stopped. Velocity is decreasing.

Since acceleration is the rate at which velocity changes with time, it is expressed in units of velocity per unit of time. For example, if t is in seconds and s is in meters, then velocity units are meters per second (m/sec), and acceleration units are meters per second per second, which would be written as

$$(\text{m/sec})/\text{sec} \quad \text{or} \quad \text{m/sec}^2$$

Example 2 Let $s = s(t) = t^3 - 6t^2$ be the position function of a particle moving on a coordinate line, where t is in seconds and s is in meters. Then the instantaneous velocity, speed, and acceleration are given by

$$v(t) = \frac{ds}{dt} = 3t^2 - 12t, \quad |v(t)| = |3t^2 - 12t|, \quad a(t) = \frac{dv}{dt} = 6t - 12$$

At time $t = 1$, the instantaneous position, velocity, speed, and acceleration of the particle are

$$s(1) = -5, \quad v(1) = -9, \quad |v(1)| = 9, \quad a(1) = -6$$

Thus, the particle is 5 meters to the left of the origin and moving with a velocity of -9 m/sec (i.e., moving in the negative direction with a speed of 9 m/sec), and has an acceleration of -6 m/sec^2. Since the acceleration and velocity have the same sign at $t = 1$, the particle is speeding up at that instant. At time $t = 4$, the instantaneous position, velocity, speed, and acceleration of the particle are

$$s(4) = -32, \quad v(4) = 0, \quad |v(4)| = 0, \quad a(4) = 12$$

Thus, the particle is 32 meters to the left of the origin, and is stopped, but is accelerating at the rate of 12 m/sec^2. The position versus time curve and the graphs of velocity and acceleration versus time are shown in Figure 4.11.2. ◀

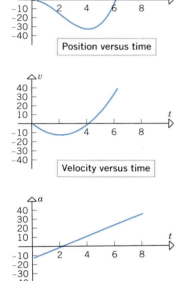

Position versus time

Velocity versus time

Acceleration versus time

Figure 4.11.2

The next example gives a schematic way of describing the motion of a particle moving on a coordinate line.

Example 3 The position function of a particle moving on a coordinate line is given by $s(t) = 2t^3 - 21t^2 + 60t + 3$, where s is in feet and t is in seconds. Describe the motion of the particle for $t \geq 0$.

Solution. The velocity and acceleration at time t are

$$v(t) = s'(t) = 6t^2 - 42t + 60 = 6(t - 2)(t - 5)$$

$$a(t) = v'(t) = 12t - 42 = 12(t - \tfrac{7}{2})$$

At each instant we can determine the direction of motion from the sign of $v(t)$ and whether the particle is speeding up or slowing down from the signs of $v(t)$ and $a(t)$ together (Figures 4.11.3a and 4.11.3b). The motion of the particle is described schematically by the curved line in Figure 4.11.3c. At time $t = 0$ the particle is at the point $s(0) = 3$ moving right with velocity $v(0) = 60$ ft/sec, but slowing down with acceleration $a(0) = -42$ ft/sec^2. The particle continues

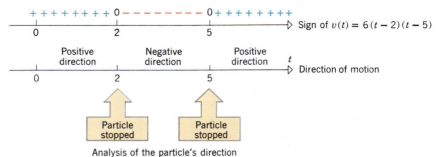

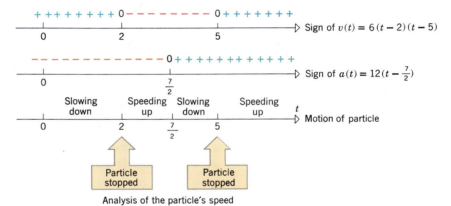

Analysis of the particle's direction

Figure 4.11.3a

Analysis of the particle's speed

Figure 4.11.3b

moving right until time $t = 2$, when it stops at the point $s(2) = 55$, reverses direction, and begins to speed up with an acceleration of $a(2) = -18$ ft/sec^2. At time $t = \frac{7}{2}$ the particle begins to slow down, but continues moving left until time $t = 5$, when it stops at the point $s(5) = 28$, reverses direction again, and begins to speed up with acceleration $a(5) = 18$ ft/sec^2. The particle then continues moving right thereafter with increasing speed.   ◄

REMARK.   The curved line in Figure 4.11.3c is descriptive only. The actual path of the particle is back and forth on the coordinate line.

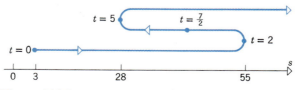

Figure 4.11.3c

► Exercise Set 4.11

1. The graphs of three position functions are shown in Figure 4.11.4. In each case determine the signs of the velocity and acceleration, then determine whether the particle is speeding up or slowing down.

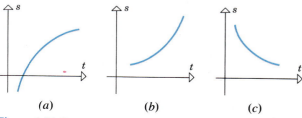

    (a)   (b)   (c)

Figure 4.11.4

2. The graphs of three velocity functions are shown in Figure 4.11.5. In each case determine the sign of the acceleration, then determine whether the particle is speeding up or slowing down.

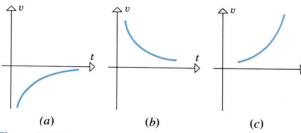

    (a)   (b)   (c)

Figure 4.11.5

3. The position function of a particle moving on a coordinate line is shown in Figure 4.11.6.
    (a) Is the particle moving left or right at time t_0?
    (b) Is the acceleration positive or negative at time t_0?
    (c) Is the particle speeding up or slowing down at time t_0?
    (d) Is the particle speeding up or slowing down at time t_1?

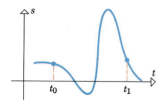

Figure 4.11.6

4. Match the graphs of the position functions shown in Figure 4.11.7 with their velocity functions shown in Figure 4.11.8.

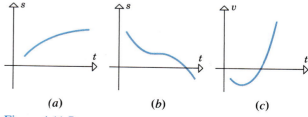

    (a)   (b)   (c)

Figure 4.11.7

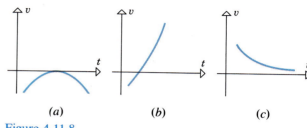

    (a)   (b)   (c)

Figure 4.11.8

5. Figure 4.11.9 shows the velocity versus time graph for a test run of the Grand Prix GTP. Using this graph, estimate
    (a) the acceleration at 60 mi/hr (in units of ft/sec²)
    (b) the time at which the maximum acceleration occurs.

[Data from *Car and Driver*, October 1990.]

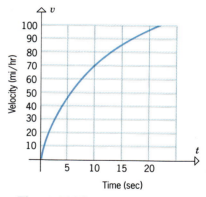

Figure 4.11.9

6. Figure 4.11.10 shows the position versus time graph for an elevator that ascends 40 m from one stop to the next.
   (a) Estimate the velocity when the elevator is halfway up.
   (b) Sketch rough graphs of the velocity versus time curve and the acceleration versus time curve.

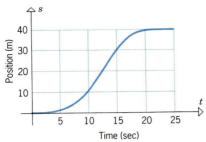

Figure 4.11.10

In Exercises 7–22, s is the position in feet, and t is the time in seconds, for a particle moving on a coordinate line.

7. (a) Let $s(t) = t^3 - 6t^2$. Make a table showing the position, velocity, speed, and acceleration at times $t = 1$, $t = 2$, $t = 3$, $t = 4$, and $t = 5$.
   (b) At each of these times specify the direction of motion, if any, and whether the particle is speeding up, slowing down, or neither.

8. Let $s = \dfrac{100}{t^2 + 12}$ for $t \geq 0$. Find the maximum speed of the particle and the direction of motion of the particle when it has this speed.

9. Let $s = 5t^2 - 22t$.
   (a) Find the maximum speed of the particle during the time interval $1 \leq t \leq 3$.
   (b) When, during the time interval $1 \leq t \leq 3$, is the particle farthest from the origin? What is its position at that instant?

In Exercises 10–15, describe the motion of the particle for $t \geq 0$ (as in Example 3) and make a sketch as in Figure 4.11.3c.

10. $s = -3t + 2$.

11. $s = 1 + 6t - t^2$.

12. $s = t^3 - 6t^2 + 9t + 1$.

13. $s = t^3 - 9t^2 + 24t$.

14. $s = t + \dfrac{9}{t + 1}$.

15. $s = \begin{cases} \cos t, & 0 \leq t \leq 2\pi \\ 1, & t > 2\pi. \end{cases}$

16. Let $s = t^3 - 6t^2 + 1$.
    (a) Find s and v when $a = 0$.
    (b) Find s and a when $v = 0$.

17. Let $s = 4t^{3/2} - 3t^2$ for $t > 0$.
    (a) Find s and v when $a = 0$.
    (b) Find s and a when $v = 0$.

18. Let $s = \sqrt{2t^2 + 1}$. Find $\lim\limits_{t \to +\infty} v$.

19. (a) Use the chain rule to show that for a particle in rectilinear motion $a = v(dv/ds)$.
    (b) Let $s = \sqrt{3t + 7}$, $t \geq 0$. Find a formula for v in terms of s and use the equation in part (a) to find the acceleration when $s = 5$.

20. If $s = t/(t^2 + 5)$ is the position function of a moving particle for $t \geq 0$, at what instant of time will the particle start to reverse its direction of motion, and where is it at that instant?

21. Suppose that the position functions of two particles, P_1 and P_2, in motion along the same line are
    $$s_1 = \tfrac{1}{2}t^2 - t + 3 \text{ and } s_2 = -\tfrac{1}{4}t^2 + t + 1$$
    respectively, for $t \geq 0$.
    (a) Prove that P_1 and P_2 do not collide.
    (b) How close can P_1 and P_2 get to one another?
    (c) During what intervals of time are they moving in opposite directions?

22. Let $s_A = 15t^2 + 10t + 20$ and $s_B = 5t^2 + 40t$, $t \geq 0$, be the position functions of cars A and B that are moving along parallel straight lanes of a highway.
    (a) How far is car A ahead of car B when $t = 0$?
    (b) At what instant of time are the cars next to one another?
    (c) At what instant of time do they have the same velocity? Which car is ahead at this instant?

23. Figure 4.11.11 shows the velocity versus distance graph for a 222 Remington Magnum 55 grain pointed soft point bullet.
    (a) Use the graph to estimate the value of dv/ds when the velocity is 2000 ft/sec.
    (b) Use the result in part (a) and the chain rule to approximate the acceleration when the velocity is 2000 ft/sec. [*Hint:* See Exercise 19.]

[Data from the *Shooter's Bible*, No. 82, Stoeger Publishing Co., 1991.]

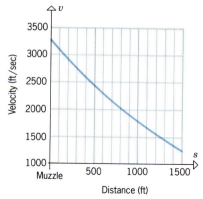

Figure 4.11.11

$$a_{ave} = \frac{\text{change in velocity}}{\text{time elapsed}} = \frac{v(t_1) - v(t_0)}{t_1 - t_0}$$

(a) Interpret v_{ave} geometrically on the graph of $s(t)$.

(b) Interpret a_{ave} geometrically on the graph of $v(t)$.

(c) Find the average velocity and acceleration over the time interval $2 \le t \le 4$ for a particle with position function $s(t) = t^3 - 3t^2$.

25. (a) Show that the average velocity v_{ave} over $[t_0, t_1]$ approaches the instantaneous velocity $v(t_0)$ as t_1 approaches t_0. [See Exercise 24 for definitions of v_{ave} and a_{ave}.]

(b) Show that the average acceleration a_{ave} over $[t_0, t_1]$ approaches the instantaneous acceleration $a(t_0)$ as t_1 approaches t_0.

24. For a particle moving on a coordinate line, the *average velocity* v_{ave} and *average acceleration* a_{ave} over a time interval $[t_0, t_1]$ are defined by

$$v_{ave} = \frac{\text{change in position}}{\text{time elapsed}} = \frac{s(t_1) - s(t_0)}{t_1 - t_0}$$

26. Prove that a particle is speeding up if the velocity and acceleration have the same sign, and slowing down if they have opposite signs. [*Hint:* Let $r(t) = |v(t)| = \sqrt{v^2(t)}$, and find $r'(t)$.]

▶ **SUPPLEMENTARY EXERCISES** [C] 10, 11, 16

1. For the hollow cylinder shown, assume that R and r are increasing at a rate of 2 m/sec, and h is decreasing at a rate of 3 m/sec. At what rate is the volume changing at the instant when $R = 7$ m, $r = 4$ m, and $h = 5$ m?

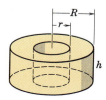

In Exercises 4–9, find the minimum value m and the maximum value M of f on the indicated interval (if they exist) and state where these extreme values occur.

4. $f(x) = 1/x$; $[-2, -1]$.

5. $f(x) = x^3 - x^4$; $[-1, \frac{3}{2}]$.

6. $f(x) = x^2(x - 2)^{1/3}$; $(0, 3]$.

7. $f(x) = 2x/(x^2 + 3)$; $(0, 2]$.

8. $f(x) = 2x^5 - 5x^4 + 7$; $(-1, 3)$.

9. $f(x) = -|x^2 - 2x|$; $[1, 3]$.

10. Use Newton's Method to approximate the smallest positive solution of $\sin x + \cos x = 0$.

11. Use Newton's Method to approximate all three solutions of $x^3 - 4x + 1 = 0$.

2. The vessel shown is filled at the rate of 4 ft³/min. How fast is the fluid level rising at the instant when the level is 1 ft?

3. A ball is dropped from a point 10 ft away from a light at the top of a 48-ft pole as shown. When the ball has dropped 16 ft, its velocity (downward) is 32 ft/sec. At what rate is its shadow moving along the ground at that instant?

In Exercises 12–19, sketch the graph of f. Use symmetry, where possible, and show all relative extrema, inflection points, and asymptotes.

12. $f(x) = (x^2 - 3)^2$.

13. $f(x) = \dfrac{1}{1 + x^2}$.

14. $f(x) = \dfrac{2x}{1 + x}$.

15. $f(x) = \dfrac{x^3 - 2}{x}$.

16. $f(x) = (1 + x)^{2/3}(3 - x)^{1/3}$.

17. $f(x) = 2\cos^2 x$, $0 \le x \le \pi$.

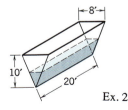

Ex. 2

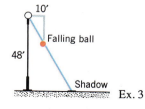

Ex. 3

18. $f(x) = x - \tan x$, $0 \le x \le 2\pi$.

19. $f(x) = \dfrac{3x}{(x + 8)^2}$.

20. Use implicit differentiation to show that a function defined implicitly by $\sin x + \cos y = 2y$ has a critical point wherever $\cos x = 0$. Then use either the first or second derivative test to classify these critical points as relative maxima or minima.

21. Find the equations of the tangent lines at all inflection points of the graph of

$$f(x) = x^4 - 6x^3 + 12x^2 - 8x + 3$$

In Exercises 22–24, find all critical points and use the first derivative test to classify them.

22. $f(x) = x^{1/3}(x - 7)^2$.

23. $f(x) = 2 \sin x - \cos 2x$, $0 \le x \le 2\pi$.

24. $f(x) = 3x - (x - 1)^{3/2}$.

In Exercises 25–27, find all critical points and use the second derivative test (if possible) to classify them.

25. $f(x) = x^{-1/2} + \frac{1}{9}x^{1/2}$. **26.** $f(x) = x^2 + 8/x$.

27. $f(x) = \sin^2 x - \cos x$, $0 \le x \le 2\pi$.

28. Find two nonnegative numbers whose sum is 20 and such that (a) the sum of their squares is a maximum, and (b) the product of the square of one and the cube of the other is a maximum.

29. Find the dimensions of the rectangle of maximum area that can be inscribed inside the ellipse $(x/4)^2 + (y/3)^2 = 1$.

30. Find the coordinates of the point on the curve $2y^2 = 5(x + 1)$ that is nearest to the origin. [*Note:* All points $P(x, y)$ on the curve satisfy $x \ge -1$.]

31. A church window consists of a blue semicircular section surmounting a clear rectangular section as shown. The blue glass lets through half as much light per unit area as the clear glass. Find the radius r of the window that admits the most light if the perimeter of the entire window is to be P feet.

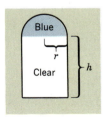

32. The cost c (in dollars per hour) to run an ocean liner at a constant speed v (in miles per hour) is given by $c = a + bv^n$, where a, b, and n are positive constants with $n > 1$. Find the speed needed to make the cheapest 3000-mi run.

33. A soup can in the shape of a right-circular cylinder of radius r and height h is to have a prescribed volume V. The top and bottom are cut from squares as shown. If the shaded corners are wasted, but there is no other waste, find the ratio r/h for the can requiring the least material (including waste).

34. If a calculator factory produces x calculators per day, the total daily cost (in dollars) incurred is $0.25x^2 + 35x + 25$. If they are sold for $50 - \frac{1}{2}x$ dollars each, find the value of x that maximizes the daily profit.

In Exercises 35–37, determine if all hypotheses of Rolle's Theorem are satisfied on the stated interval. If not, state which hypotheses fail; if so, find all values of c guaranteed in the conclusion of the theorem.

35. $f(x) = \sqrt{4 - x^2}$ on $[-2, 2]$.

36. $f(x) = x^{2/3} - 1$ on $[-1, 1]$.

37. $f(x) = \sin(x^2)$ on $[0, \sqrt{\pi}]$.

In Exercises 38–41, determine if all hypotheses of the Mean-Value Theorem are satisfied on the stated interval. If not, state which hypotheses fail; if so, find all values of c guaranteed in the conclusion of the theorem.

38. $f(x) = |x - 1|$ on $[-2, 2]$.

39. $f(x) = \sqrt{x}$ on $[0, 4]$.

40. $f(x) = \dfrac{x + 1}{x - 1}$ on $[2, 3]$.

41. $f(x) = \begin{cases} 3 - x^2, & x \le 1 \\ 2/x, & x > 1 \end{cases}$ on $[0, 2]$.

5
Integration

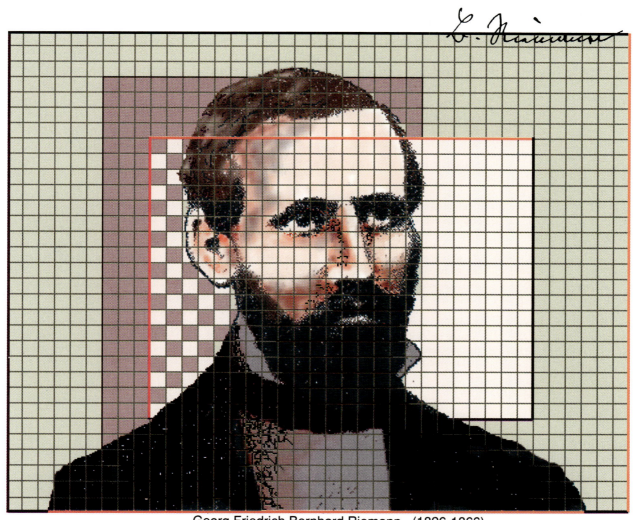

Georg Friedrich Bernhard Riemann   (1826-1866)

■ 5.1   INTRODUCTION

In this chapter we shall study the second major problem of calculus:

> **THE AREA PROBLEM.**   *Given a function f that is continuous and nonnegative on an interval* [a, b], *find the area between the graph of f and the interval* [**a, b**] *on the x-axis* (Figure 5.1.1).

   Area formulas for basic geometric figures such as rectangles, polygons, and circles date back to the earliest written records of mathematics. The first real advance beyond the elementary level of area computation was made by the Greek mathematician, Archimedes,* who devised an ingenious but cumbersome method for obtaining areas, called the *method of exhaustion.* Using this technique Archimedes was able to obtain areas of parabolic regions and spirals. By the early seventeenth century several mathematicians had learned to obtain such areas more simply by calculating appropriate limits. However, both the method of exhaustion and its successor lacked generality. For each different problem one had to devise special procedures or formulas that worked, and more often than not these were difficult or impossible to obtain. The major breakthrough in solving the general area problem was made independently by Newton and Leibniz when they discovered that areas could be obtained by reversing the process of differentiation. This major discovery, which marked the real beginning of calculus, was circulated by Newton in 1669 and then published in 1711 in a paper entitled, *De Analysi per Aequationes Numero Terminorum Infinitas* (*On Analysis by Means of Equations with Infinitely Many Terms*). Independently, Leibniz discovered the same result around 1673 and stated it in an unpublished manuscript dated November 11, 1675.

---

* ARCHIMEDES (287 B.C.–212 B.C.). Greek mathematician and scientist. Born in Syracuse, Sicily, Archimedes was the son of the astronomer Pheidias and possibly related to Heiron II, king of Syracuse. Most of the facts about his life come from the Roman biographer, Plutarch, who inserted a few tantalizing pages about him in the massive biography of the Roman soldier, Marcellus. In the words of one writer, "the account of Archimedes is slipped like a tissue-thin shaving of ham in a bull-choking sandwich."

Archimedes ranks with Newton and Gauss as one of the three greatest mathematicians who ever lived, and he is certainly the greatest mathematician of antiquity. His mathematical work is so modern in spirit and technique that it is barely distinguishable from that of a seventeenth-century mathematician, yet it was all done without benefit of algebra or a convenient number system. Among his mathematical achievements, Archimedes developed a general method (exhaustion) for finding areas and volumes, and he used the method to find areas bounded by parabolas and spirals and to find volumes of cylinders, paraboloids, and segments of spheres. He gave a procedure for approximating π and bounded its value between $3\frac{10}{71}$ and $3\frac{1}{7}$. In spite of the limitations of the Greek numbering system, he devised methods for finding square roots and invented a method based on the Greek myriad (10,000) for representing numbers as large as 1 followed by 80 million billion zeros.

Of all his mathematical work, Archimedes was most proud of his discovery of the method for finding the volume of a sphere—he showed that the volume of a sphere is two-thirds the volume of the smallest cylinder that can contain it. At his request, the figure of a sphere and cylinder was engraved on his tombstone.

In addition to mathematics, Archimedes worked extensively in mechanics and hydrostatics. Nearly every schoolchild knows Archimedes as the absent-minded scientist who, on realizing that

(continued on page 321)

In the remainder of this introductory section, we will discover the fascinating solution of the area problem for ourselves. As with most mathematical discoveries, our initial work will be intuitive; in later sections of this chapter these ideas will be made more precise.

Assume that we are interested in finding the area of the region in Figure 5.1.1. As it turns out, it will actually be easier for us to solve a whole class of area problems simultaneously. Instead of limiting the discussion to the case where the right-hand endpoint of the interval is b, we will allow this endpoint to be any real number x greater than or equal to a (Figure 5.1.2). As indicated in the figure, we will denote the desired area by $A(x)$. We have used the function notation here to emphasize that the area of the region depends on the value of the right-hand endpoint x.

Isaac Newton once said,

> *If I have seen farther than others, it is because I have stood on the shoulders of giants.*

We are fortunate here in being able to stand on the shoulders of two giants, Newton and Leibniz, for it was their idea to find $A(x)$ by first finding the *derivative* of $A(x)$, then using the known derivative $A'(x)$ to determine $A(x)$ itself. Thus, we consider the problem of finding

$$A'(x) = \lim_{h \to 0} \frac{A(x + h) - A(x)}{h} \tag{1}$$

For simplicity, consider the case where $h > 0$. The numerator on the right side of (1) is the difference of two areas: the area between a and $x + h$ minus the area between a and x (Figure 5.1.3a). If we let c be the midpoint between x

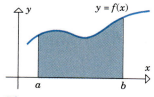

Figure 5.1.1

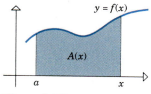

Figure 5.1.2

(*continued from page 320*)

a floating object displaces its weight of liquid, leaped from his bath and ran naked through the streets of Syracuse shouting, "Eureka, Eureka!"—(meaning, "I have found it!"). Archimedes actually created the discipline of hydrostatics and used it to find equilibrium positions for various floating bodies. He laid down the fundamental postulates of mechanics, discovered the laws of levers, and calculated centers of gravity for various flat surfaces and solids. In the excitement of discovering the mathematical laws of the lever, he is said to have declared, "Give me a place to stand and I will move the earth."

Although Archimedes was apparently more interested in pure mathematics than its applications, he was an engineering genius. During the second Punic war, when Syracuse was attacked by the Roman fleet under the command of Marcellus, it was reported by Plutarch that Archimedes' military inventions held the fleet at bay for three years. He invented super catapults that showered the Romans with rocks weighing a quarter ton or more, and fearsome mechanical devices with iron "beaks and claws" that reached over the city walls, grasped the ships, and spun them against the rocks. After the first repulse, Marcellus called Archimedes a "geometrical Briareus (a hundred-armed mythological monster) who uses our ships like cups to ladle water from the sea."

Eventually the Roman army was victorious and contrary to Marcellus' specific orders the 75-year-old Archimedes was killed by a Roman soldier. According to one report of the incident, the soldier cast a shadow across the sand in which Archimedes was working on a mathematical problem. When the annoyed Archimedes yelled, "Don't disturb my circles," the soldier flew into a rage and cut the old man down.

With his death the Greek gift of mathematics passed into oblivion, not to be fully resurrected again until the sixteenth century. Unfortunately, there is no known accurate likeness or statue of this great man.

and $x + h$, then this difference of areas can be approximated by the area of a rectangle with base h and height $f(c)$ (Figure 5.1.3b). Thus,

$$\frac{A(x + h) - A(x)}{h} \approx \frac{f(c) \cdot h}{h} = f(c) \tag{2}$$

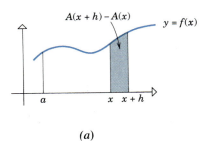
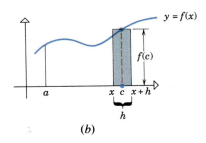

Figure 5.1.3

(a)   *(b)*

It seems plausible from Figure 5.1.3b that the error in approximation (2) will approach zero as $h \to 0$. If we accept this to be so, then it follows from (1) and (2) that

$$A'(x) = \lim_{h \to 0} \frac{A(x + h) - A(x)}{h} = \lim_{h \to 0} f(c) \tag{3}$$

Since c is the midpoint between x and $x + h$, it follows that $c \to x$ as $h \to 0$. But we have assumed f to be a continuous function, so $f(c) \to f(x)$ as $c \to x$. Therefore,

$$\lim_{h \to 0} f(c) = f(x)$$

Thus, it follows from (3) that

$$A'(x) = f(x) \tag{4}$$

This is the result we were looking for; it tells us that the derivative of the area function $A(x)$ is the function whose graph forms the upper boundary of the region in Figure 5.1.2. The following example illustrates how this formula can be used to find the area under a parabola, a problem that cannot be solved without calculus.

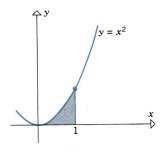

Figure 5.1.4

Example 1   We will use Formula (4) to find the area of the region under the graph of the parabola $y = x^2$ over the interval $[0, 1]$. (See Figure 5.1.4.)

As in the derivation of (4), we will start by considering the more general problem of finding the area $A(x)$ under $y = x^2$ over the interval $[0, x]$. (See Figure 5.1.5.) Since the function whose graph forms the upper boundary of the region is $f(x) = x^2$, it follows from (4) that

$$A'(x) = x^2 \tag{5}$$

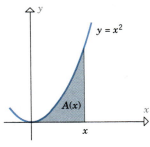

Figure 5.1.5

Thus, to find $A(x)$ we must look for a function whose derivative is x^2. This is called an ***antidifferentiation*** problem because we are trying to find $A(x)$ by "undoing" a differentiation. By simply guessing we see that

$$A(x) = \frac{1}{3}x^3$$

is one solution to (5). But this is not the only solution, since it follows from Theorem 4.10.3 that

$$A(x) = \frac{1}{3}x^3 + C \tag{6}$$

also satisfies (5) for any real value of C. We still have some work to do since this formula involves an unknown constant C which must be determined. This is where our decision to solve the area problem for a general right-hand endpoint helps. If we consider the case where $x = 0$, then the interval $[0, x]$ reduces to a single point. If we agree that the area above a single point should be taken as zero, then it follows on substituting $x = 0$ in (6) that

$$A(0) = 0 + C = 0 \quad \text{or} \quad C = 0$$

so (6) simplifies to

$$A(x) = \frac{1}{3}x^3 \tag{7}$$

which is the formula for the area under $y = x^2$ over the interval $[0, x]$. For the area over the interval $[0, 1]$ we set $x = 1$ in (7); this yields

$$A(1) = \frac{1}{3} \text{ (square units)} \quad \blacktriangleleft$$

We note that our success in finding the area in the foregoing example hinged on our ability to *guess* at a function $A(x)$ satisfying $A'(x) = x^2$. Had we not been able to find such a function, we could not have found the area. One of the goals in subsequent sections is to develop a systematic approach to finding functions from their derivatives, where possible.

We conclude this introductory section by noting that there are numerous mathematical gaps in our work here. For example, we used the term "area" freely, assuming its meaning to be understood and its properties known. Although we calculated the area under a parabola, we never stated precisely what is meant by the area under a parabola or, more generally, by the area under a curve $y = f(x)$. Until the latter part of the nineteenth century, mathematicians found their intuitive concept of area to be perfectly satisfactory. However, in the late 1800s they began to encounter problems that could only be resolved by defining the concept of area precisely and developing its properties in a mathematically rigorous manner. Research in this field has continued to the present day. One of the goals in this chapter is to give more precision to the concept of area.

■ 5.2 ANTIDERIVATIVES; THE INDEFINITE INTEGRAL

In this section we shall develop some basic results that will ultimately help us to obtain systematic procedures for finding a function from its derivative.

5.2.1 DEFINITION. A function F is called an ***antiderivative*** of a function f if $F'(x) = f(x)$ on some interval.

☐ **ANTIDERIVATIVES**

Example 1 The functions

$$\frac{1}{3}x^3, \quad \frac{1}{3}x^3 + 2, \quad \frac{1}{3}x^3 - \pi, \quad \frac{1}{3}x^3 + C \quad (C \text{ any constant})$$

are antiderivatives of $f(x) = x^2$ since the derivative of each is x^2. ◀

This example shows that a function can have many antiderivatives. In fact, if $F(x)$ is any antiderivative of $f(x)$ and C is any constant, then

$$F(x) + C$$

is also an antiderivative of $f(x)$ since

$$\frac{d}{dx}[F(x) + C] = \frac{d}{dx}[F(x)] + \frac{d}{dx}[C] = f(x) + 0 = f(x)$$

It is reasonable to ask if there are other antiderivatives of f that cannot be obtained by adding a constant to F. Provided we consider only values of x in an *interval I*, the answer is *no*. To see this, let $G(x)$ be any other antiderivative of $f(x)$; then

$$\frac{d}{dx}[F(x)] = \frac{d}{dx}[G(x)] = f(x)$$

so that by Theorem 4.10.3, F and G differ only by a constant C on I, that is,

$$G(x) = F(x) + C$$

for x in I. The following theorem summarizes these observations.

5.2.2 THEOREM. *If $F(x)$ is any antiderivative of $f(x)$, then for any value of C, the function $F(x) + C$ is also an antiderivative of $f(x)$; moreover, on any interval, every antiderivative of $f(x)$ is expressible in the form $F(x)$ plus a constant.*

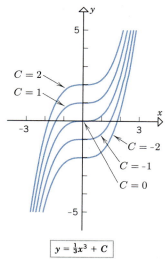

$y = \frac{1}{3}x^3 + C$

Figure 5.2.1

REMARK. We warn the reader against confusing derivatives and antiderivatives. For the function $f(x) = x^2$, the derivative is $f'(x) = 2x$, but the antiderivatives are the functions of the form $F(x) = \frac{1}{3}x^3 + C$. Because any two antiderivatives of a function f differ by a constant on an interval, the graphs of the antiderivatives of a function f over an interval form a family of curves that are vertical translations of one another. For example, Figure 5.2.1 shows some of the antiderivatives of x^2.

Example 2 For all x,

$$\frac{d}{dx}[\sin x] = \cos x$$

so $F(x) = \sin x$ is an antiderivative of $f(x) = \cos x$ on the interval $(-\infty, +\infty)$. Thus, every antiderivative of $\cos x$ on this interval is expressible in the form

$$\sin x + C$$

where C is a constant. ◄

☐ **THE INDEFINITE INTEGRAL**

The process of finding antiderivatives is called *antidifferentiation* or *integration*. If there is some function F such that

$$\frac{d}{dx}[F(x)] = f(x)$$

then the functions of the form $F(x) + C$ are the antiderivatives of $f(x)$. We denote this by writing

$$\int f(x)\, dx = F(x) + C \tag{1}$$

The symbol $\int$ is called an *integral sign*, and $f(x)$ is called the *integrand*. Statement (1) is read, "the *indefinite integral* of $f(x)$ equals $F(x)$ plus C." The adjective "indefinite" is used because the right side of (1) is not a definite function, but rather a whole set of possible functions; the constant C is called the *constant of integration*.

We observed in Example 1 that the antiderivatives of $f(x) = x^2$ are the functions of the form $F(x) = \frac{1}{3}x^3 + C$. Thus,

$$\int x^2\, dx = \frac{1}{3}x^3 + C$$

☐ **SOME MATTERS OF NOTATION**

The symbol dx in the differentiation operation

$$\frac{d}{dx}[\quad]$$

and in the antidifferentiation operation*

$$\int [\quad] \, dx$$

serves to identify the independent variable. If an independent variable other than x is used, the notation must be adjusted appropriately. Thus,

$$\frac{d}{dt}[F(t)] = f(t) \quad \text{and} \quad \int f(t) \, dt = F(t) + C$$

are equivalent statements.

For simplicity, the dx is sometimes absorbed into the integrand. For example,

$$\int 1 \, dx \quad \text{can be written as} \quad \int dx$$

$$\int \frac{1}{x^2} \, dx \quad \text{can be written as} \quad \int \frac{dx}{x^2}$$

 **INTEGRATION FORMULAS**

Every differentiation formula

$$\frac{d}{dx}[f(x)] = f'(x)$$

is equivalent to a corresponding integration formula

$$\int f'(x) \, dx = f(x) + C \tag{2}$$

Example 3

DERIVATIVE FORMULA	EQUIVALENT INTEGRATION FORMULA
$\dfrac{d}{dx}[x^3] = 3x^2$	$\displaystyle\int 3x^2 \, dx = x^3 + C$
$\dfrac{d}{dt}[\tan t] = \sec^2 t$	$\displaystyle\int \sec^2 t \, dt = \tan t + C$
$\dfrac{d}{du}[u^{3/2}] = \dfrac{3}{2} u^{1/2}$	$\displaystyle\int \dfrac{3}{2} u^{1/2} \, du = u^{3/2} + C$

◀

---

* This notation was devised by Leibniz. In his early papers Leibniz used the notation "omn." (an abbreviation for the Latin word "omnes") to denote integration. Then on October 29, 1675 he wrote, "It will be useful to write ∫ for omn., thus ∫ ℓ for omn. ℓ . . . ." Two or three weeks later he refined the notation further and wrote ∫[ ] dx rather than ∫ alone. This notation is so useful and so powerful that its development by Leibniz must be regarded as a major milestone in the history of mathematics and science.

Antidifferentiation is often educated guesswork. By looking only at the derivative of a function we try to guess the function itself. It simplifies this guessing process if we keep in mind that every differentiation formula produces a companion integration formula. For example, the differentiation formula

$$\frac{d}{dx}[\sin x] = \cos x$$

gives rise to the integration formula

$$\int \cos x \, dx = \sin x + C$$

Similarly,

$$\frac{d}{dx}[-\cos x] = \sin x \quad \text{yields} \quad \int \sin x \, dx = -\cos x + C$$

and

$$\frac{d}{dx}\left[\frac{x^{r+1}}{r+1}\right] = x^r \quad \text{yields} \quad \int x^r \, dx = \frac{x^{r+1}}{r+1} + C \quad (r \neq -1)$$

Some basic integration formulas are given in Table 5.2.1.

Table 5.2.1

DIFFERENTIATION FORMULA	INTEGRATION FORMULA
1. $\dfrac{d}{dx}[x] = 1$	$\displaystyle\int dx = x + C$
2. $\dfrac{d}{dx}\left[\dfrac{x^{r+1}}{r+1}\right] = x^r \quad (r \neq -1)$	$\displaystyle\int x^r \, dx = \dfrac{x^{r+1}}{r+1} + C \quad (r \neq -1)$
3. $\dfrac{d}{dx}[\sin x] = \cos x$	$\displaystyle\int \cos x \, dx = \sin x + C$
4. $\dfrac{d}{dx}[-\cos x] = \sin x$	$\displaystyle\int \sin x \, dx = -\cos x + C$
5. $\dfrac{d}{dx}[\tan x] = \sec^2 x$	$\displaystyle\int \sec^2 x \, dx = \tan x + C$
6. $\dfrac{d}{dx}[-\cot x] = \csc^2 x$	$\displaystyle\int \csc^2 x \, dx = -\cot x + C$
7. $\dfrac{d}{dx}[\sec x] = \sec x \tan x$	$\displaystyle\int \sec x \tan x \, dx = \sec x + C$
8. $\dfrac{d}{dx}[-\csc x] = \csc x \cot x$	$\displaystyle\int \csc x \cot x \, dx = -\csc x + C$

Example 4 From the second integration formula in Table 5.2.1 we obtain

$$\int x^2 \, dx = \frac{x^3}{3} + C \qquad \boxed{r = 2}$$

$$\int x^3 \, dx = \frac{x^4}{4} + C \quad \boxed{r = 3}$$

$$\int \frac{1}{x^5} \, dx = \int x^{-5} \, dx = \frac{x^{-5+1}}{-5+1} + C = -\frac{1}{4x^4} + C \quad \boxed{r = -5}$$

$$\int \sqrt{x} \, dx = \int x^{\frac{1}{2}} \, dx = \frac{x^{\frac{1}{2}+1}}{\frac{1}{2}+1} + C = \tfrac{2}{3}x^{\frac{3}{2}} + C \quad \boxed{r = \frac{1}{2}}$$

$$= \tfrac{2}{3}(\sqrt{x})^3 + C \quad \blacktriangleleft$$

□ PROPERTIES OF THE INDEFINITE INTEGRAL

If we differentiate an antiderivative of $f(x)$, we obtain $f(x)$ back again. Thus,

$$\frac{d}{dx}\left[\int f(x) \, dx\right] = f(x) \tag{3}$$

This result is helpful for proving the following basic properties of antiderivatives.

5.2.3 THEOREM.

(a) *A constant factor can be moved through an integral sign; that is,*

$$\int cf(x) \, dx = c \int f(x) \, dx$$

(b) *An antiderivative of a sum is the sum of the antiderivatives; that is,*

$$\int [f(x) + g(x)] \, dx = \int f(x) \, dx + \int g(x) \, dx$$

Proof. To prove (a) we must show that $c\int f(x) \, dx$ is an antiderivative of $cf(x)$, and to prove (b) we must show that $\int f(x) \, dx + \int g(x) \, dx$ is an antiderivative of $f(x) + g(x)$. But, these conclusions follow from (3):

$$\frac{d}{dx}\left[c\int f(x) \, dx\right] = c\frac{d}{dx}\left[\int f(x) \, dx\right] = cf(x)$$

$$\frac{d}{dx}\left[\int f(x) \, dx + \int g(x) \, dx\right] = \frac{d}{dx}\left[\int f(x) \, dx\right] + \frac{d}{dx}\left[\int g(x) \, dx\right]$$

$$= f(x) + g(x) \quad \blacksquare$$

Example 5 Evaluate

(a) $\displaystyle\int 4\cos x \, dx$ (b) $\displaystyle\int (x + x^2) \, dx$

Solution (a).

$$\int 4 \cos x \, dx = 4 \int \cos x \, dx = 4 (\sin x + C) = 4 \sin x + 4C$$

Theorem 5.2.3(a)   Table 5.2.1

Since C is an arbitrary constant, so is $4C$. However, no purpose is served by keeping the arbitrary constant in this form, so we will replace it by a single letter, say $K = 4C$, and write

$$\int 4 \cos x \, dx = 4 \sin x + K$$

Solution (b).

$$\int (x + x^2) \, dx = \int x \, dx + \int x^2 \, dx \qquad \text{Theorem 5.2.3(b)}$$

$$= \left[\frac{x^2}{2} + C_1 \right] + \left[\frac{x^3}{3} + C_2 \right] \qquad \text{Table 5.2.1}$$

$$= \frac{x^2}{2} + \frac{x^3}{3} + C_1 + C_2$$

Since C_1 and C_2 are arbitrary constants, so is $C_1 + C_2$. If we denote this arbitrary constant by the single letter C, we obtain

$$\int (x + x^2) \, dx = \frac{x^2}{2} + \frac{x^3}{3} + C \qquad \blacktriangleleft$$

Part (b) of Theorem 5.2.3 can be extended to more than two functions. More precisely,

$$\int [f_1(x) + f_2(x) + \cdots + f_n(x)] \, dx$$

$$= \int f_1(x) \, dx + \int f_2(x) \, dx + \cdots + \int f_n(x) \, dx$$

In addition, we have left it as an exercise to show that

$$\int [f(x) - g(x)] \, dx = \int f(x) \, dx - \int g(x) \, dx$$

Example 6

$$\int (3x^6 - 2x^2 + 7x + 1) \, dx = 3 \int x^6 \, dx - 2 \int x^2 \, dx + 7 \int x \, dx + \int 1 \, dx$$

$$= \frac{3x^7}{7} - \frac{2x^3}{3} + \frac{7x^2}{2} + x + C \qquad \blacktriangleleft$$

Sometimes an integrand must be rewritten in a different form before the integration can be performed.

Example 7   Evaluate $\int \dfrac{\cos x}{\sin^2 x}\, dx$.

Solution.

$$\int \frac{\cos x}{\sin^2 x}\, dx = \int \frac{1}{\sin x}\frac{\cos x}{\sin x}\, dx = \int \csc x \cot x\, dx = -\csc x + C \; .$$

Formula 8 in Table 5.2.1   ◀

Example 8   Evaluate $\int \dfrac{t^2 - 2t^4}{t^4}\, dt$.

Solution.

$$\int \frac{t^2 - 2t^4}{t^4}\, dt = \int \left(\frac{1}{t^2} - 2\right) dt = \int (t^{-2} - 2)\, dt$$

$$= \frac{t^{-1}}{-1} - 2t + C = -\frac{1}{t} - 2t + C \quad ◀$$

In many problems one is interested in finding a function whose derivative satisfies specified conditions. The following example illustrates a geometric problem of this type.

Example 9   Suppose that a point moves along some unknown curve $y = f(x)$ in the xy-plane in such a way that at each point (x, y) on the curve, the tangent line has slope x^2. Find an equation for the curve given that it passes through the point $(2, 1)$.

Solution.   We know that $dy/dx = x^2$, so

$$y = \int x^2\, dx = \frac{1}{3} x^3 + C \tag{4}$$

Since the curve passes through $(2, 1)$, a specific value for C can be found by using the fact that $y = 1$ if $x = 2$. Substituting these values in (4) yields

$$1 = \frac{1}{3}(2^3) + C \quad \text{or} \quad C = -\frac{5}{3}$$

so the curve is $y = \frac{1}{3}x^3 - \frac{5}{3}$.   ◀

REMARK.   Had we not specified that the curve in the foregoing example pass through a specific point [namely $(2, 1)$], then any curve of the form $y = \frac{1}{3}x^3 + C$ would have been a possible solution. The condition that the curve pass through a specific point singled out a unique curve from the family of curves suggested in Figure 5.2.1.

▶ Exercise Set 5.2

In Exercises 1–30, evaluate the integrals and check your results by differentiating the answers.

1. $\int x^8 \, dx.$

2. $\int \frac{1}{x^6} \, dx.$

3. $\int x^{5/7} \, dx.$

4. $\int \sqrt[3]{x^2} \, dx.$

5. $\int \frac{4}{\sqrt{t}} \, dt.$

6. $\int \frac{1}{2x^3} \, dx.$

7. $\int x^3 \sqrt{x} \, dx.$

8. $\int (u^3 - 2u + 7) \, du.$

9. $\int (x^{-3} + \sqrt{x} - 3x^{1/4} + x^2) \, dx.$

10. $\int (x^{2/3} - 4x^{-1/5} + 4) \, dx.$

11. $\int \left(\frac{7}{y^{3/4}} - \sqrt[3]{y} + 4\sqrt{y} \right) dy.$

12. $\int (2 + y^2)^2 \, dy.$

13. $\int x(1 + x^3) \, dx.$

14. $\int (1 + x^2)(2 - x) \, dx.$

15. $\int x^{1/3}(2 - x)^2 \, dx.$

16. $\int \frac{1 - 2t^3}{t^3} \, dt.$

17. $\int \frac{x^5 + 2x^2 - 1}{x^4} \, dx.$

18. $\int \left[\frac{1}{t^2} - \cos t \right] dt.$

19. $\int [4 \sin x + 2 \cos x] \, dx.$

20. $\int [4 \sec^2 x + \csc x \cot x] \, dx.$

21. $\int \sec x (\sec x + \tan x) \, dx.$

22. $\int [\sqrt{\theta} - \csc^2 \theta] \, d\theta.$

23. $\int \sec x (\tan x + \cos x) \, dx.$

24. $\int \frac{dy}{\csc y}.$

25. $\int \frac{\sin x}{\cos^2 x} \, dx.$

26. $\int \frac{\sin 2x}{\cos x} \, dx.$

27. $\int [1 + \sin^2 \theta \csc \theta] \, d\theta.$

28. $\int \left[\phi + \frac{2}{\sin^2 \phi} \right] d\phi.$

29. $\int \frac{\cos^3 \theta - 5}{\cos^2 \theta} \, d\theta.$

30. $\int \frac{x^2 \sin x + 2 \sin x}{2 + x^2} \, dx.$

31. Find the antiderivative $F(x)$ of $f(x) = \sqrt[3]{x}$ that satisfies $F(1) = 2$.

32. Find a function f such that $f'(x) + \sin x = 0$ and $f(0) = 2$.

33. Find the general form of a function whose second derivative is $\sqrt{x}$. [*Hint:* Solve the equation $f''(x) = \sqrt{x}$ for $f(x)$ by integrating both sides twice.]

34. Find a function f such that $f''(x) = x + \cos x$ and such that $f(0) = 1$ and $f'(0) = 2$. [*Hint:* Integrate both sides of the equation twice.]

In Exercises 35–37, find an equation of the curve that satisfies the given conditions.

35. At each point (x, y) on the curve the slope is $2x + 1$; the curve passes through the point $(-3, 0)$.

36. At each point (x, y) on the curve the slope equals the square of the distance between the point and the y-axis; the point $(-1, 2)$ is on the curve.

37. At each point (x, y) on the curve $d^2y/dx^2 = 6x$; the line $y = 5 - 3x$ is tangent to the curve at the point where $x = 1$.

38. Suppose that a uniform metal rod 50 cm long is insulated laterally, and the temperatures at the exposed ends are maintained at 25° C and 85° C, respectively. Assume that an x-axis is chosen as in Figure 5.2.2 and that the temperature $T(x)$ at each point x satisfies the equation

$$\frac{d^2 T}{dx^2} = 0$$

Find $T(x)$ for $0 \le x \le 50$.

Figure 5.2.2

In Exercises 39–42, find the derivative and state a corresponding integration formula.

39. $\frac{d}{dx} [\sqrt{x^3 + 5}].$

40. $\frac{d}{dx} \left[\frac{x}{x^2 + 3} \right].$

41. $\dfrac{d}{dx} [\sin (2\sqrt{x})].$

42. $\dfrac{d}{dx} [\sin x - x \cos x].$

43. Show that

$$F(x) = \tfrac{1}{6}(3x + 4)^2 \quad \text{and} \quad G(x) = \tfrac{3}{2}x^2 + 4x$$

are both antiderivatives of the same function by
(a) differentiation
(b) using Theorem 5.2.2.

44. Work Exercise 43 using

$$F(x) = \frac{x^2}{x^2 + 5} \quad \text{and} \quad G(x) = -\frac{5}{x^2 + 5}$$

45. Find $f(x)$ if

$$\int f(x) \, dx = 5x^3 - 3x + C$$

46. Find $g(t)$ if

$$\int g(t) \, dt = \frac{1}{\sqrt{4 - t^2}} + C$$

In Exercises 47 and 48, use a trigonometric identity to help evaluate the integral.

47. $\displaystyle\int \tan^2 x \, dx.$ **48.** $\displaystyle\int \cot^2 x \, dx.$

49. Prove:

$$\int [f(x) - g(x)] \, dx = \int f(x) \, dx - \int g(x) \, dx.$$

50. (a) Show that

$$F(x) = \begin{cases} x, & x > 0 \\ -x, & x < 0 \end{cases}$$

and

$$F_1(x) = \begin{cases} x + 2, & x > 0 \\ -x + 3, & x < 0 \end{cases}$$

are both antiderivatives of

$$f(x) = \begin{cases} 1, & x > 0 \\ -1, & x < 0 \end{cases}$$

but that $F_1(x) - F(x)$ is not a constant.
(b) Does this violate Theorem 5.2.2? Explain.

■ 5.3 INTEGRATION BY SUBSTITUTION

*In this section we shall discuss a technique, called **substitution**, which can often be used to transform complicated integration problems into simpler ones.*

□ **u-SUBSTITUTION**

The method of substitution hinges on the following formula in which u stands for a differentiable function of x.

$$\int \left[f(u) \frac{du}{dx} \right] dx = \int f(u) \, du \tag{1}$$

To justify this formula, let F be an antiderivative of f, so that

$$\frac{d}{du} [F(u)] = f(u)$$

or, equivalently,

$$\int f(u) \, du = F(u) + C \tag{2}$$

If u is a differentiable function of x, the chain rule implies that

$$\frac{d}{dx}[F(u)] = \frac{d}{du}[F(u)] \cdot \frac{du}{dx} = f(u)\frac{du}{dx}$$

or, equivalently,

$$\int \left[f(u)\frac{du}{dx} \right] dx = F(u) + C \tag{3}$$

Formula (1) follows from (2) and (3). ▌

The following example illustrates how Formula (1) is used.

Example 1   Evaluate $\int (x^2 + 1)^{50} \cdot 2x \, dx$.

Solution.   If we let $u = x^2 + 1$, then $du/dx = 2x$, so the given integral can be written as

$$\int (x^2 + 1)^{50} \cdot 2x \, dx = \int \left[u^{50}\frac{du}{dx} \right] dx = \int u^{50}\, du$$

$$\underbrace{\qquad}_{\int \left[f(u)\frac{du}{dx} \right] dx} \qquad \underbrace{\qquad}_{\int f(u)\, du}$$

$$= \frac{u^{51}}{51} + C = \frac{(x^2 + 1)^{51}}{51} + C \qquad ◀$$

In general, suppose that we are interested in evaluating

$$\int h(x) \, dx$$

It follows from (1) that if we can express this integral in the form

$$\int h(x) \, dx = \int f(g(x))g'(x) \, dx$$

then the substitution $u = g(x)$ and $du/dx = g'(x)$ will yield

$$\int h(x) \, dx = \int \left[f(u)\frac{du}{dx} \right] dx = \int f(u) \, du$$

With a "good" choice of $u = g(x)$, the integral on the right will be easier to evaluate than the original.

In practice, this substitution process is carried out as follows:

Integration by Substitution

Step 1. Make a choice for u, say $u = g(x)$.

Step 2. Compute $du/dx = g'(x)$.

Step 3. Make the substitution $u = g(x)$, $du = g'(x)\ dx$.

At this stage, the *entire* integral must be in terms of u; no x's should remain. If this is not the case, try a different choice of u.

Step 4. Evaluate the resulting integral.

Step 5. Replace u by $g(x)$, so the final answer is in terms of x.

Example 2 Evaluate $\displaystyle\int \sin^2 x \cos x\ dx$.

Solution. If we let $u = \sin x$, then

$$du/dx = \cos x \quad \text{so} \quad du = \cos x\ dx$$

Thus,

$$\int \sin^2 x \cos x\ dx = \int u^2\ du = \frac{u^3}{3} + C = \frac{\sin^3 x}{3} + C \quad \blacktriangleleft$$

Example 3 Evaluate $\displaystyle\int \frac{\cos \sqrt{x}}{2\sqrt{x}}\ dx$.

Solution. If we let $u = \sqrt{x}$, then

$$\frac{du}{dx} = \frac{1}{2\sqrt{x}} \quad \text{so} \quad du = \frac{1}{2\sqrt{x}}\ dx$$

Thus,

$$\int \frac{\cos \sqrt{x}}{2\sqrt{x}}\ dx = \int \cos u\ du = \sin u + C = \sin \sqrt{x} + C \quad \blacktriangleleft$$

Example 4 Evaluate $\displaystyle\int 3x^2 \sqrt{x^3 + 1}\ dx$.

Solution. If we let

$$u = x^3 + 1$$

then $du/dx = 3x^2$, so $du = 3x^2\, dx$. Thus,

$$\int 3x^2\sqrt{x^3 + 1}\, dx = \int \sqrt{u}\, du = \int u^{1/2}\, du$$

$$= \frac{u^{3/2}}{3/2} + C = \frac{2}{3}(x^3 + 1)^{3/2} + C \qquad \blacktriangleleft$$

Example 5 The easiest substitutions occur when the integrand is the derivative of a known function, except for a constant added to or subtracted from the independent variable. For example,

$$\int \sin (x + 9)\, dx = \int \sin u\, du = -\cos u + C = -\cos (x + 9) + C$$

$$\boxed{\begin{array}{l} u = x + 9 \\ du = 1 \cdot dx = dx \end{array}}$$

$$\int (x - 8)^{23}\, dx = \int u^{23}\, du = \frac{u^{24}}{24} + C = \frac{(x - 8)^{24}}{24} + C \qquad \blacktriangleleft$$

$$\boxed{\begin{array}{l} u = x - 8 \\ du = 1 \cdot dx = dx \end{array}}$$

Another easy u-substitution occurs when the integrand is the derivative of a known function, except for a constant that multiplies or divides the independent variable. The following example illustrates two ways to evaluate such integrals.

Example 6 Evaluate $\displaystyle\int \cos 5x\, dx$.

Solution.

$$\int \cos 5x\, dx = \int (\cos u) \cdot \frac{1}{5}\, du = \frac{1}{5}\int \cos u\, du$$

$$\boxed{\begin{array}{l} u = 5x \\ du = 5\, dx \text{ or } dx = \frac{1}{5}\, du \end{array}}$$

$$= \frac{1}{5}\sin u + C = \frac{1}{5}\sin 5x + C$$

Alternative Solution. There is a variation of the preceding method that some people prefer. The substitution $u = 5x$ requires $du = 5\, dx$. If there were a factor of 5 in the integrand, then we could group the 5 and dx together to form the du required by the substitution. Since there is no factor of 5, we shall insert one and compensate by putting a factor of $1/5$ in front of the integral.

The computations are as follows:

$$\int \cos 5x \; dx = \frac{1}{5} \int \cos 5x \cdot 5 \; dx = \frac{1}{5} \int \cos u \; du$$

$$u = 5x$$
$$du = 5 \; dx$$

$$= \frac{1}{5} \sin u + C = \frac{1}{5} \sin 5x + C \quad \blacktriangleleft$$

Example 7

$$\int \frac{dx}{(\frac{1}{3}x - 8)^5} = \int \frac{3 \; du}{u^5} = 3 \int u^{-5} \; du$$

$$u = \frac{1}{3}x - 8$$
$$du = \frac{1}{3} \; dx \text{ or } dx = 3 \; du$$

$$= -\frac{3}{4} u^{-4} + C = -\frac{3}{4} \left(\frac{1}{3}x - 8\right)^{-4} + C \quad \blacktriangleleft$$

Example 8   With the help of Theorem 5.2.3, a complicated integral can sometimes be computed by expressing it as a sum of simpler integrals. For example,

$$\int (x + \sec^2 \pi x) \; dx = \int x \; dx + \int \sec^2 \pi x \; dx$$

$$= \frac{x^2}{2} + \int \sec^2 \pi x \; dx = \frac{x^2}{2} + \frac{1}{\pi} \int \sec^2 u \; du$$

$$u = \pi x$$
$$du = \pi \; dx \text{ or } dx = \frac{1}{\pi} \; du$$

$$= \frac{x^2}{2} + \frac{1}{\pi} \tan u + C = \frac{x^2}{2} + \frac{1}{\pi} \tan \pi x + C \quad \blacktriangleleft$$

Example 9   Evaluate $\int t^4 \sqrt[3]{3 - 5t^5} \; dt$.

Solution.   After some possible false starts most readers would eventually hit on the following substitution:

$$\int t^4 \sqrt[3]{3 - 5t^5} \; dt = -\frac{1}{25} \int \sqrt[3]{u} \; du = -\frac{1}{25} \int u^{1/3} \; du$$

$$u = 3 - 5t^5$$
$$du = -25t^4 \; dt \text{ or } -\frac{1}{25} \; du = t^4 \; dt$$

$$= -\frac{1}{25} \frac{u^{4/3}}{4/3} + C = -\frac{3}{100} (3 - 5t^5)^{4/3} + C \quad \blacktriangleleft$$

Example 10 Evaluate $\int x^2 \sqrt{x-1}\,dx$.

Solution. Let

$$u = x - 1 \quad \text{so that} \quad du = dx \tag{4}$$

From the first equality in (4)

$$x^2 = (u+1)^2 = u^2 + 2u + 1$$

so that

$$\int x^2 \sqrt{x-1}\,dx = \int (u^2 + 2u + 1)\sqrt{u}\,du$$

$$= \int (u^{5/2} + 2u^{3/2} + u^{1/2})\,du$$

$$= \frac{2}{7}u^{7/2} + \frac{4}{5}u^{5/2} + \frac{2}{3}u^{3/2} + C$$

$$= \frac{2}{7}(x-1)^{7/2} + \frac{4}{5}(x-1)^{5/2} + \frac{2}{3}(x-1)^{3/2} + C \qquad \blacktriangleleft$$

REMARK. Not every function can be integrated in terms of familiar functions using u-substitution. For example, you will not find any u-substitutions that will integrate the following in terms of functions encountered thus far in this text (try):

$$\int \frac{1}{x}\,dx, \quad \int \frac{dx}{\sqrt{1-x^2}}, \quad \int \sin(x^2)\,dx$$

□ **INTEGRATION USING SYMBOLIC ALGEBRA SYSTEMS**

Recent advances in programs called *symbolic manipulators* or *symbolic algebra systems* now make it possible to evaluate many indefinite integrals on computers and calculators. For example, the following integral was evaluated in a matter of seconds using a symbolic manipulator.

$$\int \frac{x}{\sqrt{1+x}}\,dx = -2\sqrt{1+x} + \frac{2}{3}\sqrt{(1+x)^3} + C$$

In spite of the existence of symbolic manipulators, a basic level of competence in the evaluation of indefinite integrals is necessary. Just as one would not want to rely on a calculator to compute $2 + 2$, so one would not want to rely on a symbolic manipulator to integrate $f(x) = x^2$. Moreover, many of the techniques that we will develop for evaluating integrals are applicable to other kinds of mathematical problems.

▶ Exercise Set 5.3

1. Evaluate the integrals by making the indicated substitutions.

(a) $\int 2x(x^2 + 1)^{23} dx; \ u = x^2 + 1$

(b) $\int \cos^3 x \sin x \ dx; \ u = \cos x$

(c) $\int \frac{1}{\sqrt{x}} \sin \sqrt{x} \ dx; \ u = \sqrt{x}$

(d) $\int \frac{3x \ dx}{\sqrt{4x^2 + 5}}; \ u = 4x^2 + 5.$

2. Evaluate the integrals by making the indicated substitutions.

(a) $\int \sec^2 (4x + 1) \ dx; \ u = 4x + 1$

(b) $\int y\sqrt{1 + 2y^2} \ dy; \ u = 1 + 2y^2$

(c) $\int \sqrt{\sin \pi\theta} \cos \pi\theta \ d\theta; \ u = \sin \pi\theta$

(d) $\int (2x + 7)(x^2 + 7x + 3)^{4/5} \ dx;$
$u = x^2 + 7x + 3.$

3. Evaluate the integrals by making the indicated substitutions.

(a) $\int \cot x \csc^2 x \ dx; \ u = \cot x$

(b) $\int (1 + \sin t)^9 \cos t \ dt; \ u = 1 + \sin t$

(c) $\int x^2\sqrt{1 + x} \ dx; \ u = 1 + x$

(d) $\int [\csc (\sin x)]^2 \cos x \ dx; \ u = \sin x.$

In Exercises 4–29, evaluate the integrals.

4. $\int (3x - 1)^5 \ dx.$ **5.** $\int x(2 - x^2)^3 \ dx.$

6. $\int \sin 3x \ dx.$ **7.** $\int \cos 8x \ dx.$

8. $\int \sec^2 5x \ dx.$ **9.** $\int \sec 4x \tan 4x \ dx.$

10. $\int \sqrt{3t + 1} \ dt.$ **11.** $\int t\sqrt{7t^2 + 12} \ dt.$

12. $\int \frac{x}{\sqrt{4 - 5x^2}} \ dx.$ **13.** $\int \frac{x^2}{\sqrt{x^3 + 1}} \ dx.$

14. $\int \frac{1}{(1 - 3x)^2} \ dx.$ **15.** $\int \frac{x}{(4x^2 + 1)^3} \ dx.$

16. $\int x \cos (3x^2) \ dx.$ **17.** $\int \frac{\sin (5/x)}{x^2} \ dx.$

18. $\int \frac{\sec^2 (\sqrt{x})}{\sqrt{x}} \ dx.$ **19.** $\int x^2 \sec^2 (x^3) \ dx.$

20. $\int \cos^3 2t \sin 2t \ dt.$ **21.** $\int \sin^5 3t \cos 3t \ dt.$

22. $\int \frac{\sin 2\theta}{(5 + \cos 2\theta)^3} \ d\theta.$

23. $\int \cos 4\theta \ \sqrt{2 - \sin 4\theta} \ d\theta.$

24. $\int \tan^3 5x \sec^2 5x \ dx.$

25. $\int \sec^3 2x \tan 2x \ dx.$

26. $\int [\sin (\sin \theta)] \cos \theta \ d\theta.$

27. $\int [\sec^2 (\cos 3\theta)] \sin 3\theta \ d\theta.$

28. $\int \sqrt[n]{a + bx} \ dx \quad (b \neq 0).$

29. $\int \sin^n (a + bx) \cos (a + bx) \ dx \quad (n > 0, b \neq 0).$

In Exercises 30–36, evaluate the integrals. These are a little trickier than those in the preceding exercises.

30. $\int (4x^2 - 12x + 9)^{2/3} \ dx.$

31. $\int x\sqrt{x - 3} \ dx.$

32. $\int x^2\sqrt{2 - x} \ dx.$ **33.** $\int \frac{y \ dy}{\sqrt{y + 1}}.$

34. $\int \sin^3 2\theta \ d\theta.$

[*Hint:* Use the identity $\sin^2 x + \cos^2 x = 1.$]

35. $\int \tan^2 3\theta \ d\theta.$

[*Hint:* Use a trigonometric identity.]

36. $\int \sqrt{1 + x^{-2/3}} \ dx \quad (x > 0).$

37. Evaluate the integral in Example 10 by making the substitution $u = \sqrt{x} - 1$. Check your answer against the one obtained in the example.

38. (a) Evaluate the integral $\int \sin x \cos x \, dx$ by two methods: first by letting $u = \sin x$, then by letting $u = \cos x$.
(b) Explain why the two apparently different answers obtained in part (a) are really equivalent.

39. (a) Evaluate $\int (5x - 1)^2 \, dx$ by two methods: first square and integrate, then let $u = 5x - 1$.
(b) Explain why the two apparently different answers obtained in part (a) are really equivalent.

40. Find a function f such that $f'(x) = \sqrt{3x + 1}$ and $f(1) = 5$.

41. Find a function f such that $f'(x) = 6 - 5 \sin 2x$ and $f(0) = 3$.

In Exercises 42–45, express the given integral in terms of the function f. [*Hint:* $\int f'(u) \, du = f(u) + C$.]

42. $\int f'(5x) \, dx$.

43. $\int f'(3x + 2) \, dx$.

44. $\int x f'(3x^2) \, dx$.

45. $\int \frac{1}{x^2} f'(2/x) \, dx$.

5.4 SIGMA NOTATION

*In this section we shall digress from the main theme of this chapter to introduce a notation that can be used to write lengthy sums in a compact form. This material will be helpful when we discuss area in the next section. The notation uses the uppercase Greek letter Σ (sigma) and is called **sigma notation** or **summation notation**.*

☐ **SIGMA NOTATION**

To illustrate how sigma notation works, consider the sum

$$1^2 + 2^2 + 3^2 + 4^2 + 5^2$$

in which each term is of the form k^2, where k is one of the integers from 1 to 5. In sigma notation this sum can be written as

$$\sum_{k=1}^{5} k^2$$

which is read, "the summation of k^2, where k runs from 1 to 5." The notation tells us to form the sum of the terms that result when we substitute successive integers for k in the expression k^2, starting with $k = 1$ and ending with $k = 5$.

More generally, if $f(k)$ is a function of k, and a and b are integers such that $a \le b$, then

$$\sum_{k=a}^{b} f(k) \tag{1}$$

denotes the sum of the terms that result when we substitute successive integers for k, starting with $k = a$ and ending with $k = b$.

Example 1

$$\sum_{k=4}^{8} k^3 = 4^3 + 5^3 + 6^3 + 7^3 + 8^3$$

$$\sum_{k=1}^{5} 2k = 2 \cdot 1 + 2 \cdot 2 + 2 \cdot 3 + 2 \cdot 4 + 2 \cdot 5 = 2 + 4 + 6 + 8 + 10$$

$$\sum_{k=0}^{5} (2k + 1) = 1 + 3 + 5 + 7 + 9 + 11$$

$$\sum_{k=0}^{5} (-1)^k (2k + 1) = 1 - 3 + 5 - 7 + 9 - 11$$

$$\sum_{k=-3}^{1} k^3 = (-3)^3 + (-2)^3 + (-1)^3 + 0^3 + 1^3 = -27 - 8 - 1 + 0 + 1$$

$$\sum_{k=1}^{3} k \sin\left(\frac{k\pi}{5}\right) = \sin\frac{\pi}{5} + 2 \sin\frac{2\pi}{5} + 3 \sin\frac{3\pi}{5} \quad \blacktriangleleft$$

The numbers a and b in (1) are called, respectively, the *lower* and *upper limits of summation;* and the letter k is called the *index of summation.* It is not essential to use k as the index of summation; any letter will do. For example,

$$\sum_{i=1}^{6} \frac{1}{i}, \quad \sum_{j=1}^{6} \frac{1}{j}, \quad \text{and} \quad \sum_{n=1}^{6} \frac{1}{n}$$

all denote the sum

$$1 + \frac{1}{2} + \frac{1}{3} + \frac{1}{4} + \frac{1}{5} + \frac{1}{6}$$

If the upper and lower limits of summation are the same, then the "sum" in (1) reduces to one term. For example,

$$\sum_{k=2}^{2} k^3 = 2^3 \quad \text{and} \quad \sum_{i=1}^{1} \frac{1}{i + 2} = \frac{1}{1 + 2} = \frac{1}{3}$$

In the sums

$$\sum_{i=1}^{5} 2, \quad \sum_{k=3}^{6} 7, \quad \text{and} \quad \sum_{j=0}^{2} x^3$$

the expression to the right of the Σ sign does not involve the index of summation. In such cases, we take all the terms in the sum to be the same, with one term for each allowable value of the summation index. Thus,

$$\sum_{i=1}^{5} 2 = 2 + 2 + 2 + 2 + 2$$

$$\sum_{k=3}^{6} 7 = 7 + 7 + 7 + 7$$

$$\sum_{j=0}^{2} x^3 = x^3 + x^3 + x^3$$

A sum can be written in more than one way with sigma notation by changing the limits of summation. For example, the sum of the first five positive even integers can be written in the following ways:

$$\sum_{k=1}^{5} 2k = 2 + 4 + 6 + 8 + 10$$

$$\sum_{k=0}^{4} (2k + 2) = 2 + 4 + 6 + 8 + 10$$

$$\sum_{k=2}^{6} (2k - 2) = 2 + 4 + 6 + 8 + 10$$

□ **CHANGING THE INDEX OF SUMMATION**

On occasion we shall want to change the sigma notation for a given sum to a sigma notation with different limits of summation. The following example illustrates a method for doing this.

Example 2 Express

$$\sum_{k=3}^{7} 5^{k-2}$$

in sigma notation so that the lower limit of summation is 0 rather than 3.

Solution.   If we define a new summation index j by means of the formula

$$j = k - 3 \tag{2}$$

then j runs from 0 up to 4 as k runs from 3 up to 7. From (2), $k = j + 3$, so

$$\sum_{k=3}^{7} 5^{k-2} = \sum_{j=0}^{4} 5^{(j+3)-2} = \sum_{j=0}^{4} 5^{j+1}$$

As a check, the reader can verify that

$$\sum_{j=0}^{4} 5^{j+1} \quad \text{and} \quad \sum_{k=3}^{7} 5^{k-2}$$

both denote the sum $5 + 5^2 + 5^3 + 5^4 + 5^5$.   ◄

REMARK.   In the solution of Example 2 the summation index was changed from k to j. If it is desirable to keep the same symbol for the summation index, we can change the j back to k *at the very end* and express the final result as

$$\sum_{k=0}^{4} 5^{k+1} \quad \text{instead of} \quad \sum_{j=0}^{4} 5^{j+1}$$

When we want to represent a general sum we shall use letters with subscripts. For example, a general sum with five terms might be written as

$$a_1 + a_2 + a_3 + a_4 + a_5$$

or in sigma notation as

$$\sum_{k=1}^{5} a_k, \quad \sum_{j=1}^{5} a_j, \quad \text{or} \quad \sum_{m=1}^{5} a_m$$

A general sum with n terms might be written as

$$b_1 + b_2 + \cdots + b_n$$

or in sigma notation as

$$\sum_{k=1}^{n} b_k, \quad \sum_{j=1}^{n} b_j, \quad \text{or} \quad \sum_{m=1}^{n} b_m$$

☐ **PROPERTIES OF SIGMA NOTATION**

The following properties of sigma notation will help to manipulate sums.

5.4.1   THEOREM.

(a) $\displaystyle\sum_{k=1}^{n} (a_k + b_k) = \sum_{k=1}^{n} a_k + \sum_{k=1}^{n} b_k$

(b) $\displaystyle\sum_{k=1}^{n} (a_k - b_k) = \sum_{k=1}^{n} a_k - \sum_{k=1}^{n} b_k$

(c) $\displaystyle\sum_{k=1}^{n} ca_k = c \sum_{k=1}^{n} a_k$

We shall prove parts (a) and (c) and leave (b) as an exercise.

Proof (a).

$$\sum_{k=1}^{n} (a_k + b_k) = (a_1 + b_1) + (a_2 + b_2) + \cdots + (a_n + b_n)$$

$$= (a_1 + a_2 + \cdots + a_n) + (b_1 + b_2 + \cdots + b_n)$$

$$= \sum_{k=1}^{n} a_k + \sum_{k=1}^{n} b_k \quad \blacksquare$$

Proof (c).

$$\sum_{k=1}^{n} ca_k = ca_1 + ca_2 + \cdots + ca_n$$

$$= c(a_1 + a_2 + \cdots + a_n) = c \sum_{k=1}^{n} a_k \quad \blacksquare$$

REMARK. Loosely phrased, this theorem states: *Sigma of a sum equals the sum of the sigmas; sigma of a difference equals the difference of the sigmas; and a constant factor can be moved through a sigma sign.*

□ **SUMMATION FORMULAS**   The following formulas will be used in our later work.

5.4.2 THEOREM.

(a) $$\sum_{k=1}^{n} k = 1 + 2 + 3 + \cdots + n = \frac{n(n + 1)}{2}$$

(b) $$\sum_{k=1}^{n} k^2 = 1^2 + 2^2 + 3^2 + \cdots + n^2 = \frac{n(n + 1)(2n + 1)}{6}$$

(c) $$\sum_{k=1}^{n} k^3 = 1^3 + 2^3 + 3^3 + \cdots + n^3 = \left[\frac{n(n + 1)}{2}\right]^2$$

We shall prove parts (a) and (b) and leave part (c) as an exercise.

Proof (a).   If we write the terms of

$$\sum_{k=1}^{n} k = 1 + 2 + 3 + \cdots + (n - 2) + (n - 1) + n \tag{3}$$

in the opposite order, we obtain

$$\sum_{k=1}^{n} k = n + (n - 1) + (n - 2) + \cdots + 3 + 2 + 1 \tag{4}$$

Adding (3) and (4) term by term yields

$$2 \sum_{k=1}^{n} k = \underbrace{(n + 1) + (n + 1) + (n + 1) + \cdots + (n + 1)}_{n \text{ terms}} = n(n + 1)$$

Thus,

$$\sum_{k=1}^{n} k = \frac{n(n + 1)}{2} \quad \blacksquare$$

Proof (b).   This proof begins with a trick. Since

$$(k + 1)^3 - k^3 = k^3 + 3k^2 + 3k + 1 - k^3 = 3k^2 + 3k + 1$$

we obtain

$$\sum_{k=1}^{n} [(k + 1)^3 - k^3] = \sum_{k=1}^{n} (3k^2 + 3k + 1) \tag{5}$$

Writing out the left side of (5) yields

$$[2^3 - 1^3] + [3^3 - 2^3] + [4^3 - 3^3] + \cdots + [(n + 1)^3 - n^3] \tag{6}$$

Observe that each term in (6) cancels part of the next term, so that the entire sum collapses like a folding telescope, leaving only $-1^3 + (n + 1)^3$. Thus, (5) can be rewritten as

$$-1 + (n + 1)^3 = \sum_{k=1}^{n} (3k^2 + 3k + 1) \tag{7}$$

or, from Theorem 5.4.1,

$$-1 + (n + 1)^3 = 3 \sum_{k=1}^{n} k^2 + 3 \sum_{k=1}^{n} k + \sum_{k=1}^{n} 1 \tag{8}$$

But

$$\sum_{k=1}^{n} 1 = \underbrace{1 + 1 + \cdots + 1}_{n \text{ terms}} = n$$

and by part (a) of this theorem

$$\sum_{k=1}^{n} k = \frac{n(n + 1)}{2}$$

Thus, (8) can be written as

$$-1 + (n + 1)^3 = 3 \sum_{k=1}^{n} k^2 + 3 \frac{n(n + 1)}{2} + n$$

Therefore,

$$\sum_{k=1}^{n} k^2 = \frac{1}{3} \left[(n + 1)^3 - 3 \frac{n(n + 1)}{2} - (n + 1) \right]$$

$$= \frac{n + 1}{6} [2(n + 1)^2 - 3n - 2]$$

$$= \frac{n + 1}{6} (2n^2 + n) = \frac{n(n + 1)(2n + 1)}{6} \qquad \blacksquare$$

Example 3   Evaluate $\sum_{k=1}^{30} k(k + 1)$.

Solution.

$$\sum_{k=1}^{30} k(k + 1) = \sum_{k=1}^{30} (k^2 + k) = \sum_{k=1}^{30} k^2 + \sum_{k=1}^{30} k$$

$$= \frac{30(31)(61)}{6} + \frac{30(31)}{2} = 9920 \qquad \boxed{\text{Theorem 5.4.2}(a),(b)} \qquad \blacktriangleleft$$

REMARK. In formulas such as

$$\sum_{k=1}^{n} k^2 = \frac{n(n + 1)(2n + 1)}{6}$$

or

$$1^2 + 2^2 + \cdots + n^2 = \frac{n(n + 1)(2n + 1)}{6}$$

the left side of the equality is said to express the sum in **open form** and the right side is said to express it in **closed form;** the open form just indicates the terms to be added, while the closed form gives their sum.

Example 4   Express $\sum_{k=1}^{n} (3 + k)^2$ in closed form.

Solution.

$$\sum_{k=1}^{n} (3 + k)^2 = \sum_{k=1}^{n} (9 + 6k + k^2) = \sum_{k=1}^{n} 9 + 6 \sum_{k=1}^{n} k + \sum_{k=1}^{n} k^2$$

$$= 9n + 6 \frac{n(n + 1)}{2} + \frac{n(n + 1)(2n + 1)}{6}$$

$$= \frac{1}{3} n^3 + \frac{7}{2} n^2 + \frac{73}{6} n \quad \blacktriangleleft$$

▶ Exercise Set 5.4

1. Evaluate

(a) $\sum_{k=1}^{3} k^3$

(b) $\sum_{j=2}^{6} (3j - 1)$

(c) $\sum_{i=-4}^{1} (i^2 - i)$

(d) $\sum_{n=0}^{5} 1.$

2. Evaluate

(a) $\sum_{k=1}^{4} k \sin \frac{k\pi}{2}$

(b) $\sum_{j=0}^{5} (-1)^j$

(c) $\sum_{i=7}^{20} \pi$

(d) $\sum_{m=3}^{5} 2^{m+1}.$

In Exercises 3–16, express in sigma notation, but do not evaluate.

3. $1 + 2 + 3 + \cdots + 10.$

4. $3 \cdot 1 + 3 \cdot 2 + 3 \cdot 3 + \cdots + 3 \cdot 20.$

5. $1 \cdot 2 + 2 \cdot 3 + 3 \cdot 4 + \cdots + 49 \cdot 50.$

6. $1 + 2 + 2^2 + 2^3 + 2^4.$

7. $2 + 4 + 6 + 8 + \cdots + 20.$

8. $1 + 3 + 5 + 7 + \cdots + 15.$

9. $1 - 3 + 5 - 7 + 9 - 11.$

10. $1 - \frac{1}{2} + \frac{1}{3} - \frac{1}{4} + \frac{1}{5}.$

11. $-1 + \frac{1}{2} - \frac{1}{3} + \frac{1}{4} - \frac{1}{5}.$

12. $1 + \cos \frac{\pi}{7} + \cos \frac{2\pi}{7} + \cos \frac{3\pi}{7}.$

13. $\sin \frac{\pi}{8} + \sin \frac{3\pi}{8} + \sin \frac{5\pi}{8} + \sin \frac{7\pi}{8}.$

14. $2 + 4 + 8 + 16 + 32.$

15. $\frac{1}{2} + \frac{2}{3} + \frac{3}{4} + \frac{4}{5} + \frac{5}{6}.$

16. $15 + 24 + 35 + \cdots + (n^2 - 1).$

17. Express in sigma notation.

(a) $a_1 - a_2 + a_3 - a_4 + a_5$

(b) $-b_0 + b_1 - b_2 + b_3 - b_4 + b_5$

(c) $a_0 + a_1 x + a_2 x^2 + \cdots + a_n x^n$

(d) $a^5 + a^4 b + a^3 b^2 + a^2 b^3 + ab^4 + b^5.$

In Exercises 18–25, use Theorem 5.4.2 to evaluate the sums.

18. $\displaystyle\sum_{k=1}^{100} k.$ **19.** $\displaystyle\sum_{k=3}^{100} k.$ **20.** $\displaystyle\sum_{k=1}^{100} (7k + 1).$

21. $\displaystyle\sum_{k=1}^{20} k^2.$ **22.** $\displaystyle\sum_{k=4}^{20} k^2.$ **23.** $\displaystyle\sum_{k=1}^{6} (4k^3 - 2k + 1).$

24. $\displaystyle\sum_{k=1}^{6} (k - k^3).$ **25.** $\displaystyle\sum_{k=1}^{30} k(k - 2)(k + 2).$

In Exercises 26–30, express the sum in closed form.

26. (a) $\displaystyle\sum_{k=1}^{n} (4k - 3)$ (b) $\displaystyle\sum_{k=1}^{n-1} k^2.$

27. $\displaystyle\sum_{k=1}^{n} \frac{3k}{n}.$ **28.** $\displaystyle\sum_{k=1}^{n-1} \frac{k^2}{n}.$

29. $\displaystyle\sum_{k=1}^{n-1} \frac{k^3}{n^2}.$ **30.** $\displaystyle\sum_{k=1}^{n} \left(\frac{5}{n} - \frac{2k}{n}\right).$

In Exercises 31–35, the limit of a function of n is given. Express the function of n in closed form, then find the limit. [*Note:* Although n assumes only integer values, the limits can be calculated using the same techniques that we have been using for functions of a real-valued variable x. Functions of integer-valued variables will be studied in more detail later.]

31. $\displaystyle\lim_{n \to +\infty} \frac{1 + 2 + 3 + \cdots + n}{n^2}.$

32. $\displaystyle\lim_{n \to +\infty} \frac{1^2 + 2^2 + 3^2 + \cdots + n^2}{n^3}.$

33. $\displaystyle\lim_{n \to +\infty} \sum_{k=1}^{n} \frac{5k}{n^2}.$ **34.** $\displaystyle\lim_{n \to +\infty} \sum_{k=1}^{n-1} \frac{2k^2}{n^3}.$

35. $\displaystyle\lim_{n \to +\infty} \sum_{k=1}^{n-1} \left(\frac{9}{n} - \frac{k}{n^2}\right).$

36. Show that
$$1 \cdot 2 + 2 \cdot 3 + \cdots + n(n + 1) = \tfrac{1}{3}n(n + 1)(n + 2)$$

37. Show that the sum of the first n consecutive positive odd integers is n^2.

When each term of a sum cancels part of the next term, leaving only portions of the first and last terms at the end, the sum is said to *telescope*. In Exercises 38–43, evaluate the telescoping sum.

38. $\displaystyle\sum_{k=1}^{50} \left(\frac{1}{k} - \frac{1}{k + 1}\right).$ **39.** $\displaystyle\sum_{k=5}^{17} (3^k - 3^{k-1}).$

40. $\displaystyle\sum_{k=1}^{100} (2^{k+1} - 2^k).$ **41.** $\displaystyle\sum_{k=2}^{20} \left(\frac{1}{k^2} - \frac{1}{(k - 1)^2}\right).$

42. $\displaystyle\sum_{k=1}^{n} (a_k - a_{k+1}).$ **43.** $\displaystyle\sum_{k=1}^{n} (a_k - a_{k-1}).$

44. (a) Show that
$$\frac{1}{1 \cdot 2} + \frac{1}{2 \cdot 3} + \frac{1}{3 \cdot 4} + \cdots + \frac{1}{n(n + 1)} = \frac{n}{n + 1}$$

$$\left[\textit{Hint: } \frac{1}{n(n + 1)} = \frac{1}{n} - \frac{1}{n + 1}.\right]$$

(b) Use the result in part (a) to find
$$\lim_{n \to +\infty} \sum_{k=1}^{n} \frac{1}{k(k + 1)}$$

45. (a) Show that
$$\frac{1}{1 \cdot 3} + \frac{1}{3 \cdot 5} + \cdots + \frac{1}{(2n - 1)(2n + 1)} = \frac{n}{2n + 1}$$

$$\left[\textit{Hint: } \frac{1}{(2n - 1)(2n + 1)} = \frac{1}{2}\left(\frac{1}{2n - 1} - \frac{1}{2n + 1}\right).\right]$$

(b) Use the result in part (a) to find
$$\lim_{n \to +\infty} \sum_{k=1}^{n} \frac{1}{(2k - 1)(2k + 1)}$$

46. Evaluate

(a) $\displaystyle\sum_{j=0}^{m} m$ (b) $\displaystyle\sum_{n=4}^{4} 5$

(c) $\displaystyle\sum_{k=1}^{n} x$ (d) $\displaystyle\sum_{i=1}^{n} i^2 c.$

47. Evaluate

(a) $\displaystyle\sum_{k=1}^{n} n$ (b) $\displaystyle\sum_{i=0}^{0} (-3)$

(c) $\displaystyle\sum_{k=1}^{n} kx$ (d) $\displaystyle\sum_{k=m}^{n} c \quad (n \geq m).$

48. Express $1 + 2 + 2^2 + 2^3 + 2^4 + 2^5$ in sigma notation with
(a) $j = 0$ as the lower limit of summation
(b) $j = 1$ as the lower limit of summation
(c) $j = 2$ as the lower limit of summation.

49. Express
$$\sum_{k=4}^{18} k(k - 3)$$

in sigma notation with
(a) $k = 0$ as the lower limit of summation
(b) $k = 5$ as the lower limit of summation.

50. Express

$$\sum_{k=5}^{9} k2^{k+4}$$

in sigma notation with
(a) $k = 1$ as the lower limit of summation
(b) $k = 13$ as the upper limit of summation.

51. Simplify

$$\sum_{k=11}^{28} (k - 10) \sin\left(\frac{\pi}{k - 10}\right)$$

by changing the limits of summation.

52. Which of the following are valid identities?

(a) $\sum_{i=1}^{n} a_i b_i = \sum_{i=1}^{n} a_i \sum_{i=1}^{n} b_i$

(b) $\sum_{i=1}^{n} \frac{a_i}{b_i} = \sum_{i=1}^{n} a_i \bigg/ \sum_{i=1}^{n} b_i$

(c) $\sum_{i=1}^{n} a_i^2 = \left(\sum_{i=1}^{n} a_i\right)^2.$

53. By writing out the sums, determine whether the following are valid identities.

(a) $\int\left[\sum_{i=1}^{n} f_i(x)\right] dx = \sum_{i=1}^{n}\left[\int f_i(x)\, dx\right]$

(b) $\frac{d}{dx}\left[\sum_{i=1}^{n} f_i(x)\right] = \sum_{i=1}^{n}\left[\frac{d}{dx}[f_i(x)]\right].$

54. Let

$$S = \sum_{k=0}^{n} ar^k$$

Show that $S - rS = a - ar^{n+1}$ and hence that

$$\sum_{k=0}^{n} ar^k = \frac{a - ar^{n+1}}{1 - r} \quad (r \neq 1)$$

(A sum of this form is called a **geometric sum**.)

55. Use Exercise 54 to evaluate

(a) $\sum_{k=1}^{20} 3^k$ (b) $\sum_{k=5}^{30} 2^k$ (c) $\sum_{k=0}^{100} (-1)^{k+1}\frac{1}{2^k}.$

56. Use Exercise 54 to express $\sum_{k=1}^{n} \sin^k \theta$ in closed form.

57. Evaluate

$$\sum_{i=1}^{4}\left(\sum_{j=1}^{5} (i + j)\right)$$

58. Let $\bar{x}$ denote the arithmetic average of the n numbers $x_1, x_2, \ldots, x_n$. Use Theorem 5.4.1 to prove that

$$\sum_{i=1}^{n} (x_i - \bar{x}) = 0$$

59. Prove part (b) of Theorem 5.4.1.

60. Prove part (c) of Theorem 5.4.2. [*Hint:* Begin with the difference $(k + 1)^4 - k^4$ and follow the steps used to prove part (b) of the theorem.]

5.5 AREAS AS LIMITS

Art Matrix / Cornell National Supercomputer Facility

Figure 5.5.1

So far we have used the term "area" freely, assuming its meaning is to be self-evident. However, to work with areas mathematically we must ultimately define the concept of area precisely. It is natural to hope that we might be able to define the area of an arbitrary region in the plane. Surprisingly, there is no way to do this. It became clear in the late nineteenth century that there are regions in the plane of such complexity that any attempt to assign them areas in the traditional sense would ultimately lead to mathematical inconsistencies (the blue region in Figure 5.5.1, for example). Although we shall not be concerned with such regions, it is important to know that they exist. In this section we shall show how to compute areas using limits. Although we shall continue to use our intuitive concept of area in this section, the ideas we develop here will form the basis for a precise definition of area in the next section.

5.5.1 PROBLEM. *Find the area of a region R bounded below by the x-axis, on the sides by the lines x = a and x = b, and above by a curve y = f(x), where f is continuous on [a, b] and f(x) ≥ 0 for all x in [a, b] (Figure 5.5.2).*

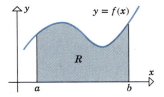

Figure 5.5.2

AREAS AS LIMITS USING INSCRIBED RECTANGLES

In Section 5.1 we showed how to obtain the area of this kind of region using antiderivatives. We shall now begin to lay the groundwork for the proof of that result by showing how the area of a region R like that in Figure 5.5.2 can be obtained as a limit. Choose an arbitrary positive integer n and divide the interval $[a, b]$ into n subintervals of width $(b - a)/n$ by introducing points

$$x_1, x_2, \ldots, x_{n-1}$$

equally spaced between a and b (Figure 5.5.3).

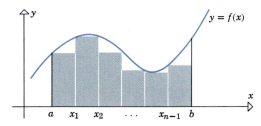

Figure 5.5.3

Next, draw vertical lines through the points a, x_1, x_2, ..., x_{n-1}, b to divide the region R into n strips of uniform width. If we approximate each of these strips by a rectangle *inscribed* under the curve $y = f(x)$ (Figure 5.5.4), then the union of these rectangles will form a region R_n, which we can view as an approximation to the entire region R. The area of this approximating region can be calculated by adding the areas of its component rectangles. Moreover, if we allow n to increase, the widths of the rectangles will get smaller, so that

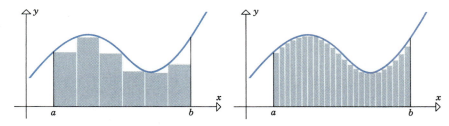

Figure 5.5.4

the approximation of R by R_n will get better as the smaller rectangles fill in more of the gaps under the curve (Figure 5.5.4). Thus, we can consider the *exact* area of R to be the limit of the areas of the approximating regions as n goes to plus infinity; that is,

$$A = \text{area } (R) = \lim_{n \to +\infty} [\text{area } (R_n)] \tag{1}$$

REMARK. There is a difference between writing $\lim_{n \to +\infty}$ and writing $\lim_{x \to +\infty}$, where n represents a positive integer, and x has no such restriction. Later we will study limits of the type $\lim_{n \to +\infty}$ in detail, but for now let it suffice to say that the computational techniques we have used for limits of the type $\lim_{x \to +\infty}$ will also work for $\lim_{n \to +\infty}$.

For computational purposes, (1) can be written in a more useful form. If we denote the heights of the inscribed rectangles by $h_1, h_2, \ldots, h_n$ and use the fact that each rectangle has a base of length $(b - a)/n$, then

$$\text{area } (R_n) = h_1 \cdot \frac{b - a}{n} + h_2 \cdot \frac{b - a}{n} + \cdots + h_n \cdot \frac{b - a}{n} \tag{2}$$

Because f is assumed to be continuous on $[a, b]$, it follows from the Extreme-Value Theorem that f assumes a minimum value on each of the n closed subintervals

$$[a, x_1], [x_1, x_2], \ldots, [x_{n-1}, b]$$

If these minimum values occur at the points $c_1, c_2, \ldots, c_n$, then the heights of the inscribed rectangles are

$$h_1 = f(c_1), h_2 = f(c_2), \ldots, h_n = f(c_n)$$

(Figure 5.5.5), so (2) can be written as

$$\text{area } (R_n) = f(c_1) \cdot \frac{b - a}{n} + f(c_2) \cdot \frac{b - a}{n} + \cdots + f(c_n) \cdot \frac{b - a}{n} \tag{3}$$

Finally, it is customary to use the notation

$$\Delta x = \frac{b - a}{n}$$

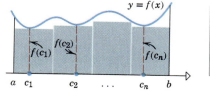

The height of each inscribed rectangle is the minimum value of $f(x)$ on the subinterval.

Figure 5.5.5

for the base dimension of the rectangles, so that (3) becomes

$$\text{area } (R_n) = f(c_1) \cdot \Delta x + f(c_2) \cdot \Delta x + \cdots + f(c_n) \cdot \Delta x$$

or, in sigma notation,

$$\text{area } (R_n) = \sum_{k=1}^{n} f(c_k) \cdot \Delta x$$

With this notation (1) becomes

$$A = \lim_{n \to +\infty} \sum_{k=1}^{n} f(c_k) \cdot \Delta x \tag{4}$$

Example 1   Use inscribed rectangles to find the area under the line $y = x$ over the interval $[1, 2]$.

Solution.   If we subdivide the interval $[1, 2]$ into n equal parts, then each part will have length

$$\Delta x = \frac{b - a}{n} = \frac{2 - 1}{n} = \frac{1}{n}$$

and, as illustrated in Figure 5.5.6, the points of subdivision will be

$$x_1 = 1 + \Delta x = 1 + \frac{1}{n}$$

$$x_2 = 1 + 2\,\Delta x = 1 + \frac{2}{n}$$

$$x_3 = 1 + 3\,\Delta x = 1 + \frac{3}{n}$$

$$\vdots$$

$$x_{n-1} = 1 + (n - 1)\,\Delta x = 1 + \frac{n - 1}{n}$$

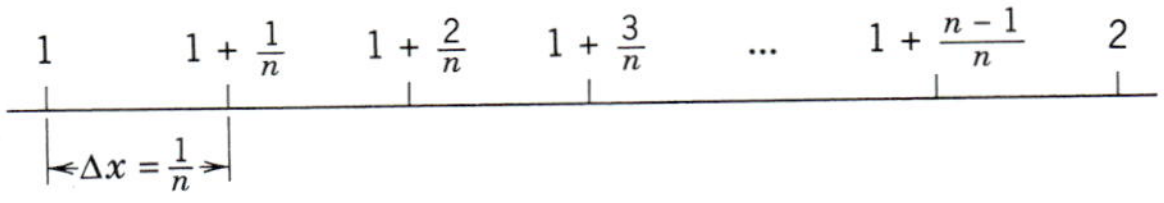

Figure 5.5.6

Because $y = f(x) = x$ is increasing, the minimum value for $f(x)$ on each sub-interval occurs at the left endpoint (Figure 5.5.7), so

$$c_1 = 1$$

$$c_2 = x_1 = 1 + \frac{1}{n}$$

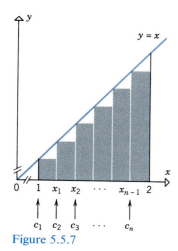

Figure 5.5.7

$$c_3 = x_2 = 1 + \frac{2}{n}$$

$$c_4 = x_3 = 1 + \frac{3}{n}$$

$$\vdots$$

$$c_n = x_{n-1} = 1 + \frac{n-1}{n}$$

Thus, the inscribed rectangles have areas

$$f(c_1) \cdot \Delta x = c_1 \cdot \Delta x = 1 \cdot \frac{1}{n}$$

$$f(c_2) \cdot \Delta x = c_2 \cdot \Delta x = \left(1 + \frac{1}{n}\right) \cdot \frac{1}{n}$$

$$f(c_3) \cdot \Delta x = c_3 \cdot \Delta x = \left(1 + \frac{2}{n}\right) \cdot \frac{1}{n}$$

$$f(c_4) \cdot \Delta x = c_4 \cdot \Delta x = \left(1 + \frac{3}{n}\right) \cdot \frac{1}{n}$$

$$\vdots$$

$$f(c_n) \cdot \Delta x = c_n \cdot \Delta x = \left(1 + \frac{n-1}{n}\right) \cdot \frac{1}{n}$$

and the sum is

$$\sum_{k=1}^{n} f(c_k) \cdot \Delta x$$

$$= \left[1 + \left(1 + \frac{1}{n}\right) + \left(1 + \frac{2}{n}\right) + \left(1 + \frac{3}{n}\right) + \cdots + \left(1 + \frac{n-1}{n}\right)\right] \cdot \frac{1}{n}$$

$$= \left[n + \left(\frac{1}{n} + \frac{2}{n} + \frac{3}{n} + \cdots + \frac{n-1}{n}\right)\right] \cdot \frac{1}{n}$$

$$= 1 + \frac{1}{n^2}[1 + 2 + 3 + \cdots + (n-1)]$$

$$= 1 + \frac{1}{n^2} \cdot \frac{(n-1)(n)}{2} \qquad \boxed{\text{Theorem 5.4.2}(a) \text{ with } n-1 \text{ substituted for } n}$$

$$= \frac{3}{2} - \frac{1}{2n}$$

Thus, from (4) the area is

$$A = \lim_{n \to +\infty} \sum_{k=1}^{n} f(c_k) \cdot \Delta x = \lim_{n \to +\infty} \left(\frac{3}{2} - \frac{1}{2n}\right) = \frac{3}{2} - 0 = \frac{3}{2}$$

The region whose area we have computed is a trapezoid with height $h = 1$ and bases $b_1 = 1$ and $b_2 = 2$. From plane geometry, the area of this trapezoid is $A = \frac{1}{2}h(b_1 + b_2) = \frac{1}{2}(1)(1 + 2) = \frac{3}{2}$, which agrees with the result that we obtained here.   ◀

Example 2   Use inscribed rectangles to find the area under the curve $y = 9 - x^2$ over the interval $[0, 3]$.

Solution.   If we divide the interval $[0, 3]$ into n subintervals of equal length, then each subinterval will have length

$$\Delta x = \frac{b - a}{n} = \frac{3 - 0}{n} = \frac{3}{n}$$

and the points of subdivision will be (Figure 5.5.8)

$$x_1 = 0 + \Delta x = \frac{3}{n}$$

$$x_2 = 0 + 2\,\Delta x = 2\left(\frac{3}{n}\right)$$

$$x_3 = 0 + 3\,\Delta x = 3\left(\frac{3}{n}\right)$$

$$\vdots$$

$$x_{n-1} = 0 + (n - 1)\,\Delta x = (n - 1)\left(\frac{3}{n}\right)$$

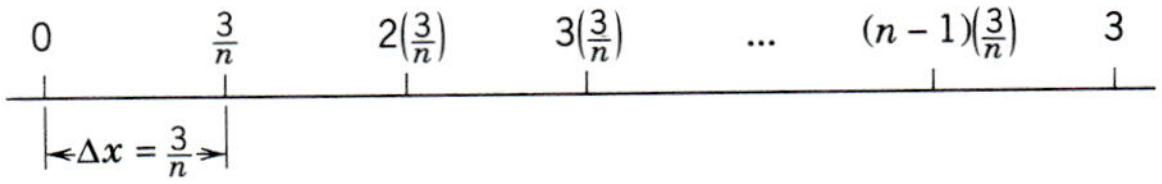

$$
\begin{array}{ccccccc}
0 & \frac{3}{n} & 2\left(\frac{3}{n}\right) & 3\left(\frac{3}{n}\right) & \cdots & (n-1)\left(\frac{3}{n}\right) & 3
\end{array}
$$

$$\left|\leftarrow \Delta x = \tfrac{3}{n} \rightarrow\right|$$

Figure 5.5.8

Because $y = f(x) = 9 - x^2$ is decreasing on $[0, 3]$, the minimum value of $f(x)$ on each subinterval occurs at the right endpoint (Figure 5.5.9), so

$$c_1 = x_1 = \frac{3}{n}$$

$$c_2 = x_2 = 2 \cdot \frac{3}{n}$$

$$c_3 = x_3 = 3 \cdot \frac{3}{n}$$

$$\vdots$$

$$c_{n-1} = x_{n-1} = (n - 1) \cdot \frac{3}{n}$$

$$c_n = 3$$

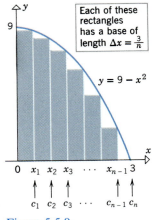

Each of these rectangles has a base of length $\Delta x = \frac{3}{n}$

$y = 9 - x^2$

Figure 5.5.9

In short, $c_k = k \cdot (3/n)$ for $k = 1, 2, \ldots, n$. Thus, the kth inscribed rectangle has area

$$f(c_k) \cdot \Delta x = (9 - c_k{}^2) \cdot \Delta x = \left[9 - k^2 \left(\frac{9}{n^2} \right) \right] \cdot \frac{3}{n} = \frac{27}{n} - \frac{27k^2}{n^3}$$

and the sum of these areas is

$$\sum_{k=1}^{n} f(c_k) \cdot \Delta x = \sum_{k=1}^{n} \left(\frac{27}{n} - \frac{27k^2}{n^3} \right) = \sum_{k=1}^{n} \frac{27}{n} - \sum_{k=1}^{n} \frac{27k^2}{n^3}$$

$$= n \cdot \frac{27}{n} - \frac{27}{n^3} \sum_{k=1}^{n} k^2 = 27 - \frac{27}{n^3} \cdot \frac{n(n + 1)(2n + 1)}{6}$$

Theorem 5.4.2(*b*)

Thus, from (4)

$$A = \lim_{n \to +\infty} \sum_{k=1}^{n} f(c_k) \cdot \Delta x = \lim_{n \to +\infty} \left[27 - \frac{27}{n^3} \cdot \frac{n(n + 1)(2n + 1)}{6} \right]$$

$$= 27 - \lim_{n \to +\infty} \frac{27}{6} \left(1 + \frac{1}{n} \right) \left(2 + \frac{1}{n} \right) = 27 - \frac{27}{6} (1)(2) = 18 \quad \blacktriangleleft$$

□ **AREAS AS LIMITS USING CIRCUMSCRIBED RECTANGLES**

It may already have occurred to the reader that we could have used *circumscribed* rectangles rather than inscribed rectangles in these examples. Indeed, if the *maximum* values of $f(x)$ in successive subintervals occur at the points $d_1, d_2, \ldots, d_n$, then the sum

$$\sum_{k=1}^{n} f(d_k) \cdot \Delta x$$

is an approximation by circumscribed rectangles to the area under the curve (Figure 5.5.10), and the exact area A is

$$A = \lim_{n \to +\infty} \sum_{k=1}^{n} f(d_k) \cdot \Delta x \tag{5}$$

since the error (represented by the overlap) diminishes as the rectangles get thinner.

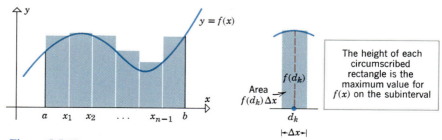

Figure 5.5.10

Example 3 Use circumscribed rectangles to find the area under the line $y = x$ over the interval $[1, 2]$.

Solution. As in Example 1, the points

$$x_1 = 1 + \frac{1}{n}$$

$$x_2 = 1 + \frac{2}{n}$$

$$x_3 = 1 + \frac{3}{n}$$

$$\vdots$$

$$x_{n-1} = 1 + \frac{n-1}{n}$$

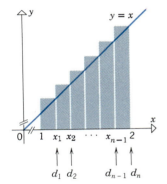

Figure 5.5.11

divide the interval $[1, 2]$ into n subintervals of length $\Delta x = 1/n$. Because $y = f(x) = x$ is increasing, the maximum value for $f(x)$ on each subinterval occurs at the right-hand endpoint (Figure 5.5.11), so that $d_1 = x_1 = 1 + 1/n$, $d_2 = x_2 = 1 + 2/n$, $d_3 = x_3 = 1 + 3/n, \ldots, d_n = 1 + n/n = 2$. Thus, the kth circumscribed rectangle has area

$$f(d_k) \cdot \Delta x = d_k \cdot \Delta x = \left(1 + \frac{k}{n}\right) \cdot \Delta x = \left(1 + \frac{k}{n}\right) \cdot \frac{1}{n}$$

and the sum of the areas of the rectangles is

$$\sum_{k=1}^{n} f(d_k) \cdot \Delta x = \sum_{k=1}^{n} \left[\left(1 + \frac{k}{n}\right) \cdot \frac{1}{n}\right] = \frac{1}{n} \sum_{k=1}^{n} 1 + \frac{1}{n^2} \sum_{k=1}^{n} k$$

$$= \frac{1}{n} \cdot n + \frac{1}{n^2} \left[\frac{n(n+1)}{2}\right] = \frac{3}{2} + \frac{1}{2n}$$

Theorem 5.4.2(a)

From (5), the area under the curve is

$$A = \lim_{n \to +\infty} \sum_{k=1}^{n} f(d_k) \cdot \Delta x = \lim_{n \to +\infty} \left(\frac{3}{2} + \frac{1}{2n}\right) = \frac{3}{2}$$

This result agrees with that obtained in Example 1 using inscribed rectangles. ◀

REMARK. Although we shall not do it, it can be proved that, in general, the method of inscribed rectangles [Formula (4)] and the method of circumscribed rectangles [Formula (5)] both yield the same value for A.

▶ Exercise Set 5.5   C 18, 19, 20, 21

In Exercises 1–4, divide the interval $[a, b]$ into $n = 4$ subintervals of equal length, then compute (a) the sum of the areas of the inscribed rectangles and (b) the sum of the areas of the circumscribed rectangles.

1. $y = 3x + 1$; $a = 2$, $b = 6$.

2. $y = 1/x$; $a = 1$, $b = 9$.

3. $y = \cos x$; $a = -\dfrac{\pi}{2}$, $b = \dfrac{\pi}{2}$.

4. $y = 2x - x^2$; $a = 1$, $b = 2$.

In Exercises 5–10, use inscribed rectangles to find the area under the curve $y = f(x)$ over the interval $[a, b]$.

5. $y = \frac{1}{2}x$; $a = 1$, $b = 4$.

6. $y = -x + 5$; $a = 0$, $b = 5$.

7. $y = x^2$; $a = 0$, $b = 1$.

8. $y = 4 - \frac{1}{4}x^2$; $a = 0$, $b = 3$.

9. $y = x^3$; $a = 2$, $b = 6$.

10. $y = 1 - x^3$; $a = -3$, $b = -1$.

11. Use circumscribed rectangles to find the area of the region in Exercise 5.

12. Use circumscribed rectangles to find the area of the region in Exercise 6.

13. Use circumscribed rectangles to find the area of the region in Exercise 7.

14. Use inscribed rectangles to find the area under $y = mx$ over the interval $[a, b]$, where $m > 0$ and $a \geq 0$.

15. (a) Show that the area under $y = x^3$ over the interval $[0, b]$ is $b^4/4$.
     (b) Find a formula for the area under $y = x^3$ over the interval $[a, b]$, where $a \geq 0$.

16. Find the area between the curve $y = \sqrt{x}$ and the interval $0 \leq y \leq 1$ on the y-axis.

17. Assuming that $a > 0$, find the area under $y = x$ over the interval $[a, b]$ by using
     (a) inscribed rectangles
     (b) circumscribed rectangles
     (c) an area formula from plane geometry.

In Exercises 18–21, use a computer or calculator to obtain an approximate value for the area under the curve with $n = 10$, 20, and 50 subintervals by using (a) inscribed rectangles and (b) circumscribed rectangles. [*Note:* A programmable calculator is best for this problem. If your calculator is not programmable, then just do the case for $n = 10$.]

18. $y = 1/x$; $[1, 2]$.

19. $y = 1/x^2$; $[1, 3]$.

20. $y = \sqrt{x}$; $[0, 4]$.

21. $y = \sin x$; $[0, \pi/2]$.

5.6 THE DEFINITE INTEGRAL

In mathematics and science there are a variety of concepts, such as length, volume, density, probability, work, and others, whose properties are remarkably similar to properties of area. In this section we shall give a precise mathematical definition of "area," and we shall introduce the notion of a "definite integral," which is the unifying thread that relates these concepts. Our work in this section will focus on ideas rather than computations. In the next section we shall develop the tools that are necessary for efficient calculations.

☐ **DEFINITION OF AREA**

The first goal in this section is to give a mathematical definition of *area*. We will begin by reviewing some of the intuitive ideas about area that were obtained in the last section. Recall that we gave two equivalent ways of viewing the area under a continuous curve $y = f(x)$ over an interval $[a, b]$ as a limit:

$$A = \lim_{n \to +\infty} \sum_{k=1}^{n} f(c_k)\, \Delta x \qquad \boxed{\text{Inscribed rectangles}} \qquad (1)$$

and

$$A = \lim_{n \to +\infty} \sum_{k=1}^{n} f(d_k)\, \Delta x \qquad \boxed{\text{Circumscribed rectangles}} \qquad (2)$$

However, these are not the only possible limit formulas for the area A. Instead of choosing the height of the rectangle on the kth subinterval to be the minimum value or the maximum value of f, we can construct a rectangle of some intermediate height. For this purpose, let us select an *arbitrary* point in each subinterval; call these points

$$x_1^*, x_2^*, \ldots, x_n^*$$

Since $f(c_k)$ and $f(d_k)$ are, respectively, the smallest and largest values of f on the kth subinterval, it follows that

$$f(c_k) \leq f(x_k^*) \leq f(d_k)$$

Thus, the areas of the inscribed rectangles, the circumscribed rectangles, and the rectangles of height $f(x_k^*)$ satisfy

$$f(c_k) \cdot \Delta x \leq f(x_k^*) \cdot \Delta x \leq f(d_k) \cdot \Delta x$$

(Figure 5.6.1).

Therefore, the sums of these areas satisfy

$$\sum_{k=1}^{n} f(c_k)\, \Delta x \leq \sum_{k=1}^{n} f(x_k^*)\, \Delta x \leq \sum_{k=1}^{n} f(d_k)\, \Delta x \qquad (3)$$

As $n \to +\infty$, the two outside sums in (3) approach A [see (1) and (2)], so the inside sum must also approach A since it is "squeezed" between the outer sums. Thus, for all possible choices of $x_1^*, x_2^*, \ldots, x_n^*$ we obtain

$$A = \lim_{n \to +\infty} \sum_{k=1}^{n} f(x_k^*)\, \Delta x$$

which is a more general limit formula for the area A of which (1) and (2) are special cases.

Although rectangles with equal widths are convenient computationally, they are not essential; we can just as well express the area A as the limit of a sum of rectangular areas with different widths (Figure 5.6.2).

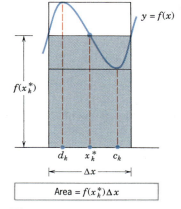

Area = $f(x_k^*)\Delta x$

Figure 5.6.1

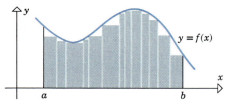

Figure 5.6.2

However, there is a complication here that did not occur before. For rectangles of equal width we have

$$\Delta x = \frac{b - a}{n}$$

so it follows that $\Delta x \to 0$ as $n \to +\infty$. In other words, we can ensure that the width of *every* rectangle decreases to zero by letting the number of rectangles increase to infinity. For rectangles whose widths vary in size this is not the case. For example, suppose we were to continually divide only the left half of the interval $[a, b]$ into more and more subintervals, but were always to leave the right half of the interval alone (Figure 5.6.3).

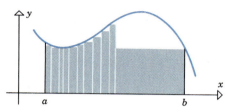

Figure 5.6.3

With this construction the number of rectangles increases to infinity, but the sum of the areas of the rectangles does not approach the area under the curve because the "error" on the right half of the interval never decreases. To remedy this problem, we need to ensure that the widths of *all* rectangles decrease to zero as the number of rectangles increases to infinity.

Let us suppose that the interval $[a, b]$ is divided into n subintervals whose widths are

$$\Delta x_1, \Delta x_2, \ldots, \Delta x_n$$

The subintervals are said to form a **partition** of the interval $[a, b]$, and the largest of the subinterval widths is called the **mesh size** of the partition. The mesh size is denoted by the symbol

$$\max \Delta x_k$$

which is read, "the maximum of the Δx_k's." For example, Figure 5.6.4 shows a partition of the interval $[0, 6]$ into four subintervals with a mesh size of 2.

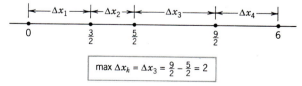

Figure 5.6.4

If an interval $[a, b]$ is partitioned into n subintervals, and if x_k^* is an arbitrary point in the kth subinterval, then

$$f(x_k^*)\, \Delta x_k$$

is the area of a rectangle of height $f(x_k^*)$ and width Δx_k, so

$$\sum_{k=1}^{n} f(x_k^*)\, \Delta x_k$$

is the sum of the shaded rectangular areas in Figure 5.6.5.

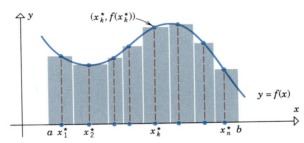

Figure 5.6.5

If we now increase n in such a way that

$$\max \Delta x_k \to 0$$

then the width of *every* rectangle tends to zero because no width exceeds the maximum; thus, (4) approaches the exact area under the curve. We denote this by writing

$$A = \lim_{\max \Delta x_k \to 0} \sum_{k=1}^{n} f(x_k^*)\, \Delta x_k$$

Motivated by this formula, we make the following definition.

5.6.1   DEFINITION (*Area Under a Curve*).   If the function f is continuous on $[a, b]$ and if $f(x) \geq 0$ for all x in $[a, b]$, then the ***area*** under the curve $y = f(x)$ over the interval $[a, b]$ is defined by

$$A = \lim_{\max \Delta x_k \to 0} \sum_{k=1}^{n} f(x_k^*)\, \Delta x_k$$

DEFINITE INTEGRALS OF CONTINUOUS FUNCTIONS WITH NONNEGATIVE VALUES

The limit in Definition 5.6.1 is so important that there is some special notation for it. We write

$$\int_a^b f(x)\, dx = \lim_{\max \Delta x_k \to 0} \sum_{k=1}^{n} f(x_k^*)\, \Delta x_k \tag{5}$$

The expression on the left side of this equation is called the ***definite integral of f from a to b,*** and the numbers b and a are called the ***upper and lower limits of integration,*** respectively. The reason for the use of the integral sign will become clear in the next section where we shall establish a relationship between the definite integral and the indefinite integral studied earlier.

It follows from Definition 5.6.1 and (5) that the area A under the curve $y = f(x)$ over the interval $[a, b]$ can be expressed in definite integral notation as

$$A = \begin{bmatrix} \text{area under} \\ y = f(x) \\ \text{over } [a, b] \end{bmatrix} = \int_a^b f(x)\, dx \qquad (6)$$

One of our goals is to develop efficient methods for evaluating definite integrals; however, the definite integrals in the following example can be evaluated using area formulas from plane geometry.

Example 1 Use appropriate area formulas from plane geometry to evaluate the following definite integrals.

$$\text{(a)} \quad \int_2^4 (x - 1)\, dx \qquad \text{(b)} \quad \int_0^1 \sqrt{1 - x^2}\, dx$$

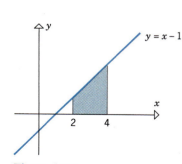

Solution (a). From (6), the integral represents the area under the graph of $y = x - 1$ over the interval $[2, 4]$. (See Figure 5.6.6.) This region is a trapezoid whose parallel sides have lengths 1 and 3 and whose height (the distance between the parallel sides) is 2. Thus,

$$\int_2^4 (x - 1)\, dx = \frac{1}{2}(1 + 3) \cdot 2 = 4$$

Figure 5.6.6

(See the inside text cover for a review of area formulas from plane geometry.)

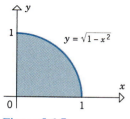

Solution (b). From (6), the integral represents the area under the graph of $y = \sqrt{1 - x^2}$ over the interval $[0, 1]$. This is the region in the first quadrant bounded by the coordinate axes and the circle $x^2 + y^2 = 1$. (See Figure 5.6.7.) The area of this region is one-fourth of the area enclosed by the entire circle, so

$$\int_0^1 \sqrt{1 - x^2}\, dx = \frac{1}{4}(\pi \cdot 1^2) = \frac{\pi}{4} \quad \blacktriangleleft$$

Figure 5.6.7

The sum

$$\sum_{k=1}^n f(x_k^*)\, \Delta x_k$$

which appears in (5), is called a ***Riemann* sum*** in honor of the German mathematician Bernhard Riemann, who first formulated many of the concepts about definite integrals. Thus, (5) defines the definite integral as a limit of Riemann sums. It is proved in advanced courses that for nonnegative continuous functions this limit must exist. As a result, we need not worry about the existence of the limit that defines the area A in Definition 5.6.1.

Although we defined the definite integral as a notation for area, limits of Riemann sums occur in a wide variety of applications that have no immediate connection with area problems. In many of these applications the functions involved assume both positive and negative values and have discontinuities, so we will want to extend the concept of a definite integral to allow for such functions.

☐ **DEFINITE INTEGRALS OF CONTINUOUS FUNCTIONS WITH NONPOSITIVE VALUES**

The simplest generalization of the definite integral is to continuous functions whose values are all nonpositive. If $f(x) \geq 0$ for all x in $[a, b]$, then it follows that $-f(x) \leq 0$ for all x in $[a, b]$, so $-f$ is a function whose values are all nonpositive on $[a, b]$. We define

$$\int_a^b -f(x)\ dx = -\int_a^b f(x)\ dx \tag{7}$$

Geometrically, the quantity in (7) is the negative of the area under $y = f(x)$ over $[a, b]$. (See Figure 5.6.8.)

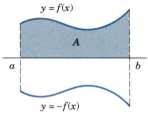

Figure 5.6.8

Example 2 Evaluate $\displaystyle\int_2^4 (1 - x)\ dx$.

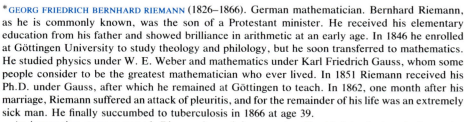

---

*GEORG FRIEDRICH BERNHARD RIEMANN (1826–1866). German mathematician. Bernhard Riemann, as he is commonly known, was the son of a Protestant minister. He received his elementary education from his father and showed brilliance in arithmetic at an early age. In 1846 he enrolled at Göttingen University to study theology and philology, but he soon transferred to mathematics. He studied physics under W. E. Weber and mathematics under Karl Friedrich Gauss, whom some people consider to be the greatest mathematician who ever lived. In 1851 Riemann received his Ph.D. under Gauss, after which he remained at Göttingen to teach. In 1862, one month after his marriage, Riemann suffered an attack of pleuritis, and for the remainder of his life was an extremely sick man. He finally succumbed to tuberculosis in 1866 at age 39.

An interesting story surrounds Riemann's work in geometry. For his introductory lecture prior to becoming an associate professor, Riemann submitted three possible topics to Gauss. Gauss surprised Riemann by choosing the topic Riemann liked the least, the foundations of geometry. The lecture was like a scene from a movie. The old and failing Gauss, a giant in his day, watching intently as his brilliant and youthful protégé skillfully pieced together portions of the old man's own work into a complete and beautiful system. Gauss is said to have gasped with delight as the lecture neared its end, and on the way home he marveled at his student's brilliance. Gauss died shortly thereafter. The results presented by Riemann that day eventually evolved into a fundamental tool that Einstein used some 50 years later to develop relativity theory.

In addition to his work in geometry, Riemann made major contributions to the theory of complex functions and mathematical physics. The notion of the definite integral, as it is presented in most basic calculus courses, is due to him. Riemann's early death was a great loss to mathematics, for his mathematical work was brilliant and of fundamental importance.

Solution. The integrand is negative over the interval $[2, 4]$ (Figure 5.6.9), so from Formula (7) and part (a) of Example 1 we obtain

$$\int_2^4 (1 - x) \, dx = \int_2^4 - (x - 1) \, dx = -\int_2^4 (x - 1) \, dx = -4 \quad \blacktriangleleft$$

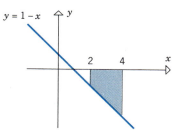

Figure 5.6.9

☐ DEFINITE INTEGRALS OF CONTINUOUS FUNCTIONS WITH POSITIVE AND NEGATIVE VALUES

Thus far we have defined the definite integral for nonnegative continuous functions and nonpositive continuous functions on an interval $[a, b]$. We will now extend the definition of the definite integral to include functions that are continuous and have both positive and negative values on $[a, b]$. It can be proved that for such functions the limit

$$\lim_{\max \Delta x_k \to 0} \sum_{k=1}^n f(x_k^*) \, \Delta x_k$$

always exists, and thus we define

$$\int_a^b f(x) \, dx = \lim_{\max \Delta x_k \to 0} \sum_{k=1}^n f(x_k^*) \, \Delta x_k \tag{8}$$

To motivate a geometric interpretation of this integral, consider the Riemann sum shown in Figure 5.6.10.

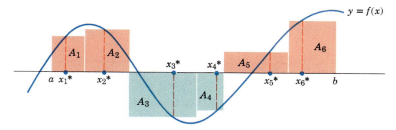

Figure 5.6.10

On those intervals where $f(x_k^*)$ is positive, the product $f(x_k^*) \, \Delta x_k$ is the area A_k of the rectangle with height $f(x_k^*)$ and base Δx_k, and on those intervals where $f(x_k^*)$ is negative, the product $f(x_k^*) \, \Delta x_k$ is the negative $-A_k$ of such an area.

For the particular Riemann sum in Figure 5.6.10 we have

$$\sum_{k=1}^{6} f(x_k^*) \, \Delta x_k = f(x_1^*) \, \Delta x_1 + f(x_2^*) \, \Delta x_2 + \cdots + f(x_6^*) \, \Delta x_6$$

$$= A_1 + A_2 - A_3 - A_4 + A_5 + A_6$$

$$= (A_1 + A_2 + A_5 + A_6) - (A_3 + A_4)$$

Geometrically, this Riemann sum is the difference of two areas—the total area of the rectangles above the x-axis minus the total area of the rectangles below the x-axis. If we were to allow the number of subdivisions to increase so that max $\Delta x_k \to 0$, then the rectangles would begin to fill out the regions between the curve $y = f(x)$ and the interval $[a, b]$ with less and less error (Figure 5.6.11). This suggests that for the function in the figure the limit of the Riemann sums would also be the difference of two areas, namely

$$\int_a^b f(x) \, dx = (A_I + A_{III}) - A_{II} = \begin{bmatrix} \text{area above} \\ [a, b] \end{bmatrix} - \begin{bmatrix} \text{area below} \\ [a, b] \end{bmatrix}$$

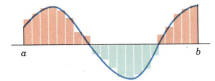 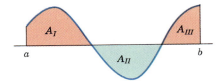

Figure 5.6.11

The area above an interval $[a, b]$ but below $y = f(x)$ minus the area below $[a, b]$ but above $y = f(x)$ will be called the *net signed area* between $y = f(x)$ and the interval $[a, b]$. More precisely, we make the following definition.

5.6.2   DEFINITION.   If the function f is continuous on $[a, b]$, and can assume both positive and negative values, then the *net signed area* A between $y = f(x)$ and the interval $[a, b]$ is defined by

$$\int_a^b f(x) \, dx = \lim_{\max \Delta x_k \to 0} \sum_{k=1}^{n} f(x_k^*) \, \Delta x_k$$

The net signed area between $y = f(x)$ and $[a, b]$ can be positive, negative, or zero; it is positive when there is more area above the interval than below, negative when there is more area below than above, and zero when the area above and below are equal.

Example 3   Use an appropriate area formula from plane geometry to evaluate the definite integral

$$\int_0^2 (x - 1) \, dx$$

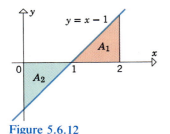

Figure 5.6.12

Solution. From Definition 5.6.2, the integral represents the net signed area between the line $y = x - 1$ and the interval $[0, 2]$. (See Figure 5.6.12.) Calculating the areas of the two triangular regions in the figure and subtracting yields

$$\int_0^2 (x - 1)\, dx = A_1 - A_2 = \frac{1}{2} - \frac{1}{2} = 0 \quad \blacktriangleleft$$

DEFINITE INTEGRALS OF FUNCTIONS WITH DISCONTINUITIES

We noted earlier without proof that if f is continuous on $[a, b]$, then the limit of the Riemann sums must exist. Thus, in Definitions 5.6.1 and 5.6.2, the existence of a numerical value for the area and the net signed area was not an issue because we assumed continuity of f. However, for functions with discontinuities, the limit of the Riemann sums may or may not exist depending on the number and the nature of the discontinuities. We make the following definition.

5.6.3 DEFINITION. If the function f is defined on the closed interval $[a, b]$, then f is called ***Riemann integrable*** on $[a, b]$ or more simply ***integrable*** on $[a, b]$ if the limit

$$\lim_{\max \Delta x_k \to 0} \sum_{k=1}^{n} f(x_k^*)\, \Delta x_k$$

exists. If f is integrable on $[a, b]$, then we define the definite integral of f from a to b by

$$\int_a^b f(x)\, dx = \lim_{\max \Delta x_k \to 0} \sum_{k=1}^{n} f(x_k^*)\, \Delta x_k$$

In our discussion up to now, and in Definition 5.6.3, we assumed the lower limit of integration to be less than the upper limit of integration. The following definition extends the concept of the definite integral to allow for equal limits of integration and for a lower limit of integration that is greater than the upper limit of integration.

5.6.4 DEFINITION.

(a) If a is in the domain of f, we define

$$\int_a^a f(x)\, dx = 0$$

(b) If f is integrable on $[a, b]$, then we define

$$\int_b^a f(x)\, dx = -\int_a^b f(x)\, dx$$

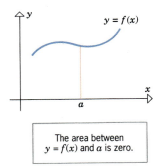

The area between
$y = f(x)$ and a is zero.

Figure 5.6.13

REMARK. Part (a) of this definition is consistent with the intuitive idea that the area between a point on the x-axis and a curve $y = f(x)$ should be zero (Figure 5.6.13). Part (b) of the definition is simply a convenient convention; it states that interchanging the limits of integration reverses the sign of the integral.

Example 4

(a) $\displaystyle\int_1^1 x^2 \, dx = 0$

(b) $\displaystyle\int_1^0 \sqrt{1 - x^2} \, dx = -\int_0^1 \sqrt{1 - x^2} \, dx = -\frac{\pi}{4}$ ◀

Example 1(b)

The problem of determining precisely which functions are integrable is quite complex and beyond the scope of this text. However, we will discuss a few results about integrability that are useful to know. We start with a definition.

5.6.5 DEFINITION. A function f is said to be **bounded** on an interval $[a, b]$ if there is a positive number M such that

$$-M \le f(x) \le M$$

for all x in $[a, b]$. Geometrically, this means that the graph of f on the interval $[a, b]$ lies between the lines $y = -M$ and $y = M$.

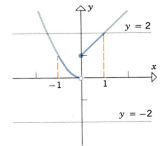

Figure 5.6.14

Example 5 The function

$$f(x) = \begin{cases} x^2, & x < 0 \\ x + 1, & x \ge 0 \end{cases}$$

is bounded on the interval $[-1, 1]$, since its graph on this interval lies between the lines $y = -2$ and $y = 2$ on this interval (Figure 5.6.14). However, the function

$$g(x) = \begin{cases} 1/x, & x \ne 0 \\ 0, & x = 0 \end{cases}$$

is not bounded on the interval $[-1, 1]$ (or on any closed interval containing the origin), since its graph on this interval cannot be enclosed between two horizontal lines (Figure 5.6.15). In general, a function that approaches $+\infty$ or $-\infty$ somewhere in an interval is not bounded on that interval. ◀

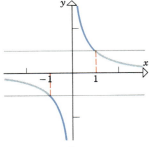

Figure 5.6.15

The following theorem, which we state without proof, lists three of the most important facts about integrability.

> **5.6.6** THEOREM.   *Let f be a function that is defined at all points in the interval $[a, b]$.*
>
> (a) *If f is continuous on $[a, b]$, then f is integrable on $[a, b]$.*
> (b) *If f is bounded on $[a, b]$ and has only finitely many points of discontinuity on $[a, b]$, then f is integrable on $[a, b]$.*
> (c) *If f is not bounded on $[a, b]$, then f is not integrable on $[a, b]$.*

Example 6   The function f in Example 5 is integrable on the interval $[-1, 1]$ because it is bounded and has only finitely many points of discontinuity on this interval (one discontinuity). However, the function g in Example 5 is not integrable on $[-1, 1]$ because it is not bounded on this interval.   ◄

■ OPTIONAL

Limits of Riemann sums are somewhat different from the kinds of limits discussed in Chapter 2. Loosely phrased, the expression

$$\lim_{\max \Delta x_k \to 0} \sum_{k=1}^{n} f(x_k^*) \, \Delta x_k = L$$

is intended to convey the idea that we can force the Riemann sums to be as close as we please to L, regardless of how $x_1^*, x_2^*, \ldots, x_n^*$ are chosen, by making the mesh size of the partition sufficiently small. This idea is captured by the following definition.

> **5.6.7** DEFINITION.   We shall write
>
> $$\lim_{\max \Delta x_k \to 0} \sum_{k=1}^{n} f(x_k^*) \, \Delta x_k = L \qquad (9)$$
>
> if given any number $\epsilon > 0$, there is a number $\delta > 0$ such that
>
> $$\left| \sum_{k=1}^{n} f(x_k^*) \, \Delta x_k - L \right| < \epsilon$$
>
> whenever $\max \Delta x_k < \delta$ and regardless of how $x_1^*, x_2^*, \ldots, x_n^*$ are chosen.

It can be shown that when a number L satisfying (9) exists, it is unique (Exercise 35). We denote this number by

$$\lim_{\max \Delta x_k \to 0} \sum_{k=1}^{n} f(x_k^*) \, \Delta x_k = \int_{a}^{b} f(x) \, dx \qquad (10)$$

REMARK.   Some writers use the symbol $\|\Delta\|$ rather than $\max \Delta x_k$ for the mesh size of the partition, in which case (10) would be written as

$$\lim_{\|\Delta\| \to 0} \sum_{k=1}^{n} f(x_k^*) \, \Delta x_k = \int_{a}^{b} f(x) \, dx$$

▶ **Exercise Set 5.6** 🅲 29, 30, 31

In Exercises 1–4, find the value of

(a) $\sum_{k=1}^{n} f(x_k^*) \, \Delta x_k$ (b) $\max \Delta x_k$.

1. $f(x) = x + 1$; $a = 0$, $b = 4$; $n = 3$;
    $\Delta x_1 = 1$, $\Delta x_2 = 1$; $\Delta x_3 = 2$;
    $x_1^* = \frac{1}{3}$, $x_2^* = \frac{3}{2}$, $x_3^* = 3$.

2. $f(x) = \cos x$; $a = 0$, $b = 2\pi$; $n = 4$;
    $\Delta x_1 = \pi/2$, $\Delta x_2 = 3\pi/4$, $\Delta x_3 = \pi/2$, $\Delta x_4 = \pi/4$;
    $x_1^* = \pi/4$, $x_2^* = \pi$, $x_3^* = 3\pi/2$, $x_4^* = 7\pi/4$.

3. $f(x) = 4 - x^2$; $a = -3$, $b = 4$; $n = 4$;
    $\Delta x_1 = 1$, $\Delta x_2 = 2$, $\Delta x_3 = 1$, $\Delta x_4 = 3$;
    $x_1^* = -\frac{5}{2}$, $x_2^* = -1$, $x_3^* = \frac{1}{4}$, $x_4^* = 3$.

4. $f(x) = x^3$; $a = -3$, $b = 3$; $n = 4$;
    $\Delta x_1 = 2$, $\Delta x_2 = 1$, $\Delta x_3 = 1$, $\Delta x_4 = 2$;
    $x_1^* = -2$, $x_2^* = 0$, $x_3^* = 0$, $x_4^* = 2$.

5. Verify that relationship (3) holds when
    $$f(x) = 1/x; \quad a = 1, \ b = 9; \ n = 4;$$
    $$\Delta x_1 = \Delta x_2 = \Delta x_3 = \Delta x_4 = 2;$$
    $$x_1^* = 2, \ x_2^* = 4, \ x_3^* = 6, \ x_4^* = 8.$$

In Exercises 6–8, use the given values of a and b to express the following limits as definite integrals. (Do not evaluate the integrals.)

6. $\displaystyle\lim_{\max \Delta x_k \to 0} \sum_{k=1}^{n} (x_k^*)^3 \, \Delta x_k$; $a = 1$, $b = 2$.

7. $\displaystyle\lim_{\max \Delta x_k \to 0} \sum_{k=1}^{n} 4x_k^*(1 - 3x_k^*) \, \Delta x_k$; $a = -3$, $b = 3$.

8. $\displaystyle\lim_{\max \Delta x_k \to 0} \sum_{k=1}^{n} (\sin^2 x_k^*) \, \Delta x_k$; $a = 0$, $b = \pi/2$.

In Exercises 9–11, express the definite integrals as limits. (Do not evaluate.)

9. $\displaystyle\int_1^2 2x \, dx$.

10. $\displaystyle\int_{-\pi/2}^{\pi/2} (1 + \cos x) \, dx$.

11. $\displaystyle\int_0^1 \frac{x}{x + 1} \, dx$.

12. Show on a sketch the area that is represented by
    (a) $\displaystyle\int_1^4 \sqrt{x} \, dx$ (b) $\displaystyle\int_1^3 \frac{1}{x} \, dx$
    (c) $\displaystyle\int_{-1}^2 \sqrt{9 - x^2} \, dx$ (d) $\displaystyle\int_0^{\pi/2} \sin x \, dx$.

13. Given the areas shown in Figure 5.6.16, find
    (a) $\displaystyle\int_a^b f(x) \, dx$ (b) $\displaystyle\int_b^c f(x) \, dx$
    (c) $\displaystyle\int_a^c f(x) \, dx$ (d) $\displaystyle\int_a^d f(x) \, dx$.

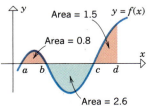

Area = 1.5 $y = f(x)$

Area = 0.8

Area = 2.6   Figure 5.6.16

In Exercises 14–27, evaluate the definite integrals by using appropriate area formulas from plane geometry, where needed.

14. (a) $\displaystyle\int_0^3 x \, dx$ (b) $\displaystyle\int_{-2}^{-1} x \, dx$
    (c) $\displaystyle\int_{-1}^4 x \, dx$ (d) $\displaystyle\int_{-5}^5 x \, dx$.

15. Let $f(x) = 1 - \frac{1}{2}x$.
    (a) $\displaystyle\int_0^2 f(x) \, dx$ (b) $\displaystyle\int_{-1}^1 f(x) \, dx$
    (c) $\displaystyle\int_2^3 f(x) \, dx$ (d) $\displaystyle\int_0^3 f(x) \, dx$.

16. $\displaystyle\int_{-10}^{-5} 6 \, dx$.   17. $\displaystyle\int_0^2 (2 - 4x) \, dx$.

18. $\displaystyle\int_0^3 |x - 2| \, dx$.   19. $\displaystyle\int_{-1}^2 |2x - 3| \, dx$.

20. $\displaystyle\int_0^2 \sqrt{4 - x^2} \, dx$.   21. $\displaystyle\int_{-1}^1 \sqrt{1 - x^2} \, dx$.

22. $\displaystyle\int_{-3}^0 (2 + \sqrt{9 - x^2}) \, dx$.

23. $\displaystyle\int_0^{10} \sqrt{10x - x^2} \, dx$. [*Hint:* Complete the square.]

24. $\displaystyle\int_{-\pi/3}^{\pi/3} \sin x \, dx$.   25. $\displaystyle\int_0^{\pi} \cos x \, dx$.

26. $\displaystyle\int_0^3 f(x) \, dx$, where
    $$f(x) = \begin{cases} 2x, & x \le 1 \\ 2, & x > 1 \end{cases}$$

27. $\int_{-2}^{2} f(x)\, dx$, where

$$f(x) = \begin{cases} 3, & x \le 0 \\ x + 3, & x > 0 \end{cases}$$

28. Suppose that

$$\sum_{k=1}^{n} f(x_k^*)\, \Delta x_k$$

is a Riemann sum for the function f on the interval $[a, b]$.

(a) Show that if the subintervals all have the same width Δx, then

$$\sum_{k=1}^{n} f(x_k^*)\, \Delta x_k = [f(x_1^*) + f(x_2^*) + \cdots + f(x_n^*)]\, \Delta x$$

(b) Show that if the subintervals all have the same width Δx and $x_1^*, x_2^*, \ldots, x_n^*$ are the midpoints of the subintervals, then

$$x_k^* = a - \tfrac{1}{2} \Delta x + k\, \Delta x$$

Since a definite integral is the limit of its Riemann sums, its value can be approximated to any degree of accuracy by a Riemann sum with sufficiently many subintervals. If the subintervals have equal width and $x_1^*, x_2^*, \ldots, x_n^*$ are chosen as the midpoints of the subintervals, then the Riemann sum is called the ***midpoint approximation*** of the definite integral. In Exercises 29–31, find the midpoint approximation of the integral using $n = 10$, 20, and 50 subintervals. (See Exercise 28.) [*Note:* A programmable calculator is best for this problem. If your calculator is not programmable, then just do the case for $n = 10$.]

29. $\int_{1}^{2} \frac{1}{x}\, dx$.

30. $\int_{0}^{2} \sin \sqrt{x}\, dx$.

31. $\int_{0}^{1} \frac{4}{x^2 + 1}\, dx$. [*Note:* The exact value is π.]

32. Prove that the function

$$f(x) = \begin{cases} 1 & \text{if } x \text{ is rational} \\ 0 & \text{if } x \text{ is irrational} \end{cases}$$

is not integrable on any closed interval $[a, b]$.

In Exercises 33 and 34, use Theorem 5.6.6 to determine whether the function is integrable on the given interval.

33. $f(x) = \begin{cases} 1/x^2, & x > 0 \\ 0, & x = 0 \end{cases}$ on $[0, 1]$.

34. $f(x) = \begin{cases} \sin 1/x, & x \ne 0 \\ 0, & x = 0 \end{cases}$ on $[-1, 1]$.

Exercises 35–37 are for readers who have read the optional material at the end of this section.

35. (Optional) Prove: If f is integrable on $[a, b]$, then the value of $\int_a^b f(x)\, dx$ is unique.

36. (Optional) Prove: If f is integrable on $[a, b]$ and c is any constant, then cf is integrable on $[a, b]$ and

$$\int_a^b cf(x)\, dx = c \int_a^b f(x)\, dx$$

37. (Optional) Prove: If f and g are integrable on $[a, b]$, then $f + g$ is integrable on $[a, b]$ and

$$\int_a^b [f(x) + g(x)]\, dx = \int_a^b f(x)\, dx + \int_a^b g(x)\, dx$$

■ 5.7 THE FIRST FUNDAMENTAL THEOREM OF CALCULUS

> *In the previous section we defined the concept of a definite integral but did not give any general methods for evaluating them. In this section we shall give a method for using antiderivatives to evaluate definite integrals.*

☐ **THE FIRST FUNDAMENTAL THEOREM OF CALCULUS**

The following theorem is the basic tool for evaluating definite integrals.

5.7.1   THEOREM (*The First Fundamental Theorem of Calculus*).   *If f is continuous on $[a, b]$ and if F is an antiderivative of f on $[a, b]$, then*

$$\int_a^b f(x)\, dx = F(b) - F(a) \tag{1}$$

Proof.   Let $x_1, x_2, \ldots, x_{n-1}$ be any points in $[a, b]$ such that

$$a < x_1 < x_2 < \cdots < x_{n-1} < b$$

These points divide $[a, b]$ into n subintervals

$$[a, x_1], [x_1, x_2], \ldots, [x_{n-1}, b] \tag{2}$$

whose lengths, as usual, we denote by

$$\Delta x_1, \Delta x_2, \ldots, \Delta x_n \tag{3}$$

We can write $F(b) - F(a)$ as a telescoping sum:

$$F(b) - F(a) = [F(x_1) - F(a)] + [F(x_2) - F(x_1)]$$
$$+ [F(x_3) - F(x_2)] + \cdots + [F(b) - F(x_{n-1})] \tag{4}$$

By hypothesis,

$$F'(x) = f(x) \tag{5}$$

for all x in $[a, b]$, so F satisfies the hypotheses of the Mean-Value Theorem (Theorem 4.10.2) on each subinterval in (2). Hence, (4) can be rewritten as

$$F(b) - F(a) = F'(x_1^*)(x_1 - a) + F'(x_2^*)(x_2 - x_1)$$
$$+ F'(x_3^*)(x_3 - x_2) + \cdots + F'(x_n^*)(b - x_{n-1}) \tag{6}$$

where $x_1^*, x_2^*, \ldots, x_n^*$ are points in successive subintervals.
  Using (3) and (5) we can rewrite (6) as

$$F(b) - F(a) = f(x_1^*)\, \Delta x_1 + f(x_2^*)\, \Delta x_2 + f(x_3^*)\, \Delta x_3 + \cdots + f(x_n^*)\, \Delta x_n$$

or in sigma notation,

$$F(b) - F(a) = \sum_{k=1}^{n} f(x_k^*) \, \Delta x_k \tag{7}$$

Let us now increase n in such a way that max $\Delta x_k \to 0$. Since f is assumed to be continuous, the right side of (7) approaches $\int_a^b f(x) \, dx$, by Theorem 5.6.6(a). However, the left side of (7) is a constant that is independent of n; thus,

$$F(b) - F(a) = \lim_{\max \Delta x_k \to 0} \sum_{k=1}^{n} f(x_k^*) \, \Delta x_k = \int_a^b f(x) \, dx \quad \blacksquare$$

The difference $F(b) - F(a)$ is commonly denoted by $F(x)]_a^b$ so that (1) can be written as

$$\int_a^b f(x) \, dx = F(x) \Big]_a^b \tag{8}$$

Some other common notations are

$$\int_a^b f(x) \, dx = \left[F(x) \right]_a^b \quad \text{and} \quad \int_a^b f(x) \, dx = F(x) \Big]_{x=a}^b$$

The latter notation emphasizes that the limits of integration refer to the variable x. This can be important in problems where more than one variable occurs.

Evaluate $\int_1^2 x \, dx$.

The function $F(x) = \frac{1}{2}x^2$ is an antiderivative of $f(x) = x$; thus, from (8)

$$\int_1^2 x \, dx = \frac{1}{2}x^2 \Big]_1^2 = \frac{1}{2}(2)^2 - \frac{1}{2}(1)^2 = 2 - \frac{1}{2} = \frac{3}{2} \quad \blacktriangleleft$$

□ **THE RELATIONSHIP BETWEEN DEFINITE AND INDEFINITE INTEGRALS**

When applying the First Fundamental Theorem of Calculus, it does not matter which antiderivative of f is used, for if F is any antiderivative of f on $[a, b]$, then all others have the form

$$F(x) + C \quad \boxed{\text{Theorem 5.2.2}}$$

Thus,

$$[F(x) + C]_a^b = [F(b) + C] - [F(a) + C]$$

$$= F(b) - F(a) = F(x)]_a^b$$

$$= \int_a^b f(x) \, dx \tag{9}$$

which shows that all antiderivatives of f on $[a, b]$ yield the same value for $\int_a^b f(x)\,dx$. Since

$$\int f(x)\,dx = F(x) + C$$

it follows from (9) that

$$\int_a^b f(x)\,dx = \left[\int f(x)\,dx\right]_a^b \tag{10}$$

which relates the definite and indefinite integrals of f.

Example 2 Use the First Fundamental Theorem of Calculus to find the area under the curve $y = \cos x$ over the interval $[0, \pi/2]$. (See Figure 5.7.1.)

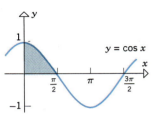

Figure 5.7.1

Solution. Since $\cos x \geq 0$ for $0 \leq x \leq \pi/2$, the area is

$$A = \int_0^{\pi/2} \cos x\,dx = \left[\int \cos x\,dx\right]_0^{\pi/2}$$

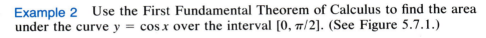

From (10)

$$= \sin x\,\Big]_0^{\pi/2} = \sin\frac{\pi}{2} - \sin 0 = 1 \quad \blacktriangleleft$$

REMARK. In the third equality of this example we took the constant of integration for the indefinite integral to be $C = 0$. This is justified because we can select any antiderivative of f on $[a, b]$, in particular the one for which $C = 0$.

☐ PROPERTIES OF THE DEFINITE INTEGRAL

The following properties of definite integrals follow from the definition of a definite integral. We shall omit the proofs.

5.7.2 THEOREM. *If f and g are integrable on $[a, b]$ and if c is a constant, then cf, $f + g$, and $f - g$ are integrable on $[a, b]$ and*

(a) $\displaystyle\int_a^b cf(x)\,dx = c\int_a^b f(x)\,dx$

(b) $\displaystyle\int_a^b [f(x) + g(x)]\,dx = \int_a^b f(x)\,dx + \int_a^b g(x)\,dx$

(c) $\displaystyle\int_a^b [f(x) - g(x)]\,dx = \int_a^b f(x)\,dx - \int_a^b g(x)\,dx$

Part (b) of this theorem can be extended to more than two functions. More precisely,

$$\int_a^b [f_1(x) + f_2(x) + \cdots + f_n(x)]\, dx$$

$$= \int_a^b f_1(x)\, dx + \int_a^b f_2(x)\, dx + \cdots + \int_a^b f_n(x)\, dx \tag{11}$$

REMARK. The bracket notation $[F(x)]_a^b$ has some properties in common with the definite integral. We leave it as an exercise to show that

$$[F(x) + G(x)]_a^b = F(x)]_a^b + G(x)]_a^b$$
$$[F(x) - G(x)]_a^b = F(x)]_a^b - G(x)]_a^b$$
$$[cF(x)]_a^b = c[F(x)]_a^b$$

Example 3 Evaluate $\displaystyle\int_0^3 (x^3 - 4x + 1)\, dx$.

Solution.

$$\int_0^3 (x^3 - 4x + 1)\, dx = \int_0^3 x^3\, dx - 4\int_0^3 x\, dx + \int_0^3 dx$$

$$= \left[\frac{x^4}{4} - 4\cdot\frac{x^2}{2} + x\right]_0^3$$

$$= \left(\frac{81}{4} - 18 + 3\right) - (0) = \frac{21}{4} \quad \blacktriangleleft$$

Formula (1) in Theorem 5.7.1 is applicable in the cases where $a = b$ or $b < a$. The following example illustrates this.

Example 4

(a) $\displaystyle\int_1^1 x^2\, dx = \frac{x^3}{3}\bigg]_1^1 = \frac{1}{3} - \frac{1}{3} = 0$

(b) $\displaystyle\int_4^0 x\, dx = \frac{x^2}{2}\bigg]_4^0 = \left[\frac{0}{2} - \frac{16}{2}\right] = -8$

This is consistent with the result that would be obtained by first reversing the limits of integration in accordance with Definition 5.6.4(b):

$$\int_4^0 x\, dx = -\int_0^4 x\, dx = -\frac{x^2}{2}\bigg]_0^4 = -\left[\frac{16}{2} - \frac{0}{2}\right] = -8 \quad \blacktriangleleft$$

□ **MORE PROPERTIES OF**
THE DEFINITE INTEGRAL

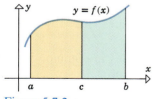

Figure 5.7.2

If f is continuous and nonnegative on $[a, b]$ and if c is a point between a and b, then it is evident that the area under $y = f(x)$ over the interval $[a, b]$ can be split into two parts, the area under the curve from a to c plus the area under the curve from c to b (Figure 5.7.2), that is,

$$\int_a^b f(x)\, dx = \int_a^c f(x)\, dx + \int_c^b f(x)\, dx$$

This is a special case of the following theorem about definite integrals, which we state without proof.

> **5.7.3 THEOREM.** *If f is integrable on a closed interval containing the three points a, b, and c, then*
>
> $$\int_a^b f(x)\, dx = \int_a^c f(x)\, dx + \int_c^b f(x)\, dx \qquad (12)$$
>
> *no matter how the points are ordered.*

This theorem is helpful when the formula for the integrand changes between the limits of integration.

Example 5 Evaluate $\displaystyle\int_0^6 f(x)\, dx$ if

$$f(x) = \begin{cases} x^2, & x < 2 \\ 3x - 2, & x \geq 2 \end{cases}$$

Solution.

$$\int_0^6 f(x)\, dx = \int_0^2 f(x)\, dx + \int_2^6 f(x)\, dx = \int_0^2 x^2\, dx + \int_2^6 (3x - 2)\, dx$$

$$= \frac{x^3}{3}\Big]_0^2 + \left[\frac{3x^2}{2} - 2x\right]_2^6 = \left(\frac{8}{3} - 0\right) + (42 - 2) = \frac{128}{3} \quad \blacktriangleleft$$

Example 6 Evaluate $\displaystyle\int_{-1}^2 |x|\, dx$.

Solution. Since $|x| = x$ when $x \geq 0$ and $|x| = -x$ when $x \leq 0$,

$$\int_{-1}^2 |x|\, dx = \int_{-1}^0 |x|\, dx + \int_0^2 |x|\, dx$$

$$= \int_{-1}^0 (-x)\, dx + \int_0^2 x\, dx$$

$$= -\frac{x^2}{2}\Big]_{-1}^0 + \frac{x^2}{2}\Big]_0^2 = \frac{1}{2} + 2 = \frac{5}{2} \quad \blacktriangleleft$$

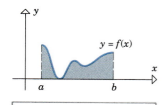

Net signed area ≥ 0

Figure 5.7.3

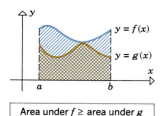

Area under $f \geq$ area under g

Figure 5.7.4

If f is a nonnegative continuous function on $[a, b]$, then the graph of $y = f(x)$ does not cross over the x-axis, so it is evident intuitively that the net signed area between $y = f(x)$ and the x-axis must be greater than or equal to zero (Figure 5.7.3). Similarly, if f and g are nonnegative continuous functions on $[a, b]$, and if $f(x) \geq g(x)$ for all x in $[a, b]$, then the area under $y = f(x)$ must be greater than or equal to the area under $y = g(x)$ over $[a, b]$ since the graph of $y = f(x)$ does not cross over the graph of $y = g(x)$ (Figure 5.7.4). The following theorem, which we state without proof, generalizes these observations to any integrable functions.

5.7.4 THEOREM.

(a) *If f is integrable on $[a, b]$ and $f(x) \geq 0$ for all x in $[a, b]$, then*

$$\int_a^b f(x) \, dx \geq 0$$

(b) *If f and g are integrable on $[a, b]$ and $f(x) \geq g(x)$ for all x in $[a, b]$, then*

$$\int_a^b f(x) \, dx \geq \int_a^b g(x) \, dx$$

REMARK. This theorem remains true if $\geq$ is replaced by $>$ throughout.

Part *(b)* of Theorem 5.7.4 states that one can integrate both sides of an inequality relating integrable functions without altering the sense of the inequality.

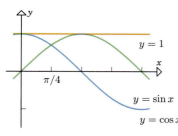

$y = 1$

$\pi/4$

$y = \sin x$

$y = \cos x$

Figure 5.7.5

Example 7 As suggested by Figure 5.7.5, the following inequalities hold on the interval $[0, \pi/4]$:

$$\sin x \leq \cos x \leq 1$$

Thus, from part *(b)* of Theorem 5.7.4, we can conclude that

$$\int_0^{\pi/4} \sin x \, dx \leq \int_0^{\pi/4} \cos x \, dx \leq \int_0^{\pi/4} dx$$

As a check, let us evaluate the integrals; this yields

$$-\cos x \, \Big]_0^{\pi/4} \leq \sin x \, \Big]_0^{\pi/4} \leq x \, \Big]_0^{\pi/4}$$

or

$$1 - \frac{1}{\sqrt{2}} \leq \frac{1}{\sqrt{2}} \leq \frac{\pi}{4}$$

We leave it for the reader to verify that these inequalities are correct (using a calculator, if necessary). ◀

We conclude this section by noting that Theorem 5.7.1 is only applicable to those continuous functions that have antiderivatives. We have not yet addressed the question of which functions actually have antiderivatives. This matter will be taken up in Section 5.9, at which time we will show that all continuous functions have antiderivatives.

► Exercise Set 5.7   C 41, 42, 43, 44

In Exercises 1–22, evaluate the definite integrals using the First Fundamental Theorem of Calculus.

1. $\int_2^3 x^3 \, dx.$

2. $\int_{-1}^1 x^4 \, dx.$

3. $\int_{-1}^2 x(1 + x^3) \, dx.$

4. $\int_{-3}^0 (x^2 - 4x + 7) \, dx.$

5. $\int_1^2 (t^2 - 2t + 8) \, dt.$

6. $\int_0^1 (x^5 - x^3 + 2x) \, dx.$

7. $\int_1^3 \frac{1}{x^2} \, dx.$

8. $\int_1^2 \frac{1}{x^6} \, dx.$

9. $\int_1^2 \left(\frac{1}{x^3} - \frac{2}{x^2} + x^{-4} \right) dx.$

10. $\int_{-2}^{-1} \left(u^{-4} + 3u^{-2} - \frac{1}{u^5} \right) du.$

11. $\int_1^9 \sqrt{x} \, dx.$

12. $\int_1^4 x^{-3/5} \, dx.$

13. $\int_4^9 2y\sqrt{y} \, dy.$

14. $\int_1^8 (5x^{2/3} - 4x^{-2}) \, dx.$

15. $\int_1^4 \left(\frac{3}{\sqrt{x}} - 5\sqrt{x} - x^{-3/2} \right) dx.$

16. $\int_4^9 (4y^{-1/2} + 2y^{1/2} + y^{-5/2}) \, dy.$

17. $\int_{-\pi/2}^{\pi/2} \sin \theta \, d\theta.$

18. $\int_0^{\pi/4} \sec^2 \theta \, d\theta.$

19. $\int_{-\pi/4}^{\pi/4} \cos x \, dx.$

20. $\int_0^1 (x - \sec x \tan x) \, dx.$

21. $\int_{\pi/6}^{\pi/2} \left(x + \frac{2}{\sin^2 x} \right) dx.$

22. $\int_a^{4a} (a^{1/2} - x^{1/2}) \, dx$ (a a positive constant).

In Exercises 23–28, use Theorem 5.7.3 to evaluate the integrals.

23. $\int_0^2 |2x - 3| \, dx.$

24. $\int_1^5 |x - 2| \, dx.$

25. $\int_0^{3\pi/4} |\cos x| \, dx.$

26. $\int_{-1}^2 \sqrt{2 + |x|} \, dx.$

27. $\int_{-2}^3 f(x) \, dx,$ where $f(x) = \begin{cases} -x, & x \ge 0 \\ x^2, & x < 0. \end{cases}$

28. $\int_0^4 f(x) \, dx,$ where $f(x) = \begin{cases} \sqrt{x}, & 0 \le x < 1 \\ 1/x^2, & x \ge 1. \end{cases}$

29. Find $\int_{-1}^2 [x + 2f(x)] \, dx$ if $\int_{-1}^2 f(x) \, dx = 3.$

30. Find $\int_1^4 [3f(x) - g(x)] \, dx$ if

$$\int_1^4 f(x) \, dx = 2 \quad \text{and} \quad \int_1^4 g(x) \, dx = 10.$$

31. Find $\int_1^5 f(x) \, dx$ if

$$\int_0^1 f(x) \, dx = -2 \quad \text{and} \quad \int_0^5 f(x) \, dx = 1.$$

32. Find $\int_3^{-2} f(x) \, dx$ if

$$\int_{-2}^1 f(x) \, dx = 2 \quad \text{and} \quad \int_1^3 f(x) \, dx = -6.$$

In Exercises 33–38, determine whether the value of the integral is positive or negative *without* using the First Fundamental Theorem of Calculus.

33. $\int_2^3 \frac{\sqrt{x}}{1 - x} \, dx.$

34. $\int_{-3}^{-1} \frac{x^4}{\sqrt{3 - x}} \, dx.$

35. $\int_0^4 \dfrac{x^2}{3 - \cos x} \, dx.$ **36.** $\int_{-2}^2 \dfrac{x^3 - 9}{|x| + 1} \, dx.$

37. $\int_2^0 x^2 \sin \sqrt{x} \, dx.$ **38.** $\int_0^{-1} \sqrt[3]{x^2 - 2} \, dx.$

39. Use the fact that

$$\frac{x}{x^4 + 1} < \frac{1}{x^3} \quad \text{for} \quad 1 \le x \le 2$$

to show that

$$\int_1^2 \frac{x}{x^4 + 1} \, dx < \frac{3}{8}.$$

40. Use the fact that

$$\frac{1}{\sqrt{x^3 + 1}} < \frac{1}{x^{3/2}} \quad \text{for} \quad 4 \le x \le 9$$

to show that

$$\int_4^9 \frac{1}{\sqrt{x^3 + 1}} \, dx < \frac{1}{3}.$$

41. Use the inequality

$$\frac{1}{x^{1.01}} \le \frac{1}{x} \le \frac{1}{x^{0.99}}$$

which holds for $x \ge 1$, to find numbers m and M such that $m < M$ and

$$m \le \int_1^3 \frac{1}{x} \, dx \le M.$$

[*Note:* A calculator will be helpful.]

42. Repeat Exercise 41 using the inequality

$$\frac{1}{x^{1.001}} \le \frac{1}{x} \le \frac{1}{x^{0.999}}$$

which holds for $x \ge 1$.

In the directions preceding Exercise 29 of Section 5.6 we defined the *midpoint approximation* of a definite integral. In Exercises 43 and 44, find the midpoint approximation of the integral using $n = 20$ subintervals, then find the exact value of the integral using the First Fundamental Theorem of Calculus.

43. $\int_1^3 \dfrac{1}{x^2} \, dx.$

44. $\int_0^{\pi/2} \sin x \, dx.$

In Exercises 45 and 46, find the limit by evaluating an appropriate definite integral. [*Hint:* Interpret the expression as a limit of Riemann sums in which the interval [0, 1] is divided into subintervals of equal width.]

45. $\displaystyle\lim_{n \to +\infty} \dfrac{\sqrt{1} + \sqrt{2} + \sqrt{3} + \cdots + \sqrt{n}}{n^{3/2}}.$

46. $\displaystyle\lim_{n \to +\infty} \dfrac{1^4 + 2^4 + 3^4 + \cdots + n^4}{n^5}.$

47. Find the area under the curve $y = x^2 + 1$ over the interval [0, 3]. Make a sketch of the region.

48. Find the area above the x-axis, but below the curve $y = (1 - x)(x - 2)$. Make a sketch of the region.

49. Find the area under the curve $y = 3 \sin x$ over the interval $[0, 2\pi/3]$. Sketch the region.

50. Find the area below the interval $[-2, -1]$, but above the curve $y = x^3$. Make a sketch of the region.

51. Find the total area that is between the curve $y = x^2 - 3x - 10$ and the interval $[-3, 8]$. Make a sketch of the region. [*Hint:* Find the portion of area above the interval and the portion of area below the interval separately.]

52. Prove:
(a) $[F(x) + G(x)]_a^b = F(x)]_a^b + G(x)]_a^b$
(b) $[F(x) - G(x)]_a^b = F(x)]_a^b - G(x)]_a^b$
(c) $[cF(x)]_a^b = c[F(x)]_a^b.$

53. Assuming the functions involved are all integrable on $[a, b]$, and $c_1, c_2, \ldots, c_k$ are constants, which of the following are always valid?

(a) $\displaystyle\int_a^b f(x)g(x) \, dx = \int_a^b f(x) \, dx \int_a^b g(x) \, dx$

(b) $\displaystyle\int_a^b [c_1 f(x) + c_2 g(x)] \, dx$

$$= c_1 \int_a^b f(x) \, dx + c_2 \int_a^b g(x) \, dx$$

(c) $\displaystyle\int_a^b \left(\sum_{k=1}^n c_k f_k(x) \right) dx = \sum_{k=1}^n \left[c_k \int_a^b f_k(x) \, dx \right]$

(d) $\displaystyle\int_a^b [f(x)]^n \, dx = \left[\int_a^b f(x) \, dx \right]^n$

(e) $\displaystyle\int_a^b \sqrt{f(x)} \, dx = \sqrt{\int_a^b f(x) \, dx}.$

54. Express Equation (11) in sigma notation.

■ 5.8 EVALUATING DEFINITE INTEGRALS BY SUBSTITUTION

In this section we shall discuss methods of evaluating definite integrals for which a substitution is needed.

We shall consider two methods for evaluating a definite integral

$$\int_a^b h(x)\, dx$$

by substitution.

Method 1

First evaluate the indefinite integral

$$\int h(x)\, dx$$

by substitution, as discussed in Section 5.3, then use the relationship

$$\int_a^b h(x)\, dx = \left[\int h(x)\, dx \right]_a^b$$

to evaluate the definite integral.

Method 2

Avoid the indefinite integral altogether by first expressing the definite integral in the form

$$\int_a^b h(x)\, dx = \int_a^b f(g(x))g'(x)\, dx \tag{1}$$

then making the substitution

$$u = g(x) \quad \text{and} \quad du = g'(x)\, dx$$

directly into the definite integral (1). However, to do this we must change the x-limits of integration to corresponding u-limits of integration: Since $u = g(x)$, it follows that

$$u = g(a) \quad \text{if} \quad x = a$$

and

$$u = g(b) \quad \text{if} \quad x = b$$

Thus, if (1) is expressed in terms of u, we obtain

$$\int_a^b h(x)\ dx = \int_{g(a)}^{g(b)} f(u)\ du$$

With a good choice for the substitution, the new definite integral involving u may be easier to evaluate than the original.

Example 1   Use the two methods above to evaluate $\int_0^2 2x(x^2 + 1)^3\ dx$.

Method 1
   If we let

$$u = x^2 + 1 \quad \text{so that} \quad du = 2x\ dx \tag{2}$$

then we obtain

$$\int 2x(x^2 + 1)^3\ dx = \int u^3\ du = \frac{u^4}{4} + C = \frac{(x^2 + 1)^4}{4} + C$$

Thus,

$$\int_0^2 2x(x^2 + 1)^3\ dx = \left[\int 2x(x^2 + 1)^3\ dx \right]_{x=0}^2 = \frac{(x^2 + 1)^4}{4} \Big]_{x=0}^2$$

$$= \frac{625}{4} - \frac{1}{4} = 156$$

Method 2
   For the substitution in (2) we have

$$u = 1 \quad \text{if} \quad x = 0$$
$$u = 5 \quad \text{if} \quad x = 2$$

Thus,

$$\int_0^2 2x(x^2 + 1)^3\ dx = \int_1^5 u^3\ du = \frac{u^4}{4} \Big]_{u=1}^5 = \frac{625}{4} - \frac{1}{4} = 156$$

which agrees with the result obtained by Method 1.   ◀

   The choice of methods for evaluating a definite integral by substitution is purely a matter of taste. However, since Method 2 requires some thought to obtain the limits of integration, we shall give a few more examples using that method.

Example 2   Evaluate $\int_0^{\pi/4} \cos(\pi - x)\ dx$.

Solution.   Let

$$u = \pi - x \quad \text{so that} \quad du = -dx$$

With this substitution we have

$$u = \pi \quad \text{if} \quad x = 0$$
$$u = 3\pi/4 \quad \text{if} \quad x = \pi/4$$

so

$$\int_0^{\pi/4} \cos(\pi - x)\, dx = \int_\pi^{3\pi/4} \cos u\,(-du)$$

$$= -\int_\pi^{3\pi/4} \cos u\, du = -\sin u \Big]_\pi^{3\pi/4}$$

$$= -[\sin(3\pi/4) - \sin(\pi)]$$

$$= -[1/\sqrt{2} - 0] = -1/\sqrt{2} \quad \blacktriangleleft$$

Example 3 Evaluate $\int_0^{\pi/8} \sin^5 2x \cos 2x\, dx$.

Solution. Let $u = \sin 2x$ so that

$$du = 2\cos 2x\, dx \quad \text{or} \quad \tfrac{1}{2} du = \cos 2x\, dx$$

With this substitution we have

$$u = \sin(0) = 0 \quad \text{if} \quad x = 0$$
$$u = \sin(\pi/4) = 1/\sqrt{2} \quad \text{if} \quad x = \pi/8$$

so

$$\int_0^{\pi/8} \sin^5 2x \cos 2x\, dx = \frac{1}{2}\int_0^{1/\sqrt{2}} u^5\, du = \frac{1}{2}\cdot\frac{u^6}{6}\Big]_0^{1/\sqrt{2}}$$

$$= \frac{1}{2}\left[\frac{1}{6(\sqrt{2})^6} - 0\right] = \frac{1}{96} \quad \blacktriangleleft$$

We conclude this section with the theorem that justifies the substitution method used in the foregoing examples.

5.8.1 THEOREM. *If g' is continuous on $[a, b]$ and f is continuous and has an antiderivative on an interval containing the values of $g(x)$ for $a \le x \le b$, then*

$$\int_a^b f(g(x))g'(x)\, dx = \int_{g(a)}^{g(b)} f(u)\, du$$

provided the integrals exist.

Proof. Let $u = g(x)$ and let F be an antiderivative of f on an interval containing the values of $g(x)$ for $a \le x \le b$. Then by the chain rule

$$\frac{d}{dx} F(g(x)) = \frac{d}{dx} F(u) = \frac{dF}{du}\frac{du}{dx} = f(u)\frac{du}{dx} = f(g(x))g'(x)$$

for each x in $[a, b]$. Thus, $F(g(x))$ is an antiderivative of $f(g(x))g'(x)$ on $[a, b]$. Therefore, by the First Fundamental Theorem of Calculus (Theorem 5.7.1)

$$\int_a^b f(g(x))g'(x)\, dx = F(g(x))\Big]_a^b$$

$$= F(g(b)) - F(g(a))$$

$$= \int_{g(a)}^{g(b)} f(u)\, du \qquad ∎$$

▶ Exercise Set 5.8

1. In each part express the integral in terms of the variable u, but do not evaluate.

(a) $\displaystyle\int_0^2 (x + 1)^7\, dx; \quad u = x + 1$

(b) $\displaystyle\int_{-1}^2 x\sqrt{8 - x^2}\, dx; \quad u = 8 - x^2$

(c) $\displaystyle\int_{-1}^1 \sin(\pi\theta)\, d\theta; \quad u = \pi\theta$

(d) $\displaystyle\int_0^{\pi/4} \tan^2 x \sec^2 x\, dx; \quad u = \tan x$

(e) $\displaystyle\int_0^1 x^3\sqrt{x^2 + 3}\, dx; \quad u = x^2 + 3$

(f) $\displaystyle\int_0^3 (x + 2)(x - 3)^{20}\, dx; \quad u = x - 3.$

In Exercises 2–11, evaluate the integrals two ways: first by a u-substitution in the definite integral, and then by a u-substitution in the corresponding indefinite integral.

2. $\displaystyle\int_1^2 (4x - 2)^3\, dx.$

3. $\displaystyle\int_0^1 (2x + 1)^4\, dx.$

4. $\displaystyle\int_1^2 (4 - 3x)^8\, dx.$

5. $\displaystyle\int_{-1}^0 (1 - 2x)^3\, dx.$

6. $\displaystyle\int_{-5}^0 x\sqrt{4 - x}\, dx.$

7. $\displaystyle\int_0^8 x\sqrt{1 + x}\, dx.$

8. $\displaystyle\int_0^{\pi/6} 2\cos 3x\, dx.$

9. $\displaystyle\int_0^{\pi/2} 4\sin(x/2)\, dx.$

10. $\displaystyle\int_{1-\pi}^{1+\pi} \sec^2(\tfrac{1}{4}x - \tfrac{1}{4})\, dx.$

11. $\displaystyle\int_{-2}^{-1} \frac{x}{(x^2 + 2)^3}\, dx.$

In Exercises 12–29, evaluate the integrals by any method.

12. $\displaystyle\int_0^1 \sqrt[3]{a + bx}\, dx \quad (b \ne 0).$

13. $\displaystyle\int_0^1 \frac{du}{\sqrt{3u + 1}}.$

14. $\displaystyle\int_1^2 \sqrt{5x - 1}\, dx.$

15. $\displaystyle\int_{-1}^1 \frac{x^2\, dx}{\sqrt{x^3 + 9}}.$

16. $\displaystyle\int_{-1}^0 6t^2(t^3 + 1)^{19}\, dt.$

17. $\displaystyle\int_1^3 \frac{x + 2}{\sqrt{x^2 + 4x + 7}}\, dx.$

18. $\displaystyle\int_1^2 \frac{dx}{x^2 - 6x + 9}.$

19. $\displaystyle\int_{-3\pi/4}^{-\pi/4} \sin x \cos x\, dx.$

20. $\displaystyle\int_0^{\pi/4} \sqrt{\tan x}\,\sec^2 x\, dx.$

21. $\displaystyle\int_0^{\sqrt{\pi}} 5x\cos(x^2)\, dx.$

22. $\displaystyle\int_0^{2\pi/t} t^2\sin tx\, dx \quad (t \text{ a positive constant}).$

23. $\displaystyle\int_{\pi^2}^{4\pi^2} \frac{1}{\sqrt{x}}\sin\sqrt{x}\, dx.$

24. $\displaystyle\int_{-\pi/4}^{\pi} \sin\theta\cos\theta\, d\theta.$

25. $\displaystyle\int_0^{\pi/2} \sin^2 3x \cos 3x\, dx.$

26. $\displaystyle\int_0^{\pi/4} \frac{\cos 2x}{\sqrt{7 - 3\sin 2x}}\, dx.$

27. $\displaystyle\int_{\pi/12}^{\pi/9} \sec^2 3\theta\, d\theta.$

28. $\displaystyle\int_{-1}^4 \frac{x\, dx}{\sqrt{5 + x}}.$

29. $\displaystyle\int_0^1 \frac{y^2\, dy}{\sqrt{4 - 3y}}.$

30. Find the area under the curve $y = 1/(3x + 1)^2$ over the interval $[0, 1]$.

31. Find the area under the curve $y = 3 \cos 2x$ over the interval $[0, \pi/8]$.

> In Exercises 32–35, make the indicated substitution and evaluate the resulting definite integral by using an appropriate area formula from plane geometry.

32. $\displaystyle\int_0^{5/3} \sqrt{25 - 9x^2}\, dx$; let $u = 3x$.

33. $\displaystyle\int_{-3}^{1} \sqrt{3 - 2x - x^2}\, dx$; let $u = x + 1$ after completing the square.

34. $\displaystyle\int_0^{2} x\sqrt{16 - x^4}\, dx$; let $u = x^2$.

35. $\displaystyle\int_{\pi/3}^{\pi/2} \sin\theta\sqrt{1 - 4\cos^2\theta}\, d\theta$; let $u = 2\cos\theta$.

36. Find $\displaystyle\int_0^{3} f(3x)\, dx$ if $\displaystyle\int_0^{9} f(x)\, dx = 5$.

37. Find $\displaystyle\int_{1/2}^{1} \frac{1}{x^2} f(1/x)\, dx$ if $\displaystyle\int_1^{2} f(x)\, dx = 3$.

38. Find $\displaystyle\int_{-2}^{0} xf(x^2)\, dx$ if $\displaystyle\int_0^{4} f(x)\, dx = 1$.

39. Find $\displaystyle\int_0^{1} f(3x + 1)\, dx$ if $\displaystyle\int_1^{4} f(x)\, dx = 5$.

40. Prove: If m and n are positive integers, then

$$\int_0^{1} x^m(1 - x)^n\, dx = \int_0^{1} x^n(1 - x)^m\, dx$$

[*Hint:* Do not evaluate; use a substitution.]

41. Prove: If n is a positive integer, then

$$\int_0^{\pi/2} \sin^n x\, dx = \int_0^{\pi/2} \cos^n x\, dx$$

[*Hint:* Do not evaluate; use a trigonometric identity and a substitution.]

42. Let f be a function that is integrable on $[-a, a]$.
  (a) Prove: If $f(-x) = -f(x)$ for all x in $[-a, a]$, then

$$\int_{-a}^{a} f(x)\, dx = 0$$

Give a geometric explanation of this result.

  (b) Prove: If $f(-x) = f(x)$ for all x in $[-a, a]$, then

$$\int_{-a}^{a} f(x)\, dx = 2\int_0^{a} f(x)\, dx$$

Give a geometric explanation of this result.

> In Exercises 43–46, use the results in Exercise 42 to express the value of the integral in terms of k.

43. $\displaystyle\int_{-2}^{2} \frac{x^4}{\sqrt{x^2 + 3}}\, dx$ if $\displaystyle\int_0^{2} \frac{x^4}{\sqrt{x^2 + 3}}\, dx = k$.

44. $\displaystyle\int_{-3}^{3} \frac{x^3}{x^2 + 5}\, dx$ if $\displaystyle\int_0^{3} \frac{x^3}{x^2 + 5}\, dx = k$.

45. $\displaystyle\int_{-1}^{1} \tan^3 x\, dx$ if $\displaystyle\int_0^{1} \tan^3 x\, dx = k$.

46. $\displaystyle\int_0^{1} \frac{\cos x}{x^2 + 1}\, dx$ if $\displaystyle\int_{-1}^{1} \frac{\cos x}{x^2 + 1}\, dx = k$.

47. Find the limit

$$\lim_{n \to +\infty} \sum_{k=1}^{n} \frac{\sin\left(\dfrac{k\pi}{n}\right)}{n}$$

by evaluating an appropriate definite integral over the interval $[0, 1]$.

48. Show that if f and g are continuous functions, then

$$\int_0^{t} f(t - x)g(x)\, dx = \int_0^{t} f(x)g(t - x)\, dx$$

49. (a) Let $I = \displaystyle\int_0^{a} \frac{f(x)}{f(x) + f(a - x)}\, dx$. Show that $I = \dfrac{a}{2}$. [*Hint:* Let $u = a - x$, and then express the integrand as the sum of two fractions.]

  (b) Use the result of part (a) to find

$$\int_0^{3} \frac{\sqrt{x}}{\sqrt{x} + \sqrt{3 - x}}\, dx.$$

  (c) Use the result of part (a) to find

$$\int_0^{\pi/2} \frac{\sin x}{\sin x + \cos x}\, dx.$$

50. Let $I = \displaystyle\int_{-1}^{1} \frac{1}{1 + x^2}\, dx$. Show that the substitution $x = 1/u$ results in

$$I = -\int_{-1}^{1} \frac{1}{1 + u^2}\, du = -I$$

so $2I = 0$, or $I = 0$, which is false (why?). Explain.

■   **5.9**   THE MEAN-VALUE THEOREM FOR INTEGRALS;
           THE SECOND FUNDAMENTAL THEOREM OF CALCULUS

> *The First Fundamental Theorem of Calculus tells us how to evaluate the definite integral of a continuous function if we can find an antiderivative for that function. We have not yet addressed the question of which functions actually have antiderivatives; that is the purpose of the Second Fundamental Theorem of Calculus, which we shall discuss in this section.*

☐ **THE MEAN-VALUE THEOREM FOR INTEGRALS**

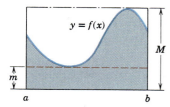

Figure 5.9.1

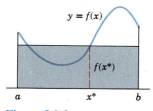

Figure 5.9.2

Our first objective in this section is to obtain a result called the *Mean-Value Theorem for Integrals*. As we shall see, this theorem is not only the cornerstone for the Second Fundamental Theorem of Calculus, but can also be used to define the notion of "average value" for continuous functions.

Let f be a continuous nonnegative function on $[a, b]$, and let m and M be the minimum and maximum values of $f(x)$ on this interval. Consider the rectangles of heights m and M over the interval $[a, b]$ (Figure 5.9.1). It is clear geometrically from this figure that the area

$$A = \int_a^b f(x)\,dx$$

under $y = f(x)$ is at least as large as the area of the rectangle of height m and no larger than the area of the rectangle of height M. It seems reasonable, therefore, that there is a rectangle over the interval $[a, b]$ of some appropriate height $f(x^*)$ between m and M whose area is precisely A; that is,

$$\int_a^b f(x)\,dx = f(x^*)(b - a)$$

(Figure 5.9.2). This is a special case of the following result.

5.9.1   THEOREM (*The Mean-Value Theorem for Integrals*).   *If f is continuous on a closed interval $[a, b]$, then there is at least one number x^* in $[a, b]$ such that*

$$\int_a^b f(x)\,dx = f(x^*)(b - a) \qquad (1)$$

Proof.   By the Extreme-Value Theorem (4.6.4), f assumes a maximum value M and a minimum value m on $[a, b]$. Thus, for all x in $[a, b]$,

$$m \le f(x) \le M$$

and from Theorem 5.7.4(*b*)

$$\int_a^b m \, dx \le \int_a^b f(x) \, dx \le \int_a^b M \, dx$$

or

$$m(b - a) \le \int_a^b f(x) \, dx \le M(b - a)$$

or

$$m \le \frac{1}{b - a} \int_a^b f(x) \, dx \le M \tag{2}$$

Since (2) states that

$$\frac{1}{b - a} \int_a^b f(x) \, dx \tag{3}$$

is a number between m and M, and since $f(x)$ assumes the values m and M on $[a, b]$, it follows from the Intermediate-Value Theorem (2.7.9) that $f(x)$ must assume the value (3) at some point x^* in $[a, b]$, that is,

$$\frac{1}{b - a} \int_a^b f(x) \, dx = f(x^*) \quad \text{or} \quad \int_a^b f(x) \, dx = f(x^*)(b - a) \quad \blacksquare$$

Example 1 Since $f(x) = x^2$ is continuous on the interval $[1, 4]$, the Mean-Value Theorem for Integrals guarantees that there is a number x^* in $[1, 4]$ such that

$$\int_1^4 x^2 \, dx = f(x^*)(4 - 1) = (x^*)^2(4 - 1) = 3(x^*)^2$$

But

$$\int_1^4 x^2 \, dx = \frac{x^3}{3} \Big]_1^4 = 21$$

so that

$$3(x^*)^2 = 21 \quad \text{or} \quad (x^*)^2 = 7 \quad \text{or} \quad x^* = \pm\sqrt{7}$$

Thus, $x^* = \sqrt{7} \approx 2.65$ is the number in the interval $[1, 4]$ whose existence is guaranteed by the Mean-Value Theorem for Integrals. ◄

□ **AVERAGE VALUE**

The number $f(x^*)$ in Theorem 5.9.1 is closely related to the familiar notion of an *arithmetic average*. To see this, divide the interval $[a, b]$ into n subintervals of equal length

$$\Delta x = \frac{b - a}{n} \tag{4}$$

and choose arbitrary points $x_1^*, x_2^*, \ldots, x_n^*$ in successive subintervals. Then the arithmetic average of the numbers $f(x_1^*), f(x_2^*), \ldots, f(x_n^*)$ is

$$\text{ave} = \frac{1}{n}[f(x_1^*) + f(x_2^*) + \cdots + f(x_n^*)]$$

or from (4)

$$\text{ave} = \frac{1}{b-a}[f(x_1^*)\,\Delta x + f(x_2^*)\,\Delta x + \cdots + f(x_n^*)\,\Delta x]$$

$$= \frac{1}{b-a}\sum_{k=1}^{n} f(x_k^*)\,\Delta x$$

Taking the limit as $n \to +\infty$ yields

$$\lim_{n \to +\infty} \frac{1}{b-a}\sum_{k=1}^{n} f(x_k^*)\,\Delta x = \frac{1}{b-a}\int_a^b f(x)\,dx$$

Since this equation describes what happens when we compute the average of "more and more" values of $f(x)$, we are led to the following definition.

5.9.2 DEFINITION. If f is integrable on $[a, b]$, then the **average value** (or **mean value**) of $f(x)$ on $[a, b]$ is defined to be

$$f(x)_{\text{ave}} = \frac{1}{b-a}\int_a^b f(x)\,dx$$

Example 2 Find the average value of $f(x) = x^2$ on the interval $[1, 4]$.

Solution.

$$f(x)_{\text{ave}} = \frac{1}{b-a}\int_a^b f(x)\,dx = \frac{1}{4-1}\int_1^4 x^2\,dx = \frac{1}{3}(21) = 7 \quad \blacktriangleleft$$

REMARK. In light of Definition 5.9.2, the quantity $f(x^*)$ in the Mean-Value Theorem for Integrals (5.9.1) is just the average value of $f(x)$ over $[a, b]$. Thus, the Mean-Value Theorem for Integrals can be stated in the form

$$\int_a^b f(x)\,dx = (b-a)f(x)_{\text{ave}}$$

☐ **DUMMY VARIABLES**

Before turning to the question of "existence" of antiderivatives, it will be helpful to consider some notational matters. Sometimes it is convenient to use a letter other than x for the variable of integration in a definite integral. For example,

$$\int_a^b f(t)\ dt, \qquad \int_a^b f(u)\ du, \qquad \int_a^b f(y)\ dy$$

These integrals all have the same value. As an example, the following integrals have the same limits of integration but different variables of integration:

$$\int_1^3 x^2\ dx = \frac{x^3}{3}\Bigg]_{x=1}^3 = \frac{27}{3} - \frac{1}{3} = \frac{26}{3}$$

$$\int_1^3 t^2\ dt = \frac{t^3}{3}\Bigg]_{t=1}^3 = \frac{27}{3} - \frac{1}{3} = \frac{26}{3}$$

$$\int_1^3 u^2\ du = \frac{u^3}{3}\Bigg]_{u=1}^3 = \frac{27}{3} - \frac{1}{3} = \frac{26}{3}$$

In general, we have the following result:

> *The value of a definite integral is unaffected if we change the letter used for the variable of integration, but do not change the limits of integration.*

Because the letter used for the variable of integration has no effect on the final value of the definite integral, it is sometimes called a *dummy variable*.

DEFINITE INTEGRALS WITH A VARIABLE UPPER LIMIT OF INTEGRATION

In the work to follow we shall consider definite integrals of the form

$$\int_a^x \underline{\hspace{2cm}}$$

where the upper limit x is allowed to vary. For such integrals we shall use a letter different from x (often t) for the variable of integration; thus, we would write

$$\int_a^x f(t)\ dt \quad \text{rather than} \quad \int_a^x f(x)\ dx$$

This avoids using x in two different ways (as a variable of integration and as a limit of integration), which can cause errors.

Example 3 Evaluate $\displaystyle\int_2^x t^2\ dt$.

Solution.

$$\int_2^x t^2\ dt = \frac{t^3}{3}\Bigg]_{t=2}^x = \frac{x^3}{3} - \frac{8}{3} \qquad \blacktriangleleft$$

Note that the final expression in the foregoing example is a function of x alone. In general, an expression of the form

$$\int_a^x f(t)\, dt$$

represents a function of x—the variable t does not enter into the final result.

☐ **THE SECOND**
FUNDAMENTAL
THEOREM OF CALCULUS

In Section 5.1 we showed informally that if f is a nonnegative continuous function, and if $A(x)$ is the area under the curve $y = f(x)$ over the interval $[a, x]$, then

$$A'(x) = f(x) \tag{5}$$

[See Figure 5.1.2 and Formula (4) in Section 5.1.] If we express $A(x)$ as the definite integral

$$A(x) = \int_a^x f(t)\, dt$$

then it follows from (5) that

$$\frac{d}{dx}\left[\int_a^x f(t)\, dt\right] = f(x)$$

The following theorem shows that this result holds for all continuous functions.

5.9.3   THEOREM (*The Second Fundamental Theorem of Calculus*).   *Let f be a continuous function on an interval I, and let a be any point in I. If F is defined by*

$$F(x) = \int_a^x f(t)\, dt \tag{6}$$

then $F'(x) = f(x)$ at each point x in the interval I.

REMARK.   In this theorem, if x is an endpoint of the interval I, then $F'(x)$ must be interpreted as the derivative from the left or right, as appropriate.

Proof.   If x is not an endpoint of the interval I, then it follows from the definition of a derivative that

$$F'(x) = \lim_{h \to 0} \frac{F(x + h) - F(x)}{h}$$

$$= \lim_{h \to 0} \frac{1}{h}\left[\int_a^{x+h} f(t)\, dt - \int_a^x f(t)\, dt\right]$$

$$= \lim_{h \to 0} \frac{1}{h}\left[\int_a^{x+h} f(t)\, dt + \int_x^a f(t)\, dt\right]$$

$$= \lim_{h \to 0} \frac{1}{h}\int_x^{x+h} f(t)\, dt \qquad \boxed{\text{Theorem 5.7.3}}$$

Applying the Mean-Value Theorem for Integrals (5.9.1) to the last expression, we obtain

$$F'(x) = \lim_{h \to 0} \frac{1}{h} [f(t^*) \cdot h] = \lim_{h \to 0} f(t^*) \tag{7}$$

where t^* is some number between x and $x + h$. Because t^* is between x and $x + h$, it follows that $t^* \to x$ as $h \to 0$. Thus, $f(t^*) \to f(x)$ as $h \to 0$, since f is assumed continuous at x. Therefore, from (7) $F'(x) = f(x)$.

If x is an endpoint of the interval I, then the two-sided limits in the proof must be replaced by the appropriate one-sided limits, but otherwise the arguments are identical. ∎

The result in Theorem 5.9.3 can be expressed by the formula

$$\frac{d}{dx} \left[\int_a^x f(t)\, dt \right] = f(x) \tag{8}$$

In words, this formula states:

Where the integrand is continuous, the derivative of a definite integral with respect to its upper limit is equal to the integrand evaluated at the upper limit.

Example 4   Since $f(x) = x^3$ is a continuous function, it follows from Formula (8) that

$$\frac{d}{dx} \left[\int_1^x t^3\, dt \right] = x^3$$

As a check, let us evaluate the integral, then differentiate:

$$\int_1^x t^3\, dt = \frac{t^4}{4} \bigg]_{t=1}^x = \frac{x^4}{4} - \frac{1}{4}$$

Differentiating this function yields x^3 as above.   ◄

Example 5   Since

$$f(x) = \frac{\sin x}{x}$$

is continuous on any interval that does not contain the origin, it follows from (8) that on the interval $(0, +\infty)$ we have

$$\frac{d}{dx} \left[\int_1^x \frac{\sin t}{t}\, dt \right] = \frac{\sin x}{x}$$

Unlike the previous example, there is no way to evaluate the integral in terms of familiar functions, so Formula (8) provides the only simple method for finding the derivative.   ◄

**EXISTENCE OF
ANTIDERIVATIVES FOR
CONTINUOUS FUNCTIONS**

If f is a continuous function on an interval I and a is any point in I, then it follows from (8) that the function

$$F(x) = \int_a^x f(t)\, dt \qquad (9)$$

is an antiderivative of f on I. Thus, thanks to the Second Fundamental Theorem of Calculus, we now know that every function that is continuous on an interval has an antiderivative on that interval.

It should be noted that unlike the indefinite integral

$$\int f(x)\, dx$$

which represents an arbitrary antiderivative of f (because of the constant of integration), the function defined by (9) is not an arbitrary antiderivative of f; it is the specific antiderivative whose value at $x = a$ is zero because

$$F(a) = \int_a^a f(t)\, dt = 0$$

Example 6 The function defined by

$$F(x) = \int_1^x t^3\, dt$$

is that antiderivative of $f(x) = x^3$ whose value at $x = 1$ is zero. As a check, in Example 4 we found that

$$F(x) = \frac{x^4}{4} - \frac{1}{4}$$

so $F(1) = 0$, as expected.   ◄

**FUNCTIONS DEFINED BY
INTEGRALS**

Although we know that continuous functions have antiderivatives, there are two possible impediments that can prevent us from applying the formula

$$\int_a^b f(x)\, dx = F(b) - F(a) \qquad (10)$$

even if f is continuous:

- We may not be clever enough to find a usable formula for an antiderivative F.
- An antiderivative F may not be expressible in terms of familiar functions.

To illustrate the second situation, suppose we are interested in evaluating the integral

$$\int_1^3 \frac{1}{x}\, dx \qquad (11)$$

Because $f(x) = 1/x$ is continuous on any interval that does not contain the origin, the function f is integrable on the interval $[1, 3]$, and we are guaranteed by the Second Fundamental Theorem of Calculus that there is some anti-derivative for f on this interval. One such antiderivative is

$$F(x) = \int_1^x \frac{1}{t}\, dt \tag{12}$$

However, this formula is of little direct help in evaluating (11) because applying (10) to (11) with this antiderivative yields

$$\int_1^3 \frac{1}{x}\, dx = F(3) - F(1) = \int_1^3 \frac{1}{t}\, dt - \int_1^1 \frac{1}{t}\, dt = \int_1^3 \frac{1}{t}\, dt$$

which is a true but valueless equation for the purpose of obtaining a numerical value for (11).

Although we might try looking for an antiderivative formula for $1/x$ that does not involve an integral, we would not have any success since it can be proved that no antiderivative of $1/x$ can be expressed in terms of finitely many poly-nomials, rational functions, trigonometric functions, or any other kinds of func-tions encountered thus far in this text. In short, our repertoire of basic functions is too limited, at present, to produce an antiderivative of $1/x$ that is simpler than (12). However, this problem does not totally preclude the possibility of obtaining a numerical value for (11), since this integral can be approximated to any degree of accuracy using Riemann sums. In Chapter 9 we shall discuss a variety of methods for approximating the value of a definite integral.

Since (12) cannot be expressed in terms of familiar functions, this formula actually defines a new function for us. This function is of special importance and will be studied in detail in Chapter 7. Many of the most important functions in scientific applications arise as integrals from the Second Fundamental Theo-rem of Calculus.

▶ **Exercise Set 5.9**

In Exercises 1–7, find the average value of the function over the given interval.

1. $f(x) = 3x$; $[1, 3]$.

2. $f(x) = x^2$; $[-1, 2]$.

3. $f(x) = \sin x$; $[0, \pi]$.

4. $f(x) = \cos 2x$; $[0, \pi]$.

5. $f(x) = \sqrt{3x + 1}$; $[0, 5]$.

6. $f(x) = 1/\sqrt{x}$; $[1, 4]$.

7. $f(x) = \sqrt{4 - x^2}$; $[-2, 2]$. [*Hint:* Evaluate the inte-gral by using a formula from geometry.]

8. (a) Find f_{ave} of $f(x) = 2x$ over $[0, 4]$.
   (b) Find a point x^* in $[0, 4]$ such that $f(x^*) = f_{ave}$.
   (c) Sketch the graph of $f(x) = 2x$ over $[0, 4]$ and construct a rectangle over the interval whose area is the same as the area under the graph of f over the interval.

9. (a) Find f_{ave} of $f(x) = x^2$ over $[0, 2]$.
   (b) Find a point x^* in $[0, 2]$ such that $f(x^*) = f_{ave}$.
   (c) Sketch the graph of $f(x) = x^2$ over $[0, 2]$ and construct a rectangle over the interval whose area is the same as the area under the graph of f over the interval.

In Exercises 10–13, find the average value of $f(x)$ over the interval and find all values of x^* described in the Mean-Value Theorem for Integrals.

10. $f(x) = 1/x^2$; $[1, 3]$.

11. $f(x) = \sqrt{x}$; $[0, 9]$.

12. $f(x) = \sin x$; $[-\pi, \pi]$.

13. $f(x) = \alpha x + \beta$ $(\alpha \neq 0)$; $[x_0, x_1]$.

14. Electricity is supplied to homes in the form of *alternating current*, which means that the voltage has a sinusoidal waveform described by

$$V = V_p \sin(2\pi ft)$$

where V_p is the amplitude (or peak voltage) measured in volts, t is time in seconds, and f is the frequency in cycles per second (Figure 5.9.3). (A cycle is the electrical term for one period of the waveform.) Alternating current voltmeters read what is called the *rms* (*root-mean-square*) value of V. This is the square root of the average value of V^2 over one cycle. Show that

$$V_{\text{rms}} = \frac{V_p}{\sqrt{2}}$$

[*Hint:* One cycle corresponds to $0 \leq t \leq 1/f$. Use the identity $\sin^2\theta = \frac{1}{2}(1 - \cos 2\theta)$ to help evaluate the integral.]

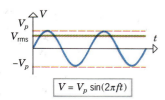

$V = V_p \sin(2\pi ft)$

Figure 5.9.3

15. The standard electrical outlets in many countries supply alternating current with an rms voltage of 120 volts at a frequency of 60 cycles/second. Use the result in Exercise 14 to find the peak voltage at such an outlet.

16. Let $s(t)$ be the position function of a particle moving on a coordinate line. In Section 4.11 (Exercise 24) we defined the average velocity v_{ave} over a time interval $[t_0, t_1]$ by

$$v_{\text{ave}} = \frac{\text{change in position}}{\text{time elapsed}} = \frac{s(t_1) - s(t_0)}{t_1 - t_0}$$

(a) Show that

$$v_{\text{ave}} = \frac{1}{t_1 - t_0} \int_{t_0}^{t_1} v(t)\, dt$$

so that v_{ave} can also be interpreted as the average value of the velocity function over the time interval $[t_0, t_1]$.

(b) Use the result in part (a) to compute the average velocity over the time interval $[0, 5]$ for a particle moving on a coordinate line with velocity function $v(t) = 32t$.

17. For a particle moving on a coordinate line, show that the average acceleration a_{ave} over a time interval $[t_0, t_1]$ (as defined in Exercise 24 of Section 4.11) can be written as

$$a_{\text{ave}} = \frac{1}{t_1 - t_0} \int_{t_0}^{t_1} a(t)\, dt$$

which is the average of the acceleration function over the time interval $[t_0, t_1]$.

18. Consider a particle moving on a coordinate line. Use Exercises 16 and 17 to find

(a) v_{ave} for $1 \leq t \leq 4$ if $v(t) = 3t^3 + 2$

(b) a_{ave} for $2 \leq t \leq 9$ if $a(t) = t^{1/2}$

(c) v_{ave} for $t_0 \leq t \leq t_1$ if $v(t) = 32t + v_0$

                            (v_0 constant).

19. Water is run at a constant rate of 1 ft³/min to fill a cylindrical tank of radius 3 ft and height 5 ft. Assuming the tank is initially empty, find the average force on the bottom over the time period required to fill the tank (density of water = 62.4 lb/ft³).

20. Prove: If $f(x) = k$ is constant on $[a, b]$, then $f_{\text{ave}} = k$ on $[a, b]$.

21. (a) Prove: If f is continuous on $[a, b]$, then

$$\int_a^b [f(x) - f_{\text{ave}}]\, dx = 0$$

(b) Does there exist a constant $c \neq f_{\text{ave}}$ such that

$$\int_a^b [f(x) - c]\, dx = 0?$$

22. Define $F(x)$ by

$$F(x) = \int_{\pi/4}^{x} \cos 2t\, dt$$

(a) Use the Second Fundamental Theorem of Calculus to find $F'(x)$.

(b) Check the result in part (a) by first integrating and then differentiating.

23. Define $F(x)$ by

$$F(x) = \int_1^x (t^3 + 1)\, dt$$

(a) Use the Second Fundamental Theorem of Calculus to find $F'(x)$.

(b) Check the result in part (a) by first integrating and then differentiating.

In Exercises 24–27, use the Second Fundamental Theorem of Calculus to find the derivative.

24. $\dfrac{d}{dx} \displaystyle\int_0^x \dfrac{dt}{1 + \sqrt{t}}.$ **25.** $\dfrac{d}{dx} \displaystyle\int_1^x \sin\left(\sqrt{t}\right)\, dt.$

26. $\dfrac{d}{dx} \displaystyle\int_0^x \dfrac{t}{\cos t}\, dt.$ **27.** $\dfrac{d}{dx} \displaystyle\int_0^x |t|\, dt.$

In Exercises 28–31, express the antiderivatives as integrals.

28. The antiderivative of $1/(1 + x^2)$ on the interval $(-\infty, +\infty)$ whose value at $x = 1$ is 0.

29. The antiderivative of $1/(x - 1)$ on the interval $(1, +\infty)$ whose value at $x = 2$ is 0.

30. The antiderivative of $1/(x - 1)$ on the interval $(-\infty, 1)$ whose value at $x = -3$ is 0.

31. The antiderivative of $1/(x - 1)$ on the interval $(-\infty, 1)$ whose value at $x = 0$ is 0.

32. (a) Over what open interval does the formula

$$F(x) = \int_1^x \frac{1}{t^2 - 9}\, dt$$

represent an antiderivative of

$$f(x) = \frac{1}{x^2 - 9}?$$

(b) Find a point where the graph of F crosses the x-axis.

33. (a) Over what open interval does the formula

$$F(x) = \int_1^x \frac{dt}{t}$$

represent an antiderivative of $f(x) = 1/x$?

(b) Find a point where the graph of F crosses the x-axis.

34. Let $F(x) = \displaystyle\int_2^x \sqrt{3t^2 + 1}\, dt$. Find

(a) $F(2)$ (b) $F'(2)$ (c) $F''(2)$.

35. Let $F(x) = \displaystyle\int_0^x \frac{\cos t}{t^2 + 3}\, dt$. Find

(a) $F(0)$ (b) $F'(0)$ (c) $F''(0)$.

36. Let $F(x) = \displaystyle\int_0^x \frac{t - 3}{t^2 + 7}\, dt$ for $-\infty < x < +\infty$.

(a) Find the value of x where F attains its minimum value.

(b) Find open intervals over which F is only increasing or only decreasing.

(c) Find open intervals over which F is only concave up or only concave down.

In Exercises 37–40, determine the domain of $F(x)$ and specify all values of x for which $F(x)$ is positive, negative, or zero. (Do not perform the integration.)

37. $F(x) = \displaystyle\int_1^x \frac{t^4}{t^2 + 3}\, dt.$ **38.** $F(x) = \displaystyle\int_2^x \frac{2 - t}{t^2 + 1}\, dt.$

39. $F(x) = \displaystyle\int_{-1}^x \sqrt{4 - t^2}\, dt.$

40. $F(x) = \displaystyle\int_{-3}^x t^2\sqrt{2 - t}\, dt.$

In Exercises 41–43, express $F(x)$ in a piecewise form that does not employ integrals.

41. $F(x) = \displaystyle\int_{-1}^x |t|\, dt.$

42. $F(x) = \displaystyle\int_0^x f(t)\, dt$, where $f(x) = \begin{cases} x, & 0 \le x \le 2 \\ 2, & x > 2. \end{cases}$

43. $F(x) = \displaystyle\int_{-1}^x f(t)\, dt$, where $f(x) = \begin{cases} x^2, & x \le 0 \\ 2x, & x > 0. \end{cases}$

44. Use the Second Fundamental Theorem of Calculus and the chain rule to show that

$$\frac{d}{dx} \int_a^{g(x)} f(t)\, dt = f(g(x))g'(x)$$

In Exercises 45 and 46, use the result in Exercise 44 to perform the differentiation.

45. $\dfrac{d}{dx} \displaystyle\int_1^{x^3} \frac{1}{t}\, dt.$ **46.** $\dfrac{d}{dx} \displaystyle\int_3^{\sin x} \frac{1}{1 + t^2}\, dt.$

47. Prove that the function

$$F(x) = \int_0^x \frac{1}{1 + t^2}\, dt + \int_0^{1/x} \frac{1}{1 + t^2}\, dt$$

is constant on the interval $(0, +\infty)$.

48. Use Exercise 44 and Theorem 5.7.3 to show that

$$\frac{d}{dx} \int_{h(x)}^{g(x)} f(t)\, dt = f(g(x))g'(x) - f(h(x))h'(x)$$

49. Use the result in Exercise 48 to perform the following differentiations:

(a) $\dfrac{d}{dx} \displaystyle\int_{x^2}^{x^3} \sin^2 t\, dt$ (b) $\dfrac{d}{dx} \displaystyle\int_{-x}^{x} \dfrac{1}{1+t}\, dt.$

50. Prove that the function

$$F(x) = \int_{x}^{3x} \frac{1}{t}\, dt$$

is constant on the interval $(0, +\infty)$ by using Exercise 48 to find $F'(x)$.

51. Prove: If f is continuous on an open interval I and b is any point in I, then at each point in I

$$\frac{d}{dx} \int_{x}^{b} f(t)\, dt = -f(x)$$

52. Prove: If f is continuous on an open interval I and a is any point in I, then

$$F(x) = \int_{a}^{x} f(t)\, dt$$

is continuous on I.

▶ SUPPLEMENTARY EXERCISES

In Exercises 1–10, evaluate the integrals and check your results by differentiation.

1. $\displaystyle\int \left[\frac{1}{x^3} + \frac{1}{\sqrt{x}} - 5\sin x \right] dx.$

2. $\displaystyle\int \frac{2t^4 - t + 2}{t^3}\, dt.$ **3.** $\displaystyle\int \frac{(\sqrt{x} + 2)^8}{\sqrt{x}}\, dx.$

4. $\displaystyle\int x^3 \cos(2x^4 - 1)\, dx.$

5. $\displaystyle\int \frac{x \sin\sqrt{2x^2 - 5}}{\sqrt{2x^2 - 5}}\, dx.$

6. $\displaystyle\int \sqrt{\cos\theta}\, \sin(2\theta)\, d\theta.$

7. $\displaystyle\int \sqrt{x}(3 + \sqrt[3]{x^4})\, dx.$

8. $\displaystyle\int \frac{x^{1/3}\, dx}{x^{8/3} + 2x^{4/3} + 1}.$

9. $\displaystyle\int \sec^2(\sin 5t)\cos 5t\, dt.$

10. $\displaystyle\int \frac{\cot^2 x}{\sin^2 x}\, dx.$

11. Evaluate $\int y(y^2 + 2)^2\, dy$ two ways: (a) by multiplying out and integrating term by term; and (b) by using the substitution $u = y^2 + 2$. Show that your answers differ by a constant.

In Exercises 12–17, evaluate the definite integral by making the indicated substitution and changing the x-limits of integration to u-limits.

12. $\displaystyle\int_{1}^{0} \sqrt[5]{1 - 2x}\, dx,\ u = 1 - 2x.$

13. $\displaystyle\int_{0}^{\pi/2} \sin^4 x \cos x\, dx,\ u = \sin x.$

14. $\displaystyle\int_{0}^{-3} \frac{x\, dx}{\sqrt{x^2 + 16}},\ u = x^2 + 16.$

15. $\displaystyle\int_{2}^{5} \frac{x - 2}{\sqrt{x - 1}}\, dx,\ u = x - 1.$

16. $\displaystyle\int_{\pi/6}^{\pi/4} \frac{\sin 2x\, dx}{\sqrt{1 - \frac{3}{2}\cos 2x}},\ u = 1 - \frac{3}{2}\cos 2x.$

17. $\displaystyle\int_{1}^{4} \frac{1}{\sqrt{x}} \cos\left(\frac{\pi\sqrt{x}}{2}\right) dx,\ u = \frac{\pi\sqrt{x}}{2}.$

In Exercises 18 and 19, evaluate $\int_{-2}^{2} f(x)\, dx.$

18. $f(x) = \begin{cases} x^3 & \text{for } x \ge 0 \\ -x & \text{for } x < 0. \end{cases}$

19. $f(x) = |2x - 1|.$

In Exercises 20–22, solve for x.

20. $\displaystyle\int_{1}^{x} \frac{1}{\sqrt{t}}\, dt = 3.$ **21.** $\displaystyle\int_{0}^{x} \frac{1}{(3t + 1)^2}\, dt = \frac{1}{6}.$

22. $\displaystyle\int_{2}^{x} (4t - 1)\, dt = 9.$

23. Evaluate:

(a) $\displaystyle\sum_{i=3}^{6} 5$

(b) $\displaystyle\sum_{i=n}^{n+3} 2$

(c) $\displaystyle\sum_{i=n}^{n+3} n$

(d) $\displaystyle\sum_{k=1}^{3} \left(\frac{k-1}{k+3}\right)$

(e) $\displaystyle\sum_{k=2}^{4} \frac{6}{k^2}$

(f) $\displaystyle\sum_{n=4}^{4} (2n+1)$

(g) $\displaystyle\sum_{k=0}^{4} \sin(k\pi/4)$

(h) $\displaystyle\sum_{k=1}^{4} \sin^k(\pi/4)$.

24. Express in sigma notation and evaluate:
(a) $3 \cdot 1 + 4 \cdot 2 + 5 \cdot 3 + \cdots + 102 \cdot 100$
(b) $200 + 198 + \cdots + 4 + 2$.

25. Express in sigma notation, first starting with $k = 1$, and then with $k = 2$. (Do not evaluate.)

(a) $\dfrac{1}{4} - \dfrac{4}{9} + \dfrac{9}{16} - \cdots - \dfrac{64}{81} + \dfrac{81}{100}$

(b) $\dfrac{\pi^2}{1} - \dfrac{\pi^3}{2} + \dfrac{\pi^4}{3} - \cdots + \dfrac{\pi^{12}}{11}$.

In Exercises 26–29, use the partition of $[a, b]$ into n subintervals of equal length, and find a closed form for the sum of the areas of (a) the inscribed rectangles and (b) the circumscribed rectangles. (c) Use your answer in either part (a) or part (b) to find the area under the curve $y = f(x)$ over the interval $[a, b]$. (Check your answer by integration.)

26. $f(x) = 6 - 2x; a = 1, b = 3$.

27. $f(x) = 16 - x^2; a = 0, b = 4$.

28. $f(x) = x^2 + 2; a = 1, b = 4$.

29. $f(x) = 6; a = -1, b = 1$.

30. The function $f(x) = \sqrt{x}$ is continuous on $[0, 4]$, so by Theorem 5.6.6 the integral $\displaystyle\int_0^4 \sqrt{x}\, dx$ exists. Find

its value by using Definition 5.6.3. Use subintervals of unequal length by taking the points $4k^2/n^2$ $(k = 1, 2, \ldots, n - 1)$ as the points of subdivision of $[0, 4]$, and let x_k^* be the right-hand endpoint of the kth subinterval $(k = 1, 2, \ldots, n)$.

31. Given that

$$\int_1^5 P(x)\, dx = -1, \quad \int_3^5 P(x)\, dx = 3,$$

and

$$\int_3^5 Q(x)\, dx = 4$$

evaluate the following:

(a) $\displaystyle\int_3^5 [2P(x) + Q(x)]\, dx$

(b) $\displaystyle\int_5^1 P(t)\, dt$

(c) $\displaystyle\int_{-3}^{-5} Q(-x)\, dx$

(d) $\displaystyle\int_3^1 P(x)\, dx$.

32. Suppose that f is continuous and $x^2 \le f(x) \le 6$ for all x in $[-1, 2]$. Find values of A and B such that

$$A \le \int_{-1}^{2} f(x)\, dx \le B$$

In Exercises 33–36, find the average value of $f(x)$ over the indicated interval and all values of x^* described in the Mean-Value Theorem for Integrals.

33. $f(x) = 3x^2; [-2, -1]$.

34. $f(x) = \dfrac{x}{\sqrt{x^2 + 9}}; [0, 4]$.

35. $f(x) = 2 + |x|; [-3, 1]$.

36. $f(x) = \sin^2 x; [0, \pi]$.
[*Hint:* $\sin^2 x = \frac{1}{2}(1 - \cos 2x)$.]

6

Applications of the Definite Integral

Gottfried Wilhelm Leibniz   (1646-1716)

■ **6.1** AREA BETWEEN TWO CURVES

> *In this section we shall discuss methods for calculating the area between curves in the plane.*

☐ **AREA BETWEEN** $y = f(x)$
AND $y = g(x)$

FIRST AREA PROBLEM. *Suppose that f and g are continuous functions on an interval $[a, b]$ and*

$$f(x) \geq g(x) \quad for \quad a \leq x \leq b$$

[This means that the curve $y = f(x)$ can touch but not cross the curve $y = g(x)$ over $[a, b]$.] Find the area A of the region bounded above by $y = f(x)$, below by $y = g(x)$, and on the sides by the lines $x = a$ and $x = b$ (Figure 6.1.1a).

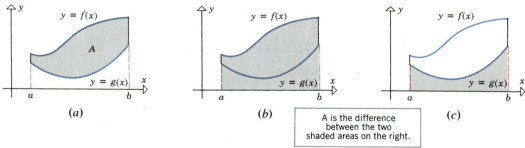

(*a*) (*b*) (*c*)

A is the difference between the two shaded areas on the right.

Figure 6.1.1

If f and g are nonnegative on $[a, b]$, then as illustrated in Figure 6.1.1b, we have

$$A = [\text{area under } f] - [\text{area under } g]$$

or equivalently

$$A = \int_a^b f(x)\, dx - \int_a^b g(x)\, dx = \int_a^b [f(x) - g(x)]\, dx$$

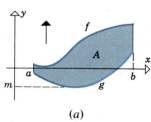

(*a*)

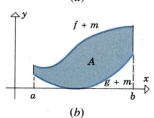

(*b*)

Figure 6.1.2

This formula can be extended to the case where g has negative values by translating the curves $y = f(x)$ and $y = g(x)$ upward until both are above the x-axis. To do this, let $-m$ be the minimum value of $g(x)$ on $[a, b]$ (Figure 6.1.2a). Since $g(x) \geq -m$, it follows that

$$g(x) + m \geq 0$$

so the functions $g + m$ and $f + m$ are nonnegative on $[a, b]$ (Figure 6.1.2b). It is intuitively clear that the area of a region is unchanged by translation, so the area A between f and g is the same as the area between $f + m$ and $g + m$. Thus,

$$A = [\text{area under } f + m] - [\text{area under } g + m]$$

or equivalently

$$A = \int_a^b [f(x) + m] \, dx - \int_a^b [g(x) + m] \, dx = \int_a^b [f(x) - g(x)] \, dx$$

which is the same formula we obtained above.

The foregoing discussion can be summarized as follows:

6.1.1 AREA FORMULA. If f and g are continuous functions on the interval $[a, b]$, and if $f(x) \geq g(x)$ for all x in $[a, b]$, then the area of the region bounded above by $y = f(x)$, below by $y = g(x)$, on the left by the line $x = a$, and on the right by the line $x = b$ is

$$A = \int_a^b [f(x) - g(x)] \, dx \qquad (1)$$

When the region is complicated, it may require some careful thought to determine the integrand and limits of integration in (1). Here is a systematic procedure that you can follow to set up this formula.

Step 1. Sketch the region and then draw a vertical line segment through the region at an arbitrary point x, connecting the top and bottom boundaries (Figure 6.1.3a).

Step 2. The top endpoint of the line segment sketched in Step 1 will be $f(x)$, the bottom one $g(x)$, and the length of the line segment will be $f(x) - g(x)$. This is the integrand in (1).

Step 3. To determine the limits of integration, imagine moving the line segment left and then right. The leftmost position at which the line segment intersects the region is $x = a$ and the rightmost is $x = b$ (Figures 6.1.3b and 6.1.3c).

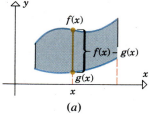 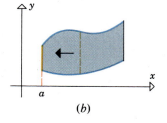 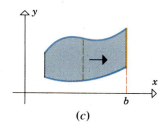

(a) (b) (c)

Figure 6.1.3

REMARK. In Step 1 above, it is not necessary to make an extremely accurate sketch of the region; sufficient accuracy to determine which curve is the upper boundary and which curve is the lower boundary is all that is required.

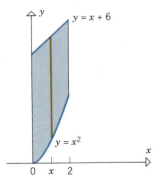

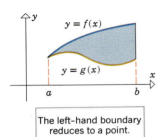

Figure 6.1.4

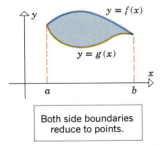

The left-hand boundary reduces to a point.

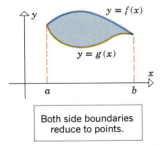

y = f(x)

y = g(x)

Both side boundaries reduce to points.

Figure 6.1.5

Example 1   Find the area of the region bounded above by $y = x + 6$, bounded below by $y = x^2$, and bounded on the sides by the lines $x = 0$ and $x = 2$.

Solution.   The region and a vertical line segment through it are shown in Figure 6.1.4. The line segment extends from $f(x) = x + 6$ on the top to $g(x) = x^2$ on the bottom. If the line segment is moved through the region, its leftmost position will be $x = 0$ and its rightmost will be $x = 2$. Thus, from (1)

$$A = \int_0^2 [(x + 6) - x^2] \, dx = \left[\frac{x^2}{2} + 6x - \frac{x^3}{3} \right]_0^2 = \frac{34}{3} - 0 = \frac{34}{3} \blacktriangleleft$$

Sometimes the upper curve $y = f(x)$ intersects the lower curve $y = g(x)$ at either the left-hand boundary $x = a$, the right-hand boundary $x = b$, or both. When this happens the side of the region where the upper and lower curves intersect reduces to a point, rather than a vertical line segment (Figure 6.1.5).

Example 2   Find the area of the region enclosed between the curves $y = x^2$ and $y = x + 6$.

Solution.   A sketch of the region (Figure 6.1.6) shows that the lower boundary is $y = x^2$ and the upper boundary is $y = x + 6$. At the endpoints of the region, the upper and lower boundaries have the same y-coordinates; thus, to find the endpoints we equate

$$y = x^2 \quad \text{and} \quad y = x + 6 \tag{2}$$

This yields

$$x^2 = x + 6 \quad \text{or} \quad x^2 - x - 6 = 0 \quad \text{or} \quad (x + 2)(x - 3) = 0$$

from which we obtain

$$x = -2 \quad \text{and} \quad x = 3$$

Although the y-coordinates of the endpoints are not essential to our solution, they may be obtained from (2) by substituting $x = -2$ and $x = 3$ in either equation. This yields $y = 4$ and $y = 9$, so the upper and lower boundaries intersect at $(-2, 4)$ and $(3, 9)$.

From (1) with $f(x) = x + 6$, $g(x) = x^2$, $a = -2$, and $b = 3$, we obtain the area

$$A = \int_{-2}^3 [(x + 6) - x^2] \, dx = \left[\frac{x^2}{2} + 6x - \frac{x^3}{3} \right]_{-2}^3$$

$$= \frac{27}{2} - \left(-\frac{22}{3} \right) = \frac{125}{6} \blacktriangleleft$$

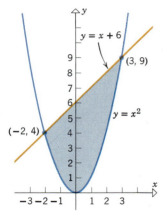

Figure 6.1.6

Example 3   Find the area of the region enclosed by $x = y^2$ and $y = x - 2$.

Solution.   To make an accurate sketch of the region we need to know where the curves $x = y^2$ and $y = x - 2$ intersect. In Example 2 we found intersections by equating the expressions for y. Here it is easier to rewrite the latter equation

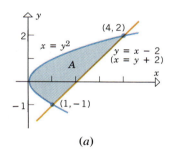

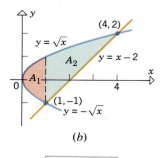

The total area A is $A_1 + A_2$.

Figure 6.1.7

as $x = y + 2$ and equate the expressions for x, namely,

$$x = y^2 \quad \text{and} \quad x = y + 2 \tag{3}$$

This yields

$$y^2 = y + 2 \quad \text{or} \quad y^2 - y - 2 = 0 \quad \text{or} \quad (y + 1)(y - 2) = 0$$

from which we obtain $y = -1$, $y = 2$. Substituting these values in either equation in (3) we see that the corresponding x values are $x = 1$ and $x = 4$, respectively, so the points of intersection are $(1, -1)$ and $(4, 2)$ (Figure 6.1.7*a*).

To apply Formula (1), the equations of the boundaries must be written so that y is expressed explicitly as a function of x. The upper boundary can be written as $y = \sqrt{x}$ (rewrite $x = y^2$ as $y = \pm\sqrt{x}$ and choose the $+$ for the upper portion of the curve). The lower portion of the boundary consists of two parts: $y = -\sqrt{x}$ for $0 \le x \le 1$ and $y = x - 2$ for $1 \le x \le 4$ (Figure 6.1.7*b*). Because of this change in the formula for the lower boundary, it is necessary to divide the region into two parts and find the area of each part separately.

From (1) with $f(x) = \sqrt{x}$, $g(x) = -\sqrt{x}$, $a = 0$, and $b = 1$, we obtain

$$A_1 = \int_0^1 [\sqrt{x} - (-\sqrt{x})]\, dx = 2 \int_0^1 \sqrt{x}\, dx$$

$$= 2 \left[\frac{2}{3} x^{3/2} \right]_0^1 = \frac{4}{3} - 0 = \frac{4}{3}$$

From (1) with $f(x) = \sqrt{x}$, $g(x) = x - 2$, $a = 1$, and $b = 4$, we obtain

$$A_2 = \int_1^4 [\sqrt{x} - (x - 2)]\, dx = \int_1^4 (\sqrt{x} - x + 2)\, dx$$

$$= \left[\frac{2}{3} x^{3/2} - \frac{1}{2} x^2 + 2x \right]_1^4 = \left(\frac{16}{3} - 8 + 8 \right) - \left(\frac{2}{3} - \frac{1}{2} + 2 \right) = \frac{19}{6}$$

Thus, the area of the entire region is

$$A = A_1 + A_2 = \tfrac{4}{3} + \tfrac{19}{6} = \tfrac{9}{2} \quad \blacktriangleleft$$

☐ **AREA BETWEEN $x = v(y)$ AND $x = w(y)$**

Sometimes it is possible to avoid splitting a region into parts by integrating with respect to y rather than x. We shall now show how this can be done.

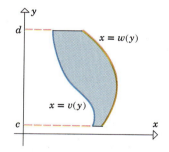

Figure 6.1.8

SECOND AREA PROBLEM. *Suppose that w and v are continuous functions on an interval $[c, d]$ and that*

$$w(y) \ge v(y) \quad \text{for} \quad c \le y \le d$$

[This means that the curve $x = w(y)$ does not cross to the left of the curve $x = v(y)$ over $[c, d]$.] Find the area A of the region bounded on the left by $x = v(y)$, on the right by $x = w(y)$, and above and below by the lines $y = d$ and $y = c$ (Figure 6.1.8).

Proceeding as in the derivation of (1), but with the roles of x and y reversed, yields the following result.

6.1.2 AREA FORMULA.   If w and v are continuous functions and if $w(y) \geq v(y)$ for all y in $[c, d]$, then the area of the region bounded on the left by $x = v(y)$, on the right by $x = w(y)$, below by $y = c$, and above by $y = d$ is

$$A = \int_c^d [w(y) - v(y)]\, dy \tag{4}$$

The procedure for finding the integrand and limits of integration in (4) is similar to that used for (1): Draw a horizontal line segment through the region from boundary to boundary at an arbitrary point y (Figure 6.1.9). The right-hand endpoint of the line segment will be $w(y)$, the left-hand endpoint $v(y)$, and the length of the line segment will be $w(y) - v(y)$, which is the integrand in (4). To determine the limits of integration, imagine moving the line segment down and then up (Figure 6.1.9). The lowest position at which the line segment intersects the region is $y = c$ and the highest is $y = d$.

In Example 3, where we integrated with respect to x to find the area of the region enclosed by $x = y^2$ and $y = x - 2$, we had to split the region into parts and evaluate two integrals. In the next example we shall see that by integrating with respect to y no splitting of the region is necessary.

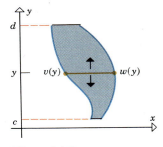

Figure 6.1.9

Example 4   Find the area of the region enclosed by $x = y^2$ and $y = x - 2$, integrating with respect to y.

Solution.   From Figure 6.1.7 the left boundary is $x = y^2$, the right boundary is $y = x - 2$, and the region extends over the interval $-1 \leq y \leq 2$. However, to apply (4) the equations for the boundaries must be written so that x is expressed explicitly as a function of y. Thus, we rewrite $y = x - 2$ as $x = y + 2$. It now follows from (4) that

$$A = \int_{-1}^{2} [(y + 2) - y^2]\, dy = \left[\frac{y^2}{2} + 2y - \frac{y^3}{3} \right]_{-1}^{2} = \frac{9}{2}$$

which agrees with the result obtained in Example 3.   ◀

REMARK.   The choice between Formulas (1) and (4) is generally dictated by the shape of the region, and one would usually choose the formula that requires the least amount of splitting. However, if the integral(s) resulting by one method are difficult to evaluate, then the other method might be preferable, even if it requires more splitting.

▶ Exercise Set 6.1   Ⓒ *39, 40*

In Exercises 1–4, find the area of the shaded region.

1.

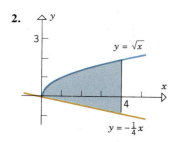

2.

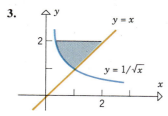

3.

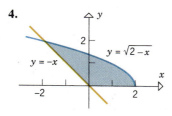

4.

5. Find the area of the region enclosed by the curves $y = x^2$ and $y = 4x$ by integrating
(a) with respect to x
(b) with respect to y.

6. Find the area of the region enclosed by the curves $y^2 = 4x$ and $y = 2x - 4$ by integrating
(a) with respect to x
(b) with respect to y.

7. Find the area of the region enclosed by the curves $y^2 = 2x$ and $y = 2x - 2$ by integrating
(a) with respect to x
(b) with respect to y.

In Exercises 8–27, sketch the region enclosed by the curves and find its area by any method.

8. $y = x^3$, $y = x$, $x = 0$, $x = 1/2$.
9. $y = x^2$, $y = \sqrt{x}$, $x = 1/4$, $x = 1$.
10. $y = x^3 - 4x$, $y = 0$, $x = 0$, $x = 2$.
11. $y = \cos 2x$, $y = 0$, $x = \pi/4$, $x = \pi/2$.
12. $y = x^3 - 4x^2 + 3x$, $y = 0$, $x = 0$, $x = 3$.
13. $x = y^2 - 4y$, $x = 0$, $y = 0$, $y = 4$.
14. $x = \sin y$, $x = 0$, $y = \pi/4$, $y = 3\pi/4$.
15. $y = \sec^2 x$, $y = 2$, $x = -\pi/4$, $x = \pi/4$.
16. $x^2 = y$, $x = y - 2$.
17. $y = x^2 + 4$, $x + y = 6$.
18. $y = x^3$, $y = -x$, $y = 8$.
19. $y^2 = -x$, $y = x - 6$, $y = -1$, $y = 4$.
20. $y = x$, $y = 4x$, $y = -x + 2$.
21. $y = 2 + |x - 1|$, $y = -\frac{1}{5}x + 7$.
22. $y = x^3 - 4x$, $y = 0$, $x = -2$, $x = 2$.
23. $x = y^3 - y$, $x = 0$.
24. $y = x^3 - 2x^2$, $y = 2x^2 - 3x$, $x = 0$, $x = 3$.
25. $y = \sin x$, $y = \cos x$, $x = 0$, $x = 2\pi$.
26. $y = \sqrt{x + 2}$, $y = x$, $y = 0$.
27. $y = 1/x^2$, $y = x$, $y = 4$.
28. Find the area between the curve $y = \sin x$ and the line segment joining the points $(0, 0)$ and $(5\pi/6, 1/2)$ on the curve.
29. Find the area enclosed by the curve $y = \sqrt{x}$, the tangent to the curve at $x = 4$, and the y-axis.
30. Let A be the area enclosed by $y = 1/x^2$, $y = 0$, $x = 1$, and $x = b$ ($b > 1$).
(a) Find A.   (b) Find $\lim_{b \to +\infty} A$.
31. Let A be the area enclosed by $y = 1/\sqrt{x}$, $y = 0$, $x = 1$, and $x = b$ ($b > 1$).
(a) Find A.   (b) Find $\lim_{b \to +\infty} A$.

32. Find a vertical line $x = k$ that divides the area enclosed by $x = \sqrt{y}$, $x = 2$, and $y = 0$ into two equal parts.

33. Find a horizontal line $y = k$ that divides the area between $y = x^2$ and $y = 9$ into two equal parts.

34. (a) Find the area of the region enclosed by the parabola $y = 2x - x^2$ and the x-axis.
(b) Find the value of m so that the line $y = mx$ divides the region in part (a) into two regions of equal area.

35. Show that the area of the ellipse in Figure 6.1.10 is πab. [*Hint:* Use a formula from geometry to help evaluate the definite integral.]

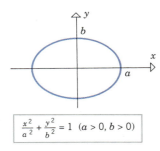

$$\frac{x^2}{a^2} + \frac{y^2}{b^2} = 1 \;\; (a > 0, b > 0)$$

Figure 6.1.10

36. Find the area that is enclosed between the curve $x^{1/2} + y^{1/2} = a^{1/2}$ and the coordinate axes.

37. Suppose that f and g are integrable on $[a, b]$, but neither $f(x) \geq g(x)$ nor $g(x) \geq f(x)$ holds for all x in $[a, b]$ [i.e., the curves $y = f(x)$ and $y = g(x)$ are intertwined].
(a) What is the geometric significance of the integral
$$\int_a^b [f(x) - g(x)]\, dx?$$
(b) What is the geometric significance of the integral
$$\int_a^b |f(x) - g(x)|\, dx?$$

38. A rectangle with edges parallel to the coordinate axes has one vertex at the origin and the diagonally opposite vertex on the curve $y = kx^m$ at the point where $x = b$ ($b > 0$, $k > 0$, and $m \geq 0$). Show that the fraction of the area of the rectangle that lies between the curve and the x-axis depends on m but not on k or b.

In Exercises 39 and 40, find the area to at least four decimal places. Use Newton's Method (Section 4.9) to approximate the x-coordinates of the points of intersection.

39. Approximate the area of the region that lies below the curve $y = \sin x$ and above the line $y = 0.2x$ where $x \geq 0$.

40. Approximate the area of the region enclosed by the graphs of $y = x^2$ and $y = \cos x$.

■ 6.2 VOLUMES BY SLICING; DISKS AND WASHERS

In this section we shall use definite integrals to find volumes of three-dimensional solids.

☐ **CYLINDERS**

A right-circular cylinder (Figure 6.2.1, center) can be generated by translating a plane circular disk along a line perpendicular to the disk. In general, we define a **right cylinder** to be any solid that can be generated by translating a plane region along a line or **axis** perpendicular to the region. All of the solids in Figure 6.2.1 are right cylinders, the center one being a **right circular cylinder**. Observe that all cross sections of a right cylinder taken perpendicular to the axis are identical in size and shape.

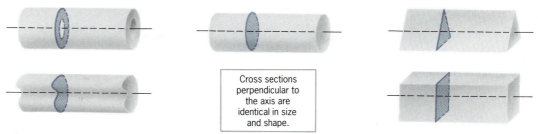

Cross sections perpendicular to the axis are identical in size and shape.

Figure 6.2.1

☐ THE METHOD OF SLICING

h

A

Volume = $A \cdot h$

Figure 6.2.2

If a right cylinder is generated by moving a plane region of area A through a distance h (Figure 6.2.2), then the volume V of the cylinder is *defined* to be

$$V = A \cdot h$$

that is, *the volume is the cross-sectional area times the height.*

Volumes of solids that are neither right cylinders nor composed of finitely many right cylinders can be obtained by a technique called "slicing." To illustrate the idea, suppose that a solid S extends along the x-axis and is bounded on the left and right by planes perpendicular to the x-axis at $x = a$ and $x = b$ (Figure 6.2.3). Because the solid S is not assumed to be a right cylinder, its cross sections perpendicular to the x-axis can vary from point to point; we will denote by $A(x)$ the area of the cross section at x (Figure 6.2.3).

Let us divide the interval $[a, b]$ into n subintervals with widths

$$\Delta x_1, \Delta x_2, \ldots, \Delta x_n$$

by inserting points

$$x_1, x_2, \ldots, x_{n-1}$$

between a and b, and let us pass a plane perpendicular to the x-axis through each of these points. As illustrated in Figure 6.2.4, these planes cut the solid S into n slices

$$S_1, S_2, \ldots, S_n$$

Consider a typical slice S_k. In general, this slice may not be a right cylinder because its cross section can vary. However, if the slice is very thin, the cross section will not vary much. Therefore, if we choose an arbitrary point x_k^* in

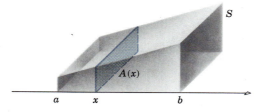

S

$A(x)$

a x b

Figure 6.2.3

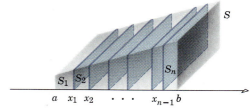

S

S_1 S_2 S_n

a x_1 x_2 $\cdots$ x_{n-1} b

Figure 6.2.4

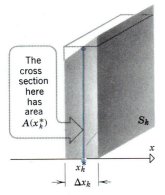

The cross section here has area $A(x_k^*)$

Figure 6.2.5

the kth subinterval, each cross section of slice S_k will be approximately the same as the cross section at x_k^*, and we can approximate slice S_k by a right cylinder of thickness Δx_k and cross-sectional area $A(x_k^*)$ (Figure 6.2.5).

Thus, the volume V_k of slice S_k is approximately the volume of this cylinder, namely,

$$V_k \approx A(x_k^*)\, \Delta x_k$$

and the volume V of the entire solid is approximately

$$V = V_1 + V_2 + \cdots + V_n \approx \sum_{k=1}^{n} A(x_k^*)\, \Delta x_k \tag{1}$$

If we now increase the number of slices in such a way that max $\Delta x_k \to 0$, then the slices will become thinner and thinner and our approximations will get better and better. Thus, intuition suggests that approximation (1) will approach the exact value of the volume V as max $\Delta x_k \to 0$, that is,

$$V = \lim_{\max \Delta x_k \to 0} \sum_{k=1}^{n} A(x_k^*)\, \Delta x_k \tag{2}$$

Since the right side of (2) is just the definite integral

$$\int_a^b A(x)\, dx$$

we are led to the following result.

VOLUMES BY CROSS SECTIONS PERPENDICULAR TO THE x-AXIS

6.2.1   VOLUME FORMULA.   Let S be a solid bounded by two parallel planes perpendicular to the x-axis at $x = a$ and $x = b$. If, for each x in $[a, b]$, the cross-sectional area of S perpendicular to the x-axis is $A(x)$, then the volume of the solid is

$$V = \int_a^b A(x)\, dx \tag{3}$$

provided $A(x)$ is integrable.

There is a similar result for cross sections perpendicular to the y-axis.

VOLUMES BY CROSS SECTIONS PERPENDICULAR TO THE y-AXIS

6.2.2   VOLUME FORMULA.   Let S be a solid bounded by two parallel planes perpendicular to the y-axis at $y = c$ and $y = d$. If, for each y in $[c, d]$, the cross-sectional area of S perpendicular to the y-axis is $A(y)$, then the volume of the solid is

$$V = \int_c^d A(y)\, dy \tag{4}$$

provided $A(y)$ is integrable.

Example 1 Derive the formula for the volume of a right pyramid whose altitude is h and whose base is a square with sides of length a.

Solution. As illustrated in Figure 6.2.6a, we introduce a rectangular coordinate system so that the y-axis passes through the apex, and the x-axis passes through the base and is parallel to a side of the base.

At any point y in the interval $[0, h]$ on the y-axis, the cross section perpendicular to the y-axis is a square. If s denotes the length of a side of this square, then by similar triangles (Figure 6.2.6b)

$$\frac{\frac{1}{2}s}{\frac{1}{2}a} = \frac{h - y}{h} \quad \text{or} \quad s = \frac{a}{h}(h - y)$$

Thus, the area $A(y)$ of the cross section at y is

$$A(y) = s^2 = \frac{a^2}{h^2}(h - y)^2$$

and by (4) the volume is

$$V = \int_0^h A(y)\, dy = \int_0^h \frac{a^2}{h^2}(h - y)^2\, dy = \frac{a^2}{h^2}\int_0^h (h - y)^2\, dy$$

$$= \frac{a^2}{h^2}\left[-\frac{1}{3}(h - y)^3\right]_{y=0}^h = \frac{a^2}{h^2}\left[0 + \frac{1}{3}h^3\right] = \frac{1}{3}a^2 h \qquad \blacktriangleleft$$

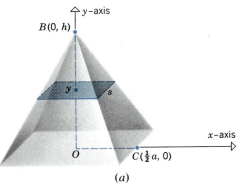

(a)   (b)

Figure 6.2.6

☐ **VOLUMES OF SOLIDS OF REVOLUTION**

Let f be nonnegative and continuous on $[a, b]$, and let R be the region bounded above by the graph of f, below by the x-axis, and on the sides by the lines $x = a$ and $x = b$ (Figure 6.2.7a). When this region is revolved about the x-axis, it generates a solid having circular cross sections (Figure 6.2.7b). Since the cross section at x has radius $f(x)$, the cross-sectional area is

$$A(x) = \pi[f(x)]^2$$

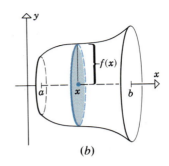

(a)   (b)

Figure 6.2.7

Therefore, from (3), the volume of the solid is

☐ **VOLUMES BY DISKS PERPENDICULAR TO THE *x*-AXIS**

$$V = \int_a^b \pi[f(x)]^2 \, dx \qquad (5)$$

Because the cross sections are circular or disk shaped, the application of this formula is called the ***method of disks.***

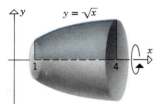

Figure 6.2.8

Example 2 Find the volume of the solid that is obtained when the region under the curve $y = \sqrt{x}$ over the interval $[1, 4]$ is revolved about the *x*-axis (Figure 6.2.8).

Solution. From (5), the volume is

$$V = \int_a^b \pi[f(x)]^2 \, dx = \int_1^4 \pi x \, dx = \frac{\pi x^2}{2} \Big]_1^4 = 8\pi - \frac{\pi}{2} = \frac{15\pi}{2} \qquad \blacktriangleleft$$

Example 3 Derive the formula for the volume of a sphere of radius r.

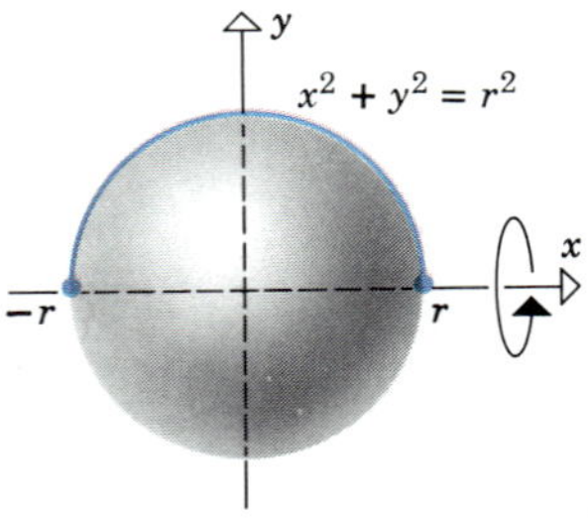

Figure 6.2.9

Solution. As indicated in Figure 6.2.9, a sphere of radius r can be generated by revolving the upper half of the circle

$$x^2 + y^2 = r^2$$

about the *x*-axis. Since the upper half of this circle is the graph of the equation $y = f(x) = \sqrt{r^2 - x^2}$, it follows from (5) that the volume of the sphere is

$$V = \int_a^b \pi[f(x)]^2 \, dx = \int_{-r}^r \pi(r^2 - x^2) \, dx$$

$$= \pi\left[r^2 x - \frac{x^3}{3}\right]_{x=-r}^r = \frac{4}{3}\pi r^3 \qquad \blacktriangleleft$$

We shall now consider more general solids of revolution. Suppose that f and g are nonnegative continuous functions such that

$$g(x) \le f(x) \quad \text{for} \quad a \le x \le b$$

and let R be the region enclosed between the graphs of these functions and the lines $x = a$ and $x = b$ (Figure 6.2.10*a*). When this region is revolved about the x-axis, it generates a solid having annular or washer-shaped cross sections (Figure 6.2.10*b*). Since the cross section at x has inner radius $g(x)$ and outer radius $f(x)$, its area is

$$A(x) = \pi[f(x)]^2 - \pi[g(x)]^2 = \pi([f(x)]^2 - [g(x)]^2)$$

Therefore, from (3), the volume of the solid is

□ **VOLUMES BY WASHERS PERPENDICULAR TO THE x-AXIS**

$$V = \int_a^b \pi([f(x)]^2 - [g(x)]^2) \, dx \qquad (6)$$

The application of this formula is called the ***method of washers.***

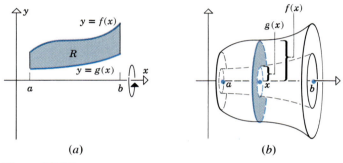

(a)   *(b)*

Figure 6.2.10

Example 4 Find the volume of the solid generated when the region between the graphs of $f(x) = \frac{1}{2} + x^2$ and $g(x) = x$ over the interval $[0, 2]$ is revolved about the x-axis (Figure 6.2.11).

Solution. From (6) the volume is

$$V = \int_a^b \pi([f(x)]^2 - [g(x)]^2) \, dx = \int_0^2 \pi([\tfrac{1}{2} + x^2]^2 - x^2) \, dx$$

$$= \int_0^2 \pi \left(\frac{1}{4} + x^4 \right) dx = \pi \left[\frac{x}{4} + \frac{x^5}{5} \right]_0^2 = \frac{69\pi}{10} \qquad \blacktriangleleft$$

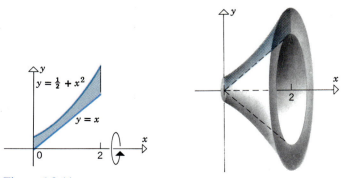

Figure 6.2.11

VOLUMES BY DISKS PERPENDICULAR TO THE y-AXIS

The methods of disks and washers have analogs for regions revolved about the y-axis. If the region of Figure 6.2.12 is revolved about the y-axis, the cross sections of the resulting solid taken perpendicular to the y-axis are disks, and it follows from Formula 6.2.2 that the volume is

$$V = \int_c^d \pi [u(y)]^2 \, dy \tag{7}$$

(Verify.)

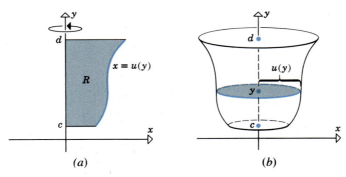

Figure 6.2.12

VOLUMES BY WASHERS PERPENDICULAR TO THE y-AXIS

Also, if the region of Figure 6.2.13a is revolved about the y-axis, the cross sections taken perpendicular to the y-axis are washers, and it follows from Formula 6.2.2 that the volume of the solid in Figure 6.2.13b is

$$V = \int_c^d \pi ([u(y)]^2 - [v(y)]^2) \, dy \tag{8}$$

(Verify.)

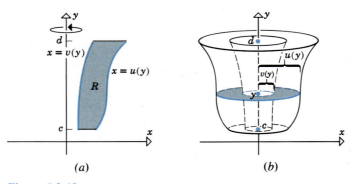

Figure 6.2.13

Example 5 Find the volume of the solid generated when the region enclosed by $y = \sqrt{x}$, $y = 2$, and $x = 0$ is revolved about the y-axis (Figure 6.2.14).

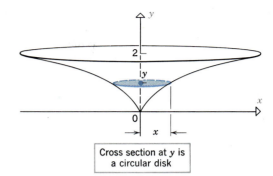

Figure 6.2.14

Solution. The cross sections taken perpendicular to the y-axis are disks, so we shall apply (7). But first we must rewrite $y = \sqrt{x}$ as $x = y^2$. Thus, from (7) with $u(y) = y^2$, the volume is

$$V = \int_c^d \pi[u(y)]^2\, dy = \int_0^2 \pi y^4\, dy = \left.\frac{\pi y^5}{5}\right]_0^2 = \frac{32\pi}{5} \quad \blacktriangleleft$$

▶ Exercise Set 6.2

In Exercises 1–4, find the volume of the solid that results when the shaded region is revolved about the indicated axis.

1.
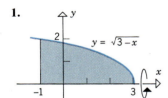

2.
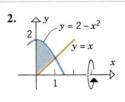

3.
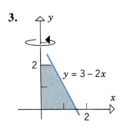

4.
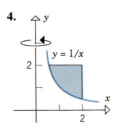

In Exercises 5–16, find the volume of the solid that results when the region enclosed by the given curves is revolved about the x-axis.

5. $y = x^2,\ x = 0,\ x = 2,\ y = 0$.

6. $y = \sec x,\ x = \pi/4,\ x = \pi/3,\ y = 0$.

7. $y = 1 + x^3,\ x = 1,\ x = 2,\ y = 0$.

8. $y = 1/x,\ x = 1,\ x = 4,\ y = 0$.

9. $y = 9 - x^2,\ y = 0$.

10. $y = \sqrt{\cos x},\ x = \pi/4,\ x = \pi/2,\ y = 0$.

11. $y = x^2,\ y = 4x$.

12. $y = x^2,\ y = 9$.

13. $y = \sin x,\ y = \cos x,\ x = 0,\ x = \pi/4$. [*Hint:* Use the identity $\cos 2x = \cos^2 x - \sin^2 x$.]

14. $y = x^2 + 1,\ y = x + 3$.

15. $y = \sqrt{x},\ y = x$.

16. $y = x^2,\ y = x^3$.

In Exercises 17–28, find the volume of the solid that results when the region enclosed by the given curves is revolved about the y-axis.

17. $y = x^3,\ x = 0,\ y = 1$.

18. $x = 1 - y^2,\ x = 0$.

19. $x = \sqrt{1 + y},\ x = 0,\ y = 3$.

20. $x = \sqrt{\cos y},\ y = 0,\ y = \pi/2,\ x = 0$.

21. $x = \csc y,\ y = \pi/4,\ y = 3\pi/4,\ x = 0$.

22. $y = 2/x,\ y = 1,\ y = 3,\ x = 0$.

23. $x = \sqrt{9 - y^2},\ y = 1,\ y = 3,\ x = 0$.

24. $y = x^2 - 1$, $x = 2$, $y = 0$.

25. $y = 1 + x^3$, $x = 1$, $y = 9$.

26. $y = x^2$, $x = y^2$.

27. $x = y^2$, $x = y + 2$.

28. $x = 1 - y^2$, $x = 2 + y^2$, $y = -1$, $y = 1$.

29. Find the volume of the solid that results when the region enclosed by the semicircle $y = \sqrt{25 - x^2}$ and the line $y = 3$ is revolved about the x-axis.

30. Find the volume of the torus that results when the region enclosed by the circle of radius r with center at $(h, 0)$, $h > r$, is revolved about the y-axis. [*Hint:* Use an appropriate formula from plane geometry to help evaluate the definite integral.]

31. Let $V(b)$ be the volume of the solid that results when the region enclosed by $y = 1/x$, $y = 0$, $x = 1$, and $x = b$ $(b > 1)$ is revolved about the x-axis.
    (a) Find $V(b)$.
    (b) Find $\lim_{b \to +\infty} V(b)$.

32. Let h, k, and m be constants such that $h > 0$, $k > 0$, and $m \geq 0$, and let S be the solid that results when the region between the curve $y = kx^m$ and the interval $[0, h]$ on the x-axis is revolved about the x-axis. Let V be the volume of the solid, and let B be the area of the cross section of the solid at $x = h$.
    (a) Show that $V = hB$ when $m = 0$.
    (b) Show that $V = \frac{1}{2}hB$ when $m = \frac{1}{2}$.
    (c) Show that $V = \frac{1}{3}hB$ when $m = 1$.
    (d) Find m so that $V = \frac{1}{n} hB$, where n is a positive integer.

33. Let V be the volume of the solid that results when the region enclosed by $y = \sqrt{x}$, $y = 0$, and $x = b$ $(b > 0)$ is revolved about the x-axis. Find the value of b for which $V = 2$.

34. Let V be the volume of the solid that results when the region enclosed by $y = 1/x$, $y = 0$, $x = 2$, and $x = b$ $(0 < b < 2)$ is revolved about the x-axis. Find the value of b for which $V = 3$.

35. Find the volume of the solid that results when the region enclosed by $y = \sqrt{x}$, $y = 0$, and $x = 9$ is revolved about the line $x = 9$.

36. Find the volume of the solid that results when the region in Exercise 35 is revolved about the line $y = 3$.

37. Find the volume of the solid that results when the region enclosed by $x = y^2$ and $x = y$ is revolved about the line $y = -1$.

38. Find the volume of the solid that results when the region in Exercise 37 is revolved about the line $x = -1$.

39. Find the volume of the solid that results when the region above the x-axis and below the ellipse

$$\frac{x^2}{a^2} + \frac{y^2}{b^2} = 1 \qquad (a > 0, b > 0)$$

is revolved about the x-axis.

40. Find the volume of the solid generated when the region enclosed by $y = \sqrt{x}$, $y = 6 - x$, and $y = 0$ is revolved about the x-axis. [*Hint:* Split the solid into two parts.]

41. Find the volume of the solid generated when the region enclosed by $y = \sqrt{x + 1}$, $y = \sqrt{2x}$, and $y = 0$ is revolved about the x-axis. [*Hint:* Split the solid into two parts.]

42. Let R_1, R_2, R_3, and R_4 be the regions indicated in Figure 6.2.15. Express the following volumes as definite integrals:
    (a) The volume of the solid generated when R_1 is revolved about the x-axis.
    (b) The volume of the solid generated when R_2 is revolved about the x-axis.
    (c) The volume of the solid generated when R_3 is revolved about the y-axis.
    (d) The volume of solid generated when R_4 is revolved about the y-axis.

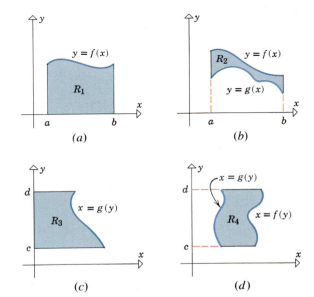

Figure 6.2.15

43. Derive the formula for the volume of a right-circular cone with radius r and height h.

44. A hole of radius $r/2$ is drilled through the center of a sphere of radius r. Find the volume of the remaining solid.

45. A cylindrical hole is drilled all the way through the center of a sphere (Figure 6.2.16). Show that the volume of the remaining solid depends only on the length L of the hole, and not on the size of the sphere.

Figure 6.2.16

46. (a) A vat, shaped like a hemisphere of radius r ft, is filled with a fluid to a depth of h ft. Find the volume of the fluid.

   (b) If fluid enters a hemispherical vat with a radius of 10 ft at a rate of $\frac{1}{2}$ ft³/min, how fast will the fluid be rising when the depth is 5 ft?

47. A cocktail glass with a bowl shaped like a hemisphere of diameter 8 cm contains a cherry with a diameter of 2 cm (Figure 6.2.17). If the glass is filled to a depth of h cm, what is the volume of liquid it contains? [*Hint:* First consider the case where the cherry is partially submerged, then the case where it is totally submerged.]

Figure 6.2.17

48. A *general cylinder* is any solid that can be generated by translating a plane region along a line passing through but not contained in the region. Derive a formula for the volume of a general cylinder with base area A and height h (Figure 6.2.18).

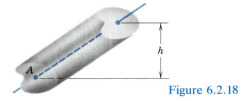

Figure 6.2.18

49. A nose cone for a space reentry vehicle is designed so that a cross section, taken x ft from the tip and perpendicular to the axis of symmetry, is a circle of radius $\frac{1}{4}x^2$ ft. Find the volume of the nose cone given that its length is 20 ft.

50. A certain solid is 1 ft high, and a horizontal cross section taken x ft above the bottom of the solid is an annulus of inner radius x^2 and outer radius $\sqrt{x}$. Find the volume of the solid.

51. The base of a certain solid is the circle $x^2 + y^2 = 9$ and each cross section perpendicular to the x-axis is an equilateral triangle with one side across the base. Find the volume of the solid.

52. The base of a certain solid is the region enclosed by $y = \sqrt{x}$, $y = 0$, and $x = 4$. Every cross section perpendicular to the x-axis is a semicircle with its diameter across the base. Find the volume of the solid.

53. The base of a certain solid is the region enclosed by $y = \sin x$, $y = 0$, $x = \pi/4$, and $x = 3\pi/4$; and every cross section perpendicular to the x-axis is a square with one side across the base. Find the volume of the solid. [*Hint:* To help with the integration, use the identity $\sin^2 x = \frac{1}{2}(1 - \cos 2x)$.]

54. The base of a certain solid is the region enclosed by $y = 1/x$, $y = 0$, $x = 1$, and $x = 3$. Every cross section perpendicular to the x-axis is an isosceles right triangle with its hypotenuse across the base. Find the volume of the solid.

55. A wedge is cut from a right-circular cylinder of radius r by two planes, one perpendicular to the axis of the cylinder and the other making an angle θ with the first. Find the volume of the wedge by slicing perpendicular to the y-axis that is shown in Figure 6.2.19.

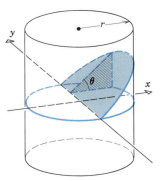

Figure 6.2.19

56. Find the volume of the wedge described in Exercise 55 by slicing perpendicular to the x-axis.

57. Two right-circular cylinders of radius r have axes that intersect at right angles. Find the volume of the solid common to the two cylinders. [*Hint:* One-eighth of the solid is sketched in Figure 6.2.20.]

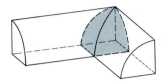

Figure 6.2.20

58. In 1635 Bonaventura Cavalieri, a student of Galileo, stated the following result, called *Cavalieri's principle*: *If two solids have the same height, and if the areas of their cross sections taken parallel to and at equal distances from their bases are always equal, then the solids have the same volume.* Prove this result.

59. Suppose that an open container of unspecified shape contains water that is evaporating into the air (Figure 6.2.21). Let $V(t)$ and $h(t)$ be the volume and depth of the water, respectively, at time t, and let $A(h)$ be the area of the water surface when the depth is h.

Assume that the volume of water decreases at a rate that is proportional to the area of the water surface, that is,

$$\frac{dV}{dt} = -kA(h)$$

where k is a positive constant.

(a) Show that

$$\frac{dh}{dt} = -k$$

that is, the depth decreases at a constant rate regardless of the shape of the container. [*Hint:* At any depth h, $V = \int_0^h A(x)\,dx$.]

(b) If at time $t = 0$ the depth is h_0, how long will it take for all of the water to evaporate? [*Hint:* Solve the equation in part (a) for h in terms of t.]

Figure 6.2.21

■ 6.3 VOLUMES BY CYLINDRICAL SHELLS

The methods discussed so far for computing volumes of solids depend on our ability to compute cross-sectional areas of the solid. In this section we shall develop an alternative technique for computing volumes that can sometimes be applied when it is inconvenient or difficult to determine the cross-sectional areas or if the integration is too difficult.

□ **CYLINDRICAL SHELLS**

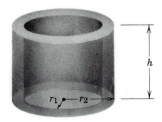

Figure 6.3.1

A *cylindrical shell* is a solid enclosed by two concentric right-circular cylinders (Figure 6.3.1). The volume V of a cylindrical shell having inner radius r_1, outer radius r_2, and height h can be written as

$$V = [\text{area of cross section}] \cdot [\text{height}] = (\pi r_2^2 - \pi r_1^2)h$$

$$= \pi(r_2 + r_1)(r_2 - r_1)h = 2\pi \cdot [\tfrac{1}{2}(r_1 + r_2)] \cdot h \cdot (r_2 - r_1)$$

But $\tfrac{1}{2}(r_1 + r_2)$ is the average radius of the shell and $r_2 - r_1$ is its thickness, so

□ **VOLUME OF A CYLINDRICAL SHELL**

$$V = 2\pi \cdot [\text{average radius}] \cdot [\text{height}] \cdot [\text{thickness}] \qquad (1)$$

We shall now show how this formula can be used to find the volume of a solid of revolution.

Let R be a plane region bounded above by a continuous curve $y = f(x)$, bounded below by the x-axis, and bounded on the left and right, respectively, by the lines $x = a$ and $x = b$. Let S be the solid generated by revolving the region R about the y-axis (Figure 6.3.2).

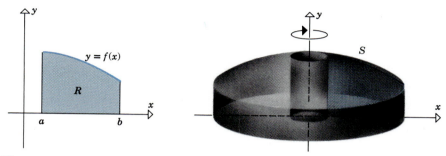

Figure 6.3.2

To find the volume of S, let us divide the interval $[a, b]$ into n subintervals with widths

$$\Delta x_1, \Delta x_2, \ldots, \Delta x_n$$

by inserting points

$$x_1, x_2, \ldots, x_{n-1}$$

between a and b, and let us draw a vertical line through each of these points to divide the region R into n strips $R_1, R_2, \ldots, R_n$ (Figure 6.3.3a).

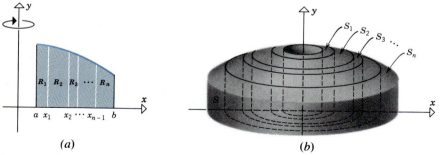

(a)   (b)

Figure 6.3.3

These strips, when revolved about the y-axis, generate solids $S_1, S_2, \ldots, S_n$. As illustrated in Figure 6.3.3b, these solids are nested one inside the other and together form the entire solid S. Thus, the volume of the solid S can be obtained by adding together the volumes of solids $S_1, S_2, \ldots, S_n$:

$$V(S) = V(S_1) + V(S_2) + \cdots + V(S_n) \tag{2}$$

Consider a typical strip R_k and the solid S_k that it generates (Figure 6.3.4). Although solid S_k resembles a cylindrical shell, it will not, in general, be a

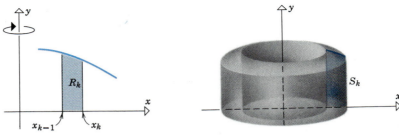

Figure 6.3.4

cylindrical shell because it can have a curved upper surface. However, if the interval width

$$\Delta x_k = x_k - x_{k-1}$$

is small, we can obtain a good approximation to the region R_k by a rectangle of width Δx_k and height $f(x_k^*)$, where

$$x_k^* = \frac{x_k + x_{k-1}}{2}$$

is the midpoint of the interval $[x_{k-1}, x_k]$ (Figure 6.3.5a). This rectangle, when revolved about the y-axis, generates a cylindrical shell, which is a good approximation to the solid S_k (Figure 6.3.5b).

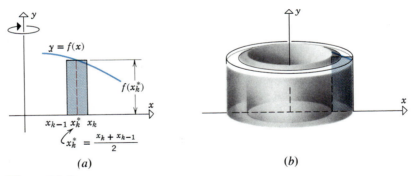

(a)   (b)

Figure 6.3.5

This approximating cylindrical shell has thickness Δx_k, height $f(x_k^*)$, and average radius x_k^*, so by (1)

$$V(S_k) \approx \text{[volume of approximating cylindrical shell]} = 2\pi x_k^* f(x_k^*) \, \Delta x_k$$

Thus, from (2), the volume, $V(S)$, of the entire solid S is approximately

$$\sum_{k=1}^{n} 2\pi x_k^* f(x_k^*) \, \Delta x_k \tag{3}$$

If we now divide $[a, b]$ into more and more subintervals in such a way that $\max \Delta x_k \to 0$, then intuition suggests that our approximations will tend to get better and (3) will approach the exact value of the volume, that is,

$$V(S) = \lim_{\max \Delta x_k \to 0} \sum_{k=1}^{n} 2\pi x_k^* f(x_k^*) \, \Delta x_k \tag{4}$$

Because the right side of (4) is just the definite integral

$$\int_a^b 2\pi x f(x) \, dx$$

we are led to the following result.

☐ **CYLINDRICAL SHELLS CENTERED ON THE *y*-AXIS**

6.3.1 VOLUME FORMULA. Let R be a plane region bounded above by a continuous curve $y = f(x)$, below by the x-axis, and on the left and right, respectively, by the lines $x = a$ and $x = b$. Then the volume of the solid generated by revolving R about the y-axis is given by

$$V = \int_a^b 2\pi x f(x) \, dx \tag{5}$$

Example 1 Use cylindrical shells to find the volume of the solid generated when the region enclosed between $y = \sqrt{x}$, $x = 1$, $x = 4$, and the x-axis is revolved about the y-axis (Figure 6.3.6).

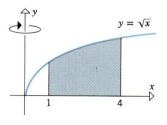

Cutaway view of the solid

Figure 6.3.6

Solution. Since $f(x) = \sqrt{x}$, $a = 1$, and $b = 4$, Formula (5) yields

$$V = \int_1^4 2\pi x \sqrt{x} \, dx = 2\pi \int_1^4 x^{3/2} \, dx$$

$$= \left[2\pi \cdot \frac{2}{5} x^{5/2} \right]_1^4 = \frac{4\pi}{5} [32 - 1] = \frac{124\pi}{5} \quad \blacktriangleleft$$

☐ **VARIATIONS OF THE METHOD OF CYLINDRICAL SHELLS**

The method of cylindrical shells is applicable in a variety of situations that do not fit the conditions required by Formula (5). For example, the solid may be generated by a revolution about the y-axis or some other line, or the region may be enclosed between two curves rather than a curve and a coordinate axis. Rather than develop separate formulas for each of these situations, we shall discuss a general way of thinking about the method of cylindrical shells that will enable the reader to adapt the method to a variety of situations. Consider Formula (5). At each point x in $[a, b]$ the vertical line through x cuts the region R in a line segment that we can view as the vertical "cross section" of R at x (Figure 6.3.7a). When the region R is revolved about the y-axis, the vertical cross section at x generates the *surface* of a right-circular cylinder having height $f(x)$ and radius x (Figure 6.3.7b). The area of this surface is

$$2\pi x f(x)$$

(see Figure 6.3.7c), which is precisely the integrand in (5). Thus, 6.3.1 can be paraphrased as follows:

The volume V by cylindrical shells is the integral of the surface area generated by an arbitrary cross section of R taken parallel to the axis about which R is revolved.

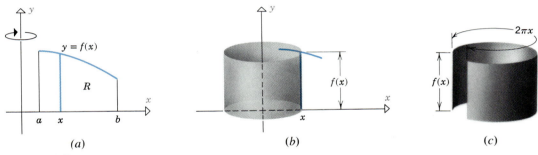

(a)　　　　　**(b)**　　　　　**(c)**

Figure 6.3.7

The following examples illustrate how this result can be used to compute volumes by cylindrical shells in situations where Formula (5) is not applicable.

Example 2　Use cylindrical shells to find the volume of the solid generated when the region R in the first quadrant enclosed between $y = x$ and $y = x^2$ is revolved about the y-axis (Figure 6.3.8).

This solid looks like a bowl with a cone-shaped interior.

Figure 6.3.8

Solution. At each x in $[0, 1]$ the cross section of R parallel to the y-axis generates a cylindrical surface of height $x - x^2$ and radius x. Since the area of this surface is

$$2\pi x(x - x^2)$$

the volume of the solid is

$$V = \int_0^1 2\pi x(x - x^2)\, dx = 2\pi \int_0^1 (x^2 - x^3)\, dx$$

$$= 2\pi \left[\frac{x^3}{3} - \frac{x^4}{4} \right]_0^1 = 2\pi \left[\frac{1}{3} - \frac{1}{4} \right] = \frac{\pi}{6} \quad ◀$$

Example 3 Use cylindrical shells to find the volume of the solid generated when the region R under $y = x^2$ over the interval $[0, 2]$ is revolved about the x-axis (Figure 6.3.9).

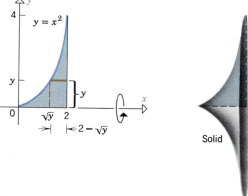

Solid

Figure 6.3.9

Solution. At each y in the interval $0 \le y \le 4$, the cross section of R parallel to the x-axis generates a cylindrical surface of height $2 - \sqrt{y}$ and radius y. Since the area of this surface is

$$2\pi y(2 - \sqrt{y})$$

the volume of the solid is

$$V = \int_0^4 2\pi y(2 - \sqrt{y})\, dy = 2\pi \int_0^4 (2y - y^{3/2})\, dy$$

$$= 2\pi \left[y^2 - \frac{2}{5} y^{5/2} \right]_0^4 = \frac{32\pi}{5} \quad ◀$$

The volume in the foregoing example can also be obtained by the method of disks [Formula (5) of Section 6.2]. The computations are

$$V = \int_0^2 \pi(x^2)^2\, dx = \int_0^2 \pi x^4\, dx = \frac{\pi x^5}{5} \Big]_0^2 = \frac{32}{5}\pi$$

► Exercise Set 6.3

In Exercises 1–4, use cylindrical shells to find the volume of the solid generated when the shaded region is revolved about the indicated axis.

1.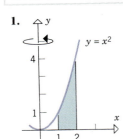

2.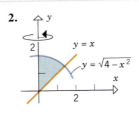

3.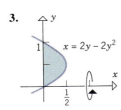

4.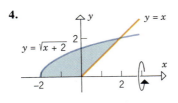

In Exercises 5–12, use cylindrical shells to find the volume of the solid generated when the region enclosed by the given curves is revolved about the y-axis.

5. $y = x^3$, $x = 1$, $y = 0$.

6. $y = \sqrt{x}$, $x = 4$, $x = 9$, $y = 0$.

7. $y = 1/x$, $y = 0$, $x = 1$, $x = 3$.

8. $y = \cos(x^2)$, $x = 0$, $x = \frac{1}{2}\sqrt{\pi}$, $y = 0$.

9. $y = 2x - 1$, $y = -2x + 3$, $x = 2$.

10. $x = y^2$, $y = x^2$.

11. $x^2 + y^3 = 4$, $x = 0$, $x = 4$, $y = 0$.

12. $y = 2x - x^2$, $y = 0$.

In Exercises 13–16, use cylindrical shells to find the volume of the solid generated when the region enclosed by the given curves is revolved about the x-axis.

13. $y^2 = x$, $y = 1$, $x = 0$.

14. $x = 2y$, $y = 2$, $y = 3$, $x = 0$.

15. $y = x^2$, $x = 1$, $y = 0$.

16. $xy = 4$, $x + y = 5$.

17. **(a)** Show by differentiation that

$$\int x \sin x \, dx = \sin x - x \cos x + C$$

(b) Find the volume of the solid generated when the region enclosed by $y = \sin x$ and $y = 0$ for $0 \le x \le \pi$ is revolved about the y-axis.

18. **(a)** Show by differentiation that

$$\int x \cos x \, dx = \cos x + x \sin x + C$$

(b) Find the volume of the solid generated when the region enclosed by $y = \cos x$, $y = 0$, and $x = 0$ for $0 \le x \le \pi/2$ is revolved about the y-axis.

19. **(a)** Use cylindrical shells to find the volume of the solid generated when the region under the curve $y = x^3 - 3x^2 + 2x$ over $[0, 1]$ is revolved about the y-axis.

(b) For this problem, is the method of cylindrical shells easier or harder than the method of slicing discussed in the last section? Explain.

20. Use cylindrical shells to find the volume of the solid that is generated when the region that is enclosed by $y = 1/x^3$, $x = 1$, $x = 2$, $y = 0$ is revolved about the line $x = -1$.

21. Use cylindrical shells to find the volume of the solid that is generated when the region that is enclosed by $y = x^3$, $y = 1$, $x = 0$ is revolved about the line $y = 1$.

22. Let R_1 and R_2 be regions of the form shown in Figure 6.3.10. Use cylindrical shells to find a formula for the volume of the solid that results when
(a) region R_1 is revolved about the y-axis
(b) region R_2 is revolved about the x-axis.

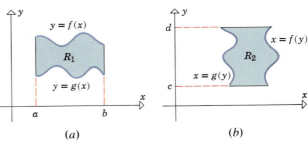

Figure 6.3.10

23. Use cylindrical shells to find the volume of the cone generated when the triangle with vertices $(0, 0)$, $(0, r)$, $(h, 0)$, where $r > 0$ and $h > 0$, is revolved about the x-axis.

24. The region enclosed between the curve $y^2 = kx$ and the line $x = \frac{1}{4}k$ is revolved about the line $x = \frac{1}{2}k$. Use cylindrical shells to find the volume of the resulting solid. (Assume $k > 0$.)

25. A round hole of radius a is drilled through the center of a solid sphere of radius r. Use cylindrical shells to find the volume of the portion removed. (Assume $r > a$.)

26. Use cylindrical shells to find the volume of the torus obtained by revolving the circle

$$x^2 + y^2 = a^2$$

about the line $x = b$, where $b > a$. [*Hint:* It may help in the integration to think of an integral as an area.]

27. Let V_x and V_y be the volumes of the solids that result when the region enclosed by $y = 1/x$, $y = 0$, $x = \frac{1}{2}$, and $x = b$ ($b > \frac{1}{2}$) is revolved about the x-axis and y-axis, respectively. Is there a value of b for which $V_x = V_y$?

■ 6.4 LENGTH OF A PLANE CURVE

In this section we shall use definite integrals to find arc lengths of plane curves. To start, we shall consider only curves that are graphs of functions. In a later section, we shall extend our results to more general curves.

□ **ARC LENGTH**

If f' is continuous on an interval, we shall say that $y = f(x)$ is a *smooth curve* (or f is a *smooth function*) on that interval. We shall restrict our discussion of arc length to smooth curves in order to eliminate some complications that would otherwise occur.

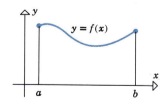

Figure 6.4.1

6.4.1 ARC LENGTH PROBLEM. *Suppose f is a smooth function on the interval $[a, b]$. Find the arc length L of the curve $y = f(x)$ over the interval $[a, b]$ (Figure 6.4.1).*

In order to solve Problem 6.4.1 we must first define the term "arc length" precisely. For motivation, consider the graph of a smooth curve $y = f(x)$ over an interval $[a, b]$, and as shown in Figure 6.4.2, divide the interval $[a, b]$ into

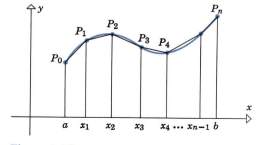

Figure 6.4.2

n subintervals with widths

$$\Delta x_1, \Delta x_2, \ldots, \Delta x_n$$

by inserting points

$$x_1, x_2, \ldots, x_{n-1}$$

between a and b. Let $P_0, P_1, \ldots, P_n$ be the points on the curve whose x-coordinates are $a, x_1, x_2, \ldots, x_{n-1}, b$ and join these points with straight-line segments. These segments form a **polygonal path** that we can regard as an approximation to the curve $y = f(x)$. Intuition suggests that the length of the approximating polygonal path will approach the length of the curve if we increase the number of points in such a way that the lengths of the line segments in the polygonal path approach zero.

To examine this idea more closely, let us isolate a typical subinterval, say the kth (Figure 6.4.3). As suggested by this figure, the length L_k of the kth line segment in the polygonal path is given by

$$L_k = \sqrt{(\Delta x_k)^2 + [f(x_k) - f(x_{k-1})]^2} \tag{1}$$

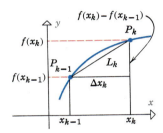

Figure 6.4.3

By the Mean-Value Theorem (4.10.2) there is a point x_k^* between x_{k-1} and x_k such that

$$\frac{f(x_k) - f(x_{k-1})}{x_k - x_{k-1}} = f'(x_k^*) \quad \text{or} \quad f(x_k) - f(x_{k-1}) = f'(x_k^*)\,\Delta x_k$$

Thus, (1) can be rewritten as

$$L_k = \sqrt{1 + [f'(x_k^*)]^2}\,\Delta x_k$$

(verify) which means that the length of the *entire* polygonal path is

$$\sum_{k=1}^{n} L_k = \sum_{k=1}^{n} \sqrt{1 + [f'(x_k^*)]^2}\,\Delta x_k$$

If we now increase the number of subintervals in such a way that max $\Delta x_k \to 0$, then the length of the polygonal path will approach the arc length L of the curve $y = f(x)$ over $[a, b]$. Thus,

$$L = \lim_{\max \Delta x_k \to 0} \sum_{k=1}^{n} \sqrt{1 + [f'(x_k^*)]^2}\,\Delta x_k \tag{2}$$

Since the right side of (2) is just the definite integral

$$\int_a^b \sqrt{1 + [f'(x)]^2}\, dx$$

we are led to the following result.

6.4.2 ARC LENGTH FORMULAS. If f is a smooth function on $[a, b]$, then the **arc length** L of the curve $y = f(x)$ from $x = a$ to $x = b$ is defined by

$$L = \int_a^b \sqrt{1 + [f'(x)]^2}\, dx = \int_a^b \sqrt{1 + \left(\frac{dy}{dx}\right)^2}\, dx \qquad (3)$$

Similarly, for a curve expressed in the form $x = g(y)$, where g' is continuous on $[c, d]$, the arc length L from $y = c$ to $y = d$ is defined by

$$L = \int_c^d \sqrt{1 + [g'(y)]^2}\, dy = \int_c^d \sqrt{1 + \left(\frac{dx}{dy}\right)^2}\, dy \qquad (4)$$

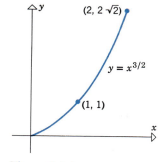

Figure 6.4.4

Example 1 Find the arc length of the curve $y = x^{3/2}$ from $(1, 1)$ to $(2, 2\sqrt{2})$ (Figure 6.4.4) using

    (a)   Formula (3)   (b)   Formula (4)

Solution (a).   Since $f(x) = x^{3/2}$,

$$f'(x) = \frac{3}{2} x^{1/2}$$

Thus, from (3), the arc length from $x = 1$ to $x = 2$ is

$$L = \int_1^2 \sqrt{1 + \frac{9}{4} x}\, dx$$

To evaluate this integral we make the u-substitution

$$u = 1 + \frac{9}{4} x, \quad du = \frac{9}{4}\, dx$$

and change the x-limits of integration ($x = 1, x = 2$) to the corresponding u-limits ($u = \frac{13}{4}, u = \frac{22}{4}$):

$$L = \frac{4}{9} \int_{13/4}^{22/4} u^{1/2}\, du = \frac{8}{27} u^{3/2} \Big]_{13/4}^{22/4} = \frac{8}{27} \left[\left(\frac{22}{4}\right)^{3/2} - \left(\frac{13}{4}\right)^{3/2} \right]$$

$$= \frac{22\sqrt{22} - 13\sqrt{13}}{27}$$

Solution (b). Solving $y = x^{3/2}$ for x in terms of y, we obtain $x = y^{2/3}$. Hence $g(y) = y^{2/3}$ and

$$g'(y) = \frac{2}{3} y^{-1/3}$$

Thus, from (4), the arc length from $y = 1$ to $y = 2\sqrt{2}$ is

$$L = \int_1^{2\sqrt{2}} \sqrt{1 + \frac{4}{9} y^{-2/3}} \, dy = \frac{1}{3} \int_1^{2\sqrt{2}} y^{-1/3} \sqrt{9 y^{2/3} + 4} \, dy$$

To evaluate this integral we make the u-substitution

$$u = 9 y^{2/3} + 4, \quad du = 6 y^{-1/3} \, dy$$

and change the y-limits of integration ($y = 1, y = 2\sqrt{2}$) to the corresponding u-limits ($u = 13, u = 22$). This gives

$$L = \frac{1}{18} \int_{13}^{22} u^{1/2} \, du = \frac{1}{27} u^{3/2} \Big]_{13}^{22}$$

$$= \frac{1}{27} [(22)^{3/2} - (13)^{3/2}]$$

$$= \frac{22\sqrt{22} - 13\sqrt{13}}{27}$$

This result agrees with that in part (a); however, the integration here is more tedious. In problems where there is a choice between using (3) or (4), it is often the case that one of the formulas leads to a simpler integral than the other. ◀

► Exercise Set 6.4 Ⓒ *13, 14*

1. Find the arc length of the curve $y = 2x$ from $(1, 2)$ to $(2, 4)$ using
   (a) Formula (3)   (b) Formula (4)
   (c) the Theorem of Pythagoras.

2. Find the arc length of the curve $y = mx + b$ from $x = k_1$ to $x = k_2$ ($m \neq 0, k_2 > k_1$) using
   (a) Formula (3)   (b) Formula (4)
   (c) the Theorem of Pythagoras.

3. Find the arc length of the curve $y = 3x^{3/2} - 1$ from $x = 0$ to $x = 1$.

4. Find the arc length of the curve $x = \frac{1}{3}(y^2 + 2)^{3/2}$ from $y = 0$ to $y = 1$.

5. Find the arc length of the curve $y = x^{2/3}$ from $x = 1$ to $x = 8$.

6. Find the arc length of the curve $y = \frac{x^4}{16} + \frac{1}{2x^2}$ from $x = 2$ to $x = 3$.

7. Find the arc length of the curve $24xy = y^4 + 48$ from $y = 2$ to $y = 4$.

8. Find the arc length of the curve $x = \frac{1}{8}y^4 + \frac{1}{4}y^{-2}$ from $y = 1$ to $y = 4$.

9. Consider the curve $y = x^{2/3}$.
   (a) Sketch the portion of the curve between $x = -1$ and $x = 8$.
   (b) Explain why (3) cannot be used to find the arc length of the curve sketched in part (a).
   (c) Find the arc length of the curve sketched in part (a).

10. Find the arc length in the second quadrant of the curve $x^{2/3} + y^{2/3} = a^{2/3}$ from the point $x = -a$ to $x = -\frac{1}{8}a$ $(a > 0)$.

11. Let $y = f(x)$ be a smooth curve and suppose $f'(x) \geq 0$ on the closed interval $[a, b]$.
   (a) Prove: There are numbers m and M such that $m \leq f'(x) \leq M$ for all x in $[a, b]$.
   (b) Prove: The arc length L of $y = f(x)$ over the closed interval $[a, b]$ satisfies the inequalities
   $$(b - a)\sqrt{1 + m^2} \leq L \leq (b - a)\sqrt{1 + M^2}$$

12. Use the result of Exercise 11 to show that the arc length L of $y = \sin x$ over the interval $0 \leq x \leq \pi/4$ satisfies

$$\frac{\pi}{4}\sqrt{\frac{3}{2}} \leq L \leq \frac{\pi}{4}\sqrt{2}$$

In Exercises 13 and 14, use the midpoint approximation (discussed preceding Exercise 29 of Section 5.6) with $n = 20$ subintervals to approximate the arc length of the curve over the given interval.

13. $y = x^2$; $[0, 2]$.

14. $y = \sin x$; $[0, \pi]$.

■ **6.5 AREA OF A SURFACE OF REVOLUTION**

In this section we shall apply the definite integral to the problem of finding the area of a surface of revolution.

☐ **DEFINITION OF SURFACE AREA**

6.5.1 SURFACE AREA PROBLEM. *Let f be a smooth, nonnegative function on $[a, b]$. Find the area of the surface generated by revolving the portion of the curve $y = f(x)$ between $x = a$ and $x = b$ about the x-axis* (Figure 6.5.1).

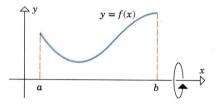
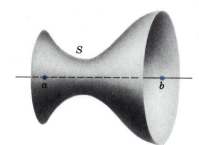

Figure 6.5.1

In order to solve this problem we must first define the term "surface area" precisely. For motivation, let us approximate the curve $y = f(x)$ by a polygonal path of straight-line segments connecting the points on the curve that have x-coordinates

$$a, x_1, x_2, \ldots, x_{n-1}, b$$

(Figure 6.5.2a). As usual, let

$$\Delta x_1, \Delta x_2, \ldots, \Delta x_n$$

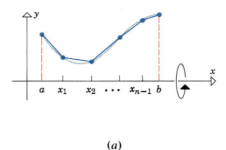
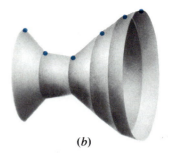

(a)

(b)

Figure 6.5.2

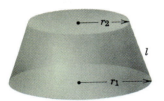

Figure 6.5.3

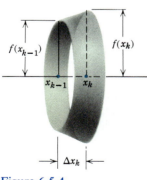

Figure 6.5.4

be the widths of the subintervals determined by these x-coordinates. If these widths are small, then the surface generated by revolving the polygonal path about the x-axis will have approximately the same area as the surface generated by revolving the curve $y = f(x)$ about the x-axis (Figure 6.5.2b). Observe that the surface generated by the polygonal path is made of parts, each of which is a frustum of a cone. Thus, the lateral area of each of these parts can be obtained from the formula

$$S = \pi(r_1 + r_2)l \tag{1}$$

for the lateral area S of a frustum of slant height l and base radii r_1 and r_2 (Figure 6.5.3). A derivation of this formula is discussed in Exercise 17. Intuition suggests that the area of the approximating surface will approach the desired area of the surface if we increase the number of subdivisions in such a way that the lengths of the line segments in the polygonal path approach zero.

To examine this idea more closely, let us isolate a typical section of the approximating surface, say the kth (Figure 6.5.4). Applying (1), we obtain as the lateral area S_k of this kth section

$$S_k = \pi[f(x_{k-1}) + f(x_k)]\sqrt{(\Delta x_k)^2 + [f(x_k) - f(x_{k-1})]^2} \tag{2}$$

By the Mean-Value Theorem (4.10.2) there is a point x_k^* between x_{k-1} and x_k such that

$$\frac{f(x_k) - f(x_{k-1})}{x_k - x_{k-1}} = f'(x_k^*) \quad \text{or} \quad f(x_k) - f(x_{k-1}) = f'(x_k^*)\,\Delta x_k$$

Thus, (2) can be rewritten as

$$S_k = \pi[f(x_{k-1}) + f(x_k)]\sqrt{1 + [f'(x_k^*)]^2}\,\Delta x_k \tag{3}$$

Since the arithmetic average of two numbers lies between those numbers, $\frac{1}{2}[f(x_{k-1}) + f(x_k)]$ is between $f(x_{k-1})$ and $f(x_k)$. Thus, because f is continuous on the interval $[x_{k-1}, x_k]$, the Intermediate-Value Theorem (2.7.9) implies that there exists a point x_k^{**} in this interval such that

$$\frac{1}{2}[f(x_{k-1}) + f(x_k)] = f(x_k^{**})$$

Thus, (3) can be rewritten as

$$S_k = 2\pi f(x_k^{**})\sqrt{1 + [f'(x_k^*)]^2}\,\Delta x_k$$

so that the area of the *entire* polygonal surface is

$$\sum_{k=1}^{n} S_k = \sum_{k=1}^{n} 2\pi f(x_k^{**})\sqrt{1 + [f'(x_k^*)]^2}\,\Delta x_k$$

If we now increase the number of subintervals in such a way that max $\Delta x_k \to 0$, then the area of the approximating polygonal surface will approach the exact surface area S. Thus,

$$S = \lim_{\max \Delta x_k \to 0} \sum_{k=1}^{n} 2\pi f(x_k^{**})\sqrt{1 + [f'(x_k^*)]^2}\,\Delta x_k \tag{4}$$

If it were the case that $x_k^{**} = x_k^*$, then the right side of (4) would be the definite integral

$$\int_a^b 2\pi f(x)\sqrt{1 + [f'(x)]^2}\,dx \tag{5}$$

However, it is proved in advanced calculus that this is true even if $x_k^* \neq x_k^{**}$ because of the continuity of f and f'. In light of (4) and (5) we are led to the following results.

6.5.2 SURFACE AREA FORMULAS. Let f be a smooth, nonnegative function on $[a, b]$. Then the **surface area S** generated by revolving the portion of the curve $y = f(x)$ between $x = a$ and $x = b$ about the x-axis is

$$S = \int_a^b 2\pi f(x)\sqrt{1 + [f'(x)]^2}\,dx \tag{6}$$

For a curve expressed in the form $x = g(y)$, where g' is continuous on $[c, d]$, and $g(y) \geq 0$ for $c \leq y \leq d$, the surface area S generated by revolving the portion of the curve from $y = c$ to $y = d$ about the y-axis is given by

$$S = \int_c^d 2\pi g(y)\sqrt{1 + [g'(y)]^2}\,dy \tag{7}$$

Example 1 Find the surface area generated by revolving the curve

$$y = \sqrt{1 - x^2}, \quad 0 \leq x \leq \tfrac{1}{2}$$

about the x-axis (Figure 6.5.5).

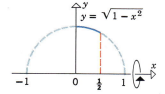

Figure 6.5.5

Solution.   Since $f(x) = \sqrt{1 - x^2}$,

$$f'(x) = -\frac{x}{\sqrt{1 - x^2}}$$

Thus, (6) yields

$$S = \int_0^{1/2} 2\pi\sqrt{1 - x^2}\,\sqrt{1 + \frac{x^2}{1 - x^2}}\,dx = \int_0^{1/2} 2\pi\,dx = 2\pi x\Big]_0^{1/2} = \pi \qquad \blacktriangleleft$$

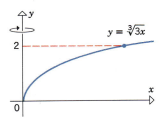

$y = \sqrt[3]{3x}$

Figure 6.5.6

Example 2   Find the surface area generated by revolving the curve

$$y = \sqrt[3]{3x}, \quad 0 \leq y \leq 2$$

about the y-axis (Figure 6.5.6).

Solution.   We shall apply (7) after rewriting $y = \sqrt[3]{3x}$ as $x = g(y) = \frac{1}{3}y^3$. Thus,

$$g'(y) = y^2$$

so that from (7) we obtain

$$S = \int_0^2 2\pi\left(\frac{1}{3}y^3\right)\sqrt{1 + y^4}\,dy = \frac{2\pi}{3}\int_0^2 y^3\sqrt{1 + y^4}\,dy \qquad (8)$$

The u-substitution

$$u = 1 + y^4, \quad du = 4y^3\,dy$$

yields

$$\int y^3\sqrt{1 + y^4}\,dy = \frac{1}{4}\int \sqrt{u}\,du = \frac{1}{4}\cdot\frac{2}{3}u^{3/2} + C = \frac{1}{6}(1 + y^4)^{3/2} + C$$

Thus, from (8),

$$S = \frac{2\pi}{3}\left[\frac{1}{6}(1 + y^4)^{3/2}\right]_0^2 = \frac{\pi}{9}(17^{3/2} - 1) \qquad \blacktriangleleft$$

▶ **Exercise Set 6.5**

In Exercises 1–6, find the area of the surface generated by revolving the given curve about the x-axis.

1. $y = 7x, \; 0 \leq x \leq 1$.

2. $y = \sqrt{x}, \; 1 \leq x \leq 4$.

3. $y = \sqrt{4 - x^2}, \; -1 \leq x \leq 1$.

4. $x = \sqrt[3]{y}, \; 1 \leq y \leq 8$.

5. $y = \sqrt{x} - \frac{1}{3}x^{3/2}, \; 1 \leq x \leq 3$.

6. $y = \frac{1}{3}x^3 + \frac{1}{4}x^{-1}, \; 1 \leq x \leq 2$.

In Exercises 7–12, find the area of the surface generated by revolving the given curve about the y-axis.

7. $x = 9y + 1, \; 0 \leq y \leq 2$.

8. $x = y^3, \; 0 \leq y \leq 1$.

9. $x = \sqrt{9 - y^2}, \; -2 \leq y \leq 2$.

10. $x = 2\sqrt{1 - y}, \; -1 \leq y \leq 0$.

11. $8xy^2 = 2y^6 + 1$, $1 \leq y \leq 2$.

12. $x = |y - 11|$, $0 \leq y \leq 2$.

13. The lateral area S of a right-circular cone with height h and base radius r is $S = \pi r \sqrt{r^2 + h^2}$. Obtain this result using (6).

14. Show that the area of the surface of a sphere of radius r is $4\pi r^2$. [*Hint:* Revolve the semicircle $y = \sqrt{r^2 - x^2}$ about the x-axis.]

15. The portion of the surface of a sphere between two parallel planes that cut the sphere and are h units apart is called a **zone** of altitude h. Show that the area of the surface of a zone depends only on the radius r of the sphere and the altitude h of the zone, and not on the location of the zone. [*Hint:* Let $-r \leq a \leq r - h$. Revolve the portion of the curve $y = \sqrt{r^2 - x^2}$ for $a \leq x \leq a + h$ about the x-axis.]

16. Assume that f is smooth and $f(x) \geq 0$ for $a \leq x \leq b$. Derive a formula for the surface area generated when the curve $y = f(x)$, $a \leq x \leq b$ is revolved about the line $y = -k$ $(k > 0)$.

17. (**Formula for Surface Area of a Frustum**)
    (a) If a cone of slant height l and base radius r is cut along a lateral edge and laid flat, it becomes a sector of a circle of radius l (Figure 6.5.7). Use the formula $\frac{1}{2}l^2\theta$ for the area of a sector with radius l and central angle θ (in radians) to show that the lateral surface area of the cone is $\pi r l$.

(b) Use the result in part (a) to obtain Formula (1) for the lateral surface area of a frustum.

 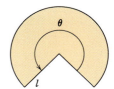

Figure 6.5.7

18. Let $y = f(x)$ be a smooth curve on the interval $[a, b]$ and assume that $f(x) \geq 0$ for $a \leq x \leq b$. By the Extreme-Value Theorem (4.6.4), the function f has a maximum value K and a minimum value k on $[a, b]$. Prove: If L is the arc length of the curve $y = f(x)$ between $x = a$ and $x = b$ and if S is the area of the surface that is generated by revolving this curve about the x-axis, then
$$2\pi k L \leq S \leq 2\pi K L$$

19. Let $y = f(x)$ be a smooth curve on $[a, b]$ and assume that $f(x) \geq 0$ for $a \leq x \leq b$. Let A be the area under the curve $y = f(x)$ between $x = a$ and $x = b$ and let S be the area of the surface obtained when this section of curve is revolved about the x-axis.
    (a) Prove that $2\pi A \leq S$.
    (b) For what functions f is $2\pi A = S$?

6.6 APPLICATION OF INTEGRATION TO RECTILINEAR MOTION

In Section 4.11 we used the derivative to define the notions of instantaneous velocity and acceleration for a particle moving along a line. In this section we shall resume the study of such motion, using the integration tools developed in the previous chapter.

□ **FINDING POSITION AND VELOCITY BY INTEGRATION**

Recall from Section 4.11 that if the position function of a particle moving on a coordinate line is $s(t)$, then its instantaneous velocity and instantaneous acceleration are given by the formulas

$$v(t) = s'(t) = \frac{ds}{dt} \quad \text{and} \quad a(t) = v'(t) = \frac{dv}{dt} = \frac{d^2s}{dt^2}$$

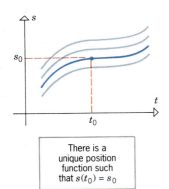

There is a
unique position
function such
that $s(t_0) = s_0$

Figure 6.6.1

respectively. It follows from these formulas that $s(t)$ is an antiderivative of $v(t)$ and $v(t)$ is an antiderivative of $a(t)$, that is,

$$s(t) = \int v(t)\, dt \quad \text{and} \quad v(t) = \int a(t)\, dt \qquad (1\text{--}2)$$

Thus, if the velocity function of a particle is known, then its position function can be obtained from (1) by integration provided there is sufficient additional information to determine the constant of integration. In particular, we can determine the constant of integration if we know the position s_0 of the particle at some time t_0, since this additional information determines a unique antiderivative of $v(t)$. (See Figure 6.6.1.) Similarly, if the acceleration function of a particle is known, then its velocity function can be obtained from (2) by integration if we know the velocity v_0 of the particle at some time t_0. (See Figure 6.6.2.)

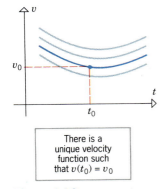

There is a
unique velocity
function such
that $v(t_0) = v_0$

Figure 6.6.2

Example 1 Find the position function of a particle moving with velocity $v(t) = \cos \pi t$ along a straight line, assuming the particle is at $s = 4$ when $t = 0$.

Solution. The position function is

$$s(t) = \int v(t)\, dt = \int \cos \pi t\, dt = \frac{1}{\pi} \sin \pi t + C$$

Since $s = 4$ when $t = 0$, it follows that

$$4 = s(0) = \frac{1}{\pi} \sin 0 + C = C$$

Thus,

$$s(t) = \frac{1}{\pi} \sin \pi t + 4 \qquad \blacktriangleleft$$

☐ **MOTION NEAR THE EARTH'S SURFACE**

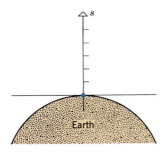

Figure 6.6.3

It is a fact of physics that an object moving on a vertical line near the earth's surface and subject only to the force of gravity moves with constant acceleration.* This constant, denoted by the letter g, is approximately 32 ft/sec² or 9.8 m/sec², depending on whether distance is measured in feet or meters.

From this simple principle it is possible to describe completely the vertical motion of a particle moving freely near the earth's surface, provided the initial position and velocity of the particle are known. To see how, assume that a coordinate line has been superimposed on the line of motion with the origin at the surface of the earth and the positive direction upward (Figure 6.6.3). More-

*Strictly speaking, the acceleration changes with the distance from the earth's center. However, near the surface of the earth the acceleration is approximately constant.

over, let us assume that time is measured in seconds, distance is measured in feet or meters, and that

the position at time $t = 0$ is s_0

the velocity at time $t = 0$ is v_0

where the initial displacement s_0 and initial velocity v_0 are known. Recall that a particle speeds up when its velocity and acceleration have the same sign and slows down when they have opposite signs. Because we have chosen the positive direction to be up, and because a particle moving up (positive velocity) slows down under the force of gravity and a particle moving down (negative velocity) speeds up under the force of gravity, it follows that the acceleration a of the particle must always be negative, that is,

$$a(t) = -g$$

Since the acceleration of the particle is constant, we obtain

$$v(t) = \int a(t) \, dt = \int -g \, dt = -gt + C_1 \tag{3}$$

To determine the constant of integration we use the fact that the velocity is v_0 when $t = 0$. Thus,

$$v_0 = v(0) = -g \cdot 0 + C_1 = C_1$$

Substituting this in (3) yields

$$v(t) = -gt + v_0 \tag{4}$$

Since v_0 is a constant, it follows that

$$s(t) = \int v(t) \, dt = \int (-gt + v_0) \, dt = -\frac{1}{2} gt^2 + v_0 t + C_2 \tag{5}$$

To determine the constant C_2 we use the fact that the position is s_0 when $t = 0$. Thus,

$$s_0 = s(0) = -\frac{1}{2} g \cdot 0 + v_0 \cdot 0 + C_2 = C_2$$

Substituting this in (5) yields

$$s(t) = -\frac{1}{2} gt^2 + v_0 t + s_0 \tag{6}$$

Example 2 A rock, initially at rest, is dropped from a height of 400 ft. Assuming gravity is the only force acting, how long does it take for the rock to hit the ground and what is its speed at the time of impact?

Solution.   Since distance is in feet, we take $g = 32$ ft/sec^2. Initially, we have $s_0 = 400$ and $v_0 = 0$, so from (6)

$$s(t) = -16t^2 + 400$$

Impact occurs when $s(t) = 0$. Solving this equation for t, we obtain

$$-16t^2 + 400 = 0$$
$$t^2 = 25$$
$$t = \pm 5$$

where t is in seconds. Since the rock is released at $t = 0$, the time of impact is positive, so we can discard the negative solution and conclude that it takes 5 sec for the rock to hit the ground. Substituting $t = 5$ and $v_0 = 0$ in (4), we obtain the velocity at time of impact,

$$v(5) = -32(5) + 0 = -160 \text{ ft/sec}$$

Thus, the speed at impact is $|v(5)| = 160$ ft/sec.   ◀

Example 3   A ball is thrown directly upward from a point 8 m above the ground with an initial velocity of 49 m/sec. Assuming gravity is the only force acting on the ball after its release, how high will the ball travel?

Solution.   Since distance is in meters, we take $g = 9.8$ m/sec^2. Initially, $s_0 = 8$ and $v_0 = 49$, so from (4) and (6)

$$v(t) = -9.8t + 49$$
$$s(t) = -4.9t^2 + 49t + 8$$

The ball will rise until $v(t) = 0$, that is, until $-9.8t + 49 = 0$ or $t = 5$. At this instant the height above the ground will be

$$s(5) = -4.9(5)^2 + 49(5) + 8 = 130.5 \text{ m}$$ ◀

☐ **DISPLACEMENT**

We conclude this section with an application of the First Fundamental Theorem of Calculus (Theorem 5.7.1) to rectilinear motion. It follows from the First Fundamental Theorem of Calculus and (1) that

$$\int_{t_1}^{t_2} v(t)\, dt = s(t) \Big]_{t_1}^{t_2} = s(t_2) - s(t_1) \tag{7}$$

Since $s(t_1)$ is the position of the particle at time t_1 and $s(t_2)$ is the position at time t_2, the difference, $s(t_2) - s(t_1)$, is the change in position or ***displacement*** of the particle during the time interval $[t_1, t_2]$. Assuming a horizontal line of motion, a positive displacement means that the particle is farther to the right at time t_2 than at time t_1; a negative displacement means that the particle is farther to the left. In the case where $v(t) \geq 0$ throughout the time interval $[t_1, t_2]$, the particle moves in the positive direction only; thus, the displacement $s(t_2) - s(t_1)$ is the same as the distance traveled by the particle (Figure 6.6.4).

In the case where $v(t) \leq 0$ throughout the time interval $[t_1, t_2]$, the particle moves in the negative direction only; thus, the displacement $s(t_2) - s(t_1)$ is the negative of the distance traveled by the particle (Figure 6.6.5). In the case where $v(t)$ assumes both positive and negative values during the time interval $[t_1, t_2]$, the particle moves back and forth and the displacement is the distance traveled in the positive direction minus the distance traveled in the negative direction.

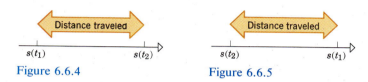

Figure 6.6.4   Figure 6.6.5

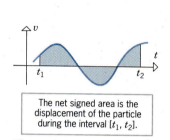

The net signed area is the displacement of the particle during the interval $[t_1, t_2]$.

Figure 6.6.6

Geometrically, (7) can be interpreted as the net signed area between the velocity versus time curve and the interval $[t_1, t_2]$, so that this net signed area represents the displacement of the particle (Figure 6.6.6).

☐ **DISTANCE TRAVELED**

If we want to find the total distance traveled in this case (distance traveled in the positive direction *plus* the distance traveled in the negative direction), we must integrate the absolute value of the velocity function (which is the speed of the particle), that is,

$$\begin{bmatrix} \text{total distance} \\ \text{traveled during} \\ \text{time interval} \\ [t_1, t_2] \end{bmatrix} = \int_{t_1}^{t_2} |v(t)| \, dt \tag{8}$$

Example 4   A particle moves on a coordinate line so that its velocity at time t is $v(t) = t^2 - 2t$ m/sec. Find

(a)   the displacement of the particle during the time interval $0 \leq t \leq 3$;

(b)   the distance traveled by the particle during the time interval $0 \leq t \leq 3$.

Solution (a).   From (7) the displacement is

$$\int_0^3 v(t) \, dt = \int_0^3 (t^2 - 2t) \, dt = \left[\frac{t^3}{3} - t^2 \right]_0^3 = 0$$

Thus, the particle is at the same position at time $t = 3$ as at $t = 0$.

Solution (b). The velocity can be written as $v(t) = t^2 - 2t = t(t - 2)$, from which we see that $v(t) \leq 0$ for $0 \leq t \leq 2$ and $v(t) \geq 0$ for $2 \leq t \leq 3$. Thus, it follows from (8) that the distance traveled is

$$\int_0^3 |v(t)| \, dt = \int_0^2 -v(t) \, dt + \int_2^3 v(t) \, dt$$

$$= \int_0^2 -(t^2 - 2t) \, dt + \int_2^3 (t^2 - 2t) \, dt$$

$$= -\left[\frac{t^3}{3} - t^2\right]_0^2 + \left[\frac{t^3}{3} - t^2\right]_2^3 = \frac{4}{3} + \frac{4}{3} = \frac{8}{3} \text{ m} \quad \blacktriangleleft$$

▶ Exercise Set 6.6

In Exercises 1–8, use the given information to find the position function of the particle.

1. $v(t) = 2t - 3$; $s(1) = 5$.

2. $v(t) = 3t^2$; $s(0) = 0$.

3. $v(t) = t^3 - 2t^2 + 1$; $s(0) = 1$.

4. $v(t) = 1 + \sin t$; $s(0) = -3$.

5. $a(t) = 4$; $v(0) = 1$; $s(0) = 0$.

6. $a(t) = t^2 - 3t + 1$; $v(0) = 0$; $s(0) = 0$.

7. $a(t) = 4 \cos 2t$; $v(0) = -1$; $s(0) = -3$.

8. $a(t) = 1/\sqrt{2t + 3}$, $t \geq 0$; $v(3) = 1$; $s(3) = 0$.

9. In each part use the given information to find the position, velocity, speed, and acceleration at time $t = 1$.
   (a) $v = \sin \frac{1}{2}\pi t$; $s = 0$ when $t = 0$.
   (b) $a = -3t$; $s = 1$ and $v = 0$ when $t = 0$.

10. A car traveling 60 mi/hr along a straight road decelerates at a constant rate of 10 ft/sec².
   (a) How long will it take until the speed is 45 mi/hr? [*Note:* 60 mi/hr = 88 ft/sec.]
   (b) How far will the car travel before coming to a stop?

11. A car traveling 60 mi/hr skids 180 ft after its brakes are applied. Find the acceleration of the car, assuming that it is constant.

12. A particle moving along a straight line is accelerating at a constant rate of 3 m/sec². Find the initial velocity if the particle moves 40 m in the first 4 sec.

In Exercises 13–24, assume that the only force acting is the earth's gravity, and apply Formulas (4) and (6) to solve the problems.

13. A projectile is launched vertically upward from ground level with an initial velocity of 112 ft/sec.
   (a) Find the velocity at $t = 3$ sec and $t = 5$ sec.
   (b) How high will the projectile rise?
   (c) Find the speed of the projectile when it hits the ground.

14. A projectile fired downward from a height of 112 ft reaches the ground in 2 sec. What is its initial velocity?

15. A projectile is fired vertically upward from ground level with an initial velocity of 16 ft/sec.
   (a) How long will it take for the projectile to hit the ground?
   (b) How long will the projectile be moving upward?

16. A rock is dropped from the top of the Washington Monument, which is 555 ft high.
   (a) How long will it take for the rock to hit the ground?
   (b) What is the speed of the rock at impact?

17. A helicopter pilot drops a package when the helicopter is 200 ft above the ground and rising at a speed of 20 ft/sec.
  (a) How long will it take for the package to hit the ground?
  (b) What will be its speed at impact?

18. A stone is thrown downward with an initial speed of 96 ft/sec from a height of 112 ft.
  (a) How long will it take for the stone to hit the ground?
  (b) What will be its speed at impact?

19. A projectile is fired vertically upward with an initial velocity of 49 m/sec from a tower 150 m high.
  (a) How long will it take for the projectile to reach its maximum height?
  (b) What is the maximum height?
  (c) How long will it take for the projectile to pass its starting point on the way down?
  (d) What is the velocity when it passes the starting point on the way down?
  (e) How long will it take for the projectile to hit the ground?
  (f) What will be its speed at impact?

20. A man drops a stone from a bridge. How high is the bridge if
  (a) the stone hits the water 4 sec later
  (b) the sound of the splash reaches the man 4 sec later? (Take 1080 ft/sec as the speed of sound.)

21. A projectile fired upward from ground level is to reach a height of 1000 ft. What must its initial velocity be?

22. A stone is released from rest from a point 40 ft above the ground. Find a formula for v in terms of s.

23. A projectile is fired upward from ground level. Find the initial velocity if it hits the ground 8 sec later. How high does it go?

24. Show that $v^2 = v_0{}^2 - 2g(s - s_0)$.

25. The graph of a velocity function over the interval $[t_1, t_2]$ is shown in Figure 6.6.7.
  (a) Is the acceleration positive or is it negative?

  (b) Is the acceleration increasing or is it decreasing?
  (c) Is the displacement positive or is it negative?

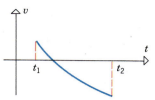

Figure 6.6.7

In Exercises 26–31, $v(t)$ is the velocity (meters/second) of a particle moving on a coordinate line. Find the displacement and the distance traveled by the particle during the given time interval.

26. $v(t) = 2t - 4;\ 0 \le t \le 4$.

27. $v(t) = t^2 + t - 2;\ 0 \le t \le 2$.

28. $v(t) = |t - 3|;\ 0 \le t \le 5$.

29. $v(t) = \cos t;\ 0 \le t \le \pi$.

30. $v(t) = 3 \sin t;\ \pi/4 \le t \le \pi$.

31. $v(t) = t^3 - 3t^2 + 2t;\ 0 \le t \le 3$.

In Exercises 32–35, $a(t)$ is the acceleration in m/sec² of a particle moving on a coordinate line, and v_0 is its velocity at time $t = 0$. Find the displacement and the distance traveled by the particle during the given time interval.

32. $a(t) = -2;\ v_0 = 3;\ 1 \le t \le 4$.

33. $a(t) = t - 2;\ v_0 = 0;\ 1 \le t \le 5$.

34. $a(t) = \sin t;\ v_0 = 1;\ \dfrac{\pi}{4} \le t \le \dfrac{\pi}{2}$.

35. $a(t) = \dfrac{1}{\sqrt{5t + 1}};\ v_0 = 2;\ 0 \le t \le 3$.

36. Suppose that the velocity of a car moving along a straight road increases from 5 to 50 ft/sec in 20 sec. If you know that the acceleration did not increase, but could have decreased, what is the minimum distance traveled by the car during this time period?

■ 6.7 WORK

> *In this section we shall use the integration tools developed in the previous chapter to study the concept of work that arises in physics and engineering.*

□ **WORK DONE BY A CONSTANT FORCE**

If an object moves a distance d along a line while subjected to a *constant* force F applied in the direction of motion, then physicists and engineers define the ***work*** W done on the object by the force F to be

$$W = F \cdot d$$
$$[\text{work}] = [\text{force}] \cdot [\text{distance}]$$

(1)

Common units for measuring force are ***pounds*** (lb) in the British engineering system and ***dynes*** (D) or ***newtons*** (N) in the metric system. (One dyne is the force required to give a mass of 1 g an acceleration of 1 cm/sec^2, and one newton is the force required to give a mass of 1 kg an acceleration of 1 m/sec^2.) The most common units of work are foot-pounds (ft·lb), dyne-centimeters (D·cm), and newton-meters (N·m). One newton-meter is also called a ***joule*** (J). One foot-pound is approximately 1.36 J.

Example 1   An object moves 5 ft along a line while subjected to a constant force of 100 lb in its direction of motion. The work done is

$$W = F \cdot d = 100 \cdot 5 = 500 \text{ ft} \cdot \text{lb}$$

An object moves 25 m along a line while subjected to a constant force of 4 N in its direction of motion. The work done is

$$W = F \cdot d = 4 \cdot 25 = 100 \text{ N} \cdot \text{m} = 100 \text{ J} \quad \blacktriangleleft$$

□ **WORK DONE BY A VARIABLE FORCE**

Calculus is needed when it is required to calculate the work done by a *variable* force. We shall consider the following problem.

> **6.7.1**   PROBLEM.   *Suppose an object moves in the positive direction along a coordinate line while subject to a force $F(x)$, in the direction of motion, whose magnitude depends on the coordinate x. Find the work done by the force when the object moves over an interval $[a, b]$.*

For example, Figure 6.7.1 shows a block subjected to the force of a compressed spring. As the block moves from a to b the spring expands and the

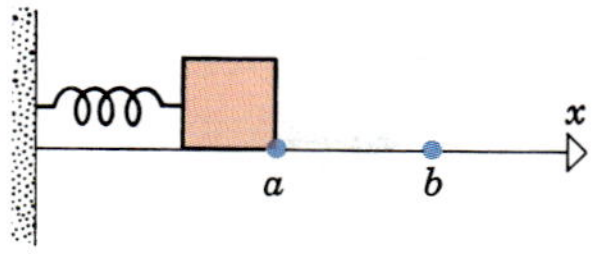

Figure 6.7.1

force it applies diminishes. Thus, the force $F(x)$ applied by the spring varies with x.

Before we can solve Problem 6.7.1 we must define precisely what is meant by the work done by a variable force. To motivate this definition, let us divide the interval $[a, b]$ into n subintervals with widths

$$\Delta x_1, \Delta x_2, \ldots, \Delta x_n$$

by inserting points

$$x_1, x_2, \ldots, x_{n-1}$$

between a and b. If we denote by W_k the work done when the object moves across the kth subinterval, then total work W done when the object moves across the entire interval $[a, b]$ will be

$$W = W_1 + W_2 + \cdots + W_n$$

Let us estimate W_k. If the kth subinterval is short and if $F(x)$ is continuous, then the force will not vary much over this subinterval; it will be almost constant. We can approximate this nearly constant force by $F(x_k^*)$, where x_k^* is any point in the kth subinterval. Thus, from (1), the work done over the kth subinterval is approximately

$$W_k \approx F(x_k^*) \, \Delta x_k$$

and the work W done over the entire interval $[a, b]$ is approximately

$$\sum_{k=1}^{n} F(x_k^*) \, \Delta x_k$$

If we now increase the number of subintervals in such a way that $\max \Delta x_k \to 0$, then intuition suggests that our approximations will tend to get better; hence,

$$W = \lim_{\max \Delta x_k \to 0} \sum_{k=1}^{n} F(x_k^*) \, \Delta x_k$$

Since the limit on the right is just the definite integral

$$\int_a^b F(x) \, dx$$

we are led to the following definition.

6.7.2 DEFINITION. If an object moves in the positive direction over the interval $[a, b]$ while subjected to a variable force $F(x)$ in the direction of motion, then the **work** done by the force is

$$W = \int_a^b F(x)\, dx \qquad (2)$$

Hooke's law [Robert Hooke (1635–1703), English physicist] states that under appropriate conditions a spring stretched x units beyond its equilibrium position pulls back with a force

$$F(x) = kx$$

where k is a constant (called the **spring constant**). The value of k depends on such factors as the thickness of the spring, the material used in its composition, and the units of force and distance.

Example 2 A spring whose natural length is 2.4 m exerts a force of 5 N when stretched 1 m beyond its natural length.

(a) Find the spring constant k.

(b) How much work is required to stretch the spring from its natural length to a length of 4.2 m?

Solution (a). From Hooke's law,

$$F(x) = kx$$

From the data, $F(x) = 5$ N when $x = 1$ m, so

$$5 = k \cdot 1$$

Thus, the spring constant is $k = 5$. This means that the force $F(x)$ required to stretch the spring x m is

$$F(x) = 5x \qquad (3)$$

Solution (b). Place the spring along a coordinate line as shown in Figure 6.7.2. We want to find the work W required to stretch the spring over the interval from $x = 0$ to $x = 1.8$. From (2) and (3) the work W required is

$$W = \int_a^b F(x)\, dx = \int_0^{1.8} 5x\, dx = \left. \frac{5x^2}{2} \right]_0^{1.8} = 8.1 \text{ J} \quad \blacktriangleleft$$

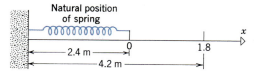

Figure 6.7.2

Example 3   A cylindrical water tank of radius 10 ft and height 30 ft is half filled with water. How much work is required to pump all the water over the upper rim of the tank?

Solution.   Introduce a coordinate line as shown in Figure 6.7.3. Imagine the water to be divided into n thin layers with thicknesses

$$\Delta x_1, \Delta x_2, \ldots, \Delta x_n$$

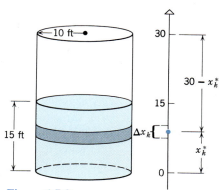

Figure 6.7.3

The force required to move the kth layer equals the weight of the layer, which can be found by multiplying its volume by the weight density of water (62.4 lb/ft³ or 9810 N/m³). Since the kth layer is a cylinder of radius $r = 10$ ft and height Δx_k, the force required to move it is

$$\begin{bmatrix} \text{force to move} \\ \text{the } k\text{th layer} \end{bmatrix} = (\pi r^2 \, \Delta x_k) \cdot [\text{weight density of water}]$$

$$= (\pi (10)^2 \, \Delta x_k)(62.4) = 6240\pi \, \Delta x_k$$

Because the kth layer has a finite thickness, the upper and lower surfaces are at different distances from the origin. However, if the layer is thin, the difference in these distances is small, and we can reasonably assume that the entire layer is concentrated at a single distance x_k^* from the origin (Figure 6.7.3). With this assumption, the work W_k required to pump the kth layer over the rim will be approximately

$$W_k \approx \underbrace{(30 - x_k^*)}_{\text{Distance}} \cdot \underbrace{6240\pi \, \Delta x_k}_{\text{Force}}$$

and the work W required to pump all n layers will be approximately

$$W = \sum_{k=1}^{n} W_k \approx \sum_{k=1}^{n} (30 - x_k^*)(6240\pi) \, \Delta x_k$$

To find the *exact* value of the work we take the limit as $\max \Delta x_k \to 0$. This yields

$$W = \lim_{\max \Delta x_k \to 0} \sum_{k=1}^{n} (30 - x_k^*)(6240\pi) \, \Delta x_k = \int_0^{15} (30 - x)(6240\pi) \, dx$$

$$= 6240\pi \left(30x - \frac{x^2}{2} \right) \Bigg]_0^{15} = 2{,}106{,}000\pi \text{ ft} \cdot \text{lb} \quad \blacktriangleleft$$

▶ Exercise Set 6.7   Ⓒ 9

1. Find the work done when
   (a) a constant force of 30 lb along the x-axis moves an object from $x = -2$ to $x = 5$ ft
   (b) a variable force of $F(x) = 1/x^2$ lb along the x-axis moves an object from $x = 1$ to $x = 6$ ft.

2. A spring whose natural length is 15 cm exerts a force of 45 N when stretched to a length of 20 cm.
   (a) Find the spring constant (in newtons/meter).
   (b) Find the work that is done in stretching the spring 3 cm beyond its natural length.
   (c) Find the work done in stretching the spring from a length of 20 cm to a length of 25 cm.

3. A spring exerts a force of 100 N when it is stretched 0.2 m beyond its natural length. How much work is required to stretch the spring 0.8 m beyond its natural length?

4. Assume that a force of 6 N is required to compress a spring from a natural length of 4 m to a length of $3\frac{1}{2}$ m. Find the work required to compress the spring from its natural length to a length of 2 m. (Hooke's law applies to compression as well as extension.)

5. Assume that 10 ft·lb of work is required to stretch a spring 1 ft beyond its natural length. What is the spring constant?

6. A cylindrical tank of radius 5 ft and height 9 ft is two-thirds filled with water. Find the work required to pump all the water over the upper rim.

7. Solve Exercise 6 assuming that the tank is two-thirds filled with a liquid that weighs ρ lb/ft^3.

8. A cone-shaped water reservoir is 20 ft in diameter across the top and 15 ft deep. If the reservoir is filled to a depth of 10 ft, how much work is required to pump all the water to the top of the reservoir?

9. The vat shown in Figure 6.7.4 contains water to a depth of 2 m. Find the work required to pump all the water to the top of the vat. [Use 9810 N/m^3 as the weight density of water.]

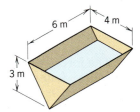

Figure 6.7.4

10. A cylindrical tank lying on its side (Figure 6.7.5) is filled with a liquid weighing 50 lb/ft^3. Find the work required to pump all the liquid to a level 1 ft above the top of the tank.

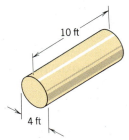

Figure 6.7.5

11. A swimming pool is built in the shape of a rectangular parallelepiped 10 ft deep, 15 ft wide, and 20 ft long.
    (a) If the pool is filled to 1 ft below the top, how much work is required to pump all the water into a drain at the top edge of the pool?
    (b) A one-horsepower motor can do 550 ft·lb of work per second. What size motor is required to empty the pool in one hour?

12. A water tower in the shape of a hemisphere of radius 10 ft is to be filled by pumping water over the top edge from a lake 200 ft below the top. How much work is required to fill the tank? [*Hint:* Solve without calculus.]

13. A 100-ft length of steel chain weighing 15 lb/ft is dangling from a pulley. How much work is required to wind the chain onto the pulley?

14. A rocket weighing 3 tons is filled with 40 tons of liquid fuel. In the initial part of the flight, fuel is burned off at a constant rate of 2 tons per 1000 ft of vertical height. How much work is done in lifting the rocket to 3000 ft?

15. (**Satellite Problem**) The weight of an object is the force exerted on it by the earth's gravity. Thus, if a person weighs 100 lb on the surface of the earth, the earth's gravity is pulling on that person with a force of 100 lb. It is a fundamental law of physics that the force the earth exerts on an object varies inversely as the square of its distance from the earth's center. Thus an object's weight $F(x)$ is related to its distance x from the earth's center by a formula of the form

$$F(x) = k/x^2$$

where k is a constant of proportionality depending on the mass of the object and the units of force and distance.
    (a) Assuming that the earth is a sphere of radius 4000 mi, find the constant k in the formula above for a satellite that weighs 6000 lb on the earth's surface.
    (b) How much work must be performed to lift this satellite to an orbital position 1000 mi above the earth's surface?

16. (**Coulomb's Law**) It follows from Coulomb's law in physics that two like electrostatic charges repel each other with a force inversely proportional to the square of the distance between them. Suppose that two charges A and B repel with a force of k N when they are positioned at points $A(-a, 0)$ and $B(a, 0)$, where a is measured in meters. Find the work W required to move charge A along the x-axis to the origin if charge B remains stationary.

6.8 FLUID PRESSURE AND FORCE

In this section we shall use the definite integral to calculate the force exerted by a liquid on a submerged surface.

☐ **FLUID PRESSURE**

If a flat surface of area A is submerged horizontally at a depth h in a container of fluid, then the weight of the fluid above exerts a force F on the surface given by

$$F = \rho h A \tag{1}$$

Figure 6.8.1

where ρ is the **weight density** of the fluid (weight per unit volume). The weight density of water is approximately 62.4 lb/ft³, for example. It is a physical fact that the force F in (1) does not depend on the shape or size of the container. Thus, if the three containers in Figure 6.8.1 have bases of the same area and are filled with a fluid to the same depth, h, then each container will have the same fluid force on its base.

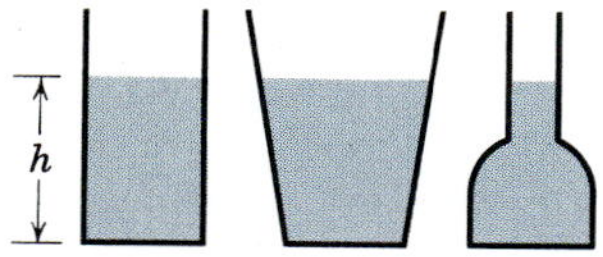

Pressure is defined to be force per unit area. Thus, from (1), the pressure p exerted at each point of a flat surface of area A submerged horizontally at a depth h is given by

$$p = \frac{F}{A} = \rho h \qquad (2)$$

In the British engineering system weight and force are commonly measured in pounds (lb) and area in square feet (ft^2), in which case weight density would be measured in units of pounds per cubic foot (lb/ft^3) and pressure in units of pounds per square foot (lb/ft^2).

In the metric system weight and force are commonly measured in newtons (N) and area in square meters (m^2), in which case weight density can be measured in units of newtons per cubic meter (N/m^3) and pressure in units of newtons per square meter (N/m^2). One newton per square meter is also called a *pascal* (Pa). One Pa is approximately 0.02089 lb/ft^2. The weight density of water is approximately 9810 N/m^3.

Example 1   If a flat circular plate of radius $r = 2$ m is submerged horizontally in water so that the top surface is at a depth of 3 m, then the force on the top surface of the plate is

$$F = \rho h A = \rho h (\pi r^2) = (9810)(3)(4\pi) = 117720\pi \text{ N}$$

and the pressure at each point on the plate surface is

$$p = \rho h = (9810)(3) = 29430 \text{ N/m}^2 = 29430 \text{ Pa} \qquad \blacktriangleleft$$

□ **PASCAL'S PRINCIPLE**

Pascal's **principle** in physics states that *fluid pressure is the same in all directions*. Thus, if a flat surface is submerged vertically (or at any angle at all), the pressure at a point of depth h is exactly the same as the pressure at a point of depth h on a horizontal surface (Figure 6.8.2). However, (1) cannot be applied

---

*BLAISE PASCAL (1623–1662). French mathematician and scientist. Pascal's mother died when he was three years old and his father, a highly educated magistrate, personally provided the boy's early education. Although Pascal showed an inclination for science and mathematics, his father refused to tutor him in those subjects until he mastered Latin and Greek. Pascal's sister and primary biographer claimed that he independently discovered the first thirty-two propositions of Euclid without ever reading a book on geometry. (However, it is generally agreed that the story is apocryphal.) Nevertheless, the precocious Pascal published a highly respected essay on conic sections by the time he was sixteen years old. Descartes, who read the essay, thought it so brilliant that he could not believe that it was written by such a young man. By age 18 his health began to fail and until his death he was in frequent pain. However, his creativity was unimpaired.

Pascal's contributions to physics include the discovery that air pressure decreases with altitude and the principle of fluid pressure that bears his name. However, the originality of his work is questioned by some historians. Pascal made major contributions to a branch of mathematics called "projective geometry," and he helped to develop probability theory through a series of letters with Fermat.

In 1646, Pascal's health problems resulted in a deep emotional crisis that led him to become increasingly concerned with religious matters. Although born a Catholic, he converted to a religious doctrine called Jansenism and spent most of his final years writing on religion and philosophy.

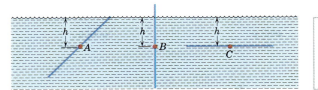

By Pascal's Principle
the fluid pressure
at points A, B,
and C is the same.

Figure 6.8.2

to obtain the total force acting on a flat surface that is not submerged horizontally, because the depth h can vary from point to point on the surface. It is in such problems that calculus comes into play.

Suppose we want to compute the total fluid force against a submerged vertical surface. As shown in Figure 6.8.3, let us introduce a vertical x-axis whose positive direction is downward and whose origin is at any convenient point.

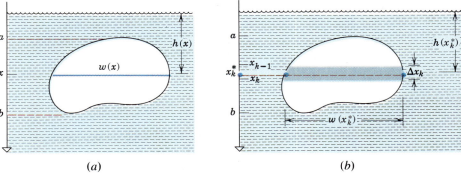

(a)   (b)

Figure 6.8.3

As indicated in Figure 6.8.3a, let us assume that the submerged portion of the surface extends from $x = a$ to $x = b$ on the x-axis. Also, suppose that a point x on the axis lies $h(x)$ units below the surface and that the cross section of the plate at x has width $w(x)$.

Next, divide the interval $[a, b]$ into n subintervals with lengths

$$\Delta x_1, \Delta x_2, \ldots, \Delta x_n$$

and in each subinterval choose an arbitrary point x_k^*. Following a familiar pattern, we approximate the section of plate along the kth subinterval by a rectangle of length $w(x_k^*)$ and width Δx_k (Figure 6.8.3b). Because the upper and lower edges of this rectangle are at different depths, (1) cannot be used to calculate the force on this rectangle. However, if Δx_k is small, the difference in depth between the upper and lower edges is small, and we can reasonably assume that the entire rectangle is concentrated at a single depth $h(x_k^*)$ below the surface. With this assumption, (1) can be used to *approximate* the force F_k on the kth rectangle. We obtain

$$F_k \approx \rho \underbrace{h(x_k^*)}_{\text{Depth}} \cdot \underbrace{w(x_k^*) \, \Delta x_k}_{\text{Area of rectangle}}$$

Thus, the total force F on the plate is approximately

$$F = \sum_{k=1}^{n} F_k \approx \sum_{k=1}^{n} \rho h(x_k^*) w(x_k^*) \, \Delta x_k$$

To find the *exact* value of the force we take the limit as $\max \Delta x_k \to 0$. This yields

$$F = \lim_{\max \Delta x_k \to 0} \sum_{k=1}^{n} \rho h(x_k^*) w(x_k^*) \, \Delta x_k = \int_a^b \rho h(x) w(x) \, dx$$

which suggests the following result.

6.8.1   FORMULA FOR FLUID FORCE.   Assume that a flat surface is immersed vertically in a liquid of weight density ρ and that the submerged portion extends from $x = a$ to $x = b$ on a vertical x-axis. For $a \leq x \leq b$, let $w(x)$ be the width of the surface at x and let $h(x)$ be the depth of the point x. Then the total *fluid force* on the surface is

$$F = \int_a^b \rho h(x) w(x) \, dx \tag{3}$$

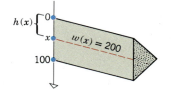

Figure 6.8.4

Example 2   The face of a dam is a vertical rectangle of height 100 ft and width 200 ft (Figure 6.8.4). Find the total fluid force exerted on the face when the water surface is level with the top of the dam.

Solution.   Introduce an x-axis with its origin at the water surface as shown in Figure 6.8.4. At a point x on this axis, the width of the dam in feet is $w(x) = 200$ and the depth in feet is $h(x) = x$. Thus, from (3) with $\rho = 62.4$ lb/ft^3 (the weight density of water) we obtain as the total force on the face

$$F = \int_0^{100} (62.4)(x)(200) \, dx = 12{,}480 \int_0^{100} x \, dx$$

$$= 12{,}480 \left. \frac{x^2}{2} \right]_0^{100} = 62{,}400{,}000 \text{ lb} \quad \blacktriangleleft$$

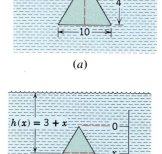

(a)

(b)

Figure 6.8.5

Example 3   A plate in the form of an isosceles triangle with base 10 ft and altitude 4 ft is submerged vertically in oil as shown in Figure 6.8.5a. Find the fluid force F against the plate surface if the oil has weight density $\rho = 30$ lb/ft^3.

Solution.   Introduce an x-axis as shown in Figure 6.8.5b. By similar triangles, the width of the plate, in feet, at a depth of $h(x) = (3 + x)$ ft satisfies

$$\frac{w(x)}{10} = \frac{x}{4} \quad \text{so} \quad w(x) = \frac{5}{2} x$$

Thus, it follows from (3) that the force on the plate is

$$F = \int_a^b \rho h(x) w(x) \, dx = \int_0^4 (30)(3 + x) \left(\frac{5}{2}x\right) dx$$

$$= 75 \int_0^4 (3x + x^2) \, dx = 75 \left[\frac{3x^2}{2} + \frac{x^3}{3}\right]_0^4 = 3400 \text{ lb} \quad \blacktriangleleft$$

▶ Exercise Set 6.8

1. A flat square plate with 3-ft sides and negligible thickness is submerged horizontally in a liquid. Find the force and pressure on a surface of the plate if
   (a) the liquid is water and the plate is at a depth of 5 ft;
   (b) the liquid has weight density 40 lb/ft³ and the plate is at a depth of 10 ft.

In Exercises 2–7, the flat surfaces shown are submerged vertically in water. Find the fluid force against the surface.

2.

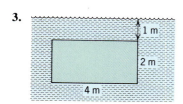

3.

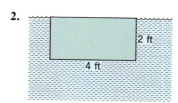

4.

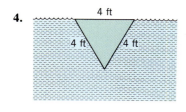

5.

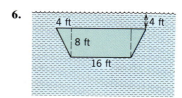

6.

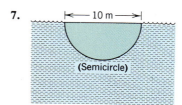

7.

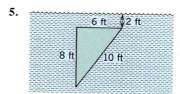

8. An oil tank is shaped like a right-circular cylinder of diameter 4 ft. Find the total fluid force against one end when the axis is horizontal and the tank is half filled with oil of weight density 50 lb/ft³.

9. A square plate of side a ft is dipped in a liquid of weight density ρ lb/ft³. Find the fluid force on the plate if a vertex is at the surface and a diagonal is perpendicular to the surface.

10. Figure 6.8.6 shows a dam whose face is an inclined rectangle. Find the fluid force on the face when the water is level with the top of this dam.

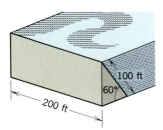

Figure 6.8.6

11. Figure 6.8.7 shows a rectangular swimming pool whose bottom is an inclined plane. Find the fluid force on the bottom when the pool is filled to the top.

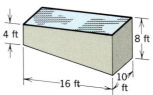

Figure 6.8.7

12. An observation window on a submarine is a square with 2-ft sides. Using ρ_0 for the weight density of seawater, find the fluid force on the window when the submarine has descended so that the window is vertical and its top is at a depth of h feet.

13. (a) Show: If the submarine in Exercise 12 descends vertically at a constant rate, then the fluid force on the window increases at a constant rate.

(b) At what rate is the force on the window increasing if the submarine is descending vertically at 20 ft/min?

▶ SUPPLEMENTARY EXERCISES

In Exercises 1–3, set up, but do not evaluate, an integral or sum of integrals that gives the area of the region R. (Set up the integral with respect to x or y as directed.)

1. R is the region in the first quadrant enclosed by $y = x^2$, $y = 2 + x$, and $x = 0$.
   (a) Integrate with respect to x.
   (b) Integrate with respect to y.

2. R is enclosed by $x = 4y - y^2$ and $y = \frac{1}{2}x$.
   (a) Integrate with respect to x.
   (b) Integrate with respect to y.

3. R is enclosed by $x = 9$ and $x = y^2$.
   (a) Integrate with respect to x.
   (b) Integrate with respect to y.

In Exercises 4–9, set up, but do not evaluate, an integral or sum of integrals that gives the stated volume. (Set up the integral with respect to x or y as directed.)

4. The volume generated by revolving the region in Exercise 1 about the x-axis.
   (a) Integrate with respect to x.
   (b) Integrate with respect to y.

5. The volume generated by revolving the region in Exercise 1 about the y-axis.

   (a) Integrate with respect to x.
   (b) Integrate with respect to y.

6. The volume generated by revolving the region in Exercise 2 about the x-axis.
   (a) Integrate with respect to x.
   (b) Integrate with respect to y.

7. The volume generated by revolving the region in Exercise 2 about the y-axis.
   (a) Integrate with respect to x.
   (b) Integrate with respect to y.

8. The volume generated by revolving the region in Exercise 3 about the x-axis.
   (a) Integrate with respect to x.
   (b) Integrate with respect to y.

9. The volume generated by revolving the region in Exercise 3 about the y-axis.
   (a) Integrate with respect to x.
   (b) Integrate with respect to y.

In Exercises 10 and 11, find (a) the area of the region described; and (b) the volume generated by revolving the region about the indicated line.

10. The region in the first quadrant enclosed by $y = \sin x$, $y = \cos x$, and $x = 0$; revolved about the x-axis. [*Hint:* $\cos^2 x - \sin^2 x = \cos 2x$.]

11. The region enclosed by the x-axis, the y-axis, and $x = \sqrt{4 - y}$; revolved about the y-axis. .

12. Set up a sum of definite integrals that represents the total shaded area between the curves $y = f(x)$ and $y = g(x)$ below.

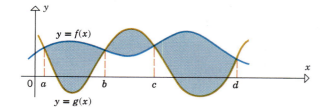

13. Find the *total* area bounded between $y = x^3$ and $y = x$ over the interval $[-1, 2]$. (See preceding exercise.)

14. Find the volume of the solid whose base is the region bounded between the curves $y = x$ and $y = x^2$, and whose cross sections perpendicular to the x-axis are squares.

15. Find the volume of the solid whose base is the triangular region with vertices $(0, 0)$, $(a, 0)$, and $(0, b)$, where $a > 0$ and $b > 0$, and whose cross sections perpendicular to the y-axis are semicircles.

In Exercises 16 and 17, find the volume generated by revolving the region described about the axis indicated.

16. The region bounded above by the curve $y = \cos x^2$, on the left by the y-axis, and below by the x-axis; revolved about the y-axis.

17. The region bounded by $y = \sqrt{x}$, $x = 4$, and $y = 0$; revolved about
(a) the line $x = 4$ (b) the line $y = 2$.

18. An American football has the shape of the solid generated by revolving the region bounded between the x-axis and the parabola $y = 4R(x^2 - \frac{1}{2}L^2)/L^2$ about the x-axis. Find its volume.

In Exercises 19–22, find the arc length of the indicated curve.

19. $8y^2 = x^3$ between $(0, 0)$ and $(2, 1)$.

20. $y = \frac{1}{3}(x^2 + 2)^{3/2}$, $0 \le x \le 3$.

21. $y = \frac{1}{10}x^5 + \frac{1}{6}x^{-3}$, $1 \le x \le 2$.

22. $y = \dfrac{1}{3}x^3 + \dfrac{1}{4x}$, $1 \le x \le 2$.

In Exercises 23–28, find the area of the surface generated by revolving the given curve about the indicated axis.

23. $y = x^3$ between $(1, 1)$ and $(2, 8)$; x-axis.

24. $y^2 = 12x$ between $(0, 0)$ and $(3, 6)$; x-axis.

25. $y = \frac{2}{3}x^{3/2} - \frac{1}{2}x^{1/2}$ between $(0, 0)$ and $(9, 33/2)$; y-axis.

26. The curve in Exercise 25 revolved about the line $x = 9$.

27. $3y = \sqrt{x}\,(3 - x)$ between $(0, 0)$ and $(3, 0)$; y-axis.

28. $y = \sqrt{2x - x^2}$ between $(\frac{1}{2}, \sqrt{3}/2)$ and $(1, 1)$; x-axis.

29. Find the work done in stretching a spring from 8 in. to 10 in. if its natural length is 6 in., and a force of 2 lb is needed to hold it at a length of 10 in.

30. Find the spring constant if 180 in·lb of work are required to stretch a spring 3 in. from its natural length.

31. A 250-lb weight is suspended from a ledge by a uniform 40-ft cable weighing 30 lb. How much work is required to bring the weight up to the ledge?

32. A tank in the shape of a right-circular cone has a 6-ft diameter at the top and a height of 5 ft. It is filled with a liquid of weight density 64 lb/ft³. How much work can be done by the liquid if it runs out of the bottom of the tank?

33. A vessel has the shape obtained by revolving about the y-axis the part of the parabola $y = 2(x^2 - 4)$ lying below the x-axis. If x and y are in feet, how much work is required to pump all the water in the full vessel to a point 4 ft above its top?

34. Two like magnetic poles repel each other with a force $F = k/x^2$ newtons, where k is a constant. Express the work needed to move them along a line from D meters apart to $D/3$ meters apart.

In Exercises 35 and 36, the flat surface shown is submerged vertically in a liquid of weight density ρ lb/ft³. Find the fluid force against the surface.

35.

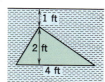

36.

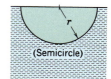

7
Logarithm and Exponential Functions

Maria Gaetana Agnesi   (1718-1799)

■ 7.1 INVERSE FUNCTIONS

In spite of their seeming simplicity, the indefinite integrals

$$\int \frac{1}{x}\, dx, \quad \int \tan x\, dx, \quad and \quad \int \sec x\, dx$$

cannot be evaluated in terms of finitely many polynomials, rational functions, or trigonometric functions. It is the primary purpose of this chapter to define new functions that will enable us to evaluate these and other important integrals. In this initial section we shall discuss some preliminary results that will be essential to our work in this chapter and the next.

□ **INVERSE FUNCTIONS**

The functions $f(x) = 2x$ and $g(x) = \frac{1}{2}x$ have the property that each cancels out the effect of the other in the sense that

$$f(g(x)) = f(\tfrac{1}{2}x) = 2(\tfrac{1}{2}x) = x$$
$$g(f(x)) = g(2x) = \tfrac{1}{2}(2x) = x$$

Similarly, the functions $f(x) = x^{1/3}$ and $g(x) = x^3$ cancel the effect of one another since

$$f(g(x)) = f(x^3) = (x^3)^{1/3} = x$$
$$g(f(x)) = g(x^{1/3}) = (x^{1/3})^3 = x$$

Pairs of functions that cancel the effect of one another are of such importance that there is some terminology associated with them.

7.1.1 DEFINITION.   If the functions f and g satisfy the two conditions

$$f(g(x)) = x \text{ for every } x \text{ in the domain of } g$$
$$g(f(x)) = x \text{ for every } x \text{ in the domain of } f$$

then we say that *f is an inverse of g* and *g is an inverse of f*. We also say that *f and g are inverse functions*.

Example 1   In the terminology of the foregoing definition, the functions $f(x) = 2x$ and $g(x) = \frac{1}{2}x$ are inverse functions, as are $f(x) = x^{1/3}$ and $g(x) = x^3$.   ◄

It can be shown that a function cannot have two different inverses. Thus, if f has an inverse, we are entitled to talk about *the* inverse of f. The inverse of f is commonly denoted by f^{-1} (read, "f inverse"); thus,

$$f(f^{-1}(x)) = x \quad \text{and} \quad f^{-1}(f(x)) = x \tag{1}$$

WARNING.   The symbol f^{-1} does not mean $1/f$.

Example 2 To emphasize that

$$f(x) = 2x \quad \text{and} \quad g(x) = \tfrac{1}{2}x$$

are inverse functions, we can write

$$f(x) = 2x \quad \text{and} \quad f^{-1}(x) = \tfrac{1}{2}x$$

or

$$g(x) = \tfrac{1}{2}x \quad \text{and} \quad g^{-1}(x) = 2x \quad \blacktriangleleft$$

☐ **SYMMETRY ABOUT THE LINE** $y = x$

There is an important relationship between the graphs of f and f^{-1} that can be seen by graphing the inverse functions in Example 1 (see Figure 7.1.1). Figure 7.1.1 shows that the graphs of $y = 2x$ and $y = \tfrac{1}{2}x$ are reflections ("mirror images") of one another about the line $y = x$. The same is true of the graphs of $y = x^3$ and $y = x^{1/3}$. This is not accidental; we shall see below that all pairs of inverse functions have this property.

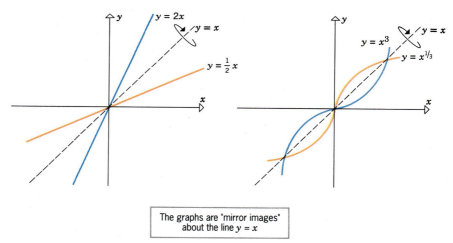

The graphs are "mirror images" about the line $y = x$

Figure 7.1.1

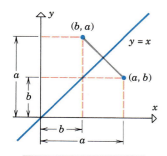

The line $y = x$ is the perpendicular bisector of the line segment joining (a, b) and (b, a).

Figure 7.1.2

It can be proved using plane geometry that the points (a, b) and (b, a) are symmetric about the line $y = x$ in the sense that this line is the perpendicular bisector of the line segment joining the points (Figure 7.1.2).

It follows from this result that interchanging the x and y coordinates of a point reflects that point about the line $y = x$ and that interchanging the x and y variables in an equation reflects the graph of that equation about the line $y = x$.

☐ **GRAPHS OF INVERSE FUNCTIONS**

Example 3 Interchanging the x and y variables in the equation

$$y = 2x$$

yields $x = 2y$, which can be rewritten as

$$y = \frac{1}{2}x$$

Thus, the graphs of $y = 2x$ and $y = \frac{1}{2}x$ are symmetric about the line $y = x$, which is consistent with Figure 7.1.1*a*. Similarly, interchanging x and y in the equation

$$y = x^3$$

yields $x = y^3$, which can be rewritten as

$$y = \sqrt[3]{x}$$

Thus, the graphs of $y = x^3$ and $y = \sqrt[3]{x}$ are symmetric about the line $y = x$, which is consistent with Figure 7.1.1*b*.   ◄

In general, we have the following theorem.

7.1.2   THEOREM.   *If a function f has an inverse, then the graphs of $y = f(x)$ and $y = f^{-1}(x)$ are symmetric about the line $y = x$.*

□ EXISTENCE OF INVERSE FUNCTIONS

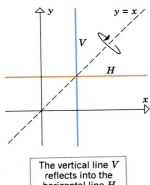

The vertical line V reflects into the horizontal line H and conversely.

Figure 7.1.3

Given a function f, we shall be interested in two important questions:

1.   Does f have an inverse?
2.   If so, how do we find it?

Let us consider the first of these questions. If f has an inverse, then the graph of $y = f^{-1}(x)$ cannot be cut more than once by any *vertical* line (see 2.3.3). But this graph is the reflection about $y = x$ of the graph of $y = f(x)$ and a reflection about $y = x$ causes horizontal lines to become vertical lines (Figure 7.1.3); hence the graph of $y = f(x)$ cannot be cut more than once by any *horizontal* line. This suggests the following result.

7.1.3   THE HORIZONTAL LINE TEST.   A function f has an inverse if and only if no horizontal line intersects its graph more than once.

Example 4   The function $f(x) = x^2$ has no inverse because it does not pass the horizontal line test (Figure 7.1.4).   ◄

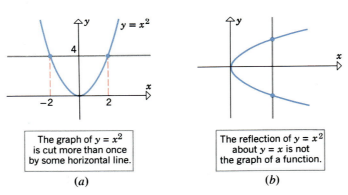

The graph of $y = x^2$ is cut more than once by some horizontal line.

(*a*)

The reflection of $y = x^2$ about $y = x$ is not the graph of a function.

(*b*)

Figure 7.1.4

□ ONE-TO-ONE FUNCTIONS

Referring to Figure 7.1.5, it is evident that if the graph of a function f has multiple intersections with a horizontal line, then those intersections correspond to points on the x-axis at which f has the same value. Thus, to state that the graph of a function f is cut at most once by any horizontal line is equivalent to stating that f does not have the same value at two distinct points in its domain. A function f with this property is said to be *one-to-one*.

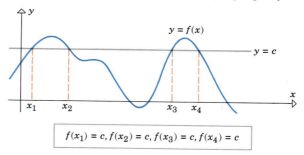

$$f(x_1) = c, f(x_2) = c, f(x_3) = c, f(x_4) = c$$

Figure 7.1.5

7.1.4   DEFINITION.   A function f is **one-to-one** if its graph is cut at most once by any horizontal line, or equivalently, if f does not have the same value at two distinct points in its domain.

Using the terminology of this definition, we can rephrase 7.1.3 as follows.

7.1.5   THEOREM.   *A function f has an inverse if and only if it is one-to-one.*

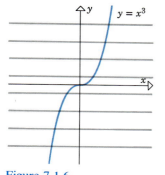

Figure 7.1.6

Example 5   We saw in Example 4 that the function $f(x) = x^2$ has no inverse (hence is not one-to-one) because there are horizontal lines that cut its graph more than once. From an algebraic viewpoint, $f(x) = x^2$ is not one-to-one because f can have the same value at two distinct points; for example, $f(-2) = 4$ and $f(2) = 4$ (Figure 7.1.4a).   ◄

Example 6   The function $f(x) = x^3$ is one-to-one because two different numbers cannot have the same cube, that is, if $x_1 \neq x_2$, then $x_1^3 \neq x_2^3$. Geometrically, no horizontal line cuts the graph of $f(x) = x^3$ more than once (Figure 7.1.6).   ◄

□ INVERSES OF INCREASING AND DECREASING FUNCTIONS

The following theorem describes a major class of functions with inverses.

7.1.6   THEOREM.   *If the domain of f is an interval, and if f is either an increasing function or a decreasing function on that interval, then f has an inverse.*

Proof.   If f is increasing and $x_1 < x_2$ are distinct points in its domain, then $f(x_1) < f(x_2)$ (Definition 4.2.1), so that $f(x_1) \neq f(x_2)$. Thus, f is one-to-one and therefore has an inverse by Theorem 7.1.5. The proof for decreasing functions is similar.   ∎

Recall from Theorem 4.2.2 and the subsequent discussion that f is increasing on an interval if $f'(x) > 0$ on the interval and is decreasing if $f'(x) < 0$. This result makes it easy to apply Theorem 7.1.6 to differentiable functions.

Example 7   Show that $f(x) = x^5 + 7x^3 + 4x + 1$ has an inverse.

Solution.   The function f is increasing on $(-\infty, +\infty)$ because

$$f'(x) = 5x^4 + 21x^2 + 4 > 0$$

for all x. Thus, f has an inverse by Theorem 7.1.6.   ◀

☐ **FINDING A FORMULA FOR f^{-1}**

We now turn to the problem of actually finding a formula for the inverse of a one-to-one function f. If we let

$$y = f^{-1}(x) \tag{2}$$

then it follows that

$$f(y) = f(f^{-1}(x))$$

or

$$f(y) = x \tag{3}$$

Observe that (3) expresses x explicitly as a function of y and (2) expresses y explicitly as a function of x. Thus, if we start with Equation (3) and solve this equation for y as a function of x, we will obtain (2), from which we can read off the formula for f^{-1}. Thus, we have the following procedure for finding a formula for the inverse of a one-to-one function f.

> **Step 1.**   Interchange x and y in the equation $y = f(x)$ to produce the equation $x = f(y)$.
>
> **Step 2.**   Solve the equation $x = f(y)$ for y as a function of x.
>
> **Step 3.**   The resulting equation in Step 2 will be $y = f^{-1}(x)$, the right side of which is the formula for f^{-1}.

Example 8   Find the inverse of $f(x) = 4x - 5$.

Solution.   We first introduce a dependent variable y; this yields

$$y = 4x - 5$$

Interchanging x and y in this equation, then solving for y we obtain

$$x = 4y - 5$$
$$y = \tfrac{1}{4}(x + 5)$$

from which it follows that

$$f^{-1}(x) = \tfrac{1}{4}(x + 5) \tag{4}$$

The graphs of f and f^{-1} are shown in Figure 7.1.7. As expected, they are reflections of one another about $y = x$.   ◄

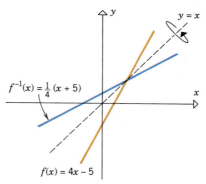

Figure 7.1.7

□ **DOMAIN AND RANGE OF INVERSE FUNCTIONS**

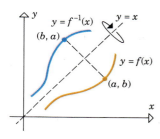

Figure 7.1.8

If f is a one-to-one function and (a, b) is a point on the graph of $y = f(x)$, then a is in the domain of f and b is in the range of f, since $b = f(a)$. As illustrated in Figure 7.1.8, reversing the x and y coordinates of this point yields the point (b, a) on the graph of $y = f^{-1}(x)$; thus, b is in the domain of f^{-1} and a is in the range of f^{-1}. This suggests the following relationships:

$$\text{range of } f^{-1} = \text{domain of } f$$
$$\text{domain of } f^{-1} = \text{range of } f \tag{5}$$

In retrospect, our computations in Example 8 were incomplete because we did not check that the computations produced the correct domain for f^{-1}. However, no harm was done because the range of $f(x) = 4x - 5$ is $(-\infty, +\infty)$ (verify), so the domain of f^{-1} is $(-\infty, +\infty)$, which is precisely the natural domain of (4). However, as a general rule you must determine the domain when calculating a formula for f^{-1}.

Example 9   Find the inverse of $f(x) = \sqrt{3x - 2}$.

Solution.   We first introduce a dependent variable y; this yields

$$y = \sqrt{3x - 2}$$

Interchanging x and y in this equation, then solving for y we obtain

$$x = \sqrt{3y - 2}$$
$$x^2 = 3y - 2$$
$$y = \tfrac{1}{3}(x^2 + 2)$$

from which it follows that

$$f^{-1}(x) = \tfrac{1}{3}(x^2 + 2)$$

The range of $f(x) = \sqrt{3x - 2}$ is $[0, +\infty)$ (Figure 7.1.9), so this interval is also the domain of f^{-1}. Thus, the inverse of f is

$$f^{-1}(x) = \tfrac{1}{3}(x^2 + 2), \quad x \geq 0 \quad ◄$$

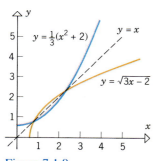

Figure 7.1.9

☐ SOME COMPLICATIONS IN FINDING INVERSES

The key step in finding the inverse of a one-to-one function f is solving the equation $x = f(y)$ for y as a function of x. If this step cannot be performed, then an explicit formula for f^{-1} cannot be produced, even though f may have an inverse. There are two reasons why you may not be able to solve the equation $x = f(y)$ for y in terms of x:

- The equation may be so complicated that it is beyond your ability to solve it.
- For some equations it is mathematically impossible to solve $x = f(y)$ for y as a function of x using finitely many algebraic operations.

The problem of determining when a function has an inverse can be quite complicated and is studied in more detail in advanced courses.

Example 10   In Example 7 we proved that $f(x) = x^5 + 7x^3 + 4x + 1$ has an inverse. However, to find a formula for f^{-1} we would have to solve the equation

$$x = y^5 + 7y^3 + 4y + 1$$

for y in terms of x, which is complicated. Thus, we have no formula for f^{-1} even though we know the inverse exists.   ◀

☐ CONTINUITY OF INVERSE FUNCTIONS

Because the graphs of f and f^{-1} are reflections of one another about the line $y = x$, it is intuitively obvious that if the graph of f has no breaks, then neither will the graph of f^{-1}. This suggests the following result, which we state without proof.

> **7.1.7   THEOREM.**   *If a function f is continuous and has an inverse, then f^{-1} is also continuous.*

☐ DIFFERENTIABILITY OF INVERSE FUNCTIONS

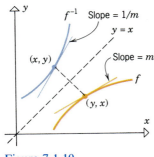

Figure 7.1.10

Because the graphs of f and f^{-1} are reflections of one another about the line $y = x$, it is intuitively obvious that if the graph of f has no corners, then neither will the graph of f^{-1}. However, since reflection about $y = x$ converts horizontal lines into vertical lines, a point where the graph of f has a horizontal tangent line will be reflected into a point where f^{-1} has a vertical tangent line, and this will be a point of nondifferentiability for f^{-1}.

A relationship between the derivatives of f and f^{-1} can be obtained as follows. Let f^{-1} be differentiable at x, and let (x, y) be the point on the graph of f^{-1} corresponding to this value of x (Figure 7.1.10). This point is the reflection about the line $y = x$ of the point (y, x), which lies on the graph of f. Because f^{-1} is differentiable at x, the tangent line to the graph of f at the point (y, x) is not horizontal; suppose that the equation of this tangent line is

$$y = mx + b \tag{6}$$

where $m \neq 0$. The reflection of this tangent line about the line $y = x$ is the tangent line at (x, y) to the graph of f^{-1}; this line has the equation

$$x = my + b$$

which we obtained by interchanging x and y in (6). Solving this equation for y yields

$$y = \frac{1}{m} x - \frac{b}{m}$$

which tells us that the slope of the tangent line to $y = f^{-1}(x)$ at (x, y) and the slope of the tangent line to $y = f(x)$ at (y, x) are reciprocals. Stating this relationship in terms of derivatives, we obtain

$$(f^{-1})'(x) = \frac{1}{f'(y)} \tag{7}$$

But (x, y) lies on the graph of f^{-1}, so $y = f^{-1}(x)$; thus, (7) can be rewritten as

$$(f^{-1})'(x) = \frac{1}{f'(f^{-1}(x))}$$

which is the relationship between the derivative of f and the derivative of f^{-1}. In summary, we have the following result.

7.1.8 THEOREM. *Suppose that the function f has an inverse and that the value of $f^{-1}(x)$ varies over an interval on which f has a nonzero derivative as x varies over an interval I. Then f^{-1} is differentiable on I and the derivative of f^{-1} is given by the formula*

$$(f^{-1})'(x) = \frac{1}{f'(f^{-1}(x))} \tag{8}$$

Formula (8) can be expressed in a less forbidding form by letting

$$y = f^{-1}(x) \quad \text{so that} \quad x = f(y)$$

Thus,

$$\frac{dy}{dx} = (f^{-1})'(x) \quad \text{and} \quad \frac{dx}{dy} = f'(y) = f'(f^{-1}(x))$$

Substituting these expressions in (8) yields the following alternative version of that formula:

$$\frac{dy}{dx} = \frac{1}{dx/dy} \tag{9}$$

If an explicit formula can be obtained for the inverse of a function, then the differentiability and the derivative of the inverse can generally be deduced from that formula. However, if no explicit formula for the inverse can be obtained, then Theorem 7.1.8 is the primary mathematical tool for establishing differentiability of the inverse. Once the differentiability has been established, a derivative formula for the inverse can be obtained either by implicit differentiation or by using Formulas (8) or (9). The following example illustrates this.

Example 11   In Example 7 we saw that the function $f(x) = x^5 + 7x^3 + 4x + 1$ has an inverse.

(a)   Show that f^{-1} is differentiable on the interval $(-\infty, +\infty)$.

(b)   Find the derivative of f^{-1} by using Formula (9).

(c)   Find the derivative of f^{-1} by implicit differentiation.

Solution (a).   Let I denote the interval $(-\infty, +\infty)$. We must show that as x varies over I, the value of $f^{-1}(x)$ varies over an interval on which f has a nonzero derivative. But this is so because the derivative of f is

$$f'(x) = 5x^4 + 21x^2 + 4$$

which is positive (hence, nonzero) for all x.

Solution (b).   If we let $y = f^{-1}(x)$, then

$$f(y) = f(f^{-1}(x)) = x$$

Thus, for the given function f, we have

$$x = f(y) = y^5 + 7y^3 + 4y + 1 \tag{10}$$

from which it follows that

$$\frac{dx}{dy} = 5y^4 + 21y^2 + 4$$

so

$$\frac{dy}{dx} = \frac{1}{dx/dy} = \frac{1}{5y^4 + 21y^2 + 4} \tag{11}$$

Because (10) is too complicated to solve for y in terms of x, we must leave (11) in terms of y.

Solution (c).   Differentiating (10) implicitly with respect to x yields

$$\frac{d}{dx}[x] = \frac{d}{dx}[y^5 + 7y^3 + 4y + 1]$$

$$1 = 5y^4 \frac{dy}{dx} + 21y^2 \frac{dy}{dx} + 4 \frac{dy}{dx}$$

$$1 = (5y^4 + 21y^2 + 4) \frac{dy}{dx}$$

$$\frac{dy}{dx} = \frac{1}{5y^4 + 21y^2 + 4}$$

which agrees with (11).   ◄

► Exercise Set 7.1

1. In (a)–(d), determine whether f and g are inverse functions.
   (a) $f(x) = 4x$, $g(x) = \frac{1}{4}x$
   (b) $f(x) = 3x + 1$, $g(x) = 3x - 1$
   (c) $f(x) = \sqrt[3]{x - 2}$, $g(x) = x^3 + 2$
   (d) $f(x) = x^4$, $g(x) = \sqrt[4]{x}$.

In Exercises 2–14, determine whether the function has an inverse.

2. $f(x) = 1 - x$.   3. $f(x) = 3x + 2$.

4. $f(x) = x^2 - 2x + 1$.   5. $f(x) = 2 - x - x^2$.

6. $f(x) = x^3 - 3x + 2$.   7. $f(x) = x^3 - x - 1$.

8. $f(x) = x^3 - 3x^2 + 3x - 1$.

9. $f(x) = x^3 + 3x^2 + 3x + 1$.

10. $f(x) = x^5 + 8x^3 + 2x - 1$.

11. $f(x) = 2x^5 + x^3 + 3x + 2$.

12. $f(x) = x + \dfrac{1}{x}$, $x > 0$.

13. $f(x) = \sin x$, $-\pi/2 < x < \pi/2$.

14. $f(x) = \tan x$, $-\pi/2 < x < \pi/2$.

In Exercises 15–25, find $f^{-1}(x)$.

15. $f(x) = x^5$.   16. $f(x) = 6x$.

17. $f(x) = 7x - 6$.   18. $f(x) = \dfrac{x + 1}{x - 1}$.

19. $f(x) = 3x^3 - 5$.   20. $f(x) = \sqrt[5]{4x + 2}$.

21. $f(x) = \sqrt[3]{2x - 1}$.

22. $f(x) = 5/(x^2 + 1)$, $x \geq 0$.

23. $f(x) = 3/x^2$, $x < 0$.

24. $f(x) = \begin{cases} 2x, & x \leq 0 \\ x^2, & x > 0. \end{cases}$

25. $f(x) = \begin{cases} 5/2 - x, & x < 2 \\ 1/x, & x \geq 2. \end{cases}$

In Exercises 26–32, use Formula (9) to find the derivative of f^{-1}, and check your work by differentiating implicitly.

26. $f(x) = 2x^3 + 5x + 3$.

27. $f(x) = 5x^3 + x - 7$.   28. $f(x) = 1/x^2$, $x > 0$.

29. $f(x) = \tan 2x$, $-\pi/4 < x < \pi/4$.

30. $f(x) = 5x - \sin 2x$.

31. $f(x) = 2x^5 + x^3 + 1$.

32. $f(x) = x^7 + 2x^5 + x^3$.

33. Let $f(x) = x^2$, $x > 1$ and $g(x) = \sqrt{x}$.
    (a) Show that $f(g(x)) = x$, $x > 1$, and $g(f(x)) = x$, $x > 1$.
    (b) Show that f and g are *not* inverses of one another by showing that the graphs of $y = f(x)$ and $y = g(x)$ are not reflections of one another about $y = x$.
    (c) Do parts (a) and (b) contradict Definition 7.1.1? Explain.

34. Let $f(x) = ax^2 + bx + c$, $a > 0$. Find f^{-1} if
    (a) $x \geq -b/(2a)$ (b) $x \leq -b/(2a)$.

In Exercises 35–39, find $f^{-1}(x)$ and its domain.

35. $f(x) = (x + 2)^4$, $x \geq 0$.

36. $f(x) = \sqrt{x + 3}$.   37. $f(x) = -\sqrt{3 - 2x}$.

38. $f(x) = 3x^2 + 5x - 2$, $x \geq 0$.

39. $f(x) = x - 5x^2$, $x \geq 1$.

40. Prove that if $a^2 + bc \neq 0$, then the graph of
$$f(x) = \frac{ax + b}{cx - a}$$
is symmetric about the line $y = x$.

41. (a) Show that $f(x) = (3 - x)/(1 - x)$ is its own inverse.
    (b) What does the result in part (a) tell you about the graph of f?

42. Suppose that a line of nonzero slope m intersects the x-axis at $(x_0, 0)$. Find an equation for the reflection of this line about $y = x$.

43. (a) Show that $f(x) = x^3 - 3x^2 + 2x$ is not one-to-one on $(-\infty, +\infty)$.
    (b) Find the largest value of k such that f is one-to-one on the interval $(-k, k)$.

44. (a) Show that $f(x) = x^4 - 2x^3$ is not one-to-one on $(-\infty, +\infty)$.
    (b) Find the smallest value of k such that f is one-to-one on the interval $[k, +\infty)$.

45. Let $f(x) = 2x^3 + 5x + 3$. Find x if $f^{-1}(x) = 1$.

46. Let $f(x) = \dfrac{x^3}{x^2 + 1}$. Find x if $f^{-1}(x) = 2$.

In Exercises 47–50, a function f and the coordinates of a point on the graph of $y = f^{-1}(x)$ are given. Find the slope of the tangent line to the graph of $y = f^{-1}(x)$ at the given point.

47. $f(x) = x^3 + x$; $(10, 2)$.

48. $f(x) = x^5 + 2x^3 + x + 4$; $(0, -1)$.

49. $f(x) = \sin 2x$, $0 \le x \le \pi/4$; $(\frac{1}{2}, \pi/12)$.

50. $f(x) = x^3 - \dfrac{2}{x}$; $(-1, 1)$.

51. Let $f(x) = \displaystyle\int_1^x \sqrt[3]{1 + t^2}\, dt$.

    (a) Without integrating, show that f is one-to-one on the interval $(-\infty, +\infty)$.

(b) Find $\dfrac{d}{dx}[f^{-1}(x)]\Big|_{x=0}$.

52. (a) Prove: If f and g are one-to-one, then so is the composition $f \circ g$.

    (b) Prove: If f and g are one-to-one, then

$$(f \circ g)^{-1} = g^{-1} \circ f^{-1}$$

53. Sketch the graph of a function that is one-to-one on $(-\infty, +\infty)$, yet not increasing on $(-\infty, +\infty)$ and not decreasing on $(-\infty, +\infty)$.

54. Prove: A one-to-one function f cannot have two different inverses.

55. Let $F(x) = f(2g(x))$ where $f(x) = x^4 + x^3 + 1$ for $0 \le x \le 2$, and $g(x) = f^{-1}(x)$. Find $F'(3)$.

■ **7.2   LOGARITHMS AND IRRATIONAL EXPONENTS (AN OVERVIEW)**

In this section we shall review some basic facts about logarithms and rational exponents, and we shall set the groundwork for a definition of irrational exponents that will be given in a later section. Our work here will be intuitive and informal. In later sections we shall make the ideas discussed here mathematically precise.

Many of the examples and exercises in this section require the use of a scientific calculator that has the capability of evaluating logarithms and powers of real numbers. Henceforth, we shall assume that you have access to such a calculator and know how to use it to compute powers and logarithms. Consult your calculator manual, if necessary.

☐ **IRRATIONAL EXPONENTS**   In algebra, integer exponents and rational exponents of a positive number b are defined by

$$b^n = b \times b \times \cdots \times b \qquad \boxed{n \text{ factors}}$$

$$b^{-n} = \frac{1}{b^n}$$

$$b^0 = 1$$

$$b^{p/q} = \sqrt[q]{b^p} = (\sqrt[q]{b})^p$$

$$b^{-p/q} = \frac{1}{b^{p/q}}$$

The next natural step is to extend the concept of an exponent to include irrational numbers, thereby giving meaning to such expressions as

$$2^\pi, \quad 3^{\sqrt{2}}, \quad \text{and} \quad \pi^{-\sqrt{7}}$$

Later, we will give a rigorous definition of irrational exponents, but we can motivate the concept by considering the example of 2^π. Using the decimal representation of π,

$$3.1415926\ldots$$

we can generate a succession of rational numbers

$$3, \quad 3.1, \quad 3.14, \quad 3.141, \quad 3.1415, \quad 3.14159, \ldots$$

that get closer and closer to π. The corresponding rational powers of 2,

$$2^3, \quad 2^{3.1}, \quad 2^{3.14}, \quad 2^{3.141}, \quad 2^{3.1415}, \quad 2^{3.14159}, \ldots$$

whose approximate values (obtained with a calculator) are shown in Table 7.2.1, get closer and closer to 2^π. Rounded to four decimal places, the value of 2^π is 8.8250.

Table 7.2.1

x	2^x
3	8.000000
3.1	8.574188
3.14	8.815241
3.141	8.821353
3.1415	8.824411
3.14159	8.824962
3.141592	8.824974

There is an underlying assumption in this discussion that as x gets closer and closer to π, the value of the function 2^x gets closer and closer to 2^π. Stated another way, we are assuming that the function 2^x is continuous at $x = \pi$. In general, we will want the definition of irrational exponents to be such that for every positive constant b the function b^x is continuous, and we will want the familiar properties of exponents, such as $b^m \cdot b^n = b^{m+n}$, to continue to hold.

☐ **THE GRAPH OF b^x**

Depending on whether

$$0 < b < 1, \quad b = 1, \quad \text{or} \quad b > 1$$

the graph of $y = b^x$ will have one of the shapes shown in Figure 7.2.1a. Because b^x is not yet defined for irrational values of x, each of the graphs consists of densely packed dots separated by holes at the irrational values of x. After we define irrational exponents so that b^x is continuous, the holes will be filled in and the graphs will look like those in Figure 7.2.1b.

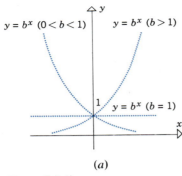

Figure 7.2.1

REMARK.   We have limited the discussion of b^x to the case where $b > 0$ since negative values of b can lead to imaginary numbers; for example, if $b = -2$ and $x = 1/2$, then $b^x = (-2)^{1/2} = \sqrt{-2}$.

☐ **REVIEW OF LOGARITHMS**

Recall from algebra that a logarithm is an exponent. More precisely, if b is a positive number other than 1 and x is a positive number, then

$$\log_b x$$

(read, "the *logarithm to the base b of x*") represents that power to which b must be raised to produce x. Thus,

$$\log_{10} 100 = 2$$

because 10 must be raised to the second power to produce 100. Similarly,

$$\log_2 8 = 3 \qquad \text{since} \qquad 2^3 = 8$$

$$\log_{10} \tfrac{1}{1000} = -3 \quad \text{since} \qquad 10^{-3} = \tfrac{1}{1000}$$

$$\log_{10} 1 = 0 \qquad \text{since} \qquad 10^0 = 1$$

$$\log_3 81 = 4 \qquad \text{since} \qquad 3^4 = 81$$

In general,

$$y = \log_b x \quad \text{and} \quad x = b^y$$

are equivalent statements. It follows from these equations that

$$\log_b b^y = y \quad \text{and} \quad b^{\log_b x} = x$$

Restating the first of these equations in terms of x rather than y, we have

$$\log_b b^x = x \tag{1}$$

$$b^{\log_b x} = x \tag{2}$$

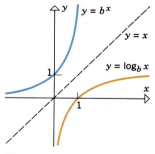

Figure 7.2.2

which implies that

$$f(x) = b^x \quad \text{and} \quad g(x) = \log_b x$$

are inverse functions. Thus, the graphs of $y = b^x$ and $y = \log_b x$ are reflections of one another about the line $y = x$. (Figure 7.2.2 illustrates the case where $b > 1$.)

REMARK.　It is important to note that for every base b, $\log_b x$ is only defined for $x > 0$. This is because $y = \log_b x$ is equivalent to $x = b^y$, and $b^y > 0$ for all real values of y (since b is positive).

The reader should already be familiar with the properties of logarithms listed in the following theorem.

7.2.1 THEOREM.

(a)　$\log_b 1 = 0$ $\qquad\qquad$ (b)　$\log_b b = 1$

(c)　$\log_b ac = \log_b a + \log_b c$ $\qquad$ (d)　$\log_b \dfrac{a}{c} = \log_b a - \log_b c$

(e)　$\log_b a^r = r \log_b a$ $\qquad\qquad$ (f)　$\log_b \dfrac{1}{c} = -\log_b c$

We shall prove (a) and (c) and leave the remaining proofs as exercises.

Proof (a).　Since $b^0 = 1$, it follows that $\log_b 1 = 0$.

Proof (c).　Let

$$x = \log_b a \quad \text{and} \quad y = \log_b c \tag{3}$$

so

$$b^x = a \quad \text{and} \quad b^y = c$$

Therefore,

$$ac = b^x b^y = b^{x+y} \quad \text{or equivalently} \quad \log_b ac = x + y$$

Thus, from (3)

$$\log_b ac = \log_b a + \log_b c \quad \blacksquare$$

☐ **THE NUMBER e**
NATURAL LOGARITHMS

The most important logarithms are *common logarithms,* which have base 10, and *natural logarithms,* which have a certain irrational base that is denoted by the letter e. To five decimal places the value of e is

$$e \approx 2.71828$$

Figure 7.2.3

This number was discovered by the Swiss mathematician Leonard Euler (see p. 71), who showed that $y = e$ is a horizontal asymptote of the graph of

$$y = \left(1 + \frac{1}{x}\right)^x \tag{4}$$

(Figure 7.2.3). Euler suggested the applicability of e to logarithms in an unpublished paper he wrote in 1728.

The fact that $y = e$ is a horizontal asymptote of (4) is expressed by the limit

$$e = \lim_{x \to +\infty} \left(1 + \frac{1}{x}\right)^x \tag{5}$$

Later, after we have defined the number e precisely, we will prove (5) and the following related result.

$$e = \lim_{x \to 0} (1 + x)^{1/x} \tag{6}$$

Most scientific calculators have a special key for evaluating e; however, the reader may want to try using (5) to approximate e by evaluating (4) for a succession of increasing values of x. Some sample calculations are shown in Table 7.2.2.

Table 7.2.2

x	$1 + \frac{1}{x}$	$\left(1 + \frac{1}{x}\right)^x$
1	2	2.000000
10	1.1	2.593742
100	1.01	2.704814
1,000	1.001	2.716924
10,000	1.0001	2.718146
100,000	1.00001	2.718268
1,000,000	1.000001	2.718280

NOTATION FOR COMMON AND NATURAL LOGARITHMS

It is standard to drop the subscript on common logarithms and write $\log x$ rather than $\log_{10} x$. It is also standard to write $\ln x$ (read, "ell en of x") rather than $\log_e x$ for natural logarithms. For example,

$\ln 1 = 0$ Since $e^0 = 1$

$\ln e = 1$ Since $e^1 = e$

$\ln \dfrac{1}{e} = -1$ Since $e^{-1} = \dfrac{1}{e}$

$\ln (e^2) = 2$ ◄

REMARK.   It should be noted that some calculators, computer programs, and mathematical literature use "log" rather than "ln" to denote the natural logarithm, so it is a good idea to check the convention being followed.

☐ **CALCULATOR APPROXIMATIONS OF LOGARITHMS**

Only the simplest of logarithms can be evaluated by inspection. Generally, a calculator will be required to obtain approximate numerical values of $\log_{10} x$ or $\ln x$. Figure 7.2.4 shows a table of values for $\ln x$ obtained with a calculator and the graph of $y = \ln x$ obtained from the data in the table.

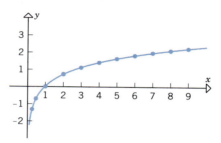

x	$y = \ln x$
0.25	−1.39
0.50	−0.69
1	0
2	0.69
3	1.10
4	1.39
5	1.61
6	1.79
7	1.95
8	2.08
9	2.20

Figure 7.2.4

☐ **EXPANDING AND CONDENSING LOGARITHMIC EXPRESSIONS**

The results in Theorem 7.2.1 are commonly used in two ways:

• To expand a single logarithm into sums, differences, and multiples of logarithms.

• To condense sums, differences, and multiples of logarithms into a single logarithm.

Example 1   Express

$$\log \frac{xy^5}{\sqrt{z}}$$

in terms of sums, differences, and multiples of $\log x$, $\log y$, and $\log z$.

Solution.

$$\log \frac{xy^5}{\sqrt{z}} = \log xy^5 - \log \sqrt{z}$$

$$= \log x + \log y^5 - \log z^{1/2}$$

$$= \log x + 5 \log y - \tfrac{1}{2} \log z \quad \blacktriangleleft$$

Example 2   In each part rewrite the expression as a single logarithm.

(a) $5 \log 2 + \log 3 - \log 8$

(b) $\tfrac{1}{3} \ln x - \ln (x^2 - 1) + 2 \ln (x + 3)$

Solution (a).

$$5 \log 2 + \log 3 - \log 8 = \log 2^5 + \log 3 - \log 8$$
$$= \log 32 + \log 3 - \log 8$$
$$= \log (32 \cdot 3) - \log 8$$
$$= \log \frac{32 \cdot 3}{8} = \log 12$$

Solution (b).

$$\tfrac{1}{3} \ln x - \ln (x^2 - 1) + 2 \ln (x + 3) = \ln x^{1/3} - \ln (x^2 - 1) + \ln (x + 3)^2$$
$$= \ln \frac{\sqrt[3]{x}(x + 3)^2}{x^2 - 1} \qquad ◄$$

REMARK.   Expressions of the form

$$\log_b (u + v), \quad \log_b (u - v), \quad \log_b u \cdot \log_b v, \quad \frac{\log_b u}{\log_b v}$$

have no useful simplifications in terms of $\log_b u$ and $\log_b v$. In particular,

$$\log_b (u + v) \neq \log_b u + \log_b v$$

$$\log_b (u - v) \neq \log_b u - \log_b v$$

$$\log_b u \cdot \log_b v \neq \log_b (uv)$$

$$\frac{\log_b u}{\log_b v} \neq \log_b \left(\frac{u}{v} \right)$$

☐ **SOLVING EQUATIONS OF THE FORM** $\log_b f(x) = k$

Equations of the form $\log_b f(x) = k$ can be solved by converting them to exponential form.

Example 3   Solve for x.

(a) $\log x = 2$ (b) $\ln (x + 1) = 5$

Solution (a).   Rewriting the equation in exponential form yields

$$x = 10^2 = 100$$

Solution (b).   Rewriting the equation in exponential form yields

$$x + 1 = e^5 \quad \text{or} \quad x = e^5 - 1$$

If desired, a calculator can be used to obtain a decimal approximation of this exact solution; to four decimal places this approximation is $x \approx 147.4132$.   ◄

☐ **SOLVING EQUATIONS OF THE FORM** $b^{f(x)} = k$

Equations of the form $b^{f(x)} = k$ can be solved by taking the logarithm of both sides. Usually common logarithms or natural logarithms are used.

Example 4 Solve for x.

(a) $5^x = 7$ (b) $e^x = 9$

Solution (a).   Taking the common logarithm of both sides yields

$$\log 5^x = \log 7$$

$$x \log 5 = \log 7$$

$$x = \frac{\log 7}{\log 5} \approx \frac{0.845098}{0.698970} \approx 1.2091$$

Alternatively, taking the natural logarithm of both sides yields

$$\ln 5^x = \ln 7$$

$$x \ln 5 = \ln 7$$

$$x = \frac{\ln 7}{\ln 5} \approx \frac{1.945910}{1.609438} \approx 1.2091$$

Solution (b).   Because the equation involves a power of e, natural logarithms will result in simpler computations than common logarithms; taking the natural logarithm of both sides yields

$$\ln (e^x) = \ln 9$$

$$x \ln e = \ln 9$$

$$x = \ln 9 \approx 2.1972 \quad \blacktriangleleft$$

Example 5 A satellite has a radioisotope power supply for which the power output in watts is given by the equation

$$P = 75e^{-t/125} \tag{7}$$

where t is the time in days that the supply is used.

(a)   What will the power output be after one year of use?
(b)   How long will it take until the satellite is operating at half power?

Solution (a).   Substituting $t = 365$ yields

$$P = 75e^{-365/125} = 75e^{-2.92} \approx 4.05$$

Thus, the power output will be approximately 4.05 watts after one year.

Solution (b).   To find the initial power output, we substitute $t = 0$ into (7); this yields

$$P = 75e^0 = 75 \text{ watts}$$

Thus, it follows from (7) that the satellite will be operating at half power when t satisfies

$$37.5 = 75e^{-t/125}$$

Dividing through by 75, then taking the natural logarithm of both sides yields

$$\frac{1}{2} = e^{-t/125}$$

$$\ln\left(\frac{1}{2}\right) = \ln(e^{-t/125})$$

$$-\ln 2 = -t/125 \quad \boxed{\text{See Theorem 7.2.1}(f).}$$

$$t = 125\ln 2 \approx 86.64$$

Thus, the satellite will be operating at half power on the 86th day.   ◄

OTHER TYPES OF LOGARITHMIC AND EXPONENTIAL EQUATIONS

The methods used in the previous examples can be adapted to solve other types of logarithmic and exponential equations.

Example 6   Solve $\dfrac{e^x - e^{-x}}{2} = 1$ for x.

Solution.   Multiplying both sides of the given equation by 2 yields

$$e^x - e^{-x} = 2$$

or equivalently

$$e^x - \frac{1}{e^x} = 2$$

Multiplying through by e^x yields

$$e^{2x} - 1 = 2e^x \quad \text{or} \quad e^{2x} - 2e^x - 1 = 0$$

This is really a quadratic equation in disguise; this can be seen by rewriting the equation as

$$(e^x)^2 - 2e^x - 1 = 0$$

and letting

$$u = e^x \qquad\qquad\qquad (8)$$

to obtain

$$u^2 - 2u - 1 = 0$$

Solving for u by the quadratic formula yields

$$u = \frac{2 \pm \sqrt{4+4}}{2} = \frac{2 \pm \sqrt{8}}{2} = 1 \pm \sqrt{2}$$

Thus, from (8)

$$e^x = 1 \pm \sqrt{2}$$

But e^x cannot be negative, so we discard the negative value $1 - \sqrt{2}$; thus,

$$e^x = 1 + \sqrt{2}$$
$$\ln e^x = \ln (1 + \sqrt{2})$$
$$x = \ln (1 + \sqrt{2}) \approx 0.881 \quad \blacktriangleleft$$

▶ Exercise Set 7.2   Ⓒ 3, 4, 7, 8, 39, 41–50

In Exercises 1 and 2, simplify the expression without using a calculator.

1. (a) $-8^{2/3}$ (b) $(-8)^{2/3}$ (c) $8^{-2/3}$.
2. (a) 2^{-4} (b) $4^{1.5}$ (c) $9^{-0.5}$.

In Exercises 3 and 4, use a calculator to obtain an approximate value of the expression. Round the answer to four decimal places.

3. (a) $2^{1.57}$ (b) $5^{-2.1}$.
4. (a) $\sqrt[5]{24}$ (b) $\sqrt[8]{0.6}$.

In Exercises 5 and 6, find the exact value of the logarithm without using a calculator.

5. (a) $\log_2 16$ (b) $\log_2 (\tfrac{1}{32})$
    (c) $\log_4 4$ (d) $\log_9 3$.
6. (a) $\log_{10} (0.001)$ (b) $\log_{10} (10^4)$
    (c) $\ln (e^3)$ (d) $\ln (\sqrt{e})$.

In Exercises 7 and 8, use a calculator to obtain an approximate value of the logarithm. Round the answer to four decimal places.

7. (a) $\log 23.2$ (b) $\ln 0.74$.
8. (a) $\log 0.3$ (b) $\ln \pi$.

In Exercises 9 and 10, use properties of logarithms (Theorem 7.2.1) to help expand the logarithm in terms of r, s, and t, where $r = \ln a$, $s = \ln b$, and $t = \ln c$.

9. (a) $\ln a^2 \sqrt{bc}$ (b) $\ln \dfrac{b}{a^3 c}$.
10. (a) $\ln \dfrac{\sqrt[3]{c}}{ab}$ (b) $\ln \sqrt{\dfrac{ab^3}{c^2}}$.

In Exercises 11 and 12, expand the logarithm in terms of sums, differences, and multiples of simpler logarithms.

11. (a) $\log (10x \sqrt{x-3})$ (b) $\ln \dfrac{x^2 \sin^3 x}{\sqrt{x^2 + 1}}$.
12. (a) $\log \dfrac{\sqrt[3]{x+2}}{\cos 5x}$ (b) $\ln \sqrt{\dfrac{x^2 + 1}{x^3 + 5}}$.

In Exercises 13–15, rewrite the expression as a single logarithm.

13. $4 \log 2 - \log 3 + \log 16$.
14. $\tfrac{1}{2} \log x - 3 \log (\sin 2x) + 2$.
15. $2 \ln (x + 1) + \tfrac{1}{3} \ln x - \ln (\cos x)$.

In Exercises 16–25, solve for x without using a calculator.

16. $\log_{10} (1 + x) = 3$.   17. $\log_{10} (\sqrt{x}) = -1$.
18. $\ln (x^2) = 4$.   19. $\ln (1/x) = -2$.
20. $\log_3 (3^x) = 7$.   21. $\log_5 (5^{2x}) = 8$.
22. $\log_{10} x^2 + \log_{10} x = 30$.
23. $\log_{10} x^{3/2} - \log_{10} \sqrt{x} = 5$.
24. $\ln 4x - 3 \ln (x^2) = \ln 2$.
25. $\ln (1/x) + \ln (2x^3) = \ln 3$.

In Exercises 26–35, solve for x without using a calculator. Use the natural logarithm when logarithms are needed.

26. $3^x = 2$.   27. $5^{-2x} = 3$.
28. $3e^{-2x} = 5$.   29. $2e^{3x} = 7$.
30. $e^x - 2xe^x = 0$.   31. $xe^{-x} + 2e^{-x} = 0$.

32. $e^{2x} - e^x = 6.$

33. $e^{-2x} - 3e^{-x} = -2.$

34. $\dfrac{e^x}{1 - e^x} = 2.$

35. $\dfrac{1}{1 - 3e^{2x}} = 4.$

36. In each part, sketch the graph of $f(x)$.
(a) $f(x) = 2^x$
(b) $f(x) = (1/2)^x$
(c) $f(x) = 3^x$
(d) $f(x) = (1/3)^x$.

37. Find the fallacy in the following "proof" that $\frac{1}{8} > \frac{1}{4}$: Multiply both sides of the inequality $3 > 2$ by $\log\frac{1}{2}$ to get

$$3\log\tfrac{1}{2} > 2\log\tfrac{1}{2}$$
$$\log(\tfrac{1}{2})^3 > \log(\tfrac{1}{2})^2$$
$$\log\tfrac{1}{8} > \log\tfrac{1}{4}$$
$$\tfrac{1}{8} > \tfrac{1}{4}$$

38. (a) Prove part (b) of Theorem 7.2.1.
(b) Prove part (d) of Theorem 7.2.1.
(c) Prove part (e) of Theorem 7.2.1.
(d) Prove part (f) of Theorem 7.2.1.

39. (a) Prove the following change of base formula for logarithms:

$$\log_a c = \frac{\log_b c}{\log_b a}$$

(b) Use a calculator and the change of base formula in part (a) to find the values of $\log_2 7.35$ and $\log_5 0.6$ rounded to four decimal-place accuracy (use either common or natural logarithms).

40. (a) Use the change of base formula that is given in Exercise 39(a) to prove that $(\log_a b)(\log_b a) = 1$.
(b) Find the exact value of $(\log_2 81)(\log_3 32)$ without using a calculator.

41. If equipment in the satellite of Example 5 requires 15 watts to operate correctly, what is the operational lifetime of the power supply?

42. The equation $Q = 12e^{-0.055t}$ gives the mass Q in grams of radioactive Potassium-42 that will remain from some initial quantity after t hours of radioactive decay.
(a) How many grams were there initially?
(b) How many grams remain after 4 hours?
(c) How long will it take to reduce the amount of radioactive Potassium-42 to half of the initial amount?

43. The acidity of a substance is measured by its pH value, which is defined by the formula

$$\text{pH} = -\log[H^+]$$

where the symbol $[H^+]$ denotes the concentration

of hydrogen ions measured in moles per liter. Distilled water has a pH of 7; a substance is called *acidic* if it has pH < 7 and *basic* if it has pH > 7. Find the pH of each of the following substances and state whether it is acidic or basic.

	SUBSTANCE	$[H^+]$
(a)	Arterial blood	3.9×10^{-8} mol/L
(b)	Tomatoes	6.3×10^{-5} mol/L
(c)	Milk	4.0×10^{-7} mol/L
(d)	Coffee	1.2×10^{-6} mol/L

44. Use the definition of pH in Exercise 43 to find $[H^+]$ in a solution having a pH equal to
(a) 2.44
(b) 8.06.

45. The perceived loudness L of a sound in decibels (dB) is related to its intensity I in watts/square meter (W/m²) by the equation

$$L = 10\log(I/I_0)$$

where $I_0 = 10^{-12}$ W/m². Damage to the average ear occurs at 90 dB or greater. Find the decibel level of each of the following sounds and state whether it will cause ear damage.

	SOUND	I
(a)	Jet aircraft (from 500 ft)	1.0×10^2 W/m²
(b)	Amplified rock music	1.0 W/m²
(c)	Garbage disposal	1.0×10^{-4} W/m²
(d)	TV (middle volume from 10 ft)	3.2×10^{-5} W/m²

In Exercises 46–48, use the definition of the decibel level of a sound (see Exercise 45).

46. If one sound is three times as intense as another, how much greater is its decibel level?

47. According to one source, the noise inside a moving automobile is about 70 dB, while an electric blender generates 93 dB. Find the ratio of the intensity of the noise of the blender to that of the automobile.

48. Suppose that the decibel level of an echo is $\frac{2}{3}$ the decibel level of the original sound. If each echo results in another echo, how many echoes will be heard from a 120-dB sound given that the average human ear can hear a sound as low as 10 dB?

49. On the *Richter scale,* the magnitude M of an earthquake is related to the released energy E in joules (J) by the equation

$$\log E = 4.4 + 1.5M$$

(a) Find the energy E of the 1906 San Francisco earthquake that registered $M = 8.2$ on the Richter scale.

(b) If the released energy of one earthquake is 10 times that of another, how much greater is its magnitude on the Richter scale?

50. Suppose that the magnitudes of two earthquakes differ by 1 on the Richter scale. Find the ratio of the released energy of the larger earthquake to that of the smaller earthquake. [*Note:* See Exercise 49 for terminology.]

51. Find $\lim\limits_{x \to 0} (1 - 2x)^{1/x}$. [*Hint:* Let $t = -2x$.]

52. Find $\lim\limits_{x \to +\infty} (1 + 3/x)^x$. [*Hint:* Let $t = 3/x$.]

■ 7.3 THE NATURAL LOGARITHM

In the previous section we gave an informal discussion of irrational exponents and natural logarithms. In this section we shall put those concepts on a sounder mathematical footing.

☐ **THE ROLE OF THE NATURAL LOGARITHM IN CALCULUS**

The natural logarithm plays a special role in calculus that we can motivate by differentiating the function

$$f(x) = \log_b x$$

where b is an arbitrary base. For purposes of this introductory discussion we shall assume that $\log_b x$ is continuous for $x > 0$. The proof of this will be given later. Using the definition of a derivative we obtain

$$\frac{d}{dx}[\log_b x] = \lim_{h \to 0} \frac{\log_b (x + h) - \log_b x}{h}$$

$$= \lim_{h \to 0} \frac{1}{h} \log_b \left(\frac{x + h}{x} \right) \qquad \boxed{\text{Theorem 7.2.1}(d)}$$

$$= \lim_{h \to 0} \frac{1}{h} \log_b \left(1 + \frac{h}{x} \right)$$

If we now let $t = h/x$, then $h = tx$ and $t \to 0$ as $h \to 0$, so

$$\frac{d}{dx}[\log_b x] = \lim_{t \to 0} \frac{1}{tx} \log_b (1 + t)$$

$$= \frac{1}{x} \lim_{t \to 0} \frac{1}{t} \log_b (1 + t) \qquad \boxed{\begin{array}{l}1/x \text{ does not vary with} \\ t, \text{ so it can be moved} \\ \text{through the limit sign.}\end{array}}$$

$$= \frac{1}{x} \lim_{t \to 0} \log_b (1 + t)^{1/t} \qquad \boxed{\text{Theorem 7.2.1}(e)}$$

$$= \frac{1}{x} \log_b \left[\lim_{t \to 0} (1 + t)^{1/t} \right] \qquad \boxed{\begin{array}{l}\text{See the remark}\\\text{following}\\\text{Theorem 2.7.5.}\end{array}}$$

$$= \frac{1}{x} \log_b e \qquad \boxed{\begin{array}{l}\text{Formula (6)}\\\text{of Section 7.2}\end{array}}$$

Thus,

$$\frac{d}{dx} [\log_b x] = \frac{1}{x} \log_b e$$

In the special case where $b = e$, we have $\log_b x = \log_e x = \ln x$, so this formula becomes

$$\frac{d}{dx} [\ln x] = \frac{1}{x} \ln e$$

But $\ln e = 1$, so this simplifies to

$$\frac{d}{dx} [\ln x] = \frac{1}{x} \tag{1}$$

Thus, among all possible choices for the base b, the one that yields the simplest derivative formula for $\log_b x$ is $b = e$. This is one of the main reasons why the natural logarithm plays such a special role in calculus.

☐ **OBJECTIVES**

Our primary objective in this section is to give precise definitions of irrational exponents, the number e, and the natural logarithm. We could define irrational exponents as limits of rational exponents, but such a definition would be cumbersome for obtaining properties relating to continuity and differentiability. For this reason, mathematicians have developed an alternative approach that is "less natural" but ultimately easier to work with. The basic idea is to give a definition of the natural logarithm that does not make use of exponents, then use the natural logarithm to define irrational exponents. To ensure that there is no circular reasoning, we will have to proceed carefully in three steps:

1. Define $\ln x$ in such a way that irrational exponents are not used.
2. Define e using properties of $\ln x$.
3. Define irrational exponents using properties of e and $\ln x$.

☐ **DEFINITION OF THE NATURAL LOGARITHM**

Let us consider how we might define $\ln x$ without using irrational exponents. Motivated by Formula (1) and Theorem 7.2.1(a), we will want our definition of $\ln x$ to be such that

$$\ln 1 = 0 \quad \text{and} \quad \frac{d}{dx} [\ln x] = \frac{1}{x}$$

Thus, for positive values of x, $\ln x$ can be viewed as that antiderivative of $1/x$ that has a value of 0 when $x = 1$. However, it follows from the Second Fundamental Theorem of Calculus (Theorem 5.9.3) that

$$F(x) = \int_1^x \frac{1}{t}\, dt$$

is a function with these two properties, since

$$\frac{d}{dx}[F(x)] = \frac{d}{dx}\left[\int_1^x \frac{1}{t}\, dt\right] = \frac{1}{x} \quad \text{and} \quad F(1) = \int_1^1 \frac{1}{t}\, dt = 0$$

This suggests the following definition of $\ln x$, which requires no use of irrational exponents.

7.3.1 DEFINITION.   The **natural logarithm** function is defined by the formula

$$\ln x = \int_1^x \frac{1}{t}\, dt, \quad x > 0 \tag{2}$$

☐ **DERIVATIVES AND INTEGRALS INVOLVING $\ln x$**

The restriction in (2) that x be positive is necessary to ensure that the integrand is continuous as required by the Second Fundamental Theorem of Calculus (Theorem 5.9.3). It follows from this definition that

$$\frac{d}{dx}[\ln x] = \frac{1}{x}, \quad x > 0 \tag{3}$$

If $u(x) > 0$, and if the function u is differentiable at x, then it follows from the chain rule [Formula (2) in Section 3.5] that

$$\frac{d}{dx}[\ln u] = \frac{1}{u} \cdot \frac{du}{dx} \tag{4}$$

Example 1   Find $\dfrac{d}{dx}[\ln (x^2 + 1)]$.

Solution.   From (4) with $u = x^2 + 1$,

$$\frac{d}{dx}[\ln (x^2 + 1)] = \frac{1}{x^2 + 1} \cdot \frac{d}{dx}[x^2 + 1] = \frac{1}{x^2 + 1} \cdot 2x = \frac{2x}{x^2 + 1} \quad ◀$$

Example 2   Find $\dfrac{d}{dx}[\ln |x|]$.

Solution.   We shall consider the cases where $x > 0$ and $x < 0$ separately. (The case where $x = 0$ is excluded because $\ln 0$ is undefined.) If $x > 0$, then $|x| = x$, so that

$$\frac{d}{dx}\,[\ln |x|] = \frac{d}{dx}\,[\ln x] = \frac{1}{x}$$

If $x < 0$, then $|x| = -x$, so that from (4) with $u = -x$,

$$\frac{d}{dx}\,[\ln |x|] = \frac{d}{dx}\,[\ln (-x)] = \frac{1}{(-x)} \cdot \frac{d}{dx}\,[-x] = \frac{1}{x}$$

Thus,

$$\frac{d}{dx}\,[\ln |x|] = \frac{1}{x} \qquad \text{if } x \neq 0 \qquad \blacktriangleleft \tag{5}$$

Formula (5) states that the function $\ln |x|$ is an antiderivative of $1/x$ everywhere except at $x = 0$. In contrast, the function $\ln x$ is an antiderivative of $1/x$ only for $x > 0$. The companion integration formula for (5) is

$$\int \frac{1}{x}\,dx = \ln |x| + C \tag{6}$$

Example 3   From (5) and the chain rule,

$$\frac{d}{dx}\,[\ln |\sin x|] = \frac{1}{\sin x} \cdot \frac{d}{dx}\,[\sin x] = \frac{\cos x}{\sin x} = \cot x \qquad \blacktriangleleft$$

Example 4   Evaluate $\displaystyle\int \frac{3x^2}{x^3 + 5}\,dx.$

Solution.   Make the substitution

$$u = x^3 + 5, \quad du = 3x^2\,dx$$

so that

$$\int \frac{3x^2}{x^3 + 5}\,dx = \int \frac{1}{u}\,du = \ln |u| + C = \ln |x^3 + 5| + C \qquad \blacktriangleleft$$

Formula (6)

REMARK.   The foregoing example illustrates an important point: Any integral of the form

$$\int \frac{g'(x)}{g(x)}\,dx$$

(where the numerator of the integrand is the derivative of the denominator) can be evaluated by the u-substitution $u = g(x)$, $du = g'(x) \, dx$, since this substitution yields

$$\int \frac{g'(x)}{g(x)} \, dx = \int \frac{du}{u} = \ln |u| + C = \ln |g(x)| + C$$

Example 5 Evaluate $\int \tan x \, dx$.

Solution.

$$\int \tan x \, dx = \int \frac{\sin x}{\cos x} \, dx = -\int \frac{1}{u} \, du = -\ln |u| + C = -\ln |\cos x| + C \quad \blacktriangleleft$$

$$u = \cos x$$
$$du = -\sin x \, dx$$

☐ **PROPERTIES OF THE NATURAL LOGARITHM**

The following theorem shows that $\ln x$, as defined in Definition 7.3.1, has the basic properties of a logarithm, thereby justifying the use of the term "logarithm" in its name.

7.3.2 THEOREM. *For any positive numbers a and c and any rational number r,*

(a) $\ln 1 = 0$ (b) $\ln ac = \ln a + \ln c$

(c) $\ln \dfrac{a}{c} = \ln a - \ln c$ (d) $\ln a^r = r \ln a$

(e) $\ln \dfrac{1}{c} = -\ln c$

Proof (a). From (2) with $x = 1$,

$$\ln 1 = \int_1^1 \frac{1}{t} \, dt = 0$$

Proof (b). Consider the function $f(x) = \ln ax$. Treating a as constant and differentiating with respect to x, we obtain

$$f'(x) = \frac{d}{dx} [\ln ax] = \frac{1}{ax} \cdot \frac{d}{dx} [ax] = \frac{1}{ax} \cdot a = \frac{1}{x}$$

which shows that $\ln ax$ and $\ln x$ have the same derivative on $(0, +\infty)$. Thus, by Theorem 4.10.3 and the remark that follows that theorem, there is a constant k such that

$$\ln ax - \ln x = k \tag{7}$$

If we let $x = 1$ in this equation and use the fact that $\ln 1 = 0$, we obtain

$$\ln a = k$$

so that (7) can be written as

$$\ln ax - \ln x = \ln a$$

In particular, if $x = c$, we obtain

$$\ln ac - \ln c = \ln a \quad \text{or} \quad \ln ac = \ln a + \ln c$$

Proofs (c) and (e). Using parts (a) and (b), we can write

$$0 = \ln 1 = \ln\left(c \cdot \frac{1}{c}\right) = \ln c + \ln \frac{1}{c}$$

so that

$$\ln \frac{1}{c} = -\ln c$$

which proves (e). To prove (c) we write

$$\ln \frac{a}{c} = \ln\left(a \cdot \frac{1}{c}\right) = \ln a + \ln \frac{1}{c} = \ln a - \ln c$$

Proof (d). The functions $\ln x^r$ and $r \ln x$ have the same derivative on $(0, +\infty)$, since

$$\frac{d}{dx}[\ln x^r] = \frac{1}{x^r} \cdot \frac{d}{dx}[x^r] = \frac{1}{x^r} \cdot rx^{r-1} = \frac{r}{x}$$

and

$$\frac{d}{dx}[r \ln x] = r\frac{d}{dx}[\ln x] = \frac{r}{x}$$

Thus, there is a constant k such that

$$\ln x^r - r \ln x = k$$

If we let $x = 1$ in this equation and use the fact that $\ln 1 = 0$, we obtain $k = 0$ (verify), so that this equation can be written as

$$\ln x^r - r \ln x = 0 \quad \text{or} \quad \ln x^r = r \ln x$$

Letting $x = a$ completes the proof. ∎

Example 6 Property (d) in Theorem 7.3.2 yields the following useful result:

$$\ln \sqrt[n]{a} = \ln a^{1/n} = \frac{1}{n} \ln a \quad \blacktriangleleft$$

When possible, the properties of logarithms in Theorem 7.3.2 should be used to convert products, quotients, and exponents into sums, differences, and constant multiples *before* differentiating a function involving natural logarithms.

Example 7

$$\frac{d}{dx}\left[\ln\left(\frac{x^2 \sin x}{\sqrt{1 + x}}\right)\right] = \frac{d}{dx}\left[2 \ln x + \ln(\sin x) - \frac{1}{2}\ln(1 + x)\right]$$

$$= \frac{2}{x} + \frac{\cos x}{\sin x} - \frac{1}{2(1 + x)}$$

$$= \frac{2}{x} + \cot x - \frac{1}{2 + 2x} \quad \blacktriangleleft$$

☐ **THE GRAPH OF ln x**

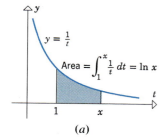

$y = \frac{1}{t}$

Area $= \displaystyle\int_1^x \frac{1}{t}\, dt = \ln x$

(a)

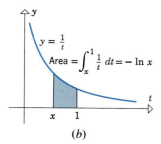

$y = \frac{1}{t}$

Area $= \displaystyle\int_x^1 \frac{1}{t}\, dt = -\ln x$

(b)

Figure 7.3.1

We now turn to the problem of graphing the function $\ln x$. For this purpose it will be helpful to interpret the values of $\ln x$ as areas. If $x > 1$, then

$$\ln x = \int_1^x \frac{1}{t}\, dt$$

can be viewed as the area under the curve $y = 1/t$ between the points 1 and x on the t-axis (Figure 7.3.1a). On the other hand, if $0 < x < 1$, then by writing

$$\ln x = -\int_x^1 \frac{1}{t}\, dt$$

we can view $\ln x$ as the *negative* of the area under the curve $y = 1/t$ between the points x and 1 on the t-axis (Figure 7.3.1b). Therefore,

$$\ln x > 0 \quad \text{if} \quad x > 1$$
$$\ln x < 0 \quad \text{if} \quad 0 < x < 1$$
$$\ln x = 0 \quad \text{if} \quad x = 1$$

Since $\ln x$ is differentiable for $x > 0$, it is continuous on the interval $(0, +\infty)$. The first derivative

$$\frac{d}{dx}[\ln x] = \frac{1}{x}$$

is positive for all x in $(0, +\infty)$, so that $\ln x$ is *increasing* on this interval, and the second derivative

$$\frac{d^2}{dx^2}[\ln x] = \frac{d}{dx}\left[\frac{1}{x}\right] = -\frac{1}{x^2}$$

is negative for all x in $(0, +\infty)$, so that the graph of $\ln x$ is *concave down* on this interval.

To complete the picture of the graph, we must investigate the behavior of $\ln x$ as $x \to 0^+$ and $x \to +\infty$. For this purpose we shall need an estimate of $\ln 2$. To obtain such an estimate, suppose that $1 \le t \le 2$, from which it follows that

$$\frac{1}{2} \le \frac{1}{t} \le 1$$

Thus, from the definition of the natural logarithm and Theorem 5.7.4(*b*) we obtain

$$\int_1^2 \frac{1}{2}\, dt \le \int_1^2 \frac{1}{t}\, dt \le \int_1^2 1\, dt$$

from which it follows on integrating that

$$\frac{1}{2} \le \ln 2 \le 1 \tag{8}$$

7.3.3   THEOREM.

(*a*) $\displaystyle \lim_{x \to +\infty} \ln x = +\infty$

(*b*) $\displaystyle \lim_{x \to 0^+} \ln x = -\infty$

Proof (a).   We shall show that for any positive number N (no matter how large) the values of $\ln x$ eventually exceed N as $x \to +\infty$. It will then follow that $\displaystyle \lim_{x \to +\infty} \ln x = +\infty$.

Since $\ln x$ is an increasing function, it follows from (8) that for $x > 2^{2N}$ we must have

$$\ln x > \ln 2^{2N} = 2N \ln 2 \ge 2N\left(\frac{1}{2}\right) = N$$

Therefore, $\displaystyle \lim_{x \to +\infty} \ln x = +\infty$.

Proof (b).   If we let $v = 1/x$, then $v \to +\infty$ as $x \to 0^+$, so that from part (*a*) of this theorem and Theorem 7.3.2(*e*) we obtain

$$\lim_{x \to 0^+} \ln x = \lim_{x \to 0^+} \left(-\ln \frac{1}{x}\right) = \lim_{v \to +\infty} (-\ln v) = -\lim_{v \to +\infty} \ln v = -\infty \quad \blacksquare$$

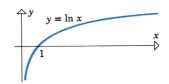

Figure 7.3.2

From the foregoing theorem and the fact that $y = \ln x$ is increasing and concave down on $(0, +\infty)$, the graph of $\ln x$ is as shown in Figure 7.3.2.

REMARK.   In the previous section, where we were working informally, we used the notations $\log_e x$ and $\ln x$ interchangeably. After we have defined irrational exponents (which will happen in the next section), we will be able to give a precise definition of $\log_e x$ and we will show at that time that $\log_e x$ and $\ln x$ are the same. This explains why the graph in Figure 7.3.2 is consistent with that in Figure 7.2.4.

□ **LOGARITHMIC DIFFERENTIATION**

We conclude this section by discussing a technique, called *logarithmic differentiation*, that is useful for differentiating functions that are composed of products, quotients, and powers.

Example 8 The derivative of

$$y = \frac{x^2 \sqrt[3]{7x - 14}}{(1 + x^2)^4} \tag{9}$$

is messy to calculate directly. However, if we first take ln of both sides and use properties of the natural logarithm, we can write

$$\ln y = 2 \ln x + \frac{1}{3} \ln (7x - 14) - 4 \ln (1 + x^2)$$

Differentiating both sides with respect to x yields

$$\frac{1}{y} \frac{dy}{dx} = \frac{2}{x} + \frac{7/3}{7x - 14} - \frac{8x}{1 + x^2}$$

Thus, on solving for dy/dx and using (9) we obtain

$$\frac{dy}{dx} = \frac{x^2 \sqrt[3]{7x - 14}}{(1 + x^2)^4} \left[\frac{2}{x} + \frac{1}{3x - 6} - \frac{8x}{1 + x^2} \right] \qquad \blacktriangleleft$$

Since $\ln y$ is defined only for $y > 0$, results obtained by logarithmic differentiation are valid only at points where this condition is satisfied. The derivative formula obtained in the last example is valid for $x > 2$ since $y > 0$ for such x [see (9)]. As the next example shows, logarithmic differentiation can sometimes be used with functions having negative values by first taking absolute values.

Example 9 The derivative of

$$y = \frac{x \sqrt[3]{x - 5}}{1 + \sin^3 x}$$

can be obtained by writing

$$\ln |y| = \ln \left| \frac{x \sqrt[3]{x - 5}}{1 + \sin^3 x} \right|$$

$$\ln |y| = \ln |x| + \frac{1}{3} \ln |x - 5| - \ln |1 + \sin^3 x|$$

$$\frac{1}{y} \frac{dy}{dx} = \frac{1}{x} + \frac{1}{3(x - 5)} - \frac{3 \sin^2 x \cos x}{1 + \sin^3 x}$$

$$\frac{dy}{dx} = \frac{x \sqrt[3]{x - 5}}{1 + \sin^3 x} \left[\frac{1}{x} + \frac{1}{3(x - 5)} - \frac{3 \sin^2 x \cos x}{1 + \sin^3 x} \right] \qquad \blacktriangleleft$$

▶ Exercise Set 7.3   © 56, 74, 75

In Exercises 1–6, find the domain of $f(x)$.

1. $f(x) = \ln(3x + 2)$.
2. $f(x) = \ln(1 - 2x)$.
3. $f(x) = \ln(4 - x^2)$.
4. $f(x) = \ln|5 - 3x|$.
5. $f(x) = \sqrt{1 + \ln x}$.
6. $f(x) = \ln(\ln x)$.
7. (a) Find the domain of $\ln(x^2)$.
   (b) For what values of x does $\ln(x^2) = 2\ln x$?

In Exercises 8–28, find dy/dx.

8. $y = 4\ln x$.
9. $y = \ln 2x$.
10. $y = \ln(x^3)$.
11. $y = (\ln x)^2$.
12. $y = \ln(\sin x)$.
13. $y = \ln|\tan x|$.
14. $y = \ln(2 + \sqrt{x})$.
15. $y = \ln\left(\dfrac{x}{1 + x^2}\right)$.
16. $y = \ln(\ln x)$.
17. $y = \ln|x^3 - 7x^2 - 3|$.
18. $y = x^3\ln x$.
19. $y = \sqrt{\ln x}$.
20. $y = \cos(\ln x)$.
21. $y = \sin\left(\dfrac{5}{\ln x}\right)$.
22. $y = \sqrt{1 + \ln^2 x}$.
23. $y = x^3\ln(3 - 2x)$.
24. $y = x[\ln(x^2 - 2x)]^3$.
25. $y = (x^2 + 1)[\ln(x^2 + 1)]^2$.
26. $y = \dfrac{\ln x}{1 + \ln x}$.
27. $y = \dfrac{x^2}{1 + \ln x}$.
28. $y = \ln\left|\dfrac{1 - \cos\pi x}{1 + \cos\pi x}\right|$.

29. Find dy/dx by implicit differentiation if $y + \ln xy = 1$.
30. Find dy/dx by implicit differentiation if $y = \ln(x\tan y)$.

In Exercises 31–42, evaluate the indefinite integrals.

31. $\displaystyle\int \dfrac{dx}{2x}$.
32. $\displaystyle\int \dfrac{5x^4}{x^5 + 1}\,dx$.
33. $\displaystyle\int \dfrac{x^2}{x^3 - 4}\,dx$.
34. $\displaystyle\int \dfrac{t + 1}{t}\,dt$.
35. $\displaystyle\int \dfrac{\sec^2 x}{\tan x}\,dx$.
36. $\displaystyle\int \cot x\,dx$.
37. $\displaystyle\int \dfrac{\sin 3\theta}{1 + \cos 3\theta}\,d\theta$.
38. $\displaystyle\int \dfrac{dx}{x\ln x}$.
39. $\displaystyle\int \dfrac{x^3}{x^2 + 1}\,dx$.
40. $\displaystyle\int \dfrac{1}{x}\cos(\ln x)\,dx$.
41. $\displaystyle\int \dfrac{1}{y}(\ln y)^3\,dy$.
42. $\displaystyle\int \dfrac{dx}{\sqrt{x}(1 - 2\sqrt{x})}$.

In Exercises 43–46, evaluate the definite integral.

43. $\displaystyle\int_0^1 \dfrac{1}{3x + 2}\,dx$.
44. $\displaystyle\int_1^4 \dfrac{3}{1 - 2x}\,dx$.
45. $\displaystyle\int_{-1}^0 \dfrac{x}{x^2 + 5}\,dx$.
46. $\displaystyle\int_1^4 \dfrac{1}{\sqrt{x}(1 + \sqrt{x})}\,dx$.

In Exercises 47–50, use Theorem 7.3.2 to help perform the indicated differentiation.

47. $\dfrac{d}{dx}\left[\ln\dfrac{\cos x}{\sqrt{4 - 3x^2}}\right]$.
48. $\dfrac{d}{dx}\left[\ln\sqrt{\dfrac{x - 1}{x + 1}}\right]$.
49. $\dfrac{d}{dx}\left[\ln(\sqrt{x}\sqrt[3]{x} + 3\sqrt[5]{3x} - 2)\right]$.
50. $\dfrac{d}{dx}\left[\ln\left(\dfrac{\sqrt{x}\sqrt[3]{x} + 1}{\sin x\sec x}\right)\right]$.

In Exercises 51–54, obtain dy/dx by logarithmic differentiation.

51. $y = x\sqrt[3]{1 + x^2}$.
52. $y = \sqrt[5]{\dfrac{x - 1}{x + 1}}$.
53. $y = \dfrac{(x^2 - 8)^{1/3}\sqrt{x^3 + 1}}{x^6 - 7x + 5}$.
54. $y = \dfrac{\sin x\cos x\tan^3 x}{\sqrt{x}}$.

55. (a) Make an appropriate u-substitution to show that
$$\int_a^{ab}\dfrac{1}{t}\,dt = \int_1^b\dfrac{1}{t}\,dt$$
   (b) Use this equality to obtain another proof of $\ln(ab) = \ln a + \ln b$.

56. If the interval $[1, x]$ on the t-axis is divided into n equal subintervals and A_1 and A_2 denote, respectively, the total areas of the inscribed and circumscribed rectangles shown in Figure 7.3.3, then

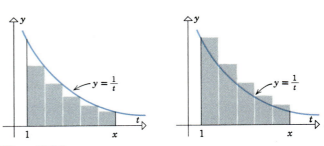

Figure 7.3.3

$$A_1 \leq \int_1^x \frac{1}{t} \, dt \leq A_2$$

Use this to find upper and lower estimates for $\ln 2$ using $n = 10$ subintervals. Compare your answer with a calculator value of $\ln 2$.

In Exercises 57–59, use natural logarithms and the change of base formula in Exercise 39, Section 7.2, to help solve the problems.

57. Show that $\dfrac{d}{dx} [\log_b x] = \dfrac{1}{\ln b} \cdot \dfrac{1}{x}$.

58. Find $\dfrac{d}{dx} [\log_x e]$.   **59.** Find $\dfrac{d}{dx} [\log_x 2]$.

In Exercises 60 and 61, interpret the expression as a limit of Riemann sums in which the interval $[0, 1]$ is divided into n subintervals of equal width, then find the limit by evaluating an appropriate definite integral.

60. $\displaystyle \lim_{n \to +\infty} \sum_{k=1}^n \frac{1}{n+k}$.   **61.** $\displaystyle \lim_{n \to +\infty} \sum_{k=1}^n \frac{k}{n^2+k^2}$.

62. Find $\displaystyle \lim_{h \to 0} \frac{\ln(1+h)}{h}$. [*Hint:* Let $f(x) = \ln x$ and consider $f'(1)$ directly from the definition of the derivative.]

63. Find the minimum value of $x^2 - \ln x$.

64. Find the minimum value of $\sqrt{x} \ln x$.

65. Find the relative extrema of $\dfrac{\ln^2 x}{x}$.

66. Prove: $\ln x < \sqrt{x}$. [*Hint:* Show that $\sqrt{x} - \ln x$ is always positive.]

67. Prove: $\ln x \leq x - 1$.
[*Hint:* Show that $x - 1 - \ln x \geq 0$.]

68. Prove: $\ln x \geq 1 - 1/x$.
[*Hint:* Show that $1 - 1/x - \ln x \leq 0$.]

69. *Boyle's law* in physics states that under appropriate conditions the pressure p exerted by a gas is related to its volume v by $pv = c$, where c is a constant depending on the units and various physical factors. Prove: When such a gas increases in volume from v_0 to v_1 the average pressure it exerts is

$$p_{\text{ave}} = \frac{c}{v_1 - v_0} \ln \frac{v_1}{v_0}$$

70. Consider the region enclosed by the curve $y = 1/x$, the x-axis, and the lines $x = 1$ and $x = b$, where $b > 1$.
(a) Let A be the area of the region. Find A and $\displaystyle \lim_{b \to +\infty} A$.
(b) Let V be the volume of the solid generated when the region is revolved about the x-axis. Find V and $\displaystyle \lim_{b \to +\infty} V$.

71. Find the area of the region enclosed by $y = \tan x$, $y = 0$, and $x = \pi/3$. [*Hint:* See Example 5.]

72. Find the volume of the solid that is generated when the region enclosed by $y = 1/x^2$, $y = 4$, and $x = 3$ is revolved about the y-axis.

73. Find the volume of the solid that is generated when the region enclosed by $y = 1/\sqrt{x}$, $y = 0$, $x = 1$, and $x = 4$ is revolved about the x-axis.

In Exercises 74 and 75, use Newton's Method (Section 4.9) to approximate the solutions of the equation. Show as many decimal places as your calculator displays.

74. $\sin x = \ln x$.   **75.** $\ln x = x - 2$.

■ 7.4 IRRATIONAL EXPONENTS; EXPONENTIAL FUNCTIONS

The primary goal of this section is to define irrational exponents. As discussed in Section 7.2, we want this definition to be such that the familiar properties of rational exponents continue to hold for irrational exponents, and for all positive values of b we want the function $f(x) = b^x$ to be continuous.

There will be three steps in defining irrational exponents:

- First we will give a precise definition of the number e. (This number was introduced informally in Section 7.2.)
- Second, we will define e^k for both rational and irrational values of the exponent k.
- Finally, for every positive real number b we will use the definition of e^k to define b^k for irrational values of the exponent k.

☐ **DEFINITION OF** e

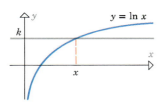

Figure 7.4.1

We saw in the previous section that $\ln x$ is an increasing, continuous function on the interval $(0, +\infty)$ and that its values vary from $-\infty$ to $+\infty$ as x varies over this interval. This being the case, it follows that the curve $y = \ln x$ crosses every horizontal line exactly once and that the intersection occurs over the interval $(0, +\infty)$. (See Figure 7.4.1.) Thus, we have the following result.

> **7.4.1 THEOREM.**   *For every real number k there is a unique real number x such that $\ln x = k$. Moreover, the number x is positive.*

It follows as a special case of this theorem that there is a unique positive real number x such that $\ln x = 1$. We *define* this number to be e (Figure 7.4.2); thus,

$$\ln e = 1 \tag{1}$$

(This agrees with the value of $\ln e$ obtained informally in Section 7.2.)

☐ **DEFINITION OF** e^k **FOR REAL VALUES OF** k

It follows from Theorem 7.3.2(d) that if k is a *rational* number, then

$$\ln e^k = k \ln e$$

so from (1)

$$\ln e^k = k \tag{2}$$

In words, the natural logarithm of e^k is k. But, it follows from Theorem 7.4.1 that e^k is the only real number whose natural logarithm is k. Thus, for rational values of k, we can actually *define* e^k to be that real number whose natural logarithm is k (Figure 7.4.3) (as opposed to using the more familiar definitions stated at the beginning of Section 7.2).

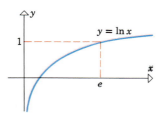

Figure 7.4.2

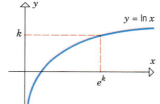

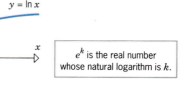

e^k is the real number whose natural logarithm is k.

Figure 7.4.3

Motivated by (2), which is valid for rational values of k, we make the following definition that holds for all real values of k, both rational and irrational.

7.4.2 DEFINITION.   If k is any real number, then e^k is defined to be that real number whose natural logarithm is k; that is,

$$\ln e^k = k$$

The function $f(x) = e^x$, called the ***natural exponential function,*** has special importance because of the following theorem, which states that the natural logarithm and exponential functions are inverse functions.

7.4.3 THEOREM.

(a) $\ln e^x = x \quad for \ -\infty < x < +\infty$

(b) $e^{\ln x} = x \quad for \ x > 0$

Proof (a).

$$\ln e^x = x \ln e = x$$

Proof (b).   Let $y = e^{\ln x}$. Then

$$\ln y = \ln(e^{\ln x}) = (\ln x)(\ln e) = \ln x$$

Since y and x have the same natural logarithm, it follows that $y = x$ or, equivalently, $e^{\ln x} = x$. ▐

Because $\ln x$ and e^x are inverse functions, it follows that the graph of $y = e^x$ is the reflection of the graph $y = \ln x$ about the line $y = x$ (Figure 7.4.4). Moreover, it follows from Theorem 7.1.7 that e^x is a continuous function since $\ln x$ is a continuous function.

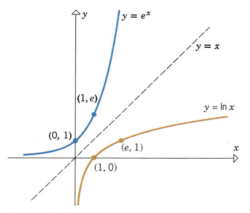

Figure 7.4.4

Figure 7.4.5 shows a table of values of e^x obtained with a calculator and a graph of $y = e^x$ obtained using those values.

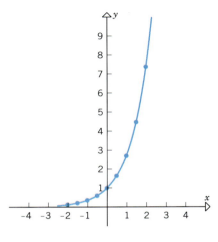

x	$y = e^x$
-2	0.1353353
-1.5	0.2231302
-1	0.3678794
-0.5	0.6065307
0	1
0.5	1.648721
1	2.718282
1.5	4.481689
2	7.389056
2.5	12.18249
3	20.08554
3.5	33.11545
4	54.59815

Figure 7.4.5

□ **EQUIVALENCE OF THE INTEGRAL AND EXPONENT DEFINITIONS OF ln x**

In words, part (b) of Theorem 7.4.3 states that $\ln x$ is that power to which e must be raised to produce x. This implies that

$$\log_e x = \ln x$$

which shows that the exponent and integral definitions of the natural logarithm are equivalent.

□ **DEFINITION OF b^k FOR REAL VALUES OF k**

It follows from Theorem 7.3.2(d) that if b is a positive real number and k is a *rational* number, then

$$\ln b^k = k \ln b$$

from which it follows that

$$e^{\ln b^k} = e^{k \ln b}$$

From Theorem 7.4.3(b) this can be rewritten as

$$b^k = e^{k \ln b} \tag{3}$$

Motivated by this formula, which is valid for rational values of k, we make the following definition.

7.4.4 DEFINITION.   If k is any real number and b is a positive real number, then b^k is defined by the formula

$$b^k = e^{k \ln b} \tag{4}$$

Example 1   We leave it for the reader to check the following computations with a calculator:

$$2^{\pi} = e^{\pi \ln 2} \approx 8.82498$$
$$7^{-\sqrt{2}} = e^{-\sqrt{2} \ln 7} \approx 0.0638044$$
$$\sqrt{5}^{\sqrt{3}} = e^{\sqrt{3} \ln \sqrt{5}} \approx 4.03019 \quad \blacktriangleleft$$

☐ **CONTINUITY OF b^x**

If $b > 0$, then it follows from (4) that the function $f(x) = b^x$ can be expressed as

$$f(x) = b^x = e^{x \ln b} \tag{5}$$

If $b \neq 1$, then (5) is called the *exponential function with base b*. Some examples are

$$2^x, \quad (1/3)^x, \quad \pi^x, \quad \text{and} \quad \sqrt{5}^x$$

REMARK.   If $b = 1$, then $b^x = 1^x = 1$ is a constant function. This is not usually classified as an exponential function.

One of our original goals was to define irrational exponents in such a way that b^x is continuous. To see that this is so, observe that (5) expresses b^x as the composition of e^x and $x \ln b$, both of which are continuous functions. Since the composition of continuous functions is continuous (Theorem 2.7.6), it follows that $f(x) = b^x$ is a continuous function.

☐ **SOME PROPERTIES OF EXPONENTIAL FUNCTIONS**

The following theorem, whose proof is left as an exercise, states that the exponential function with base b and the logarithm function with base b are inverses. Theorem 7.4.3 is a special case of this result.

7.4.5   THEOREM.

(a) $\log_b (b^x) = x \quad$ *for* $-\infty < x < +\infty$

(b) $b^{\log_b x} = x \qquad$ *for* $x > 0$

Because $\log_b x$ and b^x are inverse functions, it follows that their graphs are reflections of one another about the line $y = x$. (See Figure 7.4.6 for the case where $b > 1$.) Moreover, since b^x is continuous, so is $\log_b x$ by Theorem 7.1.7.

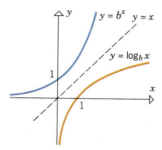

Figure 7.4.6

□ **LAWS OF REAL EXPONENTS**

One of the goals of this section was to define irrational exponents so that the familiar properties of rational exponents continue to hold for all real exponents. The following theorem shows that Definition 7.4.4 achieves that goal.

7.4.6 THEOREM. *If a and b are positive real numbers, then for all real numbers k and l the following laws of exponents hold.*

(a) $a^0 = 1$ (b) $a^1 = a$

(c) $a^k \cdot a^l = a^{k+l}$ (d) $a^{-l} = \dfrac{1}{a^l}$

(e) $\dfrac{a^k}{a^l} = a^{k-l}$ (f) $a^k \cdot b^k = (ab)^k$

(g) $\dfrac{a^k}{b^k} = \left(\dfrac{a}{b}\right)^k$ (h) $(a^k)^l = a^{kl}$

Proof. We shall prove (c) and leave the rest as exercises. We can write

$$\ln(a^k \cdot a^l) = \ln a^k + \ln a^l = k \ln a + l \ln a$$
$$= (k + l) \ln a = \ln(a^{k+l})$$

Thus, we have $a^k \cdot a^l = a^{k+l}$ since the two sides have the same natural logarithm. ∎

□ **SOME LIMITS OF EXPONENTIAL FUNCTIONS**

The following limits, which are discussed in the exercises, are evident from the graph of e^x (Figure 7.4.4).

7.4.7 THEOREM.

(a) $e^x > 0$ *for all x*

(b) $\displaystyle\lim_{x \to +\infty} e^x = +\infty$

(c) $\displaystyle\lim_{x \to -\infty} e^x = 0$

□ **DERIVATIVES OF EXPONENTIAL FUNCTIONS**

The following theorem provides the derivative formula for e^x.

7.4.8 THEOREM. *The function e^x is differentiable on $(-\infty, +\infty)$ and*

$$\frac{d}{dx}[e^x] = e^x \tag{6}$$

Proof. First we will show that $g(x) = e^x$ is a differentiable function on the interval $(-\infty, +\infty)$, then we will derive Formula (6). The function $g(x) = e^x$ is the inverse of the function

$$f(x) = \ln x$$

As x varies over the interval $(-\infty, +\infty)$, the value of $g(x) = e^x$ varies over the interval $(0, +\infty)$, on which f is differentiable and has a *nonzero* derivative [since $f'(x) = 1/x$]. Thus, it follows from Theorem 7.1.8 that g is differentiable on the interval $(-\infty, +\infty)$.

To obtain (6) let

$$y = e^x$$

which can be rewritten as

$$x = \ln y$$

Differentiating this equation implicitly with respect to x yields

$$\frac{d}{dx}[x] = \frac{d}{dx}[\ln y]$$

$$1 = \frac{1}{y}\frac{dy}{dx}$$

$$\frac{dy}{dx} = y$$

$$\frac{dy}{dx} = e^x \qquad \blacksquare$$

If u is a differentiable function of x, then it follows from (6) and the chain rule that

$$\frac{d}{dx}[e^u] = e^u \cdot \frac{du}{dx} \tag{7}$$

Example 2

$$\frac{d}{dx}[e^{-2x}] = e^{-2x} \cdot \frac{d}{dx}[-2x] = -2e^{-2x}$$

$$\frac{d}{dx}[e^{x^3}] = e^{x^3} \cdot \frac{d}{dx}[x^3] = 3x^2e^{x^3}$$

$$\frac{d}{dx}[e^{\cos x}] = e^{\cos x} \cdot \frac{d}{dx}[\cos x] = -(\sin x)e^{\cos x}$$

Because e^x and $x \ln b$ are differentiable on $(-\infty, +\infty)$, it follows from the chain rule (Theorem 3.5.2) that the composition

$$b^x = e^{x \ln b}$$

is differentiable on $(-\infty, +\infty)$ and

$$\frac{d}{dx}[b^x] = \frac{d}{dx}[e^{x \ln b}] = e^{x \ln b} \cdot \frac{d}{dx}[x \ln b] = e^{x \ln b} \cdot \ln b = b^x \ln b$$

In summary,

$$\frac{d}{dx}[b^x] = b^x \ln b \tag{8}$$

Example 3

$$\frac{d}{dx}[\pi^x] = \pi^x \ln \pi \qquad \blacktriangleleft$$

If u is a differentiable function of x, then it follows from (8) and the chain rule that

$$\frac{d}{dx}[b^u] = b^u \ln b \cdot \frac{du}{dx} \tag{9}$$

Example 4   From (9) with $b = 2$ and $u = \sin x$

$$\frac{d}{dx}[2^{\sin x}] = (2^{\sin x})(\ln 2) \cdot \frac{d}{dx}[\sin x] = (2^{\sin x})(\ln 2)(\cos x) \qquad \blacktriangleleft$$

The following theorem extends a familiar derivative formula so that it applies to arbitrary exponents.

7.4.9   THEOREM.   *If r is any real exponent, then*

$$\frac{d}{dx}[x^r] = rx^{r-1}, \quad x > 0$$

Proof.   From Definition 7.4.4

$$x^r = e^{r \ln x}$$

so that

$$\frac{d}{dx}[x^r] = \frac{d}{dx}[e^{r \ln x}] = e^{r \ln x} \cdot \frac{d}{dx}[r \ln x] = x^r \cdot \frac{r}{x} = rx^{r-1} \qquad \blacksquare$$

Example 5

$$\frac{d}{dx}[x^\pi] = \pi x^{\pi - 1} \quad \blacktriangleleft$$

REMARK. It is important to distinguish between the derivative formulas for b^x (variable exponent) and x^b (variable base). Compare, for example, the derivative of x^π in Example 5 and the derivative of π^x in Example 3.

☐ **INTEGRALS OF EXPONENTIAL FUNCTIONS**

Associated with derivatives (6) and (8) are the companion integration formulas

$$\int e^u \, du = e^u + C \tag{10}$$

$$\int b^u \, du = \frac{b^u}{\ln b} + C \tag{11}$$

Example 6

$$\int 2^x \, dx = \frac{2^x}{\ln 2} + C \quad \blacktriangleleft$$

Example 7 To evaluate

$$\int e^{5x} \, dx$$

let $u = 5x$ so that $du = 5 \, dx$ or $dx = \frac{1}{5} \, du$, which yields

$$\int e^{5x} \, dx = \frac{1}{5} \int e^u \, du = \frac{1}{5} e^u + C = \frac{1}{5} e^{5x} + C \quad \blacktriangleleft$$

Example 8

$$\int e^{-x} \, dx = - \int e^u \, du = -e^u + C = -e^{-x} + C \quad \blacktriangleleft$$

$$u = -x$$
$$du = -dx$$

Example 9

$$\int x^2 e^{x^3} \, dx = \int \frac{1}{3} e^u \, du = \frac{1}{3} e^u + C = \frac{1}{3} e^{x^3} + C \quad \blacktriangleleft$$

$$u = x^3$$
$$du = 3x^2 \, dx$$

Example 10   Evaluate

$$\int_0^{\ln 3} e^x(1 + e^x)^{1/2} \, dx$$

Solution.   Make the *u*-substitution

$$u = 1 + e^x, \quad du = e^x \, dx$$

and change the *x*-limits of integration ($x = 0, x = \ln 3$) to the *u*-limits ($u = 1 + e^0 = 2, u = 1 + e^{\ln 3} = 1 + 3 = 4$).

$$\int_0^{\ln 3} e^x(1 + e^x)^{1/2} \, dx = \int_2^4 u^{1/2} \, du = \frac{2}{3} u^{3/2} \Big]_2^4 = \frac{2}{3}[4^{3/2} - 2^{3/2}]$$

$$= \frac{16}{3} - \frac{4\sqrt{2}}{3}$$

REMARK.   For ease of printing, the exponential function is often denoted by $\exp x$ rather than e^x. For example, in this notation the statements

$$e^{x_1 + x_2} = e^{x_1}e^{x_2} \qquad\qquad \exp(x_1 + x_2) = \exp x_1 \cdot \exp x_2$$

$$e^{\ln x} = x \qquad \text{would be written as} \qquad \exp(\ln x) = x$$

$$\ln(e^x) = x \qquad\qquad \ln(\exp x) = x$$

▶ **Exercise Set 7.4**   [C] 91, 92, 93

---

1. Simplify the expression and state the values of x for which your simplification is valid.
   (a) $e^{-\ln x}$ (b) $e^{\ln x^2}$
   (c) $\ln(e^{-x^2})$ (d) $\ln(1/e^x)$
   (e) $\exp(3 \ln x)$ (f) $\ln(xe^x)$
   (g) $\ln(e^{x - \sqrt[3]{x}})$ (h) $e^{x - \ln x}$.

2. Let $f(x) = e^{-2x}$. Find the simplest exact value of $f(\ln 3)$.

3. Let $f(x) = e^x + 3e^{-x}$. Find the simplest exact value of $f(\ln 2)$.

In Exercises 4 and 5, use Definition 7.4.4 to express the given quantity as a power of e.

4. (a) 3^π (b) $2^{\sqrt{2}}$.
5. (a) π^{-x} (b) $x^{2x}, x > 0$.

In Exercises 6–21, find dy/dx.

6. $y = e^{7x}$. 7. $y = e^{-5x^2}$.
8. $y = e^{1/x}$. 9. $y = x^3 e^x$.
10. $y = \sin(e^x)$. 11. $y = \dfrac{e^x - e^{-x}}{e^x + e^{-x}}$.

12. $y = \dfrac{e^x}{\ln x}$. 13. $y = e^{x \tan x}$.
14. $y = \exp(\sqrt{1 + 5x^3})$. 15. $y = e^{(x - e^{3x})}$.
16. $y = \ln(\cos e^x)$. 17. $y = \ln(1 - xe^{-x})$.
18. $y = \sqrt{1 + e^x}$.
19. $y = e^{ax} \cos bx$ (a, b constant).
20. $y = \dfrac{a}{1 + be^{-x}}$ (a, b constant).
21. $y = e^{\ln(x^3 + 1)}$.

In Exercises 22–26, find $f'(x)$ by Formula (9) and then by logarithmic differentiation.

22. $f(x) = 2^x$. 23. $f(x) = 3^{-x}$.
24. $f(x) = \pi^{\sin x}$. 25. $f(x) = \pi^{x \tan x}$.
26. $f(x) = (\sqrt{2})^{x \ln x}$.

27. (a) Explain why Formula (8) cannot be used to find $(d/dx)[x^x]$.
    (b) Find this derivative by logarithmic differentiation.

In Exercises 28–33, find dy/dx by logarithmic differentiation.

28. $y = x^{\sin x}$.

29. $y = (x^3 - 2x)^{\ln x}$.

30. $y = (x^2 + 3)^{\ln x}$.

31. $y = (\ln x)^{\tan x}$.

32. $y = (1 + x)^{1/x}$.

33. $y = x^{(e^x)}$.

34. Show that $y = e^{3x}$ and $y = e^{-3x}$ both satisfy the equation $y'' - 9y = 0$.

35. Show that for any constants A and B, the function $y = Ae^{2x} + Be^{-4x}$ satisfies the equation

$$y'' + 2y' - 8y = 0$$

36. Show that for any constants A and k, the function $y = Ae^{kt}$ satisfies the equation $dy/dt = ky$.

37. Let $f(x) = e^{kx}$ and $g(x) = e^{-kx}$. Find
  (a) $f^{(n)}(x)$ (b) $g^{(n)}(x)$.

38. Find dy/dt if $y = e^{-\lambda t}(A \sin \omega t + B \cos \omega t)$, where A, B, λ, and ω are constants.

39. Find $f'(x)$ if

$$f(x) = \frac{1}{\sqrt{2\pi}\sigma} \exp\left[-\frac{1}{2}\left(\frac{x - \mu}{\sigma}\right)^2 \right]$$

where μ and σ are constants and $\sigma \neq 0$.

In Exercises 40–67, evaluate the integrals.

40. $\displaystyle\int \frac{dx}{e^x}$.

41. $\displaystyle\int e^{-5x}\, dx$.

42. $\displaystyle\int e^{\tan x} \sec^2 x \, dx$.

43. $\displaystyle\int e^{\sin x} \cos x \, dx$.

44. $\displaystyle\int x^3 e^{x^4}\, dx$.

45. $\displaystyle\int x^2 e^{-2x^3}\, dx$.

46. $\displaystyle\int \frac{e^x + e^{-x}}{e^x - e^{-x}}\, dx$.

47. $\displaystyle\int \frac{e^x}{1 + e^x}\, dx$.

48. $\displaystyle\int \sqrt{e^x}\, dx$.

49. $\displaystyle\int e^{2t}\sqrt{1 + e^{2t}}\, dt$.

50. $\displaystyle\int (x + 3) \exp(x^2 + 6x)\, dx$.

51. $\displaystyle\int \cos x \exp(\sin x)\, dx$.

52. $\displaystyle\int e^x \sin(1 + e^x)\, dx$.   **53.** $\displaystyle\int e^{-x}\sec^2(2 - e^{-x})\, dx$.

54. $\displaystyle\int 2^{5x}\, dx$.

55. $\displaystyle\int \pi^{\sin x} \cos x \, dx$.

56. $\displaystyle\int \left[e x^2 + \left(\frac{1}{2}\ln 2\right) \sin x \right] dx$.

57. $\displaystyle\int \left(x \ln 3 - 4\pi e^2 \cos x \right) dx$.

58. $\displaystyle\int e^{2\ln x}\, dx$.

59. $\displaystyle\int \left[\ln(e^x) + \ln(e^{-x}) \right] dx$.

60. $\displaystyle\int \frac{dy}{\sqrt{y}e^{\sqrt{y}}}$.

61. $\displaystyle\int \frac{e^{\sqrt{y}}}{\sqrt{y}}\, dy$.

62. $\displaystyle\int_0^{\ln 2} e^{-3x}\, dx$.

63. $\displaystyle\int_0^{\ln 5} e^x(3 - 4e^x)\, dx$.

64. $\displaystyle\int_1^{\sqrt{2}} x 4^{-x^2}\, dx$.

65. $\displaystyle\int_1^2 (3 - e^x)\, dx$.

66. $\displaystyle\int_0^e \frac{dx}{x + e}$.

67. $\displaystyle\int_{-\ln 3}^{\ln 3} \frac{e^x}{e^x + 4}\, dx$.

68. Show that the equation $e^{1/x} - e^{-1/x} = 0$ has no solution.

69. Prove: If $f'(x) = f(x)$ for all x in $(-\infty, +\infty)$, then $f(x)$ has the form $f(x) = ke^x$ for some constant k. [*Hint:* Let $g(x) = e^{-x} f(x)$ and find $g'(x)$.]

70. A particle moves along the x-axis so that its x-coordinate at time t is given by $x = ae^{kt} + be^{-kt}$. Show that its acceleration is proportional to x.

71. Find the intersections of the curves $y = 2^x$ and $y = 3^{x+1}$.

72. Find a point on the graph of $y = e^{3x}$ at which the tangent line passes through the origin.

73. Find $f'(x)$ if $f(x) = x^e$.

74. Show that
  (a) $y = xe^{-x}$ satisfies the equation
     $xy' = (1 - x)y$
  (b) $y = xe^{-x^2/2}$ satisfies the equation
     $xy' = (1 - x^2)y$.

75. The equilibrium constant k of a balanced chemical reaction changes with the absolute temperature T according to the law

$$k = k_0 \exp\left(-\frac{q}{2}\frac{T - T_0}{T_0 T} \right)$$

where k_0, q, and T_0 are constants. Find the rate of change of k with respect to T.

76. Find a function $y = f(x)$ such that $e^y - e^{-y} = x$.

77. Evaluate

$$\int \frac{e^{2x}}{e^x + 3}\, dx$$

In Exercises 78–81, find $f^{-1}(x)$.

78. $f(x) = e^{2x+1}$.   79. $f(x) = e^{1/x}$.

80. $f(x) = 4 \ln(x + 1)$.   81. $f(x) = 1 - \ln(3x)$.

82. Find

$$\lim_{n \to +\infty} \frac{e^{1/n} + e^{2/n} + e^{3/n} + \cdots + e^{n/n}}{n}$$

[*Hint*: Interpret this as a limit of Riemann sums in which the interval $[0, 1]$ is divided into n subintervals of equal width.]

83. Find the maximum value of $x^3 e^{-2x}$.

84. Prove: $e^x \geq 1 + x$.
[*Hint*: Show that $1 + x - e^x \leq 0$.]

85. Find the area of the region enclosed by $y = e^x$, $y = 3$, and $x = 0$.

86. Find the positive value of k such that the area under the graph of $y = e^{2x}$ over the interval $[0, k]$ is 3 square units.

87. Find the volume of the solid that is generated when the region enclosed by $y = e^x$, $y = 0$, $x = 0$, and $x = \ln 3$ is revolved about the x-axis.

88. (a) Show by differentiation that $xe^x - e^x$ is an antiderivative of xe^x.
     (b) Find the volume of the solid that is generated when the region enclosed by $y = e^x$, $y = 0$, $x = 0$, and $x = 2$ is revolved about the y-axis.

89. Evaluate $\int_1^5 \ln x \, dx$. [*Hint*: Interpret the integral as an area and integrate with respect to y.]

90. Evaluate $\int_1^3 x \ln x \, dx$. [*Hint*: Relate the integral to the volume of a solid by the method of shells and use the method of washers to find the volume.]

91. Find all values of x for which $e^x < 2 - x^2$. State your answer to at least five decimal places. [*Hint*: Use Newton's Method (Section 4.9).]

92. Find the three solutions of $x^2 = 2^x$. [*Hint*: You should be able to find two of them by inspection. Use Newton's Method (Section 4.9) to approximate the third one to at least five decimal places.]

93. In physics, the **spectral radiancy** S of an ideal radiator at a constant temperature is given by **Planck's radiation law**

$$S = \frac{af^3}{e^{bf} - 1}$$

where a and b are positive constants and f is the frequency of the radiant photons.

(a) Show that the frequency at which S attains its maximum value satisfies the equation

$$3e^{-bf} + bf - 3 = 0$$

(b) Assuming that $f > 0$, use Newton's Method (Section 4.9) to approximate the value of bf that satisfies the equation in part (a). Express your answer to at least five decimal places. [*Hint*: Let $x = bf$.]

(c) For a radiator with

$$b = \frac{4.8043 \times 10^{-11}}{T}$$

where T is the temperature in kelvins, express the frequency at which S attains its maximum value as a function of the temperature T.

94. Prove that $\lim\limits_{x \to +\infty} e^x = +\infty$ by showing that for any $N > 0$, there is a point x_0 such that $e^x > N$ whenever $x > x_0$.

95. Prove that $\lim\limits_{x \to -\infty} e^x = 0$ by showing that for any $\epsilon > 0$ there is a point x_0 such that $0 < e^x < \epsilon$ whenever $x < x_0$.

96. Sketch the graph of $y = x^{1/\ln x}$. [*Hint*: First find a simpler form for $x^{1/\ln x}$.]

97. Find the minimum value of x^x, $x > 0$.

98. Find the maximum value of $x^{1/x}$, $x > 0$.

99. (a) Give a geometric argument to show that

$$\frac{1}{x + 1} < \int_x^{x+1} \frac{1}{t} \, dt < \frac{1}{x}, \quad x > 0$$

(b) Use the result in part (a) to prove that

$$\frac{1}{x + 1} < \ln\left(1 + \frac{1}{x}\right) < \frac{1}{x}, \quad x > 0$$

(c) Use the result in part (b) to prove that

$$e^{\frac{x}{x+1}} < \left(1 + \frac{1}{x}\right)^x < e, \quad x > 0$$

and hence that

$$\lim_{x \to +\infty} \left(1 + \frac{1}{x}\right)^x = e$$

(d) Use the inequality in part (c) to prove that

$$\left(1 + \frac{1}{x}\right)^x < e < \left(1 + \frac{1}{x}\right)^{x+1}, \quad x > 0$$

7.5 GRAPHS OF EQUATIONS INVOLVING EXPONENTIALS AND LOGARITHMS

In this section we shall discuss techniques for graphing equations involving exponential and logarithmic functions.

□ **OBTAINING GRAPHS USING PROPERTIES OF EXPONENTS AND LOGARITHMS**

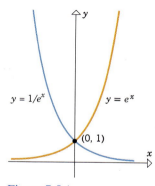

Figure 7.5.1

Graphs of equations involving exponents and logarithms can sometimes be obtained by making appropriate reflections and translations of known graphs.

Example 1 Sketch the graph of $y = 1/e^x$.

Solution. The given equation can be rewritten as

$$y = e^{-x}$$

which is the equation that results on replacing x by $-x$ in $y = e^x$. Thus, the graphs of $y = 1/e^x$ and $y = e^x$ are reflections of one another about the y-axis (Figure 7.5.1). ◀

Example 1 is a special case of a general result about exponential functions: Replacing x by $-x$ in the equation $y = b^x$ yields

$$y = b^{-x} = \frac{1}{b^x} = \left(\frac{1}{b}\right)^x$$

from which it follows that *exponential functions with reciprocal bases have graphs that are reflections of one another about the y-axis.*

Example 2 Sketch the graph of

$$y = \ln\left(\frac{1}{x-1}\right)$$

Solution. The equation can be rewritten as

$$y = -\ln(x-1) \quad \text{or equivalently} \quad -y = \ln(x-1)$$

Since the last equation results by first replacing y by $-y$ in the equation $y = \ln x$, then replacing x by $x - 1$, the graph of the given equation can be obtained by first reflecting the graph of $y = \ln x$ about the x-axis, then translating the resulting graph 1 unit to the right (Figure 7.5.2). ◀

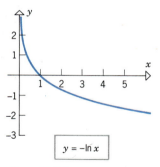
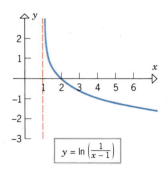

Figure 7.5.2

□ **EXPONENTIAL AND LOGARITHMIC GROWTH**

Recall the following limits, which were stated in Theorems 7.3.3 and 7.4.7.

$$\lim_{x \to +\infty} e^x = +\infty \tag{1}$$

$$\lim_{x \to -\infty} e^x = 0 \tag{2}$$

$$\lim_{x \to +\infty} \ln x = +\infty \tag{3}$$

$$\lim_{x \to 0^+} \ln x = -\infty \tag{4}$$

The following limits are consequences of (1) and (2):

$$\lim_{x \to +\infty} e^{-x} = \lim_{x \to +\infty} \frac{1}{e^x} = 0 \tag{5}$$

$$\lim_{x \to -\infty} e^{-x} = \lim_{x \to -\infty} \frac{1}{e^x} = +\infty \tag{6}$$

Formulas (1) and (3) tell us that e^x and $\ln x$ both increase toward $+\infty$ as $x \to +\infty$. However, the two functions increase in very different ways. As illustrated in Table 7.5.1, the increase in e^x is extremely rapid, while the increase in $\ln x$ is extremely slow. For example, an increase in x from 1 to 10 produces an increase in e^x of more than 22,000 units but produces less than a 3-unit increase in $\ln x$.

Mathematicians often use integer powers of x as a "measuring stick" for describing how rapidly a function grows. For example, it is shown in the exercises (Exercises 35 and 37) that if n is any positive integer power of x, then

$$\lim_{x \to +\infty} \frac{e^x}{x^n} = +\infty \tag{7}$$

$$\lim_{x \to +\infty} \frac{\ln x}{x^n} = 0 \tag{8}$$

Table 7.5.1

x	e^x	$\ln x$
1	2.718282	0
2	7.389056	0.693147
3	20.08554	1.098612
4	54.59815	1.386294
5	148.4132	1.609438
6	403.4287	1.791760
7	1096.633	1.945910
8	2980.958	2.079442
9	8103.084	2.197225
10	22026.47	2.302585

Comparing (1) and (7) tells us that e^x increases so rapidly that no matter how large we choose the integer n, division by x^n does not prevent the growth toward $+\infty$. Thus, we say that *e^x increases more rapidly than any positive power of x.*

The following alternative forms of (7) and (8) can be obtained by taking reciprocals

$$\lim_{x \to +\infty} \frac{x^n}{e^x} = 0 \tag{9}$$

$$\lim_{x \to +\infty} \frac{x^n}{\ln x} = +\infty \tag{10}$$

and the limit

$$\lim_{x \to 0^+} x^n \ln x = 0 \tag{11}$$

can be deduced from (8).

□ **GRAPHS INVOLVING EXPONENTIAL AND LOGARITHMIC FUNCTIONS**

Example 3   Sketch the graph of $y = e^{-x^2/2}$.

Solution.

Symmetries: Replacing x by $-x$ does not change the equation, so the graph is symmetric about the y-axis.

x-intercepts: Setting $y = 0$ yields the equation $e^{-x^2/2} = 0$, which has no solutions since all powers of e have positive values; thus, there are no x-intercepts.

y-intercepts: Setting $x = 0$ yields the y-intercept $y = 1$.

Vertical asymptotes: None, since $e^{-x^2/2}$ is a continuous function.

Horizontal asymptotes: Since $x^2/2 \to +\infty$ as $x \to +\infty$ or $x \to -\infty$, it follows from (5) that

$$\lim_{x \to +\infty} e^{-x^2/2} = \lim_{x \to -\infty} e^{-x^2/2} = 0$$

Thus, $y = 0$ is a horizontal asymptote.

Derivatives:

$$\frac{dy}{dx} = e^{-x^2/2}\, \frac{d}{dx}\left[-\frac{x^2}{2}\right] = -xe^{-x^2/2}$$

$$\frac{d^2y}{dx^2} = -x\,\frac{d}{dx}[e^{-x^2/2}] + e^{-x^2/2}\,\frac{d}{dx}[-x]$$

$$= x^2 e^{-x^2/2} - e^{-x^2/2}$$

$$= (x^2 - 1)e^{-x^2/2}$$

Intervals of increase and decrease: Since $e^{-x^2/2} > 0$ for all x, the sign of dy/dx is the same as that of $-x$.

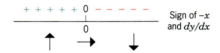

This analysis reveals a relative maximum at $x = 0$.

Concavity: Since $e^{-x^2/2} > 0$ for all x, the sign of d^2y/dx^2 is the same as that of $x^2 - 1$.

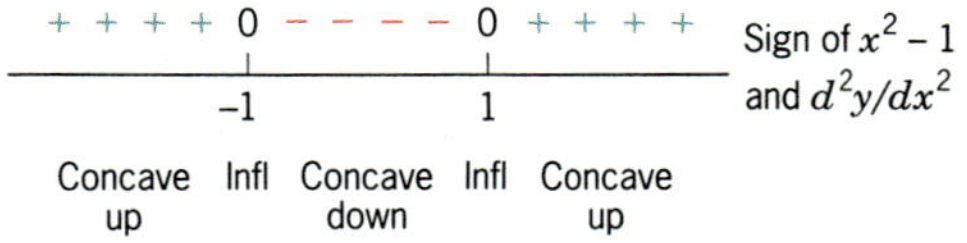

```
+ + + + 0 - - - - 0 + + + +    Sign of x² – 1
        |         |            and d²y/dx²
       -1         1

Concave  Infl  Concave  Infl  Concave
  up          down              up
```

This analysis reveals inflection points at $x = 1$ and $x = -1$.

Combining the foregoing information yields the graph in Figure 7.5.3.   ◀

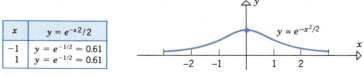

x	$y = e^{-x^2/2}$
-1	$y = e^{-1/2} \approx 0.61$
1	$y = e^{-1/2} \approx 0.61$

Figure 7.5.3

Example 4   Sketch the graph of $y = \dfrac{\ln x}{x}$.

Solution.

Domain: Since $\ln x$ is not defined for $x \leq 0$, the graph lies entirely to the right of the y-axis.

Symmetries: None.

x-intercepts: Setting $y = 0$ yields the equation

$$\frac{\ln x}{x} = 0$$

This yields a single x-intercept of $x = 1$, since this is the only value of x for which $\ln x = 0$.

y-intercepts: There are no y-intercepts since $\ln x$ is undefined at $x = 0$.

Vertical asymptotes: It follows from (11) with $n = 1$ that

$$\lim_{x \to 0^+} \frac{\ln x}{x} = -\infty$$

so the graph has a vertical asymptote at $x = 0$.

Horizontal asymptotes: It follows from (8) with $n = 1$ that

$$\lim_{x \to +\infty} \frac{\ln x}{x} = 0$$

so $y = 0$ is a horizontal asymptote.

Derivatives:

$$\frac{dy}{dx} = \frac{x\left(\dfrac{1}{x}\right) - (\ln x)(1)}{x^2} = \frac{1 - \ln x}{x^2}$$

$$\frac{d^2y}{dx^2} = \frac{x^2\left(-\dfrac{1}{x}\right) - (1 - \ln x)(2x)}{x^4} = \frac{2x \ln x - 3x}{x^4} = \frac{2 \ln x - 3}{x^3}$$

Intervals of increase and decrease: Since $x^2 > 0$ for $x > 0$, the sign of dy/dx is the same as that of $1 - \ln x$. But

$$\ln e = 1$$
$$\ln x > 1 \quad \text{if} \quad x > e$$
$$\ln x < 1 \quad \text{if} \quad x < e$$

so the signs for $1 - \ln x$ are as shown in the following diagram.

This analysis reveals a relative maximum at $x = e$.

Concavity: Since we are only concerned with positive values of x, the sign of d^2y/dx^2 will be determined by the sign of the numerator, $2 \ln x - 3$. Thus, $d^2y/dx^2 > 0$ if

$$2 \ln x - 3 > 0$$

$$\ln x > \tfrac{3}{2}$$

$$e^{\ln x} > e^{3/2} \qquad \text{Because } e^x \text{ is an increasing function}$$

$$x > e^{3/2}$$

Similarly, $2 \ln x - 3 < 0$ if $x < e^{3/2}$, and $2 \ln x - 3 = 0$ if $x = e^{3/2}$. Thus, the signs of $2 \ln x - 3$ are as shown in the following diagram.

```
0 – – – – – – – 0 + + + + +        Sign of 2 ln x – 3
└──────────────┴─────────          and d²y/dx²
              e^{3/2}
        Concave    Infl   Concave
        down              up
```

This analysis reveals an inflection point at $x = e^{3/2}$.

Combining the foregoing information yields the graph in Figure 7.5.4.

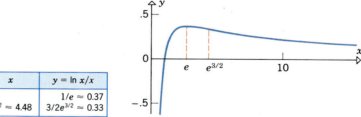

x	$y = \ln x/x$
e	$1/e \approx 0.37$
$e^{3/2} \approx 4.48$	$3/2e^{3/2} \approx 0.33$

Figure 7.5.4

▶ **Exercise Set 7.5**   Ⓒ *26, 27, 28, 29, 30, 44, 45, 46*

In Exercises 1–8, find the limits.

1. (a) $\lim\limits_{x \to +\infty} 4e^{3x}$ (b) $\lim\limits_{x \to -\infty} 4e^{3x}$.

2. (a) $\lim\limits_{x \to +\infty} 3e^{-2x}$ (b) $\lim\limits_{x \to -\infty} 3e^{-2x}$.

3. (a) $\lim\limits_{x \to +\infty} (e^x + e^{-x})$ (b) $\lim\limits_{x \to -\infty} (e^x + e^{-x})$.

4. (a) $\lim\limits_{x \to +\infty} (e^x - e^{-x})$ (b) $\lim\limits_{x \to -\infty} (e^x - e^{-x})$.

5. (a) $\lim\limits_{x \to +\infty} (1 - e^{-x^2})$ (b) $\lim\limits_{x \to -\infty} (1 - e^{-x^2})$.

6. (a) $\lim\limits_{x \to +\infty} e^{1/x}$ (b) $\lim\limits_{x \to -\infty} e^{1/x}$.

7. (a) $\lim\limits_{x \to 0^+} e^{1/x}$ (b) $\lim\limits_{x \to 0^-} e^{1/x}$.

8. (a) $\lim\limits_{x \to 0^+} e^{-1/x^2}$ (b) $\lim\limits_{x \to 0^-} e^{-1/x^2}$.

In Exercises 9–12, sketch the graph. Avoid point-plotting.

9. $y = e^x - 1$.   **10.** $y = e^{x-1}$.

11. $y = 1 - e^{-x}$.   **12.** $y = e^{1-x}$.

13. Let $f(x) = e^{|x|}$.
   (a) Is f continuous at $x = 0$?
   (b) Is f differentiable at $x = 0$?
   (c) Sketch the graph of f.

14. In each part determine whether the limit exists. If so, find it.

(a) $\lim\limits_{x \to +\infty} e^x \cos x$ (b) $\lim\limits_{x \to -\infty} e^x \cos x$.

15. Sketch the graphs of e^x, $-e^x$, and $e^x \cos x$, all in the same figure, and label any points of intersection.

In Exercises 16–19, find the limits.

16. $\lim\limits_{x \to +\infty} \dfrac{e^x + e^{-x}}{e^x - e^{-x}}$.   **17.** $\lim\limits_{x \to -\infty} \dfrac{e^x + e^{-x}}{e^x - e^{-x}}$.

18. $\lim\limits_{x \to +\infty} \dfrac{2 + e^x}{1 + 3e^x}$.   **19.** $\lim\limits_{x \to +\infty} e^{(-e^x)}$.

20. One of the fundamental functions of mathematical statistics is

$$f(x) = \frac{1}{\sqrt{2\pi}\sigma} \exp\left[-\frac{1}{2}\left(\frac{x - \mu}{\sigma}\right)^2\right]$$

where μ and σ are constants such that $\sigma > 0$ and $-\infty < \mu < +\infty$.

(a) Locate the inflection points and relative extreme points.

(b) Find $\lim\limits_{x \to +\infty} f(x)$ and $\lim\limits_{x \to -\infty} f(x)$.

(c) Sketch the graph of f.

21. Prove that

$$\lim_{h \to 0} \frac{e^h - 1}{h} = 1$$

by expressing the derivative of e^x at $x = 0$ as a limit.

In Exercises 22–25, use the method of Exercise 21 to find the limits.

22. $\lim\limits_{x \to 0} \dfrac{e^{2x} - e^x}{x}$.   **23.** $\lim\limits_{x \to 0} \dfrac{1 - e^{-x}}{x}$.

24. $\lim\limits_{x \to a} \dfrac{e^x - e^a}{x - a}$.   **25.** $\lim\limits_{x \to +\infty} x(e^{1/x} - 1)$.

In Exercises 26–30, Formulas (7) and (9) will be helpful.

26. (a) Find $\lim\limits_{x \to +\infty} xe^{-2x}$ and $\lim\limits_{x \to -\infty} xe^{-2x}$.

(b) Sketch the graph of $y = xe^{-2x}$ and label all relative extrema and inflection points.

27. (a) Find $\lim\limits_{x \to +\infty} xe^x$ and $\lim\limits_{x \to -\infty} xe^x$.

(b) Sketch the graph of $y = xe^x$ and label all relative extrema and inflection points.

28. (a) Find $\lim\limits_{x \to +\infty} x^2 e^{2x}$ and $\lim\limits_{x \to -\infty} x^2 e^{2x}$.

(b) Sketch the graph of $y = x^2 e^{2x}$ and label all relative extrema and inflection points.

29. (a) Find $\lim\limits_{x \to +\infty} x^2/e^{2x}$ and $\lim\limits_{x \to -\infty} x^2/e^{2x}$.

(b) Sketch the graph of $y = x^2/e^{2x}$ and label all relative extrema and inflection points.

30. (a) Find $\lim\limits_{x \to +\infty} e^x/x$, $\lim\limits_{x \to -\infty} e^x/x$, $\lim\limits_{x \to 0^+} e^x/x$, and $\lim\limits_{x \to 0^-} e^x/x$.

(b) Sketch the graph of $y = e^x/x$ and label all relative extrema and inflection points.

31. Let c_1, c_2, k_1, and k_2 be positive constants. In each part make a sketch that shows the relative positions of the curves $y = c_1 e^{k_1 x}$ and $y = c_2 e^{k_2 x}$ when

(a) $c_1 = c_2$ and $k_1 < k_2$

(b) $c_1 < c_2$ and $k_1 = k_2$

(c) $c_1 < c_2$ and $k_2 < k_1$.

32. Repeat Exercise 31 for the curves $y = c_1 e^{-k_1 x}$ and $y = c_2 e^{-k_2 x}$.

33. Find the area of the region enclosed by the curve $y = e^{-x}$ and the line through the points $(0, 1)$ and $(1, 1/e)$.

34. A rectangle has one side along the x-axis, one vertex at the origin, and another vertex on the curve $y = e^{-x}$, $x > 0$. Find the maximum area that the rectangle can have.

35. (a) Use the inequality $\ln x < \sqrt{x}$ (see Exercise 66, Section 7.3), to show that

$$1/x > e^{-\sqrt{x}} \quad (x > 0)$$

(b) If n is a positive integer, then from part (a)

$$1/x^n > e^{-n\sqrt{x}} \quad \text{so} \quad e^x/x^n > e^{x - n\sqrt{x}}$$

Use this result to show that $\lim\limits_{x \to +\infty} e^x/x^n = +\infty$.

36. Use Formula (7) to deduce that $\lim\limits_{x \to +\infty} x^n/e^x = 0$.

37. Given that $\ln x < \sqrt{x}$ (see Exercise 66, Section 7.3), use the Squeezing Theorem (Theorem 2.8.2) to prove that if n is any positive integer, then

$$\lim_{x \to +\infty} \frac{\ln x}{x^n} = 0$$

[*Hint*: If $x > 1$, then $0 < \ln x < \sqrt{x}$.]

38. Use Formula (8) to deduce that $\lim\limits_{x \to +\infty} x^n/\ln x = +\infty$.

39. Use Formula (8) to deduce that $\lim\limits_{x \to 0^+} x^n \ln x = 0$ where n is a positive integer. [*Hint*: Let $x = 1/t$.]

In Exercises 40–43, sketch the graph. Avoid point-plotting.

40. $y = \ln|x|$.

41. $y = \ln\dfrac{1}{x}$.

42. $y = \ln\sqrt{x}$.

43. $y = \ln(x - 1)$.

In Exercises 44–46, sketch the graph. Formulas (8) and (11) will be helpful.

44. $y = x\ln x$.   **45.** $y = \ln x/x^2$.   **46.** $y = x^2\ln x$.

47. Show that $x^e \le e^x$ for all $x > 0$, with equality only for $x = e$. [*Hint:* From Example 4, $(\ln x)/x \le 1/e$.]

■ **7.6   THE HYPERBOLIC FUNCTIONS**

In this section we shall study certain combinations of e^x and e^{-x}, called **hyperbolic functions.** *These functions have numerous engineering applications and arise naturally in many mathematical problems. It will become evident as we progress that the hyperbolic functions have many properties in common with the trigonometric functions. This similarity is reflected in the names of the hyperbolic functions.*

□ **DEFINITIONS OF HYPERBOLIC FUNCTIONS**

7.6.1   DEFINITION.   The **hyperbolic sine** and **hyperbolic cosine** functions, denoted by **sinh** and **cosh,** respectively, are defined by

$$\sinh x = \frac{e^x - e^{-x}}{2} \quad \text{and} \quad \cosh x = \frac{e^x + e^{-x}}{2}$$

REMARK.   We note that sinh rhymes with "cinch" and cosh rhymes with "gosh."

The graph of $\cosh x = \frac{1}{2}e^x + \frac{1}{2}e^{-x}$ can be obtained by separately graphing $\frac{1}{2}e^x$ and $\frac{1}{2}e^{-x}$ and then adding the y-coordinates together at each point (Figure 7.6.1*a*). This graphing technique, called **addition of ordinates,** can also be used to graph $\sinh x = \frac{1}{2}e^x - \frac{1}{2}e^{-x}$ (Figure 7.6.1*b*).

As an illustration of how hyperbolic functions occur in physical problems, consider a homogeneous flexible cable hanging suspended between two points (e.g., an electrical transmission line suspended between two poles). The cable forms a curve called a **catenary** (from the Latin "catena" meaning chain). If, as in Figure 7.6.2, a coordinate system is introduced in such a way that the low point of the cable lies on the y-axis, then it can be shown using principles of physics that the equation of the curve formed by the cable is

$$y = a\cosh\left(\frac{x}{a}\right)$$

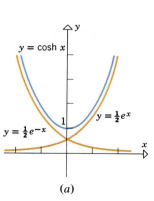

(a)

(b)

Figure 7.6.1

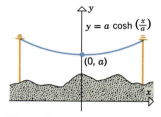

Figure 7.6.2

where a depends on the tension and physical properties of the cable.

The remaining hyperbolic functions, **hyperbolic tangent, hyperbolic cotangent, hyperbolic secant,** and **hyperbolic cosecant** are defined in terms of sinh and cosh as follows.

$$\tanh x = \frac{\sinh x}{\cosh x} = \frac{e^x - e^{-x}}{e^x + e^{-x}}$$

$$\coth x = \frac{\cosh x}{\sinh x} = \frac{e^x + e^{-x}}{e^x - e^{-x}}$$

$$\operatorname{sech} x = \frac{1}{\cosh x} = \frac{2}{e^x + e^{-x}}$$

$$\operatorname{csch} x = \frac{1}{\sinh x} = \frac{2}{e^x - e^{-x}}$$

The graphs of these functions are shown in Figure 7.6.3 (see Exercises 39–42).

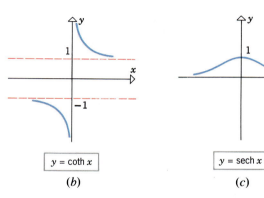

$y = \tanh x$

(a)

$y = \coth x$

(b)

$y = \operatorname{sech} x$

(c)

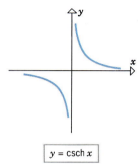

$y = \operatorname{csch} x$

(d)

Figure 7.6.3

☐ **HYPERBOLIC IDENTITIES**   The hyperbolic functions satisfy various identities similar to the identities for the trigonometric functions. The most fundamental of these is

$$\cosh^2 x - \sinh^2 x = 1 \tag{1}$$

which can be proved by writing

$$\cosh^2 x - \sinh^2 x = \left(\frac{e^x + e^{-x}}{2}\right)^2 - \left(\frac{e^x - e^{-x}}{2}\right)^2$$

$$= \frac{1}{4}(e^{2x} + 2e^0 + e^{-2x}) - \frac{1}{4}(e^{2x} - 2e^0 + e^{-2x})$$

$$= 1$$

If we divide (1) by $\cosh^2 x$, we obtain

$$1 - \tanh^2 x = \text{sech}^2 x \tag{2}$$

and if we divide (1) by $\sinh^2 x$, we obtain

$$\coth^2 x - 1 = \text{csch}^2 x \tag{3}$$

The addition formulas for sinh and cosh are

$$\sinh (x + y) = \sinh x \cosh y + \cosh x \sinh y \tag{4a}$$
$$\cosh (x + y) = \cosh x \cosh y + \sinh x \sinh y \tag{4b}$$

These are easier to prove than the corresponding identities for the trigonometric functions. For example, to prove (4a) we use the relations

$$\cosh x + \sinh x = e^x \tag{5a}$$
$$\cosh x - \sinh x = e^{-x} \tag{5b}$$

which follow immediately from Definition 7.6.1. From (5a) and (5b) it follows that

$$\sinh (x + y) = \frac{e^{(x+y)} - e^{-(x+y)}}{2} = \frac{e^x e^y - e^{-x}e^{-y}}{2}$$

$$= \frac{1}{2} [(\cosh x + \sinh x)(\cosh y + \sinh y)$$

$$- (\cosh x - \sinh x)(\cosh y - \sinh y)]$$

$$= \sinh x \cosh y + \cosh x \sinh y$$

which proves (4a). The proof of (4b) is similar.

If we let $x = y$ in (4a) and (4b), we obtain the following analogs of the trigonometric double-angle formulas:

$$\sinh 2x = 2 \sinh x \cosh x \qquad (6a)$$
$$\cosh 2x = \cosh^2 x + \sinh^2 x \qquad (6b)$$

By using (1) and (6b) the reader should be able to show that

$$\cosh 2x = 2 \sinh^2 x + 1 \qquad (7a)$$
$$\cosh 2x = 2 \cosh^2 x - 1 \qquad (7b)$$

To obtain subtraction formulas for sinh and cosh we shall use the addition formulas and the relations

$$\cosh(-x) = \cosh x \qquad (8a)$$
$$\sinh(-x) = -\sinh x \qquad (8b)$$

which follow immediately from Definition 7.6.1. If we replace y by $-y$ in (4a) and (4b) and then apply (8a) and (8b), we obtain

$$\sinh(x - y) = \sinh x \cosh y - \cosh x \sinh y \qquad (9a)$$
$$\cosh(x - y) = \cosh x \cosh y - \sinh x \sinh y \qquad (9b)$$

☐ **DERIVATIVE AND INTEGRAL FORMULAS**

Derivative formulas for $\sinh x$ and $\cosh x$ follow easily from Definition 7.6.1. For example,

$$\frac{d}{dx}[\cosh x] = \frac{d}{dx}\left[\frac{e^x + e^{-x}}{2}\right] = \frac{1}{2}\left(\frac{d}{dx}[e^x] + \frac{d}{dx}[e^{-x}]\right)$$

$$= \frac{e^x - e^{-x}}{2} = \sinh x$$

Similarly,

$$\frac{d}{dx}[\sinh x] = \cosh x$$

The derivatives of the remaining hyperbolic functions can be obtained by first expressing them in terms of sinh and cosh. For example,

$$\frac{d}{dx}[\tanh x] = \frac{d}{dx}\left[\frac{\sinh x}{\cosh x}\right] = \frac{\cosh x \dfrac{d}{dx}[\sinh x] - \sinh x \dfrac{d}{dx}[\cosh x]}{\cosh^2 x}$$

$$= \frac{\cosh^2 x - \sinh^2 x}{\cosh^2 x} = \frac{1}{\cosh^2 x} = \operatorname{sech}^2 x$$

Derivations of differentiation formulas for the remaining hyperbolic functions are exercises. For reference we summarize the generalized derivative formulas.

$$\frac{d}{dx}[\sinh u] = \cosh u \frac{du}{dx} \qquad \frac{d}{dx}[\coth u] = -\operatorname{csch}^2 u \frac{du}{dx}$$

$$\frac{d}{dx}[\cosh u] = \sinh u \frac{du}{dx} \qquad \frac{d}{dx}[\operatorname{sech} u] = -\operatorname{sech} u \tanh u \frac{du}{dx}$$

$$\frac{d}{dx}[\tanh u] = \operatorname{sech}^2 u \frac{du}{dx} \qquad \frac{d}{dx}[\operatorname{csch} u] = -\operatorname{csch} u \coth u \frac{du}{dx}$$

REMARK.   Except for a difference in the pattern of signs, these formulas are similar to the formulas for the derivatives of the trigonometric functions.

These differentiation formulas produce the following companion integration formulas.

$$\int \sinh u \; du = \cosh u + C \qquad \int \operatorname{csch}^2 u \; du = -\coth u + C$$

$$\int \cosh u \; du = \sinh u + C \qquad \int \operatorname{sech} u \tanh u \; du = -\operatorname{sech} u + C$$

$$\int \operatorname{sech}^2 u \; du = \tanh u + C \qquad \int \operatorname{csch} u \coth u \; du = -\operatorname{csch} u + C$$

Example 1

$$\frac{d}{dx}[\cosh(x^3)] = \sinh(x^3) \cdot \frac{d}{dx}[x^3] = 3x^2 \sinh(x^3)$$

$$\frac{d}{dx}[\ln(\tanh x)] = \frac{1}{\tanh x} \cdot \frac{d}{dx}[\tanh x] = \frac{\operatorname{sech}^2 x}{\tanh x} \qquad \blacktriangleleft$$

Example 2

$$\int \sinh^5 x \cosh x \; dx = \frac{1}{6} \sinh^6 x + C \qquad \boxed{\begin{array}{l} u = \sinh x \\ du = \cosh x \; dx \end{array}}$$

$$\int \tanh x \; dx = \int \frac{\sinh x}{\cosh x} \; dx$$

$$= \ln|\cosh x| + C \qquad \boxed{\begin{array}{l} u = \cosh x \\ du = \sinh x \; dx \end{array}}$$

$$= \ln(\cosh x) + C$$

We are justified in writing $|\cosh x| = \cosh x$ because $\cosh x$ is positive for all x. In fact, $\cosh x \geq 1$ for all x. This is geometrically clear from Figure 7.6.1a and can be proved using (1) (Exercise 41). $\blacktriangleleft$

If t is any real number, then the point $(\cos t, \sin t)$ lies on the circle $x^2 + y^2 = 1$ because

$$\cos^2 t + \sin^2 t = 1$$

(Figure 7.6.4). For this reason sine and cosine are called **circular functions**. Analogously, for any real number t the point $(\cosh t, \sinh t)$ lies on the curve $x^2 - y^2 = 1$ because

$$\cosh^2 t - \sinh^2 t = 1$$

(Figure 7.6.5). As we shall see later, this curve is called a *hyperbola* and accordingly sinh and cosh are called **hyperbolic functions**.

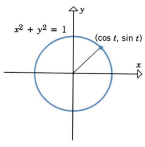

Figure 7.6.4

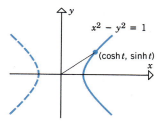

Figure 7.6.5

▶ Exercise Set 7.6   Ⓒ 48, 54, 55

1. In each part, a value for one of the hyperbolic functions is given at x_0. Find the values of the remaining five hyperbolic functions at x_0.
    (a) $\sinh x_0 = -2$
    (b) $\cosh x_0 = \frac{5}{4}, x_0 < 0$
    (c) $\tanh x_0 = -\frac{4}{5}$
    (d) $\coth x_0 = 2$
    (e) $\operatorname{sech} x_0 = \frac{15}{17}, x_0 > 0$
    (f) $\operatorname{csch} x_0 = -1$.

In Exercises 2–14, establish the identities.

2. $\cosh (x + y) = \cosh x \cosh y + \sinh x \sinh y$.

3. $\cosh 2x = 2 \sinh^2 x + 1$.

4. $\cosh 2x = 2 \cosh^2 x - 1$.

5. $\cosh (-x) = \cosh x$.

6. $\sinh (-x) = -\sinh x$.

7. $\tanh (x + y) = \dfrac{\tanh x + \tanh y}{1 + \tanh x \tanh y}$.

8. $\tanh (x - y) = \dfrac{\tanh x - \tanh y}{1 - \tanh x \tanh y}$.

9. $\tanh 2x = \dfrac{2 \tanh x}{1 + \tanh^2 x}$.

10. $\cosh \dfrac{1}{2} x = \sqrt{\dfrac{1}{2} (\cosh x + 1)}$.

11. $\sinh \dfrac{1}{2} x = \pm \sqrt{\dfrac{1}{2} (\cosh x - 1)}$.

12. $\sinh x + \sinh y = 2 \sinh \left(\dfrac{x + y}{2} \right) \cosh \left(\dfrac{x - y}{2} \right)$.

13. $\cosh x + \cosh y = 2 \cosh \left(\dfrac{x + y}{2} \right) \cosh \left(\dfrac{x - y}{2} \right)$.

14. $\cosh 3x = 4 \cosh^3 x - 3 \cosh x$.

15. Derive the differentiation formula for
    (a) $\sinh x$
    (b) $\coth x$
    (c) $\operatorname{sech} x$
    (d) $\operatorname{csch} x$.

In Exercises 16–25, find dy/dx.

16. $y = \cosh (x^4)$.

17. $y = \sinh (4x - 8)$.

18. $y = \ln (\tanh 2x)$.

19. $y = \coth (\ln x)$.

20. $y = \operatorname{sech} (e^{2x})$.

21. $y = \operatorname{csch} (1/x)$.

22. $y = \sinh^3 (2x)$.

23. $y = \sqrt{4x + \cosh^2 (5x)}$.

24. $y = \sinh (\cos 3x)$.

25. $y = x^3 \tanh^2 (\sqrt{x})$.

In Exercises 26–35, evaluate the integral.

26. $\displaystyle \int \cosh (2x - 3)\, dx$.

27. $\displaystyle \int \sinh^6 x \cosh x\, dx$.

28. $\int \text{csch}^2 (3x)\, dx.$ **29.** $\int \sqrt{\tanh x}\ \text{sech}^2 x\, dx.$

30. $\int \coth^2 x\ \text{csch}^2 x\, dx.$ **31.** $\int \tanh x\, dx.$

32. $\int \dfrac{e^x - e^{-x}}{e^x + e^{-x}}\, dx.$ **33.** $\int \tanh x\ \text{sech}^3 x\, dx.$

34. $\int \dfrac{\sinh 2x}{3 + 5 \cosh 2x}\, dx.$ **35.** $\int \dfrac{\cosh (\sqrt{x})}{\sqrt{x}}\, dx.$

36. In each part, find the exact numerical value:
    (a) $\sinh (\ln 3)$ (b) $\cosh (-\ln 2)$
    (c) $\tanh (2 \ln 5)$ (d) $\sinh (-3 \ln 2)$.

37. In each part, express as a rational function:
    (a) $\cosh (\ln x)$ (b) $\sinh (\ln x)$
    (c) $\tanh (2 \ln x)$ (d) $\cosh (-\ln x)$.

38. Prove the following facts about $\tanh x$:
    (a) $-1 < \tanh x < 1$ for all x
    (b) $\lim\limits_{x \to +\infty} \tanh x = 1$
    (c) $\lim\limits_{x \to -\infty} \tanh x = -1$.

39. Determine the intervals on which $y = \tanh x$ is positive, negative, increasing, decreasing, concave up, and concave down. Then use these results and Exercise 38 to obtain the graph of $y = \tanh x$.

40. Use the results found in Exercises 38 and 39 to help obtain the graph of $y = \coth x$.

41. (a) Prove: $\cosh x \geq 1$ for all x.
    (b) Prove: $0 < \text{sech}\, x \leq 1$ for all x.
    (c) Obtain the graph of $y = \text{sech}\, x$.

42. (a) Show that $\lim\limits_{x \to +\infty} \text{csch}\, x = \lim\limits_{x \to -\infty} \text{csch}\, x = 0$.
    (b) Show that
$$\lim\limits_{x \to 0^+} \text{csch}\, x = +\infty \quad \text{and} \quad \lim\limits_{x \to 0^-} \text{csch}\, x = -\infty$$
    (c) Obtain the graph of $y = \text{csch}\, x$.

43. Prove: $(\sinh x + \cosh x)^n = \sinh nx + \cosh nx$.

44. (a) It is sometimes said that $\sinh x$ and $\cosh x$ "behave like $\frac{1}{2} e^x$" for x large and positive. Give an informal justification of this statement.
    (b) Make a sketch showing the graphs of $y = \sinh x$, $y = \cosh x$, and $y = \frac{1}{2} e^x$, all in the same figure.

45. Find
    (a) $\lim\limits_{x \to +\infty} \dfrac{\cosh x}{e^x}$

    (b) $\lim\limits_{x \to +\infty} \dfrac{\sinh ax}{e^x}$ $(a > 0,\ \text{constant})$.

[*Hint:* In part (b) consider the cases $a > 1$, $a = 1$, and $0 < a < 1$ separately.]

46. Show that
$$\int_{-a}^{a} e^{tx}\, dx = \frac{2 \sinh at}{t}$$
where t is a nonzero constant.

47. Find the area enclosed by $y = \sinh 2x$, $y = 0$, and $x = \ln 3$.

48. Use Newton's Method to approximate the positive value of the constant a such that the area enclosed by $y = \cosh ax$, $y = 0$, $x = 0$, and $x = 1$ is 2 square units. Express your answer to at least five decimal places.

49. Find the volume of the solid that is generated when the region enclosed by $y = \cosh 2x$, $y = \sinh 2x$, $x = 0$, and $x = 5$ is revolved about the x-axis.

50. Find the volume of the solid that is generated when the region enclosed by $y = \text{sech}\, x$, $y = 0$, $x = 0$, and $x = \ln 2$ is revolved about the x-axis.

51. Find the arc length of $y = \cosh x$ between $x = 0$ and $x = \ln 2$.

52. Find the arc length of the catenary $y = a \cosh (x/a)$ between $x = 0$ and $x = x_1$ $(x_1 > 0)$.

53. A cable is suspended between two poles as shown in Figure 7.6.2. The equation of the curve formed by the cable is $y = a \cosh (x/a)$, where a is a positive constant. Suppose that the x-coordinates of the points of support are $x = -b$ and $x = b$, where $b > 0$.
    (a) Show that the length L of the cable is given by
$$L = 2a \sinh \frac{b}{a}$$
    (b) Show that the sag S (the vertical distance between the highest and lowest points on the cable) is given by
$$S = a \cosh \frac{b}{a} - a$$

Exercises 54 and 55 refer to the hanging cable described in Exercise 53.

54. Assuming that the cable is 120 ft long and the poles are 100 ft apart, approximate the sag in the cable by using Newton's Method to approximate a. Express your final answer to the nearest tenth of a foot. [*Hint:* First let $u = 50/a$.]

55. Assuming that the poles are 400 ft apart and the sag in the cable is 30 ft, approximate the length of the cable by using Newton's Method to approximate a. Express your final answer to the nearest tenth of a foot. [*Hint:* First let $u = 200/a$.]

■ 7.7 FIRST-ORDER DIFFERENTIAL EQUATIONS AND APPLICATIONS

*In this section we shall begin to study equations that involve an unknown function and its derivatives. These are called **differential equations**. Such equations describe many of the fundamental principles in science and engineering, and their study constitutes a major field of mathematics. Our work in this section is limited to the most basic types of differential equations. In the final chapter of this text we shall go a little further into this topic.*

☐ **TERMINOLOGY**

Some examples of differential equations are

$$\frac{dy}{dx} = 3y \tag{1}$$

$$\frac{d^2y}{dx^2} - 6\frac{dy}{dx} + 8y = 0 \tag{2}$$

$$y' - y = e^{2x} \tag{3}$$

$$\frac{d^3y}{dt^3} - t\frac{dy}{dt} + (t^2 - 1)y = e^t \tag{4}$$

In the first three equations, $y = y(x)$ is an unknown function of x and in the last equation $y = y(t)$ is an unknown function of t. The **order** of a differential equation is the order of the highest derivative that appears in the equation. Thus, (1) and (3) are first-order equations, (2) is second order, and (4) is third order.

A function $y = y(x)$ is a **solution** of a differential equation if the equation is satisfied when $y(x)$ and its derivatives are substituted. For example,

$$y = e^{2x} \tag{5}$$

is a solution of the equation

$$\frac{dy}{dx} - y = e^{2x} \tag{6}$$

since

$$\frac{dy}{dx} - y = 2e^{2x} - e^{2x} = e^{2x}$$

Similarly,

$$y = e^x + e^{2x} \tag{7}$$

is also a solution of (6). More generally,

$$y = Ce^x + e^{2x} \tag{8}$$

is a solution of (6) for any constant C (verify). This solution is of special importance since it can be proved that *all* solutions of (6) can be obtained by substituting values for the arbitrary constant C. For example, $C = 0$ yields solution (5) and $C = 1$ yields solution (7). A solution of a differential equation from which all other solutions can be derived by substituting values for arbitrary constants is called the **general solution** of the equation. Usually, the general solution of an nth-order equation contains n arbitrary constants. Thus, (8), which is the general solution of the first-order equation (6), contains one arbitrary constant.

Often, solutions of differential equations are expressed as implicitly defined functions. For example,

$$\ln y = xy + C \tag{9}$$

defines a solution of

$$\frac{dy}{dx} = \frac{y^2}{1 - xy} \tag{10}$$

for any value of the constant C, since implicit differentiation of (9) yields

$$\frac{1}{y}\frac{dy}{dx} = x\frac{dy}{dx} + y$$

or

$$\frac{dy}{dx} - xy\frac{dy}{dx} = y^2$$

from which (10) follows.

☐ **FIRST-ORDER SEPARABLE EQUATIONS**

A first-order differential equation is called **separable** if it is expressible in the form

$$\frac{dy}{dx} = \frac{g(x)}{h(y)}$$

To solve such an equation we rewrite it in the differential form

$$h(y)\,dy = g(x)\,dx \tag{11}$$

and integrate both sides, thereby obtaining the general solution

$$\int h(y)\,dy = \int g(x)\,dx + C$$

where C is an arbitrary constant.

In (11), the x and y variables are "separated" from each other, hence the term *separable* differential equation.

Example 1   Solve the equation $\dfrac{dy}{dx} = \dfrac{x}{y^2}$.

Solution. Changing to differential form and integrating yields

$$y^2 \, dy = x \, dx$$

$$\int y^2 \, dy = \int x \, dx$$

$$\frac{y^3}{3} = \frac{x^2}{2} + C$$

This expresses the solution implicitly. If desired, we can solve explicitly for y to obtain

$$y = (\tfrac{3}{2}x^2 + 3C)^{1/3} \qquad \text{or} \qquad y = (\tfrac{3}{2}x^2 + K)^{1/3}$$

where $K \, (= 3C)$ is an arbitrary constant. ◀

Example 2 Solve the equation $x(y - 1) \dfrac{dy}{dx} = y$.

Solution. Separating the variables and integrating yields

$$x(y - 1) \, dy = y \, dx$$

$$\frac{y - 1}{y} \, dy = \frac{dx}{x}$$

$$\int \frac{y - 1}{y} \, dy = \int \frac{dx}{x}$$

$$\int \left(1 - \frac{1}{y}\right) dy = \int \frac{dx}{x}$$

$$y - \ln|y| = \ln|x| + C$$

or, using properties of logarithms,

$$y = \ln|xy| + C$$

This gives a solution implicitly as a function of x; in this case there is no simple formula for y explicitly as a function of x. ◀

☐ **INITIAL-VALUE PROBLEMS**

When a physical problem leads to a differential equation, there are usually conditions in the problem that determine specific values for the arbitrary constants in the general solution of the equation. For a first-order equation, a condition that specifies the value of the unknown function $y(x)$ at some point $x = x_0$ is called an *initial condition*. A first-order differential equation together with one initial condition constitutes a *first-order initial-value problem.*

Example 3 Solve the initial-value problem

$$\frac{dy}{dx} = -4xy^2, \quad y(0) = 1$$

Solution. We first solve the differential equation:

$$\frac{dy}{y^2} = -4x \, dx$$

$$\int \frac{dy}{y^2} = \int -4x \, dx$$

$$-\frac{1}{y} = -2x^2 + C_1$$

or on multiplying by -1, taking reciprocals, and writing C in place of $-C_1$,

$$y = \frac{1}{2x^2 + C} \tag{12}$$

The initial condition, $y(0) = 1$, requires that $y = 1$ when $x = 0$. Substituting these values in (12) yields $C = 1$. Thus, the solution of the initial-value problem is

$$y = \frac{1}{2x^2 + 1} \qquad \blacktriangleleft$$

☐ **FIRST-ORDER LINEAR EQUATIONS**

Not every first-order differential equation is separable. For example, it is impossible to separate the variables in the equation

$$\frac{dy}{dx} + x^2 y = e^x$$

However, this equation can be solved by a different method that we shall now consider.

A first-order differential equation is called ***linear*** if it is expressible in the form

$$\frac{dy}{dx} + p(x)y = q(x) \tag{13}$$

where the functions $p(x)$ and $q(x)$ may or may not be constant. Some examples are

$$\frac{dy}{dx} + x^2 y = e^x \qquad \boxed{p(x) = x^2, \ q(x) = e^x}$$

$$y' - 3e^x y = 0 \qquad \boxed{p(x) = -3e^x, \ q(x) = 0}$$

$$\frac{dy}{dx} + 5y = 2 \qquad \boxed{p(x) = 5, \ q(x) = 2}$$

$$\frac{dy}{dx} + (\sin x)y + x^3 = 0 \qquad \boxed{p(x) = \sin x, \ q(x) = -x^3}$$

One procedure for solving (13) is based on the observation that if we define $\rho = \rho(x)$ by

$$\rho = e^{\int p(x)\, dx}$$

then

$$\frac{d\rho}{dx} = e^{\int p(x)\, dx} \cdot \frac{d}{dx} \int p(x)\, dx = \rho p(x)$$

Thus,

$$\frac{d}{dx}(\rho y) = \rho \frac{dy}{dx} + \frac{d\rho}{dx} y = \rho \frac{dy}{dx} + \rho p(x)y \qquad (14)$$

If (13) is multiplied through by ρ, it becomes

$$\rho \frac{dy}{dx} + \rho p(x)y = \rho q(x)$$

or from (14),

$$\frac{d}{dx}(\rho y) = \rho q(x)$$

This equation can be solved by integrating both sides to obtain

$$\rho y = \int \rho q(x)\, dx + C$$

or

$$y = \frac{1}{\rho}\left[\int \rho q(x)\, dx + C \right]$$

To summarize, (13) can be solved in three steps:

Step 1. Calculate

$$\rho = e^{\int p(x)\, dx}$$

This is called the ***integrating factor***. Since any ρ will suffice, we can take the constant of integration to be zero in this step.

Step 2. Multiply both sides of (13) by ρ and express the result as

$$\frac{d}{dx}(\rho y) = \rho q(x)$$

Step 3. Integrate both sides of the equation obtained in Step 2 and then solve for y. Be sure to include a constant of integration in this step.

Example 4   Solve the equation

$$\frac{dy}{dx} - 4xy = x \tag{15}$$

Solution.   Since $p(x) = -4x$, the integrating factor is

$$\rho = e^{\int(-4x)\,dx} = e^{-2x^2}$$

If we multiply (15) by ρ, we obtain

$$\frac{d}{dx}(e^{-2x^2}y) = xe^{-2x^2}$$

Integrating both sides of this equation yields

$$e^{-2x^2}y = \int xe^{-2x^2}\,dx = -\frac{1}{4}e^{-2x^2} + C$$

and then multiplying both sides by e^{2x^2} yields

$$y = -\frac{1}{4} + Ce^{2x^2} \qquad \blacktriangleleft$$

Example 5   Solve the equation

$$x\frac{dy}{dx} - y = x \quad (x > 0)$$

Solution.   To put the equation in form (13), we divide through by x to obtain

$$\frac{dy}{dx} - \frac{1}{x}y = 1 \tag{16}$$

Since $p(x) = -1/x$, the integrating factor is

$$\rho = e^{\int-(1/x)\,dx} = e^{-\ln|x|} = \frac{1}{|x|} = \frac{1}{x}$$

(The absolute value was dropped because of the assumption that $x > 0$.) If we multiply (16) by ρ, we obtain

$$\frac{d}{dx}\left(\frac{1}{x}y\right) = \frac{1}{x}$$

Integrating both sides of this equation yields

$$\frac{1}{x}y = \int\frac{1}{x}\,dx = \ln x + C$$

or

$$y = x\ln x + Cx \qquad \blacktriangleleft$$

☐ **APPLICATIONS**

We conclude this section with some applications of first-order differential equations.

Example 6 (*Geometry*) Find a curve in the xy-plane that passes through $(0, 3)$ and whose tangent line at a point (x, y) has slope $2x/y^2$.

Solution. Since the slope of the tangent line is dy/dx, we have

$$\frac{dy}{dx} = \frac{2x}{y^2} \tag{17}$$

and, since the curve passes through $(0, 3)$, we have the initial condition

$$y(0) = 3 \tag{18}$$

Equation (17) is separable and can be written as

$$y^2 \, dy = 2x \, dx$$

so

$$\int y^2 \, dy = \int 2x \, dx \qquad \text{or} \qquad \tfrac{1}{3}y^3 = x^2 + C$$

From the initial condition, (18), it follows that $C = 9$, and so the curve has the equation

$$\tfrac{1}{3}y^3 = x^2 + 9 \qquad \text{or} \qquad y = (3x^2 + 27)^{1/3} \qquad \blacktriangleleft$$

Example 7 (*Mixing Problems*) At time $t = 0$, a tank contains 4 lb of salt dissolved in 100 gal of water. Suppose that brine containing 2 lb of salt per gallon of water is allowed to enter the tank at a rate of 5 gal/min and that the mixed solution is drained from the tank at the same rate. Find the amount of salt in the tank after 10 min.

Solution. Let $y(t)$ be the amount of salt (in pounds) at time t. We are interested in finding $y(10)$, the amount of salt at time $t = 10$. We will begin by finding an expression for dy/dt, the rate of change of the amount of salt in the tank at time t. Clearly,

$$\frac{dy}{dt} = \text{rate in} - \text{rate out} \tag{19}$$

where *rate in* is the rate at which salt enters the tank and *rate out* is the rate at which salt leaves the tank. But

$$\text{rate in} = (2 \text{ lb/gal}) \cdot (5 \text{ gal/min}) = 10 \text{ lb/min}$$

At time t, the mixture contains $y(t)$ lb of salt in 100 gal of water; thus, the concentration of salt at time t is $y(t)/100$ lb/gal and

$$\text{rate out} = \left(\frac{y(t)}{100} \text{ lb/gal} \right) \cdot (5 \text{ gal/min}) = \frac{y(t)}{20} \text{ lb/min}$$

Therefore, (19) can be written as

$$\frac{dy}{dt} = 10 - \frac{y}{20}$$

or

$$\frac{dy}{dt} + \frac{y}{20} = 10 \tag{20}$$

which is a first-order linear differential equation. We also have the initial condition

$$y(0) = 4 \tag{21}$$

since the tank contains 4 lb of salt at time $t = 0$.

Multiplying both sides of (20) by the integrating factor

$$\rho = e^{\int (1/20)\,dt} = e^{t/20}$$

yields

$$\frac{d}{dt}(e^{t/20}y) = 10e^{t/20}$$

so

$$e^{t/20}y = \int 10e^{t/20}\,dt = 200e^{t/20} + C$$

or

$$y(t) = 200 + Ce^{-t/20}$$

From the initial condition, (21), it follows that

$$4 = 200 + C \quad \text{or} \quad C = -196$$

so

$$y(t) = 200 - 196e^{-t/20}$$

Thus, after 10 min ($t = 10$), the amount of salt in the tank is

$$y(10) = 200 - 196e^{-0.5} \approx 81.1 \text{ lb}$$

☐ **EXPONENTIAL GROWTH**   Many quantities increase or decrease with time in proportion to the amount of the quantity present. Some examples are human population, bacteria in a culture, drug concentration in the bloodstream, radioactivity, and the values of certain kinds of investments. We shall show how differential equations can be used to study the growth and decay of such quantities.

7.7.1   DEFINITION.   A quantity is said to have an *exponential growth (decay) model* if at each instant of time its rate of increase (decrease) is proportional to the amount of the quantity present.

Consider a quantity with an exponential growth or decay model and denote by $y(t)$ the amount of the quantity present at time t. We assume that $y(t) > 0$ for all t. Since the rate of change of $y(t)$ is proportional to the amount present, it follows that $y(t)$ satisfies

$$\frac{dy}{dt} = ky \tag{22}$$

where k is a constant of proportionality.

The constant k in (22) is called the **growth constant** if $k > 0$ and the **decay constant** if $k < 0$. If $k > 0$, then $dy/dt > 0$, so that y is increasing with time (a growth model). If $k < 0$, then $dy/dt < 0$, so that y is decreasing with time (a decay model).

REMARK. The constant k in (22) is sometimes called the **growth rate** (a negative growth rate meaning that y is decreasing). Strictly speaking, this is not correct since the growth rate of y is not k, but rather $dy/dt \; (= ky)$. Although it is rarely done, it would be more accurate to call k the **relative growth rate**, since it follows from (22) that

$$k = \frac{dy/dt}{y}$$

However, it is so common to call k the growth rate that we shall use this standard terminology in this section. The growth rate k is usually expressed as a percentage; thus, a growth rate of 3% means $k = 0.03$ and a growth rate of -500% means $k = -5$.

If we are given a quantity y with an exponential growth or decay model, and if we know the amount y_0 present at some initial time $t = 0$, then we can find the amount present at any time t by solving the initial-value problem

$$\frac{dy}{dt} = ky, \quad y(0) = y_0 \tag{23}$$

This differential equation is both separable and linear and thus can be solved by either of the methods we have studied in this section. We shall treat it as a linear equation. Rewriting the differential equation as

$$\frac{dy}{dt} - ky = 0$$

and multiplying through by the integrating factor

$$\rho = e^{\int -k\, dt} = e^{-kt}$$

yields

$$\frac{d}{dt}(e^{-kt}y) = 0$$

After integrating,

$$e^{-kt}y = C \quad \text{or} \quad y = Ce^{kt}$$

From the initial condition, $y(0) = y_0$, it follows that $C = y_0$; thus, the solution of (23) is

$$y(t) = y_0e^{kt} \qquad (24)$$

Exponential models have proved useful in studies of population growth. Although populations (e.g., people, bacteria, and flowers) grow in discrete steps, we can apply the results of this section if we are willing to approximate the population graph by a continuous curve $y = y(t)$ (Figure 7.7.1).

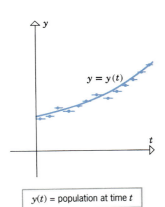

$y(t) = $ population at time t

Figure 7.7.1

Example 8 (*Population Growth*)   According to United Nations data, the world population at the beginning of 1990 was approximately 5.3 billion and growing at a rate of about 2% per year. Assuming an exponential growth model, estimate the world population at the beginning of the year 2015.

Solution.   Let

$t = $ time elapsed from the beginning of 1990 (in years)

$y = $ world population (in billions)

Since the beginning of 1990 corresponds to $t = 0$, it follows from the given data that

$y_0 = y(0) = 5.3$ (billion)

Since the growth rate is 2% ($k = 0.02$), it follows from (24) that the world population at time t will be

$$y(t) = y_0e^{kt} = 5.3e^{0.02t} \qquad (25)$$

Since the beginning of the year 2015 corresponds to an elapsed time of $t = 25$ years ($2015 - 1990 = 25$), it follows from (25) that the world population by the year 2015 will be

$$y(25) = 5.3e^{0.02(25)} = 5.3e^{0.5} \approx 5.3(1.6487) = 8.7381 \text{ (billion)}$$

which is a population of approximately 8.7 billion.   ◄

☐ **DOUBLING AND HALVING TIME**

If a quantity has an exponential growth model, then the time required for it to double in size is called the *doubling time*. Similarly, if a quantity has an exponential decay model, then the time required for it to reduce in value by half is called the *halving time*. As it turns out, doubling and halving times depend only on the growth rate and not on the amount present initially. To see why, suppose y has an exponential growth model so that

$$y = y_0e^{kt} \quad (k > 0)$$

At any fixed time t_1 let

$$y_1 = y_0 e^{kt_1} \tag{26}$$

be the value of y, and let T denote the amount of time required for y to double in size. Thus, at time $t_1 + T$ the value of y will be $2y_1$, so

$$2y_1 = y_0 e^{k(t_1+T)} = y_0 e^{kt_1} e^{kT}$$

or, from (26),

$$2y_1 = y_1 e^{kT}$$

Thus,

$$2 = e^{kT} \quad \text{and} \quad \ln 2 = kT$$

Therefore, the doubling time T is

Doubling Time

$$T = \frac{1}{k} \ln 2 \tag{27}$$

which does not depend on y_0 or t_1. We leave it as an exercise to show that the halving time for a quantity with an exponential decay model ($k < 0$) is

Halving Time

$$T = -\frac{1}{k} \ln 2 \tag{28}$$

Example 9   It follows from (27) that at the current 2% annual growth rate, the doubling time for the world population is

$$T = \frac{1}{0.02} \ln 2 \approx \frac{1}{0.02} (0.6931) = 34.655$$

or approximately 35 years. Thus, with a continued 2% annual growth rate the population of 5.3 billion in 1990 will double to 10.6 billion by the year 2025 and will double again to 21.2 billion by 2060.   ◀

Radioactive elements continually undergo a process of disintegration called *radioactive decay.* It is a physical fact that at each instant of time the rate of decay is proportional to the amount of the element present. Consequently, the amount of any radioactive element has an exponential decay model. For radioactive elements, halving time is called *half-life.*

Example 10 (*Radioactive Decay*)   The radioactive element carbon-14 has a half-life of 5750 years. If 100 grams of this element are present initially, how much will be left after 1000 years?

Solution. From (28) the decay constant is

$$k = -\frac{1}{T}\ln 2 \approx -\frac{1}{5750}(0.6931) \approx -0.00012$$

Thus, if we take $t = 0$ to be the present time, then $y_0 = y(0) = 100$, so that (24) implies that the amount of carbon-14 after 1000 years will be

$$y(1000) = 100e^{-0.00012(1000)} = 100e^{-0.12} \approx 100(0.88692) = 88.692$$

Thus, about 88.69 grams of carbon-14 will remain. ◀

▶ Exercise Set 7.7   Ⓒ 27, 29–32, 34–41, 46

In Exercises 1–6, solve the given separable differential equation. Where convenient, express the solution explicitly as a function of x.

1. $\dfrac{dy}{dx} = \dfrac{y}{x}$.

2. $\dfrac{dy}{dx} = \dfrac{x^3}{(1 + x^4)y}$.

3. $\sqrt{1 + x^2}\, y' + x(1 + y) = 0$.

4. $3\tan y - \dfrac{dy}{dx}\sec x = 0$.

5. $e^{-y}\sin x - y'\cos^2 x = 0$.

6. $\dfrac{dy}{dx} = 1 - y + x^2 - yx^2$.

In Exercises 7–12, solve the given first-order linear differential equation. Where convenient, express the solution explicitly as a function of x.

7. $\dfrac{dy}{dx} + 3y = e^{-2x}$.

8. $\dfrac{dy}{dx} - \dfrac{5}{x}y = x$ $(x > 0)$.

9. $y' + y = \cos(e^x)$.   **10.** $2\dfrac{dy}{dx} + 4y = 1$.

11. $x^2 y' + 3xy + 2x^5 = 0$ $(x > 0)$.

12. $\dfrac{dy}{dx} + y - \dfrac{1}{1 + e^x} = 0$.

In Exercises 13–18, solve the initial-value problems.

13. $\dfrac{dy}{dx} - xy = x$, $y(0) = 3$.

14. $2y\dfrac{dy}{dx} = 3x^2(x^3 + 1)^{-1/2}$, $y(2) = 1$.

15. $\dfrac{dy}{dt} + y = 2$, $y(0) = 1$.

16. $y' - xe^y = 2e^y$, $y(0) = 0$.

17. $y^2 t\dfrac{dy}{dt} - t + 1 = 0$, $y(1) = 3$ $(t > 0)$.

18. $y'\cosh x + y\sinh x = \cosh^2 x$, $y(0) = \frac{1}{4}$.

In Exercises 19–22, find an equation of the curve in the xy-plane that passes through the given point and whose tangent at (x, y) has the given slope.

19. $(1, -1)$; slope $= \dfrac{y^2}{3\sqrt{x}}$.

20. $(1, 1)$; slope $= \dfrac{3x^2}{2y}$.   **21.** $(2, 0)$; slope $= xe^y$.

22. $(1, -1)$; slope $= 2y + 3, y > -\frac{3}{2}$.

The following discussion is needed for Exercises 23 and 24. Suppose that a tank containing a liquid is vented to the air at the top and has an outlet at the bottom through which the liquid can drain. It follows from *Torricelli's law* in physics that if the outlet is opened at time $t = 0$, then at each instant the depth of the liquid $h(t)$ and the area $A(h)$ of the liquid's surface are related by

$$A(h)\frac{dh}{dt} = -k\sqrt{h}$$

where k is a positive constant that depends on such factors as the viscosity of the liquid and the cross-sectional area of the outlet. Use this result in Exercises 23 and 24, assuming that h is in feet, $A(h)$ is in square feet, and t is in seconds. A calculator will be useful.

23. Suppose that the cylindrical tank in Figure 7.7.2 is filled to a depth of 4 ft at time $t = 0$ and that the constant in Torricelli's law is $k = 0.025$.
  (a) Find $h(t)$.
  (b) How many minutes will it take for the tank to drain completely?

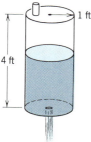

Figure 7.7.2

24. Follow the directions of Exercise 23 for the cylindrical tank in Figure 7.7.3, assuming that the tank is filled to a depth of 4 ft at time $t = 0$ and that the constant in Torricelli's law is $k = 0.025$.

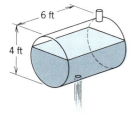

Figure 7.7.3

25. A particle moving along the x-axis encounters a resisting force that results in an acceleration of $a = dv/dt = -0.04v^2$. Given that $x = 0$ cm and $v = 50$ cm/sec at time $t = 0$, find the velocity v and position x as a function of t for $t \geq 0$.

26. A particle moving along the x-axis encounters a resisting force that results in an acceleration of $a = dv/dt = -0.02\sqrt{v}$. Given that $x = 0$ cm and $v = 9$ cm/sec at time $t = 0$, find the velocity v and position x as a function of t for $t \geq 0$.

27. A rocket, fired upward from rest at time $t = 0$, has an initial mass of m_0 (including its fuel). Assuming that the fuel is consumed at a constant rate k, the mass m of the rocket, while fuel is being burned, will be given by $m = m_0 - kt$. It can be shown that if air resistance is neglected and the fuel gases are expelled at a constant speed c relative to the rocket,

then the velocity v of the rocket will satisfy the equation

$$m \frac{dv}{dt} = ck - mg$$

where g is the acceleration due to gravity.
  (a) Find $v(t)$ keeping in mind that the mass m is a function of t.
  (b) Suppose that the fuel accounts for 80% of the initial mass of the rocket and that all of the fuel is consumed in 100 sec. Find the velocity of the rocket in meters/second at the instant the fuel is exhausted. (Use $g = 9.8$ m/sec² and $c = 2500$ m/sec.)

28. An object of mass m is dropped from rest at time $t = 0$. If we assume that the only forces acting on the object are the constant force due to gravity and a retarding force of air resistance, which is proportional to the velocity $v(t)$ of the object, then the velocity will satisfy the equation

$$m \frac{dv}{dt} = mg - kv$$

where k is a positive constant and g is the acceleration due to gravity.
  (a) Find $v(t)$.
  (b) Use the result in part (a) to find $\lim_{t \to +\infty} v(t)$.
  (c) Find the distance $x(t)$ that the object has fallen at time t given that $x = 0$ when $t = 0$.

29. An object of constant mass is projected upward from the surface of the earth. Neglecting air resistance, the gravitational force exerted by the earth on the object results in an acceleration of

$$a = dv/dt = -gR^2/x^2$$

where x is the distance of the object from the center of the earth, R is the radius of the earth, and g is the acceleration due to gravity at the surface of the earth. From the chain rule $dv/dt = v \, dv/dx$, so v and x must satisfy the equation

$$v \, dv/dx = -gR^2/x^2$$

  (a) Find v^2 in terms of x given that $v = v_0$ when $x = R$.
  (b) Use the result in part (a) to show that the velocity cannot reach zero if $v_0 \geq \sqrt{2gR}$.
  (c) The minimum value of v_0 that is required for the object not to fall back to the earth is called the **escape velocity**. Assuming that $g = 32$ ft/sec² and $R = 3960$ mi, show that the escape velocity is approximately 6.9 mi/sec.

30. A bullet of mass m, fired straight up with an initial velocity of v_0, is slowed by the force of gravity and a drag force of air resistance kv^2, where g is the constant acceleration due to gravity and k is a positive constant. As the bullet moves upward, its velocity v satisfies the equation

$$m\frac{dv}{dt} = -(kv^2 + mg)$$

(a) Show that if x is the position of the bullet at time t (Figure 7.7.4), then

$$mv\frac{dv}{dx} = -(kv^2 + mg)$$

(b) Express x in terms of v given that $x = 0$ when $v = v_0$.

(c) Assuming that $v_0 = 988$ m/sec, $g = 9.8$ m/sec^2, $m = 3.56 \times 10^{-3}$ kg, and $k = 7.3 \times 10^{-6}$, use the result in part (b) to find out how high the bullet rises. [*Hint:* Find the velocity of the bullet at its highest point.]

Figure 7.7.4

31. At time $t = 0$, a tank contains 25 lb of salt dissolved in 50 gal of water. Then brine containing 4 lb of salt per gallon of water is allowed to enter the tank at a rate of 2 gal/min and the mixed solution is drained from the tank at the same rate.
    (a) How much salt is in the tank at an arbitrary time t?
    (b) How much salt is in the tank after 25 min?

32. A tank initially contains 200 gal of pure water. Then at time $t = 0$ brine containing 5 lb of salt per gallon of water is allowed to enter the tank at a rate of 10 gal/min and the mixed solution is drained from the tank at the same rate.
    (a) How much salt is in the tank at an arbitrary time t?
    (b) How much salt is in the tank after 30 min?

33. A tank with a 1000-gal capacity initially contains 500 gal of brine containing 50 lb of salt. At time $t = 0$, pure water is added at a rate of 20 gal/min and the mixed solution is drained off at a rate of 10 gal/min. How much salt is in the tank when it reaches the point of overflowing?

34. The number of bacteria in a certain culture grows exponentially at a rate of 1% per hour. Assuming that 10,000 bacteria are present initially, find
    (a) the number of bacteria present at any time t
    (b) the number of bacteria present after 5 hr
    (c) the time required for the number of bacteria to reach 45,000.

35. Polonium-210 is a radioactive element with a half-life of 140 days. Assume that a sample weighs 10 mg initially.
    (a) Find a formula for the amount that will remain after t days.
    (b) How much will remain after 10 weeks?

36. In a certain chemical reaction a substance decomposes at a rate proportional to the amount present. Tests show that under appropriate conditions 15,000 grams will reduce to 5000 grams in 10 hours.
    (a) Find a formula for the amount that will remain from a 15,000-gram sample after t hours.
    (b) How long will it take for 50% of an initial sample of y_0 grams to decompose?

37. One hundred fruit flies are placed in a breeding container that can support a population of at most 5000 flies. If the population grows exponentially at a rate of 2% per day, how long will it take for the container to reach capacity?

38. In 1960 the American scientist W. F. Libby won the Nobel prize for his discovery of carbon dating, a method for determining the age of certain fossils. Carbon dating is based on the fact that nitrogen is converted to radioactive carbon-14 by cosmic radiation in the upper atmosphere. This radioactive carbon is absorbed by plant and animal tissue through the life processes while the plant or animal lives. However, when the plant or animal dies the absorp-

tion process stops and the amount of carbon-14 decreases through radioactive decay. Suppose that tests on a fossil show that 70% of its carbon-14 has decayed. Estimate the age of the fossil, assuming a half-life of 5750 years for carbon-14.

39. Forty percent of a radioactive substance decays in 5 years. Find the half-life of the substance.

40. Assume that if the temperature is constant, then the atmospheric pressure p varies with the altitude h (above sea level) in such a way that $dp/dh = kp$, where k is a constant.
(a) Find a formula for p in terms of k, h, and the atmospheric pressure p_0 at sea level.
(b) Given that p measures 15 lb/in² at sea level and 12 lb/in² at 5000 ft above sea level, find the pressure at 10,000 ft (assuming temperature is constant).

41. The town of Grayrock had a population of 10,000 in 1980 and 12,000 in 1990.
(a) Assuming an exponential growth model, estimate the population in 2000.
(b) What is the doubling time for the town's population?

42. Prove: If a quantity A has an exponential growth or decay model and A has values A_1 and A_2 at times t_1 and t_2, respectively, then the growth rate k is

$$k = \frac{1}{t_1 - t_2} \ln\left(\frac{A_1}{A_2}\right)$$

43. Newton's law of cooling states that the rate at which an object cools is proportional to the difference in temperature between the object and the surrounding medium. Show that if C is the constant temperature of a surrounding medium, then the temperature $T(t)$ at time t of a cooling object is given by

$$T(t) = (T_0 - C)e^{kt} + C$$

where T_0 is the temperature of the object at $t = 0$ and k is a negative constant.

44. A liquid with an initial temperature of 200° is enclosed in a metal container that is held at a constant temperature of 80°. If the liquid cools to 120° in 30 min, what will the temperature be after 1 hr? (Use the result of Exercise 43.)

45. Suppose P dollars is invested at an annual interest rate of $r \times 100\%$. If the accumulated interest is cred-

ited to the account at the end of the year, then the interest is said to be *compounded annually;* if it is credited at the end of each six-month period, then it is said to be *compounded semiannually;* and if it is credited at the end of each three-month period, then it is said to be *compounded quarterly.* The more frequently the interest is compounded, the better it is for the investor since more of the interest is itself earning interest.
(a) Show that if interest is compounded n times a year at equally spaced intervals, then the value A of the investment after t years is

$$A = P\left(1 + \frac{r}{n}\right)^{nt}$$

(b) One can imagine interest to be compounded each day, each hour, each minute, and so forth. Carried to the limit one can conceive of interest compounded at each instant of time; this is called **continuous compounding.** Thus, from part (a), the value A of P dollars after t years when invested at an annual rate of $r \times 100\%$, compounded continuously, is

$$A = \lim_{n \to +\infty} P\left(1 + \frac{r}{n}\right)^{nt}$$

Use the fact that $\lim_{x \to 0} (1 + x)^{1/x} = e$ to prove that $A = Pe^{rt}$.

(c) Use the result in part (b) to show that money invested at continuous compound interest increases at a rate proportional to the amount present.

46. (a) If $1000 is invested at 8% per year compounded continuously (Exercise 45), what will the investment be worth after 5 years?
(b) If it is desired that an investment at 8% per year compounded continuously should have a value of $10,000 after 10 years, how much should be invested now?
(c) How long does it take for an investment at 8% per year compounded continuously to double in value?

47. Derive Formula (28) for halving time.

48. Let a quantity have an exponential growth model with growth rate k. How long does it take for the quantity to triple in size?

▶ SUPPLEMENTARY EXERCISES $\boxed{C}$ 69, 77

1. In each part determine whether f and g are inverse functions.
   (a) $f(x) = mx$ $g(x) = 1/(mx)$
   (b) $f(x) = 3/(x + 1)$ $g(x) = (3 - x)/x$
   (c) $f(x) = x^3 - 8$ $g(x) = \sqrt[3]{x} + 2$
   (d) $f(x) = x^3 - 1$ $g(x) = \sqrt[3]{x} + 1$
   (e) $f(x) = \sqrt{e^x}$ $g(x) = 2 \ln x$

In Exercises 2–6, find $f^{-1}(x)$ if it exists.

2. $f(x) = 8x^3 - 1$.

3. $f(x) = x^2 - 2x + 1$.

4. $f(x) = x^2 - 2x + 1$, $x \geq 1$.

5. $f(x) = (e^x)^2 + 1$.

6. $f(x) = \exp(x^2) + 1$.

7. Let $f(x) = (ax + b)/(cx + d)$. What conditions on a, b, c, d guarantee that f^{-1} exists? Find $f^{-1}(x)$.

8. Show that $f(x) = (x + 2)/(x - 1)$ is its own inverse.

9. Find the largest open interval containing the origin on which f is one-to-one.
   (a) $f(x) = |2x - 5|$ (b) $f(x) = x^2 + 4x$
   (c) $f(x) = \cos(x - 2\pi/3)$.

In Exercises 10–13, find $f^{-1}(x)$, and then use Formula (8) of Section 7.1 to obtain $(f^{-1})'(x)$. Check your work by differentiating $f^{-1}(x)$ directly.

10. $f(x) = x^3 - 8$.

11. $f(x) = 3/(x + 1)$.

12. $f(x) = mx + b$ $(m \neq 0)$.

13. $f(x) = \sqrt{e^x}$.

14. Prove that the line $y = x$ is the perpendicular bisector of the line segment joining (a, b) and (b, a).

15. If $r = \ln 2$ and $s = \ln 3$, express the following in terms of r and s:
   (a) $\ln(1/12)$ (b) $\ln(9/\sqrt{8})$ (c) $\ln(\sqrt[4]{8/3})$.

16. Simplify:
   (a) $e^{2 - \ln x}$ (b) $\exp(\ln x^2 - 2 \ln y)$
   (c) $\ln[x^3 \exp(-x^2)]$.

17. Solve for x in terms of $\ln 3$ and $\ln 5$:
   (a) $25^x = 3^{1-x}$ (b) $\sinh x = \frac{1}{4} \cosh x$.

18. Express the following as a rational function of x:
   $3 \ln(e^{2x}(e^x)^3) + 2 \exp(\ln 1)$.

19. If $\sinh x = -3/5$, find
   (a) $\cosh x$ (b) $\tanh x$ (c) $\sinh(2x)$.

20. Express each of the following as a power of e:
   (a) 2^e (b) $(\sqrt{2})^\pi$.

In Exercises 21–42, find dy/dx. When appropriate, use implicit or logarithmic differentiation.

21. $y = 1/\sqrt{e^x}$.   22. $y = 1/e^{\sqrt{x}}$.

23. $y = x/\ln x$.   24. $y = e^x \ln(1/x)$.

25. $y = x/e^{\ln x}$.   26. $y = \ln\sqrt{x^2 + 2x}$.

27. $y = \ln(10^x/\sin x)$.   28. $y = \cos(e^{-2x})$.

29. $y = e^{\tan x} e^{4 \ln x}$.   30. $y = \ln \left| \dfrac{a + x}{a - x} \right|$.

31. $y = \ln|x + \sqrt{x^2 + a^2}|$.

32. $y = \ln|\tan 3x + \sec 3x|$.

33. $y = [\exp(x^2)]^3$.

34. $y = \ln(x^3/\sqrt{5 + \sin x})$.

35. $y = \sqrt{\ln(\sqrt{x})}$.   36. $y = e^{5x} + (5x)^e$.

37. $y = \pi^x x^\pi$.   38. $y = 4(e^x)^3/\sqrt{\exp(5x)}$.

39. $y = \sinh[\tanh(5x)]$.

40. $x^4 + e^{xy} - y^2 = 20$.

41. $y = e^{3x}(1 + e^{-x})^2$.   42. $y = (\cosh x)^{x^3}$.

43. Show that the function $y = e^{ax} \sin bx$ satisfies the equation $y'' - 2ay' + (a^2 + b^2)y = 0$ for any real constants a and b.

44. Assuming that u and v are differentiable functions of x, use logarithmic differentiation to derive a formula for $d(u^v)/dx$ in terms of u, v, du/dx, and dv/dx. What formula results when the base u is constant? When the exponent v is constant?

45. Find dy if
   (a) $y = e^{-x}$ (b) $y = \ln(1 + x)$ (c) $y = 2^{x^2}$.

46. If $y = 3^{2x}5^{7x}$, show that dy/dx is proportional to y.

47. Use the chain rule and the Second Fundamental Theorem of Calculus to find the derivative.
   (a) $\dfrac{d}{dx}\left(\displaystyle\int_0^{\ln x} \dfrac{dt}{\sqrt{4 + e^t}} \right)$

   (b) $\dfrac{d}{dx}\left(\displaystyle\int_1^{e^{5x}} \sqrt{\ln u + u}\; du \right)$.

   [*Hint*: See Exercise 44, Section 5.9.]

48. Suppose $F(-\pi) = 0$ and that $y = F(x)$ satisfies

$$dy/dx = (3 \sin^2 2x + 2 \cos^2 3x)^{1/2}$$

Express $F(x)$ as an integral.

49. If $y = Ce^{kt}$ (C, k constant) and $Y = \ln y$, show that the graph of Y versus t is a straight line.

50. Evaluate $\int e^{2x}(4 + e^{2x}) \, dx$ two ways: (a) using the substitution $u = 4 + e^{2x}$, and (b) expanding the integrand. Verify that the antiderivatives in parts (a) and (b) differ by a constant.

In Exercises 51–66, evaluate the indicated integral.

51. $\displaystyle \int \frac{e^x}{1 + e^x} \, dx.$

52. $\displaystyle \int \frac{1 + e^x}{e^x} \, dx.$

53. $\displaystyle \int x^e \, dx.$

54. $\displaystyle \int \frac{x^2}{5 - 2x^3} \, dx.$

55. $\displaystyle \int \frac{4x^2 - 3x}{x^3} \, dx.$

56. $\displaystyle \int \frac{(\ln x^2)^2}{x} \, dx.$

57. $\displaystyle \int \frac{\sec x \tan x}{2 \sec x - 1} \, dx.$

58. $\displaystyle \int \frac{e^{5x}}{3 + e^{5x}} \, dx.$

59. $\displaystyle \int (\cos 2x) \exp(\sin 2x) \, dx.$

60. $\displaystyle \int \tanh(3x + 1) \, dx.$

61. $\displaystyle \int \operatorname{sech}^2 x \tanh x \, dx.$

62. $\displaystyle \int e^{2x}(4 + e^{-3x}) \, dx.$

63. $\displaystyle \int_e^{e^2} \frac{dx}{x \ln x}.$

64. $\displaystyle \int_0^1 \frac{dx}{\sqrt{e^x}}.$

65. $\displaystyle \int_0^{\pi/4} \frac{2 \tan x}{\cos^2 x} \, dx.$

66. $\displaystyle \int_1^4 \frac{dx}{\sqrt{x} e^{\sqrt{x}}}.$

67. Show that $\int e^{kx} \, dx = (e^{kx}/k) + C$ for any nonzero constant k.

68. In each part find the indicated limit by interpreting the expression as a derivative and evaluating the derivative.

(a) $\displaystyle \lim_{h \to 0} \frac{10^h - 1}{h}$

(b) $\displaystyle \lim_{h \to 0} \frac{e^{(3+h)^2} - e^9}{h}$

(c) $\displaystyle \lim_{h \to 0} \frac{\ln(e^2 + h) - 2}{h}$

(d) $\displaystyle \lim_{x \to 1} \frac{2^x - 2}{x - 1}.$

69. Sketch $y = x^3 e^{-x}$, showing all relative extrema and inflection points.

70. Use the second derivative test to determine the nature of the critical points of $f(x) = x^2 e^{-x}$.

71. Show that if $y = f(x)$ satisfies $y' = e^y + 2y + x$, then every critical point of y is a relative minimum.

72. Sketch the curve $y = e^{-x/2} \sin 2x$ on $[-\pi/2, 3\pi/2]$. Show all x-intercepts and points where $y = \pm e^{-x/2}$.

73. Let A denote the area under the curve $y = e^{-2x}$ over the interval $[0, b]$. Express A as a function of b, and then find the value of b for which $A = \frac{1}{4}$. What is the limiting value of A as $b \to +\infty$?

74. Let R be the region bounded by the curve $y = 4(2x - 1)^{-1/2}$, the x-axis, and the lines $x = 1$ and $x = 3$. Find the volume of the solid generated by revolving R about the x-axis.

75. Find the surface area generated when the curve $y = \cosh x$, $0 \le x \le 1$, is revolved about the x-axis. [*Hint:* Equation (7b) of Section 7.6 will help to evaluate the integral.]

76. Suppose that a crystal dissolves at a rate proportional to the amount *un*dissolved. If 9 g are undissolved initially and 6 g remain undissolved after 1 min, how many grams remain undissolved after 3 min?

77. The population of the United States was 205 million in 1970. Assuming an annual growth rate of 1.8%, find (a) the population in the year 2000 and (b) the year in which the population will reach 1 billion.

8

Inverse Trigonometric and Hyperbolic Functions

Isaac Barrow   (1630-1677)

■ 8.1 INVERSE TRIGONOMETRIC FUNCTIONS

Because the six basic trigonometric functions are periodic, each of their values is repeated infinitely many times. As a result, none of these functions are one-to-one and consequently none of them have inverses. In this section we shall impose restrictions on the domains of the trigonometric functions that yield one-to-one functions, called the "inverse trigonometric functions." We shall develop the algebraic properties of these functions in this section and in the next section we shall obtain their derivatives and discuss some of their applications to calculus.

□ **INVERSE SINE**

Figure 8.1.1 shows the graph of $y = \sin x$ with the segment of the graph over the interval $[-\pi/2, \pi/2]$ emphasized in solid color. This emphasized segment is the graph of a function f that has the same values as $\sin x$, but whose domain is the interval $[-\pi/2, \pi/2]$. Since no horizontal line cuts the graph of f more than once, this function has an inverse, which we call the ***inverse sine*** function and denote by $\sin^{-1}$. Because f has domain $[-\pi/2, \pi/2]$ and range $[-1, 1]$, the inverse sine function has domain $[-1, 1]$ and range $[-\pi/2, \pi/2]$. These ideas are summarized in the following definition.

8.1.1   DEFINITION.   For each x in the interval $[-1, 1]$ we define $\sin^{-1} x$ to be that number y in the interval $[-\pi/2, \pi/2]$ such that $\sin y = x$; that is, for the stated restrictions on x and y the equations

$$y = \sin^{-1} x \quad \text{and} \quad \sin y = x$$

are equivalent.

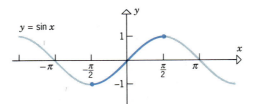

Figure 8.1.1

REMARK.   The symbol $\sin^{-1} x$ is *never* used to denote $1/\sin x$. If desired, $1/\sin x$ can be rewritten as $(\sin x)^{-1}$ or $\csc x$. In the older literature, $\sin^{-1} x$ is called the ***arcsine*** function and is denoted by $\arcsin x$. We shall not use this terminology or notation.

If we think of y as an angle in radian measure, then the algebraic requirement in Definition 8.1.1 that y lie in the interval $[-\pi/2, \pi/2]$ imposes the geometric requirement that $\sin^{-1} x$ be an angle that terminates in the first or fourth quadrant or on an axis adjacent to those quadrants. Thus, from a geometric viewpoint, $\sin^{-1} x$ is an angle whose sine is x and which terminates in the first or fourth quadrant (or on an adjacent axis).

Example 1 Find

(a) $\sin^{-1}(\tfrac{1}{2})$ (b) $\sin^{-1}(-1/\sqrt{2})$ (c) $\sin^{-1}(-1)$

Solution (a). Let $y = \sin^{-1}(\tfrac{1}{2})$. From Definition 8.1.1 this equation is equivalent to

$$\sin y = \tfrac{1}{2}, \quad -\pi/2 \le y \le \pi/2$$

Viewing y as an angle in radian measure, it follows that y terminates in the first quadrant since $\sin y > 0$. Thus, we are looking for an angle terminating in the first quadrant whose sine is $1/2$. The desired angle is $y = \pi/6$, so $\sin^{-1}(\tfrac{1}{2}) = \pi/6$.

Solution (b). Let $y = \sin^{-1}(-1/\sqrt{2})$. This is equivalent to

$$\sin y = -1/\sqrt{2}, \quad -\pi/2 \le y \le \pi/2$$

Viewing y as an angle in radian measure, it follows that y terminates in the fourth quadrant since $\sin y < 0$. Thus, we are looking for an angle terminating in the fourth quadrant whose sine is $-1/\sqrt{2}$. The desired angle is $y = -\pi/4$, so $\sin^{-1}(-1/\sqrt{2}) = -\pi/4$.

Solution (c). Let $y = \sin^{-1}(-1)$. This is equivalent to

$$\sin y = -1$$

from which it follows that $y = -\pi/2$. Thus, $\sin^{-1}(-1) = -\pi/2$. ◀

Because $\sin^{-1} x$ is the inverse of $\sin x$ restricted to the interval $[-\pi/2, \pi/2]$, it follows that

$$\sin^{-1}(\sin x) = x \quad \text{if} \quad -\frac{\pi}{2} \le x \le \frac{\pi}{2} \tag{1a}$$

$$\sin(\sin^{-1} x) = x \quad \text{if} \quad -1 \le x \le 1 \tag{1b}$$

Moreover, the graph of $y = \sin^{-1} x$ can be obtained by reflecting the graph of $y = \sin x$, $-\pi/2 \le x \le \pi/2$ (which is the solid color curve in Figure 8.1.1), about the line $y = x$ (Figure 8.1.2). Observe that $\sin^{-1} x$ is a continuous function since it is the inverse of a continuous function (Theorem 7.1.7).

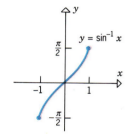

Figure 8.1.2

☐ **INVERSE COSINE**

The *inverse cosine* function, denoted by $\cos^{-1}$, is defined to be the inverse of the restricted cosine function

$$f(x) = \cos x, \quad 0 \leq x \leq \pi$$

whose graph is shown in solid color in Figure 8.1.3. Because f has domain $[0, \pi]$ and range $[-1, 1]$, the inverse cosine function has domain $[-1, 1]$ and range $[0, \pi]$. These ideas are summarized in the following definition.

8.1.2 DEFINITION. For each x in the interval $[-1, 1]$ we define $\cos^{-1} x$ to be that number y in the interval $[0, \pi]$ such that $\cos y = x$; that is, for the stated restrictions on x and y, the equations

$$y = \cos^{-1} x \quad \text{and} \quad \cos y = x$$

are equivalent.

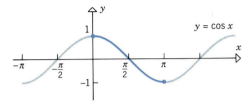

Figure 8.1.3

If we think of y as an angle in radian measure, then the algebraic requirement in Definition 8.1.2 that y lie in the interval $[0, \pi]$ imposes the geometric requirement that y be an angle that terminates in the first or second quadrant or on an axis adjacent to those quadrants. Thus, from a geometric viewpoint, $\cos^{-1} x$ is an angle whose cosine is x and which terminates in the first or second quadrant or on an adjacent axis.

Example 2 Find $\cos^{-1}(-\sqrt{3}/2)$.

Solution. Let $y = \cos^{-1}(-\sqrt{3}/2)$. This is equivalent to

$$\cos y = -\sqrt{3}/2$$

Viewing y as an angle in radian measure, it follows that y terminates in the second quadrant since $\cos y < 0$. Thus, we are looking for an angle terminating in the second quadrant whose cosine is $-\sqrt{3}/2$. The desired angle is $y = 5\pi/6$, so $\cos^{-1}(-\sqrt{3}/2) = 5\pi/6$. ◀

Because $\cos^{-1} x$ is the inverse of $\cos x$ restricted to the interval $[0, \pi]$, it follows that

$$\cos^{-1}(\cos x) = x \quad \text{if} \quad 0 \leq x \leq \pi \tag{2a}$$

$$\cos(\cos^{-1} x) = x \quad \text{if} \quad -1 \leq x \leq 1 \tag{2b}$$

Moreover, the graph of $y = \cos^{-1} x$ can be obtained by reflecting the graph of $y = \cos x$, $0 \le x \le \pi$ (which is the solid color curve in Figure 8.1.3), about the line $y = x$ (Figure 8.1.4). Observe that $\cos^{-1} x$ is a continuous function because it is the inverse of a continuous function.

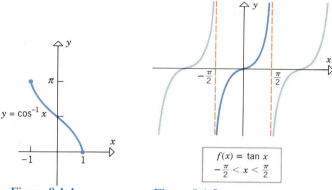

Figure 8.1.4   Figure 8.1.5

□ **INVERSE TANGENT**

The *inverse tangent* function, denoted by $\tan^{-1}$, is defined to be the inverse of the restricted tangent function

$$f(x) = \tan x, \quad -\frac{\pi}{2} < x < \frac{\pi}{2}$$

whose graph is shown in solid color in Figure 8.1.5. Because f has domain $(-\pi/2, \pi/2)$ and range $(-\infty, +\infty)$, the inverse tangent function has domain $(-\infty, +\infty)$ and range $(-\pi/2, \pi/2)$. These ideas are summarized in the following definition.

8.1.3 DEFINITION. For each x in the interval $(-\infty, +\infty)$ we define $\tan^{-1} x$ to be that number y in the interval $(-\pi/2, \pi/2)$ such that $\tan y = x$; that is, for the stated restrictions on x and y, the equations

$$y = \tan^{-1} x \quad \text{and} \quad \tan y = x$$

are equivalent.

If we think of y as an angle in radian measure, then the algebraic requirement in Definition 8.1.3 that y lie in the interval $(-\pi/2, \pi/2)$ imposes the geometric requirement that y be an angle that terminates in the first or fourth quadrant or on the positive x-axis. Thus, from a geometric viewpoint, $\tan^{-1} x$ is an angle whose tangent is x and which terminates in the first or fourth quadrant or on the positive x-axis.

Because $\tan^{-1} x$ is the inverse of $\tan x$ restricted to the interval $(-\pi/2, \pi/2)$, it follows that

$$\tan^{-1} (\tan x) = x \quad \text{if} \quad -\pi/2 < x < \pi/2 \qquad (3a)$$
$$\tan (\tan^{-1} x) = x \quad \text{if} \quad -\infty < x < +\infty \qquad (3b)$$

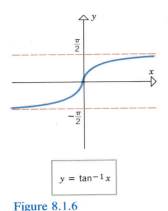

$y = \tan^{-1} x$

Figure 8.1.6

Moreover, the graph of $y = \tan^{-1} x$ can be obtained by reflecting the graph of $y = \tan x$, $-\pi/2 < x < \pi/2$ (which is the solid color curve in Figure 8.1.5), about the line $y = x$ (Figure 8.1.6). Observe that $\tan^{-1} x$ is a continuous function because it is the inverse of a continuous function.

As evidenced by Figure 8.1.6, the graph of $y = \tan^{-1} x$ has two asymptotes: $y = \pi/2$ as x approaches $+\infty$ and $y = -\pi/2$ as x approaches $-\infty$. Expressing these observations as limits we obtain

$$\lim_{x \to +\infty} \tan^{-1} x = \pi/2 \tag{4}$$

$$\lim_{x \to -\infty} \tan^{-1} x = -\pi/2 \tag{5}$$

☐ **INVERSE SECANT**

The *inverse secant* function, denoted by $\sec^{-1}$, is defined to be the inverse of the restricted secant function

$$f(x) = \sec x, \quad 0 \le x < \pi/2 \quad \text{or} \quad \pi \le x < 3\pi/2$$

whose graph is shown in solid color in Figure 8.1.7. Because f has domain $[0, \pi/2) \cup [\pi, 3\pi/2)$ and range $(-\infty, -1] \cup [1, +\infty)$, the inverse secant function has domain $(-\infty, -1] \cup [1, +\infty)$ and range $[0, \pi/2) \cup [\pi, 3\pi/2)$. These ideas are summarized in the following definition.

8.1.4 DEFINITION. For each x in the set $(-\infty, -1] \cup [1, +\infty)$, we define $\sec^{-1} x$ to be that number y in the set $[0, \pi/2) \cup [\pi, 3\pi/2)$ such that $\sec y = x$; that is, for the stated restrictions on x and y, the equations

$$y = \sec^{-1} x \quad \text{and} \quad \sec y = x$$

are equivalent.

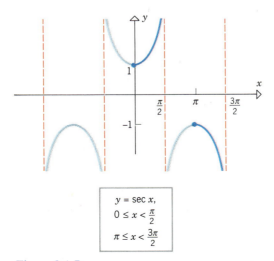

$y = \sec x,$
$0 \le x < \dfrac{\pi}{2}$
$\pi \le x < \dfrac{3\pi}{2}$

Figure 8.1.7

If we think of y as an angle in radian measure, then the algebraic requirement in Definition 8.1.4 that y lie in the set $[0, \pi/2) \cup [\pi, 3\pi/2)$ imposes the geometric requirement that y be an angle that terminates in the first or third quadrant or on the x-axis. Thus, from a geometric viewpoint, $\sec^{-1} x$ is an angle whose secant is x and which terminates in the first or third quadrant or on the x-axis.

REMARK. There is no universal agreement among mathematicians about the definition of $\sec^{-1} x$. For example, some writers define $\sec^{-1} x$ by restricting x so that $0 \leq x < \pi/2$ or $\pi/2 < x \leq \pi$. Although there are some advantages to this definition, our definition will produce simpler derivative and integration formulas in the next section.

Because $\sec^{-1} x$ is the inverse of $\sec x$ restricted to the set $[0, \pi/2) \cup [\pi, 3\pi/2)$, it follows that

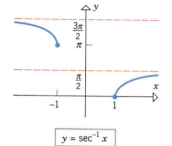

$$\sec^{-1}(\sec x) = x \quad \text{if} \quad 0 \leq x < \pi/2 \quad \text{or} \quad \pi \leq x < 3\pi/2 \tag{6a}$$

$$\sec(\sec^{-1} x) = x \quad \text{if} \quad x \leq -1 \quad \text{or} \quad x \geq 1 \tag{6b}$$

$y = \sec^{-1} x$

Figure 8.1.8

Moreover, the graph of $y = \sec^{-1} x$ can be obtained by reflecting the graph of $y = \sec x$, $0 \leq x < \pi/2$ or $\pi \leq x < 3\pi/2$ (which is the solid color curve in Figure 8.1.7), about the line $y = x$ (Figure 8.1.8).

☐ **INVERSE COTANGENT AND COSECANT**

The *inverse cotangent* and *inverse cosecant* functions are of lesser importance, so we shall summarize their properties briefly in Table 8.1.1.

Table 8.1.1

FUNCTION	DOMAIN (x VALUE)	RANGE (y VALUE)	EQUIVALENT STATEMENTS
$\cot^{-1} x$	$(-\infty, +\infty)$	$(0, \pi)$	$y = \cot^{-1} x$ and $x = \cot y$
$\csc^{-1} x$	$(-\infty, -1] \cup [1, +\infty)$	$(-\pi, -\pi/2] \cup (0, \pi/2]$	$y = \csc^{-1} x$ and $x = \csc y$

We leave it for the reader to show that the graphs of $y = \cot^{-1} x$ and $y = \csc^{-1} x$ are as shown in Figures 8.1.9 and 8.1.10.

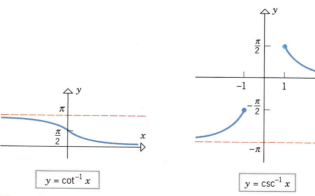

$y = \cot^{-1} x$

Figure 8.1.9

$y = \csc^{-1} x$

Figure 8.1.10

☐ **IDENTITIES INVOLVING INVERSE TRIGONOMETRIC FUNCTIONS**

In the definitions of the six inverse trigonometric functions, the restrictions on the domains are designed not only to produce one-to-one functions, but also to ensure that certain "natural" identities hold for the inverse trigonometric functions. For example, if α and β are acute complementary angles, then from basic trigonometry, $\sin \alpha$ and $\cos \beta$ are equal (Figure 8.1.11). Let us write $x = \sin \alpha = \cos \beta$ so that

$$\alpha = \sin^{-1} x \quad \text{and} \quad \beta = \cos^{-1} x$$

Since $\alpha + \beta = \pi/2$ we obtain the identity

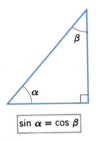

$$\sin^{-1} x + \cos^{-1} x = \frac{\pi}{2} \tag{7}$$

sin α = cos β

Figure 8.1.11

(Because α and β were assumed to be nonnegative acute angles, this derivation is only valid for $0 \le x \le 1$; for a derivation that is valid for all x in $[-1, 1]$ see Exercise 25.) Similarly, we can obtain the identities

$$\tan^{-1} x + \cot^{-1} x = \frac{\pi}{2} \tag{8}$$

$$\sec^{-1} x + \csc^{-1} x = \frac{\pi}{2} \tag{9}$$

☐ **SIMPLIFYING EXPRESSIONS INVOLVING INVERSE TRIGONOMETRIC FUNCTIONS**

In the next section we will encounter functions that are compositions of trigonometric and inverse trigonometric functions. The following examples illustrate techniques for simplifying such functions.

Example 3   Simplify the function $\cos (\sin^{-1} x)$.

Solution.   The idea is to express cosine in terms of sine in order to take advantage of the simplification $\sin (\sin^{-1} x) = x$. Thus, we start with the identity

$$\cos^2 y = 1 - \sin^2 y$$

and substitute $y = \sin^{-1} x$ to obtain

$$\cos^2 (\sin^{-1} x) = 1 - \sin^2 (\sin^{-1} x)$$

or on taking square roots

$$\left| \cos (\sin^{-1} x) \right| = \sqrt{1 - \sin^2 (\sin^{-1} x)}$$

or

$$\left| \cos (\sin^{-1} x) \right| = \sqrt{1 - x^2}$$

Since $-\pi/2 \le \sin^{-1} x \le \pi/2$, it follows that $\cos(\sin^{-1} x)$ is nonnegative; thus, we can drop the absolute value sign and write

$$\cos(\sin^{-1} x) = \sqrt{1 - x^2} \tag{10}$$

Alternative Solution. In the case where $0 < \sin^{-1} x < \pi/2$, an alternative geometric solution is possible. We let

$$y = \sin^{-1} x$$

and construct a right triangle with an acute angle of y as shown in Figure 8.1.12. It follows from the definition of y that $\sin y = x$, so we can assume that the side opposite the angle y has length x and the hypotenuse has length 1 since these values yield

$$\sin y = \frac{\text{opposite side}}{\text{hypotenuse}} = \frac{x}{1} = x$$

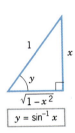

$y = \sin^{-1} x$

Figure 8.1.12

By the Theorem of Pythagoras, the side that is adjacent to the angle y has length $\sqrt{1 - x^2}$, so

$$\cos y = \frac{\text{adjacent side}}{\text{hypotenuse}} = \frac{\sqrt{1 - x^2}}{1} = \sqrt{1 - x^2}$$

or

$$\cos(\sin^{-1} x) = \sqrt{1 - x^2}$$

which agrees with (10). ◄

Example 4 Simplify the function $\sec^2(\tan^{-1} x)$.

Solution. The object is to express secant in terms of tangent to take advantage of the simplification $\tan(\tan^{-1} x) = x$. Thus, if we let $y = \tan^{-1} x$ in the identity

$$\sec^2 y = 1 + \tan^2 y$$

we obtain

$$\sec^2(\tan^{-1} x) = 1 + \tan^2(\tan^{-1} x)$$

or

$$\sec^2(\tan^{-1} x) = 1 + x^2 \tag{11}$$

Alternative Solution. In the case where $0 < \tan^{-1} x < \pi/2$, we can let

$$y = \tan^{-1} x$$

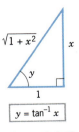

$y = \tan^{-1} x$

Figure 8.1.13

and construct a right triangle with an acute angle of y as shown in Figure 8.1.13. It follows from the definition of y that $\tan y = x$, so we can assume that the side opposite the angle y has length x and the side adjacent to y has length 1 since these values yield

$$\tan y = \frac{\text{opposite side}}{\text{adjacent side}} = \frac{x}{1} = x$$

By the Theorem of Pythagoras the hypotenuse has length $\sqrt{1 + x^2}$, so it follows from the triangle that

$$\sec(\tan^{-1} x) = \sec y = \frac{\text{hypotenuse}}{\text{adjacent side}} = \frac{\sqrt{1 + x^2}}{1} = \sqrt{1 + x^2}$$

or

$$\sec^2(\tan^{-1} x) = 1 + x^2$$

which agrees with (11).   ◀

▶ Exercise Set 8.1   Ⓒ *27, 31–41*

1. Find the exact value of
   (a) $\sin^{-1}(-1)$
   (b) $\cos^{-1}(-1)$
   (c) $\tan^{-1}(-1)$
   (d) $\cot^{-1}(1)$
   (e) $\sec^{-1}(1)$
   (f) $\csc^{-1}(1)$.

2. Find the exact value of
   (a) $\sin^{-1}(\frac{1}{2}\sqrt{3})$
   (b) $\cos^{-1}(\frac{1}{2})$
   (c) $\tan^{-1}(1)$
   (d) $\cot^{-1}(-1)$
   (e) $\sec^{-1}(-2)$
   (f) $\csc^{-1}(-2)$.

3. Given that $\theta = \sin^{-1}(-\frac{1}{2}\sqrt{3})$, find the exact values of $\cos\theta$, $\tan\theta$, $\cot\theta$, $\sec\theta$, and $\csc\theta$.

4. Given that $\theta = \cos^{-1}(\frac{1}{2})$, find the exact values of $\sin\theta$, $\tan\theta$, $\cot\theta$, $\sec\theta$, and $\csc\theta$.

5. Given that $\theta = \tan^{-1}(\frac{3}{4})$, find the exact values of $\sin\theta$, $\cos\theta$, $\cot\theta$, $\sec\theta$, and $\csc\theta$.

6. Make a table that lists the six inverse trigonometric functions together with their domains and ranges.

7. Find the exact value of
   (a) $\sin^{-1}\sin(\pi/7)$
   (b) $\sin^{-1}(\sin\pi)$
   (c) $\sin^{-1}(\sin 5\pi/7)$
   (d) $\sin^{-1}(\sin 630)$.

8. Find the exact value of
   (a) $\cos^{-1}(\cos\pi/7)$
   (b) $\cos^{-1}(\cos\pi)$
   (c) $\cos^{-1}(\cos 12\pi/7)$
   (d) $\cos^{-1}(\cos 200)$.

9. For which values of x is it true that
   (a) $\cos^{-1}(\cos x) = x$
   (b) $\cos(\cos^{-1}x) = x$
   (c) $\tan^{-1}(\tan x) = x$
   (d) $\tan(\tan^{-1}x) = x$
   (e) $\csc^{-1}(\csc x) = x$
   (f) $\csc(\csc^{-1}x) = x$?

In Exercises 10–15, find the exact value of the given quantity.

10. $\sec[\sin^{-1}(-\frac{3}{4})]$.
11. $\sin[2\cos^{-1}(\frac{3}{5})]$.
12. $\tan^{-1}[\sin(-\pi/2)]$.
13. $\sin^{-1}[\cot(\pi/4)]$.
14. $\sin[\sin^{-1}(\frac{2}{3}) + \cos^{-1}(\frac{1}{3})]$.
15. $\tan[2\sec^{-1}(\frac{3}{2})]$.

16. Prove:

$$\tan^{-1}x + \tan^{-1}y = \tan^{-1}\left(\frac{x + y}{1 - xy}\right)$$

provided $-\pi/2 < \tan^{-1}x + \tan^{-1}y < \pi/2$. [*Hint:* Use an identity for $\tan(\alpha + \beta)$.]

17. Use the result in Exercise 16 to show that
   (a) $\tan^{-1}\frac{1}{2} + \tan^{-1}\frac{1}{3} = \pi/4$
   (b) $2\tan^{-1}\frac{1}{3} + \tan^{-1}\frac{1}{7} = \pi/4$.

Following Example 3 in this section, we derived the identity $\cos(\sin^{-1}x) = \sqrt{1 - x^2}$ by considering a right triangle with an acute angle of $\sin^{-1}x$ (see Figure 8.1.12). In Exercises 18 and 19, complete the identity by constructing an appropriate triangle.

18. (a) $\sin(\cos^{-1}x) = ?$ (b) $\tan(\cos^{-1}x) = ?$
   (c) $\csc(\tan^{-1}x) = ?$ (d) $\sin(\tan^{-1}x) = ?$

19. (a) $\cos(\tan^{-1}x) = ?$ (b) $\tan(\cot^{-1}x) = ?$
   (c) $\sin(\sec^{-1}x) = ?$ (d) $\cot(\csc^{-1}x) = ?$

20. Use Figure 8.1.14 to deduce the results in parts (a) and (b).

(a) $\tan^{-1}1 + \tan^{-1}2 + \tan^{-1}3 = \pi$.
[*Hint:* Consider angles A, B, and C.]

(b) $\tan^{-1}1 + \tan^{-1}\frac{1}{2} + \tan^{-1}\frac{1}{3} = \pi/2$.
[*Hint:* Consider angles α, β, and γ.]

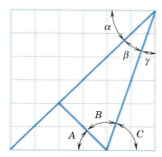

Figure 8.1.14

21. Sketch the graphs of

(a) $y = \sin^{-1}2x$ (b) $y = \tan^{-1}\frac{1}{2}x$.

22. Sketch the graphs of

(a) $y = \cos^{-1}\frac{1}{3}x$ (b) $y = 2\cot^{-1}2x$.

23. Prove:

(a) $\sin^{-1}(-x) = -\sin^{-1}x$

(b) $\tan^{-1}(-x) = -\tan^{-1}x$.

24. Prove:

(a) $\cos^{-1}(-x) = \pi - \cos^{-1}x$

(b) $\sec^{-1}(-x) = \pi + \sec^{-1}x$, if $x \geq 1$.

25. In the text we proved that

$$\sin^{-1}x + \cos^{-1}x = \frac{\pi}{2}$$

for $0 \leq x \leq 1$. Use this result together with Exercises 23(a) and 24(a) to prove this identity for $-1 \leq x \leq 1$.

26. Prove:

(a) $\cot^{-1}x = \tan^{-1}\frac{1}{x}$,   if $x > 0$

(b) $\sec^{-1}x = \cos^{-1}\frac{1}{x}$,   if $x \geq 1$

(c) $\csc^{-1}x = \sin^{-1}\frac{1}{x}$,   if $x \geq 1$.

27. Most scientific calculators have keys for the values of only $\sin^{-1}x$, $\cos^{-1}x$, and $\tan^{-1}x$. The formulas in Exercise 26 show how a calculator can be used to obtain values of $\cot^{-1}x$, $\sec^{-1}x$, and $\csc^{-1}x$ for positive values of x. Use these formulas and a cal-culator to find numerical values for each of the following inverse trigonometric functions. Express your answers in degrees, rounded to the nearest tenth of a degree.

(a) $\cot^{-1}0.7$ (b) $\sec^{-1}1.2$ (c) $\csc^{-1}2.3$.

In Exercises 28–30, solve the given equation for x in terms of k using the corresponding inverse trigono-metric function. Note that x may not be in the range of the inverse function.

28. $\cos x = k$,   if $0 < k < 1$ and $3\pi/2 < x < 2\pi$.

29. $\tan x = k$,   if $k < 0$ and $\pi/2 < x < \pi$.

30. $\sin 2x = k$,   if $0 < k < 1$ and $0 < x < \pi/2$.
[*Hint:* Consider two cases: $0 < 2x < \pi/2$ and $\pi/2 < 2x < \pi$.]

In Exercises 31–36, use a calculator to approximate the solution of the equation. Where radians are used, express your answer to four decimal places, and where degrees are used, express it to the nearest tenth of a degree. [*Note:* In each part, the solution is not in the range of the corresponding inverse trigonometric function.]

31. $\sin x = 0.37$, $\pi/2 < x < \pi$.

32. $\cos x = -0.85$, $\pi < x < 3\pi/2$.

33. $\tan x = 3.16$, $-\pi < x < -\pi/2$.

34. $\sin \theta = -0.61$, $180° < \theta < 270°$.

35. $\cos \theta = 0.23$, $-90° < \theta < 0°$.

36. $\tan \theta = -0.45$, $90° < \theta < 180°$.

37. An Earth-observing satellite has horizon sensors that can measure the angle θ shown in Figure 8.1.15. Let R be the radius of Earth (assumed spherical) and h the distance between the satellite and Earth's surface.

(a) Show that $\sin \theta = \dfrac{R}{R + h}$.

(b) Find θ, to the nearest degree, for a satellite that is 10,000 km from Earth's surface (use $R = 6378$ km).

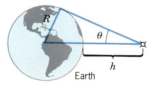

Figure 8.1.15

38. The number of hours of daylight on a given day at a given point on Earth's surface depends on the latitude λ of the point, the angle γ through which Earth has moved in its orbital plane during the time period from the vernal equinox (March 21), and the angle of inclination ι of Earth's axis of rotation measured from ecliptic north ($\iota \approx 23.55°$). The number of hours of daylight h can be approximated by the formula

$$h = \begin{cases} 24, & D \geq 1 \\ 12 + \frac{2}{15} \sin^{-1} D, & |D| < 1 \\ 0, & D \leq -1 \end{cases}$$

where

$$D = \frac{\sin \iota \sin \gamma \tan \lambda}{\sqrt{1 - \sin^2 \iota \sin^2 \gamma}}$$

and $\sin^{-1} D$ is in degree measure. Given that Fairbanks, Alaska, is located at a latitude of $\lambda = 65°$ N and that $\gamma = 90°$ on June 20 and $\gamma = 270°$ on December 20, approximate

(a) the maximum number of daylight hours at Fairbanks to one decimal place;

(b) the minimum number of daylight hours at Fairbanks to one decimal place.

[*Note:* This problem was adapted from *TEAM, A Path to Applied Mathematics*, The Mathematical Association of America, Washington, D.C., 1985.]

39. A soccer player kicks a ball with an initial speed of 14 m/sec at an angle θ with the horizontal (Figure 8.1.16). The ball lands 18 m down the field. If air resistance is neglected, then the ball will have a parabolic trajectory and the horizontal range R will be given by

$$R = \frac{v^2}{g} \sin 2\theta$$

where v is the initial speed of the ball and g is the acceleration due to gravity. Using $g = 9.8$ m/sec^2, approximate two values of θ, to the nearest degree, at which the ball could have been kicked. Which angle results in the shorter time of flight? Why?

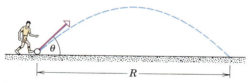

Figure 8.1.16

40. The *law of cosines* states that

$$c^2 = a^2 + b^2 - 2ab \cos \theta$$

where a, b, and c are the lengths of the sides of a triangle and θ is the angle formed by sides a and b. Find θ, to the nearest degree, for the triangle with $a = 2$, $b = 3$, and $c = 4$.

41. An airplane is flying at a constant height of 3000 ft above water at a speed of 400 ft/sec. The pilot is to release a survival package so that it lands in the water at a sighted point P. If air resistance is neglected, then the package will follow a parabolic trajectory whose equation relative to the coordinate system in Figure 8.1.17 is

$$y = 3000 - \frac{g}{2v^2} x^2$$

where g is the acceleration due to gravity and v is the speed of the airplane. Using $g = 32$ ft/sec^2, find the "line of sight" angle θ, to the nearest degree, that will result in the package hitting the target point.

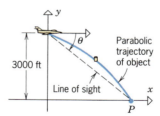

Figure 8.1.17

42. A camera is positioned x feet from the base of a missile launching pad. If a missile of length a feet is launched vertically, show that when the base of the missile is b feet above the camera lens, the angle θ subtended at the lens by the missile is

$$\theta = \cot^{-1} \frac{x}{a + b} - \cot^{-1} \frac{x}{b}$$

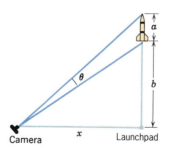

43. In the computer programming language, BASIC, the instruction ATN(X) causes the computer to calculate $\tan^{-1}(X)$. According to the programming manual for a popular microcomputer, this instruction can be used to calculate $\sin^{-1}(X)$ and $\cos^{-1}(X)$ by means of the formulas

(a) $\sin^{-1}(X) = \text{ATN}\left(\dfrac{X}{\sqrt{1 - X^2}}\right)$

(b) $\cos^{-1}(X) = -\text{ATN}\left(\dfrac{X}{\sqrt{1 - X^2}}\right) + 1.5708.$

Derive these results.

■ **8.2   DERIVATIVES AND INTEGRALS INVOLVING INVERSE TRIGONOMETRIC FUNCTIONS**

> *In this section we shall obtain the derivatives of the inverse trigonometric functions and the related integration formulas.*

□ **DERIVATIVE FORMULAS**

8.2.1   THEOREM.

$$\frac{d}{dx}[\sin^{-1}x] = \frac{1}{\sqrt{1 - x^2}} \tag{1}$$

$$\frac{d}{dx}[\cos^{-1}x] = -\frac{1}{\sqrt{1 - x^2}} \tag{2}$$

$$\frac{d}{dx}[\tan^{-1}x] = \frac{1}{1 + x^2} \tag{3}$$

$$\frac{d}{dx}[\cot^{-1}x] = -\frac{1}{1 + x^2} \tag{4}$$

$$\frac{d}{dx}[\sec^{-1}x] = \frac{1}{x\sqrt{x^2 - 1}} \tag{5}$$

$$\frac{d}{dx}[\csc^{-1}x] = -\frac{1}{x\sqrt{x^2 - 1}} \tag{6}$$

Proof (1).   First, we will show that $g(x) = \sin^{-1}x$ is differentiable on the interval $(-1, 1)$, then we will derive Formula (1). The function $g(x) = \sin^{-1}x$ is the inverse of the function

$$f(x) = \sin x, \quad -\frac{\pi}{2} \le x \le \frac{\pi}{2}$$

As x varies over the interval $(-1, 1)$, the value of $g(x) = \sin^{-1}x$ varies over the interval $(-\pi/2, \pi/2)$ on which f is differentiable and has a *nonzero* derivative (verify). Thus, it follows from Theorem 7.1.8 that g is differentiable on the interval $(-1, 1)$.

To obtain (1) let

$$y = \sin^{-1} x$$

which can be rewritten as

$$x = \sin y$$

Differentiating this equation implicitly with respect to x yields

$$\frac{d}{dx}[x] = \frac{d}{dx}[\sin y]$$

$$1 = \cos y \cdot \frac{dy}{dx}$$

$$\frac{dy}{dx} = \frac{1}{\cos y} = \frac{1}{\cos(\sin^{-1} x)} \tag{7}$$

As illustrated in Example 3 of Section 8.1,

$$\cos(\sin^{-1} x) = \sqrt{1 - x^2}$$

Thus, (7) simplifies to

$$\frac{dy}{dx} = \frac{1}{\sqrt{1 - x^2}}$$

which establishes Formula (1).

Proof (3).   As in the proof of (1), the differentiability of $g(x) = \tan^{-1} x$ follows from Theorem 7.1.8. (We omit the details.) To obtain (3), let

$$y = \tan^{-1} x$$

which can be rewritten as

$$x = \tan y$$

Differentiating this equation implicitly with respect to x yields

$$\frac{d}{dx}[x] = \frac{d}{dx}[\tan y]$$

$$1 = \sec^2 y \cdot \frac{dy}{dx}$$

$$\frac{dy}{dx} = \frac{1}{\sec^2 y} = \frac{1}{\sec^2(\tan^{-1} x)} \tag{8}$$

As shown in Example 4 of Section 8.1,

$$\sec^2(\tan^{-1} x) = 1 + x^2$$

Thus, (8) simplifies to

$$\frac{dy}{dx} = \frac{1}{1 + x^2}$$

which establishes Formula (3).

Proof (5).   As in the proof of (1), the differentiability of $g(x) = \sec^{-1} x$ follows from Theorem 7.1.8. (We omit the details.) To obtain (5), let

$$y = \sec^{-1} x$$

which can be rewritten as

$$x = \sec y \qquad\qquad (9)$$

Differentiating this equation implicitly with respect to x yields

$$\frac{d}{dx}[x] = \frac{d}{dx}[\sec y]$$

$$1 = \sec y \tan y \cdot \frac{dy}{dx}$$

$$\frac{dy}{dx} = \frac{1}{\sec y \tan y} = \frac{1}{\sec(\sec^{-1} x)\tan(\sec^{-1} x)} \qquad (10)$$

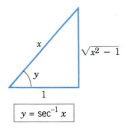

Figure 8.2.1

We can simplify (10) using the methods of Examples 3 and 4 of Section 8.1. We do this now for $x > 1$, leaving the case $x < -1$ for the reader. As illustrated in Figure 8.2.1, construct a right triangle with an acute angle of $y = \sec^{-1} x$. From (9), we can assume that the side adjacent to y has length 1 and that the hypotenuse has length x (>1). It then follows from the Theorem of Pythagoras that the side opposite to y has length $\sqrt{x^2 - 1}$. From the triangle we obtain

$$\sec(\sec^{-1} x) = \sec y = x$$
$$\tan(\sec^{-1} x) = \tan y = \sqrt{x^2 - 1}$$

Thus, (10) simplifies to

$$\frac{dy}{dx} = \frac{1}{x\sqrt{x^2 - 1}}$$

which establishes Formula (5).

Proofs (2), (4), *and* (6).   Formulas (2), (4), and (6) follow from Formulas (1), (3), and (5) above and from identities (7), (8), and (9) in Section 8.1. For example,

$$\frac{d}{dx}[\cos^{-1} x] = \frac{d}{dx}\left[\frac{\pi}{2} - \sin^{-1} x\right] = -\frac{d}{dx}[\sin^{-1} x] = -\frac{1}{\sqrt{1 - x^2}}$$

We leave the rest of the details for the reader.   ▮

If u is a differentiable function of x, then applying the chain rule to (1)–(6) yields the following generalized derivative formulas:

$$\frac{d}{dx}[\sin^{-1} u] = \frac{1}{\sqrt{1 - u^2}}\frac{du}{dx} \tag{1a}$$

$$\frac{d}{dx}[\cos^{-1} u] = -\frac{1}{\sqrt{1 - u^2}}\frac{du}{dx} \tag{2a}$$

$$\frac{d}{dx}[\tan^{-1} u] = \frac{1}{1 + u^2}\frac{du}{dx} \tag{3a}$$

$$\frac{d}{dx}[\cot^{-1} u] = -\frac{1}{1 + u^2}\frac{du}{dx} \tag{4a}$$

$$\frac{d}{dx}[\sec^{-1} u] = \frac{1}{u\sqrt{u^2 - 1}}\frac{du}{dx} \tag{5a}$$

$$\frac{d}{dx}[\csc^{-1} u] = -\frac{1}{u\sqrt{u^2 - 1}}\frac{du}{dx} \tag{6a}$$

Example 1   Find dy/dx if $y = \sin^{-1}(x^3)$.

Solution.   From (1a),

$$\frac{dy}{dx} = \frac{1}{\sqrt{1 - (x^3)^2}}(3x^2) = \frac{3x^2}{\sqrt{1 - x^6}} \quad \blacktriangleleft$$

Example 2   Find dy/dx if $y = \sec^{-1}(e^x)$.

Solution.   From (5a),

$$\frac{dy}{dx} = \frac{1}{e^x\sqrt{(e^x)^2 - 1}}(e^x) = \frac{1}{\sqrt{e^{2x} - 1}} \quad \blacktriangleleft$$

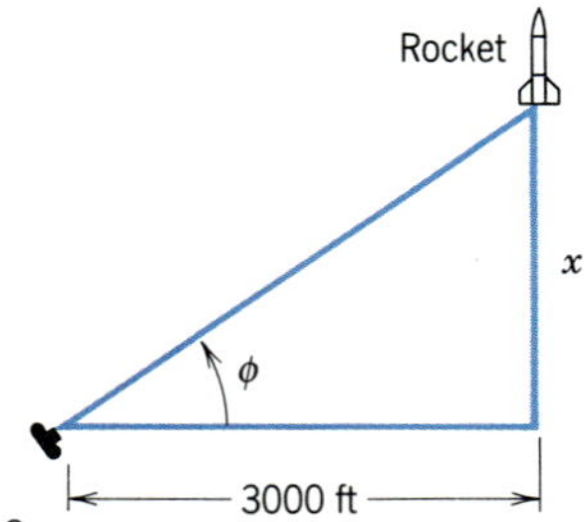

Rocket

x

ϕ

3000 ft

Camera

Figure 8.2.2

Example 3   If the rocket shown in Figure 8.2.2 is rising vertically at 880 ft/sec when it is 4000 ft up, how fast must the camera elevation angle change at that instant to keep the rocket in sight?

Solution.   We solved this related rates problem in Example 3 of Section 4.1 by differentiating both sides of the equation

$$\tan \phi = \frac{x}{3000} \tag{11}$$

implicitly with respect to the time t, after which we solved for $d\phi/dt$, obtaining

$$\frac{d\phi}{dt} = \frac{1}{3000 \sec^2 \phi}\frac{dx}{dt} \tag{12}$$

We then used the values of dx/dt and $\sec\phi$ at the given instant to find the value of $d\phi/dt$ at that instant.

Now that we can differentiate $\tan^{-1}$, an alternative solution is possible. Solving (11) for ϕ yields

$$\phi = \tan^{-1}\left(\frac{x}{3000}\right)$$

so

$$\frac{d\phi}{dt} = \frac{1}{1 + \left(\dfrac{x}{3000}\right)^2} \cdot \frac{1}{3000}\frac{dx}{dt} \tag{13}$$

We are given that $dx/dt = 880$ when $x = 4000$, so it follows from (13) that

$$\left.\frac{d\phi}{dt}\right|_{x=4000} = \frac{1}{1 + \left(\dfrac{4000}{3000}\right)^2} \cdot \frac{880}{3000} = \frac{66}{625} \approx 0.11 \text{ radian/second}$$

which agrees with the result obtained in Example 3 of Section 4.1. ◀

INTEGRATION FORMULAS

Differentiation formulas (1)–(6) yield useful integration formulas. Those most commonly needed are

$$\int \frac{du}{\sqrt{1 - u^2}} = \sin^{-1} u + C \tag{14}$$

$$\int \frac{du}{1 + u^2} = \tan^{-1} u + C \tag{15}$$

$$\int \frac{du}{u\sqrt{u^2 - 1}} = \sec^{-1} u + C \tag{16}$$

Example 4 Evaluate $\displaystyle\int \frac{dx}{1 + 3x^2}$.

Solution. Substituting

$$u = \sqrt{3}\, x, \quad du = \sqrt{3}\, dx$$

yields

$$\int \frac{dx}{1 + 3x^2} = \frac{1}{\sqrt{3}}\int \frac{du}{1 + u^2} = \frac{1}{\sqrt{3}}\tan^{-1} u + C$$

$$= \frac{1}{\sqrt{3}}\tan^{-1}(\sqrt{3}\, x) + C \quad ◀$$

Example 5   Evaluate $\displaystyle\int \frac{e^x}{\sqrt{1-e^{2x}}}\,dx$.

Solution.   Substituting

$$u = e^x, \quad du = e^x\,dx$$

yields

$$\int \frac{e^x}{\sqrt{1-e^{2x}}}\,dx = \int \frac{du}{\sqrt{1-u^2}} = \sin^{-1}u + C = \sin^{-1}(e^x) + C \quad \blacktriangleleft$$

Example 6   Evaluate $\displaystyle\int \frac{dx}{a^2 + x^2}$, where $a \neq 0$ is a constant.

Solution.   If we had a 1 in place of a^2, we could use (15). Thus, we look for a u-substitution that will enable us to replace the a^2 with a 1. If we let

$$x = au, \quad dx = a\,du$$

then

$$\int \frac{dx}{a^2 + x^2} = \int \frac{a\,du}{a^2 + a^2 u^2} = \frac{1}{a}\int \frac{du}{1 + u^2}$$

$$= \frac{1}{a}\tan^{-1}u + C = \frac{1}{a}\tan^{-1}\frac{x}{a} + C \quad \blacktriangleleft$$

The method of Example 6 leads to the following generalizations of (14), (15), and (16) for $a > 0$:

$$\int \frac{du}{\sqrt{a^2 - u^2}} = \sin^{-1}\frac{u}{a} + C \tag{17}$$

$$\int \frac{du}{a^2 + u^2} = \frac{1}{a}\tan^{-1}\frac{u}{a} + C \tag{18}$$

$$\int \frac{du}{u\sqrt{u^2 - a^2}} = \frac{1}{a}\sec^{-1}\frac{u}{a} + C \tag{19}$$

Example 7   Evaluate $\displaystyle\int \frac{dx}{\sqrt{2 - x^2}}$.

Solution.   Applying (17) with $u = x$ and $a = \sqrt{2}$ yields

$$\int \frac{dx}{\sqrt{2 - x^2}} = \sin^{-1}\frac{x}{\sqrt{2}} + C \quad \blacktriangleleft$$

▶ Exercise Set 8.2 $\boxed{C}$ *43, 45, 49, 50*

In Exercises 1–11, find dy/dx.

1. (a) $y = \sin^{-1}(\tfrac{1}{3}x)$ (b) $y = \cos^{-1}(2x + 1)$.

2. (a) $y = \tan^{-1}(x^2)$ (b) $y = \cot^{-1}(\sqrt{x})$.

3. (a) $y = \sec^{-1}(x^7)$ (b) $y = \csc^{-1}(e^x)$.

4. (a) $y = (\tan x)^{-1}$ (b) $y = \dfrac{1}{\tan^{-1}x}$.

5. (a) $y = \sin^{-1}\left(\dfrac{1}{x}\right)$ (b) $y = \cos^{-1}(\cos x)$.

6. (a) $y = \ln(\cos^{-1}x)$ (b) $y = \sqrt{\cot^{-1}x}$.

7. (a) $y = e^x \sec^{-1}x$ (b) $y = x^2(\sin^{-1}x)^3$.

8. (a) $y = \sin^{-1}x + \cos^{-1}x$
    (b) $y = \sec^{-1}x + \csc^{-1}x$.

9. (a) $y = \tan^{-1}\left(\dfrac{1-x}{1+x}\right)$
    (b) $y = (1 + x\csc^{-1}x)^{10}$.

10. (a) $y = \sin^{-1}(e^{-3x})$ (b) $y = \tan^{-1}(xe^{2x})$.

11. (a) $y = \tan^{-1}\sqrt{\dfrac{1-x}{1+x}}$ (b) $y = \sin^{-1}(x^2 \ln x)$.

In Exercises 12–13, find dy/dx by implicit differentiation.

12. $x^3 + x\tan^{-1}y = e^y$.

13. $\sin^{-1}(xy) = \cos^{-1}(x - y)$.

In Exercises 14–27, evaluate the integral.

14. $\displaystyle\int_0^{1/\sqrt{2}} \dfrac{dx}{\sqrt{1 - x^2}}$.

15. $\displaystyle\int_{-1}^{1} \dfrac{dx}{1 + x^2}$.

16. $\displaystyle\int_{\sqrt{2}}^{2} \dfrac{dx}{x\sqrt{x^2 - 1}}$.

17. $\displaystyle\int_{-\sqrt{2}}^{-2/\sqrt{3}} \dfrac{dx}{x\sqrt{x^2 - 1}}$.

18. $\displaystyle\int \dfrac{dx}{\sqrt{1 - 4x^2}}$.

19. $\displaystyle\int \dfrac{dx}{1 + 16x^2}$.

20. $\displaystyle\int \dfrac{dx}{x\sqrt{9x^2 - 1}}$.

21. $\displaystyle\int \dfrac{e^x}{1 + e^{2x}}\,dx$.

22. $\displaystyle\int_{\ln 2}^{\ln(2/\sqrt{3})} \dfrac{e^{-x}\,dx}{\sqrt{1 - e^{-2x}}}$.

23. $\displaystyle\int_1^3 \dfrac{dx}{\sqrt{x}(x + 1)}$.

24. $\displaystyle\int \dfrac{t}{t^4 + 1}\,dt$.

25. $\displaystyle\int \dfrac{\sec^2 x\,dx}{\sqrt{1 - \tan^2 x}}$.

26. $\displaystyle\int \dfrac{\sin\theta}{\cos^2\theta + 1}\,d\theta$.

27. $\displaystyle\int \dfrac{dx}{x\sqrt{1 - (\ln x)^2}}$.

28. Derive integration formulas (17) and (19).

In Exercises 29–34, use (17), (18), and (19) to evaluate the integrals.

29. (a) $\displaystyle\int \dfrac{dx}{\sqrt{9 - x^2}}$ (b) $\displaystyle\int \dfrac{dx}{5 + x^2}$

    (c) $\displaystyle\int \dfrac{dx}{x\sqrt{x^2 - \pi}}$.

30. (a) $\displaystyle\int \dfrac{e^x}{4 + e^{2x}}\,dx$ (b) $\displaystyle\int \dfrac{dx}{\sqrt{9 - 4x^2}}$

    (c) $\displaystyle\int \dfrac{dy}{y\sqrt{5y^2 - 3}}$.

31. $\displaystyle\int_0^1 \dfrac{x}{\sqrt{4 - 3x^4}}\,dx$.   **32.** $\displaystyle\int_1^2 \dfrac{1}{\sqrt{x}\sqrt{4 - x}}\,dx$.

33. $\displaystyle\int_0^{2/\sqrt{3}} \dfrac{1}{4 + 9x^2}\,dx$.   **34.** $\displaystyle\int_1^{\sqrt{3}} \dfrac{x}{3 + x^4}\,dx$.

35. Find the area of the region enclosed by the graphs of $y = 1/\sqrt{1 - 9x^2}$, $y = 0$, $x = 0$, and $x = 1/6$.

36. Find the area of the region enclosed by the graphs of $y = \sin^{-1}x$, $x = 0$, and $y = \pi/2$.

37. Find the volume of the solid generated when the region bounded by $x = 2$, $x = -2$, $y = 0$, and $y = 1/\sqrt{4 + x^2}$ is revolved about the x-axis.

38. (a) Find the volume V of the solid generated when the region bounded by $y = 1/(1 + x^4)$, $y = 0$, $x = 1$, and $x = b$ ($b > 1$) is revolved about the y-axis.

    (b) Find $\displaystyle\lim_{b \to +\infty} V$.

39. Evaluate $\displaystyle\int_0^1 \sin^{-1}x\,dx$. [*Hint:* Interpret the integral as the area of a region in the xy-plane, and integrate with respect to y.]

40. In Exercise 46 of Section 8.1, how far from the launching pad should the camera be positioned to maximize the angle θ subtended at the lens by the missile?

41. Given points $A(2, 1)$ and $B(5, 4)$, find the point P in the interval $[2, 5]$ on the x-axis that maximizes angle APB.

42. An aircraft is flying at an altitude of 4 mi at a speed of 800 mi/hr in a direction away from a tracking station on the ground. How fast is the angle of elevation changing when the aircraft is over a point 10 mi from the station?

43. A lighthouse is located 3 mi off a straight shore. If the light revolves at 2 revolutions per minute, how fast is the beam moving along the coastline at a point 2 mi down the coast?

44. A 25-ft ladder is leaning against a vertical wall. If the bottom of the ladder slides away from the base of the wall at the rate of 4 ft/sec, how fast is the angle between the ladder and the wall changing when the top of the ladder is 20 ft above the ground?

45. The lower edge of a painting, 10 ft in height, is 2 ft above an observer's eye level. Assuming that the best view is obtained when the angle subtended at the observer's eye by the painting is maximum, how far from the wall should the observer stand?

46. Use Theorem 4.10.3 to prove that
   (a) $2 \sin^{-1} \sqrt{x} = \sin^{-1}(2x - 1) + \pi/2$
       for $0 < x < 1$
   (b) $\sin^{-1}(\tanh x) = \tan^{-1}(\sinh x)$.

47. Use Theorem 4.10.2 (the Mean-Value Theorem) to prove that
$$\frac{x}{1 + x^2} < \tan^{-1}x < x \quad (x > 0)$$

48. Find $\displaystyle\lim_{n \to +\infty} \sum_{k=1}^{n} \frac{n}{n^2 + k^2}$. [*Hint:* Interpret this as the

limit of a Riemann sum in which the interval $[0, 1]$ is divided into n subintervals of equal width.]

49. A student wants to find the area enclosed by the graphs of $y = 1/\sqrt{1 - x^2}$, $y = 0$, $x = 0$, and $x = 0.8$.
   (a) Show that the exact area is $\sin^{-1} 0.8$.
   (b) The student uses a calculator to approximate the result in part (a) to two decimal places and obtains an incorrect answer of 53.13. What was the student's error? Find the correct approximation.

Exercise 50 requires separation of variables covered in Section 7.7.

---

50. Refer to Exercise 30, Section 7.7.
   (a) Solve the differential equation
$$m \frac{dv}{dt} = -(kv^2 + mg)$$
       for v in terms of t given that $v = v_0$ when $t = 0$.
   (b) Find, to the nearest tenth of a second, how long it takes the bullet to reach its highest point. Use the numerical values of v_0, g, m, and k given in part (b) of Exercise 30, Section 7.7.

■ **8.3 INVERSE HYPERBOLIC FUNCTIONS**

In this section we shall discuss inverses of hyperbolic functions. We shall be interested primarily in the integration formulas that these functions produce.

□ **HYPERBOLIC SINE**

Because the hyperbolic sine function is increasing on the interval $(-\infty, +\infty)$ (Figure 7.6.1b), it follows that $\sinh x$ is one-to-one on this interval and consequently has an inverse function, $\sinh^{-1} x$. Thus, for all real values of x and y

$$\sinh^{-1}(\sinh x) = x$$
$$\sinh(\sinh^{-1}x) = x$$

Moreover,

$$y = \sinh^{-1}x \quad \text{and} \quad \sinh y = x$$

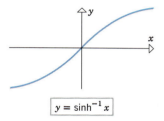

$y = \sinh^{-1} x$

Figure 8.3.1

are equivalent statements. The graph of $\sinh^{-1} x$ can be obtained by reflecting the graph of $\sinh x$ about the line $y = x$ (Figure 8.3.1).

Because $\sinh x$ is expressible in terms of e^x, one might suspect that $\sinh^{-1} x$ is expressible in terms of natural logarithms. To see that this is so, we write $y = \sinh^{-1} x$ in the alternative form

$$x = \sinh y = \frac{e^y - e^{-y}}{2}$$

so that

$$e^y - 2x - e^{-y} = 0$$

or, on multiplying through by e^y

$$e^{2y} - 2xe^y - 1 = 0$$

Applying the quadratic formula yields

$$e^y = \frac{2x \pm \sqrt{4x^2 + 4}}{2} = x \pm \sqrt{x^2 + 1}$$

Since $e^y > 0$, the solution involving the minus sign is extraneous and must be discarded. Thus,

$$e^y = x + \sqrt{x^2 + 1}$$

Taking natural logarithms yields

$$y = \ln (x + \sqrt{x^2 + 1})$$

or

$$\sinh^{-1} x = \ln (x + \sqrt{x^2 + 1}) \tag{1}$$

This formula and a calculator can be used to evaluate $\sinh^{-1} x$. Moreover, we can use (1) to obtain the derivative formula for $\sinh^{-1} x$ by writing

$$\frac{d}{dx} [\sinh^{-1} x] = \frac{d}{dx} [\ln (x + \sqrt{x^2 + 1})]$$

$$= \frac{1}{x + \sqrt{x^2 + 1}} \left(1 + \frac{x}{\sqrt{x^2 + 1}}\right)$$

$$= \frac{\sqrt{x^2 + 1} + x}{(x + \sqrt{x^2 + 1})(\sqrt{x^2 + 1})}$$

or, on canceling,

$$\frac{d}{dx} [\sinh^{-1} x] = \frac{1}{\sqrt{x^2 + 1}} \tag{2}$$

HYPERBOLIC COSINE

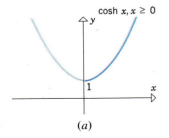

cosh x, $x \geq 0$

(a)

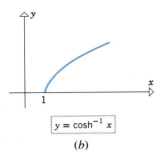

$y = \cosh^{-1} x$

(b)

Figure 8.3.2

Derivatives of the remaining inverse hyperbolic functions are obtained in a similar manner with the exception of $\cosh^{-1}$ and sech^{-1}. As evidenced by Figures 7.6.1a and 7.6.3, $\cosh x$ and $\operatorname{sech} x$ are not one-to-one, so we must restrict the domains of these functions appropriately before we can obtain inverses. For example, if we let

$$f(x) = \cosh x, \quad x \geq 0$$

then f is one-to-one (Figure 8.3.2a) and consequently has an inverse. We call this inverse $\cosh^{-1} x$. The graph of $y = \cosh^{-1} x$ is the reflection of the graph of f about the line $y = x$ (Figure 8.3.2b). The domain of $\cosh^{-1}$ is the interval $[1, +\infty)$, the range is $[0, +\infty)$, and

$$\cosh^{-1}(\cosh x) = x \quad \text{if} \quad x \geq 0$$
$$\cosh(\cosh^{-1} x) = x \quad \text{if} \quad x \geq 1$$

Moreover, if $x \geq 1$ and $y \geq 0$, then the statements

$$y = \cosh^{-1} x \quad \text{and} \quad x = \cosh y$$

are equivalent.

By following the derivations (1) and (2), the reader should be able to show that if $x > 1$, then

$$\cosh^{-1} x = \ln(x + \sqrt{x^2 - 1}) \quad \text{and} \quad \frac{d}{dx}[\cosh^{-1} x] = \frac{1}{\sqrt{x^2 - 1}}$$

SUMMARY OF THE HYPERBOLIC FUNCTIONS

For reference, we summarize the definitions, logarithmic expressions, and derivatives of the inverse hyperbolic functions.

8.3.1   DEFINITION.

$y = \sinh^{-1} x$ is equivalent to $x = \sinh y$ for all x, y

$y = \cosh^{-1} x$ is equivalent to $x = \cosh y$ if $\begin{cases} x \geq 1 \\ y \geq 0 \end{cases}$

$y = \tanh^{-1} x$ is equivalent to $x = \tanh y$ if $\begin{cases} -1 < x < 1 \\ -\infty < y < +\infty \end{cases}$

$y = \coth^{-1} x$ is equivalent to $x = \coth y$ if $\begin{cases} |x| > 1 \\ y \neq 0 \end{cases}$

$y = \operatorname{sech}^{-1} x$ is equivalent to $x = \operatorname{sech} y$ if $\begin{cases} 0 < x \leq 1 \\ y \geq 0 \end{cases}$

$y = \operatorname{csch}^{-1} x$ is equivalent to $x = \operatorname{csch} y$ if $\begin{cases} x \neq 0 \\ y \neq 0 \end{cases}$

8.3.2 THEOREM.

$$\sinh^{-1} x = \ln(x + \sqrt{x^2 + 1}) \qquad (-\infty < x < +\infty)$$

$$\cosh^{-1} x = \ln(x + \sqrt{x^2 - 1}) \qquad (x \geq 1)$$

$$\tanh^{-1} x = \frac{1}{2} \ln \frac{1 + x}{1 - x} \qquad (-1 < x < 1)$$

$$\coth^{-1} x = \frac{1}{2} \ln \frac{x + 1}{x - 1} \qquad (|x| > 1)$$

$$\operatorname{sech}^{-1} x = \ln \left(\frac{1 + \sqrt{1 - x^2}}{x} \right) \qquad (0 < x \leq 1)$$

$$\operatorname{csch}^{-1} x = \ln \left(\frac{1}{x} + \frac{\sqrt{1 + x^2}}{|x|} \right) \qquad (x \neq 0)$$

8.3.3 THEOREM.

$$\frac{d}{dx} [\sinh^{-1} x] = \frac{1}{\sqrt{1 + x^2}}$$

$$\frac{d}{dx} [\cosh^{-1} x] = \frac{1}{\sqrt{x^2 - 1}} \qquad (x > 1)$$

$$\frac{d}{dx} [\tanh^{-1} x] = \frac{1}{1 - x^2} \qquad (|x| < 1)$$

$$\frac{d}{dx} [\coth^{-1} x] = \frac{1}{1 - x^2} \qquad (|x| > 1)$$

$$\frac{d}{dx} [\operatorname{sech}^{-1} x] = -\frac{1}{x\sqrt{1 - x^2}} \qquad (0 < x < 1)$$

$$\frac{d}{dx} [\operatorname{csch}^{-1} x] = -\frac{1}{|x|\sqrt{1 + x^2}} \qquad (x \neq 0)$$

REMARK. Care must be exercised in differentiating $\tanh^{-1}x$ and $\coth^{-1}x$. In both cases the expression for the derivative is

$$\frac{1}{1 - x^2} \tag{3}$$

However, the domain of $\tanh^{-1}x$ is the interval $-1 < x < 1$ (verify this), so that the derivative formula for $\tanh^{-1}x$ applies only when $|x| < 1$. Similarly, the derivative formula for $\coth^{-1}x$ applies only on the domain of $\coth^{-1}x$, that is, when $|x| > 1$. This distinction becomes important when we integrate (3); for then we must write

$$\int \frac{1}{1 - x^2}\, dx = \begin{cases} \tanh^{-1}x + C & \text{if } |x| < 1 \\ \coth^{-1}x + C & \text{if } |x| > 1 \end{cases}$$

The following integration formulas follow from Theorem 8.3.3.

8.3.4 THEOREM.

$$\int \frac{du}{\sqrt{1 + u^2}} = \sinh^{-1}u + C$$

$$\int \frac{du}{\sqrt{u^2 - 1}} = \cosh^{-1}u + C \qquad (u > 1)$$

$$\int \frac{du}{1 - u^2} = \begin{cases} \tanh^{-1}u + C & \text{if } |u| < 1 \\ \coth^{-1}u + C & \text{if } |u| > 1 \end{cases}$$

$$\int \frac{du}{u\sqrt{1 - u^2}} = -\text{sech}^{-1}|u| + C$$

$$\int \frac{du}{u\sqrt{1 + u^2}} = -\text{csch}^{-1}|u| + C$$

The first three integration formulas follow immediately from the corresponding differentiation formulas. The last two require additional work (see Exercises 27 and 28). By using Theorem 8.3.2 these formulas can also be expressed in terms of the natural logarithm. In particular, we leave it as an exercise to show that the third formula can be written as

$$\int \frac{du}{1 - u^2} = \frac{1}{2} \ln \left| \frac{1 + u}{1 - u} \right| + C \tag{4}$$

If u is a differentiable function of x, then applying the chain rule to the formulas in Theorem 8.3.3 yields the following generalized derivative formulas:

$$\frac{d}{dx}[\sinh^{-1} u] = \frac{1}{\sqrt{1 + u^2}}\frac{du}{dx}$$

$$\frac{d}{dx}[\cosh^{-1} u] = \frac{1}{\sqrt{u^2 - 1}}\frac{du}{dx} \qquad (u > 1)$$

$$\frac{d}{dx}[\tanh^{-1} u] = \frac{1}{1 - u^2}\frac{du}{dx} \qquad (|u| < 1)$$

$$\frac{d}{dx}[\coth^{-1} u] = \frac{1}{1 - u^2}\frac{du}{dx} \qquad (|u| > 1)$$

$$\frac{d}{dx}[\operatorname{sech}^{-1} u] = -\frac{1}{u\sqrt{1 - u^2}}\frac{du}{dx} \qquad (0 < u < 1)$$

$$\frac{d}{dx}[\operatorname{csch}^{-1} u] = -\frac{1}{|u|\sqrt{1 + u^2}}\frac{du}{dx} \qquad (u \neq 0)$$

▶ **Exercise Set 8.3** C 22, 23, 24, 35

1. (a) Prove: $\cosh^{-1} x = \ln(x + \sqrt{x^2 - 1})$, $x \geq 1$.
   (b) Use part (a) to obtain the derivative of $\cosh^{-1} x$.

2. (a) Prove: $\tanh^{-1} x = \dfrac{1}{2}\ln\dfrac{1 + x}{1 - x}$, $-1 < x < 1$.

   (b) Use part (a) to obtain the derivative of $\tanh^{-1} x$.

3. (a) Prove:

   $$\operatorname{sech}^{-1} x = \cosh^{-1}(1/x), \quad 0 < x \leq 1$$
   $$\coth^{-1} x = \tanh^{-1}(1/x), \quad |x| > 1$$
   $$\operatorname{csch}^{-1} x = \sinh^{-1}(1/x), \quad x \neq 0$$

   (b) Use part (a) and the derivatives of $\sinh^{-1} x$, $\cosh^{-1} x$, and $\tanh^{-1} x$ to find the derivatives of $\operatorname{sech}^{-1} x$, $\coth^{-1} x$, and $\operatorname{csch}^{-1} x$.
   (c) Use part (a) and the logarithmic expressions for $\sinh^{-1} x$, $\cosh^{-1} x$, and $\tanh^{-1} x$ to derive the logarithmic expressions for $\operatorname{sech}^{-1} x$, $\coth^{-1} x$, and $\operatorname{csch}^{-1} x$.

4. Without referring to the text, state the domains of the six inverse hyperbolic functions.

In Exercises 5 and 6, find the logarithmic equivalent of each of the given expressions.

5. (a) $\cosh^{-1}(3)$ (b) $\sinh^{-1}(-2)$.
6. (a) $\tanh^{-1}(3/4)$ (b) $\coth^{-1}(-5/4)$.

In Exercises 7–15, find dy/dx.

7. (a) $y = \sinh^{-1}(\frac{1}{3}x)$
   (b) $y = \cosh^{-1}(2x + 1)$.
8. (a) $y = \tanh^{-1}(x^2)$ (b) $y = \coth^{-1}(\sqrt{x})$.
9. (a) $y = \operatorname{sech}^{-1}(x^7)$ (b) $y = \operatorname{csch}^{-1}(e^x)$.
10. (a) $y = (\tanh^{-1} x)^2$ (b) $y = \dfrac{1}{\tanh^{-1} x}$.
11. (a) $y = \sinh^{-1}\left(\dfrac{1}{x}\right)$
    (b) $y = \cosh^{-1}(\cosh x)$.
12. (a) $y = \ln(\cosh^{-1} x)$ (b) $y = \sqrt{\coth^{-1} x}$.
13. (a) $y = e^x \operatorname{sech}^{-1} x$ (b) $y = x^2(\sinh^{-1} x)^3$.
14. (a) $y = \sinh^{-1}(\tanh x)$
    (b) $y = \cosh^{-1}(\sinh^{-1} x)$.

15. (a) $y = \tanh^{-1}\left(\dfrac{1-x}{1+x}\right)$

(b) $y = (1 + x \operatorname{csch}^{-1} x)^{10}$.

In Exercises 16–21, evaluate the integral.

16. $\displaystyle\int \frac{dx}{\sqrt{1 + 9x^2}}$.

17. $\displaystyle\int \frac{dx}{\sqrt{x^2 - 2}}$ $(x > \sqrt{2})$.

18. $\displaystyle\int \frac{dx}{\sqrt{9x^2 - 25}}$ $(x > 5/3)$.

19. $\displaystyle\int \frac{dx}{\sqrt{1 - e^{2x}}}$.   **20.** $\displaystyle\int \frac{\sin\theta\, d\theta}{\sqrt{1 + \cos^2\theta}}$.

21. $\displaystyle\int \frac{dx}{x\sqrt{1 + x^6}}$.

In Exercises 22–24, use a calculator to obtain a numerical value for the integral.

22. $\displaystyle\int_0^{1/2} \frac{dx}{1 - x^2}$.   **23.** $\displaystyle\int_2^3 \frac{dx}{1 - x^2}$.

24. $\displaystyle\int_0^{\sqrt{3}} \frac{dt}{\sqrt{t^2 + 1}}$.

25.   Sketch the graphs of
(a) $\tanh^{-1} x$ (b) $\operatorname{csch}^{-1} x$.

26.   Sketch the graphs of
(a) $\coth^{-1} x$ (b) $\operatorname{sech}^{-1} x$.

27.   Show that
$$\frac{d}{dx}[\operatorname{sech}^{-1}|x|] = -\frac{1}{x\sqrt{1 - x^2}}$$

28.   Show that
$$\frac{d}{dx}[\operatorname{csch}^{-1}|x|] = -\frac{1}{x\sqrt{1 + x^2}}$$

29.   Derive integration formula (4).

30.   Let a be a positive constant. Derive an integration formula for
(a) $\displaystyle\int \frac{du}{\sqrt{a^2 + u^2}}$ (b) $\displaystyle\int \frac{du}{\sqrt{u^2 - a^2}}$

(c) $\displaystyle\int \frac{du}{a^2 - u^2}$.

[*Hint:* See Example 6 of Section 8.2.]

31.   Our derivation of the differentiation formula for $\sinh^{-1} x$ used the logarithmic expression for this function. Show that the derivative can also be ob-

tained by the method we used in Section 8.2 to obtain the derivative formula for $\sin^{-1} x$.

32.   Find
(a) $\displaystyle\lim_{x \to +\infty} \sinh^{-1} x$ (b) $\displaystyle\lim_{x \to +\infty} \coth^{-1} x$

(c) $\displaystyle\lim_{x \to 0^+} \operatorname{csch}^{-1} x$ (d) $\displaystyle\lim_{x \to +\infty} (\cosh^{-1} x - \ln x)$.

33.   Show that
$$\int \frac{1}{\sqrt{x^2 - 1}}\, dx = -\cosh^{-1}(-x) + C$$
if $x < -1$.

34.   Use the result in Exercise 33 to show that
$$\int \frac{1}{\sqrt{x^2 - 1}}\, dx = \ln|x + \sqrt{x^2 - 1}| + C$$
if $x < -1$.

Exercises 35 and 36 require separation of variables covered in Section 7.7.

35.   A thin flexible chain of length 2 ft is held stretched out on the top of a table with $\frac{1}{4}$ ft of the chain hanging over the edge (Figure 8.3.3). The chain is released and allowed to slide off the table. Assuming that the surface of the table is frictionless, the velocity v of the chain will satisfy the differential equation
$$v\frac{dv}{dx} = \frac{g}{L}x$$
where L is the total length of the chain, g is the acceleration due to gravity (32 ft/sec²), and x is the distance from the top of the table to the hanging end of the chain.
(a)   Solve the differential equation for v in terms of x.
(b)   Use the result in part (a) and the fact that $v = dx/dt$ to find x in terms of the time t. Approximate the time it will take for the chain to slip completely off the table. Express your answer to the nearest tenth of a second.

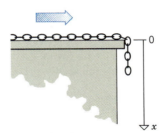

Figure 8.3.3

36. An object of mass m, released from rest at time $t = 0$, from a point above the surface of the earth is acted on by the force of gravity and a drag force that is proportional to the square of the velocity $v(t)$. The differential equation of motion is

$$m \frac{dv}{dt} = mg - kv^2$$

where g is the constant acceleration due to gravity and k is a positive constant.

(a) Solve the differential equation for $v(t)$.

(b) Use the result in part (a) to find $\lim_{t \to +\infty} v(t)$.

(c) Find the distance $x(t)$ of the object from its starting point.

▶ SUPPLEMENTARY EXERCISES

In Exercises 1–3, find the exact value.

1. (a) $\cos^{-1}(-1/2)$ (b) $\cot^{-1}[\cot(3/4)]$
(c) $\cos[\sin^{-1}(4/5)]$ (d) $\cos[\sin^{-1}(-4/5)]$.

2. (a) $\tan^{-1}(-1)$ (b) $\csc^{-1}(-2/\sqrt{3})$
(c) $\cos^{-1}[\cos(-\pi/3)]$ (d) $\sin[-\sec^{-1}(2/\sqrt{3})]$.

3. (a) $\sin^{-1}(1/\sqrt{2})$ (b) $\sin^{-1}[\sin(5\pi/4)]$
(c) $\tan(\sec^{-1}5)$ (d) $\tan^{-1}[\cot(\pi/6)]$.

4. Use a double-angle formula to convert the given expression to an algebraic function of x.
(a) $\sin(2\csc^{-1}x)$, $|x| \geq 1$
(b) $\cos(2\sin^{-1}x)$, $|x| \leq 1$
(c) $\sin(2\tan^{-1}x)$.

5. Simplify:
(a) $\cos[\cos^{-1}(4/5) + \sin^{-1}(5/13)]$
(b) $\sin[\sin^{-1}(4/5) + \cos^{-1}(5/13)]$
(c) $\tan[\tan^{-1}(1/3) + \tan^{-1}(2)]$.

6. If $u = \mathrm{csch}^{-1}(-5/12)$, find $\coth u$, $\sinh u$, $\cosh u$, and $\sinh(2u)$.

7. If $u = \tanh^{-1}(-3/5)$, find $\cosh u$, $\sinh u$, and $\cosh(2u)$.

In Exercises 8 and 9, sketch the graph of f.

8. (a) $f(x) = 3\sin^{-1}(x/2)$
(b) $f(x) = \cos^{-1}x - \pi/2$.

9. (a) $f(x) = 2\tan^{-1}(-3x)$
(b) $f(x) = \cos^{-1}x + \sin^{-1}x$.

In Exercises 10–23, find dy/dx, using implicit or logarithmic differentiation where convenient.

10. $y = \sin^{-1}(e^x) + 2\tan^{-1}(3x)$.

11. $y = \dfrac{1}{\sec^{-1}x^2}$.

12. $y = x\sin^{-1}x + \sqrt{1 - x^2}$.

13. $y = \cosh^{-1}(\sec x)$.   **14.** $\tan^{-1}y = \sin^{-1}x$.

15. $y = x\tanh^{-1}(\ln x)$.   **16.** $y = \tan^{-1}\left(\dfrac{2x}{1 - x^2}\right)$.

17. $y = \sqrt{\sin^{-1}3x}$.   **18.** $y = (\sin^{-1}2x)^{-1}$.

19. $y = \exp(\sec^{-1}x)$.   **20.** $y = (\tan^{-1}x)/\ln x$.

21. $y = \pi^{\sin^{-1}x}$.   **22.** $y = (\sinh^{-1}x)^{\pi}$.

23. $y = \tanh^{-1}\left(\dfrac{1}{\coth x}\right)$.

24. Let $f(x) = \tan^{-1}x + \tan^{-1}(1/x)$.
(a) By considering $f'(x)$, show that $f(x) = C_1$ on $(-\infty, 0)$ and $f(x) = C_2$ on $(0, +\infty)$, where C_1 and C_2 are constants.
(b) Find C_1 and C_2 in part (a).

25. Show that $y = \tan^{-1}x$ satisfies $y'' = -2\sin y \cos^3 y$.

In Exercises 26–35, evaluate the integral.

26. $\displaystyle\int \frac{dx}{\sqrt{9 - 4x^2}}$.   **27.** $\displaystyle\int \frac{e^x\, dx}{1 - e^{2x}}$.

28. $\displaystyle\int \frac{\cot x\, dx}{\sqrt{1 - \sin^2 x}}$.   **29.** $\displaystyle\int \frac{dx}{x\sqrt{(\ln x)^2 - 1}}$, $x > e$.

30. $\displaystyle\int \frac{dx}{e^x\sqrt{1 - e^{-2x}}}$.   **31.** $\displaystyle\int_{2\sqrt{3}/9}^{2/3} \frac{dx}{3x\sqrt{9x^2 - 1}}$.

32. $\displaystyle\int_0^{\sqrt{2}} \frac{x\, dx}{4 + x^4}$.   **33.** $\displaystyle\int \frac{dx}{x^{1/2} + x^{3/2}}$.

34. $\displaystyle\int \frac{x^2\, dx}{\sqrt{1 + x^6}}$.   **35.** $\displaystyle\int_{1/4}^{1/2} \frac{dx}{\sqrt{x}\sqrt{1 - x}}$.

36. The hypotenuse of a right triangle is growing at a rate of a cm/sec and one leg is decreasing at a rate of b cm/sec. How fast is the acute angle between the hypotenuse and the other leg changing at the instant when both legs are 1 cm?

37. For $f(x) = \tan^{-1}(2x) - \tan^{-1}x$ find the value(s) of x at which $f(x)$ assumes its maximum value on the interval $[0, +\infty)$.

38. Find the area between the curve $y = 1/(9 + x^2)$, the x-axis, and the lines $x = \pm\sqrt{3}$.

9
Techniques of Integration

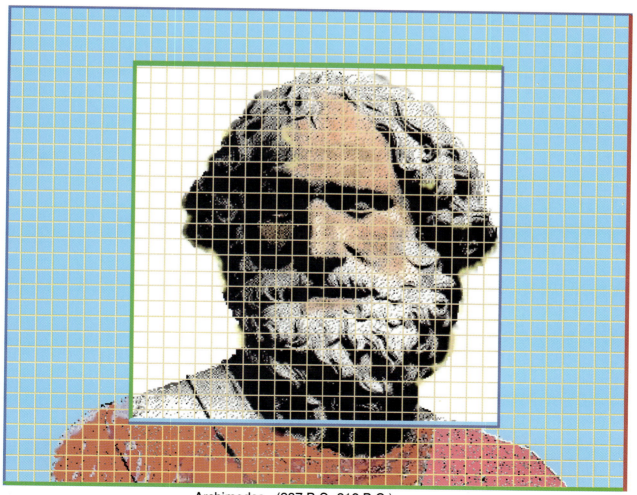

Archimedes   (287 B.C.-212 B.C.)

■ **9.1 INTEGRATION USING TABLES**

> *Although we can now integrate a wide variety of functions, there remain many important kinds of indefinite integrals that we cannot yet evaluate. The purpose of this chapter is to develop some additional techniques of integration and also to systematize the procedure of integration.*

☐ **A REVIEW OF FAMILIAR INTEGRATION FORMULAS**

In Section 5.3 we showed how to evaluate indefinite integrals by the method of substitution. The underlying idea was to make an appropriate u-substitution that would transform the given integral into a more basic integral that could be evaluated easily. The following is a list of basic integrals that we have encountered thus far.

CONSTANTS, POWERS, EXPONENTIALS | See Sections 5.2, 7.3, and 7.4. |

1. $\displaystyle\int du = u + C$

2. $\displaystyle\int a\, du = a \int du = au + C$

3. $\displaystyle\int u^r\, du = \frac{u^{r+1}}{r+1} + C,\ r \neq -1$

4. $\displaystyle\int \frac{du}{u} = \ln|u| + C$

5. $\displaystyle\int e^u\, du = e^u + C$

6. $\displaystyle\int a^u\, du = \frac{a^u}{\ln a} + C,\ a > 0$

TRIGONOMETRIC FUNCTIONS | See Sections 5.2 and 7.3. |

7. $\displaystyle\int \sin u\, du = -\cos u + C$

8. $\displaystyle\int \cos u\, du = \sin u + C$

9. $\displaystyle\int \sec^2 u\, du = \tan u + C$

10. $\displaystyle\int \csc^2 u\, du = -\cot u + C$

11. $\displaystyle\int \sec u \tan u\, du = \sec u + C$

12. $\displaystyle\int \csc u \cot u\, du = -\csc u + C$

13. $\displaystyle\int \tan u\, du = -\ln|\cos u| + C$

14. $\displaystyle\int \cot u\, du = \ln|\sin u| + C$

HYPERBOLIC FUNCTIONS | See Section 7.6. |

15. $\displaystyle\int \sinh u\, du = \cosh u + C$

16. $\displaystyle\int \cosh u\, du = \sinh u + C$

17. $\displaystyle\int \operatorname{sech}^2 u\, du = \tanh u + C$

18. $\displaystyle\int \operatorname{csch}^2 u\, du = -\coth u + C$

19. $\displaystyle\int \text{sech } u \text{ tanh } u \, du = -\text{sech } u + C$

20. $\displaystyle\int \text{csch } u \text{ coth } u \, du = -\text{csch } u + C$

ALGEBRAIC FUNCTIONS | See Sections 8.2 and 8.3. |

21. $\displaystyle\int \frac{du}{\sqrt{a^2 - u^2}} = \sin^{-1}\frac{u}{a} + C$ **22.** $\displaystyle\int \frac{du}{a^2 + u^2} = \frac{1}{a}\tan^{-1}\frac{u}{a} + C$

23. $\displaystyle\int \frac{du}{u\sqrt{u^2 - a^2}} = \frac{1}{a}\sec^{-1}\frac{u}{a} + C$

24. $\displaystyle\int \frac{du}{\sqrt{a^2 + u^2}} = \ln(u + \sqrt{u^2 + a^2}) + C$

25. $\displaystyle\int \frac{du}{\sqrt{u^2 - a^2}} = \ln|u + \sqrt{u^2 - a^2}| + C$

26. $\displaystyle\int \frac{du}{a^2 - u^2} = \frac{1}{2a}\ln\left|\frac{a + u}{a - u}\right| + C$

27. $\displaystyle\int \frac{du}{u\sqrt{a^2 - u^2}} = -\frac{1}{a}\ln\left|\frac{a + \sqrt{a^2 - u^2}}{u}\right| + C$

28. $\displaystyle\int \frac{du}{u\sqrt{a^2 + u^2}} = -\frac{1}{a}\ln\left|\frac{a + \sqrt{a^2 + u^2}}{u}\right| + C$

REMARK. Formulas 24–28 are generalizations of those in Section 8.3. Readers who did not cover that section can ignore those formulas for now; we shall develop alternative methods for evaluating them in this chapter.

☐ **USING TABLES OF INTEGRALS**

Mathematicians have formulated extensive tables of integral formulas, of which the foregoing list is only a small sample. A more extensive list, which we shall refer to as *The Endpaper Table of Integrals,* appears in the endpapers of this text. More comprehensive tables are published in such reference books as the *CRC Standard Mathematical Tables and Formulae*, CRC Press, Inc., 1991. In later sections of this chapter we shall show how some of the formulas in integral tables are derived; however, in those applications where answers rather than derivations are needed, integrals are often evaluated by looking them up in tables. The examples that follow illustrate some techniques for using integral tables.

As with most tables of integrals, the entries in The Endpaper Table of Integrals are grouped according to various characteristics of the integrand. To use the table, find a characteristic of the integrand that matches one of those in the table headings, then locate the appropriate integral formula under that heading. The following example illustrates this procedure.

Example 1 Use The Endpaper Table of Integrals to evaluate

$$\int x^2 \sqrt{7 + 3x} \, dx$$

Solution.   The integral involves an expression of the form

$$\sqrt{a + bu}$$

where $a = 7$, $b = 3$, and $u = x$, so we are led to the section of the table with the heading *Integrals Containing* $\sqrt{a + bu}$. Under that heading we find that the integral in Formula 15 matches the given integral with the above values of a, b, and u. Since $u = x$, it follows that $du = dx$, so Formula 15 yields

$$\int x^2 \sqrt{7 + 3x}\, dx = \frac{2}{2835}(135x^2 - 252x + 392)(7 + 3x)^{3/2} + C \qquad \blacktriangleleft$$

Sometimes, a given integral may not match any of the forms in a table but can be made to match by an appropriate *u*-substitution. The following example illustrates this case.

Example 2   Use The Endpaper Table of Integrals to help evaluate

$$\int \sqrt{x - 4x^2}\, dx$$

Solution.   The integral does not involve any expressions that match the table headings precisely, but if the factor multiplying x^2 had been a 1 rather than a 4, we would have had a match with

$$2au - u^2$$

by taking $a = \frac{1}{2}$ and $u = x$. This suggests that we consider making the *u*-substitution

$$u = 2x$$

since this substitution has the effect of converting $4x^2$ to u^2. With this substitution we have

$$x = \tfrac{1}{2}u \quad \text{and} \quad dx = \tfrac{1}{2}\, du$$

so the given integral becomes

$$\int \sqrt{x - 4x^2}\, dx = \frac{1}{2} \int \sqrt{\tfrac{1}{2}u - u^2}\, du$$

Thus, from Formula 50 with $a = \frac{1}{4}$ we obtain

$$\int \sqrt{x - 4x^2}\, dx = \frac{1}{2}\left[\frac{u - \frac{1}{4}}{2} \sqrt{\tfrac{1}{2}u - u^2} + \frac{1}{32} \sin^{-1}\left(\frac{u - \frac{1}{4}}{\frac{1}{4}} \right) \right] + C$$

$$= \frac{1}{2}\left[\frac{2x - \frac{1}{4}}{2} \sqrt{x - 4x^2} + \frac{1}{32} \sin^{-1}\left(\frac{2x - \frac{1}{4}}{\frac{1}{4}} \right) \right] + C$$

$$= \frac{8x - 1}{16} \sqrt{x - 4x^2} + \frac{1}{64} \sin^{-1}(8x - 1) + C \qquad \blacktriangleleft$$

Example 3 Use The Endpaper Table of Integrals to help evaluate

$$\int e^{\pi x} \sin^{-1}(e^{\pi x})\, dx$$

Solution. The integral does not involve any expressions that match the table headings; however, if we make the substitution

$$u = e^{\pi x} \quad \text{so that} \quad du = \pi e^{\pi x}\, dx$$

we obtain

$$\int e^{\pi x} \sin^{-1}(e^{\pi x})\, dx = \frac{1}{\pi} \int \sin^{-1} u\, du$$

Thus, from Formula 90

$$\int e^{\pi x} \sin^{-1}(e^{\pi x})\, dx = \frac{1}{\pi} [u \sin^{-1} u + \sqrt{1 - u^2}] + C$$

$$= \frac{1}{\pi} [e^{\pi x} \sin^{-1}(e^{\pi x}) + \sqrt{1 - e^{2\pi x}}] + C \quad \blacktriangleleft$$

Example 4 Find the volume of the solid generated when the region in the first quadrant that is enclosed by the *x*-axis, the *y*-axis, the line $x = 1$, and the curve $y = \sqrt{x^2 + 2x + 5}$ is revolved about the *y*-axis (Figure 9.1.1).

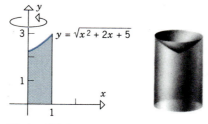

Figure 9.1.1

Solution. Using cylindrical shells we obtain

$$V = \int_0^1 2\pi x \sqrt{x^2 + 2x + 5}\, dx$$

[See Formula (5) of Section 6.3.] The Endpaper Table of Integrals has forms involving

$$\sqrt{u^2 \pm a^2} \tag{1}$$

but none involving expressions of the form

$$\sqrt{u^2 + bu + c}$$

so we shall begin by completing the square in an attempt to introduce expressions of form (1). We obtain

$$V = \int_0^1 2\pi x \sqrt{(x+1)^2 + 4} \, dx \tag{2}$$

We now make the substitution

$$u = x + 1, \quad du = dx$$

With this substitution we have

$$u = 1 \quad \text{if} \quad x = 0$$
$$u = 2 \quad \text{if} \quad x = 1$$

so (2) yields

$$V = \int_1^2 2\pi(u-1)\sqrt{u^2+4} \, du$$

$$= 2\pi \int_1^2 u\sqrt{u^2+4} \, du - 2\pi \int_1^2 \sqrt{u^2+4} \, du \tag{3}$$

The first integral in (3) could be evaluated by making the substitution $t = u^2 + 4$, but we will use Formula (29) in The Endpaper Table of Integrals with $a = 2$; this yields

$$2\pi \int_1^2 u\sqrt{u^2+4} \, du = \frac{2\pi}{3}(u^2+4)^{3/2}\Big]_1^2 = \frac{32\pi\sqrt{2}}{3} - \frac{10\pi\sqrt{5}}{3} \tag{4}$$

and the second integral in (3) can be evaluated using Formula (28) in The Endpaper Table of Integrals with $a = 2$; this yields

$$2\pi \int_1^2 \sqrt{u^2+4} \, du = 2\pi \left[\frac{u}{2}\sqrt{u^2+4} + 2\ln|u + \sqrt{u^2+4}| \right]_1^2$$

$$= 4\pi\sqrt{2} - \pi\sqrt{5} + 4\pi \ln\left(\frac{2 + 2\sqrt{2}}{1 + \sqrt{5}}\right) \tag{5}$$

Thus, it follows from (3), (4), and (5) that

$$V = \frac{20\pi\sqrt{2}}{3} - \frac{7\pi\sqrt{5}}{3} - 4\pi \ln\left(\frac{2 + 2\sqrt{2}}{1 + \sqrt{5}}\right) \approx 8.2 \quad \blacktriangleleft$$

☐ **INTEGRATION AND NEW TECHNOLOGY**

Methods for obtaining approximate numerical values of *definite* integrals date back to the earliest days of calculus; some of these will be studied later in this chapter. However, in the last few years significant strides have been made in evaluating *indefinite* integrals symbolically on computers and calculators, and we can anticipate that it will be only a matter of time until this technology makes tables of integral formulas obsolete just as it has done for tables of logarithms and trigonometric functions.

▶ Exercise Set 9.1 Ⓒ 53–66

In Exercises 1–24, use The Endpaper Table of Integrals to evaluate the integral.

1. $\int \dfrac{3x}{4x - 1}\, dx.$

2. $\int \dfrac{x}{(2 - 3x)^2}\, dx.$

3. $\int \dfrac{1}{x(2x + 5)}\, dx.$

4. $\int \dfrac{1}{x^2(1 - 5x)}\, dx.$

5. $\int x\sqrt{2x - 3}\, dx.$

6. $\int \dfrac{x}{\sqrt{2 - x}}\, dx.$

7. $\int \dfrac{1}{x\sqrt{4 - 3x}}\, dx.$

8. $\int \dfrac{1}{x\sqrt{3x - 4}}\, dx.$

9. $\int \dfrac{1}{5 - x^2}\, dx.$

10. $\int \dfrac{1}{x^2 - 9}\, dx.$

11. $\int \sqrt{x^2 - 3}\, dx.$

12. $\int \dfrac{\sqrt{x^2 + 5}}{x^2}\, dx.$

13. $\int \dfrac{x^2}{\sqrt{x^2 + 4}}\, dx.$

14. $\int \dfrac{1}{x^2\sqrt{x^2 - 2}}\, dx.$

15. $\int \sqrt{9 - x^2}\, dx.$

16. $\int \dfrac{\sqrt{4 - x^2}}{x^2}\, dx.$

17. $\int \dfrac{\sqrt{3 - x^2}}{x}\, dx.$

18. $\int \dfrac{1}{x\sqrt{6x - x^2}}\, dx.$

19. $\int \sin 3x \sin 2x\, dx.$

20. $\int \sin 2x \cos 5x\, dx.$

21. $\int x^3 \ln x\, dx.$

22. $\int \dfrac{\ln x}{\sqrt{x}}\, dx.$

23. $\int e^{-2x} \sin 3x\, dx.$

24. $\int e^x \cos 2x\, dx.$

In Exercises 25–36, make the indicated u-substitution and use The Endpaper Table of Integrals to evaluate the integral.

25. $\int \dfrac{e^{4x}}{(4 - 3e^{2x})^2}\, dx,\; u = e^{2x}.$

26. $\int \dfrac{\cos 2x}{(\sin 2x)(3 - \sin 2x)}\, dx,\; u = \sin 2x.$

27. $\int \dfrac{1}{\sqrt{x}(9x + 4)}\, dx,\; u = 3\sqrt{x}.$

28. $\int \dfrac{\cos 4x}{9 + \sin^2 4x}\, dx,\; u = \sin 4x.$

29. $\int \dfrac{1}{\sqrt{9x^2 - 4}}\, dx,\; u = 3x.$

30. $\int x\sqrt{2x^4 + 3}\, dx,\; u = \sqrt{2}x^2.$

31. $\int \dfrac{x^5}{\sqrt{5 - 9x^4}}\, dx,\; u = 3x^2.$

32. $\int \dfrac{1}{x^2\sqrt{3 - 4x^2}}\, dx,\; u = 2x.$

33. $\int \dfrac{\sin^2(\ln x)}{x}\, dx,\; u = \ln x.$

34. $\int e^{-2x} \cos^2(e^{-2x})\, dx,\; u = e^{-2x}.$

35. $\int xe^{-2x}\, dx,\; u = -2x.$

36. $\int \ln(5x - 1)\, dx,\; u = 5x - 1.$

In Exercises 37–48, make a u-substitution and use The Endpaper Table of Integrals to evaluate the integral.

37. $\int \dfrac{\sin 3x}{(\cos 3x)(\cos 3x + 1)^2}\, dx.$

38. $\int \dfrac{\ln x}{x\sqrt{4 \ln x - 1}}\, dx.$

39. $\int \dfrac{x}{16x^4 - 1}\, dx.$

40. $\int \dfrac{e^x}{3 - 4e^{2x}}\, dx.$

41. $\int e^x\sqrt{3 - 4e^{2x}}\, dx.$

42. $\int \dfrac{\sqrt{4 - 9x^2}}{x^2}\, dx.$

43. $\int \sqrt{5x - 9x^2}\, dx.$

44. $\int \dfrac{1}{x\sqrt{x - 5x^2}}\, dx.$

45. $\int x \sin 3x\, dx.$

46. $\int \cos \sqrt{x}\, dx.$

47. $\int e^{-\sqrt{x}}\, dx.$

48. $\int x \ln(2 - 3x^2)\, dx.$

In Exercises 49–52, complete the square, then make a u-substitution to evaluate the integral.

49. $\int \dfrac{1}{x^2 + 4x - 5}\, dx.$

50. $\int \sqrt{3 - 2x - x^2}\, dx.$

51. $\int \dfrac{x}{\sqrt{5 + 4x - x^2}}\, dx.$

52. $\int \dfrac{x}{x^2 + 6x + 13}\, dx.$

In Exercises 53 and 54, solve for x. Express your answer to at least three decimal places.

53. $\displaystyle\int_2^x \dfrac{1}{t(4 - t)}\, dt = 0.5,\; 0 < x < 4.$

54. $\displaystyle\int_1^x \dfrac{1}{t\sqrt{2t - 1}}\, dt = 1,\; x > \tfrac{1}{2}.$

In Exercises 55–58, find the area of the region enclosed by the curves. Express your answer to at least three decimal places.

55. $y = \sqrt{25 - x^2}$, $y = 0$, $x = 0$, $x = 4$.

56. $y = \sqrt{9x^2 - 4}$, $y = 0$, $x = 2$.

57. $y = \dfrac{1}{25 - 16x^2}$, $y = 0$, $x = 0$, $x = 1$.

58. $y = \sqrt{x}\,\ln x$, $y = 0$, $x = 4$.

In Exercises 59–62, find the volume of the solid generated when the region enclosed by the curves is revolved about the y-axis. Express your answer to at least three decimal places.

59. $y = \cos x$ and $y = 0$ for $0 \leq x \leq \pi/2$.

60. $y = \sqrt{x - 4}$, $y = 0$, $x = 8$.

61. $y = e^{-x}$, $y = 0$, $x = 0$, $x = 3$.

62. $y = \ln x$, $y = 0$, $x = 5$.

In Exercises 63 and 64, find the arc length of the curve. Express your answer to at least three decimal places.

63. $y = 2x^2$ for $0 \leq x \leq 2$.

64. $y = 3 \ln x$ for $1 \leq x \leq 3$.

In Exercises 65 and 66, find the area of the surface generated by revolving the curve about the x-axis. Express your answer to at least three decimal places.

65. $y = \sin x$ for $0 \leq x \leq \pi$.

66. $y = 1/x$ for $1 \leq x \leq 4$.

■ 9.2 INTEGRATION BY PARTS

In this section, we shall develop a technique that will help us to evaluate a wide variety of integrals that do not fit any of the basic integration formulas.

□ **DERIVATION OF THE FORMULAS FOR INTEGRATION BY PARTS**

If f and g are differentiable functions, then by the rule for differentiating products

$$\frac{d}{dx}[f(x)g(x)] = f(x)g'(x) + g(x)f'(x)$$

Integrating both sides we obtain

$$\int \frac{d}{dx}[f(x)g(x)]\, dx = \int f(x)g'(x)\, dx + \int g(x)f'(x)\, dx$$

or

$$f(x)g(x) + C = \int f(x)g'(x)\, dx + \int g(x)f'(x)\, dx$$

or

$$\int f(x)g'(x)\, dx = f(x)g(x) - \int g(x)f'(x)\, dx + C$$

Since the integral on the right will produce another constant of integration, there is no need to keep the C in this last equation; thus, we obtain

$$\int f(x)g'(x)\,dx = f(x)g(x) - \int f'(x)g(x)\,dx \tag{1}$$

which is called the formula for **integration by parts**. By using this formula we can sometimes reduce a hard integration problem to an easier one.

In practice, it is usual to rewrite (1) by letting

$$u = f(x) \quad du = f'(x)\,dx$$
$$v = g(x) \quad dv = g'(x)\,dx$$

This yields the following formula.

Integration by Parts for Indefinite Integrals

$$\int u\,dv = uv - \int v\,du \tag{2a}$$

For definite integrals the corresponding formula is

Integration by Parts for Definite Integrals

$$\int_a^b u\,dv = uv\Big]_a^b - \int_a^b v\,du \tag{2b}$$

REMARK. It is important to keep in mind that the variables u and v in these formulas are functions of x, even though it is not indicated explicitly. Moreover, the limits of integration in (2b) are limits on the variable x. Some people find it helpful to emphasize this by writing (2b) as

$$\int_{x=a}^{x=b} u\,dv = uv\Big]_{x=a}^{x=b} - \int_{x=a}^{x=b} v\,du \tag{2c}$$

Example 1 Evaluate $\int xe^x\,dx$.

Solution. To apply (2a) we must write the integral in the form

$$\int u\,dv$$

One way to do this is to let

$$u = x \quad \text{and} \quad dv = e^x\,dx$$

so that

$$du = dx \quad \text{and} \quad v = \int e^x \, dx = e^x$$

Thus, from (2a)

$$\int \underbrace{x}_{u} \underbrace{e^x \, dx}_{dv} = \underbrace{x}_{u} \underbrace{e^x}_{v} - \int \underbrace{e^x}_{v} \underbrace{dx}_{du}$$

or

$$\int xe^x \, dx = xe^x - e^x + C \qquad \blacktriangleleft$$

REMARK. In the calculation of v from dv above, we omitted the constant of integration and wrote $v = \int e^x \, dx = e^x$. Had we included a constant of integration and written $v = \int e^x \, dx = e^x + C_1$, the constant C_1 would have eventually canceled out [Exercise 56(a)]. This is always the case in integration by parts [Exercise 56(b)], so we shall usually omit the constant when calculating v from dv.

To use integration by parts successfully, the choice of u and dv must be made so that the new integral is easier than the original. For example, had we decided above to let

$$u = e^x \qquad dv = x \, dx$$

$$du = e^x \, dx \qquad v = \int x \, dx = \frac{x^2}{2}$$

then we would have obtained

$$\int xe^x \, dx = \int u \, dv = uv - \int v \, du = \frac{x^2}{2} e^x - \frac{1}{2} \int x^2 e^x \, dx$$

For this choice of u and dv the new integral is actually more complicated than the original. It is difficult to give hard and fast rules for choosing u and dv. It is a matter of experience that comes with lots of practice.

The next example shows that it is sometimes necessary to use integration by parts more than once in the same problem.

Example 2 Evaluate $\int x^2 e^{-x} \, dx$.

Solution. Let

$$u = x^2 \qquad dv = e^{-x} \, dx$$

$$du = 2x \, dx \qquad v = \int e^{-x} \, dx = -e^{-x}$$

so that

$$\int x^2 e^{-x}\, dx = \int u\, dv = uv - \int v\, du = -x^2 e^{-x} + 2 \int x e^{-x}\, dx \qquad (3)$$

The last integral is similar to the original except that we have replaced x^2 by x. Another integration by parts applied to $\int x e^{-x}\, dx$ will complete the problem. We let

$$u = x \qquad dv = e^{-x}\, dx$$

$$du = dx \qquad v = \int e^{-x}\, dx = -e^{-x}$$

so that

$$\int x e^{-x}\, dx = \int u\, dv = uv - \int v\, du$$

$$= -x e^{-x} + \int e^{-x}\, dx$$

$$= -x e^{-x} - e^{-x} + C_1$$

Substituting in (3) we obtain

$$\int x^2 e^{-x}\, dx = -x^2 e^{-x} + 2(-x e^{-x} - e^{-x} + C_1)$$

$$= -x^2 e^{-x} - 2x e^{-x} - 2e^{-x} + 2C_1$$

$$= -(x^2 + 2x + 2)e^{-x} + C$$

where $C = 2C_1$. ◀

Example 3 Evaluate $\int \ln x\, dx$.

Solution. Let

$$u = \ln x \qquad dv = dx$$

$$du = \frac{1}{x}\, dx \qquad v = \int dx = x$$

so that

$$\int \ln x\, dx = \int u\, dv = uv - \int v\, du = x \ln x - \int x\left(\frac{1}{x}\right) dx$$

$$= x \ln x - \int dx = x \ln x - x + C \qquad ◀$$

The next example illustrates how integration by parts can be used to integrate the inverse trigonometric functions.

Example 4   Evaluate $\displaystyle\int_0^1 \tan^{-1} x \, dx$.

Solution.   Let

$$u = \tan^{-1} x \qquad dv = dx$$

$$du = \frac{1}{1+x^2} \, dx \qquad v = \int dx = x$$

Thus,

$$\int_0^1 \tan^{-1} x \, dx = \int_0^1 u \, dv = uv \Big]_0^1 - \int_0^1 v \, du$$

$$= x \tan^{-1} x \Big]_0^1 - \int_0^1 \frac{x}{1+x^2} \, dx$$

> The limits of integration refer to x; that is, $x = 0$ and $x = 1$.

But

$$\int_0^1 \frac{x}{1+x^2} \, dx = \frac{1}{2} \int_0^1 \frac{2x}{1+x^2} \, dx = \frac{1}{2} \ln(1+x^2) \Big]_0^1 = \frac{1}{2} \ln 2$$

so

$$\int_0^1 \tan^{-1} x \, dx = x \tan^{-1} x \Big]_0^1 - \frac{1}{2} \ln 2 = \left(\frac{\pi}{4} - 0\right) - \frac{1}{2} \ln 2$$

$$= \frac{\pi}{4} - \ln \sqrt{2} \qquad \blacktriangleleft$$

Example 5   Evaluate $\displaystyle\int e^x \cos x \, dx$.

Solution.   Let

$$u = e^x \qquad dv = \cos x \, dx$$

$$du = e^x \, dx \qquad v = \int \cos x \, dx = \sin x$$

Thus,

$$\int e^x \cos x \, dx = \int u \, dv = uv - \int v \, du$$

$$= e^x \sin x - \int e^x \sin x \, dx \qquad (4)$$

Since the integral $\int e^x \sin x \, dx$ is similar in form to the original integral $\int e^x \cos x \, dx$, it seems that nothing has been accomplished. However, let us integrate this new integral by parts; we let

$$u = e^x \qquad dv = \sin x \, dx$$

$$du = e^x \, dx \qquad v = \int \sin x \, dx = -\cos x$$

Thus,

$$\int e^x \sin x \, dx = \int u \, dv = uv - \int v \, du$$

$$= -e^x \cos x + \int e^x \cos x \, dx$$

Substituting in (4) yields

$$\int e^x \cos x \, dx = e^x \sin x - \left[-e^x \cos x + \int e^x \cos x \, dx \right]$$

or

$$\int e^x \cos x \, dx = e^x \sin x + e^x \cos x - \int e^x \cos x \, dx$$

which is an equation we can solve for the unknown integral. We obtain

$$2 \int e^x \cos x \, dx = e^x \sin x + e^x \cos x$$

or

$$\int e^x \cos x \, dx = \frac{1}{2} e^x \sin x + \frac{1}{2} e^x \cos x + C \qquad \blacktriangleleft$$

☐ **REDUCTION FORMULAS**   If n is a positive integer ($n \geq 2$), then

$$\int \sin^n x \, dx \quad \text{and} \quad \int \cos^n x \, dx$$

can be evaluated by using **reduction formulas**. These are formulas that express the given integral in terms of a similar integral involving a *lower* power. For example, we can obtain a reduction formula for $\int \cos^n x \, dx$ by writing $\cos^n x$ as $\cos^{n-1} x \cdot \cos x$ and letting

$$u = \cos^{n-1} x \qquad\qquad dv = \cos x \, dx$$

$$du = (n - 1) \cos^{n-2} x (-\sin x) \, dx \qquad v = \int \cos x \, dx = \sin x$$

$$= -(n - 1) \cos^{n-2} x \sin x \, dx$$

so that

$$\int \cos^n x \, dx = \int \cos^{n-1} x \cos x \, dx = \int u \, dv = uv - \int v \, du$$

$$= \cos^{n-1} x \sin x + (n-1) \int \sin^2 x \cos^{n-2} x \, dx$$

$$= \cos^{n-1} x \sin x + (n-1) \int (1 - \cos^2 x) \cos^{n-2} x \, dx$$

$$= \cos^{n-1} x \sin x + (n-1) \int \cos^{n-2} x \, dx - (n-1) \int \cos^n x \, dx$$

Transposing the last term on the right to the left side yields

$$n \int \cos^n x \, dx = \cos^{n-1} x \sin x + (n-1) \int \cos^{n-2} x \, dx$$

or

$$\int \cos^n x \, dx = \frac{1}{n} \cos^{n-1} x \sin x + \frac{n-1}{n} \int \cos^{n-2} x \, dx \qquad (5)$$

This reduction formula reduces the exponent by 2. Thus, if we apply it repeatedly, we can eventually express $\int \cos^n x \, dx$ in terms of

$$\int \cos x \, dx = \sin x + C$$

if n is odd, or

$$\int \cos^0 x \, dx = \int dx = x + C$$

if n is even.

Example 6   Evaluate

$$\text{(a)} \quad \int \cos^3 x \, dx \qquad \text{(b)} \quad \int \cos^4 x \, dx$$

Solution (a).   From (5) with $n = 3$

$$\int \cos^3 x \, dx = \frac{1}{3} \cos^2 x \sin x + \frac{2}{3} \int \cos x \, dx$$

$$= \frac{1}{3} \cos^2 x \sin x + \frac{2}{3} \sin x + C$$

Solution (b). From (5) with $n = 4$

$$\int \cos^4 x \, dx = \frac{1}{4} \cos^3 x \sin x + \frac{3}{4} \int \cos^2 x \, dx$$

and from (5) with $n = 2$

$$\int \cos^2 x \, dx = \frac{1}{2} \cos x \sin x + \frac{1}{2} \int dx = \frac{1}{2} \cos x \sin x + \frac{1}{2} x + C_1$$

so that

$$\int \cos^4 x \, dx = \frac{1}{4} \cos^3 x \sin x + \frac{3}{4} \left(\frac{1}{2} \cos x \sin x + \frac{1}{2} x + C_1 \right)$$

$$= \frac{1}{4} \cos^3 x \sin x + \frac{3}{8} \cos x \sin x + \frac{3}{8} x + C$$

where $C = \frac{3}{4} C_1$ ◄

We leave it as an exercise to derive the following companion formula to (5).

$$\int \sin^n x \, dx = -\frac{1}{n} \sin^{n-1} x \cos x + \frac{n-1}{n} \int \sin^{n-2} x \, dx \qquad (6)$$

► Exercise Set 9.2

In Exercises 1–28, evaluate the integral.

1. $\int x e^{-x} \, dx$.

2. $\int x e^{3x} \, dx$.

3. $\int x^2 e^x \, dx$.

4. $\int x^2 e^{-2x} \, dx$.

5. $\int x \sin 2x \, dx$.

6. $\int x \cos 3x \, dx$.

7. $\int x^2 \cos x \, dx$.

8. $\int x^2 \sin x \, dx$.

9. $\int \sqrt{x} \ln x \, dx$.

10. $\int x \ln x \, dx$.

11. $\int (\ln x)^2 \, dx$.

12. $\int \frac{\ln x}{\sqrt{x}} \, dx$.

13. $\int \ln (2x + 3) \, dx$.

14. $\int \ln (x^2 + 4) \, dx$.

15. $\int \sin^{-1} x \, dx$.

16. $\int \cos^{-1} (2x) \, dx$.

17. $\int \tan^{-1} (2x) \, dx$.

18. $\int x \tan^{-1} x \, dx$.

19. $\int e^x \sin x \, dx$.

20. $\int e^{-3\theta} \sin 3\theta \, d\theta$.

21. $\int e^{ax} \sin bx \, dx$.

22. $\int e^{2x} \cos 3x \, dx$.

23. $\int \sin (\ln x) \, dx$.

24. $\int \cos (\ln x) \, dx$.

25. $\int x \sec^2 x \, dx$.

26. $\int x \tan^2 x \, dx$.

27. $\int x^3 e^{x^2} \, dx$.

28. $\int \frac{x e^x}{(x + 1)^2} \, dx$.

In Exercises 29–41, evaluate the definite integral.

29. $\int_0^1 xe^{-5x}\, dx.$

30. $\int_0^2 xe^{2x}\, dx.$

31. $\int_1^e x^2 \ln x\, dx.$

32. $\int_{\sqrt{e}}^e \frac{\ln x}{x^2}\, dx.$

33. $\int_{-2}^2 \ln(x+3)\, dx.$

34. $\int_0^{1/2} \sin^{-1} x\, dx.$

35. $\int_2^4 \sec^{-1}\sqrt{\theta}\, d\theta.$

36. $\int_1^2 x\sec^{-1} x\, dx.$

37. $\int_0^{\pi/2} x\sin 4x\, dx.$

38. $\int_0^\pi (x + x\cos x)\, dx.$

39. $\int_1^3 \sqrt{x}\tan^{-1}\sqrt{x}\, dx.$

40. $\int_0^2 \ln(x^2+1)\, dx.$

41. $\int_0^1 \frac{x^3}{\sqrt{x^2+1}}\, dx.$

42. Solve Exercise 41 without using integration by parts. [*Hint:* Let $u = \sqrt{x^2+1}$.]

43. (a) Find the area of the region enclosed by $y = \ln x$, the line $x = e$, and the x-axis.
(b) Find the volume of the solid generated when the region in part (a) is revolved about the x-axis.

44. Find the area of the region between $y = x\sin x$ and $y = x$ for $0 \le x \le \pi/2$.

45. Find the volume of the solid generated when the region between $y = \sin x$ and $y = 0$ for $0 \le x \le \pi$ is revolved about the y-axis.

46. Find the volume of the solid generated when the region enclosed between $y = \cos x$ and $y = 0$ for $0 \le x \le \pi/2$ is revolved about the y-axis.

47. Use reduction formula (6) to evaluate
(a) $\int \sin^3 x\, dx$ (b) $\int_0^{\pi/4} \sin^4 x\, dx.$

48. Use reduction formula (5) to evaluate
(a) $\int \cos^5 x\, dx$ (b) $\int_0^{\pi/2} \cos^6 x\, dx.$

49. Use reduction formula (5) to help evaluate
(a) $\int \cos^3 5x\, dx$ (b) $\int x\cos^4(x^2)\, dx.$

[*Hint:* First make a substitution.]

50. Use reduction formula (6) to help evaluate
(a) $\int \sin^4 2x\, dx$ (b) $\int \frac{\sin^3\sqrt{x}}{\sqrt{x}}\, dx.$

[*Hint:* First make a substitution.]

51. Derive reduction formula (6).

In Exercises 52 and 53, derive the reduction formula in part (a) and use it to evaluate the integral in (b).

52. (a) $\int \sec^n x\, dx$
$$= \frac{\sec^{n-2} x\tan x}{n-1} + \frac{n-2}{n-1}\int \sec^{n-2} x\, dx$$
(b) $\int \sec^4 x\, dx.$

53. (a) $\int x^n e^x\, dx = x^n e^x - n\int x^{n-1} e^x\, dx$
(b) $\int x^3 e^x\, dx.$

54. Use the reduction formula in part (a) of Exercise 53 to help evaluate
(a) $\int x^2 e^{3x}\, dx$ (b) $\int_0^1 xe^{-\sqrt{x}}\, dx.$

[*Hint:* First make a substitution.]

55. Let f be a function whose second derivative is continuous on $[-1, 1]$. Show that
$$\int_{-1}^1 xf''(x)\, dx = f'(1) + f'(-1) + f(-1) - f(1)$$

56. (a) In Example 1, let
$$u = x \qquad dv = e^x\, dx$$
$$du = dx \qquad v = \int e^x\, dx = e^x + C_1$$
and show that the constant C_1 cancels out, thus leading to the same solution we obtained by omitting C_1.
(b) Show that
$$uv - \int v\, du = u(v + C_1) - \int (v + C_1)\, du$$
thereby justifying the omission of the constant of integration when calculating v in integration by parts.

57. Use integration by parts on $\int \frac{1}{x}\, dx$ with $u = 1/x$ and $dv = dx$. Explain.

◼ 9.3 INTEGRATING POWERS OF SINE AND COSINE

In this section we shall study methods for evaluating integrals of the form

$$\int \sin^m x \cos^n x \, dx$$

where m and n are nonnegative integers.

Integrals of the form

$$\int \sin^m x \, dx \quad \text{and} \quad \int \cos^n x \, dx$$

can be evaluated using reduction formulas (5) and (6) derived in the previous section. (See Example 6 of that section.) In the first three examples that follow we shall give alternative methods for evaluating such integrals. In some cases these alternative methods will result in simpler forms of the antiderivatives than those that result from the reduction formulas. The methods that we shall describe use the trigonometric identities

$$\sin^2 x = \frac{1}{2}(1 - \cos 2x) \tag{1a}$$

$$\cos^2 x = \frac{1}{2}(1 + \cos 2x) \tag{1b}$$

which follow from the double-angle formulas

$$\cos 2x = 1 - 2\sin^2 x \quad \text{and} \quad \cos 2x = 2\cos^2 x - 1$$

Example 1 Show that

$$\int \sin^2 x \, dx = \frac{1}{2}x - \frac{1}{4}\sin 2x + C \tag{2}$$

$$\int \cos^2 x \, dx = \frac{1}{2}x + \frac{1}{4}\sin 2x + C \tag{3}$$

Solution. It follows from (1a) and (1b) that

$$\int \sin^2 x \, dx = \frac{1}{2}\int(1 - \cos 2x)\, dx = \frac{1}{2}x - \frac{1}{4}\sin 2x + C$$

$$\int \cos^2 x \, dx = \frac{1}{2}\int(1 + \cos 2x)\, dx = \frac{1}{2}x + \frac{1}{4}\sin 2x + C \quad ◀$$

Example 2   Show that

$$\int \sin^4 x \; dx = \frac{3}{8} x - \frac{1}{4} \sin 2x + \frac{1}{32} \sin 4x + C \qquad (4)$$

$$\int \cos^4 x \; dx = \frac{3}{8} x + \frac{1}{4} \sin 2x + \frac{1}{32} \sin 4x + C \qquad (5)$$

Solution.

$$\int \cos^4 x \; dx = \int (\cos^2 x)^2 \; dx = \int \left[\frac{1}{2} (1 + \cos 2x) \right]^2 dx$$

$$= \frac{1}{4} \int (1 + 2 \cos 2x + \cos^2 2x) \; dx$$

To finish, we apply (1b) again and write

$$\cos^2 2x = \frac{1}{2} (1 + \cos 4x) = \frac{1}{2} + \frac{1}{2} \cos 4x$$

which gives

$$\int \cos^4 x \; dx = \frac{1}{4} \int \left(\frac{3}{2} + 2 \cos 2x + \frac{1}{2} \cos 4x \right) dx$$

$$= \frac{3}{8} x + \frac{1}{4} \sin 2x + \frac{1}{32} \sin 4x + C$$

A similar procedure, using (1a), will yield (4).   ◀

Example 3   Show that

$$\int \sin^3 x \; dx = -\cos x + \frac{1}{3} \cos^3 x + C \qquad (6)$$

$$\int \cos^3 x \; dx = \sin x - \frac{1}{3} \sin^3 x + C \qquad (7)$$

Solution.

$$\int \sin^3 x \; dx = \int \sin^2 x \sin x \; dx$$

$$= \int (1 - \cos^2 x) \sin x \; dx$$

$$= \int \sin x \; dx - \int \cos^2 x \sin x \; dx$$

> To integrate, let
> $u = \cos x$, $du = -\sin x \; dx$.

$$= -\cos x + \frac{1}{3}\cos^3 x + C$$

$$\int \cos^3 x \, dx = \int \cos^2 x \cos x \, dx$$

$$= \int (1 - \sin^2 x) \cos x \, dx$$

$$= \int \cos x \, dx - \int \sin^2 x \cos x \, dx$$

$$= \sin x - \frac{1}{3}\sin^3 x + C \quad \blacktriangleleft$$

To integrate, let
$u = \sin x, \, du = \cos x \, dx.$

If m and n are both positive integers, then the integral

$$\int \sin^m x \cos^n x \, dx$$

can be evaluated by one of three procedures, depending on whether m and n are odd or even. The procedures are outlined in Table 9.3.1.

Table 9.3.1

$\int \sin^m x \cos^n x \, dx$	PROCEDURE	RELEVANT IDENTITIES
n odd	Substitute $u = \sin x$	$\cos^2 x = 1 - \sin^2 x$
m odd	Substitute $u = \cos x$	$\sin^2 x = 1 - \cos^2 x$
$\begin{cases} m \text{ even} \\ n \text{ even} \end{cases}$	Use identities to reduce the powers on $\sin x$ and $\cos x$	$\begin{cases} \sin^2 x = \frac{1}{2}(1 - \cos 2x) \\ \cos^2 x = \frac{1}{2}(1 + \cos 2x) \end{cases}$

Example 4   Evaluate $\int \sin^4 x \cos^5 x \, dx$.

Solution.   Since $n = 5$ is odd, we shall follow the first procedure in Table 9.3.1 and make the substitution

$$u = \sin x \quad du = \cos x \, dx$$

To form the du, we shall first split off a factor of $\cos x$.

$$\int \sin^4 x \cos^5 x \, dx = \int \sin^4 x \cos^4 x \cos x \, dx$$

$$= \int \sin^4 x (1 - \sin^2 x)^2 \cos x \, dx$$

$$= \int u^4 (1 - u^2)^2 \, du$$

$$= \int (u^4 - 2u^6 + u^8) \, du$$

$$= \frac{1}{5} u^5 - \frac{2}{7} u^7 + \frac{1}{9} u^9 + C$$

$$= \frac{1}{5} \sin^5 x - \frac{2}{7} \sin^7 x + \frac{1}{9} \sin^9 x + C \qquad \blacktriangleleft$$

Example 5   Evaluate $\int \sin^3 x \cos^2 x \, dx$.

Solution.   Since $m = 3$ is odd, we shall follow the second procedure in Table 9.3.1 and make the substitution

$$u = \cos x \quad du = -\sin x \, dx$$

To form the du, we shall split off a factor of $\sin x$.

$$\int \sin^3 x \cos^2 x \, dx = \int \sin^2 x \cos^2 x \sin x \, dx$$

$$= \int (1 - \cos^2 x) \cos^2 x \sin x \, dx$$

$$= -\int (1 - u^2) u^2 \, du$$

$$= \int (u^4 - u^2) \, du$$

$$= \frac{1}{5} u^5 - \frac{1}{3} u^3 + C$$

$$= \frac{1}{5} \cos^5 x - \frac{1}{3} \cos^3 x + C \qquad \blacktriangleleft$$

Example 6   Evaluate $\int \sin^4 x \cos^4 x \, dx$.

Solution.   Since $m = n = 4$, we shall follow the third procedure in Table 9.3.1.

$$\int \sin^4 x \cos^4 x \, dx = \int (\sin^2 x)^2 (\cos^2 x)^2 \, dx$$

$$= \int (\tfrac{1}{2}[1 - \cos 2x])^2 (\tfrac{1}{2}[1 + \cos 2x])^2 \, dx$$

$$= \frac{1}{16} \int (1 - \cos^2 2x)^2 \, dx$$

$$= \frac{1}{16} \int \sin^4 2x \, dx$$

To finish, we shall let $u = 2x$, $du = 2 \, dx$ and then use (4).

$$\int \sin^4 x \cos^4 x \, dx = \frac{1}{32} \int \sin^4 u \, du$$

$$= \frac{1}{32} \left(\frac{3}{8} u - \frac{1}{4} \sin 2u + \frac{1}{32} \sin 4u \right) + C$$

$$= \frac{3}{128} x - \frac{1}{128} \sin 4x + \frac{1}{1024} \sin 8x + C \quad \blacktriangleleft$$

Integrals of the form

$$\int \sin mx \cos nx \, dx, \quad \int \sin mx \sin nx \, dx, \quad \int \cos mx \cos nx \, dx$$

can be found using the product to sum formulas from trigonometry (see 28a, 28b, and 28c of Section 1 in Appendix B).

Example 7 Evaluate $\int \sin 7x \cos 3x \, dx$.

Solution. Since

$$\sin 7x \cos 3x = \frac{1}{2} (\sin 4x + \sin 10x)$$

we can write

$$\int \sin 7x \cos 3x \, dx = \frac{1}{2} \int (\sin 4x + \sin 10x) \, dx$$

$$= -\frac{1}{8} \cos 4x - \frac{1}{20} \cos 10x + C \quad \blacktriangleleft$$

▶ Exercise Set 9.3

In Exercises 1–30, perform the indicated integration.

1. $\int \cos^5 x \sin x \, dx$.

2. $\int \sin^4 3x \cos 3x \, dx$.

3. $\int \sin ax \cos ax \, dx \quad (a \neq 0)$.

4. $\int \cos^2 3x \, dx$.

5. $\int \sin^2 5\theta \, d\theta$.

6. $\int \cos^3 at \, dt \quad (a \neq 0)$.

7. $\int \cos^4 \left(\frac{x}{4} \right) \, dx$.

8. $\int \sin^5 x \, dx$.

9. $\int \cos^5 \theta \, d\theta$.

10. $\int \sin^3 x \cos^3 x \, dx$.

11. $\int \sin^2 2t \cos^3 2t \, dt$.

12. $\int \sin^4 x \cos^5 x \, dx$.

13. $\int \cos^4 x \sin^3 x \, dx$.

14. $\int \sin^3 2x \cos^2 2x \, dx$.

15. $\int \sin^5 \theta \cos^4 \theta \, d\theta$.

16. $\int \cos^{1/5} x \sin x \, dx$.

17. $\int \sin^2 x \cos^2 x \, dx$.

18. $\displaystyle\int \sin^2 x \cos^4 x \, dx.$

19. $\displaystyle\int \sin x \cos 2x \, dx.$

20. $\displaystyle\int \sin 3\theta \cos 2\theta \, d\theta.$

21. $\displaystyle\int \sin x \cos \left(\frac{x}{2}\right) dx.$

22. $\displaystyle\int \sin ax \cos bx \, dx$ $(a > 0, b > 0, a \neq b).$

23. $\displaystyle\int \frac{\sin x}{\cos^8 x} \, dx.$

24. $\displaystyle\int \sqrt{\cos \theta} \sin \theta \, d\theta.$

25. $\displaystyle\int_0^{\pi/4} \cos^3 x \, dx.$

26. $\displaystyle\int_{-\pi}^{\pi} \cos^2 5\theta \, d\theta.$

27. $\displaystyle\int_0^{\pi/3} \sin^4 3x \cos^3 3x \, dx.$

28. $\displaystyle\int_0^{\pi/2} \sin^2 \frac{x}{2} \cos^2 \frac{x}{2} \, dx.$

29. $\displaystyle\int_0^{\pi/6} \sin 2x \cos 4x \, dx.$

30. $\displaystyle\int_0^{2\pi} \sin^2 kx \, dx$ $(k \neq 0).$

31. Let m, n be distinct nonnegative integers. Prove:

(a) $\displaystyle\int_0^{2\pi} \sin mx \cos nx \, dx = 0$

(b) $\displaystyle\int_0^{2\pi} \cos mx \cos nx \, dx = 0$

(c) $\displaystyle\int_0^{2\pi} \sin mx \sin nx \, dx = 0.$

32. The region bounded below by the x-axis and above

by the portion of $y = \sin x$ from $x = 0$ to $x = \pi$ is revolved about the x-axis. Find the volume of the resulting solid.

33. Find the volume of the solid that results when the region enclosed by $y = \cos x$, $y = \sin x$, $x = 0$, and $x = \pi/4$ is revolved about the x-axis.

34. (a) Use Formula (6) in Section 9.2 to show that

$$\int_0^{\pi/2} \sin^n x \, dx = \frac{n-1}{n} \int_0^{\pi/2} \sin^{n-2} x \, dx$$

(b) Use this result to derive the **Wallis sine formulas:**

$$\int_0^{\pi/2} \sin^n x \, dx = \frac{\pi}{2} \cdot \frac{1 \cdot 3 \cdot 5 \cdots (n-1)}{2 \cdot 4 \cdot 6 \cdots n} \quad \left(\begin{array}{c} n \text{ even} \\ \text{and} \geq 2 \end{array}\right)$$

$$\int_0^{\pi/2} \sin^n x \, dx = \frac{2 \cdot 4 \cdot 6 \cdots (n-1)}{3 \cdot 5 \cdot 7 \cdots n} \quad \left(\begin{array}{c} n \text{ odd} \\ \text{and} \geq 3 \end{array}\right)$$

35. Use the Wallis formulas in Exercise 34 to evaluate

(a) $\displaystyle\int_0^{\pi/2} \sin^3 x \, dx$ (b) $\displaystyle\int_0^{\pi/2} \sin^4 x \, dx$

(c) $\displaystyle\int_0^{\pi/2} \sin^5 x \, dx$ (d) $\displaystyle\int_0^{\pi/2} \sin^6 x \, dx.$

36. Use Formula (5) in Section 9.2 and the method of Exercise 34 to derive the **Wallis cosine formulas:**

$$\int_0^{\pi/2} \cos^n x \, dx = \frac{2 \cdot 4 \cdot 6 \cdots (n-1)}{3 \cdot 5 \cdot 7 \cdots n} \quad \left(\begin{array}{c} n \text{ odd} \\ \text{and} \geq 3 \end{array}\right)$$

$$\int_0^{\pi/2} \cos^n x \, dx = \frac{\pi}{2} \cdot \frac{1 \cdot 3 \cdot 5 \cdots (n-1)}{2 \cdot 4 \cdot 6 \cdots n} \quad \left(\begin{array}{c} n \text{ even} \\ \text{and} \geq 2 \end{array}\right)$$

37. Derive (6) and (7) using reduction formulas and appropriate trigonometric identities.

9.4   INTEGRATING POWERS OF SECANT AND TANGENT

In this section we shall discuss methods for evaluating integrals of the form

$$\int \tan^m x \sec^n x \, dx$$

where m and n are nonnegative integers.

We shall begin with the integrals

$$\int \tan x \, dx \qquad \text{and} \qquad \int \sec x \, dx$$

The first integral is evaluated by writing

$$\int \tan x \, dx = \int \frac{\sin x}{\cos x} \, dx$$

from which it follows with $u = \cos x$ and $du = -\sin x \, dx$ that

$$\int \tan x = -\ln |\cos x| + C \tag{1a}$$

or since $-\ln |\cos x| = \ln (1/|\cos x|) = \ln |\sec x|$,

$$\int \tan x \, dx = \ln |\sec x| + C \tag{1b}$$

The second integral requires a trick. We write

$$\int \sec x \, dx = \int \sec x \left(\frac{\sec x + \tan x}{\sec x + \tan x} \right) dx$$

$$= \int \frac{\sec^2 x + \sec x \tan x}{\sec x + \tan x} \, dx$$

$$= \int \frac{du}{u} \qquad \boxed{\begin{array}{l} u = \sec x + \tan x \\ du = (\sec^2 x + \sec x \tan x) \, dx \end{array}}$$

$$= \ln |u| + C$$

from which it follows that

$$\int \sec x \, dx = \ln |\sec x + \tan x| + C \tag{2}$$

Higher powers of secant and tangent can be evaluated using the reduction formulas

$$\int \sec^n x \, dx = \frac{\sec^{n-2} x \tan x}{n - 1} + \frac{n - 2}{n - 1} \int \sec^{n-2} x \, dx \tag{3}$$

$$\int \tan^m x \, dx = \frac{\tan^{m-1} x}{m - 1} - \int \tan^{m-2} x \, dx \tag{4}$$

Formula (3) was discussed in Exercise 52 of Section 9.2, and Formula (4) can be obtained from the identity

$$1 + \tan^2 x = \sec^2 x$$

by writing

$$\int \tan^m x \, dx = \int \tan^{m-2} x \, \tan^2 x \, dx = \int \tan^{m-2} x \, (\sec^2 x - 1) \, dx$$

$$= \int \tan^{m-2} x \, \sec^2 x \, dx - \int \tan^{m-2} x \, dx$$

$$= \frac{\tan^{m-1} x}{m-1} - \int \tan^{m-2} x \, dx \qquad \boxed{\begin{array}{l} u = \tan x \\ du = \sec^2 x \, dx \end{array}}$$

Example 1   Evaluate $\int \sec^3 x \, dx$.

Solution.   From (3) with $n = 3$,

$$\int \sec^3 x \, dx = \frac{\sec x \tan x}{2} + \frac{1}{2} \int \sec x \, dx$$

$$= \frac{1}{2} \sec x \tan x + \frac{1}{2} \ln |\sec x + \tan x| + C \qquad \blacktriangleleft$$

Example 2   Evaluate $\int \tan^5 x \, dx$.

Solution.   We shall use Formula (4) twice.

$$\int \tan^5 x \, dx = \frac{\tan^4 x}{4} - \int \tan^3 x \, dx$$

$$= \frac{\tan^4 x}{4} - \left[\frac{\tan^2 x}{2} - \int \tan x \, dx \right]$$

$$= \frac{1}{4} \tan^4 x - \frac{1}{2} \tan^2 x - \ln |\cos x| + C \qquad \blacktriangleleft$$

If m and n are positive integers, then the integral

$$\int \tan^m x \, \sec^n x \, dx$$

can be evaluated by one of the three procedures in Table 9.4.1.

Table 9.4.1

$\int \tan^m x \, \sec^n x \, dx$	PROCEDURE	RELEVANT IDENTITIES
n even	Substitute $u = \tan x$	$\sec^2 x = \tan^2 x + 1$
m odd	Substitute $u = \sec x$	$\tan^2 x = \sec^2 x - 1$
$\begin{cases} m \text{ even} \\ n \text{ odd} \end{cases}$	Reduce to powers of $\sec x$ alone	$\tan^2 x = \sec^2 x - 1$

Example 3 Evaluate $\int \tan^2 x \sec^4 x \, dx$.

Solution. Since $n = 4$ is even, we shall follow the first procedure in Table 9.4.1 and make the substitution

$$u = \tan x, \quad du = \sec^2 x \, dx$$

To form the du, we shall split off a factor of $\sec^2 x$.

$$\int \tan^2 x \sec^4 x \, dx = \int \tan^2 x \sec^2 x \sec^2 x \, dx$$

$$= \int \tan^2 x (\tan^2 x + 1) \sec^2 x \, dx$$

$$= \int u^2(u^2 + 1) \, du$$

$$= \frac{1}{5} u^5 + \frac{1}{3} u^3 + C = \frac{1}{5} \tan^5 x + \frac{1}{3} \tan^3 x + C \quad \blacktriangleleft$$

Example 4 Evaluate $\int \tan^3 x \sec^3 x \, dx$.

Solution. Since $m = 3$ is odd, we shall follow the second procedure in Table 9.4.1 and make the substitution

$$u = \sec x, \quad du = \sec x \tan x \, dx$$

To form the du, we shall split off the product $\sec x \tan x$.

$$\int \tan^3 x \sec^3 x \, dx = \int \tan^2 x \sec^2 x (\sec x \tan x) \, dx$$

$$= \int (\sec^2 x - 1) \sec^2 x (\sec x \tan x) \, dx$$

$$= \int (u^2 - 1)u^2 \, du$$

$$= \frac{1}{5} u^5 - \frac{1}{3} u^3 + C = \frac{1}{5} \sec^5 x - \frac{1}{3} \sec^3 x + C \quad \blacktriangleleft$$

Example 5 Evaluate $\int \tan^2 x \sec x \, dx$.

Solution. Since $m = 2$ and $n = 1$, we shall follow the third procedure in Table 9.4.1.

$$\int \tan^2 x \sec x \, dx = \int (\sec^2 x - 1) \sec x \, dx = \int \sec^3 x \, dx - \int \sec x \, dx$$

See Example 1.

$$= \frac{1}{2} \sec x \tan x + \frac{1}{2} \ln |\sec x + \tan x| - \ln |\sec x + \tan x| + C$$

$$= \frac{1}{2} \sec x \tan x - \frac{1}{2} \ln |\sec x + \tan x| + C \quad \blacktriangleleft$$

Example 6   Instead of using reduction formula (3), the first procedure of Table 9.4.1 can be used to integrate an even power of $\sec x$. For example,

$$\int \sec^6 x \, dx = \int \sec^4 x \sec^2 x \, dx = \int (\sec^2 x)^2 \sec^2 x \, dx$$

$$= \int (\tan^2 x + 1)^2 \sec^2 x \, dx = \int (u^2 + 1)^2 \, du$$

To integrate, let $u = \tan x$, $du = \sec^2 x \, dx$.

$$= \int (u^4 + 2u^2 + 1) \, du = \frac{1}{5} u^5 + \frac{2}{3} u^3 + u + C$$

$$= \frac{1}{5} \tan^5 x + \frac{2}{3} \tan^3 x + \tan x + C \quad \blacktriangleleft$$

REMARK.   With the aid of the identity $1 + \cot^2 x = \csc^2 x$ some of the techniques in this section can be adapted to treat integrals of the form

$$\int \cot^m x \csc^n x \, dx$$

▶ Exercise Set 9.4

In Exercises 1–34, perform the indicated integration.

1. $\int \sec^2 (3x + 1) \, dx$.

2. $\int \tan 5x \, dx$.

3. $\int e^{-2x} \tan (e^{-2x}) \, dx$.

4. $\int \cot 3x \, dx$.

5. $\int \sec 2x \, dx$.

6. $\int \frac{\sec (\sqrt{x})}{\sqrt{x}} \, dx$.

7. $\int \tan^2 x \sec^2 x \, dx$.

8. $\int \tan^5 x \sec^4 x \, dx$.

9. $\int \tan^3 4x \sec^4 4x \, dx$.

10. $\int \tan^4 \theta \sec^4 \theta \, d\theta$.

11. $\int \sec^5 x \tan^3 x \, dx$.

12. $\int \tan^5 \theta \sec \theta \, d\theta$.

13. $\int \tan^4 x \sec x \, dx$.

14. $\int \tan^2 \frac{x}{2} \sec^3 \frac{x}{2} \, dx$.

15. $\int \tan 2t \sec^3 2t \, dt$.

16. $\int \tan x \sec^5 x \, dx$.

17. $\int \sec^4 x \, dx$.

18. $\int \sec^5 x \, dx$.

19. $\int \sec^6 (\pi x) \, dx$.

20. $\int \tan^3 4x \, dx$.

21. $\int \tan^4 x \, dx$.

22. $\int \tan^7 \theta \, d\theta$.

23. $\int x \tan^2 (x^2) \sec^2 (x^2) \, dx$.

24. $\int \tan^2 (1 - 2x) \sec (1 - 2x) \, dx$.

25. $\displaystyle\int \cot^3 x \csc^3 x \, dx.$ **26.** $\displaystyle\int \cot^2 3t \sec 3t \, dt.$

27. $\displaystyle\int \cot^3 x \, dx.$ **28.** $\displaystyle\int \csc^4 x \, dx.$

29. $\displaystyle\int \sqrt{\tan x} \sec^4 x \, dx.$ **30.** $\displaystyle\int \tan x \sec^{3/2} x \, dx.$

31. $\displaystyle\int_0^{\pi/6} \tan^2 2x \, dx.$ **32.** $\displaystyle\int_0^{\pi/6} \sec^3 \theta \tan \theta \, d\theta.$

33. $\displaystyle\int_0^{\pi/2} \tan^5 \frac{x}{2} \, dx.$ **34.** $\displaystyle\int_{\pi/4}^{\pi/2} \csc^3 x \cot x \, dx.$

35. Find the arc length of the curve $y = \ln(\cos x)$ over the interval $[0, \pi/4]$.

36. Find the volume of the solid generated when the region enclosed by $y = \tan x$, $y = 1$, and $x = 0$ is revolved about the x-axis.

37. (a) Show that

$$\int \csc x \, dx = -\ln|\csc x + \cot x| + C$$

(b) Show that the result in part (a) can also be written as

$$\int \csc x \, dx = \ln|\csc x - \cot x| + C$$

and

$$\int \csc x \, dx = \ln|\tan \tfrac{1}{2} x| + C$$

38. Rewrite $\sin x + \cos x$ in the form

$$A \sin(x + \phi)$$

and use your result together with Exercise 37 to evaluate

$$\int \frac{dx}{\sin x + \cos x}$$

39. Use the method of Exercise 38 to evaluate

$$\int \frac{dx}{a \sin x + b \cos x} \qquad (a, b \text{ not both zero})$$

40. Use integration by parts and Formula (2) to evaluate $\int \sec^3 x \, dx$.

9.5 TRIGONOMETRIC SUBSTITUTIONS

In this section we shall show how to evaluate integrals that contain expressions of the form

$$\sqrt{a^2 - x^2}, \quad \sqrt{x^2 + a^2}, \quad and \quad \sqrt{x^2 - a^2}$$

($a > 0$) by making substitutions involving trigonometric functions. We shall also show how the method of completing the square can sometimes be used to help evaluate integrals involving expressions of the form $ax^2 + bx + c$.

The basic idea for evaluating an integral that involves one of the radicals described above is to make a substitution that will eliminate the radical. For example, to eliminate the radical in the expression

$$\sqrt{a^2 - x^2}$$

we can make the substitution

$$x = a \sin \theta, \quad -\pi/2 \le \theta \le \pi/2 \tag{1}$$

which yields

$$\sqrt{a^2 - x^2} = \sqrt{a^2 - a^2 \sin^2 \theta} = \sqrt{a^2(1 - \sin^2 \theta)}$$

$$= a\sqrt{\cos^2 \theta} = a |\cos \theta| = a \cos \theta$$

$$\boxed{\cos \theta \ge 0 \text{ since } -\pi/2 \le \theta \le \pi/2}$$

The purpose of the restriction on θ in (1) is twofold. First, it enables us to replace $|\cos \theta|$ by $\cos \theta$, thereby simplifying the resulting calculations, and second, the interval to which θ is restricted is the range of the function $\sin^{-1}$ (see Section 8.1), so (1) can be rewritten as

$$\theta = \sin^{-1}\left(\frac{x}{a}\right)$$

where needed. The following table lists the trigonometric substitutions that we will be using in this section.

EXPRESSION IN THE INTEGRAND	SUBSTITUTION	RESTRICTION ON θ	TRIGONOMETRIC IDENTITY NEEDED FOR SIMPLIFICATION
$\sqrt{a^2 - x^2}$	$x = a \sin \theta$	$-\pi/2 \le \theta \le \pi/2$	$a^2 - a^2 \sin^2 \theta = a^2 \cos^2 \theta$
$\sqrt{a^2 + x^2}$	$x = a \tan \theta$	$-\pi/2 < \theta < \pi/2$	$a^2 + a^2 \tan^2 \theta = a^2 \sec^2 \theta$
$\sqrt{x^2 - a^2}$	$x = a \sec \theta$	$\begin{cases} 0 \le \theta < \pi/2 & \text{(if } x \ge a) \\ \pi \le \theta < 3\pi/2 & \text{(if } x \le -a) \end{cases}$	$a^2 \sec^2 \theta - a^2 = a^2 \tan^2 \theta$

Example 1 Evaluate

$$\int \frac{dx}{x^2\sqrt{4 - x^2}}$$

Solution. To eliminate the radical we make the substitution

$$x = 2 \sin \theta, \quad -\pi/2 \le \theta \le \pi/2$$

so that

$$\frac{dx}{d\theta} = 2 \cos \theta \quad \text{or} \quad dx = 2 \cos \theta \, d\theta$$

This yields

$$\int \frac{dx}{x^2\sqrt{4 - x^2}} = \int \frac{2 \cos \theta \, d\theta}{(2 \sin \theta)^2 \sqrt{4 - 4 \sin^2 \theta}}$$

$$= \int \frac{2 \cos \theta \, d\theta}{(2 \sin \theta)^2 (2 \cos \theta)} = \frac{1}{4} \int \frac{d\theta}{\sin^2 \theta}$$

$$= \frac{1}{4} \int \csc^2 \theta \, d\theta = -\frac{1}{4} \cot \theta + C$$

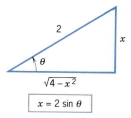

Figure 9.5.1

To complete the solution we must express $\cot\theta$ in terms of x. This can be done using trigonometric identities or more simply by displaying the substitution $x = 2\sin\theta$ ($\sin\theta = x/2$) as in Figure 9.5.1. From the figure we obtain

$$\cot\theta = \frac{\sqrt{4-x^2}}{x}$$

so that

$$\int \frac{dx}{x^2\sqrt{4-x^2}} = -\frac{1}{4}\cot\theta + C = -\frac{1}{4}\frac{\sqrt{4-x^2}}{x} + C \quad \blacktriangleleft$$

Example 2 Evaluate

$$\int \frac{dx}{\sqrt{x^2+a^2}}$$

Solution. To eliminate the radical we make the substitution

$$x = a\tan\theta, \quad -\frac{\pi}{2} < \theta < \frac{\pi}{2} \tag{2}$$

so that

$$\frac{dx}{d\theta} = a\sec^2\theta \quad \text{or} \quad dx = a\sec^2\theta\,d\theta$$

This yields

$$\int \frac{dx}{\sqrt{x^2+a^2}} = \int \frac{a\sec^2\,d\theta}{\sqrt{a^2\tan^2\theta+a^2}} = \int \frac{a\sec^2\theta\,d\theta}{a\,|\sec\theta|} = \int \frac{a\sec^2\theta\,d\theta}{a\sec\theta}$$

$$\boxed{\sec\theta > 0 \text{ since } -\pi/2 < \theta < \pi/2}$$

$$= \int \sec\theta\,d\theta = \ln|\sec\theta + \tan\theta| + C$$

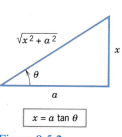

Figure 9.5.2

To express the solution in terms of x, we can represent (2) by the triangle in Figure 9.5.2, from which we obtain

$$\sec\theta = \frac{\sqrt{x^2+a^2}}{a} \quad \text{and} \quad \tan\theta = \frac{x}{a}$$

so

$$\int \frac{dx}{\sqrt{x^2+a^2}} = \ln\left|\frac{\sqrt{x^2+a^2}}{a} + \frac{x}{a}\right| + C$$

or if preferred we can rewrite the expression on the right as

$$\ln\left|\sqrt{x^2+a^2} + x\right| - \ln a + C$$

and combine the constant $\ln a$ with the constant of integration to obtain

$$\int \frac{dx}{\sqrt{x^2 + a^2}} = \ln|\sqrt{x^2 + a^2} + x| + C'$$

Moreover, $\sqrt{x^2 + a^2} + x > 0$ for all x, so we can drop the absolute value sign and write

$$\int \frac{dx}{\sqrt{x^2 + a^2}} = \ln(\sqrt{x^2 + a^2} + x) + C' \tag{3}$$

◄

Example 3   Evaluate

$$\int \frac{\sqrt{x^2 - 25}}{x} \, dx$$

Solution.   To eliminate the radical, we make the substitution

$$x = 5 \sec \theta, \quad 0 \le \theta < \pi/2 \quad \text{or} \quad \pi \le \theta < 3\pi/2 \tag{4}$$

so that

$$\frac{dx}{d\theta} = 5 \sec \theta \tan \theta \quad \text{or} \quad dx = 5 \sec \theta \tan \theta \, d\theta$$

Thus,

$$\int \frac{\sqrt{x^2 - 25}}{x} \, dx = \int \frac{\sqrt{25 \sec^2 \theta - 25}}{5 \sec \theta} (5 \sec \theta \tan \theta) \, d\theta$$

$$= \int \frac{5 |\tan \theta|}{5 \sec \theta} (5 \sec \theta \tan \theta) \, d\theta$$

$$= 5 \int \tan^2 \theta \, d\theta \qquad \boxed{\begin{array}{l} \tan \theta \ge 0 \text{ since} \\ 0 \le \theta < \pi/2 \quad \text{or} \quad \pi \le \theta < 3\pi/2 \end{array}}$$

$$= 5 \int (\sec^2 \theta - 1) \, d\theta$$

$$= 5 \tan \theta - 5\theta + C$$

To express the solution in terms of x, we can represent (4) by the triangle in Figure 9.5.3, from which we obtain

$$\tan \theta = \frac{\sqrt{x^2 - 25}}{5}$$

so that

$$\int \frac{\sqrt{x^2 - 25}}{x} \, dx = \sqrt{x^2 - 25} - 5 \sec^{-1}\left(\frac{x}{5}\right) + C \qquad ◄$$

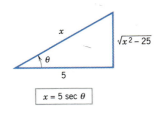

$x = 5 \sec \theta$

Figure 9.5.3

The integral in the next example will arise frequently in later sections.

Example 4 Evaluate

$$\int_{-a}^{a} \sqrt{a^2 - x^2}\, dx \quad (a > 0)$$

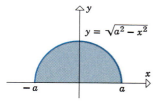

$y = \sqrt{a^2 - x^2}$

$-a \qquad a$

Figure 9.5.4

Solution. This integral can be evaluated by the substitution $x = a \sin \theta$ (verify); however, a better approach is to observe that the integral represents the area of a semicircle of radius a (Figure 9.5.4). Thus,

$$\int_{-a}^{a} \sqrt{a^2 - x^2}\, dx = \frac{1}{2}\pi a^2 \qquad \blacktriangleleft$$

☐ **INTEGRALS INVOLVING**
$ax^2 + bx + c$

Integrals that involve a quadratic expression $ax^2 + bx + c$, where $a \neq 0$ and $b \neq 0$, can often be evaluated by first completing the square, then making an appropriate substitution. The following examples illustrate this idea.

Example 5 Evaluate

$$\int \frac{dx}{x^2 - 2x + 5}$$

Solution. Completing the square yields

$$x^2 - 2x + 5 = (x^2 - 2x + 1) + 5 - 1 = (x - 1)^2 + 4$$

Thus,

$$\int \frac{dx}{x^2 - 2x + 5} = \int \frac{dx}{(x-1)^2 + 4} = \int \frac{du}{u^2 + 4}$$

$$\begin{array}{l} u = x - 1 \\ du = dx \end{array}$$

$$= \frac{1}{2} \tan^{-1} \frac{u}{2} + C = \frac{1}{2} \tan^{-1} \left(\frac{x - 1}{2} \right) + C \qquad \blacktriangleleft$$

Formula (22)
of Section 9.1

Example 6 Evaluate

$$\int \frac{dx}{\sqrt{5 - 4x - 2x^2}}$$

Solution. Completing the square yields

$$5 - 4x - 2x^2 = 5 - 2(x^2 + 2x) = 5 - 2(x^2 + 2x + 1) + 2$$
$$= 5 - 2(x + 1)^2 + 2 = 7 - 2(x + 1)^2$$

Thus,

$$\int \frac{dx}{\sqrt{5 - 4x - 2x^2}} = \int \frac{dx}{\sqrt{7 - 2(x + 1)^2}}$$

$$= \int \frac{du}{\sqrt{7 - 2u^2}} \qquad \boxed{\begin{array}{l} u = x + 1 \\ du = dx \end{array}}$$

$$= \frac{1}{\sqrt{2}} \int \frac{du}{\sqrt{(7/2) - u^2}}$$

$$= \frac{1}{\sqrt{2}} \sin^{-1}\left(\frac{u}{\sqrt{7/2}}\right) + C \qquad \boxed{\begin{array}{l} \text{Formula (21),} \\ \text{Section 9.1} \\ \text{with } a = \sqrt{7/2} \end{array}}$$

$$= \frac{1}{\sqrt{2}} \sin^{-1}(\sqrt{2/7}\, u) + C$$

$$= \frac{1}{\sqrt{2}} \sin^{-1}(\sqrt{2/7}(x + 1)) + C \qquad ◀$$

Example 7   Evaluate

$$\int \frac{x}{x^2 - 4x + 8}\, dx$$

Solution.   Completing the square yields

$$x^2 - 4x + 8 = (x^2 - 4x + 4) + 8 - 4 = (x - 2)^2 + 4$$

Thus, the substitution

$$u = x - 2, \quad du = dx$$

yields

$$\int \frac{x}{x^2 - 4x + 8}\, dx = \int \frac{x}{(x - 2)^2 + 4}\, dx = \int \frac{u + 2}{u^2 + 4}\, du$$

$$= \int \frac{u}{u^2 + 4}\, du + 2 \int \frac{du}{u^2 + 4}$$

$$= \frac{1}{2} \int \frac{2u}{u^2 + 4}\, du + 2 \int \frac{du}{u^2 + 4}$$

$$= \frac{1}{2} \ln(u^2 + 4) + 2\left(\frac{1}{2}\right) \tan^{-1} \frac{u}{2} + C$$

$$= \frac{1}{2} \ln[(x - 2)^2 + 4] + \tan^{-1}\left(\frac{x - 2}{2}\right) + C \qquad ◀$$

▶ Exercise Set 9.5

In Exercises 1–30, perform the indicated integration.

1. $\displaystyle\int \sqrt{4 - x^2}\, dx.$

2. $\displaystyle\int \sqrt{1 - 4x^2}\, dx.$

3. $\displaystyle\int \frac{x^2}{\sqrt{9 - x^2}}\, dx.$

4. $\displaystyle\int \frac{dx}{x^2\sqrt{16 - x^2}}.$

5. $\displaystyle\int \frac{dx}{(4 + x^2)^2}.$

6. $\displaystyle\int \frac{dx}{(x^2 + 1)^{3/2}}.$

7. $\displaystyle\int \frac{\sqrt{x^2 - 9}}{x}\, dx.$

8. $\displaystyle\int \frac{dx}{x^2\sqrt{x^2 - 16}}.$

9. $\displaystyle\int \frac{x^3}{\sqrt{2 - x^2}}\, dx.$

10. $\displaystyle\int x^3\sqrt{5 - x^2}\, dx.$

11. $\displaystyle\int \frac{dx}{(3 + x^2)^{3/2}}.$

12. $\displaystyle\int \frac{x^2}{\sqrt{5 + x^2}}\, dx.$

13. $\displaystyle\int \frac{dx}{x^2\sqrt{4x^2 - 9}}.$

14. $\displaystyle\int \frac{\sqrt{1 + t^2}}{t}\, dt.$

15. $\displaystyle\int \frac{dx}{(1 - x^2)^{3/2}}.$

16. $\displaystyle\int \frac{dx}{x^2\sqrt{x^2 + 25}}.$

17. $\displaystyle\int \frac{dx}{\sqrt{x^2 - 1}}.$

18. $\displaystyle\int \frac{dx}{1 + 2x^2 + x^4}.$

19. $\displaystyle\int \frac{dx}{x^2\sqrt{9 - 4x^2}}.$

20. $\displaystyle\int \frac{x^2}{\sqrt{x^2 - 25}}\, dx.$

21. $\displaystyle\int \frac{dx}{(9x^2 - 1)^{3/2}}.$

22. $\displaystyle\int \frac{\cos\theta}{\sqrt{2 - \sin^2\theta}}\, d\theta.$

23. $\displaystyle\int e^x\sqrt{1 - e^{2x}}\, dx.$

24. $\displaystyle\int_0^{1/3} \frac{dx}{(4 - 9x^2)^2}.$

25. $\displaystyle\int_0^4 x^3\sqrt{16 - x^2}\, dx.$

26. $\displaystyle\int_{\sqrt{2}}^2 \frac{\sqrt{2x^2 - 4}}{x}\, dx.$

27. $\displaystyle\int_{\sqrt{2}}^2 \frac{dx}{x^2\sqrt{x^2 - 1}}.$

28. $\displaystyle\int_{-1/\sqrt{2}}^{1/\sqrt{2}} (1 - 2x^2)^{3/2}\, dx.$

29. $\displaystyle\int_1^3 \frac{dx}{x^4\sqrt{x^2 + 3}}.$

30. $\displaystyle\int_0^3 \frac{x^3}{(3 + x^2)^{5/2}}\, dx.$

31. The integral

$$\int \frac{x}{x^2 + 4}\, dx$$

can be evaluated either by a trigonometric substitution or by the substitution $u = x^2 + 4$. Do it both ways and show that the results are equivalent.

32. By integrating, prove that the area of a circle of radius r is πr^2. [*Hint:* $x^2 + y^2 = r^2$ is the equation of such a circle.]

33. Find the arc length of the curve $y = \ln x$ from $x = 1$ to $x = 2$.

34. Find the arc length of the curve $y = x^2$ from $x = 0$ to $x = 1$.

35. Find the area of the surface generated when the curve in Exercise 34 is revolved about the x-axis.

36. Find the volume of the solid generated when the region enclosed by $x = y(1 - y^2)^{1/4}$, $y = 0$, $y = 1$, and $x = 0$ is revolved about the y-axis.

In Exercises 37 and 38 the trigonometric substitutions $x = a \sec\theta$ and $x = a \tan\theta$ lead to difficult integrals; for such integrals it is possible to use the **hyperbolic substitutions**

$$x = a \sinh u \text{ for integrals involving } \sqrt{x^2 + a^2}$$
$$x = a \cosh u \text{ for integrals involving } \sqrt{x^2 - a^2}$$

These substitutions are useful because in each case the hyperbolic identity

$$a^2 \cosh^2 u - a^2 \sinh^2 u = a^2$$

removes the radical.

37. (a) Evaluate

$$\int \frac{dx}{\sqrt{x^2 + 9}}$$

using the hyperbolic substitution that is suggested above.

(b) Evaluate the integral in part (a) by a trigonometric substitution and show that the results in parts (a) and (b) agree.

38. Follow the directions of Exercise 37 for the integral

$$\int \sqrt{x^2 - 1}\, dx, \quad x \ge 1$$

In Exercises 39–53, perform the integrations.

39. $\displaystyle\int \frac{dx}{x^2 - 4x + 13}.$

40. $\displaystyle\int \frac{dx}{\sqrt{2x - x^2}}.$

41. $\displaystyle\int \frac{dx}{\sqrt{8 + 2x - x^2}}.$

42. $\displaystyle\int \frac{dx}{16x^2 + 16x + 5}.$

43. $\displaystyle\int \frac{dx}{\sqrt{x^2 - 6x + 10}}.$

44. $\displaystyle\int \frac{x}{x^2 + 6x + 10}\, dx.$

45. $\int \sqrt{3 - 2x - x^2} \, dx.$

46. $\int \dfrac{e^x}{\sqrt{1 + e^x + e^{2x}}} \, dx.$ **47.** $\int \dfrac{dx}{2x^2 + 4x + 7}.$

48. $\int \dfrac{\cos \theta}{\sin^2 \theta - 6 \sin \theta + 12} \, d\theta.$

49. $\int \dfrac{2x + 5}{x^2 + 2x + 5} \, dx.$ **50.** $\int \dfrac{2x + 3}{4x^2 + 4x + 5} \, dx.$

51. $\int \dfrac{x + 3}{\sqrt{x^2 + 2x + 2}} \, dx.$

52. $\int_1^2 \dfrac{dx}{\sqrt{4x - x^2}}.$ **53.** $\int_0^1 \sqrt{x(4 - x)} \, dx.$

■ 9.6 INTEGRATING RATIONAL FUNCTIONS; PARTIAL FRACTIONS

Recall that a rational function is the quotient of two polynomials. In general, rational functions can be quite difficult to integrate. However, in this section we shall give a method that, in theory, can be used to express any rational function as a sum of simple rational functions that can be integrated by the methods studied in earlier sections.

□ **PARTIAL FRACTIONS**

In algebra one learns to combine two or more fractions into a single fraction by finding a common denominator; for example,

$$\frac{2}{x - 4} + \frac{3}{x + 1} = \frac{2(x + 1) + 3(x - 4)}{(x - 4)(x + 1)} = \frac{5x - 10}{x^2 - 3x - 4} \tag{1}$$

However, the left side of (1) is easier to integrate than the right side:

$$\int \frac{5x - 10}{x^2 - 3x - 4} \, dx = \int \frac{2}{x - 4} \, dx + \int \frac{3}{x + 1} \, dx$$

$$= 2 \ln |x - 4| + 3 \ln |x + 1| + C$$

Thus, for purposes of integration it would be desirable to have a method for obtaining the left side of (1) starting with the right side. To see how this can be done, let us factor the denominator on the right side of (1) and *assume* that there exist unknown constants A and B such that

$$\frac{5x - 10}{(x - 4)(x + 1)} = \frac{A}{x - 4} + \frac{B}{x + 1} \tag{2}$$

The terms on the right side of (2) are called **partial fractions** (because they each constitute *part* of the expression on the left side), and the entire right side of (2) is called the **partial fraction decomposition** of the left side. Observe that the denominators of the partial fractions are the factors of the denominator on the left side.

To find the constants A and B, we first multiply (2) through by $(x - 4)(x + 1)$ to clear fractions:

$$5x - 10 = A(x + 1) + B(x - 4) \tag{3}$$

This equation is actually an identity that holds for all real values of x; thus, it holds for any values of x that we choose to substitute. In particular, we can substitute $x = 4$ (which makes the second term on the right zero) and $x = -1$ (which makes the first term on the right zero). These substitutions yield

$$A = 2 \quad \text{and} \quad B = 3$$

which agrees with (1).

As an alternative method for finding A and B, we can multiply out on the right side of (3) and collect like powers of x:

$$5x - 10 = (A + B)x + (A - 4B) \tag{4}$$

Equating coefficients of like powers of x on the two sides yields

$$A + B = 5$$
$$A - 4B = -10$$

which is a system of two linear equations in the unknowns A and B. Solving this system yields $A = 2$ and $B = 3$ as before. This method is justified because (4) is an identity holding for all x, and two polynomials are equal for all x if and only if their corresponding coefficients are equal (Exercise 62).

The critical step in the foregoing work was the (correct) guess that the partial fraction decomposition had form (2). Once we knew the form of the decomposition, it was a straightforward matter to find A and B. One of the goals of this section is to explain how to obtain the form of a partial fraction decomposition.

There is a theorem in advanced algebra which states that every rational function $P(x)/Q(x)$ in which the *degree of the numerator is less than the degree of the denominator* can be expressed as a sum

$$\frac{P(x)}{Q(x)} = F_1(x) + F_2(x) + \cdots + F_n(x) \tag{5}$$

where $F_1(x), F_2(x), \ldots, F_n(x)$ are rational functions of the form

$$\frac{A}{(ax + b)^k} \quad \text{or} \quad \frac{Ax + B}{(ax^2 + bx + c)^k}$$

The terms $F_1(x), F_2(x), \ldots, F_n(x)$ on the right side of (5) are called *partial fractions,* and the entire right side is called the *partial fraction decomposition* of the left side. We will see below that the number of terms of each type depends on the factors of $Q(x)$; however, first let us review some results about factoring polynomials.

In theory, every polynomial $Q(x)$ with real coefficients can be factored into a product of linear and quadratic factors with real coefficients. For example, the polynomial

$$Q(x) = x^3 - 3x^2 + x - 3$$

factors into

$$Q(x) = (x - 3)(x^2 + 1)$$

The quadratic factor $x^2 + 1$ cannot be further decomposed into linear factors without using imaginary numbers $[x^2 + 1 = (x - i)(x + i)]$. Such quadratic factors are said to be *irreducible.* Although it is theoretically possible to factor every polynomial into linear and irreducible quadratic factors with real coefficients, it can be difficult to do so in many cases.

☐ **FINDING THE FORM OF A PARTIAL FRACTION DECOMPOSITION**

The first step in finding the partial fraction decomposition of a rational function $P(x)/Q(x)$ *whose numerator has smaller degree than the denominator* is to factor $Q(x)$ completely into linear and irreducible quadratic factors, and then collect all repeated factors so that $Q(x)$ is expressed as a product of distinct factors of the form

$$(ax + b)^m \quad \text{and} \quad (ax^2 + bx + c)^m$$

From these factors we can determine the form of the partial fraction decomposition using two rules which we shall now discuss.

☐ **LINEAR FACTORS**

If all of the factors of $Q(x)$ are linear, then the partial fraction decomposition of $P(x)/Q(x)$ can be determined by using the following rule:

LINEAR FACTOR RULE.   For each factor of the form $(ax + b)^m$, the partial fraction decomposition contains the following sum of m partial fractions:

$$\frac{A_1}{ax + b} + \frac{A_2}{(ax + b)^2} + \cdots + \frac{A_m}{(ax + b)^m}$$

where $A_1, A_2, \ldots, A_m$ are constants to be determined.

Example 1   Evaluate

$$\int \frac{dx}{x^2 + x - 2}$$

Solution.   The integrand can be written as

$$\frac{1}{x^2 + x - 2} = \frac{1}{(x - 1)(x + 2)}$$

which is a rational function whose numerator has smaller degree than the denominator. According to the linear factor rule, the factor $x - 1$ introduces one term (since $m = 1$) of the form

$$\frac{A}{x - 1}$$

and the factor $x + 2$ introduces one term of the form

$$\frac{B}{x + 2}$$

so that the partial fraction decomposition is

$$\frac{1}{(x-1)(x+2)} = \frac{A}{x-1} + \frac{B}{x+2} \tag{6}$$

where A and B are constants to be determined so that (6) becomes an *identity*. To find these constants we can multiply both sides of (6) by $(x-1)(x+2)$ to obtain

$$1 = A(x+2) + B(x-1) \tag{7}$$

As previously discussed, we can find A and B by substituting the values of x that make the terms on the right side of (7) zero, namely $x = -2$ and $x = 1$. Substituting $x = -2$ yields

$$1 = -3B \quad \text{or} \quad B = -\frac{1}{3}$$

and substituting $x = 1$ yields

$$1 = 3A \quad \text{or} \quad A = \frac{1}{3}$$

Thus, (6) becomes

$$\frac{1}{(x-1)(x+2)} = \frac{1/3}{x-1} + \frac{-1/3}{x+2}$$

and

$$\int \frac{dx}{(x-1)(x+2)} = \frac{1}{3} \int \frac{dx}{x-1} - \frac{1}{3} \int \frac{dx}{x+2}$$

$$= \frac{1}{3} \ln|x-1| - \frac{1}{3} \ln|x+2| + C$$

$$= \frac{1}{3} \ln\left|\frac{x-1}{x+2}\right| + C$$

Alternative Solution. The constants A and B in (7) can also be determined by collecting like powers of x,

$$1 = (A+B)x + (2A-B)$$

and then equating the coefficients of the like powers of x on both sides to obtain

$$A + B = 0$$
$$2A - B = 1$$

The solution of this system of linear equations is $A = 1/3$ and $B = -1/3$, which agrees with the results obtained above.   ◄

When $Q(x)$ has only distinct linear factors, as in the foregoing example, then the unknown constants in the partial fraction decomposition can all be found by substituting appropriate values of x. However, if there are repeated linear factors, then it is not as easy to obtain all of the constants by substituting values of x. For such problems one can either find all of the constants by equating like powers of x and solving the resulting system of linear equations or one can use a combination of the two methods, first finding as many constants as possible by substituting values, then finding the rest by equating appropriate like powers of x. The following example illustrates the latter procedure.

Example 2 Evaluate

$$\int \frac{2x + 4}{x^3 - 2x^2} \, dx$$

Solution. The integrand can be rewritten as

$$\frac{2x + 4}{x^3 - 2x^2} = \frac{2x + 4}{x^2(x - 2)}$$

Although x^2 is a quadratic factor, it is *not* irreducible since $x^2 = xx$. Thus, by the linear factor rule, x^2 introduces two terms (since $m = 2$) of the form

$$\frac{A}{x} + \frac{B}{x^2}$$

and the factor $x - 2$ introduces one term (since $m = 1$) of the form

$$\frac{C}{x - 2}$$

so the partial fraction decomposition is

$$\frac{2x + 4}{x^2(x - 2)} = \frac{A}{x} + \frac{B}{x^2} + \frac{C}{x - 2} \tag{8}$$

Multiplying by $x^2(x - 2)$ yields

$$2x + 4 = Ax(x - 2) + B(x - 2) + Cx^2 \tag{9}$$

which, after multiplying out and collecting like powers of x, becomes

$$2x + 4 = (A + C)x^2 + (-2A + B)x - 2B \tag{10}$$

To determine A, B, and C we shall use a combination of the methods illustrated in Example 1. Let $x = 0$ and $x = 2$ in (9) to obtain

$$B = -2 \quad \text{and} \quad C = 2$$

Equating corresponding coefficients of x^2 in (10) gives

$$A + C = 0, \quad \text{so} \quad A = -C = -2$$

and (8) becomes

$$\frac{2x + 4}{x^2(x - 2)} = \frac{-2}{x} + \frac{-2}{x^2} + \frac{2}{x - 2}$$

Thus,

$$\int \frac{2x + 4}{x^2(x - 2)} \, dx = -2 \int \frac{dx}{x} - 2 \int \frac{dx}{x^2} + 2 \int \frac{dx}{x - 2}$$

$$= -2 \ln |x| + \frac{2}{x} + 2 \ln |x - 2| + C$$

$$= 2 \ln \left| \frac{x - 2}{x} \right| + \frac{2}{x} + C \quad \blacktriangleleft$$

☐ **QUADRATIC FACTORS**   If some of the factors of $Q(x)$ are irreducible quadratics, then the contribution of those factors to the partial fraction decomposition of $P(x)/Q(x)$ can be determined from the following rule:

QUADRATIC FACTOR RULE.   For each factor of the form $(ax^2 + bx + c)^m$, the partial fraction decomposition contains the following sum of m partial fractions:

$$\frac{A_1 x + B_1}{ax^2 + bx + c} + \frac{A_2 x + B_2}{(ax^2 + bx + c)^2} + \cdots + \frac{A_m x + B_m}{(ax^2 + bx + c)^m}$$

where $A_1, A_2, \ldots, A_m, B_1, B_2, \ldots, B_m$ are constants to be determined.

Example 3   Evaluate

$$\int \frac{x^2 + x - 2}{3x^3 - x^2 + 3x - 1} \, dx$$

Solution.   The denominator in the integrand can be factored by grouping:

$$\frac{x^2 + x - 2}{3x^3 - x^2 + 3x - 1} = \frac{x^2 + x - 2}{x^2(3x - 1) + (3x - 1)} = \frac{x^2 + x - 2}{(3x - 1)(x^2 + 1)}$$

By the linear factor rule, the factor $3x - 1$ introduces one term:

$$\frac{A}{3x - 1}$$

and by the quadratic factor rule, the factor $x^2 + 1$ introduces one term:

$$\frac{Bx + C}{x^2 + 1}$$

Thus, the partial fraction decomposition is

$$\frac{x^2 + x - 2}{(3x - 1)(x^2 + 1)} = \frac{A}{3x - 1} + \frac{Bx + C}{x^2 + 1} \tag{11}$$

Multiplying by $(3x - 1)(x^2 + 1)$ yields

$$x^2 + x - 2 = A(x^2 + 1) + (Bx + C)(3x - 1) \tag{12}$$

To determine A, B, and C, we multiply out and collect like terms:

$$x^2 + x - 2 = (A + 3B)x^2 + (-B + 3C)x + (A - C) \tag{13}$$

Equating corresponding coefficients gives

$$
\begin{aligned}
A + 3B \quad\quad &= \quad 1 \\
- \quad B + 3C &= \quad 1 \\
A \quad\quad - \quad C &= -2
\end{aligned}
$$

To solve this system, subtract the third equation from the first to eliminate A. Then use the resulting equation together with the second equation to solve for B and C. Finally, determine A from the first or third equation. This yields (verify)

$$A = -\frac{7}{5}, \quad B = \frac{4}{5}, \quad C = \frac{3}{5}$$

Thus, (11) becomes

$$\frac{x^2 + x - 2}{(3x - 1)(x^2 + 1)} = \frac{-\frac{7}{5}}{3x - 1} + \frac{\frac{4}{5}x + \frac{3}{5}}{x^2 + 1}$$

and

$$\int \frac{x^2 + x - 2}{(3x - 1)(x^2 + 1)} \, dx$$

$$= -\frac{7}{5} \int \frac{dx}{3x - 1} + \frac{4}{5} \int \frac{x}{x^2 + 1} \, dx + \frac{3}{5} \int \frac{dx}{x^2 + 1}$$

$$= -\frac{7}{15} \ln|3x - 1| + \frac{2}{5} \ln(x^2 + 1) + \frac{3}{5} \tan^{-1} x + C \quad \blacktriangleleft$$

REMARK.   The partial fraction decomposition in the foregoing example can also be obtained by a combination of the two methods we have discussed. Substituting $x = 1/3$ in (12) yields $A = -7/5$, then equating the corresponding constant terms and the corresponding coefficients of x^2 in (13) yields equations that express B and C in terms of the known value of A. (Verify that this procedure yields the same results as those obtained in the example.)

Example 4   Evaluate

$$\int \frac{3x^4 + 4x^3 + 16x^2 + 20x + 9}{(x + 2)(x^2 + 3)^2} \, dx$$

Solution. Observe first that the degree of the numerator (which is 4) is less than the degree of the denominator (which is 5), so the method of partial fractions is applicable. By the linear factor rule, the factor $x + 2$ introduces one term:

$$\frac{A}{x + 2}$$

and by the quadratic factor rule, the factor $(x^2 + 3)^2$ introduces two terms (since $m = 2$):

$$\frac{Bx + C}{x^2 + 3} + \frac{Dx + E}{(x^2 + 3)^2}$$

Thus, the partial fraction decomposition of the integrand is

$$\frac{3x^4 + 4x^3 + 16x^2 + 20x + 9}{(x + 2)(x^2 + 3)^2} = \frac{A}{x + 2} + \frac{Bx + C}{x^2 + 3} + \frac{Dx + E}{(x^2 + 3)^2} \tag{14}$$

Multiplying by $(x + 2)(x^2 + 3)^2$ yields

$$3x^4 + 4x^3 + 16x^2 + 20x + 9$$
$$= A(x^2 + 3)^2 + (Bx + C)(x^2 + 3)(x + 2) + (Dx + E)(x + 2) \tag{15}$$

which, after multiplying out and collecting like powers of x, becomes

$$3x^4 + 4x^3 + 16x^2 + 20x + 9$$
$$= (A + B)x^4 + (2B + C)x^3 + (6A + 3B + 2C + D)x^2$$
$$+ (6B + 3C + 2D + E)x + (9A + 6C + 2E) \tag{16}$$

Equating corresponding coefficients in (16) yields the following system of five linear equations in five unknowns:

$$\begin{aligned}
A + B &= 3 \\
2B + C &= 4 \\
6A + 3B + 2C + D &= 16 \\
6B + 3C + 2D + E &= 20 \\
9A + 6C + 2E &= 9
\end{aligned} \tag{17}$$

This system is tedious to solve, but the work can be reduced by first substituting $x = -2$ in (15), which yields $A = 1$. Substituting this known value of A in (17) yields the simpler system

$$\begin{aligned}
B &= 2 \\
2B + C &= 4 \\
3B + 2C + D &= 10 \\
6B + 3C + 2D + E &= 20 \\
6C + 2E &= 0
\end{aligned} \tag{18}$$

This system can be solved by starting at the top and working down, first substituting $B = 2$ in the second equation to get $C = 0$, then substituting the known values of B and C in the third equation to get $D = 4$, and so forth. This yields

$$A = 1, \quad B = 2, \quad C = 0, \quad D = 4, \quad E = 0$$

Thus, (14) becomes

$$\frac{3x^4 + 4x^3 + 16x^2 + 20x + 9}{(x + 2)(x^2 + 3)^2} = \frac{1}{x + 2} + \frac{2x}{x^2 + 3} + \frac{4x}{(x^2 + 3)^2}$$

and so

$$\int \frac{3x^4 + 4x^3 + 16x^2 + 20x + 9}{(x + 2)(x^2 + 3)^2} \, dx$$

$$= \int \frac{dx}{x + 2} + \int \frac{2x}{x^2 + 3} \, dx + 4 \int \frac{x}{(x^2 + 3)^2} \, dx$$

$$= \ln|x + 2| + \ln(x^2 + 3) - \frac{2}{x^2 + 3} + C \quad \blacktriangleleft$$

☐ **INTEGRATING IMPROPER RATIONAL FUNCTIONS**

As noted earlier, partial fraction decomposition applies only to rational functions in which the degree of the numerator is less than the degree of the denominator; these are called *proper rational functions*. Rational functions in which the degree of the numerator is greater than or equal to the degree of the denominator are called *improper rational functions*. The following example shows that improper rational functions can be integrated by first performing a long division, and then working with the remainder term.

Example 5 Evaluate

$$\int \frac{3x^4 + 3x^3 - 5x^2 + x - 1}{x^2 + x - 2} \, dx$$

Solution. Since the integrand is an improper rational function, we cannot use a partial fraction decomposition directly. However, if we perform the long division

$$
\begin{array}{r}
3x^2 + 1 \\
x^2 + x - 2 \overline{)\ 3x^4 + 3x^3 - 5x^2 + x - 1} \\
\underline{3x^4 + 3x^3 - 6x^2} \\
x^2 + x - 1 \\
\underline{x^2 + x - 2} \\
1
\end{array}
$$

we can write the integrand as the quotient plus the remainder over the divisor, that is,

$$\frac{3x^4 + 3x^3 - 5x^2 + x - 1}{x^2 + x - 2} = (3x^2 + 1) + \frac{1}{x^2 + x - 2}$$

Thus,

$$\int \frac{3x^4 + 3x^3 - 5x^2 + x - 1}{x^2 + x - 2} \, dx = \int (3x^2 + 1) \, dx + \int \frac{dx}{x^2 + x - 2}$$

The second integral on the right now involves a proper rational function and can thus be evaluated by a partial fraction decomposition. Using the result of Example 1 we obtain

$$\int \frac{3x^4 + 3x^3 - 5x^2 + x - 1}{x^2 + x - 2} \, dx = x^3 + x + \frac{1}{3} \ln \left| \frac{x - 1}{x + 2} \right| + C \qquad \blacktriangleleft$$

■ FACTORING POLYNOMIALS—OPTIONAL DISCUSSION

The method of partial fractions depends on our ability to carry out the necessary factorization. This is not always easy to do. The following results, usually proved in algebra courses, are helpful to know.

9.6.1 THEOREM (*Factor Theorem*).   *If $p(x)$ is a polynomial and r is a solution of the equation $p(x) = 0$, then $x - r$ is a factor of $p(x)$.*

9.6.2 THEOREM. *Let*

$$p(x) = a_0 x^n + a_1 x^{n-1} + \cdots + a_{n-1}x + a_n$$

be a polynomial of degree n with integer coefficients.
(a) *If r is an integer solution of $p(x) = 0$, then the constant term a_n is an integer multiple of r.*
(b) *If c/d is a rational solution of $p(x) = 0$, and if c/d is expressed in lowest terms, then the constant term a_n is an integer multiple of c; and the leading coefficient a_0 is an integer multiple of d.*

Example 6   For the equation

$$x^3 + x^2 - 10x + 8 = 0$$

it follows from part (a) of the preceding theorem that the only possible integer solutions are $\pm 1, \pm 2, \pm 4, \pm 8$. By substitution, or by using synthetic division, the reader can show that 1, 2, -4 are solutions and the rest are not. It follows that

$$x^3 + x^2 - 10x + 8 = (x - 1)(x - 2)(x + 4) \qquad \blacktriangleleft$$

Example 7   For the equation

$$2x^3 + x^2 - 6x - 3 = 0$$

the only possible numerators for rational solutions are ± 1, ± 3, and the only possible denominators are ± 1, ± 2; thus, the only possible rational solutions are

$$\pm 1, \ \pm 3, \ \pm \tfrac{1}{2}, \ \pm \tfrac{3}{2}$$

By substitution, or by using synthetic division, the reader can show that $-\tfrac{1}{2}$ is a solution, but the rest are not. It follows by division that

$$2x^3 + x^2 - 6x - 3 = (x + \tfrac{1}{2})(2x^2 - 6) = 2(x + \tfrac{1}{2})(x + \sqrt{3})(x - \sqrt{3})$$

Therefore, the solutions of the given equation are $x = -\tfrac{1}{2}$, $x = -\sqrt{3}$, and $x = \sqrt{3}$. ◀

▶ Exercise Set 9.6   © 54, 55

In Exercises 1–8, write out the form of the partial fraction decomposition. (Do not find the numerical values of the coefficients.)

1. $\dfrac{3x - 1}{(x - 2)(x + 5)}$.

2. $\dfrac{5}{x(x^2 - 9)}$.

3. $\dfrac{2x - 3}{x^3 - x^2}$.

4. $\dfrac{x^2}{(x + 2)^3}$.

5. $\dfrac{1 - 5x^2}{x^3(x^2 + 1)}$.

6. $\dfrac{2x}{(x - 1)(x^2 + 5)}$.

7. $\dfrac{4x^3 - x}{(x^2 + 5)^2}$.

8. $\dfrac{1 - 3x^4}{(x - 2)(x^2 + 1)^2}$.

In Exercises 9–46, perform the integrations.

9. $\displaystyle\int \dfrac{dx}{x^2 + 3x - 4}$.

10. $\displaystyle\int \dfrac{dx}{x^2 + 8x + 7}$.

11. $\displaystyle\int \dfrac{x}{x^2 - 5x + 6}\,dx$.

12. $\displaystyle\int \dfrac{5x - 4}{x^2 - 4x}\,dx$.

13. $\displaystyle\int \dfrac{11x + 17}{2x^2 + 7x - 4}\,dx$.

14. $\displaystyle\int \dfrac{5x - 5}{3x^2 - 8x - 3}\,dx$.

15. $\displaystyle\int \dfrac{dx}{(x - 1)(x + 2)(x - 3)}$.

16. $\displaystyle\int \dfrac{dx}{x(x^2 - 1)}$.

17. $\displaystyle\int \dfrac{2x^2 - 9x - 9}{x^3 - 9x}\,dx$.

18. $\displaystyle\int \dfrac{2x^2 + 4x - 8}{x^3 - 4x}\,dx$.

19. $\displaystyle\int \dfrac{x^2 + 2}{x + 2}\,dx$.

20. $\displaystyle\int \dfrac{x^2 - 4}{x - 1}\,dx$.

21. $\displaystyle\int \dfrac{3x^2 - 10}{x^2 - 4x + 4}\,dx$.

22. $\displaystyle\int \dfrac{x^2}{x^2 - 3x + 2}\,dx$.

23. $\displaystyle\int \dfrac{x^3}{x^2 - 3x + 2}\,dx$.

24. $\displaystyle\int \dfrac{x^3}{x^2 - x - 6}\,dx$.

25. $\displaystyle\int \dfrac{x^5 + 2x^2 + 1}{x^3 - x}\,dx$.

26. $\displaystyle\int \dfrac{2x^5 - x^3 - 1}{x^3 - 4x}\,dx$.

27. $\displaystyle\int \dfrac{2x^2 + 3}{x(x - 1)^2}\,dx$.

28. $\displaystyle\int \dfrac{3x^2 - x + 1}{x^3 - x^2}\,dx$.

29. $\displaystyle\int \dfrac{x^2 + x - 16}{(x + 1)(x - 3)^2}\,dx$.

30. $\displaystyle\int \dfrac{2x^2 - 2x - 1}{x^3 - x^2}\,dx$.

31. $\displaystyle\int \dfrac{x^2}{(x + 2)^3}\,dx$.

32. $\displaystyle\int \dfrac{2x^2 + 3x + 3}{(x + 1)^3}\,dx$.

33. $\displaystyle\int \dfrac{2x^2 - 1}{(4x - 1)(x^2 + 1)}\,dx$.

34. $\displaystyle\int \dfrac{dx}{x(x^2 + x + 1)}$.

35. $\displaystyle\int \dfrac{dx}{x^4 - 16}$.

36. $\displaystyle\int \dfrac{dx}{x^3 + x}$.

37. $\displaystyle\int \dfrac{x^3 + 3x^2 + x + 9}{(x^2 + 1)(x^2 + 3)}\,dx$.

38. $\displaystyle\oint \dfrac{x^3 + x^2 + x + 2}{(x^2 + 1)(x^2 + 2)}\,dx$.

39. $\displaystyle\int \dfrac{x^3 - 3x^2 + 2x - 3}{x^2 + 1}\,dx$.

40. $\displaystyle\int \dfrac{x^4 + 6x^3 + 10x^2 + x}{x^2 + 6x + 10}\,dx$.

41. $\displaystyle\int \dfrac{x^2 + 1}{(x^2 + 2x + 3)^2}\,dx$.

42. $\displaystyle\int \dfrac{x^5 + x^4 + 4x^3 + 4x^2 + 4x + 4}{(x^2 + 2)^3}\,dx$.

43. $\displaystyle\int \dfrac{\cos \theta}{\sin^2 \theta + 4 \sin \theta - 5}\,d\theta$.

44. $\int \dfrac{e^t}{e^{2t} - 4}\, dt.$ **45.** $\int \dfrac{dx}{1 + e^x}.$

46. $\int \dfrac{\sec^2 \theta}{\tan^3 \theta - \tan^2 \theta}\, d\theta.$

47. (a) Find constants a and b such that

$$x^4 + 1 = (x^2 + ax + 1)(x^2 + bx + 1)$$

(b) Use the result in part (a) to show that

$$\int_0^1 \frac{x}{x^4 + 1}\, dx = \frac{\pi}{8}$$

48. Find the area of the region that is enclosed by $y = (x - 3)/(x^3 + x^2)$, $y = 0$, $x = 1$, and $x = 2$.

49. Find the volume of the solid generated when the region enclosed by $y = x^2/(9 - x^2)$, $y = 0$, $x = 0$, and $x = 2$ is revolved about the x-axis.

In Exercises 50–53, solve the differential equations.

---

50. $\dfrac{dy}{dx} = y^2 + y.$ **51.** $\dfrac{dy}{dx} = y^2 - 5y + 6.$

52. $\dfrac{dy}{dt} = t^2 y^2 - 4t^2 y.$

53. $t(t - 1)\dfrac{dy}{dt} - (y^2 + y) = 0.$

54. In some chemical reactions in which two reacting substances combine to form a resultant substance, the concentration of the resultant changes with time at a rate that is proportional to the product of the concentrations of the reacting substances. If x is the concentration of the resultant at any time $t \geq 0$, and a and b are the concentrations of the reactants at time $t = 0$, then x satisfies the differential equation

$$\frac{dx}{dt} = k(a - x)(b - x) \qquad (a \neq b)$$

(a) Assuming that $x = 0$ when $t = 0$, solve the equation for t in terms of x, a, b, and k.

(b) Assuming that the concentrations are measured in moles/liter, time is measured in minutes, and the values of the constants are $a = 0.10$, $b = 0.06$, and $k = 0.3$, find the length of time that it will take for the concentration of the resultant to reach 0.02 mol/L. Express your answer to the nearest tenth of a minute.

55. The differential equation

$$\frac{dy}{dt} = ay - by^2 \qquad (a > 0, b > 0)$$

which is called the **logistic equation,** arises in the study of human population growth.

(a) By solving the equation, show that its general solution is

$$y = \frac{a}{b + Ce^{-at}}$$

where C is an arbitrary constant.

(b) Find $\lim\limits_{t \to +\infty} y(t)$ and $\lim\limits_{t \to -\infty} y(t)$.

56. (a) Use the result in part (a) of Exercise 55 to find the solution of the logistic equation that satisfies the initial condition $y(0) = \frac{1}{2}$ in the case where $a = 2$ and $b = 1$.

(b) Sketch the graph of the solution in part (a).

In Exercises 57–60, use Theorems 9.6.1 and 9.6.2.

---

57. Find all rational solutions, if any, and use your results to factor the polynomial into a product of linear and irreducible quadratic factors.

(a) $x^3 - 6x^2 + 11x - 6 = 0$

(b) $x^3 - 3x^2 + x - 20 = 0$

(c) $x^4 - 5x^3 + 7x^2 - 5x + 6 = 0.$

58. Find all rational solutions, if any, and use your results to factor the polynomial into a product of linear and irreducible quadratic factors.

(a) $8x^3 + 4x^2 - 2x - 1 = 0$

(b) $6x^4 - 7x^3 + 6x^2 - 1 = 0$

(c) $9x^4 - 56x^3 + 57x^2 + 98x - 24 = 0.$

59. Evaluate

$$\int \frac{dx}{x^4 - 3x^3 - 7x^2 + 27x - 18}$$

60. Evaluate

$$\int \frac{dx}{16x^3 - 4x^2 + 4x - 1}$$

61. Use Theorem 9.6.2 to prove:

(a) $\sqrt{2}$ is irrational

(b) If a is a positive integer, then $\sqrt{a}$ is either an integer or is irrational.

62. (a) Prove: $a_0 x^n + a_1 x^{n-1} + \cdots + a_n \equiv 0$ if and only if $a_0 = a_1 = \cdots = a_n = 0$.

(b) Use the result in part (a) to prove:

$$a_0 x^n + a_1 x^{n-1} + \cdots + a_n$$
$$\equiv b_0 x^n + b_1 x^{n-1} + \cdots + b_n$$

if and only if $a_0 = b_0$, $a_1 = b_1, \ldots, a_n = b_n$.

[*Note:* The symbol $\equiv$ means that the two sides are equal for all values of x.]

■ **9.7 MISCELLANEOUS SUBSTITUTIONS**

In this section we shall consider some integrals that do not fit into any of the categories previously studied.

□ **INTEGRALS INVOLVING RATIONAL EXPONENTS**

Integrals involving rational powers of x can often be simplified by substituting

$$u = x^{1/n}$$

where n is the least common multiple of the denominators of the exponents. The effect of this substitution is to replace fractional exponents with integer exponents, which are easier to work with.

Example 1 Evaluate

$$\int \frac{\sqrt{x}}{1 + \sqrt[3]{x}}\, dx$$

Solution. The integrand involves $x^{1/2}$ and $x^{1/3}$, so we make the substitution

$$u = x^{1/6}$$

or

$$x = u^6 \quad \text{and} \quad dx = 6u^5\, du$$

Thus,

$$\int \frac{\sqrt{x}}{1 + \sqrt[3]{x}}\, dx = \int \frac{(u^6)^{1/2}}{1 + (u^6)^{1/3}}\,(6u^5)\, du = 6\int \frac{u^8}{1 + u^2}\, du$$

By long division

$$\frac{u^8}{1 + u^2} = u^6 - u^4 + u^2 - 1 + \frac{1}{1 + u^2}$$

Thus,

$$\int \frac{\sqrt{x}}{1 + \sqrt[3]{x}}\, dx = 6\int \left(u^6 - u^4 + u^2 - 1 + \frac{1}{1 + u^2}\right) du$$

$$= \frac{6}{7}u^7 - \frac{6}{5}u^5 + 2u^3 - 6u + 6\tan^{-1}u + C$$

$$= \frac{6}{7}x^{7/6} - \frac{6}{5}x^{5/6} + 2x^{1/2} - 6x^{1/6} + 6\tan^{-1}(x^{1/6}) + C \quad ◀$$

Example 2 Evaluate

$$\int \frac{dx}{2 + 2\sqrt{x}}$$

Solution. The integrand contains $\sqrt{x} = x^{1/2}$, so we make the substitution

$$u = x^{1/2}$$

or

$$x = u^2 \quad \text{and} \quad dx = 2u \, du$$

This yields

$$\int \frac{dx}{2 + 2\sqrt{x}} = \int \frac{2u}{2 + 2u} \, du$$

$$= \int \left(1 - \frac{1}{1 + u}\right) du \qquad \boxed{\text{Long division}}$$

$$= u - \ln|1 + u| + C = \sqrt{x} - \ln|1 + \sqrt{x}| + C \qquad \blacktriangleleft$$

The following example illustrates a variation on the above idea.

Example 3 Evaluate $\displaystyle\int \sqrt{1 + e^x} \, dx$.

Solution. The substitution

$$u^2 = 1 + e^x$$

will eliminate the square root. To express dx in terms of du, it is helpful to solve this equation for x and then differentiate. We obtain

$$e^x = u^2 - 1$$

$$x = \ln(u^2 - 1)$$

$$\frac{dx}{du} = \frac{2u}{u^2 - 1}, \quad dx = \frac{2u}{u^2 - 1} \, du$$

Thus,

$$\int \sqrt{1 + e^x} \, dx = \int u \left(\frac{2u}{u^2 - 1}\right) du$$

$$= \int \frac{2u^2}{u^2 - 1} \, du$$

$$= \int \left(2 + \frac{2}{u^2 - 1}\right) du \qquad \boxed{\text{Long division}}$$

$$= 2u + \int \left(\frac{1}{u - 1} - \frac{1}{u + 1}\right) du \qquad \boxed{\text{Partial fractions}}$$

$$= 2u + \ln|u - 1| - \ln|u + 1| + C$$

$$= 2u + \ln\left|\frac{u - 1}{u + 1}\right| + C$$

$$= 2\sqrt{1 + e^x} + \ln\left|\frac{\sqrt{1 + e^x} - 1}{\sqrt{1 + e^x} + 1}\right| + C \qquad \blacktriangleleft$$

☐ **INTEGRALS CONTAINING RATIONAL EXPRESSIONS IN sin x AND cos x**

Functions such as

$$\frac{\sin x + 3 \cos^2 x}{\cos x + 4 \sin x}, \qquad \frac{\sin x}{1 + \cos x - \cos^2 x}, \qquad \frac{3 \sin^5 x}{1 + 4 \sin x}$$

are called *rational expressions in sin x and cos x.* Such expressions consist of finitely many sums, differences, products, and quotients of $\sin x$ and $\cos x$. If an integrand is a rational expression in $\sin x$ and $\cos x$, then the substitution

$$u = \tan (x/2), \quad -\pi < x < \pi \tag{1}$$

will transform the integrand into a rational function of u (which can then be integrated by methods already discussed). To see this, observe that

$$\cos (x/2) = \frac{1}{\sec (x/2)} = \frac{1}{\sqrt{1 + \tan^2 (x/2)}} = \frac{1}{\sqrt{1 + u^2}}$$

$$\sin (x/2) = \tan (x/2) \cos (x/2) = u \left(\frac{1}{\sqrt{1 + u^2}} \right)$$

Therefore,

$$\sin x = 2 \sin (x/2) \cos (x/2) = 2u \left(\frac{1}{\sqrt{1 + u^2}} \right) \left(\frac{1}{\sqrt{1 + u^2}} \right)$$

or

$$\sin x = \frac{2u}{1 + u^2} \tag{2}$$

Also,

$$\cos x = 1 - 2 \sin^2 \left(\frac{x}{2} \right) = 1 - \frac{2u^2}{1 + u^2}$$

so

$$\cos x = \frac{1 - u^2}{1 + u^2} \tag{3}$$

Moreover, from (1)

$$\frac{x}{2} = \tan^{-1} u \quad \text{so that} \quad \frac{dx}{du} = \frac{2}{1 + u^2}$$

or

$$dx = \frac{2}{1 + u^2} \, du \tag{4}$$

It follows from (2), (3), and (4) that substitution (1) will yield an integrand that is a rational function of u.

Example 4 Evaluate

$$\int \frac{dx}{1 + \sin x}$$

Solution. Making substitution (1), it follows from (2) and (4) that

$$\int \frac{dx}{1 + \sin x} = \int \frac{1}{1 + \left(\dfrac{2u}{1 + u^2}\right)} \left(\frac{2}{1 + u^2}\right) du = \int \frac{2}{1 + 2u + u^2} du$$

$$= \int \frac{2}{(1 + u)^2} du = -\frac{2}{1 + u} + C = -\frac{2}{1 + \tan\left(\dfrac{x}{2}\right)} + C \quad \blacktriangleleft$$

REMARK. While substitution (1) is a useful tool, it can lead to cumbersome partial fraction decompositions. Consequently, this method should be used only after looking for simpler methods.

▶ **Exercise Set 9.7**

In Exercises 1–26, perform the integrations.

1. $\int x\sqrt{x - 2}\, dx.$

2. $\int_0^8 \frac{x}{\sqrt{x + 1}}\, dx.$

3. $\int_4^8 \frac{\sqrt{x - 4}}{x}\, dx.$

4. $\int_0^9 \frac{\sqrt{x}}{x + 9}\, dx.$

5. $\int_0^4 \frac{1}{3 + \sqrt{x}}\, dx.$

6. $\int \frac{x^5}{\sqrt{x^3 + 1}}\, dx.$

7. $\int x^5 \sqrt{x^3 + 1}\, dx.$

8. $\int \frac{1}{x\sqrt{x^3 - 1}}\, dx.$

9. $\int \frac{dx}{\sqrt{x} + \sqrt[3]{x}}.$

10. $\int \frac{dx}{x - x^{3/5}}.$

11. $\int \frac{dv}{v(1 - v^{1/4})}.$

12. $\int \frac{x^{2/3}}{x + 1}\, dx.$

13. $\int \frac{dt}{t^{1/2} - t^{1/3}}.$

14. $\int \frac{1 + \sqrt{x}}{1 - \sqrt{x}}\, dx.$

15. $\int \frac{x^3}{\sqrt{1 + x^2}}\, dx.$

16. $\int \frac{x}{(x + 3)^{1/5}}\, dx.$

17. $\int \sin \sqrt{x}\, dx.$

18. $\int e^{\sqrt{x}}\, dx.$

19. $\int \frac{1}{\sqrt{e^x + 1}}\, dx.$

20. $\int_0^{\ln 2} \sqrt{e^x - 1}\, dx.$

21. $\int \frac{dx}{1 + \sin x + \cos x}.$

22. $\int \frac{dx}{2 + \sin x}.$

23. $\int_{\pi/2}^{\pi} \frac{d\theta}{1 - \cos \theta}.$

24. $\int \frac{dx}{4 \sin x - 3 \cos x}.$

25. $\int \frac{\cos x}{2 - \cos x}\, dx.$

26. $\int \frac{dx}{\sin x + \tan x}.$

27. (a) Use the substitution $u = \tan(x/2)$ to show that

$$\int \sec x\, dx = \ln \left| \frac{1 + \tan \frac{1}{2}x}{1 - \tan \frac{1}{2}x} \right| + C$$

(b) Use the result in part (a) to show that

$$\int \sec x\, dx = \ln \left| \tan \left(\frac{\pi}{4} + \frac{x}{2} \right) \right| + C$$

(c) Show that this agrees with (2) of Section 9.4.

28. (a) Use the substitution $u = \tan(x/2)$ to show that

$$\int \csc x\, dx = \frac{1}{2} \ln \left[\frac{1 - \cos x}{1 + \cos x} \right] + C$$

(b) Show that this agrees with Exercise 37(a) of Section 9.4.

29. Develop a substitution that can be used to integrate rational functions of $\sinh x$ and $\cosh x$ and use your substitution to evaluate

$$\int \frac{dx}{2 \cosh x + \sinh x}$$

without expressing the integrand in terms of e^x and e^{-x}.

In Exercises 30–33, use the substitution $x = 1/u$ to help evaluate the integral. (Assume that $x > 0$.)

30. $\displaystyle \int \frac{\sqrt{4 - x^2}}{x^4}\, dx.$

31. $\displaystyle \int \frac{1}{x^2\sqrt{3 - x^2}}\, dx.$

32. $\displaystyle \int \frac{1}{x^2\sqrt{x^2 + 1}}\, dx.$

33. $\displaystyle \int \frac{\sqrt{x^2 - 5}}{x^4}\, dx.$

■ 9.8 NUMERICAL INTEGRATION; SIMPSON'S RULE

> *The most direct way to evaluate a definite integral is to find an antiderivative of the integrand and apply the First Fundamental Theorem of Calculus. If an antiderivative cannot be found, then the value of the integral can be approximated using methods that we shall study in this section.*

☐ **RIEMANN SUM APPROXIMATIONS**

The simplest approximations of definite integrals use Riemann sums. Recall from the definition of a definite integral (Definition 5.6.2) that

$$\int_a^b f(x)\, dx = \lim_{\max \Delta x_k \to 0} \sum_{k=1}^n f(x_k^*)\, \Delta x_k$$

In the special case where the interval $[a, b]$ is divided into n subintervals of equal width Δx, we have

$$\Delta x_k = \Delta x = \frac{b - a}{n}$$

so the above formula simplifies to

$$\int_a^b f(x)\, dx = \lim_{n \to +\infty} \sum_{k=1}^n f(x_k^*)\, \Delta x$$

This suggests that the approximation

$$\int_a^b f(x)\, dx \approx \sum_{k=1}^n f(x_k^*)\, \Delta x = \Delta x \sum_{k=1}^n f(x_k^*) = \frac{b - a}{n} \sum_{k=1}^n f(x_k^*)$$

should be good for large values of n. This approximation can be written in expanded form as

$$\int_a^b f(x)\, dx \approx \left(\frac{b - a}{n} \right) \left[f(x_1^*) + f(x_2^*) + \cdots + f(x_n^*) \right] \tag{1}$$

where $x_1^*, x_2^*, \ldots, x_n^*$ denote arbitrary points in successive subintervals. Different versions of this approximation result from different choices of these points. Two possibilities are to choose them as the left endpoints of the subintervals or the right endpoints of the subintervals. To obtain the resulting formulas for these choices, let

$$y_0 = f(a), \quad y_1 = f(x_1), \quad y_2 = f(x_2), \ldots, \quad y_{n-1} = f(x_{n-1}), \quad y_n = f(b)$$

denote the values of f at the endpoints of the subintervals (Figure 9.8.1a); then, as illustrated in Figures 9.8.1b and 9.8.1c, the left-hand and right-hand endpoint approximations resulting from (1) are

Left-hand Endpoint Approximation

$$\int_a^b f(x)\, dx \approx \left(\frac{b - a}{n}\right)[y_0 + y_1 + \cdots + y_{n-1}] \tag{2}$$

Right-hand Endpoint Approximation

$$\int_a^b f(x)\, dx \approx \left(\frac{b - a}{n}\right)[y_1 + y_2 + \cdots + y_n] \tag{3}$$

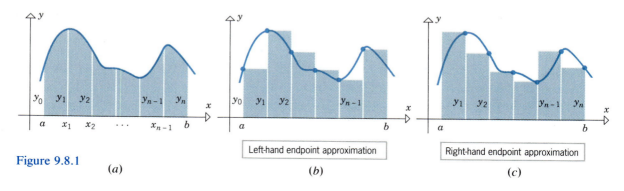

Figure 9.8.1

(a)   Left-hand endpoint approximation (b)   Right-hand endpoint approximation (c)

Another variation of (1) results from choosing $x_1^*, x_2^*, \ldots, x_n^*$ to be the midpoints of the subintervals. If we denote these midpoints by $m_1, m_2, \ldots, m_n$ and the values of f at these midpoints by

$$y_{m_1}, y_{m_2}, \ldots, y_{m_n}$$

then, as illustrated in Figure 9.8.2a, the approximation resulting from (1) when $x_1^*, x_2^*, \ldots, x_n^*$ are the midpoints is

Midpoint Approximation

$$\int_a^b f(x)\, dx \approx \left(\frac{b - a}{n}\right)[y_{m_1} + y_{m_2} + \cdots + y_{m_n}]$$

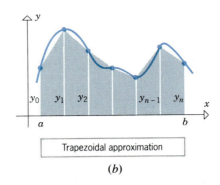

Midpoint approximation

Trapezoidal approximation

(a)   (b)

Figure 9.8.2

The left-hand and right-hand endpoint approximations are rarely used; however, if we take the average of the left-hand and right-hand endpoint approximations, we obtain a result, called the *trapezoidal approximation*, which is commonly used:

> **Trapezoidal Approximation**
>
> $$\int_a^b f(x)\, dx \approx \left(\frac{b-a}{2n}\right)[y_0 + 2y_1 + \cdots + 2y_{n-1} + y_n] \qquad (4)$$

The name, *trapezoidal approximation*, can be explained by considering the case where $f(x) \geq 0$ on $[a, b]$, so that $\int_a^b f(x)\, dx$ represents the area under $f(x)$ over $[a, b]$. Geometrically, the trapezoidal approximation formula results if we approximate this area by the sum of the trapezoidal areas shown in Figure 9.8.2b (Exercise 41).

Example 1   In the tables at the top of the next page we have approximated

$$\ln 2 = \int_1^2 \frac{1}{x}\, dx$$

using the midpoint approximation and the trapezoidal approximation. In each case we used $n = 10$ subdivisions of the interval $[1, 2]$, so that

$$\frac{b-a}{n} = \frac{2-1}{10} = 0.1 \quad \text{and} \quad \frac{b-a}{2n} = \frac{2-1}{20} = 0.05 \qquad \blacktriangleleft$$

REMARK.   In Example 1 we rounded the numerical values to nine places to the right of the decimal point; we shall follow this procedure throughout this section. If your calculator cannot produce this many places, then you will have to make the appropriate adjustments. What is important here is that you understand the principles involved.

Midpoint Approximation

i	MIDPOINT m_i	$y_{m_i} = f(m_i) = 1/m_i$
1	1.05	0.952380952
2	1.15	0.869565217
3	1.25	0.800000000
4	1.35	0.740740741
5	1.45	0.689655172
6	1.55	0.645161290
7	1.65	0.606060606
8	1.75	0.571428571
9	1.85	0.540540541
10	1.95	0.512820513
		6.928353603

$$\int_1^2 \frac{1}{x}\,dx \approx (0.1)(6.928353603) = 0.692835360$$

Trapezoidal Approximation

i	ENDPOINT x_i	$y_i = f(x_i) = 1/x_i$	MULTIPLIER w_i	$w_i y_i$
0	1.0	1.000000000	1	1.000000000
1	1.1	0.909090909	2	1.818181818
2	1.2	0.833333333	2	1.666666667
3	1.3	0.769230769	2	1.538461538
4	1.4	0.714285714	2	1.428571429
5	1.5	0.666666667	2	1.333333333
6	1.6	0.625000000	2	1.250000000
7	1.7	0.588235294	2	1.176470588
8	1.8	0.555555556	2	1.111111111
9	1.9	0.526315789	2	1.052631579
10	2.0	0.500000000	1	0.500000000
				13.875428063

$$\int_1^2 \frac{1}{x}\,dx \approx (0.05)(13.875428063) = 0.693771403$$

COMPARISON OF THE MIDPOINT AND TRAPEZOIDAL APPROXIMATIONS

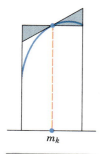

m_k

The shaded triangles have equal areas.

Figure 9.8.3

The value of ln 2 rounded to nine decimal places is

$$\ln 2 = \int_1^2 \frac{1}{x}\,dx \approx 0.693147181 \tag{5}$$

so that the midpoint approximation in Example 1 produced a more accurate result than the trapezoidal approximation (verify). To see why this should be so, we need to look at the midpoint approximation from another viewpoint. [For simplicity in the explanations, we shall assume that $f(x) \geq 0$, but the conclusions will be true without this assumption.] For differentiable functions, the midpoint approximation is sometimes called the *tangent line approximation* because over each subinterval the area of the rectangle used in the midpoint approximation is equal to the area of the trapezoid whose upper boundary is the tangent line to $y = f(x)$ at the midpoint of the interval (Figure 9.8.3). The equality of these areas follows from the fact that the shaded triangles in Figure 9.8.3 are congruent.

Let us denote the midpoint and trapezoidal approximations of

$$\int_a^b f(x)\,dx$$

with n subintervals by M_n and T_n, respectively. We define the *errors* in these approximations to be

$$E_M = \int_a^b f(x)\,dx - M_n \quad \text{and} \quad E_T = \int_a^b f(x)\,dx - T_n$$

The quantities $|E_M|$ and $|E_T|$, which describe the sizes of the errors without regard to sign, are also important; these are called the *magnitudes* of the errors.

In Figure 9.8.4a we have isolated a subinterval of $[a, b]$ on which the graph of a function f is concave down, and we have shaded the areas that represent the magnitudes of the errors in the midpoint and trapezoidal approximations over the subinterval. In Figure 9.8.4b we show a succession of four illustrations which make it evident that the magnitude of the error from the midpoint approximation is less than that from the trapezoidal approximation. If the graph of f were concave up, analogous figures would lead to the same conclusion. (This argument, due to Frank Buck, appeared in *The College Mathematics Journal*, Vol. 16, No. 1, 1985.)

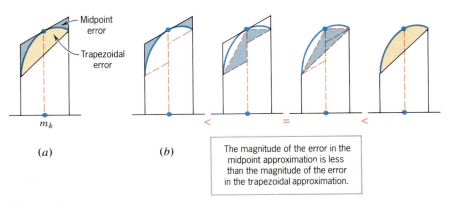

The magnitude of the error in the midpoint approximation is less than the magnitude of the error in the trapezoidal approximation.

Figure 9.8.4

Figure 9.8.4a also suggests that on a subinterval where the graph is concave down, the midpoint approximation is larger than the value of the integral and the trapezoidal approximation is smaller. On an interval where the graph is concave up it is the other way around. In summary, we have the following result, which we state without formal proof.

9.8.1   THEOREM.   *Let f be continuous on $[a, b]$, and let $|E_M|$ and $|E_T|$ be the magnitudes of the errors that result from the midpoint and trapezoidal approximations of $\int_a^b f(x)\, dx$ using n subintervals.*

(a)   *If the graph of f is either concave up or concave down on (a, b), then $|E_M| < |E_T|$, that is, the magnitude of the error from the midpoint approximation is less than that from the trapezoidal approximation.*

(b)   *If the graph of f is concave down on (a, b), then*

$$T_n < \int_a^b f(x)\, dx < M_n$$

(c)   *If the graph of f is concave up on (a, b), then*

$$M_n < \int_a^b f(x)\, dx < T_n$$

Example 2   We observed earlier that the midpoint approximation of $\ln 2$ obtained in Example 1 was more accurate than the trapezoidal approximation. This is consistent with part (a) of Theorem 9.8.1, since $f(x) = 1/x$ is continuous on $[1, 2]$ and concave up on $(1, 2)$. Moreover, a comparison of the two approximations to (5) shows that the midpoint approximation is smaller than $\ln 2$ and the trapezoidal approximation is larger. This is consistent with part (c) of Theorem 9.8.1.   ◀

REMARK.   Do not erroneously conclude that the midpoint approximation is always better than the trapezoidal approximation; for functions with inflection points, the trapezoidal approximation can be more accurate.

□ **SIMPSON'S RULE**

Intuition suggests that we might improve on the midpoint and trapezoidal approximations by replacing the linear upper boundaries of the approximating strips in Figure 9.8.2 by curved upper boundaries chosen to fit the shape of the curve $y = f(x)$ more closely. This is the idea behind **Simpson's* rule**, which uses parabolic curves of the form

$$y = ax^2 + bx + c \tag{6}$$

to approximate sections of the curve $y = f(x)$. [Recall from Section 1.6 that (6) is the equation of a parabola with axis of symmetry parallel to the y-axis.]

To simplify the description of Simpson's rule we shall assume that $f(x) \geq 0$ on $[a, b]$ so that we can interpret $\int_a^b f(x)\,dx$ as an area. However, the method is valid without this assumption. The heart of Simpson's rule is the formula

$$A = \frac{h}{3}[Y_0 + 4Y_1 + Y_2] \tag{7}$$

which gives the area under the curve

$$y = ax^2 + bx + c$$

over an arbitrary interval of width $2h$. In this formula Y_0, Y_1, and Y_2 represent the y values at the left-hand endpoint, the midpoint m, and the right-hand endpoint of the interval (Figure 9.8.5).

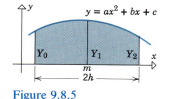

Figure 9.8.5

---

*THOMAS SIMPSON (1710–1761). English mathematician. Simpson was the son of a weaver. He was trained to follow in his father's footsteps and had little formal education in his early life. His interest in science and mathematics was aroused in 1724, when he witnessed an eclipse of the sun and received two books from a peddler, one on astrology and the other on arithmetic. Simpson quickly absorbed their contents and soon became a successful local fortune teller. His improved financial situation enabled him to give up weaving and marry his landlady, an older woman. Then in 1733 some mysterious "unfortunate incident" forced him to move. He settled in Derby, where he taught in an evening school and worked at weaving during the day. In 1736 he moved to London and published his first mathematical work in a periodical called the *Ladies' Diary* (of which he later became the editor). In 1737 he published a successful calculus textbook that enabled him to give up weaving completely and concentrate on textbook writing and teaching. His fortunes improved further in 1740 when one Robert Heath accused him of plagiarism. The publicity was marvelous, and Simpson proceeded to dash off a succession of best-selling textbooks: *Algebra* (ten editions plus translations), *Geometry* (twelve editions plus translations), *Trigonometry* (five editions plus translations), and numerous others.

It is interesting to note that Simpson did not discover the rule that bears his name. It was a well-known result by Simpson's time.

To derive (7), observe that the left-hand endpoint of the interval is $m - h$ and the right-hand endpoint is $m + h$, so the area A under $y = ax^2 + bx + c$ over this interval is

$$A = \int_{m-h}^{m+h} (ax^2 + bx + c) \, dx$$

$$= \frac{a}{3} x^3 + \frac{b}{2} x^2 + cx \bigg]_{m-h}^{m+h}$$

$$= \frac{a}{3} [(m + h)^3 - (m - h)^3] + \frac{b}{2} [(m + h)^2 - (m - h)^2]$$

$$+ c[(m + h) - (m - h)]$$

or on simplifying

$$A = \frac{h}{3} [a(6m^2 + 2h^2) + b(6m) + 6c] \tag{8}$$

But the values of $y = ax^2 + bx + c$ at the left-hand endpoint, the midpoint, and the right-hand endpoint are, respectively,

$$Y_0 = a(m - h)^2 + b(m - h) + c$$
$$Y_1 = am^2 + bm + c$$
$$Y_2 = a(m + h)^2 + b(m + h) + c$$

from which it follows that

$$Y_0 + 4Y_1 + Y_2 = a(6m^2 + 2h^2) + b(6m) + 6c \tag{9}$$

Thus, (7) follows from (8) and (9).

Simpson's rule is obtained by dividing the interval $[a, b]$ into an *even* number of subintervals of equal width h and applying Formula (7) to approximate the area under $y = f(x)$ over successive pairs of subintervals. The sum of these approximations then serves as an estimate of $\int_a^b f(x) \, dx$. More precisely, let $[a, b]$ be divided into n subintervals of width $h = (b - a)/n$ (n even) and let

$$y_0, y_1, \ldots, y_n$$

be the values of $y = f(x)$ at the subinterval endpoints

$$a = x_0, x_1, \ldots, x_n = b$$

By (7) the area under $y = f(x)$ over the first two subintervals is approximately

$$\frac{h}{3} [y_0 + 4y_1 + y_2]$$

and the area over the second pair of subintervals is approximately

$$\frac{h}{3} [y_2 + 4y_3 + y_4]$$

and the area over the last pair of subintervals is approximately

$$\frac{h}{3}[y_{n-2} + 4y_{n-1} + y_n]$$

Adding all the approximations, collecting terms, and replacing h by $(b - a)/n$ yields

Simpson's Rule

$$\int_a^b f(x)\, dx \approx \left(\frac{b - a}{3n}\right)[y_0 + 4y_1 + 2y_2 + 4y_3 + 2y_4 + \cdots$$

$$+ 2y_{n-2} + 4y_{n-1} + y_n]$$

We shall denote the Simpson's rule approximation with n subintervals by S_n and define the error in this approximation to be

$$E_S = \int_a^b f(x)\, dx - S_n$$

Example 3 In the following table we have approximated

$$\ln 2 = \int_1^2 \frac{1}{x}\, dx$$

by Simpson's rule using $n = 10$ subdivisions so that

$$\frac{b - a}{3n} = \frac{2 - 1}{3(10)} = \frac{1}{30}$$

Simpson's Rule

i	ENDPOINT x_i	$y_i = f(x_i) = 1/x_i$	MULTIPLIER w_i	$w_i y_i$
0	1.0	1.000000000	1	1.000000000
1	1.1	0.909090909	4	3.636363636
2	1.2	0.833333333	2	1.666666666
3	1.3	0.769230769	4	3.076923076
4	1.4	0.714285714	2	1.428571428
5	1.5	0.666666667	4	2.666666668
6	1.6	0.625000000	2	1.250000000
7	1.7	0.588235294	4	2.352941176
8	1.8	0.555555556	2	1.111111112
9	1.9	0.526315789	4	2.105263156
10	2.0	0.500000000	1	0.500000000
				20.794506918

$$\int_1^2 \frac{1}{x}\, dx \approx (\tfrac{1}{30})(20.794506918) = 0.693150231$$

Observe, by comparing this result to (5), that Simpson's rule produced a more accurate approximation of $\ln 2$ than either of the methods in Example 1. ◄

□ **ERROR ESTIMATES**

With all the methods studied in this section, there are two sources of error, the *intrinsic* or *truncation error* due to the approximation formula, and the *roundoff* error introduced in the calculations. In general, increasing n reduces the truncation error but increases the roundoff error, since more computations are required for larger n. In practical applications, it is important to know how large n must be taken to ensure that a specified degree of accuracy is obtained. The analysis of roundoff error is complicated and will not be considered here. However, the following theorems, which are proved in books on **numerical analysis,** provide estimates of the truncation errors in the midpoint, trapezoidal, and Simpson's rule approximations.

9.8.2   THEOREM (*Midpoint and Trapezoidal Error Estimates*).   *If f'' is continuous on $[a, b]$ and if K_2 is the maximum value of $|f''(x)|$ on $[a, b]$, then for n subdivisions of $[a, b]$*

(a) $\displaystyle |E_M| \leq \frac{(b - a)^3 K_2}{24n^2}$ (10)

(b) $\displaystyle |E_T| \leq \frac{(b - a)^3 K_2}{12n^2}$ (11)

9.8.3   THEOREM (*Simpson Error Estimate*).   *If $f^{(4)}$ is continuous on $[a, b]$ and if K_4 is the maximum value of $|f^{(4)}(x)|$ on $[a, b]$, then for n subdivisions of $[a, b]$*

$$|E_S| \leq \frac{(b - a)^5 K_4}{180n^4} \qquad (12)$$

The quantities on the right sides of (10), (11), and (12) (and any numbers larger than these) are called **upper bounds** on the magnitudes of the errors.

Example 4   Find an upper bound on the magnitude of the error that results from approximating

$$\ln 2 = \int_1^2 \frac{1}{x}\, dx$$

using $n = 10$ subintervals by

(a)   the trapezoidal approximation,

(b)   the midpoint approximation,

(c)   and Simpson's rule.

Solution.   We shall apply Formulas (10), (11), and (12) with

$$f(x) = \frac{1}{x}, \quad a = 1, \quad b = 2, \quad \text{and} \quad n = 10$$

We have

$$f'(x) = -\frac{1}{x^2}, \quad f''(x) = \frac{2}{x^3}, \quad f'''(x) = -\frac{6}{x^4}, \quad f^{(4)}(x) = \frac{24}{x^5}$$

Thus,

$$|f''(x)| = \left|\frac{2}{x^3}\right| = \frac{2}{x^3} \tag{13}$$

$$|f^{(4)}(x)| = \left|\frac{24}{x^5}\right| = \frac{24}{x^5} \tag{14}$$

where we have dropped the absolute values because $f''(x)$ and $f^{(4)}(x)$ have positive values for $1 \leq x \leq 2$. Since (13) and (14) are continuous and decreasing on $[1, 2]$, both functions have their maximum values at $x = 1$; for (13) this maximum value is 2 and for (14) it is 24, so we can take $K_2 = 2$ in (10) and (11) and $K_4 = 24$ in (12). This yields

$$|E_T| \leq \frac{(b-a)^3 K_2}{12n^2} = \frac{1^3 \cdot 2}{12 \cdot 10^2} \approx 0.001666667$$

$$|E_M| \leq \frac{(b-a)^3 K_2}{24n^2} = \frac{1^3 \cdot 2}{24 \cdot 10^2} \approx 0.000833333$$

$$|E_S| \leq \frac{(b-a)^5 K_4}{180n^4} = \frac{1^5 \cdot 24}{180 \cdot 10^4} \approx 0.000013333 \quad ◀$$

Table 9.8.1 shows that the estimates in the foregoing example are consistent with the computations in Examples 1 and 3. In the table we have obtained approximate values of $|E_T|$, $|E_M|$, and $|E_S|$ by computing the absolute value of the difference between the value of ln 2 (accurate to nine decimal places) and the approximations obtained in Examples 1 and 3. Observe that these values of $|E_T|$, $|E_M|$, and $|E_S|$ satisfy the upper bounds obtained in Example 4 and, in fact, are considerably smaller than the upper bounds. It is quite common that the actual errors in the approximations are substantially smaller than the upper bounds.

Table 9.8.1

ln 2 (NINE DECIMAL PLACES)	APPROXIMATION	ABSOLUTE VALUE OF THE DIFFERENCE
0.693147181	$T_{10} = 0.693771403$	$|E_T| \approx 0.000624222$
0.693147181	$M_{10} = 0.692835360$	$|E_M| \approx 0.000311821$
0.693147181	$S_{10} = 0.693150231$	$|E_S| \approx 0.000003050$

Example 5 How many subintervals might be used in approximating

$$\ln 2 = \int_1^2 \frac{1}{x} \, dx$$

by Simpson's rule to ensure that the magnitude of the error is less than 10^{-6}?

Solution.   We want to find an even value of n (the number of subintervals) such that $|E_S| < 10^{-6}$, which, from (12), can be achieved by choosing n so that

$$\frac{(b-a)^5 K_4}{180n^4} < 10^{-6}$$

Using the values of a, b, and K_4 obtained in Example 4, this inequality can be written as

$$\frac{24}{180n^4} < 10^{-6} \quad \text{or} \quad \frac{2}{15n^4} < 10^{-6}$$

which can be rewritten as

$$n^4 > \frac{2 \times 10^6}{15} \quad \text{or} \quad n > \sqrt[4]{\frac{2 \times 10^6}{15}}$$

The right side of this inequality is slightly larger than 19.1 (verify), so the smallest even integer satisfying the inequality is $n = 20$.   ◀

In cases where it is difficult to find the values of K_2 and K_4 required in Formulas (10), (11), and (12), these constants may be replaced by any larger constants if such constants are easier to find. For example, if $K_2 < K$, then

$$|E_T| \le \frac{(b-a)^3 K_2}{12n^2} < \frac{(b-a)^3 K}{12n^2} \tag{15}$$

so the right side of (15) is also an upper bound on the value of $|E_T|$ (although it is larger and therefore less desirable than the upper bound using K_2). The following example illustrates this idea.

Example 6   How many subintervals might be used in approximating

$$\int_0^1 \cos(x^2)\, dx$$

by the midpoint approximation to ensure that the magnitude of the error is less than 5×10^{-4}?

Solution.   We want to find a value of n such that $|E_M| < 5 \times 10^{-4}$. From (10) using $a = 0$ and $b = 1$ we obtain

$$|E_M| \le \frac{K_2}{24n^2} \tag{16}$$

We have

$$f(x) = \cos(x^2)$$

so

$$f'(x) = -2x \sin(x^2)$$

$$f''(x) = -4x^2 \cos(x^2) - 2 \sin(x^2) = -(4x^2 \cos(x^2) + 2 \sin(x^2))$$

so

$$|f''(x)| = |4x^2 \cos(x^2) + 2 \sin(x^2)| \tag{17}$$

It is tedious to find the maximum value of (17) on [0, 1], so we will use a larger value instead. Such a value can be obtained by observing that

$$|4x^2 \cos(x^2) + 2 \sin(x^2)| \le |4x^2 \cos(x^2)| + |2 \sin(x^2)| \qquad \boxed{\begin{array}{c} \text{Theorem 1.2.6} \\ \text{(the triangle inequality)} \end{array}}$$

$$= 4x^2 |\cos(x^2)| + 2 |\sin(x^2)| \qquad \boxed{\begin{array}{l} |\cos(x^2)| \le 1, \\ |\sin(x^2)| \le 1, \\ x^2 \le 1 \quad \text{(on [0, 1])} \end{array}}$$

$$\le 4 \cdot 1 + 2 \cdot 1$$

$$= 6$$

Thus, $K_2 \le 6$, so from (16)

$$|E_M| \le \frac{K_2}{24n^2} \le \frac{6}{24n^2} = \frac{1}{4n^2}$$

It follows that $|E_M| < 5 \times 10^{-4}$ if

$$\frac{1}{4n^2} < 5 \times 10^{-4}$$

or equivalently

$$n > \sqrt{500} > 22.3$$

Thus, $n = 23$ subintervals will ensure the desired accuracy (as will any larger number of subintervals). ◀

□ **A COMPARISON OF THE THREE METHODS**

Of the three methods studied in this section, Simpson's rule generally produces more accurate results than the midpoint or trapezoidal approximations for the same amount of work. To make this plausible, let us express (10), (11), and (12) in terms of the subinterval width

$$\Delta x = \frac{b - a}{n}$$

We obtain

$$|E_M| \le \frac{1}{24} K_2(b - a)(\Delta x)^2 \tag{18}$$

$$|E_T| \le \frac{1}{12} K_2(b - a)(\Delta x)^2 \tag{19}$$

$$|E_S| \le \frac{1}{180} K_4(b - a)(\Delta x)^4 \tag{20}$$

(verify). Thus, for Simpson's rule the upper bound on the magnitude of the error is proportional to $(\Delta x)^4$, whereas it is proportional to $(\Delta x)^2$ for the midpoint and trapezoidal approximations. Thus, reducing the interval width by a factor of 10, for example, reduces the error bound by a factor of $1/100$ for the midpoint and trapezoidal approximations but reduces it by a factor of $1/10,000$ for Simpson's rule. This suggests that the accuracy of Simpson's rule improves much more rapidly than that of the other approximations as n increases.

As a final note, observe that if $f(x)$ is a polynomial of degree 3 or less, then $f^{(4)}(x) = 0$ for all x, so $K_4 = 0$ in (12) and consequently $|E_S| = 0$. Thus, Simpson's rule gives exact results for polynomials of degree 3 or less. Similarly, the midpoint and trapezoidal approximations give exact results for polynomials of degree 1 or less. (You should also be able to see that this is so geometrically.)

▶ Exercise Set 9.8   [C] 1–30, 33–40

In Exercises 1–6, use $n = 10$ subdivisions to approximate the value of the integral by (a) the midpoint rule, (b) the trapezoidal rule, and (c) Simpson's rule. In each case find the exact value of the integral and approximate the magnitude of the error. Express your answers to at least four decimal places.

1. $\displaystyle\int_0^3 \sqrt{x + 1}\, dx.$

2. $\displaystyle\int_1^4 \frac{1}{\sqrt{x}}\, dx.$

3. $\displaystyle\int_0^\pi \sin x\, dx.$

4. $\displaystyle\int_0^1 \cos x\, dx.$

5. $\displaystyle\int_1^3 e^{-x}\, dx.$

6. $\displaystyle\int_{-1}^1 \frac{1}{2x + 3}\, dx.$

In Exercises 7–12, use inequalities (10), (11), and (12) to find upper bounds on the magnitudes of the errors in parts (a), (b), and (c) of the indicated exercise.

7. Exercise 1.   8. Exercise 2.

9. Exercise 3.   10. Exercise 4.

11. Exercise 5.   12. Exercise 6.

In Exercises 13–18, use inequalities (10), (11), and (12) to find a value for n to ensure that the magnitude of the error will be less than the given value if n subdivisions are used to approximate the integral by (a) the midpoint rule, (b) the trapezoidal rule, and (c) Simpson's rule.

13. Exercise 1; 5×10^{-4}. 14. Exercise 2; 5×10^{-4}.

15. Exercise 3; 10^{-3}.   16. Exercise 4; 10^{-3}.

17. Exercise 5; 10^{-6}.   18. Exercise 6; 10^{-6}.

In Exercises 19–24, use $n = 10$ subdivisions to approximate the value of the integral by (a) the midpoint rule, (b) the trapezoidal rule, and (c) Simpson's rule. Express your answers to at least four decimal places.

19. $\displaystyle\int_0^1 e^{-x^2}\, dx.$

20. $\displaystyle\int_0^2 \frac{x}{\sqrt{1 + x^3}}\, dx.$

21. $\displaystyle\int_1^2 \sqrt{1 + x^3}\, dx.$

22. $\displaystyle\int_0^\pi \frac{1}{2 - \sin x}\, dx.$

23. $\displaystyle\int_0^2 \sin(x^2)\, dx.$

24. $\displaystyle\int_1^3 \sqrt{\ln x}\, dx.$

In Exercises 25 and 26, the exact value of the integral is π (verify). Use $n = 10$ subdivisions to approximate the integral by (a) the midpoint rule, (b) the trapezoidal rule, and (c) Simpson's rule. In each case approximate the magnitude of the error. Express your answers to at least four decimal places.

25. $\displaystyle\int_0^1 \frac{4}{1 + x^2}\, dx.$

26. $\displaystyle\int_0^2 \sqrt{4 - x^2}\, dx.$

27. In Example 5 we showed that taking $n = 20$ subdivisions ensures that the magnitude of the error in approximating

$$\ln 2 = \int_1^2 \frac{1}{x}\, dx$$

by Simpson's rule is less than 10^{-6}. Use Simpson's

rule with $n = 14$ subdivisions to approximate this integral, and show that the magnitude of the resulting error is less than 10^{-6} by comparing your result to the approximation in expression (5), which is accurate to nine decimal places. Does this contradict the fact that we obtained $n = 20$ in Example 5? Explain.

28. In parts (a) and (b), determine whether an approximation of the integral by the trapezoidal rule would be less than or would be greater than the exact value of the integral.

    (a) $\displaystyle\int_{1}^{2} e^{-x^2}\, dx$ (b) $\displaystyle\int_{0}^{0.5} e^{-x^2}\, dx$

In Exercises 29 and 30, find a value for n to ensure that the magnitude of the error in approximating the integral by the midpoint rule will be less than 10^{-4}.

29. $\displaystyle\int_{0}^{2} x \sin x\, dx.$ 30. $\displaystyle\int_{0}^{1} e^{\cos x}\, dx.$

In Exercises 31 and 32, show that inequalities (10) and (11) are of no value in finding an upper bound on the magnitude of the error in approximating the integral by either the midpoint rule or the trapezoidal rule.

31. $\displaystyle\int_{0}^{1} \sqrt{x}\, dx.$ 32. $\displaystyle\int_{0}^{1} \sin \sqrt{x}\, dx.$

In Exercises 33 and 34, use Simpson's rule with $n = 10$ subdivisions to approximate the length of the curve. Round your answer to three decimal places.

33. $y = \sin x,\ 0 \le x \le \pi.$
34. $y = 1/x,\ 1 \le x \le 3.$

Numerical integration methods can be used in problems where only measured or experimentally determined values of the integrand are available. In Exercises 35–40, use Simpson's rule to estimate the value of the integral.

35. A graph of the speed v versus time t for a test run of an Infiniti G20 automobile is shown in Figure 9.8.6. Estimate the speeds at $t = 0, 5, 10, 15$, and 20 sec from the graph, convert to feet/second using 1 mi/hr $= 22/15$ ft/sec, and use these speeds to approximate the number of feet traveled during the first 20 sec. Round your answer to the nearest foot. [*Hint:* Distance traveled $= \int_{0}^{20} v(t)\, dt.$] [Data from *Road and Track*, October 1990.]

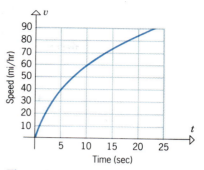

Figure 9.8.6

36. A graph of the acceleration a versus time t for an object moving on a straight line is shown in Figure 9.8.7. Estimate the accelerations at $t = 0, 1, 2, \ldots, 8$ sec from the graph and use them to approximate the change in velocity from $t = 0$ to $t = 8$ sec. Round your answer to the nearest tenth centimeter/second. [*Hint:* Change in velocity $= \int_{0}^{8} a(t)\, dt.$]

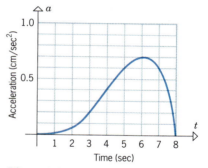

Figure 9.8.7

37. Table 9.8.2 gives the speeds, in miles/second, at various times for a test rocket that was fired upward from the surface of Earth. Use these values to approximate the number of miles traveled during the first 180 sec. Round your answer to the nearest tenth of a mile. [*Hint:* Distance traveled $= \int_{0}^{180} v(t)\, dt.$]

Table 9.8.2

TIME t (sec)	SPEED v (mi/sec)
0	0.00
30	0.03
60	0.08
90	0.16
120	0.27
150	0.42
180	0.65

38. Table 9.8.3 gives the speeds of a bullet at various distances from the muzzle of a rifle. Use these values to approximate the number of seconds for the bullet to travel 1800 ft. Express your answer to the nearest hundreth of a second. [*Hint:* If v is the speed of the bullet and x is the distance traveled, then $v = dx/dt$ so that $dt/dx = 1/v$ and $t = \int_0^{1800} (1/v)\, dx$.]

Table 9.8.3

DISTANCE x (ft)	SPEED v (ft/sec)
0	3100
300	2908
600	2725
900	2549
1200	2379
1500	2216
1800	2059

39. Measurements of a pottery shard recovered from an archaeological dig reveal that the shard came from a pot with a flat bottom and circular cross sections (Figure 9.8.8). The figure shows interior radius measurements of the shard made every 4 cm from the bottom of the pot to the top. Use those values to approximate the interior volume of the pot to the nearest tenth of a liter (1 L = 1000 cm³). [*Hint:* Use 6.2.2 (volume by cross sections) to set up an appropriate integral for the volume.]

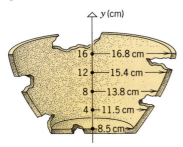

Figure 9.8.8

40. Engineers want to construct a straight and level road 600 ft long and 75 ft wide by cutting through an intervening hill (Figure 9.8.9). Heights of the hill above the centerline of the proposed road, as obtained at various points from a contour map of the region, are shown in Table 9.8.4. To estimate the construction costs, the engineers need to know the volume of earth that must be removed. Approximate this volume, rounded to the nearest cubic foot. [*Hint:* First, set up an integral for the cross-sectional area of the cut along the centerline of the road, then assume that the height of the hill does not vary between the centerline and edges of the road.]

Table 9.8.4

HORIZONTAL DISTANCE x (ft)	HEIGHT h (ft)
0	0
100	7
200	16
300	24
400	25
500	16
600	0

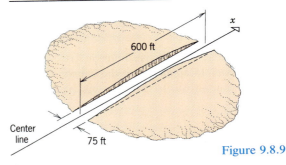

Figure 9.8.9

41. Derive the trapezoidal rule by summing the areas of the trapezoids in Figure 9.8.2(b).

▶ SUPPLEMENTARY EXERCISES   [C] 78–83

In Exercises 1–64, evaluate the integrals.

1. $\int x \cos 2x \, dx.$

2. $\int x \cos x^2 \, dx.$

3. $\int \tan^3 x \sec x \, dx.$

4. $\int \sin^3 x \cos^2 x \, dx.$

5. $\int \tan^2 3t \sec^2 3t \, dt.$

6. $\int \cot 2x \csc^3 2x \, dx.$

7. $\int \frac{\sin^2 x \, dx}{1 + \cos x}.$

8. $\int \frac{\sin 2x \, dx}{\cos x \,(1 + \cos x)}.$

9. $\int x^2 \cos^2 x \, dx.$

10. $\displaystyle\int \sin^2 2x \cos^2 2x \, dx.$

11. $\displaystyle\int \sec^5 x \sin x \, dx.$ **12.** $\displaystyle\int \tan^5 2x \, dx.$

13. $\displaystyle\int \sin^4 x \cos^2 x \, dx.$ **14.** $\displaystyle\int \frac{dx}{\sec^4 x}.$

15. $\displaystyle\int_0^{\pi/4} \sin 5x \sin 3x \, dx.$

16. $\displaystyle\int_{-\pi/10}^0 \sin 2x \cos 3x \, dx.$

17. $\displaystyle\int_0^1 \sin^2 \pi x \, dx.$ **18.** $\displaystyle\int_0^{\pi/3} \sin^3 3x \, dx.$

19. $\displaystyle\int_0^{\sqrt{\pi/2}} x \sec^2 (x^2) \, dx.$

20. $\displaystyle\int_0^{\pi/4} \frac{\sec^2 x \, dx}{\sqrt{1 + 3\tan x}}.$

21. $\displaystyle\int \frac{\sin (\cot^{-1} x) \, dx}{1 + x^2}.$ **22.** $\displaystyle\int \frac{e^{\tan 3x} \, dx}{\cos^2 3x}.$

23. $\displaystyle\int e^x \sec (e^x) \, dx.$ **24.** $\displaystyle\int x \sec^2 3x \, dx.$

25. $\displaystyle\int \frac{e^{2x}}{\sqrt{e^{2x} + 1}} \, dx.$ **26.** $\displaystyle\int \frac{e^{2x}}{e^{2x} + 1} \, dx.$

27. $\displaystyle\int e^{3x} \sin 2x \, dx.$ **28.** $\displaystyle\int \ln (a^2 + x^2) \, dx.$

29. $\displaystyle\int_1^2 \sin^{-1} (x/2) \, dx.$ **30.** $\displaystyle\int \frac{\cos 2\pi x}{e^{2\pi x}} \, dx.$

31. $\displaystyle\int \sin (3 \ln x) \, dx.$ **32.** $\displaystyle\int x^3 e^{-x^2} \, dx.$

33. $\displaystyle\int \frac{x \, dx}{\sqrt{x^2 - 9}}.$ **34.** $\displaystyle\int_1^2 \frac{\sqrt{4x^2 - 1}}{x} \, dx.$

35. $\displaystyle\int_1^3 \frac{\sqrt{9 - x^2}}{x} \, dx.$ **36.** $\displaystyle\int_0^{\pi/6} \frac{\cos 3x}{\sqrt{4 - \sin^2 3x}} \, dx.$

37. $\displaystyle\int \frac{x^2 \, dx}{\sqrt{2x + 3}}.$ **38.** $\displaystyle\int \frac{1}{\sqrt{x}(x + 9)} \, dx.$

39. $\displaystyle\int \frac{dt}{\sqrt{3 - 4t - 4t^2}}.$ **40.** $\displaystyle\int \frac{dx}{\sqrt{6x - x^2}}.$

41. $\displaystyle\int \frac{dx}{x^2 \sqrt{a^2 - x^2}}.$ **42.** $\displaystyle\int \frac{x^3}{(x^2 + 4)^{1/3}} \, dx.$

43. $\displaystyle\int \sqrt{a^2 - x^2} \, dx.$ **44.** $\displaystyle\int x\sqrt{a^2 - x^2} \, dx.$

45. $\displaystyle\int \frac{x - 2}{\sqrt{4x - x^2}} \, dx.$ **46.** $\displaystyle\int_1^3 \frac{dx}{x^2 - 2x + 5}.$

47. $\displaystyle\int \frac{dx}{2x^2 + 3x + 1}.$ **48.** $\displaystyle\int \frac{dx}{(x^2 + 4)^2}.$

49. $\displaystyle\int \frac{x + 1}{x^3 + x^2 - 6x} \, dx.$ **50.** $\displaystyle\int \frac{x^3 + 1}{x - 2} \, dx.$

51. $\displaystyle\int \frac{x^2 - 1}{x^3 - 3x} \, dx.$ **52.** $\displaystyle\int \frac{x - 3}{x^3 - 1} \, dx.$

53. $\displaystyle\int \frac{2x^2 + 5}{x^4 - 1} \, dx.$

54. $\displaystyle\int \frac{x^4 - x^3 - x - 1}{x^3 - x^2} \, dx.$

55. $\displaystyle\int \frac{dx}{(x^2 + 4)(x - 3)}.$ **56.** $\displaystyle\int \frac{x \, dx}{(x + 1)^3}.$

57. $\displaystyle\int \frac{3x^2 + 12x + 2}{(x^2 + 4)^2} \, dx.$

58. $\displaystyle\int \frac{(4x + 2) \, dx}{x^4 + 2x^3 + x^2}.$

59. $\displaystyle\int \frac{x \, dx}{x^2 + 2x + 5}.$ **60.** $\displaystyle\int \frac{6x \, dx}{(x^2 + 9)^3}.$

61. $\displaystyle\int \frac{dx}{\sqrt{3 - 2x^2}}.$ **62.** $\displaystyle\int \frac{1 + t}{\sqrt{t}} \, dt.$

63. $\displaystyle\int \frac{\sqrt{t} \, dt}{1 + t}.$ **64.** $\displaystyle\int \frac{\sqrt{1 - x^2}}{x^2} \, dx.$

Exercises 65–70 relate to Section 9.7. Evaluate the integrals.

65. $\displaystyle\int_0^1 \frac{x^{2/3}}{1 + x^{1/3}} \, dx.$ **66.** $\displaystyle\int \frac{dx}{x^{1/2} + x^{1/4}}.$

67. $\displaystyle\int \frac{dx}{1 - \tan x}.$ **68.** $\displaystyle\int \frac{dx}{3 \cos x + 5}.$

69. $\displaystyle\int \frac{dx}{\sin x - \tan x}.$ **70.** $\displaystyle\int_0^{\pi/3} \frac{dx}{5 \sec x - 3}.$

71. Use partial fractions to show that
$$\int \frac{dx}{x^2 - a^2} = \frac{1}{2a} \ln \left| \frac{x - a}{x + a} \right| + C \quad (a \neq 0)$$

72. Find the arc length of (a) the parabola $y = x^2/2$ from $(0, 0)$ to $(2, 2)$ and (b) the curve $y = \ln (\sec x)$ from $(0, 0)$ to $(\pi/4, \frac{1}{2} \ln 2)$.

73. Let R be the region bounded by the curve $y = 1/(4 + x^2)$ and the lines $x = 0$, $y = 0$, and $x = 2$. Find (a) the area of R, (b) the volume of the solid obtained by revolving R about the x-axis, and (c) the volume of the solid obtained by revolving R about the y-axis.

74. Derive the following reduction formulas for $a \neq 0$:

(a) $\displaystyle \int x^n e^{ax} \, dx = \frac{x^n e^{ax}}{a} - \frac{n}{a} \int x^{n-1} e^{ax} \, dx$

(b) $\displaystyle \int x^n \sin ax \, dx$

$$= \frac{-x^n \cos ax}{a} + \frac{n}{a} \int x^{n-1} \cos ax \, dx$$

$$\int x^n \cos ax \, dx$$

$$= \frac{x^n \sin ax}{a} - \frac{n}{a} \int x^{n-1} \sin ax \, dx$$

(c) $\displaystyle \int \sin^n ax \cos^m ax \, dx = -\frac{\sin^{n-1} ax \cos^{m+1} ax}{a(m+n)}$

$$+ \frac{n-1}{m+n} \int \sin^{n-2} ax \cos^m ax \, dx$$

$$= \frac{\sin^{n+1} ax \cos^{m-1} ax}{a(m+n)}$$

$$+ \frac{m-1}{m+n} \int \sin^n ax \cos^{m-2} ax \, dx.$$

75. Use Exercise 74 to evaluate the following integrals.

(a) $\displaystyle \int x^3 e^{2x} \, dx$ (b) $\displaystyle \int_0^{\pi/10} x^2 \sin 5x \, dx$

(c) $\displaystyle \int \sin^2 x \cos^4 x \, dx$

76. Evaluate the following integrals assuming that $a \neq 0$.

(a) $\displaystyle \int x^n \ln ax \, dx \quad (n \neq -1)$

(b) $\displaystyle \int \sec^n ax \tan ax \, dx \quad (n \geq 1)$

77. Find $\int (\sin^3 \theta / \cos^5 \theta) \, d\theta$ two ways: (a) letting $u = \cos \theta$ and (b) expressing the integrand in terms of $\sec \theta$ and $\tan \theta$. Show that your answers differ by a constant.

In Exercises 78–81, approximate the integral using the given value of n and (a) the trapezoidal rule, (b) Simpson's rule. Use a calculator and express the answer to four decimal places.

78. $\displaystyle \int_0^1 \sqrt{x} \, dx, \ n = 4.$ **79.** $\displaystyle \int_{-4}^2 e^{-x} \, dx, \ n = 6.$

80. $\displaystyle \int_0^4 \sinh x \, dx, \ n = 4.$ **81.** $\displaystyle \int_4^{5.2} \ln x \, dx, \ n = 6.$

In Exercises 82 and 83, use Simpson's rule with $n = 10$ to approximate the given integral. Use a calculator and express the answer to five decimal places.

82. $\displaystyle \int_0^2 \cos(\sinh x) \, dx.$ **83.** $\displaystyle \int_1^2 \sin(\ln x) \, dx.$

84. (a) Show that if $f(x)$ is continuous for $0 \leq x \leq 1$, then

$$\int_0^\pi x f(\sin x) \, dx = \frac{\pi}{2} \int_0^\pi f(\sin x) \, dx$$

[*Hint:* Let $x = \pi - u$.]

(b) Use the result in part (a) to find

$$\int_0^\pi \frac{x \sin x}{2 - \sin^2 x} \, dx$$

In Exercises 85–94, evaluate the integrals.

85. $\displaystyle \int \frac{1}{e^{ax} + 1} \, dx, \ a \neq 0.$ **86.** $\displaystyle \int \frac{(x-2)^3}{\sqrt{4x - x^2}} \, dx.$

87. $\displaystyle \int \frac{\sqrt{1+x} + \sqrt{1-x}}{\sqrt{1+x} - \sqrt{1-x}} \, dx.$

88. $\displaystyle \int (\cos^{32} x \sin^{30} x - \cos^{30} x \sin^{32} x) \, dx.$

89. $\displaystyle \int \frac{\sqrt{x+1}}{(x-1)^{5/2}} \, dx. \ [\textit{Hint:} \text{ Let } x - 1 = 1/u.]$

90. $\displaystyle \int \frac{1}{x^{10} + x} \, dx. \ [\textit{Hint:} \text{ Rewrite the denominator as}$
$x^{10}(1 + x^{-9})$ and let $u = 1 + x^{-9}$.]

91. $\displaystyle \int \frac{1}{x(3x^5 + 2)} \, dx. \ [\textit{Hint:} \text{ See Exercise 90.}]$

92. $\displaystyle \int \frac{3x^6 - 2}{x(2x^6 + 5)} \, dx. \ [\textit{Hint:} \text{ Rewrite as the sum of two}$
integrals and see Exercise 90.]

93. $\displaystyle \int \sqrt{x - \sqrt{x^2 - 4}} \, dx.$

[*Hint:* $\frac{1}{2}(\sqrt{x+2} - \sqrt{x-2})^2 = ?]$

94. $\displaystyle \int_0^1 \sqrt{1 + \sqrt{1 - x^2}} \, dx.$

[*Hint:* $\frac{1}{2}(\sqrt{1+x} + \sqrt{1-x})^2 = ?]$

10

Improper Integrals;
L'Hôpital's Rule

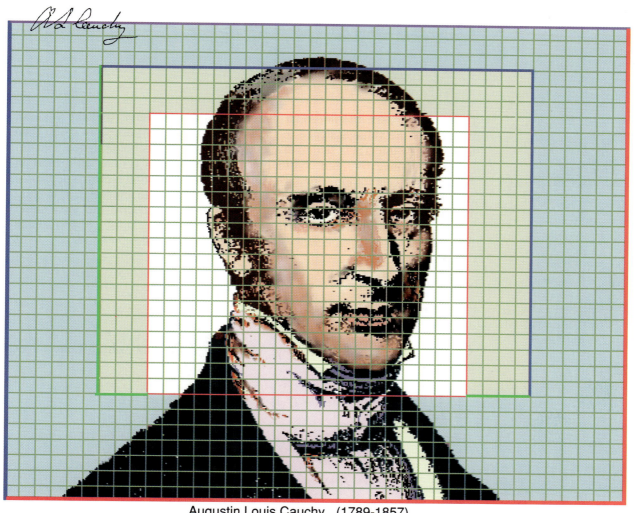

Augustin Louis Cauchy   (1789-1857)

■ 10.1   IMPROPER INTEGRALS

> *In this section we shall extend the concept of the definite integral to include*
>
> • *integrals over infinite intervals;*
> • *integrals in which the integrand becomes infinite within the interval of integration.*
>
> *These are called **improper integrals**.*

□ **INTEGRALS OVER INFINITE INTERVALS**

In the definition of $\int_a^b f(x)\,dx$, it is assumed that the interval $[a, b]$ is finite. If f is continuous on the interval $[a, +\infty)$, then we define the improper integral $\int_a^{+\infty} f(x)\,dx$ as a limit in the following way:

$$\int_a^{+\infty} f(x)\,dx = \lim_{l \to +\infty} \int_a^l f(x)\,dx \tag{1}$$

If this limit exists, the improper integral is said to *converge,* and the value of the limit is the value assigned to the integral. If the limit does not exist, then the improper integral is said to *diverge,* in which case it is not assigned a value.

If f is nonnegative and continuous on $[a, +\infty)$, then (1) has an important geometric interpretation. For each value of $l > a$, the definite integral $\int_a^l f(x)\,dx$ represents the area under the curve $y = f(x)$ over the finite interval $[a, l]$ (Figure 10.1.1a). As we let l approach $+\infty$, this area tends toward the area under $y = f(x)$ over the entire interval $[a, +\infty)$ (Figure 10.1.1b). Thus, $\int_a^{+\infty} f(x)\,dx$ can be regarded as the area under $y = f(x)$ over the interval $[a, +\infty)$.

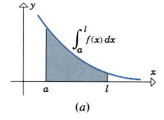

(a)

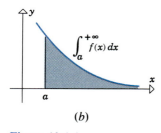

(b)

Figure 10.1.1

Example 1   Find $\displaystyle\int_1^{+\infty} \frac{dx}{x^2}$.

Solution.   We begin by replacing the infinite upper limit with a finite upper limit l:

$$\int_1^l \frac{dx}{x^2} = -\frac{1}{x}\bigg]_1^l = -\frac{1}{l} - (-1) = 1 - \frac{1}{l}$$

Thus,

$$\int_1^{+\infty} \frac{dx}{x^2} = \lim_{l \to +\infty} \int_1^l \frac{dx}{x^2} = \lim_{l \to +\infty}\left(1 - \frac{1}{l}\right) = 1$$

so the given integral converges to 1.   ◄

Example 2   Evaluate $\displaystyle\int_1^{+\infty}\frac{dx}{x}$.

Solution.

$$\int_1^{+\infty}\frac{dx}{x}=\lim_{l\to+\infty}\int_1^l\frac{dx}{x}=\lim_{l\to+\infty}\left[\ln|x|\right]_1^l=\lim_{l\to+\infty}\ln|l|=+\infty$$

Thus, the integral diverges.   ◀

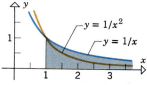

Figure 10.1.2

It is worthwhile to reflect on the results of Examples 1 and 2. Why is it that $\int_1^{+\infty}dx/x^2$ converges while $\int_1^{+\infty}dx/x$ diverges when, in fact, the graphs of $1/x$ and $1/x^2$ look similar over the interval $[1,+\infty)$ (Figure 10.1.2)? The explanation is that $1/x^2$ approaches zero more rapidly than $1/x$ as $x\to+\infty$, so that area accumulates under the curve $y=1/x^2$ less rapidly than under the curve $y=1/x$. When we calculate

$$\lim_{l\to+\infty}\int_1^l\frac{dx}{x}\quad\text{and}\quad\lim_{l\to+\infty}\int_1^l\frac{dx}{x^2}$$

the difference is enough that the first limit is infinite while the second is finite.

If f is continuous on the interval $(-\infty,b]$, then we define the improper integral $\int_{-\infty}^b f(x)\,dx$ as a limit in the following way:

$$\int_{-\infty}^b f(x)\,dx=\lim_{l\to-\infty}\int_l^b f(x)\,dx$$

As before, the improper integral is said to **converge** if the limit exists and **diverge** if it does not. If f is nonnegative, then the integral represents the area under $y=f(x)$ over the interval $(-\infty,b]$.

Example 3   Evaluate $\displaystyle\int_{-\infty}^0 e^x\,dx$.

Solution.   We begin by replacing the infinite lower limit with a finite lower limit l:

$$\int_l^0 e^x\,dx=e^x\bigg]_l^0=e^0-e^l=1-e^l$$

Thus,

$$\int_{-\infty}^0 e^x\,dx=\lim_{l\to-\infty}\int_l^0 e^x\,dx=\lim_{l\to-\infty}(1-e^l)=1\quad◀$$

If the two improper integrals $\int_{-\infty}^0 f(x)\,dx$ and $\int_0^{+\infty}f(x)\,dx$ both converge, then we say that $\int_{-\infty}^{+\infty}f(x)\,dx$ **converges** and we define

$$\int_{-\infty}^{+\infty}f(x)\,dx=\int_{-\infty}^0 f(x)\,dx+\int_0^{+\infty}f(x)\,dx \tag{2}$$

If either integral on the right side of (2) diverges, then we say that $\int_{-\infty}^{+\infty} f(x)\,dx$ *diverges*. If f is nonnegative, then the integral $\int_{-\infty}^{+\infty} f(x)\,dx$ represents the area under $y = f(x)$ over the interval $(-\infty, +\infty)$.

Example 4 Evaluate $\displaystyle\int_{-\infty}^{+\infty} \frac{dx}{1 + x^2}$.

Solution.

$$\int_0^{+\infty} \frac{dx}{1 + x^2} = \lim_{l \to +\infty} \int_0^l \frac{dx}{1 + x^2} = \lim_{l \to +\infty} \left[\tan^{-1} x \right]_0^l = \lim_{l \to +\infty} (\tan^{-1} l) = \frac{\pi}{2}$$

$$\int_{-\infty}^0 \frac{dx}{1 + x^2} = \lim_{l \to -\infty} \int_l^0 \frac{dx}{1 + x^2} = \lim_{l \to -\infty} \left[\tan^{-1} x \right]_l^0 = \lim_{l \to -\infty} (-\tan^{-1} l) = \frac{\pi}{2}$$

Thus,

$$\int_{-\infty}^{+\infty} \frac{dx}{1 + x^2} = \int_{-\infty}^0 \frac{dx}{1 + x^2} + \int_0^{+\infty} \frac{dx}{1 + x^2} = \frac{\pi}{2} + \frac{\pi}{2} = \pi \quad \blacktriangleleft$$

REMARK.   In (2), the decision to split the integral $\int_{-\infty}^{+\infty} f(x)\,dx$ at $x = 0$ is arbitrary. We can just as well make the split at any other point $x = c$ without affecting the convergence, the divergence, or the value of the integral; that is,

$$\int_{-\infty}^{+\infty} f(x)\,dx = \int_{-\infty}^c f(x)\,dx + \int_c^{+\infty} f(x)\,dx$$

(We omit the proof.)

☐ **INTEGRALS WHOSE INTEGRANDS BECOME INFINITE**

Recall from Theorem 5.6.6(c) that if a function f is not bounded on an interval $[a, b]$, then f is not integrable on $[a, b]$; that is, the limit

$$\lim_{\max \Delta x_k \to 0} \sum_{k=1}^n f(x_k^*)\,\Delta x_k = \int_a^b f(x)\,dx$$

does not exist. For example, the integral

$$\int_0^3 \frac{1}{(x - 2)^{2/3}}\,dx$$

does not exist since the integrand approaches $+\infty$ at the point $x = 2$ in the interval of integration. It is possible, however, to extend the concept of an integral to include functions that are not bounded by defining the integral of such a function as a limit of integrals of bounded functions in an appropriate way. These limits are also called *improper integrals*. We shall first consider the case where the integrand of f approaches $+\infty$ or $-\infty$ at one of the endpoints of integration.

If f is continuous on the interval $[a, b]$, but $f(x) \to +\infty$ or $f(x) \to -\infty$ as x approaches b from the left, then we define the improper integral $\int_a^b f(x)\,dx$ as

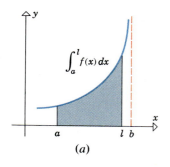

(a)

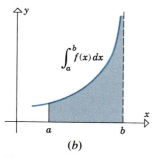

(b)

Figure 10.1.3

a limit in the following way:

$$\int_a^b f(x)\ dx = \lim_{l \to b^-} \int_a^l f(x)\ dx \tag{3}$$

To visualize this definition geometrically, consider the case where f is non-negative on the interval $[a, b)$. For each number l satisfying $a \leq l < b$, the integral $\int_a^l f(x)\ dx$ represents the area under $y = f(x)$ over the interval $[a, l]$ (Figure 10.1.3a). As we let l approach b from the left, these areas tend to fill out the entire area under $y = f(x)$ over the interval $[a, b)$ (Figure 10.1.3b). As before, the improper integral $\int_a^b f(x)\ dx$ is said to **converge** or **diverge** depending on whether the limit in (3) does or does not exist.

Example 5 Evaluate $\displaystyle\int_0^1 \frac{dx}{\sqrt{1 - x}}$.

Solution. The integral is improper because the integrand approaches $+\infty$ as x approaches the upper limit 1 from the left. From (3),

$$\int_0^1 \frac{dx}{\sqrt{1 - x}} = \lim_{l \to 1^-} \int_0^l \frac{dx}{\sqrt{1 - x}} = \lim_{l \to 1^-} \left[-2\sqrt{1 - x} \right]_0^l$$

$$= \lim_{l \to 1^-} \left[-2\sqrt{1 - l} + 2 \right] = 2 \quad \blacktriangleleft$$

If f is continuous on the interval $(a, b]$, but $f(x) \to +\infty$ or $f(x) \to -\infty$ as x approaches a from the right, then we define the improper integral $\int_a^b f(x)\ dx$ as

$$\int_a^b f(x)\ dx = \lim_{l \to a^+} \int_l^b f(x)\ dx$$

The improper integral **converges** or **diverges** depending on whether the limit does or does not exist.

Example 6 The integral

$$\int_1^2 \frac{dx}{1 - x}$$

is improper because the integrand approaches $-\infty$ as x approaches the lower limit 1 from the right. The integral diverges because

$$\int_1^2 \frac{dx}{1 - x} = \lim_{l \to 1^+} \int_l^2 \frac{dx}{1 - x} = \lim_{l \to 1^+} \left[-\ln|1 - x| \right]_l^2$$

$$= \lim_{l \to 1^+} \left[-\ln|-1| + \ln|1 - l| \right]$$

$$= \lim_{l \to 1^+} \ln|1 - l| = -\infty \quad \blacktriangleleft$$

Let f be continuous on the interval $[a, b]$ with the exception that at some point c satisfying $a < c < b$, $f(x)$ becomes infinite (tends to $+\infty$ or $-\infty$) as x approaches c from the left or the right. If the two improper integrals $\int_a^c f(x)\,dx$ and $\int_c^b f(x)\,dx$ both converge, then we say that the improper integral $\int_a^b f(x)\,dx$ *converges,* and we define

$$\int_a^b f(x)\,dx = \int_a^c f(x)\,dx + \int_c^b f(x)\,dx \tag{4}$$

If either integral on the right side of (4) diverges, then we say that $\int_a^b f(x)\,dx$ *diverges.*

Example 7 Evaluate $\int_1^4 \dfrac{dx}{(x-2)^{2/3}}$.

Solution. The integrand approaches $+\infty$ as $x \to 2$, so we use (4) to write

$$\int_1^4 \frac{dx}{(x-2)^{2/3}} = \int_1^2 \frac{dx}{(x-2)^{2/3}} + \int_2^4 \frac{dx}{(x-2)^{2/3}} \tag{5}$$

But

$$\int_1^2 \frac{dx}{(x-2)^{2/3}} = \lim_{l \to 2^-} \int_1^l \frac{dx}{(x-2)^{2/3}} = \lim_{l \to 2^-} [3(l-2)^{1/3} - 3(1-2)^{1/3}] = 3$$

$$\int_2^4 \frac{dx}{(x-2)^{2/3}} = \lim_{l \to 2^+} \int_l^4 \frac{dx}{(x-2)^{2/3}} = \lim_{l \to 2^+} [3(4-2)^{1/3} - 3(l-2)^{1/3}] = 3\sqrt[3]{2}$$

Thus, from (5),

$$\int_1^4 \frac{dx}{(x-2)^{2/3}} = 3 + 3\sqrt[3]{2} \quad \blacktriangleleft$$

WARNING. It is sometimes tempting to apply the First Fundamental Theorem of Calculus directly to an improper integral without taking the appropriate limits. To illustrate what can go wrong with this procedure, suppose we ignore the fact that the integral

$$\int_0^2 \frac{dx}{(x-1)^2} \tag{6}$$

is improper and write

$$\int_0^2 \frac{dx}{(x-1)^2} = -\frac{1}{x-1}\Big]_0^2 = -1 - (1) = -2$$

This result is clearly nonsense because the integrand is never negative and consequently the integral cannot be negative! To evaluate (6) correctly we should write

$$\int_0^2 \frac{dx}{(x-1)^2} = \int_0^1 \frac{dx}{(x-1)^2} + \int_1^2 \frac{dx}{(x-1)^2} \tag{7}$$

But

$$\int_0^1 \frac{dx}{(x-1)^2} = \lim_{l \to 1^-} \int_0^l \frac{dx}{(x-1)^2} = \lim_{l \to 1^-} \left[-\frac{1}{l-1} - 1 \right] = +\infty$$

so that (6) diverges.

▶ Exercise Set 10.1   C 63, 64, 65, 66

In Exercises 1–32, evaluate the integrals that converge.

1. $\int_0^{+\infty} e^{-x}\, dx.$

2. $\int_1^{+\infty} \frac{dx}{x^3}.$

3. $\int_1^{+\infty} \frac{dx}{\sqrt{x}}.$

4. $\int_{-1}^{+\infty} \frac{x}{1+x^2}\, dx.$

5. $\int_4^{+\infty} \frac{2}{x^2-1}\, dx.$

6. $\int_0^{+\infty} xe^{-x^2}\, dx.$

7. $\int_e^{+\infty} \frac{1}{x \ln^3 x}\, dx.$

8. $\int_2^{+\infty} \frac{1}{x\sqrt{\ln x}}\, dx.$

9. $\int_a^{+\infty} \frac{x\, dx}{(x^2+1)^2}.$

10. $\int_0^{+\infty} \frac{dx}{a^2+b^2x^2}, \quad a > 0, \ b > 0.$

11. $\int_{-\infty}^0 \frac{dx}{(2x-1)^3}.$

12. $\int_{-\infty}^2 \frac{dx}{x^2+4}.$

13. $\int_{-\infty}^0 e^{3x}\, dx.$

14. $\int_{-\infty}^0 \frac{e^x\, dx}{3-2e^x}.$

15. $\int_{-\infty}^{+\infty} x^3\, dx.$

16. $\int_{-\infty}^{+\infty} \frac{x}{\sqrt{x^2+2}}\, dx.$

17. $\int_{-\infty}^{+\infty} \frac{x}{(x^2+3)^2}\, dx.$

18. $\int_{-\infty}^{+\infty} \frac{e^{-t}}{1+e^{-2t}}\, dt.$

19. $\int_3^4 \frac{dx}{(x-3)^2}.$

20. $\int_0^8 \frac{dx}{\sqrt[3]{x}}.$

21. $\int_0^{\pi/2} \tan x\, dx.$

22. $\int_0^9 \frac{dx}{\sqrt{9-x}}.$

23. $\int_0^1 \frac{dx}{\sqrt{1-x^2}}.$

24. $\int_{-3}^1 \frac{x\, dx}{\sqrt{9-x^2}}.$

25. $\int_0^{\pi/6} \frac{\cos x}{\sqrt{1-2\sin x}}\, dx.$

26. $\int_0^{\pi/4} \frac{\sec^2 x}{1-\tan x}\, dx.$

27. $\int_0^3 \frac{dx}{x-2}.$

28. $\int_{-2}^2 \frac{dx}{x^2}.$

29. $\int_{-1}^8 x^{-1/3}\, dx.$

30. $\int_0^4 \frac{dx}{(x-2)^{2/3}}.$

31. $\int_0^{+\infty} \frac{1}{x^2}\, dx.$

32. $\int_1^{+\infty} \frac{dx}{x\sqrt{x^2-1}}.$

In Exercises 33–36, make the given substitution and evaluate the resulting integral.

33. $\int_0^{+\infty} \frac{e^{-\sqrt{x}}}{\sqrt{x}}\, dx; \ u = \sqrt{x}.$

34. $\int_0^{+\infty} \frac{dx}{\sqrt{x}(x+4)}; \ u = \sqrt{x}.$

35. $\int_0^{+\infty} \frac{e^{-x}}{\sqrt{1-e^{-x}}}\, dx; \ u = 1 - e^{-x}.$

36. $\int_0^{+\infty} \frac{e^{-x}}{\sqrt{1-e^{-2x}}}\, dx; \ u = e^{-x}.$

In Exercises 37 and 38, find the value of a.

37. $\int_0^{+\infty} e^{-ax}\, dx = 5.$

38. $\int_0^{+\infty} \frac{1}{x^2+a^2}\, dx = 1, \quad a > 0.$

In Exercises 39 and 40, evaluate the integral given that $\int_0^{+\infty} e^{-x^2} \, dx = \frac{1}{2}\sqrt{\pi}$.

39. $\displaystyle\int_0^{+\infty} \frac{e^{-x}}{\sqrt{x}} \, dx.$

40. $\displaystyle\int_0^{+\infty} e^{-a^2 x^2} \, dx, \ a > 0.$

41. Evaluate $\displaystyle\int_0^{+\infty} \frac{\sin x}{\sqrt{x}} \, dx$ given that

$$\int_0^{+\infty} \sin(x^2) \, dx = \frac{1}{2}\sqrt{\frac{\pi}{2}}$$

42. The *average speed*, $\bar{v}$, of the molecules of an ideal gas is given by

$$\bar{v} = \frac{4}{\sqrt{\pi}}\left(\frac{M}{2RT}\right)^{3/2}\int_0^{+\infty} v^3 e^{-Mv^2/(2RT)} \, dv$$

and the *root-mean-square speed*, v_{rms}, by

$$v_{\text{rms}}^2 = \frac{4}{\sqrt{\pi}}\left(\frac{M}{2RT}\right)^{3/2}\int_0^{+\infty} v^4 e^{-Mv^2/(2RT)} \, dv$$

where v is the molecular speed, T is the gas temperature, M is the molecular weight of the gas, and R is the gas constant.

(a) Show that $\bar{v} = \sqrt{\dfrac{8RT}{\pi M}}$ given that

$$\int_0^{+\infty} x^3 e^{-a^2 x^2} \, dx = \frac{1}{2a^4}$$

(b) Show that $v_{\text{rms}} = \sqrt{3RT/M}$ given that

$$\int_0^{+\infty} x^4 e^{-a^2 x^2} \, dx = \frac{3\sqrt{\pi}}{8a^5}, \ a > 0$$

43. (a) It is possible for an improper integral to diverge without becoming infinite. Show that $\int_0^{+\infty} \cos x \, dx$ does this.

(b) Show that $\displaystyle\int_0^{+\infty} \frac{\cos\sqrt{x}}{\sqrt{x}} \, dx$ diverges.

(c) Show that $\displaystyle\int_0^1 \frac{\cos(1/x)}{x^2} \, dx$ diverges.

44. Given that $I = \int_0^{\pi/2} \ln(\tan x) \, dx$ converges, show that the value of the integral is zero. [*Hint:* Use the substitution $x = \pi/2 - u$ to show that $I = -I$.]

45. Evaluate $\displaystyle\int_0^{+\infty} e^{-x} \cos x \, dx.$

46. For what values of p does $\displaystyle\int_0^{+\infty} e^{px} \, dx$ converge?

47. Show that $\displaystyle\int_0^1 \frac{dx}{x^p}$ converges if $p < 1$ and diverges if $p \geq 1$.

48. Show that $\displaystyle\int_1^{+\infty} \frac{dx}{x^p}$ converges if $p > 1$ and diverges if $p \leq 1$.

49. It can be shown that if f and g are continuous and $0 \leq f(x) \leq g(x)$ for all $x \geq a$, then $\int_a^{+\infty} f(x) \, dx$ converges if $\int_a^{+\infty} g(x) \, dx$ converges. Moreover,

$$\int_a^{+\infty} f(x) \, dx \leq \int_a^{+\infty} g(x) \, dx$$

and $\int_a^{+\infty} g(x) \, dx$ is an upper bound for $\int_a^{+\infty} f(x) \, dx$. Use this result to obtain an upper bound for each of the following improper integrals.

(a) $\displaystyle\int_2^{+\infty} \frac{x}{x^5 + 1} \, dx$ (b) $\displaystyle\int_1^{+\infty} e^{-x^2} \, dx$

50. It can be shown that if f and g are continuous and $0 \leq f(x) \leq g(x)$ for all $x \geq a$, then $\int_a^{+\infty} g(x) \, dx$ diverges if $\int_a^{+\infty} f(x) \, dx$ diverges. Use this result to show that each of the following improper integrals diverges.

(a) $\displaystyle\int_2^{+\infty} \frac{\sqrt{x^3 + 1}}{x} \, dx$ (b) $\displaystyle\int_0^{+\infty} \frac{e^x}{2x + 1} \, dx.$

51. Find the area of the region between the x-axis and the curve $y = e^{-3x}$ for $x \geq 0$.

52. Find the area of the region between the x-axis and the curve $y = 8/(x^2 - 4)$ for $x \geq 3$.

53. Let R be the region to the right of $x = 1$ that is bounded by the x-axis and the curve $y = 1/x$.

(a) Show that the solid obtained by revolving R about the x-axis has a finite volume.

(b) Show that the solid of part (a) has an infinite surface area. [*Hint:* See Exercise 50.] [*Note:* It has been suggested that by filling this solid with paint and letting it seep through to the surface one could paint an infinite surface area with a finite amount of paint!]

54. Suppose that the region between the x-axis and the curve $y = e^{-x}$ for $x \geq 0$ is revolved about the x-axis.

(a) Find the volume of the solid that is generated.

(b) Find the surface area of the solid.

55. In electromagnetic theory, the magnetic potential at a point on the axis of a circular coil is given by

$$u = \frac{2\pi NIr}{k}\int_a^{+\infty} \frac{dx}{(r^2 + x^2)^{3/2}}$$

where N, I, r, k, and a are constants. Find u.

56. Sketch the region whose area is $\displaystyle\int_0^{+\infty} \frac{dx}{1+x^2}$, and use your sketch to show that

$$\int_0^{+\infty} \frac{dx}{1+x^2} = \int_0^1 \sqrt{\frac{1-y}{y}}\, dy$$

57. (**Satellite Problem**) In Exercise 15 of Section 6.7, we determined the work required to lift a 6000-lb satellite to a specified orbital position. The result in part (a) of that problem will be needed here.
  (a) Find a definite integral that represents the work required to lift a 6000-lb satellite to a position l mi above the earth's surface.
  (b) Find a definite integral that represents the work required to lift a 6000-lb satellite an "infinite distance" above the earth's surface. Evaluate the integral. [*Note:* The result obtained here is sometimes called the work required to "escape" the earth's gravity.]

58. The *Laplace transform* of a function $f(t)$, denoted by $\mathcal{L}\{f(t)\}$, is defined by

$$\mathcal{L}\{f(t)\} = \int_0^{+\infty} e^{-st} f(t)\, dt$$

where s is regarded as a constant for the integration. The Laplace transform has the effect of "transforming" the function $f(t)$ into a function of s. Show that

$$\mathcal{L}\{e^{2t}\} = \frac{1}{s-2} \quad \text{if } s > 2$$

In Exercises 59–62, find $\mathcal{L}\{f(t)\}$ as defined in Exercise 58.

59. $f(t) = 1,\ s > 0.$

60. $f(t) = \begin{cases} 0, & t < 3 \\ 1, & t \geq 3 \end{cases},\ s > 0.$

61. $f(t) = \sin t,\ s > 0.$ **62.** $f(t) = \cos t,\ s > 0.$

In Exercises 63 and 64, transform the given improper integral into a proper integral by making the stated u-substitution, then approximate the proper integral by Simpson's rule (Section 9.8) with $n = 10$ subdivisions. Round your answer to three decimal places.

63. $\displaystyle\int_0^1 \frac{\cos x}{\sqrt{x}}\, dx;\ u = \sqrt{x}.$

64. $\displaystyle\int_0^1 \frac{\sin x}{\sqrt{1-x}}\, dx;\ u = \sqrt{1-x}.$

An improper integral over an infinite interval can be approximated by first replacing the infinite limit(s) of integration by finite limit(s), then using a numerical integration technique, such as Simpson's rule, to approximate the integral with finite limit(s). This is illustrated in Exercises 65 and 66.

65. (a) It can be shown that

$$\int_0^{+\infty} \frac{1}{x^6+1}\, dx = \frac{\pi}{3}$$

Approximate this integral by applying Simpson's rule with $n = 20$ subdivisions to the integral

$$\int_0^K \frac{1}{x^6+1}\, dx$$

with $K = 4$. Round your answer to three decimal places and compare it to $\pi/3$ rounded to three decimal places.
  (b) Use the result in Exercise 49 and the fact that $1/(x^6+1) < 1/x^6$ for $x \geq 4$ to show that

$$\int_0^{+\infty} \frac{1}{x^6+1}\, dx = \int_0^4 \frac{1}{x^6+1}\, dx + E$$

where $0 < E < 2 \times 10^{-4}$. [*Note: E* is the error in part (a) that results from replacing $+\infty$ by $K = 4$. The other two sources of error in the approximation are the error from Simpson's rule and the roundoff error.]

66. (a) It can be shown that

$$\int_0^{+\infty} e^{-x^2}\, dx = \frac{1}{2}\sqrt{\pi}$$

Approximate this integral by applying Simpson's rule with $n = 10$ subdivisions to the integral

$$\int_0^K e^{-x^2}\, dx$$

with $K = 3$. Round your answer to four decimal places and compare it to $\frac{1}{2}\sqrt{\pi}$ rounded to four decimal places.
  (b) Use the result in Exercise 49 and the fact that $e^{-x^2} < xe^{-x^2}$ for $x \geq 3$ to show that

$$\int_0^{+\infty} e^{-x^2}\, dx = \int_0^3 e^{-x^2}\, dx + E$$

where $0 < E < 7 \times 10^{-5}$. [See the note in Exercise 65.]

■ **10.2** L'HÔPITAL'S RULE (INDETERMINATE FORMS OF TYPE 0/0)

In this section we shall develop an important new technique for finding limits of functions.

□ **L'HÔPITAL'S RULE**

In each of the limits

$$\lim_{x \to 2} \frac{x^2 - 4}{x - 2} \quad \text{and} \quad \lim_{x \to 0} \frac{\sin x}{x} \tag{1}$$

the numerator and denominator both approach zero. It is customary to describe such limits as *indeterminate forms of type* **0/0**. As we shall see, a limit of this type can have any real number whatsoever as its value or can diverge to $+\infty$ or $-\infty$. The value of such a limit, if it converges, is not generally evident by inspection, so the term "indeterminate" is used to convey the idea that the limit cannot be determined without some additional work.

In Example 6 of Section 2.5 we evaluated the first limit in (1) by canceling the common factor $x - 2$ from the numerator and denominator, and in Theorem 2.8.3 we resorted to a rather intricate geometric argument to obtain the second limit in (1). Because geometric arguments and the technique of canceling factors apply only to a limited range of problems, it is desirable to have a general method for handling indeterminate forms. This is provided by *L'Hôpital's* rule,* which we now discuss.

10.2.1 THEOREM (*L'Hôpital's Rule for Form* **0/0**). *Let* lim *stand for one of the limits* $\lim_{x \to a}$, $\lim_{x \to a^+}$, $\lim_{x \to a^-}$, $\lim_{x \to +\infty}$, *or* $\lim_{x \to -\infty}$, *and suppose that* $\lim f(x) = 0$ *and* $\lim g(x) = 0$. *If* $\lim [f'(x)/g'(x)]$ *has a finite value L, or if this limit is* $+\infty$ *or* $-\infty$, *then*

$$\lim \frac{f(x)}{g(x)} = \lim \frac{f'(x)}{g'(x)}$$

---

*GUILLAUME FRANCOIS ANTOINE DE L'HÔPITAL (1661–1704). French mathematician. L'Hôpital, born to parents of the French high nobility, held the title of Marquis de Sainte-Mesme Comte d'Autrement. He showed mathematical talent quite early and at age 15 solved a difficult problem about cycloids posed by Pascal. As a young man he served briefly as a cavalry officer, but resigned because of nearsightedness. In his own time he gained fame as the author of the first textbook ever published on differential calculus, L'Analyse des Infiniment Petits pour l'Intelligence des Lignes Courbes (1696). L'Hôpital's rule appeared for the first time in that book. Actually, L'Hôpital's rule and most of the material in the calculus text were due to John Bernoulli, who was L'Hôpital's teacher. L'Hôpital dropped his plans for a book on integral calculus when Leibniz informed him that he intended to write such a text. L'Hôpital was apparently generous and personable, and his many contacts with major mathematicians provided the vehicle for disseminating major discoveries in calculus throughout Europe.

REMARK.   There are some hypotheses implicit in this theorem. For example, in the case where $x \to a$, the statement

$$\lim_{x \to a} \frac{f'(x)}{g'(x)} = L$$

requires that f'/g' be defined in some open interval I containing a (except possibly at a). This implies that f and g are differentiable and $g'(x) \neq 0$ in I (except possibly at a). Similar hypotheses are implicit in the other cases.

In essence, L'Hôpital's rule enables us to replace one limit problem with another that may be simpler. In each of the following examples we shall employ the following three-step process:

> **Step 1.**   Check that $\lim f(x)/g(x)$ is an indeterminate form. If it is not, then L'Hôpital's rule cannot be used.
>
> **Step 2.**   Differentiate f and g separately.
>
> **Step 3.**   Find $\lim f'(x)/g'(x)$. If this limit is finite, $+\infty$, or $-\infty$, then it is equal to $\lim f(x)/g(x)$.

Example 1   Use L'Hôpital's rule to evaluate

(a) $\displaystyle \lim_{x \to 2} \frac{x^2 - 4}{x - 2}$ (b) $\displaystyle \lim_{x \to 0} \frac{\sin 2x}{x}$

Solution (a).   Since

$$\lim_{x \to 2} (x^2 - 4) = 0 \quad \text{and} \quad \lim_{x \to 2} (x - 2) = 0$$

the given limit is an indeterminate form of type 0/0. Thus, L'Hôpital's rule applies and we can write

$$\lim_{x \to 2} \frac{x^2 - 4}{x - 2} = \lim_{x \to 2} \frac{\dfrac{d}{dx}[x^2 - 4]}{\dfrac{d}{dx}[x - 2]} = \lim_{x \to 2} \frac{2x}{1} = 4$$

Observe that this agrees with the result obtained in Example 6 of Section 2.5 by factoring.

Solution (b).   Since

$$\lim_{x \to 0} \sin 2x = 0 \quad \text{and} \quad \lim_{x \to 0} x = 0$$

the given limit is an indeterminate form of type 0/0. Thus, L'Hôpital's rule applies and we can write

$$\lim_{x \to 0} \frac{\sin 2x}{x} = \lim_{x \to 0} \frac{\dfrac{d}{dx}[\sin 2x]}{\dfrac{d}{dx}[x]} = \lim_{x \to 0} \frac{2 \cos 2x}{1} = 2$$

Observe that this agrees with the result obtained in Example 4 of Section 2.8 by substitution.   ◄

REMARK.   To be rigorous, in each of the foregoing examples the first equality is not justified until the limit on the right is shown to exist. However, for simplicity we shall usually arrange the computations as shown when applying L'Hôpital's rule.

Example 2   Evaluate $\displaystyle\lim_{x \to \pi/2} \frac{1 - \sin x}{\cos x}$.

Solution.   Since

$$\lim_{x \to \pi/2} (1 - \sin x) = \lim_{x \to \pi/2} \cos x = 0$$

the given limit is an indeterminate form of type 0/0. Thus, by L'Hôpital's rule

$$\lim_{x \to \pi/2} \frac{1 - \sin x}{\cos x} = \lim_{x \to \pi/2} \frac{\dfrac{d}{dx}[1 - \sin x]}{\dfrac{d}{dx}[\cos x]} = \lim_{x \to \pi/2} \frac{-\cos x}{-\sin x} = \frac{0}{-1} = 0 \qquad ◄$$

Example 3   Evaluate $\displaystyle\lim_{x \to 0} \frac{e^x - 1}{x^3}$.

Solution.   Since

$$\lim_{x \to 0} (e^x - 1) = \lim_{x \to 0} x^3 = 0$$

the given limit is an indeterminate form of type 0/0. Thus, by L'Hôpital's rule

$$\lim_{x \to 0} \frac{e^x - 1}{x^3} = \lim_{x \to 0} \frac{\dfrac{d}{dx}[e^x - 1]}{\dfrac{d}{dx}[x^3]} = \lim_{x \to 0} \frac{e^x}{3x^2} = +\infty \qquad ◄$$

As the following example shows, it is sometimes necessary to apply L'Hôpital's rule more than once in the same problem.

Example 4 Evaluate $\displaystyle\lim_{x\to 0}\frac{1-\cos x}{x^2}$.

Solution. Since

$$\lim_{x\to 0}(1-\cos x)=\lim_{x\to 0}x^2=0$$

the given limit is an indeterminate form of type 0/0. Thus, by L'Hôpital's rule

$$\lim_{x\to 0}\frac{1-\cos x}{x^2}=\lim_{x\to 0}\frac{\sin x}{2x}$$

However, the new limit is also an indeterminate form of type 0/0, so we apply L'Hôpital's rule again. This yields

$$\lim_{x\to 0}\frac{1-\cos x}{x^2}=\lim_{x\to 0}\frac{\sin x}{2x}=\lim_{x\to 0}\frac{\cos x}{2}=\frac{1}{2}\quad\blacktriangleleft$$

WARNING. When applying L'Hôpital's rule to $\lim f(x)/g(x)$, the derivatives of $f(x)$ and $g(x)$ are taken separately to yield the new limit, $\lim[f'(x)/g'(x)]$. Do not make the mistake of differentiating $f(x)/g(x)$ according to the quotient rule.

Example 5 Evaluate $\displaystyle\lim_{x\to 0}\frac{e^x}{x^2}$.

Solution. We have

$$\lim_{x\to 0}e^x=1\quad\text{and}\quad\lim_{x\to 0}x^2=0$$

so the given problem is not an indeterminate form of type 0/0 and consequently L'Hôpital's rule cannot be applied. By inspection

$$\lim_{x\to 0}\frac{e^x}{x^2}=+\infty\quad\blacktriangleleft$$

WARNING. Applying L'Hôpital's rule to limits that are not indeterminate forms can lead to erroneous results. As an illustration, two applications of L'Hôpital's rule in the last example would have led to the *incorrect* conclusion that the limit is 1/2.

Example 6 Evaluate $\displaystyle\lim_{x\to+\infty}\frac{x^{-4/3}}{\sin(1/x)}$.

Solution. Since

$$\lim_{x\to+\infty}x^{-4/3}=\lim_{x\to+\infty}\sin(1/x)=0$$

the given limit is an indeterminate form of type 0/0. Thus, by L'Hôpital's rule

$$\lim_{x \to +\infty} \frac{x^{-4/3}}{\sin(1/x)} = \lim_{x \to +\infty} \frac{-\frac{4}{3}x^{-7/3}}{(-1/x^2)\cos(1/x)} = \lim_{x \to +\infty} \frac{\frac{4}{3}x^{-1/3}}{\cos(1/x)} = \frac{0}{1} = 0 \quad \blacktriangleleft$$

Example 7   Evaluate $\displaystyle\lim_{x \to 0^-} \frac{\tan x}{x^2}$.

Solution.   Since

$$\lim_{x \to 0^-} \tan x = \lim_{x \to 0^-} x^2 = 0$$

the given limit is an indeterminate form of type 0/0. Thus, by L'Hôpital's rule

$$\lim_{x \to 0^-} \frac{\tan x}{x^2} = \lim_{x \to 0^-} \frac{\sec^2 x}{2x} = -\infty \quad \blacktriangleleft$$

■ **OPTIONAL**

□ **THEORY BEHIND L'HÔPITAL'S RULE**

The proof of L'Hôpital's rule depends on the following result, called the *Extended Mean-Value Theorem* or sometimes the *Cauchy* Mean-Value Theorem*.

> **10.2.2**   THEOREM (*Extended Mean-Value Theorem*).   *Let the functions f and g be differentiable on (a, b) and continuous on $[a, b]$. If $g'(x) \neq 0$ for all x in (a, b), then there is at least one point c in (a, b) such that*
>
> $$\frac{f'(c)}{g'(c)} = \frac{f(b) - f(a)}{g(b) - g(a)} \tag{2}$$

Proof.   Observe first that $g(b) - g(a) \neq 0$, since otherwise it would follow from the Mean-Value Theorem (4.10.2) that $g'(x) = 0$ at some point x in (a, b), contradicting our hypothesis. For convenience, introduce a new function F defined by

$$F(x) = [f(b) - f(a)]g(x) - [g(b) - g(a)]f(x) \tag{3}$$

---

*AUGUSTIN LOUIS CAUCHY (1789–1857). French mathematician. Cauchy's early education was acquired from his father, a barrister and master of the classics. Cauchy entered L'Ecole Polytechnique in 1805 to study engineering, but because of poor health, was advised to concentrate on mathematics. His major mathematical work began in 1811 with a series of brilliant solutions to some difficult outstanding problems. In 1814 he wrote a treatise on integrals that was to become the basis for modern complex variable theory; in 1816 there followed a classic paper on wave propagation in liquids that won a prize from the French Academy; and in 1822 he wrote a paper that formed the basis of modern elasticity theory. Cauchy's mathematical contributions for the next 35 years were brilliant and staggering in quantity, over 700 papers filling 26 modern volumes. Cauchy's work initiated the era of modern analysis. He brought to mathematics standards of precision and rigor undreamed of by Leibniz and Newton.

(*continued on next page*)

It follows from our assumptions about f and g that F is continuous on $[a, b]$ and differentiable on (a, b). Moreover, $F(b) = F(a)$ (verify), so that the Mean-Value Theorem (4.10.2) implies that there is at least one point c in (a, b) where $F'(c) = 0$. Thus, from (3)

$$[f(b) - f(a)]g'(c) - [g(b) - g(a)]f'(c) = 0$$

or

$$\frac{f'(c)}{g'(c)} = \frac{f(b) - f(a)}{g(b) - g(a)} \quad \blacksquare$$

Note that in the special case where $g(x) = x$, Formula (2) reduces to

$$f'(c) = \frac{f(b) - f(a)}{b - a}$$

which is precisely the conclusion of the Mean-Value Theorem (4.10.2). Thus, the Extended Mean-Value Theorem is, in fact, an extension of the Mean-Value Theorem.

☐ **PROOF OF L'HÔPITAL'S RULE**

Since there are numerous cases of L'Hôpital's rule, we shall prove just one of them. The techniques used in the proof of this one case are typical of those used in the other cases.

Proof of Theorem 10.2.1.   We shall consider only the case where L is finite and lim is the two-sided limit $\lim\limits_{x \to a}$, with a finite.
By hypothesis,

$$\lim_{x \to a} \frac{f'(x)}{g'(x)} = L \tag{4}$$

As noted in the remark following the statement of Theorem 10.2.1, (4) implies that there is an interval $(a, r]$ extending to the right of a and an interval $[l, a)$ extending to the left of a on which $f'(x)$ and $g'(x)$ are defined and $g'(x) \neq 0$. For convenience, define two new functions F and G by

$$F(x) = \begin{cases} f(x), & x \neq a \\ 0, & x = a \end{cases} \quad \text{and} \quad G(x) = \begin{cases} g(x), & x \neq a \\ 0, & x = a \end{cases}$$

(continued)

Cauchy's life was inextricably tied to the political upheavals of the time. A strong partisan of the Bourbons, he left his wife and children in 1830 to follow the Bourbon king Charles X into exile. For his loyalty he was made a baron by the ex-king. Cauchy eventually returned to France, but refused to accept a university position until the government waived its requirement that he take a loyalty oath.

It is difficult to get a clear picture of the man. Devoutly Catholic, he sponsored charitable work for unwed mothers, criminals, and relief for Ireland. Yet other aspects of his life cast him in an unfavorable light. The Norwegian mathematician Abel described him as, "mad, infinitely Catholic, and bigoted." Some writers praise his teaching, yet others say he rambled incoherently and, according to a report of the day, he once devoted an entire lecture to extracting the square root of seventeen to ten decimal places by a method well known to his students. In any event, Cauchy is undeniably one of the greatest minds in the history of science.

Both F and G are continuous on $[l, r]$; they are continuous on the intervals $[l, a)$ and $(a, r]$ because on these intervals $F(x) = f(x)$, $G(x) = g(x)$, and f and g are continuous (since they are differentiable). The continuity of F and G at a follows from the fact that $\lim\limits_{x \to a} f(x)/g(x)$ is an indeterminate form of type 0/0, so

$$\lim_{x \to a} F(x) = \lim_{x \to a} f(x) = 0 \quad \text{and} \quad \lim_{x \to a} G(x) = \lim_{x \to a} g(x) = 0$$

Thus, F and G satisfy the hypotheses of the Extended Mean-Value Theorem (10.2.2) on $[l, a]$ and $[a, r]$. Moreover, the definitions of F and G imply that

$$F'(c) = f'(c) \quad \text{and} \quad G'(c) = g'(c) \tag{5}$$

at any point c (different from a) in one of these intervals. If we choose a point $x \neq a$ in one of the intervals $[l, a]$ or $[a, r]$ and apply the Extended Mean-Value Theorem to $[x, a]$ (or $[a, x]$), we conclude that there is a number c between a and x such that

$$\frac{F(x) - F(a)}{G(x) - G(a)} = \frac{F'(c)}{G'(c)} \tag{6}$$

From (5) together with the fact that $F(a) = G(a) = 0$ and $F(x) = f(x)$, $G(x) = g(x)$, we can rewrite (6) as

$$\frac{f(x)}{g(x)} = \frac{f'(c)}{g'(c)}$$

Thus,

$$\lim_{x \to a} \frac{f(x)}{g(x)} = \lim_{x \to a} \frac{f'(c)}{g'(c)} \tag{7}$$

Since c is between a and x, it follows that $c \to a$ as $x \to a$. This fact together with (4) yields

$$\lim_{x \to a} \frac{f'(c)}{g'(c)} = \lim_{c \to a} \frac{f'(c)}{g'(c)} = L$$

Thus, from (7)

$$\lim_{x \to a} \frac{f(x)}{g(x)} = L \quad \blacksquare$$

▶ Exercise Set 10.2

In Exercises 1–32, find the limits.

1. $\lim\limits_{x \to 1} \dfrac{\ln x}{x - 1}$.

2. $\lim\limits_{x \to 0} \dfrac{\sin 2x}{\sin 5x}$.

3. $\lim\limits_{x \to 0} \dfrac{e^x - 1}{\sin x}$.

4. $\lim\limits_{x \to 3} \dfrac{x - 3}{3x^2 - 13x + 12}$.

5. $\lim\limits_{\theta \to 0} \dfrac{\tan \theta}{\theta}$.

6. $\lim\limits_{t \to 0} \dfrac{t e^t}{1 - e^t}$.

7. $\lim\limits_{x \to 1} \dfrac{\ln x}{\tan \pi x}$.

8. $\lim\limits_{x \to c} \dfrac{x^{1/3} - c^{1/3}}{x - c}$.

9. $\lim\limits_{x \to \pi^+} \dfrac{\sin x}{x - \pi}$.

10. $\lim\limits_{x \to 0^+} \dfrac{\sin x}{x^2}$.

11. $\lim\limits_{x \to \frac{1}{2}\pi^-} \dfrac{\cos x}{\sqrt{\frac{1}{2}\pi - x}}$.

12. $\lim\limits_{x \to 0^+} \dfrac{1 - \cos x}{x^3}$.

13. $\lim\limits_{x \to 0} \dfrac{e^x + e^{-x} - 2}{1 - \cos 2x}$.

14. $\lim\limits_{x \to 0} \dfrac{2 \cosh x - 2}{1 - \cos 2x}$.

15. $\lim\limits_{x \to 0} \dfrac{\sin 3x}{\sinh 2x}$.

16. $\lim\limits_{x \to 0} \dfrac{1 - e^{-2x}}{x^2 + 3x}$.

17. $\lim\limits_{x \to \pi/2} \dfrac{\sin 2x}{4x^2 - \pi^2}$.

18. $\lim\limits_{x \to 2} \dfrac{\ln (5x - 9)}{x^3 - 8}$.

19. $\lim\limits_{x \to 0} \dfrac{x - \ln (x + 1)}{1 - \cos 2x}$.

20. $\lim\limits_{x \to 0} \dfrac{x - \tan^{-1} x}{x^3}$.

21. $\lim\limits_{x \to 0} \dfrac{2 - x^2 - 2 \cos x}{x^4}$.

22. $\lim\limits_{x \to +\infty} \dfrac{\ln (1 + 3/x)}{\sin (2/x)}$.

23. $\lim\limits_{x \to 0} \dfrac{x - \sin x}{x^3}$.

24. $\lim\limits_{x \to -1} \dfrac{x^2 - 1}{\ln (3x + 4)}$.

25. $\lim\limits_{x \to 0} \dfrac{e^{ax} - e^{bx}}{x}$.

26. $\lim\limits_{x \to 0} \dfrac{a^x - 1}{x}$, $\quad a > 0$.

27. $\lim\limits_{x \to 0} \dfrac{x - \tan x}{\sin x - x}$.

28. $\lim\limits_{x \to \pi} \dfrac{\sin^2 x}{1 + \cos 3x}$.

29. $\lim\limits_{x \to +\infty} \dfrac{\frac{1}{2}\pi - \tan^{-1} x}{\ln \left(1 + \dfrac{1}{x^2}\right)}$.

30. $\lim\limits_{x \to 0^+} \dfrac{\tan x}{\tan 2x}$.

31. $\lim\limits_{x \to 0^+} \dfrac{\ln (\cos x)}{\ln (\cos 3x)}$.

32. $\lim\limits_{\theta \to 0} \dfrac{\sin^2 \theta - \sin (\theta^2)}{\theta^4}$.

33. (a) Find the error in the following calculation:

$$\lim_{x \to 1} \frac{x^3 - x^2 + x - 1}{x^3 - x^2} = \lim_{x \to 1} \frac{3x^2 - 2x + 1}{3x^2 - 2x}$$

$$= \lim_{x \to 1} \frac{6x - 2}{6x - 2} = 1$$

   (b) Find the correct answer.

34. Find $\lim\limits_{x \to 1} \dfrac{x^4 - 4x^3 + 6x^2 - 4x + 1}{x^4 - 3x^3 + 3x^2 - x}$.

35. Find all values of k and l such that

$$\lim_{x \to 0} \frac{k + \cos lx}{x^2} = -4$$

36. Find $\lim\limits_{x \to 0} \dfrac{(2 + x) \ln (1 - x)}{(1 - e^x) \cos x}$.

37. Find $\lim\limits_{x \to 1} \sqrt{\dfrac{\ln x}{x^4 - 1}}$.

$\left[\textit{Hint: } \text{First find } \lim\limits_{x \to 1} \dfrac{\ln x}{x^4 - 1} . \right]$

38. Find

$$\lim_{x \to 0^+} \frac{x}{\sqrt{1 - e^{-3x^2}}} \quad \text{and} \quad \lim_{x \to 0^-} \frac{x}{\sqrt{1 - e^{-3x^2}}}$$

$\left[\textit{Hint: } \text{First find } \lim\limits_{x \to 0} \dfrac{x^2}{1 - e^{-3x^2}} . \right]$

39. Find $\lim\limits_{x \to +\infty} x \ln \left(\dfrac{x + 1}{x - 1} \right)$.

[*Hint:* Rewrite the problem as an indeterminate form of type 0/0.]

40. (a) Explain why L'Hôpital's rule does not apply to the problem

$$\lim_{x \to 0} \frac{x^2 \sin (1/x)}{\sin x}$$

   (b) Find the limit.

41. Find $\lim\limits_{x \to 0^+} \dfrac{x \sin (1/x)}{\sin x}$ if it exists.

42. (a) Given that $k \neq 0$ and $x > 0$, show that

$$\int_1^x \frac{1}{t^{1-k}} \, dt = \frac{x^k - 1}{k}$$

   (b) Use the result in part (a) to make a guess at the value of the limit

$$\lim_{k \to 0} \frac{x^k - 1}{k}$$

   (c) Use L'Hôpital's rule to substantiate your guess. [This exercise was motivated by an article by Henry C. Finlayson, which appeared in the *American Math. Monthly*, Vol 94, No. 5, May 1987, p. 450.]

43. In Figure 10.2.1, let $T(\theta)$ be the area of the right triangle ABC, and $S(\theta)$ the area of the segment of the circle formed by chord AB and the arc of the unit circle subtended by the central angle θ. Find $\lim\limits_{\theta \to 0^+} T(\theta)/S(\theta)$.

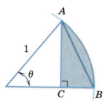

Figure 10.2.1

■ 10.3 OTHER INDETERMINATE FORMS

$$\left(\frac{\infty}{\infty},\ 0\cdot\infty,\ 0^0,\ \infty^0,\ 1^\infty,\ \infty-\infty\right)$$

> *In this section we shall continue our study of indeterminate forms and introduce another version of L'Hôpital's rule.*

□ **SOME NOTATION**

In Figure 10.3.1 we have illustrated four ways in which a function can become "infinite at a."

Sometimes we shall want to indicate that a function becomes infinite at a point a without actually specifying which of these four possibilities occurs. To do so, we shall write

$$\lim_{x\to a} f(x) = \infty$$

In situations where

$$\lim_{x\to +\infty} f(x) = +\infty \quad \text{or} \quad \lim_{x\to +\infty} f(x) = -\infty$$

but the sign is of no consequence, we shall indicate that one of these two possibilities occurs by writing

$$\lim_{x\to +\infty} f(x) = \infty$$

Similarly, we shall write

$$\lim_{x\to -\infty} f(x) = \infty$$

to mean that

$$\lim_{x\to -\infty} f(x) = +\infty \quad \text{or} \quad \lim_{x\to -\infty} f(x) = -\infty$$

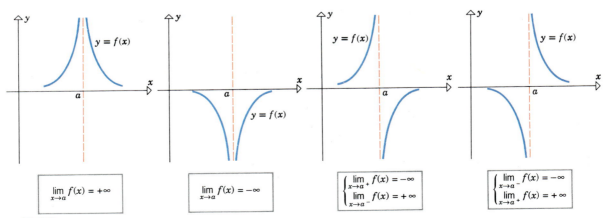

Figure 10.3.1

☐ **INDETERMINATE FORMS OF TYPE ∞/∞**

An *indeterminate form of type* ∞/∞ is a limit $\lim f(x)/g(x)$ in which $\lim f(x) = \infty$ and $\lim g(x) = \infty$. Some examples are

$$\lim_{x \to 0^+} \frac{\ln x}{\csc x}$$ Numerator → −∞, denominator → +∞

$$\lim_{x \to +\infty} \frac{x}{e^x}$$ Numerator → +∞, denominator → +∞

$$\lim_{x \to 0} \frac{1 + 1/x}{\cot x}$$ Numerator → ∞, denominator → ∞

The following version of L'Hôpital's rule, usually proved in advanced courses, is applicable to problems like these.

10.3.1   THEOREM (*L'Hôpital's Rule for Form* ∞/∞).   *Let* lim *stand for one of the limits* $\lim_{x \to a}$, $\lim_{x \to a^+}$, $\lim_{x \to a^-}$, $\lim_{x \to +\infty}$, *or* $\lim_{x \to -\infty}$, *and suppose that* $\lim f(x) = \infty$ *and* $\lim g(x) = \infty$. *If* $\lim [f'(x)/g'(x)]$ *has a finite value* L, *or if this limit is* $+\infty$ *or* $-\infty$, *then*

$$\lim \frac{f(x)}{g(x)} = \lim \frac{f'(x)}{g'(x)}$$

REMARK.   The remark about hypotheses which follows Theorem 10.2.1 is also applicable here.

Example 1   Evaluate $\lim\limits_{x \to +\infty} \dfrac{x}{e^x}$.

Solution.

$$\lim_{x \to +\infty} x = \lim_{x \to +\infty} e^x = +\infty$$

so that the given limit is an indeterminate form of type ∞/∞. Thus, by L'Hôpital's rule

$$\lim_{x \to +\infty} \frac{x}{e^x} = \lim_{x \to +\infty} \frac{\dfrac{d}{dx}[x]}{\dfrac{d}{dx}[e^x]} = \lim_{x \to +\infty} \frac{1}{e^x} = 0 \qquad \blacktriangleleft$$

Example 2   Evaluate $\lim\limits_{x \to 0^+} \dfrac{\ln x}{\csc x}$.

Solution.

$$\lim_{x \to 0^+} \ln x = -\infty \quad \text{and} \quad \lim_{x \to 0^+} \csc x = +\infty$$

so that the given limit is an indeterminate form of type ∞/∞. Thus, by L'Hôpital's rule

$$\lim_{x \to 0^+} \frac{\ln x}{\csc x} = \lim_{x \to 0^+} \frac{1/x}{-\csc x \cot x} \tag{1}$$

This last limit is again an indeterminate form of type ∞/∞. Moreover, any additional applications of L'Hôpital's rule will yield powers of $1/x$ in the numerator and expressions involving $\csc x$ and $\cot x$ in the denominator; thus, repeated application of L'Hôpital's rule simply produces new indeterminate forms. We must try something else. The last limit in (1) can be rewritten as

$$\lim_{x \to 0^+} \left(-\frac{\sin x}{x} \tan x \right) = -\lim_{x \to 0^+} \frac{\sin x}{x} \cdot \lim_{x \to 0^+} \tan x = -(1)(0) = 0$$

Thus,

$$\lim_{x \to 0^+} \frac{\ln x}{\csc x} = 0 \quad \blacktriangleleft$$

□ **INDETERMINATE FORMS OF TYPE $0 \cdot \infty$**

A limit of a product, $\lim f(x)g(x)$, is called an **indeterminate form of type $0 \cdot \infty$** if $\lim f(x) = 0$ and $\lim g(x) = \infty$. Frequently, limit problems of this type can be converted to the form $0/0$ by writing

$$f(x)g(x) = \frac{f(x)}{1/g(x)}$$

or to the form ∞/∞ by writing

$$f(x)g(x) = \frac{g(x)}{1/f(x)}$$

The limit can then be evaluated by L'Hôpital's rule.

Example 3   Evaluate $\lim\limits_{x \to 0^+} x \ln x$.

Solution.   Since

$$\lim_{x \to 0^+} x = 0 \quad \text{and} \quad \lim_{x \to 0^+} \ln x = -\infty$$

the given problem is an indeterminate form of type $0 \cdot \infty$. We shall convert the problem to the form ∞/∞ and apply L'Hôpital's rule as follows:

$$\lim_{x \to 0^+} x \ln x = \lim_{x \to 0^+} \frac{\ln x}{1/x} = \lim_{x \to 0^+} \frac{1/x}{-1/x^2} = \lim_{x \to 0^+} (-x) = 0 \quad \blacktriangleleft$$

In this example we could have converted the problem to form $0/0$ by writing

$$x \ln x = \frac{x}{1/\ln x}$$

However, this is less desirable than the first choice because of the relatively complicated derivative of $1/\ln x$.

WARNING. It is tempting to argue that an indeterminate form of type $0\cdot\infty$ has value 0 since "zero times anything is zero." However, this is fallacious since $0\cdot\infty$ is not a product of numbers, but rather a statement about limits. The following example gives an indeterminate form of type $0\cdot\infty$ whose value is not zero.

Example 4 Evaluate $\displaystyle\lim_{x\to\pi/4} (1 - \tan x) \sec 2x$.

Solution. The given problem is an indeterminate form of type $0\cdot\infty$. We convert it to type 0/0 and apply L'Hôpital's rule as follows:

$$\lim_{x\to\pi/4} (1 - \tan x) \sec 2x = \lim_{x\to\pi/4} \frac{1 - \tan x}{1/\sec 2x} = \lim_{x\to\pi/4} \frac{1 - \tan x}{\cos 2x}$$

$$= \lim_{x\to\pi/4} \frac{-\sec^2 x}{-2 \sin 2x} = \frac{-2}{-2} = 1 \quad \blacktriangleleft$$

☐ **INDETERMINATE FORMS OF TYPE 0^0, ∞^0, 1^∞**

Limits of the form

$$\lim_{x\to a} f(x)^{g(x)}$$

give rise to *indeterminate forms of the types 0^0, ∞^0, and 1^∞*. (The meaning of these symbols should be clear.) All three types are treated by first introducing a dependent variable

$$y = f(x)^{g(x)}$$

and then calculating

$$\lim_{x\to a} \ln y = \lim_{x\to a} [\ln (f(x)^{g(x)})] = \lim_{x\to a} [g(x) \ln f(x)]$$

Once the value of $\displaystyle\lim_{x\to a} \ln y$ is known, it is a simple matter to determine

$$\lim_{x\to a} y = \lim_{x\to a} f(x)^{g(x)}$$

as our examples will show.

In the next example we shall derive the two limit formulas for e stated in Section 7.2. [See Formulas (5) and (6) in that section.]

Example 5 Show that

(a) $\displaystyle\lim_{x\to 0} (1 + x)^{1/x} = e$ (b) $\displaystyle\lim_{x\to +\infty} \left(1 + \frac{1}{x}\right)^x = e$

Solution (a). Since

$$\lim_{x\to 0} (1 + x) = 1 \quad \text{and} \quad \lim_{x\to 0} \frac{1}{x} = \infty$$

the given limit is an indeterminate form of type 1^∞. As discussed above, we introduce a dependent variable

$$y = (1 + x)^{1/x}$$

and take the natural logarithm of both sides:

$$\ln y = \ln (1 + x)^{1/x} = \frac{1}{x} \ln (1 + x) = \frac{\ln (1 + x)}{x}$$

The limit

$$\lim_{x \to 0} \ln y = \lim_{x \to 0} \frac{\ln (1 + x)}{x}$$

is an indeterminate form of type 0/0, so by L'Hôpital's rule,

$$\lim_{x \to 0} \ln y = \lim_{x \to 0} \frac{\ln (1 + x)}{x} = \lim_{x \to 0} \frac{1/(1 + x)}{1} = 1 \qquad (2)$$

To complete the calculation, we shall use the known limit, $\lim_{x \to 0} \ln y = 1$, to find the unknown limit, $\lim_{x \to 0} y = \lim_{x \to 0} (1 + x)^{1/x}$. We argue as follows: As $x \to 0$, we have

$$\ln y \to 1 \qquad \boxed{\text{From (2)}}$$

$$e^{\ln y} \to e^1 \qquad \boxed{\text{Since } e^x \text{ is a continuous function}}$$

$$y \to e \qquad \boxed{\text{Simplification}}$$

Therefore,

$$\lim_{x \to 0} (1 + x)^{1/x} = e \qquad (3)$$

Solution (b).   The formula in this part can be derived by the same procedure used in part (a). Alternatively, we can simply let $x = 1/h$ in (3) to obtain

$$\lim_{h \to +\infty} \left(1 + \frac{1}{h}\right)^h = e$$

which is the same as the stated formula except for the notation used for the variable.   ◄

□ **INDETERMINATE FORMS OF TYPE** $\infty - \infty$

If a limit problem such as

$$\lim [f(x) - g(x)] \quad \text{or} \quad \lim [f(x) + g(x)]$$

leads to one of the expressions

$$(+\infty) - (+\infty), \quad (-\infty) - (-\infty)$$

$$(+\infty) + (-\infty), \quad (-\infty) + (+\infty)$$

then one term tends to make the expression large and the other tends to make it small, resulting in an indeterminate form. These are called ***indeterminate forms of type*** $\infty - \infty$. Such forms can sometimes be treated by combining the two terms into one and manipulating the result into one of the previous forms.

Example 6 Evaluate $\displaystyle\lim_{x \to 0^+} \left(\dfrac{1}{x} - \dfrac{1}{\sin x}\right)$.

Solution. Since

$$\lim_{x \to 0^+} \frac{1}{x} = +\infty \quad \text{and} \quad \lim_{x \to 0^+} \frac{1}{\sin x} = +\infty$$

the given limit is an indeterminate form of type $\infty - \infty$. Combining terms yields

$$\lim_{x \to 0^+} \left(\frac{1}{x} - \frac{1}{\sin x}\right) = \lim_{x \to 0^+} \left(\frac{\sin x - x}{x \sin x}\right)$$

which is an indeterminate form of type $0/0$. Applying L'Hôpital's rule twice yields

$$\lim_{x \to 0^+} \left(\frac{\sin x - x}{x \sin x}\right) = \lim_{x \to 0^+} \frac{\cos x - 1}{\sin x + x \cos x}$$

$$= \lim_{x \to 0^+} \frac{-\sin x}{\cos x + \cos x - x \sin x}$$

$$= \frac{0}{2} = 0 \quad \blacktriangleleft$$

Example 7 Evaluate $\displaystyle\lim_{x \to 0^+} (\cot x - \ln x)$.

Solution. Since

$$\lim_{x \to 0^+} \cot x = +\infty \quad \text{and} \quad \lim_{x \to 0^+} \ln x = -\infty$$

we are dealing with a problem of the type $(+\infty) - (-\infty)$. This is not an indeterminate form. The first term tends to make the limit large and because of the subtraction, the second term also tends to make the limit large. Thus,

$$\lim_{x \to 0^+} (\cot x - \ln x) = +\infty \quad \blacktriangleleft$$

REMARK. In general, limits of the type $0/\infty$, $\infty/0$, 0^∞, $\infty \cdot \infty$, $+\infty + (+\infty)$, $+\infty - (-\infty)$, $-\infty + (-\infty)$, $-\infty - (+\infty)$ are not indeterminate forms (Exercise 45).

▶ **Exercise Set 10.3** C 47, 61, 62

In Exercises 1–42, find the limits.

1. $\lim\limits_{x \to +\infty} \dfrac{\ln x}{x}$.

2. $\lim\limits_{x \to +\infty} \dfrac{e^{3x}}{x^2}$.

3. $\lim\limits_{x \to 0^+} \dfrac{\cot x}{\ln x}$.

4. $\lim\limits_{x \to 0^+} \dfrac{1 - \ln x}{e^{1/x}}$.

5. $\lim\limits_{x \to +\infty} \dfrac{x \ln x}{x + \ln x}$.

6. $\lim\limits_{x \to +\infty} \dfrac{x^3 - 2x + 1}{4x^3 + 2}$.

7. $\lim\limits_{x \to +\infty} \dfrac{x^{100}}{e^x}$.

8. $\lim\limits_{x \to 0^+} \dfrac{\ln (\sin x)}{\ln (\tan x)}$.

9. $\lim\limits_{x \to +\infty} xe^{-x}$.

10. $\lim\limits_{x \to \pi^-} (x - \pi) \tan \tfrac{1}{2}x$.

11. $\lim\limits_{x \to +\infty} x \sin \dfrac{\pi}{x}$.

12. $\lim\limits_{x \to 0^+} \tan x \ln x$.

13. $\lim\limits_{x \to +\infty} x(e^{\sin (2/x)} - 1)$.

14. $\lim\limits_{x \to 1} x^{1/(1-x)}$.

15. $\lim\limits_{x \to +\infty} (1 - 3/x)^x$.

16. $\lim\limits_{x \to 0} (1 + 2x)^{-3/x}$.

17. $\lim\limits_{x \to 0} (e^x + x)^{1/x}$.

18. $\lim\limits_{x \to +\infty} (1 + a/x)^{bx}$.

19. $\lim\limits_{x \to +\infty} (1 + 1/x^2)^x$.

20. $\lim\limits_{x \to +\infty} \left(\dfrac{x + 1}{x + 2}\right)^x$.

21. $\lim\limits_{x \to +\infty} (1 + 1/x)^{x^2}$.

22. $\lim\limits_{x \to 0} (1 + \sin 2x)^{1/x}$.

23. $\lim\limits_{x \to 1} (2 - x)^{\tan (\pi/2)x}$.

24. $\lim\limits_{x \to +\infty} [\cos (2/x)]^{x^2}$.

25. $\lim\limits_{x \to 0^+} x^{\sin x}$.

26. $\lim\limits_{x \to 0^+} x^x$.

27. $\lim\limits_{x \to 0^+} (\sin x)^{3/\ln x}$.

28. $\lim\limits_{x \to 0^+} (e^{2x} - 1)^{1/\ln x}$.

29. $\lim\limits_{x \to (1/2)\pi^-} (\tan x)^{\cos x}$.

30. $\lim\limits_{x \to +\infty} (\ln x)^{1/x}$.

31. $\lim\limits_{x \to +\infty} (1 + x^2)^{1/\ln x}$.

32. $\lim\limits_{x \to +\infty} (3^x + 5^x)^{1/x}$.

33. $\lim\limits_{\theta \to 0} \left(\dfrac{1}{1 - \cos \theta} - \dfrac{2}{\sin^2 \theta}\right)$.

34. $\lim\limits_{x \to 0} \left(\dfrac{1}{x^2} - \dfrac{\cos 3x}{x^2}\right)$.

35. $\lim\limits_{x \to 0} (\csc x - 1/x)$.

36. $\lim\limits_{x \to 0} \left(\dfrac{1}{x} - \dfrac{1}{e^x - 1}\right)$.

37. $\lim\limits_{x \to 0} (\cot x - \csc x)$.

38. $\lim\limits_{x \to +\infty} [\ln x - \ln (1 + x)]$.

39. $\lim\limits_{x \to +\infty} [x - \ln (x^2 + 1)]$.

40. $\lim\limits_{x \to +\infty} [x - \ln (1 + 2e^x)]$.

41. $\lim\limits_{x \to 0^+} \dfrac{\cot x}{\cot 2x}$.

42. $\lim\limits_{x \to (1/2)\pi^-} \dfrac{4 \tan x}{1 + \sec x}$.

43. Show that for any positive integer n

(a) $\lim\limits_{x \to +\infty} \dfrac{x^n}{e^x} = 0$ (b) $\lim\limits_{x \to +\infty} \dfrac{e^x}{x^n} = +\infty$.

44. Show that for any positive integer n

(a) $\lim\limits_{x \to +\infty} \dfrac{\ln x}{x^n} = 0$ (b) $\lim\limits_{x \to +\infty} \dfrac{x^n}{\ln x} = +\infty$.

45. Limits of the type

$$0/\infty, \ \infty/0, \ 0^\infty, \ \infty \cdot \infty, \ +\infty + (+\infty),$$

$$+\infty - (-\infty), \ -\infty + (-\infty), \ -\infty - (+\infty)$$

are *not* indeterminate forms. Find the following limits by inspection.

(a) $\lim\limits_{x \to 0^+} \dfrac{x}{\ln x}$ (b) $\lim\limits_{x \to +\infty} \dfrac{x^3}{e^{-x}}$

(c) $\lim\limits_{x \to (1/2)\pi^-} (\cos x)^{\tan x}$ (d) $\lim\limits_{x \to 0^+} (\ln x) \cot x$

(e) $\lim\limits_{x \to (1/2)\pi^-} \left(\dfrac{1}{\frac{1}{2}\pi - x} + \tan x\right)$

(f) $\lim\limits_{x \to 0^+} \left(\dfrac{1}{x} - \ln x\right)$

(g) $\lim\limits_{x \to -\infty} (x + x^3)$ (h) $\lim\limits_{x \to +\infty} \left(\ln \left(\dfrac{1}{x}\right) - e^x\right)$.

46. Find $\lim\limits_{x \to +\infty} (e^x - x^2)$.

47. Sketch the graph of x^x, $x > 0$.

48. Sketch the graph of $(1/x) \tan x$, $-\pi/2 < x < \pi/2$.

In Exercises 49–51, evaluate the improper integral.

49. $\displaystyle\int_0^1 \ln x \, dx$.

50. $\displaystyle\int_1^{+\infty} \dfrac{\ln x}{x^2} \, dx$.

51. $\displaystyle\int_0^{+\infty} xe^{-3x} \, dx$.

52. Find the area between the x-axis and the curve $y = (\ln x - 1)/x^2$ for $x \geq e$.

53. Find the volume of the solid that is generated when the region between the x-axis and the curve $y = e^{-x}$ for $x \geq 0$ is revolved about the y-axis.

54. (a) Show that $\int_1^{+\infty} e^{t^2}\, dt = +\infty$. [*Hint:* Don't try to carry out the integration. Instead, compare the sizes of $\int_1^l e^t\, dt$ and $\int_1^l e^{t^2}\, dt$.]

(b) Use L'Hôpital's rule to evaluate

$$\lim_{x \to +\infty} \frac{1}{x} \int_1^x e^{t^2}\, dt$$

55. (a) Show that

$$\int_0^{+\infty} \sqrt{1 + t^3}\, dt = +\infty$$

[*Hint:* $\sqrt{1 + t^3} \geq t^{3/2}$ for $t \geq 0$.]

(b) Use L'Hôpital's rule to evaluate

$$\lim_{x \to +\infty} \frac{\int_0^{2x} \sqrt{1 + t^3}\, dt}{x^{5/2}}$$

56. (a) Show that

$$\int_0^{+\infty} \sqrt{3 + e^{-2x}}\, dx = +\infty$$

[*Hint:* $\sqrt{3 + e^{-2x}} > \sqrt{3}$ for $x \geq 0$.]

(b) Let f_{ave} denote the average value of the function $f(x) = \sqrt{3 + e^{-2x}}$ on the interval $[0, a]$ where $a > 0$. Find $\lim_{a \to +\infty} f_{\text{ave}}$.

In Exercises 57–60, verify that L'Hôpital's rule is of no help in finding the limit, then find the limit by some other method.

57. $\displaystyle\lim_{x \to +\infty} \frac{x + \sin 2x}{x}$.

58. $\displaystyle\lim_{x \to +\infty} \frac{2x - \sin x}{3x + \sin x}$.

59. $\displaystyle\lim_{x \to +\infty} \frac{x(2 + \sin x)}{x + 1}$.

60. $\displaystyle\lim_{x \to +\infty} \frac{x(2 + \sin x)}{x^2 + 1}$.

61. (a) Show that if k is a positive constant, then

$$\lim_{x \to +\infty} x(k^{1/x} - 1) = \ln k$$

[*Hint:* Let $t = 1/x$.]

(b) If n is a positive integer, then it follows from part (a) with $x = n$ that the approximation

$$n(\sqrt[n]{k} - 1) \approx \ln k$$

should be good when n is large. Use this result and the square root key on a calculator to approximate $\ln 0.3$ and $\ln 2$ with $n = 1024$, then compare the values obtained with values of the logarithms generated directly from the calcu-

lator. [*Hint:* The nth roots for which n is a power of 2 can be obtained as successive square roots.]

62. (a) Show that $\displaystyle\lim_{x \to \pi/2} (\pi/2 - x) \tan x = 1$.

(b) Show that

$$\lim_{x \to \pi/2} \left(\frac{1}{\pi/2 - x} - \tan x\right) = 0$$

(c) It follows from part (b) that the approximation

$$\tan x \approx \frac{1}{\pi/2 - x}$$

should be good for values of x near $\pi/2$. Use a calculator to find $\tan x$ and $1/(\pi/2 - x)$ for $x = 1.57$; compare the results.

63. Find $\displaystyle\lim_{x \to +\infty} xf(t/\sqrt{x})$ if $\displaystyle\lim_{x \to 0^+} f(x) = \lim_{x \to 0^+} f'(x) = 0$ and $\displaystyle\lim_{x \to 0^+} f''(x) = a$. [*Hint:* Let $u = t/\sqrt{x}$.]

64. Let f and g be differentiable functions for which $\displaystyle\lim_{x \to +\infty} f(x) = \lim_{x \to +\infty} g(x) = +\infty$, $\displaystyle\lim_{x \to +\infty} f'(x) = a$, and $\displaystyle\lim_{x \to +\infty} g'(x) = b$ where $a \neq 0$ and $b \neq 0$. Find

$$\lim_{x \to +\infty} \frac{\ln f(x)}{\ln g(x)}$$

In Exercises 65 and 66, find the Laplace transform of $f(t)$. (See Exercise 58 of Section 10.1 for the definition of the Laplace transform.)

65. $f(t) = t,\ s > 0$.　　**66.** $f(t) = t^2,\ s > 0$.

67. There is a myth that circulates among beginning calculus students which states that all indeterminate forms of types 0^0, ∞^0, and 1^∞ have value 1 because "anything to the zero power is 1" and "1 to any power is 1." The fallacy is that 0^0, ∞^0, and 1^∞ are not powers of numbers, but rather descriptions of limits. The following examples, which were transmitted to me by the late Professor Jack Staib of Drexel University, show that such indeterminate forms can have any positive real value:

(a) $\displaystyle\lim_{x \to 0^+} [x^{(\ln a)/(1 + \ln x)}] = 0^0 = a$

(b) $\displaystyle\lim_{x \to +\infty} [x^{(\ln a)/(1 + \ln x)}] = \infty^0 = a$

(c) $\displaystyle\lim_{x \to 0} [(x + 1)^{(\ln a)/x}] = 1^\infty = a$.

Prove these results.

68. The *Gamma function,* $\Gamma(x)$, is defined as

$$\Gamma(x) = \int_0^{+\infty} t^{x-1}e^{-t}\,dt$$

It can be shown that this improper integral converges if and only if $x > 0$.

(a) Find $\Gamma(1)$.

(b) Prove: $\Gamma(x + 1) = x\Gamma(x)$ for all $x > 0$. [*Hint:* Use integration by parts.]

(c) Use the results in parts (a) and (b) to find $\Gamma(2)$, $\Gamma(3)$, and $\Gamma(4)$.

(d) Show that $\Gamma(\tfrac{1}{2}) = \sqrt{\pi}$ given that

$$\int_0^{+\infty} e^{-x^2}\,dx = \tfrac{1}{2}\sqrt{\pi}$$

(e) Use the results in parts (b) and (d) to show that $\Gamma(\tfrac{3}{2}) = \tfrac{1}{2}\sqrt{\pi}$ and $\Gamma(\tfrac{5}{2}) = \tfrac{3}{4}\sqrt{\pi}$.

69. Refer to the Gamma function defined in Exercise 68 to show that

(a) $\int_0^1 (\ln x)^n\,dx = (-1)^n\,\Gamma(n + 1), \quad n > 0$. [*Hint:* Let $t = -\ln x$.]

(b) $\int_0^{+\infty} e^{-x^n}\,dx = \Gamma(1/n + 1), \quad n > 0$. [*Hint:* Let $t = x^n$. Use the result in part (b) of Exercise 68.]

▶ SUPPLEMENTARY EXERCISES

In Exercises 1–14, evaluate the integrals that converge.

1. $\displaystyle\int_{-\infty}^{+\infty} \frac{dx}{x^2 + 4}$.

2. $\displaystyle\int_4^6 \frac{dx}{4 - x}$.

3. $\displaystyle\int_0^1 \frac{x\,dx}{\sqrt{1 - x^2}}$.

4. $\displaystyle\int_0^5 \frac{dx}{\sqrt{25 - x^2}}$.

5. $\displaystyle\int_{-1}^1 \frac{dx}{\sqrt{x^2}}$.

6. $\displaystyle\int_0^{\pi/2} \sec^2 x\,dx$.

7. $\displaystyle\int_{-\infty}^{+\infty} xe^{-x^2}\,dx$.

8. $\displaystyle\int_{-\infty}^0 xe^x\,dx$.

9. $\displaystyle\int_0^{\pi/2} \cot x\,dx$.

10. $\displaystyle\int_0^{+\infty} \frac{dx}{x^5}$.

11. $\displaystyle\int_e^{+\infty} \frac{dx}{x(\ln x)^2}$.

12. $\displaystyle\int_0^1 \sqrt{x}\,\ln x\,dx$.

13. $\displaystyle\int_0^{+\infty} \frac{dx}{x^2 + 2x + 2}$.

14. $\displaystyle\int_0^4 \frac{e^{-\sqrt{x}}}{\sqrt{x}}\,dx$.

15. For what values of n does $\displaystyle\int_0^1 x^n \ln x\,dx$ converge? What is its value?

In Exercises 16–31, find the limit.

16. $\displaystyle\lim_{x\to 1} \frac{\ln x}{x - 1}$.

17. $\displaystyle\lim_{x\to 0} \frac{xe^{3x} - x}{1 - \cos 2x}$.

18. $\displaystyle\lim_{x\to +\infty} \frac{\ln(\ln x)}{\sqrt{x}}$.

19. $\displaystyle\lim_{x\to 0^+} x^2 e^{1/x}$.

20. $\displaystyle\lim_{x\to +\infty} (\sqrt{x^2 + x} - x)$.

21. $\displaystyle\lim_{x\to 0^-} x^2 e^{1/x}$.

22. $\displaystyle\lim_{x\to 0^-} (1 - x)^{2/x}$.

23. $\displaystyle\lim_{\theta\to 0} \left(\frac{\csc\theta}{\theta} - \frac{1}{\theta^2}\right)$.

24. $\displaystyle\lim_{x\to 0} \frac{x - \tan^{-1}x}{x^4}$.

25. $\displaystyle\lim_{x\to 2} \frac{x - 1 - e^{x-2}}{1 - \cos 2\pi x}$.

26. $\displaystyle\lim_{x\to 0} \frac{9^x - 3^x}{x}$.

27. $\displaystyle\lim_{x\to 0} \frac{\int_0^x \sin t^2\,dt}{\sin x^2}$.

28. $\displaystyle\lim_{x\to +\infty} x^{1/x}$.

29. $\displaystyle\lim_{x\to +\infty} \frac{(\ln x)^3}{x}$.

30. $\displaystyle\lim_{x\to +\infty} \left(\frac{x}{x - 3}\right)^x$.

31. $\displaystyle\lim_{x\to 0^+} (1 + x)^{\ln x}$.

32. Find the area of the region in the first quadrant between the curves $y = e^{-x}$ and $y = e^{-2x}$.

In Exercises 33 and 34, find (a) the area of the region R, and (b) the volume obtained by revolving R about the x-axis.

33. The region R bounded between the x-axis and the curve $y = x^{-2/3}, x \geq 8$.

34. The region R bounded between the x-axis and the curve $y = x^{-1/3}, 0 \leq x \leq 1$.

35. Find the total arc length of the graph of the function $f(x) = \sqrt{x - x^2} - \sin^{-1}\sqrt{x}$. [*Hint:* First find the domain of f.]

11
Infinite Series

Brook Taylor   (1685-1731)

◾ 11.1 SEQUENCES

This chapter is concerned with the study of "infinite series," which, loosely speaking, are sums with infinitely many terms. The material in this chapter has far-reaching applications in engineering and science and is the cornerstone for many branches of mathematics. In this initial section we shall develop some preliminary results that are important in their own right.

☐ **DEFINITION OF A SEQUENCE**

In everyday language, we use the term "sequence" to suggest a succession of objects or events given in a specified order. *Informally* speaking, the term "sequence" in mathematics is used to describe an unending succession of numbers. Some possibilities are

$$1, 2, 3, 4, \ldots$$
$$2, 4, 6, 8, \ldots$$
$$1, \tfrac{1}{2}, \tfrac{1}{3}, \tfrac{1}{4}, \ldots$$
$$1, -1, 1, -1, \ldots$$

In each case, the three dots are used to suggest that the sequence continues indefinitely, following the obvious pattern. The numbers in a sequence are called the *terms* of the sequence. The terms may be described according to the positions they occupy. Thus, a sequence has a *first term*, a *second term*, a *third term*, and so forth. Because a sequence continues indefinitely, there is no last term.

The most common way to specify a sequence is to give a formula for the terms. To illustrate the idea, we have listed the terms in the sequence

$$2, 4, 6, 8, \ldots$$

together with their term numbers:

TERM NUMBER	1	2	3	4	...
TERM	2	4	6	8	...

There is a clear relationship here; each term is twice its term number. Thus, for each positive integer n the nth term in the sequence $2, 4, 6, 8, \ldots$ is given by the formula $2n$. This is denoted by writing

$$2, 4, 6, 8, \ldots, 2n, \ldots$$

or more compactly in *bracket notation* as

$$\{2n\}_{n=1}^{+\infty}$$

From the bracket notation, the terms in the sequence can be generated by successively substituting the integer values $n = 1, 2, 3, \ldots$ into the formula $2n$.

Example 1 List the first five terms of the sequence $\{2^n\}_{n=1}^{+\infty}$

Solution. Substituting $n = 1, 2, 3, 4, 5$ into the formula 2^n yields

$$2^1, \ 2^2, \ 2^3, \ 2^4, \ 2^5, \ldots$$

or, equivalently,

$$2, \ 4, \ 8, \ 16, \ 32, \ldots \quad \blacktriangleleft$$

Example 2 Express the following sequences in bracket notation.

(a) $\dfrac{1}{2}, \dfrac{2}{3}, \dfrac{3}{4}, \dfrac{4}{5}, \ldots$ \qquad (b) $\dfrac{1}{2}, \dfrac{1}{4}, \dfrac{1}{8}, \dfrac{1}{16}, \ldots$

(c) $1, \ -1, \ 1, \ -1, \ldots$ \qquad (d) $\dfrac{1}{2}, \ -\dfrac{2}{3}, \dfrac{3}{4}, \ -\dfrac{4}{5}, \ldots$

(e) $1, \ 3, \ 5, \ 7, \ldots$

Solution.

(a) Begin by comparing terms and term numbers:

TERM NUMBER	1	2	3	4	$\ldots$
TERM	$\dfrac{1}{2}$	$\dfrac{2}{3}$	$\dfrac{3}{4}$	$\dfrac{4}{5}$	$\ldots$

In each term, the numerator is the same as the term number, and the denominator is one greater than the term number. Thus, the nth term is $n/(n + 1)$ and the sequence can be written as

$$\left\{ \frac{n}{n + 1} \right\}_{n=1}^{+\infty}$$

(b) Observe that the sequence can be rewritten as

$$\frac{1}{2}, \ \frac{1}{2^2}, \ \frac{1}{2^3}, \ \frac{1}{2^4}, \ldots$$

and construct a table comparing terms and term numbers:

TERM NUMBER	1	2	3	4	$\ldots$
TERM	$\dfrac{1}{2}$	$\dfrac{1}{2^2}$	$\dfrac{1}{2^3}$	$\dfrac{1}{2^4}$	$\ldots$

From the table we see that the nth term is $1/2^n$, so the sequence can be written as

$$\left\{ \frac{1}{2^n} \right\}_{n=1}^{+\infty}$$

(c)   Observe first that $(-1)^r$ is either 1 or -1 according to whether r is an even integer or an odd integer. In the sequence $1, -1, 1, -1, \ldots$ the odd-numbered terms are 1's and the even-numbered terms are -1's. Thus, a formula for the nth term can be obtained by raising -1 to a power that will be even when n is odd and odd when n is even. This is accomplished by the formula $(-1)^{n+1}$, so the sequence can be written as

$$\{(-1)^{n+1}\}_{n=1}^{+\infty}$$

(d)   Combining the results in parts (a) and (c), we can write this sequence as

$$\left\{(-1)^{n+1}\,\frac{n}{n+1}\right\}_{n=1}^{+\infty}$$

(e)   Begin by comparing terms and term numbers:

TERM NUMBER	1	2	3	4	...
TERM	1	3	5	7	...

From the table we see that each term is one less than twice the term number. Thus, the nth term is $2n - 1$, and the sequence can be written as

$$\{2n - 1\}_{n=1}^{+\infty} \qquad \blacktriangleleft$$

Frequently we shall want to write down a sequence without specifying the numerical values of the terms. We do this by writing

$$a_1, a_2, \ldots, a_n, \ldots$$

or in bracket notation

$$\{a_n\}_{n=1}^{+\infty}$$

or sometimes simply

$$\{a_n\}$$

(There is nothing special about the letters a and n; any other letters may be used.)

In the beginning of this section we stated informally that a sequence is an unending succession of numbers. However, this is not a satisfactory mathematical definition because the word "succession" is itself an undefined term. The time has come to formally define the term "sequence." When we write a sequence such as

$$2, 4, 6, 8, \ldots, 2n, \ldots$$

in bracket notation

$$\{2n\}_{n=1}^{+\infty} \tag{1}$$

we are specifying a rule that tells us how to associate a numerical value, namely $2n$, with each positive integer n. Stated another way, $\{2n\}_{n=1}^{+\infty}$ may be regarded as a formula for a function whose independent variable n ranges over the positive integers. Indeed, we could rewrite (1) in functional notation as

$$f(n) = 2n, \quad n = 1, 2, 3, \ldots$$

From this point of view, the notation

$$2, 4, 6, 8, \ldots, 2n, \ldots$$

represents a listing of the function values

$$f(1), f(2), f(3), \ldots, f(n), \ldots$$

This suggests the following definition.

11.1.1 DEFINITION. A *sequence* (or *infinite sequence*) is a function whose domain is the set of positive integers.

□ **GRAPHS OF SEQUENCES**

Because sequences are functions, we may inquire about the graph of a sequence. For example, the graph of the sequence

$$\left\{\frac{1}{n}\right\}_{n=1}^{+\infty}$$

is the graph of the equation

$$y = \frac{1}{n}, \quad n = 1, 2, 3, \ldots$$

Because the right side of this equation is defined only for positive integer values of n, the graph consists of a succession of isolated points (Figure 11.1.1a). This is in marked distinction to the graph of

$$y = \frac{1}{x}, \quad x \geq 1$$

which is a continuous curve (Figure 11.1.1b).

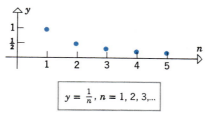

$$y = \frac{1}{n}, n = 1, 2, 3, \ldots$$

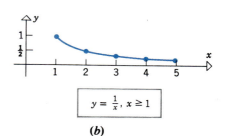

$$y = \frac{1}{x}, x \geq 1$$

Figure 11.1.1   **(a)**   **(b)**

In Figure 11.1.2 we have sketched the graphs of four sequences:

$$\{n + 1\}_{n=1}^{+\infty}, \quad \{(-1)^{n+1}\}_{n=1}^{+\infty}, \quad \left\{\frac{n}{n + 1}\right\}_{n=1}^{+\infty}, \quad \left\{1 + \left(-\frac{1}{2}\right)^n\right\}_{n=1}^{+\infty}$$

Each of these sequences behaves differently as n gets larger and larger. In the sequence $\{n + 1\}$, the terms grow without bound; in the sequence $\{(-1)^{n+1}\}$ the terms oscillate between 1 and -1; in the sequence $\left\{\dfrac{n}{n + 1}\right\}$ the terms increase toward a "limit" of 1; and finally, in the sequence $\{1 + (-\frac{1}{2})^n\}$ the terms also tend toward a "limit" of 1, but do so in an oscillatory fashion.

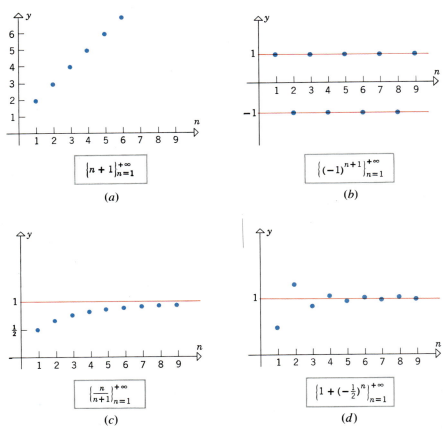

Figure 11.1.2

□ LIMIT OF A SEQUENCE   Let us try to make the term "limit" more precise. To say that a sequence $\{a_n\}_{n=1}^{+\infty}$ approaches a limit L as n gets large is intended to mean that eventually the terms in the sequence become arbitrarily close to the number L. Thus, if we choose *any* positive number ϵ, the terms in the sequence will eventually be within ϵ units of L. Geometrically, this means that if we sketch the lines $y = L + \epsilon$ and $y = L - \epsilon$, the terms in the sequence will eventually be trapped within the band between these lines, and thus be within ϵ units of L (Figure 11.1.3).

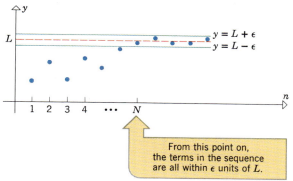

Figure 11.1.3

The following definition expresses this idea precisely.

11.1.2 DEFINITION. A sequence $\{a_n\}_{n=1}^{+\infty}$ is said to have the *limit* L if given any $\epsilon > 0$ there is a positive integer N such that $|a_n - L| < \epsilon$ when $n \geq N$.

If a sequence $\{a_n\}_{n=1}^{+\infty}$ has a limit L, we say that the sequence *converges* to L and write

$$\lim_{n \to +\infty} a_n = L$$

A sequence that does not have a finite limit is said to *diverge*.

Example 3 Figure 11.1.2 suggests that the sequences $\{n + 1\}_{n=1}^{+\infty}$ and $\{(-1)^{n+1}\}_{n=1}^{+\infty}$ diverge, while the sequences

$$\left\{ \frac{n}{n + 1} \right\}_{n=1}^{+\infty} \quad \text{and} \quad \left\{ 1 + \left(-\frac{1}{2} \right)^n \right\}_{n=1}^{+\infty}$$

converge to 1; that is,

$$\lim_{n \to +\infty} \frac{n}{n + 1} = 1 \quad \text{and} \quad \lim_{n \to +\infty} \left(1 + \left(-\frac{1}{2} \right)^n \right) = 1 \quad \blacktriangleleft$$

Many familiar properties of limits apply to limits of sequences.

11.1.3 THEOREM. *Suppose that the sequences $\{a_n\}$ and $\{b_n\}$ converge to limits L_1 and L_2, respectively, and c is a constant. Then*

(a) $\displaystyle \lim_{n \to +\infty} c = c$

(b) $\displaystyle \lim_{n \to +\infty} ca_n = c \lim_{n \to +\infty} a_n = cL_1$

(c) $\displaystyle \lim_{n \to +\infty} (a_n + b_n) = \lim_{n \to +\infty} a_n + \lim_{n \to +\infty} b_n = L_1 + L_2$

(d) $\displaystyle \lim_{n \to +\infty} (a_n - b_n) = \lim_{n \to +\infty} a_n - \lim_{n \to +\infty} b_n = L_1 - L_2$

(e) $\displaystyle \lim_{n \to +\infty} (a_n b_n) = \lim_{n \to +\infty} a_n \cdot \lim_{n \to +\infty} b_n = L_1 L_2$

(f) $\displaystyle \lim_{n \to +\infty} \left(\frac{a_n}{b_n} \right) = \frac{\displaystyle \lim_{n \to +\infty} a_n}{\displaystyle \lim_{n \to +\infty} b_n} = \frac{L_1}{L_2}$ *(if $L_2 \neq 0$)*

(We omit the proof.)

Example 4   In each part, determine whether the given sequence converges or diverges. If it converges, find the limit.

(a) $\displaystyle \left\{ \frac{n}{2n + 1} \right\}_{n=1}^{+\infty}$ (b) $\displaystyle \left\{ (-1)^{n+1} \frac{n}{2n + 1} \right\}_{n=1}^{+\infty}$

(c) $\displaystyle \left\{ (-1)^{n+1} \frac{1}{n} \right\}_{n=1}^{+\infty}$ (d) $\{8 - 2n\}_{n=1}^{+\infty}$ (e) $\displaystyle \left\{ \frac{n}{e^n} \right\}_{n=1}^{+\infty}$

Solution.

(a)   Dividing numerator and denominator by n yields

$$\lim_{n \to +\infty} \frac{n}{2n + 1} = \lim_{n \to +\infty} \frac{1}{\left(2 + \dfrac{1}{n} \right)} = \frac{\displaystyle \lim_{n \to +\infty} 1}{\displaystyle \lim_{n \to +\infty} \left(2 + \dfrac{1}{n} \right)} = \frac{\displaystyle \lim_{n \to +\infty} 1}{\displaystyle \lim_{n \to +\infty} 2 + \lim_{n \to +\infty} \dfrac{1}{n}}$$

$$= \frac{1}{2 + 0} = \frac{1}{2}$$

Thus, $\displaystyle \left\{ \frac{n}{2n + 1} \right\}_{n=1}^{+\infty}$ converges to $\dfrac{1}{2}$.

(b)   From part (a),

$$\lim_{n \to +\infty} \frac{n}{2n + 1} = \frac{1}{2}$$

Thus, since $(-1)^{n+1}$ oscillates between $+1$ and -1, the product $(-1)^{n+1} \dfrac{n}{2n + 1}$ oscillates between positive and negative values, with the odd-numbered terms approaching $\frac{1}{2}$ and the even-numbered terms approaching $-\frac{1}{2}$. Therefore, the sequence $\left\{(-1)^{n+1} \dfrac{n}{2n + 1}\right\}_{n=1}^{+\infty}$ approaches no limit—it diverges.

(c) Since $\lim\limits_{n \to +\infty} 1/n = 0$, the product $(-1)^{n+1}(1/n)$ oscillates between positive and negative values, with the odd-numbered terms approaching 0 through positive values and the even-numbered terms approaching 0 through negative values. Thus,

$$\lim_{n \to +\infty} (-1)^{n+1} \frac{1}{n} = 0$$

so the sequence converges to 0.

(d) $\lim\limits_{n \to +\infty} (8 - 2n) = -\infty$, so the sequence $\{8 - 2n\}_{n=1}^{+\infty}$ diverges.

(e) We want to find $\lim\limits_{n \to +\infty} n/e^n$, which is an indeterminate form of type ∞/∞.

Unfortunately, we cannot apply L'Hôpital's rule directly since e^n and n are not differentiable functions (n assumes only integer values). However, we can apply L'Hôpital's rule to the related problem $\lim\limits_{x \to +\infty} x/e^x$ to obtain

$$\lim_{x \to +\infty} \frac{x}{e^x} = \lim_{x \to +\infty} \frac{1}{e^x} = 0$$

We conclude from this that $\lim\limits_{n \to +\infty} n/e^n = 0$ since the values of n/e^n and x/e^x are the same when x is a positive integer. ◄

Example 5 Show that $\lim\limits_{n \to +\infty} \sqrt[n]{n} = 1$.

Solution. With the aid of L'Hôpital's rule, $\lim\limits_{n \to +\infty} \dfrac{1}{n} \ln n = 0$. Thus,

$$\lim_{n \to +\infty} \sqrt[n]{n} = \lim_{n \to +\infty} n^{1/n} = \lim_{n \to +\infty} e^{(1/n)\ln n} = e^0 = 1 \quad ◄$$

Exercise Set 11.1

In Exercises 1–18, show the first five terms of the sequence, determine whether the sequence converges, and if so find the limit. (When writing out the terms of the sequence, you need not find numerical values; leave the terms in the first form you obtain.)

1. $\left\{\dfrac{n}{n + 2}\right\}_{n=1}^{+\infty}$.

2. $\left\{\dfrac{n^2}{2n + 1}\right\}_{n=1}^{+\infty}$.

3. $\{2\}_{n=1}^{+\infty}$.

4. $\left\{\ln\left(\dfrac{1}{n}\right)\right\}_{n=1}^{+\infty}$.

5. $\left\{\dfrac{\ln n}{n}\right\}_{n=1}^{+\infty}$.

6. $\left\{n \sin \dfrac{\pi}{n}\right\}_{n=1}^{+\infty}$.

7. $\{1 + (-1)^n\}_{n=1}^{+\infty}$.

8. $\left\{\dfrac{(-1)^{n+1}}{n^2}\right\}_{n=1}^{+\infty}$.

9. $\left\{(-1)^n \dfrac{2n^3}{n^3 + 1}\right\}_{n=1}^{+\infty}$.

10. $\left\{\dfrac{n}{2^n}\right\}_{n=1}^{+\infty}$.

11. $\left\{\dfrac{(n + 1)(n + 2)}{2n^2}\right\}_{n=1}^{+\infty}$.

12. $\left\{\dfrac{\pi^n}{4^n}\right\}_{n=1}^{+\infty}$.

13. $\{\cos{(3/n)}\}_{n=1}^{+\infty}$.

14. $\left\{\cos{\dfrac{\pi n}{2}}\right\}_{n=1}^{+\infty}$.

15. $\{n^2 e^{-n}\}_{n=1}^{+\infty}$.

16. $\{\sqrt{n^2 + 3n} - n\}_{n=1}^{+\infty}$.

17. $\left\{\left(\dfrac{n+3}{n+1}\right)^n\right\}_{n=1}^{+\infty}$.

18. $\left\{\left(1 - \dfrac{2}{n}\right)^n\right\}_{n=1}^{+\infty}$.

In Exercises 19–26, express the sequence in the notation $\{a_n\}_{n=1}^{+\infty}$, determine whether the sequence converges, and if so find its limit.

19. $\dfrac{1}{2}, \dfrac{3}{4}, \dfrac{5}{6}, \dfrac{7}{8}, \ldots$

20. $0, \dfrac{1}{2^2}, \dfrac{2}{3^2}, \dfrac{3}{4^2}, \ldots$

21. $\dfrac{1}{3}, \dfrac{1}{9}, \dfrac{1}{27}, \dfrac{1}{81}, \ldots$

22. $-1, 2, -3, 4, -5, \ldots$

23. $\left(1 - \dfrac{1}{2}\right), \left(\dfrac{1}{2} - \dfrac{1}{3}\right), \left(\dfrac{1}{3} - \dfrac{1}{4}\right), \left(\dfrac{1}{4} - \dfrac{1}{5}\right), \ldots$

24. $3, \dfrac{3}{2}, \dfrac{3}{2^2}, \dfrac{3}{2^3}, \ldots$

25. $(\sqrt{2} - \sqrt{3}), (\sqrt{3} - \sqrt{4}), (\sqrt{4} - \sqrt{5}), \ldots$

26. $\dfrac{1}{3^5}, -\dfrac{1}{3^6}, \dfrac{1}{3^7}, -\dfrac{1}{3^8}, \ldots$

27. Let $\{a_n\}$ be the sequence for which $a_1 = \sqrt{6}$ and $a_{n+1} = \sqrt{6 + a_n}$ for $n \geq 1$.
   (a) Find the first three terms of the sequence.
   (b) It can be shown that the sequence $\{a_n\}$ converges. Assuming this to be so, find its limit L. [*Hint:* $\lim\limits_{n \to +\infty} a_n = \lim\limits_{n \to +\infty} a_{n+1} = L$.]

28. Let a_1 and k be any positive real numbers and let $\{a_n\}$ be the sequence for which $a_{n+1} = \frac{1}{2}(a_n + k/a_n)$ for $n \geq 1$. Assuming that this sequence converges, find its limit using the hint in Exercise 27(b).

29. The **Fibonacci sequence** is defined by $a_{n+2} = a_n + a_{n+1}$ for $n \geq 1$, where $a_1 = a_2 = 1$.
   (a) Find the first eight terms of the sequence.
   (b) Find $\lim\limits_{n \to +\infty} (a_{n+1}/a_n)$ assuming that this limit exists.
   [*Hint:* $\lim\limits_{n \to +\infty} (a_{n+1}/a_n) = \lim\limits_{n \to +\infty} (a_{n+2}/a_{n+1})$.]

30. The nth term a_n of the sequence $1, 2, 1, 4, 1, 6, \ldots$ is best written in the form

$$a_n = \begin{cases} 1, & \text{if } n \text{ is odd} \\ n, & \text{if } n \text{ is even} \end{cases}$$

since a single formula applicable to all terms would

be too complicated to be useful. By considering even and odd terms separately, find a similar expression for the nth term of the sequence

   (a) $1, \dfrac{1}{2^2}, 3, \dfrac{1}{2^4}, 5, \dfrac{1}{2^6}, \ldots$

   (b) $1, \dfrac{1}{3}, \dfrac{1}{3}, \dfrac{1}{5}, \dfrac{1}{5}, \dfrac{1}{7}, \dfrac{1}{7}, \dfrac{1}{9}, \dfrac{1}{9}, \ldots$

31. Consider the sequence $\{a_n\}_{n=1}^{+\infty}$ where

$$a_n = \dfrac{1}{n^2} + \dfrac{2}{n^2} + \cdots + \dfrac{n}{n^2}$$

   (a) Write out the first four terms of the sequence.
   (b) Find the limit of the sequence. [*Hint:* Sum up the terms in the formula for a_n.]

32. Follow the directions in Exercise 31 with

$$a_n = \dfrac{1^2}{n^3} + \dfrac{2^2}{n^3} + \cdots + \dfrac{n^2}{n^3}$$

33. If we accept the fact that the sequence $\{1/n\}_{n=1}^{+\infty}$ converges to the limit $L = 0$, then according to Definition 11.1.2, for every $\epsilon > 0$, there exists an integer N such that $|a_n - L| = |(1/n) - 0| < \epsilon$ when $n \geq N$. In each part, find the smallest possible value of N for the given value of ϵ.
   (a) $\epsilon = 0.5$ (b) $\epsilon = 0.1$ (c) $\epsilon = 0.001$.

34. If we accept the fact that the sequence

$$\left\{\dfrac{n}{n+1}\right\}_{n=1}^{+\infty}$$

converges to the limit $L = 1$, then according to Definition 11.1.2, for every $\epsilon > 0$ there exists an integer N such that

$$|a_n - L| = \left|\dfrac{n}{n+1} - 1\right| < \epsilon$$

when $n \geq N$. In each part, find the smallest value of N for the given value of ϵ.
   (a) $\epsilon = 0.25$ (b) $\epsilon = 0.1$ (c) $\epsilon = 0.001$.

35. Prove:
   (a) The sequence $\{1/n\}_{n=1}^{+\infty}$ converges to 0.
   (b) The sequence $\left\{\dfrac{n}{n+1}\right\}_{n=1}^{+\infty}$ converges to 1.

36. Consider the sequence $\{a_n\}_{n=1}^{+\infty}$ whose nth term is

$$a_n = \sum_{k=0}^{n-1} \dfrac{1}{1 + k/n} \cdot \dfrac{1}{n}$$

Show that $\lim\limits_{n \to +\infty} a_n = \ln 2$. [*Hint:* Interpret $\lim\limits_{n \to +\infty} a_n$ as a definite integral.]

37. (a) Show that a polygon with n equal sides inscribed in a circle of radius r has perimeter $p_n = 2rn \sin(\pi/n)$.

   (b) By finding the limit of the sequence $\{p_n\}_{n=1}^{+\infty}$, derive the formula for the circumference of the circle.

38. Find $\lim_{n \to +\infty} r^n$, where r is a real number. [*Hint:* Consider the cases $|r| < 1$, $|r| > 1$, $r = 1$, and $r = -1$ separately.]

39. Find the limit of the sequence $\{(2^n + 3^n)^{1/n}\}_{n=1}^{+\infty}$.

■ 11.2 MONOTONE SEQUENCES

Sometimes the critical information about a sequence is whether it converges or not, with the limit being of lesser interest. In this section we shall discuss results that are used to study convergence of sequences.

□ **TERMINOLOGY**

We begin with some terminology.

11.2.1 DEFINITION. A sequence $\{a_n\}$ is called

increasing if	$a_1 < a_2 < a_3 < \cdots < a_n < \cdots$
nondecreasing if	$a_1 \le a_2 \le a_3 \le \cdots \le a_n \le \cdots$
decreasing if	$a_1 > a_2 > a_3 > \cdots > a_n > \cdots$
nonincreasing if	$a_1 \ge a_2 \ge a_3 \ge \cdots \ge a_n \ge \cdots$

A sequence that is either nondecreasing or nonincreasing is called **monotone,** and a sequence that is increasing or decreasing is called **strictly monotone.** Observe that a strictly monotone sequence is monotone, but not conversely. (Why?)

Example 1

$$\frac{1}{2}, \frac{2}{3}, \frac{3}{4}, \ldots, \frac{n}{n+1}, \ldots \qquad \text{is increasing}$$

$$1, \frac{1}{2}, \frac{1}{3}, \ldots, \frac{1}{n}, \ldots \qquad \text{is decreasing}$$

$$1, 1, 2, 2, 3, 3, \ldots \qquad \text{is nondecreasing}$$

$$1, 1, \frac{1}{2}, \frac{1}{2}, \frac{1}{3}, \frac{1}{3}, \ldots \qquad \text{is nonincreasing}$$

All four of these sequences are monotone, but the sequence

$$1, -\frac{1}{2}, \frac{1}{3}, -\frac{1}{4}, \ldots, (-1)^{n+1}\frac{1}{n}, \ldots$$

is not. The first and second sequences are strictly monotone.   ◀

In order for a sequence to be increasing, *all* pairs of successive terms, a_n and a_{n+1}, must satisfy $a_n < a_{n+1}$, or equivalently, $a_n - a_{n+1} < 0$. More generally, monotone sequences can be classified as follows:

DIFFERENCE BETWEEN SUCCESSIVE TERMS	CLASSIFICATION
$a_n - a_{n+1} < 0$	Increasing
$a_n - a_{n+1} > 0$	Decreasing
$a_n - a_{n+1} \leq 0$	Nondecreasing
$a_n - a_{n+1} \geq 0$	Nonincreasing

Frequently, one can *guess* whether a sequence is increasing, decreasing, nondecreasing, or nonincreasing after writing out some of the initial terms. However, to be certain that the guess is correct, a precise mathematical proof is needed. The following example illustrates a method for doing this.

Example 2 Show that

$$\frac{1}{2}, \frac{2}{3}, \frac{3}{4}, \ldots, \frac{n}{n+1}, \ldots$$

is an increasing sequence.

Solution. It is intuitively clear that the sequence is increasing. To prove that this is so, let

$$a_n = \frac{n}{n+1}$$

We can obtain a_{n+1} by replacing n by $n+1$ in this formula. This yields

$$a_{n+1} = \frac{n+1}{(n+1)+1} = \frac{n+1}{n+2}$$

Thus, for $n \geq 1$

$$a_n - a_{n+1} = \frac{n}{n+1} - \frac{n+1}{n+2} = \frac{n^2 + 2n - n^2 - 2n - 1}{(n+1)(n+2)}$$

$$= -\frac{1}{(n+1)(n+2)} < 0$$

This proves that the sequence is increasing. ◄

If a_n and a_{n+1} are any successive terms in an increasing sequence, then $a_n < a_{n+1}$. If the terms in the sequence are all positive, then we can divide both sides of this inequality by a_n to obtain $1 < a_{n+1}/a_n$ or equivalently $a_{n+1}/a_n > 1$. More generally, monotone sequences with *positive* terms can be classified as follows:

RATIO OF SUCCESSIVE TERMS	CLASSIFICATION
$a_{n+1}/a_n > 1$	Increasing
$a_{n+1}/a_n < 1$	Decreasing
$a_{n+1}/a_n \geq 1$	Nondecreasing
$a_{n+1}/a_n \leq 1$	Nonincreasing

Example 3 Show that the sequence in Example 2 is increasing by examining the ratio of successive terms.

Solution. As shown in the solution of Example 2,

$$a_n = \frac{n}{n+1} \quad \text{and} \quad a_{n+1} = \frac{n+1}{n+2}$$

Thus,

$$\frac{a_{n+1}}{a_n} = \frac{(n+1)/(n+2)}{n/(n+1)} = \frac{n+1}{n+2} \cdot \frac{n+1}{n} = \frac{n^2 + 2n + 1}{n^2 + 2n} \tag{1}$$

Since the numerator in (1) exceeds the denominator, the ratio exceeds 1, that is, $a_{n+1}/a_n > 1$ for $n \geq 1$. This proves that the sequence is increasing. ◀

In our subsequent work we will encounter sequences involving *factorials*. The reader will recall that if n is a positive integer, then $n!$ (n factorial) is the product of the first n positive integers, that is, $n! = 1 \cdot 2 \cdot 3 \cdots n$. Furthermore, it is agreed that $0! = 1$.

Example 4 Show that the sequence

$$\frac{e}{2!}, \frac{e^2}{3!}, \frac{e^3}{4!}, \ldots, \frac{e^n}{(n+1)!}, \ldots$$

is decreasing.

Solution. We shall examine the ratio of successive terms. Since

$$a_n = \frac{e^n}{(n+1)!}$$

it follows on replacing n by $n + 1$ that

$$a_{n+1} = \frac{e^{n+1}}{[(n+1)+1]!} = \frac{e^{n+1}}{(n+2)!}$$

Thus,

$$\frac{a_{n+1}}{a_n} = \frac{e^{n+1}/(n+2)!}{e^n/(n+1)!} = \frac{e^{n+1}}{e^n} \cdot \frac{(n+1)!}{(n+2)!} = \frac{e}{n+2}$$

For $n \geq 1$, it follows that $n + 2 \geq 3 > e$ ($\approx 2.718...$), so

$$\frac{a_{n+1}}{a_n} = \frac{e}{n + 2} < 1$$

for $n \geq 1$. This proves that the sequence is decreasing. ◀

The following example illustrates still a third technique for determining whether a sequence is increasing or decreasing.

Example 5 In Examples 2 and 3 we proved that the sequence

$$\frac{1}{2}, \frac{2}{3}, \frac{3}{4}, \ldots, \frac{n}{n + 1}, \ldots$$

is increasing by considering the difference and ratio of successive terms. Alternatively, we can proceed as follows. Let

$$f(x) = \frac{x}{x + 1}$$

so the nth term in the given sequence is $a_n = f(n)$. The function f is increasing for $x \geq 1$ since

$$f'(x) = \frac{(x + 1)(1) - x(1)}{(x + 1)^2} = \frac{1}{(x + 1)^2} > 0$$

Thus,

$$a_n = f(n) < f(n + 1) = a_{n+1}$$

which proves that the given sequence is increasing. ◀

In general, if $f(n) = a_n$ is the nth term of a sequence, and if f is differentiable for $x \geq 1$, then we have the following results:

DERIVATIVE OF f FOR $x \geq 1$	CLASSIFICATION OF THE SEQUENCE WITH nTH TERM $a_n = f(n)$
$f'(x) > 0$	Increasing
$f'(x) < 0$	Decreasing
$f'(x) \geq 0$	Nondecreasing
$f'(x) \leq 0$	Nonincreasing

We omit the proof.

□ **CONVERGENCE OF MONOTONE SEQUENCES**

The following two theorems, whose optional proofs are discussed at the end of this section, show that a monotone sequence either converges or it becomes infinite—divergence by oscillation cannot occur.

11.2.2   THEOREM.   *If* $a_1 \leq a_2 \leq a_3 \leq \cdots \leq a_n \leq \cdots$ *is a nondecreasing sequence, then there are two possibilities:*

(a)   *There is a constant M such that* $a_n \leq M$ *for all n, in which case the sequence converges to a limit L satisfying* $L \leq M$.

(b)   *No such constant exists, in which case* $\lim\limits_{n \to +\infty} a_n = +\infty$.

11.2.3   THEOREM.   *If* $a_1 \geq a_2 \geq a_3 \geq \cdots \geq a_n \geq \cdots$ *is a nonincreasing sequence, then there are two possibilities:*

(a)   *There is a constant M such that* $a_n \geq M$ *for all n, in which case the sequence converges to a limit L satisfying* $L \geq M$.

(b)   *No such constant exists, in which case* $\lim\limits_{n \to +\infty} a_n = -\infty$.

It should be noted that these results do not give a method for obtaining limits; they tell us only whether a limit exists. To prove these theorems we need a preliminary result that takes us to the very foundations of the real number system. In this text we have not been concerned with a logical development of the real numbers; our approach has been to accept the familiar properties of real numbers and to work with them. Indeed, we have not even attempted to define the term "real number." However, by the late nineteenth century, the study of limits and functions in calculus necessitated a precise axiomatic formulation of the real numbers in much the same way that Euclidean geometry is developed from axioms. While we will not attempt to pursue this development, we shall have need for the following axiom about real numbers.

11.2.4   AXIOM (*The Completeness Axiom*).   *If S is a nonempty set of real numbers, and if there is some real number that is greater than or equal to every number in S, then there is a smallest real number that is greater than or equal to every number in S.*

For example, let S be the set of numbers in the interval $(1, 3)$. It is true that there exists a number u greater than or equal to every number in S; some examples are $u = 10$, $u = 100$, and $u = 3.2$. The smallest number u that is greater than or equal to every number in S is $u = 3$.

There is an alternative phrasing of the completeness axiom that is useful to know. Let us call a number u an **upper bound** for a set S if u is greater than or equal to every number in S; and if S has a smallest upper bound, call it the **least upper bound** of S. Using this terminology the Completeness Axiom states:

11.2.5   AXIOM (*The Completeness Axiom, Alternative Form*).   *If a nonempty set S of real numbers has an upper bound, then S has a least upper bound.*

Example 6 As shown in Examples 2 and 3, the sequence $\left\{\dfrac{n}{n+1}\right\}_{n=1}^{+\infty}$ is increasing (hence nondecreasing). Since

$$a_n = \frac{n}{n+1} < 1, \quad n = 1, 2, \ldots$$

the terms in the sequence have $M = 1$ as an upper bound. By Theorem 11.2.2 the sequence must converge to a limit $L \leq M$. This is indeed the case since

$$\lim_{n \to +\infty} \frac{n}{n+1} = \lim_{n \to +\infty} \frac{1}{1 + \dfrac{1}{n}} = 1 \quad \blacktriangleleft$$

Because the limit of a sequence $\{a_n\}$ describes the behavior of the terms as n gets *large*, one can alter or even delete a *finite* number of terms in a sequence without affecting either the convergence or the value of the limit. That is, the original and the modified sequence will both converge or both diverge, and in the case of convergence both will have the same limit. We omit the proof.

Example 7 Show that the sequence

$$\left\{\frac{5^n}{n!}\right\}_{n=1}^{+\infty}$$

converges.

Solution. It is tedious to determine convergence directly from the limit

$$\lim_{n \to +\infty} \frac{5^n}{n!}$$

Thus, we shall proceed indirectly. If we let

$$a_n = \frac{5^n}{n!}$$

then

$$a_{n+1} = \frac{5^{n+1}}{(n+1)!}$$

so that

$$\frac{a_{n+1}}{a_n} = \frac{5^{n+1}/(n+1)!}{5^n/n!} = \frac{5^{n+1}}{5^n} \cdot \frac{n!}{(n+1)!} = \frac{5}{n+1}$$

For $n = 1$, 2, and 3, the value of the ratio a_{n+1}/a_n is greater than 1, so $a_{n+1} > a_n$. Thus,

$$a_1 < a_2 < a_3 < a_4$$

For $n = 4$ the value of a_{n+1}/a_n is 1, so

$$a_4 = a_5$$

For $n \geq 5$ the value of a_{n+1}/a_n is less than 1, so

$$a_5 > a_6 > a_7 > a_8 > \cdots$$

Thus, if we discard the first four terms of the given sequence (which will not affect convergence), the resulting sequence will be decreasing. Moreover, each term in the sequence is positive, so by Theorem 11.2.3 with $M = 0$ the sequence converges to some limit that is ≥ 0. ◄

■ OPTIONAL

Proof of Theorem 11.2.2.

(a) Assume there exists a number M such that $a_n \leq M$ for $n = 1, 2, \ldots$. Then M is an upper bound for the set of terms in the sequence. By the Completeness Axiom there is a least upper bound for the terms, call it L. Now let ϵ be any positive number. Since L is the least upper bound for the terms, $L - \epsilon$ is not an upper bound for the terms, which means that there is at least one term a_N such that

$$a_N > L - \epsilon$$

Moreover, since $\{a_n\}$ is a nondecreasing sequence, we must have

$$a_n \geq a_N > L - \epsilon \tag{2}$$

when $n \geq N$. But a_n cannot exceed L since L is an upper bound for the terms. This observation together with (2) tells us that $L \geq a_n > L - \epsilon$ for $n \geq N$, so all terms from the Nth on are within ϵ units of L. This is exactly the requirement to have

$$\lim_{n \to +\infty} a_n = L$$

Finally, $L \leq M$ since M is an upper bound for the terms and L is the least upper bound. This proves part (a).

(b) If there is no number M such that $a_n \leq M$ for $n = 1, 2, \ldots$, then no matter how large we choose M, there is a term a_N such that

$$a_N > M$$

and, since the sequence is nondecreasing,

$$a_n \geq a_N > M$$

when $n \geq N$. Thus, the terms in the sequence become arbitrarily large as n increases. That is,

$$\lim_{n \to +\infty} a_n = +\infty \quad ∎$$

The proof of Theorem 11.2.3 will be omitted since it is similar to the proof of 11.2.2.

▶ Exercise Set 11.2

In Exercises 1–6, determine whether the given sequence $\{a_n\}$ is monotone by examining $a_n - a_{n+1}$. If so, classify it as increasing, decreasing, nonincreasing, or nondecreasing.

1. $\left\{\dfrac{1}{n}\right\}_{n=1}^{+\infty}.$

2. $\left\{1 - \dfrac{1}{n}\right\}_{n=1}^{+\infty}.$

3. $\left\{\dfrac{n}{2n+1}\right\}_{n=1}^{+\infty}.$

4. $\left\{\dfrac{n}{4n-1}\right\}_{n=1}^{+\infty}.$

5. $\{n - 2^n\}_{n=1}^{+\infty}.$

6. $\{n - n^2\}_{n=1}^{+\infty}.$

In Exercises 7–18, determine whether the given sequence $\{a_n\}$ is monotone by examining a_{n+1}/a_n. If so, classify it as increasing, decreasing, nonincreasing, or nondecreasing.

7. $\left\{\dfrac{n}{2n+1}\right\}_{n=1}^{+\infty}.$

8. $\left\{\dfrac{n}{2^n}\right\}_{n=1}^{+\infty}.$

9. $\{ne^{-n}\}_{n=1}^{+\infty}.$

10. $\left\{\dfrac{n^2}{3^n}\right\}_{n=1}^{+\infty}.$

11. $\left\{\dfrac{2^n}{n!}\right\}_{n=1}^{+\infty}.$

12. $\left\{\dfrac{e^n}{n!}\right\}_{n=1}^{+\infty}.$

13. $\left\{\dfrac{n!}{3^n}\right\}_{n=1}^{+\infty}.$

14. $\left\{\dfrac{n^2}{n!}\right\}_{n=1}^{+\infty}.$

15. $\left\{\dfrac{10^n}{(2n)!}\right\}_{n=1}^{+\infty}.$

16. $\left\{\dfrac{2^n}{1+2^n}\right\}_{n=1}^{+\infty}.$

17. $\left\{\dfrac{n^n}{n!}\right\}_{n=1}^{+\infty}.$

18. $\left\{\dfrac{10^n}{2^{(n^2)}}\right\}_{n=1}^{+\infty}.$

In Exercises 19–24, use differentiation to show that the sequence is strictly monotone and classify it as increasing or decreasing.

19. $\left\{\dfrac{n}{2n+1}\right\}_{n=1}^{+\infty}.$

20. $\left\{3 - \dfrac{1}{n}\right\}_{n=1}^{+\infty}.$

21. $\left\{\dfrac{1}{n + \ln n}\right\}_{n=1}^{+\infty}.$

22. $\{ne^{-2n}\}_{n=1}^{+\infty}.$

23. $\left\{\dfrac{\ln (n + 2)}{n + 2}\right\}_{n=1}^{+\infty}.$

24. $\{\tan^{-1} n\}_{n=1}^{+\infty}.$

In Exercises 25–30, show that the sequence is monotone and apply Theorem 11.2.2 or 11.2.3 to determine whether it converges.

25. $\left\{\dfrac{n}{5^n}\right\}_{n=1}^{+\infty}.$

26. $\left\{\dfrac{2^n}{(n+1)!}\right\}_{n=1}^{+\infty}.$

27. $\left\{n - \dfrac{1}{n}\right\}_{n=1}^{+\infty}.$

28. $\left\{\cos \dfrac{\pi}{2n}\right\}_{n=1}^{+\infty}.$

29. $\left\{2 + \dfrac{1}{n}\right\}_{n=1}^{+\infty}.$

30. $\left\{\dfrac{4n-1}{5n+2}\right\}_{n=1}^{+\infty}.$

31. Find the limit (if it exists) of the sequence.

(a) $1, -1, 1, \dfrac{1}{2}, \dfrac{1}{3}, \dfrac{1}{4}, \dfrac{1}{5}, \ldots$

(b) $-\dfrac{1}{2}, 0, 0, 0, 1, 2, 3, 4, \ldots.$

32. (a) Is $\{100^n/n!\}_{n=1}^{+\infty}$ a monotone sequence?

(b) Is it a convergent sequence? Justify your answer.

33. Show that $\left\{\dfrac{3^n}{1 + 3^{2n}}\right\}_{n=1}^{+\infty}$ is a decreasing sequence.

34. Show that $\left\{\dfrac{1 \cdot 3 \cdot 5 \cdots (2n - 1)}{n!}\right\}_{n=1}^{+\infty}$ is an increasing sequence.

35. Let $\{a_n\}$ be the sequence for which $a_1 = \sqrt{2}$ and $a_{n+1} = \sqrt{2 + a_n}$ for $n \geq 1$.

(a) Find the first three terms of the sequence.

(b) Show that $a_n < 2$ for $n \geq 1$.

(c) Show that $a_{n+1}^2 - a_n^2 = (2 - a_n)(1 + a_n)$ for $n \geq 1$.

(d) Use the results in parts (b) and (c) to show that $\{a_n\}$ is an increasing sequence. [*Hint*: If x and y are positive real numbers such that $x^2 - y^2 > 0$, then it follows by factoring that $x - y > 0$.]

(e) Show that $\{a_n\}$ converges and find its limit L.

36. Let $\{a_n\}$ be the sequence for which $a_1 = \sqrt{3}$ and $a_{n+1} = \sqrt{3a_n}$ for $n \geq 1$.

(a) Find the first three terms of the sequence.

(b) Show that $a_n < 3$ for $n \geq 1$.

(c) Show that $a_{n+1}^2 - a_n^2 = a_n(3 - a_n)$ for $n \geq 1$.

(d) Use the hint in Exercise 35(d) to show that $\{a_n\}$ is an increasing sequence.

(e) Show that $\{a_n\}$ converges and find its limit L.

37. (a) Show that if $\{a_n\}_{n=1}^{+\infty}$ is a nonincreasing sequence, then $\{-a_n\}_{n=1}^{+\infty}$ is a nondecreasing sequence.

(b) Use part (a) and Theorem 11.2.2 to help prove Theorem 11.2.3.

38. (a) Deduce the inequalities

$$\int_1^n \ln x \, dx < \ln n! < \int_1^{n+1} \ln x \, dx$$

from Figure 11.2.1 for $n \geq 2$ by comparing appropriate areas.

(b) Use the result in part (a) to show that

$$\frac{n^n}{e^{n-1}} < n! < \frac{(n+1)^{n+1}}{e^n}, \quad n > 1$$

39. Use the Squeezing Theorem (Theorem 2.8.2) and the result in Exercise 38(b) to show that

$$\lim_{n \to +\infty} \frac{\sqrt[n]{n!}}{n} = \frac{1}{e}$$

40. Use the left inequality in Exercise 38(b) to show that $\lim_{n \to +\infty} \sqrt[n]{n!} = +\infty$.

41. (a) Show that $\left\{ \dfrac{n^n}{n!e^n} \right\}_{n=1}^{+\infty}$ is a decreasing sequence.

[*Hint:* From Exercise 99(d) of Section 7.4 we have $(1 + 1/x)^x < e$ for $x > 0$.]

(b) Does the sequence converge? Explain.

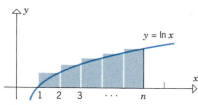

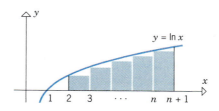

Figure 11.2.1

11.3 INFINITE SERIES

The purpose of this section is to discuss sums that contain infinitely many terms. The most familiar examples of such sums occur in the decimal representation of real numbers. For example, when we write $\frac{1}{3}$ in the decimal form $\frac{1}{3} = 0.3333 \ldots$ we mean

$$\frac{1}{3} = 0.3 + 0.03 + 0.003 + 0.0003 + \cdots$$

which suggests that the decimal representation of $\frac{1}{3}$ can be viewed as a sum of infinitely many real numbers.

☐ **SUMS OF INFINITE SERIES**

Our first objective is to define what is meant by the "sum" of infinitely many real numbers. We begin with some terminology.

11.3.1 DEFINITION. An *infinite series* is an expression of the form

$$u_1 + u_2 + u_3 + \cdots + u_k + \cdots$$

or in sigma notation

$$\sum_{k=1}^{\infty} u_k$$

The numbers $u_1, u_2, u_3, \ldots$ are called the *terms* of the series.

Since it is physically impossible to add infinitely many numbers together, sums of infinite series are defined and computed by an indirect limiting process. To motivate the basic idea, consider the infinite decimal

$$0.3333\ldots \qquad (1)$$

which can be viewed as the infinite series

$$0.3 + 0.03 + 0.003 + 0.0003 + \cdots$$

or equivalently

$$\frac{3}{10} + \frac{3}{10^2} + \frac{3}{10^3} + \frac{3}{10^4} + \cdots \qquad (2)$$

Since (1) is the decimal expansion of $\frac{1}{3}$, any reasonable definition for the sum of an infinite series should yield $\frac{1}{3}$ for the sum of (2). To obtain such a definition, consider the following sequence of (finite) sums:

$$s_1 = \frac{3}{10} = 0.3$$

$$s_2 = \frac{3}{10} + \frac{3}{10^2} = 0.33$$

$$s_3 = \frac{3}{10} + \frac{3}{10^2} + \frac{3}{10^3} = 0.333$$

$$s_4 = \frac{3}{10} + \frac{3}{10^2} + \frac{3}{10^3} + \frac{3}{10^4} = 0.3333$$

$$\vdots$$

The sequence of numbers $s_1, s_2, s_3, s_4, \ldots$ can be viewed as a succession of approximations to the "sum" of the infinite series, which we want to be $\frac{1}{3}$. As we progress through the sequence, more and more terms of the infinite series are used, and the approximations get better and better, suggesting that the desired sum of $\frac{1}{3}$ might be the *limit* of this sequence of approximations. To see that this is so, we must calculate the limit of the general term in the sequence of approximations, namely

$$s_n = \frac{3}{10} + \frac{3}{10^2} + \cdots + \frac{3}{10^n} \qquad (3)$$

The problem of calculating

$$\lim_{n \to +\infty} s_n = \lim_{n \to +\infty} \left\{ \frac{3}{10} + \frac{3}{10^2} + \cdots + \frac{3}{10^n} \right\}$$

is complicated by the fact that both the last term and the number of terms in the sum change with n. It is best to rewrite such limits in a closed form in which the number of terms does not vary, if possible. (See the remark following

Example 3 in Section 5.4.) To do this, we multiply both sides of (3) by $\frac{1}{10}$ to obtain

$$\frac{1}{10} s_n = \frac{3}{10^2} + \frac{3}{10^3} + \cdots + \frac{3}{10^n} + \frac{3}{10^{n+1}} \tag{4}$$

and then subtract (4) from (3) to obtain

$$s_n - \frac{1}{10} s_n = \frac{3}{10} - \frac{3}{10^{n+1}}$$

$$\frac{9}{10} s_n = \frac{3}{10} \left(1 - \frac{1}{10^n} \right)$$

$$s_n = \frac{1}{3} \left(1 - \frac{1}{10^n} \right)$$

Since $1/10^n \to 0$ as $n \to +\infty$, it follows that

$$\lim_{n \to +\infty} s_n = \lim_{n \to +\infty} \frac{1}{3} \left(1 - \frac{1}{10^n} \right) = \frac{1}{3}$$

which we denote by writing

$$\frac{1}{3} = \frac{3}{10} + \frac{3}{10^2} + \frac{3}{10^3} + \cdots + \frac{3}{10^n} + \cdots$$

Motivated by the foregoing example, we are now ready to define the general concept of the "sum" of an infinite series

$$\sum_{k=1}^{\infty} u_k$$

We begin with some terminology: Let s_n denote the sum of the first n terms of the series. Thus,

$$s_1 = u_1$$

$$s_2 = u_1 + u_2$$

$$s_3 = u_1 + u_2 + u_3$$

$$\vdots$$

$$s_n = u_1 + u_2 + u_3 + \cdots + u_n = \sum_{k=1}^{n} u_k$$

The number s_n is called the **nth partial sum** of the series and the sequence $\{s_n\}_{n=1}^{+\infty}$ is called the **sequence of partial sums.**

WARNING.   In everyday English the words "sequence" and "series" are often used interchangeably. However, this is not so in mathematics—mathematically, a sequence is a *succession* and a series is a *sum*. It is essential that you keep this distinction in mind.

As n increases, the partial sum $s_n = u_1 + u_2 + \cdots + u_n$ includes more and more terms of the series. Thus, if s_n tends toward a limit as $n \to +\infty$, it is reasonable to view this limit as the sum of *all* the terms in the series. This suggests the following definition.

11.3.2 DEFINITION. Let $\{s_n\}$ be the sequence of partial sums of the series $\sum\limits_{k=1}^{\infty} u_k$. If the sequence $\{s_n\}$ converges to a limit S, then the series is said to *converge* and S is called the *sum* of the series. We denote this by writing

$$S = \sum_{k=1}^{\infty} u_k$$

If the sequence of partial sums diverges, then the series is said to *diverge*. A divergent series has no sum.

Example 1   Determine whether the series

$$1 - 1 + 1 - 1 + 1 - 1 + \cdots$$

converges or diverges. If it converges, find the sum.

Solution. It is tempting to conclude that the sum of the series is zero by arguing that the positive and negative terms cancel one another. However, this is *not correct;* the problem is that algebraic operations that hold for finite sums do not carry over to infinite series in all cases. Later, we shall discuss conditions under which familiar algebraic operations can be applied to infinite series, but for this example we turn directly to Definition 11.3.2. The partial sums are

$$s_1 = 1$$
$$s_2 = 1 - 1 = 0$$
$$s_3 = 1 - 1 + 1 = 1$$
$$s_4 = 1 - 1 + 1 - 1 = 0$$

and so forth. Thus, the sequence of partial sums is

$$1, 0, 1, 0, 1, 0, \ldots$$

Since this is a divergent sequence, the given series diverges and consequently has no sum.   ◀

☐ **GEOMETRIC SERIES**

The series encountered thus far are examples of *geometric series*. A geometric series is one of the form

$$a + ar + ar^2 + ar^3 + \cdots + ar^{k-1} + \cdots \quad (a \neq 0)$$

where each term is obtained by multiplying the previous one by a constant r. The multiplier r is called the *ratio* for the series. Some examples of geometric series are

$$1 + 2 + 4 + 8 + \cdots + 2^{k-1} + \cdots \qquad \boxed{a = 1, r = 2}$$

$$\frac{3}{10} + \frac{3}{10^2} + \frac{3}{10^3} + \cdots + \frac{3}{10^{k-1}} + \cdots \qquad \boxed{a = \frac{3}{10}, r = \frac{1}{10}}$$

$$\frac{1}{2} - \frac{1}{4} + \frac{1}{8} - \frac{1}{16} + \cdots + (-1)^{k+1}\frac{1}{2^k} + \cdots \qquad \boxed{a = \frac{1}{2}, r = -\frac{1}{2}}$$

$$1 + 1 + 1 + \cdots + 1 + \cdots \qquad \boxed{a = 1, r = 1}$$

$$1 - 1 + 1 - 1 + \cdots + (-1)^{k+1} + \cdots \qquad \boxed{a = 1, r = -1}$$

The following theorem is the fundamental result on convergence of geometric series.

11.3.3   THEOREM.   *A geometric series*

$$a + ar + ar^2 + \cdots + ar^{k-1} + \cdots \qquad (a \neq 0)$$

converges if $|r| < 1$ and diverges if $|r| \geq 1$. If the series converges, then the sum is

$$\frac{a}{1 - r} = a + ar + ar^2 + \cdots + ar^{k-1} + \cdots$$

Proof.   Let us treat the case $|r| = 1$ first. If $r = 1$, then the series is

$$a + a + a + \cdots + a + \cdots$$

so the nth partial sum is $s_n = na$ and $\lim_{n \to +\infty} s_n = \lim_{n \to +\infty} na = \pm\infty$ (the sign depending on whether a is positive or negative). This proves divergence. If $r = -1$, the series is

$$a - a + a - a + \cdots$$

so the sequence of partial sums is

$$a, 0, a, 0, a, 0, \ldots$$

which diverges.
Now let us consider the case where $|r| \neq 1$. The nth partial sum of the series is

$$s_n = a + ar + ar^2 + \cdots + ar^{n-1} \tag{5}$$

Multiplying both sides of (5) by r yields

$$rs_n = ar + ar^2 + \cdots + ar^{n-1} + ar^n \tag{6}$$

and subtracting (6) from (5) gives

$$s_n - rs_n = a - ar^n$$

or

$$(1 - r)s_n = a - ar^n \tag{7}$$

Since $r \neq 1$ in the case we are considering, this can be rewritten as

$$s_n = \frac{a - ar^n}{1 - r} = \frac{a}{1 - r} - \frac{ar^n}{1 - r} \tag{8}$$

If $|r| < 1$, then $\lim\limits_{n \to +\infty} r^n = 0$, so $\{s_n\}$ converges. From (8)

$$\lim_{n \to +\infty} s_n = \frac{a}{1 - r}$$

If $|r| > 1$, then either $r > 1$ or $r < -1$. In the case $r > 1$, $\lim\limits_{n \to +\infty} r^n = +\infty$, and in the case $r < -1$, r^n oscillates between positive and negative values that grow in magnitude, so $\{s_n\}$ diverges in both cases. ∎

Example 2 The series

$$5 + \frac{5}{4} + \frac{5}{4^2} + \cdots + \frac{5}{4^{k-1}} + \cdots$$

is a geometric series with $a = 5$ and $r = \frac{1}{4}$. Since $|r| = \frac{1}{4} < 1$, the series converges and the sum is

$$\frac{a}{1 - r} = \frac{5}{1 - \frac{1}{4}} = \frac{20}{3} \quad \blacktriangleleft$$

Example 3 Find the rational number represented by the repeating decimal

$$0.784784784 \ldots$$

Solution. We can write

$$0.784784784 \ldots = 0.784 + 0.000784 + 0.000000784 + \cdots$$

so the given decimal is the sum of a geometric series with $a = 0.784$ and $r = 0.001$. Thus,

$$0.784784784 \ldots = \frac{a}{1 - r} = \frac{0.784}{1 - 0.001} = \frac{0.784}{0.999} = \frac{784}{999} \quad \blacktriangleleft$$

Example 4 Determine whether the series

$$\sum_{k=1}^{\infty} \frac{1}{k(k + 1)} = \frac{1}{1 \cdot 2} + \frac{1}{2 \cdot 3} + \frac{1}{3 \cdot 4} + \frac{1}{4 \cdot 5} + \cdots$$

converges or diverges. If it converges, find the sum.

Solution. The *n*th partial sum of the series is

$$s_n = \sum_{k=1}^{n} \frac{1}{k(k+1)} = \frac{1}{1 \cdot 2} + \frac{1}{2 \cdot 3} + \frac{1}{3 \cdot 4} + \cdots + \frac{1}{n(n+1)}$$

To calculate $\lim_{n \to +\infty} s_n$ we shall rewrite s_n in closed form. This may be accomplished by using the method of partial fractions to obtain (verify)

$$\frac{1}{k(k+1)} = \frac{1}{k} - \frac{1}{k+1}$$

from which it follows that

$$s_n = \sum_{k=1}^{n} \left(\frac{1}{k} - \frac{1}{k+1} \right)$$

$$= \left(1 - \frac{1}{2} \right) + \left(\frac{1}{2} - \frac{1}{3} \right) + \left(\frac{1}{3} - \frac{1}{4} \right) + \cdots + \left(\frac{1}{n} - \frac{1}{n+1} \right)$$

$$= 1 + \left(-\frac{1}{2} + \frac{1}{2} \right) + \left(-\frac{1}{3} + \frac{1}{3} \right) + \cdots + \left(-\frac{1}{n} + \frac{1}{n} \right) - \frac{1}{n+1} \qquad (9)$$

The sum in (9) is an example of a ***telescoping sum,*** which means that each term cancels part of the next term, thereby collapsing the sum (like a folding telescope) into only two terms. After the cancellation, (9) can be written as

$$s_n = 1 - \frac{1}{n+1}$$

so

$$\lim_{n \to +\infty} s_n = \lim_{n \to +\infty} \left(1 - \frac{1}{n+1} \right) = 1$$

and therefore

$$1 = \sum_{k=1}^{\infty} \frac{1}{k(k+1)} \qquad \blacktriangleleft$$

☐ **HARMONIC SERIES**

One of the most famous and important of all diverging series is the ***harmonic series,***

$$\sum_{k=1}^{\infty} \frac{1}{k} = 1 + \frac{1}{2} + \frac{1}{3} + \frac{1}{4} + \frac{1}{5} + \cdots$$

which arises in connection with the overtones produced by a vibrating musical string. It is not immediately evident that this series diverges. However, the divergence will become apparent when we examine the partial sums in detail.

Because the terms in the series are all positive, the partial sums

$$s_1 = 1, \ s_2 = 1 + \frac{1}{2}, \ s_3 = 1 + \frac{1}{2} + \frac{1}{3}, \ s_4 = 1 + \frac{1}{2} + \frac{1}{3} + \frac{1}{4}, \dots$$

form an increasing sequence

$$s_1 < s_2 < s_3 < \cdots < s_n < \cdots$$

Thus, by Theorem 11.2.2 we can prove divergence by demonstrating that there is no constant M that is greater than or equal to *every* partial sum. To this end, we shall consider some selected partial sums, namely $s_2, s_4, s_8, s_{16}, s_{32}, \dots$. Note that the subscripts are successive powers of 2, so that these are the partial sums of the form s_{2^n}. These partial sums satisfy the inequalities

$$s_2 = 1 + \frac{1}{2} > \frac{1}{2} + \frac{1}{2} = \frac{2}{2}$$

$$s_4 = s_2 + \frac{1}{3} + \frac{1}{4} > s_2 + \left(\frac{1}{4} + \frac{1}{4}\right) = s_2 + \frac{1}{2} > \frac{3}{2}$$

$$s_8 = s_4 + \frac{1}{5} + \frac{1}{6} + \frac{1}{7} + \frac{1}{8} > s_4 + \left(\frac{1}{8} + \frac{1}{8} + \frac{1}{8} + \frac{1}{8}\right) = s_4 + \frac{1}{2} > \frac{4}{2}$$

$$s_{16} = s_8 + \frac{1}{9} + \frac{1}{10} + \frac{1}{11} + \frac{1}{12} + \frac{1}{13} + \frac{1}{14} + \frac{1}{15} + \frac{1}{16}$$

$$> s_8 + \left(\frac{1}{16} + \frac{1}{16} + \frac{1}{16} + \frac{1}{16} + \frac{1}{16} + \frac{1}{16} + \frac{1}{16} + \frac{1}{16}\right) = s_8 + \frac{1}{2} > \frac{5}{2}$$

$$\vdots$$

$$s_{2^n} > \frac{n+1}{2}$$

Now if M is any constant, we can certainly find a positive integer n such that $(n + 1)/2 > M$. But for this n

$$s_{2^n} > \frac{n+1}{2} > M$$

so that no constant M is greater than or equal to *every* partial sum of the harmonic series. This proves divergence.

▶ Exercise Set 11.3

1. In each part, find the first four partial sums; find a closed form for the nth partial sum; and determine whether the series converges (if so, give the sum).

    (a) $\displaystyle\sum_{k=1}^{\infty} \frac{2}{5^{k-1}}$ (b) $\displaystyle\sum_{k=1}^{\infty} \frac{1}{(k+1)(k+2)}$

    (c) $\displaystyle\sum_{k=1}^{\infty} \frac{2^{k-1}}{4}$.

In Exercises 2–16, determine whether the series converges or diverges. If it converges, find the sum.

2. $\displaystyle\sum_{k=1}^{\infty} \frac{1}{5^k}$.

3. $\displaystyle\sum_{k=1}^{\infty} \left(-\frac{3}{4}\right)^{k-1}$.

4. $\displaystyle\sum_{k=1}^{\infty} \left(\frac{2}{3}\right)^{k+2}$.

5. $\displaystyle\sum_{k=1}^{\infty} (-1)^{k-1} \frac{7}{6^{k-1}}$.

6. $\displaystyle\sum_{k=1}^{\infty} 4^{k-1}.$

7. $\displaystyle\sum_{k=1}^{\infty} \left(-\frac{3}{2}\right)^{k+1}.$

8. $\displaystyle\sum_{k=1}^{\infty} \left(\frac{1}{k+3} - \frac{1}{k+4}\right).$

9. $\displaystyle\sum_{k=1}^{\infty} \frac{1}{(k+2)(k+3)}.$

10. $\displaystyle\sum_{k=1}^{\infty} \left(\frac{1}{2^k} - \frac{1}{2^{k+1}}\right).$

11. $\displaystyle\sum_{k=1}^{\infty} \frac{1}{9k^2 + 3k - 2}.$

12. $\displaystyle\sum_{k=2}^{\infty} \frac{1}{k^2 - 1}.$

13. $\displaystyle\sum_{k=1}^{\infty} \frac{4^{k+2}}{7^{k-1}}.$

14. $\displaystyle\sum_{k=1}^{\infty} (e/\pi)^{k-1}.$

15. $\displaystyle\sum_{k=1}^{\infty} (-1/2)^k.$

16. $\displaystyle\sum_{k=3}^{\infty} \frac{5}{k-2}.$

In Exercises 17–22, express the repeating decimal as a fraction.

17. $0.4444\ldots.$

18. $0.9999\ldots.$

19. $5.373737\ldots.$

20. $0.159159159\ldots.$

21. $0.782178217821\ldots.$

22. $0.451141414\ldots.$

23. Find a closed form for the nth partial sum of the series

$$\ln\frac{1}{2} + \ln\frac{2}{3} + \ln\frac{3}{4} + \cdots + \ln\frac{n}{n+1} + \cdots$$

and determine whether the series converges.

24. A ball is dropped from a height of 10 m. Each time it strikes the ground it bounces vertically to a height that is $\frac{3}{4}$ of the previous height. Find the total distance the ball will travel if it is allowed to bounce indefinitely.

25. Show: $\displaystyle\sum_{k=2}^{\infty} \ln(1 - 1/k^2) = -\ln 2.$

26. Show: $\displaystyle\sum_{k=1}^{\infty} \frac{\sqrt{k+1} - \sqrt{k}}{\sqrt{k^2 + k}} = 1.$

27. Show: $\displaystyle\sum_{k=1}^{\infty} \left(\frac{1}{k} - \frac{1}{k+2}\right) = \frac{3}{2}.$

28. (a) Find A and B such that

$$\frac{6^k}{(3^{k+1} - 2^{k+1})(3^k - 2^k)} = \frac{2^k A}{3^k - 2^k} + \frac{2^k B}{3^{k+1} - 2^{k+1}}$$

(b) Use the result in part (a) to help show that

$$\sum_{k=1}^{\infty} \frac{6^k}{(3^{k+1} - 2^{k+1})(3^k - 2^k)} = 2$$

[This problem appeared in the Forty-Fifth Annual William Lowell Putnam Mathematical Competition.]

29. Show: $\dfrac{1}{1 \cdot 3} + \dfrac{1}{3 \cdot 5} + \dfrac{1}{5 \cdot 7} + \cdots = \dfrac{1}{2}.$

30. Show: $\dfrac{1}{1 \cdot 3} + \dfrac{1}{2 \cdot 4} + \dfrac{1}{3 \cdot 5} + \cdots = \dfrac{3}{4}.$

31. Use geometric series to show that

(a) $\displaystyle\sum_{k=0}^{\infty} (-1)^k x^k = \frac{1}{1+x}$ if $-1 < x < 1$

(b) $\displaystyle\sum_{k=0}^{\infty} (x-3)^k = \frac{1}{4-x}$ if $2 < x < 4$

(c) $\displaystyle\sum_{k=0}^{\infty} (-1)^k x^{2k} = \frac{1}{1+x^2}$ if $-1 < x < 1.$

In Exercises 32–35, find all values of x for which the series converges, and for these values find its sum.

32. $x - x^3 + x^5 - x^7 + x^9 - \cdots.$

33. $\dfrac{1}{x^2} + \dfrac{2}{x^3} + \dfrac{4}{x^4} + \dfrac{8}{x^5} + \dfrac{16}{x^6} + \cdots.$

34. $e^{-x} + e^{-2x} + e^{-3x} + e^{-4x} + e^{-5x} + \cdots.$

35. $\sin x - \frac{1}{2}\sin^2 x + \frac{1}{4}\sin^3 x - \frac{1}{8}\sin^4 x + \cdots.$

36. Prove the following decimal equality assuming that $a_n \neq 9$:

$$0.a_1 a_2 \ldots a_n 9999\ldots = 0.a_1 a_2 \ldots (a_n + 1)0000\ldots$$

37. Let a_1 be any real number and define

$$a_{n+1} = \tfrac{1}{2}(a_n + 1) \text{ for } n = 1, 2, 3, \ldots$$

Show that the sequence $\{a_n\}_{n=1}^{+\infty}$ converges and find its limit. [*Hint:* Express a_n in terms of a_1.]

38. Lines L_1 and L_2 form an angle θ, $0 < \theta < \pi/2$, at their point of intersection P (Figure 11.3.1). A point P_0 is chosen that is on L_1 and a units from P. Starting from P_0 a zig-zag path is constructed by successively going back and forth between L_1 and L_2 along a perpendicular from one line to the other. Find the following sums in terms of θ:

(a) $P_0 P_1 + P_1 P_2 + P_2 P_3 + \cdots$

(b) $P_0 P_1 + P_2 P_3 + P_4 P_5 + \cdots$

(c) $P_1 P_2 + P_3 P_4 + P_5 P_6 + \cdots.$

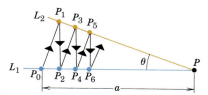

Figure 11.3.1

39. As shown in Figure 11.3.2, suppose that an angle θ is bisected using a straightedge and compass to produce ray R_1, then the angle between R_1 and the initial side is bisected to produce ray R_2. Thereafter, rays $R_3, R_4, R_5, \ldots$ are constructed in succession by bisecting the angle between the previous two rays. Show that the sequence of angles that these rays make with the initial side has a limit of $\theta/3$. [This problem is based on *Trisection of an Angle in an Infinite Number of Steps* by Eric Kincannon, which appeared in *The College Mathematics Journal*, Vol. 21, No. 5, November 1990.]

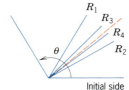

Initial side   **Figure 11.3.2**

40. In his *Treatise on the Configurations of Qualities and Motions* (written in the 1350s), the French Bishop of Lisieux, Nicole Oresme, used a geometric method to show that

$$\sum_{k=1}^{\infty} \frac{k}{2^k} = \frac{1}{2} + \frac{2}{4} + \frac{3}{8} + \frac{4}{16} + \cdots = 2$$

In Figure 11.3.3a each term in the series is represented by the area of a rectangle. In Figure 11.3.3b the configuration in part (a) has been divided into rectangles with areas $A_1, A_2, A_3, \ldots$. Show that $A_1 + A_2 + A_3 + \cdots = 2$. [For convenience, the horizontal and vertical scales in Figure 11.3.3 are different.]

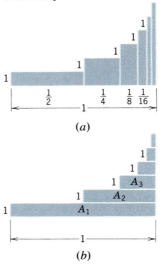

(a)

(b)   **Figure 11.3.3**

41. The great Swiss mathematician, Leonhard Euler, (biography on p. 71) made occasional errors in his pioneering work on infinite series. For example, Euler deduced that

$$\tfrac{1}{2} = 1 - 1 + 1 - 1 + \cdots$$

and

$$-1 = 1 + 2 + 4 + 8 + \cdots$$

by substituting $x = -1$ and $x = 2$ in the formula

$$\frac{1}{1 - x} = 1 + x + x^2 + x^3 + \cdots$$

What was the error in his reasoning?

11.4 CONVERGENCE; THE INTEGRAL TEST

In the previous section we found sums of series and investigated convergence by first writing the nth partial sum s_n in closed form and then examining its limit. However, it is relatively rare that the nth partial sum of a series can be written in closed form; for most series, convergence or divergence is determined by using convergence tests, some of which we shall introduce in this section. Once it is established that a series converges, the sum of the series can generally be approximated to any degree of accuracy by a partial sum with sufficiently many terms.

☐ **THE DIVERGENCE TEST**

Our first theorem states that the terms of an infinite series must tend toward zero if the series is to converge.

11.4.1 THEOREM. *If the series Σu_k converges, then $\lim\limits_{k \to +\infty} u_k = 0$.*

Proof. The term u_k can be written as

$$u_k = s_k - s_{k-1} \tag{1}$$

where s_k is the sum of the first k terms and s_{k-1} is the sum of the first $k - 1$ terms. If S denotes the sum of the series, then $\lim\limits_{k \to +\infty} s_k = S$, and since $(k - 1) \to +\infty$ as $k \to +\infty$, we also have $\lim\limits_{k \to +\infty} s_{k-1} = S$. Thus, from (1)

$$\lim_{k \to +\infty} u_k = \lim_{k \to +\infty} (s_k - s_{k-1}) = S - S = 0 \quad ■$$

The following result is just an alternative phrasing of the above theorem and needs no additional proof.

11.4.2 THEOREM (*The Divergence Test*). *If $\lim\limits_{k \to +\infty} u_k \neq 0$, then the series Σu_k diverges.*

Example 1 The series

$$\sum_{k=1}^{\infty} \frac{k}{k + 1} = \frac{1}{2} + \frac{2}{3} + \frac{3}{4} + \cdots + \frac{k}{k + 1} + \cdots$$

diverges since

$$\lim_{k \to +\infty} \frac{k}{k + 1} = \lim_{k \to +\infty} \frac{1}{1 + 1/k} = 1 \neq 0 \quad ◀$$

WARNING. The converse of Theorem 11.4.1 is false. To prove that a series converges it does not suffice to show that $\lim\limits_{k \to +\infty} u_k = 0$, since this property may hold for divergent as well as convergent series. For example, the kth term of the divergent harmonic series $1 + 1/2 + 1/3 + \cdots + 1/k + \cdots$ approaches zero as $k \to +\infty$, and the kth term of the convergent geometric series $1/2 + 1/2^2 + \cdots + 1/2^k + \cdots$ tends to zero as $k \to +\infty$.

□ **ALGEBRAIC PROPERTIES OF INFINITE SERIES**

For brevity, the proof of the following result is left for the exercises.

11.4.3 THEOREM.

(a) *If Σu_k and Σv_k are convergent series, then $\Sigma(u_k + v_k)$ and $\Sigma(u_k - v_k)$ are convergent series and the sums of these series are related by*

$$\sum_{k=1}^{\infty} (u_k + v_k) = \sum_{k=1}^{\infty} u_k + \sum_{k=1}^{\infty} v_k$$

$$\sum_{k=1}^{\infty} (u_k - v_k) = \sum_{k=1}^{\infty} u_k - \sum_{k=1}^{\infty} v_k$$

(b) *If c is a nonzero constant, then the series Σu_k and $\Sigma c u_k$ both converge or both diverge. In the case of convergence, the sums are related by*

$$\sum_{k=1}^{\infty} c u_k = c \sum_{k=1}^{\infty} u_k$$

(c) *Convergence or divergence is unaffected by deleting a finite number of terms from the beginning of a series; that is, for any positive integer K, the series*

$$\sum_{k=1}^{\infty} u_k = u_1 + u_2 + u_3 + \cdots$$

and

$$\sum_{k=K}^{\infty} u_k = u_K + u_{K+1} + u_{K+2} + \cdots$$

both converge or both diverge.

REMARK. Do not read too much into part (c) of this theorem. Although the convergence is not affected when a finite number of terms is deleted from the beginning of a convergent series, the *sum* of the series is changed by the removal of these terms.

Example 2 Find the sum of the series

$$\sum_{k=1}^{\infty} \left(\frac{3}{4^k} - \frac{2}{5^{k-1}} \right)$$

Solution. The series

$$\sum_{k=1}^{\infty} \frac{3}{4^k} = \frac{3}{4} + \frac{3}{4^2} + \frac{3}{4^3} + \cdots$$

is a convergent geometric series ($a = \frac{3}{4}, r = \frac{1}{4}$), and the series

$$\sum_{k=1}^{\infty} \frac{2}{5^{k-1}} = 2 + \frac{2}{5} + \frac{2}{5^2} + \frac{2}{5^3} + \cdots$$

is also a convergent geometric series ($a = 2, r = \frac{1}{5}$). Thus, from Theorems 11.4.3(a) and 11.3.3 the given series converges and

$$\sum_{k=1}^{\infty} \left(\frac{3}{4^k} - \frac{2}{5^{k-1}} \right) = \sum_{k=1}^{\infty} \frac{3}{4^k} - \sum_{k=1}^{\infty} \frac{2}{5^{k-1}} = \frac{\frac{3}{4}}{1 - \frac{1}{4}} - \frac{2}{1 - \frac{1}{5}}$$

$$= 1 - \frac{5}{2} = -\frac{3}{2} \quad \blacktriangleleft$$

Example 3 The series

$$\sum_{k=1}^{\infty} \frac{5}{k} = 5 + \frac{5}{2} + \frac{5}{3} + \cdots + \frac{5}{k} + \cdots$$

diverges by part (b) of Theorem 11.4.3, since

$$\sum_{k=1}^{\infty} \frac{5}{k} = \sum_{k=1}^{\infty} 5 \left(\frac{1}{k} \right)$$

which shows that each term is a constant times the corresponding term of the divergent harmonic series. $\blacktriangleleft$

Example 4 The series

$$\sum_{k=10}^{\infty} \frac{1}{k} = \frac{1}{10} + \frac{1}{11} + \frac{1}{12} + \cdots$$

diverges by part (c) of Theorem 11.4.3, since this series results by deleting the first nine terms from the divergent harmonic series. $\blacktriangleleft$

□ **CONVERGENCE TESTS**

If an infinite series $u_1 + u_2 + u_3 + \cdots + u_k + \cdots$ has *nonnegative terms*, then the partial sums $s_1 = u_1$, $s_2 = u_1 + u_2$, $s_3 = u_1 + u_2 + u_3, \ldots$ form a nondecreasing sequence, that is,

$$s_1 \leq s_2 \leq s_3 \leq \cdots \leq s_n \leq \cdots$$

If there is a finite constant M such that $s_n \leq M$ for all n, then according to Theorem 11.2.2, the sequence of partial sums will converge to a limit S, satisfying $S \leq M$. If no such constant exists, then $\lim\limits_{n \to +\infty} s_n = +\infty$. This yields the following theorem.

11.4.4 THEOREM. *If Σu_k is a series with nonnegative terms, and if there is a constant M such that*

$$s_n = u_1 + u_2 + \cdots + u_n \leq M$$

for every n, then the series converges and the sum S satisfies $S \leq M$. If no such M exists, then the series diverges.

☐ **THE INTEGRAL TEST**

If we have a series with positive terms, say

$$\sum_{k=1}^{\infty} \frac{1}{k^2}$$

and if we form the improper integral

$$\int_{1}^{+\infty} \frac{1}{x^2}\, dx$$

whose integrand is obtained by replacing the summation index k by x, then there is a relationship between convergence of the series and convergence of the improper integral.

11.4.5 THEOREM (*The Integral Test*). *Let Σu_k be a series with positive terms, and let $f(x)$ be the function that results when k is replaced by x in the formula for u_k. If f is decreasing and continuous for $x \geq 1$, then*

$$\sum_{k=1}^{\infty} u_k \quad and \quad \int_{1}^{+\infty} f(x)\, dx$$

both converge or both diverge.

We shall defer the proof to the end of the section and proceed with some examples.

Example 5 Determine whether

$$\sum_{k=1}^{\infty} \frac{1}{k^2}$$

converges or diverges.

Solution. If we replace k by x in the formula for u_k, we obtain the function

$$f(x) = \frac{1}{x^2}$$

which satisfies the hypotheses of the integral test. (Verify.) Since

$$\int_{1}^{+\infty} \frac{1}{x^2}\, dx = \lim_{l \to +\infty} \int_{1}^{l} \frac{dx}{x^2} = \lim_{l \to +\infty} \left[-\frac{1}{x} \right]_{1}^{l} = \lim_{l \to +\infty} \left[1 - \frac{1}{l} \right] = 1$$

the integral converges and consequently the series converges. ◄

REMARK. In the above example, do *not* erroneously conclude that $\sum_{k=1}^{\infty} \frac{1}{k^2} = 1$ from the fact that $\int_{1}^{+\infty} \frac{1}{x^2}\, dx = 1$. (In advanced texts it is proved that the sum of this series is $\pi^2/6$; and, in fact, the sum of the first two terms alone exceeds 1.)

Example 6 The integral test provides another way to demonstrate divergence of the harmonic series $\sum\limits_{k=1}^{\infty} \dfrac{1}{k}$. If we replace k by x in the formula for u_k, we obtain the function $f(x) = 1/x$, which satisfies the hypotheses of the integral test. (Verify.) Since

$$\int_1^{+\infty} \frac{1}{x} \, dx = \lim_{l \to +\infty} \int_1^l \frac{1}{x} \, dx = \lim_{l \to +\infty} [\ln l - \ln 1] = +\infty$$

the integral diverges and consequently so does the series.   ◀

Example 7 Determine whether the series

$$\frac{1}{e} + \frac{2}{e^4} + \frac{3}{e^9} + \cdots + \frac{k}{e^{k^2}} + \cdots$$

converges or diverges.

Solution. If we replace k by x in the formula for u_k, we obtain the function

$$f(x) = \frac{x}{e^{x^2}} = xe^{-x^2}$$

For $x \geq 1$, this function has positive values and is continuous. Moreover, for $x \geq 1$ the derivative

$$f'(x) = e^{-x^2} - 2x^2 e^{-x^2} = e^{-x^2}(1 - 2x^2)$$

is negative, so that f is decreasing for $x \geq 1$. Thus, the hypotheses of the integral test are met. But

$$\int_1^{+\infty} xe^{-x^2} \, dx = \lim_{l \to +\infty} \int_1^l xe^{-x^2} \, dx = \lim_{l \to +\infty} \left[-\frac{1}{2} e^{-x^2} \right]_1^l$$

$$= \left(-\frac{1}{2} \right) \lim_{l \to +\infty} \left[e^{-l^2} - e^{-1} \right] = \frac{1}{2e}$$

Thus, the improper integral and the series converge.   ◀

☐ *p*-SERIES

The harmonic series and the series in Example 5 are special cases of a class of series called *p-series* or *hyperharmonic series*. A *p*-series is an infinite series of the form

$$\sum_{k=1}^{\infty} \frac{1}{k^p} = 1 + \frac{1}{2^p} + \frac{1}{3^p} + \cdots + \frac{1}{k^p} + \cdots$$

where $p > 0$. Examples of *p*-series are

$$\sum_{k=1}^{\infty} \frac{1}{k} = 1 + \frac{1}{2} + \frac{1}{3} + \cdots + \frac{1}{k} + \cdots \qquad \boxed{p = 1}$$

$$\sum_{k=1}^{\infty} \frac{1}{k^2} = 1 + \frac{1}{2^2} + \frac{1}{3^2} + \cdots + \frac{1}{k^2} + \cdots \qquad \boxed{p = 2}$$

$$\sum_{k=1}^{\infty} \frac{1}{\sqrt{k}} = 1 + \frac{1}{\sqrt{2}} + \frac{1}{\sqrt{3}} + \cdots + \frac{1}{\sqrt{k}} + \cdots \qquad \boxed{p = \tfrac{1}{2}}$$

The following theorem tells when a *p*-series converges.

11.4.6   THEOREM (*Convergence of p-Series*).

$$\sum_{k=1}^{\infty} \frac{1}{k^p} = 1 + \frac{1}{2^p} + \frac{1}{3^p} + \cdots + \frac{1}{k^p} + \cdots$$

converges if $p > 1$ *and diverges if* $0 < p \leq 1$.

Proof.   To establish this result when $p \neq 1$, we shall use the integral test.

$$\int_{1}^{+\infty} \frac{1}{x^p} \, dx = \lim_{l \to +\infty} \int_{1}^{l} x^{-p} \, dx$$

$$= \lim_{l \to +\infty} \frac{x^{1-p}}{1 - p} \Big]_{1}^{l}$$

$$= \lim_{l \to +\infty} \left[\frac{l^{1-p}}{1 - p} - \frac{1}{1 - p} \right]$$

If $p > 1$, then $1 - p < 0$, so $l^{1-p} \to 0$ as $l \to +\infty$. Thus, the integral converges [its value is $-1/(1 - p)$] and consequently the series also converges. For $0 < p < 1$, it follows that $1 - p > 0$ and $l^{1-p} \to +\infty$ as $l \to +\infty$, so the integral and the series diverge. The case $p = 1$ is the harmonic series, which was previously shown to diverge.   ∎

Example 8

$$1 + \frac{1}{\sqrt[3]{2}} + \frac{1}{\sqrt[3]{3}} + \cdots + \frac{1}{\sqrt[3]{k}} + \cdots$$

diverges since it is a *p*-series with $p = \tfrac{1}{3} < 1$.   ◀

□ **PROOF OF THE INTEGRAL TEST**

We conclude this section by proving Theorem 11.4.5 (the integral test).

Proof.   Let $f(x)$ satisfy the hypotheses of the theorem. Since

$$f(1) = u_1, \; f(2) = u_2, \ldots, f(n) = u_n, \ldots$$

the rectangles in Figures 11.4.1*a* and 11.4.1*b* have areas $u_1, u_2, \ldots, u_n$ as

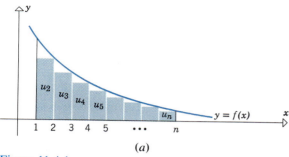
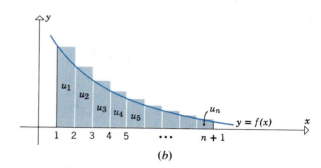

(a)   (b)

Figure 11.4.1

indicated. In Figure 11.4.1a, the total area of the rectangles is less than the area under the curve from $x = 1$ to $x = n$, so

$$u_2 + u_3 + \cdots + u_n < \int_1^n f(x)\,dx$$

Therefore,

$$u_1 + u_2 + u_3 + \cdots + u_n < u_1 + \int_1^n f(x)\,dx \tag{2}$$

In Figure 11.4.1b, the total area of the rectangles is greater than the area under the curve from $x = 1$ to $x = n + 1$, so that

$$\int_1^{n+1} f(x)\,dx < u_1 + u_2 + \cdots + u_n \tag{3}$$

If we let $s_n = u_1 + u_2 + \cdots + u_n$ be the nth partial sum of the series, then (2) and (3) yield

$$\int_1^{n+1} f(x)\,dx < s_n < u_1 + \int_1^n f(x)\,dx \tag{4}$$

If the integral $\int_1^{+\infty} f(x)\,dx$ converges to a finite value L, then from the right-hand inequality in (4)

$$s_n < u_1 + \int_1^n f(x)\,dx < u_1 + \int_1^{+\infty} f(x)\,dx = u_1 + L$$

Thus, each partial sum is less than the finite constant $u_1 + L$, and the series converges by Theorem 11.4.4. On the other hand, if the integral $\int_1^{+\infty} f(x)\,dx$ diverges, then

$$\lim_{n \to +\infty} \int_1^{n+1} f(x)\,dx = +\infty$$

so that from the left-hand inequality in (4), $\lim_{n \to +\infty} s_n = +\infty$. This means that the series also diverges. ∎

REMARK. If the summation index in a series Σu_k does not begin with $k = 1$, a variation of the integral test may still apply. It can be shown that

$$\sum_{k=K}^{\infty} u_k \quad \text{and} \quad \int_{K}^{+\infty} f(x)\, dx$$

both converge or both diverge provided the hypotheses of Theorem 11.4.5 hold for $x \geq K$.

▶ Exercise Set 11.4   C 37, 38, 39, 40

In Exercises 1–4, use Theorem 11.4.3 to find the sum of the series.

1. $\displaystyle\sum_{k=1}^{\infty} \left[\frac{1}{2^k} + \frac{1}{4^k} \right].$

2. $\displaystyle\sum_{k=1}^{\infty} \left[\frac{1}{5^k} - \frac{1}{k(k+1)} \right].$

3. $\displaystyle\sum_{k=2}^{\infty} \left[\frac{1}{k^2-1} - \frac{7}{10^{k-1}} \right].$

4. $\displaystyle\sum_{k=1}^{\infty} \left[\frac{7}{3^k} + \frac{6}{(k+3)(k+4)} \right].$

5. In each part, determine whether the given p-series converges or diverges.

(a) $\displaystyle\sum_{k=1}^{\infty} \frac{1}{k^3}$

(b) $\displaystyle\sum_{k=1}^{\infty} \frac{1}{\sqrt{k}}$

(c) $\displaystyle\sum_{k=1}^{\infty} k^{-1}$

(d) $\displaystyle\sum_{k=1}^{\infty} k^{-2/3}$

(e) $\displaystyle\sum_{k=1}^{\infty} k^{-4/3}$

(f) $\displaystyle\sum_{k=1}^{\infty} \frac{1}{\sqrt[4]{k}}$

(g) $\displaystyle\sum_{k=1}^{\infty} \frac{1}{\sqrt[3]{k^5}}$

(h) $\displaystyle\sum_{k=1}^{\infty} \frac{1}{k^{\pi}}.$

In Exercises 6–8, use the divergence test to show that the series diverges.

6. (a) $\displaystyle\sum_{k=1}^{\infty} \frac{k+1}{k+2}$

(b) $\displaystyle\sum_{k=1}^{\infty} \ln k.$

7. (a) $\displaystyle\sum_{k=1}^{\infty} \frac{k^2+k+3}{2k^2+1}$

(b) $\displaystyle\sum_{k=1}^{\infty} \left(1 + \frac{1}{k} \right)^k.$

8. (a) $\displaystyle\sum_{k=1}^{\infty} \cos k\pi$

(b) $\displaystyle\sum_{k=1}^{\infty} \frac{e^k}{k}.$

In Exercises 9–30, determine whether the series converges or diverges.

9. $\displaystyle\sum_{k=1}^{\infty} \frac{1}{k+6}.$

10. $\displaystyle\sum_{k=1}^{\infty} \frac{3}{5k}.$

11. $\displaystyle\sum_{k=1}^{\infty} \frac{1}{5k+2}.$

12. $\displaystyle\sum_{k=1}^{\infty} \frac{k}{1+k^2}.$

13. $\displaystyle\sum_{k=1}^{\infty} \frac{1}{1+9k^2}.$

14. $\displaystyle\sum_{k=1}^{\infty} \frac{1}{(4+2k)^{3/2}}.$

15. $\displaystyle\sum_{k=1}^{\infty} \frac{1}{\sqrt{k+5}}.$

16. $\displaystyle\sum_{k=1}^{\infty} \frac{1}{\sqrt[k]{e}}.$

17. $\displaystyle\sum_{k=1}^{\infty} \frac{1}{\sqrt[3]{2k-1}}.$

18. $\displaystyle\sum_{k=3}^{\infty} \frac{\ln k}{k}.$

19. $\displaystyle\sum_{k=1}^{\infty} \frac{k}{\ln(k+1)}.$

20. $\displaystyle\sum_{k=1}^{\infty} k e^{-k^2}.$

21. $\displaystyle\sum_{k=1}^{\infty} \frac{1}{(k+1)[\ln(k+1)]^2}.$

22. $\displaystyle\sum_{k=1}^{\infty} \frac{k^2+1}{k^2+3}.$

23. $\displaystyle\sum_{k=1}^{\infty} \left(1 + \frac{1}{k} \right)^{-k}.$

24. $\displaystyle\sum_{k=1}^{\infty} \frac{1}{\sqrt{k^2+1}}.$

25. $\displaystyle\sum_{k=1}^{\infty} \frac{\tan^{-1} k}{1+k^2}.$

26. $\displaystyle\sum_{k=1}^{\infty} \operatorname{sech}^2 k.$

27. $\displaystyle\sum_{k=5}^{\infty} 7k^{-p} \quad (p > 1).$

28. $\displaystyle\sum_{k=1}^{\infty} 7(k+5)^{-p} \quad (p \leq 1).$

29. $\displaystyle\sum_{k=1}^{\infty} k^2 \sin^2 \left(\frac{1}{k} \right).$

30. $\displaystyle\sum_{k=1}^{\infty} k^2 e^{-k^3}.$

31. Prove: $\displaystyle\sum_{k=2}^{\infty} \frac{1}{k(\ln k)^p}$ converges if $p > 1$ and diverges if $p \leq 1$.

32. Prove: $\displaystyle\sum_{k=3}^{\infty} \frac{1}{k(\ln k)[\ln(\ln k)]^p}$ converges if $p > 1$ and diverges if $p \leq 1$.

33. Prove: If Σu_k converges and Σv_k diverges, then $\Sigma(u_k + v_k)$ diverges and $\Sigma(u_k - v_k)$ diverges. [*Hint:* Assume that $\Sigma(u_k + v_k)$ converges and use Theorem

11.4.3 to obtain a contradiction. Similarly, for $\Sigma(u_k - v_k)$.]

34. Find examples to show that $\Sigma(u_k + v_k)$ and $\Sigma(u_k - v_k)$ may converge or may diverge if Σu_k and Σv_k both diverge.

35. With the help of Exercise 33, determine whether the given series in parts (a)–(d) converge or diverge.

(a) $\displaystyle\sum_{k=1}^{\infty} \left[\left(\frac{2}{3}\right)^{k-1} + \frac{1}{k} \right]$

(b) $\displaystyle\sum_{k=1}^{\infty} \left[\frac{k^2}{1 + k^2} + \frac{1}{k(k + 1)} \right]$

(c) $\displaystyle\sum_{k=1}^{\infty} \left[\frac{1}{3k + 2} + \frac{1}{k^{3/2}} \right]$

(d) $\displaystyle\sum_{k=2}^{\infty} \left[\frac{1}{k(\ln k)^2} - \frac{1}{k^2} \right]$.

36. If the sum S of a convergent series $\Sigma_{k=1}^{\infty} u_k$ of positive terms is approximated by its nth partial sum s_n, then the error in the approximation is $S - s_n$. Let $f(x)$ be the function that results when k is replaced by x in the formula for u_k. Show that if f is decreasing for $x \geq n$, then

$$\int_{n+1}^{+\infty} f(x)\, dx < S - s_n < \int_{n}^{+\infty} f(x)\, dx$$

The result in Exercise 36 can be written as

$$s_n + \int_{n+1}^{+\infty} f(x)\, dx < S < s_n + \int_{n}^{+\infty} f(x)\, dx$$

which provides an upper and lower bound on the sum S of the series. Use this result in part (a) of Exercises 37 and 38.

37. (a) Find the partial sum s_{10} of the series $\Sigma_{k=1}^{\infty} 1/k^3$, and use it to obtain an upper and lower bound on the sum S of the series. Express your results to four decimal places.

(b) Use the right-hand inequality in Exercise 36 to find a value of n to ensure that the error in approximating S by s_n is less than 10^{-3}.

38. (a) Find the partial sum s_6 of the series $\Sigma_{k=1}^{\infty} 1/k^4$, and use it to obtain an upper and lower bound on the sum S of the series. Express your results to five decimal places. [*Note:* Compare your bounds with the exact value of S, which is $\pi^4/90$.]

(b) Use the right-hand inequality in Exercise 36 to find a value of n to ensure that the error in approximating S by s_n is less than 10^{-5}.

39. Let s_n be the nth partial sum of the divergent series $\Sigma_{k=1}^{\infty} 1/k$.

(a) Use inequality (4) to show that for $n \geq 2$

$$\ln(n + 1) < s_n < 1 + \ln n$$

Then find integer upper and lower bounds for $s_{1,000,000}$.

(b) Find a value of n to ensure that $s_n > 100$.

40. Let s_n be the nth partial sum of the divergent series $\Sigma_{k=1}^{\infty} 1/\sqrt{k}$.

(a) Use inequality (4) to show that for $n \geq 2$

$$2\sqrt{n + 1} - 2 < s_n < 2\sqrt{n} - 1$$

Then find integer upper and lower bounds for $s_{10,000}$.

(b) Find a value of n to ensure that $s_n > 100$.

11.5 ADDITIONAL CONVERGENCE TESTS

In this section we shall develop some additional convergence tests for series with positive terms.

□ **THE COMPARISON TEST**

Our first result is not only a practical test for convergence, but is also a theoretical tool that will be used to develop other convergence tests.

11.5.1   THEOREM (*The Comparison Test*).   *Let Σa_k and Σb_k be series with nonnegative terms and suppose*

$$a_1 \le b_1, a_2 \le b_2, a_3 \le b_3, \ldots, a_k \le b_k, \ldots$$

(a)   *If the "bigger series" Σb_k converges, then the "smaller series" Σa_k also converges.*

(b)   *On the other hand, if the "smaller series" Σa_k diverges, then the "bigger series" Σb_k also diverges.*

Proof (a).   Suppose that the series Σb_k converges and its sum is B. Then for all n

$$b_1 + b_2 + \cdots + b_n \le \sum_{k=1}^{\infty} b_k = B$$

From our hypothesis it follows that

$$a_1 + a_2 + \cdots + a_n \le b_1 + b_2 + \cdots + b_n$$

so that

$$a_1 + a_2 + \cdots + a_n \le B$$

Thus, each partial sum of the series Σa_k is less than or equal to B, so that Σa_k converges by Theorem 11.4.4.

Proof (b).   This part is really just an alternative phrasing of part (a). If Σa_k diverges, then Σb_k must diverge since convergence of Σb_k would imply convergence of Σa_k, contrary to the hypothesis.   ▮

☐ **THE RATIO TEST**

Since the comparison test requires a little ingenuity to use, we shall wait until the next section before applying it. For now, we shall use the comparison test to develop some other tests that are easier to apply.

11.5.2   THEOREM (*The Ratio Test*).   *Let Σu_k be a series with positive terms and suppose*

$$\lim_{k \to +\infty} \frac{u_{k+1}}{u_k} = \rho$$

(a)   *If $\rho < 1$, the series converges.*

(b)   *If $\rho > 1$ or $\rho = +\infty$, the series diverges.*

(c)   *If $\rho = 1$, the series may converge or diverge, so that another test must be tried.*

Proof (a). Assume that $\rho < 1$, and let $r = \frac{1}{2}(1 + \rho)$. Thus, $\rho < r < 1$, since r is the midpoint between 1 and ρ. It follows that the number

$$\epsilon = r - \rho \tag{1}$$

is positive. Since

$$\rho = \lim_{k \to +\infty} \frac{u_{k+1}}{u_k}$$

it follows that for k sufficiently large, say $k \geq K$, the ratios u_{k+1}/u_k are within ϵ units of ρ. Thus, we will have

$$\frac{u_{k+1}}{u_k} < \rho + \epsilon \quad \text{when} \quad k \geq K$$

or on substituting (1)

$$\frac{u_{k+1}}{u_k} < r \quad \text{when} \quad k \geq K$$

that is,

$$u_{k+1} < r u_k \quad \text{when} \quad k \geq K$$

This yields the inequalities

$$u_{K+1} < r u_K$$
$$u_{K+2} < r u_{K+1} < r^2 u_K$$
$$u_{K+3} < r u_{K+2} < r^3 u_K$$
$$u_{K+4} < r u_{K+3} < r^4 u_K \tag{2}$$
$$\vdots$$

But $|r| < 1$ (why?), so

$$r u_K + r^2 u_K + r^3 u_K + \cdots$$

is a convergent geometric series. From the inequalities in (2) and the comparison test it follows that

$$u_{K+1} + u_{K+2} + u_{K+3} + \cdots$$

must also be a convergent series. Thus, $u_1 + u_2 + u_3 + \cdots + u_k + \cdots$ converges by Theorem 11.4.3(c).

Proof (b). Assume that $\rho > 1$. Thus,

$$\epsilon = \rho - 1 \tag{3}$$

is a positive number. Since

$$\rho = \lim_{k \to +\infty} \frac{u_{k+1}}{u_k}$$

it follows that for k sufficiently large, say $k \geq K$, the ratio u_{k+1}/u_k is within ϵ units of ρ. Thus,

$$\frac{u_{k+1}}{u_k} > \rho - \epsilon \quad \text{when} \quad k \geq K$$

or on substituting (3)

$$\frac{u_{k+1}}{u_k} > 1 \quad \text{when} \quad k \geq K$$

that is,

$$u_{k+1} > u_k \quad \text{when} \quad k \geq K$$

This yields the inequalities

$$
\begin{aligned}
u_{K+1} &> u_K \\
u_{K+2} &> u_{K+1} > u_K \\
u_{K+3} &> u_{K+2} > u_K \\
u_{K+4} &> u_{K+3} > u_K \\
&\vdots
\end{aligned}
\tag{4}
$$

Since $u_K > 0$, it follows from the inequalities in (4) that $\lim_{k \to +\infty} u_k \neq 0$, so $u_1 + u_2 + \cdots + u_k + \cdots$ diverges by Theorem 11.4.2. The proof in the case where $\rho = +\infty$ is omitted.

Proof (c). The series

$$\sum_{k=1}^{\infty} \frac{1}{k} \quad \text{and} \quad \sum_{k=1}^{\infty} \frac{1}{k^2}$$

both have $\rho = 1$ (verify). Since the first is the divergent harmonic series and the second is a convergent p-series, the ratio test does not distinguish between convergence and divergence when $\rho = 1$. ∎

Example 1 The series

$$\sum_{k=1}^{\infty} \frac{1}{k!}$$

converges by the ratio test since

$$\rho = \lim_{k \to +\infty} \frac{u_{k+1}}{u_k} = \lim_{k \to +\infty} \frac{1/(k+1)!}{1/k!} = \lim_{k \to +\infty} \frac{k!}{(k+1)!} = \lim_{k \to +\infty} \frac{1}{k+1} = 0$$

so $\rho < 1$. ◀

Example 2 The series

$$\sum_{k=1}^{\infty} \frac{k}{2^k}$$

converges by the ratio test since

$$\rho = \lim_{k \to +\infty} \frac{u_{k+1}}{u_k} = \lim_{k \to +\infty} \frac{k+1}{2^{k+1}} \cdot \frac{2^k}{k} = \frac{1}{2} \lim_{k \to +\infty} \frac{k+1}{k} = \frac{1}{2}$$

so $\rho < 1$. ◄

Example 3 The series

$$\sum_{k=1}^{\infty} \frac{k^k}{k!}$$

diverges by the ratio test since

$$\rho = \lim_{k \to +\infty} \frac{u_{k+1}}{u_k} = \lim_{k \to +\infty} \frac{(k+1)^{k+1}}{(k+1)!} \cdot \frac{k!}{k^k}$$

$$= \lim_{k \to +\infty} \frac{(k+1)^k}{k^k}$$

$$= \lim_{k \to +\infty} \left(1 + \frac{1}{k}\right)^k = e \qquad \boxed{\begin{array}{l} \text{See Section 7.2,} \\ \text{Formula (5).} \end{array}}$$

Since $\rho = e > 1$, the series diverges. ◄

Example 4 Determine whether the series

$$1 + \frac{1}{3} + \frac{1}{5} + \frac{1}{7} + \cdots + \frac{1}{2k-1} + \cdots$$

converges or diverges.

Solution. The ratio test is of no help since

$$\rho = \lim_{k \to +\infty} \frac{u_{k+1}}{u_k} = \lim_{k \to +\infty} \frac{1}{2(k+1)-1} \cdot \frac{2k-1}{1} = \lim_{k \to +\infty} \frac{2k-1}{2k+1} = 1$$

However, the integral test proves that the series diverges since

$$\int_1^{+\infty} \frac{dx}{2x-1} = \lim_{l \to +\infty} \int_1^l \frac{dx}{2x-1} = \lim_{l \to +\infty} \frac{1}{2} \ln (2x-1) \Big]_1^l = +\infty \qquad ◄$$

Example 5   The series

$$\frac{2!}{4} + \frac{4!}{4^2} + \frac{6!}{4^3} + \cdots + \frac{(2k)!}{4^k} + \cdots$$

diverges since

$$\rho = \lim_{k \to +\infty} \frac{u_{k+1}}{u_k} = \lim_{k \to +\infty} \frac{[2(k+1)]!}{4^{k+1}} \cdot \frac{4^k}{(2k)!} = \lim_{k \to +\infty} \left(\frac{(2k+2)!}{(2k)!} \cdot \frac{1}{4} \right)$$

$$= \frac{1}{4} \lim_{k \to +\infty} (2k+2)(2k+1) = +\infty \quad \blacktriangleleft$$

☐ **THE ROOT TEST**

Sometimes the following result is easier to apply than the ratio test.

***

11.5.3   THEOREM (*The Root Test*).   *Let Σu_k be a series with positive terms and suppose*

$$\rho = \lim_{k \to +\infty} \sqrt[k]{u_k} = \lim_{k \to +\infty} (u_k)^{1/k}$$

(a)   *If $\rho < 1$, the series converges.*
(b)   *If $\rho > 1$, or $\rho = +\infty$, the series diverges.*
(c)   *If $\rho = 1$, the series may converge or diverge, so that another test must be tried.*

***

Since the proof of the root test is similar to the proof of the ratio test, we shall omit it.

Example 6   The series

$$\sum_{k=2}^{\infty} \left(\frac{4k-5}{2k+1} \right)^k$$

diverges by the root test since

$$\rho = \lim_{k \to +\infty} (u_k)^{1/k} = \lim_{k \to +\infty} \frac{4k-5}{2k+1} = 2 > 1 \quad \blacktriangleleft$$

Example 7   The series

$$\sum_{k=1}^{\infty} \frac{1}{(\ln (k+1))^k}$$

converges by the root test, since

$$\lim_{k \to +\infty} (u_k)^{1/k} = \lim_{k \to +\infty} \frac{1}{\ln (k+1)} = 0 < 1 \quad \blacktriangleleft$$

☐ **COMMENTS ON NOTATION**

We conclude this section with a remark about notation. Until now we have written most of our infinite series in the form

$$\sum_{k=1}^{\infty} u_k \tag{5}$$

with the summation index beginning at 1. If the summation index begins at some other integer, it is always possible to rewrite the series in form (5). Thus, for example, the series

$$\sum_{k=0}^{\infty} \frac{2^k}{k!} = 1 + 2 + \frac{2^2}{2!} + \frac{2^3}{3!} + \cdots \tag{6}$$

can be written as

$$\sum_{k=1}^{\infty} \frac{2^{k-1}}{(k-1)!} = 1 + 2 + \frac{2^2}{2!} + \frac{2^3}{3!} + \cdots \tag{7}$$

However, for purposes of applying the convergence tests, it is not necessary that the series have form (5). For example, we can apply the ratio test to (6) without converting to the more complicated form (7). Doing so yields

$$\rho = \lim_{k \to +\infty} \frac{u_{k+1}}{u_k} = \lim_{k \to +\infty} \frac{2^{k+1}}{(k+1)!} \cdot \frac{k!}{2^k} = \lim_{k \to +\infty} \frac{2}{k+1} = 0$$

which shows that the series converges since $\rho < 1$.

▶ **Exercise Set 11.5**   Ⓒ *41, 42, 43, 44*

In Exercises 1–6, apply the ratio test. According to the test, does the series converge, does the series diverge, or are the results inconclusive?

1. $\sum_{k=1}^{\infty} \frac{3^k}{k!}$.

2. $\sum_{k=1}^{\infty} \frac{4^k}{k^2}$.

3. $\sum_{k=2}^{\infty} \frac{1}{5k}$.

4. $\sum_{k=1}^{\infty} k \left(\frac{1}{2} \right)^k$.

5. $\sum_{k=1}^{\infty} \frac{k!}{k^3}$.

6. $\sum_{k=1}^{\infty} \frac{k}{k^2 + 1}$.

In Exercises 7–10, apply the root test. According to the test, does the series converge, does the series diverge, or are the results inconclusive?

7. $\sum_{k=1}^{\infty} \left(\frac{3k+2}{2k-1} \right)^k$.

8. $\sum_{k=1}^{\infty} \left(\frac{k}{100} \right)^k$.

9. $\sum_{k=1}^{\infty} \frac{k}{5^k}$.

10. $\sum_{k=1}^{\infty} (1 + e^{-k})^k$.

In Exercises 11–32, use any appropriate test to determine whether the series converges.

11. $\sum_{k=1}^{\infty} \frac{2^k}{k^3}$.

12. $\sum_{k=1}^{\infty} \frac{1}{k^2}$.

13. $\sum_{k=0}^{\infty} \frac{7^k}{k!}$.

14. $\sum_{k=1}^{\infty} \frac{1}{2k+1}$.

15. $\sum_{k=1}^{\infty} \frac{k^2}{5^k}$.

16. $\sum_{k=1}^{\infty} \frac{k! \, 10^k}{3^k}$.

17. $\sum_{k=1}^{\infty} k^{50} e^{-k}$.

18. $\sum_{k=1}^{\infty} \frac{k^2}{k^3 + 1}$.

19. $\sum_{k=1}^{\infty} k \left(\frac{2}{3} \right)^k$.

20. $\sum_{k=1}^{\infty} k^k$.

21. $\sum_{k=2}^{\infty} \frac{1}{k \ln k}$.

22. $\sum_{k=1}^{\infty} \frac{2^k}{k^3 + 1}$.

23. $\sum_{k=1}^{\infty} \left(\frac{4}{7k-1} \right)^k$.

24. $\sum_{k=1}^{\infty} \frac{(k!)^2 2^k}{(2k+2)!}$.

25. $\sum_{k=0}^{\infty} \frac{(k!)^2}{(2k)!}$.

26. $\sum_{k=1}^{\infty} \frac{1}{k^2 + 25}$.

27. $\sum_{k=1}^{\infty} \frac{1}{1 + \sqrt{k}}$.

28. $\sum_{k=1}^{\infty} \frac{k!}{k^k}$.

29. $\sum_{k=1}^{\infty} \frac{\ln k}{e^k}$.

30. $\sum_{k=1}^{\infty} \frac{k!}{e^{k^2}}$.

31. $\sum_{k=0}^{\infty} \frac{(k + 4)!}{4!k!4^k}$.

32. $\sum_{k=1}^{\infty} \left(\frac{k}{k + 1}\right)^{k^2}$.

In Exercises 33–35, show that the series converges.

33. $1 + \dfrac{1 \cdot 2}{1 \cdot 3} + \dfrac{1 \cdot 2 \cdot 3}{1 \cdot 3 \cdot 5} + \dfrac{1 \cdot 2 \cdot 3 \cdot 4}{1 \cdot 3 \cdot 5 \cdot 7} + \cdots$.

34. $1 + \dfrac{1 \cdot 3}{3!} + \dfrac{1 \cdot 3 \cdot 5}{5!} + \dfrac{1 \cdot 3 \cdot 5 \cdot 7}{7!} + \cdots$.

35. $\dfrac{2!}{1} + \dfrac{3!}{1 \cdot 4} + \dfrac{4!}{1 \cdot 4 \cdot 7} + \dfrac{5!}{1 \cdot 4 \cdot 7 \cdot 10} + \cdots$.

36. For which positive values of α does $\sum_{k=1}^{\infty} \alpha^k/k^\alpha$ converge?

37. (a) Show: $\lim\limits_{k \to +\infty} (\ln k)^{1/k} = 1$. [*Hint:* Let $y = (\ln x)^{1/x}$ and find $\lim\limits_{x \to +\infty} \ln y$.]

    (b) Use the result in part (a) and the root test to show that $\sum_{k=1}^{\infty} (\ln k)/3^k$ converges.

    (c) Show that the series converges using the ratio test.

38. Prove: $\lim\limits_{k \to +\infty} k!/k^k = 0$. [*Hint:* Use Theorem 11.4.1.]

39. Prove: $\lim\limits_{k \to +\infty} \dfrac{a^k}{k!} = 0$ for every real number a. [See the hint in Exercise 38.]

40. If the sum S of a convergent series $\sum_{k=1}^{\infty} u_k$ of positive terms is approximated by the nth partial sum s_n, then the error in the approximation is defined as

$$S - s_n = \sum_{k=n+1}^{\infty} u_k = u_{n+1} + u_{n+2} + u_{n+3} + \cdots$$

Let $r_k = u_{k+1}/u_k$.

(a) Prove: If r_k is *decreasing* for all $k \geq n + 1$ and if $r_{n+1} < 1$, then

$$S - s_n < \frac{u_{n+1}}{1 - r_{n+1}}$$

[*Hint:* Show that the sum of the series $\sum_{k=n+1}^{\infty} u_k$ is less than the sum of the convergent geometric series

$$u_{n+1} + r_{n+1}u_{n+1} + r_{n+1}^2 u_{n+1} + \cdots$$

(b) Prove: If r_k is *increasing* for all $k \geq n + 1$ and if $\lim\limits_{k \to +\infty} r_k = \rho < 1$, then

$$S - s_n < \frac{u_{n+1}}{1 - \rho}$$

[*Hint:* Show that the sum of the series $\sum_{k=n+1}^{\infty} u_k$ is less than the sum of the convergent geometric series

$$u_{n+1} + \rho u_{n+1} + \rho^2 u_{n+1} + \cdots$$

In Exercises 41–44, use the results in Exercise 40. For each exercise do the following:

(a) Compute the stated partial sum and find an upper bound on the error in approximating S by the partial sum. Express your results to five decimal places.

(b) Find a value of n to ensure that s_n will approximate S with an error that is less than 10^{-5}.

41. $S = \sum_{k=1}^{\infty} \dfrac{1}{k!}$; s_5.

42. $S = \sum_{k=1}^{\infty} \dfrac{k}{3^k}$; s_8.

43. $S = \sum_{k=1}^{\infty} \dfrac{1}{k2^k}$; s_7.

44. $S = \sum_{k=1}^{\infty} \dfrac{1}{\sqrt{k}3^k}$; s_4.

11.6 APPLYING THE COMPARISON TEST

In this section we shall discuss procedures for applying the comparison test and we shall state an alternative version of this test which is easier to work with. Before starting, we remind the reader that the comparison test applies only to series with positive terms.

☐ **SOME USEFUL INFORMAL PRINCIPLES**

There are two basic steps required to apply the comparison test to a series Σu_k of positive terms:

- Guess at whether the series Σu_k converges or diverges.
- Find a series that proves the guess to be correct. Thus, if the guess is divergence, we must find a divergent series whose terms are "smaller" than the corresponding terms of Σu_k, and if the guess is convergence, we must find a convergent series whose terms are "bigger" than the corresponding terms of Σu_k.

To help with the guessing process in the first step we have formulated some principles that sometimes *suggest* whether a series is likely to converge or diverge. We have called these "informal principles" because they are not intended as formal theorems. In fact, we shall not guarantee that they *always* work. However, they work often enough to be useful as a starting point for the comparison test.

11.6.1 INFORMAL PRINCIPLE. *Constant terms in the denominator of u_k can usually be deleted without affecting the convergence or divergence of the series.*

Example 1 Use the above principle to help guess whether the following series converge or diverge.

$$\text{(a)} \quad \sum_{k=1}^{\infty} \frac{1}{2^k + 1} \qquad \text{(b)} \quad \sum_{k=5}^{\infty} \frac{1}{\sqrt{k} - 2} \qquad \text{(c)} \quad \sum_{k=1}^{\infty} \frac{1}{(k + \frac{1}{2})^3}$$

Solution.

(a) Deleting the constant 1 suggests that

$$\sum_{k=1}^{\infty} \frac{1}{2^k + 1} \quad \text{behaves like} \quad \sum_{k=1}^{\infty} \frac{1}{2^k}$$

The modified series is a convergent geometric series, so the given series is likely to converge.

(b) Deleting the -2 suggests that

$$\sum_{k=5}^{\infty} \frac{1}{\sqrt{k} - 2} \quad \text{behaves like} \quad \sum_{k=5}^{\infty} \frac{1}{\sqrt{k}}$$

The modified series is a portion of a divergent p-series ($p = \frac{1}{2}$), so the given series is likely to diverge.

(c) Deleting the $\frac{1}{2}$ suggests that

$$\sum_{k=1}^{\infty} \frac{1}{(k + \frac{1}{2})^3} \quad \text{behaves like} \quad \sum_{k=1}^{\infty} \frac{1}{k^3}$$

The modified series is a convergent p-series ($p = 3$), so the given series is likely to converge. ◄

> **11.6.2   INFORMAL PRINCIPLE.**   *If a polynomial in k appears as a factor in the numerator or denominator of u_k, all but the highest power of k in the polynomial may usually be deleted without affecting the convergence or divergence of the series.*

Example 2   Use the above principle to help guess whether the following series converge or diverge.

$$\text{(a)} \quad \sum_{k=1}^{\infty} \frac{1}{\sqrt{k^3 + 2k}} \qquad \text{(b)} \quad \sum_{k=1}^{\infty} \frac{6k^4 - 2k^3 + 1}{k^5 + k^2 - 2k}$$

Solution (a).   Deleting the term $2k$ suggests that

$$\sum_{k=1}^{\infty} \frac{1}{\sqrt{k^3 + 2k}} \quad \text{behaves like} \quad \sum_{k=1}^{\infty} \frac{1}{\sqrt{k^3}} = \sum_{k=1}^{\infty} \frac{1}{k^{3/2}}$$

Since the modified series is a convergent p-series ($p = \frac{3}{2}$), the given series is likely to converge.

Solution (b).   Deleting all but the highest powers of k in the numerator and also in the denominator suggests that

$$\sum_{k=1}^{\infty} \frac{6k^4 - 2k^3 + 1}{k^5 + k^2 - 2k} \quad \text{behaves like} \quad \sum_{k=1}^{\infty} \frac{6k^4}{k^5} = \sum_{k=1}^{\infty} 6 \left(\frac{1}{k} \right)$$

Since each term in the modified series is a constant times the corresponding term in the divergent harmonic series, the given series is likely to diverge.   ◄

Once it is decided whether a series is likely to converge or diverge, the second step in applying the comparison test is to produce a series with which the given series can be compared to substantiate the guess. Let us consider the case of convergence first. To prove Σa_k converges by the comparison test, we must find a *convergent* series Σb_k such that

$$a_k \leq b_k$$

for all k. Frequently, b_k is derived from the formula for a_k by either increasing the numerator of a_k, or decreasing the denominator of a_k, or both.

Example 3   Use the comparison test to determine whether

$$\sum_{k=1}^{\infty} \frac{1}{2k^2 + k}$$

converges or diverges.

Solution.   Using Principle 11.6.2, the given series behaves like the series

$$\sum_{k=1}^{\infty} \frac{1}{2k^2} = \frac{1}{2} \sum_{k=1}^{\infty} \frac{1}{k^2}$$

which is a constant times a convergent p-series. Thus, the given series is likely to converge. To prove the convergence, observe that when we discard the k from the denominator of $1/(2k^2 + k)$, the denominator decreases and the ratio increases, so that

$$\frac{1}{2k^2 + k} < \frac{1}{2k^2}$$

for $k = 1, 2, \ldots$. Since

$$\sum_{k=1}^{\infty} \frac{1}{2k^2} = \frac{1}{2} \sum_{k=1}^{\infty} \frac{1}{k^2}$$

converges, so does $\displaystyle\sum_{k=1}^{\infty} \frac{1}{2k^2 + k}$ by the comparison test. ◀

Example 4 Use the comparison test to determine whether

$$\sum_{k=1}^{\infty} \frac{1}{2k^2 - k}$$

converges or diverges.

Solution. Using Principle 11.6.2, the series behaves like the convergent series

$$\sum_{k=1}^{\infty} \frac{1}{2k^2} = \frac{1}{2} \sum_{k=1}^{\infty} \frac{1}{k^2}$$

Thus, the given series is likely to converge. However, if we discard the k from the denominator of $1/(2k^2 - k)$, the denominator increases and the ratio decreases, so that

$$\frac{1}{2k^2 - k} > \frac{1}{2k^2}$$

Unfortunately, this inequality is in the wrong direction to prove convergence of the given series. A different approach is needed; we must do something to decrease the denominator, not increase it. We accomplish this by replacing k by k^2 to obtain

$$\frac{1}{2k^2 - k} \leq \frac{1}{2k^2 - k^2} = \frac{1}{k^2}$$

Since $\displaystyle\sum_{k=1}^{\infty} \frac{1}{k^2}$ is a convergent p-series, the given series converges by the comparison test. ◀

To prove that a series Σa_k diverges by the comparison test, we must produce a divergent series Σb_k of positive terms such that $a_k \geq b_k$ for all k.

Example 5 Use the comparison test to determine whether

$$\sum_{k=1}^{\infty} \frac{1}{k - \frac{1}{4}}$$

converges or diverges.

Solution. Using Principle 11.6.1, the series behaves like the divergent harmonic series

$$\sum_{k=1}^{\infty} \frac{1}{k}$$

Thus, the given series is likely to diverge. Since

$$\frac{1}{k - \frac{1}{4}} > \frac{1}{k} \quad \text{for } k = 1, 2, \ldots$$

and since $\sum_{k=1}^{\infty} \frac{1}{k}$ diverges, the given series diverges by the comparison test. ◀

Example 6 Use the comparison test to determine whether

$$\sum_{k=1}^{\infty} \frac{1}{\sqrt{k} + 5}$$

converges or diverges.

Solution. Using Principle 11.6.1, the series behaves like the divergent p-series

$$\sum_{k=1}^{\infty} \frac{1}{\sqrt{k}}$$

Thus, the given series is likely to diverge. For $k \geq 25$ we have

$$\frac{1}{\sqrt{k} + 5} \geq \frac{1}{\sqrt{k} + \sqrt{k}} = \frac{1}{2\sqrt{k}}$$

and since

$$\sum_{k=25}^{\infty} \frac{1}{2\sqrt{k}}$$

diverges (why?), the series

$$\sum_{k=25}^{\infty} \frac{1}{\sqrt{k} + 5}$$

diverges by the comparison test; consequently, the given series diverges by Theorem 11.4.3(*c*). ◀

□ **THE LIMIT COMPARISON TEST**

As the last four examples show, some tricky work with inequalities may be required to apply the comparison test. Fortunately, there is an alternative version of this test, which avoids this complication. We shall state the result, but defer its proof to the end of the section so that we can proceed immediately to some examples.

11.6.3 THEOREM (*The Limit Comparison Test*). *Let Σa_k and Σb_k be series with positive terms and suppose*

$$\rho = \lim_{k \to +\infty} \frac{a_k}{b_k}$$

If ρ is finite and $\rho \neq 0$, then the series both converge or both diverge.

To illustrate how this test works we shall reconsider the problems in Examples 4 and 5.

Example 7 Use the limit comparison test to determine whether the following series converge or diverge:

(a) $\displaystyle\sum_{k=1}^{\infty} \frac{1}{2k^2 - k}$ (b) $\displaystyle\sum_{k=1}^{\infty} \frac{1}{k - \frac{1}{4}}$

Solution (a). As in Example 4, we guess that the given series behaves like the convergent series

$$\sum_{k=1}^{\infty} \frac{1}{2k^2} \tag{1}$$

Thus, from Theorem 11.6.3 with

$$a_k = \frac{1}{2k^2 - k} \quad \text{and} \quad b_k = \frac{1}{2k^2}$$

we obtain

$$\rho = \lim_{k \to +\infty} \frac{a_k}{b_k} = \lim_{k \to +\infty} \frac{2k^2}{2k^2 - k} = \lim_{k \to +\infty} \frac{2}{2 - 1/k} = 1$$

Since (1) converges, so does the given series since ρ is finite and positive.

Solution (b). As in Example 5, we guess that the given series behaves like the divergent series

$$\sum_{k=1}^{\infty} \frac{1}{k} \tag{2}$$

Thus, from Theorem 11.6.3 with

$$a_k = \frac{1}{k - \frac{1}{4}} \quad \text{and} \quad b_k = \frac{1}{k}$$

we obtain

$$\rho = \lim_{k \to +\infty} \frac{a_k}{b_k} = \lim_{k \to +\infty} \frac{k}{k - \frac{1}{4}} = \lim_{k \to +\infty} \frac{1}{1 - \frac{1}{4k}} = 1$$

Since (2) diverges, so does the given series since ρ is finite and nonzero.   ◀

Example 8   Use the limit comparison test to determine whether

$$\sum_{k=1}^{\infty} \frac{3k^3 - 2k^2 + 4}{k^5 - k^3 + 2}$$

converges or diverges.

Solution.   From Principle 11.6.2, the series behaves like

$$\sum_{k=1}^{\infty} \frac{3k^3}{k^5} = \sum_{k=1}^{\infty} \frac{3}{k^2} \tag{3}$$

which converges since it is a constant times a convergent *p*-series. Thus, the given series is likely to converge. To substantiate this, we apply the limit comparison test to series (3) and the given series. We obtain

$$\rho = \lim_{k \to +\infty} \frac{\dfrac{3k^3 - 2k^2 + 4}{k^5 - k^3 + 2}}{\dfrac{3}{k^2}} = \lim_{k \to +\infty} \frac{3k^5 - 2k^4 + 4k^2}{3k^5 - 3k^3 + 6} = 1$$

Since $\rho \neq 0$, the given series converges because series (3) converges.   ◀

Unlike the comparison test, the limit comparison test does not require any tricky manipulation of inequalities. However, the test only applies when $0 < \rho < +\infty$. We note, though, that if $\rho = 0$ or $\rho = +\infty$, conclusions about convergence or divergence may be drawn in certain cases (see Exercise 44).

REMARK.   As a practical matter you should generally try the limit comparison test before the comparison test because it is easier to apply.

■ OPTIONAL

We conclude this section with a proof of the limit comparison test.

Proof of Theorem 11.6.3.   We must show that Σb_k converges when Σa_k converges and conversely. The idea of the proof is to apply the comparison test to Σa_k and suitable multiples of Σb_k. To this end, let

$$\epsilon = \frac{\rho}{2} \tag{4}$$

so that $\epsilon > 0$ because $\rho > 0$ by hypothesis. From the assumption that

$$\rho = \lim_{k \to +\infty} \frac{a_k}{b_k}$$

it follows that for k sufficiently large, say $k \geq K$, the ratio a_k/b_k will be within ϵ units of ρ. Thus,

$$\rho - \epsilon < \frac{a_k}{b_k} < \rho + \epsilon \quad \text{when} \quad k \geq K$$

or on substituting (4)

$$\frac{1}{2}\rho < \frac{a_k}{b_k} < \frac{3}{2}\rho \quad \text{when} \quad k \geq K$$

or

$$\frac{1}{2}\rho b_k < a_k < \frac{3}{2}\rho b_k \quad \text{when} \quad k \geq K \tag{5}$$

If $\displaystyle\sum_{k=1}^{\infty} a_k$ converges, then $\displaystyle\sum_{k=K}^{\infty} a_k$ converges by Theorem 11.4.3(c). It follows from the comparison test and the left-hand inequality in (5) that $\displaystyle\sum_{k=K}^{\infty} \frac{1}{2}\rho b_k$ converges. Thus, $\displaystyle\sum_{k=K}^{\infty} b_k$ converges [Theorem 11.4.3(b)] and $\displaystyle\sum_{k=1}^{\infty} b_k$ converges [Theorem 11.4.3(c)].

Conversely, if $\displaystyle\sum_{k=1}^{\infty} b_k$ converges, then $\displaystyle\sum_{k=K}^{\infty} \frac{3}{2}\rho b_k$ converges [Theorem 11.4.3(b), (c)], so $\displaystyle\sum_{k=K}^{\infty} a_k$ converges by the comparison test and the right-hand inequality in (5). Thus, $\displaystyle\sum_{k=1}^{\infty} a_k$ converges. ∎

▶ Exercise Set 11.6

In Exercises 1–6, prove that the series converges by the comparison test.

1. $\displaystyle\sum_{k=1}^{\infty} \frac{1}{3^k + 5}.$

2. $\displaystyle\sum_{k=1}^{\infty} \frac{2}{k^4 + k}.$

3. $\displaystyle\sum_{k=1}^{\infty} \frac{1}{5k^2 - k}.$

4. $\displaystyle\sum_{k=1}^{\infty} \frac{k}{8k^3 + 2k^2 - 1}.$

5. $\displaystyle\sum_{k=1}^{\infty} \frac{2^k - 1}{3^k + 2k}.$

6. $\displaystyle\sum_{k=1}^{\infty} \frac{5\sin^2 k}{k!}.$

In Exercises 7–12, prove that the series diverges by the comparison test.

7. $\displaystyle\sum_{k=1}^{\infty} \frac{3}{k - \frac{1}{4}}.$

8. $\displaystyle\sum_{k=1}^{\infty} \frac{1}{\sqrt{k} + 8}.$

9. $\displaystyle\sum_{k=1}^{\infty} \frac{9}{\sqrt{k} + 1}.$

10. $\displaystyle\sum_{k=2}^{\infty} \frac{k + 1}{k^2 - k}.$

11. $\displaystyle\sum_{k=1}^{\infty} \frac{k^{4/3}}{8k^2 + 5k + 1}.$

12. $\displaystyle\sum_{k=1}^{\infty} \frac{k^{-1/2}}{2 + \sin^2 k}.$

In Exercises 13–18, use the limit comparison test to determine whether the series converges or diverges.

13. $\displaystyle\sum_{k=1}^{\infty} \frac{4k^2 - 2k + 6}{8k^7 + k - 8}$.   **14.** $\displaystyle\sum_{k=1}^{\infty} \frac{1}{9k + 6}$.

15. $\displaystyle\sum_{k=1}^{\infty} \frac{5}{3^k + 1}$.

16. $\displaystyle\sum_{k=1}^{\infty} \frac{k(k + 3)}{(k + 1)(k + 2)(k + 5)}$.

17. $\displaystyle\sum_{k=1}^{\infty} \frac{1}{\sqrt[3]{8k^2 - 3k}}$.   **18.** $\displaystyle\sum_{k=1}^{\infty} \frac{1}{(2k + 3)^{17}}$.

In Exercises 19–34, use any method to determine whether the series converges or diverges. In some cases, you may have to use tests from earlier sections.

19. $\displaystyle\sum_{k=1}^{\infty} \frac{1}{k^3 + 2k + 1}$.   **20.** $\displaystyle\sum_{k=1}^{\infty} \frac{1}{(3 + k)^{2/5}}$.

21. $\displaystyle\sum_{k=1}^{\infty} \frac{1}{9k - 2}$.   **22.** $\displaystyle\sum_{k=1}^{\infty} \frac{\ln k}{k}$.

23. $\displaystyle\sum_{k=1}^{\infty} \frac{\sqrt{k}}{k^3 + 1}$.   **24.** $\displaystyle\sum_{k=1}^{\infty} \frac{4}{2 + 3^k k}$.

25. $\displaystyle\sum_{k=1}^{\infty} \frac{1}{\sqrt{k(k + 1)}}$.   **26.** $\displaystyle\sum_{k=1}^{\infty} \frac{2 + (-1)^k}{5^k}$.

27. $\displaystyle\sum_{k=1}^{\infty} \frac{2 + \sqrt{k}}{(k + 1)^3 - 1}$.   **28.** $\displaystyle\sum_{k=1}^{\infty} \frac{4 + |\cos k|}{k^3}$.

29. $\displaystyle\sum_{k=1}^{\infty} \frac{1}{4 + 2^{-k}}$.   **30.** $\displaystyle\sum_{k=1}^{\infty} \frac{\sqrt{k} \ln k}{k^3 + 1}$.

31. $\displaystyle\sum_{k=1}^{\infty} \frac{\tan^{-1} k}{k^2}$.   **32.** $\displaystyle\sum_{k=1}^{\infty} \frac{5^k + k}{k! + 3}$.

33. $\displaystyle\sum_{k=1}^{\infty} \frac{\ln k}{k\sqrt{k}}$.   **34.** $\displaystyle\sum_{k=1}^{\infty} \frac{\cos (1/k)}{k^2}$.

35. Use the limit comparison test to show that the series

$\displaystyle\sum_{k=1}^{\infty} (1 - \cos (1/k))$ converges.

$\left[\textit{Hint:} \text{ Compare with the series } \displaystyle\sum_{k=1}^{\infty} 1/k^2. \right]$

36. Use the limit comparison test to show that the series

$\displaystyle\sum_{k=1}^{\infty} \sin (\pi/k)$ diverges.

$\left[\textit{Hint:} \text{ Compare with the series } \displaystyle\sum_{k=1}^{\infty} \pi/k. \right]$

37. Use the comparison test to determine whether $\displaystyle\sum_{k=1}^{\infty} \frac{\ln k}{k^2}$ converges or diverges. [*Hint:* $\ln x < \sqrt{x}$ by Exercise 66 of Section 7.3.]

38. Determine whether $\displaystyle\sum_{k=2}^{\infty} \frac{1}{(\ln k)^2}$ converges or diverges. [*Hint:* See the hint in Exercise 37.]

39. Let a, b, and p be positive constants. For which values of p does the series $\displaystyle\sum_{k=1}^{\infty} \frac{1}{(a + bk)^p}$ converge?

40. (a) Show that $k^k \geq k!$ and use this result to prove that the series $\displaystyle\sum_{k=1}^{\infty} k^{-k}$ converges by the comparison test.
(b) Prove convergence using the root test.

41. Use the limit comparison test to investigate convergence of $\displaystyle\sum_{k=1}^{\infty} \frac{(k + 1)^2}{(k + 2)!}$.

42. Use the limit comparison test to investigate convergence of the series $1 + \frac{1}{3} + \frac{1}{5} + \frac{1}{7} + \cdots$

43. Prove that $\sum_{k=1}^{\infty} 1/k!$ converges by comparison with a suitable geometric series.

44. Let Σa_k and Σb_k be series with positive terms. Prove:
(a) If $\displaystyle\lim_{k \to +\infty} (a_k/b_k) = 0$ and Σb_k converges, then Σa_k converges.
(b) If $\displaystyle\lim_{k \to +\infty} (a_k/b_k) = +\infty$ and Σb_k diverges, then Σa_k diverges.

■ 11.7 ALTERNATING SERIES; CONDITIONAL CONVERGENCE

> *So far our emphasis has been on series with positive terms. In this section we shall discuss series containing both positive and negative terms.*

□ **ALTERNATING SERIES**   Of special importance are series whose terms are alternately positive and negative. These are called **alternating series**. Some examples are

$$1 - 1 + 1 - 1 + \cdots + (-1)^{k+1} + \cdots$$

$$1 - \frac{1}{2} + \frac{1}{3} - \frac{1}{4} + \cdots + (-1)^{k+1}\frac{1}{k} + \cdots$$

$$-1 + 2 - 3 + 4 - \cdots + (-1)^k\, k + \cdots$$

$$-1 + \frac{1}{2!} - \frac{1}{3!} + \frac{1}{4!} - \cdots + (-1)^k\frac{1}{k!} + \cdots$$

In general, an alternating series has one of two possible forms:

$$\sum_{k=1}^{\infty} (-1)^{k+1}\, a_k = a_1 - a_2 + a_3 - a_4 + \cdots \tag{1}$$

or

$$\sum_{k=1}^{\infty} (-1)^{k}\, a_k = -a_1 + a_2 - a_3 + a_4 - \cdots \tag{2}$$

where the a_k's are assumed to be positive in both cases.

The following theorem is the key result on convergence of alternating series.

11.7.1   THEOREM (*Alternating Series Test*).   *An alternating series of either form*
(1) *or form* (2) *converges if the following two conditions are satisfied:*

(*a*) $a_1 > a_2 > a_3 > \cdots > a_k > \cdots$

(*b*) $\displaystyle\lim_{k \to +\infty} a_k = 0$

Proof.   Our proof of this theorem will use the following result about sequences which is intuitively obvious, but which we will not actually prove:

> *If the even-numbered terms of a sequence converge to a limit L, and if the odd-numbered terms of the sequence converge to the same limit L, then the entire sequence converges to the limit L.*

To prove convergence of an alternating series, we shall show that its even partial sums and odd partial sums converge to the same limit S; it will then follow from the foregoing result that the entire sequence of partial sums converges to S, which will imply convergence of the series.

We shall consider the case

$$a_1 - a_2 + a_3 - a_4 + \cdots + (-1)^{k+1} a_k + \cdots$$

The other case is left as an exercise. Consider the partial sums

$$s_2 = a_1 - a_2$$
$$s_4 = (a_1 - a_2) + (a_3 - a_4)$$
$$s_6 = (a_1 - a_2) + (a_3 - a_4) + (a_5 - a_6)$$
$$s_8 = (a_1 - a_2) + (a_3 - a_4) + (a_5 - a_6) + (a_7 - a_8)$$
$$\vdots$$

By condition (a), each of the differences appearing in the parentheses is positive, so

$$s_2 < s_4 < s_6 < s_8 < \cdots$$

Moreover, the terms in this sequence are all less than or equal to a_1 since we can write

$$s_2 = a_1 - a_2$$
$$s_4 = a_1 - (a_2 - a_3) - a_4$$
$$s_6 = a_1 - (a_2 - a_3) - (a_4 - a_5) - a_6$$
$$s_8 = a_1 - (a_2 - a_3) - (a_4 - a_5) - (a_6 - a_7) - a_8$$
$$\vdots$$

Thus, the sequence

$$s_2, s_4, s_6, s_8, \ldots, s_{2n}, \ldots$$

is increasing and has an upper bound (namely, a_1), so it converges to some limit S by Theorem 11.2.2; that is,

$$\lim_{n \to +\infty} s_{2n} = S \tag{3}$$

We shall show next that the sequence

$$s_1, s_3, s_5, \ldots, s_{2n-1}, \ldots$$

also has limit S. Since the $(2n)$-th term in the given alternating series is $-a_{2n}$, it follows that $s_{2n} - s_{2n-1} = -a_{2n}$, which can be written as

$$s_{2n-1} = s_{2n} + a_{2n} \tag{4}$$

But $2n \to +\infty$ as $n \to +\infty$, so $\lim\limits_{n \to +\infty} a_{2n} = 0$ by condition (b). Thus, from (3) and (4)

$$\lim_{n \to +\infty} s_{2n-1} = \lim_{n \to +\infty} s_{2n} + \lim_{n \to +\infty} a_{2n} = S + 0 = S \tag{5}$$

From (3) and (5), the sequence $s_1, s_2, s_3, \ldots, s_n, \ldots$ converges to S, so the series converges. ∎

Example 1 The series

$$1 - \frac{1}{2} + \frac{1}{3} - \frac{1}{4} + \cdots + (-1)^{k+1} \frac{1}{k} + \cdots$$

is called the **alternating harmonic series**. Since

$$a_k = \frac{1}{k} > \frac{1}{k+1} = a_{k+1}$$

and

$$\lim_{k \to +\infty} a_k = \lim_{k \to +\infty} \frac{1}{k} = 0$$

this series converges by the alternating series test. ◀

Example 2 Determine whether the alternating series

$$\sum_{k=1}^{\infty} (-1)^{k+1} \frac{k+3}{k(k+1)}$$

converges or diverges.

Solution. Condition (*b*) of the alternating series test is satisfied since

$$\lim_{k \to +\infty} a_k = \lim_{k \to +\infty} \frac{k+3}{k(k+1)} = \lim_{k \to +\infty} \frac{\dfrac{1}{k} + \dfrac{3}{k^2}}{1 + \dfrac{1}{k}} = 0$$

To see if condition (*a*) is met, we must determine whether the sequence

$$\{a_k\}_{k=1}^{+\infty} = \left\{ \frac{k+3}{k(k+1)} \right\}_{k=1}^{+\infty}$$

is decreasing. Since

$$\frac{a_{k+1}}{a_k} = \frac{k+4}{(k+1)(k+2)} \cdot \frac{k(k+1)}{k+3} = \frac{k^2 + 4k}{k^2 + 5k + 6}$$

$$= \frac{k^2 + 4k}{(k^2 + 4k) + (k+6)} < 1$$

we have $a_k > a_{k+1}$, so the series converges by the alternating series test. ◀

REMARK.   If an alternating series violates condition (*b*) of the alternating series test, then the series must diverge by the divergence test (11.4.2). However, if condition (*b*) is satisfied, but condition (*a*) is not, the series can either converge or diverge.*

Figure 11.7.1 provides some insight into the way in which an alternating series

$$a_1 - a_2 + a_3 - a_4 + \cdots + (-1)^{k+1} a_k + \cdots$$

converges to its sum *S* when the conditions of the alternating series test are satisfied.

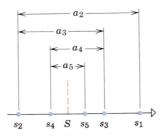

Figure 11.7.1

In the figure we have plotted the successive partial sums on a horizontal. Because

$$a_1 > a_2 > a_3 > a_4 > \cdots$$

and

$$\lim_{k \to +\infty} a_k = 0$$

the successive partial sums oscillate in smaller and smaller steps, closing in on the sum *S*. It is of interest to note that the even-numbered partial sums are less than or equal to *S* and the odd-numbered partial sums are greater than *S*. Thus, the sum *S* falls between any two successive partial sums; that is, for any positive integer *n*

$$s_n < S < s_{n+1} \quad \text{or} \quad s_{n+1} < S < s_n$$

depending on whether *n* is even or odd. In either case,

$$|S - s_n| < |s_{n+1} - s_n| \tag{6}$$

*The interested reader will find some nice examples in an article by R. Lariviere, "On a Convergence Test for Alternating Series," *Mathematics Magazine*, Vol. 29, 1956, p. 88.

But $s_{n+1} - s_n = \pm a_{n+1}$ (the sign depending on whether n is even or odd), so (6) yields

$$|S - s_n| < a_{n+1} \tag{7}$$

Since $|S - s_n|$ represents the magnitude of error that results when we approximate the sum of the entire series by the sum of the first n terms, (7) tells us that this error is less than the magnitude of the $(n + 1)$-st term in the series. The reader can check that (7) also holds for alternating series of the form

$$-a_1 + a_2 - a_3 + a_4 - \cdots + (-1)^k a_k + \cdots$$

In summary, we have the following result.

11.7.2 THEOREM. *If an alternating series satisfies the conditions of the alternating series test, and if the sum S of the series is approximated by the nth partial sum s_n, thereby resulting in an error of $S - s_n$, then*

$$|S - s_n| < a_{n+1}$$

Moreover, the sign of the error is the same as that of the coefficient of a_{n+1} in the series.

Example 3   As shown in Example 1, the alternating harmonic series

$$1 - \frac{1}{2} + \frac{1}{3} - \frac{1}{4} + \cdots + (-1)^{k+1} \frac{1}{k} + \cdots$$

satisfies the conditions of the alternating series test; hence, the series has a sum S, which we know must lie between any two successive partial sums. In particular, it must lie between

$$s_7 = 1 - \frac{1}{2} + \frac{1}{3} - \frac{1}{4} + \frac{1}{5} - \frac{1}{6} + \frac{1}{7} = \frac{319}{420}$$

and

$$s_8 = 1 - \frac{1}{2} + \frac{1}{3} - \frac{1}{4} + \frac{1}{5} - \frac{1}{6} + \frac{1}{7} - \frac{1}{8} = \frac{533}{840}$$

so

$$\frac{533}{840} < S < \frac{319}{420} \tag{8}$$

Later in this chapter we shall show that the sum S of the alternating harmonic series is ln 2. If we accept this to be so for now, it follows from (8) that

$$\frac{533}{840} < \ln 2 < \frac{319}{420}$$

or with the help of a calculator (verify)

$$0.6345 < \ln 2 < 0.7596$$

The value of ln 2, rounded to four decimal places, is 0.6931, which is consistent with these inequalities. It follows from Theorem 11.7.2 that

$$|\ln 2 - s_7| = \left| \ln 2 - \frac{319}{420} \right| < a_8 = \frac{1}{8}$$

and

$$|\ln 2 - s_8| = \left| \ln 2 - \frac{533}{840} \right| < a_9 = \frac{1}{9}$$

We leave it for the reader to verify that these error estimates are correct by using a calculator.   ◄

ABSOLUTE AND CONDITIONAL CONVERGENCE

The series

$$1 - \frac{1}{2} - \frac{1}{2^2} + \frac{1}{2^3} + \frac{1}{2^4} - \frac{1}{2^5} - \frac{1}{2^6} + \cdots$$

does not fit in any of the categories studied so far—it has mixed signs, but is not alternating. We shall now develop some convergence tests that can be applied to such series.

11.7.3 DEFINITION.   A series

$$\sum_{k=1}^{\infty} u_k = u_1 + u_2 + \cdots + u_k + \cdots$$

is said to *converge absolutely* if the series of absolute values

$$\sum_{k=1}^{\infty} |u_k| = |u_1| + |u_2| + \cdots + |u_k| + \cdots$$

converges.

Example 4   The series

$$1 - \frac{1}{2} - \frac{1}{2^2} + \frac{1}{2^3} + \frac{1}{2^4} - \frac{1}{2^5} - \frac{1}{2^6} + \cdots$$

converges absolutely since the series of absolute values

$$1 + \frac{1}{2} + \frac{1}{2^2} + \frac{1}{2^3} + \frac{1}{2^4} + \frac{1}{2^5} + \frac{1}{2^6} + \cdots$$

is a convergent geometric series. On the other hand, the alternating harmonic series

$$1 - \frac{1}{2} + \frac{1}{3} - \frac{1}{4} + \frac{1}{5} - \cdots$$

does not converge absolutely since the series of absolute values

$$1 + \frac{1}{2} + \frac{1}{3} + \frac{1}{4} + \frac{1}{5} + \cdots$$

diverges.   ◄

Absolute convergence is of importance because of the following theorem.

11.7.4 THEOREM. *If the series*

$$\sum_{k=1}^{\infty} |u_k| = |u_1| + |u_2| + \cdots + |u_k| + \cdots$$

converges, then so does the series

$$\sum_{k=1}^{\infty} u_k = u_1 + u_2 + \cdots + u_k + \cdots$$

In other words, if a series converges absolutely, then it converges.

Proof. Our proof is based on a trick. We shall show that the series

$$\sum_{k=1}^{\infty} (u_k + |u_k|) \tag{9}$$

converges. Since $\Sigma|u_k|$ is assumed to converge, it will then follow from Theorem 11.4.3(a) that Σu_k converges, since

$$\sum_{k=1}^{\infty} u_k = \sum_{k=1}^{\infty} [(u_k + |u_k|) - |u_k|]$$

For all k, the value of $u_k + |u_k|$ is either 0 or $2|u_k|$, depending on whether u_k is negative or not. Thus, for all values of k

$$0 \leq u_k + |u_k| \leq 2|u_k| \tag{10}$$

But $\Sigma 2|u_k|$ is a convergent series since it is a constant times the convergent series $\Sigma|u_k|$. Thus, from (10), series (9) converges by the comparison test.   ■

Example 5 In Example 4 we showed that

$$1 - \frac{1}{2} - \frac{1}{2^2} + \frac{1}{2^3} + \frac{1}{2^4} - \frac{1}{2^5} - \frac{1}{2^6} + \cdots$$

converges absolutely. It follows from Theorem 11.7.4 that the series converges. ◄

Example 6 Show that the series

$$\sum_{k=1}^{\infty} \frac{\cos k}{k^2}$$

converges.

Solution. Since $|\cos k| \leq 1$ for all k,

$$\left| \frac{\cos k}{k^2} \right| \leq \frac{1}{k^2}$$

Thus,

$$\sum_{k=1}^{\infty} \left| \frac{\cos k}{k^2} \right|$$

converges by the comparison test, and consequently

$$\sum_{k=1}^{\infty} \frac{\cos k}{k^2}$$

converges. ◄

If $\Sigma |u_k|$ *diverges*, no conclusion can be drawn about the convergence or divergence of Σu_k. For example, consider the two series

$$1 - \frac{1}{2} + \frac{1}{3} - \frac{1}{4} + \cdots + (-1)^{k+1}\frac{1}{k} + \cdots \tag{11}$$

$$-1 - \frac{1}{2} - \frac{1}{3} - \frac{1}{4} - \cdots - \frac{1}{k} - \cdots \tag{12}$$

Series (11), the alternating harmonic series, converges, whereas series (12), being a constant times the harmonic series, diverges. Yet in each case the series of absolute values is

$$1 + \frac{1}{2} + \frac{1}{3} + \cdots + \frac{1}{k} + \cdots$$

which diverges. A series such as (11), which is convergent, but not absolutely convergent, is called *conditionally convergent*.

□ **THE RATIO TEST FOR ABSOLUTE CONVERGENCE**

The following version of the ratio test is useful for investigating absolute convergence.

11.7.5 THEOREM (*Ratio Test for Absolute Convergence*). *Let Σu_k be a series with nonzero terms and suppose*

$$\lim_{k \to +\infty} \frac{|u_{k+1}|}{|u_k|} = \rho$$

(a) *If $\rho < 1$, the series Σu_k converges absolutely and therefore converges.*
(b) *If $\rho > 1$ or if $\rho = +\infty$, then the series Σu_k diverges.*
(c) *If $\rho = 1$, no conclusion about convergence or absolute convergence can be drawn from this test.*

The proof is discussed in the exercises.

Example 7 The series

$$\sum_{k=1}^{\infty} (-1)^k \frac{2^k}{k!}$$

converges absolutely since

$$\rho = \lim_{k \to +\infty} \frac{|u_{k+1}|}{|u_k|} = \lim_{k \to +\infty} \frac{2^{k+1}}{(k+1)!} \cdot \frac{k!}{2^k} = \lim_{k \to +\infty} \frac{2}{k+1} = 0 < 1 \qquad ◄$$

Example 8 We proved earlier (Theorem 11.3.3) that a geometric series

$$a + ar + ar^2 + \cdots + ar^{k-1} + \cdots$$

converges if $|r| < 1$ and diverges if $|r| \geq 1$. However, a stronger statement can be made—the series converges *absolutely* if $|r| < 1$. This follows from Theorem 11.7.5 since

$$\rho = \lim_{k \to +\infty} \frac{|u_{k+1}|}{|u_k|} = \lim_{k \to +\infty} \frac{|ar^k|}{|ar^{k-1}|} = \lim_{k \to +\infty} |r| = |r|$$

so that $\rho < 1$ if $|r| < 1$. ◄

The following review is included as a ready reference to convergence tests.

Review of Convergence Tests

NAME	STATEMENT	COMMENTS
Divergence Test (11.4.2)	If $\lim\limits_{k \to +\infty} u_k \neq 0$, then Σu_k diverges.	If $\lim\limits_{k \to +\infty} u_k = 0$, Σu_k may or may not converge.
Integral Test (11.4.5)	Let Σu_k be a series with positive terms, and let $f(x)$ be the function that results when k is replaced by x in the formula for u_k. If f is decreasing and continuous for $x \geq 1$, then $$\sum_{k=1}^{\infty} u_k \quad \text{and} \quad \int_1^{+\infty} f(x)\,dx$$ both converge or both diverge.	Use this test when $f(x)$ is easy to integrate. This test only applies to series that have positive terms.
Comparison Test (11.5.1)	Let Σa_k and Σb_k be series with nonnegative terms such that $$a_1 \leq b_1,\ a_2 \leq b_2, \ldots, a_k \leq b_k, \ldots$$ If Σb_k converges, then Σa_k converges, and if Σa_k diverges, then Σb_k diverges.	Use this test as a last resort; other tests are often easier to apply. This test only applies to series with nonnegative terms.
Ratio Test (11.5.2)	Let Σu_k be a series with positive terms and suppose $$\lim_{k \to +\infty} \frac{u_{k+1}}{u_k} = \rho$$ (a) Series converges if $\rho < 1$. (b) Series diverges if $\rho > 1$ or $\rho = +\infty$. (c) No conclusion if $\rho = 1$.	Try this test when u_k involves factorials or kth powers.
Root Test (11.5.3)	Let Σu_k be a series with positive terms such that $$\rho = \lim_{k \to +\infty} \sqrt[k]{u_k}$$ (a) Series converges if $\rho < 1$. (b) Series diverges if $\rho > 1$ or $\rho = +\infty$. (c) No conclusion if $\rho = 1$.	Try this test when u_k involves kth powers.
Limit Comparison Test (11.6.3)	Let Σa_k and Σb_k be series with positive terms such that $$\rho = \lim_{k \to +\infty} \frac{a_k}{b_k}$$ If $0 < \rho < +\infty$, then both series converge or both diverge.	This is easier to apply than the comparison test, but still requires some skill in choosing the series Σb_k for comparison.
Alternating Series Test (11.7.1)	The series $$a_1 - a_2 + a_3 - a_4 + \cdots$$ $$-a_1 + a_2 - a_3 + a_4 - \cdots$$ converge if (a) $a_1 > a_2 > a_3 > \cdots$ (b) $\lim\limits_{k \to +\infty} a_k = 0$	This test applies only to alternating series. It is assumed that $a_k > 0$ for all k.

Review of Convergence Tests (Continued)

NAME	STATEMENT	COMMENTS
Ratio Test for Absolute Convergence (11.7.5)	Let Σu_k be a series with nonzero terms such that $$\rho = \lim_{k \to +\infty} \frac{|u_{k+1}|}{|u_k|}$$ (a) Series converges absolutely if $\rho < 1$. (b) Series diverges if $\rho > 1$ or $\rho = +\infty$. (c) No conclusion if $\rho = 1$.	The series need not have positive terms and need not be alternating to use this test.

▶ Exercise Set 11.7   Ⓒ 38–46, 56

In Exercises 1–6, use the alternating series test to determine whether the series converges or diverges.

1. $\displaystyle\sum_{k=1}^{\infty} \frac{(-1)^{k+1}}{2k+1}$.

2. $\displaystyle\sum_{k=1}^{\infty} (-1)^{k+1} \frac{k}{3^k}$.

3. $\displaystyle\sum_{k=1}^{\infty} (-1)^{k+1} \frac{k+1}{3k+1}$.

4. $\displaystyle\sum_{k=1}^{\infty} (-1)^{k+1} \frac{k+4}{k^2+k}$.

5. $\displaystyle\sum_{k=1}^{\infty} (-1)^{k+1} e^{-k}$.

6. $\displaystyle\sum_{k=3}^{\infty} (-1)^k \frac{\ln k}{k}$.

In Exercises 7–12, use the ratio test for absolute convergence to determine whether the series converges absolutely or diverges.

7. $\displaystyle\sum_{k=1}^{\infty} \left(-\frac{3}{5}\right)^k$.

8. $\displaystyle\sum_{k=1}^{\infty} (-1)^{k+1} \frac{2^k}{k!}$.

9. $\displaystyle\sum_{k=1}^{\infty} (-1)^{k+1} \frac{3^k}{k^2}$.

10. $\displaystyle\sum_{k=1}^{\infty} (-1)^k \left(\frac{k}{5^k}\right)$.

11. $\displaystyle\sum_{k=1}^{\infty} (-1)^k \left(\frac{k^3}{e^k}\right)$.

12. $\displaystyle\sum_{k=1}^{\infty} (-1)^{k+1} \frac{k^k}{k!}$.

In Exercises 13–30, classify the series as absolutely convergent, conditionally convergent, or divergent.

13. $\displaystyle\sum_{k=1}^{\infty} \frac{(-1)^{k+1}}{3k}$.

14. $\displaystyle\sum_{k=1}^{\infty} \frac{(-1)^{k+1}}{k^{4/3}}$.

15. $\displaystyle\sum_{k=1}^{\infty} \frac{(-4)^k}{k^2}$.

16. $\displaystyle\sum_{k=1}^{\infty} \frac{(-1)^{k+1}}{k!}$.

17. $\displaystyle\sum_{k=1}^{\infty} \frac{\cos k\pi}{k}$.

18. $\displaystyle\sum_{k=3}^{\infty} \frac{(-1)^k \ln k}{k}$.

19. $\displaystyle\sum_{k=1}^{\infty} (-1)^{k+1} \left(\frac{k+2}{3k-1}\right)^k$.

20. $\displaystyle\sum_{k=1}^{\infty} \frac{(-1)^{k+1}}{k^2+1}$.

21. $\displaystyle\sum_{k=1}^{\infty} (-1)^{k+1} \frac{k+2}{k(k+3)}$.

22. $\displaystyle\sum_{k=1}^{\infty} \frac{(-1)^{k+1}k^2}{k^3+1}$.

23. $\displaystyle\sum_{k=1}^{\infty} \sin \frac{k\pi}{2}$.

24. $\displaystyle\sum_{k=1}^{\infty} \frac{\sin k}{k^3}$.

25. $\displaystyle\sum_{k=2}^{\infty} \frac{(-1)^k}{k \ln k}$.

26. $\displaystyle\sum_{k=1}^{\infty} \frac{(-1)^k}{\sqrt{k(k+1)}}$.

27. $\displaystyle\sum_{k=2}^{\infty} \left(-\frac{1}{\ln k}\right)^k$.

28. $\displaystyle\sum_{k=1}^{\infty} \frac{(-1)^{k+1}}{\sqrt{k+1} + \sqrt{k}}$.

29. $\displaystyle\sum_{k=2}^{\infty} \frac{(-1)^k(k^2+1)}{k^3+2}$.

30. $\displaystyle\sum_{k=1}^{\infty} \frac{k \cos k\pi}{k^2+1}$.

In Exercises 31–34, the series satisfies the conditions of the alternating series test. For the stated value of n, use Theorem 11.7.2 to find an upper bound on the magnitude of the error that results if the sum of the series is approximated by the nth partial sum.

31. $\displaystyle\sum_{k=1}^{\infty} \frac{(-1)^{k+1}}{k}$; $n = 7$.

32. $\displaystyle\sum_{k=1}^{\infty} \frac{(-1)^{k+1}}{k!}$; $n = 5$.

33. $\displaystyle\sum_{k=1}^{\infty} \frac{(-1)^{k+1}}{\sqrt{k}}$; $n = 99$.

34. $\displaystyle\sum_{k=1}^{\infty} \frac{(-1)^{k+1}}{(k+1)\ln(k+1)}$; $n = 3$.

In Exercises 35–38, the series satisfies the conditions of the alternating series test. Use Theorem 11.7.2 to find a value of n for which the nth partial sum is ensured to approximate the sum of the series to the stated accuracy.

35. $\displaystyle\sum_{k=1}^{\infty} \frac{(-1)^{k+1}}{k}$; $|\text{error}| < 0.0001$.

36. $\displaystyle\sum_{k=1}^{\infty} \frac{(-1)^{k+1}}{k!}$; $|\text{error}| < 0.00001$.

37. $\displaystyle\sum_{k=1}^{\infty} \frac{(-1)^{k+1}}{\sqrt{k}}$; $|\text{error}| < 0.005$.

38. $\displaystyle\sum_{k=1}^{\infty} \frac{(-1)^{k+1}}{(k+1)\ln(k+1)}$; $|\text{error}| < 0.1$.

In Exercises 39 and 40, use Theorem 11.7.2 to find an upper bound on the magnitude of the error that results if s_{10} is used to approximate the sum of the given *geometric* series. Compute s_{10} rounded to four decimal places and compare this value with the exact sum of the series.

39. $\dfrac{3}{4} - \dfrac{3}{8} + \dfrac{3}{16} - \dfrac{3}{32} + \cdots$.

40. $1 - \dfrac{2}{3} + \dfrac{4}{9} - \dfrac{8}{27} + \cdots$.

In Exercises 41–44, the series satisfies the conditions of the alternating series test. Use Theorem 11.7.2 to find a value of n for which s_n is ensured to approximate the sum of the series with an error that is less than 10^{-4} in magnitude. Compute s_n rounded to five decimal places and compare this value with the sum of the series.

41. $\sin 1 = 1 - \dfrac{1}{3!} + \dfrac{1}{5!} - \dfrac{1}{7!} + \cdots$.

42. $\cos 1 = 1 - \dfrac{1}{2!} + \dfrac{1}{4!} - \dfrac{1}{6!} + \cdots$.

43. $\ln \tfrac{3}{2} = \dfrac{1}{1\cdot 2} - \dfrac{1}{2\cdot 2^2} + \dfrac{1}{3\cdot 2^3} - \dfrac{1}{4\cdot 2^4} + \cdots$.

44. $\dfrac{\pi}{16} = \dfrac{1}{1^5 + 4\cdot 1} - \dfrac{1}{3^5 + 4\cdot 3}$
$\qquad\qquad + \dfrac{1}{5^5 + 4\cdot 5} - \dfrac{1}{7^5 + 4\cdot 7} + \cdots$.

45. For the series $\dfrac{\pi^2}{12} = 1 - \dfrac{1}{2^2} + \dfrac{1}{3^2} - \dfrac{1}{4^2} + \cdots$

(a) use Theorem 11.7.2 to find a value of n for which s_n is ensured to approximate the sum of the series with an error that is less than 5×10^{-3} in magnitude

(b) compute s_{10} and show that the magnitude of the error is less than 5×10^{-3}, thus showing that the value of n obtained from Theorem 11.7.2 is a conservative estimate.

46. For the series $\dfrac{\pi}{4} = 1 - \tfrac{1}{3} + \tfrac{1}{5} - \tfrac{1}{7} + \cdots$

(a) use Theorem 11.7.2 to find a value of n for which s_n is ensured to approximate the sum of the series with an error that is less than 10^{-2} in magnitude

(b) compute s_{26} and show that the magnitude of the error is less than 10^{-2}, thus showing that the value of n obtained from Theorem 11.7.2 is a conservative estimate.

47. Prove: If Σa_k converges absolutely, then Σa_k^2 converges.

48. Show that the converse of the result in Exercise 47 is false by finding a series for which Σa_k^2 converges, but $\Sigma|a_k|$ diverges.

49. Prove Theorem 11.7.1 for series of the form

$$-a_1 + a_2 - a_3 + a_4 - \cdots + (-1)^k a_k + \cdots$$

50. Prove Theorem 11.7.5. [*Hint:* Theorem 11.7.4 will help in part (a). For part (b), it may help to review the proof of Theorem 11.5.2.]

51. The sum of an absolutely convergent series is independent of the order in which the terms are added, but the terms of a conditionally convergent series can be rearranged to converge to any given value, or even diverge. For example, let S be the sum of the conditionally convergent alternating harmonic series,

$$S = 1 - \frac{1}{2} + \frac{1}{3} - \frac{1}{4} + \frac{1}{5} - \frac{1}{6} + \cdots$$

Rearrange the terms in this series to get

$$\left(1 - \frac{1}{2} - \frac{1}{4}\right) + \left(\frac{1}{3} - \frac{1}{6} - \frac{1}{8}\right)$$
$$+ \left(\frac{1}{5} - \frac{1}{10} - \frac{1}{12}\right) + \cdots$$

Show that this rearrangement results in a series that converges to $S/2$. [*Hint:* Add the first two terms within each pair of parentheses.]

52. Based on the discussion in Exercise 51, rearrange the terms in the convergent series

$$1 - \frac{1}{\sqrt{2}} + \frac{1}{\sqrt{3}} - \frac{1}{\sqrt{4}} + \frac{1}{\sqrt{5}} - \frac{1}{\sqrt{6}} + \cdots$$

as

$$\left(1 + \frac{1}{\sqrt{3}} - \frac{1}{\sqrt{2}}\right) + \left(\frac{1}{\sqrt{5}} + \frac{1}{\sqrt{7}} - \frac{1}{\sqrt{4}}\right) + \cdots$$

$$= \sum_{k=1}^{\infty} \left(\frac{1}{\sqrt{4k-3}} + \frac{1}{\sqrt{4k-1}} - \frac{1}{\sqrt{2k}}\right)$$

Show that this rearrangement results in a series that diverges to $+\infty$. [*Hint:* Note that $1/\sqrt{4k-3} > 1/\sqrt{4k}$ and $1/\sqrt{4k-1} > 1/\sqrt{4k}$. Show that the kth term in the rearranged series is greater than $(1 - 1/\sqrt{2})/\sqrt{k}$.]

In Exercises 53–55, use parts (*a*) and (*b*) of Theorem 11.4.3 and the fact that the sum of an absolutely convergent series is independent of the order in which the terms are added.

53. Given: $\dfrac{\pi^2}{6} = 1 + \dfrac{1}{2^2} + \dfrac{1}{3^2} + \dfrac{1}{4^2} + \cdots$.

Show: $\dfrac{\pi^2}{8} = 1 + \dfrac{1}{3^2} + \dfrac{1}{5^2} + \dfrac{1}{7^2} + \cdots$.

54. Given: $\dfrac{\pi^4}{90} = 1 + \dfrac{1}{2^4} + \dfrac{1}{3^4} + \dfrac{1}{4^4} + \cdots$.

Show: $\dfrac{\pi^4}{96} = 1 + \dfrac{1}{3^4} + \dfrac{1}{5^4} + \dfrac{1}{7^4} + \cdots$.

55. Given: $\dfrac{\pi^2}{6} = 1 + \dfrac{1}{2^2} + \dfrac{1}{3^2} + \dfrac{1}{4^2} + \cdots$.

Show: $\dfrac{\pi^2}{12} = 1 - \dfrac{1}{2^2} + \dfrac{1}{3^2} - \dfrac{1}{4^2} + \cdots$.

56. A small bug moves back and forth along a straight line as follows: it walks D units, stops and reverses direction, walks $D/2$ units, stops and reverses direction, walks $D/3$ units, stops and reverses direction, walks $D/4$ units, stops and reverses direction, and so forth, until it stops for the 1000th time. Given that $D = 180$ cm, find upper and lower bounds for
(a) the final distance between the bug and its starting point; [*Hint:* Use Theorem 11.7.2 and the fact that $1 - \frac{1}{2} + \frac{1}{3} - \frac{1}{4} + \cdots = \ln 2$.]
(b) the total distance traveled by the bug. [*Hint:* Use inequality (4) in Section 11.4.]

■ 11.8 POWER SERIES

In previous sections we studied series with constant terms. In this section we shall consider series whose terms involve variables. Such series are of fundamental importance in many branches of mathematics and the physical sciences.

□ **POWER SERIES IN** *x*

If $c_0, c_1, c_2, \ldots$ are constants and x is a variable, then a series of the form

$$\sum_{k=0}^{\infty} c_k x^k = c_0 + c_1 x + c_2 x^2 + \cdots + c_k x^k + \cdots$$

is called a ***power series in x.*** Some examples are

$$\sum_{k=0}^{\infty} x^k = 1 + x + x^2 + x^3 + \cdots$$

$$\sum_{k=0}^{\infty} \frac{x^k}{k!} = 1 + x + \frac{x^2}{2!} + \frac{x^3}{3!} + \cdots$$

$$\sum_{k=0}^{\infty} (-1)^k \frac{x^{k+1}}{k+1} = x - \frac{x^2}{2} + \frac{x^3}{3} - \frac{x^4}{4} + \cdots$$

$$\sum_{k=0}^{\infty} (-1)^k \frac{x^{2k}}{(2k)!} = 1 - \frac{x^2}{2!} + \frac{x^4}{4!} - \frac{x^6}{6!} + \cdots$$

$$\sum_{k=0}^{\infty} (-1)^k \frac{x^{2k+1}}{(2k+1)!} = x - \frac{x^3}{3!} + \frac{x^5}{5!} - \frac{x^7}{7!} + \cdots$$

If a numerical value is substituted for x in a power series $\Sigma c_k x^k$, then we obtain a series of constants that may either converge or diverge. This leads to the following basic problem.

A FUNDAMENTAL PROBLEM.   *For what values of x does a given power series, $\Sigma c_k x^k$, converge?*

The following theorem is the fundamental result on convergence of power series. The proof can be found in most advanced calculus texts.

11.8.1   THEOREM.   *For any power series in x, exactly one of the following is true:*

(a)   The series converges only for $x = 0$.

(b)   The series converges absolutely for all real values of x.

(c)   The series converges absolutely for all x in some finite open interval $(-R, R)$, and diverges if $x < -R$ or $x > R$ (Figure 11.8.1). At the points $x = R$ and $x = -R$ the series may converge absolutely, converge conditionally, or diverge, depending on the particular series.

Series diverges   Series converges absolutely   Series diverges

$-R$ 0 R

Figure 11.8.1

☐ **RADIUS AND INTERVAL OF CONVERGENCE**

In case (c), where the power series converges absolutely for $|x| < R$ and diverges for $|x| > R$, we call R the *radius of convergence*. In case (a), where the series converges only for $x = 0$, we define the radius of convergence to be $R = 0$; and in case (b), where the series converges absolutely for all x, we define the radius of convergence to be $R = +\infty$. The set of all values of x for which a power series converges is called the *interval of convergence*.

Example 1   Find the interval of convergence and radius of convergence of the power series

$$\sum_{k=0}^{\infty} x^k = 1 + x + x^2 + \cdots + x^k + \cdots$$

Solution.   For every x, the given series is a geometric series with ratio $r = x$. Thus, by Example 8 of Section 11.7, the series converges absolutely if $-1 < x < 1$ and diverges if $|x| \geq 1$. Therefore, the interval of convergence is $(-1, 1)$ and the radius of convergence is $R = 1$.   ◀

Example 2   Find the interval of convergence and radius of convergence of

$$\sum_{k=0}^{\infty} \frac{x^k}{k!}$$

Solution.   We shall apply the ratio test for absolute convergence (11.7.5). For every real number x,

$$\rho = \lim_{k \to +\infty} \left| \frac{u_{k+1}}{u_k} \right| = \lim_{k \to +\infty} \left| \frac{x^{k+1}}{(k+1)!} \cdot \frac{k!}{x^k} \right| = \lim_{k \to +\infty} \left| \frac{x}{k+1} \right| = 0$$

Since $\rho < 1$ for all x, the series converges absolutely for all x. Thus, the interval of convergence is $(-\infty, +\infty)$ and the radius of convergence is $R = +\infty$.   ◀

REMARK.   There is a useful consequence of Example 2. Since

$$\sum_{k=0}^{\infty} \frac{x^k}{k!}$$

converges for all x, Theorem 11.4.1 implies that for all values of x

$$\lim_{k \to +\infty} \frac{x^k}{k!} = 0 \tag{1}$$

We shall need this result later.

Example 3   Find the interval of convergence and radius of convergence of

$$\sum_{k=0}^{\infty} k! x^k$$

Solution.   If $x = 0$, the series has only one nonzero term and therefore converges. If $x \neq 0$, the ratio test yields

$$\rho = \lim_{k \to +\infty} \left| \frac{u_{k+1}}{u_k} \right| = \lim_{k \to +\infty} \left| \frac{(k+1)! x^{k+1}}{k! x^k} \right| = \lim_{k \to +\infty} |(k+1)x| = +\infty$$

Therefore, the series converges if $x = 0$, but diverges for all other x. Consequently, the interval of convergence is the single point $x = 0$ and the radius of convergence is $R = 0$.   ◀

Example 4   Find the interval of convergence and radius of convergence of

$$\sum_{k=0}^{\infty} \frac{(-1)^k x^k}{3^k (k+1)}$$

Solution. Since $|(-1)^k| = |(-1)^{k+1}| = 1$, we obtain

$$\rho = \lim_{k \to +\infty} \left| \frac{u_{k+1}}{u_k} \right| = \lim_{k \to +\infty} \left| \frac{x^{k+1}}{3^{k+1}(k+2)} \cdot \frac{3^k(k+1)}{x^k} \right|$$

$$= \lim_{k \to +\infty} \left[\frac{|x|}{3} \cdot \left(\frac{k+1}{k+2} \right) \right]$$

$$= \frac{|x|}{3} \lim_{k \to +\infty} \left(\frac{1+1/k}{1+2/k} \right) = \frac{|x|}{3}$$

The ratio test for absolute convergence implies that the series converges absolutely if $|x| < 3$ and diverges if $|x| > 3$. The ratio test fails to provide any information when $|x| = 3$, so the cases $x = -3$ and $x = 3$ need separate analyses. Substituting $x = -3$ in the given series yields

$$\sum_{k=0}^{\infty} \frac{(-1)^k(-3)^k}{3^k(k+1)} = \sum_{k=0}^{\infty} \frac{(-1)^k(-1)^k 3^k}{3^k(k+1)} = \sum_{k=0}^{\infty} \frac{1}{k+1}$$

which is the divergent harmonic series $1 + \frac{1}{2} + \frac{1}{3} + \frac{1}{4} + \cdots$. Substituting $x = 3$ in the given series yields

$$\sum_{k=0}^{\infty} \frac{(-1)^k 3^k}{3^k(k+1)} = \sum_{k=0}^{\infty} \frac{(-1)^k}{k+1} = 1 - \frac{1}{2} + \frac{1}{3} - \frac{1}{4} + \cdots$$

which is the conditionally convergent alternating harmonic series. Thus, the interval of convergence for the given series is $(-3, 3]$ and the radius of convergence is $R = 3$. ◀

Example 5 Find the interval and radius of convergence of the series

$$\sum_{k=0}^{\infty} (-1)^k \frac{x^{2k}}{(2k)!}$$

Solution. Since $|(-1)^k| = |(-1)^{k+1}| = 1$, we have

$$\rho = \lim_{k \to +\infty} \left| \frac{u_{k+1}}{u_k} \right| = \lim_{k \to +\infty} \left| \frac{x^{2(k+1)}}{[2(k+1)]!} \cdot \frac{(2k)!}{x^{2k}} \right| = \lim_{k \to +\infty} \left| \frac{x^{2k+2}}{(2k+2)!} \frac{(2k)!}{x^{2k}} \right|$$

$$= \lim_{k \to +\infty} \left| \frac{x^2}{(2k+2)(2k+1)} \right| = x^2 \lim_{k \to +\infty} \frac{1}{(2k+2)(2k+1)} = x^2 \cdot 0 = 0$$

Thus, $\rho < 1$ for all x, which means that the interval of convergence is $(-\infty, +\infty)$ and the radius of convergence is $R = +\infty$. ◀

☐ **POWER SERIES IN** $x - a$ In addition to power series in x, we shall be interested in series of the form

$$\sum_{k=0}^{\infty} c_k(x-a)^k = c_0 + c_1(x-a) + c_2(x-a)^2 + \cdots + c_k(x-a)^k + \cdots$$

where $c_0, c_1, c_2, \ldots$ and a are constants. Such a series is called a ***power series in*** $x - a$. Some examples are

$$\sum_{k=0}^{\infty} \frac{(x-1)^k}{k+1} = 1 + \frac{(x-1)}{2} + \frac{(x-1)^2}{3} + \frac{(x-1)^3}{4} + \cdots$$

$$\sum_{k=0}^{\infty} \frac{(-1)^k(x+3)^k}{k!} = 1 - (x+3) + \frac{(x+3)^2}{2!} - \frac{(x+3)^3}{3!} + \cdots$$

The first is a power series in $x - 1$ ($a = 1$) and the second a power series in $x + 3$ ($a = -3$).

The convergence properties of a power series $\Sigma c_k(x - a)^k$ may be obtained from Theorem 11.8.1 by substituting $X = x - a$ to obtain

$$\sum_{k=0}^{\infty} a_k X^k$$

which is a power series in X. There are three possibilities for this series: it converges only when $X = 0$, or equivalently only when $x = a$; it converges absolutely for all values of X, or equivalently for all values of x; or finally, it converges absolutely for all X satisfying

$$-R < X < R \tag{2}$$

and diverges when $X < -R$ or $X > R$. But (2) can be written as

$$-R < x - a < R$$

or

$$a - R < x < a + R$$

Thus, we are led to the following result.

11.8.2 THEOREM. *For a power series $\Sigma c_k(x - a)^k$, exactly one of the following is true:*

(a) *The series converges only for $x = a$.*

(b) *The series converges absolutely for all real values of x.*

(c) *The series converges absolutely for all x in some finite open interval $(a - R, a + R)$ and diverges if $x < a - R$ or $x > a + R$ (Figure 11.8.2). At the points $x = a - R$ and $x = a + R$, the series may converge absolutely, converge conditionally, or diverge, depending on the particular series.*

Series diverges   Series converges absolutely   Series diverges

$a - R$ a $a + R$

Figure 11.8.2

In cases (a), (b), and (c) of Theorem 11.8.2 the series is said to have **radius of convergence** 0, $+\infty$, and R, respectively. The set of all values of x for which the series converges is called the **interval of convergence**.

Example 6   Find the interval of convergence and radius of convergence of the series

$$\sum_{k=1}^{\infty} \frac{(x-5)^k}{k^2}$$

Solution.   We apply the ratio test for absolute convergence.

$$\rho = \lim_{k \to +\infty} \left| \frac{u_{k+1}}{u_k} \right| = \lim_{k \to +\infty} \left| \frac{(x-5)^{k+1}}{(k+1)^2} \cdot \frac{k^2}{(x-5)^k} \right|$$

$$= \lim_{k \to +\infty} \left[|x-5| \left(\frac{k}{k+1} \right)^2 \right]$$

$$= |x-5| \lim_{k \to +\infty} \left(\frac{1}{1+1/k} \right)^2 = |x-5|$$

Thus, the series converges absolutely if $|x-5| < 1$, or $-1 < x-5 < 1$, or $4 < x < 6$. The series diverges if $x < 4$ or $x > 6$.

To determine the convergence behavior at the endpoints $x = 4$ and $x = 6$, we substitute these values in the given series. If $x = 6$, the series becomes

$$\sum_{k=1}^{\infty} \frac{1^k}{k^2} = \sum_{k=1}^{\infty} \frac{1}{k^2} = 1 + \frac{1}{2^2} + \frac{1}{3^2} + \frac{1}{4^2} + \cdots$$

which is a convergent *p*-series ($p = 2$). If $x = 4$, the series becomes

$$\sum_{k=1}^{\infty} \frac{(-1)^k}{k^2} = -1 + \frac{1}{2^2} - \frac{1}{3^2} + \frac{1}{4^2} - \cdots$$

Since this series converges absolutely, the interval of convergence for the given series is $[4, 6]$. The radius of convergence is $R = 1$ (Figure 11.8.3).   ◀

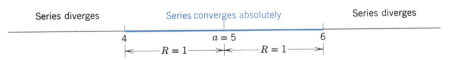

Figure 11.8.3

REMARK.   The ratio test should never be used for testing for convergence at the endpoints of the interval of convergence, since the ratio ρ is always 1 at those points (why?).

▶ **Exercise Set 11.8**

In Exercises 1–24, find the radius of convergence and the interval of convergence.

1. $\displaystyle\sum_{k=0}^{\infty} \frac{x^k}{k+1}$.

2. $\displaystyle\sum_{k=0}^{\infty} 3^k x^k$.

3. $\displaystyle\sum_{k=0}^{\infty} \frac{(-1)^k x^k}{k!}$.

4. $\displaystyle\sum_{k=0}^{\infty} \frac{k!}{2^k} x^k$.

5. $\displaystyle\sum_{k=1}^{\infty} \frac{5^k}{k^2} x^k$.

6. $\displaystyle\sum_{k=2}^{\infty} \frac{x^k}{\ln k}$.

7. $\displaystyle\sum_{k=1}^{\infty} \frac{x^k}{k(k+1)}$.

8. $\displaystyle\sum_{k=0}^{\infty} \frac{(-2)^k x^{k+1}}{k+1}$.

9. $\displaystyle\sum_{k=1}^{\infty} (-1)^{k-1} \frac{x^k}{\sqrt{k}}$.

10. $\displaystyle\sum_{k=0}^{\infty} \frac{(-1)^k x^{2k}}{(2k)!}$.

11. $\displaystyle\sum_{k=0}^{\infty} (-1)^k \frac{x^{2k+1}}{(2k+1)!}$.

12. $\displaystyle\sum_{k=1}^{\infty} (-1)^k \frac{x^{3k}}{k^{3/2}}$.

13. $\displaystyle\sum_{k=0}^{\infty} \frac{3^k}{k!} x^k$.

14. $\displaystyle\sum_{k=2}^{\infty} (-1)^{k+1} \frac{x^k}{k(\ln k)^2}$.

15. $\displaystyle\sum_{k=0}^{\infty} \frac{x^k}{1+k^2}$.

16. $\displaystyle\sum_{k=0}^{\infty} \frac{(x-3)^k}{2^k}$.

17. $\displaystyle\sum_{k=1}^{\infty} (-1)^{k+1} \frac{(x+1)^k}{k}$.

18. $\displaystyle\sum_{k=0}^{\infty} (-1)^k \frac{(x-4)^k}{(k+1)^2}$.

19. $\displaystyle\sum_{k=0}^{\infty} \left(\frac{3}{4}\right)^k (x+5)^k$.

20. $\displaystyle\sum_{k=1}^{\infty} \frac{(2k+1)!}{k^3} (x-2)^k$.

21. $\displaystyle\sum_{k=1}^{\infty} (-1)^k \frac{(x+1)^{2k+1}}{k^2+4}$.

22. $\displaystyle\sum_{k=1}^{\infty} \frac{(\ln k)(x-3)^k}{k}$.

23. $\displaystyle\sum_{k=0}^{\infty} \frac{\pi^k (x-1)^{2k}}{(2k+1)!}$.

24. $\displaystyle\sum_{k=0}^{\infty} \frac{(2x-3)^k}{4^{2k}}$.

In Exercises 25–27, show the first four terms of the series, and find the radius of convergence.

25. $\displaystyle\sum_{k=1}^{\infty} \frac{1 \cdot 2 \cdot 3 \cdots k}{1 \cdot 4 \cdot 7 \cdots (3k-2)} x^k$.

26. $\displaystyle\sum_{k=1}^{\infty} (-1)^k \frac{1 \cdot 2 \cdot 3 \cdots k}{1 \cdot 3 \cdot 5 \cdots (2k-1)} x^{2k+1}$.

27. $\displaystyle\sum_{k=1}^{\infty} \frac{1 \cdot 3 \cdot 5 \cdots (2k-1)}{(2k-2)!} x^k$.

28. Use the root test to find the interval of convergence of $\displaystyle\sum_{k=2}^{\infty} \frac{x^k}{(\ln k)^k}$.

29. Find the interval of convergence of $\displaystyle\sum_{k=0}^{\infty} \frac{(x-a)^k}{b^k}$, where $b > 0$.

30. Find the radius of convergence of the power series $\displaystyle\sum_{k=0}^{\infty} \frac{(pk)!}{(k!)^p} x^k$, where p is a positive integer.

31. Find the radius of convergence of the power series $\displaystyle\sum_{k=0}^{\infty} \frac{(k+p)!}{k!(k+q)!} x^k$, where p and q are positive integers.

32. Prove: If $\displaystyle\lim_{k \to +\infty} |c_k|^{1/k} = L$, where $L \neq 0$, then $1/L$ is the radius of convergence of the power series $\sum_{k=0}^{\infty} c_k x^k$.

33. Prove: If the power series $\sum_{k=0}^{\infty} c_k x^k$ has radius of convergence R, then the series $\sum_{k=0}^{\infty} c_k x^{2k}$ has radius of convergence $\sqrt{R}$.

34. Prove: If the interval of convergence of the series $\sum_{k=0}^{\infty} c_k (x-a)^k$ is $(a-R, a+R]$, then the series converges conditionally at $a+R$.

11.9 TAYLOR AND MACLAURIN SERIES

One of the early applications of calculus was the computation of approximate numerical values for functions such as $\sin x$, $\ln x$, and e^x. One common method for obtaining such values is to approximate the function by a polynomial, then use the polynomial to compute the desired numerical values. In this section we shall discuss procedures for approximating functions by polynomials, and in the next section we shall investigate the errors in these approximations. In this section we shall also see how polynomial approximations lead naturally to the important problem of finding a power series that converges to a specified function.

☐ **APPROXIMATING FUNCTIONS BY POLYNOMIALS**

The problem of primary interest in this section can be phrased informally as follows:

> **PROBLEM.** *Given a function f and a point a on the x-axis, find a polynomial of specified degree that best approximates the function f in the "vicinity" of the point a.*

As stated, the problem is somewhat vague in that we have not specified any requirements on f such as continuity, differentiability, and so forth, and it is not at all evident what we mean by the "best approximation in the vicinity of a point." However, the problem is suggestive enough to get us started, and we shall resolve the ambiguities as we progress.

Suppose that we are interested in approximating a function f in the vicinity of the point $a = 0$ by a polynomial

$$p(x) = c_0 + c_1x + \cdots + c_nx^n \tag{1}$$

Because $p(x)$ has $n + 1$ coefficients, it seems reasonable that we should be able to impose $n + 1$ conditions on this polynomial to achieve a good approximation to $f(x)$. Because the point $a = 0$ is the center of interest, our strategy will be to choose the coefficients of $p(x)$ so that the value of p and its first n derivatives are the same as the value of f and its first n derivatives at $a = 0$. By forcing this high degree of "match" at $a = 0$, it is reasonable to hope that $f(x)$ and $p(x)$ will remain close over some interval (possibly quite small) centered at $a = 0$. Thus, we shall assume that f can be differentiated n times at 0, and we shall try to find the coefficients in (1) such that

$$f(0) = p(0), \quad f'(0) = p'(0), \quad f''(0) = p''(0), \quad \ldots, \quad f^{(n)}(0) = p^{(n)}(0) \tag{2}$$

The first n derivatives of (1) are

$$p(x) = c_0 + c_1x + c_2x^2 + c_3x^3 + \cdots + c_nx^n$$
$$p'(x) = c_1 + 2c_2x + 3c_3x^2 + \cdots + nc_nx^{n-1}$$
$$p''(x) = 2c_2 + 3\cdot 2c_3x + \cdots + n(n-1)c_nx^{n-2}$$
$$p'''(x) = 3\cdot 2c_3 + \cdots + n(n-1)(n-2)c_nx^{n-3}$$
$$\vdots$$
$$p^{(n)}(x) = n(n-1)(n-2)\cdots(1)c_n$$

from which it follows that

$$p(0) = c_0$$
$$p'(0) = c_1$$
$$p''(0) = 2c_2 = 2!c_2$$
$$p'''(0) = 3\cdot 2c_3 = 3!c_3$$
$$\vdots$$
$$p^{(n)}(0) = n(n-1)(n-2)\cdots(1)c_n = n!c_n$$

Thus, to satisfy the conditions in (2) we must have

$$f(0) = c_0$$

$$f'(0) = c_1$$

$$f''(0) = 2!c_2$$

$$f'''(0) = 3!c_3$$

$$\vdots$$

$$f^{(n)}(0) = n!c_n$$

which yields the following values for the coefficients of $p(x)$:

$$c_0 = f(0), \quad c_1 = f'(0), \quad c_2 = \frac{f''(0)}{2!}, \quad c_3 = \frac{f'''(0)}{3!}, \quad \ldots, c_n = \frac{f^{(n)}(0)}{n!}$$

☐ **MACLAURIN POLYNOMIALS**

The polynomial that results by using these coefficients in (1) is called the *nth Maclaurin* polynomial for f*. In summary, we have the following definition.

11.9.1 DEFINITION. If f can be differentiated n times at 0, then we define the ***nth Maclaurin polynomial for f*** to be

$$p_n(x) = f(0) + f'(0)x + \frac{f''(0)}{2!}x^2 + \frac{f'''(0)}{3!}x^3 + \cdots + \frac{f^{(n)}(0)}{n!}x^n \qquad (3)$$

This polynomial has the property that its value and the values of its first n derivatives match the value of $f(x)$ and its first n derivatives when $x = 0$.

Example 1 Find the Maclaurin polynomials p_0, p_1, p_2, p_3, and p_n for e^x.

Solution. Let $f(x) = e^x$. Thus,

$$f'(x) = f''(x) = f'''(x) = \cdots = f^{(n)}(x) = e^x$$

---

*COLIN MACLAURIN (1698–1746). Scottish mathematician. Maclaurin's father, a minister, died when the boy was only six months old, and his mother when he was nine years old. He was then raised by an uncle who was also a minister.

Maclaurin entered Glasgow University as a divinity student, but transferred to mathematics after one year. He received his Master's degree at age 17 and, in spite of his youth, began teaching at Marischal College in Aberdeen, Scotland.

Maclaurin met Isaac Newton during a visit to London in 1719 and from that time on he became Newton's disciple. During that era, some of Newton's analytic methods were bitterly attacked by major mathematicians and much of Maclaurin's important mathematical work resulted from his efforts to defend Newton's ideas geometrically. Maclaurin's work, *A Treatise of Fluxions* (1742), was the first systematic formulation of Newton's methods. The treatise was so carefully done that it was a standard of mathematical rigor in calculus until the work of Cauchy in 1821.

Maclaurin was an outstanding experimentalist. He devised numerous ingenious mechanical devices, made important astronomical observations, performed actuarial computations for insurance societies, and helped to improve maps of the islands around Scotland.

and

$$f(0) = f'(0) = f''(0) = f'''(0) = \cdots = f^{(n)}(0) = e^0 = 1$$

Therefore,

$$p_0(x) = f(0) = 1$$

$$p_1(x) = f(0) + f'(0)x = 1 + x$$

$$p_2(x) = f(0) + f'(0)x + \frac{f''(0)}{2!}x^2 = 1 + x + \frac{x^2}{2!} = 1 + x + \frac{1}{2}x^2$$

$$p_3(x) = f(0) + f'(0)x + \frac{f''(0)}{2!}x^2 + \frac{f'''(0)}{3!}x^3$$

$$= 1 + x + \frac{x^2}{2!} + \frac{x^3}{3!} = 1 + x + \frac{1}{2}x^2 + \frac{1}{6}x^3$$

$$p_n(x) = f(0) + f'(0)x + \frac{f''(0)}{2!}x^2 + \cdots + \frac{f^{(n)}(0)}{n!}x^n$$

$$= 1 + x + \frac{x^2}{2!} + \cdots + \frac{x^n}{n!} \quad \blacktriangleleft$$

Figure 11.9.1 shows the graphs of e^x and its first four Maclaurin polynomials. Note that the graphs of e^x and $p_3(x)$ are virtually indistinguishable over the interval from $-\frac{1}{2}$ to $\frac{1}{2}$. (In the next section we shall investigate in detail the accuracy of Maclaurin polynomial approximations.)

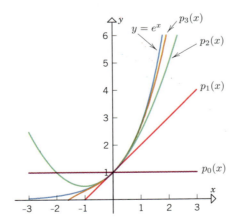

Figure 11.9.1

Example 2   Find the nth Maclaurin polynomial for $\ln(x + 1)$.

Solution.   Let $f(x) = \ln(x + 1)$ and arrange the computations as follows:

$$f(x) = \ln(x + 1) \qquad\qquad f(0) = \ln 1 = 0$$

$$f'(x) = \frac{1}{x+1} \qquad\qquad f'(0) = 1$$

$$f''(x) = -\frac{1}{(x+1)^2} \qquad\qquad f''(0) = -1$$

$$f'''(x) = \frac{2}{(x+1)^3} \qquad\qquad f'''(0) = 2$$

$$f^{(4)}(x) = -\frac{3 \cdot 2}{(x+1)^4} \qquad\qquad f^{(4)}(0) = -3!$$

$$f^{(5)}(x) = \frac{4 \cdot 3 \cdot 2}{(x+1)^5} \qquad\qquad f^{(5)}(0) = 4!$$

$$\vdots \qquad\qquad\qquad\qquad \vdots$$

$$f^{(n)}(x) = (-1)^{n+1}\frac{(n-1)!}{(x+1)^n} \qquad f^{(n)}(0) = (-1)^{n+1}(n-1)!$$

Substituting these values in (3) yields

$$p_n(x) = x - \frac{x^2}{2} + \frac{x^3}{3} - \cdots + (-1)^{n+1}\frac{x^n}{n}$$

The graphs of $f(x) = \ln(x+1)$, $p_1(x)$, $p_2(x)$, $p_3(x)$, and $p_4(x)$ are shown in Figure 11.9.2. Observe that the approximations tend to deteriorate rapidly as x moves away from the origin, but that the higher-degree Maclaurin polynomials remain good approximations over larger intervals. ◀

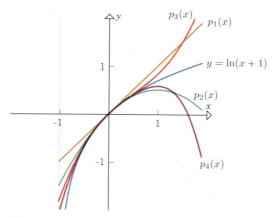

Figure 11.9.2

Example 3 In the Maclaurin polynomials for $\sin x$, only the odd powers of x appear explicitly. To see this, let $f(x) = \sin x$; thus,

$$f(x) = \sin x \qquad f(0) = 0$$
$$f'(x) = \cos x \qquad f'(0) = 1$$
$$f''(x) = -\sin x \qquad f''(0) = 0$$
$$f'''(x) = -\cos x \qquad f'''(0) = -1$$

Since $f^{(4)}(x) = \sin x = f(x)$, the pattern 0, 1, 0, -1 will repeat over and over as we evaluate successive derivatives at 0. Therefore, the successive Maclaurin polynomials for $\sin x$ are

$$p_1(x) = 0 + x = x$$

$$p_2(x) = 0 + x + 0 = x$$

$$p_3(x) = 0 + x + 0 - \frac{x^3}{3!} = x - \frac{x^3}{3!}$$

$$p_4(x) = 0 + x + 0 - \frac{x^3}{3!} + 0 = x - \frac{x^3}{3!}$$

$$p_5(x) = 0 + x + 0 - \frac{x^3}{3!} + 0 + \frac{x^5}{5!} = x - \frac{x^3}{3!} + \frac{x^5}{5!}$$

$$p_6(x) = 0 + x + 0 - \frac{x^3}{3!} + 0 + \frac{x^5}{5!} + 0 = x - \frac{x^3}{3!} + \frac{x^5}{5!}$$

$$p_7(x) = 0 + x + 0 - \frac{x^3}{3!} + 0 + \frac{x^5}{5!} + 0 - \frac{x^7}{7!} = x - \frac{x^3}{3!} + \frac{x^5}{5!} - \frac{x^7}{7!}$$

$$\vdots$$

In general, the Maclaurin polynomials for $\sin x$ are

$$p_{2n+1}(x) = p_{2n+2}(x) = x - \frac{x^3}{3!} + \frac{x^5}{5!} - \frac{x^7}{7!} + \cdots + (-1)^n \frac{x^{2n+1}}{(2n+1)!}$$

$$(n = 0, 1, 2, \ldots)$$

The graphs of $f(x) = \sin x$, $p_1(x)$, $p_3(x)$, $p_5(x)$, and $p_7(x)$ are shown in Figure 11.9.3. Observe that $p_7(x)$ is such a good approximation to $\sin x$ over the interval $[-\pi, \pi]$ that its graph is nearly indistinguishable from that of $\sin x$ over this interval. ◀

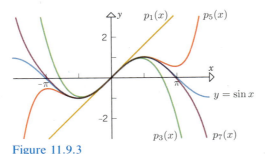

Figure 11.9.3

Example 4   In the Maclaurin polynomials for $\cos x$, only the even powers of x appear explicitly. Using computations similar to those in Example 3, the reader should be able to show that the successive Maclaurin polynomials for $\cos x$ are

$$p_0(x) = p_1(x) = 1$$

$$p_2(x) = p_3(x) = 1 - \frac{x^2}{2!}$$

$$p_4(x) = p_5(x) = 1 - \frac{x^2}{2!} + \frac{x^4}{4!}$$

$$p_6(x) = p_7(x) = 1 - \frac{x^2}{2!} + \frac{x^4}{4!} - \frac{x^6}{6!}$$

$$p_8(x) = p_9(x) = 1 - \frac{x^2}{2!} + \frac{x^4}{4!} - \frac{x^6}{6!} + \frac{x^8}{8!}$$

In general, the Maclaurin polynomials for $\cos x$ are

$$p_{2n}(x) = p_{2n+1}(x) = 1 - \frac{x^2}{2!} + \frac{x^4}{4!} - \cdots + (-1)^n \frac{x^{2n}}{(2n)!}$$

$$(n = 0, 1, 2, \ldots)$$

(See Figure 11.9.4.) ◀

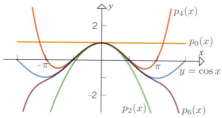

Figure 11.9.4

☐ **TAYLOR POLYNOMIALS**

Because the Maclaurin polynomials for a function f are obtained by matching their values and the values of their derivatives with those of f at the point $a = 0$, the accuracy of the Maclaurin polynomials as approximations to f is best in the vicinity of the origin. Generally, the farther x is from the origin, the less accurate the Maclaurin polynomial is as an approximation to f. If we are interested in a polynomial approximation to f that has its best accuracy in the vicinity of a general point a (possibly different from 0), then we should choose the approximating polynomial p so that its value and the values of its derivatives match those of f at the point $x = a$. The computations are simplest if $p(x)$ is expressed in powers of $x - a$, that is,

$$p(x) = c_0 + c_1(x - a) + c_2(x - a)^2 + \cdots + c_n(x - a)^n \qquad (4)$$

We leave it as an exercise for the reader to calculate the first n derivatives of $p(x)$ and show that

$$p(a) = c_0, \quad p'(a) = c_1, \quad p''(a) = 2!c_2, \quad p'''(a) = 3!c_3, \quad \ldots, \quad p^{(n)}(a) = n!c_n$$

Thus, if we want the values of $p(x)$ and its first n derivatives to match the values of $f(x)$ and its first n derivatives at $x = a$, we must have

$$c_0 = f(a), \quad c_1 = f'(a), \quad c_2 = \frac{f''(a)}{2!}, \quad c_3 = \frac{f'''(a)}{3!}, \quad \ldots, \quad c_n = \frac{f^{(n)}(a)}{n!}$$

Substituting these values in (4) we obtain a polynomial called the *nth Taylor* polynomial about x = a for f.*

> **11.9.2 DEFINITION.** If f can be differentiated n times at a, then we define the **nth Taylor polynomial for f about x = a** to be
>
> $$p_n(x) = f(a) + f'(a)(x - a) + \frac{f''(a)}{2!}(x - a)^2$$
>
> $$+ \frac{f'''(a)}{3!}(x - a)^3 + \cdots + \frac{f^{(n)}(a)}{n!}(x - a)^n \quad (5)$$

REMARK.   Observe that with $a = 0$, the nth Taylor polynomial for f is the nth Maclaurin polynomial for f; that is, the Maclaurin polynomials are special cases of the Taylor polynomials.

Example 5   Find the Taylor polynomials $p_1(x)$, $p_2(x)$, and $p_3(x)$ for $\sin x$ about $x = \pi/3$.

Solution.   Let $f(x) = \sin x$. Thus,

$$f(x) = \sin x \qquad f(\pi/3) = \sin\frac{\pi}{3} = \frac{\sqrt{3}}{2}$$

$$f'(x) = \cos x \qquad f'(\pi/3) = \cos\frac{\pi}{3} = \frac{1}{2}$$

$$f''(x) = -\sin x \qquad f''(\pi/3) = -\sin\frac{\pi}{3} = -\frac{\sqrt{3}}{2}$$

---

* BROOK TAYLOR (1685–1731). English mathematician. Taylor was born of well-to-do parents. Musicians and artists were entertained frequently in the Taylor home, which undoubtedly had a lasting influence on young Brook. In later years, Taylor published a definitive work on the mathematical theory of perspective and obtained major mathematical results about the vibrations of strings. There also exists an unpublished work, *On Musick*, that was intended to be part of a joint paper with Isaac Newton.

Taylor's life was scarred with unhappiness, illness, and tragedy. Because his first wife was not rich enough to suit his father, the two men argued bitterly and parted ways. Subsequently, his wife died in childbirth. Then, after he remarried, his second wife also died in childbirth, though his daughter survived.

Taylor's most productive period was from 1714 to 1719, during which time he wrote on a wide range of subjects—magnetism, capillary action, thermometers, perspective, and calculus. In his final years, Taylor devoted his writing efforts to religion and philosophy. According to Taylor, the results that bear his name were motivated by coffeehouse conversations about works of Newton on planetary motion and works of Halley (''Halley's comet'') on roots of polynomials.

Taylor's writing style was so terse and hard to understand that he never received credit for many of his innovations.

$$f'''(x) = -\cos x \qquad f'''(\pi/3) = -\cos\frac{\pi}{3} = -\frac{1}{2}$$

Substituting in (5) with $a = \pi/3$ yields

$$p_1(x) = f(\pi/3) + f'(\pi/3)\left(x - \frac{\pi}{3}\right) = \frac{\sqrt{3}}{2} + \frac{1}{2}\left(x - \frac{\pi}{3}\right)$$

$$p_2(x) = f(\pi/3) + f'(\pi/3)\left(x - \frac{\pi}{3}\right) + \frac{f''(\pi/3)}{2!}\left(x - \frac{\pi}{3}\right)^2$$

$$= \frac{\sqrt{3}}{2} + \frac{1}{2}\left(x - \frac{\pi}{3}\right) - \frac{\sqrt{3}}{2\cdot 2!}\left(x - \frac{\pi}{3}\right)^2$$

$$p_3(x) = f(\pi/3) + f'(\pi/3)\left(x - \frac{\pi}{3}\right) + \frac{f''(\pi/3)}{2!}\left(x - \frac{\pi}{3}\right)^2 + \frac{f'''(\pi/3)}{3!}\left(x - \frac{\pi}{3}\right)^3$$

$$= \frac{\sqrt{3}}{2} + \frac{1}{2}\left(x - \frac{\pi}{3}\right) - \frac{\sqrt{3}}{2\cdot 2!}\left(x - \frac{\pi}{3}\right)^2 - \frac{1}{2\cdot 3!}\left(x - \frac{\pi}{3}\right)^3 \quad \blacktriangleleft$$

☐ **SIGMA NOTATION FOR TAYLOR AND MACLAURIN POLYNOMIALS**

Frequently, it is convenient to express the defining formula for the Taylor polynomial in sigma notation. To do this, we use the notation $f^{(k)}(a)$ to denote the kth derivative of f at $x = a$, and we make the added convention that $f^{(0)}(a)$ denotes $f(a)$. This enables us to write

$$\sum_{k=0}^{n} \frac{f^{(k)}(a)}{k!}(x - a)^k = f(a) + f'(a)(x - a)$$

$$+ \frac{f''(a)}{2!}(x - a)^2 + \cdots + \frac{f^{(n)}(a)}{n!}(x - a)^n$$

In particular, the nth Maclaurin polynomial for $f(x)$ may be written as

$$\sum_{k=0}^{n} \frac{f^{(k)}(0)}{k!}x^k = f(0) + f'(0)x + \frac{f''(0)}{2!}x^2 + \cdots + \frac{f^{(n)}(0)}{n!}x^n \qquad (6)$$

☐ **TAYLOR AND MACLAURIN SERIES**

If we assume that the function f has derivatives of all orders at $x = 0$, then the nth Maclaurin polynomial of f can be regarded as a partial sum of the infinite series

$$\sum_{k=0}^{\infty} \frac{f^{(k)}(0)}{k!}x^k = f(0) + f'(0)x + \frac{f''(0)}{2!}x^2 + \cdots + \frac{f^{(k)}(0)}{k!}x^k + \cdots$$

(A similar conclusion can be drawn for the Taylor polynomials about $x = a$.) As n increases, the Maclaurin polynomials that form the partial sums have more and more derivatives whose values match those of f at the origin, so for each fixed x in the vicinity of the origin, it is reasonable to expect that the partial sums will become better and better approximations of $f(x)$ and that they might actually converge to $f(x)$. Should this happen, we would have

$$f(x) = \lim_{n\to+\infty} \sum_{k=0}^{n} \frac{f^{(k)}(0)}{k!}x^k = \sum_{k=0}^{\infty} \frac{f^{(k)}(0)}{k!}x^k$$

which expresses $f(x)$ as the sum of a power series in the vicinity of the origin. A similar conclusion can be drawn for the Taylor polynomials at any point $x = a$. We make the following definitions.

11.9.3   DEFINITION.   If f has derivatives of all orders at a, then we define the *Taylor series for f about x = a* to be

$$\sum_{k=0}^{\infty} \frac{f^{(k)}(a)}{k!} (x - a)^k = f(a) + f'(a)(x - a)$$

$$+ \frac{f''(a)}{2!} (x - a)^2 + \cdots + \frac{f^{(k)}(a)}{k!} (x - a)^k + \cdots \quad (7)$$

11.9.4   DEFINITION.   If f has derivatives of all orders at 0, then we define the *Maclaurin series for f* to be

$$\sum_{k=0}^{\infty} \frac{f^{(k)}(0)}{k!} x^k = f(0) + f'(0)x + \frac{f''(0)}{2!} x^2 + \cdots + \frac{f^{(k)}(0)}{k!} x^k + \cdots \quad (8)$$

Observe that the Maclaurin series for f is just the Taylor series for f about $a = 0$.

Example 6   In Example 1, we found the nth Maclaurin polynomial for the function e^x to be

$$\sum_{k=0}^{n} \frac{x^k}{k!} = 1 + x + \frac{x^2}{2!} + \cdots + \frac{x^n}{n!}$$

Thus, the Maclaurin series for e^x is

$$\sum_{k=0}^{\infty} \frac{x^k}{k!} = 1 + x + \frac{x^2}{2!} + \frac{x^3}{3!} + \cdots + \frac{x^n}{n!} + \cdots$$

Similarly, from Example 3 the Maclaurin series for $\sin x$ is

$$\sum_{k=0}^{\infty} (-1)^k \frac{x^{2k+1}}{(2k + 1)!} = x - \frac{x^3}{3!} + \frac{x^5}{5!} - \frac{x^7}{7!} + \cdots$$

and from Example 4 the Maclaurin series for $\cos x$ is

$$\sum_{k=0}^{\infty} (-1)^k \frac{x^{2k}}{(2k)!} = 1 - \frac{x^2}{2!} + \frac{x^4}{4!} - \frac{x^6}{6!} + \cdots \quad \blacktriangleleft$$

Example 7   Find the Taylor series about $x = 1$ for $1/x$.

Solution.   Let $f(x) = 1/x$ so that

$$f(x) = \frac{1}{x} \qquad\qquad f(1) = 1$$

$$f'(x) = -\frac{1}{x^2} \qquad\qquad f'(1) = -1$$

$$f''(x) = \frac{2}{x^3} \qquad\qquad f''(1) = 2!$$

$$f'''(x) = -\frac{3 \cdot 2}{x^4} \qquad\qquad f'''(1) = -3!$$

$$f^{(4)}(x) = \frac{4 \cdot 3 \cdot 2}{x^5} \qquad\qquad f^{(4)}(1) = 4!$$

$$\vdots \qquad\qquad\qquad \vdots$$

$$f^{(k)}(x) = (-1)^k \frac{k!}{x^{k+1}} \qquad f^{(k)}(1) = (-1)^k k!$$

$$\vdots \qquad\qquad\qquad \vdots$$

Thus, substituting in (7) with $a = 1$ yields

$$\sum_{k=0}^{\infty} \frac{(-1)^k k!}{k!}(x - 1)^k = \sum_{k=0}^{\infty} (-1)^k (x - 1)^k$$

$$= 1 - (x - 1) + (x - 1)^2 - (x - 1)^3 + \cdots \quad \blacktriangleleft$$

We conclude this section by emphasizing again that we have no guarantee that the Maclaurin and Taylor series of a function f converge, or if they do converge, that the sum is $f(x)$. Convergence questions will be addressed in the next section.

▶ Exercise Set 11.9

In Exercises 1–12, find the fourth Maclaurin polynomial ($n = 4$) for the given function.

1. e^{-2x}.

2. $\dfrac{1}{1 + x}$.

3. $\sin 2x$.

4. $e^x \cos x$.

5. $\tan x$.

6. $x^3 - x^2 + 2x + 1$.

7. xe^x.

8. $\tan^{-1} x$.

9. $\sec x$.

10. $\sqrt{1 + x}$.

11. $\ln(3 + 2x)$.

12. $\sinh x$.

In Exercises 13–22, find the third Taylor polynomial ($n = 3$) about $x = a$ for the given function.

13. e^x; $a = 1$.

14. $\ln x$; $a = 1$.

15. $\sqrt{x}$; $a = 4$.

16. $x^4 + x - 3$; $a = -2$.

17. $\cos x$; $a = \dfrac{\pi}{4}$.

18. $\tan x$; $a = \dfrac{\pi}{3}$.

19. $\sin \pi x$; $a = -\dfrac{1}{3}$.

20. $\csc x$; $a = \dfrac{\pi}{2}$.

21. $\tan^{-1} x$; $a = 1$.

22. $\cosh x$; $a = \ln 2$.

In Exercises 23–31, find the Maclaurin series for the given function. Express your answer in sigma notation.

23. e^{-x}.   **24.** e^{ax}.   **25.** $\dfrac{1}{1+x}$.

26. xe^x.   **27.** $\ln(1+x)$.   **28.** $\sin \pi x$.

29. $\cos\left(\dfrac{x}{2}\right)$.   **30.** $\sinh x$.   **31.** $\cosh x$.

In Exercises 32–39, find the Taylor series about $x = a$ for the given function. Express your answer in sigma notation.

32. $\dfrac{1}{x}$; $a = 3$.   **33.** $\dfrac{1}{x}$; $a = -1$.

34. e^x; $a = 2$.   **35.** $\ln x$; $a = 1$.

36. $\cos x$; $a = \pi/2$.   **37.** $\sin \pi x$; $a = 1/2$.

38. $\dfrac{1}{x+2}$; $a = 3$.   **39.** $\sinh x$; $a = \ln 4$.

40. Prove: The value of $p_n(x) = \displaystyle\sum_{k=0}^{n} \dfrac{f^{(k)}(a)}{k!}(x-a)^k$ and its first n derivatives match the value of $f(x)$ and its first n derivatives at $x = a$.

■ 11.10 TAYLOR FORMULA WITH REMAINDER; CONVERGENCE OF TAYLOR SERIES

In this section we shall analyze the error that results when a function f is approximated by a Taylor or Maclaurin polynomial. This work will enable us to investigate convergence properties of Taylor and Maclaurin series.

□ **TAYLOR'S THEOREM**

If we approximate a function f by its nth Taylor polynomial p_n, then the error at a point x is represented by the difference $f(x) - p_n(x)$. This difference is commonly called the **nth remainder** and is denoted by

$$R_n(x) = f(x) - p_n(x)$$

The following theorem gives an important explicit formula for this remainder.

11.10.1 THEOREM (*Taylor's Theorem*). *Suppose that a function f can be differentiated n + 1 times at each point in an interval containing the point a, and let*

$$p_n(x) = f(a) + f'(a)(x-a) + \frac{f''(a)}{2!}(x-a)^2 + \cdots + \frac{f^{(n)}(a)}{n!}(x-a)^n$$

be the nth Taylor polynomial about x = a for f. Then for each x in the interval, there is at least one point c between a and x such that

$$R_n(x) = f(x) - p_n(x) = \frac{f^{(n+1)}(c)}{(n+1)!}(x-a)^{n+1} \tag{1}$$

In the foregoing theorem, the statement that c is "between" a and x means that c is in the interval (a, x) if $a < x$, or in (x, a) if $x < a$, or $c = a$ if $x = a$.

Rewriting (1) as

$$f(x) = p_n(x) + \frac{f^{(n+1)}(c)}{(n+1)!}(x-a)^{n+1}$$

then substituting the expression for $p_n(x)$ stated in Theorem 11.10.1 yields the following result, called ***Taylor's formula with remainder***.

$$f(x) = f(a) + f'(a)(x-a) + \frac{f''(a)}{2!}(x-a)^2 + \cdots$$

$$+ \frac{f^{(n)}(a)}{n!}(x-a)^n + \frac{f^{(n+1)}(c)}{(n+1)!}(x-a)^{n+1} \qquad (2)$$

□ **LAGRANGE'S FORM OF THE REMAINDER**

This formula expresses $f(x)$ as the sum of its nth Taylor polynomial about $x = a$ plus a remainder or error term,

$$R_n(x) = \frac{f^{(n+1)}(c)}{(n+1)!}(x-a)^{n+1} \qquad (3)$$

in which c is some unspecified number between a and x. The value of c depends on a, x, and n. Historically, (3) was not discovered by Taylor, but rather by Joseph Louis Lagrange.* For this reason, (3) is frequently called ***Lagrange's form of the remainder***. Other formulas for the remainder can be found in ad-

---

*JOSEPH LOUIS LAGRANGE (1736–1813). French–Italian mathematician and astronomer. Lagrange, the son of a public official, was born in Turin, Italy. (Baptismal records list his name as Giuseppe Lodovico Lagrangia.) Although his father wanted him to be a lawyer, Lagrange was attracted to mathematics and astronomy after reading a memoir by the astronomer Halley. At age 16 he began to study mathematics on his own and by age 19 was appointed to a professorship at the Royal Artillery School in Turin. The following year Lagrange sent Euler solutions to some famous problems using new methods that eventually blossomed into a branch of mathematics called calculus of variations. These methods and Lagrange's applications of them to problems in celestial mechanics were so monumental that by age 25 he was regarded by many of his contemporaries as the greatest living mathematician.

In 1776, on the recommendation of Euler, he was chosen to succeed Euler as the director of the Berlin Academy. During his stay in Berlin, Lagrange distinguished himself not only in celestial mechanics, but also in algebraic equations and the theory of numbers. After twenty years in Berlin, he moved to Paris at the invitation of Louis XVI. He was given apartments in the Louvre and treated with great honor, even during the revolution.

Napoleon was a great admirer of Lagrange and showered him with honors—count, senator, and Legion of Honor. The years Lagrange spent in Paris were devoted primarily to didactic treatises summarizing his mathematical conceptions. One of Lagrange's most famous works is a memoir, *Mécanique Analytique*, in which he reduced the theory of mechanics to a few general formulas from which all other necessary equations could be derived.

It is an interesting historical fact that Lagrange's father speculated unsuccessfully in several financial ventures, so his family was forced to live quite modestly. Lagrange himself stated that if his family had money, he would not have made mathematics his vocation.

In spite of his fame, Lagrange was always a shy and modest man. On his death, he was buried with honor in the Pantheon.

vanced calculus texts. The proof of Taylor's Theorem will be deferred until the end of the section.

For Maclaurin polynomials ($a = 0$), (2) and (3) become

$$f(x) = f(0) + f'(0)x + \frac{f''(0)}{2!}x^2 + \cdots + \frac{f^{(n)}(0)}{n!}x^n + \frac{f^{(n+1)}(c)}{(n+1)!}x^{n+1} \quad (4)$$

$$R_n(x) = \frac{f^{(n+1)}(c)}{(n+1)!}x^{n+1} \quad \text{(where } c \text{ is between 0 and } x\text{)} \quad (5)$$

Example 1 From Example 1 of the previous section, the nth Maclaurin polynomial for $f(x) = e^x$ is

$$1 + x + \frac{x^2}{2!} + \cdots + \frac{x^n}{n!}$$

Since $f^{(n+1)}(x) = e^x$, it follows that

$$f^{(n+1)}(c) = e^c$$

Thus, from Taylor's formula with remainder (4)

$$e^x = 1 + x + \frac{x^2}{2!} + \cdots + \frac{x^n}{n!} + \frac{e^c}{(n+1)!}x^{n+1} \quad (6)$$

where c is between 0 and x. Note that (6) is valid for all real values of x since the hypotheses of Taylor's Theorem are satisfied on the interval $(-\infty, +\infty)$ (verify). ◀

Example 2 In Example 5 of the previous section, we found the Taylor polynomial of degree 3 about $x = \pi/3$ for $f(x) = \sin x$ to be

$$\frac{\sqrt{3}}{2} + \frac{1}{2}\left(x - \frac{\pi}{3}\right) - \frac{\sqrt{3}}{4}\left(x - \frac{\pi}{3}\right)^2 - \frac{1}{12}\left(x - \frac{\pi}{3}\right)^3$$

Since $f^{(4)}(x) = \sin x$ (verify), it follows that $f^{(4)}(c) = \sin c$. Thus, from Taylor's formula with remainder (2) in the case $a = \pi/3$ and $n = 3$ we obtain

$$\sin x = \frac{\sqrt{3}}{2} + \frac{1}{2}\left(x - \frac{\pi}{3}\right) - \frac{\sqrt{3}}{4}\left(x - \frac{\pi}{3}\right)^2 - \frac{1}{12}\left(x - \frac{\pi}{3}\right)^3 + \frac{\sin c}{4!}\left(x - \frac{\pi}{3}\right)^4$$

where c is some number between $\pi/3$ and x. ◀

□ CONVERGENCE OF TAYLOR SERIES

In the previous section we raised the problem of finding those values of x, if any, for which the Taylor series for f about $x = a$ converges to $f(x)$, that is,

$$f(x) = f(a) + f'(a)(x - a) + \frac{f''(a)}{2!}(x - a)^2$$

$$+ \cdots + \frac{f^{(k)}(a)}{k!}(x - a)^k + \cdots \quad (7)$$

To solve this problem, let us write (7) in sigma notation:

$$f(x) = \sum_{k=0}^{\infty} \frac{f^{(k)}(a)}{k!}(x - a)^k$$

This equality is equivalent to

$$f(x) = \lim_{n \to +\infty} \sum_{k=0}^{n} \frac{f^{(k)}(a)}{k!}(x - a)^k$$

or

$$\lim_{n \to +\infty} \left[f(x) - \sum_{k=0}^{n} \frac{f^{(k)}(a)}{k!}(x - a)^k \right] = 0 \quad (8)$$

But the bracketed expression in (8) is $R_n(x)$, since it is the difference between $f(x)$ and its nth Taylor polynomial about $x = a$. Thus, (8) can be written as

$$\lim_{n \to +\infty} R_n(x) = 0$$

which leads us to the following result.

11.10.2 THEOREM. *The equality*

$$f(x) = \sum_{k=0}^{\infty} \frac{f^{(k)}(a)}{k!}(x - a)^k$$

holds if and only if $\lim_{n \to +\infty} R_n(x) = 0$.

To paraphrase this theorem, the *Taylor series for f converges to f(x) at precisely those points where the remainder approaches zero.*

Example 3 Show that the Maclaurin series for e^x converges to e^x for all x.

Solution. We want to show that

$$e^x = 1 + x + \frac{x^2}{2!} + \frac{x^3}{3!} + \cdots + \frac{x^n}{n!} + \cdots \quad (9)$$

holds for all x. From (6),

$$e^x = 1 + x + \frac{x^2}{2!} + \cdots + \frac{x^n}{n!} + R_n(x)$$

where

$$R_n(x) = \frac{e^c}{(n + 1)!} x^{n+1}$$

and c is between 0 and x. Thus, we must show that for all x

$$\lim_{n \to +\infty} R_n(x) = \lim_{n \to +\infty} \frac{e^c}{(n + 1)!} x^{n+1} = 0 \tag{10}$$

(Keep in mind that c is a function of x, not a constant.) Our proof of this result will hinge on the limit

$$\lim_{n \to +\infty} \frac{x^{n+1}}{(n + 1)!} = 0 \tag{11}$$

which follows from Formula (1) of Section 11.8. We shall consider three cases: $x > 0$, $x < 0$, and $x = 0$. If $x > 0$, then

$$0 < c < x$$

from which it follows that

$$0 < e^c < e^x$$

and consequently

$$0 < \frac{e^c}{(n + 1)!} x^{n+1} < \frac{e^x}{(n + 1)!} x^{n+1} \tag{12}$$

From (11) we obtain

$$\lim_{n \to +\infty} \frac{e^x}{(n + 1)!} x^{n+1} = e^x \lim_{n \to +\infty} \frac{x^{n+1}}{(n + 1)!} = e^x \cdot 0 = 0$$

Thus, from (12) and the Squeezing Theorem (2.8.2), result (10) follows.

In the case $x < 0$, we have $c < 0$ since c is between 0 and x. Thus, $0 < e^c < 1$ and consequently

$$0 < e^c \left| \frac{x^{n+1}}{(n + 1)!} \right| < \left| \frac{x^{n+1}}{(n + 1)!} \right|$$

or

$$0 < \left| \frac{e^c}{(n + 1)!} x^{n+1} \right| < \left| \frac{x^{n+1}}{(n + 1)!} \right|$$

or

$$0 < |R_n(x)| < \left| \frac{x^{n+1}}{(n+1)!} \right|$$

From (11) and the Squeezing Theorem it follows that $\lim\limits_{n \to +\infty} |R_n(x)| = 0$, and so $\lim\limits_{n \to +\infty} R_n(x) = 0$.

Convergence in the case $x = 0$ is obvious since (9) reduces to

$$e^0 = 1 + 0 + 0 + 0 + \cdots \quad \blacktriangleleft$$

Example 4 Show that the Maclaurin series for $\sin x$ converges to $\sin x$ for all x.

Solution. We must show that

$$\sin x = x - \frac{x^3}{3!} + \frac{x^5}{5!} - \frac{x^7}{7!} + \cdots$$

holds for all x. Let $f(x) = \sin x$, so that for all x either

$$f^{(n+1)}(x) = \pm \cos x \quad \text{or} \quad f^{(n+1)}(x) = \pm \sin x$$

In all of these cases $|f^{(n+1)}(x)| \leq 1$, so that for all possible values of c

$$|f^{(n+1)}(c)| \leq 1$$

Hence,

$$0 \leq |R_n(x)| = \left| \frac{f^{(n+1)}(c)}{(n+1)!} x^{n+1} \right| \leq \frac{|x|^{n+1}}{(n+1)!} \tag{13}$$

From (11) with $|x|$ replacing x,

$$\lim_{n \to +\infty} \frac{|x|^{n+1}}{(n+1)!} = 0$$

so from (13) and the Squeezing Theorem

$$\lim_{n \to +\infty} |R_n(x)| = 0$$

and consequently

$$\lim_{n \to +\infty} R_n(x) = 0$$

for all x. $\blacktriangleleft$

Example 5 Find the Taylor series for $\sin x$ about $x = \pi/2$ and show that the series converges to $\sin x$ for all x.

Solution.   Let $f(x) = \sin x$. Thus,

$$f(x) \quad = \sin x \qquad f\left(\frac{\pi}{2}\right) \quad = \sin \frac{\pi}{2} = 1$$

$$f'(x) \quad = \cos x \qquad f'\left(\frac{\pi}{2}\right) \quad = \cos \frac{\pi}{2} = 0$$

$$f''(x) \quad = -\sin x \qquad f''\left(\frac{\pi}{2}\right) \quad = -\sin \frac{\pi}{2} = -1$$

$$f'''(x) \quad = -\cos x \qquad f'''\left(\frac{\pi}{2}\right) \quad = -\cos \frac{\pi}{2} = 0$$

Since $f^{(4)}(x) = \sin x$, the pattern $1, 0, -1, 0$ will repeat over and over as we evaluate successive derivatives at $\pi/2$. Thus, the Taylor series representation about $x = \pi/2$ for $\sin x$ is

$$\sin x = 1 - \frac{1}{2!}\left(x - \frac{\pi}{2}\right)^2 + \frac{1}{4!}\left(x - \frac{\pi}{2}\right)^4 - \frac{1}{6!}\left(x - \frac{\pi}{2}\right)^6 + \cdots \qquad (14)$$

which we are trying to show is valid for all x. As shown in Example 4, $|f^{(n+1)}(c)| \leq 1$, so

$$0 \leq |R_n(x)| = \left|\frac{f^{(n+1)}(c)}{(n+1)!}\left(x - \frac{\pi}{2}\right)^{n+1}\right| \leq \frac{\left|x - \frac{\pi}{2}\right|^{n+1}}{(n+1)!}$$

From (11) with $|x - \pi/2|$ replacing x it follows that

$$\lim_{n \to +\infty} \frac{\left|x - \frac{\pi}{2}\right|^{n+1}}{(n+1)!} = 0$$

so that by the same argument given in Example 4, $\lim_{n \to +\infty} R_n(x) = 0$ for all x.

This shows that (14) is valid for all x.   ◀

REMARK.   It can be shown that the Taylor series for e^x, $\sin x$, and $\cos x$ about any point $x = a$ converges to these functions for all x.

☐ **CONSTRUCTING MACLAURIN SERIES BY SUBSTITUTION**

Sometimes Maclaurin series can be obtained by substituting in other Maclaurin series.

Example 6   Using the Maclaurin series

$$e^x = 1 + x + \frac{x^2}{2!} + \frac{x^3}{3!} + \frac{x^4}{4!} + \cdots \qquad -\infty < x < +\infty$$

we can derive the Maclaurin series for e^{-x} by substituting $-x$ for x to obtain

$$e^{-x} = 1 + (-x) + \frac{(-x)^2}{2!} + \frac{(-x)^3}{3!} + \frac{(-x)^4}{4!} + \cdots \qquad -\infty < -x < +\infty$$

or

$$e^{-x} = 1 - x + \frac{x^2}{2!} - \frac{x^3}{3!} + \frac{x^4}{4!} - \cdots \qquad -\infty < x < +\infty$$

From the Maclaurin series for e^x and e^{-x} we can obtain the Maclaurin series for $\cosh x$ by writing

$$\cosh x = \frac{1}{2}(e^x + e^{-x}) = \frac{1}{2}\left(\left[1 + x + \frac{x^2}{2!} + \frac{x^3}{3!} + \frac{x^4}{4!} + \cdots\right]\right.$$
$$\left. + \left[1 - x + \frac{x^2}{2!} - \frac{x^3}{3!} + \frac{x^4}{4!} + \cdots\right]\right)$$

or

$$\cosh x = 1 + \frac{x^2}{2!} + \frac{x^4}{4!} + \cdots \qquad -\infty < x < +\infty \qquad \blacktriangleleft$$

REMARK. We could have derived the foregoing Maclaurin series directly. However, sometimes indirect methods are useful when it is messy to calculate the higher derivatives required for a Maclaurin series. There is, however, a loose thread in the logic of Example 6. We have produced a power series in x that converges to $\cosh x$ for all x. But isn't it conceivable that we have produced a power series in x *different* from the Maclaurin series? In Section 11.12 we shall show that if a power series in $x - a$ converges to $f(x)$ on some interval containing a, then the series must be the Taylor series for f about $x = a$. Thus, we are assured the series obtained for $\cosh x$ is, in fact, the Maclaurin series.

Example 7 Using the Maclaurin series

$$\frac{1}{1 - x} = 1 + x + x^2 + x^3 + \cdots \qquad -1 < x < 1$$

we can derive the Maclaurin series for $1/(1 - 2x^2)$ by substituting $2x^2$ for x to obtain

$$\frac{1}{1 - 2x^2} = 1 + (2x^2) + (2x^2)^2 + (2x^2)^3 + \cdots \qquad -1 < 2x^2 < 1$$

or

$$\frac{1}{1 - 2x^2} = 1 + 2x^2 + 4x^4 + 8x^6 + \cdots = \sum_{k=0}^{\infty} 2^k x^{2k} \qquad -1 < 2x^2 < 1$$

Since $2x^2 \geq 0$ for all x, the convergence condition $-1 < 2x^2 < 1$ can be written in the equivalent form $0 \leq 2x^2 < 1$ or $0 \leq x^2 < 1/2$ or $-1/\sqrt{2} < x < 1/\sqrt{2}$. $\qquad \blacktriangleleft$

☐ **BINOMIAL SERIES**

If m is a real number, then the Maclaurin series for $(1 + x)^m$ is called the *binomial series;* it is given by (verify)

$$1 + mx + \frac{m(m - 1)}{2!}x^2 + \frac{m(m - 1)(m - 2)}{3!}x^3 + \cdots$$

REMARK.   If m is a nonnegative integer, then $f(x) = (1 + x)^m$ is a polynomial of degree m, so

$$f^{(m+1)}(0) = f^{(m+2)}(0) = f^{(m+3)}(0) = \cdots = 0$$

and the binomial series reduces to the familiar binomial expansion

$$(1 + x)^m = 1 + mx + \frac{m(m-1)}{2!} x^2 + \frac{m(m-1)(m-2)}{3!} x^3 + \cdots + x^m$$

which is valid for $-\infty < x < +\infty$.

It is proved in advanced calculus that if m is not a nonnegative integer, then the binomial series converges to $(1 + x)^m$ if $|x| < 1$. Thus, for such values of x

$$(1 + x)^m = 1 + mx + \frac{m(m-1)}{2!} x^2 + \cdots$$
$$+ \frac{m(m-1)(m-2)\cdots(m-k+1)}{k!} x^k + \cdots \qquad (15)$$

or in sigma notation

$$(1 + x)^m = 1 + \sum_{k=1}^{\infty} \frac{m(m-1)\cdots(m-k+1)}{k!} x^k \quad \text{if } |x| < 1 \qquad (16)$$

Example 8   Express $1/\sqrt{1 + x}$ as a binomial series.

Solution.   Substituting $m = -\frac{1}{2}$ in (15) yields

$$\frac{1}{\sqrt{1+x}} = 1 - \frac{1}{2}x + \frac{(-\frac{1}{2})(-\frac{1}{2}-1)}{2!} x^2 + \frac{(-\frac{1}{2})(-\frac{1}{2}-1)(-\frac{1}{2}-2)}{3!} x^3$$

$$+ \cdots + \frac{(-\frac{1}{2})(-\frac{3}{2})(-\frac{5}{2})\cdots(-\frac{1}{2}-k+1)}{k!} x^k + \cdots$$

$$= 1 - \frac{1}{2}x + \frac{1 \cdot 3}{2^2 \cdot 2!} x^2 - \frac{1 \cdot 3 \cdot 5}{2^3 \cdot 3!} x^3 + \cdots$$

$$+ (-1)^k \frac{1 \cdot 3 \cdot 5 \cdots (2k-1)}{2^k k!} x^k + \cdots \qquad \blacktriangleleft$$

For reference, we have listed in Table 11.10.1 the Maclaurin series for a number of important functions, and we have indicated the interval over which the series converges to the function. It should be noted that the intervals of convergence stated for $\ln(1 + x)$ and $\tan^{-1} x$ are somewhat difficult to obtain directly. However, these intervals can be obtained by indirect methods that we shall study in the last section of this chapter.

Table 11.10.1

MACLAURIN SERIES	INTERVAL OF VALIDITY
$\dfrac{1}{1-x} = \displaystyle\sum_{k=0}^{\infty} x^k = 1 + x + x^2 + x^3 + \cdots$	$-1 < x < 1$
$e^x = \displaystyle\sum_{k=0}^{\infty} \dfrac{x^k}{k!} = 1 + x + \dfrac{x^2}{2!} + \dfrac{x^3}{3!} + \dfrac{x^4}{4!} + \cdots$	$-\infty < x < +\infty$
$\sin x = \displaystyle\sum_{k=0}^{\infty} (-1)^k \dfrac{x^{2k+1}}{(2k+1)!} = x - \dfrac{x^3}{3!} + \dfrac{x^5}{5!} - \dfrac{x^7}{7!} + \cdots$	$-\infty < x < +\infty$
$\cos x = \displaystyle\sum_{k=0}^{\infty} (-1)^k \dfrac{x^{2k}}{(2k)!} = 1 - \dfrac{x^2}{2!} + \dfrac{x^4}{4!} - \dfrac{x^6}{6!} + \cdots$	$-\infty < x < +\infty$
$\ln(1+x) = \displaystyle\sum_{k=0}^{\infty} (-1)^k \dfrac{x^{k+1}}{k+1} = x - \dfrac{x^2}{2} + \dfrac{x^3}{3} - \dfrac{x^4}{4} + \cdots$	$-1 < x \leq 1$
$\tan^{-1} x = \displaystyle\sum_{k=0}^{\infty} (-1)^k \dfrac{x^{2k+1}}{2k+1} = x - \dfrac{x^3}{3} + \dfrac{x^5}{5} - \dfrac{x^7}{7} + \cdots$	$-1 \leq x \leq 1$
$\sinh x = \displaystyle\sum_{k=0}^{\infty} \dfrac{x^{2k+1}}{(2k+1)!} = x + \dfrac{x^3}{3!} + \dfrac{x^5}{5!} + \dfrac{x^7}{7!} + \cdots$	$-\infty < x < +\infty$
$\cosh x = \displaystyle\sum_{k=0}^{\infty} \dfrac{x^{2k}}{(2k)!} = 1 + \dfrac{x^2}{2!} + \dfrac{x^4}{4!} + \dfrac{x^6}{6!} + \cdots$	$-\infty < x < +\infty$
$(1+x)^m = 1 + \displaystyle\sum_{k=1}^{\infty} \dfrac{m(m-1)\cdots(m-k+1)}{k!} x^k$	$-1 < x < 1$

■ **OPTIONAL**

We conclude this section with a proof of Taylor's Theorem.

Proof of Theorem 11.10.1. By hypothesis, f can be differentiated $n + 1$ times at each point in an interval containing the point a. Choose any point b in this interval. To be specific, we shall assume that $b > a$. (The cases $b < a$ and $b = a$ are left to the reader.) Let $p_n(x)$ be the nth Taylor polynomial for $f(x)$ about $x = a$ and define

$$h(x) = f(x) - p_n(x) \tag{17}$$
$$g(x) = (x - a)^{n+1} \tag{18}$$

Because $f(x)$ and $p_n(x)$ have the same value and the same first n derivatives at $x = a$, it follows that

$$h(a) = h'(a) = h''(a) = \cdots = h^{(n)}(a) = 0 \tag{19}$$

Also, we leave it for the reader to show that

$$g(a) = g'(a) = g''(a) = \cdots = g^{(n)}(a) = 0 \tag{20}$$

and that $g(x)$ and its first n derivatives are nonzero when $x \neq a$.

It is straightforward to check that h and g satisfy the hypotheses of the Extended Mean-Value Theorem (10.2.2) on the interval $[a, b]$, so that there is a point c_1 with $a < c_1 < b$ such that

$$\frac{h(b) - h(a)}{g(b) - g(a)} = \frac{h'(c_1)}{g'(c_1)} \tag{21}$$

or, from (19) and (20),

$$\frac{h(b)}{g(b)} = \frac{h'(c_1)}{g'(c_1)} \tag{22}$$

If we now apply the Extended Mean-Value Theorem to h' and g' over the interval $[a, c_1]$, we may deduce that there is a point c_2 with $a < c_2 < c_1 < b$ such that

$$\frac{h'(c_1) - h'(a)}{g'(c_1) - g'(a)} = \frac{h''(c_2)}{g''(c_2)}$$

or, from (19) and (20),

$$\frac{h'(c_1)}{g'(c_1)} = \frac{h''(c_2)}{g''(c_2)}$$

which, when combined with (22), yields

$$\frac{h(b)}{g(b)} = \frac{h''(c_2)}{g''(c_2)}$$

It should now be clear that if we continue in this way, applying the Extended Mean-Value Theorem to the successive derivatives of h and g, we shall eventually obtain a relationship of the form

$$\frac{h(b)}{g(b)} = \frac{h^{(n+1)}(c_{n+1})}{g^{(n+1)}(c_{n+1})} \tag{23}$$

where $a < c_{n+1} < b$. However, $p_n(x)$ is a polynomial of degree n, so that its $(n + 1)$-st derivative is zero. Thus, from (17)

$$h^{(n+1)}(c_{n+1}) = f^{(n+1)}(c_{n+1}) \tag{24}$$

Also, from (18), the $(n + 1)$-st derivative of $g(x)$ is the constant $(n + 1)!$, so that

$$g^{(n+1)}(c_{n+1}) = (n + 1)! \tag{25}$$

Substituting (24) and (25) into (23) yields

$$\frac{h(b)}{g(b)} = \frac{f^{(n+1)}(c_{n+1})}{(n + 1)!}$$

Letting $c = c_{n+1}$, and using (17) and (18), it follows that

$$f(b) - p_n(b) = \frac{f^{(n+1)}(c)}{(n + 1)!}(b - a)^{n+1}$$

But this is precisely Formula (1) in Taylor's Theorem (Theorem 11.10.1), with the exception that the variable here is b rather than x. Thus, to finish we need only replace b by x. ▮

▶ Exercise Set 11.10

For the functions in Exercises 1–14, find Lagrange's form of the remainder for the given values of a and n.

1. e^{2x}; $a = 0$; $n = 5$.   **2.** $\cos x$; $a = 0$; $n = 8$.

3. $\dfrac{1}{x + 1}$; $a = 0$; $n = 4$.   **4.** $\tan x$; $a = 0$; $n = 2$.

5. xe^x; $a = 0$; $n = 3$.

6. $\ln(1 + x)$; $a = 0$; $n = 5$.

7. $\tan^{-1} x$; $a = 0$; $n = 2$.   **8.** $\sinh x$; $a = 0$; $n = 6$.

9. $\sqrt{x}$; $a = 4$; $n = 3$.   **10.** $\dfrac{1}{x}$; $a = 1$; $n = 5$.

11. $\sin x$; $a = \dfrac{\pi}{6}$; $n = 4$.   **12.** $\cos \pi x$; $a = \dfrac{1}{2}$; $n = 2$.

13. $\dfrac{1}{(1 + x)^2}$; $a = -2$; $n = 5$.

14. $\csc x$; $a = \dfrac{\pi}{2}$; $n = 1$.

In Exercises 15–18, find Lagrange's form of the remainder $R_n(x)$ when $a = 0$.

15. $f(x) = \dfrac{1}{1 - x}$.   **16.** $f(x) = e^{-x}$.

17. $f(x) = e^{2x}$.   **18.** $f(x) = \ln(1 + x)$.

19. Prove: The Maclaurin series for $\cos x$ converges to $\cos x$ for all x.

20. Prove: The Taylor series for $\sin x$ about $x = \pi/4$ converges to $\sin x$ for all x.

21. Prove: The Taylor series for e^x about $x = 1$ converges to e^x for all x.

22. (a) Prove: The Maclaurin series for $\ln(1 + x)$ converges to $\ln(1 + x)$ if $0 \le x \le 1$. [*Remark:* The

Maclaurin series actually converges to $\ln(1 + x)$ on the interval $(-1, 1]$, but the Lagrange form of the remainder is not strong enough to establish this fact.]

(b) Use $x = 1$ in the Maclaurin series for $\ln(1 + x)$, Table 11.10.1, to show that

$$\ln 2 = 1 - \frac{1}{2} + \frac{1}{3} - \frac{1}{4} + \cdots$$

23. Prove: The Taylor series for e^x about any point $x = a$ converges to e^x for all x.

24. Prove: The Taylor series for $\sin x$ about any point $x = a$ converges to $\sin x$ for all x.

25. Prove: The Taylor series for $\cos x$ about any point $x = a$ converges to $\cos x$ for all x.

26. Derive (15). (Do not try to prove convergence.)

In Exercises 27–50, use the Maclaurin series in Table 11.10.1 to obtain the Maclaurin series for the given function. In each case give the first four terms of the series and specify the interval on which the series converges to the function.

27. e^{-2x}.   **28.** $x^2 e^x$.

29. xe^{-x}.   **30.** e^{x^2}.

31. $\sin 2x$.   **32.** $\cos 2x$.

33. $x^2 \cos x$.   **34.** $\sin(x^2)$.

35. $\sin^2 x$. [*Hint:* $\sin^2 x = \frac{1}{2}(1 - \cos 2x)$.]

36. $\cos^2 x$. [*Hint:* $\cos^2 x = \frac{1}{2}(1 + \cos 2x)$.]

37. $\ln(1 - x^2)$.   **38.** $\ln(1 + 2x)$.

39. $\dfrac{1}{1 - 4x^2}$.   **40.** $\dfrac{x}{1 - x}$.

41. $\dfrac{x^2}{1 + 3x}$.

42. $\dfrac{x}{1 + x^2}$.

43. $x \sinh 2x$.

44. $\cosh(x^2)$.

45. $\sqrt{1 + 3x}$.

46. $\sqrt{1 + x^2}$.

47. $\dfrac{1}{(1 - 2x)^2}$.

48. $\dfrac{x}{(1 + 2x)^3}$.

49. $\dfrac{x}{\sqrt{1 - x^2}}$.

50. $x(1 - x^2)^{3/2}$.

51. Use the Maclaurin series for $1/(1 - x)$ to express $1/x$ in powers of $x - 1$. Find the interval of convergence. $\left[\textit{Hint:} \dfrac{1}{x} = \dfrac{1}{1 + (x - 1)} . \right]$

52. Show that the series

$$1 - \frac{x}{2!} + \frac{x^2}{4!} - \frac{x^3}{6!} + \cdots$$

converges to the function

$$f(x) = \begin{cases} \cos \sqrt{x}, & x \geq 0 \\ \cosh \sqrt{-x}, & x < 0 \end{cases}$$

[*Hint:* Use the Maclaurin series for $\cos x$ and $\cosh x$ to obtain series for $\cos \sqrt{x}$ where $x \geq 0$, and $\cosh \sqrt{-x}$ where $x \leq 0$.]

53. If m is any real number, and k is a nonnegative integer, then we define the *binomial coefficients* $\dbinom{m}{k}$ by the formulas $\dbinom{m}{0} = 1$ and

$$\binom{m}{k} = \frac{m(m - 1)(m - 2) \cdots (m - k + 1)}{k!}$$

for $k \geq 1$.

Express Formula (16) in terms of binomial coefficients.

In Exercises 54–57, use the known Maclaurin series for e^x, $\sin x$, and $\cos x$ to help find the sum of the given series.

54. $2 + \dfrac{4}{2!} + \dfrac{8}{3!} + \dfrac{16}{4!} + \cdots$.

55. $\pi - \dfrac{\pi^3}{3!} + \dfrac{\pi^5}{5!} - \dfrac{\pi^7}{7!} + \cdots$.

56. $1 - \dfrac{e^2}{2!} + \dfrac{e^4}{4!} - \dfrac{e^6}{6!} + \cdots$.

57. $1 - \ln 3 + \dfrac{(\ln 3)^2}{2!} - \dfrac{(\ln 3)^3}{3!} + \cdots$.

58. (a) Use the Maclaurin series for $\dfrac{1}{1 - x}$ to help find the Maclaurin series for the function

$$f(x) = \frac{x}{1 - x^2}$$

(b) Use the result in part (a) to help find $f^{(5)}(0)$ and $f^{(6)}(0)$.

59. (a) Use the Maclaurin series for $\cos x$ to find the Maclaurin series for $f(x) = x^2 \cos 2x$.

(b) Use the result in part (a) to help find $f^{(5)}(0)$.

60. The purpose of this exercise is to show that the Taylor series of a function f may possibly converge to a value different from $f(x)$ for certain x. Let

$$f(x) = \begin{cases} e^{-1/x^2}, & x \neq 0 \\ 0, & x = 0 \end{cases}$$

(a) Use the definition of a derivative to show that $f'(0) = 0$.

(b) With some difficulty it can be shown that $f^{(n)}(0) = 0$ for $n \geq 2$. Accepting this fact, show that the Maclaurin series of f converges for all x, but converges to $f(x)$ only at the point $x = 0$.

■ **11.11 COMPUTATIONS USING TAYLOR SERIES**

In this section we shall show how Taylor and Maclaurin series can be used to obtain approximate values for trigonometric functions and logarithms.

□ **AN UPPER BOUND ON
THE REMAINDER**

In the previous section we showed that for all x

$$e^x = 1 + x + \frac{x^2}{2!} + \frac{x^3}{3!} + \frac{x^4}{4!} + \cdots + \frac{x^k}{k!} + \cdots$$

In particular, if we let $x = 1$, we obtain the following expression for e as the sum of an infinite series

$$e = 1 + 1 + \frac{1}{2!} + \frac{1}{3!} + \frac{1}{4!} + \cdots + \frac{1}{k!} + \cdots$$

Thus, we can approximate e to any degree of accuracy using an appropriate partial sum

$$e \approx 1 + 1 + \frac{1}{2!} + \frac{1}{3!} + \cdots + \frac{1}{n!} \tag{1}$$

The value of n in this formula is at our disposal—the larger we choose n the more accurate the approximation. This is illustrated in Table 11.11.1, where we have used (1) to approximate e for various values of n. To nine decimal-place accuracy, the value of e is 2.718281828. This level of accuracy is obtained with $n = 13$.

Table 11.11.1

n	$1 + 1 + \dfrac{1}{2!} + \dfrac{1}{3!} + \cdots + \dfrac{1}{n!}$
0	1.000000000
1	2.000000000
3	2.500000000
4	2.666666667
5	2.708333333
10	2.718281526
11	2.718281801

In practical applications where approximate values of e are needed, one would be interested in approximating e to a specified degree of accuracy, in which case it would be important to determine how large n should be chosen to attain the desired accuracy. For example, how large should n be taken in (1) to ensure an error that is less than 0.00005? We shall now show how Lagrange's remainder formula can be used to answer such questions.

According to Taylor's Theorem, the absolute value of the error that results when $f(x)$ is approximated by its nth Taylor polynomial $p_n(x)$ about $x = a$ is

$$|R_n(x)| = |f(x) - p_n(x)| = \left| \frac{f^{(n+1)}(c)}{(n+1)!}(x - a)^{n+1} \right| \tag{2}$$

In this formula, c is an unknown number between a and x, so that the value of $f^{(n+1)}(c)$ usually cannot be determined. However, it is frequently possible to determine an upper bound on the size of $|f^{(n+1)}(c)|$; that is, one can often find a constant M such that $|f^{(n+1)}(c)| \le M$. For such an M, it follows from (2) that

$$|R_n(x)| = |f(x) - p_n(x)| \le \frac{M}{(n+1)!} |x - a|^{n+1} \tag{3}$$

which gives an upper bound on the magnitude of the error $R_n(x)$. In the examples to follow we shall give applications of this formula, but first it will be helpful to introduce some terminology relating to the accuracy of approximations.

An approximation is said to be **accurate to n decimal places** if the magnitude of the error is less than 0.5×10^{-n}. For example,

DESCRIPTION	MAGNITUDE OF THE ERROR IS LESS THAN	
1 decimal-place accuracy	0.05	$= 0.5 \times 10^{-1}$
2 decimal-place accuracy	0.005	$= 0.5 \times 10^{-2}$
3 decimal-place accuracy	0.0005	$= 0.5 \times 10^{-3}$
4 decimal-place accuracy	0.00005	$= 0.5 \times 10^{-4}$

☐ **APPROXIMATING e**

Example 1   Use (1) to approximate e to four decimal-place accuracy.

Solution.   It follows from Taylor's formula with remainder that

$$e^x = 1 + x + \frac{x^2}{2!} + \cdots + \frac{x^n}{n!} + \frac{e^c}{(n+1)!} x^{n+1}$$

where c is between 0 and x (see Example 1, Section 11.10). Thus, in the case $x = 1$ we obtain

$$e = 1 + 1 + \frac{1}{2!} + \cdots + \frac{1}{n!} + \frac{e^c}{(n+1)!}$$

where c is between 0 and 1. This tells us that the magnitude of the error in approximation (1) is

$$|R_n| = \left| \frac{e^c}{(n+1)!} \right| = \frac{e^c}{(n+1)!} \tag{4}$$

where $0 < c < 1$. Since $c < 1$, it follows that

$$e^c < e^1 = e$$

so that from (4)

$$|R_n| < \frac{e}{(n+1)!} \tag{5}$$

This inequality provides an upper bound on the magnitude of the error R_n. Unfortunately, this inequality is not very useful since the right side involves the quantity e, which we are trying to estimate. However, if we use the fact that $e < 3$, then we can replace (5) with the following less precise but more useful result:

$$|R_n| < \frac{3}{(n + 1)!} \tag{5a}$$

It follows that if we choose n so that

$$|R_n| < \frac{3}{(n + 1)!} < 0.5 \times 10^{-4} = 0.00005 \tag{6}$$

then approximation (1) will be accurate to four decimal places. An appropriate value for n may be found by trial and error. For example, using a calculator one can evaluate $3/(n + 1)!$ for $n = 0, 1, 2, \ldots$ until a value of n satisfying (6) is obtained. We leave it for the reader to show that $n = 8$ is the first positive integer satisfying (6). Thus, to four decimal-place accuracy

$$e \approx 1 + 1 + \frac{1}{2!} + \frac{1}{3!} + \frac{1}{4!} + \frac{1}{5!} + \frac{1}{6!} + \frac{1}{7!} + \frac{1}{8!} \approx 2.7183 \tag{7}$$

◀

REMARK. It should be noted that there are two types of errors that result when computing with series. The first, called *truncation error,* is the error that results when the entire series is approximated by a partial sum. The second kind of error, called *roundoff error,* results when decimal approximations are used. For example, (7) involves a truncation error of at most 0.5×10^{-4} (four decimal-place accuracy). However, to obtain the numerical value in (7) we used a calculator to evaluate the left side, resulting in roundoff error due to the limitations of the calculator. The problem of controlling roundoff error is surprisingly difficult and is studied in courses in a branch of mathematics called *numerical analysis.* As a rule of thumb, to achieve n decimal-place accuracy in a final result, one should use more than $n + 1$ decimal places in each intermediate computation, then round off to n decimal places at the end. However, even this procedure may occasionally not produce n decimal-place accuracy. For purposes of this text, we recommend that you perform all intermediate calculations with the maximum number of decimal places that your calculator will allow, then round off the final result.

☐ **APPROXIMATING TRIGONOMETRIC FUNCTIONS**

Example 2   Use the Maclaurin series for $\sin x$ to approximate $\sin 3°$ to five decimal-place accuracy.

Solution.   In the Maclaurin series

$$\sin x = x - \frac{x^3}{3!} + \frac{x^5}{5!} - \frac{x^7}{7!} + \cdots \tag{8}$$

the angle x is assumed to be in radians (because the differentiation formulas for the trigonometric functions were derived with this assumption). Since $3° = \pi/60$ radians, it follows from (8) that

$$\sin 3° = \sin \frac{\pi}{60} = \left(\frac{\pi}{60}\right) - \frac{\left(\frac{\pi}{60}\right)^3}{3!} + \frac{\left(\frac{\pi}{60}\right)^5}{5!} - \frac{\left(\frac{\pi}{60}\right)^7}{7!} + \cdots \tag{9}$$

We must now decide how many terms in this series must be kept in order to obtain five decimal-place accuracy. We shall consider two possible approaches, one using Lagrange's remainder formula, and the other exploiting the fact that (9) satisfies the conditions of the alternating series test.

If we let $f(x) = \sin x$, then the magnitude of the error that results when $\sin x$ is approximated by its nth Maclaurin polynomial is

$$|R_n| = \left| \frac{f^{(n+1)}(c)}{(n+1)!} x^{n+1} \right|$$

where c is between 0 and x. Since $f^{(n+1)}(c)$ is either $\pm \sin c$ or $\pm \cos c$, it follows that $|f^{(n+1)}(c)| \leq 1$, so

$$|R_n| \leq \frac{|x|^{n+1}}{(n+1)!}$$

In particular, if $x = \pi/60$, then

$$|R_n| \leq \frac{\left(\frac{\pi}{60}\right)^{n+1}}{(n+1)!}$$

Thus, for five decimal-place accuracy, we must choose n so that

$$\frac{\left(\frac{\pi}{60}\right)^{n+1}}{(n+1)!} < 0.5 \times 10^{-5} = 0.000005$$

By trial and error with the help of a calculator or computer, the reader can check that $n = 3$ is the smallest n that works. Thus, in (9) we need only keep terms up to the third power for five decimal-place accuracy, that is,

$$\sin 3° \approx \left(\frac{\pi}{60}\right) - \frac{\left(\frac{\pi}{60}\right)^3}{3!} \approx 0.05234 \tag{10}$$

An alternative approach to determining n uses the fact that (9) satisfies the conditions of the alternating series test, Theorem 11.7.1. (Verify.) Thus, by Theorem 11.7.2, if we use only those terms up to and including

$$\pm \frac{\left(\dfrac{\pi}{60}\right)^m}{m!} \qquad \boxed{m \text{ is an odd positive integer.}}$$

then the magnitude of the error will be at most

$$\frac{\left(\dfrac{\pi}{60}\right)^{m+2}}{(m+2)!}$$

Thus, for five decimal-place accuracy, we look for the first positive odd integer m such that

$$\frac{\left(\dfrac{\pi}{60}\right)^{m+2}}{(m+2)!} < 0.5 \times 10^{-5} = 0.000005$$

By trial and error, $m = 3$ is the first such integer. Thus, to five decimal-place accuracy

$$\sin 3° \approx \left(\frac{\pi}{60}\right) - \frac{\left(\dfrac{\pi}{60}\right)^3}{3!} \approx 0.05234 \tag{11}$$

which is the same as our earlier result. ◄

To approximate the value of a function f at a point x_0 using a Taylor series, two factors enter into the selection of the point $x = a$ for the series. First, the point $x = a$ must be selected so that f and its derivatives can be evaluated at a, since those values are needed to find Taylor series. Second, it is desirable to choose a as close as possible to x_0 since the "rate of convergence" of a Taylor series is usually most rapid close to a; that is, fewer terms are required in a partial sum to achieve a given level of accuracy.

Example 3 Approximate $\sin 92°$ to five decimal-place accuracy.

Solution. We will use the Taylor series for $\sin x$ with $a = \pi/2 \, (= 90°)$, since $\sin x$ and its derivatives can be readily evaluated at this point and this value of a is reasonably close to the point $x = 92° = \frac{23}{45}\pi$ where we want to make the approximation. From Example 5 in Section 11.10, the Taylor series for $\sin x$ about $x = \pi/2$ is

$$\sin x = 1 - \frac{1}{2!}\left(x - \frac{\pi}{2}\right)^2 + \frac{1}{4!}\left(x - \frac{\pi}{2}\right)^4 - \frac{1}{6!}\left(x - \frac{\pi}{2}\right)^6 + \cdots \tag{12}$$

Substituting $x = \frac{23}{45}\pi$ in (12) yields

$$\sin 92° = \sin \frac{23}{45}\pi = 1 - \frac{1}{2!}\left(\frac{\pi}{90}\right)^2 + \frac{1}{4!}\left(\frac{\pi}{90}\right)^4 - \frac{1}{6!}\left(\frac{\pi}{90}\right)^6 + \cdots \tag{13}$$

If we let $f(x) = \sin x$, then the magnitude of the error in approximating $\sin x$ by its nth Taylor polynomial about $a = \pi/2$ is

$$|R_n| = \left| \frac{f^{(n+1)}(c)}{(n+1)!} \left(x - \frac{\pi}{2} \right)^{n+1} \right|$$

where c is between $\pi/2$ and x. As was the case in the previous example, $|f^{(n+1)}(c)| \leq 1$, so

$$|R_n| \leq \frac{\left| x - \dfrac{\pi}{2} \right|^{n+1}}{(n+1)!}$$

In particular, if $x = \frac{23}{45}\pi$, then

$$|R_n| \leq \frac{\left(\dfrac{\pi}{90} \right)^{n+1}}{(n+1)!}$$

For five decimal-place accuracy, we must choose n so that

$$\frac{\left(\dfrac{\pi}{90} \right)^{n+1}}{(n+1)!} < 0.5 \times 10^{-5} = 0.000005$$

By trial and error, the smallest such integer is $n = 3$. Thus, in (13) we need only keep terms up to the third power for five decimal-place accuracy, that is,

$$\sin 92° \approx 1 + 0 \cdot \left(\frac{\pi}{90} \right) - \frac{1}{2!} \left(\frac{\pi}{90} \right)^2 + 0 \cdot \left(\frac{\pi}{90} \right)^3$$

$$= 1 - \frac{1}{2!} \left(\frac{\pi}{90} \right)^2 \approx 0.99939$$

As in Example 2, we could also have obtained this result by exploiting the fact that (13) is an alternating series. ◄

□ **APPROXIMATING LOGARITHMS**

The Maclaurin series

$$\ln (1 + x) = x - \frac{x^2}{2} + \frac{x^3}{3} - \frac{x^4}{4} + \cdots \quad -1 < x \leq 1 \tag{14}$$

is the starting point for the approximation of natural logarithms. Unfortunately, the usefulness of this series is limited because of its slow convergence and the restriction $-1 < x \leq 1$. However, if we replace x by $-x$ in this series, we obtain

$$\ln (1 - x) = -x - \frac{x^2}{2} - \frac{x^3}{3} - \frac{x^4}{4} - \cdots \quad -1 \leq x < 1 \tag{15}$$

and on subtracting (15) from (14) we obtain

$$\ln\left(\frac{1+x}{1-x}\right) = 2\left(x + \frac{x^3}{3} + \frac{x^5}{5} + \frac{x^7}{7} + \cdots\right) \quad -1 < x < 1 \tag{16}$$

Series (16), first obtained by James Gregory* in 1668, can be used to compute the natural logarithm of any positive number y by letting

$$y = \frac{1+x}{1-x}$$

or equivalently

$$x = \frac{y-1}{y+1} \tag{17}$$

and noting that $-1 < x < 1$. For example, to compute $\ln 2$ we let $y = 2$ in (17), which yields $x = 1/3$. Substituting this value in (16) gives

$$\ln 2 = 2\left[(1/3) + \frac{(1/3)^3}{3} + \frac{(1/3)^5}{5} + \frac{(1/3)^7}{7} + \cdots\right]$$

Adding the four terms shown and then rounding to four decimal places at the end yields

$$\ln 2 \approx 0.6931$$

It is of interest to note that in the case $x = 1$, series (14) yields

$$\ln 2 = 1 - \frac{1}{2} + \frac{1}{3} - \frac{1}{4} + \frac{1}{5} - \cdots$$

This result is noteworthy because it is the first time we have been able to obtain the sum of the alternating harmonic series. However, this series converges too slowly to be of any computational value.

☐ **APPROXIMATING** π

If we let $x = 1$ in the Maclaurin series

$$\tan^{-1} x = x - \frac{x^3}{3} + \frac{x^5}{5} - \frac{x^7}{7} + \cdots \quad -1 \leq x \leq 1 \tag{18}$$

---

*JAMES GREGORY (1638–1675). Scottish mathematician and astronomer. Gregory, the son of a minister, was famous in his time as the inventor of the Gregorian reflecting telescope, so named in his honor. Although he is not generally ranked with the great mathematicians, much of his work relating to calculus was studied by Leibniz and Newton and undoubtedly influenced some of their discoveries. There is a manuscript, discovered posthumously, which shows that Gregory had anticipated Taylor series well before Taylor.

we obtain

$$\frac{\pi}{4} = \tan^{-1} 1 = 1 - \frac{1}{3} + \frac{1}{5} - \frac{1}{7} + \cdots ,$$

or

$$\pi = 4\left[1 - \frac{1}{3} + \frac{1}{5} - \frac{1}{7} + \cdots \right]$$

This famous series, obtained by Leibniz in 1674, converges too slowly to be of computational importance. A more practical procedure for approximating π uses the identity

$$\frac{\pi}{4} = \tan^{-1}\frac{1}{2} + \tan^{-1}\frac{1}{3} \tag{19}$$

By using this identity and series (18) to approximate $\tan^{-1}\frac{1}{2}$ and $\tan^{-1}\frac{1}{3}$, the value of π can be effectively approximated to any degree of accuracy.

▶ Exercise Set 11.11   Ⓒ 1–14, 19, 21

1. Use inequality (5a) to find a value of n to ensure that (1) will approximate e to
   (a) five decimal-place accuracy.
   (b) ten decimal-place accuracy.

In Exercises 2–8, apply Lagrange's form of the remainder.

2. Use $x = -1$ in the Maclaurin series for e^x to approximate $1/e$ to three decimal-place accuracy.

3. Use $x = \frac{1}{2}$ in the Maclaurin series for e^x to approximate $\sqrt{e}$ to four decimal-place accuracy.

4. Use the Maclaurin series for $\sin x$ to approximate $\sin 4°$ to five decimal-place accuracy.

5. Use the Maclaurin series for $\cos x$ to approximate $\cos(\pi/20)$ to four decimal-place accuracy.

6. Use an appropriate Taylor series for $\sin x$ to approximate $\sin 85°$ to four decimal-place accuracy.

7. Use an appropriate Taylor series for $\cos x$ to approximate $\cos 58°$ to four decimal-place accuracy.

8. Use an appropriate Taylor series for $\sin x$ to approximate $\sin 35°$ to four decimal-place accuracy.

9. Use the first two terms of series (16) to approximate ln 1.25. Round your answer to three decimal places.

10. Use the first six terms of series (16) to approximate ln 3. Round your answer to four decimal places.

11. Use the Maclaurin series for $\tan^{-1} x$ to approximate $\tan^{-1} 0.1$ to three decimal-place accuracy. [*Hint:* Use the fact that (18) is an alternating series.]

12. Use the Maclaurin series for $\sinh x$ to approximate $\sinh 0.5$ to three decimal-place accuracy.

13. Use the Maclaurin series for $\cosh x$ to approximate $\cosh 0.1$ to four decimal-place accuracy.

14. Use an appropriate Taylor series for $\sqrt[3]{x}$ to approximate $\sqrt[3]{28}$ to three decimal-place accuracy.

15. Find an interval of values for x containing $x = 0$ over which $\sin x$ can be approximated by $x - x^3/3!$ with three decimal-place accuracy ensured.

16. Find an interval of values for x over which e^x can be approximated by $1 + x + x^2/2!$ with three decimal-place accuracy ensured.

17. Find an upper bound on the magnitude of the error in the approximation $\cos x \approx 1 - x^2/2! + x^4/4!$ if $-0.2 \le x \le 0.2$.

18. Find an upper bound on the magnitude of the error in the approximation $\ln(1 + x) \approx x$ if $|x| < 0.01$.

19. In each part find the number of terms of the stated series that are required to ensure that the sum of the terms approximates $\ln 2$ to six decimal-place accuracy.

(a) The alternating harmonic series
(b) Series (16).

20. Prove identity (19).

21. Approximate $\tan^{-1}\frac{1}{2}$ and $\tan^{-1}\frac{1}{3}$ to three decimal-place accuracy, and then use identity (19) to approximate π.

■ 11.12 DIFFERENTIATION AND INTEGRATION OF POWER SERIES

In this section we shall show that if a power series in $x - a$ converges in some interval and has a sum of $f(x)$ in that interval, then the power series must be the Taylor series about $x = a$ for the function f. This result is important for many reasons, but it is important for computational purposes because it will enable us to find Taylor series without using the defining formula for such series. This is often necessary in cases where successive derivatives of f are prohibitively complicated to compute. In this section we shall also consider conditions under which a Taylor series can be differentiated or integrated term by term.

☐ **POWER SERIES REPRESENTATIONS OF FUNCTIONS**

If a function f is expressed as the sum of a power series for all x in some interval, then we shall say that the power series *represents* f on the interval or that the series is a *power series representation of f* on the interval. For example, we know from our study of geometric series that

$$\frac{1}{1 - x} = 1 + x + x^2 + x^3 + \cdots \quad -1 < x < 1$$

Thus, the series $1 + x + x^2 + x^3 + \cdots$ represents the function

$$f(x) = \frac{1}{1 - x}$$

on the interval $(-1, 1)$.

The following theorems show that if a function f is represented by a power series on an interval, then differentiating the series term by term produces a power series representation of f' on the interval, and integrating the series term by term produces a power series representation of the integral of f on the interval.

11.12.1 THEOREM (*Differentiation of Power Series*). *If a function f is repre-sented by a power series, say*

$$f(x) = \sum_{k=0}^{\infty} c_k(x - a)^k$$

where the series has a nonzero radius of convergence R, then:

(a) *The series of differentiated terms*

$$\sum_{k=0}^{\infty} \frac{d}{dx}[c_k(x - a)^k] = \sum_{k=1}^{\infty} kc_k(x - a)^{k-1}$$

has radius of convergence R.

(b) *The function f is differentiable on the interval $(a - R, a + R)$, and for every x in this interval*

$$f'(x) = \sum_{k=0}^{\infty} \frac{d}{dx}[c_k(x - a)^k]$$

To paraphrase this theorem informally, *a power series representation of a function can be differentiated term by term on any open interval within the interval of convergence.*

Example 1 To illustrate this theorem, we shall use the Maclaurin series

$$\sin x = x - \frac{x^3}{3!} + \frac{x^5}{5!} - \frac{x^7}{7!} + \cdots \qquad -\infty < x < +\infty$$

$$\cos x = 1 - \frac{x^2}{2!} + \frac{x^4}{4!} - \frac{x^6}{6!} + \cdots \qquad -\infty < x < +\infty$$

$$e^x = 1 + x + \frac{x^2}{2!} + \frac{x^3}{3!} + \frac{x^4}{4!} + \cdots \qquad -\infty < x < +\infty$$

to obtain the familiar derivative formulas

$$\frac{d}{dx}[\sin x] = \cos x \quad \text{and} \quad \frac{d}{dx}[e^x] = e^x$$

Differentiating the Maclaurin series for $\sin x$ and e^x term by term yields

$$\frac{d}{dx}[\sin x] = \frac{d}{dx}\left[x - \frac{x^3}{3!} + \frac{x^5}{5!} - \frac{x^7}{7!} + \cdots\right] = 1 - 3\frac{x^2}{3!} + 5\frac{x^4}{5!} - 7\frac{x^6}{7!} + \cdots$$

$$= 1 - \frac{x^2}{2!} + \frac{x^4}{4!} - \frac{x^6}{6!} + \cdots$$

$$= \cos x$$

$$\frac{d}{dx}[e^x] = \frac{d}{dx}\left[1 + x + \frac{x^2}{2!} + \frac{x^3}{3!} + \cdots\right] = 1 + 2\frac{x}{2!} + 3\frac{x^2}{3!} + 4\frac{x^3}{4!} + \cdots$$

$$= 1 + x + \frac{x^2}{2!} + \frac{x^3}{3!} + \cdots$$

$$= e^x \quad \blacktriangleleft$$

11.12.2   THEOREM (***Integration of Power Series***).   *If a function f is represented by a power series, say*

$$f(x) = \sum_{k=0}^{\infty} c_k(x - a)^k$$

where the series has a nonzero radius of convergence R, then:

(*a*)   *The series of integrated terms*

$$\sum_{k=0}^{\infty}\left[\int c_k(x - a)^k\, dx\right] = \sum_{k=0}^{\infty} \frac{c_k}{k + 1}(x - a)^{k+1}$$

has radius of convergence R.

(*b*)   *The function f is continuous on the interval* $(a - R, a + R)$ *and for all x in this interval*

$$\int f(x)\, dx = \sum_{k=0}^{\infty}\left[\int c_k(x - a)^k\, dx\right] + C$$

(*c*)   *For all* α *and* β *in the interval* $(a - R, a + R)$, *the series*

$$\sum_{k=0}^{\infty}\left[\int_{\alpha}^{\beta} c_k(x - a)^k\, dx\right]$$

converges absolutely and

$$\int_{\alpha}^{\beta} f(x)\, dx = \sum_{k=0}^{\infty}\left[\int_{\alpha}^{\beta} c_k(x - a)^k\, dx\right]$$

To paraphrase this theorem informally, *a power series representation of a function can be integrated term by term on any interval within the interval of convergence.*

REMARK.   Note that in part (*b*) a separate constant of integration is not introduced for each term in the series; rather, a single constant C is added to the entire series.

Example 2 To illustrate part (*b*) of the foregoing theorem, we shall use the Maclaurin series for $\sin x$ and $\cos x$ to obtain the familiar integration formula

$$\int \cos x \, dx = \sin x + C$$

Integrating the Maclaurin series for $\cos x$ term by term yields

$$
\begin{aligned}
\int \cos x \, dx &= \int \left[1 - \frac{x^2}{2!} + \frac{x^4}{4!} - \frac{x^6}{6!} + \cdots \right] dx \\
&= \left[x - \frac{x^3}{3(2!)} + \frac{x^5}{5(4!)} - \frac{x^7}{7(6!)} + \cdots \right] + C \\
&= \left[x - \frac{x^3}{3!} + \frac{x^5}{5!} - \frac{x^7}{7!} + \cdots \right] + C \\
&= \sin x + C \quad \blacktriangleleft
\end{aligned}
$$

Example 3 The integral

$$\int_0^1 e^{-x^2} \, dx$$

cannot be evaluated directly because there is no elementary antiderivative of e^{-x^2}. However, it is possible to approximate the integral by some numerical technique such as Simpson's rule. Still, another possibility is to represent e^{-x^2} by its Maclaurin series and then integrate term by term in accordance with part (*c*) of Theorem 11.12.2. This produces a series that converges to the integral.

The simplest way to obtain the Maclaurin series for e^{-x^2} is to replace x by $-x^2$ in the Maclaurin series

$$e^x = 1 + x + \frac{x^2}{2!} + \frac{x^3}{3!} + \frac{x^4}{4!} + \cdots$$

to obtain

$$e^{-x^2} = 1 - x^2 + \frac{x^4}{2!} - \frac{x^6}{3!} + \frac{x^8}{4!} - \cdots$$

Therefore,

$$
\begin{aligned}
\int_0^1 e^{-x^2} \, dx &= \int_0^1 \left[1 - x^2 + \frac{x^4}{2!} - \frac{x^6}{3!} + \frac{x^8}{4!} - \cdots \right] dx \\
&= \left[x - \frac{x^3}{3} + \frac{x^5}{5(2!)} - \frac{x^7}{7(3!)} + \frac{x^9}{9(4!)} - \cdots \right]_0^1 \\
&= 1 - \frac{1}{3} + \frac{1}{5 \cdot 2!} - \frac{1}{7 \cdot 3!} + \frac{1}{9 \cdot 4!} - \cdots \quad (1)
\end{aligned}
$$

Thus, we have found a series that converges to the value of the integral $\int_0^1 e^{-x^2}\, dx$. Using the first three terms in this series we obtain the approximation

$$\int_0^1 e^{-x^2}\, dx \approx 1 - \frac{1}{3} + \frac{1}{10} = \frac{23}{30} \approx 0.767$$

Since series (1) satisfies the conditions of the alternating series test, it follows from Theorem 11.7.2 that the magnitude of the error in this approximation is at most $1/(7 \cdot 3!) = 1/42 \approx 0.0238$. Greater accuracy can be obtained by using more terms in the series. ◀

POWER SERIES REPRESENTATIONS MUST BE TAYLOR SERIES

The following theorem shows that if a function f is represented by a power series in $x - a$ on an interval, then that series must be the Taylor series for f; thus, Taylor series are the only power series that can represent functions on an interval.

11.12.3 THEOREM. *If*

$$f(x) = c_0 + c_1(x - a) + c_2(x - a)^2 + \cdots + c_n(x - a)^n + \cdots$$

for all x in some open interval containing a, then the series is the Taylor series for f about a.

Proof. By repeated application of Theorem 11.12.1(b) we obtain

$$f(x) = c_0 + c_1(x - a) + c_2(x - a)^2 + c_3(x - a)^3 + c_4(x - a)^4 + \cdots$$
$$f'(x) = c_1 + 2c_2(x - a) + 3c_3(x - a)^2 + 4c_4(x - a)^3 + \cdots$$
$$f''(x) = 2!c_2 + (3 \cdot 2)c_3(x - a) + (4 \cdot 3)c_4(x - a)^2 + \cdots$$
$$f'''(x) = 3!c_3 + (4 \cdot 3 \cdot 2)c_4(x - a) + \cdots$$
$$\vdots$$

On substituting $x = a$, all the powers of $x - a$ drop out leaving

$$f(a) = c_0 \qquad\qquad c_0 = f(a)$$
$$f'(a) = c_1 \qquad\qquad c_1 = f'(a)$$
$$f''(a) = 2!c_2 \quad \text{or} \quad c_2 = \frac{f''(a)}{2!}$$
$$f'''(a) = 3!c_3 \qquad\qquad c_3 = \frac{f'''(a)}{3!}$$
$$\vdots$$

which shows that the coefficients $c_0, c_1, c_2, c_3, \ldots$ are precisely the coefficients in the Taylor series about a for $f(x)$. ■

REMARK.   The foregoing theorem tells us that no matter how we arrive at a power series in $x - a$ converging to $f(x)$, be it by substitution, by integration, by differentiation, or by algebraic manipulation, the resulting series will be the Taylor series about a for $f(x)$.

Example 4   Find the Maclaurin series for $\tan^{-1} x$.

Solution.   We could calculate this Maclaurin series directly. However, we can also exploit Theorem 11.12.2 by first observing that

$$\int \frac{1}{1 + x^2} \, dx = \tan^{-1} x + C$$

and then integrating the Maclaurin series for $1/(1 + x^2)$ term by term. Since

$$\frac{1}{1 - x} = 1 + x + x^2 + x^3 + x^4 + \cdots \qquad -1 < x < 1$$

it follows on replacing x by $-x^2$ that

$$\frac{1}{1 + x^2} = 1 - x^2 + x^4 - x^6 + x^8 - \cdots \qquad -1 < x < 1 \qquad (2)$$

Thus,

$$\tan^{-1} x + C = \int \frac{1}{1 + x^2} \, dx = \int [1 - x^2 + x^4 - x^6 + x^8 - \cdots] \, dx$$

or

$$\tan^{-1} x = \left[x - \frac{x^3}{3} + \frac{x^5}{5} - \frac{x^7}{7} + \frac{x^9}{9} - \cdots \right] - C$$

The constant of integration may be evaluated by substituting $x = 0$ and using the condition $\tan^{-1} 0 = 0$. This gives $C = 0$, so that

$$\tan^{-1} x = x - \frac{x^3}{3} + \frac{x^5}{5} - \frac{x^7}{7} + \frac{x^9}{9} - \cdots \qquad (3)$$

We are guaranteed by Theorem 11.12.3 that the series we have produced is the Maclaurin series for $\tan^{-1} x$ and that it actually converges to $\tan^{-1} x$ for $-1 < x < 1$.   ◄

REMARK.   Theorems 11.12.1 and 11.12.2 say nothing about the behavior of the differentiated and integrated series at the endpoints $a - R$ and $a + R$. Indeed, by differentiating termwise, convergence may be lost at one or both endpoints; and by integrating termwise, convergence may be gained at one or both endpoints. As an illustration, we derived series (3) for $\tan^{-1} x$ by integrating series (2) for $1/(1 + x^2)$. We omit the proof that the series for $\tan^{-1} x$

converges to $\tan^{-1} x$ on the interval $[-1, 1]$, while the series for $1/(1 + x^2)$ converges to $1/(1 + x^2)$ only on $(-1, 1)$. Thus, convergence was gained at both endpoints by integrating.

☐ **MISCELLANEOUS TECHNIQUES FOR OBTAINING TAYLOR SERIES**

We conclude this section with some techniques for obtaining Taylor series that would be messy to obtain directly.

Example 5 Find the first three terms that occur in the Maclaurin series for $e^{-x^2} \tan^{-1} x$.

Solution. Using the series for e^{-x^2} and $\tan^{-1} x$ obtained in Examples 3 and 4 gives

$$e^{-x^2} \tan^{-1} x = \left(1 - x^2 + \frac{x^4}{2} - \cdots\right)\left(x - \frac{x^3}{3} + \frac{x^5}{5} - \cdots\right)$$

We now multiply out, following a familiar format from elementary algebra:

$$
\begin{array}{r}
1 - x^2 + \dfrac{x^4}{2} - \cdots \\
\times \quad x - \dfrac{x^3}{3} + \dfrac{x^5}{5} - \cdots \\
\hline
x - x^3 + \dfrac{x^5}{2} - \cdots \\
- \dfrac{x^3}{3} + \dfrac{x^5}{3} - \dfrac{x^7}{6} + \cdots \\
\dfrac{x^5}{5} - \dfrac{x^7}{5} + \cdots \\
\hline
x - \frac{4}{3}x^3 + \frac{31}{30}x^5 - \cdots
\end{array}
$$

Thus,

$$e^{-x^2} \tan^{-1} x = x - \frac{4}{3} x^3 + \frac{31}{30} x^5 - \cdots \qquad -1 < x < 1$$

If desired, more terms in the series can be obtained by including more terms in the factors. ◀

Example 6 Find the first three nonzero terms in the Maclaurin series for $\tan x$.

Solution. Instead of computing the series directly, we write

$$\tan x = \frac{\sin x}{\cos x} = \frac{x - \dfrac{x^3}{3!} + \dfrac{x^5}{5!} - \cdots}{1 - \dfrac{x^2}{2!} + \dfrac{x^4}{4!} - \cdots} \qquad -\frac{\pi}{2} < x < \frac{\pi}{2}$$

and follow the familiar "long division" process to obtain

$$1 - \frac{x^2}{2} + \frac{x^4}{24} - \cdots \overline{\smash{\big)}\ x - \frac{x^3}{6} + \frac{x^5}{120} - \cdots}$$

$$x + \frac{x^3}{3} + \frac{2x^5}{15} + \cdots$$

$$x - \frac{x^3}{2} + \frac{x^5}{24} - \cdots$$

$$\frac{x^3}{3} - \frac{x^5}{30} + \cdots$$

$$\frac{x^3}{3} - \frac{x^5}{6} + \cdots$$

$$\frac{2x^5}{15} + \cdots$$

Thus,

$$\tan x = x + \frac{x^3}{3} + \frac{2x^5}{15} + \cdots \qquad -\frac{\pi}{2} < x < \frac{\pi}{2} \qquad \blacktriangleleft$$

▶ **Exercise Set 11.12**   © *14–21*

In Exercises 1–4, obtain the stated results by differentiating or integrating Maclaurin series term by term.

1. (a) $\dfrac{d}{dx}[e^x] = e^x$ (b) $\displaystyle\int e^x\,dx = e^x + C.$

2. (a) $\dfrac{d}{dx}[\cos x] = -\sin x$

  (b) $\displaystyle\int \sin x\,dx = -\cos x + C.$

3. (a) $\dfrac{d}{dx}[\sinh x] = \cosh x$

  (b) $\displaystyle\int \sinh x\,dx = \cosh x + C.$

4. (a) $\dfrac{d}{dx}[\ln(1 + x)] = \dfrac{1}{1 + x}$

  (b) $\displaystyle\int \dfrac{1}{1 + x}\,dx = \ln(1 + x) + C.$

5. Derive the Maclaurin series for $1/(1 + x)^2$ by differentiating an appropriate Maclaurin series term by term.

6. By differentiating an appropriate series, show that

$$\sum_{k=1}^{\infty} kx^k = \frac{x}{(1 - x)^2} \qquad \text{for } -1 < x < 1$$

$$\left[\textit{Hint: Consider } x\,\frac{d}{dx}\left[\frac{1}{1 - x}\right]. \right]$$

7. By integrating an appropriate series, show that

$$\sum_{k=1}^{\infty} \frac{x^k}{k} = \ln\left(\frac{1}{1 - x}\right) \qquad \text{for } -1 < x < 1$$

$$\left[\textit{Hint: } \ln\left(\frac{1}{1 - x}\right) = -\ln(1 - x). \right]$$

8. Use the result of Exercise 6 to find the sum of the series

$$\frac{1}{3} + \frac{2}{3^2} + \frac{3}{3^3} + \frac{4}{3^4} + \cdots$$

9. Use the result of Exercise 7 to find the sum of the series

$$\frac{1}{4} + \frac{1}{2(4^2)} + \frac{1}{3(4^3)} + \frac{1}{4(4^4)} + \cdots$$

10. Find the sum

$$\sum_{k=0}^{\infty} \frac{k + 1}{k!} = 1 + 2 + \frac{3}{2!} + \frac{4}{3!} + \frac{5}{4!} + \cdots$$

  [*Hint:* Differentiate the Maclaurin series for xe^x.]

11. Find the sum of the series

$$2 + 6x + 12x^2 + 20x^3 + \cdots$$

  [*Hint:* Find the second derivative of the Maclaurin series for $1/(1 - x)$.]

12. Find the sum $\sum_{k=1}^{\infty} \dfrac{k^2}{4^k}$. [*Hint:* Differentiate the Maclaurin series for $1/(1-x)$, multiply by x, differentiate, and multiply by x again.]

13. Let $f(x) = \sum_{k=0}^{\infty} (-1)^k \dfrac{x^{k+1}}{k+1}$

$$= x - \frac{x^2}{2} + \frac{x^3}{3} - \frac{x^4}{4} + \cdots$$

(a) Use the ratio test to show that the series converges for all x in the interval $(-1, 1)$.

(b) Use part (a) of Theorem 11.12.1 to find a power series for $f'(x)$. What is its interval of convergence?

(c) From the series obtained in part (b), deduce that $f'(x) = 1/(1 + x)$ and hence that

$$f(x) = \ln(1 + x) \text{ for } -1 < x < 1$$

[*Remark:* The Lagrange form of the remainder can be used to show that the Maclaurin series for $\ln(1 + x)$ converges to $\ln(1 + x)$ for $x = 1$ as well.]

In Exercises 14–21, use series to approximate the value of the integral to three decimal-place accuracy.

14. $\displaystyle\int_0^1 \sin x^2 \, dx.$ 　　**15.** $\displaystyle\int_0^1 \cos \sqrt{x}\, dx.$

16. $\displaystyle\int_0^{0.1} \frac{\sin x}{x}\, dx.$ 　　**17.** $\displaystyle\int_0^{1/2} \frac{dx}{1+x^4}.$

18. $\displaystyle\int_0^{1/2} \tan^{-1} 2x^2 \, dx.$ 　　**19.** $\displaystyle\int_0^{0.1} e^{-x^3}\, dx.$

20. $\displaystyle\int_0^{0.2} \sqrt[3]{1+x^4}\, dx.$ 　　**21.** $\displaystyle\int_0^{1/2} \frac{dx}{\sqrt[4]{x^2+1}}.$

In Exercises 22–29, use any method to find the first four nonzero terms in the Maclaurin series of the given function.

22. $x^4 e^x.$ 　　**23.** $e^{-x^2}\cos x.$

24. $\dfrac{x^2}{1+x^4}.$ 　　**25.** $\dfrac{\sin x}{e^x}.$

26. $\tanh x.$ 　　**27.** $x \ln(1-x^2).$

28. $\dfrac{\ln(1+x)}{1-x}.$ 　　**29.** $x^2 e^{4x}\sqrt{1+x}.$

30. Obtain the familiar result, $\lim_{x\to 0} (\sin x)/x = 1$, by finding a power series for $(\sin x)/x$ and taking the limit term by term.

31. Use the method of Exercise 30 to find the limits.

(a) $\displaystyle\lim_{x\to 0} \frac{1-\cos x}{\sin x}$

(b) $\displaystyle\lim_{x\to 0} \frac{\ln \sqrt{1+x} - \sin 2x}{x}.$

32. (a) Use the relationship

$$\int \frac{1}{\sqrt{1-x^2}}\, dx = \sin^{-1}x + C$$

to find the first four nonzero terms in the Maclaurin series for $\sin^{-1}x$.

(b) Express the series in sigma notation.

(c) What is the radius of convergence?

33. (a) Use the relationship

$$\int \frac{1}{\sqrt{1+x^2}}\, dx = \sinh^{-1}x + C$$

to find the first four nonzero terms in the Maclaurin series for $\sinh^{-1}x$.

(b) Express the series in sigma notation.

(c) What is the radius of convergence?

34. Prove: If the power series $\sum_{k=0}^{\infty} a_k x^k$ and $\sum_{k=0}^{\infty} b_k x^k$ have the same sum on an interval $(-r, r)$, then $a_k = b_k$ for all values of k.

▶ SUPPLEMENTARY EXERCISES 　Ⓒ 26, 50, 51, 52

In Exercises 1–6, find $L = \lim_{n\to+\infty} a_n$ if it exists.

1. $a_n = (-1)^n/e^n.$ 　　**2.** $a_n = e^{1/n}.$

3. $a_n = \dfrac{1}{\sqrt{n}} - \dfrac{1}{\sqrt{n+1}}.$ 　　**4.** $a_n = \sin(\pi n).$

5. $a_n = \sin\left(\dfrac{(2n-1)\pi}{2}\right).$ 　　**6.** $a_n = \dfrac{n+1}{n(n+2)}.$

7. Which of the sequences $\{a_n\}_{n=1}^{+\infty}$ in Exercises 1–6 are (a) decreasing, (b) nondecreasing, and (c) alternating?

8. Suppose $f(x)$ satisfies

$$f'(x) > 0 \quad \text{and} \quad f(x) \le 1 - e^{-x}$$

for all $x \ge 1$. What can you conclude about the convergence of $\{a_n\}$ if $a_n = f(n)$, $n = 1, 2, \ldots$?

9. Use your knowledge of geometric series and p-series to determine all values of q for which the following series converge.

(a) $\displaystyle\sum_{k=0}^{\infty} \pi^k/q^{2k}$ (b) $\displaystyle\sum_{k=1}^{\infty} (1/k^q)^3$

(c) $\displaystyle\sum_{k=2}^{\infty} 1/(\ln q^k)$ (d) $\displaystyle\sum_{k=2}^{\infty} 1/(\ln q)^k$.

10. (a) Use a suitable test to find all values of q for which $\displaystyle\sum_{k=2}^{\infty} 1/[k (\ln k)^q]$ converges.

   (b) Why can't you use the integral test for the series

$$\sum_{k=1}^{\infty} (2 + \cos k)/k^2?$$ Test for convergence using a test that does apply.

11. Express $1.3636\ldots$ as (a) an infinite series in sigma notation, and (b) a ratio of integers.

12. In parts (a)–(d), use the comparison test to determine whether the series converges.

(a) $\displaystyle\sum_{k=1}^{\infty} \frac{2k - 1}{3k^2 - k}$ (b) $\displaystyle\sum_{k=1}^{\infty} \frac{2k + 1}{3k^2 + k}$

(c) $\displaystyle\sum_{k=1}^{\infty} \frac{2k - 1}{3k^3 - k^2}$ (d) $\displaystyle\sum_{k=1}^{\infty} \frac{2k + 1}{3k^3 + k^2}$.

13. Find the sum of the series (if it converges).

(a) $\displaystyle\sum_{k=1}^{\infty} \frac{2^k + 3^k}{6^{k+1}}$ (b) $\displaystyle\sum_{k=2}^{\infty} \ln\left(1 + \frac{1}{k}\right)$

(c) $\displaystyle\sum_{k=1}^{\infty} [k^{-1/2} - (k + 1)^{-1/2}]$.

In Exercises 14–21, determine whether the series converges or diverges. You may use the following limits without proof:

$$\lim_{k \to +\infty} (1 + 1/k)^k = e, \quad \lim_{k \to +\infty} \sqrt[k]{k} = 1, \quad \lim_{k \to +\infty} \sqrt[k]{a} = 1$$

14. $\displaystyle\sum_{k=0}^{\infty} e^{-k}$.   **15.** $\displaystyle\sum_{k=1}^{\infty} ke^{-k^2}$.

16. $\displaystyle\sum_{k=1}^{\infty} \frac{k}{k^2 + 2k + 7}$.   **17.** $\displaystyle\sum_{k=1}^{\infty} \frac{\sqrt{k}}{k^2 + 7}$.

18. $\displaystyle\sum_{k=1}^{\infty} \left(\frac{k}{k + 1}\right)^k$.   **19.** $\displaystyle\sum_{k=0}^{\infty} \frac{3^k k!}{(2k)!}$.

20. $\displaystyle\sum_{k=0}^{\infty} \frac{k^6 3^k}{(k + 1)!}$.   **21.** $\displaystyle\sum_{k=1}^{\infty} \left(\frac{5k}{2k + 1}\right)^{3k}$.

In Exercises 22–25, determine whether the given series is absolutely convergent, conditionally convergent, or divergent.

22. $\displaystyle\sum_{k=1}^{\infty} (-1)^k/e^{1/k}$.   **23.** $\displaystyle\sum_{k=0}^{\infty} (-2)^k/(3^k + 1)$.

24. $\displaystyle\sum_{k=0}^{\infty} (-1)^k/(2k + 1)$.   **25.** $\displaystyle\sum_{k=0}^{\infty} (-1)^k 3^k/2^{k+1}$.

26. Find a value of n to ensure that the nth partial sum approximates the sum of the series to the stated accuracy.

(a) $\displaystyle\sum_{k=1}^{\infty} \frac{(-1)^k}{k^2 + 1}$; $|\text{error}| < 0.0001$

(b) $\displaystyle\sum_{k=1}^{\infty} \frac{(-1)^k}{5^k + 1}$; $|\text{error}| < 0.00005$.

In Exercises 27–32, determine the radius of convergence and the interval of convergence of the given power series.

27. $\displaystyle\sum_{k=1}^{\infty} \frac{(x - 1)^k}{k\sqrt{k}}$.   **28.** $\displaystyle\sum_{k=1}^{\infty} \frac{(2x)^k}{3k}$.

29. $\displaystyle\sum_{k=1}^{\infty} \frac{(1 - x)^{2k}}{4^k k}$.   **30.** $\displaystyle\sum_{k=1}^{\infty} \frac{k^2(x + 2)^k}{(k + 1)!}$.

31. $\displaystyle\sum_{k=1}^{\infty} \frac{k!(x - 1)^k}{5^k}$.   **32.** $\displaystyle\sum_{k=1}^{\infty} \frac{(2k)! x^k}{(2k + 1)!}$.

In Exercises 33–35, find
(a) the nth Taylor polynomial for f about $x = a$ (for the stated values of n and a);
(b) Lagrange's form of $R_n(x)$ (for the stated values of n and a);
(c) an upper bound on the absolute value of the error if $f(x)$ is approximated over the given interval by the Taylor polynomial obtained in part (a).

33. $f(x) = \ln(x - 1)$; $a = 2$; $n = 3$; $[\frac{3}{2}, 2]$.

34. $f(x) = e^{x/2}$; $a = 0$; $n = 4$; $[-1, 0]$.

35. $f(x) = \sqrt{x}$; $a = 1$; $n = 2$; $[\frac{4}{9}, 1]$.

36. (a) Use the identity $a - x = a(1 - x/a)$ to find the Maclaurin series for $1/(a - x)$ from the geometric series. What is its radius of convergence?

(b) Find the Maclaurin series and radius of convergence of $1/(3 + x)$.

(c) Find the Maclaurin series and radius of convergence of $2x/(4 + x^2)$.

(d) Use partial fractions to find the Maclaurin series and radius of convergence of

$$\frac{1}{(1 - x)(2 - x)}$$

37. Use the known Maclaurin series for $\ln(1 + x)$ to find the Maclaurin series and radius of convergence of $\ln(a + x)$ for $a > 0$.

38. Use the identity $x = a + (x - a)$ and the known Maclaurin series for e^x, $\sin x$, $\cos x$, and $1/(1 - x)$ to find the Taylor series about $x = a$ for (a) e^x, (b) $\sin x$, and (c) $1/x$.

39. Use the series of Example 8 in Section 11.10 to find the Maclaurin series and radius of convergence of $1/\sqrt{9 + x}$.

In Exercises 40–45, use any method to find the first three nonzero terms of the Maclaurin series.

40. $e^{\tan x}$.

41. $\sec x$.

42. $(\sin x)/(e^x - x)$.

43. $\sqrt{\cos x}$.

44. $e^x \ln(1 - x)$.

45. $\ln(1 + \sin x)$.

46. Find a power series for $\dfrac{1 - \cos 3x}{x^2}$ and use it to evaluate $\lim\limits_{x \to 0} \dfrac{1 - \cos 3x}{x^2}$.

47. Find a power series for $\dfrac{\ln(1 - 2x)}{x}$ and use it to evaluate $\lim\limits_{x \to 0} \dfrac{\ln(1 - 2x)}{x}$.

48. How many decimal places of accuracy can be guaranteed if we approximate $\cos x$ by $1 - x^2/2$ for $-0.1 < x < 0.1$?

49. For what values of x can $\sin x$ be replaced by $x - x^3/6 + x^5/120$ with an ensured accuracy of 6×10^{-4}?

In Exercises 50–52, approximate the indicated quantity to three decimal-place accuracy.

50. $\cos(10°)$.

51. $\displaystyle\int_0^1 \frac{(1 - e^{-t/2})}{t}\, dt$.

52. $\displaystyle\int_0^1 \frac{\sin x}{\sqrt{x}}\, dx$.

53. Show that $y = \displaystyle\sum_{n=0}^{\infty} k^n x^n/n!$ satisfies $y' - ky = 0$ for any fixed k.

12
Topics in Analytic Geometry

Colin Maclaurin   (1698-1746)

■ 12.1 INTRODUCTION TO THE CONIC SECTIONS

The surface shown at the left in Figure 12.1.1 is called a ***double right-circular cone*** or sometimes simply a ***cone***. It is the surface in three-dimensional space generated by a line that revolves about a fixed ***axis*** in such a way that the line passes through a fixed point on the axis and always makes the same angle with the axis. The fixed point is called the ***vertex*** of the cone. The cone consists of two parts, called ***nappes***, that intersect at the vertex.

The curves that can be obtained as intersections of a cone and a plane are called ***conics*** or ***conic sections***, the most important of which are circles, ellipses, parabolas, and hyperbolas (Figure 12.1.1). A ***circle*** is obtained by cutting a cone with a plane that is perpendicular to the axis and does not contain the vertex. If the cutting plane is tilted slightly and intersects only one nappe, the resulting intersection is an ***ellipse***. If the cutting plane is tilted still further so that it is parallel to a line on the surface of the cone, but intersects only one nappe, the resulting intersection is a ***parabola***. Finally, if the plane intersects both nappes, but does not contain the vertex, the resulting intersection is a ***hyperbola***.

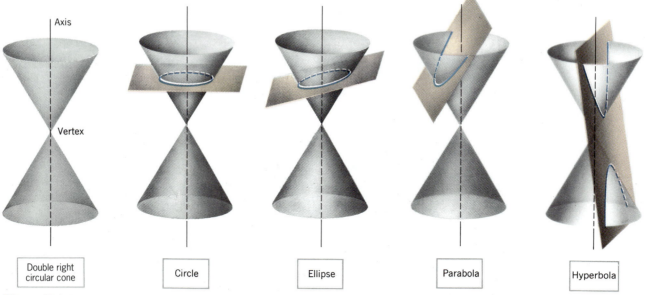

Figure 12.1.1

By choosing the cutting plane to pass through the vertex, it is possible to obtain a point, a line, or a pair of lines for the intersection (Figure 12.1.2). These are called ***degenerate conic sections***.

According to the Alexandrian geographer and astronomer Eratosthenes, the conic sections were first discovered by Menaechmus, a geometer and astronomer in Plato's academy. Although it is not known for certain what motivated the discovery of the conic sections, it is commonly believed that they resulted from the study of construction problems. Menaechmus, for example, used them

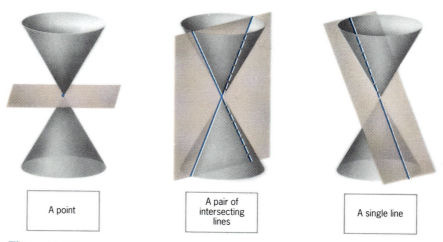

A point

A pair of
intersecting
lines

A single line

Figure 12.1.2

to solve the problem of "doubling the cube," that is, constructing a cube whose volume is twice that of a given cube. Another theory suggests that the conic sections may have originated as a result of work on sundials. With the advent of analytic geometry and calculus, conic sections gained importance in the physical sciences. In 1609 Johannes Kepler published a book known as *Astronomia Nova* in which he presented his landmark discovery that the path of each planet about the sun is an ellipse. Galileo and Newton showed that objects subject to gravitational forces can also move along paths that are parabolas or hyperbolas.

Today, properties of conic sections are used in the construction of telescopes, radar antennas, medical equipment, navigational systems, and in the determination of satellite orbits. In this chapter we shall develop the basic geometric properties and equations of conic sections. For simplicity, our working definitions of the conic sections will be based on their geometric properties rather than their interpretation as intersections of a plane with a cone. This will enable us to keep our work in a two-dimensional setting.

12.2 THE PARABOLA; TRANSLATION OF COORDINATE AXES

In this section we shall discuss properties of parabolas.

□ **DEFINITION OF A PARABOLA**

12.2.1 DEFINITION.   A *parabola* is the set of all points in the plane that are equidistant from a given line and a given point not on the line.

The given line is called the *directrix* of the parabola, and the given point the *focus* (Figure 12.2.1). A parabola is symmetric about the line that passes through the focus at right angles to the directrix. This line, called the *axis* or the *axis of symmetry* of the parabola, meets the parabola at a point called the *vertex.*

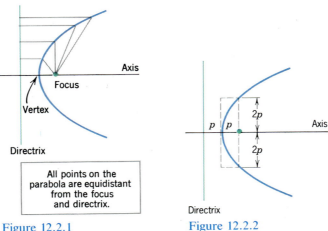

Figure 12.2.1

Figure 12.2.2

□ GEOMETRIC PROPERTIES OF A PARABOLA

It is traditional in the study of parabolas to denote the distance between the focus and the vertex by p. The vertex is equidistant from the focus and the directrix (why?), so the distance between the vertex and the directrix is also p; consequently, the distance between the focus and the directrix is $2p$ (Figure 12.2.2). As illustrated in the figure, the parabola passes through the corners of a box that extends from the vertex to the focus along the axis of symmetry and extends $2p$ units above and $2p$ units below the axis of symmetry.

□ STANDARD EQUATIONS OF A PARABOLA

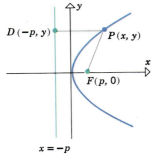

Figure 12.2.3

The equation of a parabola is simplest if the coordinate axes are positioned so that the vertex is at the origin and the axis of symmetry is along the x-axis or y-axis. The four possible such orientations are shown in Table 12.2.1. These are called the *standard positions* of a parabola, and the resulting equations are called the *standard equations* of a parabola.

To illustrate how the equations in Table 12.2.1 are obtained, let us consider the first entry in the table—the parabola with focus at $(p, 0)$, directrix $x = -p$. Let $P(x, y)$ be any point on the parabola. Since P is equidistant from the focus and directrix, the distances PF and PD in Figure 12.2.3 are equal; that is,

$$PF = PD \qquad (1)$$

where $D(-p, y)$ is the foot of the perpendicular from P to the directrix. From the distance formula, the distances PF and PD are

$$PF = \sqrt{(x - p)^2 + y^2} \quad \text{and} \quad PD = \sqrt{(x + p)^2} \qquad (2)$$

Substituting in (1) and squaring yields

$$(x - p)^2 + y^2 = (x + p)^2 \qquad (3)$$

Table 12.2.1

ORIENTATION	DESCRIPTION	STANDARD EQUATION
	• Vertex at the origin. • Parabola opens in the positive x-direction. • Symmetric about the x-axis.	$y^2 = 4px$
	• Vertex at the origin. • Parabola opens in the negative x-direction. • Symmetric about the x-axis.	$y^2 = -4px$
	• Vertex at the origin. • Parabola opens in the positive y-direction. • Symmetric about the y-axis.	$x^2 = 4py$
	• Vertex at the origin. • Parabola opens in the negative y-direction. • Symmetric about the y-axis.	$x^2 = -4py$

and after simplifying

$$y^2 = 4px \tag{4}$$

Conversely, any point $P(x, y)$ satisfying (4) also satisfies (3) (reverse the steps in the simplification); thus, from (2), $PF = PD$, which shows that P is equidistant from the focus and directrix. Therefore, each point satisfying (4) lies on the parabola.

The remaining equations in Table 12.2.1 have similar derivations; they are left as exercises.

☐ A TECHNIQUE FOR GRAPHING PARABOLAS

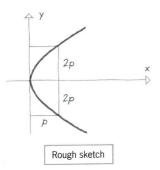

Rough sketch

Figure 12.2.4

Parabolas can be graphed from their *standard equations* using four basic steps:

- Determine whether the axis of symmetry is along the x-axis or the y-axis. This can be ascertained from the quadratic term in the equation; referring to Table 12.2.1, the axis of symmetry is along the x-axis if the equation has a y^2-term, and it is along the y-axis if it has an x^2-term.
- Determine which way the parabola opens. If the axis of symmetry is along the x-axis, then the parabola opens to the right if the coefficient of x is positive, and it opens to the left if the coefficient is negative. If the axis of symmetry is along the y-axis, then the parabola opens up if the coefficient of y is positive, and it opens down if the coefficient is negative.
- Determine the value of p and draw a box extending p units from the origin along the axis of symmetry in the direction in which the parabola opens and extending $2p$ units above and $2p$ units below the axis of symmetry.
- Using the box as a guide, sketch the parabola so that its vertex is at the origin and it passes through the corners of the box (Figure 12.2.4).

Example 1 Sketch the graphs of the parabolas

(a) $x^2 = 12y$ (b) $y^2 + 8x = 0$

showing the focus and directrix of each.

Solution (a). This equation involves x^2, so the axis of symmetry is along the y-axis, and the coefficient of y is positive, so the parabola opens upward. From the coefficient of y, we obtain $4p = 12$ or $p = 3$. Drawing a box extending $p = 3$ units up from the origin and $2p = 6$ units to the left and $2p = 6$ units to the right of the y-axis, then using the corners of the box as a guide, yields the graph in Figure 12.2.5.

The focus lies $p = 3$ units from the vertex along the axis of symmetry in the direction in which the parabola opens, so its coordinates are $(0, 3)$. The directrix is perpendicular to the axis of symmetry at a distance of $p = 3$ units from the vertex on the opposite side from the focus, so its equation is $y = -3$.

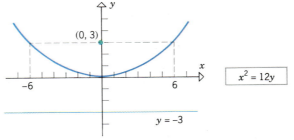

Figure 12.2.5

Solution (b). We first rewrite the equation in the standard form

$$y^2 = -8x$$

This equation involves y^2, so the axis of symmetry is along the x-axis, and the

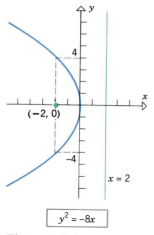

$y^2 = -8x$

Figure 12.2.6

coefficient of x is negative, so the parabola opens to the left. From the coefficient of x we obtain $4p = 8$, so $p = 2$. Drawing a box extending $p = 2$ units left from the origin and $2p = 4$ units above and $2p = 4$ units below the x-axis, then using the corners of the box as a guide, yields the graph in Figure 12.2.6. ◀

Example 2 Find an equation for the parabola that is symmetric about the y-axis, has its vertex at the origin, and passes through the point $(5, 2)$.

Solution. Since the parabola is symmetric about the y-axis and has its vertex at the origin, the equation is of the form

$$x^2 = 4py \quad \text{or} \quad x^2 = -4py$$

where the sign depends on whether the parabola opens up or down. But the parabola must open up, since it passes through the point $(5, 2)$, which lies in the first quadrant. Thus, the equation is of the form

$$x^2 = 4py \tag{5}$$

Since the parabola passes through $(5, 2)$, we must have $5^2 = 4p \cdot 2$ or $4p = \frac{25}{2}$. Therefore, (5) becomes

$$x^2 = \frac{25}{2} y \quad ◀$$

☐ **TRANSLATION OF AXES**

If a parabola has its axis of symmetry parallel to one of the coordinate axes, but its vertex is not at the origin, then the location of the focus and directrix can be found by introducing an auxiliary coordinate system with its origin at the vertex and its axes parallel to the original axes. Before we can discuss the details, we need some preliminary results about translating coordinate axes.

In Figure 12.2.7a we have translated the axes of an xy-coordinate system to obtain a new $x'y'$-coordinate system whose origin O' is at the point $(x, y) = (h, k)$.

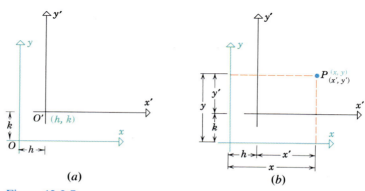

(a)   (b)

Figure 12.2.7

As a result, a point P in the plane will have both (x, y)-coordinates and (x', y')-coordinates. As suggested by Figure 12.2.7b, these coordinates are related by

$$x' = x - h, \quad y' = y - k \tag{6}$$

or, equivalently,

$$x = x' + h, \quad y = y' + k \qquad (7)$$

These are called the **translation equations.**

As an illustration, if the new origin is at $(h, k) = (4, -1)$ and the xy-coordinates of a point P are $(2, 5)$, then the $x'y'$-coordinates of P are

$$x' = x - h = 2 - 4 = -2 \quad \text{and} \quad y' = y - k = 5 - (-1) = 6$$

☐ **TRANSLATED PARABOLAS**

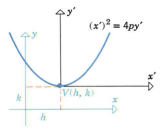

Figure 12.2.8

Let us now consider a parabola with vertex V at (h, k) in an xy-coordinate system and axis of symmetry parallel to the y-axis. If, as in Figure 12.2.8, we translate the axes so that the vertex V is at the origin of an $x'y'$-coordinate system, then in $x'y'$-coordinates the equation of the parabola will be

$$(x')^2 = 4py' \quad \text{or} \quad (x')^2 = -4py'$$

and from the translation equations (6), the corresponding equation in xy-coordinates will be

> **Parabola with Vertex (h, k) and Axis Parallel to y-axis**
>
> $$(x - h)^2 = \pm 4p(y - k) \qquad (8)$$
>
> where the $+$ sign occurs if the parabola opens in the positive y-direction and the $-$ sign if it opens in the negative y-direction.

Thus, in xy-coordinates, an equation of form (8) represents a parabola with vertex at (h, k) and axis of symmetry parallel to the y-axis. The parabola opens in the positive or negative y-direction according to whether the coefficient of p is positive or negative. Similarly,

> **Parabola with Vertex (h, k) and Axis Parallel to x-axis**
>
> $$(y - k)^2 = \pm 4p(x - h) \qquad (9)$$
>
> where the $+$ sign occurs if the parabola opens in the positive x-direction and the $-$ sign if it opens in the negative x-direction

represents a parabola with vertex at (h, k) and axis of symmetry parallel to the x-axis. The parabola opens in the positive or negative x-direction according to whether the coefficient of p is positive or negative.

Example 3 Sketch the parabola

$$(y - 3)^2 = 8(x + 4)$$

and locate the focus and directrix.

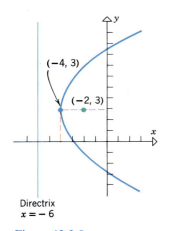

Directrix
$x = -6$

Figure 12.2.9

Solution.   From (9) with the plus sign, the equation represents a parabola with vertex at $(h, k) = (-4, 3)$ and axis of symmetry parallel to the x-axis. Moreover, $4p = 8$ or $p = 2$. Since the coefficient of p is positive, the parabola opens in the positive x-direction. This places the focus 2 units to the right of the vertex or at the point $(-2, 3)$. The directrix is 2 units to the left of the vertex (and parallel to the y-axis), so its equation is $x = -6$. The parabola is sketched in Figure 12.2.9.   ◀

Sometimes (8) and (9) appear in expanded form, in which case some algebraic manipulations are required to identify the parabola.

Example 4   Show that the curve

$$y = 6x^2 - 12x + 8$$

is a parabola.

Solution.   Because the equation involves x to the second power and y to the first power, and because the expanded form of (8) has the same property, we shall try to rewrite the equation in form (8). To do this, we first divide by the coefficient of x^2 and collect all the x-terms on one side:

$$\frac{1}{6}y - \frac{8}{6} = x^2 - 2x$$

Next, we complete the square on the x-terms by adding 1 to both sides:

$$\frac{1}{6}y - \frac{2}{6} = (x - 1)^2$$

Finally, we factor out the coefficient of the y-term to obtain

$$(x - 1)^2 = \frac{1}{6}(y - 2)$$

which has form (8) with the plus sign and

$$h = 1, \quad k = 2, \quad 4p = \frac{1}{6}, \quad p = \frac{1}{24}$$

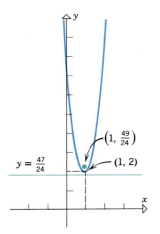

Figure 12.2.10

Thus, the curve is a parabola with vertex at $(1, 2)$ and axis of symmetry parallel to the y-axis. Since the coefficient of p is positive, the parabola opens in the positive y-direction, the focus is $\frac{1}{24}$ unit above the vertex, and the directrix $\frac{1}{24}$ unit below the vertex. Thus, the focus is at $(1, \frac{49}{24})$ and the directrix has equation $y = \frac{47}{24}$ (Figure 12.2.10).   ◀

The procedure used in Example 4 can be used to prove the following general result.

12.2.2 THEOREM. *The graph of*

$$y = Ax^2 + Bx + C \quad (A \neq 0)$$

is a parabola with axis of symmetry parallel to the y-axis; the parabola opens in the positive y-direction if A > 0 and in the negative y-direction if A < 0. The graph of

$$x = Ay^2 + By + C \quad (A \neq 0)$$

is a parabola with axis of symmetry parallel to the x-axis; the parabola opens in the positive x-direction if A > 0 and in the negative x-direction if A < 0.

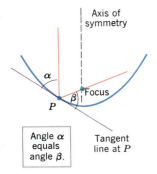

Angle α equals angle β.

Figure 12.2.11

Example 5   Find an equation for the parabola that has its vertex at $(1, 2)$ and its focus at $(4, 2)$.

Solution.   Since the focus and vertex are on a horizontal line, and since the focus is to the right of the vertex, the parabola opens to the right and its equation has the form

$$(y - k)^2 = 4p(x - h)$$

Since the vertex and focus are three units apart we have $p = 3$, and since the vertex is at $(h, k) = (1, 2)$ we obtain

$$(y - 2)^2 = 12(x - 1) \quad \blacktriangleleft$$

□ **REFLECTION PROPERTIES OF PARABOLAS**

Parabolas have important applications in the design of telescopes, radar antennas, and lighting systems. This is due to the following property of parabolas (Exercise 44).

12.2.3   THEOREM (*A Geometric Property of Parabolas*).   *The tangent line at a point P on a parabola makes equal angles with the line through P parallel to the axis of symmetry and the line through P and the focus* (Figure 12.2.11).

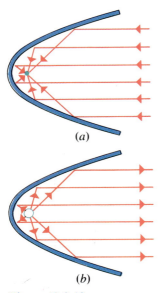

(a)

(b)

Figure 12.2.12

It is a principle of physics that when light is reflected from a point P on a surface, the angle of incidence equals the angle of reflection; that is, the angle between the incoming ray and the tangent line at P equals the angle between the outgoing ray and the tangent line at P. Therefore, if the reflecting surface has parabolic cross sections with a common focus, then it follows from Theorem 12.2.3 that all light rays entering parallel to the axis will be reflected through the focus (Figure 12.2.12*a*). In reflecting telescopes, this principle is used to reflect the (approximately) parallel rays of light from the stars or planets off a parabolic mirror to an eyepiece at the focus of the parabola. Conversely, if a light source is located at the focus of a parabolic reflector, it follows from Theorem 12.2.3 that the reflected rays will form a beam parallel to the axis (Figure 12.2.12*b*). The parabolic reflectors in flashlights and automobile headlights utilize this principle. The optical principles just discussed also apply to radar signals, which explains the parabolic shape of many radar antennas.

▶ Exercise Set 12.2

In Exercises 1–16, sketch the parabola. Show the focus, vertex, and directrix.

1. $y^2 = 6x$.

2. $y^2 = -10x$.

3. $x^2 = -9y$.

4. $x^2 = 4y$.

5. $5y^2 = 12x$.

6. $x^2 - 40y = 0$.

7. $(y - 3)^2 = 6(x - 2)$.

8. $(y + 1)^2 = -7(x - 4)$.

9. $(x + 2)^2 = -(y + 2)$.

10. $(x - \frac{1}{2})^2 = 2(y - 1)$.

11. $x^2 - 4x + 2y = 1$.

12. $y^2 - 6y - 2x + 1 = 0$.

13. $x = y^2 - 4y + 2$.

14. $y = 4x^2 + 8x + 5$.

15. $-y^2 + 2y + x = 0$.

16. $y = 1 - 4x - x^2$.

In Exercises 17–31, find an equation for the parabola satisfying the given conditions.

17. Vertex $(0, 0)$; focus $(3, 0)$.

18. Vertex $(0, 0)$; focus $(0, -4)$.

19. Vertex $(0, 0)$; directrix $x = 7$.

20. Vertex $(0, 0)$; directrix $y = \frac{1}{2}$.

21. Vertex $(0, 0)$; symmetric about the x-axis; passes through $(2, 2)$.

22. Vertex $(0, 0)$; symmetric about the y-axis; passes through $(-1, 3)$.

23. Focus $(0, -3)$; directrix $y = 3$.

24. Focus $(6, 0)$; directrix $x = -6$.

25. Axis $y = 0$; passes through $(3, 2)$ and $(2, -3)$.

26. Axis $x = 0$; passes through $(2, -1)$ and $(-4, 5)$.

27. Focus $(3, 0)$; directrix $x = 0$.

28. Vertex $(4, -5)$; focus $(1, -5)$.

29. Vertex $(1, 1)$; directrix $y = -2$.

30. Focus $(-1, 4)$; directrix $x = 5$.

31. Vertex $(5, -3)$; axis parallel to the y-axis; passes through $(9, 5)$.

32. Use the definition of a parabola (12.2.1) to find the equation of the parabola with focus $(2, 1)$ and directrix $x + y + 1 = 0$. [*Hint:* Use the result of Exercise 19, Section 1.6.]

33. (a) Find an equation for the parabola with axis parallel to the y-axis and passing through $(0, 3)$, $(2, 0)$, and $(3, 2)$.

(b) Find an equation for the parabola with axis parallel to the x-axis and passing through the points in part (a).

34. Prove: The line tangent to the parabola $x^2 = 4py$ at (x_0, y_0) is

$$y = \frac{x_0}{2p} x - y_0$$

35. Find the vertex, focus, and directrix of the parabola $y = Ax^2 + Bx + C$ $(A \neq 0)$.

36. (a) Find an equation for the following parabolic arch with base b and height h.

(b) Find the area under the arch.

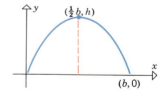

37. A parabolic arch spans a road 40 feet wide. How high is the arch if a center section of the road 20 feet wide has a minimum clearance of 12 feet?

38. Let C be a circle of radius r and L a line that does not intersect C but is in the same plane as C. Show that the centers of all circles that do not enclose C and are tangent to both C and L lie on a parabola. Specify the location of the focus and directrix of the parabola.

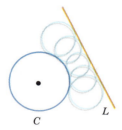

39. Prove: The vertex is the closest point on a parabola to the focus.

40. A comet moves in a parabolic orbit with the sun at its focus. When the comet is 40 million miles from the sun, the line from the sun to the comet makes an angle of 60° with the axis of the parabola. How close will the comet come to the sun? [*Hint:* See Exercise 39.]

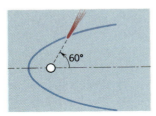

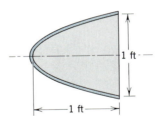

42. Derive the equation $x^2 = 4py$ in Table 12.2.1.

43. Derive the equation $y^2 = -4px$ in Table 12.2.1.

44. Prove Theorem 12.2.3. [*Hint:* Choose coordinate axes so that the parabola has the equation $x^2 = 4py$. Show that the tangent line at $P(x_0, y_0)$ intersects the y-axis at $Q(0, -y_0)$ and that the triangle whose three vertices are at P, Q, and the focus is isosceles.]

41. How far from the vertex should a light source be placed on the axis of the following parabolic reflector to produce a beam of parallel rays?

■ 12.3 THE ELLIPSE

In this section we shall discuss properties of ellipses.

☐ **DEFINITION OF AN ELLIPSE**

12.3.1 DEFINITION. An *ellipse* is the set of all points in the plane, the sum of whose distances from two fixed points is a given positive constant that is greater than the distance between the fixed points.

The two fixed points are called the *foci* (plural of "focus"), and the midpoint of the line segment joining the foci is called the *center* of the ellipse (Figure 12.3.1). To help visualize Definition 12.3.1, imagine that two ends of a string are tacked to the foci and a pencil traces a curve as it is held tight against the string (Figure 12.3.2). The resulting curve will be an ellipse since the sum of the distances to the foci is a constant, namely the total length of the string. Note that if the foci coincide, the ellipse reduces to a circle.

The line segment through the foci and across the ellipse is called the *major axis* (Figure 12.3.3), while the line segment across the ellipse, through the center, and perpendicular to the major axis is called the *minor axis*. It is traditional in the study of ellipses to denote the length of the major axis by $2a$, the length of the minor axis by $2b$, and the distance between the foci by $2c$ (Figure 12.3.4). The numbers a and b are called the *semiaxes* of the ellipse.

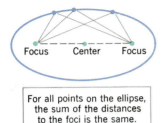

For all points on the ellipse, the sum of the distances to the foci is the same.

Figure 12.3.1

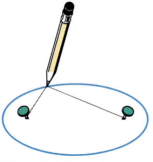

Figure 12.3.2

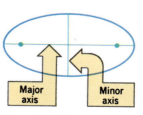

Major axis   Minor axis

Figure 12.3.3

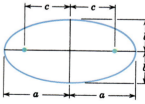

Figure 12.3.4

GEOMETRIC PROPERTIES OF AN ELLIPSE

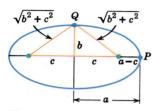

Figure 12.3.5

There is a basic relationship between the numbers a, b, and c that can be obtained by considering a point P at the end of the major axis and a point Q at the end of the minor axis. Because P and Q both lie on the ellipse, the sum of the distances from each of them to the foci will be the same. If we express this fact as an equation, we obtain (see Figure 12.3.5)

$$2\sqrt{b^2 + c^2} = (a - c) + (a + c)$$

from which it follows that

$$a = \sqrt{b^2 + c^2} \tag{1}$$

or equivalently

$$c = \sqrt{a^2 - b^2} \tag{2}$$

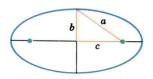

Figure 12.3.6

Formula (1) shows that the distance from a focus to an end of the minor axis is a (Figure 12.3.6), which implies that for *all* points on the ellipse the sum of the distances to the foci is $2a$. (Why?)

It also follows from (1) that $a \geq b$ with the equality holding only when $c = 0$. Geometrically, this means that the major axis of an ellipse is at least as large as the minor axis, and that the two axes have equal length only when the foci coincide, in which case the ellipse is a circle.

STANDARD EQUATIONS OF AN ELLIPSE

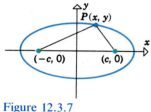

Figure 12.3.7

The equation of an ellipse is simplest if the coordinate axes are positioned so that the center of the ellipse is at the origin and the foci are on the x-axis or y-axis. The two possible such orientations are shown in Table 12.3.1. These are called the **standard positions** of an ellipse, and the resulting equations are called the **standard equations** of an ellipse.

To illustrate how the equations in Table 12.3.1 are derived, let us consider the first entry in the table. Let $P(x, y)$ be any point on the ellipse. Since the sum of the distances from P to the foci is $2a$, it follows (Figure 12.3.7) that

$$\sqrt{(x + c)^2 + y^2} + \sqrt{(x - c)^2 + y^2} = 2a$$

Table 12.3.1

ORIENTATION	DESCRIPTION	STANDARD EQUATION
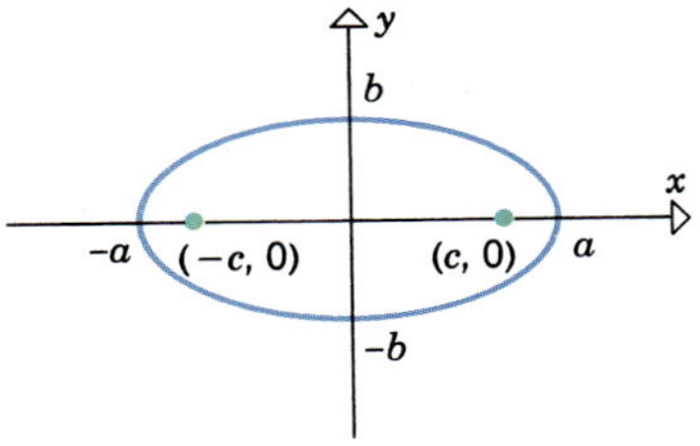	• Foci and major axis on the x-axis. • Minor axis on the y-axis. • Center at the origin. • x-intercepts: $\pm a$. • y-intercepts: $\pm b$. • $a \geq b$	$\dfrac{x^2}{a^2} + \dfrac{y^2}{b^2} = 1$
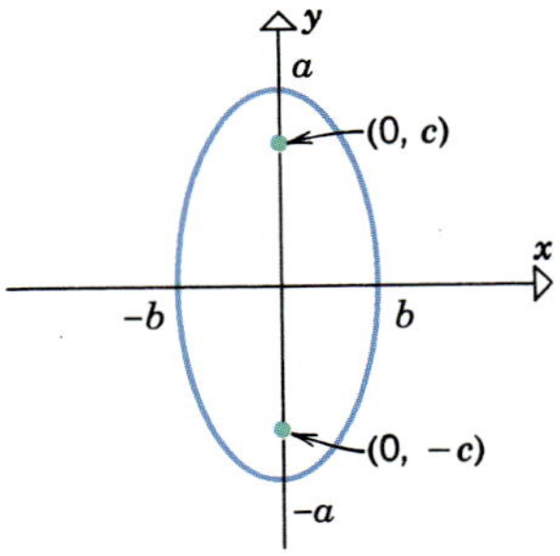	• Foci and major axis on the y-axis. • Minor axis on the x-axis. • Center at the origin. • x-intercepts: $\pm b$. • y-intercepts: $\pm a$. • $a \geq b$	$\dfrac{x^2}{b^2} + \dfrac{y^2}{a^2} = 1$

Transposing the second radical to the right side of the equation and squaring yields

$$(x + c)^2 + y^2 = 4a^2 - 4a\sqrt{(x - c)^2 + y^2} + (x - c)^2 + y^2$$

and, on simplifying,

$$\sqrt{(x - c)^2 + y^2} = a - \frac{c}{a}x$$

Squaring again and simplifying yields

$$\frac{x^2}{a^2} + \frac{y^2}{a^2 - c^2} = 1$$

which, by virtue of (1), can be written as

$$\frac{x^2}{a^2} + \frac{y^2}{b^2} = 1 \tag{3}$$

Conversely, it can be shown that any point whose coordinates satisfy (3) has $2a$ as the sum of its distances from the foci, so that such a point is on the ellipse.

The second equation in Table 12.3.1 has a similar derivation.

☐ A TECHNIQUE FOR GRAPHING ELLIPSES

Ellipses can be graphed from their *standard equations* using three basic steps:

- Determine whether the major axis is on the *x*-axis or the *y*-axis. This can be ascertained from the sizes of the denominators in the equation. Referring to Table 12.3.1, and keeping in mind that $a^2 > b^2$ (since $a > b$), the major axis is along the *x*-axis if x^2 has the larger denominator, and it is along the *y*-axis if y^2 has the larger denominator. If the denominators are equal, the ellipse is a circle.

- Determine the values of *a* and *b* and draw a box extending *a* units on each side of the center along the major axis and *b* units on each side of the center along the minor axis.

- Using the box as a guide, sketch the ellipse so that its center is at the origin and it touches the sides of the box where the sides intersect the coordinate axes (Figure 12.3.8).

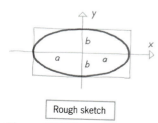

Rough sketch

Figure 12.3.8

Example 1 Sketch the graphs of the ellipses

$$\text{(a)}\quad \frac{x^2}{9} + \frac{y^2}{16} = 1 \qquad \text{(b)}\quad x^2 + 2y^2 = 4$$

showing the foci of each.

Solution (a). Since y^2 has the larger denominator, the major axis is along the *y*-axis. Moreover, since $a^2 > b^2$, we must have $a^2 = 16$ and $b^2 = 9$, so

$$a = 4 \quad \text{and} \quad b = 3$$

Drawing a box extending 4 units on each side of the origin along the *y*-axis and 3 units on each side of the origin along the *x*-axis as a guide yields the graph in Figure 12.3.9.

The foci lie *c* units on each side of the center along the major axis, where *c* is given by (2). From the values of a^2 and b^2 above, we obtain

$$c = \sqrt{a^2 - b^2} = \sqrt{16 - 9} = \sqrt{7} \approx 2.6$$

Since the foci lie on the *y*-axis in this case, their coordinates are $(0, \sqrt{7})$ and $(0, -\sqrt{7})$.

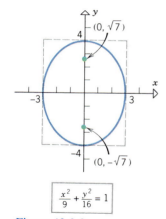

$$\frac{x^2}{9} + \frac{y^2}{16} = 1$$

Figure 12.3.9

Solution (b). We first rewrite the equation in the standard form

$$\frac{x^2}{4} + \frac{y^2}{2} = 1$$

Since x^2 has the larger denominator, the major axis lies along the *x*-axis, and we have $a^2 = 4$ and $b^2 = 2$. Drawing a box extending $a = 2$ on each side of the origin along the *x*-axis and extending $b = \sqrt{2} \approx 1.4$ units on each side of the origin along the *y*-axis as a guide yields the graph in Figure 12.3.10.

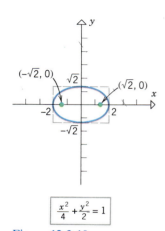

$$\frac{x^2}{4} + \frac{y^2}{2} = 1$$

Figure 12.3.10

From (2), we obtain

$$c = \sqrt{a^2 - b^2} = \sqrt{2} \approx 1.4$$

so the foci, which lie on the x-axis in this case, have coordinates $(\sqrt{2}, 0)$ and $(-\sqrt{2}, 0)$. ◀

Example 2   Find an equation for the ellipse with foci $(0, \pm 2)$ and major axis with endpoints $(0, \pm 4)$.

Solution.   From Table 12.3.1, the equation has the form

$$\frac{x^2}{b^2} + \frac{y^2}{a^2} = 1$$

and from the given information, $a = 4$ and $c = 2$. It follows from (1) that

$$b^2 = a^2 - c^2 = 16 - 4 = 12$$

so the equation of the ellipse is

$$\frac{x^2}{12} + \frac{y^2}{16} = 1 \qquad ◀$$

 **TRANSLATED ELLIPSES**

If the axes of an ellipse are parallel to the coordinate axes but the center is not at the origin, then the equation of the ellipse may be determined by translation of axes. Specifically, if the center of the ellipse is at the point with xy-coordinates (h, k), and if we translate the xy-axes so that the origin of the $x'y'$-system is at the center (h, k), then the equation of the ellipse in the $x'y'$-system will be

$$\frac{(x')^2}{a^2} + \frac{(y')^2}{b^2} = 1 \quad \text{or} \quad \frac{(x')^2}{b^2} + \frac{(y')^2}{a^2} = 1$$

depending on the orientation of the major and minor axes (Figure 12.3.11). Thus, from the translation equations $x' = x - h$, $y' = y - k$, the equation of the ellipse in the xy-system will be one of the following:

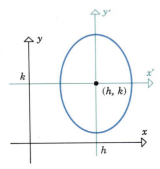

Figure 12.3.11

> *Ellipse with Center (h, k) and Major Axis Parallel to x-axis*
>
> $$\frac{(x - h)^2}{a^2} + \frac{(y - k)^2}{b^2} = 1 \qquad (a \geq b) \tag{4}$$

> *Ellipse with Center (h, k) and Major Axis Parallel to y-axis*
>
> $$\frac{(x - h)^2}{b^2} + \frac{(y - k)^2}{a^2} = 1 \qquad (a \geq b) \tag{5}$$

Sometimes (4) and (5) appear in expanded form, in which case it is necessary to complete the squares before the ellipse can be analyzed.

Example 3   Show that the curve

$$16x^2 + 9y^2 - 64x - 54y + 1 = 0$$

is an ellipse. Sketch the ellipse and show the location of the foci.

Solution.   Our objective is to rewrite the equation in one of the forms, (4) or (5). To do this, group the x-terms and y-terms and take the constant to the right side:

$$(16x^2 - 64x) + (9y^2 - 54y) = -1$$

Next, factor out the coefficients of x^2 and y^2 and complete the squares:

$$16(x^2 - 4x + 4) + 9(y^2 - 6y + 9) = -1 + 64 + 81$$

or

$$16(x - 2)^2 + 9(y - 3)^2 = 144$$

Finally, divide through by 144 to introduce a 1 on the right side:

$$\frac{(x - 2)^2}{9} + \frac{(y - 3)^2}{16} = 1$$

This is an equation of form (5), with $h = 2$, $k = 3$, $a^2 = 16$, and $b^2 = 9$. Thus, the given equation is an ellipse with center $(2, 3)$ and major axis parallel to the y-axis. Since $a = 4$, the major axis extends 4 units above and 4 units below the center, so its endpoints are $(2, 7)$ and $(2, -1)$ (Figure 12.3.12). Since $b = 3$, the minor axis extends 3 units to the left and 3 units to the right of the center, so its endpoints are $(-1, 3)$ and $(5, 3)$. Since

$$c = \sqrt{a^2 - b^2} = \sqrt{16 - 9} = \sqrt{7}$$

the foci lie $\sqrt{7}$ units above and below the center, placing them at the points $(2, 3 + \sqrt{7})$ and $(2, 3 - \sqrt{7})$.   ◄

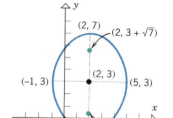

Figure 12.3.12

□ **REFLECTION PROPERTIES OF ELLIPSES**

Tangent line at P

Angles α and β are equal.

Figure 12.3.13

In the exercises the reader is asked to prove the following result.

12.3.2   THEOREM (*A Geometric Property of Ellipses*).   *A line tangent to an ellipse at a point P makes equal angles with the lines through P and the foci* (Figure 12.3.13).

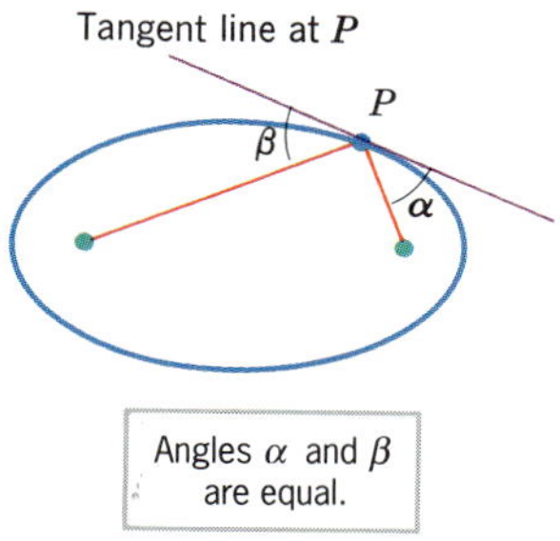

It follows from Theorem 12.3.2 that a ray of light emanating from one focus of an ellipse will be reflected through the other focus. This property of ellipses is used in "whispering galleries." Such rooms have ceilings whose cross sections are elliptical in shape with common foci. As a result, if a person standing at one focus whispers, the sound waves are reflected by the ceiling to the other focus, making it possible for a person at that focus to hear the whispered sound.

► **Exercise Set 12.3**   ☐ 48

In Exercises 1–14, sketch the ellipse. Label the foci and the ends of the major and minor axes.

1. $\dfrac{x^2}{16} + \dfrac{y^2}{9} = 1.$ **2.** $\dfrac{x^2}{4} + \dfrac{y^2}{25} = 1.$

3. $9x^2 + y^2 = 9.$ **4.** $4x^2 + 9y^2 = 36.$

5. $x^2 + 3y^2 = 2.$ **6.** $16x^2 + 4y^2 = 1.$

7. $9(x-1)^2 + 16(y-3)^2 = 144.$

8. $(x+3)^2 + 4(y-5)^2 = 16.$

9. $3(x+2)^2 + 4(y+1)^2 = 12.$

10. $\frac{1}{4}x^2 + \frac{1}{9}(y+2)^2 - 1 = 0.$

11. $x^2 + 9y^2 + 2x - 18y + 1 = 0.$

12. $9x^2 + 4y^2 + 18x - 24y + 9 = 0.$

13. $4x^2 + y^2 + 8x - 10y = -13.$

14. $5x^2 + 9y^2 - 20x + 54y = -56.$

In Exercises 15–26, find an equation for the ellipse satisfying the given conditions.

15. Ends of major axis $(\pm 3, 0)$; ends of minor axis $(0, \pm 2)$.

16. Ends of major axis $(0, \pm\sqrt{5})$; ends of minor axis $(\pm 1, 0)$.

17. Length of major axis 26; foci $(\pm 5, 0)$.

18. Length of minor axis 16; foci $(0, \pm 6)$.

19. Foci $(\pm 1, 0)$; $b = \sqrt{2}$.

20. Foci $(\pm 3, 0)$; $a = 4$.

21. $c = 2\sqrt{3}$; $a = 4$; center at the origin; foci on a coordinate axis (two answers).

22. $b = 3$; $c = 4$; center at the origin; foci on a coordinate axis (two answers).

23. Ends of major axis $(\pm 6, 0)$; passes through $(2, 3)$.

24. Center at $(0, 0)$; major and minor axes along the coordinate axes; passes through $(3, 2)$ and $(1, 6)$.

25. Foci $(1, 2)$ and $(1, 4)$; minor axis of length 2.

26. Foci $(2, 1)$ and $(2, -3)$; major axis of length 6.

27. Find an equation of the ellipse traced by a point which moves so that the sum of its distances to $(4, 1)$ and $(4, 5)$ is 12.

28. Find an equation of the ellipse traced by a point which moves so that the sum of its distances to $(0, 0)$ and $(1, 1)$ is 4.

In Exercises 29–32, find all intersections of the given curves, and make a sketch of the curves that shows the points of intersection.

29. $x^2 + 4y^2 = 40$ and $x + 2y = 8.$

30. $y^2 = 2x$ and $x^2 + 2y^2 = 12.$

31. $x^2 + 9y^2 = 36$ and $x^2 + y^2 = 20.$

32. $16x^2 + 9y^2 = 36$ and $5x^2 + 18y^2 = 45.$

33. Find two values of k such that the line $x + 2y = k$ is tangent to the ellipse $x^2 + 4y^2 = 8$. Find the points of tangency.

34. Prove: The line that is tangent to the ellipse $x^2/a^2 + y^2/b^2 = 1$ at (x_0, y_0) has the equation $xx_0/a^2 + yy_0/b^2 = 1.$

Exercises 35–38 refer to an ellipse with major and minor axes of lengths $2a$ and $2b$, respectively (Figure 12.3.14).

35. Find the area enclosed by the ellipse in Figure 12.3.14.

36. Find the volume of the solid that results when the region enclosed by the ellipse in Figure 12.3.14 is revolved about
(a) the major axis   (b) the minor axis.

37. Show that if $2c$ is the distance between the foci of the ellipse in Figure 12.3.14, then the area of the surface generated by revolving the ellipse about the major axis is

$$2\pi ab\left(\frac{b}{a} + \frac{a}{c}\sin^{-1}\frac{c}{a}\right)$$

38. Show that if $2c$ is the distance between the foci of the ellipse in Figure 12.3.14, then the area of the surface generated by revolving the ellipse about the minor axis is

$$2\pi ab\left(\frac{a}{b} + \frac{b}{c}\ln\frac{a+c}{b}\right)$$

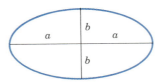

Figure 12.3.14

39. The tank of an oil truck is 18 feet long and has elliptical cross sections that are 6 feet wide and 4 feet high (Figure 12.3.15). Find a formula for the volume of oil in the tank in terms of the depth h of the oil.

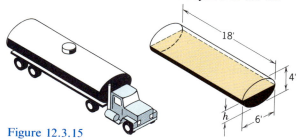

Figure 12.3.15

40. A semielliptic arch spans a highway 50 feet wide. How high is the arch if a center section of the highway 30 feet wide has a minimum clearance of 14 feet?

41. Find the area of the square that can be inscribed in the ellipse $x^2/a^2 + y^2/b^2 = 1$.

42. Suppose that you want to draw an ellipse having given values for the lengths of the major and minor axes by using the method shown in Figure 12.3.2. Assuming that the axes are drawn, explain how a compass can be used to locate the positions for the tacks.

43. The distance between a point $P(x, y)$ and the point $(4, 0)$ is $4/5$ of the distance between P and the line $x = 25/4$. Show that the set of all such points is an ellipse. Find the center and the lengths of the major and minor axes.

44. Show that the curve of intersection of a plane and a right-circular cylinder is an ellipse, assuming that the plane is neither perpendicular to nor parallel to the axis of the cylinder. [*Hint:* Let θ be the angle that the plane makes with a circular cross section of the cylinder. Introduce xy- and $x'y'$-axes as shown in Figure 12.3.16, and find x' and y' in terms of x and y.]

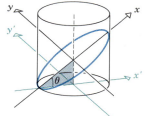

Figure 12.3.16

45. A carpenter needs to cut an elliptical hole in a sloped roof through which a circular vent pipe of diameter D is to be inserted vertically (Figure 12.3.17). The carpenter wants to draw the outline of the hole on the roof using a pencil, a piece of string, and two tacks, as illustrated in Figure 12.3.2. The center point of the ellipse is known, and common sense suggests that its major axis must be perpendicular to the drip line of the roof. The carpenter needs to determine the length L of the string and the distance T between a tack and the center point. The architect's plans show that the pitch of the roof is p (pitch = rise over run, as in Figure 12.3.17). Find T and L in terms of D and p. [*Note:* This exercise is based on an article by William H. Enos, which appeared in the *Mathematics Teacher*, Feb. 1991, p. 148.]

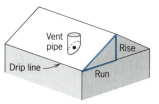

Figure 12.3.17

46. Suppose that the base of a solid is elliptical with a major axis of length 9 and a minor axis of length 4. Find the volume of the solid if
   (a) the cross-sections perpendicular to the major axis are squares
   (b) the cross-sections perpendicular to the minor axis are equilateral triangles.

47. The "flatness" of an ellipse is related to the ratio $e = c/a$, which is called the *eccentricity* of the ellipse. (Note that the number e, as used here, is not the base of the natural logarithm.)
   (a) Show that $0 < e < 1$.
   (b) Find an equation of the ellipse with eccentricity $e = 3/5$ and foci at the points $(0, -3)$ and $(0, 3)$.
   (c) What happens to the shape of an ellipse if the length of the major axis is constant but the eccentricity approaches 1?
   (d) What happens to the shape of an ellipse if the length of the major axis is constant but the eccentricity approaches 0?

48. The planets and comets in our solar system have elliptical orbits with the sun at a focus. The point closest to the sun in an orbit is called the **perihelion,** and the point farthest from the sun is called the **aphelion;** these points are at the ends of the major axis of the orbit (Figure 12.3.18).

(a) Find the eccentricity of the earth's orbit given that the ratio of the distance from the sun at the perihelion to the distance from the sun at the aphelion is 59/61. [*Note:* Eccentricity is defined in Exercise 47.]

(b) Find the distance between the earth and the sun at the perihelion given that the semimajor axis of the orbit has a length of 93 million miles.

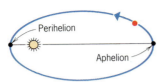

Figure 12.3.18

49. Let C_1 and C_2 be circles, in the same plane, with radii r_1 and r_2, respectively. Assume that $r_1 < r_2$ and that C_1 lies entirely inside C_2, but that C_1 and C_2 do not have the same center (Figure 12.3.19). Show that the centers of all circles that are outside C_1 and inside C_2, and are tangent to both C_1 and C_2, lie on an ellipse whose foci are the centers of C_1 and C_2. Find the length of the major axis and the location of the center of the ellipse.

Figure 12.3.19

50. Prove Theorem 12.3.2. [*Hint:* Introduce a rectangular coordinate system so that the ellipse has the equation $x^2/a^2 + y^2/b^2 = 1$, and use the result of Exercise 31 of Section 1.4.]

51. Derive the equation $x^2/b^2 + y^2/a^2 = 1$ in Table 12.3.1.

■ 12.4 THE HYPERBOLA

In this section we shall discuss properties of hyperbolas.

□ **DEFINITION OF A HYPERBOLA**

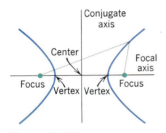

Figure 12.4.1

12.4.1 DEFINITION. A *hyperbola* is the set of all points in the plane, the difference of whose distances from two fixed points is a given positive constant that is less than the distance between the fixed points.

In this definition the "difference" of the distances is understood to mean the distance to the farther point minus the distance to the closer point. The two fixed points are called the *foci,* and the midpoint of the line segment joining the foci is called the *center* of the hyperbola (Figure 12.4.1). The line through the foci is called the *focal axis* (or *transverse axis*), and the line through the center and perpendicular to the focal axis is called the *conjugate axis.* The hyperbola intersects the focal axis at two points, called *vertices.* The two separate parts of a hyperbola are called the *branches.*

☐ ASYMPTOTES OF A HYPERBOLA

Associated with every hyperbola is a pair of lines, called the *asymptotes* of the hyperbola. These lines intersect at the center of the hyperbola and have the property that as a point P moves along the hyperbola away from the center, the distance between P and one of the asymptotes approaches zero (Figure 12.4.2).

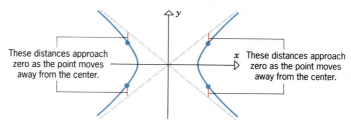

These distances approach zero as the point moves away from the center.

These distances approach zero as the point moves away from the center.

Figure 12.4.2

☐ GEOMETRIC PROPERTIES OF A HYPERBOLA

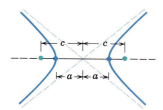

Figure 12.4.3

It is traditional in the study of hyperbolas to denote the distance between the vertices by $2a$, the distance between the foci by $2c$ (Figure 12.4.3), and to define the quantity b as

$$b = \sqrt{c^2 - a^2} \tag{1}$$

This relationship, which can also be expressed as

$$c = \sqrt{a^2 + b^2} \tag{2}$$

is pictured geometrically in Figure 12.4.4. As illustrated in that figure, the asymptotes pass through the center of the hyperbola and the corners of a box extending b units on each side of the center along the conjugate axis and a units on each side of the center along the focal axis. We shall verify this later in this section.

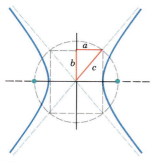

Figure 12.4.4

If V is one vertex of a hyperbola, then, as illustrated in Figure 12.4.5, the distance from V to the farther focus minus the distance from V to the closer focus is

$$[(c - a) + 2a] - (c - a) = 2a$$

Thus, for *all* points on a hyperbola, the distance to the farther focus minus the distance to the closer focus is $2a$. (Why?)

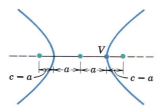

Figure 12.4.5

☐ **STANDARD EQUATIONS OF A HYPERBOLA**

The equation of a hyperbola is simplest if the coordinate axes are positioned so that the center of the hyperbola is at the origin and the foci are on the x-axis or y-axis. The two possible such orientations are shown in Table 12.4.1. These are called the **standard positions** of a hyperbola, and the resulting equations are called the **standard equations** of a hyperbola.

To illustrate how the equations in Table 12.4.1 are derived, let us consider the first entry in the table. Let $P(x, y)$ be any point on the hyperbola, so that the distance from P to the farther focus minus the distance from P to the closer focus is $2a$. Depending on which focus is farther from P, this condition leads either to the equation

$$\sqrt{(x + c)^2 + y^2} - \sqrt{(x - c)^2 + y^2} = 2a \tag{3}$$

or to

$$\sqrt{(x - c)^2 + y^2} - \sqrt{(x + c)^2 + y^2} = 2a \tag{4}$$

(See Figure 12.4.6.) Rewriting (3) yields

$$\sqrt{(x + c)^2 + y^2} = 2a + \sqrt{(x - c)^2 + y^2}$$

then squaring both sides yields

$$(x + c)^2 + y^2 = 4a^2 + 4a\sqrt{(x - c)^2 + y^2} + (x - c)^2 + y^2$$

Simplifying, then isolating the radical, then squaring again to remove the radical, ultimately yields (verify)

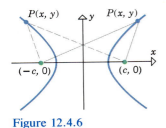

Figure 12.4.6

Table 12.4.1

ORIENTATION	DESCRIPTION	STANDARD EQUATION	ASYMPTOTE EQUATIONS
	• Foci on the x-axis. • Conjugate axis on the y-axis. • Center at the origin.	$\dfrac{x^2}{a^2} - \dfrac{y^2}{b^2} = 1$	$y = \dfrac{b}{a}x$ $y = -\dfrac{b}{a}x$
	• Foci on the y-axis. • Conjugate axis on the x-axis. • Center at the origin.	$\dfrac{y^2}{a^2} - \dfrac{x^2}{b^2} = 1$	$y = \dfrac{a}{b}x$ $y = -\dfrac{a}{b}x$

$$\frac{x^2}{a^2} - \frac{y^2}{c^2 - a^2} = 1$$

which, by virtue of (1), can be written as

$$\frac{x^2}{a^2} - \frac{y^2}{b^2} = 1 \tag{5}$$

We leave it as an exercise to show that (4) also simplifies to (5). Thus, each point $P(x, y)$ on the hyperbola satisfies (5), regardless of the branch on which it lies. Conversely, it can be shown that for any point P whose coordinates satisfy (5), the distance from P to the farther focus minus the distance from P to the closer focus is $2a$, so that such a point must lie on the hyperbola.

To prove that the asymptotes of (5) have the equations stated in Table 12.4.1, we rewrite (5) in the form

$$y^2 = \frac{b^2}{a^2}(x^2 - a^2)$$

which is equivalent to the pair of equations

$$y = \frac{b}{a}\sqrt{x^2 - a^2} \quad \text{and} \quad y = -\frac{b}{a}\sqrt{x^2 - a^2}$$

Thus, in the first quadrant, the vertical distance between the line $y = (b/a)x$ and the hyperbola can be written (Figure 12.4.7) as

$$\frac{b}{a}x - \frac{b}{a}\sqrt{x^2 - a^2}$$

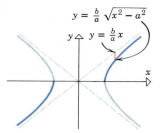

Figure 12.4.7

But this distance tends to zero as $x \to +\infty$ since

$$\lim_{x \to +\infty}\left(\frac{b}{a}x - \frac{b}{a}\sqrt{x^2 - a^2}\right) = \lim_{x \to +\infty}\frac{b}{a}(x - \sqrt{x^2 - a^2})$$

$$= \lim_{x \to +\infty}\frac{b}{a}\frac{(x - \sqrt{x^2 - a^2})(x + \sqrt{x^2 - a^2})}{x + \sqrt{x^2 - a^2}}$$

$$= \lim_{x \to +\infty}\frac{ab}{x + \sqrt{x^2 - a^2}} = 0$$

The analysis in the remaining quadrants is similar.

REMARK. There is a trick that can be used to avoid memorizing the equations of the asymptotes of a hyperbola. They can be obtained, when needed, by substituting 0 for the 1 on the right side of the hyperbola equation, and then solving for y in terms of x. For example, for the hyperbola

$$\frac{x^2}{a^2} - \frac{y^2}{b^2} = 1$$

we would write

$$\frac{x^2}{a^2} - \frac{y^2}{b^2} = 0 \quad \text{or} \quad y^2 = \frac{b^2}{a^2}x^2 \quad \text{or} \quad y = \pm\frac{b}{a}x$$

which are the equations for the asymptotes.

☐ **A TECHNIQUE FOR GRAPHING HYPERBOLAS**

Hyperbolas can be graphed from their *standard equations* using four basic steps:

- Determine whether the focal axis is on the x-axis or the y-axis. This can be ascertained from the location of the minus sign in the equation. Referring to Table 12.4.1, the focal axis is along the x-axis when the minus sign precedes the y^2-term, and it is along the y-axis when the minus sign precedes the x^2-term.
- Determine the values of a and b and draw a box extending a units on either side of the center along the focal axis and b units on either side of the center along the conjugate axis. (The squares of a and b can be read directly from the equation.)
- Draw the asymptotes along the diagonals of the box.
- Using the box and the asymptotes as a guide, sketch the graph of the hyperbola (Figure 12.4.8).

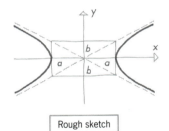

Rough sketch

Figure 12.4.8

Example 1 Sketch the graphs of the hyperbolas

$$\text{(a)} \quad \frac{x^2}{4} - \frac{y^2}{9} = 1 \qquad \text{(b)} \quad \frac{y^2}{7} - \frac{x^2}{2} = 1$$

showing their vertices, foci, and asymptotes.

Solution (a). The minus sign precedes the y^2-term, so the focal axis is along the x-axis. From the denominators in the equation we obtain

$$a^2 = 4 \quad \text{and} \quad b^2 = 9$$

Since a and b are positive, we must have $a = 2$ and $b = 3$. Recalling that the vertices lie a units on each side of the center on the focal axis, it follows that their coordinates in this case are $(2, 0)$ and $(-2, 0)$. Drawing a box extending $a = 2$ units along the x-axis on each side of the origin and $b = 3$ units on each side of the origin along the y-axis, then drawing the asymptotes along the diagonals of the box as a guide, yields the graph in Figure 12.4.9.

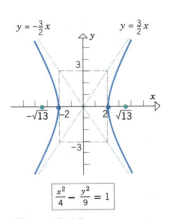

Figure 12.4.9

To obtain equations for the asymptotes, we substitute 0 for 1 in the given equation; this yields

$$\frac{x^2}{4} - \frac{y^2}{9} = 0 \quad \text{or} \quad y = \pm \frac{3}{2} x$$

The foci lie c units on each side of the center along the focal axis, where c is given by (2). From the values of a^2 and b^2 above we obtain

$$c = \sqrt{a^2 + b^2} = \sqrt{4 + 9} = \sqrt{13} \approx 3.6$$

Since the foci lie on the x-axis in this case, their coordinates are $(\sqrt{13}, 0)$ and $(-\sqrt{13}, 0)$.

Solution (b). The minus sign precedes the x^2-term, so the focal axis is along the y-axis. From the denominators in the equation we obtain $a^2 = 7$ and $b^2 = 2$, from which it follows that

$$a = \sqrt{7} \approx 2.6 \quad \text{and} \quad b = \sqrt{2} \approx 1.4$$

Thus, the vertices are at $(0, \sqrt{7})$ and $(0, -\sqrt{7})$. Drawing a box extending $a = \sqrt{7}$ units on either side of the origin along the y-axis and $b = \sqrt{2}$ units on either side of the origin along the x-axis, then drawing the asymptotes, yields the graph in Figure 12.4.10. Substituting 0 for 1 in the given equation yields

$$\frac{y^2}{7} - \frac{x^2}{2} = 0 \quad \text{or} \quad y = \pm \sqrt{\frac{7}{2}} x$$

which are the equations of the asymptotes. Also,

$$c = \sqrt{a^2 + b^2} = \sqrt{7 + 2} = 3$$

so the foci, which lie on the y-axis in this case, are at $(0, 3)$ and $(0, -3)$. ◀

Example 2 Find the equation of the hyperbola with vertices $(0, \pm 8)$ and asymptotes $y = \pm\frac{4}{3}x$.

Solution. Since the vertices are on the y-axis, the equation of the hyperbola has the form $y^2/a^2 - x^2/b^2 = 1$ and the asymptotes are

$$y = \pm \frac{a}{b} x$$

From the location of the vertices we have $a = 8$, so the given equations of the asymptotes yield

$$y = \pm \frac{a}{b} x = \pm \frac{8}{b} x = \pm \frac{4}{3} x$$

from which it follows that $b = 6$. Thus, the hyperbola has the equation

$$\frac{y^2}{64} - \frac{x^2}{36} = 1 \quad ◀$$

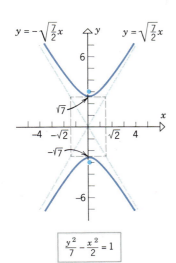

Figure 12.4.10

□ **TRANSLATED HYPERBOLAS**

If the center of a hyperbola is at the point (h, k) and its focal and conjugate axes are parallel to the coordinate axes, then its equation has one of the following forms:

> **Hyperbola with Center (h, k) and Focal Axis Parallel to x-axis**
>
> $$\frac{(x - h)^2}{a^2} - \frac{(y - k)^2}{b^2} = 1 \tag{6}$$

> **Hyperbola with Center (h, k) and Focal Axis Parallel to y-axis**
>
> $$\frac{(y - k)^2}{a^2} - \frac{(x - h)^2}{b^2} = 1 \tag{7}$$

The derivations are similar to those for the parabola and ellipse and will be omitted. As before, the equations of the asymptotes can be obtained by replacing the 1 by a 0 on the right side of (6) or (7), and then solving for y in terms of x.

Example 3 Show that the curve

$$x^2 - y^2 - 4x + 8y - 21 = 0$$

is a hyperbola. Sketch the hyperbola and show the foci, vertices, and asymptotes in the figure.

Solution. Our objective is to rewrite the equation in form (6) or (7) by completing the squares. To do this, we first group the x-terms and y-terms and take the constant to the right side:

$$(x^2 - 4x) - (y^2 - 8y) = 21$$

Next, complete the squares:

$$(x^2 - 4x + 4) - (y^2 - 8y + 16) = 21 + 4 - 16$$

or

$$(x - 2)^2 - (y - 4)^2 = 9$$

Finally, divide by 9 to introduce a 1 on the right side:

$$\frac{(x - 2)^2}{9} - \frac{(y - 4)^2}{9} = 1 \tag{8}$$

This is an equation of form (6) with $h = 2$, $k = 4$, $a^2 = 9$, and $b^2 = 9$. Thus, the equation represents a hyperbola with center $(2, 4)$ and focal axis parallel to the x-axis. Since $a = 3$, the vertices are located 3 units to the left and 3 units to the right of the center, or at the points $(-1, 4)$ and $(5, 4)$. From (2), $c = \sqrt{a^2 + b^2} = \sqrt{9 + 9} = 3\sqrt{2}$, so the foci are located $3\sqrt{2}$ units to the left and right of the center, or at the points $(2 - 3\sqrt{2}, 4)$ and $(2 + 3\sqrt{2}, 4)$.

The equations of the asymptotes may be found by substituting 0 for 1 in (8) to obtain

$$\frac{(x-2)^2}{9} - \frac{(y-4)^2}{9} = 0 \quad \text{or} \quad y - 4 = \pm(x-2)$$

which yields the asymptotes

$$y = x + 2 \quad \text{and} \quad y = -x + 6$$

With the aid of a box extending $a = 3$ units left and right of the center and $b = 3$ units above and below the center, we obtain the sketch in Figure 12.4.11.   ◄

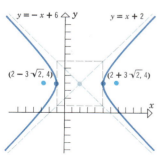

Figure 12.4.11

A hyperbola (such as that in the last example) is called *equilateral* if $a = b$. The asymptotes of an equilateral hyperbola are perpendicular. (Verify.)

☐ **HYPERBOLIC NAVIGATION**

By measuring the difference in reception times of synchronized radio signals from two widely spaced transmitters, a ship's electronic equipment can determine the difference $2a$ in its distances from the two transmitters. This information places the ship somewhere on the hyperbola whose foci are at the transmitters and whose points have $2a$ as the difference in their distances from the foci. By using two pairs of transmitters, the position of a ship can be determined as the intersection of two hyperbolas (Figure 12.4.12).

☐ **REFLECTION PROPERTIES OF HYPERBOLAS**

In the exercises, the reader is asked to prove the following result.

> **12.4.2**   THEOREM (*A Geometric Property of Hyperbolas*).   *A line tangent to a hyperbola at a point P makes equal angles with the lines through P and the foci* (Figure 12.4.13).

It follows from Theorem 12.4.2 that a ray of light emanating from one focus of a hyperbola will be reflected back along the line from the opposite focus (Figure 12.4.14). Light reflection properties of hyperbolas are used to advantage in the design of high-quality telescopes.

Figure 12.4.12

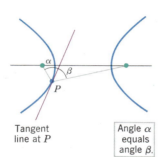

Figure 12.4.13

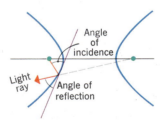

Figure 12.4.14

▶ Exercise Set 12.4

In Exercises 1–16, sketch the hyperbola. Find the co-ordinates of the vertices and foci, and find equations for the asymptotes.

1. $\dfrac{x^2}{16} - \dfrac{y^2}{4} = 1.$

2. $\dfrac{y^2}{9} - \dfrac{x^2}{25} = 1.$

3. $9y^2 - 4x^2 = 36.$

4. $16x^2 - 25y^2 = 400.$

5. $8x^2 - y^2 = 8.$

6. $3x^2 - y^2 = -9.$

7. $x^2 - y^2 = 1.$

8. $4y^2 - 4x^2 = 1.$

9. $\dfrac{(x-2)^2}{9} - \dfrac{(y-4)^2}{4} = 1.$

10. $\dfrac{(y+4)^2}{3} - \dfrac{(x-2)^2}{5} = 1.$

11. $(y+3)^2 - 9(x+2)^2 = 36.$

12. $16(x+1)^2 - 8(y-3)^2 = 16.$

13. $x^2 - 4y^2 + 2x + 8y - 7 = 0.$

14. $4x^2 - 9y^2 + 16x + 54y - 29 = 0.$

15. $16x^2 - y^2 - 32x - 6y = 57.$

16. $4y^2 - x^2 + 40y - 4x = -60.$

In Exercises 17–30, find an equation for the hyperbola satisfying the given conditions.

17. Vertices $(\pm 2, 0)$; foci $(\pm 3, 0)$.

18. Vertices $(0, \pm 3)$; foci $(0, \pm 5)$.

19. Vertices $(\pm 1, 0)$; asymptotes $y = \pm 2x$.

20. Vertices $(0, \pm 3)$; asymptotes $y = \pm x$.

21. Asymptotes $y = \pm\frac{3}{2}x$; $b = 4$.

22. Foci $(0, \pm 5)$; asymptotes $y = \pm 2x$.

23. Passes through $(5, 9)$; asymptotes $y = \pm x$.

24. Asymptotes $y = \pm\frac{3}{4}x$; $c = 5$.

25. Vertices $(\pm 2, 0)$; passes through $(4, 2)$.

26. Foci $(\pm 3, 0)$; asymptotes $y = \pm 2x$.

27. Vertices $(4, -3)$ and $(0, -3)$; foci 6 units apart.

28. Foci $(1, 8)$ and $(1, -12)$; vertices 4 units apart.

29. Vertices $(2, 4)$ and $(10, 4)$; foci 10 units apart.

30. Asymptotes $y = 2x + 1$ and $y = -2x + 3$; passes through the origin.

31. Find the equation of the hyperbola traced by a point which moves so that the difference between its distances to $(0, 0)$ and $(1, 1)$ is 1.

32. Find the equation of the hyperbola traced by a point which moves so that the difference between its distances to $(4, -3)$ and $(-2, 5)$ is 6.

In Exercises 33–36, find all intersections of the given curves, and make a sketch of the curves that shows the points of intersection.

33. $x^2 - 4y^2 = 36$ and $x - 2y - 20 = 0.$

34. $y^2 - 8x^2 = 5$ and $y - 2x^2 = 0.$

35. $3x^2 - 7y^2 = 5$ and $9y^2 - 2x^2 = 1.$

36. $x^2 - y^2 = 1$ and $x^2 + y^2 = 7.$

37. A point moves so that the product of its distances from the lines $y = mx$ and $y = -mx$ is a constant k^2. Show that the point moves along a hyperbola having these lines as asymptotes.

38. Prove: The line that is tangent to the hyperbola $x^2/a^2 - y^2/b^2 = 1$ at (x_0, y_0) has the equation $xx_0/a^2 - yy_0/b^2 = 1.$

39. A line tangent to the hyperbola $4x^2 - y^2 = 36$ intersects the y-axis at the point $(0, 4)$. Find the point(s) of tangency.

40. Let R be the region enclosed between the hyperbola $b^2x^2 - a^2y^2 = a^2b^2$ and the line $x = c$ through the focus $(c, 0)$. Find the volume of the solid generated by revolving R about
    (a) the x-axis   (b) the y-axis.

41. Find the coordinates of all points on the hyperbola $4x^2 - y^2 = 4$ where the two lines that pass through the point and the foci are perpendicular.

42. The distance between a point $P(x, y)$ and the point $(5, 0)$ is $5/3$ of the distance between P and the line $x = 9/5$. Show that the set of all such points is a hyperbola.

43. The "flatness" of a hyperbola is related to the ratio $e = c/a$, which is called the **eccentricity** of the hyperbola. (See Exercise 47 of Section 12.3 for the eccentricity of an ellipse.)
    (a) Show that $e > 1$.
    (b) Find an equation of the hyperbola with eccentricity $e = 5/3$ and foci at the points $(5, 0)$ and $(-5, 0)$.
    (c) What happens to the shape of a hyperbola if the distance between the vertices is constant but the eccentricity approaches 1?

(d) What happens to the shape of a hyperbola if the distance between the vertices is constant but the eccentricity approaches $+\infty$?

44. Let C_1 and C_2 be circles in the same plane with unequal radii r_1 and r_2, respectively. Assume that C_1 and C_2 do not intersect and that neither circle is inside the other (Figure 12.4.15). Show that the centers of all circles that are outside C_1 and C_2, and are tangent to both C_1 and C_2, lie on a branch of a hyperbola whose foci are the centers of C_1 and C_2. Where is the center of the hyperbola?

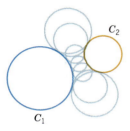

Figure 12.4.15

45. Suppose that the report of a gun is heard first by one observer and then t seconds later by a second observer in another location. Assuming that the gun and the ear levels of the observers are all in the same horizontal plane and that the speed of sound is constant, show that the gun was fired somewhere on a branch of a hyperbola whose foci coincide with the observers. [*Hint:* Let v be the speed of sound and d_1 and d_2 the horizontal distances between the gun and the observers (Figure 12.4.16).]

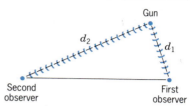

Figure 12.4.16

46. Show that Equation (3) simplifies to (5).

47. Derive the equation $y^2/a^2 - x^2/b^2 = 1$ in Table 12.4.1.

48. Prove Theorem 12.4.2. [*Hint:* Introduce coordinate axes so that the hyperbola has the equation $x^2/a^2 - y^2/b^2 = 1$, and use the result of Exercise 31 of Section 1.4.]

49. (a) Show that if an ellipse and a hyperbola have the same foci, then at each point of intersection their tangent lines are perpendicular. [*Hint:* Use the results in Exercise 38 above and Exercise 34 of Section 12.3.]

(b) Give a geometric proof of the result in part (a) using the reflection properties in Theorems 12.3.2 and 12.4.2.

12.5 ROTATION OF AXES; SECOND-DEGREE EQUATIONS

In previous sections we obtained equations of conic sections with axes parallel to the coordinate axes. In this section we shall study the equations of conics that are "tilted" relative to the coordinate axes. This will lead us to investigate rotations of coordinate axes.

QUADRATIC EQUATIONS IN x AND y

Each of the equations

$$(y - k)^2 = 4p(x - h), \qquad (x - h)^2 = 4p(y - k)$$

$$\frac{(x - h)^2}{a^2} + \frac{(y - k)^2}{b^2} = 1, \qquad \frac{(x - h)^2}{b^2} + \frac{(y - k)^2}{a^2} = 1$$

$$\frac{(x - h)^2}{a^2} - \frac{(y - k)^2}{b^2} = 1, \qquad \frac{(y - k)^2}{a^2} - \frac{(x - h)^2}{b^2} = 1$$

represents a conic section with axis or axes parallel to the coordinate axes. By squaring out the quadratic terms and simplifying, each of these equations can be rewritten in the form

$$Ax^2 + Cy^2 + Dx + Ey + F = 0 \tag{1}$$

For example, $(y - k)^2 = 4p(x - h)$ can be written as

$$y^2 - 4px - 2ky + (k^2 + 4ph) = 0$$

which is of form (1) with

$$A = 0, \quad C = 1, \quad D = -4p, \quad E = -2k, \quad F = k^2 + 4ph$$

Equation (1) is a special case of the more general equation

$$Ax^2 + Bxy + Cy^2 + Dx + Ey + F = 0 \tag{2}$$

which, if A, B, and C are not all zero, is called a **second-degree equation** or **quadratic equation** in x and y. We shall show later that the graph of any second-degree equation is a conic section (possibly a degenerate conic section). If $B = 0$, then (2) reduces to (1) and the conic section has its axis or axes parallel to the coordinate axes. However, if $B \neq 0$, then (2) contains a "cross-product" term Bxy, and the graph of the conic section represented by the equation has its axis or axes "tilted" relative to the coordinate axes. As an illustration, consider the ellipse with foci $F_1(1, 2)$ and $F_2(-1, -2)$ and such that the sum of the distances from each point $P(x, y)$ on the ellipse to the foci is 6 units. Expressing this condition as an equation, we obtain (Figure 12.5.1)

$$\sqrt{(x - 1)^2 + (y - 2)^2} + \sqrt{(x + 1)^2 + (y + 2)^2} = 6$$

Squaring both sides, then isolating the remaining radical, then squaring again, ultimately yields

$$8x^2 - 4xy + 5y^2 = 36$$

as the equation of the ellipse. This is an equation of form (2) with $A = 8$, $B = -4$, $C = 5$, $D = 0$, $E = 0$, $F = -36$.

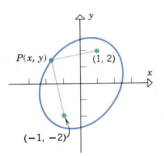

Figure 12.5.1

□ **ROTATION OF AXES**

To study conics that are tilted relative to the coordinate axes, it is frequently helpful to rotate the coordinate axes, so that the rotated coordinate axes are parallel to the axes of the conic. Before we can discuss the details, we need to develop some ideas about rotation of coordinate axes.

In Figure 12.5.2a the axes of an xy-coordinate system have been rotated about the origin through an angle θ to produce a new $x'y'$-coordinate system. As shown in the figure, each point P in the plane has coordinates (x', y') as well as coordinates (x, y). To see how the two are related, let r be the distance from the common origin to the point P, and let α be the angle shown in Figure 12.5.2b. It follows that

$$x = r \cos(\theta + \alpha), \quad y = r \sin(\theta + \alpha) \tag{3}$$

and

$$x' = r \cos \alpha, \quad y' = r \sin \alpha \tag{4}$$

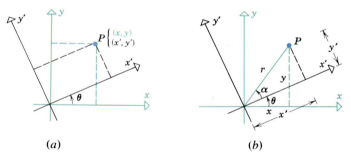

(a)   (b)

Figure 12.5.2

Using familiar trigonometric identities, the relationships in (3) can be written as

$$x = r \cos \theta \cos \alpha - r \sin \theta \sin \alpha$$
$$y = r \sin \theta \cos \alpha + r \cos \theta \sin \alpha$$

and on substituting (4) in these equations we obtain the following relationships:

> **The Rotation Equations**
>
> $$x = x' \cos \theta - y' \sin \theta$$
> $$y = x' \sin \theta + y' \cos \theta$$

(5)

Example 1   Suppose that the axes of an xy-coordinate system are rotated through an angle of $\theta = 45°$ to obtain an $x'y'$-coordinate system. Find the equation of the curve

$$x^2 - xy + y^2 - 6 = 0$$

in $x'y'$-coordinates.

Solution.   Substituting $\sin \theta = \sin 45° = 1/\sqrt{2}$ and $\cos \theta = \cos 45° = 1/\sqrt{2}$ in (5) yields the rotation equations

$$x = \frac{x'}{\sqrt{2}} - \frac{y'}{\sqrt{2}}$$

$$y = \frac{x'}{\sqrt{2}} + \frac{y'}{\sqrt{2}}$$

Substituting these into the given equation yields

$$\left(\frac{x'}{\sqrt{2}} - \frac{y'}{\sqrt{2}}\right)^2 - \left(\frac{x'}{\sqrt{2}} - \frac{y'}{\sqrt{2}}\right)\left(\frac{x'}{\sqrt{2}} + \frac{y'}{\sqrt{2}}\right) + \left(\frac{x'}{\sqrt{2}} + \frac{y'}{\sqrt{2}}\right)^2 - 6 = 0$$

or

$$\frac{x'^2 - 2x'y' + y'^2 - x'^2 + y'^2 + x'^2 + 2x'y' + y'^2}{2} = 6$$

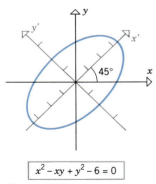

$x^2 - xy + y^2 - 6 = 0$

Figure 12.5.3

or

$$\frac{x'^2}{12} + \frac{y'^2}{4} = 1$$

which is the equation of an ellipse (Figure 12.5.3). ◄

If the rotation equations (5) are solved for x' and y' in terms of x and y, one obtains (Exercise 14):

$$\begin{align} x' &= x \cos \theta + y \sin \theta \\ y' &= -x \sin \theta + y \cos \theta \end{align} \tag{6}$$

Example 2 Find the new coordinates of the point $(2, 4)$ if the coordinate axes are rotated through an angle of $\theta = 30°$.

Solution. Using the rotation equations in (6) with $x = 2$, $y = 4$, $\cos \theta = \cos 30° = \sqrt{3}/2$, and $\sin \theta = \sin 30° = 1/2$, we obtain

$$\begin{align} x' &= 2(\sqrt{3}/2) + 4(1/2) = \sqrt{3} + 2 \\ y' &= -2(1/2) + 4(\sqrt{3}/2) = -1 + 2\sqrt{3} \end{align}$$

Thus, the new coordinates are $(\sqrt{3} + 2, -1 + 2\sqrt{3})$. ◄

☐ **ELIMINATING THE CROSS-PRODUCT TERM**

In Example 1, we were able to identify the curve $x^2 - xy + y^2 - 6 = 0$ as an ellipse because the rotation of axes eliminated the xy-term, thereby reducing the equation to a familiar form. This occurred because the new $x'y'$-axes were aligned with the axes of the ellipse. The following theorem tells how to determine an appropriate rotation of axes to eliminate the cross-product term of a second-degree equation in x and y.

12.5.1 THEOREM. *If the equation*

$$Ax^2 + Bxy + Cy^2 + Dx + Ey + F = 0 \tag{7}$$

is such that $B \neq 0$, and if an $x'y'$-coordinate system is obtained by rotating the xy-axes through an angle θ satisfying

$$\cot 2\theta = \frac{A - C}{B} \tag{8}$$

then, in $x'y'$-coordinates, Equation (7) will have the form

$$A'x'^2 + C'y'^2 + D'x' + E'y' + F' = 0$$

Proof. Substituting (5) into (7) and simplifying yields

$$A'x'^2 + B'x'y' + C'y'^2 + D'x' + E'y' + F' = 0$$

where

$$A' = A \cos^2 \theta + B \cos \theta \sin \theta + C \sin^2 \theta$$
$$B' = B (\cos^2 \theta - \sin^2 \theta) + 2(C - A) \sin \theta \cos \theta$$
$$C' = A \sin^2 \theta - B \sin \theta \cos \theta + C \cos^2 \theta \qquad (9)$$
$$D' = D \cos \theta + E \sin \theta$$
$$E' = -D \sin \theta + E \cos \theta$$
$$F' = F$$

(Verify.) To complete the proof we must show that $B' = 0$ if

$$\cot 2\theta = \frac{A - C}{B}$$

or equivalently

$$\frac{\cos 2\theta}{\sin 2\theta} = \frac{A - C}{B} \qquad (10)$$

However, by using the trigonometric double-angle formulas, we can rewrite B' in the form

$$B' = B \cos 2\theta - (A - C) \sin 2\theta$$

Thus, $B' = 0$ if θ satisfies (10).   █

REMARK. It is always possible to satisfy (8) with an angle θ in the range $0 < \theta < \pi/2$. We shall always use such a value of θ.

Example 3 Identify and sketch the curve $xy = 1$.

Solution. As a first step, we shall rotate the coordinate axes to eliminate the cross-product term. Comparing the given equation to (7), we have

$$A = 0, \quad B = 1, \quad C = 0$$

Thus, the desired angle of rotation must satisfy

$$\cot 2\theta = \frac{A - C}{B} = \frac{0 - 0}{1} = 0$$

This condition can be met by taking $2\theta = \pi/2$ or $\theta = \pi/4 = 45°$. Substituting $\cos \theta = \cos 45° = 1/\sqrt{2}$ and $\sin \theta = \sin 45° = 1/\sqrt{2}$ in (5) yields

$$x = \frac{x'}{\sqrt{2}} - \frac{y'}{\sqrt{2}}$$
$$\qquad (11)$$
$$y = \frac{x'}{\sqrt{2}} + \frac{y'}{\sqrt{2}}$$

and substituting these in the equation $xy = 1$ yields

$$\left(\frac{x'}{\sqrt{2}} - \frac{y'}{\sqrt{2}}\right)\left(\frac{x'}{\sqrt{2}} + \frac{y'}{\sqrt{2}}\right) = 1$$

or on simplifying

$$\frac{x'^2}{2} - \frac{y'^2}{2} = 1 \tag{12}$$

This is an equation of the form

$$\frac{x'^2}{a^2} - \frac{y'^2}{b^2} = 1$$

with $a = b = \sqrt{2}$. With the exception that the variables are x' and y' rather than x and y, this is the familiar equation of a hyperbola (see Table 12.4.1). This hyperbola has its focal axis on the x'-axis, its conjugate axis on the y'-axis. Moreover, $c = \sqrt{a^2 + b^2} = 2$, so the vertices, foci, and asymptotes are

VERTICES	$(x', y') = (\pm a, 0) = (\pm\sqrt{2}, 0)$
FOCI	$(x', y') = (\pm c, 0) = (\pm 2, 0)$
ASYMPTOTES	$y' = \pm\dfrac{b}{a}x'$ or $y' = \pm x'$

The vertices in xy-coordinates can be found by substituting $x' = \sqrt{2}$, $y' = 0$ and $x' = -\sqrt{2}$, $y' = 0$ in (11), and the foci may be obtained by substituting $x' = 2$, $y' = 0$ and $x' = -2$, $y' = 0$. The results are

VERTICES: $(x, y) = (1, 1)$ and $(x, y) = (-1, -1)$
FOCI: $(x, y) = (\sqrt{2}, \sqrt{2})$ and $(x, y) = (-\sqrt{2}, -\sqrt{2})$

To obtain the xy-equations of the asymptotes, it is desirable to use form (6) of the rotation equations. Substituting

$$\cos\theta = \cos 45° = 1/\sqrt{2} \quad\text{and}\quad \sin\theta = \sin 45° = 1/\sqrt{2}$$

in (6) yields

$$x' = \frac{1}{\sqrt{2}}x + \frac{1}{\sqrt{2}}y$$
$$y' = -\frac{1}{\sqrt{2}}x + \frac{1}{\sqrt{2}}y \tag{13}$$

Substituting (13) in the asymptote equations $y' = x'$ and $y' = -x'$ and simplifying shows that the asymptotes are

$$x = 0 \quad\text{and}\quad y = 0$$

The curve $xy = 1$ is sketched in Figure 12.5.4.   ◄

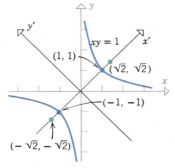

Figure 12.5.4

In problems where it is inconvenient to solve

$$\cot 2\theta = \frac{A - C}{B}$$

for θ, the values of $\sin \theta$ and $\cos \theta$ needed for the rotation equations may be obtained by first calculating $\cos 2\theta$ and then computing $\sin \theta$ and $\cos \theta$ from the identities

$$\sin \theta = \sqrt{\frac{1 - \cos 2\theta}{2}} \quad \text{and} \quad \cos \theta = \sqrt{\frac{1 + \cos 2\theta}{2}}$$

Example 4 Identify and sketch the curve

$$153x^2 - 192xy + 97y^2 - 30x - 40y - 200 = 0$$

Solution. We have $A = 153$, $B = -192$, and $C = 97$, so

$$\cot 2\theta = \frac{A - C}{B} = -\frac{56}{192} = -\frac{7}{24}$$

Since θ is to be chosen in the range $0 < \theta < \pi/2$, this relationship is represented by the triangle in Figure 12.5.5. From that triangle we obtain $\cos 2\theta = -7/25$, which implies that

$$\cos \theta = \sqrt{\frac{1 + \cos 2\theta}{2}} = \sqrt{\frac{1 - 7/25}{2}} = \frac{3}{5}$$

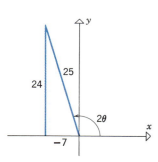

Figure 12.5.5

$$\sin \theta = \sqrt{\frac{1 - \cos 2\theta}{2}} = \sqrt{\frac{1 + 7/25}{2}} = \frac{4}{5}$$

Substituting these values in (5) yields the rotation equations

$$x = \frac{3}{5}x' - \frac{4}{5}y'$$

$$y = \frac{4}{5}x' + \frac{3}{5}y'$$

Substituting these expressions in the given equation yields

$$\frac{153}{25}(3x' - 4y')^2 - \frac{192}{25}(3x' - 4y')(4x' + 3y') + \frac{97}{25}(4x' + 3y')^2$$

$$- \frac{30}{5}(3x' - 4y') - \frac{40}{5}(4x' + 3y') - 200 = 0$$

which simplifies to

$$25x'^2 + 225y'^2 - 50x' - 200 = 0$$

or

$$x'^2 + 9y'^2 - 2x' - 8 = 0$$

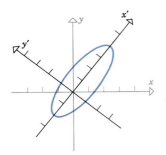

Completing the square yields

$$\frac{(x' - 1)^2}{9} + y'^2 = 1$$

This is an ellipse whose center is $(1, 0)$ in $x'y'$-coordinates (Figure 12.5.6).   ◄

Figure 12.5.6

■ OPTIONAL (THE DISCRIMINANT)

It is possible to describe the graph of a second-degree equation without rotating coordinate axes.

12.5.2 THEOREM.   *Consider a second-degree equation*

$$Ax^2 + Bxy + Cy^2 + Dx + Ey + F = 0 \qquad (14)$$

(a)   *If $B^2 - 4AC < 0$, the equation represents an ellipse, a circle, a point, or else has no graph.*

(b)   *If $B^2 - 4AC > 0$, the equation represents a hyperbola or a pair of intersecting lines.*

(c)   *If $B^2 - 4AC = 0$, the equation represents a parabola, a line, a pair of parallel lines, or else has no graph.*

The quantity $B^2 - 4AC$ in this theorem is called the ***discriminant*** of the quadratic equation. To see why Theorem 12.5.2 is true, we need a fact about the discriminant. It can be shown (Exercise 19) that if the coordinate axes are rotated through any angle θ, and if

$$A'x'^2 + B'x'y' + C'y'^2 + D'x' + E'y' + F' = 0 \qquad (15)$$

is the equation resulting from (14) after rotation, then

$$B^2 - 4AC = B'^2 - 4A'C' \qquad (16)$$

In other words, the discriminant of a quadratic equation is not altered by rotating the coordinate axes. For this reason the discriminant is said to be ***invariant*** under a rotation of coordinate axes. In particular, if we choose the angle of rotation to eliminate the cross-product term, then (15) becomes

$$A'x'^2 + C'y'^2 + D'x' + E'y' + F' = 0 \qquad (17)$$

and since $B' = 0$, (16) tells us that

$$B^2 - 4AC = -4A'C' \tag{18}$$

Part (*a*) of Theorem 12.5.2 can now be proved as follows. If $B^2 - 4AC < 0$, then from (18), $A'C' > 0$, so (17) may be divided through by $A'C'$ and written in the form

$$\frac{1}{C'}\left(x'^2 + \frac{D'}{A'}x'\right) + \frac{1}{A'}\left(y'^2 + \frac{E'}{C'}y'\right) = -\frac{F'}{A'C'}$$

By completing the squares, this equation may be rewritten in the form

$$\frac{1}{C'}(x' - h)^2 + \frac{1}{A'}(y' - k)^2 = K \tag{19}$$

But A' and C' have the same sign (since $A'C' > 0$), so by multiplying (19) through by -1, if necessary, we can assume A' and C' to be positive. Thus, (19) can be written as

$$\frac{(x' - h)^2}{(\sqrt{C'})^2} + \frac{(y' - k)^2}{(\sqrt{A'})^2} = K \tag{20}$$

If $K < 0$, this equation has no graph (the left side is nonnegative for all x' and y'); if $K = 0$, the equation is satisfied only by $x' = h$ and $y' = k$, so the graph is the single point (h, k); finally, if $K > 0$, we can rewrite (20) in the form

$$\frac{(x' - h)^2}{(\sqrt{C'}\sqrt{K})^2} + \frac{(y' - k)^2}{(\sqrt{A'}\sqrt{K})^2} = 1$$

which is an ellipse or a circle. The proofs of parts (*b*) and (*c*) require a similar kind of analysis. ▮

Example 5 Use the discriminant to identify the graph of

$$8x^2 - 3xy + 5y^2 - 7x + 6 = 0$$

Solution. We have

$$B^2 - 4AC = (-3)^2 - 4(8)(5) = -151$$

Since the discriminant is negative, the equation represents an ellipse, a point, or else has no graph. (Why can't the graph be a circle?) ◄

In cases where a quadratic equation represents a point, a pair of parallel lines, a pair of intersecting lines or has no graph, we say that equation represents a *degenerate conic section*. Thus, if we allow for possible degeneracy, it follows from Theorem 12.5.2 that *every quadratic equation has a conic section as its graph.*

▶ Exercise Set 12.5

1. Let an $x'y'$-coordinate system be obtained by rotating an xy-coordinate system through an angle of $\theta = 60°$.
   (a) Find the $x'y'$-coordinates of the point whose xy-coordinates are $(-2, 6)$.
   (b) Find an equation of the curve
      $\sqrt{3}xy + y^2 = 6$ in $x'y'$-coordinates.
   (c) Sketch the curve in part (b), showing both xy-axes and $x'y'$-axes.

2. Let an $x'y'$-coordinate system be obtained by rotating an xy-coordinate system through an angle of $\theta = 30°$.
   (a) Find the $x'y'$-coordinates of the point whose xy-coordinates are $(1, -\sqrt{3})$.
   (b) Find an equation of the curve
      $2x^2 + 2\sqrt{3}xy = 3$ in $x'y'$-coordinates.
   (c) Sketch the curve in part (b), showing both xy-axes and $x'y'$-axes.

In Exercises 3–12, rotate the coordinate axes to remove the xy-term. Then name the conic and sketch its graph.

3. $xy = -9$.

4. $x^2 - xy + y^2 - 2 = 0$.

5. $x^2 + 4xy - 2y^2 - 6 = 0$.

6. $31x^2 + 10\sqrt{3}xy + 21y^2 - 144 = 0$.

7. $x^2 + 2\sqrt{3}xy + 3y^2 + 2\sqrt{3}x - 2y = 0$.

8. $34x^2 - 24xy + 41y^2 - 25 = 0$.

9. $9x^2 - 24xy + 16y^2 - 80x - 60y + 100 = 0$.

10. $5x^2 - 6xy + 5y^2 - 8\sqrt{2}x + 8\sqrt{2}y = 8$.

11. $52x^2 - 72xy + 73y^2 + 40x + 30y - 75 = 0$.

12. $6x^2 + 24xy - y^2 - 12x + 26y + 11 = 0$.

13. Let an $x'y'$-coordinate system be obtained by rotating an xy-coordinate system through an angle θ. Prove: For every choice of θ, the equation $x^2 + y^2 = r^2$ becomes $x'^2 + y'^2 = r^2$. Give a geometric explanation of this result.

14. Derive (6) by solving the rotation equations in (5) for x' and y' in terms of x and y.

15. Let an $x'y'$-coordinate system be obtained by rotating an xy-coordinate system through an angle of 45°. Use (6) to find an equation in xy-coordinates of the curve $3x'^2 + y'^2 = 6$.

16. Let an $x'y'$-coordinate system be obtained by rotating an xy-coordinate system through an angle of 30°. Use (5) to find an equation in $x'y'$-coordinates of the curve $y = x^2$.

17. Show that the graph of the equation

$$\sqrt{x} + \sqrt{y} = 1$$

is a portion of a parabola. [*Hint*: First rationalize the equation and then perform a rotation of axes.]

18. Derive the expression for B' in (9).

19. Use (9) to prove that $B^2 - 4AC = B'^2 - 4A'C'$ for all values of θ.

20. Use (9) to prove that $A + C = A' + C'$ for all values of θ.

21. Prove: If $A = C$ in (7), then the cross-product term can be eliminated by rotating through 45°.

22. Prove: If $B \neq 0$, then the graph of $x^2 + Bxy + F = 0$ is a hyperbola if $F \neq 0$ and two intersecting lines if $F = 0$.

In Exercises 23–27, use the discriminant to identify the graph of the given equation.

23. $x^2 - xy + y^2 - 2 = 0$.

24. $x^2 + 4xy - 2y^2 - 6 = 0$.

25. $x^2 + 2\sqrt{3}xy + 3y^2 + 2\sqrt{3}x - 2y = 0$.

26. $6x^2 + 24xy - y^2 - 12x + 26y + 11 = 0$.

27. $34x^2 - 24xy + 41y^2 - 25 = 0$.

28. Each of the following represents a degenerate conic section. Where possible, sketch the graph.
    (a) $x^2 - y^2 = 0$
    (b) $x^2 + 3y^2 + 7 = 0$
    (c) $8x^2 + 7y^2 = 0$
    (d) $x^2 - 2xy + y^2 = 0$
    (e) $9x^2 + 12xy + 4y^2 - 36 = 0$
    (f) $x^2 + y^2 - 2x - 4y = -5$.

29. Prove parts (*b*) and (*c*) of Theorem 12.5.2.

▶ SUPPLEMENTARY EXERCISES [C] 29

In Exercises 1–8, identify the curve as a parabola, ellipse, or hyperbola, and give the following information:

Parabola: The coordinates of the vertex and focus; the equation of the directrix.
Ellipse: The coordinates of the center and foci; the lengths of the major and minor axes.
Hyperbola: The coordinates of the center, foci, and vertices; the equations of the asymptotes.

1. $y^2 + 12x - 6y + 33 = 0$.

2. $x^2 - 4y^2 = -1$.

3. $9x^2 + 4y^2 + 36x - 8y + 4 = 0$.

4. $6x + 8y - x^2 - 4y^2 = 12$.

5. $x^2 - 9y^2 - 4x + 18y - 14 = 0$.

6. $4y = x^2 + 2x - 7$.   7. $3x + 2y^2 - 4y - 7 = 0$.

8. $4x^2 = y^2 - 4y$.

In Exercises 9–16, find an equation for the curve described.

9. The parabola with vertex at $(1, 3)$ and directrix $x = -3$.

10. The ellipse with major axis of length 12 and foci at $(2, 7)$ and $(2, -1)$.

11. The hyperbola with foci $(0, \pm 5)$ and vertices 6 units apart.

12. The parabola with axis $y = 1$, vertex $(2, 1)$, and passing through $(3, -1)$.

13. The ellipse with foci $(\pm 3, 0)$ and such that the distances from the foci to $P(x, y)$ on the ellipse add up to 10 units.

14. The hyperbola with vertices $(0, \pm 2)$ and asymptotes $y = \pm 3x$.

15. The curve C with the property that the distance between the point $(3, 4)$ and any point $P(x, y)$ on C is equal to the distance between P and the line $y = 2$.

16. The hyperbola with vertices $(-3, 2)$ and $(1, 2)$ and perpendicular asymptotes.

In Exercises 17–20, sketch the curve whose equation is given in the stated exercise.

17. Exercise 1.   18. Exercise 2.

19. Exercise 3.   20. Exercise 6.

In Exercises 21–26, find the rotation angle θ needed to remove the xy-term; then name the conic and give its equation in $x'y'$-coordinates after the xy-term is removed.

21. $3x^2 - 2xy + 3y^2 = 4$.

22. $7x^2 - 8xy + y^2 = 9$.

23. $11x^2 + 10\sqrt{3}xy + y^2 = 4$.

24. $x^2 + 4xy + 4y^2 - 2\sqrt{5}x + \sqrt{5}y = 0$.

25. $16x^2 - 24xy + 9y^2 - 60x - 80y + 100 = 0$.

26. $73x^2 - 72xy + 52y^2 - 100 = 0$.

In Exercises 27–29, sketch the graph of the equation in the stated exercise, showing the xy- and $x'y'$-axes.

27. Exercise 21.   28. Exercise 23.

29. Exercise 24.

30. For the curve $xy + x - y = 1$,
    (a) find the rotation angle necessary to remove the xy-term
    (b) find the equation in $x'y'$-coordinates obtained by rotating the xy-coordinate axes through an angle of $\theta = \tan^{-1}(3/4)$ radians and verify the identities $A' + C' = A + C$ and $B'^2 - 4A'C' = B^2 - 4AC$.

Exercise 31 will show that the conic sections can all be obtained as curves traced by a point P moving in such a way that the ratio of the distances from P to a fixed line L and to a fixed point F not on L is a constant e. The line L is called a **directrix** of the curve, the point F the **focus** of the curve, and the number e the **eccentricity** of the curve. (The constant e, as used here, is not the base of the natural logarithm.) If $e = 1$, then from Definition 12.2.1 the curve is a parabola. It will be shown that the curve is an ellipse if $0 < e < 1$ and a hyperbola if $e > 1$. To obtain these results we shall assume that F is at the origin of an xy-coordinate system, L is the vertical line $x = k$ ($k > 0$), and S is the curve traced by a point $P(x, y)$ which moves so that

$$PF/PD = e$$

where PF is the distance between P and F and PD is the distance between P and L (see following figure).

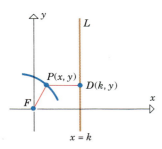

31. (a) Show that if $0 < e < 1$, then S is the ellipse given by the equation

$$\frac{(x + c)^2}{a^2} + \frac{y^2}{b^2} = 1$$

where

$$c = \frac{e^2 k}{e^2 - 1}, \quad a^2 = \frac{e^2 k^2}{(e^2 - 1)^2}, \quad b^2 = \frac{e^2 k^2}{1 - e^2}$$

(b) Show that if $e > 1$, then S is the hyperbola given by the equation

$$\frac{(x - c)^2}{a^2} - \frac{y^2}{b^2} = 1$$

where

$$c = \frac{e^2 k}{e^2 - 1}, \quad a^2 = \frac{e^2 k^2}{(e^2 - 1)^2}, \quad b^2 = \frac{e^2 k^2}{e^2 - 1}$$

Exercise 32 shows that the concepts of focus and directrix, as defined in Exercise 31, are consistent with their earlier usage.

32. Let a^2, b^2, and c be as given in Exercise 31.

(a) Show that if $0 < e < 1$, then $c^2 = a^2 - b^2$, and use this result to show that F is a focus of the ellipse.

(b) Show that if $e > 1$, then $c^2 = a^2 + b^2$, and use this result to show that F is a focus of the hyperbola.

(c) Show that in the case of an ellipse or a hyperbola $e = c/a$.

13

Polar Coordinates and Parametric Equations

Johann Bernoulli   (1667-1748)

■ 13.1 POLAR COORDINATES

*Up to now, we have specified the location of points in the plane by means of coordinates relative to a rectangular coordinate system consisting of two perpendicular coordinate axes. In this section, we shall introduce a new kind of coordinate system, called a **polar coordinate system**, that will be easier to work with for many problems.*

□ POLAR COORDINATE SYSTEMS

Figure 13.1.1

To form a polar coordinate system in a plane, we pick a fixed point O, called the *origin* or *pole,* and using the origin as an endpoint we construct a ray, called the *polar axis*. After selecting a unit of measurement, we may associate with any point P in the plane a pair of *polar coordinates* (r, θ), where r is the distance from P to the origin and θ measures the angle from the polar axis to the line segment OP (Figure 13.1.1). The number r is called the *radial distance* of P and θ is called a *polar angle* of P. In Figure 13.1.2, the points $(6, 45°)$, $(3, 225°)$, $(5, 120°)$, and $(4, 330°)$ are plotted in polar coordinate systems.

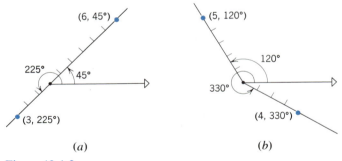

(a)   (b)

Figure 13.1.2

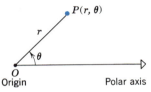

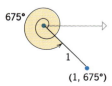

All three pairs of polar coordinates represent the same point.

Figure 13.1.3

The polar coordinates of a point are not unique. For example, the polar coordinates

$$(1, 315°), \quad (1, -45°), \quad \text{and} \quad (1, 675°)$$

all represent the same point (Figure 13.1.3). In general, if a point P has polar coordinates (r, θ), then for any integer $n = 0, 1, 2, \ldots$

$$(r, \theta + n \cdot 360°) \quad \text{and} \quad (r, \theta - n \cdot 360°)$$

are also polar coordinates of P.

In the case where P is the origin, the line segment OP in Figure 13.1.1 reduces to a point, since $r = 0$. Because there is no clearly defined polar angle in this case, we will agree that an arbitrary polar angle θ may be used. Thus, for every θ, the point $(0, \theta)$ is the origin.

☐ **RELATIONSHIP BETWEEN POLAR AND RECTANGULAR COORDINATES**

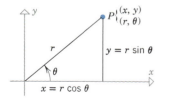

Figure 13.1.4

Frequently, it is helpful to use both polar and rectangular coordinates in the same problem. To do this, we let the positive x-axis of the rectangular coordinate system serve as the polar axis for the polar coordinate system. When this is done, each point P has both polar coordinates (r, θ) and rectangular coordinates (x, y). As suggested by Figure 13.1.4, these coordinates are related by the equations

$$x = r \cos \theta \qquad \text{and} \qquad r^2 = x^2 + y^2$$
$$y = r \sin \theta \qquad\qquad \tan \theta = \frac{y}{x} \qquad\qquad (1\text{a–b})$$

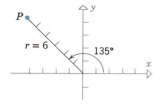

The point P has polar coordinates (6, 135°) and rectangular coordinates $(-3\sqrt{2}, 3\sqrt{2})$

Figure 13.1.5

Example 1 Find the rectangular coordinates of the point P whose polar coordinates are $(6, 135°)$.

Solution. Substituting the polar coordinates $r = 6$ and $\theta = 135°$ in (1a) yields

$$x = 6 \cos 135° = 6\left(-\sqrt{2}/2\right) = -3\sqrt{2}$$
$$y = 6 \sin 135° = 6\left(\sqrt{2}/2\right) = 3\sqrt{2}$$

Thus, the rectangular coordinates of the point P are $(-3\sqrt{2}, 3\sqrt{2})$ (Figure 13.1.5). ◀

☐ **NEGATIVE VALUES OF r**

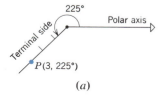

(a)

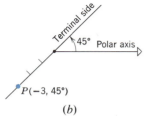

(b)

Figure 13.1.6

When we start graphing curves in polar coordinates, it will be desirable to allow negative values for r. This will require a special definition. For motivation, consider the point P with polar coordinates $(3, 225°)$. We can reach this point by rotating the polar axis 225° and then moving *forward* from the origin 3 units along the terminal side of the angle (Figure 13.1.6a). On the other hand, we can also reach the point P by rotating the polar axis 45° and then moving *backward* 3 units from the origin along the *extension* of the terminal side of the angle (Figure 13.1.6b). This suggests that the point $(3, 225°)$ might also be denoted by $(-3, 45°)$, with the minus sign serving to indicate that the point is on the extension of the angle's terminal side rather than on the terminal side itself.

Since the terminal side of the angle $\theta + 180°$ is the extension of the terminal side of the angle θ, we shall define

$$(-r, \theta) \quad \text{and} \quad (r, \theta + 180°) \qquad\qquad (2)$$

to be polar coordinates for the same point. With $r = 3$ and $\theta = 45°$ in (2) it follows that $(-3, 45°)$ and $(3, 225°)$ represent the same point.

In the exercises we ask the reader to show that relationships (1a) and (1b) remain valid when r is negative.

REMARK. For many purposes it does not matter whether polar angles are represented in degrees or radians. However, later on, when we are concerned with differentiation problems, it will be essential to measure polar angles in radians, since the derivative formulas for the trigonometric functions were derived under this assumption. From here on, we shall use radian measure.

Example 2   Find polar coordinates of the point P whose rectangular coordinates are $(-2, 2\sqrt{3})$.

Solution.   We will find polar coordinates (r, θ) of P such that $r > 0$ and $0 \le \theta < 2\pi$. From the first equation in (1b),

$$r^2 = x^2 + y^2 = (-2)^2 + (2\sqrt{3})^2 = 4 + 12 = 16$$

so $r = 4$. From the second equation in (1b),

$$\tan \theta = \frac{y}{x} = \frac{2\sqrt{3}}{-2} = -\sqrt{3}$$

From this and the fact that $(-2, 2\sqrt{3})$ lies in the second quadrant, it follows that $\theta = 2\pi/3$. Thus, $(4, 2\pi/3)$ are polar coordinates of P. All other polar coordinates of P have the form

$$\left(4, \frac{2\pi}{3} + 2n\pi \right) \quad \text{or} \quad \left(-4, \frac{5\pi}{3} + 2n\pi \right), \text{ where } n \text{ is an integer.} \quad \blacktriangleleft$$

Polar coordinates provide a new way of graphing certain equations. As an example, consider the equation

$$r = \sin \theta \tag{3}$$

By substituting values for θ at increments of $\pi/6$ ($= 30°$) and calculating r, we can construct the following table:

Table 13.1.1

θ (radians)	0	$\dfrac{\pi}{6}$	$\dfrac{\pi}{3}$	$\dfrac{\pi}{2}$	$\dfrac{2\pi}{3}$	$\dfrac{5\pi}{6}$	π	$\dfrac{7\pi}{6}$	$\dfrac{4\pi}{3}$	$\dfrac{3\pi}{2}$	$\dfrac{5\pi}{3}$	$\dfrac{11\pi}{6}$	2π
$r = \sin \theta$	0	$\dfrac{1}{2}$	$\dfrac{\sqrt{3}}{2}$	1	$\dfrac{\sqrt{3}}{2}$	$\dfrac{1}{2}$	0	$-\dfrac{1}{2}$	$-\dfrac{\sqrt{3}}{2}$	-1	$-\dfrac{\sqrt{3}}{2}$	$-\dfrac{1}{2}$	0

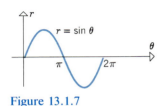

$r = \sin \theta$

Figure 13.1.7

The pairs of values listed in this table can be plotted in one of two ways. One possibility is to view each pair (r, θ) as *rectangular* coordinates of a point and plot each point in the $r\theta$-plane. This yields points on the familiar sine curve (Figure 13.1.7). Alternatively, we can view each pair (r, θ) as *polar* coordinates of a point and plot each point in a polar coordinate system (Figure 13.1.8). Note that there are 13 pairs listed in Table 13.1.1, but only 6 points plotted in Figure 13.1.8. This is because the pairs from $\theta = \pi$ on yield duplicates of the preceding points. For example, $(-1/2, 7\pi/6)$ and $(1/2, \pi/6)$ represent the same point.

The points in Figure 13.1.8 appear to lie on a circle. That this is indeed the case may be seen by expressing (3) in terms of x and y. To do this we first multiply (3) through by r to obtain

$$r^2 = r \sin \theta$$

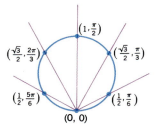

Figure 13.1.8

which can be rewritten using (1a) and (1b) as

$$x^2 + y^2 = y \quad \text{or} \quad x^2 + y^2 - y = 0$$

or on completing the square

$$x^2 + \left(y - \frac{1}{2}\right)^2 = \frac{1}{4}$$

This is a circle of radius $\frac{1}{2}$ centered at the point $(0, \frac{1}{2})$ in the xy-plane.

Example 3 Sketch the curve $r = \sin 2\theta$ in polar coordinates.

Solution. A rough sketch of the graph may be obtained without plotting points by observing how the value of r changes with θ.

- If θ increases from 0 to $\pi/4$, then 2θ increases from 0 to $\pi/2$ and $r = \sin 2\theta$ increases from 0 to 1 (Figure 13.1.9a).
- If θ increases from $\pi/4$ to $\pi/2$, then 2θ increases from $\pi/2$ to π and $r = \sin 2\theta$ decreases from 1 back to 0. Thus, the curve completes a loop (Figure 13.1.9b).
- If θ increases from $\pi/2$ to $3\pi/4$, then 2θ increases from π to $3\pi/2$ and r decreases from 0 to -1. Because r is negative, the resulting portion of the graph is in the fourth quadrant, although the values of θ are in the second quadrant (Figure 13.1.9c).
- If θ increases from $3\pi/4$ to π, then 2θ increases from $3\pi/2$ to 2π and $r = \sin 2\theta$ increases from -1 to 0; thus, the curve completes a loop in the fourth quadrant (Figure 13.1.9d).

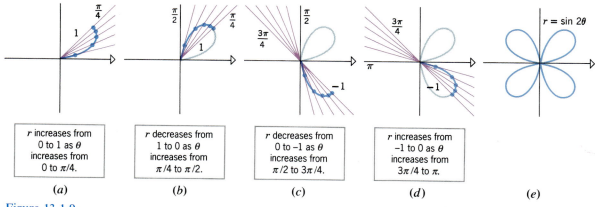

Figure 13.1.9

We leave it for the reader to continue the analysis for θ in the range π to 2π and show that the remaining portion of the curve consists of two loops in the second and third quadrants (Figure 13.1.9e). No new points are generated by allowing θ to vary past 2π or through negative values (verify), so Figure 13.1.9e is the complete graph. ◀

▶ Exercise Set 13.1

1. Plot the following points in polar coordinates.
   (a) $(3, \pi/4)$ (b) $(5, 2\pi/3)$
   (c) $(1, \pi/2)$ (d) $(4, 7\pi/6)$
   (e) $(2, 4\pi/3)$ (f) $(0, \pi)$.

2. Plot the following points in polar coordinates.
   (a) $(2, -\pi/3)$ (b) $(3/2, -7\pi/4)$
   (c) $(-3, 3\pi/2)$ (d) $(-5, -\pi/6)$
   (e) $(-6, -\pi)$ (f) $(-1, 9\pi/4)$.

3. Find rectangular coordinates of the points whose
   polar coordinates are given.
   (a) $(6, \pi/6)$ (b) $(7, 2\pi/3)$
   (c) $(8, 9\pi/4)$ (d) $(5, 0)$
   (e) $(7, 17\pi/6)$ (f) $(0, \pi)$.

4. Find rectangular coordinates of the points whose
   polar coordinates are given.
   (a) $(-8, \pi/4)$ (b) $(7, -\pi/4)$
   (c) $(-6, -5\pi/6)$ (d) $(0, -\pi)$
   (e) $(-2, -3\pi/2)$ (f) $(-5, 0)$.

5. The following points are given in rectangular coordinates. Express the points in polar coordinates with
   $r \geq 0$ and $0 \leq \theta < 2\pi$.
   (a) $(-5, 0)$ (b) $(2\sqrt{3}, -2)$
   (c) $(0, -2)$ (d) $(-8, -8)$
   (e) $(-3, 3\sqrt{3})$ (f) $(1, 1)$.

6. Express the points in Exercise 5 in polar coordinates
   with $r \geq 0$ and $-\pi < \theta \leq \pi$.

7. Express the points in Exercise 5 in polar coordinates
   with $r \leq 0$ and $0 \leq \theta < 2\pi$.

8. In each part find polar coordinates satisfying the
   stated conditions for the point whose rectangular
   coordinates are $(-\sqrt{3}, 1)$.
   (a) $r \geq 0$ and $0 \leq \theta < 2\pi$
   (b) $r \leq 0$ and $0 \leq \theta < 2\pi$
   (c) $r \geq 0$ and $-2\pi < \theta \leq 0$
   (d) $r \leq 0$ and $-\pi < \theta \leq \pi$.

In Exercises 9–20, identify the curve by transforming
to rectangular coordinates.

9. $r = 2$.   10. $r = 3$.

11. $r \sin \theta = 4$.   12. $r = 5 \sec \theta$.

13. $r = 3 \cos \theta$.   14. $r = 2 \sin \theta$.

15. $r^2 \sin 2\theta = 8$. 16. $r^2 \cos 2\theta = 9$.

17. $r = \dfrac{2}{1 + \sin \theta}$.   18. $r = \dfrac{6}{2 - \cos \theta}$.

19. $r = \dfrac{6}{3 \cos \theta + 2 \sin \theta}$.

20. $r = \sec \theta \tan \theta$.

In Exercises 21–32, express the equations in polar coordinates.

21. $x = 7$.   22. $y = -3$.

23. $x^2 + y^2 = 9$.   24. $x^2 + y^2 = 5$.

25. $x^2 + y^2 - 6y = 0$.   26. $x^2 + y^2 + 4x = 0$.

27. $x^2 = 9y$.   28. $x^2 - y^2 = 4$.

29. $4xy = 9$.

30. $(x^2 + y^2)^2 = 16(x^2 - y^2)$.

31. $(x^2 + y^2)^2 = 2xy$.   32. $x^2(x^2 + y^2) = y^2$.

33. Sketch the graph of $r^2 - 3r + 2 = 0$ in polar coor-
    dinates. [*Hint:* Factor.]

In Exercises 34–37, use the method of Example 3 to
sketch the curve in polar coordinates.

34. $r = \cos 2\theta$.   35. $r = 2(1 + \sin \theta)$.

36. $r = 4 \cos 3\theta$.   37. $r = 1 - \cos \theta$.

38. Prove that the distance between the points with polar
    coordinates (r_1, θ_1) and (r_2, θ_2) is

$$d = \sqrt{r_1^2 + r_2^2 - 2r_1r_2 \cos (\theta_1 - \theta_2)}$$

39. Prove: In polar coordinates, the equation
    $r = a \sin \theta + b \cos \theta$ represents a circle.

40. Prove: The area of the triangle whose vertices have
    polar coordinates $(0, 0)$, (r_1, θ_1), and (r_2, θ_2) is

$$A = \tfrac{1}{2}r_1r_2 \sin (\theta_2 - \theta_1)$$

(Assume that $0 \leq \theta_1 < \theta_2 \leq \pi$ and r_1 and r_2 are positive.)

41. Prove: Equations (1a) hold if r is negative.

42. Prove that

$$rr_1 \sin (\theta - \theta_1) + rr_2 \sin (\theta_2 - \theta)$$
$$+ r_1r_2 \sin (\theta_1 - \theta_2) = 0$$

is the polar equation of the line through the points
(r_1, θ_1) and (r_2, θ_2).

13.2 GRAPHS IN POLAR COORDINATES

> *In this section we shall graph some curves in polar coordinates. Since we shall be working with both rectangular xy-coordinates and polar coordinates, we shall assume throughout that the positive x-axis coincides with the polar axis.*

☐ **LINES IN POLAR COORDINATES**

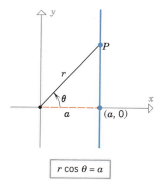

Figure 13.2.1

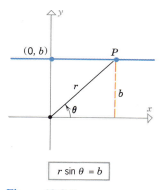

Figure 13.2.2

A line perpendicular to the x-axis and passing through the point with xy-coordinates $(a, 0)$ has the equation $x = a$. To express this equation in polar coordinates we substitute $x = r \cos \theta$ [see (1a) in the previous section]; this yields

$$r \cos \theta = a \tag{1}$$

This result makes sense geometrically since each point $P(r, \theta)$ on this line will yield the value a for $r \cos \theta$ (Figure 13.2.1).

A line parallel to the x-axis that meets the y-axis at the point with xy-coordinates $(0, b)$ has the equation $y = b$. Substituting $y = r \sin \theta$ yields

$$r \sin \theta = b \tag{2}$$

as the polar equation of this line. This makes sense geometrically since each point $P(r, \theta)$ on this line will yield the value b for $r \sin \theta$ (Figure 13.2.2).

For any constant θ_0, the equation

$$\theta = \theta_0 \tag{3}$$

is satisfied by the coordinates of all points of the form $P(r, \theta_0)$, regardless of the value of r. Thus, the equation represents the line through the origin making an angle of θ_0 (radians) with the polar axis (Figure 13.2.3).

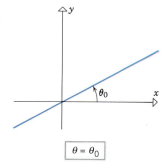

Figure 13.2.3

Example 1 Sketch the graphs of the following equations in polar coordinates.

(a) $r \cos \theta = 3$ (b) $r \sin \theta = -2$ (c) $\theta = \dfrac{3\pi}{4}$

Solution. From (1), (2), and (3), each graph is a line; they are shown in Figure 13.2.4. ◀

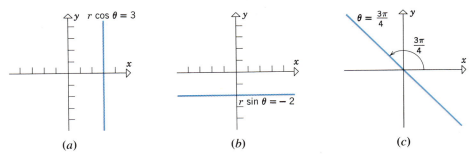

Figure 13.2.4

By substituting $x = r \cos \theta$ and $y = r \sin \theta$ in the equation $Ax + By + C = 0$, we obtain the **general polar form of a line,**

$$r(A \cos \theta + B \sin \theta) + C = 0$$

□ **CIRCLES IN POLAR COORDINATES**

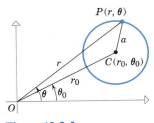

Figure 13.2.5

Let us try to find the polar equation of a circle whose radius is a and whose center has polar coordinates (r_0, θ_0) (Figure 13.2.5). If we let $P(r, \theta)$ be an arbitrary point on the circle, and if we apply the law of cosines to the triangle OCP in Figure 13.2.5, we obtain

$$r^2 - 2rr_0 \cos(\theta - \theta_0) + r_0^2 = a^2 \qquad (4)$$

This general equation is rarely used since the corresponding equation in rectangular coordinates is easier to work with. However, we will now consider some special cases of (4) that arise frequently.

A circle of radius a, centered at the origin, has an especially simple polar equation. If we let $r_0 = 0$ in (4), we obtain $r^2 = a^2$ or, since $a \geq 0$,

$$r = a \qquad (5)$$

This equation makes sense geometrically since the circle of radius a, centered at the origin, consists of all points $P(r, \theta)$ for which $r = a$, regardless of the value of θ (Figure 13.2.6a).

If a circle of radius a has its center on the x-axis and passes through the origin, then the polar coordinates of the center are either

$$(a, 0) \quad \text{or} \quad (a, \pi)$$

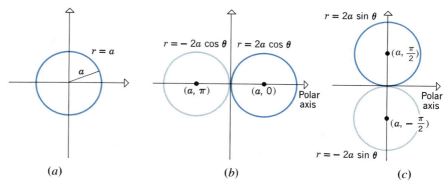

Figure 13.2.6

depending on whether the center is to the right or left of the origin (Figure 13.2.6*b*). For the circle to the right of the origin we have $r_0 = a$ and $\theta_0 = 0$, so (4) reduces to

$$r^2 - 2ra \cos \theta = 0$$

or

$$r = 2a \cos \theta \qquad\qquad (6a)$$

For the circle to the left of the origin, $r_0 = a$ and $\theta_0 = \pi$, so (4) reduces to

$$r = -2a \cos \theta \qquad\qquad (6b)$$

because $\cos(\theta - \pi) = -\cos \theta$.

If a circle of radius a passes through the origin and has its center on the y-axis, then the polar coordinates of the center are $(a, \pi/2)$ or $(a, -\pi/2)$, and (4) reduces to

$$r = 2a \sin \theta \qquad \text{or} \qquad r = -2a \sin \theta \qquad\qquad (7a\text{–}b)$$

depending on whether the center is above or below the origin (Figure 13.2.6*c*).

Equations (6a) and (7a) are interpreted geometrically in Figures 13.2.7*a* and 13.2.7*b*. In each figure part, *OPA* is a right triangle (why?), so for each point $P(r, \theta)$ on the circle, we have $r = 2a \cos \theta$ or $r = 2a \sin \theta$, depending on the position of the circle.

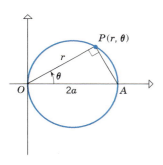

$r = 2a \cos \theta$

(*a*)

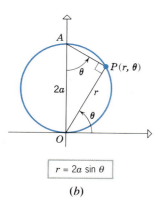

$r = 2a \sin \theta$

(*b*)

Figure 13.2.7

Example 2   Sketch the graphs of the following equations in polar coordinates.

(a) $r = 4 \cos \theta$ (b) $r = -5 \sin \theta$ (c) $r = 3$

Solution.   These equations are of forms (6a), (7b), and (5), respectively, and therefore represent circles. The graphs are shown in Figure 13.2.8.   ◄

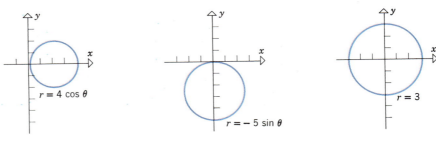

Figure 13.2.8

The points with polar coordinates (r, θ) and $(r, -\theta)$ are symmetric about the x-axis (Figure 13.2.9a). Thus, if a polar equation is unchanged on replacing θ by $-\theta$, then its graph is symmetric about the x-axis. The following tests for symmetry will be useful.

13.2.1   THEOREM (*Symmetry Tests*).

(*a*) *A curve in polar coordinates is symmetric about the x-axis if replacing θ by $-\theta$ in its equation produces an equivalent equation* (Figure 13.2.9a).

(*b*) *A curve in polar coordinates is symmetric about the y-axis if replacing θ by $\pi - \theta$ in its equation produces an equivalent equation* (Figure 13.2.9b).

(*c*) *A curve in polar coordinates is symmetric about the origin if replacing r by $-r$ in its equation produces an equivalent equation* (Figure 13.2.9c).

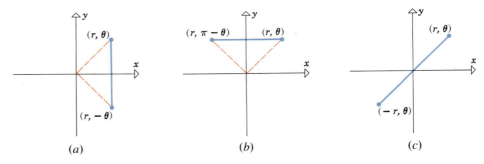

Figure 13.2.9

☐ **CARDIOIDS AND LIMAÇONS**

Equations of the form

$$r = a + b \sin \theta, \qquad r = a - b \sin \theta \tag{8a–8b}$$
$$r = a + b \cos \theta, \qquad r = a - b \cos \theta \tag{8c–8d}$$

produce polar curves called *limaçons* (from the Latin word "limax," for a slug, a snail-like creature). There are four possible shapes for a limaçon* that can be determined from the ratio a/b when $a > 0$ and $b > 0$ (Figure 13.2.10).

---

*For a detailed discussion of limaçons and their shapes, the reader is referred to the article by Jane T. Grossman and Michael P. Grossman, "Dimple or No Dimple," *Two Year College Mathematics Journal*, Vol. 13, No. 1, 1982, pp. 52–55.

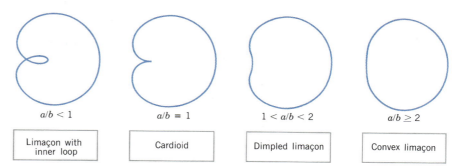

Figure 13.2.10

The position of the limaçon relative to the polar axis depends on whether $\sin \theta$ or $\cos \theta$ appears in the equation and whether the $+$ or $-$ occurs. Because of the heart-shaped appearance of the curve in the case $a = b$, limaçons of this type are called **cardioids** (from the Greek word "kardia," for heart).

Example 3 Assuming a to be a positive constant, sketch the graph of the equation $r = a(1 - \cos \theta)$ in polar coordinates.

Solution. This is an equation of form (8d) with $a = b$; therefore, it represents a cardioid. Since $\cos(-\theta) = \cos \theta$, the equation is unaltered when θ is replaced by $-\theta$; thus, the cardioid is symmetric about the x-axis. This being the case, we can obtain the entire curve by first sketching the portion of the cardioid above the x-axis and then reflecting this portion about the x-axis.

As θ varies from 0 to π, $\cos \theta$ decreases steadily from 1 to -1, and $1 - \cos \theta$ increases steadily from 0 to 2. Thus, as θ varies from 0 to π, the value of $r = a(1 - \cos \theta)$ will increase steadily from an initial value of $r = 0$ to a final value of $r = 2a$. Using this information, and by plotting the points in Table 13.2.1, we obtain the graph in Figure 13.2.11a.

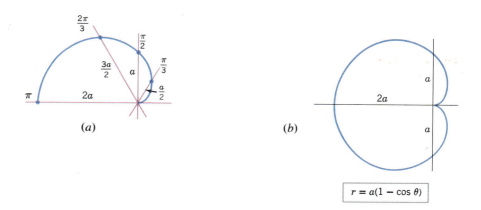

$$r = a(1 - \cos \theta)$$

Figure 13.2.11

Table 13.2.1

θ	0	$\pi/3$	$\pi/2$	$2\pi/3$	π
$r = a(1 - \cos\theta)$	0	$a/2$	a	$3a/2$	$2a$

On reflecting the curve in Figure 13.2.11a about the x-axis, we obtain the entire cardioid (Figure 13.2.11b). ◀

REMARK.   We have sketched the cardioid in Figure 13.2.11a with the line $\theta = 0$ tangent to the curve at the origin. Although it is not evident from our discussion that this is the case, we will prove in Section 13.5 that if a polar curve passes through the origin when $\theta = \theta_0$, then the line $\theta = \theta_0$ is tangent to the curve at the origin.

☐ LEMNISCATES

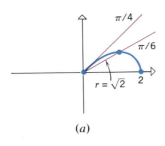

A lemniscate

Figure 13.2.12

If $a > 0$, then equations of the form

$$r^2 = a^2 \cos 2\theta, \qquad r^2 = -a^2 \cos 2\theta \tag{9a–9b}$$
$$r^2 = a^2 \sin 2\theta, \qquad r^2 = -a^2 \sin 2\theta \tag{9c–9d}$$

represent propeller-shaped curves, called **lemniscates** (from the Greek word "lemniscos" for a looped ribbon resembling the figure 8) (Figure 13.2.12). The lemniscates are centered at the origin, but the position relative to the polar axis depends on the sign preceding the a^2 and whether $\sin 2\theta$ or $\cos 2\theta$ appears in the equation.

Example 4   Sketch the curve

$$r^2 = 4 \cos 2\theta \tag{10}$$

in polar coordinates.

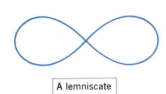

(a)

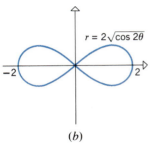

(b)

Figure 13.2.13

Solution.   The equation is of type (9a) with $a = 2$, and thus represents a lemniscate. Solving (10) for r in terms of θ yields two functions of θ:

$$r = 2\sqrt{\cos 2\theta} \quad \text{and} \quad r = -2\sqrt{\cos 2\theta} \tag{11a–b}$$

We leave it for the reader to apply the symmetry tests in Theorem 13.2.1 and show that the graph of each of these equations is symmetric about the x-axis and the y-axis. Therefore, we can obtain each graph by first sketching the portion of the graph in the range $0 \leq \theta \leq \pi/2$ and then reflecting that portion about the x- and y-axes.

We shall consider the graph of (11a) first. As θ varies from 0 to $\pi/4$, the value of $\cos 2\theta$ decreases steadily from 1 to 0, so that $r = 2\sqrt{\cos 2\theta}$ decreases steadily from 2 to 0. With this information and the points plotted from Table 13.2.2, we obtain the sketch in Figure 13.2.13a. (Note that the curve passes through the origin when $\theta = \pi/4$, so the line $\theta = \pi/4$ is tangent to the curve at the origin by the remark following Example 3.)

For θ in the range $\pi/4 < \theta < \pi/2$, the quantity $\cos 2\theta$ is negative, so there are no real values of r satisfying (11a). Thus, there are no points on the graph for such θ. The entire graph is obtained by reflecting the curve in Figure 13.2.13a about the x-axis and then reflecting the resulting curve about the y-axis (Figure 13.2.13b).

Table 13.2.2

θ	0	$\pi/6$	$\pi/4$
$r = 2\sqrt{\cos 2\theta}$	2	$\sqrt{2}$	0

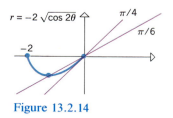

Figure 13.2.14

To graph $r = -2\sqrt{\cos 2\theta}$, we simply note that (11a) and (11b) differ only in the sign of r, so the graph of (11b) is identical to that in Figure 13.2.13b, but traced in a "diagonally opposite" manner. For example, as θ varies from 0 to $\pi/4$, the portion of the graph generated is in the third quadrant because r is negative (Figure 13.2.14). To summarize, the graph of (10) actually consists of two superimposed lemniscates, one being the graph of (11a) and the other the graph of (11b).   ◄

☐ **SPIRALS**

A curve that "winds around the origin" infinitely many times in such a way that r increases (or decreases) steadily as θ increases is called a *spiral*. The most common example is the *spiral of Archimedes,* which has an equation of the form

$$r = a\theta \quad (\theta \geq 0) \qquad \text{or} \qquad r = a\theta \quad (\theta \leq 0) \tag{12a–b}$$

In these equations, θ is in radians and a is positive.

Example 5   Sketch the curve

$$r = \theta \quad (\theta \geq 0)$$

in polar coordinates.

Solution.   This is an equation of form (12a) with $a = 1$; thus, it represents an Archimedean spiral. Since $r = 0$ when $\theta = 0$, the origin is on the curve and the polar axis is tangent to the spiral.

A reasonably accurate sketch may be obtained by plotting the intersections of the spiral with the x and y axes and noting that r increases steadily as θ increases. The intersections with the x-axis occur when

$$\theta = 0, \pi, 2\pi, 3\pi, \ldots$$

at which points r has the values

$$r = 0, \pi, 2\pi, 3\pi, \ldots$$

and the intersections with the y-axis occur when

$$\theta = \frac{\pi}{2}, \frac{3\pi}{2}, \frac{5\pi}{2}, \frac{7\pi}{2}, \ldots$$

at which points r has the values

$$r = \frac{\pi}{2}, \frac{3\pi}{2}, \frac{5\pi}{2}, \frac{7\pi}{2}, \ldots$$

The graph is shown in Figure 13.2.15.   ◄

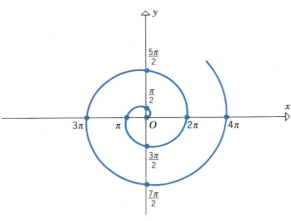

Figure 13.2.15

Starting from the origin, the spiral in Figure 13.2.15 loops counterclockwise around the origin. This is true of all Archimedean spirals of form (12a); Archimedean spirals of form (12b) loop clockwise around the origin.

□ **ROSE CURVES**

Equations of the form

$$r = a \sin n\theta \qquad \text{and} \qquad r = a \cos n\theta \tag{13a–b}$$

represent flower-shaped curves called **roses**. The rose has n equally spaced petals or loops if n is odd and $2n$ equally spaced petals if n is even (Figure 13.2.16). The orientation of the rose relative to the polar axis depends on the sign of the constant a and whether $\sin\theta$ or $\cos\theta$ appears in the equation. (A rose curve occurred in Example 3 of the previous section.)

If $n = 1$ in (13a) or (13b), then we obtain the equation of a circle, which can be regarded as a one-petal rose.

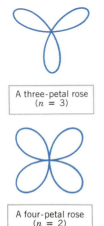

A three-petal rose
($n = 3$)

A four-petal rose
($n = 2$)

Figure 13.2.16

REMARK.   It can be shown that a rose with an even number of petals is traced out exactly once as θ varies over the interval $0 \le \theta < 2\pi$ (Example 3 in the previous section), and a rose with an odd number of petals is traced out exactly once as θ varies over the interval $0 \le \theta < \pi$ (Exercise 62).

▶ **Exercise Set 13.2**   □C 38, 42, 43, 44, 45, 46, 48

In Exercises 1–36, sketch the curve in polar coordinates and give its name.

1. $\theta = \pi/6$.

2. $\theta = -3\pi/4$.

3. $r = 5$.

4. $r = 4 \sin \theta$.

5. $r = -6 \cos \theta$.

6. $r = -3 \sin \theta$.

7. $2r = \cos \theta$.

8. $r = 1 + \sin \theta$.

9. $r = 3(1 - \sin \theta)$.

10. $r - 2 = 2 \cos \theta$.

11. $r = 4 - 4 \cos \theta$.

12. $r = -5 + 5 \sin \theta$.

13. $r = -1 - \cos \theta$.

14. $r = 1 + 2 \sin \theta$.

15. $r = 1 - 2 \cos \theta$.

16. $r = 4 + 3 \cos \theta$.

17. $r = 3 + 2 \sin \theta$.

18. $r = 3 - \cos \theta$.

19. $r = 2 + \sin \theta$.

20. $r - 5 = 3 \sin \theta$.

21. $r = 3 + 4 \cos \theta$.

22. $r = -3 - 4 \sin \theta$.

23. $r = 5 - 2 \cos \theta$.

24. $r^2 = 9 \cos 2\theta$.

25. $r^2 = -9 \cos 2\theta$.

26. $r^2 = \sin 2\theta$.

27. $r^2 = -16 \sin 2\theta$.

28. $r = 4\theta$ $(\theta \geq 0)$.

29. $r = 4\theta$ $(\theta \leq 0)$.

30. $r = 4\theta$.

31. $r = \cos 2\theta$.

32. $r = 3 \sin 2\theta$.

33. $r = \sin 3\theta$.

34. $r = 2 \cos 3\theta$.

35. $r = 9 \sin 4\theta$.

36. $r = \cos 5\theta$.

In Exercises 37–40, sketch the curve in polar coordinates.

37. $r = 4 \cos \theta + 4 \sin \theta$. [*Hint:* Convert to rectangular coordinates.]

38. $r = e^\theta$ (logarithmic spiral).

39. $r = \sin (\theta/2)$.

40. $r^2 = \theta$ (parabolic spiral).

In Exercises 41–44, the curves all have vertical or horizontal asymptotes. Find the asymptotes by expressing the rectangular coordinates x and y as functions of θ, then sketch the curve in polar coordinates.

41. $r = 4 \tan \theta$ (kappa curve). [*Hint:* Show that $x \to 4$ or $x \to -4$ and $y \to +\infty$ or $y \to -\infty$ as θ approaches an odd multiple of $\pi/2$.]

42. $r = 2 + 2 \sec \theta$ (conchoid of Nicomedes). [*Hint:* Show that $x \to 2$ and $y \to +\infty$ or $y \to -\infty$ as θ approaches an odd multiple of $\pi/2$.]

43. $r = 2 \sin \theta \tan \theta$ (cissoid). [*Hint:* See the hint in Exercise 42.]

44. $r = 1/\theta$ $(\theta > 0)$ (hyperbolic spiral). [*Hint:* Find $\lim_{\theta \to 0^+} y$ to show that there is a horizontal asymptote.]

45. $r = 1/\sqrt{\theta}$ $(\theta > 0)$. [*Hint:* Find $\lim_{\theta \to 0^+} y$ to show that there is a horizontal asymptote.]

46. $r = 1/\theta^2$ $(\theta > 0)$. [*Hint:* Find $\lim_{\theta \to 0^+} y$ to show that there is no horizontal asymptote.]

47. Find the maximum value of the y-coordinate of points on the cardioid $r = 1 + \cos \theta$ for θ in the interval $[0, \pi]$.

48. (a) Show that the maximum value of the y-coordinate of points on the curve $r = 1/\sqrt{\theta}$ for θ in the interval $(0, \pi]$ occurs when θ satisfies the equation $\tan \theta = 2\theta$.

(b) Use Newton's Method to solve the equation in part (a) for θ to at least four decimal-place accuracy.

(c) Use the result of part (b) to approximate the maximum value of y for $0 < \theta \leq \pi$.

In Exercises 49 and 50, take the *width* of a petal of a rose curve to be the dimension shown in Figure 13.2.17.

Petal width

Figure 13.2.17

49. Show that the width of a petal of the four-petal rose $r = \cos 2\theta$ is $2\sqrt{6}/9$. [*Hint:* Express y in terms of θ and find the maximum value of y for $0 \leq \theta \leq \pi/4$.]

50. The width of a petal of the rose $r = \cos n\theta$ is twice the maximum value of y for $0 \leq \theta \leq \pi/(2n)$. Show that the value of θ for which y attains this maximum value satisfies the equation $n \tan n\theta \tan \theta = 1$.

51. Find the minimum value of the x-coordinate of points on the cardioid $r = 1 + \cos \theta$ for θ in the interval $[0, \pi]$.

52. Show that the minimum value of the x-coordinate of points on the curve $r = a + b \cos \theta$, $a > 0$, $b > 0$, for θ in the interval $[0, \pi]$ is $-a^2/(4b)$ if $a < 2b$ and is $b - a$ if $a \geq 2b$.

Exercises 53–60 use the notions of focus, directrix, and eccentricity developed for conic sections in Supplementary Exercises 31 and 32 of Chapter 12.

53. Show that if the focus is at the pole of a polar coordinate system, and the directrix is perpendicular to the polar axis and k units from the pole ($k > 0$), then the equation of the conic section with eccentricity e is

$$r = \frac{ek}{1 - e \cos \theta} \quad \text{or} \quad r = \frac{ek}{1 + e \cos \theta}$$

where the first equation applies when the directrix is to the left of the pole and the second when it is to the right.

54. Show that if the focus is at the pole of a polar coordinate system, and the directrix is parallel to the polar axis and k units from the pole ($k > 0$), then the equation of the conic section with eccentricity e

is

$$r = \frac{ek}{1 - e \sin \theta} \quad \text{or} \quad r = \frac{ek}{1 + e \sin \theta}$$

where the first equation applies when the directrix is below the pole and the second when it is above.

In Exercises 55–60, find the eccentricity e and sketch a graph of the conic (see Exercises 53 and 54).

55. $r = \dfrac{3}{2 - 2 \cos \theta}$.

56. $r = \dfrac{5}{3 + 3 \sin \theta}$.

57. $r = \dfrac{3}{2 + \sin \theta}$.

58. $r = \dfrac{4}{3 - 2 \cos \theta}$.

59. $r = \dfrac{4}{2 + 3 \cos \theta}$.

60. $r = \dfrac{3}{3 - 4 \sin \theta}$.

61. What is the geometric relationship between the graphs of $r = f(\theta)$ and $r = f(\theta + \alpha)$ in polar coordinates if α is a constant?

62. Prove that a rose with an even number of petals is traced out exactly once as θ varies over the interval $0 \leq \theta < 2\pi$ and a rose with an odd number of petals is traced out exactly once as θ varies over the interval $0 \leq \theta < \pi$.

■ 13.3   AREA IN POLAR COORDINATES

In this section we shall show how to find areas of regions that are bounded by curves whose equations are given in polar coordinates.

□ **THE AREA PROBLEM IN POLAR COORDINATES**

13.3.1   PROBLEM. *Find the area of the region R between a polar curve $r = f(\theta)$ and two lines, $\theta = \alpha$ and $\theta = \beta$ (Figure 13.3.1).*

To solve this area problem, we shall make two assumptions:

1. $\alpha < \beta$.

2. $f(\theta)$ is continuous and nonnegative for $\alpha \leq \theta \leq \beta$.

As a first step toward finding the area of the region R, let $\theta_1, \theta_2, \ldots, \theta_{n-1}$ be any numbers such that

$$\alpha < \theta_1 < \theta_2 < \cdots < \theta_{n-1} < \beta$$

and construct the lines

$$\theta = \theta_1, \ \theta = \theta_2, \ldots, \ \theta = \theta_{n-1}$$

These lines divide the angle from α to β into n subangles, and they divide the region R into n subregions (Figure 13.3.2). As indicated in the figure, we shall denote the subangles by

$$\Delta\theta_1, \ \Delta\theta_2, \ldots, \ \Delta\theta_n$$

and the areas of the subregions by

$$A_1, A_2, \ldots, A_n$$

Then the area A of the entire region may be written as

$$A = A_1 + A_2 + \cdots + A_n = \sum_{k=1}^{n} A_k$$

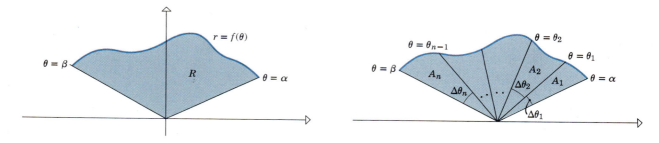

Figure 13.3.1

Figure 13.3.2

Let us now concentrate on the areas $A_1, A_2, \ldots, A_n$ of the subregions. If $\Delta\theta_k$ is not too large, we can approximate the area A_k by the area of a *sector* having central angle $\Delta\theta_k$ and radius $f(\theta_k^*)$, where θ_k^* is an arbitrary angle terminating in the kth subregion (Figure 13.3.3). Thus, from the area formula for a sector [Formula (1), Section 2.8],

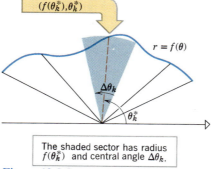

The shaded sector has radius $f(\theta_k^*)$ and central angle $\Delta\theta_k$.

Figure 13.3.3

$$A_k \approx \frac{1}{2} [f(\theta_k^*)]^2 \, \Delta \theta_k$$

and

$$A = \sum_{k=1}^{n} A_k \approx \sum_{k=1}^{n} \frac{1}{2} [f(\theta_k^*)]^2 \, \Delta \theta_k \qquad (1)$$

If we now increase n in such a way that max $\Delta \theta_k \to 0$, then the sectors will become better and better approximations to the subregions and (1) will approach the exact value of the area A; that is,

$$A = \lim_{\max \Delta \theta_k \to 0} \sum_{k=1}^{n} \frac{1}{2} [f(\theta_k^*)]^2 \, \Delta \theta_k = \int_{\alpha}^{\beta} \frac{1}{2} [f(\theta)]^2 \, d\theta$$

Thus, we have the following result.

13.3.2 AREA IN POLAR COORDINATES.   If $f(\theta)$ is continuous and nonnegative for $\alpha \le \theta \le \beta$, and if $\beta - \alpha \le 2\pi$, then the area A enclosed by the polar curve $r = f(\theta)$ and the lines $\theta = \alpha$ and $\theta = \beta$ is

$$A = \int_{\alpha}^{\beta} \frac{1}{2} [f(\theta)]^2 \, d\theta = \int_{\alpha}^{\beta} \frac{1}{2} r^2 \, d\theta \qquad (2)$$

The hardest part of applying (2) is determining the limits of integration. This can be done as follows:

Step 1. Sketch the region R whose area is to be determined.

Step 2. Draw an arbitrary "radial line" from the origin to the boundary curve $r = f(\theta)$.

Step 3. Ask, "Over what interval of values must θ vary in order for the radial line to sweep out the region R?"

Step 4. Your answer in Step 3 will determine the lower and upper limits of integration.

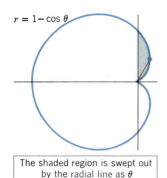

$r = 1 - \cos \theta$

The shaded region is swept out by the radial line as θ varies from 0 to $\frac{\pi}{2}$.

Figure 13.3.4

Example 1   Find the area of the region in the first quadrant within the cardioid $r = 1 - \cos \theta$.

Solution.   The region and a typical radial line are shown in Figure 13.3.4. For the radial line to sweep out the region, θ must vary from 0 to $\pi/2$. Thus, from (2) with $\alpha = 0$ and $\beta = \pi/2$,

$$A = \int_{0}^{\pi/2} \frac{1}{2} r^2 \, d\theta = \frac{1}{2} \int_{0}^{\pi/2} (1 - \cos \theta)^2 \, d\theta$$

$$= \frac{1}{2} \int_{0}^{\pi/2} (1 - 2 \cos \theta + \cos^2 \theta) \, d\theta$$

With the help of the identity $\cos^2\theta = \frac{1}{2}(1 + \cos 2\theta)$, this can be rewritten as

$$A = \frac{1}{2}\int_0^{\pi/2}\left(\frac{3}{2} - 2\cos\theta + \frac{1}{2}\cos 2\theta\right)d\theta$$

$$= \frac{1}{2}\left[\frac{3}{2}\theta - 2\sin\theta + \frac{1}{4}\sin 2\theta\right]_0^{\pi/2} = \tfrac{3}{8}\pi - 1 \quad \blacktriangleleft$$

Example 2 Find the entire area within the cardioid of Example 1.

Solution. For the radial line to sweep out the entire cardioid, θ must vary from 0 to 2π. Thus, from (2) with $\alpha = 0$ and $\beta = 2\pi$,

$$A = \int_0^{2\pi}\frac{1}{2}r^2\,d\theta = \frac{1}{2}\int_0^{2\pi}(1-\cos\theta)^2\,d\theta$$

If we proceed as in Example 1, this reduces to

$$A = \frac{1}{2}\int_0^{2\pi}\left(\frac{3}{2} - 2\cos\theta + \frac{1}{2}\cos 2\theta\right)d\theta = \frac{3\pi}{2}$$

Alternative Solution. Since the cardioid is symmetric about the x-axis, we can calculate the portion of area above the x-axis and double the result. In the portion of the cardioid above the x-axis, θ ranges from 0 to π, so that

$$A = 2\int_0^{\pi}\frac{1}{2}r^2\,d\theta = \int_0^{\pi}(1-\cos\theta)^2\,d\theta = \frac{3\pi}{2} \quad \blacktriangleleft$$

☐ **USING SYMMETRY TO AVOID NEGATIVE VALUES OF r**

It is important to keep in mind that Formula (2) requires $r = f(\theta)$ to be non-negative. For example, suppose we were interested in finding the area of the region enclosed by the rose $r = \sin 2\theta$ shown in Figure 13.1.9. Although this region is swept out by a radial line as θ varies from 0 to 2π, the value of

$$r = \sin 2\theta$$

is negative over portions of this interval (if $\pi/2 < \theta < \pi$ or $3\pi/2 < \theta < 2\pi$). Thus, Formula (2) is not applicable over the entire interval $0 \leq \theta \leq 2\pi$. The following example shows how to avoid this difficulty by calculating a portion of the area and using symmetry.

Example 3 Find the area of the region enclosed by the rose curve $r = \sin 2\theta$.

Solution. The petal in the first quadrant is swept out as θ varies over the interval $0 \leq \theta \leq \pi/2$ (Figures 13.1.9a and b), and for these values of θ, the value of $r = \sin 2\theta$ is nonnegative, so Formula (2) can be used to find the area of this petal. By symmetry, the area enclosed by one petal is one-fourth of the entire area A, so

$$A = 4 \int_0^{\pi/2} \frac{1}{2} r^2 \, d\theta = 2 \int_0^{\pi/2} \sin^2 2\theta \, d\theta$$

$$= 2 \int_0^{\pi/2} \frac{1}{2} (1 - \cos 4\theta) \, d\theta = \int_0^{\pi/2} (1 - \cos 4\theta) \, d\theta$$

$$= \left[\theta - \frac{1}{4} \sin 4\theta \right]_0^{\pi/2} = \pi/2 \quad \blacktriangleleft$$

☐ **USING NEGATIVE ANGLES AS LIMITS OF INTEGRATION**

It was assumed in our derivation of Formula (2) that $\alpha < \beta$. Sometimes, this is best achieved by using negative angles for one or both limits of integration. The following example illustrates this.

Example 4 Find the area of the region that is inside the cardioid $r = 4 + 4 \cos \theta$ and outside the circle $r = 6$.

Solution. To sketch the region, we need to know where the circle and cardioid intersect. To find these points, we equate the given expressions for r. This yields

$$4 + 4 \cos \theta = 6$$

from which we obtain $\cos \theta = \frac{1}{2}$ or

$$\theta = \frac{\pi}{3} \quad \text{and} \quad \theta = -\frac{\pi}{3}$$

The region is sketched in Figure 13.3.5. The desired area can be obtained by subtracting the shaded areas in parts (*b*) and (*c*) of Figure 13.3.5. Thus,

$$A = \int_{-\pi/3}^{\pi/3} \frac{1}{2} (4 + 4 \cos \theta)^2 \, d\theta - \int_{-\pi/3}^{\pi/3} \frac{1}{2} (6)^2 \, d\theta$$

$$= \int_{-\pi/3}^{\pi/3} \frac{1}{2} [(4 + 4 \cos \theta)^2 - 36] \, d\theta = \int_{-\pi/3}^{\pi/3} (16 \cos \theta + 8 \cos^2 \theta - 10) \, d\theta$$

$$= \left[16 \sin \theta + (4\theta + 2 \sin 2\theta) - 10\theta \right]_{-\pi/3}^{\pi/3} = 18\sqrt{3} - 4\pi$$

Alternative Solution. Using symmetry, we can calculate the portion of the area above the x-axis and double it to obtain the entire area A. This yields (verify)

$$A = 2 \int_0^{\pi/3} \frac{1}{2} [(4 + 4 \cos \theta)^2 - 36] \, d\theta = 2(9\sqrt{3} - 2\pi) = 18\sqrt{3} - 4\pi$$

which agrees with the previous result. $\quad \blacktriangleleft$

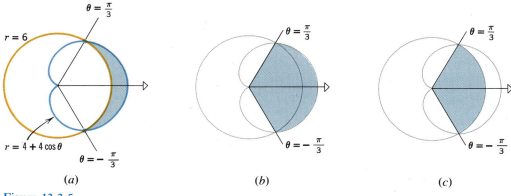

Figure 13.3.5

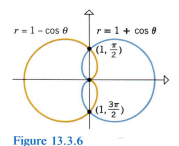

Figure 13.3.6

WARNING. In the last example we determined the intersections of the curves $r = 4 + 4\cos\theta$ and $r = 6$ by equating the right-hand sides and solving for θ. However, it may not be possible to find *all* the intersections of two polar curves, $r = f_1(\theta)$ and $r = f_2(\theta)$, by solving the equation $f_1(\theta) = f_2(\theta)$ for θ. This is because every point has infinitely many pairs of polar coordinates. It is possible, therefore, that a point of intersection may have no single pair of polar coordinates that satisfies both equations. For example, the cardioids

$$r = 1 - \cos\theta \quad \text{and} \quad r = 1 + \cos\theta \tag{3}$$

intersect at three points, the origin, the point $(1, \pi/2)$, and the point $(1, 3\pi/2)$ (Figure 13.3.6). Equating the right-hand sides of the equations in (3) yields $1 - \cos\theta = 1 + \cos\theta$ or $\cos\theta = 0$, so

$$\theta = \frac{\pi}{2} + k\pi, \quad k = 0, \pm 1, \pm 2, \ldots$$

Substituting any of these values in (3) yields $r = 1$, so that we have found only two distinct points of intersection, $(1, \pi/2)$ and $(1, 3\pi/2)$; the origin has been missed.

When calculating intersections in polar coordinates, it is a good idea to make a sketch of the curves to determine how many intersections there should be.

▶ Exercise Set 13.3

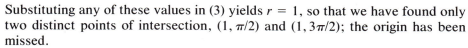

In Exercises 1–21, find the area of the region described.

1. The region in the first quadrant enclosed by the first loop of the spiral $r = \theta \, (\theta \geq 0)$ and the lines $\theta = \pi/6$ and $\theta = \pi/3$.

2. The region in the first quadrant within the cardioid $r = 1 + \sin\theta$.

3. The region that is enclosed by the cardioid $r = 2 + 2\cos\theta$.

4. The region outside the cardioid $r = 2 - 2\cos\theta$ and inside the circle $r = 4$.

5. The region inside the circle $r = 5\sin\theta$ and outside the limaçon $r = 2 + \sin\theta$.

6. The region enclosed by the inner loop of the limaçon $r = 1 + 2\cos\theta$.

7. The region inside the cardioid $r = 2 + 2\cos\theta$ and outside the circle $r = 3$.

8. The region inside the circle $r = 2a\sin\theta$.

9. The region enclosed by the curve $r^2 = \sin 2\theta$.

10. The region common to the circles $r = 4\cos\theta$ and $r = 4\sin\theta$.

11. The region enclosed by the rose $r = 4\cos 3\theta$.

12. The region inside the rose $r = 2a\cos 2\theta$ and outside the circle $r = a\sqrt{2}$.

13. The region common to the circle $r = 3\cos\theta$ and the cardioid $r = 1 + \cos\theta$.

14. The region between the loops of the limaçon $r = \frac{1}{2} + \cos\theta$.

15. The region inside the circle $r = 10$ and to the right of the line $r\cos\theta = 6$.

16. The region inside the cardioid $r = a(1 + \sin\theta)$ and outside the circle $r = a\sin\theta$.

17. The region enclosed by $r = a\sec^2\frac{1}{2}\theta$ and the rays whose polar angles are $\theta = 0$ and $\theta = \frac{1}{2}\pi$.

18. The region inside the cardioid $r = 2 + 2\cos\theta$ and to the right of the line $r\cos\theta = 3/2$.

19. The region swept out by a radial line from the origin to the curve $r = 3e^{-2\theta}$ as θ varies over the interval $0 \le \theta \le \pi$.

20. The region swept out by a radial line from the origin to the curve $r = 2/\theta$ as θ varies over the interval $1 \le \theta \le 3$.

21. The region swept out by a radial line from the origin to the curve $r = 1/\sqrt{\theta}$ as θ varies over the interval $1/9 \le \theta \le 4$.

22. (a) Find the error: The area inside the lemniscate $r^2 = a^2 \cos 2\theta$ is

$$A = \int_0^{2\pi} \frac{1}{2}r^2\,d\theta = \int_0^{2\pi} \frac{1}{2}a^2 \cos 2\theta\,d\theta$$

$$= \frac{1}{4}a^2 \sin 2\theta \Big]_0^{2\pi} = 0$$

(b) Find the correct area.

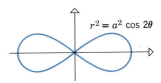

23. Find the area inside the lemniscate $r^2 = 4\cos 2\theta$ and outside the circle $r = \sqrt{2}$.

24. A radial line is drawn from the origin to the spiral $r = a\theta$ ($a > 0$ and $\theta \ge 0$). Find the area swept out during the second revolution of the radial line that was not swept out during the first revolution.

25. The graph of $x^3 - xy + y^3 = 0$, called a **folium of Descartes**, is shown in the figure below.
(a) Show that in polar coordinates its equation is

$$r = \frac{\sin\theta \cos\theta}{\cos^3\theta + \sin^3\theta}$$

(b) Find the area of the region that is enclosed by the loop.
[*Hint:* Show that $r = \sec\theta \tan\theta/(1 + \tan^3\theta)$.]

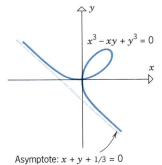

Asymptote: $x + y + 1/3 = 0$

26. Show that if $f(\theta)$ is continuous and nonnegative for $\alpha \le \theta \le \beta$ and $0 \le \beta - \alpha \le 2\pi$, then the area enclosed by $r = 2f(\theta)$ and the lines $\theta = \alpha$ and $\theta = \beta$ is four times the area enclosed by $r = f(\theta)$ and the lines $\theta = \alpha$ and $\theta = \beta$.

■ 13.4 PARAMETRIC EQUATIONS

> *Recall that the graph in the xy-plane of a function f can be cut at most once by any vertical line. Consequently, there are many important curves in the xy-plane that are not graphs of equations of the form y = f(x) (circles, for example). In this section we shall show that such curves can be obtained from pairs of functions, one used to specify the x-coordinate of a point on the graph and the other used to specify the y-coordinate.*

□ **PARAMETRIC EQUATIONS**

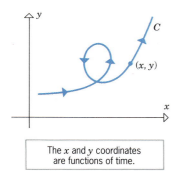

The x and y coordinates are functions of time.

Figure 13.4.1

Imagine a particle moving along some curve C in the xy-plane, which may or may not be the graph of a function (Figure 13.4.1). The x and y coordinates of the moving particle are functions of time t, say

$$x = f(t) \quad \text{and} \quad y = g(t)$$

Physicists and engineers call these the **equations of motion** of the particle and the curve C the **trajectory** of the particle.

Example 1 Suppose that a particle moves in the xy-plane with the equations of motion

$$x = \tfrac{1}{2}t^3 - 6t, \quad y = \tfrac{1}{2}t^2$$

Sketch the trajectory of the particle over the time interval $0 \le t \le 4$.

Solution. The trajectory, shown in Figure 13.4.2, was found by calculating the x and y coordinates of the particle for $t = 0, 1, 2, 3, 4$ (Table 13.4.1), plotting the points (x, y), and connecting successive points with a smooth curve. The arrows on the curve indicate the direction of motion. ◄

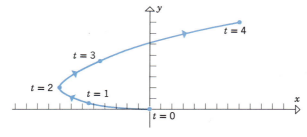

Figure 13.4.2

Table 13.4.1

t	0	1	2	3	4
$x = \frac{1}{2}t^3 - 6t$	0	$-\frac{11}{2}$	-8	$-\frac{9}{2}$	8
$y = \frac{1}{2}t^2$	0	$\frac{1}{2}$	2	$\frac{9}{2}$	8
(x, y)	$(0, 0)$	$(-\frac{11}{2}, \frac{1}{2})$	$(-8, 2)$	$(-\frac{9}{2}, \frac{9}{2})$	$(8, 8)$

Motivated by the concept of equations of motion, mathematicians often specify curves in the *xy*-plane by using a pair of equations

$$x = x(t), \quad y = y(t) \tag{1}$$

to express the coordinates of a point (x, y) on the curve as functions of an auxiliary variable t. These are called **parametric equations** for the curve, and the variable t is called a **parameter**. The parameter t should be viewed as an independent variable that varies over an interval of real values. If no restrictions on t are stated explicitly, then it is understood that t varies between $-\infty$ and $+\infty$. To indicate that t is restricted to values in an interval $[a, b]$, we will write

$$x = x(t), \quad y = y(t) \quad (a \leq t \leq b)$$

Although time is a common parameter, it is not the only possibility; other parameters will be illustrated in the examples.

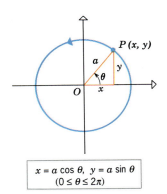

$x = a \cos \theta, \ y = a \sin \theta$
$(0 \leq \theta \leq 2\pi)$

Figure 13.4.3

REMARK.   In (1) we followed a common practice of writing the parametric equations as $x = x(t)$ and $y = y(t)$, rather than $x = f(t)$ and $y = g(t)$, to reduce the number of different letters being used. This dual use of a letter to denote both a dependent variable and a function is common and rarely leads to difficulties.

Example 2   Let C be a circle of radius a centered at the origin. If $P(x, y)$ is any point on the circle and θ is the angle measured counterclockwise from the positive x-axis to the line segment OP (Figure 13.4.3), then

$$x = a \cos \theta \quad \text{and} \quad y = a \sin \theta$$

As θ varies from 0 to 2π, the point $P(x, y)$ makes one complete revolution. Thus, the circle can be represented by the parametric equations

$$x = a \cos \theta, \quad y = a \sin \theta \quad (0 \leq \theta \leq 2\pi) \tag{2}$$

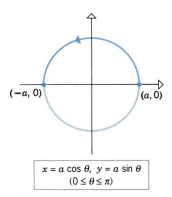

$x = a \cos \theta, \ y = a \sin \theta$
$(0 \leq \theta \leq \pi)$

Figure 13.4.4

By changing the interval over which θ is allowed to vary, we can obtain different portions of the circle. For example,

$$x = a \cos \theta, \quad y = a \sin \theta \quad (0 \leq \theta \leq \pi) \tag{3}$$

represents just the upper half of the circle (Figure 13.4.4).   ◀

☐ **ORIENTATION OF A PARAMETRIC CURVE**

A curve that is represented parametrically is traced in a certain direction as the parameter increases. This is called the *direction of increasing parameter* or sometimes the *orientation* of the curve. For example, the circle and semicircle given parametrically by (2) and (3) are traced counterclockwise as θ increases, so this is the direction of increasing parameter. Phrased another way, the circle is oriented counterclockwise. We have indicated this in Figures 13.4.3 and 13.4.4 with arrows.

Example 3 Let m denote the slope of the tangent line to the parabola $y = x^2$ at an arbitrary point (x, y). Find parametric equations for the parabola in terms of the parameter m.

Solution. Our objective is to express both x and y in terms of m. But

$$m = \frac{dy}{dx} = 2x$$

so

$$x = \frac{1}{2}\frac{dy}{dx} = \frac{1}{2}m \quad \text{and} \quad y = x^2 = \left(\frac{1}{2}m\right)^2 = \frac{1}{4}m^2$$

which leads to the parametric equations

$$x = \frac{1}{2}m, \quad y = \frac{1}{4}m^2$$

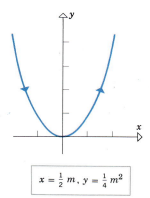

$x = \frac{1}{2}m, \ y = \frac{1}{4}m^2$

Figure 13.4.5

The direction of increasing parameter, indicated in Figure 13.4.5, is obtained by noting that x increases as the slope m increases, since $x = \frac{1}{2}m$. ◀

☐ **ELIMINATING THE PARAMETER**

Sometimes a curve given parametrically can be recognized by eliminating the parameter. The following example illustrates this.

Example 4 Sketch the curve

$$x = 2t - 3, \quad y = 6t - 7$$

Solution. Solving the first equation for t, we obtain

$$t = \frac{x + 3}{2}$$

Substituting this into the second equation yields

$$y = 6\left(\frac{x + 3}{2}\right) - 7 \quad \text{or} \quad y = 3x + 2$$

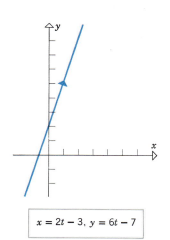

$x = 2t - 3, \ y = 6t - 7$

Figure 13.4.6

which is the straight line shown in Figure 13.4.6. The direction of increasing parameter, shown in the figure, was obtained from the original parametric equations by observing that x and y both increase as t increases. ◀

REMARK. As the foregoing example illustrates, the direction of increasing parameter is lost when the parameter is eliminated; it must be recovered from the original parametric equations.

Example 5 Sketch the curve

$$x = a \cos t, \quad y = b \sin t \quad (0 \le t \le 2\pi)$$

where $0 < b < a$.

Solution. We can eliminate t by writing

$$\frac{x}{a} = \cos t \quad \text{and} \quad \frac{y}{b} = \sin t$$

from which it follows that

$$\frac{x^2}{a^2} + \frac{y^2}{b^2} = 1$$

This is the equation of the ellipse shown in Figure 13.4.7. We leave it for the reader to show that the ellipse has counterclockwise orientation. ◄

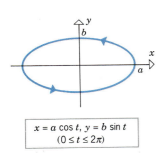

$$x = a \cos t, \ y = b \sin t$$
$$(0 \le t \le 2\pi)$$

Figure 13.4.7

If the parameter is eliminated from a pair of parametric equations, the graph of the resulting equation in x and y may possibly extend beyond the graph of the original parametric equations. For example, consider the curve with parametric equations

$$x = \cosh t, \quad y = \sinh t \tag{4}$$

We may eliminate the parameter by using the identity

$$\cosh^2 t - \sinh^2 t = 1$$

[Formula (1) of Section 7.6] to obtain

$$x^2 - y^2 = 1 \tag{5}$$

which is the equation of the hyperbola shown in Figure 13.4.8. However, the original parametric equations represent only the right branch of this hyperbola since $x = \cosh t$ is never less than 1. To exclude the unwanted left branch we would have to replace (5) by

$$x^2 - y^2 = 1, \quad x \ge 1$$

or more simply,

$$x = \sqrt{1 + y^2}$$

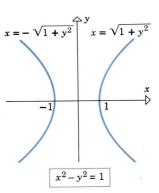

$$x^2 - y^2 = 1$$

Figure 13.4.8

To summarize, whenever the parameter is eliminated from parametric equations, it is important to check that the graph of the resulting equation does not include too much. If it does, added conditions must be imposed.

REPRESENTING GRAPHS OF FUNCTIONS PARAMETRICALLY

Although parametric equations are of special importance for curves that are not graphs of functions, curves of the form $y = f(x)$ or $x = g(y)$ may also be represented parametrically. This can always be done by introducing a parameter t that is equal to the independent variable.

Example 6   The curves $y = 4x^2 - 1$ and $x = 5y^3 - y$ can be represented parametrically as

$$x = t, \quad y = 4t^2 - 1 \quad \text{and} \quad x = 5t^3 - t, \quad y = t$$

respectively.   ◄

REMARK.   A given curve can always be represented parametrically in infinitely many different ways. For example, a parametrization of $y = 4x^2 - 1$ different from that in Example 6 can be obtained by making the substitution $t = s + 1$. This yields new parametric equations

$$x = s + 1, \quad y = 4(s + 1)^2 - 1 = 4s^2 + 8s + 3$$

Different substitutions would yield different parametric equations for the curve.

Example 7   If $r = f(\theta)$ is a polar curve, we can obtain parametric equations for the curve in terms of θ by substituting $r = f(\theta)$ in the relationships

$$x = r \cos \theta$$
$$y = r \sin \theta$$

This yields the parametric equations

$$x = f(\theta) \cos \theta, \quad y = f(\theta) \sin \theta$$

For example, the cardioid $r = 1 - \cos \theta$ can be represented by the parametric equations

$$x = (1 - \cos \theta) \cos \theta, \quad y = (1 - \cos \theta) \sin \theta$$

If we impose the added restriction, $0 \le \theta \le 2\pi$, then the point (x, y) will traverse the cardioid exactly once as θ varies over $[0, 2\pi]$ (Figure 13.4.9).   ◄

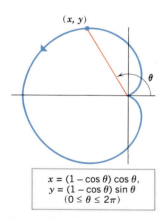

(x, y)

θ

$x = (1 - \cos \theta) \cos \theta,$
$y = (1 - \cos \theta) \sin \theta$
$(0 \le \theta \le 2\pi)$

Figure 13.4.9

TANGENT LINES TO CURVES DEFINED PARAMETRICALLY

Sometimes it is difficult to eliminate the parameter from a pair of parametric equations and express y directly in terms of x. Nevertheless, by using the chain rule, it is possible to find the derivative of y with respect to x without eliminating the parameter. We start with the relationship

$$\frac{dy}{dt} = \frac{dy}{dx}\frac{dx}{dt}$$

If $dx/dt \ne 0$, this can be rewritten as

$$\frac{dy}{dx} = \frac{dy/dt}{dx/dt} \tag{6}$$

Example 8   Given the curve

$$x = t^2, \quad y = t^3$$

find dy/dx and d^2y/dx^2 at the point $(1, 1)$ without eliminating the parameter t.

Solution.

$$\frac{dy}{dx} = \frac{dy/dt}{dx/dt} = \frac{3t^2}{2t} = \frac{3}{2}t \tag{7}$$

To find the second derivative, denote dy/dx by y' and use (6) to write

$$\frac{d^2y}{dx^2} = \frac{dy'}{dx} = \frac{dy'/dt}{dx/dt} = \frac{3/2}{2t} = \frac{3}{4t} \tag{8}$$

Note that these calculations express dy/dx and d^2y/dx^2 in terms of the parameter t. Since the point $(1, 1)$ on the curve corresponds to $t = 1$, it follows from (7) and (8) that

$$\left.\frac{dy}{dx}\right|_{t=1} = \frac{3}{2}(1) = \frac{3}{2} \quad \text{and} \quad \left.\frac{d^2y}{dx^2}\right|_{t=1} = \frac{3}{4}$$

The curve and the tangent at $(1, 1)$ are shown in Figure 13.4.10. Note that the tangent line has slope $\frac{3}{2}$ at $(1, 1)$ and that the curve is concave up at $(1, 1)$, in keeping with our calculations.   ◀

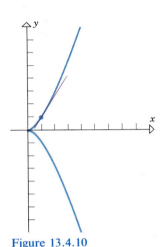

Figure 13.4.10

REMARK.   Formula (6) does not apply if $dx/dt = 0$. However, at points where $dx/dt = 0$ and $dy/dt \neq 0$, there is generally a **vertical tangent** line. Points (x, y) where both $dx/dt = 0$ and $dy/dt = 0$ are called **singular points**. The behavior of a curve at a singular point must be determined on a case-by-case basis.

☐ **ARC LENGTH OF**
CURVES DEFINED
PARAMETRICALLY

In Section 6.4 we obtained the following formula for the arc length L of a curve $y = f(x)$ from $x = a$ to $x = b$:

$$L = \int_a^b \sqrt{1 + [f'(x)]^2} \, dx = \int_a^b \sqrt{1 + \left(\frac{dy}{dx}\right)^2} \, dx \tag{9}$$

This formula is a special case of the following general result about arc length of parametric curves:

13.4.1   **THEOREM** (*Arc Length of Parametric Curves*).   *If $x'(t)$ and $y'(t)$ are continuous functions for $a \leq t \leq b$, then the parametric curve*

$$x = x(t), \quad y = y(t), \quad a \leq t \leq b$$

has arc length L given by

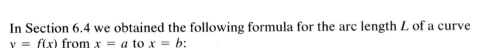

$$L = \int_a^b \sqrt{[x'(t)]^2 + [y'(t)]^2} \, dt = \int_a^b \sqrt{\left(\frac{dx}{dt}\right)^2 + \left(\frac{dy}{dt}\right)^2} \, dt \tag{10}$$

The derivation of these formulas is similar to the derivation of (9) and will be omitted. Observe that (9) follows from (10) by expressing the curve $y = f(x)$ in the parametric form

$$x = t, \quad y = f(t)$$

in which case

$$\frac{dx}{dt} = 1 \quad \text{and} \quad \frac{dy}{dt} = f'(t) = f'(x) = \frac{dy}{dx}$$

Example 9 Use (10) to find the arc length of the circle

$$x = a \cos t, \quad y = a \sin t \qquad (0 \le t \le 2\pi)$$

where $a > 0$.

Solution.

$$L = \int_0^{2\pi} \sqrt{\left(\frac{dx}{dt}\right)^2 + \left(\frac{dy}{dt}\right)^2} \, dt$$

$$= \int_0^{2\pi} \sqrt{(-a \sin t)^2 + (a \cos t)^2} \, dt$$

$$= \int_0^{2\pi} a \, dt = at \Big]_0^{2\pi} = 2\pi a \qquad \blacktriangleleft$$

■ **OPTIONAL**

If a wheel rolls along a straight line without slipping, then a point on the rim of the wheel traces a curve called a *cycloid* (Figure 13.4.11). The cycloid is of special interest because it provides the solution of two famous mathematical problems: the *brachistochrone problem* (from Greek words meaning "shortest time") and the *tautochrone problem* (from Greek words meaning "equal time").

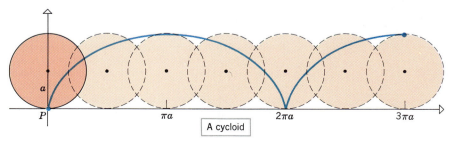

A cycloid

Figure 13.4.11

Figure 13.4.12

In June of 1696, Johann Bernoulli* posed the brachistochrone problem in the form of a challenge to other mathematicians. The problem was to determine the shape of a wire down which a bead might slide from a point P to another point Q, not directly below, in the *shortest time*. At first, one might conjecture that the wire should form a straight line, since that shape yields the shortest distance from P to Q. However, the correct answer is half of one arch of an inverted cycloid (Figure 13.4.12). In essence, this shape allows the bead to fall rapidly at first, building up sufficient initial speed to reach Q in the shortest time, even though the path does not provide the shortest distance from P to Q. This problem was solved by Newton, Leibniz, L'Hôpital, Johann Bernoulli, and his older brother Jakob Bernoulli. It was formulated and incorrectly solved years earlier by Galileo, who gave the arc of a circle as the answer.

In the tautochrone problem the object is to find the shape of a wire from P to Q such that two beads started at *any* points on the wire between P and Q reach Q in the same amount of time. Again, the answer is half of one arch of a cycloid.

We conclude this section by deriving parametric equations for a cycloid.

---

*BERNOULLI. An amazing Swiss family that included several generations of outstanding mathematicians and scientists. Nikolaus Bernoulli (1623–1708), a druggist, fled from Antwerp to escape religious persecution and ultimately settled in Basel, Switzerland. There he had three sons, Jakob I (also called Jacques or James), Nikolaus, and Johann I (also called Jean or John). The Roman numerals are used to distinguish family members with identical names.

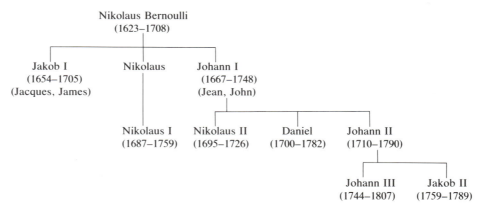

Following Newton and Leibniz, the Bernoulli brothers, Jakob I and Johann I, are considered by some to be the two most important founders of calculus. Jakob I was self-taught in mathematics. His father wanted him to study for the ministry, but he turned to mathematics and in 1686 became a professor at the University of Basel. When he started working in mathematics, he knew nothing of Newton's and Leibniz's work. He eventually became familiar with Newton's results, but because so little of Leibniz's work was published, Jakob duplicated many of Leibniz's results.

Jakob's younger brother Johann I was urged to enter into business by his father. Instead, he turned to medicine and studied mathematics under the guidance of his older brother. He eventually became a mathematics professor at Gröningen in Holland, and then, when Jakob died in 1705, Johann succeeded him as mathematics professor at Basel. Throughout their lives, Jakob I and Johann I had a mutual passion for criticizing each other's work, which frequently erupted into ugly confrontations. Leibniz tried to mediate the disputes, but Jakob, who resented Leibniz's superior intellect, accused him of siding with Johann, and thus Leibniz became entangled in the arguments. The brothers often worked on common problems that they posed as challenges to one

(*continued on next page*)

Example 10   Let P be a fixed point on the rim of a wheel of radius a. Suppose that the wheel rolls along the positive x-axis of a rectangular coordinate system and that P is initially at the origin. When the wheel rolls a given distance, the radial line from the center C to the point P rotates through an angle ϕ (Figure 13.4.13), which we shall use as our parameter.

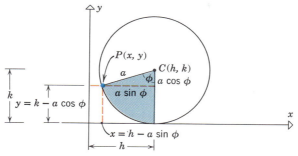

Figure 13.4.13

(It is customary to regard ϕ as a positive angle, even though it is generated by a clockwise rotation.) Our objective is to express the coordinates of $P(x, y)$ in terms of the angle ϕ. Figure 13.4.13 suggests that the coordinates of $P(x, y)$ and the coordinates of the wheel's center $C(h, k)$ are related by

$$x = h - a \sin \phi \qquad\qquad (11)$$
$$y = k - a \cos \phi$$

Since the height of the wheel's center is the radius of the wheel, we have $k = a$; and since the distance h moved by the center is the same as the circular arc length subtended by ϕ (why?), we have $h = a\phi$. Therefore, (11) yields

$$x = a\phi - a \sin \phi, \quad y = a - a \cos \phi \qquad (0 \le \phi < +\infty)$$

These are the parametric equations of the cycloid.   ◀

(continued)

another. Johann, interested in gaining fame, often used unscrupulous means to make himself appear the originator of his brother's results; Jakob occasionally retaliated. Thus, it is often difficult to determine who deserves credit for many results. However, both men made major contributions to the development of calculus. In addition to his work on calculus, Jakob helped establish fundamental principles in probability, including the Law of Large Numbers, which is a cornerstone of modern probability theory. Johann was Euler's teacher at Basel and L'Hôpital's tutor in France.

Among the other members of the Bernoulli family, Daniel, son of Johann I, is the most famous. He was a professor of mathematics at St. Petersburg Academy in Russia and subsequently a professor of anatomy and then physics at Basel. He did work in calculus and probability, but is best known for his work in physics. A basic law of fluid flow, called Bernoulli's principle, is named in his honor. He won the annual prize of the French Academy 10 times for work on vibrating strings, tides of the sea, and kinetic theory of gases.

Johann II succeeded his father as professor of mathematics at Basel. His research was on the theory of heat and sound. Nikolaus I was a mathematician and law scholar who worked on probability and series. On the recommendation of Leibniz, he was appointed professor of mathematics at Padua and then went to Basel as a professor of logic and then law. Nikolaus II was professor of jurisprudence in Switzerland and then professor of mathematics at St. Petersburg Academy. Johann III was a professor of mathematics and astronomy in Berlin and Jakob II succeeded his uncle Daniel as professor of mathematics at St. Petersburg Academy in Russia. Truly an incredible family!

▶ **Exercise Set 13.4**   C 49

1. (a) Sketch the curve $x = t$, $y = t^2$ by plotting the points corresponding to $t = -3, -2, -1, 0, 1, 2, 3$.
    (b) Show that the curve is a parabola by eliminating the parameter t.

2. Sketch the curve $x = t^3 - 4t$, $y = 2t$, $-3 \le t \le 3$ by plotting some appropriate points.

In Exercises 3–15, sketch the curve by eliminating the parameter t, and indicate the direction of increasing t.

3. $x = \cos t$, $y = \sin t$, $0 \le t \le 2\pi$.

4. $x = 1 + \cos t$, $y = 3 - \sin t$, $0 \le t \le 2\pi$.

5. $x = 3t - 4$, $y = 6t + 2$.

6. $x = t - 3$, $y = 3t - 7$, $0 \le t \le 3$.

7. $x = 2 \cos t$, $y = 5 \sin t$, $0 \le t \le 2\pi$.

8. $x = \sqrt{t}$, $y = 2t + 4$.

9. $x = 3 + 2 \cos t$, $y = 2 + 4 \sin t$, $0 \le t \le 2\pi$.

10. $x = 2 \cosh t$, $y = 4 \sinh t$.

11. $x = 4 \sin 2\pi t$, $y = 4 \cos 2\pi t$, $0 \le t \le 1$.

12. $x = \sec t$, $y = \tan t$, $\pi \le t < 3\pi/2$.

13. $x = \cos 2t$, $y = \sin t$, $-\pi/2 \le t \le \pi/2$.

14. $x = 4t + 3$, $y = 16t^2 - 9$.

15. $x = t^2$, $y = 2 \ln t$, $t \ge 1$.

In Exercises 16–25, sketch the curve.

16. $x = 3e^{-t} - 2$, $y = 4e^{-t} - 1$, $t \ge 0$.

17. $x = 3e^{-2t} - 1$, $y = e^{-t}$, $t \ge 0$.

18. $x = \sec^2 t$, $y = \tan^2 t$.

19. $x = 2 \sin^2 t$, $y = 3 \cos^2 t$.

20. $x = \sec t$, $y = \tan t$, $-\pi/2 < t < \pi/2$.

21. $x = \cos t$, $y = \sin^2 t$.

22. $x = \cos 2t$, $y = 2 \cos t$.

23. $x = \sin^2 t$, $y = 1 + \cos t$.

24. $x = \cos 3t$, $y = 2 \sin^2 3t - 1$.

25. $x = \cos(e^{-t})$, $y = \sin(e^{-t})$, $t \ge 0$.

26. Find parametric equations of the upper semicircle $y = \sqrt{a^2 - x^2}$ $(a > 0)$ in terms of the parameter t, where t is the distance between $(-a, 0)$ and any point on the semicircle.

27. Find parametric equations of the curve $y = \sqrt{x}$ in terms of the parameter t, using for t the x-coordinate of the point where the tangent line to the curve crosses the x-axis.

In Exercises 28–34, find dy/dx at the point corresponding to the given value of the parameter without eliminating the parameter.

28. $x = t^2 + 4$, $y = 8t$; $t = 2$.

29. $x = \cos t$, $y = \sin t$; $t = 3\pi/4$.

30. $x = t + 5$, $y = 5t - 7$; $t = 1$.

31. $x = \sqrt{t}$, $y = 2t + 4$; $t = 9$.

32. $x = \sec \theta$, $y = \tan \theta$; $\theta = \pi/3$.

33. $x = 4 \cos 2\pi s$, $y = 3 \sin 2\pi s$; $s = -1/4$.

34. $x = \sinh t$, $y = \cosh t$; $t = 0$.

In Exercises 35–38, find d^2y/dx^2 at the point corresponding to the given value of the parameter without eliminating the parameter.

35. $x = \frac{1}{2}t^2$, $y = \frac{1}{3}t^3$; $t = 2$.

36. $x = \cos \phi$, $y = \sin \phi$; $\phi = \pi/4$.

37. $x = \sqrt{t}$, $y = 2t + 4$; $t = 1$.

38. $x = \sec t$, $y = \tan t$; $t = \pi/3$.

In Exercises 39–47, find the arc length of the curve.

39. $x = 4t + 3$, $y = 3t - 2$, $0 \le t \le 2$.

40. $x = \cos^3 t$, $y = \sin^3 t$, $0 \le t \le \pi/2$.

41. $x = \frac{1}{3}t^3$, $y = \frac{1}{2}t^2$, $0 \le t \le 1$.

42. $x = \frac{1}{3}t^3$, $y = \frac{1}{2}t^2$, $-1 \le t \le 0$.

43. $x = \cos 2t$, $y = \sin 2t$, $0 \le t \le \pi/2$.

44. $x = e^t (\sin t + \cos t)$, $y = e^t (\cos t - \sin t)$, $1 \le t \le 4$.

45. $x = (1 + t)^2$, $y = (1 + t)^3$, $0 \le t \le 1$.

46. $x = e^t \cos t$, $y = e^t \sin t$, $0 \le t \le \pi/2$.

47. One arch of the cycloid
$$x = a(t - \sin t), \quad y = a(1 - \cos t) \qquad (a > 0).$$

48. Show that the total arc length of the ellipse $x = a \cos t$, $y = b \sin t$, $0 \le t \le 2\pi$ for $a > b > 0$ is given by

$$4a \int_0^{\pi/2} \sqrt{1 - e^2 \cos^2 t} \; dt$$

where $e = \sqrt{a^2 - b^2}/a$.

49. (a) Show that the total arc length of the ellipse $x = 2\cos t$, $y = \sin t$, $0 \le t \le 2\pi$ is given by

$$4\int_0^{\pi/2} \sqrt{1 + 3\sin^2 t}\ dt$$

(b) Use Simpson's rule with $n = 10$ subdivisions to approximate the total arc length of the ellipse in part (a). Round your answer to two decimal places.

(c) Suppose that the parametric equations in part (a) describe the path of a particle moving in the xy-plane, where t is time in seconds and x and y are in centimeters. Use Simpson's rule with $n = 10$ subdivisions to approximate the distance traveled from $t = 1.5$ sec to $t = 4.8$ sec. Round your answer to two decimal places.

50. Find parametric equations for the rose $r = 2\cos 2\theta$ using θ as the parameter.

51. Find parametric equations for the limaçon $r = 2 + 3\sin\theta$ using θ as the parameter.

52. Find the equation of the tangent line to the curve $x = 2t + 4$, $y = 8t^2 - 2t + 4$ at the point where $t = 1$.

53. Find the equation of the tangent line to the curve $x = e^t$, $y = e^{-t}$ at the point where $t = 2$.

54. Find all values of t at which the curve $x = 2\cos t$, $y = 4\sin t$ has a tangent that is
(a) horizontal   (b) vertical.

55. Find all values of the parameter t at which the curve $x = 2t^3 - 15t^2 + 24t + 7$, $y = t^2 + t + 1$ has a tangent line that is
(a) horizontal   (b) vertical.

56. Show that the curve $x = t^3 - 4t$, $y = t^2$ intersects itself at the point $(0, 4)$, and find equations for two tangent lines to the curve at the point of intersection.

57. Show that the curve with parametric equations $x = t^2 - 3t + 5$, $y = t^3 + t^2 - 10t + 9$ intersects itself at the point $(3, 1)$, and find equations for two tangent lines to the curve at the point of intersection.

58. A point traces the circle $x^2 + y^2 = 25$ so that $dx/dt = 8$ when the point reaches $(4, 3)$. Find dy/dt there.

59. Describe the curve whose parametric equations are $x = a\cos t + h$, $y = b\sin t + k$, $0 \le t \le 2\pi$.

60. A *hypocycloid* is a curve traced by a point P on the circumference of a circle that rolls inside a larger fixed circle. Suppose that the fixed circle has radius a, the rolling circle has radius b, and the fixed circle is centered at the origin. Let ϕ be the angle shown

in the following figure, and assume that the point P is at $(a, 0)$ when $\phi = 0$. Show that the hypocycloid generated is given by the parametric equations

$$x = (a - b)\cos\phi + b\cos\left(\frac{a - b}{b}\phi\right)$$

$$y = (a - b)\sin\phi - b\sin\left(\frac{a - b}{b}\phi\right)$$

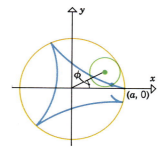

61. If $b = \tfrac{1}{4}a$ in Exercise 60, then the resulting curve is called a four-cusped hypocycloid.
(a) Sketch this curve.
(b) Show that the curve is given by the parametric equations $x = a\cos^3\phi$, $y = a\sin^3\phi$.
(c) Show that the curve is given by the equation $x^{2/3} + y^{2/3} = a^{2/3}$ in rectangular coordinates.

Exercises 62–67 require the formulas developed in the following discussion: If $x'(t)$ and $y'(t)$ are continuous functions for $a \le t \le b$, then it can be shown that the area of the surface generated by revolving the curve

$$x = x(t), \quad y = y(t) \qquad (a \le t \le b)$$

about the x-axis is

$$S = \int_a^b 2\pi y(t)\sqrt{[x'(t)]^2 + [y'(t)]^2}\ dt$$

and the area of the surface generated by revolving the curve about the y-axis is

$$S = \int_a^b 2\pi x(t)\sqrt{[x'(t)]^2 + [y'(t)]^2}\ dt$$

[The derivations are similar to those used to obtain Formulas (6) and (7) in Section 6.5.]

62. Find the area of the surface generated by revolving $x = t^2$, $y = 2t$, $0 \le t \le 4$ about the x-axis.

63. Find the area of the surface generated by revolving $x = e^t\cos t$, $y = e^t\sin t$, $0 \le t \le \pi/2$ about the x-axis.

64. Find the area of the surface generated by revolving $x = \cos^2 t$, $y = \sin^2 t$, $0 \le t \le \pi/2$ about the y-axis.

65. Find the area of the surface generated by revolving $x = t$, $y = 2t^2$, $0 \le t \le 1$ about the y-axis.

66. By revolving the semicircle $x = r \cos t$, $y = r \sin t$, $0 \le t \le \pi$ about the x-axis, show that the surface area of a sphere of radius r is $4\pi r^2$.

67. The equations

$$x = a\phi - a \sin \phi, \quad y = a - a \cos \phi \quad (0 \le \phi \le 2\pi)$$

represent one arch of a cycloid. Show that the surface area generated by revolving this curve about the x-axis is $S = 64\pi a^2/3$.

■ **13.5 TANGENT LINES AND ARC LENGTH IN POLAR COORDINATES**

In this section we shall use our results on parametric equations to derive formulas for arc length and slopes of tangent lines to polar curves.

□ **TANGENT LINES**

Let $r = f(\theta)$ be a curve in polar coordinates, for which f is a differentiable function of θ. We can obtain parametric equations for this curve in terms of θ by substituting $r = f(\theta)$ in the relationships

$$x = r \cos \theta$$
$$y = r \sin \theta$$

This yields the parametric equations

$$x = f(\theta) \cos \theta, \quad y = f(\theta) \sin \theta \tag{1}$$

If we assume the existence of a tangent line to the curve $r = f(\theta)$ at the point $P(r, \theta)$, then the slope of this tangent line is

$$m = \tan \phi = \frac{dy}{dx}$$

where ϕ is the angle of inclination. From (1) we have

$$\frac{dx}{d\theta} = -f(\theta) \sin \theta + f'(\theta) \cos \theta \tag{2}$$

$$\frac{dy}{d\theta} = f(\theta) \cos \theta + f'(\theta) \sin \theta$$

Thus, if $dx/d\theta \ne 0$, it follows from Equation (6) of Section 13.4 that

$$m = \tan \phi = \frac{dy}{dx} = \frac{dy/d\theta}{dx/d\theta} \tag{3}$$

or

$$m = \tan \phi = \frac{f(\theta) \cos \theta + f'(\theta) \sin \theta}{-f(\theta) \sin \theta + f'(\theta) \cos \theta} = \frac{r \cos \theta + \sin \theta \dfrac{dr}{d\theta}}{-r \sin \theta + \cos \theta \dfrac{dr}{d\theta}} \tag{4}$$

which is a formula for the slope of the tangent line to $r = f(\theta)$ at $P(r, \theta)$.

Formula (4) was derived by assuming that $dx/d\theta \neq 0$. If $dx/d\theta = 0$ and $dy/d\theta \neq 0$, we will agree that the curve has a **vertical tangent**. This is reasonable since $\tan \phi$ becomes infinite in this case [see (3)], indicating that $\phi = \pi/2$. Points where both $dx/d\theta = 0$ and $dy/d\theta = 0$ are called **singular points**. No general statement can be made about the behavior of a polar curve at a singular point. Each case requires its own analysis.

If $\cos \theta \neq 0$, we may divide the numerator and denominator of (4) by $\cos \theta$ to obtain the following alternative formula for the slope of a tangent line.

$$m = \tan \phi = \frac{r + \tan \theta \dfrac{dr}{d\theta}}{-r \tan \theta + \dfrac{dr}{d\theta}} \tag{5}$$

Example 1 Find the slope of the tangent line to the circle

$$r = 4 \cos \theta \tag{6}$$

at the point where $\theta = \pi/4$.

Solution. Differentiating (6) yields

$$\frac{dr}{d\theta} = -4 \sin \theta \tag{7}$$

If $\theta = \pi/4$, it follows from (6) and (7) that

$$r = 4 \cos \frac{\pi}{4} = 2\sqrt{2} \quad \text{and} \quad \frac{dr}{d\theta} = -4 \sin \frac{\pi}{4} = -2\sqrt{2}$$

Substituting these values and $\tan \theta = \tan \dfrac{\pi}{4} = 1$ in (5) yields

$$\tan \phi = \frac{2\sqrt{2} + (1)(-2\sqrt{2})}{-(2\sqrt{2})(1) + (-2\sqrt{2})} = 0$$

Thus, the circle has a horizontal tangent line when $\theta = \pi/4$ (Figure 13.5.1). ◀

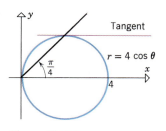

Figure 13.5.1

Example 2 At what points does the cardioid $r = 1 - \cos \theta$ have a vertical tangent line?

Solution.   A vertical tangent line will occur at points where $dx/d\theta = 0$ and $dy/d\theta \neq 0$. The cardioid is given parametrically by the equations

$$x = r \cos \theta = (1 - \cos \theta) \cos \theta$$
$$y = r \sin \theta = (1 - \cos \theta) \sin \theta \qquad (0 \leq \theta \leq 2\pi)$$

Differentiating and then simplifying, we obtain (verify)

$$\frac{dx}{d\theta} = \sin \theta (2 \cos \theta - 1) \tag{8}$$

$$\frac{dy}{d\theta} = (1 - \cos \theta)(1 + 2 \cos \theta) \tag{9}$$

From (8), $dx/d\theta = 0$ if $\sin \theta = 0$ or $\cos \theta = \frac{1}{2}$; this occurs where $\theta = 0$, π, $\pi/3$, $5\pi/3$, or 2π. But from (9), $dy/d\theta \neq 0$ if $\theta = \pi$, $\pi/3$, and $5\pi/3$, so vertical tangents occur at these points. Since $dy/d\theta = 0$ if $\theta = 0$ or $\theta = 2\pi$, these are singular points. Although we shall not prove it, the cardioid has a horizontal tangent at the origin (Figure 13.5.2). ◀

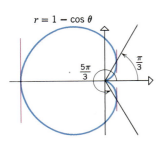

$r = 1 - \cos \theta$

Figure 13.5.2

Another way to find the tangent line to a polar curve at a point $P(r, \theta)$ is to find the angle ψ between the **radial line** OP and the tangent line (Figure 13.5.3).

By definition, ψ is measured counterclockwise from the radial line OP to the tangent line, and is selected so that

$$0 \leq \psi < \pi$$

If we choose the polar angle θ in the range $0 \leq \theta < 2\pi$, there is a simple algebraic relationship between θ, ϕ, and ψ that depends on the relative sizes of ϕ and θ. If $\phi \geq \theta$, then

$$\psi = \phi - \theta \tag{10}$$

(Figure 13.5.3*a*), and if $\phi < \theta$, then the relationship is

$$\psi = \phi + (\pi - \theta) = (\phi - \theta) + \pi \tag{11}$$

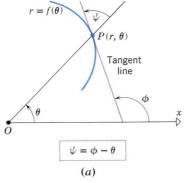

$r = f(\theta)$

$P(r, \theta)$

Tangent line

$\psi = \phi - \theta$

(*a*)

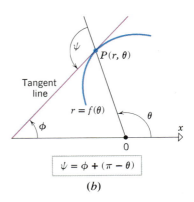

$P(r, \theta)$

Tangent line

$r = f(\theta)$

$\psi = \phi + (\pi - \theta)$

(*b*)

Figure 13.5.3

(Figure 13.5.3*b*). In either case, we may write

$$\tan \psi = \tan (\phi - \theta) \tag{12}$$

from which it follows that

$$\tan \psi = \frac{\tan \phi - \tan \theta}{1 + \tan \phi \tan \theta} \tag{13}$$

Substituting (5) into (13) yields

$$\tan \psi = \frac{\left[\left(r + \tan \theta \dfrac{dr}{d\theta}\right) \Big/ \left(-r \tan \theta + \dfrac{dr}{d\theta}\right)\right] - \tan \theta}{1 + \left[\left(r + \tan \theta \dfrac{dr}{d\theta}\right) \Big/ \left(-r \tan \theta + \dfrac{dr}{d\theta}\right)\right] \tan \theta}$$

$$= \frac{r + \tan \theta \dfrac{dr}{d\theta} + r \tan^2 \theta - \tan \theta \dfrac{dr}{d\theta}}{-r \tan \theta + \dfrac{dr}{d\theta} + r \tan \theta + \tan^2 \theta \dfrac{dr}{d\theta}}$$

$$= \frac{r(1 + \tan^2 \theta)}{(1 + \tan^2 \theta) \dfrac{dr}{d\theta}}$$

or

$$\tan \psi = \frac{r}{dr/d\theta} \tag{14}$$

Example 3   For the cardioid $r = 1 - \cos \theta$, find the angle ψ between the tangent line and radial line at the points where the cardioid crosses the y-axis.

Solution.   For the given cardioid we have

$$\tan \psi = \frac{r}{dr/d\theta} = \frac{1 - \cos \theta}{\sin \theta}$$

so that from the trigonometric identity

$$\tan \frac{\theta}{2} = \frac{1 - \cos \theta}{\sin \theta}$$

(derive) it follows that

$$\tan \psi = \tan \frac{\theta}{2}$$

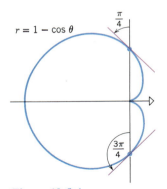

$r = 1 - \cos\theta$

Figure 13.5.4

If $0 \le \theta < 2\pi$, this implies that

$$\psi = \frac{\theta}{2}$$

since $0 \le \psi < \pi$. In particular, where the cardioid crosses the y-axis we have $\theta = \pi/2$ and $\theta = 3\pi/2$, so the angles from the radial lines to the tangent lines at these points are $\psi = \pi/4$ and $\psi = 3\pi/4$, respectively (Figure 13.5.4).   ◄

In Section 13.2 we remarked informally that if a polar curve $r = f(\theta)$ passes through the origin where $\theta = \theta_0$, then the line $\theta = \theta_0$ is tangent to the curve at the origin. To see that this is so, we need only observe that in order for the curve $r = f(\theta)$ to pass through the origin when $\theta = \theta_0$, we must have $r = 0$ for this value of θ. Substituting $r = 0$ in (14) yields

$$\tan\psi = 0$$

(provided $dr/d\theta \ne 0$), which tells us that the angle ψ between the radial line $\theta = \theta_0$ and the tangent line is zero. Thus, the tangent line coincides with the line $\theta = \theta_0$.

□ ARC LENGTH OF A POLAR CURVE

The arc length formula for a parametric curve can be used to obtain a formula for the arc length of a polar curve.

**13.5.1   THEOREM (*Arc Length of a Polar Curve*).   *If a curve has the polar equation $r = f(\theta)$, where $f'(\theta)$ is continuous for $\alpha \le \theta \le \beta$, then its arc length L from $\theta = \alpha$ to $\theta = \beta$ is*

$$L = \int_{\alpha}^{\beta} \sqrt{[f(\theta)]^2 + [f'(\theta)]^2}\, d\theta \qquad (15)$$

or equivalently

$$L = \int_{\alpha}^{\beta} \sqrt{r^2 + \left(\frac{dr}{d\theta}\right)^2}\, d\theta \qquad (16)$$

Proof.   It follows from (2) that (verify)

$$\left(\frac{dx}{d\theta}\right)^2 + \left(\frac{dy}{d\theta}\right)^2 = [f(\theta)]^2 + [f'(\theta)]^2$$

Thus, from (10) in Theorem 13.4.1 with θ in place of t, we obtain (15).   ∎

Example 4   Find the arc length of the spiral $r = e^{\theta}$ between $\theta = 0$ and $\theta = 1$.

Solution.

$$L = \int_\alpha^\beta \sqrt{r^2 + \left(\frac{dr}{d\theta}\right)^2} \, d\theta = \int_0^1 \sqrt{(e^\theta)^2 + (e^\theta)^2} \, d\theta$$

$$= \int_0^1 \sqrt{2} \, e^\theta \, d\theta = \sqrt{2} e^\theta \Big]_0^1 = \sqrt{2}(e - 1) \quad \blacktriangleleft$$

Example 5   Find the total arc length of the cardioid $r = 1 + \cos \theta$.

Solution.   The cardioid is traced out once as θ varies from $\theta = 0$ to $\theta = 2\pi$. Thus,

$$L = \int_\alpha^\beta \sqrt{r^2 + \left(\frac{dr}{d\theta}\right)^2} \, d\theta = \int_0^{2\pi} \sqrt{(1 + \cos \theta)^2 + (-\sin \theta)^2} \, d\theta$$

$$= \sqrt{2} \int_0^{2\pi} \sqrt{1 + \cos \theta} \, d\theta$$

From the identity $1 + \cos \theta = 2 \cos^2 \tfrac{1}{2}\theta$, we obtain

$$L = 2 \int_0^{2\pi} \sqrt{\cos^2 \tfrac{1}{2}\theta} \, d\theta = 2 \int_0^{2\pi} |\cos \tfrac{1}{2}\theta| \, d\theta \qquad (17)$$

Since

$$\cos \tfrac{1}{2}\theta \geq 0 \quad \text{when} \quad 0 \leq \theta \leq \pi$$

and

$$\cos \tfrac{1}{2}\theta \leq 0 \quad \text{when} \quad \pi \leq \theta \leq 2\pi$$

we may rewrite (17) as

$$L = 2 \left[\int_0^\pi \cos \tfrac{1}{2}\theta \, d\theta - \int_\pi^{2\pi} \cos \tfrac{1}{2}\theta \, d\theta \right]$$

$$= 4 \sin \tfrac{1}{2}\theta \Big]_0^\pi - 4 \sin \tfrac{1}{2}\theta \Big]_\pi^{2\pi} = 8$$

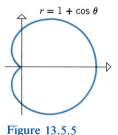

$r = 1 + \cos \theta$

Figure 13.5.5

Alternative Solution.   Since the cardioid is symmetric about the x-axis (Figure 13.5.5), the entire arc length is twice the arc length from $\theta = 0$ to $\theta = \pi$. Thus, proceeding as above, we obtain

$$L = 2\sqrt{2} \int_0^\pi \sqrt{1 + \cos \theta} \, d\theta = 4 \int_0^\pi \cos \tfrac{1}{2}\theta \, d\theta = 8 \quad \blacktriangleleft$$

Exercise Set 13.5 $\boxed{\text{C}}$ 21, 23

In Exercises 1–6, find the slope of the tangent to the curve at the point with the given value of θ.

1. $r = 2 \cos \theta$; $\theta = \pi/3$.

2. $r = 1 + \sin \theta$; $\theta = \pi/4$.

3. $r = 1/\theta$; $\theta = 2$.

4. $r = a \sec 2\theta$; $\theta = \pi/6$.

5. $r = \cos 3\theta$; $\theta = 3\pi/4$.

6. $r = 4 - 3 \sin \theta$; $\theta = \pi$.

In Exercises 7–12, find $\tan \psi$ at the point on the curve with the given value of θ.

7. $r = 3(1 - \cos \theta)$; $\theta = \pi/2$.

8. $r = 2\theta$; $\theta = 1$.

9. $r = 5 \sin \theta$; $\theta = \pi$.

10. $r = \tan \theta$; $\theta = 3\pi/4$.

11. $r = 4 \sin^2 \theta$; $\theta = 5\pi/6$.

12. $r = \sin \theta/(1 + \cos \theta)$; $\theta = \pi/3$.

In Exercises 13–19, find the arc length of the curve.

13. $r = e^{3\theta}$ from $\theta = 0$ to $\theta = 2$.

14. The entire circle $r = a$.

15. The entire circle $r = 2a \cos \theta$.

16. $r = \sin^2(\theta/2)$ from $\theta = 0$ to $\theta = \pi$.

17. $r = a\theta^2$ from $\theta = 0$ to $\theta = \pi$.

18. $r = \sin^3(\theta/3)$ from $\theta = 0$ to $\theta = \pi/2$.

19. The entire cardioid $r = a(1 - \cos \theta)$. [*Hint:* See Example 5.]

20. (a) Find the arc length L of the curve $r = e^{-a\theta}$, $a > 0$, for $0 \le \theta \le \theta_0$.
   (b) Find $\lim_{\theta_0 \to +\infty} L$.

21. (a) Show that the arc length of one petal of the rose $r = \cos n\theta$ is given by

$$2 \int_0^{\pi/(2n)} \sqrt{1 + (n^2 - 1) \sin^2 n\theta} \; d\theta$$

   (b) Use the result in part (a) and Simpson's rule with ten subdivisions to approximate the arc

length of one petal of the four-petal rose $r = \cos 2\theta$. Round your answer to two decimal places.

22. Use Theorem 13.5.1 to show that from $\theta = \alpha$ to $\theta = \beta$ the arc length of $r = 2f(\theta)$ is twice that of $r = f(\theta)$.

23. Suppose that a long thin rod with one end fixed at the pole of a polar coordinate system rotates counterclockwise at the constant rate of 0.5 rad/sec. At time $t = 0$ a bug on the rod is 10 mm from the pole and is moving outward along the rod at the constant speed of 2 mm/sec.
   (a) Find an equation of the form $r = f(\theta)$ for the path of motion of the bug, assuming that $\theta = 0$ when $t = 0$.
   (b) Find the distance the bug travels along the path in part (a) during the first 5 sec. Round your answer to the nearest tenth of a millimeter.

24. Find all points on the cardioid $r = a(1 + \cos \theta)$ where the tangent is
   (a) horizontal   (b) vertical.

25. Find all points on the limaçon $r = 1 - 2 \sin \theta$ where the tangent is horizontal.

26. Prove: If the polar curves $r = f_1(\theta)$ and $r = f_2(\theta)$ intersect at a point P, and if β is the smallest nonnegative angle between the tangent lines at P, then

$$\tan \beta = \left| \frac{\tan \psi_2 - \tan \psi_1}{1 + \tan \psi_1 \tan \psi_2} \right|$$

where ψ_1 and ψ_2 are evaluated at P.

27. Show that the curves $r = \sin 2\theta$ and $r = \cos \theta$ intersect at $(\sqrt{3}/2, \pi/6)$, and use the result of Exercise 26 to find the smallest nonnegative angle between their tangent lines at the point of intersection.

28. Find all points of intersection of the curves $r = \cos \theta$ and $r = 1 - \cos \theta$, and use the result of Exercise 26 to find the smallest nonnegative angle between the tangent lines at each point of intersection.

29. Prove that ψ is the same at each point of the curve $r = e^{a\theta}$.

30. Prove: At all points of intersection of the cardioids $r = a(1 + \cos \theta)$ and $r = b(1 - \cos \theta)$ (excluding the origin), the tangent lines are perpendicular.

▶ SUPPLEMENTARY EXERCISES   Ⓒ *22, 23, 37*

1. Find the rectangular coordinates of the points with the given polar coordinates.
   (a) $(-2, 4\pi/3)$
   (b) $(2, -\pi/2)$
   (c) $(0, -\pi)$
   (d) $(-\sqrt{2}, -\pi/4)$
   (e) $(3, \pi)$
   (f) $(1, \tan^{-1}(-\frac{4}{3}))$.

2. In parts (a)–(c), points are given in rectangular coordinates. Express them in polar coordinates in three ways:
   (i) With $r \geq 0$ and $0 \leq \theta < 2\pi$
   (ii) With $r \geq 0$ and $-\pi < \theta \leq \pi$
   (iii) With $r \leq 0$ and $0 \leq \theta < 2\pi$.
   (a) $(-\sqrt{3}, -1)$ (b) $(-3, 0)$ (c) $(1, -1)$.

3. Sketch the region in polar coordinates determined by the given inequalities.
   (a) $1 \leq r \leq 2$, $\cos \theta \leq 0$
   (b) $-1 \leq r \leq 1$, $\pi/4 \leq \theta \leq \pi/2$.

In Exercises 4–11, identify the curve by transforming to rectangular coordinates.

4. $r = 2/(1 - \cos \theta)$.
5. $r^2 \sin(2\theta) = 1$.
6. $r = \pi/2$.
7. $r = -4 \csc \theta$.
8. $r = 6/(3 - \sin \theta)$.
9. $\theta = \pi/3$.
10. $r = 2 \sin \theta + 3 \cos \theta$.
11. $r = 0$.

In Exercises 12–15, express the given equation in polar coordinates.

12. $x^2 + y^2 = kx$.
13. $x = -3$.
14. $y^2 = 4x$.
15. $y = 3x$.

In Exercises 16–23, sketch the curve in polar coordinates.

16. $r = -4 \sin 3\theta$.
17. $r = -1 - 2 \cos \theta$.
18. $r = 5 \cos \theta$.
19. $r = 4 - \sin \theta$.
20. $r = 3(\cos \theta - 1)$.
21. $r = \theta/\pi$ $(\theta \geq 0)$.
22. $r = \sqrt{2} \cos(\theta/2)$.
23. $r = e^{-\theta/\pi}$ $(\theta \geq 0)$.

In Exercises 24–26, sketch the curves in the same polar coordinate system, and find all points of intersection.

24. $r = 3 \cos \theta$, $r = 1 + \cos \theta$.
25. $r = a \cos(2\theta)$, $r = a/2$ $(a > 0)$.
26. $r = 2 \sin \theta$, $r = 2 + 2 \cos \theta$.

In Exercises 27–29, set up, but *do not evaluate*, definite integrals for the stated area and arc length.

27. (a) The area inside both the circle and cardioid in Exercise 24.
    (b) The arc length of that part of the cardioid outside the circle in Exercise 24.

28. (a) The area inside the rose and outside the circle in Exercise 25.
    (b) The arc length of that part of the rose lying inside the circle in Exercise 25.

29. (a) The area inside the circle and outside the cardioid in Exercise 26.
    (b) The arc length of that portion of the circle lying inside the cardioid in Exercise 26.

In Exercises 30–33, find the area of the region described.

30. One petal of the rose $r = a \sin 3\theta$.
31. The region outside the circle $r = a$ and inside the lemniscate $r^2 = 2a^2 \cos 2\theta$.
32. The region in part (a) of Exercise 28.
33. The region in part (a) of Exercise 29.

In Exercises 34–37:
(a) Sketch the curve and indicate the direction of increasing parameter.
(b) Use the parametric equations to find dy/dx, d^2y/dx^2, and the equation of the tangent line at the point on the curve corresponding to the parameter value t_0 (or θ_0).

34. $x = 3 - t^2$, $y = 2 + t$, $0 \leq t \leq 3$; $t_0 = 1$.
35. $x = 1 + 3 \cos \theta$, $y = -1 + 2 \sin \theta$, $0 \leq \theta \leq \pi$; $\theta_0 = \pi/2$.
36. $x = 2 \tan \theta$, $y = \sec \theta$, $-\pi/2 < \theta < \pi/2$; $\theta_0 = \pi/3$.
37. $x = 1/t$, $y = \ln t$, $1 \leq t \leq e$; $t_0 = 2$.

In Exercises 38–43, find the arc length of the curve described.

38. $x = 2t^3$, $y = 3t^2$, $-4 \leq t \leq 4$.
39. $x = \ln \cos 2t$, $y = 2t$, $0 \leq t \leq \pi/6$.

40. $x = 3 \cos t - 1$, $y = 3 \sin t + 4$, $0 \le t \le \pi$.

41. $x = 3t^2$, $y = t^3 - 3t$, $0 \le t \le 1$.

42. $x = 1 - \cos t$, $y = t - \sin t$, $-\pi \le t \le \pi$.

43. $r = e^{\theta}$, $0 \le \theta \le 2\pi$.

In Exercises 44–46, $x(t)$ and $y(t)$ describe the motion of a particle. Find the coordinates of the particle when the instantaneous direction of motion is (a) horizontal and (b) vertical.

44. $x = -2t^2$, $y = t^3 - 3t + 5$.

45. $x = 1 - 2 \sin t$, $y = t + 2 \cos t$, $0 \le t \le \pi$.

46. $x = \ln t$, $y = t^2 - 4t$, $t > 0$.

47. At what instant does the trajectory described in Exercise 46 have a point of inflection?

48. Find a set of parametric equations for
(a) the line $y = 2x + 3$
(b) the ellipse $4(x - 2)^2 + y^2 = 4$.

49. For the cardioid $r = 2(1 + \cos \theta)$, find the slope of the tangent line and the angle ψ when $\theta = \pi/2$.

50. For the circle $r = 4 \sin \theta$, show that $\psi = \theta$ if $0 \le \theta < \pi$, and find a formula for the inclination angle ϕ as a function of θ for $0 \le \theta < \pi/2$.

14

Second-Order Differential Equations

Jakob Bernoulli (1667-1748)

■ **14.1   SECOND-ORDER LINEAR HOMOGENEOUS DIFFERENTIAL EQUATIONS WITH CONSTANT COEFFICIENTS**

> *In Section 7.7 we showed how to solve first-order linear differential equations. In this section we shall show how to solve certain second-order linear differential equations.*

☐ **SECOND-ORDER LINEAR DIFFERENTIAL EQUATIONS**

Recall from Section 7.7 that a first-order linear differential equation has the form

$$\frac{dy}{dx} + p(x)y = q(x)$$

In this section we shall be concerned with the general **second-order linear differential equation**

$$\frac{d^2y}{dx^2} + p(x)\frac{dy}{dx} + q(x)y = r(x) \tag{1}$$

or in an alternative notation

$$y'' + p(x)y' + q(x)y = r(x)$$

where $p(x)$, $q(x)$, and $r(x)$ are continuous functions. If $r(x) = 0$ for all x, then (1) reduces to

$$\frac{d^2y}{dx^2} + p(x)\frac{dy}{dx} + q(x)y = 0$$

which is called the general second-order linear **homogeneous** differential equation. If $r(x)$ is not identically zero, then (1) is said to be **nonhomogeneous**. Some examples of second-order linear differential equations are

$$\frac{d^2y}{dx^2} + x^2\frac{dy}{dx} - xy = e^x \qquad \boxed{p(x) = x^2, \quad q(x) = -x, \quad r(x) = e^x}$$

$$y'' + y' - 3y = \sin x \qquad \boxed{p(x) = 1, \quad q(x) = -3, \quad r(x) = \sin x}$$

$$y'' + e^x y = 3 \qquad \boxed{p(x) = 0, \quad q(x) = e^x, \quad r(x) = 3}$$

$$\frac{d^2y}{dx^2} - \frac{dy}{dx} + 2y = 0 \qquad \boxed{p(x) = -1, \quad q(x) = 2, \quad r(x) = 0}$$

The last equation is homogeneous and the first three are nonhomogeneous.

In Section 7.7 we gave a general procedure for solving first-order linear differential equations by integration. For second-order linear differential equations the situation is more complicated and simple general procedures for solving such equations can only be given in special cases. Before we can pursue this matter further, we shall need some preliminary results.

First, some terminology. Two continuous functions f and g are said to be *linearly dependent* if one is a constant multiple of the other. If neither is a constant multiple of the other, then they are called *linearly independent*. Thus,

$$f(x) = \sin x \quad \text{and} \quad g(x) = 3 \sin x$$

are linearly dependent, but

$$f(x) = x \quad \text{and} \quad g(x) = x^2$$

are linearly independent.

The following theorem is central to the study of second-order linear differential equations.

14.1.1 THEOREM. *If $y_1 = y_1(x)$ and $y_2 = y_2(x)$ are linearly independent solutions of the homogeneous equation*

$$\frac{d^2y}{dx^2} + p(x)\frac{dy}{dx} + q(x)y = 0 \tag{2}$$

then

$$y(x) = c_1 y_1(x) + c_2 y_2(x) \tag{3}$$

is the general solution of (2) in the sense that every solution of (2) can be obtained from (3) by choosing appropriate values for the arbitrary constants c_1 and c_2; conversely, (3) is a solution of (2) for all choices of c_1 and c_2.

(Readers interested in a proof of this theorem are referred to *Elementary Differential Equations and Boundary Value Problems*, John Wiley & Sons, New York, 1988, by William E. Boyce and Richard C. DiPrima.)

REMARK. The expression on the right side of (3) is called a *linear combination* of $y_1(x)$ and $y_2(x)$. Thus, Theorem 14.1.1 tells us that once we find two linearly independent solutions of (2), we essentially know all the solutions because every other solution can be expressed as a linear combination of those two.

☐ **CONSTANT COEFFICIENTS**

For the remainder of this section we shall restrict our attention to second-order linear homogeneous equations in which $p(x)$ and $q(x)$ are constants, p and q. Such equations have the form

$$\frac{d^2y}{dx^2} + p\frac{dy}{dx} + qy = 0 \tag{4}$$

Our objective is to find two linearly independent solutions of this equation. To start, we note that the function e^{mx} has the property that its derivatives are constant multiples of the function itself. This suggests the possibility that

$$y = e^{mx} \tag{5}$$

might be a solution of (4) if the constant m is suitably chosen. Since

$$\frac{dy}{dx} = me^{mx}, \quad \frac{d^2y}{dx^2} = m^2 e^{mx} \tag{6}$$

substituting (5) and (6) into (4) yields

$$(m^2 + pm + q)e^{mx} = 0 \tag{7}$$

which is satisfied if and only if

$$m^2 + pm + q = 0 \tag{8}$$

since $e^{mx} \neq 0$ for every x.

Equation (8), which is called the **auxiliary equation** for (4), can be obtained from (4) by replacing d^2y/dx^2 by m^2, dy/dx by $m (= m^1)$, and y by $1 (= m^0)$. The solutions, m_1 and m_2, of the auxiliary equation can be obtained by factoring or by the quadratic formula. These solutions are

$$m_1 = \frac{-p + \sqrt{p^2 - 4q}}{2}, \quad m_2 = \frac{-p - \sqrt{p^2 - 4q}}{2} \tag{9}$$

Depending on whether $p^2 - 4q$ is positive, zero, or negative, these roots will be distinct and real, equal and real, or complex conjugates.* We shall consider each of these cases separately.

☐ **DISTINCT REAL ROOTS**   If m_1 and m_2 are distinct real roots, then (4) has the two solutions

$$y_1 = e^{m_1 x}, \quad y_2 = e^{m_2 x}$$

Neither of the functions $e^{m_1 x}$ and $e^{m_2 x}$ is a constant multiple of the other (Exercise 29), so the general solution of (4) in this case is

$$y(x) = c_1 e^{m_1 x} + c_2 e^{m_2 x} \tag{10}$$

Example 1   Find the general solution of $y'' - y' - 6y = 0$.

Solution.   The auxiliary equation

$$m^2 - m - 6 = 0$$

can be rewritten as

$$(m + 2)(m - 3) = 0$$

---

*Recall that the complex solutions of a polynomial equation, and in particular of a quadratic equation, occur as conjugate pairs $a + bi$ and $a - bi$.

so its roots are $m = -2$, $m = 3$. Thus, from (10) the general solution of the differential equation is

$$y = c_1 e^{-2x} + c_2 e^{3x}$$

where c_1 and c_2 are arbitrary constants. ◀

☐ **EQUAL REAL ROOTS**

If m_1 and m_2 are equal real roots, say $m_1 = m_2 \, (= m)$, then the auxiliary equation yields only one solution of (4):

$$y_1(x) = e^{mx}$$

We shall now show that

$$y_2(x) = xe^{mx} \tag{11}$$

is a second linearly independent solution. To see that this is so, note that $p^2 - 4q = 0$ in (9) since the roots are equal. Thus,

$$m = m_1 = m_2 = -p/2$$

and (11) becomes

$$y_2(x) = xe^{(-p/2)x}$$

Differentiating yields

$$y_2'(x) = \left(1 - \frac{p}{2}x\right)e^{(-p/2)x} \quad \text{and} \quad y_2''(x) = \left(\frac{p^2}{4}x - p\right)e^{-(p/2)x}$$

so

$$y_2''(x) + py_2'(x) + qy_2(x) = \left[\left(\frac{p^2}{4}x - p\right) + p\left(1 - \frac{p}{2}x\right) + qx\right]e^{(-p/2)x}$$

$$= \left[-\frac{p^2}{4} + q\right]xe^{(-p/2)x} \tag{12}$$

But $p^2 - 4q = 0$ implies that $(-p^2/4) + q = 0$, so (12) becomes

$$y_2''(x) + py_2'(x) + qy_2(x) = 0$$

which tells us that $y_2(x)$ is a solution of (4). It can be shown that

$$y_1(x) = e^{mx} \quad \text{and} \quad y_2(x) = xe^{mx}$$

are linearly independent (Exercise 29), so the general solution of (4) in this case is

$$y = c_1 e^{mx} + c_2 xe^{mx} \tag{13}$$

Example 2 Find the general solution of $y'' - 8y' + 16y = 0$.

Solution. The auxiliary equation

$$m^2 - 8m + 16 = 0$$

can be rewritten as

$$(m - 4)^2 = 0$$

so $m = 4$ is the only root. Thus, from (13) the general solution of the differential equation is

$$y = c_1 e^{4x} + c_2 x e^{4x} \quad \blacktriangleleft$$

☐ **COMPLEX ROOTS**

If the auxiliary equation has complex roots $m_1 = a + bi$ and $m_2 = a - bi$, then $y_1(x) = e^{ax} \cos bx$ and $y_2(x) = e^{ax} \sin bx$ are linearly independent solutions of (4) and

$$y = e^{ax}(c_1 \cos bx + c_2 \sin bx) \tag{14}$$

is the general solution. The proof is discussed in the exercises (Exercise 30).

Example 3 Find the general solution of $y'' + y' + y = 0$.

Solution. The auxiliary equation $m^2 + m + 1 = 0$ has roots

$$m_1 = \frac{-1 + \sqrt{1 - 4}}{2} = -\frac{1}{2} + \frac{\sqrt{3}}{2} i$$

$$m_2 = \frac{-1 - \sqrt{1 - 4}}{2} = -\frac{1}{2} - \frac{\sqrt{3}}{2} i$$

Thus, from (14) with $a = -1/2$ and $b = \sqrt{3}/2$, the general solution of the differential equation is

$$y = e^{-x/2} \left(c_1 \cos \frac{\sqrt{3}}{2} x + c_2 \sin \frac{\sqrt{3}}{2} x \right) \quad \blacktriangleleft$$

☐ **INITIAL-VALUE PROBLEMS**

When a physical problem leads to a second-order differential equation, there are usually two conditions in the problem that determine specific values for the two arbitrary constants in the general solution of the equation. Conditions that specify the value of the solution $y(x)$ and its derivative $y'(x)$ at some point $x = x_0$ are called *initial conditions*. A second-order differential equation with initial conditions is called a *second-order initial-value problem*.

Example 4 Solve the initial-value problem

$$y'' - y = 0, \quad y(0) = 1, \quad y'(0) = 0$$

Solution. We must first solve the differential equation. The auxiliary equation

$$m^2 - 1 = 0$$

has distinct real roots $m_1 = 1$, $m_2 = -1$, so from (10) the general solution is

$$y(x) = c_1 e^x + c_2 e^{-x} \tag{15}$$

and the derivative of this solution is

$$y'(x) = c_1 e^x - c_2 e^{-x} \tag{16}$$

Substituting $x = 0$ in (15) and (16) and using the initial conditions $y(0) = 1$ and $y'(0) = 0$ yields the system of equations

$$c_1 + c_2 = 1$$
$$c_1 - c_2 = 0$$

Solving this system yields $c_1 = \frac{1}{2}$, $c_2 = \frac{1}{2}$, so from (15) the solution of the initial-value problem is

$$y(x) = \tfrac{1}{2} e^x + \tfrac{1}{2} e^{-x} \quad \blacktriangleleft$$

The following summary is included as a ready reference for the solution of second-order homogeneous linear differential equations with constant coefficients.

Summary

EQUATION: $y'' + py' + qy = 0$
AUXILIARY EQUATION: $m^2 + pm + q = 0$

CASE	GENERAL SOLUTION
Distinct real roots m_1, m_2 to the auxiliary equation	$y = c_1 e^{m_1 x} + c_2 e^{m_2 x}$
Equal real roots $m_1 = m_2 \,(= m)$ to the auxiliary equation	$y = c_1 e^{mx} + c_2 x e^{mx}$
Complex roots $m_1 = a + bi$, $m_2 = a - bi$ to the auxiliary equation	$y = e^{ax}(c_1 \cos bx + c_2 \sin bx)$

▶ Exercise Set 14.1

1. Verify that the following are solutions of the differential equation $y'' - y' - 2y = 0$ by substituting these functions into the equation:

(a) e^{2x} and e^{-x}

(b) $c_1 e^{2x} + c_2 e^{-x}$ (c_1, c_2 constants).

2. Verify that the following are solutions of the differential equation $y'' + 4y' + 4y = 0$ by substituting these functions into the equation:

(a) e^{-2x} and xe^{-2x}

(b) $c_1 e^{-2x} + c_2 x e^{-2x}$ (c_1, c_2 constants).

In Exercises 3–16, find the general solution of the differential equation.

3. $y'' + 3y' - 4y = 0.$ 4. $y'' + 6y' + 5y = 0.$

5. $y'' - 2y' + y = 0.$ 6. $y'' + 6y' + 9y = 0.$

7. $y'' + 5y = 0.$ 8. $y'' + y = 0.$

9. $\dfrac{d^2y}{dx^2} - \dfrac{dy}{dx} = 0.$ 10. $\dfrac{d^2y}{dx^2} + 3\dfrac{dy}{dx} = 0.$

11. $\dfrac{d^2y}{dt^2} + 4\dfrac{dy}{dt} + 4y = 0.$

12. $\dfrac{d^2y}{dt^2} - 10\dfrac{dy}{dt} + 25y = 0.$

13. $\dfrac{d^2y}{dx^2} - 4\dfrac{dy}{dx} + 13y = 0.$

14. $\dfrac{d^2y}{dx^2} - 6\dfrac{dy}{dx} + 25y = 0.$

15. $8y'' - 2y' - 1 = 0.$ 16. $9y'' - 6y' + 1 = 0.$

In Exercises 17–22, solve the initial-value problem.

17. $y'' + 2y' - 3y = 0;$ $y(0) = 1,\ y'(0) = 5.$

18. $y'' - 6y' - 7y = 0;$ $y(0) = 5,\ y'(0) = 3.$

19. $y'' - 6y' + 9y = 0;$ $y(0) = 2,\ y'(0) = 1.$

20. $y'' + 4y' + y = 0;$ $y(0) = 5,\ y'(0) = 4.$

21. $y'' + 4y' + 5y = 0;$ $y(0) = -3,\ y'(0) = 0.$

22. $y'' - 6y' + 13y = 0;$ $y(0) = -1,\ y'(0) = 1.$

23. In each part find a second-order linear homogeneous differential equation with constant coefficients that has the given functions as solutions.
    (a) $y_1 = e^{5x},\ y_2 = e^{-2x}$
    (b) $y_1 = e^{4x},\ y_2 = xe^{4x}$
    (c) $y_1 = e^{-x}\cos 4x,\ y_2 = e^{-x}\sin 4x.$

24. Show that if e^x and e^{-x} are solutions of a second-order linear homogeneous differential equation, then so are $\cosh x$ and $\sinh x$.

25. Find all values of k for which the differential equation $y'' + ky' + ky = 0$ has a general solution of the given form.
    (a) $y = c_1 e^{ax} + c_2 e^{bx}$ (b) $y = c_1 e^{ax} + c_2 xe^{ax}$
    (c) $y = c_1 e^{ax}\cos bx + c_2 e^{ax}\sin bx.$

26. The equation
$$x^2 \frac{d^2y}{dx^2} + px\frac{dy}{dx} + qy = 0 \qquad (x > 0)$$
where p and q are constants, is called *Euler's equidimensional equation*. Show that the substitution

$x = e^z$ transforms this equation into the equation
$$\frac{d^2y}{dz^2} + (p - 1)\frac{dy}{dz} + qy = 0$$

27. Use the result in Exercise 26 to find the general solution of
    (a) $x^2\dfrac{d^2y}{dx^2} + 3x\dfrac{dy}{dx} + 2y = 0$ $(x > 0)$
    (b) $x^2\dfrac{d^2y}{dx^2} - x\dfrac{dy}{dx} - 2y = 0$ $(x > 0).$

28. Let $y(x)$ be a solution of $y'' + py' + qy = 0.$ Prove: If p and q are positive constants, then $\lim\limits_{x \to +\infty} y(x) = 0.$

29. The *Wronskian* of two differentiable functions y_1 and y_2 is denoted by $W(y_1, y_2)$ and is defined to be the function
$$W(y_1, y_2) = y_1 y_2' - y_1' y_2 = \begin{vmatrix} y_1 & y_2 \\ y_1' & y_2' \end{vmatrix}$$
The value of $W(y_1, y_2)$ at a point x is denoted by $W(y_1, y_2)(x)$ or often more simply by $W(x)$. It can be proved that two solutions, $y_1 = y_1(x)$ and $y_2 = y_2(x)$, of Equation (2) are linearly dependent if and only if $W(x) = 0$ for all x. Equivalently, the functions are linearly independent if and only if $W(x) \neq 0$ for at least one value of x. Use this result to prove that the following solutions of Equation (4) are linearly independent:
    (a) $y_1 = e^{m_1 x},\ y_2 = e^{m_2 x}$ $(m_1 \neq m_2)$
    (b) $y_1 = e^{mx},\ y_2 = xe^{mx}.$

30. Prove: If the auxiliary equation of
$$y'' + py' + qy = 0$$
has complex roots $a + bi$ and $a - bi$, then the general solution of this differential equation is
$$y(x) = e^{ax}(c_1\cos bx + c_2\sin bx)$$
[*Hint:* By substitution, verify that $y_1 = e^{ax}\cos bx$ and $y_2 = e^{ax}\sin bx$ are solutions of the differential equation. Then use Exercise 29 to prove that y_1 and y_2 are linearly independent.]

31. Suppose that the auxiliary equation of the differential equation $y'' + py' + qy = 0$ has distinct real roots μ and m.
    (a) Show that the function
$$g_\mu(x) = \frac{e^{\mu x} - e^{mx}}{\mu - m}$$
is a solution of the differential equation.

(b) Use L'Hôpital's rule to show that

$$\lim_{\mu \to m} g_\mu(x) = xe^{mx}$$

[*Note:* Can you see how the result in part (b) makes it plausible that the function $y(x) = xe^{mx}$ is a solution of $y'' + py' + qy = 0$ when m is a repeated root of the auxiliary equation?]

32. Consider the problem of solving the differential equation

$$y'' + \lambda y = 0$$

subject to the conditions $y(0) = 0$, $y(\pi) = 0$.

(a) Show that if $\lambda \leq 0$, then $y = 0$ is the only solution.

(b) Show that if $\lambda > 0$, then the solution is

$$y = c \sin \sqrt{\lambda} x$$

where c is an arbitrary constant, if

$$\lambda = 1, 2^2, 3^2, 4^2, \ldots$$

and the only solution is $y = 0$ otherwise.

■ 14.2 SECOND-ORDER LINEAR NONHOMOGENEOUS DIFFERENTIAL EQUATIONS WITH CONSTANT COEFFICIENTS; UNDETERMINED COEFFICIENTS

> *In this section we shall study techniques for solving second-order linear differential equations that are not homogeneous.*

□ **THE COMPLEMENTARY EQUATION**

The following theorem is the key result for solving a second-order *nonhomogeneous* linear differential equation with constant coefficients

$$y'' + py' + qy = r(x) \tag{1}$$

where p and q are constants and $r(x)$ is a continuous function of x.

14.2.1 THEOREM. *The general solution of* (1) *is*

$$y(x) = c_1 y_1(x) + c_2 y_2(x) + y_p(x)$$

where $c_1 y_1(x) + c_2 y_2(x)$ is the general solution of the homogeneous equation

$$y'' + py' + qy = 0 \tag{2}$$

and $y_p(x)$ is any solution of (1).

The proof is deferred to the end of the section.

REMARK. Equation (2) is called the ***complementary equation*** to (1) and $y_p(x)$ is called a ***particular solution*** of (1). Thus, Theorem 14.2.1 states that *the general solution of* (1) *is obtained by adding a particular solution of* (1) *to the general solution of the complementary equation.*

☐ **THE METHOD OF UNDETERMINED COEFFICIENTS**

Since we already know how to obtain the general solution of (2), we shall focus our attention on the problem of obtaining a particular solution of (1). In this section we shall discuss a procedure for doing this called the method of *undetermined coefficients,* and in the next section we shall discuss a second procedure called *variation of parameters.*

The first step in the method of undetermined coefficients is to make a reasonable guess about the form of the particular solution. This guess will involve one or more unknown coefficients which we shall then determine from conditions obtained by substituting the proposed solution into the differential equation.

We shall begin with some examples that illustrate the basic idea.

Example 1 Find a particular solution of the differential equation

$$y'' + 2y' - 8y = e^{3x} \tag{3}$$

Solution. It should be clear that such functions as $\sin x$, $\cos x$, $\ln x$, or x^3 are not reasonable possibilities for $y_p(x)$ because the derivatives of such functions do not yield expressions involving e^{3x}. Since the obvious choice for $y_p(x)$ is an expression involving e^{3x}, we shall *guess* that $y_p(x)$ has the form

$$y_p(x) = Ae^{3x} \tag{4}$$

where A is an unknown constant to be determined. It follows that

$$y_p'(x) = 3Ae^{3x} \tag{5}$$
$$y_p''(x) = 9Ae^{3x} \tag{6}$$

Substituting expressions (4), (5), and (6) for y, y', and y'' in (3) yields

$$9Ae^{3x} + 2(3Ae^{3x}) - 8(Ae^{3x}) = e^{3x}$$

or

$$7Ae^{3x} = e^{3x}$$

Thus, $A = \frac{1}{7}$. From this result and (4), a particular solution of (3) is

$$y_p(x) = \tfrac{1}{7}e^{3x} \quad \blacktriangleleft$$

Example 2 Find a particular solution of the differential equation

$$y'' - y' - 6y = e^{3x} \tag{7}$$

Solution. As in Example 1, a reasonable guess is $y_p(x) = Ae^{3x}$. However, if we substitute (4), (5), and (6) in (7), we obtain

$$9Ae^{3x} - 3Ae^{3x} - 6Ae^{3x} = e^{3x} \quad \text{or} \quad 0 = e^{3x}$$

which is contradictory. The problem here is that $y_p(x) = Ae^{3x}$ is a solution of the complementary equation

$$y'' - y' - 6y = 0$$

which makes it impossible for y_p to satisfy (7), no matter how A is selected. How should we proceed in this case? Experience has shown that multiplying (4) by x produces the correct form for the solution. Thus, we shall try

$$y_p(x) = Axe^{3x} \tag{8}$$

It follows that

$$y_p'(x) = 3Axe^{3x} + Ae^{3x} \tag{9}$$
$$y_p''(x) = 9Axe^{3x} + 6Ae^{3x} \tag{10}$$

Substituting expressions (8), (9), and (10) for y, y', and y'' in (7) yields

$$(9Axe^{3x} + 6Ae^{3x}) - (3Axe^{3x} + Ae^{3x}) - 6(Axe^{3x}) = e^{3x}$$

or

$$5Ae^{3x} = e^{3x}$$

Thus, $A = \frac{1}{5}$, so (8) implies that a particular solution of (7) is

$$y_p(x) = \frac{1}{5}xe^{3x} \quad \blacktriangleleft$$

REMARK. Had it turned out that Axe^{3x} was also a solution of the complementary equation, then we would have tried $y_p(x) = Ax^2e^{3x}$ as our candidate for a particular solution.

 SUMMARY

In summary, a particular solution of an equation of the form

$$y'' + py' + qy = ke^{ax}$$

can be obtained as follows:

> **Step 1.** Start with $y_p = Ae^{ax}$ as an initial guess.
>
> **Step 2.** Determine if the initial guess is a solution of the complementary equation $y'' + py' + qy = 0$.
>
> **Step 3.** If the initial guess is not a solution of the complementary equation, then $y_p = Ae^{ax}$ is the correct form of a particular solution.
>
> **Step 4.** If the initial guess is a solution of the complementary equation, then multiply it by the smallest positive integer power of x required to produce a function that is not a solution of the complementary equation. This will yield either $y_p = Axe^{ax}$ or $y_p = Ax^2e^{ax}$.

Table 14.2.1, which we provide without proof, restates the preceding procedure more compactly and also explains how to find a particular solution of (1) when $r(x)$ is a polynomial or a combination of sine and cosine functions.

Table 14.2.1

EQUATION	INITIAL GUESS FOR y_p
$y'' + py' + qy = ke^{ax}$	$y_p = Ae^{ax}$
$y'' + py' + qy = a_0 + a_1 x + \cdots + a_n x^n$	$y_p = A_0 + A_1 x + \cdots + A_n x^n$
$y'' + py' + qy = a_1 \cos bx + a_2 \sin bx$	$y_p = A_1 \cos bx + A_2 \sin bx$

MODIFICATION RULE
If any term in the initial guess is a solution of the complementary equation, then the correct form for y_p is obtained by multiplying the initial guess by the smallest positive integer power of x required so that no term is a solution of the complementary equation.

Example 3   Find the general solution of

$$y'' + y' = 4x^2 \tag{11}$$

Solution.   We begin by finding the general solution of the complementary equation

$$y'' + y' = 0$$

The auxiliary equation for this homogeneous equation is

$$m^2 + m = 0$$

which has roots $m_1 = 0$, $m_2 = -1$. Thus, the general solution, $y_c(x)$, of the complementary equation is

$$y_c(x) = c_1 e^{0x} + c_2 e^{-x} = c_1 + c_2 e^{-x}$$

Since the right-hand side of (11) is a second-degree polynomial, our initial guess for y_p is

$$y_p = A_0 + A_1 x + A_2 x^2$$

But A_0 is a solution of the complementary equation (take $c_1 = A_0$, $c_2 = 0$), so our second guess for y_p is

$$y_p = x(A_0 + A_1 x + A_2 x^2) = A_0 x + A_1 x^2 + A_2 x^3 \tag{12}$$

Since no term of this function is a solution of the complementary equation, this is the correct form for y_p. Differentiating (12) we obtain

$$y_p'(x) = A_0 + 2A_1 x + 3A_2 x^2 \tag{13}$$
$$y_p''(x) = 2A_1 + 6A_2 x \tag{14}$$

Substituting (13) and (14) in (11) yields

$$(2A_1 + 6A_2 x) + (A_0 + 2A_1 x + 3A_2 x^2) = 4x^2$$

or

$$(A_0 + 2A_1) + (2A_1 + 6A_2)x + 3A_2x^2 = 4x^2$$

To solve for A_0, A_1, and A_2, we equate corresponding coefficients on the two sides of this equation; this yields

$$A_0 + 2A_1 = 0$$
$$2A_1 + 6A_2 = 0$$
$$3A_2 = 4$$

from which we obtain

$$A_2 = \tfrac{4}{3}, \quad A_1 = -4, \quad A_0 = 8$$

Substituting these values in (12) yields the particular solution

$$y_p(x) = 8x - 4x^2 + \tfrac{4}{3}x^3$$

Thus, the general solution of (11) is

$$y(x) = y_c(x) + y_p(x) = c_1 + c_2e^{-x} + 8x - 4x^2 + \tfrac{4}{3}x^3 \qquad \blacktriangleleft$$

Example 4   Find the general solution of

$$y'' - 2y' + y = \sin 2x \qquad (15)$$

Solution.   We begin by finding the general solution of the complementary equation

$$y'' - 2y' + y = 0$$

The auxiliary equation for this homogeneous equation is

$$m^2 - 2m + 1 = 0$$

which has the repeated root $m_1 = 1$, $m_2 = 1$. Thus, the general solution of the complementary equation is

$$y_c(x) = c_1e^x + c_2xe^x \qquad (16)$$

Since $\sin 2x$ has the form $a_1 \cos bx + a_2 \sin bx$ ($a_1 = 0$, $a_2 = 1$, $b = 2$), our initial guess for y_p is

$$y_p(x) = A_1 \cos 2x + A_2 \sin 2x \qquad (17)$$

Since no term of y_p is a solution of the complementary equation [see (16)], this is the correct form for a particular solution. Differentiating (17) we obtain

$$y_p'(x) = -2A_1 \sin 2x + 2A_2 \cos 2x \qquad (18)$$
$$y_p''(x) = -4A_1 \cos 2x - 4A_2 \sin 2x \qquad (19)$$

Substituting (17), (18), and (19) in (15) yields

$$(-4A_1 \cos 2x - 4A_2 \sin 2x) - 2(-2A_1 \sin 2x + 2A_2 \cos 2x)$$
$$+ (A_1 \cos 2x + A_2 \sin 2x) = \sin 2x$$

or

$$(-3A_1 - 4A_2) \cos 2x + (4A_1 - 3A_2) \sin 2x = \sin 2x$$

To solve for A_1 and A_2 we equate the coefficients of $\sin 2x$ and $\cos 2x$ on the two sides of this equation; this yields

$$-3A_1 - 4A_2 = 0$$
$$4A_1 - 3A_2 = 1$$

from which we obtain

$$A_1 = \tfrac{4}{25}, \quad A_2 = -\tfrac{3}{25}$$

(Verify.) From this result and (17), a particular solution of (15) is

$$y_p(x) = \tfrac{4}{25} \cos 2x - \tfrac{3}{25} \sin 2x$$

Thus, the general solution of (15) is

$$y(x) = y_c(x) + y_p(x) = c_1 e^x + c_2 x e^x + \tfrac{4}{25} \cos 2x - \tfrac{3}{25} \sin 2x \qquad \blacktriangleleft$$

We conclude this section with a proof of Theorem 14.2.1.

■ OPTIONAL
_____

Proof of Theorem 14.2.1. To prove that

$$y(x) = c_1 y_1(x) + c_2 y_2(x) + y_p(x) \qquad (20)$$

is the general solution of (1) we must prove two results: first, that for all choices of c_1 and c_2 the function $y(x)$ defined by (20) satisfies (1); and second, that every solution of (1) can be obtained from (20) by choosing appropriate values for the constants c_1 and c_2.

To prove the first statement, let c_1 and c_2 have any real values. Then differentiating (20) yields

$$y'(x) = c_1 y_1'(x) + c_2 y_2'(x) + y_p'(x)$$
$$y''(x) = c_1 y_1''(x) + c_2 y_2''(x) + y_p''(x)$$

so

$$y''(x) + py'(x) + qy(x) = c_1(y_1''(x) + py_1'(x) + qy_1(x))$$
$$+ c_2(y_2''(x) + py_2'(x) + qy_2(x))$$
$$+ y_p''(x) + py_p'(x) + qy_p(x) \qquad (21)$$

But $y_1(x)$ and $y_2(x)$ satisfy (2) and $y_p(x)$ satisfies (1), so (21) reduces to

$$y''(x) + py'(x) + qy(x) = 0 + 0 + r(x) = r(x)$$

which proves that $y(x)$ satisfies (1).

To prove the second statement, let $y(x)$ be any solution of (1). We must find values of c_1 and c_2 such that

$$y(x) = c_1y_1(x) + c_2y_2(x) + y_p(x) \tag{22}$$

But $y(x) - y_p(x)$ satisfies (2) since

$$\begin{aligned}(y(x) &- y_p(x))'' + p(y(x) - y_p(x))' + q(y(x) - y_p(x)) \\ &= [y''(x) + py'(x) + qy(x)] - [y_p''(x) + py_p'(x) + qy_p(x)] \\ &= r(x) - r(x) = 0\end{aligned}$$

Thus, since $c_1y_1(x) + c_2y_2(x)$ is the general solution of (2), there exist values of c_1 and c_2 such that the solution $y(x) - y_p(x)$ can be written as

$$y(x) - y_p(x) = c_1y_1(x) + c_2y_2(x)$$

from which (22) follows. ∎

▶ Exercise Set 14.2

In Exercises 1–24, use the method of undetermined coefficients to find the general solution of the differential equation.

1. $y'' + 6y' + 5y = 2e^{3x}$. 2. $y'' + 3y' - 4y = 5e^{7x}$.

3. $y'' - 9y' + 20y = -3e^{5x}$.

4. $y'' + 7y' - 8y = 7e^x$.

5. $y'' + 2y' + y = e^{-x}$.

6. $y'' + 4y' + 4y = 4e^{-2x}$.

7. $y'' + y' - 12y = 4x^2$.

8. $y'' - 4y' - 5y = -6x^2$.

9. $y'' - 6y' = x - 1$.   10. $y'' + 3y' = 2x + 2$.

11. $y'' - x^3 + 1 = 0$.   12. $y'' + 3x^3 + x = 0$.

13. $y'' - y' - 2y = 10\cos x$.

14. $y'' - 3y' - 4y = 2\sin x$.

15. $y'' - 4y = 2\sin 2x + 3\cos 2x$.

16. $y'' - 9y = \cos 3x - \sin 3x$.

17. $y'' + y = \sin x$.

18. $y'' + 4y = \cos 2x$.

19. $y'' - 3y' + 2y = x$.

20. $y'' + 4y' + 4y = 3x + 3$.

21. $y'' + 4y' + 9y = x^2 + 3x$.

22. $y'' - y = 1 + x + x^2$.

23. $y'' + 4y = \sin x \cos x$.

24. $y'' + 4y = \cos^2 x - \sin^2 x$.

25. (a) Prove: If $y_1(x)$ is a solution of

$$y'' + p(x)y' + q(x)y = r_1(x)$$

and $y_2(x)$ is a solution of

$$y'' + p(x)y' + q(x)y = r_2(x)$$

then $y_1(x) + y_2(x)$ is a solution of

$$y'' + p(x)y' + q(x)y = r_1(x) + r_2(x)$$

(b) Use the result in part (a) to find a particular solution of the equation

$$y'' + 3y' - 4y = x + e^x$$

(c) State a generalization of the result in part (a) that is applicable to the equation

$$\begin{aligned}y'' &+ p(x)y' + q(x)y \\ &= r_1(x) + r_2(x) + \cdots + r_n(x)\end{aligned}$$

In Exercises 26–34, use the results in Exercise 25 to find the general solution of the differential equation.

26. $y'' - y' - 2y = x + e^{-x}$.

27. $y'' - y = 1 + e^x$.

28. $y'' - 4y' + 3y = 2\cos x + 4\sin x$.

29. $y'' + 4y = 1 + x + \sin x$.

30. $y'' + 2y' + y = 2 + 3x + 3e^x + 2\cos 2x$.

31. $y'' - 2y' + y = \sinh x$. [*Hint:* $\sinh x = \frac{1}{2}(e^x - e^{-x})$.]

32. $y'' + 4y' - 5y = \cosh x$.
[*Hint:* $\cosh x = \frac{1}{2}(e^x + e^{-x})$.]

33. $y'' + y = 12\cos^2 x$. [*Hint:* $\cos^2 x = \frac{1}{2}(1 + \cos 2x)$.]

34. $y'' + 2y' + y = \sin^2 x$.
[*Hint:* $\sin^2 x = \frac{1}{2}(1 - \cos 2x)$.]

35. (a) Find the general solution of
$$y'' + \mu^2 y = a\sin bx$$

where a is an arbitrary constant and μ and b are positive constants such that $\mu \neq b$.

(b) Use part (a) and Exercise 25 to find the general solution of
$$y'' + \mu^2 y = \sum_{k=1}^{n} a_k \sin k\pi x$$

where $\mu > 0$ and $\mu \neq k\pi$, $k = 1, 2, \ldots, n$.

36. Find the general solution of
$$y'' + \lambda^2 y = \sum_{k=1}^{n} a_k \cos k\pi x$$

where $\lambda > 0$ and $\lambda \neq k\pi$, $k = 1, 2, \ldots, n$. [*Hint:* See Exercise 35.]

37. Find all solutions of the equation
$$y'' - y' = 4 - 4x$$

(if any) with the property that $y'(x_0) = y''(x_0) = 0$ at some point x_0.

■ 14.3 VARIATION OF PARAMETERS

In this section* we shall consider an alternative method for finding a particular solution of a second-order linear nonhomogeneous equation that often works when the method of undetermined coefficients cannot be applied.

As in the previous section, we shall be concerned with equations of the form

$$y'' + py' + qy = r(x) \tag{1}$$

where p and q are constants and $r(x)$ is a continuous function of x. The method we shall discuss, called *variation of parameters,* is based on the (not very obvious) fact that if

$$y_c(x) = c_1 y_1(x) + c_2 y_2(x) \tag{2}$$

---

*It is assumed in this section that the reader is familiar with Cramer's rule, which is discussed in Section VI of Appendix C.

is the general solution of the complementary equation

$$y'' + py' + qy = 0$$

then it is possible to find functions $u(x)$ and $v(x)$ such that

$$y_p(x) = u(x)y_1(x) + v(x)y_2(x) \tag{3}$$

is a particular solution of (1).

To see how the functions $u(x)$ and $v(x)$ can be found, consider the derivative of (3):

$$y_p' = (uy_1' + vy_2') + (u'y_1 + v'y_2) \tag{4}$$

If we now differentiate y_p', we shall introduce second derivatives of the unknown functions $u(x)$ and $v(x)$. To avoid this complication we shall require that $u(x)$ and $v(x)$ satisfy the condition

$$u'y_1 + v'y_2 = 0 \tag{5}$$

so (4) simplifies to

$$y_p' = uy_1' + vy_2'$$

Thus,

$$y_p'' = uy_1'' + u'y_1' + vy_2'' + v'y_2' \tag{6}$$

On substituting (3), (4), and (6) in (1) and rearranging terms we obtain

$$u(y_1'' + py_1' + qy_1) + v(y_2'' + py_2' + qy_2) + u'y_1' + v'y_2' = r(x) \tag{7}$$

Since y_1 and y_2 are solutions of the complementary equation, we have

$$y_1'' + py_1' + qy_1 = 0 \quad \text{and} \quad y_2'' + py_2' + qy_2 = 0$$

so (7) simplifies to

$$u'y_1' + v'y_2' = r(x) \tag{8}$$

Equations (5) and (8) together yield two equations in the two unknowns $u'(x)$ and $v'(x)$:

$$\begin{aligned} u'y_1 + v'y_2 &= 0 \\ u'y_1' + v'y_2' &= r(x) \end{aligned} \tag{9}$$

It can be shown (see Boyce and DiPrima, *Elementary Differential Equations and Boundary Value Problems*, John Wiley & Sons, New York, 1988) that this system has a unique solution for u' and v'. Once this solution is found, u and v can be obtained by integration.

Example 1   Find the general solution of

$$y'' + y = \sec x \qquad \left(-\frac{\pi}{2} < x < \frac{\pi}{2} \right) \tag{10}$$

Solution.   We begin by finding the general solution of the complementary equation

$$y'' + y = 0$$

The auxiliary equation for this homogeneous equation is

$$m^2 + 1 = 0$$

which has roots $m_1 = i$ and $m_2 = -i$. Thus, the general solution of the complementary equation is

$$y_c(x) = c_1 \cos x + c_2 \sin x$$

Comparing this with (2) yields $y_1(x) = \cos x$ and $y_2(x) = \sin x$, so we shall look for a particular solution of the form

$$y_p(x) = u(x) \cos x + v(x) \sin x \tag{11}$$

Substituting $y_1 = \cos x$, $y_2 = \sin x$, and $r(x) = \sec x$ in (9) yields

$$\begin{aligned} u' \cos x + v' \sin x &= 0 \\ -u' \sin x + v' \cos x &= \sec x \end{aligned} \tag{12}$$

Solving this system for u' and v' by Cramer's rule we obtain

$$u' = \frac{\begin{vmatrix} 0 & \sin x \\ \sec x & \cos x \end{vmatrix}}{\begin{vmatrix} \cos x & \sin x \\ -\sin x & \cos x \end{vmatrix}} = \frac{-\sec x \sin x}{\cos^2 x + \sin^2 x} = \frac{-\tan x}{1} = -\tan x$$

$$v' = \frac{\begin{vmatrix} \cos x & 0 \\ -\sin x & \sec x \end{vmatrix}}{\begin{vmatrix} \cos x & \sin x \\ -\sin x & \cos x \end{vmatrix}} = \frac{\cos x \sec x}{\cos^2 x + \sin^2 x} = \frac{1}{1} = 1$$

Integrating u' and v' yields

$$u(x) = \int -\tan x \, dx = \ln |\cos x| \quad \text{and} \quad v(x) = \int 1 \, dx = x$$

[We set the constants of integration equal to zero because any functions $u(x)$ and $v(x)$ satisfying (12) will suffice.] Substituting these functions in (11) yields

$$y_p(x) = (\ln |\cos x|) \cos x + x \sin x$$

Thus, the general solution of (10) is

$$y(x) = y_c(x) + y_p(x) = c_1 \cos x + c_2 \sin x + (\ln |\cos x|) \cos x + x \sin x \quad \blacktriangleleft$$

▶ Exercise Set 14.3

In Exercises 1–6, use variation of parameters to find the general solution of the differential equation and check your result using undetermined coefficients.

1. $y'' + y = x^2$.

2. $y'' + 9y = 3x$.

3. $y'' + y' - 2y = 2e^x$.

4. $y'' + 5y' + 6y = e^{-x}$.

5. $y'' + 4y = \sin 2x$.

6. $y'' + 9y = \cos 3x$.

In Exercises 7–28, use variation of parameters to find the general solution of the differential equation.

7. $y'' + y = \tan x$.

8. $y'' + y = \cot x$.

9. $y'' - 2y' + y = e^x/x$.

10. $y'' - 4y' + 4y = e^{2x}/x$.

11. $y'' + y = 3 \sin^2 x$.

12. $y'' + y = 6 \cos^2 x$.

13. $y'' + y = \csc x$.

14. $y'' + 9y = 6 \sec 3x$.

15. $y'' + y = \sec x \tan x$.

16. $y'' + y = \csc x \cot x$.

17. $y'' + 2y' + y = e^{-x}/x^2$.

18. $y'' - y = x^2 e^x$.

19. $y'' + 4y' + 4y = xe^{-x}$.

20. $y'' + 4y' + 4y = xe^{2x}$.

21. $y'' + y = \sec^2 x$.

22. $y'' + y = \sec^3 x$.

23. $y'' - 2y' + y = e^x/x^2$.

24. $y'' - 2y' + y = x^3 e^x$.

25. $y'' - y = e^x \cos x$.

26. $y'' - 2y' + 2y = e^{2x} \sin x$.

27. $y'' + 2y' + y = e^{-x} \ln|x|$.

28. $y'' - 3y' + 2y = \dfrac{e^x}{1 + e^x}$.

29. Use the method of variation of parameters to show that the general solution of $y'' + y = r(x)$ is

$$y = c_1 \cos x + c_2 \sin x + g(x) \cos x + h(x) \sin x$$

where

$$g(x) = -\int r(x) \sin x \, dx$$

$$h(x) = \int r(x) \cos x \, dx$$

30. Use the method of variation of parameters to show that if y_1 and y_2 are linearly independent solutions of $y'' + py' + qy = 0$, then a particular solution of $y'' + py' + qy = r(x)$ is given by

$$y_p(x) = -y_1(x) \int \frac{y_2(x)r(x)}{W(x)} \, dx + y_2(x) \int \frac{y_1(x)r(x)}{W(x)} \, dx$$

where $W(x) = y_1(x)y_2'(x) - y_1'(x)y_2(x)$.

■ **14.4   VIBRATION OF A SPRING**

In this section we shall use second-order linear differential equations with constant coefficients to study the vibration of a spring.

We shall be interested in solving the following problem.

14.4.1   THE VIBRATING SPRING PROBLEM.   *As shown in* Figure 14.4.1, *let a mass attached to a vertical spring be allowed to settle into an equilibrium position. Then, let the spring be stretched (or compressed) by pulling (or pushing) the mass, and finally, let the mass be released, thereby causing it to undergo a vibratory motion. We shall be interested in finding a formula for the position of the mass at any time t.*

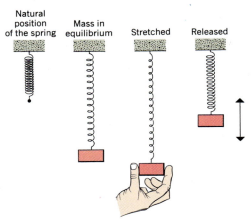

Figure 14.4.1

To solve this problem we shall need three results from physics:

HOOKE'S LAW. *If a spring is stretched (or compressed) l units beyond its natural position, then it pulls back (or pushes) with a force of magnitude*

$$F = kl$$

*where k is a positive constant, called the **spring constant**. This constant depends on such factors as the thickness of the spring, the material from which it is made, and the units of force and length; it is measured in units of force per unit of length.*

NEWTON'S SECOND LAW OF MOTION. *If an object with mass M is subjected to a force **F**, then it undergoes an acceleration **a** satisfying*

$$\mathbf{F} = M\mathbf{a}$$

WEIGHT. *The gravitational force exerted by the earth on an object is called the **weight** (more precisely, **earth weight**) of the object. It follows from Newton's Second Law of Motion that an object with mass M has a weight whose magnitude w is given by*

$$w = Mg \tag{1}$$

*where g is a constant, called the **acceleration due to gravity**. Near the surface of the earth an approximate value of g is $g \approx 32$ ft/sec^2 if length is measured in feet and time in seconds, or $g \approx 980$ cm/sec^2 if length is measured in centimeters and time in seconds.*

In any problem involving length, mass, time, and force it is important that the units of measurement be consistent. The most important systems of measurement are summarized in Table 14.4.1.

Table 14.4.1

SYSTEM OF MEASUREMENT	LENGTH	TIME	MASS	FORCE
Engineering system	foot (ft)	second (sec)	slug	pound (lb)
Mks system	meter (m)	second (sec)	kilogram (kg)	newton (N)
cgs system	centimeter (cm)	second (sec)	gram (g)	dyne

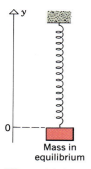

Figure 14.4.2

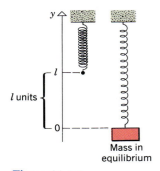

Figure 14.4.3

To solve the vibrating spring problem posed above, we introduce a coordinate axis (a y-axis) with the positive direction up and the origin at the bottom end of the spring when the mass is in its equilibrium position (Figure 14.4.2). If we take $t = 0$ to be the time at which the mass is released, then at each subsequent time t, the end of the spring has a position $y(t)$, a velocity $y'(t)$, and an acceleration $y''(t)$.* Because the positive direction of the y-axis is up, force, velocity, and acceleration are positive when directed up and negative when directed down.

Before we can solve the spring problem, it will be necessary to derive a preliminary result from the equilibrium conditions for the mass. Let us assume that the spring constant is k, the mass of the object is M, and the spring is stretched l units beyond its natural length when the mass is in equilibrium (Figure 14.4.3). When the mass is in its equilibrium position its downward weight, $-Mg$, is balanced exactly by the upward force, kl, of the spring, so the sum of these forces must be zero†; that is,

$$kl - Mg = 0$$

or

$$kl = Mg \qquad (2)$$

The basic strategy for solving the spring problem is to find the total force $F(t)$ that acts on the mass at time t. Then, since the acceleration of the mass at time t is $y''(t)$, it will follow from Newton's Second Law of Motion that

$$My''(t) = F(t)$$

or

$$y''(t) = F(t)/M \qquad (3)$$

which is a second-order linear differential equation that can be solved for the position function $y(t)$.

---

*For convenience of terminology we shall refer to $y(t)$, $y'(t)$, and $y''(t)$ as the position, velocity, and acceleration of the mass, respectively.

†We shall assume that the weight of the spring is small relative to the weight of the mass and can be neglected.

At each instant, the force $F(t)$ in (3) consists of four possible components:

$F_g(t)$ = the force of gravity (weight of the mass)

$F_s(t)$ = the force of the spring

$F_d(t)$ = the **damping force** or frictional force exerted on the mass by the surrounding medium (air, water, oil, etc.)

$F_e(t)$ = external forces due to such factors as movement in the spring support, magnetic forces acting on the mass, and so on.

☐ **UNDAMPED FREE VIBRATIONS**

The simplest case occurs when there is no damping ($F_d = 0$) and the mass is *free* of external forces ($F_e = 0$). In this case the only forces acting on the mass are the force of gravity, F_g, and the spring force, F_s. Let us try to calculate these forces when the mass is at an arbitrary point $y(t)$.

When the mass is at the point $y(t)$, the spring is stretched (or compressed) $l - y(t)$ units from its natural length (Figure 14.4.4), so by Hooke's law the spring exerts a force of

$$F_s(t) = k(l - y(t))$$

on the mass. Adding this to the force of gravity,

$$F_g(t) = -Mg$$

acting on the mass yields the total force acting on the mass:

$$F_s(t) + F_g(t) = k(l - y(t)) - Mg$$

But $kl = Mg$ from (2), so

$$F_s(t) + F_g(t) = -ky(t) \tag{4}$$

Substituting this expression for $F(t)$ in (3) yields

$$y''(t) = -\frac{k}{M}y(t)$$

or

$$y''(t) + \frac{k}{M}y(t) = 0 \tag{5}$$

which is a second-order linear differential equation with constant coefficients.

Because the mass is *released* (i.e., has zero initial velocity) at time $t = 0$, we have $y'(0) = 0$. If we assume, in addition, that the position of the mass at time $t = 0$ is $y(0) = y_0$, then we have two initial conditions that can be combined with (5) to yield an initial-value problem for $y(t)$:

$$y'' + \frac{k}{M}y = 0$$

$$y(0) = y_0, \quad y'(0) = 0 \tag{6}$$

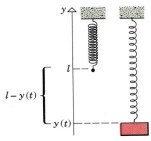

Figure 14.4.4

The auxiliary equation for the differential equation in (6) is

$$m^2 + \frac{k}{M} = 0$$

which has roots $m_1 = \sqrt{k/M}\, i$, $m_2 = -\sqrt{k/M}\, i$ (since k and M are positive), so the general solution of the differential equation is

$$y(t) = c_1 \cos(\sqrt{k/M}\, t) + c_2 \sin(\sqrt{k/M}\, t)$$

From the initial conditions $y(0) = y_0$ and $y'(0) = 0$ it follows that $c_1 = y_0$, $c_2 = 0$ (verify), so the solution of (6) is

$$y(t) = y_0 \cos(\sqrt{k/M}\, t) \tag{7}$$

This formula describes a periodic vibration with an *amplitude* of $|y_0|$ and *period* T given by

$$T = \frac{2\pi}{\sqrt{k/M}} = 2\pi \sqrt{M/k} \tag{8}$$

(Figure 14.4.5). The *frequency f* of the vibration is the number of cycles per unit time, so

$$f = \frac{1}{T} = \frac{1}{2\pi} \sqrt{k/M} \tag{9}$$

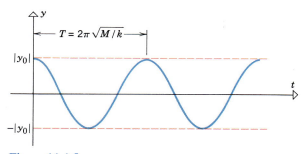

Figure 14.4.5

REMARK. In the preceding problem, the spring oscillates indefinitely with constant amplitude because there are no frictional forces to dissipate the energy in the mass–spring system.

Example 1 Suppose that the top of a spring is fixed to a ceiling and a mass attached to the bottom end stretches the spring $\frac{1}{2}$ foot. If the mass is pulled 2 feet below its equilibrium position and released, find

(a)   a formula for the position of the mass at any time t;

(b)   the amplitude, period, and frequency of the motion.

Solution (a).   The appropriate formula is (7). Although we are not given the mass M or the spring constant k, it does not matter because we are given that the mass stretches the spring $l = \frac{1}{2}$ ft and we know that $g = 32$ ft/sec^2, so (2) implies that

$$\frac{k}{M} = \frac{g}{l} = \frac{32}{\frac{1}{2}} = 64 \text{ lb/ft}$$

Also, the mass is initially 2 units *below* its equilibrium position, so $y_0 = -2$. Therefore, from (7) the formula for the position of the mass at time t is

$$y(t) = -2 \cos 8t$$

Solution (b).   From (8) and (9)

$$\text{amplitude} = |y_0| = |-2| = 2 \text{ ft}$$

$$\text{period} = T = 2\pi\sqrt{M/k} = 2\pi\sqrt{1/64} = \pi/4 \text{ sec/cycle}$$

$$\text{frequency} = f = 1/T = 4/\pi \text{ cycles/sec} \quad \blacktriangleleft$$

☐ **DAMPED FREE VIBRATIONS**

We shall now consider the solution of the spring problem in the case where damping cannot be neglected; that is, $F_d \neq 0$.

Physicists have shown that under appropriate conditions the damping force F_d is opposite to the direction of motion of the mass (i.e., tends to slow the mass down) and has a magnitude that is proportional to the speed of the mass (the greater the speed, the greater the effect of friction). This type of damping force is described by an equation of the form

$$F_d(t) = -cy'(t) \tag{10}$$

where c is a positive constant, called the *damping constant*. The damping constant depends on the viscosity of the surrounding medium; it is measured in units of force per unit of velocity (lb/ft/sec or equivalently lb · sec/ft, for example).

It follows from (4) and (10) that

$$F_s(t) + F_g(t) + F_d(t) = -ky(t) - cy'(t)$$

If we substitute this expression in (3) for $F(t)$, we obtain

$$y''(t) = -\frac{k}{M}y(t) - \frac{c}{M}y'(t)$$

or

$$y''(t) + \frac{c}{M}y'(t) + \frac{k}{M}y(t) = 0$$

Combining this equation with the initial conditions $y(0) = y_0$, $y'(0) = 0$ yields the following initial-value problem whose solution describes the motion of the mass subject to damping:

$$y'' + \frac{c}{M}y' + \frac{k}{M}y = 0$$

$$y(0) = y_0, \quad y'(0) = 0 \tag{11}$$

The form of the solution to (11) will depend on whether the auxiliary equation

$$m^2 + \frac{c}{M}m + \frac{k}{M} = 0 \tag{12}$$

has distinct real roots, equal real roots, or complex roots. We leave it for the reader to show that the roots are distinct and real if $c^2 > 4kM$, equal and real if $c^2 = 4kM$, and complex if $c^2 < 4kM$ (Exercise 22).

☐ **UNDERDAMPED VIBRATIONS**

The cases $c^2 > 4kM$ and $c^2 = 4kM$ are called **overdamped** and **critically damped**, respectively. In these cases vibration in the usual sense does not occur: the mass simply drifts slowly toward its equilibrium position without ever passing through it. We shall leave the analysis of these cases for the exercises and concentrate on the case $c^2 < 4kM$ in which true vibratory motion occurs. This is called the **underdamped case**.

In the underdamped case the roots of the auxiliary equation in (12) are

$$m_1 = -\frac{c}{2M} + \frac{\sqrt{4Mk - c^2}}{2M}i, \quad m_2 = -\frac{c}{2M} - \frac{\sqrt{4Mk - c^2}}{2M}i$$

(Verify.) For convenience, let

$$\alpha = \frac{c}{2M}, \quad \beta = \frac{\sqrt{4Mk - c^2}}{2M} \tag{13}$$

so the solution of the differential equation in (11) is

$$y(t) = e^{-\alpha t}(c_1 \cos \beta t + c_2 \sin \beta t) \tag{14}$$

It follows that

$$y'(t) = -\alpha e^{-\alpha t}(c_1 \cos \beta t + c_2 \sin \beta t) + \beta e^{-\alpha t}(-c_1 \sin \beta t + c_2 \cos \beta t)$$

so the initial conditions $y(0) = y_0$, $y'(0) = 0$ yield the equations

$$c_1 = y_0$$
$$-\alpha c_1 + \beta c_2 = 0$$

which can be solved to obtain

$$c_1 = y_0, \quad c_2 = \alpha y_0/\beta$$

Substituting these values in (14) yields the solution of (11):

$$y(t) = \frac{y_0}{\beta} e^{-\alpha t}(\beta \cos \beta t + \alpha \sin \beta t) \tag{15}$$

In the exercises (Exercise 23) we ask the reader to use the cosine addition formula to show that this solution can be rewritten in the alternative form

$$y(t) = \frac{y_0\sqrt{\alpha^2 + \beta^2}}{\beta} e^{-\alpha t} \cos(\beta t - \omega) \tag{16a}$$

where

$$\omega = \tan^{-1}\left(\frac{\alpha}{\beta}\right), \quad 0 < \omega < \pi/2 \tag{16b}$$

Since $\cos(\beta t - \omega)$ has values between $+1$ and -1, it follows from (16a) and (16b) that the graph of $y(t)$ oscillates between the curves

$$y = \frac{y_0\sqrt{\alpha^2 + \beta^2}}{\beta} e^{-\alpha t} \quad \text{and} \quad y = -\frac{y_0\sqrt{\alpha^2 + \beta^2}}{\beta} e^{-\alpha t}$$

Thus, the graph of $y(t)$ resembles a cosine curve, but with decreasing amplitude (Figure 14.4.6). Strictly speaking, the function $y(t)$ is not periodic. However, $\cos(\beta t - \omega)$ has a period of $2\pi/\beta$, so the displacement $y(t)$ reaches a relative maximum at times spaced $2\pi/\beta$ units apart. Thus, we shall define the **period T** of this motion as

$$T = \frac{2\pi}{\beta} = \frac{4M\pi}{\sqrt{4Mk - c^2}} \tag{17}$$

(Figure 14.4.6) and the **frequency** to be

$$f = \frac{1}{T} = \frac{\sqrt{4Mk - c^2}}{4M\pi} \tag{18}$$

If we think of a relative maximum as marking the end of one cycle of motion and the start of the next, then the period T is the time required for the completion of one cycle, and the frequency is the number of cycles that occur per unit time. The physically significant fact is that the frequency and period of the

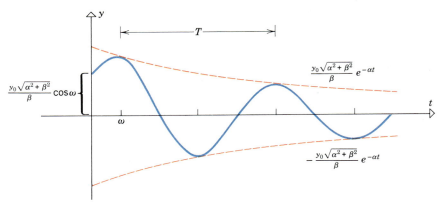

Figure 14.4.6

mass–spring system do not change as the amplitudes of the vibrations get smaller (why?). Galileo, who first observed this fact, used it to help design clocks.

Example 2 A spring with a spring constant of $k = 4$ lb/ft is attached to a ceiling and a 64-lb weight is attached to the bottom end. The weight is pushed 2 ft above its equilibrium position and released with an initial velocity of 0. Assuming that the damping constant due to air resistance is $c = 4$ lb/ft, find

(a) a formula for the position of the weight at any time t;

(b) the period and frequency of the vibration.

Solution (a). We shall first use (13) to find α and β, after which we can use either (15) or (16a) and (16b). The attached weight is 64 lb, so from (1) its mass is

$$M = \frac{w}{g} = \frac{64}{32} = 2 \text{ slugs}$$

Thus,

$$\alpha = \frac{c}{2M} = \frac{4}{4} = 1, \quad \beta = \frac{\sqrt{4Mk - c^2}}{2M} = \frac{4}{4} = 1$$

Since the weight is initially 2 ft above its equilibrium position, we have $y_0 = 2$, so from (15) the position function of the weight is

$$y(t) = 2e^{-t}(\cos t + \sin t)$$

Alternatively, we can apply (16a) and (16b), but we must first calculate ω. Since

$$\omega = \tan^{-1}\left(\frac{\alpha}{\beta}\right) = \tan^{-1}(1) = \frac{\pi}{4}$$

(16a) and (16b) yield the alternative formula

$$y(t) = 2\sqrt{2}e^{-t} \cos\left(t - \frac{\pi}{4}\right)$$

Solution (b). From (17) and (18)

$$T = \frac{2\pi}{\beta} = 2\pi \quad \text{sec/cycle}$$

$$f = \frac{1}{T} = \frac{1}{2\pi} \quad \text{cycles/sec} \quad \blacktriangleleft$$

☐ **FORCED VIBRATIONS**

If there are external forces such as movement of the spring support or magnetic forces acting on the mass, then $F_e \neq 0$ and the vibrations are said to be *forced*. The study of forced vibrations leads to important phenomena such as *resonance* and *beats,* but this is beyond the scope of this section. Readers interested in this topic are referred to *Elementary Differential Equations and Boundary Value Problems*, John Wiley & Sons, New York, 1988, by William E. Boyce and Richard C. DiPrima.

▶ Exercise Set 14.4

In this exercise set assume that the y-axis is oriented as in Figure 14.4.2.

1. A weight of 64 lb is attached to a vertical spring with spring constant $k = 8$ lb/ft. The weight is pushed 1 ft above its equilibrium position and released.
   (a) Find an initial-value problem whose solution $y(t)$ is the position function of the weight, assuming that there is no damping.
   (b) Solve the initial-value problem.
   (c) Check the solution in (b) using Formula (7).

2. A mass of 1000 g is attached to a vertical spring with spring constant $k = 25$ dyne/cm. The mass is pushed 50 cm above its equilibrium position and released.
   (a) Find an initial-value problem whose solution $y(t)$ is the position function of the mass, assuming that there is no damping.
   (b) Solve the initial-value problem.
   (c) Check the solution in (b) using Formula (7).

3. A mass attached to a vertical spring stretches the spring 5 cm. The mass is pulled 10 cm below its equilibrium position and released.
   (a) Find an initial-value problem whose solution $y(t)$ is the position function of the mass, assuming that there is no damping.
   (b) Solve the initial-value problem.
   (c) Check the solution in (b) using Formula (7).

4. A mass attached to a vertical spring stretches the spring 2 ft. The mass is pulled 4 ft below its equilibrium position and released.
   (a) Find an initial-value problem whose solution $y(t)$ is the position function of the mass, assuming that there is no damping.
   (b) Solve the initial-value problem.
   (c) Check the solution in (b) using Formula (7).

5. A weight of $\frac{1}{2}$ lb is attached to a vertical spring with spring constant $k = 1$ lb/ft. The weight is pushed 2 ft above its equilibrium position and released. Use Formulas (7), (8), and (9) to find
   (a) the position function of the weight
   (b) the amplitude of the vibration
   (c) the period of the vibration
   (d) the frequency of the vibration.

6. A mass of 2 slugs is attached to a vertical spring with spring constant $k = 4$ lb/ft. The mass is pushed

1 ft above its equilibrium position and released. Use Formulas (7), (8), and (9) to find
   (a) the position function of the mass
   (b) the amplitude of the vibration
   (c) the period of the vibration
   (d) the frequency of the vibration.

7. A mass attached to a vertical spring stretches the spring 1 in. The mass is pulled 3 in. below its equilibrium position and released. Use Formulas (7), (8), and (9) to find
   (a) the position function of the mass
   (b) the amplitude of the vibration
   (c) the period of the vibration
   (d) the frequency of the vibration.

8. A mass attached to a vertical spring stretches the spring 8 m. The mass is pulled 2 m below its equilibrium position and released. Use Formulas (7), (8), and (9) to find
   (a) the position function of the mass
   (b) the amplitude of the vibration
   (c) the period of the vibration
   (d) the frequency of the vibration.

9. A weight of 32 lb is attached to a vertical spring with spring constant $k = 8$ lb/ft. The surrounding medium has a damping constant of $c = 4$ lb · sec/ft. The weight is pulled 3 ft below its equilibrium position and released.
   (a) Find an initial-value problem whose solution $y(t)$ is the position function of the weight.
   (b) Solve the initial-value problem.
   (c) Check the solution in (b) using Formula (15).
   (d) Express the solution in the form of (16a).
   (e) Find the period of the vibration.
   (f) Find the frequency of the vibration.

10. A mass of 3 slugs is attached to a vertical spring with spring constant $k = 9$ lb/ft. The surrounding medium has a damping constant of $c = 6$ lb · sec/ft. The mass is pulled 1 ft below its equilibrium position and released.
    (a) Find an initial-value problem whose solution $y(t)$ is the position function of the mass.
    (b) Solve the initial-value problem.
    (c) Check the solution in (b) using Formula (15).
    (d) Express the solution in the form of (16a).
    (e) Find the period of the vibration.
    (f) Find the frequency of the vibration.

11. A mass of 25 g is attached to a vertical spring with spring constant $k = 3$ dyne/cm. The surrounding medium has a damping constant of $c = 10$ dyne · sec/cm. The mass is pushed 5 cm above its equilibrium position and released.
  (a) Use Formula (16a) to find the position function of the mass.
  (b) Find the period of the vibration.
  (c) Find the frequency of the vibration.

12. A mass of 490 g is attached to a vertical spring with spring constant $k = 40$ dyne/cm. The surrounding medium has a damping constant of $c = 140$ dyne · sec/cm. The mass is pushed 20 cm above its equilibrium position and released.
  (a) Use Formula (16a) to find the position function of the mass.
  (b) Find the period of the vibration.
  (c) Find the frequency of the vibration.

> If the object in Figure 14.4.1 is given an initial velocity v_0 rather than being released with initial velocity 0, then in Formulas (6) and (11) the initial conditions become $y(0) = y_0$, $y'(0) = v_0$. Use this fact in Exercises 13–16.

13. A weight of 3 lb attached to a vertical spring stretches the spring $\frac{1}{2}$ ft. While in its equilibrium position, the weight is struck to give it a downward initial velocity of 2 ft/sec. Assuming that there is no damping, find
  (a) the position function of the weight
  (b) the amplitude of the vibration
  (c) the period of the vibration
  (d) the frequency of the vibration.

14. A mass of 64 slugs attached to a vertical spring stretches the spring $1\frac{1}{2}$ in. The mass is pulled 4 in. below its equilibrium position and struck to give it a downward initial velocity of 8 ft/sec. Assuming that there is no damping, find
  (a) the position function of the weight
  (b) the amplitude of the vibration
  (c) the period of the vibration
  (d) the frequency of the vibration.

15. A weight of 4 lb is attached to a vertical spring with spring constant $k = 6\frac{1}{4}$ lb/ft. The weight is pushed $\frac{1}{3}$ ft above its equilibrium position and struck to give it a downward initial velocity of 5 ft/sec. Assuming that the damping constant is $c = \frac{1}{4}$ lb · sec/ft, find the position function of the weight.

16. A weight of 49 dynes is attached to a vertical spring with spring constant $k = \frac{1}{4}$ dyne/cm. The weight is pushed 1 cm above its equilibrium position and struck to give it an upward initial velocity of 2 cm/sec. Assuming that the damping constant is $c = \frac{1}{5}$ dyne · sec/cm, find the position function of the weight.

17. A weight of w pounds is attached to a spring, then pulled below its equilibrium position and released, thereby causing it to vibrate with a period of 3 sec. When 4 additional pounds are added, the period becomes 5 sec. Assuming there is no damping, find
  (a) the spring constant k (b) the weight w.

18. As illustrated in the following figure, let a toy cart of mass M be attached to a wall by a spring with spring constant k, and let an x-axis be introduced as shown with its origin at the point where the cart is in equilibrium. Suppose that the cart is pulled or pushed horizontally, then released. Find a differential equation for the position function of the cart if
  (a) there is no damping
  (b) there is a damping constant of c.

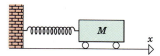

19. A cylindrical buoy of height h and radius r floats in the water with its axis vertical. By *Archimedes' principle* the water exerts an upward force on the buoy (the buoyancy force) with magnitude equal to the weight of the water displaced. Neglecting all forces except the buoyancy force and the force of gravity, determine the period with which the buoy will vibrate vertically if it is depressed slightly from its equilibrium position and released. Take ρ lb/in³ to be the density of water and δ lb/in³ to be the density of the buoy material (density = weight per unit of volume).

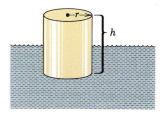

20. Suppose that the mass of the object in Figure 14.4.1 is M and the spring constant is k. Show that if the mass is displaced y_0 units from its equilibrium position and given an initial velocity of v_0 rather than being released with velocity 0, then its position function is given by

$$y(t) = y_0 \cos\left(\sqrt{\frac{k}{M}}\, t\right) + v_0 \sqrt{\frac{M}{k}} \sin\left(\sqrt{\frac{k}{M}}\, t\right)$$

21. A mass attached to a vertical spring is displaced from its equilibrium position and released, thereby causing it to vibrate with amplitude $|y_0|$ and period T (no damping).

(a) Show that the velocity of the mass has maximum magnitude $2\pi|y_0|/T$ and that the maximum occurs when the mass is at its equilibrium position.

(b) Show that the acceleration of the mass has maximum magnitude $4\pi^2|y_0|/T^2$ and that the maximum occurs when the mass is at a top or bottom point of its motion.

22. Prove that the roots of Equation (12) are distinct and real, equal and real, or complex according to whether $c^2 > 4kM$, $c^2 = 4kM$, or $c^2 < 4kM$.

23. Use the addition formula for cosine to show that Formula (15) can be rewritten as (16a).

24. Overdamped Motion

(a) Prove that in the case of overdamped motion $(c^2 > 4kM)$ the solution of (11) is

$$y(t) = \frac{y_0}{m_2 - m_1} (m_2 e^{m_1 t} - m_1 e^{m_2 t})$$

(b) Prove that $\lim_{t \to +\infty} y(t) = 0$. [*Hint:* Show that m_1 and m_2 are negative.]

25. Critically Damped Motion

(a) Prove that in the case of critically damped motion $(c^2 = 4kM)$ the solution of (11) is

$$y(t) = y_0 e^{-\alpha t}(1 + \alpha t)$$

(b) Prove that $\lim_{t \to +\infty} y(t) = 0$.

(c) Prove that $y(t) \neq 0$ for any positive value of t (assuming that the initial displacement y_0 is not zero).

Appendix

REVIEW OF SETS

In this appendix we shall review the basic notation and operations on sets.

A *set* can be viewed as a collection of objects, called *members* or *elements* of the set. In this text most of the sets of interest will consist of real numbers or points in two- or three-dimensional space. As discussed in the subsection on intervals in Section 1.1, sets can be described by listing the elements between braces or by using the notation $\{x: \underline{\hspace{1cm}}\}$.

It is common to denote sets by capital letters and general elements of sets by lowercase letters. To indicate that an element a is a member of a set A, we write

$$a \in A$$

which is read, "a is an element of A" or "a belongs to A." To indicate that the element a is *not* a member of the set A, we write

$$a \notin A$$

which is read, "a is not an element of A." or "a does not belong to A."

Example 1 Let A be the set of all rational numbers and B the set of all real values of x that satisfy the inequalities $2 < x < 4$. Then

$$\tfrac{3}{4} \in A, \quad -2 \in A, \quad \pi \notin A, \quad -\sqrt{2} \notin A$$
$$2.5 \in B, \quad \pi \in B, \quad -1 \notin B, \quad 4 \notin B \quad \blacktriangleleft$$

Sometimes sets arise that have no members. Such a set is denoted by the symbol $\varnothing$ and is called an *empty set* or a *null set*. Thus,

$$\{x:x \text{ is a real number satisfying } x^2 < 0\} = \varnothing$$

DEFINITION **1.**   Two sets A and B are said to be *equal* if they have the same elements, in which case we write $A = B$.

Example 2

$$\{x : x^2 = 1\} = \{-1, 1\}, \quad \{\pi, 0, 3\} = \{3, \pi, 0\}$$
$$\{x : x^2 < 9\} = \{x : -3 < x < 3\} \quad \blacktriangleleft$$

DEFINITION 2. If every member of a set A is also a member of set B, then we say A is a **subset** of B and write $A \subset B$.

REMARK. By convention, an empty set $\varnothing$ is a subset of every set.

Example 3

$$\{-2, 4\} \subset \{-2, 1, 0, 4\}$$
$$\{x : x \text{ is rational}\} \subset \{x : x \text{ is a real number}\}$$
$$\varnothing \subset A \text{ (for every set } A) \quad \blacktriangleleft$$

REMARK. If $A \subset B$ and $B \subset A$, then $A = B$. (Why?)

DEFINITION 3. If A and B are two given sets, then

(a) the set of all elements belonging to both A and B is denoted by $A \cap B$ and is called the **intersection** of A and B;

(b) the set of all elements belonging to A or B or both is denoted by $A \cup B$ and is called the **union** of A and B.

The union and intersection of two sets, A and B, can be visualized geometrically by viewing A and B as sets of points in the plane enclosed by circles that overlap (Figure A.1a). This is called a **Venn diagram**. The sets $A \cap B$ and $A \cup B$ are shown as shaded regions in Figures A.1b and A.1c.

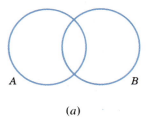

(a)

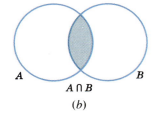

$A \cap B$

(b)

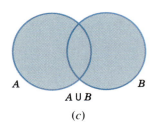

$A \cup B$

(c)

Figure A.1

Example 4   If

$$A = \{a, b, c, d, e\} \quad \text{and} \quad B = \{b, d, e, g\}$$

then

$$A \cap B = \{b, d, e\} \quad \text{and} \quad A \cup B = \{a, b, c, d, e, g\} \quad \blacktriangleleft$$

B

Appendix

TRIGONOMETRY REVIEW

■ I. TRIGONOMETRIC FUNCTIONS AND IDENTITIES

In this section we shall review the definitions and basic properties of trigonometric functions.

□ **ANGLES**

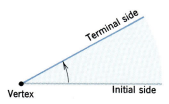

Figure B.1

We shall be concerned with angles in the plane that are generated by rotating a ray (or half-line) about its endpoint. The starting position of the ray is called the *initial side* of the angle and the final position of the ray is called the *terminal side*. The initial and terminal sides meet at a point called the *vertex* of the angle (Figure B.1).

An angle is considered *positive* if it is generated by a counterclockwise rotation and *negative* if it is generated by a clockwise rotation. In a rectangular coordinate system, an angle is said to be in *standard position* if its vertex is at the origin and its initial side is along the positive x-axis (Figure B.2). As shown in Figure B.3, an angle may be generated by making more than one complete revolution.

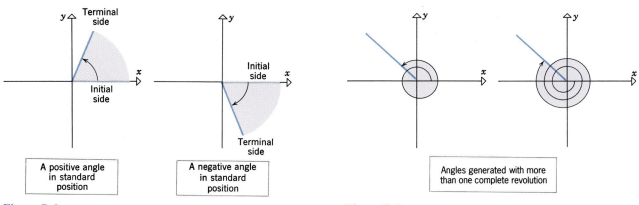

Figure B.2

Figure B.3

The size of an angle is commonly measured in *degrees*. One degree (written 1°) is the measure of an angle generated by a ray that rotates 1/360 of one revolution. Thus, there are 360° in an angle of one complete revolution, 180° in an angle of one-half revolution, 90° in an angle of one-quarter revolution, and so forth (Figure B.4).

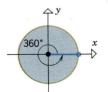 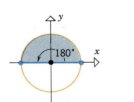 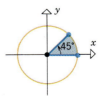 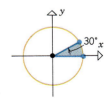

Figure B.4

Degrees are divided into sixty equal parts, called *minutes,* and minutes are divided into sixty equal parts, called *seconds*. Thus, one minute (written 1′) is 1/60 of a degree, and one second (written 1″) is 1/60 of a minute. Smaller subdivisions of a degree are expressed as fractions of a second.

In calculus, angles are measured in radians rather than degrees because it simplifies many important formulas. To define the radian measure of an angle, suppose that a circle of radius 1 is constructed so that its center is at the vertex of the angle. As the angle is generated by a ray rotating from the initial side to the terminal side, the intersection of this ray with the unit circle travels some distance d along the unit circle (Figure B.5). To distinguish between clockwise and counterclockwise rotations, we introduce the *signed distance* or *signed arc length* s traveled by the point; it is defined by

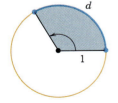

Figure B.5

$s = d$ if the rotation is counterclockwise

$s = -d$ if the rotation is clockwise

$s = 0$ if there is no rotation

The signed arc length s is called the *radian measure* of the angle generated by the rotating ray. Since a circle of radius 1 has a circumference of 2π, there are 2π radians in an angle of one complete revolution, π radians in an angle of one-half revolution, $\pi/2$ radians in an angle of one-quarter revolution, and so forth (Figure B.6).

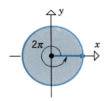 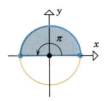 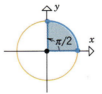 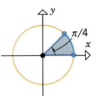 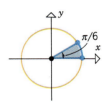

Figure B.6

The following table shows the relationship between the degree measure and radian measure of some of the more important positive angles.

DEGREES	30°	45°	60°	90°	120°	135°	150°	180°	270°	360°
RADIANS	$\dfrac{\pi}{6}$	$\dfrac{\pi}{4}$	$\dfrac{\pi}{3}$	$\dfrac{\pi}{2}$	$\dfrac{2\pi}{3}$	$\dfrac{3\pi}{4}$	$\dfrac{5\pi}{6}$	π	$\dfrac{3\pi}{2}$	2π

Observe that in the foregoing table the degree symbol was used for angles measured in degrees, but that the angles measured in radians were given as numbers with no units specified. This is standard practice in mathematics—when no units are specified for an angle, it is understood that the units are radians. Engineers are usually more specific; they use the symbol *rad* to denote radians. Thus, an angle of 2π radians can be denoted by 2π rad or by 2π.

From the fact that an angle of π radians corresponds to one of 180°, we obtain the following formulas, which are useful for converting from degrees to radians and conversely.

$$1° = \frac{\pi}{180} \text{ rad} \approx 0.01745 \text{ rad}$$

$$1 \text{ rad} = \left(\frac{180}{\pi}\right)° \approx 57° \ 17' \ 44.8''$$

Example 1

(a)   Express 146° in radians.

(b)   Express 3 radians in degrees.

Solution (a).   Since

$$1° = \frac{\pi}{180} \text{ rad}$$

it follows that

$$146° = \frac{\pi}{180} \cdot 146 \text{ rad} = \frac{73\pi}{90} \text{ rad} \approx 2.5482 \text{ rad}$$

Solution (b).   Since

$$1 \text{ rad} = \left(\frac{180}{\pi}\right)°$$

it follows that

$$3 \text{ rad} = \left(3 \cdot \frac{180}{\pi}\right)° = \left(\frac{540}{\pi}\right)° \approx 171.9° \quad \blacktriangleleft$$

☐ **THE RELATIONSHIP BETWEEN ARC LENGTH, ANGLE, AND RADIUS**

There is a theorem in plane geometry which states that for two concentric circles, the ratio of the arc lengths subtended by a central angle is equal to the corresponding ratio of the radii. Thus, for the circles in Figure B.7 we have

$$\frac{s_1}{s_2} = \frac{r_1}{r_2} \quad \text{or equivalently} \quad \frac{s_1}{r_1} = \frac{s_2}{r_2}$$

In particular, suppose that a central angle of θ radians subtends an arc of length s on a circle of radius r and an arc of length s' on a circle of radius 1. Then

$$\frac{s}{r} = \frac{s'}{1} = s' \tag{1}$$

But, by definition of radian measure, s' is the radian measure of the central angle, that is, $s' = \theta$. Substituting this in (1) yields the following fundamental relationship between the radian measure of an angle and the length of arc that it subtends on a circle of radius r centered at its vertex (Figure B.8):

$$\theta = \frac{s}{r} \tag{2}$$

This formula applies to negative as well as positive angles if s is interpreted as the signed arc length.

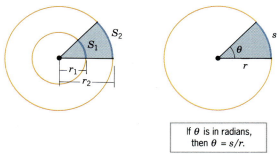

If θ is in radians, then $\theta = s/r$.

Figure B.7   Figure B.8

☐ **AREA OF A SECTOR**

The shaded region in Figure B.8 is called a *sector*. It is a theorem from plane geometry that the ratio of the area A of this sector to the area of the entire circle is the same as the ratio of the central angle of the sector to the central angle of the entire circle; thus, if the angles are in radians, we have

$$\frac{A}{\pi r^2} = \frac{\theta}{2\pi}$$

Solving for A yields the following formula for the area of a sector in terms of the radius r and the angle θ in radians:

$$A = \tfrac{1}{2}r^2\theta \tag{3}$$

☐ **TRIGONOMETRIC FUNCTIONS**

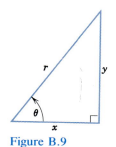

Figure B.9

The *sine, cosine, tangent, cosecant, secant,* and *cotangent* of a positive acute angle θ can be defined as ratios of the sides of a right triangle. Using the notation from Figure B.9, these definitions take the following form:

$$\sin \theta = \frac{\text{side opposite } \theta}{\text{hypotenuse}} = \frac{y}{r} \tag{4}$$

$$\cos \theta = \frac{\text{side adjacent to } \theta}{\text{hypotenuse}} = \frac{x}{r} \tag{5}$$

$$\tan \theta = \frac{\text{side opposite } \theta}{\text{side adjacent to } \theta} = \frac{y}{x} \tag{6}$$

$$\csc \theta = \frac{\text{hypotenuse}}{\text{side opposite } \theta} = \frac{r}{y} \tag{7}$$

$$\sec \theta = \frac{\text{hypotenuse}}{\text{side adjacent to } \theta} = \frac{r}{x} \tag{8}$$

$$\cot \theta = \frac{\text{side adjacent to } \theta}{\text{side opposite } \theta} = \frac{x}{y} \tag{9}$$

We shall call sin, cos, tan, csc, sec, and cot the *trigonometric functions*. Because similar triangles have proportional sides, the values of the trigonometric functions depend only on the size of θ and not on the particular right triangle used to compute the ratios. Moreover, in these definitions it does not matter whether θ is in degrees or radians.

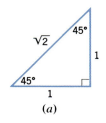

(a)

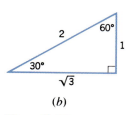

(b)

Figure B.10

Example 2 Recall from high school geometry that the two legs of a 45°–45°–90° triangle are of equal size. This fact and the Theorem of Pythagoras yield Figure B.10*a*. Recall also that the hypotenuse of a 30°–60°–90° triangle is twice the size of the shorter leg and that the shorter leg is opposite the 30° angle. These facts and the Theorem of Pythagoras yield Figure B.10*b*.

From Figure B.10 we obtain

$$\sin 45° = \frac{1}{\sqrt{2}} \quad \cos 45° = \frac{1}{\sqrt{2}} \quad \tan 45° = 1$$

$$\csc 45° = \sqrt{2} \quad \sec 45° = \sqrt{2} \quad \cot 45° = 1$$

$$\sin 30° = \frac{1}{2} \quad \cos 30° = \frac{\sqrt{3}}{2} \quad \tan 30° = \frac{1}{\sqrt{3}}$$

$$\csc 30° = 2 \quad \sec 30° = \frac{2}{\sqrt{3}} \quad \cot 30° = \sqrt{3}$$

$$\sin 60° = \frac{\sqrt{3}}{2} \quad \cos 60° = \frac{1}{2} \quad \tan 60° = \sqrt{3}$$

$$\csc 60° = \frac{2}{\sqrt{3}} \quad \sec 60° = 2 \quad \cot 60° = \frac{1}{\sqrt{3}} \qquad \blacktriangleleft$$

Since a right triangle cannot have an angle larger than 90°, Formulas (4)–(9) are not applicable if θ is obtuse. To obtain definitions of the trigonometric functions that apply to all angles, we take the following approach: Given an angle θ, introduce a coordinate system so that the angle is in standard position. Then construct a circle of arbitrary positive radius r centered at the origin, and let $P(x, y)$ be the point where the terminal side of the angle intersects the circle (Figure B.11). We make the following definition:

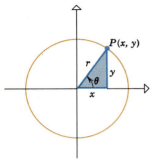

Figure B.11

DEFINITION 1.

$$\sin \theta = \frac{y}{r}, \quad \cos \theta = \frac{x}{r}, \quad \tan \theta = \frac{y}{x}$$

$$\csc \theta = \frac{r}{y}, \quad \sec \theta = \frac{r}{x}, \quad \cot \theta = \frac{x}{y}$$

The formulas in Definition 1 agree with Formulas (4)–(9). However, Definition 1 applies to all angles (positive, negative, acute, or obtuse), whereas Formulas (4)–(9) apply only to positive acute angles. Note that $\tan \theta$ and $\sec \theta$ are undefined if the terminal side of θ is on the y-axis (since $x = 0$), while $\csc \theta$ and $\cot \theta$ are undefined if the terminal side of θ is on the x-axis (since $y = 0$). The following relationships follow from Definition 1 (verify).

$$\csc \theta = \frac{1}{\sin \theta}, \quad \sec \theta = \frac{1}{\cos \theta}, \quad \cot \theta = \frac{1}{\tan \theta} \tag{10a}$$

$$\tan \theta = \frac{\sin \theta}{\cos \theta}, \quad \cot \theta = \frac{\cos \theta}{\sin \theta} \tag{10b}$$

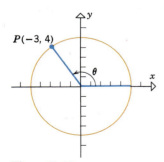

Figure B.12

Example 3   For the angle θ shown in Figure B.12 we have $x = -3$, $y = 4$, and $r = 5$, so

$$\sin \theta = \frac{4}{5}, \quad \cos \theta = -\frac{3}{5}, \quad \tan \theta = -\frac{4}{3}$$

$$\csc \theta = \frac{5}{4}, \quad \sec \theta = -\frac{5}{3}, \quad \cot \theta = -\frac{3}{4} \qquad \blacktriangleleft$$

☐ **TRIGONOMETRIC FUNCTIONS AND THE UNIT CIRCLE**

It can be proved using properties of similar triangles that the values of the trigonometric functions in Definition 1 do not depend on the radius r of the circle. In particular, we can use the circle of radius 1 (called the *unit circle*), in which case the formulas for $\sin \theta$ and $\cos \theta$ simplify to

$$x = \cos \theta, \quad y = \sin \theta$$

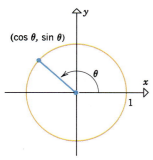

Figure B.13

Thus, we have the following geometric interpretation of $\sin\theta$ and $\cos\theta$ (Figure B.13):

THEOREM 2 (*Unit Circle Interpretation of* $\sin\theta$ *and* $\cos\theta$). *If an angle θ is placed in standard position, then its terminal side intersects the unit circle at the point* $(\cos\theta, \sin\theta)$.

Theorem 2, together with Formulas (10a) and (10b), can be used as follows to evaluate the trigonometric functions of many common angles:

- Construct the angle θ in standard position in an xy-coordinate system.
- Find the coordinates of the intersection of the terminal side of the angle and the unit circle; the x- and y-coordinates of this intersection are the values of $\cos\theta$ and $\sin\theta$, respectively.
- Use Formulas (10a) and (10b) to find the values of the remaining trigonometric functions from the values of $\cos\theta$ and $\sin\theta$.

The following example illustrates this method.

Example 4   Evaluate the trigonometric functions of $\theta = 150°$.

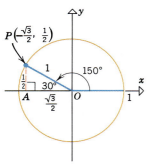

Figure B.14

Solution.   Construct a unit circle and place the angle $\theta = 150°$ in standard position (Figure B.14). Since $\angle AOP$ is $30°$ and $\triangle OAP$ is a $30°$–$60°$–$90°$ triangle, the leg AP has length $\frac{1}{2}$ (half the hypotenuse) and the leg OA has length $\sqrt{3}/2$ by the Theorem of Pythagoras. Thus, the coordinates of P are $(-\sqrt{3}/2, 1/2)$ and Theorem 2 yields

$$\sin 150° = \frac{1}{2}, \quad \cos 150° = -\frac{\sqrt{3}}{2}$$

From Formulas (10a) and (10b)

$$\tan 150° = \frac{\sin 150°}{\cos 150°} = \frac{1/2}{-\sqrt{3}/2} = -\frac{1}{\sqrt{3}}$$

$$\cot 150° = \frac{1}{\tan 150°} = -\sqrt{3}$$

$$\sec 150° = \frac{1}{\cos 150°} = -\frac{2}{\sqrt{3}}$$

$$\csc 150° = \frac{1}{\sin 150°} = 2 \quad \blacktriangleleft$$

Example 5 Evaluate the trigonometric functions at $\theta = 5\pi/6$.

Solution. Since $5\pi/6 = 150°$, this problem is equivalent to that of Example 4. From that example we obtain

$$\sin\frac{5\pi}{6} = \frac{1}{2}, \quad \cos\frac{5\pi}{6} = -\frac{\sqrt{3}}{2}, \quad \tan\frac{5\pi}{6} = -\frac{1}{\sqrt{3}}$$

$$\csc\frac{5\pi}{6} = 2, \quad \sec\frac{5\pi}{6} = -\frac{2}{\sqrt{3}}, \quad \cot\frac{5\pi}{6} = -\sqrt{3} \quad \blacktriangleleft$$

Example 6 Evaluate the trigonometric functions at $\theta = -\pi/2$.

Solution. As shown in Figure B.15, the terminal side of $\theta = -\pi/2$ intersects the unit circle at the point $(0, -1)$, so

$$\sin(-\pi/2) = -1, \quad \cos(-\pi/2) = 0$$

and from Formulas (10a) and (10b)

$$\tan(-\pi/2) = \frac{\sin(-\pi/2)}{\cos(-\pi/2)} = \frac{-1}{0} \quad \text{(undefined)}$$

$$\cot(-\pi/2) = \frac{\cos(-\pi/2)}{\sin(-\pi/2)} = \frac{0}{-1} = 0$$

$$\sec(-\pi/2) = \frac{1}{\cos(-\pi/2)} = \frac{1}{0} \quad \text{(undefined)}$$

$$\csc(-\pi/2) = \frac{1}{\sin(-\pi/2)} = \frac{1}{-1} = -1 \quad \blacktriangleleft$$

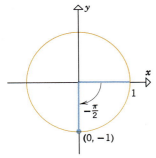

Figure B.15

The reader should be able to obtain all of the results in Table 1 by the methods illustrated in the last three examples.

Table 1

	$\theta = 0$ (0°)	$\pi/6$ (30°)	$\pi/4$ (45°)	$\pi/3$ (60°)	$\pi/2$ (90°)	$2\pi/3$ (120°)	$3\pi/4$ (135°)	$5\pi/6$ (150°)	π (180°)	$3\pi/2$ (270°)	2π (360°)
$\sin\theta$	0	1/2	$1/\sqrt{2}$	$\sqrt{3}/2$	1	$\sqrt{3}/2$	$1/\sqrt{2}$	1/2	0	-1	0
$\cos\theta$	1	$\sqrt{3}/2$	$1/\sqrt{2}$	1/2	0	$-1/2$	$-1/\sqrt{2}$	$-\sqrt{3}/2$	-1	0	1
$\tan\theta$	0	$1/\sqrt{3}$	1	$\sqrt{3}$	—	$-\sqrt{3}$	-1	$-1/\sqrt{3}$	0	—	0
$\csc\theta$	—	2	$\sqrt{2}$	$2/\sqrt{3}$	1	$2/\sqrt{3}$	$\sqrt{2}$	2	—	-1	—
$\sec\theta$	1	$2/\sqrt{3}$	$\sqrt{2}$	2	—	-2	$-\sqrt{2}$	$-2/\sqrt{3}$	-1	—	1
$\cot\theta$	—	$\sqrt{3}$	1	$1/\sqrt{3}$	0	$-1/\sqrt{3}$	-1	$-\sqrt{3}$	—	0	—

☐ **SIGNS OF THE TRIGONOMETRIC FUNCTIONS**

The signs of the trigonometric functions of an angle are determined by the quadrant in which the terminal side of the angle falls. For example, if the terminal side falls in the first quadrant, then x and y are positive in Definition 1, so all of the trigonometric functions have positive values. If the terminal side falls in the second quadrant, then x is negative and y is positive, so sin and csc are positive, but all other trigonometric functions are negative. The diagram in Figure B.16 shows which trigonometric functions are positive in the various quadrants. The reader will find it instructive to check that the results in Table 1 are consistent with Figure B.16.

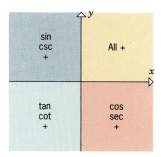

Figure B.16

☐ **TRIGONOMETRIC IDENTITIES**

A *trigonometric identity* is an equation involving trigonometric functions that is true for all angles for which both sides of the equation are defined. One of the most important identities in trigonometry can be derived by applying the Theorem of Pythagoras to the triangle in Figure B.11 to obtain

$$x^2 + y^2 = r^2$$

Dividing both sides by r^2 and using the definitions of $\sin\theta$ and $\cos\theta$ (Definition 1), we obtain the following fundamental result:

$$\sin^2\theta + \cos^2\theta = 1 \tag{11}$$

The following identities can be obtained from (11) by dividing through by $\cos^2\theta$ and $\sin^2\theta$, respectively, then applying (10a) and (10b):

$$\tan^2\theta + 1 = \sec^2\theta \tag{12}$$

$$1 + \cot^2\theta = \csc^2\theta \tag{13}$$

If (x, y) is a point on the unit circle, then the points $(-x, y)$, $(-x, -y)$, and $(x, -y)$ also lie on the unit circle (why?), and the four points are corners of a rectangle with sides parallel to the coordinate axes (Figure B.17a). Theorem 2 and parts (b), (c), and (d) of Figure B.17 suggest important identities for sine

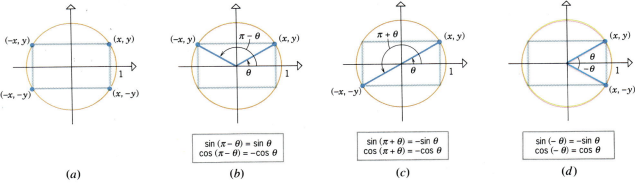

$$\sin(\pi - \theta) = \sin\theta$$
$$\cos(\pi - \theta) = -\cos\theta$$

$$\sin(\pi + \theta) = -\sin\theta$$
$$\cos(\pi + \theta) = -\cos\theta$$

$$\sin(-\theta) = -\sin\theta$$
$$\cos(-\theta) = \cos\theta$$

(a) (b) (c) (d)

Figure B.17

and cosine, and dividing those identities leads to related identities for the tangent:

$$\sin(\pi - \theta) = \sin\theta \tag{14a}$$
$$\cos(\pi - \theta) = -\cos\theta \tag{14b}$$
$$\tan(\pi - \theta) = -\tan\theta \tag{14c}$$

$$\sin(\pi + \theta) = -\sin\theta \tag{15a}$$
$$\cos(\pi + \theta) = -\cos\theta \tag{15b}$$
$$\tan(\pi + \theta) = \tan\theta \tag{15c}$$

$$\sin(-\theta) = -\sin\theta \tag{16a}$$
$$\cos(-\theta) = \cos\theta \tag{16b}$$
$$\tan(-\theta) = -\tan\theta \tag{16c}$$

Two angles in standard position that have the same terminal side must have the same values for their trigonometric functions since their terminal sides intersect the unit circle at the same point. In particular, two angles whose radian measures differ by a multiple of 2π have the same terminal side and hence have the same values for their trigonometric functions. This yields the identities

$$\sin\theta = \sin(\theta + 2\pi) = \sin(\theta - 2\pi) \tag{17a}$$
$$\cos\theta = \cos(\theta + 2\pi) = \cos(\theta - 2\pi) \tag{17b}$$

and more generally

$$\sin\theta = \sin(\theta \pm 2n\pi), \quad n = 0, 1, 2, \ldots \tag{18a}$$
$$\cos\theta = \cos(\theta \pm 2n\pi), \quad n = 0, 1, 2, \ldots \tag{18b}$$

Identities (14c), (15c), and (16c) imply that

$$\tan \theta = \tan (\theta + \pi) \tag{19a}$$

$$\tan \theta = \tan (\theta - \pi) \tag{19b}$$

Identity (19a) is just (15c) with the terms in the sum reversed, and identity (19b) follows from (14c) and (16c) (verify). These two identities state that adding or subtracting π from an angle does not affect the value of the tangent of the angle. It follows that the same is true for any multiple of π; thus,

$$\tan \theta = \tan (\theta \pm n\pi), \quad n = 0, 1, 2, \ldots \tag{19c}$$

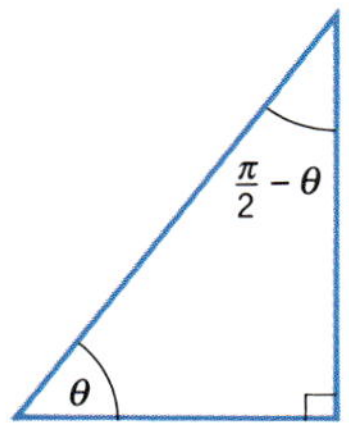

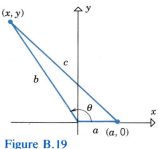

Figure B.18

Figure B.18 shows complementary angles θ and $\pi/2 - \theta$ of a right triangle. It follows from Formulas (4) and (5) that

$$\sin \theta = \frac{\text{side opposite } \theta}{\text{hypotenuse}} = \frac{\text{side adjacent to } \pi/2 - \theta}{\text{hypotenuse}} = \cos \left(\frac{\pi}{2} - \theta \right)$$

$$\cos \theta = \frac{\text{side adjacent to } \theta}{\text{hypotenuse}} = \frac{\text{side opposite } \pi/2 - \theta}{\text{hypotenuse}} = \sin \left(\frac{\pi}{2} - \theta \right)$$

which yields the identities

$$\sin \left(\frac{\pi}{2} - \theta \right) = \cos \theta \tag{20a}$$

$$\cos \left(\frac{\pi}{2} - \theta \right) = \sin \theta \tag{20b}$$

and, on dividing (20a) by (20b), we have

$$\tan \left(\frac{\pi}{2} - \theta \right) = \cot \theta \tag{20c}$$

□ **THE LAW OF COSINES**

The next theorem, called the *law of cosines,* generalizes the Theorem of Pythagoras. This result is not only important in its own right but is also the starting point for some important trigonometric identities.

THEOREM 3 (*Law of Cosines*).   *If the sides of a triangle have lengths a, b, and c, and if θ is the angle between the sides with lengths a and b, then*

$$c^2 = a^2 + b^2 - 2ab \cos \theta$$

Figure B.19

Proof.   Introduce a coordinate system so that θ is in standard position and the side of length a falls along the positive x-axis (Figure B.19). As shown in Figure

B.19, the side of length a extends from the origin to $(a, 0)$ and the side of length b extends from the origin to some point (x, y). From the definition of $\sin \theta$ and $\cos \theta$ we have

$$\sin \theta = \frac{y}{b}, \quad \cos \theta = \frac{x}{b}$$

so

$$y = b \sin \theta, \quad x = b \cos \theta \tag{21}$$

From the distance formula in Theorem 1.6.1 we obtain

$$c^2 = (x - a)^2 + (y - 0)^2$$

so that, from (21),

$$\begin{aligned}
c^2 &= (b \cos \theta - a)^2 + b^2 \sin^2 \theta \\
&= a^2 + b^2 (\cos^2 \theta + \sin^2 \theta) - 2ab \cos \theta \\
&= a^2 + b^2 - 2ab \cos \theta
\end{aligned}$$

which completes the proof.   ▮

We will now show how the law of cosines can be used to obtain the following identities, called the **addition formulas** for sine and cosine:

$$\sin (\alpha + \beta) = \sin \alpha \cos \beta + \cos \alpha \sin \beta \tag{22a}$$
$$\cos (\alpha + \beta) = \cos \alpha \cos \beta - \sin \alpha \sin \beta \tag{22b}$$

$$\sin (\alpha - \beta) = \sin \alpha \cos \beta - \cos \alpha \sin \beta \tag{23a}$$
$$\cos (\alpha - \beta) = \cos \alpha \cos \beta + \sin \alpha \sin \beta \tag{23b}$$

We will derive (23b) first. In our derivation we will assume that $0 \leq \beta < \alpha < 2\pi$ (Figure B.20). The proofs for the remaining cases will be omitted.

As shown in Figure B.20, the terminal sides of α and β intersect the unit circle at the points $P_1(\cos \alpha, \sin \alpha)$ and $P_2(\cos \beta, \sin \beta)$. If we denote the lengths of the sides of triangle OP_1P_2 by OP_1, P_1P_2, and OP_2, then $OP_1 = OP_2 = 1$ and, from the distance formula in Theorem 1.6.1,

$$\begin{aligned}
(P_1P_2)^2 &= (\cos \beta - \cos \alpha)^2 + (\sin \beta - \sin \alpha)^2 \\
&= (\sin^2 \alpha + \cos^2 \alpha) + (\sin^2 \beta + \cos^2 \beta) - 2(\cos \alpha \cos \beta + \sin \alpha \sin \beta) \\
&= 2 - 2(\cos \alpha \cos \beta + \sin \alpha \sin \beta)
\end{aligned}$$

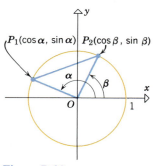

Figure B.20

But angle $P_2OP_1 = \alpha - \beta$, so that the law of cosines yields

$$\begin{aligned}
(P_1P_2)^2 &= (OP_1)^2 + (OP_2)^2 - 2(OP_1)(OP_2) \cos (\alpha - \beta) \\
&= 2 - 2 \cos (\alpha - \beta)
\end{aligned}$$

Equating the two expressions for $(P_1P_2)^2$ and simplifying, we obtain

$$\cos (\alpha - \beta) = \cos \alpha \cos \beta + \sin \alpha \sin \beta$$

which completes the derivation of (23b).

We can use (16a) and (20b) to derive (23a) as follows:

$$\sin(\alpha - \beta) = \cos\left[\frac{\pi}{2} - (\alpha - \beta)\right] = \cos\left[\left(\frac{\pi}{2} - \alpha\right) - (-\beta)\right]$$

$$= \cos\left(\frac{\pi}{2} - \alpha\right)\cos(-\beta) + \sin\left(\frac{\pi}{2} - \alpha\right)\sin(-\beta)$$

$$= \cos\left(\frac{\pi}{2} - \alpha\right)\cos\beta - \sin\left(\frac{\pi}{2} - \alpha\right)\sin\beta$$

$$= \sin\alpha\cos\beta - \cos\alpha\sin\beta$$

Identities (22a) and (22b) can be obtained from (23a) and (23b) by substituting $-\beta$ for β and using the identities

$$\sin(-\beta) = -\sin\beta, \quad \cos(-\beta) = \cos\beta$$

We leave it for the reader to derive the identities

$$\tan(\alpha + \beta) = \frac{\tan\alpha + \tan\beta}{1 - \tan\alpha\tan\beta} \tag{24a}$$

$$\tan(\alpha - \beta) = \frac{\tan\alpha - \tan\beta}{1 + \tan\alpha\tan\beta} \tag{24b}$$

Identity (24a) can be obtained by dividing (22a) by (22b) and then simplifying. Identity (24b) can be obtained from (24a) by substituting $-\beta$ for β and simplifying.

In the special case where $\alpha = \beta$, identities (22a), (22b), and (24a) yield the *double-angle formulas*

$$\sin 2\alpha = 2\sin\alpha\cos\alpha \tag{25a}$$

$$\cos 2\alpha = \cos^2\alpha - \sin^2\alpha \tag{25b}$$

$$\tan 2\alpha = \frac{2\tan\alpha}{1 - \tan^2\alpha} \tag{25c}$$

By using the identity $\sin^2\alpha + \cos^2\alpha = 1$, (25b) can be rewritten in the alternative forms

$$\cos 2\alpha = 2\cos^2\alpha - 1 \tag{26a}$$

$$\cos 2\alpha = 1 - 2\sin^2\alpha \tag{26b}$$

If we replace α by $\alpha/2$ in (26a) and (26b) and use some algebra, we obtain the *half-angle formulas*

$$\cos^2\frac{\alpha}{2} = \frac{1 + \cos\alpha}{2} \tag{27a}$$

$$\sin^2\frac{\alpha}{2} = \frac{1 - \cos\alpha}{2} \tag{27b}$$

We leave it for the exercises to derive the following *product-to-sum formulas* from (22) and (23):

$$\sin \alpha \cos \beta = \frac{1}{2}[\sin (\alpha - \beta) + \sin (\alpha + \beta)] \tag{28a}$$

$$\sin \alpha \sin \beta = \frac{1}{2}[\cos (\alpha - \beta) - \cos (\alpha + \beta)] \tag{28b}$$

$$\cos \alpha \cos \beta = \frac{1}{2}[\cos (\alpha - \beta) + \cos (\alpha + \beta)] \tag{28c}$$

We also leave it for the exercises to derive the following *sum-to-product formulas:*

$$\sin \alpha + \sin \beta = 2 \sin \frac{\alpha + \beta}{2} \cos \frac{\alpha - \beta}{2} \tag{29a}$$

$$\sin \alpha - \sin \beta = 2 \cos \frac{\alpha + \beta}{2} \sin \frac{\alpha - \beta}{2} \tag{29b}$$

$$\cos \alpha + \cos \beta = 2 \cos \frac{\alpha + \beta}{2} \cos \frac{\alpha - \beta}{2} \tag{29c}$$

$$\cos \alpha - \cos \beta = -2 \sin \frac{\alpha + \beta}{2} \sin \frac{\alpha - \beta}{2} \tag{29d}$$

☐ **FINDING AN ANGLE FROM THE VALUE OF ITS TRIGONOMETRIC FUNCTIONS**

There are numerous situations in which it is desired to find an unknown angle from a known value of one of its trigonometric functions. The following example illustrates a method for doing this.

Example 7   Find θ if $\sin \theta = \frac{1}{2}$.

Solution.   We shall begin by looking for positive angles that satisfy the given equation.

Because $\sin \theta$ is positive, the angle θ must terminate in the first or second quadrant. If it terminates in the first quadrant, then the hypotenuse of $\triangle OAP$ in Figure B.21a is double the leg AP, so

$$\theta = 30° = \frac{\pi}{6} \text{ radians}$$

If θ terminates in the second quadrant (Figure B.21b) then the hypotenuse of $\triangle OAP$ is double the leg AP, so $\angle AOP = 30°$, which implies that

$$\theta = 180° - 30° = 150° = \frac{5\pi}{6} \text{ radians}$$

Now that we have found these two solutions, all other solutions are obtained by adding or subtracting multiples of 360° (2π radians) to them. Thus, the entire

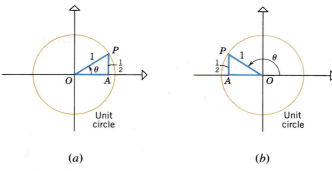

(a)   (b)

Figure B.21

set of solutions is given by the formulas

$$\theta = 30° \pm n \cdot 360°, \quad n = 0, 1, 2, \ldots$$

and

$$\theta = 150° \pm n \cdot 360°, \quad n = 0, 1, 2, \ldots$$

or in radian measure

$$\theta = \frac{\pi}{6} \pm n \cdot 2\pi, \quad n = 0, 1, 2, \ldots$$

and

$$\theta = \frac{5\pi}{6} \pm n \cdot 2\pi, \quad n = 0, 1, 2, \ldots \quad \blacktriangleleft$$

▶ Exercises   Ⓒ 7, 27, 28

In Exercises 1 and 2, express the angles in radians.

1. (a) 75° (b) 390° (c) 20° (d) 138°.

2. (a) 420° (b) 15° (c) 225° (d) 165°.

In Exercises 3 and 4, express the angles in degrees.

3. (a) $\pi/15$ (b) 1.5 (c) $8\pi/5$ (d) 3π.

4. (a) $\pi/10$ (b) 2 (c) $2\pi/5$ (d) $7\pi/6$.

5. Sketch the following angles in standard position.
 (a) 60° (b) $-150°$ (c) $9\pi/4$ (d) $-\pi$.

6. Suppose that the angle θ shown in Figure B.8 is measured in *degrees* instead of radians. Find a formula, in terms of r and θ, for
 (a) the arc length s
 (b) the area A of the sector.

7. A point rotates through an angle of 150° around a circle of radius 4 cm. Find

 (a) the distance traveled
 (b) the area of the sector swept out.

8. Find a formula for the area A of a circular sector in terms of its radius r and arc length s.

9. A right-circular cone is made from a circular piece of paper of radius R by cutting out a sector of angle θ radians and gluing the cut edges of the remaining piece together (Figure B.22). Find
 (a) the radius r of the base of the cone in terms of R and θ
 (b) the height h of the cone in terms of R and θ.

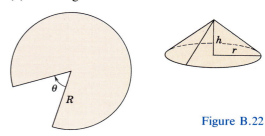

Figure B.22

10. Let r and L be the radius of the base and the slant height of a right-circular cone (Figure B.23). Show that the lateral surface area, S, of the cone is $S = \pi r L$. [*Hint:* As shown in Figure B.22, the lateral surface of the cone becomes a circular sector when cut along a line from the vertex to the base and flattened.]

Figure B.23

In Exercises 11–22, the angle θ is an acute angle of a right triangle. Solve the problems by drawing an appropriate right triangle. Do *not* use a calculator.

---

11. Find $\sin \theta$ and $\tan \theta$ given that the hypotenuse has length 5 and the side adjacent to θ has length 2.

12. Find $\cos \theta$ and $\tan \theta$ given that the hypotenuse has length 4 and the side opposite θ has length 3.

13. Find $\sin \theta$ and $\cos \theta$ given that the side adjacent to θ has length 3 and the side opposite θ has length 5.

14. Find $\csc \theta$ and $\cot \theta$ given that the hypotenuse has length 3 and the side adjacent to θ has length 1.

15. Find $\sin \theta$ and $\cos \theta$ given that $\tan \theta = 3$.

16. Find $\sin \theta$ and $\tan \theta$ given that $\cos \theta = 2/3$.

17. Find $\tan \theta$ and $\csc \theta$ given that $\sec \theta = 5/2$.

18. Find $\cot \theta$ and $\sec \theta$ given that $\csc \theta = 4$.

19. Find the length of the side adjacent to θ given that the hypotenuse has length 6 and $\cos \theta = 0.3$.

20. Find the length of the side opposite θ given that the side adjacent to θ has length 15.6 and $\cot \theta = 5.2$.

21. Find the length of the hypotenuse given that the side opposite θ has length 2.4 and $\sin \theta = 0.8$.

22. Find the length of the hypotenuse given that the side adjacent to θ has length 1.5 and $\cos \theta = 0.6$.

23. (a) Let θ be an acute angle such that $\sin \theta = a/3$. Express the remaining trigonometric functions in terms of a by using the triangle in Figure B.24.

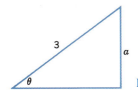

Figure B.24

(b) Let θ be an acute angle such that $\tan \theta = a/5$. Construct an appropriate triangle, and use it to express the remaining trigonometric functions in terms of a.

(c) Let θ be an acute angle such that $\sec \theta = a$. Construct an appropriate triangle, and use it to express the remaining trigonometric functions in terms of a.

24. How could you use a ruler and protractor to approximate $\sin 17°$ and $\cos 17°$?

In Exercises 25 and 26, do *not* use a calculator.

---

25. Two sides of a triangle have lengths of 3 cm and 7 cm and meet at an angle of 60°. Find the area of the triangle.

26. Let ABC be a triangle whose angles at A and B are 30° and 45°. If the side opposite the angle B has length 9, find the lengths of the remaining sides and the size of the angle C.

27. A 10-foot ladder leans against a house and makes an angle of 67° with level ground. How far is the top of the ladder above the ground? Express your answer to the nearest tenth of a foot.

28. From a point 120 feet on level ground from a building, the angle of elevation to the top of the building is 76°. Find the height of the building. Express your answer to the nearest foot.

29. An observer on level ground is at a distance d from a building. The angles of elevation to the bottom of the windows on the second and third floors are α and β, respectively. Find the distance h between the bottoms of the windows in terms of α, β, and d.

30. From a point on level ground, the angle of elevation to the top of a tower is α. From a point that is d units closer to the tower, the angle of elevation is β. Find the height h of the tower in terms of α, β, and d.

Exercises 31 and 32 refer to an arbitrary triangle ABC in which side a is opposite angle A, side b is opposite angle B, and side c is opposite angle C.

---

31. Prove: The area of a triangle ABC can be written as

$$\text{area} = \tfrac{1}{2}bc \sin A$$

Find two other similar formulas for the area.

32. Prove the *law of sines:* In any triangle, the ratios of the sides to the sines of the opposite angles are equal.

That is,

$$\frac{a}{\sin A} = \frac{b}{\sin B} = \frac{c}{\sin C}$$

In Exercises 33 and 34, the value of an angle θ is given. Find the values of all six trigonometric functions of θ without using a calculator.

33. (a) $225°$ (b) $-210°$
    (c) $5\pi/3$ (d) $-3\pi/2$.

34. (a) $330°$ (b) $-120°$
    (c) $9\pi/4$ (d) -3π.

35. Find the values of all six trigonometric functions of the angle θ shown in Figure B.25. Do *not* use a calculator.

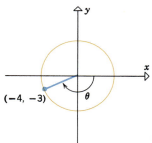

Figure B.25

36. Find the values of all six trigonometric functions of the angle θ shown in Figure B.26. Do *not* use a calculator.

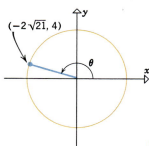

Figure B.26

37. Use identities and Table 1 to find the exact values of the expression.
    (a) $\cos 15°$ (b) $\sin 22.5°$ (c) $\sin 37.5°$.

38. Use identities and Table 1 to find the exact values of the expression.
    (a) $\sin 75°$ (b) $\tan 75°$.

In Exercises 39 and 40, do *not* use a calculator.

39. If $\cos \theta = 2/3$ and $0 < \theta < \pi/2$, find
    (a) $\sin 2\theta$ (b) $\cos 2\theta$.

40. If $\tan \alpha = 3/4$ and $\tan \beta = 2$, where $0 < \alpha < \pi/2$ and $0 < \beta < \pi/2$, find
    (a) $\sin (\alpha - \beta)$ (b) $\cos (\alpha + \beta)$.

41. Use identities (22) and (23) to express each of the following in terms of $\sin \theta$ or $\cos \theta$.
    (a) $\sin \left(\dfrac{\pi}{2} + \theta\right)$ (b) $\cos \left(\dfrac{\pi}{2} + \theta\right)$
    (c) $\sin \left(\dfrac{3\pi}{2} - \theta\right)$ (d) $\cos \left(\dfrac{3\pi}{2} + \theta\right)$.

42. Derive identities (24).

43. Derive identity
    (a) (28a)   (b) (28b)   (c) (28c).

44. If $A = \alpha + \beta$ and $\beta = \alpha - \beta$, then $\alpha = \frac{1}{2}(A + B)$ and $\beta = \frac{1}{2}(A - B)$ (verify). Use this result and identities (28) to derive identity
    (a) (29a)   (b) (29c)   (c) (29d).

45. Substitute $-\beta$ for β in identity (29a) to derive identity (29b).

In Exercises 46–56, derive the given identities.

46. $\dfrac{\cos \theta \sec \theta}{1 + \tan^2 \theta} = \cos^2 \theta$.

47. $\dfrac{\cos \theta \tan \theta + \sin \theta}{\tan \theta} = 2 \cos \theta$.

48. $2 \csc 2\theta = \sec \theta \csc \theta$.

49. $\tan \theta + \cot \theta = 2 \csc 2\theta$.

50. $\dfrac{\sin 2\theta}{\sin \theta} - \dfrac{\cos 2\theta}{\cos \theta} = \sec \theta$.

51. $\dfrac{\sin \theta + \cos 2\theta - 1}{\cos \theta - \sin 2\theta} = \tan \theta$.

52. $\sin 3\theta + \sin \theta = 2 \sin 2\theta \cos \theta$.

53. $\sin 3\theta - \sin \theta = 2 \cos 2\theta \sin \theta$.

54. $\tan \dfrac{\theta}{2} = \dfrac{1 - \cos \theta}{\sin \theta}$.

55. $\tan \dfrac{\theta}{2} = \dfrac{\sin \theta}{1 + \cos \theta}$.

56. $\cos \left(\dfrac{\pi}{3} + \theta\right) + \cos \left(\dfrac{\pi}{3} - \theta\right) = \cos \theta$.

57. Express $\sin 3\theta$ and $\cos 3\theta$ in terms of $\sin \theta$ and $\cos \theta$.

58. (a) Express $3 \sin \alpha + 5 \cos \alpha$ in the form
         $C \sin (\alpha + \phi)$

    (b) Show that a sum of the form
         $A \sin \alpha + B \cos \alpha$

      can be rewritten in the form $C \sin (\alpha + \phi)$.

59. Show that the length of the diagonal of the parallelogram in Figure B.27 is

$$d = \sqrt{a^2 + b^2 + 2ab\cos\theta}$$

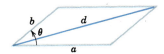

Figure B.27

60. Find all values of θ (in radians) such that
(a) $\sin\theta = 1$ (b) $\cos\theta = 1$ (c) $\tan\theta = 1$
(d) $\csc\theta = 1$ (e) $\sec\theta = 1$ (f) $\cot\theta = 1$.

61. Find all values of θ (in radians) such that
(a) $\sin\theta = 0$ (b) $\cos\theta = 0$ (c) $\tan\theta = 0$
(d) $\csc\theta$ is undefined   (e) $\sec\theta$ is undefined
(f) $\cot\theta$ is undefined.

In Exercises 62–77, find all values of θ (in radians) that satisfy the given equation. Do not use a calculator.

62. $\cos\theta = -\dfrac{1}{\sqrt{2}}$.

63. $\sin\theta = -\dfrac{1}{\sqrt{2}}$.

64. $\tan\theta = -1$.

65. $\cos\theta = \dfrac{1}{2}$.

66. $\sin\theta = -\dfrac{1}{2}$.

67. $\tan\theta = \sqrt{3}$.

68. $\tan\theta = \dfrac{1}{\sqrt{3}}$.

69. $\sin\theta = -\dfrac{\sqrt{3}}{2}$.

70. $\sin\theta = -1$.

71. $\cos\theta = -1$.

72. $\cot\theta = -1$.

73. $\cot\theta = \sqrt{3}$.

74. $\sec\theta = -2$.

75. $\csc\theta = -2$.

76. $\csc\theta = \dfrac{2}{\sqrt{3}}$.

77. $\sec\theta = \dfrac{2}{\sqrt{3}}$.

■ II. GRAPHS OF TRIGONOMETRIC FUNCTIONS

> *In this section we shall review the basic properties of the graphs of trigonometric functions.*

□ **THE GRAPHS OF THE SINE AND COSINE FUNCTIONS**

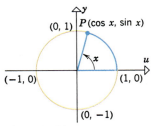

Figure B.28

Recall that an angle written without any units is understood to be measured in radians. Thus, the expression $\sin 3$ means the sine of 3 radians (*not* the sine of 3 degrees). More generally, for any real number x, the expression $\sin x$ means the sine of x radians.

To obtain the graphs of $y = \sin x$ and $y = \cos x$, it will be helpful to interpret $\sin x$ and $\cos x$ as coordinates of the intersection P of the terminal side of the angle x with the unit circle. However, because we want to use x as the angle here, we cannot work with the unit circle in an xy-coordinate system as before, since this would lead to conflicting interpretations of x. Thus, we shall consider the unit circle in a uy-coordinate system, as shown in Figure B.28.

The general shape of the graph of $y = \sin x$ (Figure B.29) can be deduced by observing the y-coordinate of P as the terminal side of the angle x rotates counterclockwise (positive x) and then clockwise (negative x). Consider the

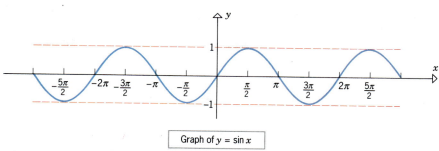

Figure B.29

counterclockwise rotation first (Figure B.30): When $x = 0$, the point P is on the u-axis, so $v = 0$. As x increases, v increases to 1 (at $\pi/2$), then decreases to 0 (at π), then decreases further to -1 (at $3\pi/2$), and then increases to 0 (at 2π). Additional counterclockwise rotations produce repetitions of these values, so the graph of $y = \sin x$ repeats every 2π units. The analysis for negative x is similar.

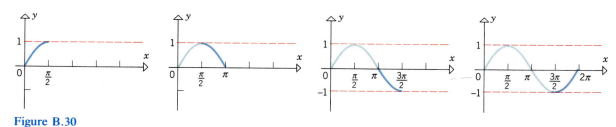

Figure B.30

The general shape of the graph of $y = \cos x$ (Figure B.31) can be obtained by observing the u-coordinate of P in Figure B.28 as x varies through positive and negative values.

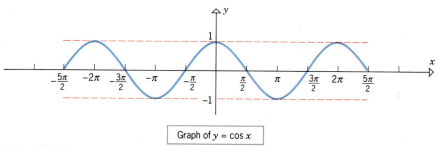

Figure B.31

□ THE GRAPH OF THE TANGENT FUNCTION

The general shape of the graph of $y = \tan x$ can also be deduced geometrically. To begin, let us assume that

$$-\pi/2 < x < \pi/2$$

and construct the angle x in standard position in a uv-coordinate system. As illustrated in Figure B.32, the intersection of the terminal side of the angle x with the vertical line L through $B(1,0)$ has a y-coordinate of $\tan x$. We will

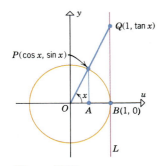

Figure B.32

prove this for nonnegative x; the proof for negative x is similar: Observe that triangles OAP and OBQ are similar, so that

$$\frac{BQ}{OB} = \frac{AP}{OA} \tag{1}$$

But $OB = 1$, $OA = \cos x$, and $AP = \sin x$, so that from (1)

$$y = BQ = \frac{\sin x}{\cos x} = \tan x$$

The general shape of the graph of $y = \tan x$ can now be deduced from Figure B.32 by observing the y-coordinate of Q as the terminal side of x rotates counterclockwise (positive x) and then clockwise (negative x). Consider the counterclockwise rotation first (Figure B.33a): When $x = 0$, the point Q is on the u-axis, so $y = 0$. As x increases toward $\pi/2$, the point Q travels up the line L getting higher and higher; thus, the value of y increases without bound from its initial value of 0. For the negative rotation, y decreases without bound from its initial value of zero (Figure B.33b).

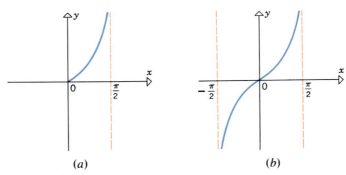

(a)   (b)

Figure B.33

Finally, from the identities

$$\tan (x + \pi) = \tan x \quad \text{and} \quad \tan (x - \pi) = \tan x$$

which were given in Section I (see 19a and 19b), we see that the graph of $y = \tan x$ repeats every π units. This fact together with Figure B.33b leads us to the complete graph of $y = \tan x$ shown in Figure B.34.

Observe that unlike the functions $\sin x$ and $\cos x$, which are defined for all values of x, the function $\tan x$ ($= \sin x/\cos x$) is undefined at the values of x where $\cos x = 0$; those values are

$$x = \pm\frac{\pi}{2}, \pm\frac{3\pi}{2}, \pm\frac{5\pi}{2}, \dots$$

☐ **AMPLITUDE AND PERIOD**

The values of $\sin x$ and $\cos x$ repeat every 2π units, and the values of $\tan x$ repeat every π units; we say that $\sin x$ and $\cos x$ have *period* 2π and that $\tan x$ has *period* π. More generally, we make the following definition.

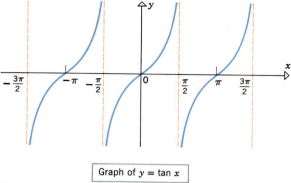

Graph of $y = \tan x$

Figure B.34

DEFINITION 4.   A function f is called *periodic* if there is a positive number p such that

$$f(x + p) = f(x) \tag{2}$$

whenever x and $x + p$ lie in the domain of f. The smallest value of p for which (2) holds is called the *fundamental period* of f or sometimes the *period* of f.

In words, (2) states that the values of f repeat every p units.

Functions of the form

$$a \sin bx \quad \text{and} \quad a \cos bx \tag{3}$$

occur frequently in applications. The graphs of such functions are considered in the following examples.

Example 1   Sketch the graphs of

   (a) $y = 3 \sin x$ (b) $y = \sin 4x$ (c) $y = 3 \sin 4x$

Solution (a).   The factor of 3 in $y = 3 \sin x$ has the effect of multiplying each y-coordinate in the graph of $y = \sin x$ by 3, thereby stretching the graph of $y = \sin x$ vertically by a factor of 3 (Figure B.35).

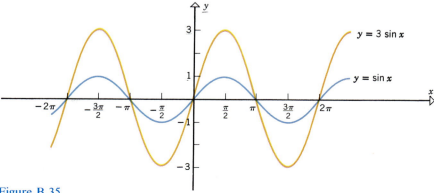

Figure B.35

Solution (b).   The graph of $y = \sin x$ repeats when x changes by 2π. Thus, the graph of $y = \sin 4x$ repeats whenever $4x$ changes by 2π, or equivalently when x changes by $\frac{1}{4}(2\pi) = \pi/2$. Thus, the factor of 4 in $y = \sin 4x$ has the effect of reducing the fundamental period by a factor of $\frac{1}{4}$ (Figure B.36).

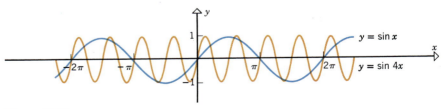

Figure B.36

Solution (c).   The graph of $y = 3 \sin 4x$ combines both the stretching effect discussed in part (a) and the period effect discussed in part (b). The graph is shown in Figure B.37.   ◄

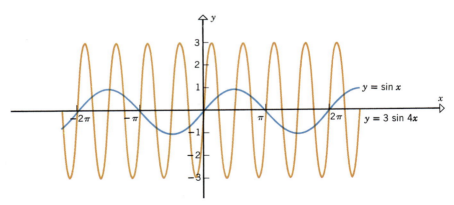

Figure B.37

Example 2 illustrates the following general result.

> **THEOREM 5.**   *If $a \neq 0$ and $b \neq 0$, then the functions $a \sin bx$ and $a \cos bx$ have fundamental period $2\pi/|b|$ and their graphs oscillate between $-a$ and a.*

We leave the proof as an exercise. The number $|a|$ is called the **amplitude** of $a \sin bx$ and $a \cos bx$.

Example 2   The graph of the function $6 \cos 7x$ has amplitude 6 and fundamental period $2\pi/7$; the function $-\frac{1}{3}\sin \frac{1}{4}x$ has amplitude $\frac{1}{3}$ and fundamental period 8π.   ◄

We leave it as an exercise to prove the following result.

THEOREM 6.   *If $b \neq 0$, then the function $\tan bx$ has fundamental period $\pi/|b|$.*

▶ Exercises

1. Find the fundamental period of

   (a) $\sin 5x$ (b) $\cos \frac{1}{3} x$ (c) $\tan \frac{1}{8} x$

   (d) $\tan 7x$ (e) $\sin \pi x$ (f) $\sec \frac{\pi}{5} x$

   (g) $\cot \pi x$ (h) $\csc \frac{x}{k}$.

In Exercises 2 and 3, find the amplitudes of the functions.

---

2. (a) $4 \sin 6x$ (b) $-8 \cos \pi x$
   (c) $\frac{1}{2} \sin x$ (d) $-\frac{1}{8} \cos 2x$.

3. (a) $5 \cos 2x$ (b) $\frac{1}{3} \sin 4x$
   (c) $-\frac{1}{2} \cos x$ (d) $-\sin \pi x$.

In Exercises 4–11, sketch the graph of the equation.

---

4. $y = \sin 3x$.   5. $y = 2 \sin 3x$.

6. $y = 2 \sin x$.   7. $y = \frac{1}{2} \cos \pi x$.

8. $y = -4 \cos x$.   9. $y = \cos(-4x)$.

10. $y = \sin(2\pi x)$.   11. $y = -2 \sin(-2x)$.

12. (a) How are the graphs of $y = \cos x$ and
         $y = \cos(-x)$ related?
    (b) How are the graphs of $y = \sin x$ and
         $y = \sin(-x)$ related?

13. Use the graphs of $\sin x$, $\cos x$, $\tan x$, and the relationships

   $$\cot x = \frac{1}{\tan x}, \quad \sec x = \frac{1}{\cos x}, \quad \csc x = \frac{1}{\sin x}$$

   to help sketch the graphs of
   (a) $\cot x$ (b) $\sec x$ (c) $\csc x$.

In Exercises 14–17, sketch the graph of the equation.

---

14. $y = 3 \cot 2x$.   15. $y = \frac{1}{4} \sec \pi x$.

16. $y = 2 \csc \frac{x}{3}$.   17. $y = 2 \cot \frac{x}{4}$.

In Exercises 18–32, sketch the graph of the equation.

---

18. $y = \sin(x + \pi)$.   19. $y = \cos(x - \pi)$.

20. $y = \cos\left(x - \frac{\pi}{4}\right)$.   21. $y = \sin\left(x + \frac{\pi}{2}\right)$.

22. $y = 2 - \cos x$.   23. $y = 1 + \sin x$.

24. $y = 2 \cos\left(4x + \frac{\pi}{2}\right)$.

25. $y = 3 \sin\left(2x + \frac{\pi}{4}\right)$.

26. $y = \tan\left(\pi x - \frac{1}{2}\right)$.   27. $y = 2 \tan(2\pi x + 1)$.

28. $y = |\tan x|$.   29. $y = |\cos x|$.

30. $y = \sin x - \cos x$.   31. $y = \sin x + \cos x$.

32. $y = 2 \sin x + 3 \cos 2x$.

33. Classify the following functions as even, odd, or neither. (See Exercise 47 in Section 2.3.)
    (a) $\sin x$ (b) $\cos x$ (c) $\tan x$
    (d) $|\sec x|$ (e) $\cot(x^2)$ (f) $3 \sec(-2x)$
    (g) $-4 \csc x$ (h) $\frac{\sin x}{x}$.

34. Prove: If $0 < x < \pi/2$, then

   $$\sin x < x < \tan x$$

   [*Hint:* In Figure B.32, compare the areas of triangle OAP, sector OBP, and triangle OBQ.]

35. Prove: For all real numbers x,

   $$|\sin x| \leq |x|$$

   [*Hint:* First consider the cases $|x| \geq \pi/2$ and $x = 0$; these are easy. Then use the left-hand inequality in Exercise 34 to complete the proof.]

36. Prove Theorem 5 of this section.

37. Prove Theorem 6 of this section.

Appendix

SUPPLEMENTARY MATERIAL

I. ONE-SIDED AND INFINITE LIMITS: A RIGOROUS APPROACH

> *In this section we shall give precise definitions of one-sided and infinite limits and illustrate how these definitions are used to prove results about limits.*

In Section 2.4 we discussed the one-sided limits

$$\lim_{x \to a^-} f(x) = L \quad \text{and} \quad \lim_{x \to a^+} f(x) = L$$

from an intuitive viewpoint. We interpreted the first statement to mean that the value of $f(x)$ approaches L as x approaches a from the left side, and the second statement to mean that the value of $f(x)$ approaches L as x approaches a from the right side. We can formalize these ideas mathematically as follows:

DEFINITION 1. Let f be defined on some open interval extending to the left from a. We shall write

$$\lim_{x \to a^-} f(x) = L$$

if given any number $\epsilon > 0$, we can find a number $\delta > 0$ such that $f(x)$ satisfies $|f(x) - L| < \epsilon$ whenever x satisfies $a - \delta < x < a$.

DEFINITION 2. Let f be defined on some open interval extending to the right from a. We shall write

$$\lim_{x \to a^+} f(x) = L$$

if given any number $\epsilon > 0$, we can find a number $\delta > 0$ such that $f(x)$ satisfies $|f(x) - L| < \epsilon$ whenever x satisfies $a < x < a + \delta$.

Before discussing some examples, let us compare Definitions 1 and 2 with the definition of $\lim\limits_{x \to a} f(x) = L$ as stated in Definition 2.6.1. Given $\epsilon > 0$, the definition of $\lim\limits_{x \to a} f(x) = L$ requires that we be able to find a $\delta > 0$ such that

$$|f(x) - L| < \epsilon \tag{1}$$

whenever x satisfies

$$0 < |x - a| < \delta$$

or, equivalently, whenever x lies in the set

$$(a - \delta, a) \cup (a, a + \delta)$$

(See Figure C.1a.)

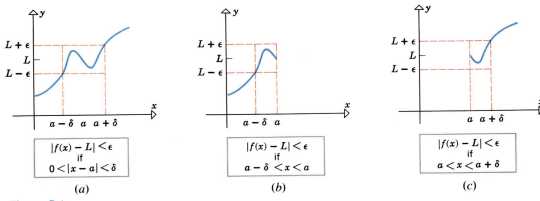

(a)   (b)   (c)

Figure C.1

On the other hand, the definition of $\lim\limits_{x \to a^-} f(x) = L$ requires only that (1) hold for x in the set $(a - \delta, a)$ (Figure C.1b), while the definition of $\lim\limits_{x \to a^+} f(x) = L$ requires only that (1) be satisfied on the set $(a, a + \delta)$ (Figure C.1c).

Example 1   Because the function $f(x) = \sqrt{x}$ is undefined for $x < 0$, the limits

$$\lim_{x \to 0} \sqrt{x} \quad \text{and} \quad \lim_{x \to 0^-} \sqrt{x}$$

are undefined. The expression

$$\lim_{x \to 0^+} \sqrt{x}$$

is the only one that makes sense. Prove that

$$\lim_{x \to 0^+} \sqrt{x} = 0$$

Solution. Let $\epsilon > 0$ be given. We must find a $\delta > 0$ such that $f(x) = \sqrt{x}$ satisfies

$$|\sqrt{x} - 0| < \epsilon \tag{2}$$

whenever x satisfies

$$0 < x < 0 + \delta$$

or equivalently

$$0 < x < \delta \tag{3}$$

To find δ, we rewrite (2) as

$$\sqrt{x} < \epsilon \tag{4}$$

(since $\sqrt{x} = |\sqrt{x}|$). We must choose δ so that (4) holds whenever (3) does. We can do this by taking

$$\delta = \epsilon^2$$

To see that this choice of δ works, assume that x satisfies (3). Since we are letting $\delta = \epsilon^2$, it follows from (3) that

$$0 < x < \epsilon^2$$

or equivalently

$$0 < \sqrt{x} < \epsilon \tag{5}$$

Thus, (4) holds since it is just the right-hand inequality in (5). This proves that

$$\lim_{x \to 0^+} \sqrt{x} = 0 \qquad \blacktriangleleft$$

Example 2 Let

$$f(x) = \begin{cases} 2x, & x < 1 \\ x^3, & x \geq 1 \end{cases}$$

Prove that

$$\lim_{x \to 1^-} f(x) = 2$$

Solution. Let $\epsilon > 0$ be given. We must find a $\delta > 0$ such that $f(x)$ satisfies

$$|f(x) - 2| < \epsilon \tag{6}$$

whenever x satisfies

$$1 - \delta < x < 1 \tag{7}$$

Because we are considering only x satisfying (7), it follows from the formula for $f(x)$ that

$$f(x) = 2x$$

Thus, we can rewrite (6) as

$$|2x - 2| < \epsilon$$

or

$$|x - 1| < \frac{\epsilon}{2}$$

or

$$1 - \frac{\epsilon}{2} < x < 1 + \frac{\epsilon}{2} \qquad (8)$$

We must choose δ so that (8) is satisfied whenever (7) is satisfied. We can do this by taking

$$\delta = \frac{\epsilon}{2}$$

To see that this choice of δ works, assume that x satisfies (7). With $\delta = \epsilon/2$, it follows from (7) that

$$1 - \frac{\epsilon}{2} < x < 1 \qquad (9)$$

and this implies that

$$1 - \frac{\epsilon}{2} < x < 1 + \frac{\epsilon}{2}$$

since ϵ is positive.

Thus, we have proved

$$\lim_{x \to 1^-} f(x) = 2 \qquad \blacktriangleleft$$

☐ **LIMITS INVOLVING INFINITY**

In Section 2.4 we discussed the limits

$$\lim_{x \to -\infty} f(x) = L \quad \text{and} \quad \lim_{x \to +\infty} f(x) = L$$

from an intuitive viewpoint. We interpreted the first statement to mean that the value of $f(x)$ approaches L as x moves indefinitely far from the origin along the negative x-axis, and we interpreted the second statement to mean that the value of $f(x)$ approaches L as x moves indefinitely far from the origin along the positive x-axis.

The phrase "x moves indefinitely far from the origin along the positive x-axis" is intended to convey the idea that given *any* positive number N (no matter how large) eventually x is larger than N (Figure C.2a). Similarly, the phrase "x moves indefinitely far from the origin along the negative x-axis" is

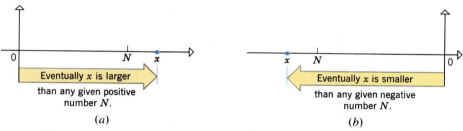

Eventually x is larger than any given positive number N.

Eventually x is smaller than any given negative number N.

(a)

(b)

Figure C.2

intended to convey the idea that given any negative number N (no matter how small) eventually x is smaller than N (Figure C.2b). Thus, when we say that $\lim\limits_{x \to +\infty} f(x) = L$, we mean that for any $\epsilon > 0$, we can find a positive number N (sufficiently large) so that $f(x)$ satisfies

$$|f(x) - L| < \epsilon$$

when x is greater than N (Figure C.3). Similarly, when we say that $\lim\limits_{x \to -\infty} f(x) = L$, we mean that for any $\epsilon > 0$, we can find a negative number N so that $f(x)$ satisfies

$$|f(x) - L| < \epsilon$$

when x is smaller than N (Figure C.4).

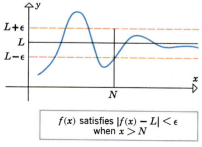

$f(x)$ satisfies $|f(x) - L| < \epsilon$ when $x > N$

Figure C.3

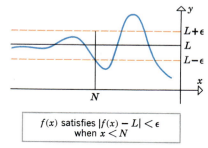

$f(x)$ satisfies $|f(x) - L| < \epsilon$ when $x < N$

Figure C.4

These ideas suggest the following formal definitions.

DEFINITION 3. Let f be a function that is defined on some infinite open interval $(x_0, +\infty)$. We shall write

$$\lim_{x \to +\infty} f(x) = L$$

if given any number $\epsilon > 0$, there corresponds a positive number N such that

$$|f(x) - L| < \epsilon$$

whenever $x > N$.

DEFINITION 4. Let f be a function that is defined on some infinite open interval $(-\infty, x_0)$. We shall write

$$\lim_{x \to -\infty} f(x) = L$$

if given any number $\epsilon > 0$, there corresponds a negative number N such that

$$|f(x) - L| < \epsilon$$

whenever x satisfies $x < N$.

Example 3 Prove that

$$\lim_{x \to +\infty} \frac{1}{x} = 0$$

Solution. We must show that for any $\epsilon > 0$ we can find a positive number N such that $f(x) = 1/x$ satisfies

$$\left| \frac{1}{x} - 0 \right| < \epsilon \tag{10}$$

whenever x satisfies

$$x > N \tag{11}$$

To find N, rewrite (10) as

$$\frac{1}{|x|} < \epsilon$$

or

$$|x| > \frac{1}{\epsilon} \tag{12}$$

Thus, we must find a positive number N such that x satisfies (12) whenever x satisfies (11). We can do this by taking

$$N = \frac{1}{\epsilon}$$

To prove that this choice of N works, assume that x satisfies (11). Since we are letting $N = 1/\epsilon$, (11) becomes

$$x > \frac{1}{\epsilon} \tag{13}$$

But $1/\epsilon$ is positive (since $\epsilon > 0$), so the x in (13) is positive. Thus, (13) can be written as

$$|x| > \frac{1}{\epsilon}$$

which is precisely inequality (12). Thus, we have proved

$$\lim_{x \to +\infty} \frac{1}{x} = 0 \quad \blacktriangleleft$$

In Section 2.4 we discussed the limits

$$\lim_{x \to a} f(x) = +\infty \qquad \lim_{x \to a} f(x) = -\infty \qquad (14)$$

$$\lim_{x \to a^+} f(x) = +\infty \qquad \lim_{x \to a^+} f(x) = -\infty \qquad (15)$$

$$\lim_{x \to a^-} f(x) = +\infty \qquad \lim_{x \to a^-} f(x) = -\infty \qquad (16)$$

$$\lim_{x \to +\infty} f(x) = +\infty \qquad \lim_{x \to +\infty} f(x) = -\infty \qquad (17)$$

$$\lim_{x \to -\infty} f(x) = +\infty \qquad \lim_{x \to -\infty} f(x) = -\infty \qquad (18)$$

from an intuitive viewpoint. Recall that each of these expressions describes a particular way in which the limit fails to exist. The $+\infty$ to the right of the equal sign indicates that the limit fails to exist because $f(x)$ is increasing without bound, and the $-\infty$ indicates that the limit fails to exist because $f(x)$ is decreasing without bound.

To make these ideas mathematically precise, we must clarify the meanings of the phrases "$f(x)$ is increasing without bound" and "$f(x)$ is decreasing without bound." We shall limit our discussion to the limits in (14); formal definitions of the remaining limits are discussed in the exercises.

When we say that $f(x)$ increases without bound as x approaches a from either side, we mean that given any positive number N (no matter how large) the value of $f(x)$ eventually exceeds N as x approaches a from either side. Figure C.5 illustrates this idea. For the curve in that figure we have

$$\lim_{x \to a} f(x) = +\infty$$

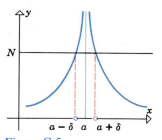

Figure C.5

We picked an arbitrary positive number N and marked it on the y-axis. As x approaches a from either side, x will eventually lie between the points

$$a - \delta \quad \text{and} \quad a + \delta$$

shown in the figure. When this occurs the value of $f(x)$ will satisfy

$$f(x) > N \qquad (19)$$

It is evident from the figure that no matter how large we make N, we shall always be able to find a sufficiently small positive number δ such that $f(x)$

satisfies (19) when x is between $a - \delta$ and $a + \delta$ (but different from a). This suggests the following definition.

DEFINITION 5.   Let f be defined in some open interval containing the number a, except that f need not be defined at a. We shall write

$$\lim_{x \to a} f(x) = +\infty$$

if given any positive number N, we can find a number $\delta > 0$ such that $f(x)$ satisfies

$$f(x) > N$$

whenever x satisfies $0 < |x - a| < \delta$.

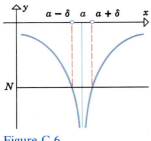

Figure C.6

When we say that $f(x)$ decreases without bound as x approaches a from either side, we mean that given any negative number N (no matter how small) the value of $f(x)$ will eventually be smaller than N as x approaches a from either side. Figure C.6 illustrates this idea. For the curve in that figure we have

$$\lim_{x \to a} f(x) = -\infty$$

We picked an arbitrary negative number N and marked it on the y-axis. As x approaches a from either side, x will eventually lie between the points

$$a - \delta \quad \text{and} \quad a + \delta$$

shown in the figure. When this occurs the value of $f(x)$ will satisfy

$$f(x) < N$$

This suggests the following definition.

DEFINITION 6.   Let f be defined in some open interval containing the number a, except that f need not be defined at a. We shall write

$$\lim_{x \to a} f(x) = -\infty$$

if given any negative number N, we can find a number $\delta > 0$ such that $f(x)$ satisfies

$$f(x) < N$$

whenever x satisfies $0 < |x - a| < \delta$.

Example 4 Prove that

$$\lim_{x \to 0} \frac{1}{x^2} = +\infty$$

Solution. We must show that given any positive number N, we can find a positive number δ such that $f(x) = 1/x^2$ satisfies

$$\frac{1}{x^2} > N \tag{20}$$

whenever x satisfies

$$0 < |x - 0| < \delta$$

or equivalently

$$0 < |x| < \delta \tag{21}$$

To find δ we shall rewrite (20) as

$$x^2 < \frac{1}{N} \tag{22}$$

Thus, we must find a positive number δ such that x satisfies (22) whenever x satisfies (21). We can do this by taking

$$\delta = \frac{1}{\sqrt{N}}$$

To prove that this choice of δ works, assume that x satisfies (21). Since we are letting $\delta = 1/\sqrt{N}$, (21) becomes

$$0 < |x| < \frac{1}{\sqrt{N}}$$

or equivalently

$$0 < |x|^2 < \frac{1}{N}$$

or since $|x|^2 = x^2$,

$$0 < x^2 < \frac{1}{N} \tag{23}$$

Thus, (22) is satisfied, since it is just the right-hand inequality in (23). This proves that

$$\lim_{x \to 0} \frac{1}{x^2} = +\infty \quad \blacktriangleleft$$

▶ Exercises

In Exercises 1–6, use Definitions 1 and 2 to prove that the stated limit is correct.

1. $\lim_{x \to 2^+} (x + 1) = 3$.

2. $\lim_{x \to 1^-} (3x + 2) = 5$.

3. $\lim_{x \to 4^+} \sqrt{x - 4} = 0$.

4. $\lim_{x \to 0^-} \sqrt{-x} = 0$.

5. $\lim_{x \to 2^+} f(x) = 2$, where $f(x) = \begin{cases} x, & x > 2 \\ 3x, & x \le 2. \end{cases}$

6. $\lim_{x \to 2^-} f(x) = 6$, where $f(x) = \begin{cases} x, & x > 2 \\ 3x, & x \le 2. \end{cases}$

In Exercises 7–14, use Definitions 3 and 4 to prove that the stated limit is correct.

7. $\lim_{x \to +\infty} \frac{1}{x^2} = 0$.

8. $\lim_{x \to -\infty} \frac{1}{x} = 0$.

9. $\lim_{x \to -\infty} \frac{1}{x + 2} = 0$.

10. $\lim_{x \to +\infty} \frac{1}{x + 2} = 0$.

11. $\lim_{x \to +\infty} \frac{x}{x + 1} = 1$.

12. $\lim_{x \to -\infty} \frac{x}{x + 1} = 1$.

13. $\lim_{x \to -\infty} \frac{4x - 1}{2x + 5} = 2$.

14. $\lim_{x \to +\infty} \frac{4x - 1}{2x + 5} = 2$.

In Exercises 15–20, use Definitions 5 and 6 to prove that the stated limit is correct.

15. $\lim_{x \to 3} \frac{1}{(x - 3)^2} = +\infty$.

16. $\lim_{x \to 3} \frac{-1}{(x - 3)^2} = -\infty$.

17. $\lim_{x \to 0} \frac{1}{|x|} = +\infty$.

18. $\lim_{x \to 1} \frac{1}{|x - 1|} = +\infty$.

19. $\lim_{x \to 0} \left(-\frac{1}{x^4} \right) = -\infty$.

20. $\lim_{x \to 0} \frac{1}{x^4} = +\infty$.

21. Define limit statements (15) and (16).

22. Use the definitions in Exercise 21 to prove

(a) $\lim_{x \to 0^+} \frac{1}{x} = +\infty$

(b) $\lim_{x \to 0^-} \frac{1}{x} = -\infty$.

23. Use the definitions in Exercise 21 to prove

(a) $\lim_{x \to 1^+} \frac{1}{1 - x} = -\infty$

(b) $\lim_{x \to 1^-} \frac{1}{1 - x} = +\infty$.

24. Define limit statements (17) and (18).

25. Use the definitions in Exercise 24 to prove

(a) $\lim_{x \to +\infty} (x + 1) = +\infty$

(b) $\lim_{x \to -\infty} (x + 1) = -\infty$.

26. Use the definitions in Exercise 24 to prove

(a) $\lim_{x \to +\infty} (x^2 - 3) = +\infty$

(b) $\lim_{x \to -\infty} (x^3 + 5) = -\infty$.

■ II. SOME PROOFS OF LIMIT THEOREMS

In this section we shall prove a number of results stated in Section 2.5, as well as some other basic properties of limits.

Our first result states that the limit of a function, if it exists, is unique; that is, there cannot be two distinct limits.

THEOREM 7.   *If $\lim_{x \to a} f(x) = L_1$ and $\lim_{x \to a} f(x) = L_2$, then $L_1 = L_2$.*

REMARK.   In this section we shall only consider the cases where a and the limits are finite (i.e., not equal to $+\infty$ or $-\infty$). Other cases are considered in the exercises.

Proof. We shall assume that $L_1 \neq L_2$ and show that this assumption leads to a contradiction. It wili then follow that $L_1 = L_2$. Let

$$\epsilon = \frac{|L_1 - L_2|}{2} \tag{1}$$

Since we are assuming $L_1 \neq L_2$, it follows that $\epsilon > 0$.
From the assumption

$$\lim_{x \to a} f(x) = L_1$$

we can find $\delta_1 > 0$ such that

$$|f(x) - L_1| < \epsilon \tag{2}$$

whenever

$$0 < |x - a| < \delta_1 \tag{3}$$

and since

$$\lim_{x \to a} f(x) = L_2 \tag{4}$$

we can find $\delta_2 > 0$ such that

$$|f(x) - L_2| < \epsilon \tag{5}$$

whenever

$$0 < |x - a| < \delta_2 \tag{6}$$

Let δ be the smaller of the numbers δ_1 and δ_2, and choose any x satisfying

$$0 < |x - a| < \delta$$

Since $\delta \leq \delta_1$ and $\delta \leq \delta_2$, this x satisfies both (3) and (6) and, therefore, $f(x)$ satisfies both (2) and (5). But

$$
\begin{aligned}
2\epsilon &= |L_1 - L_2| &&\boxed{\text{From (1)}} \\
&= |L_1 - f(x) + f(x) - L_2| \\
&= |(L_1 - f(x)) + (f(x) - L_2)| \\
&\leq |L_1 - f(x)| + |f(x) - L_2| &&\boxed{\text{The triangle inequality}} \\
&< \epsilon + \epsilon &&\boxed{\text{(2) and (5)}} \\
&= 2\epsilon
\end{aligned}
$$

Thus, we have shown that

$$2\epsilon < 2\epsilon$$

which is a contradiction. ▮

Next, we prove Formula (1) of Section 2.5.

THEOREM **8.** *For any constant k,*

$$\lim_{x \to a} k = k$$

Proof. Let $\epsilon > 0$ be given. We must find $\delta > 0$ such that

$$|k - k| < \epsilon \tag{7}$$

whenever

$$0 < |x - a| < \delta \tag{8}$$

But (7) can be written as

$$0 < \epsilon$$

which holds for all values of x. Thus, for *any* positive δ whatever, (7) will be satisfied when (8) is satisfied. ▌

Next we shall prove part (*a*) of Theorem 2.5.1.

THEOREM **9** [***Theorem* 2.5.1(*a*)].** *If* $\lim_{x \to a} f(x) = L_1$ *and* $\lim_{x \to a} g(x) = L_2$, *then*

$$\lim_{x \to a} [f(x) + g(x)] = \lim_{x \to a} f(x) + \lim_{x \to a} g(x) = L_1 + L_2$$

Proof. Let $\epsilon > 0$ be given. To prove that

$$\lim_{x \to a} [f(x) + g(x)] = L_1 + L_2$$

we must show that given $\epsilon > 0$ we can find $\delta > 0$ such that

$$|[f(x) + g(x)] - [L_1 + L_2]| < \epsilon \tag{9}$$

whenever

$$0 < |x - a| < \delta \tag{10}$$

Since

$$\lim_{x \to a} f(x) = L_1$$

we can find $\delta_1 > 0$ such that

$$|f(x) - L_1| < \frac{\epsilon}{2} \tag{11}$$

whenever

$$0 < |x - a| < \delta_1 \tag{12}$$

and since

$$\lim_{x \to a} g(x) = L_2$$

we can find $\delta_2 > 0$ such that

$$|g(x) - L_2| < \frac{\epsilon}{2} \tag{13}$$

whenever

$$0 < |x - a| < \delta_2 \tag{14}$$

Let δ be the smaller of the numbers δ_1 and δ_2. If x satisfies (10), then x will also satisfy both (12) and (14) since $\delta \leq \delta_1$ and $\delta \leq \delta_2$. Consequently, $f(x)$ and $g(x)$ will satisfy both (11) and (13). Therefore,

$$|[f(x) + g(x)] - [L_1 + L_2]| = |[f(x) - L_1] + [g(x) - L_2]|$$
$$\leq |f(x) - L_1| + |g(x) - L_2|$$
$$< \frac{\epsilon}{2} + \frac{\epsilon}{2} = \epsilon$$

Thus, for the δ we have selected, (9) is satisfied whenever (10) is satisfied. This completes the proof. ▌

Next we shall prove part (c) of Theorem 2.5.1. This proof is a little more complicated than the previous proofs.

THEOREM **10** [*Theorem* **2.5.1(c)**]. *If* $\lim_{x \to a} f(x) = L_1$ *and* $\lim_{x \to a} g(x) = L_2$, *then*

$$\lim_{x \to a} [f(x)g(x)] = \lim_{x \to a} f(x) \lim_{x \to a} g(x) = L_1 L_2$$

Proof. Let $\epsilon > 0$ be given. We must find $\delta > 0$ such that

$$|f(x)g(x) - L_1 L_2| < \epsilon \tag{15}$$

whenever

$$0 < |x - a| < \delta \tag{16}$$

To find δ we shall express (15) in a different form. We can write

$$f(x) = L_1 + [f(x) - L_1] \quad \text{and} \quad g(x) = L_2 + [g(x) - L_2]$$

When we multiply these expressions and subtract L_1L_2, we obtain

$$f(x)g(x) - L_1L_2 = L_1[g(x) - L_2] + L_2[f(x) - L_1] + [f(x) - L_1][g(x) - L_2]$$

so that (15) can be rewritten as

$$|L_1[g(x) - L_2] + L_2[f(x) - L_1] + [f(x) - L_1][g(x) - L_2]| < \epsilon \qquad (17)$$

Thus, we must find $\delta > 0$ such that (17) holds whenever (16) does.

Since ϵ is positive, the numbers

$$\sqrt{\epsilon/3}, \quad \frac{\epsilon}{3(1 + |L_1|)}, \quad \frac{\epsilon}{3(1 + |L_2|)}$$

are also positive. Therefore, since

$$\lim_{x \to a} f(x) = L_1 \quad \text{and} \quad \lim_{x \to a} g(x) = L_2$$

we can find positive numbers δ_1, δ_2, δ_3, and δ_4 such that

$$\left.\begin{array}{lll} |f(x) - L_1| < \sqrt{\epsilon/3} & \text{when} & 0 < |x - a| < \delta_1 \\[2mm] |f(x) - L_1| < \dfrac{\epsilon}{3(1 + |L_2|)} & \text{when} & 0 < |x - a| < \delta_2 \\[2mm] |g(x) - L_2| < \sqrt{\epsilon/3} & \text{when} & 0 < |x - a| < \delta_3 \\[2mm] |g(x) - L_2| < \dfrac{\epsilon}{3(1 + |L_1|)} & \text{when} & 0 < |x - a| < \delta_4 \end{array}\right\} \qquad (18)$$

Let δ be the smallest of the numbers δ_1, δ_2, δ_3, and δ_4. Then $\delta \le \delta_1$, $\delta \le \delta_2$, $\delta \le \delta_3$, and $\delta \le \delta_4$. Thus, if x satisfies (16), then x will also satisfy the four conditions on the right side of (18). Consequently, $f(x)$ and $g(x)$ will satisfy the four conditions on the left side of (18). Therefore,

$$\begin{aligned} |L_1[g(x) &- L_2] + L_2[f(x) - L_1] + [f(x) - L_1][g(x) - L_2]| \\ &\le |L_1[g(x) - L_2]| + |L_2[f(x) - L_1]| + |[f(x) - L_1][g(x) - L_2]| \\ &= |L_1|\,|g(x) - L_2| + |L_2|\,|f(x) - L_1| + |f(x) - L_1|\,|g(x) - L_2| \\ &< |L_1|\frac{\epsilon}{3(1 + |L_1|)} + |L_2|\frac{\epsilon}{3(1 + |L_2|)} + \sqrt{\epsilon/3}\sqrt{\epsilon/3} \qquad \boxed{\text{From (18)}} \\ &= \frac{\epsilon}{3}\frac{|L_1|}{1 + |L_1|} + \frac{\epsilon}{3}\frac{|L_2|}{1 + |L_2|} + \frac{\epsilon}{3} \\ &< \frac{\epsilon}{3} + \frac{\epsilon}{3} + \frac{\epsilon}{3} = \epsilon \end{aligned}$$

$$\boxed{\text{Since } \frac{|L_1|}{1 + |L_1|} < 1 \text{ and } \frac{|L_2|}{1 + |L_2|} < 1}$$

Thus, for the δ we have selected, (17) holds whenever (16) does. This completes the proof. ∎

□ **RELATIONSHIP BETWEEN ONE-SIDED AND TWO-SIDED LIMITS**

In Section 2.4 we interpreted the statement

$$\lim_{x \to a} f(x) = L$$

to mean

$$\lim_{x \to a^+} f(x) = L \quad \text{and} \quad \lim_{x \to a^-} f(x) = L$$

The following theorem shows that this interpretation is consistent with our formal limit definitions.

THEOREM 11. *If* $\lim_{x \to a^+} f(x) = \lim_{x \to a^-} f(x) = L$, *then* $\lim_{x \to a} f(x) = L$ *and conversely, if* $\lim_{x \to a} f(x) = L$, *then* $\lim_{x \to a^+} f(x) = L$ *and* $\lim_{x \to a^-} f(x) = L$.

Proof. Assume that

$$\lim_{x \to a^+} f(x) = \lim_{x \to a^-} f(x) = L \tag{19}$$

To prove

$$\lim_{x \to a} f(x) = L$$

we must show that given $\epsilon > 0$ there exists a $\delta > 0$ such that

$$|f(x) - L| < \epsilon \tag{20}$$

whenever

$$0 < |x - a| < \delta \tag{21}$$

Because of (19), there exists a number $\delta_1 > 0$ such that (20) is satisfied whenever

$$a < x < a + \delta_1 \tag{22}$$

and there exists a number $\delta_2 > 0$ such that (20) is satisfied whenever

$$a - \delta_2 < x < a \tag{23}$$

Let δ be the smaller of δ_1 and δ_2, so that $\delta \leq \delta_1$ and $\delta \leq \delta_2$. Thus, if x satisfies (21), x will satisfy both (22) and (23) (Figure C.7) and consequently $f(x)$ will satisfy (20). Thus, for the δ we have selected, (20) holds whenever (21) does. This completes this part of the proof. The proof of the converse is left as an exercise. ∎

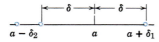

Figure C.7

► Exercises

1. Prove: For any constant k, $\lim_{x \to +\infty} k = k$.

2. Prove: For any constant k, $\lim_{x \to -\infty} k = k$.

3. Prove: If $\lim_{x \to -\infty} f(x) = L_1$ and $\lim_{x \to -\infty} g(x) = L_2$, then
$\lim_{x \to -\infty} [f(x) + g(x)] = L_1 + L_2$.

4. Prove: If $\lim_{x \to +\infty} f(x) = L_1$ and $\lim_{x \to +\infty} g(x) = L_2$, then
$\lim_{x \to +\infty} [f(x) + g(x)] = L_1 + L_2$.

5. Use Theorems 8, 9, and 10 of this section to prove:
If $\lim_{x \to a} f(x) = L_1$ and $\lim_{x \to a} g(x) = L_2$, then
$$\lim_{x \to a} [f(x) - g(x)] = \lim_{x \to a} f(x) - \lim_{x \to a} g(x)$$
$$= L_1 - L_2$$

6. Suppose $\lim_{x \to a} f(x) = +\infty$ and $\lim_{x \to a} g(x) = +\infty$.
   (a) Prove: $\lim_{x \to a} [f(x) + g(x)] = +\infty$.

(b) Is it true that $\lim_{x \to a} [f(x) - g(x)] = 0$?

7. Suppose $\lim_{x \to a} f(x) = -\infty$ and $\lim_{x \to a} g(x) = +\infty$.
   (a) Prove: $\lim_{x \to a} [f(x) - g(x)] = -\infty$.
   (b) Is it true that $\lim_{x \to a} [f(x) + g(x)] = 0$?

8. Use Theorems 8 and 10 of this section to prove: If $\lim_{x \to a} f(x) = L$ and k is a constant, then
$$\lim_{x \to a} [kf(x)] = kL$$

9. Prove: $\lim_{x \to a} f(x) = L$ if and only if
$$\lim_{x \to a} [f(x) - L] = 0$$

10. Prove: If $\lim_{x \to a} f(x) = L$, then $\lim_{x \to a} |f(x)| = |L|$.

11. Finish the proof of Theorem 11 of this section.

■ III. PROOF OF THE LIMIT PROPERTY OF CONTINUITY

> *In this section we shall prove Theorem 2.7.5 for the case of the two-sided limit* $\lim_{x \to c} f(x)$.

THEOREM 12 (*Theorem 2.7.5*).   *If* $\lim_{x \to c} g(x) = L$ *and if the function f is continuous at L, then* $\lim_{x \to c} f(g(x)) = f(L)$; *that is,* $\lim_{x \to c} f(g(x)) = f(\lim_{x \to c} g(x))$

Proof. To prove that
$$\lim_{x \to c} f(g(x)) = f(L)$$

we must show that for every $\epsilon > 0$ there is a $\delta > 0$ such that $f(g(x))$ satisfies
$$|f(g(x)) - f(L)| < \epsilon \tag{1}$$

whenever x satisfies

$$0 < |x - c| < \delta \tag{2}$$

Since f is continuous at L,

$$\lim_{x \to L} f(x) = f(L)$$

If we use u as the variable rather than x, then this limit can be written as

$$\lim_{u \to L} f(u) = f(L)$$

Thus, given $\epsilon > 0$ we can find a $\delta_1 > 0$ such that $f(u)$ satisfies

$$|f(u) - f(L)| < \epsilon \tag{3}$$

whenever u satisfies

$$|u - L| < \delta_1 \tag{4}$$

Since

$$\lim_{x \to c} g(x) = L$$

there exists $\delta > 0$ such that

$$|g(x) - L| < \delta_1 \tag{5}$$

whenever x satisfies (2). To complete the proof let

$$u = g(x)$$

and assume that x satisfies (2). This implies that $g(x)$ satisfies (5), which implies that u satisfies (4). Therefore, $f(u)$ satisfies (3), which implies that $f(g(x))$ satisfies (1). Thus, (1) is satisfied whenever (2) is satisfied. ∎

IV.   PROOF OF THE CHAIN RULE

In this section we shall prove the chain rule (Theorem 3.5.2).

We begin with a preliminary result.

THEOREM **13.**   *If f is differentiable at x and if $y = f(x)$, then*

$$\Delta y = f'(x)\, \Delta x + \epsilon\, \Delta x$$

where $\epsilon \to 0$ as $\Delta x \to 0$.

Proof. Define

$$\epsilon = \begin{cases} \dfrac{f(x + \Delta x) - f(x)}{\Delta x} - f'(x) & \text{if } \Delta x \neq 0 \\ 0 & \text{if } \Delta x = 0 \end{cases} \tag{1}$$

If $\Delta x \neq 0$, it follows from (1) that

$$\epsilon \, \Delta x = [f(x + \Delta x) - f(x)] - f'(x) \, \Delta x \tag{2}$$

But,

$$\Delta y = f(x + \Delta x) - f(x) \tag{3}$$

so (2) can be written as

$$\epsilon \, \Delta x = \Delta y - f'(x) \, \Delta x$$

or

$$\Delta y = f'(x) \, \Delta x + \epsilon \, \Delta x \tag{4}$$

If $\Delta x = 0$, then (4) still holds (why?), so (4) is valid for all values of Δx. It remains to show that $\epsilon \to 0$ as $\Delta x \to 0$. But, this follows from the assumption that f is differentiable at x, since

$$\lim_{\Delta x \to 0} \epsilon = \lim_{\Delta x \to 0} \left[\frac{f(x + \Delta x) - f(x)}{\Delta x} - f'(x) \right] = f'(x) - f'(x) = 0 \quad \blacksquare$$

We are now ready to prove the chain rule.

THEOREM **14** (***Theorem** 3.5.2*). *If g is differentiable at the point x and f is differentiable at the point $g(x)$, then the composition $f \circ g$ is differentiable at the point x. Moreover, if $y = f(g(x))$ and $u = g(x)$, then*

$$\frac{dy}{dx} = \frac{dy}{du} \cdot \frac{du}{dx}$$

Proof. Since g is differentiable at x and $u = g(x)$, it follows from Theorem 13 that

$$\Delta u = g'(x) \, \Delta x + \epsilon_1 \, \Delta x \tag{5}$$

where $\epsilon_1 \to 0$ as $\Delta x \to 0$. And since $y = f(g(x)) = f(u)$ is differentiable at $u = g(x)$, it follows from Theorem 13 that

$$\Delta y = f'(u) \, \Delta u + \epsilon_2 \, \Delta u \tag{6}$$

where $\epsilon_2 \to 0$ as $\Delta u \to 0$.

Factoring out the Δu in (6) and then substituting (5) yields

$$\Delta y = [f'(u) + \epsilon_2][g'(x) \, \Delta x + \epsilon_1 \, \Delta x]$$

or

$$\Delta y = [f'(u) + \epsilon_2][g'(x) + \epsilon_1]\, \Delta x$$

or if $\Delta x \neq 0$,

$$\frac{\Delta y}{\Delta x} = [f'(u) + \epsilon_2][g'(x) + \epsilon_1] \tag{7}$$

Since $\epsilon_1 \to 0$ and $\epsilon_2 \to 0$ as $\Delta x \to 0$, it follows from (7) and Theorem 10 that

$$\lim_{\Delta x \to 0} \frac{\Delta y}{\Delta x} = f'(u)g'(x)$$

or

$$\frac{dy}{dx} = f'(u)g'(x) = \frac{dy}{du} \cdot \frac{du}{dx} \quad \blacksquare$$

■ V. PROOFS OF KEY RESULTS

In this section we shall prove two fundamental results: Theorem 4.3.6 (the first derivative test) and Theorem 4.3.7 (the second derivative test).

THEOREM **15** (*Theorem* **4.3.6**). *Suppose f is continuous at a critical point x_0.*

(a) *If $f'(x) > 0$ on an open interval extending left from x_0 and $f'(x) < 0$ on an open interval extending right from x_0, then f has a relative maximum at x_0.*

(b) *If $f'(x) < 0$ on an open interval extending left from x_0 and $f'(x) > 0$ on an open interval extending right from x_0, then f has a relative minimum at x_0.*

(c) *If $f'(x)$ has the same sign [either $f'(x) > 0$ or $f'(x) < 0$] on an open interval extending left from x_0 and on an open interval extending right from x_0, then f does not have a relative extremum at x_0.*

Proof (a).   In accordance with the hypothesis, assume that $f'(x) > 0$ on the interval (a, x_0) and $f'(x) < 0$ on the interval (x_0, b). To prove that f has a relative maximum at x_0, we shall show that

$$f(x_0) \geq f(x) \tag{1}$$

for all x in (a, b). From Theorem 4.2.2, f is increasing on the interval $(a, x_0]$, and decreasing on the interval $[x_0, b)$, so $f(x_0) \geq f(x)$ for all x in (a, b) with equality only at $x = x_0$, which proves (1).   ∎

The proofs of parts (b) and (c) are left as exercises.

THEOREM 16 (*Theorem* **4.3.7**).   *Suppose f is twice differentiable at a stationary point x_0.*

(a)   *If $f''(x_0) > 0$, then f has a relative minimum at x_0.*
(b)   *If $f''(x_0) < 0$, then f has a relative maximum at x_0.*

Proof (a).   Using the definition of a derivative we can write

$$f''(x_0) = \lim_{h \to 0} \frac{f'(x_0 + h) - f'(x_0)}{h} \tag{2}$$

By hypothesis, $f''(x_0) > 0$, so we can use $\epsilon = \frac{1}{2} f''(x_0)$ in the definition of a limit and deduce from (2) that there exists a $\delta > 0$ such that

$$\left| \frac{f'(x_0 + h) - f'(x_0)}{h} - f''(x_0) \right| < \frac{1}{2} f''(x_0) \tag{3}$$

whenever h satisfies

$$0 < |h| < \delta \tag{4}$$

To prove that f has a relative minimum at x_0, we shall show that

$$f'(x) > 0 \quad \text{for all } x \text{ in } (x_0, x_0 + \delta) \tag{5}$$

and

$$f'(x) < 0 \quad \text{for all } x \text{ in } (x_0 - \delta, x_0) \tag{6}$$

It will then follow from the first derivative test (Theorem 15) that f has a relative minimum at x_0.

From (3) and (4) it follows that

$$\frac{1}{2} f''(x_0) < \frac{f'(x_0 + h) - f'(x_0)}{h} < \frac{3}{2} f''(x_0) \tag{7}$$

whenever h satisfies

$$0 < |h| < \delta$$

By hypothesis, x_0 is a stationary point for f, so that $f'(x_0) = 0$. From this, the fact that $f''(x_0) > 0$, and the left-hand inequality in (7), it follows that

$$0 < \frac{f'(x_0 + h)}{h} \tag{8}$$

whenever h satisfies (4).

To prove (5), let x be any point in $(x_0, x_0 + \delta)$. If we let

$$h = x - x_0 \tag{9}$$

then $h > 0$ and h satisfies (4), so that (8) holds.
Multiplying (8) by h yields

$$0 < f'(x_0 + h) \tag{10}$$

then substituting (9) in (10) yields $0 < f'(x)$, which establishes (5). To obtain (6) let x be any point in $(x_0 - \delta, x_0)$. If we let

$$h = x - x_0 \tag{11}$$

then $h < 0$ and h satisfies (4) so that (8) holds.
Multiplying (8) by h yields

$$f'(x_0 + h) < 0 \tag{12}$$

then substituting (11) in (12) yields $f'(x) < 0$, which establishes (6). ∎

The proof of part (b) is similar and is left as an exercise.

▶ Exercises

1. Prove:
(a) part (b) of Theorem 15
(b) part (c) of Theorem 15.

2. Prove part (b) of Theorem 16.

3. A function f is called ***nondecreasing*** on a given interval if

$$f(x_2) \geq f(x_1) \text{ whenever } x_2 > x_1$$

where x_1 and x_2 are points in the interval. Similarly, f is called ***nonincreasing*** on the interval if

$$f(x_2) \leq f(x_1) \text{ whenever } x_2 > x_1$$

(a) Sketch the graph of a function that is nondecreasing, yet is not increasing.
(b) Sketch the graph of a function that is nonincreasing, yet is not decreasing.
[*Remark:* Unfortunately, terminology is not always consistent in the mathematical literature. Some writers use the terms "strictly increasing" and "strictly decreasing" where we have used the terms increasing and decreasing. These writers then use the terms "increasing" and "decreasing" where we have used the terms nondecreasing and nonincreasing.]

4. Use the Mean-Value Theorem to prove each of the following. (See Exercise 3 for the definitions of nondecreasing and nonincreasing.)

(a) If $f'(x) \geq 0$ on an open interval (a, b), then f is nondecreasing on (a, b).
(b) If $f'(x) \leq 0$ on an open interval (a, b), then f is nonincreasing on (a, b).

5. Prove: If $f'(x) < g'(x)$ for all x in (a, b), then for all points x_1, x_2 in (a, b), where $x_2 > x_1$,

$$f(x_2) - f(x_1) < g(x_2) - g(x_1)$$

6. (a) Prove: If $f'(x_0) > 0$ and f' is continuous at x_0, then there is an open interval containing x_0 on which f is increasing.
(b) Prove: If $f'(x_0) < 0$ and f' is continuous at x_0, then there is an open interval containing x_0 on which f is decreasing.

7. (a) Use Exercise 6(a) to prove: If $f''(x_0) > 0$ and f'' is continuous at x_0, then there is an open interval containing x_0 on which f is concave up.
(b) Use Exercise 6(b) to prove: If $f''(x_0) < 0$ and f'' is continuous at x_0, then there is an open interval containing x_0 on which f is concave down.

8. Use the results in Exercise 6 to prove: If f is differentiable at each point of the interval $[a, b]$ and f' is continuous at a and b, and $f'(a)f'(b) < 0$, then there is a number c in (a, b) such that $f'(c) = 0$.

■ VI. CRAMER'S RULE

In this section we shall show how determinants can be used to solve systems of two linear equations in two unknowns and three linear equations in three unknowns. We assume that the reader has read the text through Section 14.6.

Recall that a ***solution*** of a system of two linear equations in two unknowns

$$a_1 x + b_1 y = k_1$$
$$a_2 x + b_2 y = k_2$$
(1)

is a value for x and a value for y that satisfy *both* equations. The graphs of these equations are lines, which we shall denote by L_1 and L_2. Since a point (x, y) lies on a line if and only if the numbers x and y satisfy the equation of the line, the solutions of the system will correspond to points of intersection of L_1 and L_2. There are three possibilities:

- The lines L_1 and L_2 are parallel and distinct (Figure C.8*a*), in which case there is no intersection and, consequently, no solution to the system.
- The lines L_1 and L_2 intersect at only one point (Figure C.8*b*), in which case the system has exactly one solution.
- The lines L_1 and L_2 coincide (Figure C.8*c*), in which case there are infinitely many points of intersection, and consequently infinitely many solutions to the system.

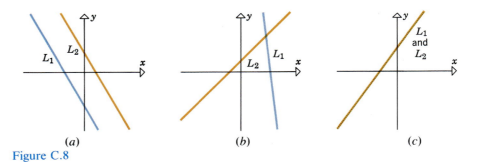

(a)   (b)   (c)

Figure C.8

Example 1   If we multiply the second equation of the system

$$x + y = 4$$
$$2x + 2y = 6$$

by $\frac{1}{2}$, it becomes evident that there is no solution since the two equations in the resulting system

$$x + y = 4$$
$$x + y = 3$$

contradict each other. Geometrically, the lines $x + y = 4$ and $2x + 2y = 6$ are distinct and parallel.   ◄

Example 2   Since the second equation of the system

$$x + y = 4$$
$$2x + 2y = 8$$

is a multiple of the first equation, it is evident that any solution of the first equation will satisfy the second equation automatically. But the first equation, $x + y = 4$, has infinitely many solutions since we may assign x an arbitrary value and determine y from the relationship $y = 4 - x$. (Some possibilities are $x = 0$, $y = 4$; $x = 1$, $y = 3$; $x = -1$, $y = 5$.) Thus, the system has infinitely many solutions. Geometrically, the lines $x + y = 4$ and $2x + 2y = 8$ coincide.   ◄

Example 3   If we add the equations of the system

$$x + y = 3$$
$$2x - y = 6$$

we obtain $3x = 9$ or $x = 3$, and if we substitute this value in the first equation we obtain $y = 0$, so that this system has the unique solution $x = 3$, $y = 0$. Geometrically, the lines $x + y = 3$ and $2x - y = 6$ intersect only at the point $(3, 0)$.   ◄

A *solution* of a system of three linear equations in three unknowns

$$a_1 x + b_1 y + c_1 z = k_1$$
$$a_2 x + b_2 y + c_2 z = k_2 \tag{2}$$
$$a_3 x + b_3 y + c_3 z = k_3$$

consists of values of x, y, and z that satisfy all three equations. The graphs of these equations are planes in 3-space, and the solutions of this system correspond to points where all three planes intersect. It follows that system (2), like system (1), has zero, one, or infinitely many solutions, depending on the relative positions of the planes (Figure C.9).

   If a system of two linear equations in two unknowns or three linear equations in three unknowns has a unique solution, the solution can be expressed as a

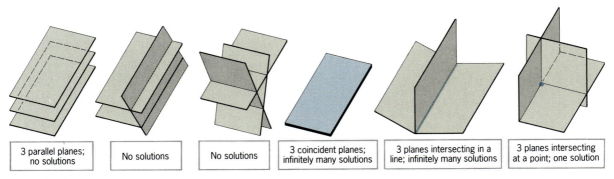

| 3 parallel planes; no solutions | No solutions | No solutions | 3 coincident planes; infinitely many solutions | 3 planes intersecting in a line; infinitely many solutions | 3 planes intersecting at a point; one solution |

Figure C.9

ratio of determinants by using the following theorems, which are cases of a general result called *Cramer's* rule*.

THEOREM 17 (*Cramer's Rule for Two Unknowns*).   *If the system of equations*

$$a_1 x + b_1 y = k_1$$
$$a_2 x + b_2 y = k_2$$

is such that $\begin{vmatrix} a_1 & b_1 \\ a_2 & b_2 \end{vmatrix} \neq 0$

then the system has a unique solution. This solution is

$$x = \frac{\begin{vmatrix} k_1 & b_1 \\ k_2 & b_2 \end{vmatrix}}{\begin{vmatrix} a_1 & b_1 \\ a_2 & b_2 \end{vmatrix}}, \quad y = \frac{\begin{vmatrix} a_1 & k_1 \\ a_2 & k_2 \end{vmatrix}}{\begin{vmatrix} a_1 & b_1 \\ a_2 & b_2 \end{vmatrix}}$$

---

*GABRIEL CRAMER (1704–1752). Swiss mathematician. Although Cramer does not rank with the great mathematicians of his time, his contributions as a disseminator of mathematical ideas have earned him a well-deserved place in the history of mathematics. The son of a physician, Cramer was born and educated in Geneva, Switzerland. At age 20 he competed for, but failed to secure, the chair of philosophy at the Académie de Calvin at Geneva. However, the awarding magistrates were sufficiently impressed with Cramer and a fellow competitor to create a new chair of mathematics for both men to share. Alternately, each assumed the full responsibility and salary associated with the chair for two or three years while the other traveled. During his travels Cramer met many of the great mathematicians and scientists of his day: the Bernoullis, Euler, Halley, D'Alembert, and others. Many of these contacts and friendships led to extensive correspondence in which information about new mathematical discoveries was transmitted. Eventually, Cramer became sole possessor of the mathematics chair and the chair of philosophy as well.

Cramer's mathematical work was primarily in geometry and probability; he had relatively little knowledge of calculus and did not use it to any great extent in his work. In 1730 he finished second to Johann I Bernoulli in a competition for a prize offered by the Paris Academy to explain properties of planetary orbits.

(*continued on next page*)

THEOREM 18 (*Cramer's Rule for Three Unknowns*). *If the system of equations*

$$a_1 x + b_1 y + c_1 z = k_1$$
$$a_2 x + b_2 y + c_2 z = k_2$$
$$a_3 x + b_3 y + c_3 z = k_3$$

is such that $\begin{vmatrix} a_1 & b_1 & c_1 \\ a_2 & b_2 & c_2 \\ a_3 & b_3 & c_3 \end{vmatrix} \neq 0$

then the system has a unique solution. This solution is

$$x = \frac{\begin{vmatrix} k_1 & b_1 & c_1 \\ k_2 & b_2 & c_2 \\ k_3 & b_3 & c_3 \end{vmatrix}}{\begin{vmatrix} a_1 & b_1 & c_1 \\ a_2 & b_2 & c_2 \\ a_3 & b_3 & c_3 \end{vmatrix}}, \quad y = \frac{\begin{vmatrix} a_1 & k_1 & c_1 \\ a_2 & k_2 & c_2 \\ a_3 & k_3 & c_3 \end{vmatrix}}{\begin{vmatrix} a_1 & b_1 & c_1 \\ a_2 & b_2 & c_2 \\ a_3 & b_3 & c_3 \end{vmatrix}}, \quad z = \frac{\begin{vmatrix} a_1 & b_1 & k_1 \\ a_2 & b_2 & k_2 \\ a_3 & b_3 & k_3 \end{vmatrix}}{\begin{vmatrix} a_1 & b_1 & c_1 \\ a_2 & b_2 & c_2 \\ a_3 & b_3 & c_3 \end{vmatrix}}$$

REMARK. There is a pattern to the formulas in these theorems. In each formula the determinant in the denominator is formed from the coefficients of the unknowns; and the determinant in the numerator differs from the determinant in the denominator in that the coefficients of the unknown being calculated are replaced by the k's. It is assumed in these theorems that the system is written so that like unknowns are aligned vertically and the constants appear by themselves on the right side of each equation.

Example 4 Use Cramer's rule to solve

$$5x - 2y = -1$$
$$2x + 3y = 3$$

(continued)

Cramer's most widely known work, *Introduction à l'analyse des lignes courbes algébriques* (1750), was a study and classification of algebraic curves; Cramer's rule appeared in the appendix. Although the rule bears his name, variations of the basic idea were formulated earlier by Leibniz (and even earlier by Chinese mathematicians). However, Cramer's superior notation helped clarify and popularize the technique.

Perhaps Cramer's most important contributions stemmed from his work as an editor of the mathematical creations of others. He edited and published the works of Jacob I Bernoulli and Leibniz.

Overwork combined with a fall from a carriage eventually led to his death in 1752. Cramer was apparently a good-natured and pleasant person, though he never married. His interests were broad. He wrote on philosophy of law and government and the history of mathematics. He served in public office, participated in artillery and fortifications activities for the government, instructed workers on techniques of cathedral repair, and undertook excavations of cathedral archives. Cramer received numerous honors for his activities.

Solution.

$$x = \frac{\begin{vmatrix} -1 & -2 \\ 3 & 3 \end{vmatrix}}{\begin{vmatrix} 5 & -2 \\ 2 & 3 \end{vmatrix}} = \frac{3}{19}; \quad y = \frac{\begin{vmatrix} 5 & -1 \\ 2 & 3 \end{vmatrix}}{\begin{vmatrix} 5 & -2 \\ 2 & 3 \end{vmatrix}} = \frac{17}{19} \quad \blacktriangleleft$$

Example 5 Use Cramer's rule to solve

$$\begin{aligned} x \quad\;\; + 2z &= 6 \\ -3x + 4y + 6z &= 30 \\ -x - 2y + 3z &= 8 \end{aligned}$$

Solution.

$$x = \frac{\begin{vmatrix} 6 & 0 & 2 \\ 30 & 4 & 6 \\ 8 & -2 & 3 \end{vmatrix}}{\begin{vmatrix} 1 & 0 & 2 \\ -3 & 4 & 6 \\ -1 & -2 & 3 \end{vmatrix}} = \frac{-10}{11}; \quad y = \frac{\begin{vmatrix} 1 & 6 & 2 \\ -3 & 30 & 6 \\ -1 & 8 & 3 \end{vmatrix}}{\begin{vmatrix} 1 & 0 & 2 \\ -3 & 4 & 6 \\ -1 & -2 & 3 \end{vmatrix}} = \frac{18}{11};$$

$$z = \frac{\begin{vmatrix} 1 & 0 & 6 \\ -3 & 4 & 30 \\ -1 & -2 & 8 \end{vmatrix}}{\begin{vmatrix} 1 & 0 & 2 \\ -3 & 4 & 6 \\ -1 & -2 & 3 \end{vmatrix}} = \frac{38}{11} \quad \blacktriangleleft$$

▶ Exercises

In Exercises 1–8, solve the system using Cramer's rule.

1. $3x - 4y = -5$
$\quad 2x + y = 4.$

2. $-x + 3y = 8$
$\quad 2x + 5y = 7.$

3. $2x_1 - 5x_2 = -2$
$\quad 4x_1 + 6x_2 = 1.$

4. $3a + 2b = 4$
$\quad -a + b = 7.$

5. $x + 2y + z = 3$
$\quad 2x + y - z = 0$
$\quad x - y + z = 6.$

6. $x - 3y + z = 4$
$\quad 2x - y = -2$
$\quad 4x - 3z = 0.$

7. $x_1 + x_2 - 2x_3 = 1$
$\quad 2x_1 - x_2 + x_3 = 2$
$\quad x_1 - 2x_2 - 4x_3 = -4.$

8. $r + s + t = 2$
$\quad r - s - 2t = 0$
$\quad -r + 2s + t = 4.$

9. Use Cramer's rule to solve the rotation equations

$$\begin{aligned} x &= x' \cos\theta - y' \sin\theta \\ y &= x' \sin\theta + y' \cos\theta \end{aligned}$$

for x' and y' in terms of x and y.

10. Solve the following system of equations for the unknown angles α, β, and γ, where $0 \le \alpha \le 2\pi$, $0 \le \beta \le 2\pi$, and $0 \le \gamma < \pi$:

$$\begin{aligned} 2\sin\alpha - \cos\beta + 3\tan\gamma &= 3 \\ 4\sin\alpha + 2\cos\beta - 2\tan\gamma &= 2 \\ 6\sin\alpha - 3\cos\beta + \tan\gamma &= 9 \end{aligned}$$

[*Hint:* First solve for $\sin\alpha$, $\cos\beta$, and $\tan\gamma$.]

Answers to Odd-Numbered Exercises

▶ Exercise Set 1.1 (Page 16)

1. (a) rational   (b) integer, rational
     (c) integer, rational   (d) rational
     (e) integer, rational   (f) irrational
     (g) rational   (h) integer, rational

3. (a) $\frac{41}{333}$ (b) $\frac{115}{9}$ (c) $\frac{20943}{550}$ (d) $\frac{537}{1250}$

5. (a) $\frac{256}{81}$ (b) worse

7.

Line	Blocks
2	3, 4
3	1, 2
4	3, 4
5	2, 4, 5
6	1, 2
7	3, 4

9. (a), (d), (f)

11. (a) all values   (b) none

13. (a) yes   (b) no

15. (a) $\{x: x \text{ is a positive odd integer}\}$
     (b) $\{x: x \text{ is an even integer}\}$
     (c) $\{x: x \text{ is irrational}\}$
     (d) $\{x: x \text{ is an integer and } 7 \le x \le 10\}$

17. (a) false   (b) true   (c) true   (d) false
     (e) true   (f) true   (g) true

19. (a) ——○→ 4   (b) ○——→ −3
     (c) ●——●→ −1 ... 7   (d) ●——●→ −3 ... 3
     (e) ●——●→ −3 ... 3   (f) ●——●→ −3 ... 3

21. (a) $[-2, 2]$ (b) $(-\infty, -2) \cup (2, +\infty)$

23. $(-\infty, \frac{10}{3})$

25. $(-\infty, -\frac{11}{2}]$

27. $(-\frac{3}{2}, \frac{1}{2}]$

29. $(-\infty, 3) \cup (4, +\infty)$

31. $(-\frac{3}{2}, 2)$

33. $(-\infty, -2] \cup (2, +\infty)$

35. $(-\infty, -3) \cup (3, +\infty)$

37. $(-\infty, -2) \cup (4, +\infty)$

39. $[4, 5]$

41. $(-8, 0) \cup (4, +\infty)$

43. $(2, +\infty)$

45. $(-\infty, -3] \cup [2, +\infty)$

47. $77 \le F \le 104$ **55.** $(-\infty, -\frac{1}{2})$

▶ Exercise Set 1.2 (Page 26)

1. (a) 7   (b) $\sqrt{2}$ (c) k^2 (d) k^2

3. $x \le 3$ **5.** all real x **7.** $x \ge 0$ or $x = -\frac{2}{3}$

9. $x \ge -5$ **13.** (a) 2   (b) 1   (c) 14
                 (d) $3 + \sqrt{2}$ (e) 7   (f) 5

15. (a) -9 (b) 7   (c) 12

17. $-\frac{5}{6}, \frac{3}{2}$ **19.** $\frac{1}{2}, \frac{5}{2}$ **21.** $-\frac{11}{10}, \frac{11}{8}$

23. $1, \frac{17}{5}$ **25.** $(-9, -3)$ **27.** $[-\frac{3}{2}, \frac{9}{2}]$

29. $(-\infty, -3) \cup (-1, +\infty)$ **31.** $(-\infty, \frac{1}{2}] \cup [\frac{9}{2}, +\infty)$

33. $(-\infty, \frac{1}{2}) \cup (\frac{3}{2}, +\infty)$ **35.** $[\frac{1}{8}, \frac{1}{2}) \cup (\frac{1}{2}, \frac{7}{8}]$

37. $(-\infty, \frac{5}{2})$ **39.** $[\frac{7}{10}, \frac{7}{2}]$ **41.** $(0, +\infty)$

43. $(-\infty, -7) \cup (-7, -\frac{3}{2})$ **45.** $x \in (-\infty, 2] \cup [3, +\infty)$

47. $-3, 9$ **55.** $\frac{1}{3}$ **57.** $\frac{13}{11}$ (Other answers are possible.)

▶ **Exercise Set 1.3 (Page 38)**

1.

(−2, 5)
(3, 4)
(4, 0)
(−2.5, −3)
(1.7, −2)
(0, −6)

3. (a)

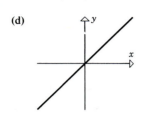

(b)

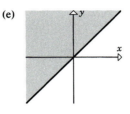

(c)

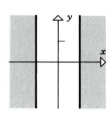

(d)

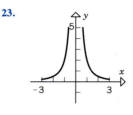

(e)

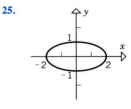

(f)

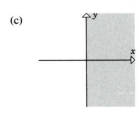

5. (a) vertical **(b)** horizontal **(c)** vertical

7. (a) yes **(b)** no **(c)** yes **(d)** yes

9. (a) origin **(b)** x-axis
   (c) none **(d)** y-axis

11. (a) $x = y^2$, $x = -y^2$
   (b) $y = x^2$, $y = -x^2$, $y = 1/x^2$, $y = -1/x^2$
   (c) $y = x^3$, $y = \sqrt[3]{x}$, $y = 1/x$, $y = -1/x$

13. (a) y-axis **(b)** origin
   (c) x-axis, y-axis, origin

23.

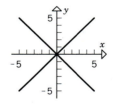

25.

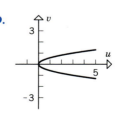

27. the union of the graphs of $x - y = 0$ and $x + y = 0$

15.

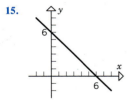

17.

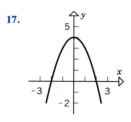

19.

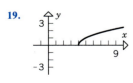

21.

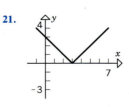

29.

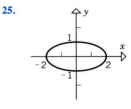

▶ **Exercise Set 1.4 (Page 46)**

1. (a) $\frac{1}{2}$ **(b)** -1 **(c)** 0 **(d)** not defined

3. (a) yes **(b)** no

5.

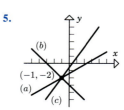

7. (c), (b), (d), (a)   **9.** (a) $\frac{3}{2}$ (b) $-\frac{3}{4}$ **23.** (a) 153°   (b) 45°   (c) 117°   (d) 89°

11. about 240 N/m   **13.** about 0.6 m/sec/°C   **25.** (a) parallel   (b) perpendicular   (c) neither

15. (a) 14   (b) $-\frac{1}{3}$ **17.** 29   **19.** $\frac{13}{7}$ **27.** (2, 0), (7, 0)

21. (a) $\dfrac{1}{\sqrt{3}}$ (b) -1 (c) $\sqrt{3}$

33. (a) $\frac{7}{19}$ (b) $\frac{57}{25}$ (c) $\frac{4}{13}$

35. (a) 20°   (b) 66°   (c) 17°   **37.** 11.09

▶ Exercise Set 1.5 **(Page 54)**

1. (a) 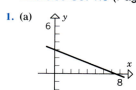   (b) 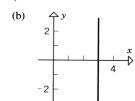   (c) 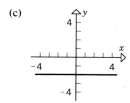   (d)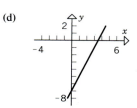

3. (a) 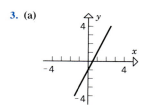   (b) 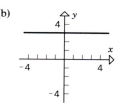   (c)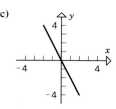

5.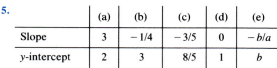

	(a)	(b)	(c)	(d)	(e)
Slope	3	$-1/4$	$-3/5$	0	$-b/a$
y-intercept	2	3	8/5	1	b

7. (a) 60°   (b) 117°   **9.** $y = -2x + 4$

11. $y = 4x + 7$ **13.** $y = -\frac{1}{5}x + 6$

15. $y = 11x - 18$

17. $y = \dfrac{1}{\sqrt{3}}x - 3$ **19.** $y = \frac{1}{2}x + 2$

21. $y = 1$ **23.** $x = 5$

25. (a) parallel   (b) perpendicular   (c) parallel
(d) perpendicular   (e) neither

27. (a) $-\frac{3}{2}$ (b) $\frac{4}{5}$ (c) $\frac{5}{2}$
(d) $-\frac{15}{2}$ (e) -4

29. (a) $F = \frac{9}{5}C + 32$ (b) $\frac{5}{9}$
(c) $-40°$ (d) 37° C

31. (a) $p = 0.098h + 1$
(b) approximately 10.2 m

33. (a) $r = -0.0125t + 0.8$ **37.** $\frac{49}{6}$
(b) 64 days

39. (a) $(4, -1)$ (b) $(1, -2)$

41. $(\frac{1}{2}, \frac{1}{4})$

▶ Exercise Set 1.6 **(Page 65)**

1. in the proof of Theorem 1.6.1

3. (a) 10   (b) (4, 5)

5. (a) $\sqrt{29}$ (b) $(-\frac{9}{5}, -5)$

11. 0   **13.** $y = -3x + 4$

15. $(-\frac{29}{8}, -\frac{23}{4})$

17. 3   **21.** 4

23. (a) (0, 0); 5   (b) (1, 4); 4
(c) $(-1, -3)$; $\sqrt{5}$ (d) $(0, -2)$; 1

25. $(x - 3)^2 + (y + 2)^2 = 16$

27. $(x + 4)^2 + (y - 8)^2 = 64$

29. $(x + 3)^2 + (y + 4)^2 = 25$

31. $(x - 1)^2 + (y - 1)^2 = 2$

33. circle; center (1, 2), radius 4

35. circle; center $(-1, 1)$, radius $\sqrt{2}$

37. the point $(-1, -1)$

39. circle; center (0, 0), radius $\frac{1}{3}$

41. no graph

43. circle; center $(-\frac{5}{4}, -\frac{1}{2})$, radius $\frac{3}{2}$

45. (a) $y = -\sqrt{16 - x^2}$
(b) $y = 2 + \sqrt{3 - 2x - x^2}$

47. (a) 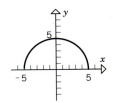   (b)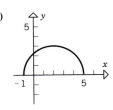

49. $y = -\frac{3}{4}x + \frac{25}{4}$

51. (a) inside
(b) largest $3\sqrt{5}$, smallest $\sqrt{5}$

53. $(1/3, \pm\sqrt{8}/3)$

55. (a) equation: $2x^2 + 2y^2 - 12x + 8y + 1 = 0$
(b) center $(3, -2)$, radius $5/\sqrt{2}$

57. 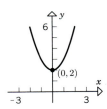   **59.** 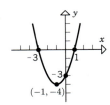   **61.** 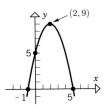   **63.**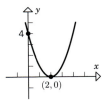

65. 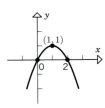   **67.** 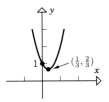   **69.**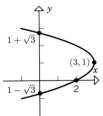

71. (a) $x = \sqrt{3 - y}$ (b) $x = 1 - \sqrt{y + 1}$

73. (a) 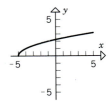   (b)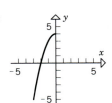

75. (a)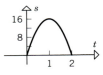

77. (a) $y = 150 - \frac{3}{2}x$ (b) $A = 150x - \frac{3}{2}x^2$
(c) 3750 ft^2

▶ **Chapter 1 Supplementary Exercises** (Page 67)

1. (a) $(-3, 5]$ (b) $[-3, 3]$
(c) $(-\infty, -\frac{1}{2}] \cup [\frac{1}{2}, +\infty)$

3. (a) $(-\infty, -\frac{1}{2}) \cup (3, +\infty)$ (b) $[1, 4]$

5. (a) $(-2, -1] \cup [1, 2)$
(b) $(-\infty, -5] \cup [-1, +\infty)$

7. (a) Take $a = -2$, $b = 1$.   (b) $a + b > 0$

9. Both legs of a right triangle are no longer than the hypotenuse.

11. (a) all points on the y-axis or the line $y = x$
(b) all points on the line $x = 1$ or the line $y = x + 1$

13. (a) all points on or outside the circle of radius 4 with center at $(1, 3)$
(b)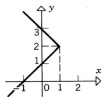

15. at $(2, 4)$ and $(-1, 1)$

17.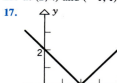

19.

21. $(x - 1)^2 + (y - 2)^2 = 25$

23. four circles: $(x - h)^2 + (y - k)^2 = 25$, where $h = 1$ or 11 and $k = 2$ or 12

25. point $(-2, -1)$ **27.** no graph

29. (a) $y = 4x/3$; 10; $(0, 0)$
   (b) $x = 3$; 8; $(3, 0)$
   (c) $y = 4$; 6; $(0, 4)$
   (d) $y = 7 - x$; $\sqrt{2}$; $(\frac{7}{2}, \frac{7}{2})$

31. $(4, -3)$ and $(-4, 3)$

33. $y = -3$ **35.** $y = -\frac{1}{2}x$

37. L: $y - 0 = (-2)(x - 1)$
   L': $y = \frac{1}{2}x - 3$; $(2, -2)$

39. L: $y - 1 = \frac{2}{3}(x - 3)$ **41.** no
   L': $3x + 2y = -8$; $(-2, -1)$

▶ **Exercise Set 2.1** (Page 81)

1. (a) 14   (b) 50   (c) 2   (d) 11
   (e) $3a^2 + 6a + 5$ (f) $27t^2 + 2$

3. (a) -8 (b) $\frac{1}{4}$ (c) 0
   (d) 6   (e) 5.8   (f) $\dfrac{1}{t^2 + 5}$

5. $(-\infty, 3) \cup (3, +\infty)$

7. $(-\infty, -\sqrt{3}] \cup [\sqrt{3}, +\infty)$

9. $(-\infty, -2) \cup [1, +\infty)$ **11.** $(-\infty, +\infty)$

13. $[5, 8]$ **15.** $(-\infty, +\infty)$

17. $(-\infty, 0) \cup (0, +\infty)$ **19.** $[0, +\infty)$

21. $x \neq \dfrac{\pi}{2} + 2k\pi$, $k = 0, \pm1, \pm2, \ldots$

23. domain $(-\infty, 3]$,   **25.** domain $[-2, 2]$,
   range $[0, +\infty)$ range $[0, 2]$

27. domain $[0, +\infty)$,   **29.** domain $(-\infty, +\infty)$,
   range $[3, +\infty)$ range $[3, +\infty)$

31. domain $(-\infty, +\infty)$,   **33.** domain $(-\infty, +\infty)$,
   range $(-\infty, +\infty)$ range $[-3, 3]$

35. domain $(-\infty, +\infty)$,
   range $[1, 3]$

37. $f(x) = \begin{cases} 2x + 1, & x < 0 \\ 4x + 1, & x \geq 0 \end{cases}$

39. $g(x) = \begin{cases} 1 - 2x, & x < 0 \\ 1, & 0 \leq x < 1 \\ 2x - 1, & x \geq 1 \end{cases}$

41. $\frac{38}{3}$ **43.** $\pm\sqrt{2}$

45. $2k\pi$, $k = 0, \pm1, \pm2, \ldots$

47. $(\frac{1}{6} + 2k)^2\pi^2$ or $(\frac{5}{6} + 2k)^2\pi^2$ for $k = 0, 1, 2, \ldots$

49. $A = \dfrac{C^2}{4\pi}$

51. (a) $S = 6x^2$ (b) $S = 6V^{2/3}$

53. $V = 4x^3 - 46x^2 + 120x$

55. $h = L(1 - \cos\theta)$

57. (a) $25°$ F   (b) $2°$ F   (c) $-15°$ F

59. $5°$ F   **61.** $\dfrac{1 - (1/x)}{1 + (1/x)} = \dfrac{x - 1}{x + 1}$, $x \neq 0$

63. $x - 1$, $x \neq -1$ or -2

65. $\sqrt{x + 1} + 1$, $x \neq -1$ **67.** x, $x \neq -3$ or 1

▶ **Exercise Set 2.2** (Page 90)

1. (a) $t^2 + 1$ (b) $t^2 + 4t + 5$
   (c) $x^2 + 4x + 5$ (d) $\dfrac{1}{x^2} + 1$
   (e) $x^2 + 2hx + h^2 + 1$ (f) $x^2 + 1$
   (g) $x + 1$, $x \geq 0$ (h) $9x^2 + 1$

3. (a) 1   (b) 12   (c) -5 (d) 4

5. (a) $x^2 + 2x + 1$ (b) $-x^2 + 2x - 1$
   (c) $2x(x^2 + 1)$ (d) $\dfrac{2x}{x^2 + 1}$
   (e) $2(x^2 + 1)$ (f) $4x^2 + 1$

7. (a) $\sqrt{x + 1} + x - 2$ (b) $\sqrt{x + 1} - x + 2$
   (c) $(x - 2)\sqrt{x + 1}$ (d) $\dfrac{\sqrt{x + 1}}{x - 2}$
   (e) $\sqrt{x - 1}$ (f) $\sqrt{x + 1} - 2$

9. (a) $\sqrt{x - 2} + \sqrt{x - 3}$
   (b) $\sqrt{x - 2} - \sqrt{x - 3}$
   (c) $\sqrt{x - 2}\sqrt{x - 3}$ (d) $\dfrac{\sqrt{x - 2}}{\sqrt{x - 3}}$
   (e) $\sqrt{\sqrt{x - 3} - 2}$ (f) $\sqrt{\sqrt{x - 2} - 3}$

11. (a) $\sqrt{1 - x^2} + \sin 3x$ (b) $\sqrt{1 - x^2} - \sin 3x$
   (c) $\sqrt{1 - x^2} \sin 3x$ (d) $\sqrt{1 - x^2}/\sin 3x$
   (e) $|\cos 3x|$ (f) $\sin(3\sqrt{1 - x^2})$

13. $(f \circ g)(x) = \dfrac{4}{x + 5}$, $x \geq 0$; $(g \circ f)(x) = \dfrac{2}{\sqrt{x^2 + 5}}$

15. (a) $4x - 15$ ·(b) $4x^2 - 20x + 25$

21. $g(x) = \sqrt{x}$, $h(x) = x + 2$ **23.** $g(x) = x^7$, $h(x) = x - 5$

25. $g(x) = |x|$, $h(x) = x^2 - 3x + 5$

27. $g(x) = x^2$, $h(x) = \sin x$

29. $g(x) = \dfrac{3}{5 + x}$, $h(x) = \cos x$

31. $g(x) = \dfrac{x}{3 + x}$, $h(x) = \tan x$

33. $f(x) = \sqrt{x}$, $g(x) = 3 - x^2$, $h(x) = \sin x$

35. $f(x) = x^2 + x + 3$ **37.** $0, \frac{3}{2}$ **39.** $g(x) = 9x^2 - 5$

43. (a) monomial, polynomial, rational, explicit algebraic
    (b) explicit algebraic
    (c) rational, explicit algebraic
    (d) polynomial, rational, explicit algebraic

45. (a) explicit algebraic
    (b) rational, explicit algebraic
    (c) monomial, polynomial, rational, explicit algebraic
    (d) explicit algebraic

▶ Exercise Set 2.3 (Page 102)

1. (a) $-4, -3, -2, 2, 3$ (b) $0, 4$
   (c) $-4 \le x \le -3, -2 \le x \le 2, x \ge 3$
   (d) $x \le -4, -3 \le x \le -2, 2 \le x \le 3$

3.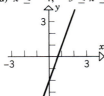

5.

7.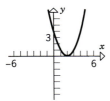

9.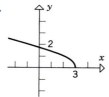

11.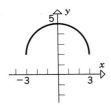

13.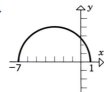

15.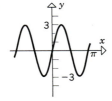

17.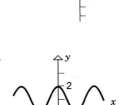

19.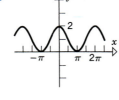

21.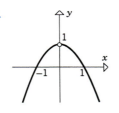

23.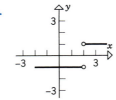

25.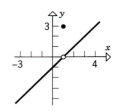

27.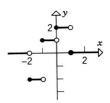

29.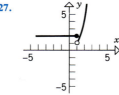

31. $f(x) = \begin{cases} x + 2, & x \le 2 \\ 3x - 2, & x > 2 \end{cases}$

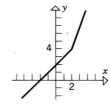

33. $g(x) = \begin{cases} 2, & x < 3 \\ 8 - 2x, & 3 \le x < 5 \\ -2, & x \ge 5 \end{cases}$

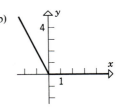

35. (a)   (b)

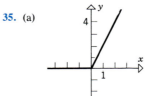

35. (c) 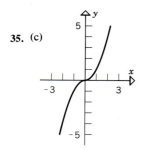 (d)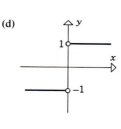

37. $A = \begin{cases} x^2, & 0 \le x \le 1 \\ 2x - 1, & x > 1 \end{cases}$

39. $g(x) = \begin{cases} 2, & x < -1 \\ 1 - x, & -1 \le x < 1 \\ \frac{1}{2}(x - 1), & x \ge 1 \end{cases}$

41. (a) 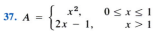 (b)

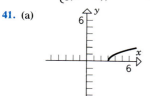

(c) 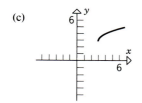 (d)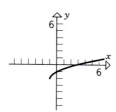

43.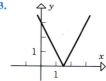

45. (a) 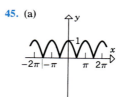 (b)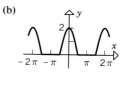

47. (a) even   (b) odd   (c) even
  (d) neither   (e) odd   (f) even

49. (a) even   (b) odd   (c) odd   (d) neither

53. (a) 1.324717957

55. 0.739085133

57. (a) both   (b) y is a function of x.
  (c) neither   (d) both

59. (a) $y = \dfrac{1}{x^2}$ (b) $y = \dfrac{1 - x}{1 + x}$ (c) $y = -x$

63. (a) both   (b) a function of x
  (c) a function of y (d) neither

▶ Exercise Set 2.4 (Page 115)

1. (a) -1 (b) 3   (c) does not exist
  (d) 1   (e) -1 (f) 3

3. (a) 1   (b) 1   (c) 1
  (d) 1   (e) $-\infty$ (f) $+\infty$

5. (a) 0   (b) 0   (c) 0
  (d) 3   (e) $+\infty$ (f) $+\infty$

7. (a) $-\infty$ (b) $+\infty$ (c) does not exist
  (d) not defined   (e) 2   (f) 0

9. (a) $-\infty$ (b) $-\infty$ (c) $-\infty$
  (d) 1   (e) 2   (f) 2

11. (a) 0   (b) 0   (c) 0
  (d) 0   (e) does not exist   (f) does not exist

13. all values except -4

▶ Exercise Set 2.5 (Page 129)

1. 7   **3.** π **5.** 36

7. $\sqrt{109}$ **9.** 14   **11.** 0

13. 8   **15.** 4   **17.** $-\frac{4}{5}$

19. $\frac{3}{2}$ **21.** 0   **23.** 0

25. $-\sqrt{5}$ **27.** $1/\sqrt{6}$ **29.** $\sqrt{3}$

31. $+\infty$ **33.** does not exist   **35.** $-\infty$

37. $+\infty$ **39.** does not exist   **41.** $+\infty$

43. $-\infty$ **45.** $-\frac{1}{7}$ **47.** -1

49. 6   **51.** $+\infty$ **53.** $+\infty$

55. $-\infty$ **57.** $\frac{1}{2}$ if $a \ne 0$, 1 if $a = 0$

59. (a) 2   (b) 2   (c) 2   **61.** 4

65. $\frac{1}{4}$ **67.** $\frac{5}{8}$

69. 0   **71.** $\frac{5}{2}$

73. $a/2$ **75.** if $r(a)$ is defined

▶ Exercise Set 2.6 **(Page 139)**

Note: There are other possible answers for Exercises 1–23, 29.

1. 0.05 **3.** 1/700 **5.** 0.05

7. 1/9000 **9.** 1 **11.** $\delta = \epsilon/3$

13. $\delta = \epsilon/2$

15. $\delta = \epsilon$ **17.** $\delta = \min(\epsilon/6, 1)$

19. $\delta = \min(\epsilon/36, \frac{1}{4})$ **21.** $\delta = \min(2\epsilon, 4)$

23. $\delta = \epsilon$ **29.** $\delta = \min(\epsilon/8, 2)$

▶ Exercise Set 2.7 **(Page 148)**

1. continuous on $(1, 2)$, $[2, 3]$, $(2, 3)$; discontinuous on $[1, 3]$, $(1, 3)$, and $[1, 2]$ at $x = 2$

3. continuous on $(1, 3)$, $(1, 2)$, $(2, 3)$; discontinuous on $[1, 3]$ at $x = 1$ and $x = 3$; on $[1, 2]$ at $x = 1$, and on $[2, 3]$ at $x = 3$

5. none **7.** none

9. $x = \pm 4$ **11.** $x = \pm 3$

13. none **15.** none

17. (a) 5 (b) $\frac{4}{3}$

21. $0, \pm 1, \pm 2, \dots$

23. (a) $x = 0$; not removable
(b) $x = -3$; removable
(c) $x = 2$; removable
  $x = -2$; not removable

33. $-1.65, 1.35$

35. (a) 2.25 (b) 2.235

▶ Exercise Set 2.8 **(Page 157)**

1. none **3.** $x = n\pi$, $n = 0, \pm 1, \pm 2, \dots$

5. $x = n\pi$, $n = 0, \pm 1, \pm 2, \dots$ **7.** none

9. $x = \dfrac{\pi}{6} + 2n\pi$ or $\dfrac{5\pi}{6} + 2n\pi$, $n = 0, \pm 1, \pm 2, \dots$

13. 1 **15.** $-\sqrt{3}/2$ **17.** 3

19. -1 **21.** 0 **23.** $\frac{7}{3}$

25. 1 **27.** 2 **29.** 0

31. $-\frac{25}{49}$ **33.** does not exist

35. 3 **37.** $\frac{1}{2}$

39. (a) 1 (b) 0 (c) 1 **41.** $-\pi$

43. (a) 0 (b) 0 **45.** 1

51. (a) 0.17365 (b) 0.17453

53. (a) 0.08749 (b) 0.08727

▶ Chapter 2 Supplementary Exercises **(Page 159)**

1. $|x| \le 2$; $\sqrt{2}, 2, 1$

3. $x \ne -2, 1$; $\frac{1}{2}$, undefined, $\frac{1}{4}$

5. all x; $-1, 3, \sqrt{3}$

7. (a) $(-6 + 6x - 2x^2)/x^2$
(b) $-9/[x(x + 3)]$
(c) $(3x^2 - 9x + 3)/(x - 3)$
(d) $(4x - 3)/(3 - x)$

9. the horizontal line $y = -\pi$; domain: all x; range: $\{-\pi\}$

11. domain: $x \ne -2$; range $y \ne -2$

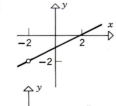

13. domain: $x \ge -\frac{1}{3}$; range: $y \le 0$

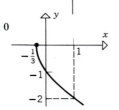

15. domain: $x \ne \pm 2$; range: $y \ne 0, \frac{1}{2}$

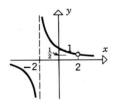

17. Some possible answers are
(a) $h(x) = x^3$, $g(x) = x^2 + 3$;
  $h(x) = x^6$, $g(x) = x + 3$
(b) $h(x) = x^2 + 1$, $g(x) = \sqrt{x}$;
  $h(x) = x^2$, $g(x) = \sqrt{x + 1}$
(c) $h(x) = 3x + 2$, $g(x) = \sin x$;
  $h(x) = 3x$, $g(x) = \sin(x + 2)$.

19. (a) -1 (b) does not exist (c) 1
(d) 0 (e) $-\infty$ (f) 0
(g) 0 (h) $-\infty$

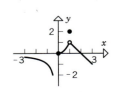

21. 2, 1, 0, does not exist, $+\infty$, does not exist

23. 5, 10, 0, 10, 0, $-\infty$, $+\infty$

25. a/b

27. does not exist

29. 0

31. $3 - k$

▶ **Exercise Set 3.1** (Page 170)

1. (a) $\frac{7}{2}$ (b) 3, $y = 3x - \frac{9}{2}$ (c)

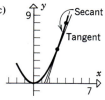

3. (a) $-\frac{1}{6}$ (b) $-\frac{1}{4}$, $y = -\frac{1}{4}x + 1$
(c)

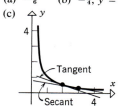

5. (b) $y = 75x - 250$ (c) $y = 3x_0{}^2 x - 2x_0{}^3$

7. (a) $2x_0 + 1$ (b) $y = 5x - 4$
(c) $y = (2x_0 + 1)x - x_0{}^2$

9. (a) 4 m/sec   (b)

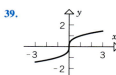

11. (a) t_0 (b) 0   (c) speeding up
(d) slowing down

13. It is a straight line with slope equal to the velocity.

15. (a) 72° F at about 4:30 P.M.
(b) 4° F/hr
(c) $-7°$ F/hr at about 9 P.M.

17. (a) first year
(b) 6 cm/yr
(c) 10 cm/yr at about age 14
(d)

19. (a) 320,000 ft   (b) 8000 ft/sec
(c) 45 ft/sec   (d) 24,000 ft/sec

21. (a) 720 ft/min   (b) 192 ft/min

23. (a) 10   (b) 4

25. (a) 3π (b) 4π

▶ **Exercise Set 3.2** (Page 183)

1. $6x$ **3.** $3x^2$ **5.** $\dfrac{1}{2\sqrt{x+1}}$

7. $-\dfrac{1}{x^2}$ **9.** $2ax$ **11.** $-\dfrac{1}{2x^{3/2}}$

13. 18; $y = 18x - 27$ **15.** 0; $y = 0$

17. $\frac{1}{6}$; $y = \frac{1}{6}x + \frac{5}{3}$ **19.** (a) $8x$ (b) 8

21. $8t + 1$ **23.** $6\lambda - 1$

25. (a) D   (b) F   (c) B
(d) C   (e) A   (f) E

27. 0.08 and 0.018 mol/L/sec

29. (a) $F \approx 200$ lb, $dF/d\theta \approx 60$ lb/rad   (b) $\mu \approx 0.3$

31.

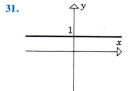

33.

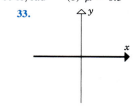

35.

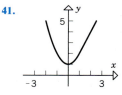

37.

39.

41.

43. $f(1) = 0$, $f'(1) = 5$

▶ **Exercise Set 3.3 (Page 197)**

1. $28x^6$

3. $24x^7 + 2$

5. 0

7. $-\frac{1}{3}(7x^6 + 2)$

9. $3ax^2 + 2bx + c$

11. $24x^{-9} + 1/\sqrt{x}$

13. $-3x^{-4} - 7x^{-8}$

15. $18x^2 - \frac{3}{2}x + 12$

17. $-15x^{-2} - 14x^{-3} + 48x^{-4} + 32x^{-5}$

19. $12x(3x^2 + 1)$

21. $-5/(5x - 3)^2$

23. $3/(2x + 1)^2$

25. $7/(x + 3)^2$

27. $\left(\dfrac{3x + 2}{x}\right)(-5x^{-6}) + (x^{-5} + 1)\left(-\dfrac{2}{x^2}\right)$

29. (a) $-\frac{37}{4}$ (b) $-\frac{23}{18}$

31. $32t$

33. $3\pi r^2$

35. $\dfrac{7 - 2t^3}{(t^3 + 7)^2}$

37. $-\dfrac{2GmM}{r^3}$

39. (a) $42x - 10$ (b) 24
  (c) $2/x^3$ (d) $700x^3 - 96x$

41. (a) $-210x^{-8} + 60x^2$ (b) $-6x^{-4}$ (c) $6a$

43. (a) 0 (b) 112 (c) 360

47. $F''(x) = xf''(x) + 2f'(x)$

49. $(1, \frac{5}{8})$, $(2, \frac{2}{3})$

51. $y = 5x + 17$

53. $a = 3, b = 2$

55. $y = 3x^2 - x - 2$

57. $\frac{1}{2}$ **59.** $2 \pm \sqrt{3}$ **61.** $-2x_0$

67. (a) $2(1 + 1/x)(x^{-3} + 7)$
  $\quad + (2x + 1)(-1/x^2)(x^{-3} + 7)$
  $\quad\quad + (2x + 1)(1 + 1/x)(-3x^{-4})$

  (b) $(-5x^{-6})(x^2 + 2x)(4 - 3x)(2x^9 + 1)$
  $\quad + x^{-5}(2x + 2)(4 - 3x)(2x^9 + 1)$
  $\quad\quad + x^{-5}(x^2 + 2x)(-3)(2x^9 + 1)$
  $\quad\quad\quad + x^{-5}(x^2 + 2x)(4 - 3x)(18x^8)$

  (c) $3(7x^6 + 2)(x^7 + 2x - 3)^2$ (d) $100x(x^2 + 1)^{49}$

69. $2(2x^3 - 5x^2 + 7x - 2)(6x^2 - 10x + 7)$

71. not differentiable at $x = 1$

73. $a = 6, b = -3$

75. (a) $x = \frac{2}{3}$ (b) $x = \pm 2$

77. (a) $n(n - 1)(n - 2) \cdots 1$ (b) 0
  (c) $a_n n(n - 1)(n - 2) \cdots 1$

83. (b) f and all its derivatives up to $f^{(n-1)}(x)$ are continuous on (a, b).

▶ **Exercise Set 3.4 (Page 204)**

1. $-2 \sin x - 3 \cos x$

3. $\dfrac{x \cos x - \sin x}{x^2}$

5. $x^3 \cos x + (3x^2 + 5) \sin x$

7. $\sec x \tan x - \sqrt{2} \sec^2 x$ **9.** $\sec^3 x + \sec x \tan^2 x$

11. $1 + 4 \csc x \cot x - 2 \csc^2 x$

13. $-\dfrac{\csc x}{1 + \csc x}$ **15.** 0

17. $\dfrac{1}{(1 + x \tan x)^2}$ **19.** $-x \cos x - 2 \sin x$

21. $-x \sin x + 5 \cos x$ **23.** $-4 \sin x \cos x$

25. (a) $x = n\pi$, $n = 0, \pm1, \pm2, \ldots$ (b) none
  (c) $x = \dfrac{\pi}{2} + n\pi$, $n = 0, \pm1, \pm2, \ldots$

27. (a) $y = x$ (b) $y = 2x - \pi/2 + 1$
  (c) $y = 2x + \pi/2 - 1$

29. 0.087 ft/degree **31.** 1.75 m/degree

33. (a) all x (b) all x
  (c) $x \neq \pi/2 + n\pi$, $n = 0, \pm1, \pm2, \ldots$
  (d) $x \neq n\pi$, $n = 0, \pm1, \pm2, \ldots$
  (e) $x \neq \pi/2 + n\pi$, $n = 0, \pm1, \pm2, \ldots$
  (f) $x \neq n\pi$, $n = 0, \pm1, \pm2, \ldots$
  (g) $x \neq \pi + 2n\pi$, $n = 0, \pm1, \pm2, \ldots$
  (h) $x \neq n\pi/2$, $n = 0, \pm1, \pm2, \ldots$
  (i) all x

35. $3, 7, 11, \ldots$ **37.** $\sec^2 y$

▶ **Exercise Set 3.5 (Page 211)**

1. $37(x^3 + 2x)^{36}(3x^2 + 2)$

3. $-2\left(x^3 - \dfrac{7}{x}\right)^{-3}\left(3x^2 + \dfrac{7}{x^2}\right)$

5. $\dfrac{24(1 - 3x)}{(3x^2 - 2x + 1)^4}$ **7.** $\dfrac{3}{4\sqrt{x}\sqrt{4 + 3\sqrt{x}}}$

9. $3x^2 \cos(x^3)$ **11.** $8x \sec^2(4x^2)$

13. $-20 \cos^4 x \sin x$ **15.** $-\dfrac{2}{x^3} \cos\left(\dfrac{1}{x^2}\right)$

17. $28x^6 \sec^2(x^7) \tan(x^7)$

19. $-\dfrac{5 \sin(5x)}{2\sqrt{\cos(5x)}}$

21. $-3[x + \csc(x^3 + 3)]^{-4}$
  $\quad \cdot [1 - 3x^2 \csc(x^3 + 3) \cot(x^3 + 3)]$

23. $\dfrac{x(10 - 3x^2)}{\sqrt{5 - x^2}}$

25. $10x^3 \sin 5x \cos 5x + 3x^2 \sin^2 5x$

27. $-x^3 \sec\left(\dfrac{1}{x}\right) \tan\left(\dfrac{1}{x}\right) + 5x^4 \sec\left(\dfrac{1}{x}\right)$

29. $\sin x \sin(\cos x)$

31. $-6\cos^2(\sin 2x)\sin(\sin 2x)\cos 2x$

33. $12(5x + 8)^{13}(x^3 + 7x)^{11}(3x^2 + 7)$
$+ 65(x^3 + 7x)^{12}(5x + 8)^{12}$

35. $\dfrac{33(x - 5)^2}{(2x + 1)^4}$

37. $-\dfrac{2(2x + 3)^2(52x^2 + 96x + 3)}{(4x^2 - 1)^9}$

39. $5[x\sin 2x + \tan^4(x^7)]^4[2x\cos 2x + \sin 2x$
$+ 28x^6\tan^3(x^7)\sec^2(x^7)]$

41. $-25x\cos(5x) - 10\sin(5x) - 2\cos(2x)$

43. $y = -x$ **45.** $y = -1$

47. $3\cot^2\theta\csc^2\theta$ **49.** $\pi(b - a)\sin 2\pi\omega$

51. (c) $f = 1/T$
(d) amplitude $= 0.6$ cm, $T = 2\pi/15$, $f = 15/(2\pi)$

53. (a) 10 lb/in^2, -2 lb/(in$^3 \cdot$ mi)
(b) -0.6 lb/(in$^2 \cdot$ sec)

55. $3x^2y^2\dfrac{dy}{dx} + 2xy^3$

57. $\left(x\dfrac{dy}{dx} + y\right)\cos(xy)$ **59.** $2x\dfrac{dx}{dt} + 2y\dfrac{dy}{dt}$

61. $\dfrac{x^2}{2\sqrt{y}}\dfrac{dy}{dt} + 2x\sqrt{y}\dfrac{dx}{dt}$

63. $\begin{cases} \cos x, & 0 < x < \pi \\ -\cos x, & -\pi < x < 0 \end{cases}$

65. (a) $-\dfrac{1}{x}\cos\dfrac{1}{x} + \sin\dfrac{1}{x}$

67. (a) 21   (b) -36 **69.** 6

71. $1/(2x)$ **73.** $\frac{2}{3}x$ **75.** $f'\big(g(h(x))\big)g'(h(x))h'(x)$

► **Exercise Set 3.6** (Page 220)

1. $\frac{2}{3}(2x - 5)^{-2/3}$ **3.** $\dfrac{9}{2(x + 2)^2}\left(\dfrac{x - 1}{x + 2}\right)^{1/2}$

5. $\frac{1}{3}x^2(5x^2 + 1)^{-5/3}(25x^2 + 9)$

7. $-\dfrac{15[\sin(3/x)]^{3/2}\cos(3/x)}{2x^2}$

9. $-\frac{2}{3}(2x - 1)^{-4/3}\sec^2[(2x - 1)^{-1/3}]$

11. $-1, \frac{2}{3}$

13. $-\dfrac{x}{y}$ **15.** $\dfrac{1 - 2xy - 3y^3}{x^2 + 9xy^2}$

17. $-\dfrac{y^2}{x^2}$ **19.** $-\dfrac{\sqrt{y}}{\sqrt{x}}$

21. $\dfrac{1 - 70x(x^2 + 3y^2)^{34}}{210y(x^2 + 3y^2)^{34}}$ **23.** $\dfrac{\frac{3}{2}x^2(x^3 + y^2)^{1/2} - y}{x - y(x^3 + y^2)^{1/2}}$

25. $\dfrac{1 - 2xy^2\cos(x^2y^2)}{2x^2y\cos(x^2y^2)}$

27. $\dfrac{1 - 3y^2\tan^2(xy^2 + y)\sec^2(xy^2 + y)}{3(2xy + 1)\tan^2(xy^2 + y)\sec^2(xy^2 + y)}$

29. $\dfrac{3y^2\sin^2(xy^2)\cos(xy^2)}{2\sqrt{1 + \sin^3(xy^2)} - 6xy\sin^2(xy^2)\cos(xy^2)}$

31. $-\frac{14}{13}$ **33.** 2   **35.** -2

37. 1   **39.** $\frac{6}{5}$ **41.** $-2x/y^5$

43. $-3/(y - x)^3$

45. $-\dfrac{\sin 2y + y(\sin^2 y + 1)}{(1 + x\sin y)^3}$

47. $\dfrac{2t^3 + 3a^2}{2a^3 - 6at}$ **49.** $-\dfrac{b^2\lambda}{a^2\omega}$

51. $-\dfrac{2y^3 + 3t^2y}{(6ty^2 + t^3)\cos t}$

53. $\dfrac{dy}{dt} = \dfrac{3\cos 3x - y^2}{2xy}\dfrac{dx}{dt}$

55. $a = \frac{1}{4}, b = \frac{5}{4}$

57. $y = x/\sqrt{3}, y = -x/\sqrt{3}$

59. (a) $(0, 0), (-5/\sqrt{2}, 0), (5/\sqrt{2}, 0)$
(b) $9x + 13y = 40$

► **Exercise Set 3.7** (Page 228)

1. (a) 5   (b) 4
(c)

3. (a) $-\frac{1}{3}$ (b) -0.5
(c)

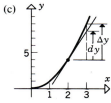

5. $dy = 3x^2\,dx$
$\Delta y = 3x^2\,\Delta x + 3x(\Delta x)^2 + (\Delta x)^3$

7. $dy = (2x - 2)\,dx$
$\Delta y = 2x\,\Delta x + (\Delta x)^2 - 2\,\Delta x$

9. $dy = (12x^2 - 14x + 2)\,dx$

11. $dy = (\cos x - x\sin x)\,dx$

13. $2x$ **15.** -1

17. 83.16   **19.** 8.0625

21. 8.9944   **23.** 2.005

25. 0.8573   **27.** 0.6947

29. 0.0225   **31.** 0.0048

33. (a) ± 2 ft^2 (b) side: $\pm 1\%$, area: $\pm 2\%$

35. (a) opposite: $\pm 0.151''$, adjacent: $\pm 0.087''$
(b) opposite: $\pm 3.0\%$, adjacent: $\pm 1.0\%$

37. $\pm 10\%$ **39.** $\pm 6\%$

41. $\pm 0.5\%$ **43.** 0.236 cm^3

45. (a) $\alpha = 1.5 \times 10^{-5}/°\text{C}$
(b) 180.1 cm

47. $x = 0.1$: 0.9091, 0.9
$x = -0.02$: 1.0204, 1.02

▶ **Chapter 3 Supplementary Exercises** (Page 230)

1. k **3.** $-2/\sqrt{9 - 4x}$ for $x < 9/4$

5. 0   **7.** $y + 1 = 5(x - 3)$

9. (a) 12   (b) -7 (c) 9
(d) $-9/4$ (e) 5   (f) 21
(g) -35 (h) -7 (i) -126
(j) -12 (k) $3/2$ (l) $-3/2$

11. $2(x - 3)^3(x^2 + 6x + 3)/(x^2 + 2x)^2$; 3, $-3 \pm \sqrt{6}$

13. $-3(3x + 1)^2(3x + 2)/x^7$; $-1/3$, $-2/3$

15. $(7x^2 + 5x + 3)x^{-1/2}(x^2 + x + 1)^{-2/3}/6$; none

17. $-2\sqrt{2}/x^3 + 2x^{-2}/5$ **19.** $\sqrt{3}$ $(z = \sin^2 2r)$

21. $2(x - 1)/x^3$ **23.** 0

25. 2 $(F(x) = 2x)$ **27.** ± 1 **29.** ± 2

31. odd integer multiples of $\pi/4$

33. $\Delta x = \pi/4$, $\Delta y = 1$, $dy = \pi/2$

35. (a) $-97/48$ (b) $1 - (\pi/90)$

37. $dh = -\pi/30$ ft

39. (a) 2000 gal/min   (b) 2500 gal/min

41. $\dfrac{dy}{dx} = \dfrac{y^2 - \cos(x + 2y)}{2\cos(x + 2y) - 2xy}$; $y = -\frac{1}{2}x$

▶ **Exercise Set 4.1** (Page 239)

1. (a) $\dfrac{dA}{dt} = 2x\dfrac{dx}{dt}$ (b) $12 \text{ ft}^2/\text{min}$

3. (a) $\dfrac{dV}{dt} = \pi\left(r^2\dfrac{dh}{dt} + 2rh\dfrac{dr}{dt}\right)$
(b) $-20\pi \text{ in}^3/\text{sec}$; decreasing

5. (a) $\dfrac{d\theta}{dt} = \dfrac{\cos^2\theta}{x^2}\left(x\dfrac{dy}{dt} - y\dfrac{dx}{dt}\right)$
(b) $-\frac{5}{16}$ radian/sec; decreasing

7. $4\pi/15 \text{ in}^2/\text{min}$ **9.** $1/\sqrt{\pi}$ mi/hr

11. $4860\pi \text{ cm}^3/\text{min}$ **13.** $\frac{5}{8}$ ft/sec

15. $\frac{8}{5} \text{ in}^2/\text{min}$ **17.** $\frac{1}{12}$ rad/sec

19. (a) 500 mi, 1716 mi
(b) 1354 mi; 27.7 mi/min

21. $9/(20\pi)$ ft/min   **23.** $125\pi \text{ ft}^3/\text{min}$

25. 250 mi/hr   **27.** $\frac{36}{25}\sqrt{69}$ ft/min

29. $8\pi/5$ km/sec   **31.** $600\sqrt{7}$ mi/hr

33. (a) $-60/7$ units/sec   (b) falling

35. -4 units/sec   **37.** $\frac{3}{2}$

39. 4.5 cm/sec; away

43. $2/(9\pi)$ cm/sec

▶ **Exercise Set 4.2** (Page 248)

1. (a) (d, f) (b) (a, d), (f, g)
(c) (a, b), (c, e) (d) (b, c), (e, g)

3. at A: negative, positive
at B: positive, negative
at C: negative, negative

5. (a) $[\frac{5}{2}, +\infty)$ (b) $(-\infty, \frac{5}{2}]$
(c) $(-\infty, +\infty)$ (d) none   (e) none

7. (a) $(-\infty, +\infty)$ (b) none
(c) $(-2, +\infty)$ (d) $(-\infty, -2)$ (e) -2

9. (a) $(-\infty, -\frac{2}{3}]$, $[\frac{2}{3}, +\infty)$
(b) $[-\frac{2}{3}, \frac{2}{3}]$ (c) $(0, +\infty)$
(d) $(-\infty, 0)$ (e) 0

11. (a) $[1, +\infty)$ (b) $(-\infty, 1]$
(c) $(-\infty, 0)$, $(\frac{2}{3}, +\infty)$
(d) $(0, \frac{2}{3})$ (e) $0, \frac{2}{3}$

13. (a) $[\pi, 2\pi]$ (b) $(0, \pi]$ (c) $(\pi/2, 3\pi/2)$
(d) $(0, \pi/2)$, $(3\pi/2, 2\pi)$ (e) $\pi/2, 3\pi/2$

15. (a) $(-\pi/2, \pi/2)$ (b) none
(c) $(0, \pi/2)$ (d) $(-\pi/2, 0)$ (e) 0

17. (a) $(-\infty, +\infty)$ (b) none
(c) $(-\infty, -2)$ (d) $(-2, +\infty)$ (e) -2

19. (a) $[-1, +\infty)$ (b) $(-\infty, -1]$
(c) $(-\infty, 0)$, $(2, +\infty)$ (d) $(0, 2)$ (e) $0, 2$

21. $(-\infty, +\infty)$

27. (a)   (b)

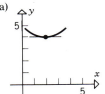

27. (c)

29. (a)

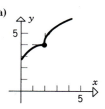

(b)

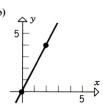

31. none

39. $f(x) = x$, $g(x) = 2x$ on $(-\infty, +\infty)$

▶ **Exercise Set 4.3** (Page 255)

1. $x = \frac{3}{2}$ (stationary)

3. $x = -3, 1$ (stationary)

5. $x = 0, \pm\sqrt{3}$ (stationary)

7. $x = \pm\sqrt{2}$ (stationary)

9. $x = 0$ (not differentiable)

11. $x = n\dfrac{\pi}{3}$, $n = 0, \pm1, \pm2, \ldots$ (stationary)

13. $x = n\dfrac{\pi}{4}$, $n = 1, 2, 3, 4, 5, 6, 7$ (stationary)

15. $x = -1$ (stationary); $x = 0$ (not differentiable)

17. (a) $x = 2$ (b) $x = 0$ (c) $x = 1, x = 3$

19. relative maximum at $x = 0$,
relative minimum at $x = \pm\sqrt{5}$

21. relative maximum at $x = \pm\frac{3}{2}$,
relative minimum at $x = -1$

23. relative max of 5 at $x = -2$

25. relative min of 0 at $x = \pi$;
relative max of 1 at $x = \pi/2, 3\pi/2$

27. none

29. relative min of 0 at $x = 1$;
relative max of $\frac{4}{27}$ at $x = \frac{1}{3}$

31. relative min of 0 at $x = 0$;
relative max of 1 at $x = -1, 1$

33. relative min of 0 at $x = 0$

35. relative min of 0 at $x = 0$

37. relative min of 0 at $x = -2, 2$;
relative max of 4 at $x = 0$

39. relative min of 0 at $x = \pm\pi/2, \pm3\pi/2$,
$\pm5\pi/2, \ldots$;
relative max of 1 at $x = 0, \pm\pi, \pm2\pi, \ldots$

41. relative min of tan (1) at $x = 0$

43. relative min of 0 at $x = \pi/2, \pi, 3\pi/2$;
relative max of 1 at $x = \pi/4, 3\pi/4, 5\pi/4, 7\pi/4$

45. 54

47. $f(x) = -x^4$ has a relative maximum at $x = 0$,
$f(x) = x^4$ has a relative minimum at $x = 0$,
$f(x) = x^3$ has neither at $x = 0$;
$f'(0) = 0$ for all three functions.

49. (a)

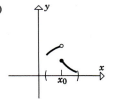

$f(x_0)$ is not an extreme value.

(b)

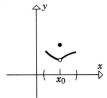

$f(x_0)$ is a relative maximum.

(c)

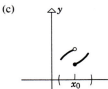

$f(x_0)$ is a relative minimum.

▶ **Exercise Set 4.4** (Page 264)

1.

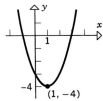

3.

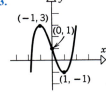

5.

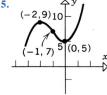

7.

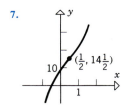

9.

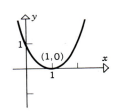

11.

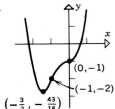

13.

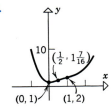

15.

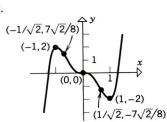

17.

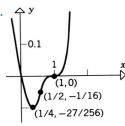

19. vertical: $x = 2$; horizontal: $y = 3$

21. vertical: $x = \pm\sqrt{5}$; horizontal: none

23. vertical: $x = -1$, $x = 3$; horizontal: $y = 1$

25. $x = -\frac{5}{2}$ **27.** $x = -\frac{1}{3}$

29.

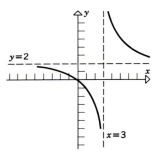

31.

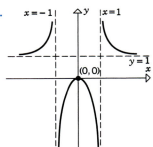

33.

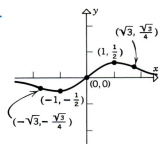

35.

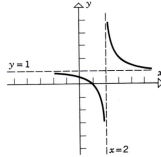

37.

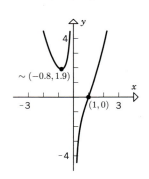

39.

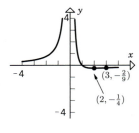

41.

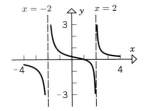

43.

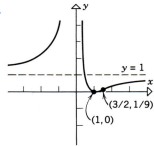

45.

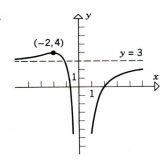

47.

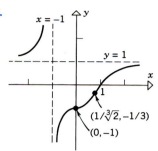

$x = -1$

$y = 1$

1

$(1/\sqrt[3]{2}, -1/3)$

$(0, -1)$

49.

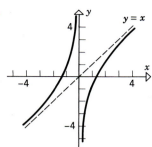

$y = x$

4

-4 4

-4

51.

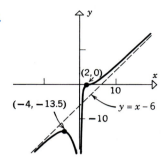

y

$(2, 0)$

10

$(-4, -13.5)$

$y = x - 6$

-10

53.

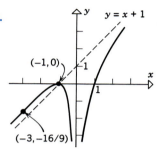

$y = x + 1$

$(-1, 0)$

1

1

$(-3, -16/9)$

55.

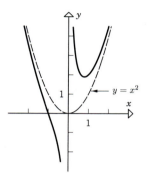

y

1

$y = x^2$

1

57.

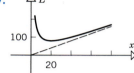

L

100

20

x

59.

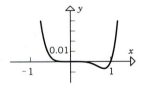

y

0.01

-1 1

x

Exercise Set 4.5

1

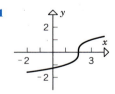

y

2

-2 3

-2

3.

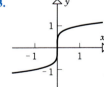

y

1

-1 1

-1

5.

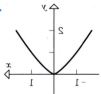

y

2

1 -1

7.

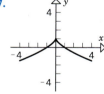

y

4

-4 4

-4

9.

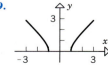

y

3

-3 3

x

11.

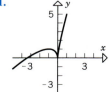

y

5

-3 3

-3

13.

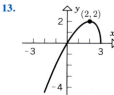

y $(2, 2)$

2

-3 3

-4

15.

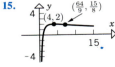

y $\left(\frac{64}{9}, \frac{15}{8}\right)$

4 $(4, 2)$

15

-4

17.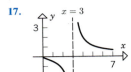

19.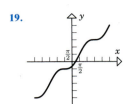

21.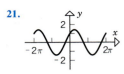

23.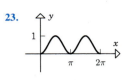

▶ **Exercise Set 4.6** (Page 278)

1. maximum value 1 when $x = 0, 1$; minimum value 0 when $x = \frac{1}{2}$

3. maximum value 27 when $x = 4$; minimum value -1 when $x = 0$

5. maximum value $3/\sqrt{5}$ when $x = 1$; minimum value $-3/\sqrt{5}$ when $x = -1$

7. maximum value 48 when $x = 8$; minimum value 0 when $x = 0, 20$

9. maximum value $1 - (\pi/4)$ when $x = -\pi/4$; minimum value $(\pi/4) - 1$ when $x = \pi/4$

11. maximum value 2 when $x = 0$; minimum value $\sqrt{3}$ when $x = \pi/6$

13. maximum value 17 when $x = -5$; minimum value 1 when $x = -3$

15. minimum $-\frac{13}{4}$, no maximum

17. maximum 1, no minimum

19. minimum 0, no maximum

21. no maximum or minimum

23. no maximum or minimum

25. maximum -4, no minimum

27. maximum value $3\sqrt{3}/2$ when $x = (\pi/6) + n\pi$ $(n = 0, \pm1, \pm2, \ldots)$ minimum value $-3\sqrt{3}/2$ when $x = (5\pi/6) + n\pi$ $(n = 0, \pm1, \pm2, \ldots)$

29. maximum $\sin(1)$, minimum $-\sin(1)$

31. maximum value 2; minimum value $-\frac{1}{4}$

33. (a) relative min of 0 at $x = a$ **35.** (b) 125 (b) none

37. $2/(3\sqrt{3})$

39. $a_0 = 9$, $a_1 = -8$, $a_2 = 2$

47. $(\frac{1}{2}, -\frac{1}{4})$ closest, $(-1, -1)$ farthest

---

▶ **Exercise Set 4.7** (Page 288)

1. $5 + 5$ **3.** (a) 1   (b) $\frac{1}{2}$

5. 500 ft by 750 ft   **7.** $10\sqrt{2}$ by $10\sqrt{2}$

9. 5 in. by $\frac{12}{5}$ in.   **11.** 2 in. square   **13.** $\frac{200}{27}$ ft^3

15. (a) Use all of the wire for the circle
(b) $\dfrac{12\pi}{\pi + 4}$ in. for the circle

17. Each side has length 4.

19. $L/12$ by $L/12$ by $L/12$

21. (a) 7000   (b) yes

23. height $L/\sqrt{3}$, radius $\sqrt{2/3}\, L$

25. height $2\sqrt{(5 - \sqrt{5})/10}\, R$
radius $\sqrt{(5 + \sqrt{5})/10}\, R$

27. $\dfrac{\pi}{3}$ **29.** $2\pi R^3/(9\sqrt{3})$

31. (a) 24   (b) \$24   (c) \$24.10
(d) $R'(x) = 10$, $P'(x) = 6 - 0.2x$

35. $p/(4 + \pi)$ **37.** (b) no cube

39. (a) $\frac{3}{4}$ (b) $\frac{3}{16}$

43. (c) $\frac{1}{4}$ mi downstream from the house

▶ **Exercise Set 4.8** (Page 296)

1. (a) 10, 10   (b) no minimum

3. 80 ft (\$1 fencing), 40 ft (\$2 fencing)

5. base: 10 cm square, height: 20 cm

7. ends: $\sqrt[3]{3V/4}$ units square
length: $\frac{4}{3}\sqrt[3]{3V/4}$ units

9. height $=$ radius $= \sqrt[3]{500/\pi}$

11. radius: $\sqrt[6]{\dfrac{450}{\pi^2}}$ cm   height: $\dfrac{30}{\pi}\sqrt[3]{\dfrac{\pi^2}{450}}$ cm

13. $(-\sqrt{2}, 1)$, $(\sqrt{2}, 1)$

15. (a) no maximum   (b) -3

17. $1/\sqrt{5}$ **19.** $(\sqrt{2}, 1/2)$ **21.** $(-1/\sqrt{3}, 3/4)$

23. height: $4R$
radius: $\sqrt{2}R$ **25.** $4(1 + 2^{2/3})^{3/2}$ ft

27. 30 cm from the weaker source

▶ Exercise Set 4.9 (Page 303)

1. 1.414213562

3. 1.817120593

5. −1.671699882

7. 1.224439550

9. 0.581138830

11. −1.452626879

13. 1.895494267

15. 4.493409458

17. (b) 3.162277660

19. −1.165373043

21. −0.474626618, 1.395336994

23. −1.220744085, 0.724491959

25. −4.098859132

27. (0.589754512, 0.347810385)

29. 171°

▶ Exercise Set 4.10 (Page 308)

1. 3   **3.** π **5.** 1

7. 1   **9.** 5/4   **11.** $-\sqrt{5}$

13. (b) $\tan x$ is not continuous on $[0, \pi]$.

29. 8   **31.** $f(x) = x^3 - 4x + 5$

▶ Exercise Set 4.11 (Page 315)

1. (a) positive, negative, slowing down
(b) positive, positive, speeding up
(c) negative, positive, slowing down

3. (a) left   (b) negative   (c) speeding up
(d) slowing down

5. (a) 6.7 ft/sec²   (b) $t = 0$ sec

7.

t	s	v	$|v|$	a	Direction; Motion
1	−5	−9	9	−6	Left; speeding up
2	−16	−12	12	0	Left; neither
3	−27	−9	9	6	Left; slowing down
4	−32	0	0	12	Stopped
5	−25	15	15	18	Right; speeding up

9. (a) 12   (b) $t = 2.2$, $s = -24.2$

11.

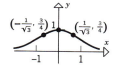

13.

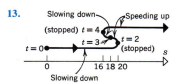

15.

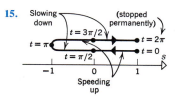

17. (a) $s = 5/16$, $v = 3/2$
(b) $s = 1$, $a = -3$

19. (b) $v = \dfrac{3}{2\sqrt{3t + 7}}$; $a = -9/500$

21. (b) 2/3 unit
(c) $0 \le t < 1$ and $t > 2$

23. (a) −1.25 ft/sec/ft
(b) −2500 ft/sec²

▶ Chapter 4 Supplementary Exercises (Page 317)

1. decreasing 39π m³/sec

3. 60 ft/sec (toward pole)

5. $m = -2 = f(-1)$; $M = 27/256 = f(3/4)$

7. no m; $M = 1/\sqrt{3} = f(\sqrt{3})$

9. $m = -3 = f(3)$; $M = 0 = f(2)$

11. −2.114907541, 0.254101688, 1.860805853

13.

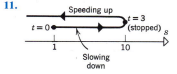

15.

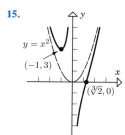

17.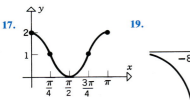

19.

21. at $(1, 2)$, $y = 2x$; at $(2, 3)$, $y = 3$

23. rel. max at $x = \pi/2$ and $3\pi/2$;
rel. min at $x = 7\pi/6$ and $11\pi/6$

25. rel. min at $x = 9$

27. rel. max at $x = 2\pi/3$ and $4\pi/3$;
rel. min at $x = \pi$

29. $2\sqrt{8}$ by $3\sqrt{2}$

31. $r = 2P/(3\pi + 8)$ ft   **33.** $r/h = \pi/8$

35. satisfied; $c = 0$ **37.** satisfied; $c = \sqrt{\pi/2}$

39. satisfied; $c = 1$ **41.** satisfied; $c = \frac{1}{2}, \sqrt{2}$

▶ Exercise Set 5.2 **(Page 331)**

1. $\frac{1}{9}x^9 + C$ **3.** $\frac{7}{12}x^{12/7} + C$

5. $8\sqrt{t} + C$ **7.** $\frac{2}{9}x^{9/2} + C$

9. $-\frac{1}{2}x^{-2} + \frac{2}{5}x^{3/2} - \frac{12}{5}x^{5/4} + \frac{1}{3}x^3 + C$

11. $28y^{1/4} - \frac{3}{4}y^{4/3} + \frac{8}{3}y^{3/2} + C$

13. $\frac{1}{2}x^2 + \frac{1}{5}x^5 + C$

15. $3x^{4/3} - \frac{12}{7}x^{7/3} + \frac{3}{10}x^{10/3} + C$

17. $\frac{1}{2}x^2 - \frac{2}{x} + \frac{1}{3x^3} + C$

19. $-4\cos x + 2\sin x + C$

21. $\tan x + \sec x + C$ **23.** $\sec x + x + C$

25. $\sec x + C$ **27.** $\theta - \cos \theta + C$

29. $\sin \theta - 5 \tan \theta + C$

31. $F(x) = \frac{3}{4}x^{4/3} + \frac{5}{4}$

33. $f(x) = \frac{4}{15}x^{5/2} + C_1x + C_2$

35. $y = x^2 + x - 6$

37. $y = x^3 - 6x + 7$

39. $\int \dfrac{3x^2}{2\sqrt{x^3 + 5}}\, dx = \sqrt{x^3 + 5} + C$

41. $\int \dfrac{\cos (2\sqrt{x})}{\sqrt{x}}\, dx = \sin (2\sqrt{x}) + C$

43. (b) $F(x) = G(x) + \frac{8}{3}$

45. $f(x) = 15x^2 - 3$ **47.** $\tan x - x + C$

▶ Exercise Set 5.3 **(Page 338)**

1. (a) $\frac{1}{24}(x^2 + 1)^{24} + C$ (b) $-\frac{1}{4}\cos^4 x + C$
(c) $-2\cos \sqrt{x} + C$ (d) $\frac{3}{4}\sqrt{4x^2 + 5} + C$

3. (a) $-\frac{1}{2}\cot^2 x + C$ (b) $\frac{1}{10}(1 + \sin t)^{10} + C$
(c) $\frac{2}{7}(1 + x)^{7/2} - \frac{4}{5}(1 + x)^{5/2} + \frac{2}{3}(1 + x)^{3/2} + C$
(d) $-\cot (\sin x) + C$

5. $-\frac{1}{8}(2 - x^2)^4 + C$ **7.** $\frac{1}{8}\sin 8x + C$

9. $\frac{1}{4}\sec 4x + C$ **11.** $\frac{1}{21}(7t^2 + 12)^{3/2} + C$

13. $\frac{2}{3}\sqrt{x^3 + 1} + C$ **15.** $-\frac{1}{16}(4x^2 + 1)^{-2} + C$

17. $\frac{1}{5}\cos (5/x) + C$ **19.** $\frac{1}{3}\tan (x^3) + C$

21. $\frac{1}{18}\sin^6 3t + C$ **23.** $-\frac{1}{6}(2 - \sin 4\theta)^{3/2} + C$

25. $\frac{1}{6}\sec^3 2x + C$ **27.** $-\frac{1}{3}\tan (\cos 3\theta) + C$

29. $\dfrac{1}{b(n + 1)}\sin^{n+1}(a + bx) + C$

31. $\frac{2}{5}(x - 3)^{5/2} + 2(x - 3)^{3/2} + C$

33. $\frac{2}{3}(y + 1)^{3/2} - 2(y + 1)^{1/2} + C$

35. $\frac{1}{3}\tan 3\theta - \theta + C$

39. (a) $\frac{25}{3}x^3 - 5x^2 + x + C$; $\frac{1}{15}(5x - 1)^3 + C$
(b) They differ by a constant.

41. $f(x) = 6x + \frac{3}{2}\cos 2x + \frac{1}{2}$

43. $\frac{1}{3}f(3x + 2) + C$ **45.** $-\frac{1}{2}f(2/x) + C$

▶ Exercise Set 5.4 **(Page 345)**

1. (a) 36   (b) 55   (c) 40   (d) 6

3. $\displaystyle\sum_{k=1}^{10} k$ **5.** $\displaystyle\sum_{k=1}^{49} k(k + 1)$

7. $\displaystyle\sum_{k=1}^{10} 2k$ **9.** $\displaystyle\sum_{k=1}^{6} (-1)^{k+1}(2k - 1)$

11. $\displaystyle\sum_{k=1}^{5} (-1)^k \frac{1}{k}$ **13.** $\displaystyle\sum_{k=1}^{4} \sin \frac{(2k - 1)\pi}{8}$

15. $\displaystyle\sum_{k=1}^{5} \frac{k}{k + 1}$

17. (a) $\displaystyle\sum_{k=1}^{5} (-1)^{k+1}a_k$ (b) $\displaystyle\sum_{k=0}^{5} (-1)^{k+1}b_k$

(c) $\displaystyle\sum_{k=0}^{n} a_k x^k$ (d) $\displaystyle\sum_{k=0}^{5} a^{5-k}b^k$

19. 5047   **21.** 2870

23. 1728   **25.** 214,365

27. $\frac{3}{2}(n + 1)$ **29.** $\frac{1}{4}(n - 1)^2$

31. $\frac{1}{2}$ **33.** $\frac{5}{2}$ **35.** $\frac{17}{2}$

39. $3^{17} - 3^4$ **41.** $-\frac{399}{400}$ **43.** $a_n - a_0$

45. (b) $\frac{1}{2}$

47. (a) n^2 (b) -3 (c) $\frac{1}{2}n(n+1)x$
(d) $(n-m+1)c$

49. (a) $\sum_{k=0}^{14} (k+4)(k+1)$ (b) $\sum_{k=5}^{19} (k-1)(k-4)$

51. $\sum_{k=1}^{18} k \sin \frac{\pi}{k}$ **53.** Both are valid.

55. (a) $\frac{3}{2}(3^{20}-1)$ (b) $2^{31}-2^5$
(c) $-\frac{2}{3}\left(1+\frac{1}{2^{101}}\right)$ **57.** 110

► **Exercise Set 5.5** (Page 355)

1. (a) 46 (b) 58

3. (a) $\dfrac{\sqrt{2}\pi}{4} \approx 1.111$

(b) $\dfrac{(\sqrt{2}+2)\pi}{4} \approx 2.682$

5. $\frac{15}{4}$ **7.** $\frac{1}{3}$ **9.** 320

11. $\frac{15}{4}$ **13.** $\frac{1}{3}$

15. (b) $\frac{1}{4}(b^4-a^4)$ **17.** $\frac{1}{2}(b^2-a^2)$

19. (a) 0.584145862, 0.623823864, 0.649145594
(b) 0.761923639, 0.712712753, 0.684701150

21. (a) 0.919403170, 0.960215997, 0.984209789
(b) 1.076482803, 1.038755813, 1.015625715

► **Exercise Set 5.6** (Page 366)

1. (a) $\frac{71}{6}$ (b) 2 **3.** (a) $-\frac{117}{16}$ (b) 3

7. $\displaystyle\int_{-3}^{3} 4x(1-3x)\,dx$

9. $\displaystyle\lim_{\max \Delta x_k \to 0} \sum_{k=1}^{n} 2x_k^* \,\Delta x_k;\ a=1,\ b=2$

11. $\displaystyle\lim_{\max \Delta x_k \to 0} \sum_{k=1}^{n} \frac{x_k^*}{x_k^*+1}\,\Delta x_k;\ a=0,\ b=1$

13. (a) 0.8 (b) -2.6 (c) -1.8 (d) -0.3
15. (a) 1 (b) 2 (c) $-\frac{1}{4}$ (d) $\frac{3}{4}$
17. -4 **19.** $\frac{13}{2}$ **21.** $\pi/2$
23. $25\pi/2$ **25.** 0 **27.** 14
29. 0.692835360, 0.693069098, 0.693134682
31. 3.142425985, 3.141800987, 3.141625987

► **Exercise Set 5.7** (Page 374)

1. $\frac{65}{4}$ **3.** $\frac{81}{10}$ **5.** $\frac{22}{3}$
7. $\frac{2}{3}$ **9.** $-\frac{1}{3}$ **11.** $\frac{52}{3}$
13. $\frac{844}{5}$ **15.** $-\frac{55}{3}$ **17.** 0
19. $\sqrt{2}$ **21.** $\pi^2/9 + 2\sqrt{3}$
23. $\frac{5}{2}$ **25.** $2-\sqrt{2}/2$
27. $-\frac{11}{6}$ **29.** $\frac{15}{2}$ **31.** 3

33. negative **35.** positive
37. negative
41. $m=1.092599583,\ M=1.104669194$
43. 0.665867079; $\frac{2}{3}$
45. $\frac{2}{3}$ **47.** 12 **49.** $\frac{9}{2}$
51. $\frac{203}{2}$ **53.** (b), (c)

► **Exercise Set 5.8** (Page 379)

1. (a) $\displaystyle\int_{1}^{3} u^7\,du$ (b) $-\dfrac{1}{2}\displaystyle\int_{7}^{4} u^{1/2}\,du$

(c) $\dfrac{1}{\pi}\displaystyle\int_{-\pi}^{\pi} \sin u\,du$ (d) $\displaystyle\int_{0}^{1} u^2\,du$

(e) $\dfrac{1}{2}\displaystyle\int_{3}^{4} (u-3)u^{1/2}\,du$ (f) $\displaystyle\int_{-3}^{0} (u+5)u^{20}\,du$

3. $\frac{121}{5}$ **5.** 10

7. $\frac{1192}{15}$ **9.** $8-4\sqrt{2}$

11. $-\frac{1}{48}$ **13.** $\frac{2}{3}$

15. $\frac{2}{3}(\sqrt{10}-2\sqrt{2})$ **17.** $2(\sqrt{7}-\sqrt{3})$
19. 0 **21.** 0
23. -4 **25.** $-\frac{1}{9}$
27. $\frac{1}{3}(\sqrt{3}-1)$ **29.** $\frac{106}{405}$
31. $3\sqrt{2}/4$ **33.** 2π
35. $\pi/8$ **37.** 3
39. $\frac{5}{3}$ **43.** $2k$
45. 0 **47.** $2/\pi$
49. (b) $\frac{3}{2}$ (c) $\pi/4$

▶ **Exercise Set 5.9** (Page 388)

1. 6 **3.** $2/\pi$

5. $\frac{14}{5}$ **7.** $\pi/2$

9. (a) $\frac{4}{3}$ (b) $2/\sqrt{3}$ (c)

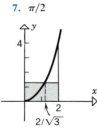

11. $f_{\text{ave}} = 2; x^* = 4$

13. $f_{\text{ave}} = \frac{1}{2}\alpha(x_0 + x_1) + \beta; x^* = \frac{1}{2}(x_0 + x_1)$

15. $120\sqrt{2}$ **19.** 1404π lb

21. (b) no **23.** (a) $x^3 + 1$

25. $\sin \sqrt{x}$ **27.** $|x|$

29. $\displaystyle\int_2^x \frac{1}{t - 1}\,dt$ **31.** $\displaystyle\int_0^x \frac{1}{t - 1}\,dt$

33. (a) $(0, +\infty)$ (b) $x = 1$

35. (a) 0 (b) $\frac{1}{3}$ (c) 0

37. $(-\infty, +\infty)$; zero if $x = 1$, positive if $x > 1$, negative if $x < 1$

39. $[-2, 2]$; zero if $x = -1$, positive if $-1 < x \le 2$, negative if $-2 \le x < -1$

41. $F(x) = \begin{cases} \frac{1}{2}(1 - x^2), & x < 0 \\ \frac{1}{2}(1 + x^2), & x \ge 0 \end{cases}$

43. $F(x) = \begin{cases} \frac{1}{3}(x^3 + 1), & x \le 0 \\ x^2 + \frac{1}{3}, & x > 0 \end{cases}$ **45.** $3/x$

49. (a) $3x^2 \sin^2(x^3) - 2x \sin^2(x^2)$ (b) $\dfrac{2}{1 - x^2}$

▶ **Chapter 5 Supplementary Exercises** (Page 391)

1. $-\frac{1}{2}x^{-2} + 2\sqrt{x} + 5\cos x + C$

3. $\frac{2}{9}(\sqrt{x} + 2)^9 + C$

5. $-\frac{1}{2}\cos\sqrt{2x^2 - 5} + C$

7. $2x^{3/2} + \frac{6}{17}x^{17/6} + C$

9. $\frac{1}{5}\tan(\sin 5t) + C$

11. (a) $\frac{1}{6}y^6 + y^4 + 2y^2 + C$
(b) $\frac{1}{6}(y^2 + 2)^3 + C$

13. $\displaystyle\int_0^1 u^4\,du = \frac{1}{5}$ **15.** $\displaystyle\int_1^4 \frac{u - 1}{\sqrt{u}}\,du = \frac{8}{3}$

17. $\displaystyle\int_{\pi/2}^{\pi} \frac{4}{\pi}\cos u\,du = -\frac{4}{\pi}$

19. $\frac{17}{2}$ **21.** $\frac{1}{3}$

23. (a) 20 (b) 8 (c) $4n$
(d) $\frac{8}{15}$ (e) $\frac{61}{24}$ (f) 9
(g) $1 + \sqrt{2}$ (h) $3(\sqrt{2} + 1)/4$

25. (a) $\displaystyle\sum_{k=1}^{9} (-1)^{k-1}\left(\frac{k}{k + 1}\right)^2 = \sum_{k=2}^{10} (-1)^k \left(\frac{k - 1}{k}\right)^2$

(b) $\displaystyle\sum_{k=1}^{11} \frac{(-\pi)^{k+1}}{k} = \sum_{k=2}^{12} (-1)^k \frac{\pi^k}{k - 1}$

27. (a) $64\left[1 - \dfrac{(n + 1)(2n + 1)}{6n^2}\right]$

(b) $64\left[1 - \dfrac{(n - 1)(2n - 1)}{6n^2}\right]$

(c) $\frac{128}{3}$

29. (a) 12 (b) 12 (c) 12

31. (a) 10 (b) 1
(c) -4 (let $u = -x$)
(d) 4

33. $f_{\text{ave}} = 7; x^* = -\sqrt{7/3}$

35. $f_{\text{ave}} = 13/4; x^* = -5/4$

▶ **Exercise Set 6.1** (Page 399)

1. $\frac{9}{2}$ **3.** 1 **5.** $\frac{32}{3}$

7. $\frac{9}{4}$ **9.** $\frac{49}{192}$ **11.** $\frac{1}{2}$

13. $\frac{32}{3}$ **15.** $\pi - 2$ **17.** $\frac{9}{2}$

19. $\frac{355}{6}$ **21.** 24 **23.** $\frac{1}{2}$

25. $4\sqrt{2}$ **27.** $\frac{11}{2}$ **29.** $\frac{2}{3}$

31. (a) $2(\sqrt{b} - 1)$ (b) $+\infty$

33. $9/\sqrt[3]{4}$ **39.** 1.180898334

▶ **Exercise Set 6.2** (Page 407)

1. 8π **3.** $13\pi/6$

5. $32\pi/5$ **7.** $373\pi/14$

9. $1296\pi/5$ **11.** $2048\pi/15$

13. $\pi/2$ **15.** $\pi/6$

17. $3\pi/5$ **19.** 8π

21. 2π **23.** $28\pi/3$

25. $58\pi/5$ **27.** $72\pi/5$

29. $256\pi/3$ **31.** (a) $\pi(1 - 1/b)$ (b) π

33. $2/\sqrt{\pi}$ **35.** $648\pi/5$ **49.** $40{,}000\pi$ ft^3 **51.** $36\sqrt{3}$

37. $\pi/2$ **39.** $\frac{4}{3}\pi ab^2$ **53.** $\frac{1}{4}(\pi + 2)$ **55.** $\frac{2}{3}r^3 \tan\theta$ **57.** $\frac{16}{3}r^3$

41. π **43.** $\frac{1}{3}\pi r^2 h$ **59.** (b) h_0/k

45. $V = \frac{1}{6}\pi L^3$

47. $V = \begin{cases} 3\pi h^2, & 0 \le h < 2 \\ \frac{1}{3}\pi(12h^2 - h^3 - 4), & 2 \le h \le 4 \end{cases}$

▶ **Exercise Set 6.3** (Page 416)

1. $15\pi/2$ **3.** $\pi/3$ **17.** (b) $2\pi^2$

5. $2; \pi/5$ **7.** 4π **19.** (a) $7\pi/30$ **21.** $9\pi/14$

9. $20\pi/3$ **11.** $3\pi\sqrt[3]{4}\,(1 + 3\sqrt[3]{3})$ **23.** $\frac{1}{3}\pi r^2 h$ **25.** $\dfrac{4\pi}{3}[r^3 - (r^2 - a^2)^{3/2}]$ **27.** $b = 1$

13. $\pi/2$ **15.** $\pi/5$

▶ **Exercise Set 6.4** (Page 420)

1. $\sqrt{5}$ **3.** $\frac{1}{243}(85\sqrt{85} - 8)$ **13.** 4.645975301

5. $\frac{1}{27}(80\sqrt{10} - 13\sqrt{13})$ **7.** $\frac{17}{6}$

9. (a)   (b) dy/dx does not exist at $x = 0$.

(c) $\frac{1}{27}(13\sqrt{13} + 80\sqrt{10} - 16)$

▶ **Exercise Set 6.5** (Page 424)

1. $35\sqrt{2}\pi$ **3.** 8π **9.** 24π **11.** $\dfrac{16{,}911\pi}{1024}$

5. $\dfrac{16\pi}{9}$ **7.** $40\pi\sqrt{82}$ **15.** $S = 2\pi rh$ **19.** (b) constant functions

---

▶ **Exercise Set 6.6** (Page 430)

1. $s(t) = t^2 - 3t + 7$ **15.** (a) 1 sec   (b) $\frac{1}{2}$ sec

3. $s(t) = \frac{1}{4}t^4 - \frac{2}{3}t^3 + t + 1$ **17.** (a) $\frac{5}{8}(1 + \sqrt{33})$ sec   (b) $20\sqrt{33}$ ft/sec

5. $s(t) = 2t^2 + t$ **19.** (a) 5 sec   (b) 272.5 m

7. $s(t) = -\cos 2t - t - 2$ (c) 10 sec   (d) -49 m/sec

9. (a) $s = 2/\pi$, $v = 1$, $|v| = 1$, $a = 0$ (e) 12.46 sec   (f) 73.1 m/sec

   (b) $s = \frac{1}{2}$, $v = -\frac{3}{2}$, $|v| = \frac{3}{2}$, $a = -3$ **21.** $80\sqrt{10}$ ft/sec   **23.** 256 ft

11. $-\frac{968}{45}$ ft/sec^2 **25.** (a) negative   (b) increasing   (c) negative

13. (a) $v(3) = 16$ ft/sec, $v(5) = -48$ ft/sec   **27.** $\frac{2}{3}$; 3   **29.** 0; 2

   (b) 196 ft   (c) 112 ft/sec   **31.** $\frac{9}{4}$; $\frac{11}{4}$ **33.** $-\frac{10}{3}$; $\frac{17}{3}$ **35.** $\frac{204}{25}$; $\frac{204}{25}$

▶ **Exercise Set 6.7** (Page 436)

1. (a) 210 ft · lb   (b) 5/6 ft · lb   **11.** (a) 926,640 ft · lb   (b) 0.468 hp

3. 160 J   **5.** 20 lb/ft   **13.** 75,000 ft · lb

7. $900\pi\rho$ ft · lb   **9.** 261,600 J   **15.** (a) 96×10^9 (b) 4,800,000 mi · lb

▶ **Exercise Set 6.8 (Page 441)**

1. (a) 2808 lb   (b) 3600 lb

3. 156,960 N

5. 6988.8 lb

7. 8.175×10^5 N

9. $\rho a^3/\sqrt{2}$

11. $14{,}976\sqrt{17}$ lb

13. (b) $80\rho_0$ lb/min

▶ **Chapter 6 Supplementary Exercises (Page 442)**

1. (a) $\displaystyle\int_0^2 (x + 2 - x^2)\, dx$

(b) $\displaystyle\int_0^2 \sqrt{y}\, dy + \int_2^4 [\sqrt{y} - (y - 2)]\, dy$

3. (a) $\displaystyle\int_0^9 2\sqrt{x}\, dx$ (b) $\displaystyle\int_{-3}^3 (9 - y^2)\, dy$

5. (a) $\displaystyle\int_0^2 2\pi x(x + 2 - x^2)\, dx$

(b) $\displaystyle\int_0^2 \pi y\, dy + \int_2^4 \pi[y - (y - 2)^2]\, dy$

7. (a) $\displaystyle\int_0^4 2\pi x[\tfrac{1}{2}x - (2 - \sqrt{4 - x})]\, dx$

(b) $\displaystyle\int_0^2 \pi[(4y - y^2)^2 - 4y^2]\, dy$

9. (a) $\displaystyle\int_0^9 4\pi x^{3/2}\, dx$

(b) $\displaystyle\int_{-3}^3 \pi(81 - y^4)\, dy$

11. (a) 16/3   (b) 8π

13. 11/4

15. $\pi a^2 b/24$

17. (a) $256\pi/15$ (b) $40\pi/3$

19. 61/27

21. 779/240

23. $\frac{1}{27}\pi(145^{3/2} - 10^{3/2})$ sq. units

25. $1017\pi/5$ sq. units

27. $28\pi\sqrt{3}/5$ sq. units

29. 3 in · lb

31. 10,600 ft · lb

33. 6656π ft · lb

35. $28\rho/3$ lb

▶ **Exercise Set 7.1 (Page 455)**

1. (a) yes (b) no (c) yes (d) no

3. yes   **5.** no   **7.** no

9. yes   **11.** yes   **13.** yes

15. $x^{1/5}$ **17.** $\frac{1}{4}(x + 6)$

19. $\sqrt[3]{\dfrac{x + 5}{3}}$ **21.** $\frac{1}{2}(x^3 + 1)$

23. $-\sqrt{\dfrac{3}{x}}$ **25.** $\begin{cases} 5/2 - x, & x > 1/2 \\ 1/x, & x \le 1/2 \end{cases}$

27. $\dfrac{1}{15y^2 + 1}$ **29.** $\dfrac{1}{2\sec^2 2y}$ **31.** $\dfrac{1}{10y^4 + 3y^2}$

33. (b)

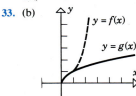

(c) No; $f(g(x)) = x$ for $x > 1$, but the domain of g is $x > 0$.

35. $x^{1/4} - 2$ for $x \ge 16$

37. $\frac{1}{2}(3 - x^2)$ for $x \le 0$

39. $\frac{1}{10}(1 + \sqrt{1 - 20x})$ for $x \le -4$

41. (b) symmetric about the line $y = x$

43. (b) $1 - \sqrt{3}/3$ **45.** 10

47. $\frac{1}{13}$ **49.** $1/\sqrt{3}$

51. (b) $1/\sqrt[3]{2}$

53. 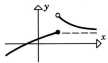   **55.** $\frac{88}{7}$

▶ **Exercise Set 7.2 (Page 465)**

1. (a) -4 (b) 4   (c) $\frac{1}{4}$

3. (a) 2.9690   (b) 0.0341

5. (a) 4   (b) -5 (c) 1   (d) $\frac{1}{2}$

7. (a) 1.3655   (b) -0.3011

9. (a) $2r + s/2 + t/2$ (b) $s - 3r - t$

11. (a) $1 + \log x + \frac{1}{2}\log (x - 3)$

(b) $2\ln |x| + 3\ln \sin x - \frac{1}{2}\ln (x^2 + 1)$

13. $\log \frac{256}{3}$

15. $\ln \dfrac{\sqrt[3]{x}(x+1)^2}{\cos x}$

17. 0.01

19. e^2

21. 4

23. 10^5

25. $\sqrt{3/2}$

27. $-\dfrac{\ln 3}{2 \ln 5}$

29. $\frac{1}{3} \ln \frac{7}{2}$

31. -2

33. $0, -\ln 2$

35. $-\ln 2$

37. $\log \frac{1}{2} < 0$, so $3 \log \frac{1}{2} < 2 \log \frac{1}{2}$

39. (b) $2.8777, -0.3174$ **41.** 201 days

43. (a) 7.4, basic   (b) 4.2, acidic
(c) 6.4, acidic   (d) 5.9, acidic

45. (a) 140 dB, damage   (b) 120 dB, damage
(c) 80 dB, no damage   (d) 75 dB, no damage

47. ~ 200

49. (a) $\sim 5 \times 10^{16}$ J
(b) ~ 0.67

51. e^{-2}

► Exercise Set 7.3 (Page 476)

1. $x > -\frac{2}{3}$

3. $-2 < x < 2$

5. $x \geq e^{-1}$

7. (a) all $x \neq 0$
(b) $x > 0$

9. $\dfrac{1}{x}$

11. $\dfrac{2 \ln x}{x}$

13. $\dfrac{\sec^2 x}{\tan x}$

15. $\dfrac{1 - x^2}{x(1 + x^2)}$

17. $\dfrac{3x^2 - 14x}{x^3 - 7x^2 - 3}$

19. $\dfrac{1}{2x \sqrt{\ln x}}$

21. $-\dfrac{5 \cos (5/\ln x)}{x(\ln x)^2}$

23. $-\dfrac{2x^3}{3 - 2x} + 3x^2 \ln (3 - 2x)$

25. $4x \ln (x^2 + 1) + 2x[\ln (x^2 + 1)]^2$

27. $\dfrac{x(1 + 2 \ln x)}{(1 + \ln x)^2}$

29. $-\dfrac{y}{x(y + 1)}$

31. $\frac{1}{2} \ln |x| + C$

33. $\frac{1}{3} \ln |x^3 - 4| + C$

35. $\ln |\tan x| + C$

37. $-\frac{1}{3} \ln (1 + \cos 3\theta) + C$

39. $\frac{1}{2}x^2 - \frac{1}{2} \ln (x^2 + 1) + C$

41. $\frac{1}{4} (\ln y)^4 + C$

43. $\frac{1}{3} \ln \frac{5}{2}$

45. $\frac{1}{2} \ln \frac{5}{6}$

47. $-\tan x + \dfrac{3x}{4 - 3x^2}$

49. $\dfrac{1}{2x} + \dfrac{1}{3(x + 3)} + \dfrac{3}{5(3x - 2)}$

51. $x \sqrt[3]{1 + x^2} \left[\dfrac{1}{x} + \dfrac{2x}{3(1 + x^2)} \right]$

53. $\left[\dfrac{(x^2 - 8)^{1/3} \sqrt{x^3 + 1}}{x^6 - 7x + 5} \right]$
$\cdot \left[\dfrac{2x}{3(x^2 - 8)} + \dfrac{3x^2}{2(x^3 + 1)} - \dfrac{6x^5 - 7}{x^6 - 7x + 5} \right]$

59. $-\dfrac{\ln 2}{x(\ln x)^2}$

61. $\frac{1}{2} \ln 2$

63. $\frac{1}{2}(1 + \ln 2)$

65. relative minimum $(1, 0)$,
relative maximum $(e^2, 4e^{-2})$

71. $\ln 2$

73. $\pi \ln 4$

75. $0.158594340, 3.146193221$

► Exercise Set 7.4 (Page 486)

1. (a) $1/x, x > 0$ (b) $x^2, x \neq 0$
(c) $-x^2, -\infty < x < +\infty$ (d) $-x, -\infty < x < +\infty$
(e) $x^3, x > 0$ (f) $x + \ln x, x > 0$
(g) $x - \sqrt[3]{x}, -\infty < x < +\infty$ (h) $e^x/x, x > 0$

3. $\frac{7}{2}$ **5.** (a) $e^{-x \ln \pi}$ (b) $e^{2x \ln x}$

7. $-10 x e^{-5x^2}$

9. $x^2 e^x (x + 3)$

11. $\dfrac{4}{(e^x + e^{-x})^2}$

13. $(x \sec^2 x + \tan x) e^{x \tan x}$

15. $(1 - 3e^{3x}) e^{(x - e^{3x})}$

17. $\dfrac{x - 1}{e^x - x}$

19. $e^{ax}(a \cos bx - b \sin bx)$

21. $3x^2$

23. $-3^{-x} \ln 3$

25. $\pi^{x \tan x} (\ln \pi)(x \sec^2 x + \tan x)$

27. (a) not of the form a^x, where a is a constant
(b) $x^x (1 + \ln x)$

29. $(x^3 - 2x)^{\ln x} \left[\dfrac{3x^2 - 2}{x^3 - 2x} \ln x + \dfrac{1}{x} \ln (x^3 - 2x) \right]$

31. $(\ln x)^{\tan x} \left[\dfrac{\tan x}{x \ln x} + (\sec^2 x) \ln (\ln x) \right]$

33. $x^{(e^x)} \left[\dfrac{e^x}{x} + e^x \ln x \right]$

37. (a) $k^n e^{kx}$ (b) $(-1)^n k^n e^{-kx}$

39. $-\dfrac{1}{\sqrt{2\pi \sigma^3}} (x - \mu) \exp \left[-\dfrac{1}{2} \left(\dfrac{x - \mu}{\sigma} \right)^2 \right]$

41. $-\frac{1}{5} e^{-5x} + C$

43. $e^{\sin x} + C$

45. $-\frac{1}{6} e^{-2x^3} + C$

47. $\ln (1 + e^x) + C$

49. $\frac{1}{3}(1 + e^{2t})^{3/2} + C$

51. $\exp (\sin x) + C$

53. $\tan (2 - e^{-x}) + C$

55. $\dfrac{\pi^{\sin x}}{\ln \pi} + C$

57. $\frac{1}{2}x^2 \ln 3 - 4\pi e^2 \sin x + C$

59. C

61. $2e^{\sqrt{y}} + C$

63. -36

65. $3 + e - e^2$

67. $\ln \frac{21}{13}$

71. $\dfrac{\ln 3}{\ln (2/3)}$

73. ex^{e-1}

75. $\dfrac{-qk_0}{2T^2} \exp \left[-(q/2) \left(\dfrac{T - T_0}{T_0 T} \right) \right]$

77. $e^x - 3 \ln (e^x + 3) + C$

79. $\dfrac{1}{\ln x}$

81. $\frac{1}{3}e^{1-x}$

83. $\frac{27}{8}e^{-3}$

85. $3 \ln 3 - 2$

87. 4π

89. $5 \ln 5 - 4$

91. $-1.315973778 < x < 0.537274449$

93. (b) 2.821439372　　(c) $f = 5.87 \times 10^{10}T$

97. $e^{-1/e}$

► Exercise Set 7.5 **(Page 494)**

1. (a) $+\infty$　(b) 0

3. (a) $+\infty$　(b) $+\infty$

5. (a) 1　(b) 1

7. (a) $+\infty$　(b) 0

9.

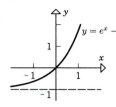

11.

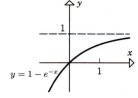

13. (a) yes　(b) no

(c)

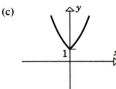

15.

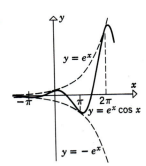

17. -1　**19.** 0　**21.** 1　**23.** 1　**25.** 1

27. (a) $\lim\limits_{x \to +\infty} xe^x = +\infty,\ \lim\limits_{x \to -\infty} xe^x = 0$

(b)

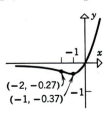

29. (a) $\lim\limits_{x \to +\infty} \dfrac{x^2}{e^{2x}} = 0,\ \lim\limits_{x \to -\infty} \dfrac{x^2}{e^{2x}} = +\infty$

(b)

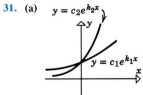

31. (a)

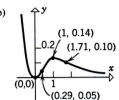

(b)

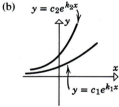

(c)

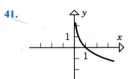

33. $\dfrac{3 - e}{2e}$

41.

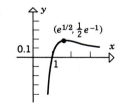

43.

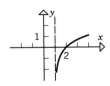

45.

▶ **Exercise Set 7.6 (Page 501)**

1.

	$\sinh x_0$	$\cosh x_0$	$\tanh x_0$	$\coth x_0$	$\operatorname{sech} x_0$	$\operatorname{csch} x_0$
(a)	-2	$\sqrt{5}$	$-2/\sqrt{5}$	$-\sqrt{5}/2$	$1/\sqrt{5}$	$-1/2$
(b)	$-3/4$	$5/4$	$-3/5$	$-5/3$	$4/5$	$-4/3$
(c)	$-4/3$	$5/3$	$-4/5$	$-5/4$	$3/5$	$-3/4$
(d)	$1/\sqrt{3}$	$2/\sqrt{3}$	$1/2$	2	$\sqrt{3}/2$	$\sqrt{3}$
(e)	$8/15$	$17/15$	$8/17$	$17/8$	$15/17$	$15/8$
(f)	-1	$\sqrt{2}$	$-1/\sqrt{2}$	$-\sqrt{2}$	$1/\sqrt{2}$	-1

17. $4\cosh(4x - 8)$

19. $-\dfrac{\operatorname{csch}^2(\ln x)}{x}$

21. $\dfrac{\operatorname{csch}(1/x)\coth(1/x)}{x^2}$

23. $\dfrac{2 + 5\cosh(5x)\sinh(5x)}{\sqrt{4x + \cosh^2(5x)}}$

25. $x^{5/2}\tanh(\sqrt{x})\operatorname{sech}^2(\sqrt{x}) + 3x^2\tanh^2(\sqrt{x})$

27. $\frac{1}{7}\sinh^7 x + C$

29. $\frac{2}{3}(\tanh x)^{3/2} + C$

31. $\ln(\cosh x) + C$

33. $-\frac{1}{3}\operatorname{sech}^3 x + C$

35. $2\sinh(\sqrt{x}) + C$

37. For $x > 0$:
(a) $\dfrac{x^2 + 1}{2x}$ (b) $\dfrac{x^2 - 1}{2x}$ (c) $\dfrac{x^4 - 1}{x^4 + 1}$ (d) $\dfrac{x^2 + 1}{2x}$

45. (a) $\frac{1}{2}$
(b) $+\infty$ if $a > 1$, $\frac{1}{2}$ if $a = 1$, 0 if $0 < a < 1$

47. $\frac{16}{9}$

49. 5π

51. $\frac{3}{4}$

55. 405.9 ft

▶ **Exercise Set 7.7 (Page 514)**

1. $y = Cx$

3. $y = Ce^{-\sqrt{1+x^2}} - 1$

5. $y = \ln(\sec x + C)$

7. $y = e^{-2x} + Ce^{-3x}$

9. $y = e^{-x}\sin(e^x) + Ce^{-x}$

11. $y = -\frac{2}{7}x^4 + Cx^{-3}$

13. $y = -1 + 4e^{x^2/2}$

15. $y = 2 - e^{-t}$

17. $y = \sqrt[3]{3t - 3\ln t + 24}$

19. $y = -3/(2\sqrt{x} + 1)$

21. $y = -\ln(3 - x^2/2)$

23. (a) $h(t) = (2 - 0.003979t)^2$
(b) about 8.4 min

25. $v = 50/(2t + 1)$, $x = 25\ln(2t + 1)$

27. (a) $v(t) = c\ln\dfrac{m_0}{m_0 - kt} - gt$
(b) 3044 m/sec

29. (a) $v^2 = 2gR^2/x + v_0{}^2 - 2gR$

31. (a) $y = 200 - 175e^{-t/25}$
(b) 136 lb

33. 25 lb

35. (a) $y = 10e^{-0.005t}$
(b) 7 mg

37. 196 days

39. 6.8 years

41. (a) 14,400
(b) 38 years

▶ **Chapter 7 Supplementary Exercises (Page 518)**

1. (a) no (b) yes (c) no (d) yes (e) yes

3. does not exist

5. $\frac{1}{2}\ln(x - 1)$

7. If $ad - bc \neq 0$, then $f^{-1}(x) = (-dx + b)/(cx - a)$.

9. (a) $(-\infty, 5/2)$ (b) $(-2, +\infty)$
(c) $(-\pi/3, 2\pi/3)$

11. $-3/x^2$

13. $2/x$

15. (a) $-(2r + s)$ (b) $2s - \frac{3}{2}r$
(c) $\frac{1}{4}(3r - s)$

17. (a) $\ln 3/(2\ln 5 + \ln 3)$
(b) $\frac{1}{2}(\ln 5 - \ln 3)$

19. (a) $\sqrt{34}/5$ (b) $-3/\sqrt{34}$
(c) $-6\sqrt{34}/25$

21. $-1/(2\sqrt{e^x})$

23. $(\ln x - 1)/(\ln x)^2$

25. 0

27. $\ln 10 - \cot x$

29. $x^4 e^{\tan x}\left(\sec^2 x + \dfrac{4}{x}\right)$

31. $(x^2 + a^2)^{-1/2}$

33. $6x\exp(3x^2)$

35. $1/(4x\sqrt{\ln\sqrt{x}})$

37. $x^{\pi-1}\pi^x(\pi + x\ln\pi)$

39. $5\cosh(\tanh(5x))\operatorname{sech}^2(5x)$

41. $3e^{3x} + 4e^{2x} + e^x$

45. (a) $dy = -e^{-x}dx$ (b) $dy = dx/(1 + x)$
(c) $dy = 2x(\ln 2)\,2^{x^2}dx$

47. (a) $\dfrac{1}{x\sqrt{4 + x}}$ (b) $5e^{5x}\sqrt{5x + e^{5x}}$

49. $Y = kt + b$, $b = \ln C$

51. $\ln(1 + e^x) + C$

53. $\dfrac{x^{e+1}}{e + 1} + C$

55. $4\ln|x| + 3/x + C$

57. $\frac{1}{2}\ln|2\sec x - 1| + C$

59. $\frac{1}{2}e^{\sin 2x} + C$

61. $\frac{1}{2} \tanh^2 x + C_1 = -\frac{1}{2} \operatorname{sech}^2 x + C_2$

63. $\ln 2$ **65.** $1/\ln 2$

69. inflection points at 0, $3 - \sqrt{3}$, and $3 + \sqrt{3}$; relative extremum at 3

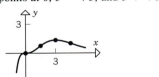

73. $A = \frac{1}{2}(1 - e^{-2b}) = \frac{1}{4}$ when $b = \ln \sqrt{2}$; $A \to \frac{1}{2}$ as $b \to +\infty$

75. $\frac{1}{2}\pi(2 + \sinh 2)$

77. (a) about 352 million   (b) the year 2058

▶ Exercise Set 8.1 (Page 530)

1. (a) $-\pi/2$ (b) π (c) $-\pi/4$
    (d) $\pi/4$ (e) 0 (f) $\pi/2$

3. $\frac{1}{2}, -\sqrt{3}, -1/\sqrt{3}, 2, -2/\sqrt{3}$ **5.** $\frac{4}{5}, \frac{3}{5}, \frac{3}{4}, \frac{5}{3}, \frac{5}{4}$

7. (a) $\pi/7$ (b) 0 (c) $2\pi/7$ (d) $201\pi - 630$

9. (a) $0 \le x \le \pi$ (b) $-1 \le x \le 1$
    (c) $-\pi/2 < x < \pi/2$ (d) $-\infty < x < +\infty$
    (e) $0 < x \le \pi/2$, $-\pi < x \le -\pi/2$ (f) $|x| \ge 1$

11. $\frac{24}{25}$ **13.** $\pi/2$ **15.** $-4\sqrt{5}$

19. (a) $\dfrac{1}{\sqrt{1 + x^2}}$ (b) $1/x$

    (c) $\dfrac{\sqrt{x^2 - 1}}{x}$ (d) $\sqrt{x^2 - 1}$

21. (a)

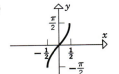

(b)

27. (a) $55.0°$ (b) $33.6°$ (c) $25.8°$

29. $x = \pi + \tan^{-1} k$

31. 2.7626 **33.** -1.8773

35. $-76.7°$ **37.** (b) $23°$

39. $32°$ or $58°$; $32°$ **41.** $29°$

▶ Exercise Set 8.2 (Page 539)

1. (a) $\dfrac{1}{\sqrt{9 - x^2}}$ (b) $-\dfrac{2}{\sqrt{1 - (2x + 1)^2}}$

3. (a) $\dfrac{7}{x\sqrt{x^{14} - 1}}$ (b) $-\dfrac{1}{\sqrt{e^{2x} - 1}}$

5. (a) $-\dfrac{1}{|x|\sqrt{x^2 - 1}}$ (b) $\begin{cases} 1, & \sin x > 0 \\ -1, & \sin x < 0 \end{cases}$

7. (a) $\dfrac{e^x}{x\sqrt{x^2 - 1}} + e^x \sec^{-1} x$

    (b) $\dfrac{3x^2(\sin^{-1} x)^2}{\sqrt{1 - x^2}} + 2x(\sin^{-1} x)^3$

9. (a) $-\dfrac{1}{x^2 + 1}$

    (b) $10(1 + x \csc^{-1} x)^9 \left(-\dfrac{1}{\sqrt{x^2 - 1}} + \csc^{-1} x \right)$

11. (a) $-\dfrac{1}{2\sqrt{1 - x^2}}$ (b) $\dfrac{x + 2x \ln x}{\sqrt{1 - x^4 \ln^2 x}}$

13. $\dfrac{y\sqrt{1 - (x - y)^2} + \sqrt{1 - x^2 y^2}}{\sqrt{1 - x^2 y^2} - x\sqrt{1 - (x - y)^2}}$

15. $\pi/2$ **17.** $-\pi/12$ **19.** $\frac{1}{4} \tan^{-1} 4x + C$

21. $\tan^{-1}(e^x) + C$ **23.** $\pi/6$

25. $\sin^{-1}(\tan x) + C$ **27.** $\sin^{-1}(\ln x) + C$

29. (a) $\sin^{-1} \left(\dfrac{x}{3} \right) + C$

    (b) $\dfrac{1}{\sqrt{5}} \tan^{-1} \left(\dfrac{x}{\sqrt{5}} \right) + C$

    (c) $\dfrac{1}{\sqrt{\pi}} \sec^{-1} \left(\dfrac{x}{\sqrt{\pi}} \right) + C$

31. $\pi/(6\sqrt{3})$ **33.** $\pi/18$

35. $\pi/18$ **37.** $\pi^2/4$

39. $\pi/2 - 1$ **41.** $1 + 2\sqrt{2}$

43. $52\pi/3$ mi/min   **45.** $2\sqrt{6}$ ft

49. (b) 0.93

▶ **Exercise Set 8.3 (Page 545)**

5. (a) $\ln(3 + \sqrt{8})$ (b) $\ln(\sqrt{5} - 2)$

7. (a) $\dfrac{1}{\sqrt{9 + x^2}}$ **9.** (a) $-\dfrac{7}{x\sqrt{1 - x^{14}}}$

  (b) $\dfrac{2}{\sqrt{(2x + 1)^2 - 1}}$ (b) $-\dfrac{1}{\sqrt{1 + e^{2x}}}$

11. (a) $-\dfrac{1}{|x|\sqrt{x^2 + 1}}$

  (b) $\begin{cases} 1, & x > 0 \\ -1, & x < 0 \end{cases}$

13. (a) $-\dfrac{e^x}{x\sqrt{1 - x^2}} + e^x \operatorname{sech}^{-1} x$

  (b) $\dfrac{3x^2(\sinh^{-1} x)^2}{\sqrt{1 + x^2}} + 2x(\sinh^{-1} x)^3$

15. (a) $-1/2x$

  (b) $10(1 + x\operatorname{csch}^{-1} x)^9\left(-\dfrac{x}{|x|\sqrt{1 + x^2}} + \operatorname{csch}^{-1} x\right)$

17. $\cosh^{-1}\left(\dfrac{x}{\sqrt{2}}\right) + C$

19. $-\operatorname{sech}^{-1}(e^x) + C$ **21.** $-\tfrac{1}{3}\operatorname{csch}^{-1}|x^3| + C$

23. -0.2028

25. (a) 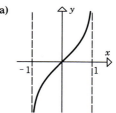 (b)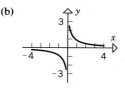

35. (a) $v = \sqrt{16x^2 - 1}$

  (b) $x = \tfrac{1}{4}\cosh 4t$; 0.7 sec

▶ **Chapter 8 Supplementary Exercises (Page 547)**

1. (a) $2\pi/3$ (b) $\tfrac{3}{4}$ (c) $\tfrac{3}{5}$ (d) $\tfrac{3}{5}$

3. (a) $\pi/4$ (b) $-\pi/4$ (c) $2\sqrt{6}$ (d) $\pi/3$

5. (a) $\tfrac{33}{65}$ (b) $\tfrac{56}{65}$ (c) 7

7. $\tfrac{5}{4}, -\tfrac{3}{4}, \tfrac{34}{16}$

9. (a)

  (b) $y = f(x) = \pi/2$ (constant), $|x| \leq 1$

11. $-2/[x(\sec^{-1} x^2)^2\sqrt{x^4 - 1}]$

13. $(\sec x \tan x)/|\tan x|$

15. $\dfrac{1}{1 - (\ln x)^2} + \tanh^{-1}(\ln x)$

17. $3/(2\sqrt{(1 - 9x^2)}\sin^{-1} 3x)$

19. $\exp(\sec^{-1} x)/(x\sqrt{x^2 - 1})$

21. $(\pi^{\sin^{-1} x})\ln \pi/\sqrt{1 - x^2}$

23. 1 (since $y = x$) **25.** Show $y' = \cos^2 y$.

27. $\tfrac{1}{2}\ln|(1 + e^x)/(1 - e^x)| + C$

29. $\cosh^{-1}(\ln x) + C, x > e$

31. $\pi/18$ **33.** $2\tan^{-1}\sqrt{x} + C$

35. $\pi/6$ ($u = \sqrt{x}$) **37.** $1/\sqrt{2}$

▶ **Exercise Set 9.1 (Page 555)**

1. $\tfrac{3}{4}x + \tfrac{3}{16}\ln|4x - 1| + C$

3. $\dfrac{1}{5}\ln\left|\dfrac{x}{2x + 5}\right| + C$ **5.** $\tfrac{1}{5}(x + 1)(2x - 3)^{3/2} + C$

7. $\dfrac{1}{2}\ln\left|\dfrac{\sqrt{4 - 3x} - 2}{\sqrt{4 - 3x} + 2}\right| + C$ **9.** $\dfrac{1}{2\sqrt{5}}\ln\left|\dfrac{x + \sqrt{5}}{x - \sqrt{5}}\right| + C$

11. $\tfrac{1}{2}x\sqrt{x^2 - 3} - \tfrac{3}{2}\ln|x + \sqrt{x^2 - 3}| + C$

13. $\tfrac{1}{2}x\sqrt{x^2 + 4} - 2\ln|x + \sqrt{x^2 + 4}| + C$

15. $\dfrac{1}{2}x\sqrt{9 - x^2} + \dfrac{9}{2}\sin^{-1}\dfrac{x}{3} + C$

17. $\sqrt{3 - x^2} - \sqrt{3}\ln\left|\dfrac{\sqrt{3} + \sqrt{3 - x^2}}{x}\right| + C$

19. $-\tfrac{1}{10}\sin 5x + \tfrac{1}{2}\sin x + C$

21. $\tfrac{1}{16}x^4(4\ln x - 1) + C$

23. $-\tfrac{1}{13}e^{-2x}(2\sin 3x + 3\cos 3x) + C$

25. $\dfrac{1}{18}\left[\dfrac{4}{4 - 3e^{2x}} + \ln|4 - 3e^{2x}|\right] + C$

27. $\dfrac{1}{3}\tan^{-1}\dfrac{3\sqrt{x}}{2} + C$

29. $\tfrac{1}{3}\ln|3x + \sqrt{9x^2 - 4}| + C$

31. $\dfrac{1}{54}\left[-\dfrac{3}{2}x^2\sqrt{5-9x^4}+\dfrac{5}{2}\sin^{-1}\left(\dfrac{3x^2}{\sqrt{5}}\right)\right]+C$

33. $\frac{1}{2}\ln x-\frac{1}{4}\sin(2\ln x)+C$

35. $-\frac{1}{4}e^{-2x}(2x+1)+C$

37. $-\dfrac{1}{3}\left[\dfrac{1}{1+\cos 3x}+\ln\left|\dfrac{\cos 3x}{1+\cos 3x}\right|\right]+C$

39. $\dfrac{1}{16}\ln\left|\dfrac{4x^2-1}{4x^2+1}\right|+C$

41. $\dfrac{1}{2}\left[e^x\sqrt{3-4e^{2x}}+\dfrac{3}{2}\sin^{-1}\dfrac{2e^x}{\sqrt{3}}\right]+C$

43. $\dfrac{1}{3}\left[\dfrac{18x-5}{12}\sqrt{5x-9x^2}+\dfrac{25}{72}\sin^{-1}\dfrac{18x-5}{5}\right]+C$

45. $\frac{1}{9}(\sin 3x-3x\cos 3x)+C$

47. $-2e^{-\sqrt{x}}(\sqrt{x}+1)+C$

49. $\dfrac{1}{6}\ln\left|\dfrac{x-1}{x+5}\right|+C$

51. $-\sqrt{5+4x-x^2}+2\sin^{-1}\dfrac{x-2}{3}+C$

53. 3.523188312   **55.** 17.59119022

57. 0.054930614   **59.** 3.586419094

61. 5.031899801   **63.** 8.409316783

65. 14.42359944

▶ **Exercise Set 9.2** (Page 563)

1. $-xe^{-x}-e^{-x}+C$

3. $x^2e^x-2xe^x+2e^x+C$

5. $-\frac{1}{2}x\cos 2x+\frac{1}{4}\sin 2x+C$

7. $x^2\sin x+2x\cos x-2\sin x+C$

9. $\frac{2}{3}x^{3/2}\ln x-\frac{4}{9}x^{3/2}+C$

11. $x(\ln x)^2-2x\ln x+2x+C$

13. $x\ln(2x+3)-x+\frac{3}{2}\ln(2x+3)+C$

15. $x\sin^{-1}x+\sqrt{1-x^2}+C$

17. $x\tan^{-1}(2x)-\frac{1}{4}\ln(1+4x^2)+C$

19. $\frac{1}{2}e^x(\sin x-\cos x)+C$

21. $\dfrac{e^{ax}}{a^2+b^2}(a\sin bx-b\cos bx)+C$

23. $\frac{1}{2}x[\sin(\ln x)-\cos(\ln x)]+C$

25. $x\tan x+\ln|\cos x|+C$ **27.** $\frac{1}{2}x^2e^{x^2}-\frac{1}{2}e^{x^2}+C$

29. $\frac{1}{25}(1-6e^{-5})$

31. $\frac{1}{9}(2e^3+1)$

33. $5\ln 5-4$

35. $\dfrac{5\pi}{6}-\sqrt{3}+1$

37. $-\pi/8$ **39.** $\frac{1}{3}(2\sqrt{3}\pi-\frac{1}{2}\pi-2+\ln 2)$

41. $\frac{4}{3}(2-\sqrt{2})$ **43.** (a) 1   (b) $\pi(e-2)$ **45.** $2\pi^2$

47. (a) $-\frac{1}{3}\sin^2 x\cos x-\frac{2}{3}\cos x+C$
(b) $\frac{1}{32}(3\pi-8)$

49. (a) $\frac{1}{15}\cos^2 5x\sin 5x+\frac{2}{15}\sin 5x+C$
(b) $\frac{1}{8}\cos^3(x^2)\sin(x^2)+\frac{3}{16}\cos(x^2)\sin(x^2)+\frac{3}{16}x^2+C$

53. (b) $x^3e^x-3x^2e^x+6xe^x-6e^x+C$

▶ **Exercise Set 9.3** (Page 569)

1. $-\frac{1}{6}\cos^6 x+C$ **3.** $\dfrac{1}{2a}\sin^2 ax+C$

5. $\frac{1}{2}\theta-\frac{1}{20}\sin 10\theta+C$

7. $\dfrac{3}{8}x+\sin\left(\dfrac{x}{2}\right)+\dfrac{1}{8}\sin x+C$

9. $\sin\theta-\frac{2}{3}\sin^3\theta+\frac{1}{5}\sin^5\theta+C$

11. $\frac{1}{6}\sin^3 2t-\frac{1}{10}\sin^5 2t+C$

13. $-\frac{1}{5}\cos^5 x+\frac{1}{7}\cos^7 x+C$

15. $-\frac{1}{5}\cos^5\theta+\frac{2}{7}\cos^7\theta-\frac{1}{9}\cos^9\theta+C$

17. $\frac{1}{8}x-\frac{1}{32}\sin 4x+C$

19. $-\frac{1}{6}\cos 3x+\frac{1}{2}\cos x+C$

21. $-\dfrac{1}{3}\cos\left(\dfrac{3x}{2}\right)-\cos\left(\dfrac{x}{2}\right)+C$

23. $\dfrac{1}{7\cos^7 x}+C$ **25.** $\dfrac{5\sqrt{2}}{12}$

27. 0   **29.** $\frac{1}{24}$ **33.** $\pi/2$

35. (a) $\frac{2}{3}$ (b) $3\pi/16$ (c) $\frac{8}{15}$ (d) $5\pi/32$

▶ **Exercise Set 9.4** (Page 574)

1. $\frac{1}{3}\tan(3x+1)+C$ **3.** $\frac{1}{2}\ln|\cos(e^{-2x})|+C$

5. $\frac{1}{2}\ln|\sec 2x+\tan 2x|+C$ **7.** $\frac{1}{3}\tan^3 x+C$

9. $\frac{1}{16}\tan^4(4x)+\frac{1}{24}\tan^6(4x)+C$

11. $\frac{1}{7}\sec^7 x-\frac{1}{5}\sec^5 x+C$

13. $\frac{1}{4}\sec^3 x\tan x-\frac{5}{8}\sec x\tan x+\frac{3}{8}\ln|\sec x+\tan x|+C$

15. $\frac{1}{6}\sec^3(2t)+C$ **17.** $\tan x+\frac{1}{3}\tan^3 x+C$

19. $\dfrac{1}{5\pi}\sec^4(\pi x)\tan(\pi x)+\dfrac{4}{15\pi}\sec^2(\pi x)\tan(\pi x)$

$\qquad\qquad\qquad\qquad+\dfrac{8}{15\pi}\tan(\pi x)+C$

21. $\frac{1}{3}\tan^3 x - \tan x + x + C$ 23. $\frac{1}{6}\tan^3(x^2) + C$

25. $-\frac{1}{5}\csc^5 x + \frac{1}{3}\csc^3 x + C$

27. $-\frac{1}{2}\csc^2 x - \ln|\sin x| + C$

29. $\frac{2}{3}\tan^{3/2} x + \frac{2}{7}\tan^{7/2} x + C$ 31. $\sqrt{3}/2 - \pi/6$

33. $-\frac{1}{2} + \ln 2$ 35. $\ln(\sqrt{2}+1)$

39. $-\dfrac{1}{\sqrt{a^2+b^2}}\ln|\csc(x+\theta)+\cot(x+\theta)| + C$, where

θ satisfies $\cos\theta = \dfrac{a}{\sqrt{a^2+b^2}}$ and $\sin\theta = \dfrac{b}{\sqrt{a^2+b^2}}$

▶ Exercise Set 9.5 (Page 581)

1. $2\sin^{-1}\left(\dfrac{x}{2}\right) + \dfrac{1}{2}x\sqrt{4-x^2} + C$

3. $\dfrac{9}{2}\sin^{-1}\left(\dfrac{x}{3}\right) - \dfrac{1}{2}x\sqrt{9-x^2} + C$

5. $\dfrac{1}{16}\tan^{-1}\dfrac{x}{2} + \dfrac{x}{8(4+x^2)} + C$

7. $\sqrt{x^2-9} - 3\sec^{-1}\dfrac{x}{3} + C$

9. $-2\sqrt{2-x^2} + \frac{1}{3}(2-x^2)^{3/2} + C$

11. $\dfrac{x}{3\sqrt{3+x^2}} + C$ 13. $\dfrac{\sqrt{4x^2-9}}{9x} + C$

15. $\dfrac{x}{\sqrt{1-x^2}} + C$ 17. $\ln|x+\sqrt{x^2-1}| + C$

19. $-\dfrac{\sqrt{9-4x^2}}{9x} + C$ 21. $-\dfrac{x}{\sqrt{9x^2-1}} + C$

23. $\frac{1}{2}\sin^{-1}(e^x) + \frac{1}{2}e^x\sqrt{1-e^{2x}} + C$

25. $\frac{2048}{15}$ 27. $\frac{1}{2}(\sqrt{3}-\sqrt{2})$

29. $\frac{1}{243}(10\sqrt{3}+18)$ 31. $\frac{1}{2}\ln(x^2+4) + C$

33. $\sqrt{5} - \sqrt{2} + \ln\left(\dfrac{2+2\sqrt{2}}{1+\sqrt{5}}\right)$

35. $\dfrac{\pi}{32}[18\sqrt{5} - \ln(2+\sqrt{5})]$

37. (a) $\sinh^{-1}\left(\dfrac{x}{3}\right) + C$

(b) $\ln\left(\dfrac{\sqrt{x^2+9}}{3} + \dfrac{x}{3}\right) + C$

39. $\frac{1}{3}\tan^{-1}\left(\dfrac{x-2}{3}\right) + C$ 41. $\sin^{-1}\left(\dfrac{x-1}{3}\right) + C$

43. $\ln(\sqrt{x^2-6x+10} + x - 3) + C$

45. $2\sin^{-1}\left(\dfrac{x+1}{2}\right) + \dfrac{1}{2}(x+1)\sqrt{3-2x-x^2} + C$

47. $\dfrac{1}{\sqrt{10}}\tan^{-1}\dfrac{\sqrt{2}(x+1)}{\sqrt{5}} + C$

49. $\ln(x^2+2x+5) + \dfrac{3}{2}\tan^{-1}\left(\dfrac{x+1}{2}\right) + C$

51. $\sqrt{x^2+2x+2} + 2\ln(\sqrt{x^2+2x+2} + x + 1) + C$

53. $\dfrac{2\pi}{3} - \dfrac{\sqrt{3}}{2}$

▶ Exercise Set 9.6 (Page 592)

1. $\dfrac{A}{x-2} + \dfrac{B}{x+5}$ 3. $\dfrac{A}{x} + \dfrac{B}{x^2} + \dfrac{C}{x-1}$

5. $\dfrac{A}{x} + \dfrac{B}{x^2} + \dfrac{C}{x^3} + \dfrac{Dx+E}{x^2+1}$ 7. $\dfrac{Ax+B}{x^2+5} + \dfrac{Cx+D}{(x^2+5)^2}$

9. $\dfrac{1}{5}\ln\left|\dfrac{x-1}{x+4}\right| + C$

11. $-2\ln|x-2| + 3\ln|x-3| + C$

13. $\frac{5}{2}\ln|2x-1| + 3\ln|x+4| + C$

15. $-\frac{1}{6}\ln|x-1| + \frac{1}{15}\ln|x+2| + \frac{1}{10}\ln|x-3| + C$

17. $\ln\left|\dfrac{x(x+3)^2}{x-3}\right| + C$

19. $\frac{1}{2}x^2 - 2x + 6\ln|x+2| + C$

21. $3x + 12\ln|x-2| - \dfrac{2}{x-2} + C$

23. $\frac{1}{2}x^2 + 3x - \ln|x-1| + 8\ln|x-2| + C$

25. $\frac{1}{3}x^3 + x + \ln\left|\dfrac{(x+1)(x-1)^2}{x}\right| + C$

27. $3\ln|x| - \ln|x-1| - \dfrac{5}{x-1} + C$

29. $\ln\dfrac{(x-3)^2}{|x+1|} + \dfrac{1}{x-3} + C$

31. $\ln|x+2| + \dfrac{4}{x+2} - \dfrac{2}{(x+2)^2} + C$

33. $-\frac{7}{34}\ln|4x-1| + \frac{6}{17}\ln(x^2+1) + \frac{3}{17}\tan^{-1}x + C$

35. $\dfrac{1}{32}\ln\left|\dfrac{x-2}{x+2}\right| - \dfrac{1}{16}\tan^{-1}\dfrac{x}{2} + C$

37. $3\tan^{-1}x + \frac{1}{2}\ln(x^2+3) + C$

39. $\frac{1}{2}x^2 - 3x + \frac{1}{2}\ln(x^2+1) + C$

41. $\dfrac{1}{\sqrt{2}}\tan^{-1}\left(\dfrac{x+1}{\sqrt{2}}\right) + \dfrac{1}{x^2+2x+3} + C$

43. $\dfrac{1}{6}\ln\left|\dfrac{\sin\theta-1}{\sin\theta+5}\right| + C$ 45. $\ln\dfrac{e^x}{1+e^x} + C$

47. (a) $\sqrt{2}$, $-\sqrt{2}$

49. $\pi(\frac{19}{5} - \frac{9}{4}\ln 5)$

51. $y = \dfrac{3 - 2Ce^x}{1 - Ce^x}$

53. $y = \dfrac{t - 1}{Ct - t + 1}$

55. (b) a/b; 0

57. (a) $(x - 1)(x - 2)(x - 3)$ (b) $(x - 4)(x^2 + x + 5)$
(c) $(x - 2)(x - 3)(x^2 + 1)$

59. $\frac{1}{8}\ln|x - 1| - \frac{1}{5}\ln|x - 2|$
$\qquad\qquad + \frac{1}{12}\ln|x - 3| - \frac{1}{120}\ln|x + 3| + C$

▶ Exercise Set 9.7 **(Page 597)**

1. $\frac{2}{5}(x - 2)^{5/2} + \frac{4}{3}(x - 2)^{3/2} + C$

3. $4 - \pi$ **5.** $4 - 6\ln\frac{5}{3}$

7. $\frac{2}{15}(x^3 + 1)^{5/2} - \frac{2}{9}(x^3 + 1)^{3/2} + C$

9. $2x^{1/2} - 3x^{1/3} + 6x^{1/6} - 6\ln(x^{1/6} + 1) + C$

11. $4\ln\dfrac{v^{1/4}}{|1 - v^{1/4}|} + C$

13. $2t^{1/2} + 3t^{1/3} + 6t^{1/6} + 6\ln|t^{1/6} - 1| + C$

15. $\frac{1}{3}(1 + x^2)^{3/2} - (1 + x^2)^{1/2} + C$

17. $-2\sqrt{x}\cos\sqrt{x} + 2\sin\sqrt{x} + C$

19. $\ln\dfrac{\sqrt{e^x + 1} - 1}{\sqrt{e^x + 1} + 1} + C$

21. $\ln\left|\tan\left(\dfrac{x}{2}\right) + 1\right| + C$ **23.** 1

25. $\dfrac{4}{\sqrt{3}}\tan^{-1}\left(\sqrt{3}\tan\dfrac{x}{2}\right) - x + C$

29. $\dfrac{2}{\sqrt{3}}\tan^{-1}\left(\dfrac{2\tanh(x/2) + 1}{\sqrt{3}}\right) + C$

31. $-\dfrac{\sqrt{3 - x^2}}{3x} + C$ **33.** $\dfrac{(x^2 - 5)^{3/2}}{15x^3} + C$

▶ Exercise Set 9.8 **(Page 610)**

1. exact value $= \frac{14}{3} \approx 4.666666667$
(a) 4.667600663, $|E_M| \approx 0.000933996$
(b) 4.664795679, $|E_T| \approx 0.001870988$
(c) 4.666651630, $|E_S| \approx 0.000015037$

3. exact value $= 2$
(a) 2.008248408, $|E_M| \approx 0.008248408$
(b) 1.983523538, $|E_T| \approx 0.016476462$
(c) 2.000109517, $|E_S| \approx 0.000109517$

5. exact value $= e^{-1} - e^{-3} \approx 0.318092373$
(a) 0.317562837, $|E_M| \approx 0.000529536$
(b) 0.319151975, $|E_T| \approx 0.001059602$
(c) 0.318095187, $|E_S| \approx 0.000002814$

7. (a) 0.002812500   (b) 0.005625000   (c) 0.000126563

9. (a) 0.012919282   (b) 0.025838564   (c) 0.000170011

11. (a) 0.001226265   (b) 0.002452530   (c) 0.000006540

13. (a) 24   (b) 34   (c) 8

15. (a) 36   (b) 51   (c) 8

17. (a) 351   (b) 496   (c) 16

19. (a) 0.747130878   (b) 0.746210796   (c) 0.746824948

21. (a) 2.129469966   (b) 2.130644002   (c) 2.129861595

23. (a) 0.809253858   (b) 0.795924733   (c) 0.805376152

25. (a) 3.142425985, $|E_M| \approx 0.000833331$
(b) 3.139925989, $|E_T| \approx 0.001666665$
(c) 3.141592614, $|E_S| \approx 0.000000040$

29. 116   **33.** 3.820   **35.** 1604 ft

37. 37.9 mi   **39.** 9.3 L

▶ Chapter 9 Supplementary Exercises **(Page 612)**

1. $\frac{1}{2}x\sin 2x + \frac{1}{4}\cos 2x + C$

3. $\frac{1}{3}\sec^3 x - \sec x + C$

5. $\frac{1}{9}\tan^3 3t + C$ **7.** $x - \sin x + C$

9. $\frac{1}{6}x^3 + \frac{1}{4}(x^2 - \frac{1}{2})\sin 2x + \frac{1}{4}x\cos 2x + C$

11. $\frac{1}{4}\sec^4 x + C$

13. $-(\sin^3 2x)/48 - (\sin 4x)/64 + x/16 + C$

15. $\frac{1}{4}$ **17.** $\frac{1}{2}$ **19.** $\frac{1}{2}$

21. $x/\sqrt{1 + x^2} + C$

23. $\ln|\sec(e^x) + \tan(e^x)| + C$

25. $\sqrt{e^{2x} + 1} + C$

27. $\frac{1}{13}e^{3x}(3\sin 2x - 2\cos 2x) + C$

29. $\frac{5}{6}\pi - \sqrt{3}$

31. $\frac{1}{10}x[\sin(3\ln x) - 3\cos(3\ln x)] + C$

33. $\sqrt{x^2 - 9} + C$ **35.** $3\ln(3 + \sqrt{8}) - \sqrt{8}$

37. $\frac{1}{5}\sqrt{2x + 3}\,(x^2 - 2x + 6) + C$

39. $\dfrac{1}{2}\sin^{-1}\left(\dfrac{2t + 1}{2}\right) + C$

41. $-\sqrt{a^2 - x^2}/(a^2 x) + C$

43. $\frac{1}{2}[x\sqrt{a^2 - x^2} + a^2\sin^{-1}(x/a)] + C$

45. $-\sqrt{4x - x^2} + C$

47. $\ln|(2x + 1)/(x + 1)| + C$

49. $-\dfrac{\ln|x|}{6} - \dfrac{2}{15}\ln|x + 3| + \dfrac{3}{10}\ln|x - 2| + C$

51. $\frac{1}{3}\ln|x^3 - 3x| + C$

53. $\dfrac{7}{4}\ln\left|\dfrac{x - 1}{x + 1}\right| - \dfrac{3}{2}\tan^{-1}x + C$

55. $\dfrac{1}{26}\left[\ln\dfrac{(x - 3)^2}{x^2 + 4} - 3\tan^{-1}\left(\dfrac{x}{2}\right)\right] + C$

57. $\dfrac{7}{8}\tan^{-1}\left(\dfrac{x}{2}\right) - \dfrac{24 + 5x}{4(4 + x^2)} + C$

59. $\frac{1}{2}\ln(x^2 + 2x + 5) - \frac{1}{2}\tan^{-1}\frac{1}{2}(x + 1) + C$

61. $\dfrac{1}{\sqrt{2}}\sin^{-1}\sqrt{\frac{2}{3}}x + C$

63. $2(\sqrt{t} - \tan^{-1}\sqrt{t}) + C$ **65.** $-\frac{7}{4} + 3\ln 2$

67. $\frac{1}{2}(x - \ln|\sin x - \cos x|) + C$

69. $\frac{1}{4}\cot^2(\frac{1}{2}x) + \frac{1}{2}\ln|\tan(\frac{1}{2}x)| + C$

73. (a) $\frac{1}{2}\tan^{-1}(1) = \pi/8$
   (b) $\pi(\pi + 2)/64$ **(c)** $\pi\ln 2$

75. (a) $e^{2x}(4x^3 - 6x^2 + 6x - 3)/8 + C$
   (b) $(\pi - 2)/125$
   (c) $\frac{1}{8}(\sin x \cos x)^3 + \frac{1}{16}x - \frac{1}{64}\sin(4x) + C$

77. (a) $\frac{1}{4}(\sec^4\theta - 2\sec^2\theta) + C_1$
   (b) $\frac{1}{4}\tan^4\theta + C_2, \; C_2 = C_1 - \frac{1}{4}$

79. (a) 58.9275   **(b)** 54.7328

81. (a) 1.8277   **(b)** 1.8278   **83.** 0.36972

85. $-\dfrac{1}{a}\ln(1 + e^{-ax}) + C$

87. $\sqrt{1 - x^2} + \ln(1 - \sqrt{1 - x^2}) + C$

89. $-\dfrac{1}{3}\left(\dfrac{x + 1}{x - 1}\right)^{3/2} + C$

91. $-\frac{1}{10}\ln|3 + 2x^{-5}| + C$

93. $\dfrac{\sqrt{2}}{3}[(x + 2)^{3/2} - (x - 2)^{3/2}] + C$

► **Exercise Set 10.1** (Page 621)

1. 1   **3.** divergent

5. $\ln\frac{5}{3}$ **7.** $\frac{1}{2}$

9. $\dfrac{1}{2(a^2 + 1)}$ **11.** $-\frac{1}{4}$

13. $\frac{1}{3}$ **15.** divergent

17. 0   **19.** divergent

21. divergent   **23.** $\pi/2$

25. 1   **27.** divergent

29. $\frac{9}{2}$ **31.** divergent

33. 2   **35.** 2

37. $\frac{1}{5}$ **39.** $\sqrt{\pi}$

41. $\sqrt{\pi}/2$ **45.** $\frac{1}{2}$

49. (a) $\dfrac{1}{24}$ with $\dfrac{x}{x^5 + 1} \le \dfrac{1}{x^4}$
   (b) $\frac{1}{2}e^{-1}$ with $e^{-x^2} \le xe^{-x^2}$

51. $\frac{1}{3}$ **53. (a)** $V = \pi$

55. $\dfrac{2\pi NI}{kr}\left(1 - \dfrac{a}{\sqrt{r^2 + a^2}}\right)$

57. (b) 24,000,000   **59.** $1/s$

61. $1/(s^2 + 1)$ **63.** 1.809

65. (a) $1.047, \; \pi/3 \approx 1.047197551$

► **Exercise Set 10.2** (Page 630)

1. 1   **3.** 1   **5.** 1

7. $1/\pi$ **9.** -1 **11.** 0

13. $\frac{1}{2}$ **15.** $\frac{3}{2}$ **17.** $-1/2\pi$

19. $\frac{1}{4}$ **21.** $-\frac{1}{12}$ **23.** $\frac{1}{6}$

25. $a - b$ **27.** 2   **29.** $+\infty$

31. $\frac{1}{9}$ **33. (b)** 2

35. $k = -1, l = \pm 2\sqrt{2}$ **37.** $\frac{1}{2}$

39. 2   **41.** does not exist   **43.** 3

► **Exercise Set 10.3** (Page 638)

1. 0   **3.** $-\infty$ **5.** $+\infty$

7. 0   **9.** 0   **11.** π

13. 2   **15.** e^{-3} **17.** e^2

19. 1   **21.** $+\infty$ **23.** $e^{2/\pi}$

25. 1   **27.** e^3 **29.** 1

31. e^2 **33.** $-\frac{1}{2}$

35. 0   **37.** 0

39. $+\infty$ **41.** 2

45. (a) 0   **(b)** $+\infty$ **(c)** 0   **(d)** $-\infty$
   (e) $+\infty$ **(f)** $+\infty$ **(g)** $-\infty$ **(h)** $-\infty$

47.

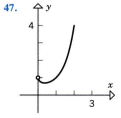

49. -1 **51.** $\frac{1}{9}$ **53.** 2π **61.** (b) $\ln 0.3 \approx -1.203265293$, $\ln 2 \approx 0.693381829$

55. $8\sqrt{2}/5$ **57.** 1 **59.** does not exist   **63.** $\frac{1}{2}at^2$ **65.** $1/s^2$

▶ Chapter 10 Supplementary Exercises (Page 640)

1. $\pi/2$ **3.** 1 **5.** 6 **21.** 0 **23.** $\frac{1}{8}$

7. 0 **9.** diverges   **11.** 1 **25.** $-1/(4\pi^2)$ **27.** 0

13. $\pi/4$ **15.** $n > -1$; $-1/(n+1)^2$ **29.** 0 **31.** 1

17. $\frac{3}{2}$ **19.** $+\infty$ **33.** (a) $+\infty$ (b) $3\pi/2$ **35.** 2

▶ Exercise Set 11.1 (Page 649)

1. $\frac{1}{3}, \frac{2}{4}, \frac{3}{5}, \frac{4}{6}, \frac{5}{7}$; converges to 1

3. $2, 2, 2, 2, 2$; converges to 2

5. $\dfrac{\ln 1}{1}, \dfrac{\ln 2}{2}, \dfrac{\ln 3}{3}, \dfrac{\ln 4}{4}, \dfrac{\ln 5}{5}$; converges to 0

7. $0, 2, 0, 2, 0$; diverges

9. $-1, \frac{16}{9}, -\frac{54}{28}, \frac{128}{65}, -\frac{250}{126}$; diverges

11. $\frac{6}{2}, \frac{12}{8}, \frac{20}{18}, \frac{30}{32}, \frac{42}{50}$; converges to $\frac{1}{2}$

13. $\cos 3, \cos \frac{3}{2}, \cos 1, \cos \frac{3}{4}, \cos \frac{3}{5}$; converges to 1

15. $e^{-1}, 4e^{-2}, 9e^{-3}, 16e^{-4}, 25e^{-5}$; converges to 0

17. $2, (\frac{5}{3})^2, (\frac{6}{4})^3, (\frac{7}{5})^4, (\frac{8}{6})^5$; converges to e^2

19. $\left\{\dfrac{2n-1}{2n}\right\}_{n=1}^{+\infty}$; converges to 1

21. $\left\{\dfrac{1}{3^n}\right\}_{n=1}^{+\infty}$; converges to 0

23. $\left\{\dfrac{1}{n} - \dfrac{1}{n+1}\right\}_{n=1}^{+\infty}$; converges to 0

25. $\{\sqrt{n+1} - \sqrt{n+2}\}_{n=1}^{+\infty}$; converges to 0

27. (a) $\sqrt{6}, \sqrt{6+\sqrt{6}}, \sqrt{6+\sqrt{6+\sqrt{6}}}$ (b) 3

29. (a) $1, 1, 2, 3, 5, 8, 13, 21$ (b) $(1+\sqrt{5})/2$

31. (a) $1, \frac{3}{4}, \frac{2}{3}, \frac{5}{8}$ (b) $\frac{1}{2}$

33. (a) 3 (b) 11 (c) 1001 **39.** 3

▶ Exercise Set 11.2 (Page 658)

1. decreasing   **3.** increasing   **25.** converges   **27.** diverges

5. decreasing   **7.** increasing   **29.** converges

9. decreasing   **11.** nonincreasing   **31.** (a) 0 (b) $+\infty$ (does not exist)

13. not monotone   **15.** decreasing

17. increasing   **19.** increasing   **35.** (a) $\sqrt{2}, \sqrt{2+\sqrt{2}}, \sqrt{2+\sqrt{2+\sqrt{2}}}$ (e) 2

21. decreasing   **23.** decreasing   **41.** (b) converges (decreasing and bounded below by 0)

▶ Exercise Set 11.3 (Page 666)

1. (a) converges to $\frac{5}{2}$ (b) converges to $\frac{1}{2}$ **21.** $\frac{869}{1111}$ **23.** diverges   **33.** $\dfrac{1}{x^2-2x}$, $|x| > 2$

  (c) diverges

3. $\frac{4}{7}$ **5.** 6 **7.** diverges   **35.** $\dfrac{2\sin x}{2+\sin x}$, $-\infty < x < +\infty$ **37.** 1

9. $\frac{1}{3}$ **11.** $\frac{1}{6}$ **13.** $\frac{448}{3}$

15. $-\frac{1}{3}$ **17.** $\frac{4}{9}$ **19.** $\frac{532}{99}$ **41.** The series converges to $1/(1-x)$ only for $-1 < x < 1$.

▶ Exercise Set 11.4 (Page 676)

1. $\frac{4}{3}$ **3.** $-\frac{1}{36}$

5. (a) converges   (b) diverges   (c) diverges

  (d) diverges   (e) converges   (f) diverges

  (g) converges   (h) converges

9. diverges **11.** diverges
13. converges **15.** diverges
17. diverges **19.** diverges
21. converges **23.** diverges
25. converges **27.** converges
29. diverges

35. (a) diverges (b) diverges
(c) diverges (d) converges
37. (a) 1.1975; $1.2016 < S < 1.2026$
(b) 23
39. (a) $13 < s_{1,000,000} < 15$
(b) 2.69×10^{43}

▶ **Exercise Set 11.5** (Page 683)

1. converges **3.** inconclusive
5. diverges **7.** diverges
9. converges **11.** diverges
13. converges **15.** converges
17. converges **19.** converges

21. diverges **23.** converges
25. converges **27.** diverges
29. converges **31.** converges
41. (a) 1.71667; error < 0.00163 (b) 8
43. (a) 0.69226; error < 0.00098 (b) 13

▶ **Exercise Set 11.6** (Page 691)

13. converges **15.** converges
17. diverges **19.** converges
21. diverges **23.** converges
25. diverges **27.** converges

29. diverges **31.** converges
33. converges **37.** converges
39. $p > 1$ **41.** converges

▶ **Exercise Set 11.7** (Page 703)

1. converges **3.** diverges
5. converges **7.** absolutely
9. diverges **11.** absolutely
13. conditionally **15.** divergent
17. conditionally **19.** absolutely
21. conditionally **23.** divergent
25. conditionally **27.** absolutely

29. conditionally **31.** 0.125
33. 0.1 **35.** $9,999$ **37.** $39,999$
39. $|\text{error}| < 0.00074$; $s_{10} \approx 0.4995$; exact sum $= 0.5$
41. $n = 4$, $s_4 \approx 0.84147$; $\sin(1) \approx 0.841470985$
43. $n = 9$, $s_9 \approx 0.40553$; $\ln\frac{3}{2} \approx 0.405465108$
45. (a) 14
(b) 0.817962176; $|\text{error}| \approx 0.004504858$

▶ **Exercise Set 11.8** (Page 710)

1. 1, $[-1, 1)$
3. $+\infty$, $(-\infty, +\infty)$
5. $\frac{1}{5}$, $[-\frac{1}{5}, \frac{1}{5}]$
7. 1, $[-1, 1]$
9. 1, $(-1, 1]$
11. $+\infty$, $(-\infty, +\infty)$
13. $+\infty$, $(-\infty, +\infty)$
15. 1, $[-1, 1]$
17. 1, $(-2, 0]$
19. $\frac{4}{3}$, $(-\frac{19}{3}, -\frac{11}{3})$

21. 1, $[-2, 0]$
23. $+\infty$, $(-\infty, +\infty)$
25. $x + \frac{1}{2}x^2 + \frac{3}{14}x^3 + \frac{3}{35}x^4 + \cdots$; $R = 3$
27. $x + \frac{3}{2}x^2 + \frac{5}{8}x^3 + \frac{7}{48}x^4 + \cdots$; $R = +\infty$
29. $(a - b, a + b)$ **31.** $+\infty$

▶ **Exercise Set 11.9** (Page 721)

1. $1 - 2x + 2x^2 - \frac{4}{3}x^3 + \frac{2}{3}x^4$
3. $2x - \frac{4}{3}x^3$
5. $x + \frac{1}{3}x^3$
7. $x + x^2 + \frac{x^3}{2!} + \frac{x^4}{3!}$
9. $1 + \frac{1}{2}x^2 + \frac{5}{24}x^4$

11. $\ln 3 + \frac{2}{3}x - \frac{2}{9}x^2 + \frac{8}{81}x^3 - \frac{4}{81}x^4$
13. $e + e(x - 1) + \frac{e}{2!}(x - 1)^2 + \frac{e}{3!}(x - 1)^3$
15. $2 + \frac{1}{4}(x - 4) - \frac{1}{64}(x - 4)^2 + \frac{1}{512}(x - 4)^3$

17. $\dfrac{\sqrt{2}}{2} - \dfrac{\sqrt{2}}{2}\left(x - \dfrac{\pi}{4}\right) - \dfrac{\sqrt{2}}{4}\left(x - \dfrac{\pi}{4}\right)^2 + \dfrac{\sqrt{2}}{12}\left(x - \dfrac{\pi}{4}\right)^3$

19. $-\dfrac{\sqrt{3}}{2} + \dfrac{\pi}{2}\left(x + \dfrac{1}{3}\right) + \dfrac{\sqrt{3}\pi^2}{4}\left(x + \dfrac{1}{3}\right)^2 - \dfrac{\pi^3}{12}\left(x + \dfrac{1}{3}\right)^3$

21. $\dfrac{\pi}{4} + \dfrac{1}{2}(x - 1) - \dfrac{1}{4}(x - 1)^2 + \dfrac{1}{12}(x - 1)^3$

23. $\displaystyle\sum_{k=0}^{\infty} (-1)^k \dfrac{x^k}{k!}$

25. $\displaystyle\sum_{k=0}^{\infty} (-1)^k x^k$

27. $\displaystyle\sum_{k=1}^{\infty} (-1)^{k+1} \dfrac{x^k}{k}$

29. $\displaystyle\sum_{k=0}^{\infty} (-1)^k \dfrac{x^{2k}}{4^k(2k)!}$

31. $\displaystyle\sum_{k=0}^{\infty} \dfrac{x^{2k}}{(2k)!}$

33. $\displaystyle\sum_{k=0}^{\infty} (-1)(x + 1)^k$

35. $\displaystyle\sum_{k=1}^{\infty} (-1)^{k+1} \dfrac{(x - 1)^k}{k}$

37. $\displaystyle\sum_{k=0}^{\infty} (-1)^k \dfrac{\pi^{2k}}{(2k)!}\left(x - \dfrac{1}{2}\right)^{2k}$

39. $\displaystyle\sum_{k=0}^{\infty} \left(\dfrac{16 + (-1)^{k+1}}{8}\right) \dfrac{(x - \ln 4)^k}{k!}$

▶ **Exercise Set 11.10** (Page 733)

1. $\dfrac{2^6 e^{2c}}{6!} x^6$

3. $-\dfrac{x^5}{(c + 1)^6}$

5. $\dfrac{(4 + c)e^c}{4!} x^4$

7. $-\dfrac{(1 - 3c^2)}{3(1 + c^2)^3} x^3$

9. $-\dfrac{5}{128 c^{7/2}} (x - 4)^4$

11. $\dfrac{\cos c}{5!}\left(x - \dfrac{\pi}{6}\right)^5$

13. $\dfrac{7}{(1 + c)^8} (x + 2)^6$

15. $\dfrac{x^{n+1}}{(1 - c)^{n+2}}$

17. $\dfrac{2^{n+1} e^{2c}}{(n + 1)!} x^{n+1}$

27. $1 - 2x + 2x^2 - \frac{4}{3}x^3 + \cdots ; (-\infty, +\infty)$

29. $x - x^2 + \dfrac{1}{2!} x^3 - \dfrac{1}{3!} x^4 + \cdots ; (-\infty, +\infty)$

31. $2x - \dfrac{2^3}{3!} x^3 + \dfrac{2^5}{5!} x^5 - \dfrac{2^7}{7!} x^7 + \cdots ; (-\infty, +\infty)$

33. $x^2 - \dfrac{1}{2!} x^4 + \dfrac{1}{4!} x^6 - \dfrac{1}{6!} x^8 + \cdots ; (-\infty, +\infty)$

35. $x^2 - \dfrac{2^3}{4!} x^4 + \dfrac{2^5}{6!} x^6 - \dfrac{2^7}{8!} x^8 + \cdots ; (-\infty, +\infty)$

37. $-x^2 - \frac{1}{2}x^4 - \frac{1}{3}x^6 - \frac{1}{4}x^8 - \cdots ; (-1, 1)$

39. $1 + 4x^2 + 16x^4 + 64x^6 + \cdots ; (-\frac{1}{2}, \frac{1}{2})$

41. $x^2 - 3x^3 + 9x^4 - 27x^5 + \cdots ; (-\frac{1}{3}, \frac{1}{3})$

43. $2x^2 + \dfrac{2^3}{3!} x^4 + \dfrac{2^5}{5!} x^6 + \dfrac{2^7}{7!} x^8 + \cdots ; (-\infty, +\infty)$

45. $1 + \frac{3}{2}x - \frac{9}{8}x^2 + \frac{27}{16}x^3 - \cdots ; (-\frac{1}{3}, \frac{1}{3})$

47. $1 + 4x + 12x^2 + 32x^3 + \cdots ; (-\frac{1}{2}, \frac{1}{2})$

49. $x + \frac{1}{2}x^3 + \frac{3}{8}x^5 + \frac{5}{16}x^7 + \cdots ; (-1, 1)$

51. $\displaystyle\sum_{k=0}^{\infty} (-1)^k (x - 1)^k, (0, 2)$

53. $\displaystyle\sum_{k=0}^{\infty} \binom{m}{k} x^k$

55. $\sin \pi = 0$

57. $e^{-\ln 3} = \frac{1}{3}$

59. (a) $x^2 - 2x^4 + \frac{2}{3}x^6 - \frac{4}{45}x^8 + \cdots$ (b) 0

▶ **Exercise Set 11.11** (Page 742)

1. (a) 9   (b) 13

3. 1.6487

5. 0.9877

7. 0.5299

9. 0.223

11. 0.100

13. 1.0050

15. $|x| < 0.569$

17. 9×10^{-8}

19. (a) 1,999,999   (b) 8

21. 3.140

▶ **Exercise Set 11.12** (Page 750)

5. $\displaystyle\sum_{k=1}^{\infty} (-1)^{k+1} k x^{k-1}$

9. $\ln \frac{4}{3}$

11. $\dfrac{2}{(1 - x)^3}$

13. (b) $\displaystyle\sum_{k=0}^{\infty} (-1)^k x^k ; (-1, 1)$

15. 0.764

17. 0.494

19. 0.100

21. 0.491

23. $1 - \frac{3}{2}x^2 + \frac{25}{24}x^4 - \frac{331}{720}x^6 + \cdots$

25. $x - x^2 + \frac{1}{3}x^3 - \frac{1}{30}x^5 + \cdots$

27. $-x^3 - \frac{1}{2}x^5 - \frac{1}{3}x^7 - \frac{1}{4}x^9 - \cdots$

29. $x^2 + \frac{9}{2}x^3 + \frac{79}{8}x^4 + \frac{683}{48}x^5 + \cdots$

31. (a) 0   (b) $-\frac{3}{2}$

33. (a) $x - \frac{1}{6}x^3 + \frac{3}{40}x^5 - \frac{5}{112}x^7$

(b) $x + \displaystyle\sum_{k=1}^{\infty} (-1)^k \dfrac{1 \cdot 3 \cdot 5 \cdots (2k - 1)}{2^k k!(2k + 1)} x^{2k+1}$ (c) 1

▶ **Chapter 11 Supplementary Exercises** (Page 751)

1. $L = 0$ **3.** $L = 0$

5. does not exist

7. (a) 2, 3, 6   (b) 4   (c) 1, 5

9. (a) $|q| > \sqrt{\pi}$ (b) $q > \frac{1}{3}$
(c) none $(p = 1)$ (d) $q > e$ or $0 < q < 1/e$

11. (a) $1 + 36 \sum_{k=1}^{\infty} (0.01)^k$ (b) $\frac{15}{11}$

13. (a) $\frac{1}{4}$ (b) diverges   (c) 1

15. converges   **17.** converges   **19.** converges

21. diverges (general term does not approach zero)

23. converges absolutely

25. diverges   **27.** $R = 1; 0 \le x \le 2$

29. $R = 2; -1 < x < 3$ **31.** $R = 0; x = 1$

33. (a) $(x - 2) - (x - 2)^2/2 + (x - 2)^3/3$
(b) $\frac{-1}{4(c-1)^4}(x - 2)^4$, c between 2 and x (c) $\frac{(\frac{1}{2})^4}{4(\frac{1}{2})^4} = \frac{1}{4}$

35. (a) $1 + (x - 1)/2 - (x - 1)^2/8$
(b) $\frac{(x - 1)^3}{16c^{5/2}}$, c between 1 and x
(c) $\frac{(5/9)^3}{16(2/3)^5} < 0.0814$

37. $\ln a + \sum_{k=0}^{\infty} (-1)^k \frac{(x/a)^{k+1}}{k + 1}$; $R = a$

39. $\frac{1}{3}\left\{1 + \sum_{k=1}^{\infty} (-1)^k \frac{1 \cdot 3 \cdots (2k - 1)}{2 \cdot 4 \cdots (2k)}(x/9)^k\right\}$;
$R = 9$

41. $1 + x^2/2 + 5x^4/24 + \cdots$

43. $1 - x^2/4 - x^4/96 + \cdots$

45. $x - \frac{1}{2}x^2 + \frac{1}{6}x^3 + \cdots$

47. $-2 - 2x - \frac{8}{3}x^3 - \cdots; -2$

49. $|x| < 1.17$ **51.** 0.444

▶ **Exercise Set 12.2** (Page 765)

1.
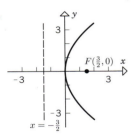

3.
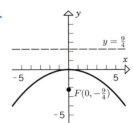

5.
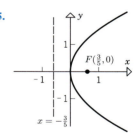

7.
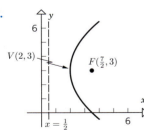

9.
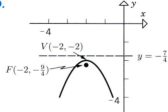

11.
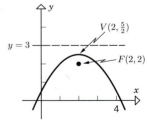

13.
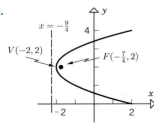

15.
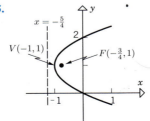

17. $y^2 = 12x$

19. $y^2 = -28x$

21. $y^2 = 2x$

23. $x^2 = -12y$

25. $y^2 = -5(x - \frac{19}{5})$

27. $y^2 = 6(x - \frac{3}{2})$

29. $(x - 1)^2 = 12(y - 1)$

31. $(x - 5)^2 = 2(y + 3)$

33. (a) $y = \frac{7}{6}x^2 - \frac{23}{6}x + 3$

(b) $x = -\frac{7}{6}y^2 + \frac{17}{6}y + 2$

35. vertex: $\left(-\dfrac{B}{2A}, \dfrac{4AC - B^2}{4A} \right)$

focus: $\left(-\dfrac{B}{2A}, \dfrac{4AC - B^2 + 1}{4A} \right)$

directrix: $y = \dfrac{4AC - B^2 - 1}{4A}$

37. 16 ft

41. $\frac{1}{16}$ ft

▶ **Exercise Set 12.3** (Page 772)

1.

3.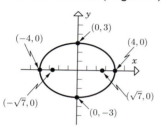

5.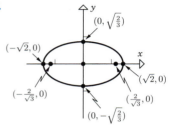

7.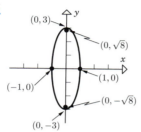

9.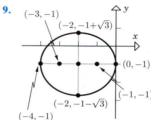

11.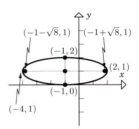

13.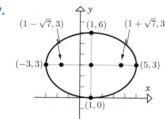

15. $\dfrac{x^2}{9} + \dfrac{y^2}{4} = 1$

17. $\dfrac{x^2}{169} + \dfrac{y^2}{144} = 1$

19. $\dfrac{x^2}{3} + \dfrac{y^2}{2} = 1$

21. $\dfrac{x^2}{16} + \dfrac{y^2}{4} = 1, \dfrac{x^2}{4} + \dfrac{y^2}{16} = 1$

23. $\dfrac{x^2}{36} + \dfrac{y^2}{81/8} = 1$

25. $(x - 1)^2 + \dfrac{(y - 3)^2}{2} = 1$

27. $\dfrac{(x - 4)^2}{32} + \dfrac{(y - 3)^2}{36} = 1$

29.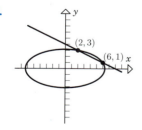

31.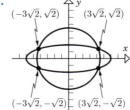

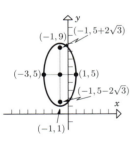

33. $k = -4$ at $(-2, -1)$ **35.** πab
 $k = 4$ at $(2, 1)$

39. $27\left[4 \sin^{-1} \dfrac{h-2}{2} + (h-2)\sqrt{4h - h^2} + 2\pi\right]$

41. $\dfrac{4a^2b^2}{a^2 + b^2}$

43. center, $(0, 0)$; major axis, 10; minor axis, 6

45. $T = \frac{1}{2}pD$, $L = D\sqrt{1 + p^2}$

47. (b) $\dfrac{x^2}{16} + \dfrac{y^2}{25} = 1$
 (c) The ellipse flattens and approaches the major axis.
 (d) The ellipse widens and approaches a circle.

49. major axis, $r_1 + r_2$; center at the midpoint of the line segment joining the centers of C_1 and C_2

▶ **Exercise Set 12.4 (Page 782)**

1.
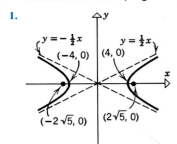

3.
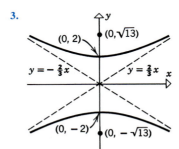

5.
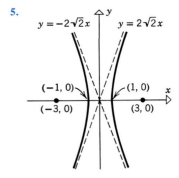

7.
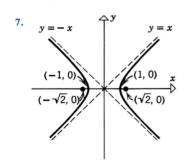

9. $y - 4 = -\frac{2}{3}(x-2)$ $y - 4 = \frac{2}{3}(x-2)$
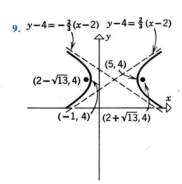

11. $y + 3 = -3(x+2)$ $y + 3 = 3(x+2)$
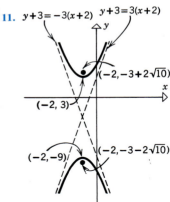

13.
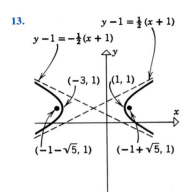

15. $y + 3 = -4(x-1)$
 $y + 3 = 4(x-1)$
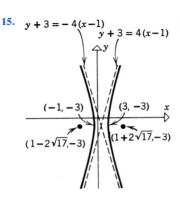

17. $\dfrac{x^2}{4} - \dfrac{y^2}{5} = 1$

19. $x^2 - \dfrac{y^2}{4} = 1$

21. $\dfrac{x^2}{64/9} - \dfrac{y^2}{16} = 1, \ \dfrac{y^2}{36} - \dfrac{x^2}{16} = 1$

23. $\dfrac{y^2}{56} - \dfrac{x^2}{56} = 1$

25. $\dfrac{x^2}{4} - \dfrac{y^2}{4/3} = 1$

27. $\dfrac{(x-2)^2}{4} - \dfrac{(y+3)^2}{5} = 1$

29. $\dfrac{(x-6)^2}{16} - \dfrac{(y-4)^2}{9} = 1$

31. $8xy - 4x - 4y + 1 = 0$

33.

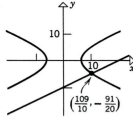

35.

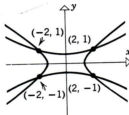

39. $(\frac{3}{2}\sqrt{13}, -9), \ (-\frac{3}{2}\sqrt{13}, -9)$

41. $(\pm 3/\sqrt{5}, 4/\sqrt{5}), \ (\pm 3/\sqrt{5}, -4/\sqrt{5})$

43. (b) $\dfrac{x^2}{9} - \dfrac{y^2}{16} = 1$

(c) The hyperbola flattens and approaches the focal axis, excluding the segment between the vertices.

(d) The hyperbola approaches the lines that are perpendicular to the focal axis at the vertices.

▶ Exercise Set 12.5 **(Page 792)**

1. (a) $(-1 + 3\sqrt{3}, \sqrt{3} + 3)$

(b) $\dfrac{x'^2}{4} - \dfrac{y'^2}{12} = 1$

(c)

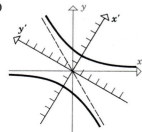

3. $\theta = 45°; \ \dfrac{y'^2}{18} - \dfrac{x'^2}{18} = 1,$

hyperbola

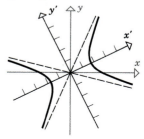

5. $\theta = \tan^{-1}\frac{1}{2};$

$\dfrac{x'^2}{3} - \dfrac{y'^2}{2} = 1,$ hyperbola

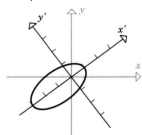

7. $\theta = 60°; \ y' = x'^2$, parabola

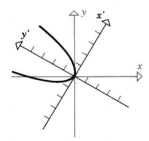

9. $\theta = \tan^{-1}\frac{3}{4};$

$y'^2 = 4(x' - 1)$, parabola

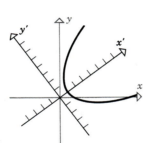

11. $\theta = \tan^{-1}\frac{3}{4};$

$\dfrac{(x'+1)^2}{4} + y'^2 = 1$, ellipse

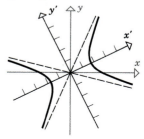

15. $x^2 + xy + y^2 = 3$

23. ellipse, point, or no graph

25. parabola, line, pair of parallel lines, or no graph

27. ellipse, point, or no graph

▶ Chapter 12 Supplementary Exercises **(Page 793)**

1. parabola: $V(-2, 3)$, $F(-5, 3)$, directrix $x = 1$

3. ellipse: $C(-2, 1)$, $F(-2, 1 \pm \sqrt{5})$, axis lengths 6 and 4

5. hyperbola: $C(2, 1)$, $F(2 \pm \sqrt{10}, 1)$, $V(2 \pm 3, 1)$, asymptotes $y - 1 = \pm(x - 2)/3$

7. parabola: $V(3, 1)$, $F(21/8, 1)$, directrix $x = 27/8$

9. $(y - 3)^2 = 16(x - 1)$ **11.** $y^2/9 - x^2/16 = 1$

13. $x^2/25 + y^2/16 = 1$ **15.** $(x - 3)^2 = 4(y - 3)$

17.

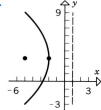

19.

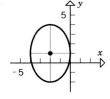

21. $\pi/4$; ellipse; $(x'/\sqrt{2})^2 + (y')^2 = 1$

23. $\pi/6$; hyperbola; $4x'^2 - y'^2 = 1$

25. $\tan^{-1}\frac{4}{3}$; parabola; $y'^2 = 4(x' - 1)$

27.

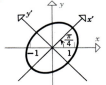

29.

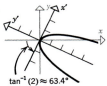

$\tan^{-1}(2) \approx 63.4°$

▶ Exercise Set 13.1 **(Page 800)**

1.

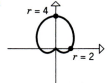

3. (a) $(3\sqrt{3}, 3)$ (b) $(-7/2, 7\sqrt{3}/2)$

(c) $(4\sqrt{2}, 4\sqrt{2})$ (d) $(5, 0)$

(e) $(-7\sqrt{3}/2, 7/2)$ (f) $(0, 0)$

5. (a) $(5, \pi)$ (b) $(4, 11\pi/6)$

(c) $(2, 3\pi/2)$ (d) $(8\sqrt{2}, 5\pi/4)$

(e) $(6, 2\pi/3)$ (f) $(\sqrt{2}, \pi/4)$

7. (a) $(-5, 0)$ (b) $(-4, 5\pi/6)$

(c) $(-2, \pi/2)$ (d) $(-8\sqrt{2}, \pi/4)$

(e) $(-6, 5\pi/3)$ (f) $(-\sqrt{2}, 5\pi/4)$

9. $x^2 + y^2 = 4$; circle   **11.** $y = 4$; line

13. $x^2 + y^2 - 3x = 0$; circle   **15.** $xy = 4$; hyperbola

17. $x^2 = 4 - 4y$; parabola

19. $3x + 2y = 6$; line

21. $r \cos \theta = 7$ **23.** $r = 3$

25. $r = 6 \sin \theta$ **27.** $r = 9 \sec \theta \tan \theta$

29. $r^2 \sin 2\theta = \frac{9}{2}$ **31.** $r^2 = \sin 2\theta$

33. The graph consists of the concentric circles $r = 1$ and $r = 2$.

35.

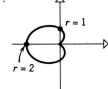

37.

► Exercise Set 13.2 **(Page 809)**

1.

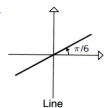

Line

3.

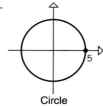

Circle

5.

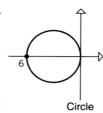

Circle

7.

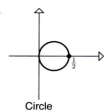

Circle

9.

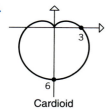

Cardioid

11.

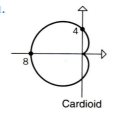

Cardioid

13.

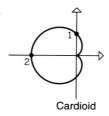

Cardioid

15.

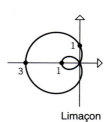

Limaçon

17.

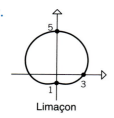

Limaçon

19.

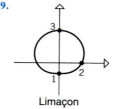

Limaçon

21.

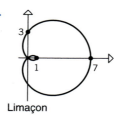

Limaçon

23.

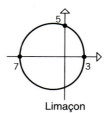

Limaçon

25.

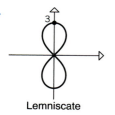

Lemniscate

27.

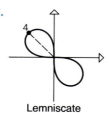

Lemniscate

29.

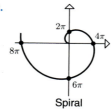

Spiral

31.

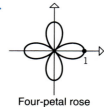

Four-petal rose

33.

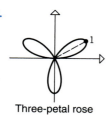

Three-petal rose

35.

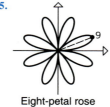

Eight-petal rose

37.

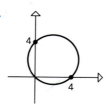

39.

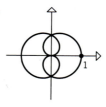

41.

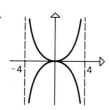

43.

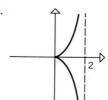

45.

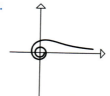

47. $\frac{3}{4}\sqrt{3}$

51. $-\frac{1}{4}$

55. $e = 1$

57. $e = \frac{1}{2}$

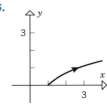

59. $e = \frac{3}{2}$

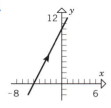

61. The graph of $r = f(\theta + \alpha)$ is the graph of $r = f(\theta)$ rotated α radians clockwise about the pole if $\alpha > 0$, $|\alpha|$ radians counterclockwise if $\alpha < 0$.

▶ **Exercise Set 13.3** (Page 815)

1. $\dfrac{7\pi^3}{1296}$

3. 6π

13. $\dfrac{5\pi}{4}$

15. $100 \cos^{-1}\left(\frac{3}{5}\right) - 48$

5. $\dfrac{8\pi}{3} + \sqrt{3}$

7. $\dfrac{9\sqrt{3}}{2} - \pi$

17. $\frac{4}{3}a^2$

19. $\frac{9}{8}(1 - e^{-4\pi})$

9. 1

11. 4π

21. $\ln 6$

23. $2\sqrt{3} - \frac{2}{3}\pi$

25. (b) $\frac{1}{6}$

▶ **Exercise Set 13.4** (Page 826)

1.

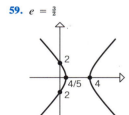

3.

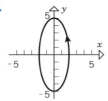

13.

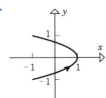

15.

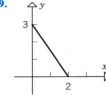

5.

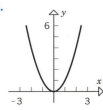

7.

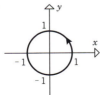

17.

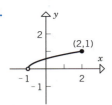

19.

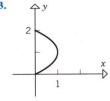

9.

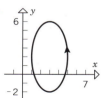

11.

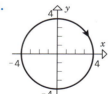

21.

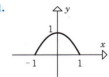

23.

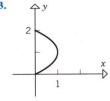

25.

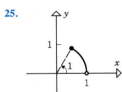

27. $x = -t,\ y = \sqrt{-t}$

29. 1

31. 12

33. 0

35. $\frac{1}{2}$

37. 4

39. 10

41. $\frac{1}{3}(2\sqrt{2} - 1)$

43. π

45. $\frac{1}{27}(80\sqrt{10} - 13\sqrt{13})$

47. $8a$

49. (b) 9.69 (c) 5.16 cm

51. $x = (2 + 3\sin\theta)\cos\theta$
$y = (2 + 3\sin\theta)\sin\theta$

53. $y = -e^{-4}x + 2e^{-2}$

55. (a) $-\frac{1}{2}$ (b) 1, 4

57. $y = 5x - 14,\ y = 6x - 17$

59. if $|a| = |b|$, a circle with center at (h, k) and radius $|a|$; if
$|a| \neq |b|$, an ellipse with center at (h, k) and vertices $(h \pm |a|, k)$
if $|a| > |b|$ and vertices $(h, k \pm |b|)$ if $|a| < |b|$

61. (a)

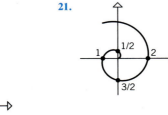

63. $\dfrac{2\sqrt{2}}{5}\pi(2e^{\pi} + 1)$

65. $\dfrac{\pi}{24}(17\sqrt{17} - 1)$

▶ **Exercise Set 13.5 (Page 834)**

1. $\dfrac{1}{\sqrt{3}}$

3. $\dfrac{\tan 2 - 2}{2\tan 2 + 1}$

5. -2

7. 1

9. 0

11. $-\dfrac{1}{2\sqrt{3}}$

13. $\dfrac{\sqrt{10}}{3}(e^6 - 1)$

15. $2\pi a$

17. $\dfrac{a}{3}[(\pi^2 + 4)^{3/2} - 8]$

19. $8a$

21. (b) 2.42

23. (a) $r = 4\theta + 10$
(b) 38.9 mm

25. $\theta = \dfrac{\pi}{2}, \dfrac{3\pi}{2},\ \sin^{-1}\frac{1}{4},\ \pi - \sin^{-1}\frac{1}{4}$

27. $\tan^{-1} 3\sqrt{3}$

▶ **Chapter 13 Supplementary Exercises (Page 835)**

1. (a) $(1, \sqrt{3})$ (b) $(0, -2)$
(c) $(0, 0)$ (d) $(-1, 1)$
(e) $(-3, 0)$ (f) $(\frac{3}{5}, -\frac{4}{5})$

3. (a)

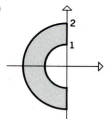

(b)

5. $2xy = 1$ (hyperbola)

7. $y = -4$ (line)

9. $y = \sqrt{3}x$ (line)

11. $x = y = 0$ (point)

13. $r\cos\theta = -3$

15. $\tan\theta = 3$

17.

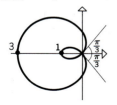

19.

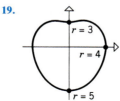

21.

23.

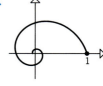

25. $(a/2, \pm\pi/6),\ (a/2, \pm\pi/3),\ (a/2, \pm 2\pi/3),$
$(a/2, \pm 5\pi/6)$

27. (a) $\displaystyle\int_0^{\pi/3}(1 + \cos\theta)^2\,d\theta + \int_{\pi/3}^{\pi/2}(3\cos\theta)^2\,d\theta$

(b) $\displaystyle 2\int_{\pi/3}^{\pi}\sqrt{2(1 + \cos\theta)}\,d\theta$

29. (a) $\int_{\pi/2}^{\pi} 2[\sin^2\theta - (1 + \cos\theta)^2]\,d\theta$

   (b) $\int_0^{\pi/2} 2\,d\theta$

31. $a^2[\sqrt{3} - \pi/3]$ 33. $4 - \pi$

35. (a) the ellipse $4(x - 1)^2 + 9(y + 1)^2 = 36$ oriented from $(4, -1)$ counterclockwise to $(-2, -1)$

   (b) $0, -2/9, \ y = 1$

37. (a) $y = -\ln x$; oriented from $(1, 0)$ to $(e^{-1}, 1)$

   (b) $-2, 4, \ y = -2(x - \frac{1}{2}) + \ln 2$

39. $\ln(2 + \sqrt{3})$

41. 4 43. $\sqrt{2}(e^{2\pi} - 1)$

45. (a) $\left(0, \dfrac{\pi}{6} + \sqrt{3}\right)$ and $\left(0, \dfrac{5\pi}{6} - \sqrt{3}\right)$

   (b) $\left(-1, \dfrac{\pi}{2}\right)$

47. $t = 1$ 49. slope $= 1, \ \ \psi = 3\pi/4$

▶ Exercise Set 14.1 **(Page 843)**

3. $y = c_1 e^x + c_2 e^{-4x}$

5. $y = c_1 e^x + c_2 x e^x$

7. $y = c_1 \cos\sqrt{5}x + c_2 \sin\sqrt{5}x$

9. $y = c_1 + c_2 e^x$ 11. $y = c_1 e^{-2t} + c_2 t e^{-2t}$

13. $y = e^{2x}(c_1 \cos 3x + c_2 \sin 3x)$

15. $y = c_1 e^{-x/4} + c_2 e^{x/2}$ 17. $y = 2e^x - e^{-3x}$

19. $y = (2 - 5x)e^{3x}$

21. $y = -e^{-2x}(3\cos x + 6\sin x)$

23. (a) $y'' - 3y' - 10y = 0$

   (b) $y'' - 8y' + 16y = 0$

   (c) $y'' + 2y' + 17y = 0$

25. (a) $k < 0$ or $k > 4$

   (b) $0, 4$ (c) $0 < k < 4$

27. (a) $y = \dfrac{1}{x}\,[c_1 \cos(\ln x) + c_2 \sin(\ln x)]$

   (b) $y = c_1 x^{1+\sqrt{3}} + c_2 x^{1-\sqrt{3}}$

▶ Exercise Set 14.2 **(Page 851)**

1. $y = c_1 e^{-x} + c_2 e^{-5x} + \frac{1}{16}e^{3x}$

3. $y = c_1 e^{4x} + c_2 e^{5x} - 3x e^{5x}$

5. $y = (c_1 + c_2 x)e^{-x} + \frac{1}{2}x^2 e^{-x}$

7. $y = c_1 e^{3x} + c_2 e^{-4x} - \frac{13}{216} - \frac{1}{18}x - \frac{1}{3}x^2$

9. $y = c_1 + c_2 e^{6x} + \frac{5}{36}x - \frac{1}{12}x^2$

11. $y = c_1 + c_2 x - \frac{1}{2}x^2 + \frac{1}{20}x^5$

13. $y = c_1 e^{-x} + c_2 e^{2x} - 3\cos x - \sin x$

15. $y = c_1 e^{-2x} + c_2 e^{2x} - \frac{3}{8}\cos 2x - \frac{1}{4}\sin 2x$

17. $y = c_1 \cos x + c_2 \sin x - \frac{1}{2}x\cos x$

19. $y = c_1 e^x + c_2 e^{2x} + \frac{3}{4} + \frac{1}{2}x$

21. $y = e^{-2x}(c_1 \cos\sqrt{5}x + c_2 \sin\sqrt{5}x) - \frac{94}{729} + \frac{19}{81}x + \frac{1}{9}x^2$

23. $y = c_1 \cos 2x + c_2 \sin 2x - \frac{1}{8}x\cos 2x$

25. (b) $-\frac{3}{16} - \frac{1}{4}x + \frac{1}{5}xe^x$

27. $y = c_1 e^{-x} + c_2 e^x - 1 + \frac{1}{2}xe^x$

29. $y = c_1 \cos 2x + c_2 \sin 2x + \frac{1}{4} + \frac{1}{4}x + \frac{1}{3}\sin x$

31. $y = (c_1 + c_2 x)e^x + \frac{1}{4}x^2 e^x + \frac{1}{8}e^{-x}$

33. $y = c_1 \cos x + c_2 \sin x + 6 - 2\cos 2x$

35. (a) $y = c_1 \cos\mu x + c_2 \sin\mu x + \dfrac{a}{\mu^2 - b^2}\sin bx$

   (b) $y = c_1 \cos\mu x + c_2 \sin\mu x + \displaystyle\sum_{k=1}^{n}\dfrac{a_k}{\mu^2 - k^2\pi^2}\sin k\pi x$

37. $x_0 = 1; \ y = 2x^2 - 4e^{x-1} + c, \ c$ arbitrary

▶ Exercise Set 14.3 **(Page 855)**

1. $y = c_1 \cos x + c_2 \sin x + x^2 - 2$

3. $y = c_1 e^x + c_2 e^{-2x} + \frac{2}{3}x e^x$

5. $y = c_1 \cos 2x + c_2 \sin 2x - \frac{1}{4}x\cos 2x$

7. $y = c_1 \cos x + c_2 \sin x - (\cos x)\ln|\sec x + \tan x|$

9. $y = c_1 e^x + c_2 x e^x + x e^x \ln|x|$

11. $y = c_1 \cos x + c_2 \sin x + \frac{3}{2} + \frac{1}{2}\cos 2x$

13. $y = c_1 \cos x + c_2 \sin x - x\cos x + (\sin x)\ln|\sin x|$

15. $y = c_1 \cos x + c_2 \sin x + x\cos x - (\sin x)\ln|\cos x|$

17. $y = c_1 e^{-x} + c_2 x e^{-x} - e^{-x}\ln|x|$

19. $y = c_1 e^{-2x} + c_2 x e^{-2x} + (x - 2)e^{-x}$

21. $y = c_1 \cos x + c_2 \sin x - 1 + (\sin x)\ln|\sec x + \tan x|$

23. $y = c_1 e^x + c_2 x e^x - e^x \ln|x|$

25. $y = c_1 e^x + c_2 e^{-x} + \frac{1}{5}e^x(2\sin x - \cos x)$

27. $y = c_1 e^{-x} + c_2 x e^{-x} + \frac{1}{2}x^2 e^{-x}\ln|x| - \frac{3}{4}x^2 e^{-x}$

▶ Exercise Set 14.4 **(Page 864)**

1. (a) $y'' + 4y = 0, \ y(0) = 1, \ y'(0) = 0$

   (b) $y = \cos 2t$

3. (a) $y'' + 196y = 0, \ y(0) = -10, \ y'(0) = 0$

   (b) $y = -10\cos 14t$

5. (a) $y = 2\cos 8t$ (b) 2
(c) $\pi/4$ (d) $4/\pi$

7. (a) $y = -\frac{1}{4}\cos 8\sqrt{6}t$ (b) $\frac{1}{4}$ (c) $\dfrac{\pi}{4\sqrt{6}}$ (d) $\dfrac{4\sqrt{6}}{\pi}$

9. (a) $y'' + 4y' + 8y = 0,\ y(0) = -3,\ y'(0) = 0$
(b) $y = -3e^{-2t}(\cos 2t + \sin 2t)$

11. (a) $y = \frac{3}{2}\sqrt{6}e^{-t/5}\cos\left(\dfrac{\sqrt{2}}{5}t - \tan^{-1}\dfrac{1}{\sqrt{2}}\right)$

(b) $\dfrac{10\pi}{\sqrt{2}}$ (c) $\dfrac{\sqrt{2}}{10\pi}$

13. (a) $y = -\frac{1}{4}\sin 8t$ (b) $\frac{1}{4}$
(c) $\pi/4$ (d) $4/\pi$

15. $y = \frac{1}{3}e^{-t}(\cos 7t - 2\sin 7t)$

17. (a) $\frac{1}{32}\pi^2$ (b) $\frac{9}{4}$

19. $T = 2\pi\sqrt{\dfrac{\delta h}{\rho g}}$

▶ **Exercises, Trigonometric Functions and Identities** (Page A17)

1. (a) $\frac{5}{12}\pi$ (b) $\frac{13}{6}\pi$ (c) $\frac{1}{9}\pi$ (d) $\frac{23}{30}\pi$

3. (a) $120°$ (b) $(270/\pi)°$ (c) $288°$ (d) $540°$

5. (a)   (b)

(c) 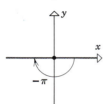   (d)

7. (a) $10\pi/3$ cm   (b) $20\pi/3$ cm²

9. (a) $\dfrac{2\pi - \theta}{2\pi}R$ (b) $\dfrac{\sqrt{4\pi\theta - \theta^2}}{2\pi}R$

11. $\sin\theta = \sqrt{21}/5,\ \tan\theta = \sqrt{21}/2$

13. $\sin\theta = 5/\sqrt{34},\ \cos\theta = 3/\sqrt{34}$

15. $\sin\theta = 3/\sqrt{10},\ \cos\theta = 1/\sqrt{10}$

17. $\tan\theta = \sqrt{21}/2,\ \csc\theta = 5/\sqrt{21}$

19. 1.8 **21.** 3

23.

	$\sin\theta$	$\cos\theta$	$\tan\theta$	$\csc\theta$	$\sec\theta$	$\cot\theta$
(a)	$\dfrac{a}{3}$	$\dfrac{\sqrt{9-a^2}}{3}$	$\dfrac{a}{\sqrt{9-a^2}}$	$\dfrac{3}{a}$	$\dfrac{3}{\sqrt{9-a^2}}$	$\dfrac{\sqrt{9-a^2}}{a}$
(b)	$\dfrac{a}{\sqrt{a^2+25}}$	$\dfrac{5}{\sqrt{a^2+25}}$	$\dfrac{a}{5}$	$\dfrac{\sqrt{a^2+25}}{a}$	$\dfrac{\sqrt{a^2+25}}{5}$	$\dfrac{5}{a}$
(c)	$\dfrac{\sqrt{a^2-1}}{a}$	$\dfrac{1}{a}$	$\sqrt{a^2-1}$	$\dfrac{a}{\sqrt{a^2-1}}$	a	$\dfrac{1}{\sqrt{a^2-1}}$

25. $\frac{21}{4}\sqrt{3}$ **27.** 9.2 ft

29. $h = (\tan\beta - \tan\alpha)d$

31. $\frac{1}{2}ac\sin B,\ \frac{1}{2}ab\sin C$

33.

	θ	$\sin\theta$	$\cos\theta$	$\tan\theta$	$\csc\theta$	$\sec\theta$	$\cot\theta$
(a)	$225°$	$-1/\sqrt{2}$	$-1/\sqrt{2}$	1	$-\sqrt{2}$	$-\sqrt{2}$	1
(b)	$-210°$	$\frac{1}{2}$	$-\sqrt{3}/2$	$-1/\sqrt{3}$	2	$-2/\sqrt{3}$	$-\sqrt{3}$
(c)	$5\pi/3$	$-\sqrt{3}/2$	$\frac{1}{2}$	$-\sqrt{3}$	$-2/\sqrt{3}$	2	$-1/\sqrt{3}$
(d)	$-3\pi/2$	1	0	—	1	—	0

35. $\sin\theta = -\frac{3}{5}$, $\cos\theta = -\frac{4}{5}$, $\tan\theta = \frac{3}{4}$, $\csc\theta = -\frac{5}{3}$, $\sec\theta = -\frac{5}{4}$, $\cot\theta = \frac{4}{3}$

37. (a) $\frac{1}{4}(\sqrt{2} + \sqrt{6})$ or $\frac{1}{2}\sqrt{2 + \sqrt{3}}$

(b) $\frac{1}{2}\sqrt{2 - \sqrt{2}}$

(c) $\frac{1}{2}\sqrt{2 - \sqrt{2 - \sqrt{3}}}$

39. (a) $4\sqrt{5}/9$ (b) $-\frac{1}{9}$

41. (a) $\cos\theta$ (b) $-\sin\theta$
(c) $-\cos\theta$ (d) $\sin\theta$

57. $\sin 3\theta = 3\sin\theta\cos^2\theta - \sin^3\theta$,
$\cos 3\theta = \cos^3\theta - 3\sin^2\theta\cos\theta$

61. (a) $\theta = \pm n\pi$, $n = 0, 1, 2, \ldots$

(b) $\theta = \frac{\pi}{2} \pm n\pi$, $n = 0, 1, 2, \ldots$

(c) $\theta = \pm n\pi$, $n = 0, 1, 2, \ldots$
(d) $\theta = \pm n\pi$, $n = 0, 1, 2, \ldots$

(e) $\theta = \frac{\pi}{2} \pm n\pi$, $n = 0, 1, 2, \ldots$

(f) $\theta = \pm n\pi$, $n = 0, 1, 2, \ldots$

63. $\theta = \frac{5\pi}{4} \pm 2n\pi$ and $\theta = \frac{7\pi}{4} \pm 2n\pi$, $n = 0, 1, 2, \ldots$

65. $\theta = \frac{\pi}{3} \pm 2n\pi$ and $\theta = \frac{5\pi}{3} \pm 2n\pi$, $n = 0, 1, 2, \ldots$

67. $\theta = \frac{\pi}{3} \pm n\pi$, $n = 0, 1, 2, \ldots$

69. $\theta = \frac{4\pi}{3} \pm 2n\pi$ and $\theta = \frac{5\pi}{3} \pm 2n\pi$, $n = 0, 1, 2, \ldots$

71. $\theta = \pi \pm 2n\pi$, $n = 0, 1, 2, \ldots$

73. $\theta = \frac{\pi}{6} \pm n\pi$, $n = 0, 1, 2, \ldots$

75. $\theta = \frac{7\pi}{6} \pm 2n\pi$ and $\theta = \frac{11\pi}{6} \pm 2n\pi$, $n = 0, 1, 2, \ldots$

77. $\theta = \frac{\pi}{6} \pm 2n\pi$ and $\theta = \frac{11\pi}{6} \pm 2n\pi$, $n = 0, 1, 2, \ldots$

▶ **Exercises, Graphs of Trigonometric Functions (Page A25)**

1. (a) $2\pi/5$ (b) 6π (c) 8π (d) $\pi/7$
(e) 2 (f) 10 (g) 1 (h) $2\pi k$

3. (a) 5 (b) $\frac{1}{3}$ (c) $\frac{1}{2}$ (d) 1

5.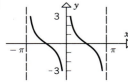

7.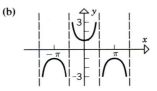

9.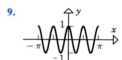

11.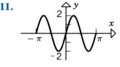

13. (a)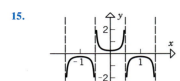

(b)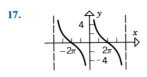

(c)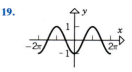

15.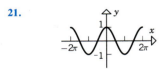

17.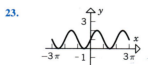

19.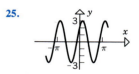

21.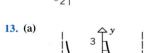

23.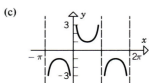

25.

27.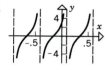

29.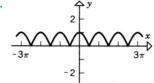

31.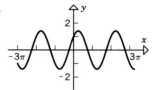

33. (a) odd   (b) even   (c) odd   (d) even
(e) even   (f) even   (g) odd   (h) even

▶ Exercises, Cramer's Rule **(Page A55)**

1. $x = 1, y = 2$

3. $x_1 = -\frac{7}{32}, x_2 = \frac{5}{16}$

5. $x = 2, y = -1, z = 3$

7. $x_1 = \frac{26}{21}, x_2 = \frac{25}{21}, x_3 = \frac{5}{7}$

9. $x' = x \cos \theta + y \sin \theta$
$ y' = -x \sin \theta + y \cos \theta$

Index

Photo Credits

☐ **CHAPTER OPENERS**

Chapter 1
Portrait: H. Josse/Art Resource
Signature: Smithsonian Institution

Chapter 2
Portrait: New York Public Library Picture Collection
Signature: Smithsonian Institution

Chapter 3
Portrait: New York Public Library Picture Collection
Signature: The Bettmann Archive

Chapter 4
Portrait: David Eugene Smith Collection, Columbia University

Chapter 5
Portrait: David Eugene Smith Collection, Columbia University
Signature: Smithsonian Institution

Chapter 6
Portrait: David Eugene Smith Collection, Columbia University
Signature: New York Public Library Picture Collection

Chapter 7
Portrait: New York Public Library Picture Collection

Chapter 8
Portrait: New York Public Library Picture Collection
Signature: The British Library

Chapter 9
Portrait: The Bettmann Archive

Chapter 10
Portrait: David Eugene Smith Collection, Columbia University
Signature: Smithsonian Institution

Chapter 11
Portrait: David Eugene Smith Collection, Columbia University
Signature: The British Library

Chapter 12
Portrait: David Eugene Smith Collection, Columbia University
Signature: Smithsonian Institution

Chapter 13
Portrait: David Eugene Smith Collection, Columbia University
Signature: Smithsonian Institution

Chapter 14
Portrait: David Eugene Smith Collection, Columbia University
Signature: The British Library

☐ **CHAPTER OPENING PORTRAITS**

The chapter opening portraits were created by Hudson River Studio from black and white engravings. They were scanned into and manipulated by computer to create the color illustrations.

Signatures for Chapter Openers 4 and 7 are simulated.